ENCYCLOPEDIA OF PHYSICS

EDITED BY

S. FLÜGGE

VOLUME VI

ELASTICITY AND PLASTICITY

WITH 254 FIGURES

Springer-Verlag Berlin Heidelberg GmbH

1958

HANDBUCH DER PHYSIK

HERAUSGEGEBEN VON

S. FLÜGGE

BAND VI

ELASTIZITÄT UND PLASTIZITÄT

MIT 254 FIGUREN

Springer-Verlag Berlin Heidelberg GmbH

1958

Alle Rechte, insbesondere das der Übersetzung in fremde Sprachen, vorbehalten.

Ohne ausdrückliche Genehmigung des Verlages ist es auch nicht gestattet, dieses Buch oder Teile daraus auf photomechanischem Wege (Photokopie, Mikrokopie) zu vervielfältigen.

ISBN 978-3-662-42801-6   ISBN 978-3-662-43081-1 (eBook)
DOI 10.1007/978-3-662-43081-1

© by Springer-Verlag Berlin Heidelberg 1958

Ursprünglich erschienen bei Springer-Verlag OHG. Berlin·Gottigen·Heidelberg 1958.

Die Wiedergabe von Gebrauchsnamen, Handelsnamen, Warenbezeichnungen usw. in diesem Werk berechtigt auch ohne besondere Kennzeichnung nicht zu der Annahme, daß solche Namen im Sinn der Warenzeichen- und Markenschutz-Gesetzgebung als frei zu betrachten wären und daher von jedermann benutzt werden dürften.

Contents.

	Page
The Classical Theory of Elasticity. By IAN NAISMITH SNEDDON, Simson Professor of Mathematics in the University of Glasgow, Glasgow (Great Britain), and Dr. DENIS STANLEY BERRY, Senior Research Assistant, University of Nottingham, Mathematics Department, Nottingham (Great Britain). (With 16 Figures)	1

A. General theory 2
    I. The analysis of strain 2
    II. The analysis of stress 8
    III. The relation between stress and strain 15

B. Torsion and flexure 27
    I. The torsion problem 27
    II. Bending of beams 38

C. Two-dimensional problems in elasto-statics 40
    I. General theory 40
    II. Airy stress function 42
    III. Complex potentials 48
    IV. Cauchy integral methods 58
    V. Fourier transform methods 72
    VI. Real potential methods 78

D. Three-dimensional problems in elasto-statics 84
    I. General theory 84
    II. BETTI's method 94
    III. The method of integral transforms 98
    IV. Applications of curvilinear coordinates 102

E. Dynamical problems 107
    I. Elastic waves 107
    II. Boundary value problems of dynamic elasticity 118

F. Thermoelasticity 123

Bibliography 126

Photoelasticity. By HERBERT T. JESSOP, T. D., M. Sc., F. Inst. P., Senior Lecturer, Department of Civil and Municipal Engineering, University College, London (Great Britain). (With 114 Figures) 127

Introduction 127

A. Theory 128
    I. History 128
    II. The optical basis of photoelasticity 130
    III. The mechanical basis of photoelasticity 136
    IV. Theory of photoelasticity 145

B. Exploration of two-dimensional stress systems 155
    I. Photoelastic equipment 155
    II. Photoelastic materials 161
    III. Exploration of two-dimensional stresses 166

Contents.

	Page
C. Exploration of three-dimensional stress systems	177
I. Frozen stress materials and techniques	177
II. The determination of the stresses	180
III. The tilting stage method of exploration of three-dimensional stresses	186
IV. The scattered light method of observation	193
D. Practical applications	202
I. Two-dimensional examples	202
II. Three-dimensional examples	215
III. Present state and possible future developments	225
Bibliography	228

The Mathematical Theories of the Inelastic Continuum. By Dr. ALFRED M. FREUDENTHAL, Professor of Civil Engineering, Columbia University, New York/N. Y. (USA), and Dr. HILDA GEIRINGER, Professor of Mathematics, Cambridge/Massachusetts (USA). (With 60 Figures) 229

First part. The inelastic continuum 229

  A. Mechanics and thermodynamics of the inelastic continuum 229
    I. The inelastic behavior of solids 229
    II. Mechanics of deformable media 234
    III. Thermodynamic considerations 243

  B. Stress-strain relations 256
    I. General formulation 256
    II. Anelastic relations 263
    III. Visco-elastic relations 269
    IV. Plastic relations 278
    V. Combined quasi-linear relations 289

  C. The visco-elastic and the visco-plastic medium 293
    I. The visco-elastic continuum 293
    II. The visco-plastic medium 301

  D. Problems of structural mechanics 308
    I. Visco-elastic structures 308
    II. Elastic-plastic structures 313

Second part. The ideal plastic body 322

  E. The basic equations 322
    I. The three-dimensional problem 322
    II. Discontinuous solutions 333
      a) Characteristics. Application to the three-dimensional problem of the perfectly plastic body 333
      b) General consideration of discontinuous solutions 340
      c) HADAMARD's theory 343
      d) Shock conditions. Stress discontinuities 346

  F. The problem of plane strain 349
    I. Plane strain, plane stress, and generalizations 349
    II. The theory of plane strain 353
      a) Differential relations 353
      b) Integration. Particular solutions 360

  G. The general plane problem 367
    I. Basic theory 367
      a) The equations 367
      b) Characteristics of the complete plane problem 372
      c) Remarks on integration. Examples 381

	Page
II. Singular solutions and various remarks	384
a) Limit line singularities and branch line singularities	384
b) Simple waves	390
c) Various remarks	396
H. Boundary-value problems	399
I. Some elastic-plastic problems	399
a) The torsion problem	399
b) The thick walled tube	408
c) Flat ring and flat sheet in plane stress. Further elastic-plastic problems	418
II. Some plastic-rigid problems	425
a) Introductory remarks	425
b) Wedge with pressure on one face	427
c) Plastic mass between rough rigid plates	429
Reference Books	432

Rheology. By Dr. Marcus Reiner, Professor of Applied Mechanics, Israel Institute of Technology, Haifa (Israel). (With 45 Figures) 434

A. Preliminaries	434
I. Introduction	434
II. The classical bodies	448
B. Macrorheology	452
I. First-order phenomena	452
II. Higher order phenomena	487
III. Strength	519
C. Microrheology	522
D. Rheometry	535
E. Addenda	542
Symbols	549
Bibliography	549

Fracture. By Dr. George R. Irwin, Superintendent of Mechanics Division, U. S. Naval Research Laboratory, Washington/D.C. (USA). (With 10 Figures) 551

I. Tensile strength of liquids	551
II. Stress and force relations in fracture	556
III. Forming and spreading of cracks	567
IV. Stress field, velocity, and division of a running crack	575
V. Effects of size upon fracturing	580
Bibliography references	589

Fatigue. By Dr. Alfred M. Freudenthal, Professor of Civil Engineering, Columbia University, New York/N. Y. (USA). (With 9 Figures) 591

I. The fatigue phenomenon	591
II. Micromechanics of progressive damage	596
III. Fatigue theories	603
IV. Cumulative damage	608
References	612

Sachverzeichnis (Deutsch-Englisch) 614

Subject Index (English-German) 628

The Classical Theory of Elasticity.

By

I. N. SNEDDON and D. S. BERRY.

With 16 Figures.

1. Introduction. The theory of elasticity is concerned with the mechanics of deformable bodies which recover their original shape upon the removal of the forces causing the deformation. The first discussions of elastic phenomena occur in the writings of HOOKE (1676) but the first real attempts to construct a theory of elasticity using the continuum approach, in which speculations on the molecular structure of the body are avoided and macroscopic phenomena are described in terms of field variables, date from the first half of the eighteen century[1]. Since that time a tremendous amount of scientific effort has been devoted to the study of the mathematical theory of elasticity and its applications to physics and engineering. The sheer volume of the published work in the subject makes it quite impossible for an author to cover the entire subject at all adequately within the compass of a single book. The present article has a much more modest aim than that: It tries to give a brief survey of certain parts of the basic theory of elasticity with sufficient discussion of special problems to give some indication of the mathematical techniques available for the solution of such problems. Even within that limited framework there are notable omissions; for example, nothing is said about such an important technological topic as the theory of elastic stability or about such a basic topic as the calculation of the elastic constants of a crystal by the theory of crystal lattices.

The consequences of most of the omissions are mitigated by the appearance recently of several excellent textbooks — by GREEN and ZERNA (1954), SOKOLNIKOFF (1956), and the publication in 1953 by P. Noordhoff Ltd. of Groningen of J. R. M. RADOK's English translation of an earlier edition of MUSKHELISHVILI's classic work (1954)[2].

The first part of the article (Division A) is concerned with the basic general theory. Although this article is devoted to the *classical* theory of elasticity, in which the strains are assumed to be infinitesimal, we have followed here the treatment of the theory of large deformations developed by GREEN and ZERNA since it throws light upon the fundamental problems of the subject.

Division B is a discussion of two very simple problems — the torsion of cylinders and the flexure of beams. This provides simple illustrations of the application of the general theory and the solutions themselves are of some importance in engineering.

The largest division of the article (Division C) is a discussion of two-dimensional problems in elasto-statics. This division naturally owes much to the recent

---

[1] For the early history of the theory of elasticity the reader is referred to I. TODHUNTER and K. PEARSON: A History of the Theory of Elasticity and of the Strength of Materials, 2 vols. Cambridge 1893.

[2] For details the reader is referred to the bibliography.

important contributions to the theory of elasticity made by Russian mathematicians and described in detail in Muskhelishvili's masterly treatise. This is followed by a division (D) on the methods developed for the solution of three-dimensional problems.

Finally there are short divisions on dynamical problems (E) and thermoelasticity (F). A physicist interested in the physical principles underlying the classical theory of elasticity might confine his attention to these divisions and to division A.

The bibliography is restricted to better known books on the subject and in addition to mathematical treatises contains works mainly interested in applications of the theory of elasticity to geophysics and engineering. Of more fundamental topics omitted from this article: The solution of special problems involving large deformations is discussed by Green and Zerna (1954) and Murnaghan (1951). The theory of shells is treated by Timoshenko (1940) and Goldenveiser (1953) and the theory of elastic stability by Timoshenko (1934). Detailed discussions of the use of variational methods are given by Leibenson (1943) and Sokolnikoff (1956). The applications of functional analysis to the approximate solutions of problems in elasticity is described by Kantorovich (1952) and Synge (1957). The use of analogue methods (such as the soap film) and of finite-difference equations in elasticity is discussed by Timoshenko and Goodier (1951). Direct experimental methods such as photoelasticity and strain gauges are the subject of separate articles in this volume.

A. General theory.

I. The analysis of strain.

In the first subdivision of the general theory we shall consider the analysis of strain and the construction of the strain tensor. As already stated in the introduction the treatment given here is based upon the treatise by Green and Zerna but, in order to achieve a measure of uniformity in this *Handbuch*, we have adopted the notation, not of Green and Zerna, but that used by Dr. Horst Tietz in Sect. D of his article "Geometrie" in Vol. II.

2. The displacement vector. Suppose that at time $t=0$ the elastic body is at rest in a configuration S described in terms of a fixed system of rectangular cartesian axes with origin O so that the coordinates of a typical point P are (x_1, x_2, x_3). We denote by $\boldsymbol{x}$ the position vector of P. We now suppose that the elastic body is deformed to a new configuration $\bar{S}$ in such a way that at time t the representative point P has moved to a point $\bar{P}$ with position vector $\bar{\boldsymbol{x}}=(\bar{x}_1, \bar{x}_2, \bar{x}_3)$ referred to a new set of axes with origin at a point $\bar{O}$ (compare Fig. 1). The position vector $\boldsymbol{u}$ of the point $\bar{P}$ with respect to the point P represents the displacement of the point P when the body moves from configuration S to configuration $\bar{S}$, and, for that reason, it is known as the *displacement vector*. If the position vector of $\bar{O}$ relative to O is denoted by $\boldsymbol{a}$ then it is a matter of simple vector algebra to show that

$$\boldsymbol{u} = \bar{\boldsymbol{x}} - \boldsymbol{x} + \boldsymbol{a}. \qquad (2.1)$$

So far we have made no assumption about the nature of the material forming the elastic body other than that it is continuous. We now make the additional assumption that the material of the body is of such a nature that the coordinates $\bar{x}_1, \bar{x}_2, \bar{x}_3$ of a typical point $\bar{P}$ in the state $\bar{S}$ are single-valued functions of the time t and the coordinates x_1, x_2, x_3 of the position P of the corresponding point in the original state S. It is also assumed that, except possibly at certain singular points, curves or surfaces, these functions possess continuous derivatives with respect to x_1, x_2, x_3 of as high an order as we wish. We write

$$\bar{x}_i = \bar{x}_i(x_1, x_2, x_3, t) \quad (i = 1, 2, 3) \tag{2.2}$$

and assume that x_i can be expressed as a function of $\bar{x}_1, \bar{x}_2, \bar{x}_3, t$ in the same way:

$$x_i = x_i(\bar{x}_1, \bar{x}_2, \bar{x}_3, t) \quad (i = 1, 2, 3). \tag{2.3}$$

The physical assumption embodied in the statements (2.2) and (2.3) should be regarded as axioms upon which our theory is based, their justification lying in the fact that they lead to theoretical results which are in agreement with the observed experimental facts. We also assume that S and $\bar{S}$ are situated in three-dimensional Euclidean space.

It follows from the relations (2.2) and (2.3) that the differentials $d\bar{x}^r$, evaluated at time t, are related to the differentials dx^r by the equations

$$\left.\begin{array}{l} d\bar{x}^r = \dfrac{\partial \bar{x}^r}{\partial x^s} dx^s, \\[6pt] dx^r = \dfrac{\partial x^r}{\partial \bar{x}^s} d\bar{x}^s. \end{array}\right\} \tag{2.4}$$

Now the square of the line element in the configuration S may be written

$$ds^2 = dx^i dx^i \tag{2.5}$$

while that in $\bar{S}$ is

$$d\bar{s}^2 = d\bar{x}^i d\bar{x}^i. \tag{2.6}$$

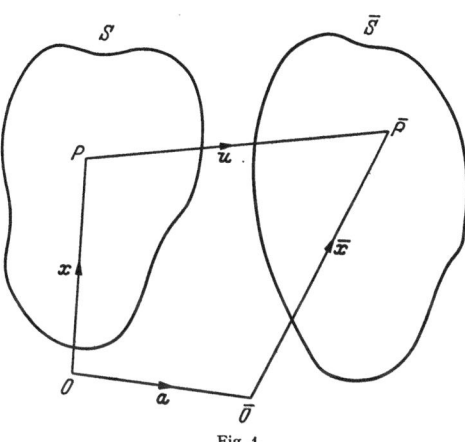

Fig. 1.

The original configuration S may be specified by means of a system of orthogonal curvilinear coordinates (q_1, q_2, q_3) and we shall suppose that this system moves continuously with the body in the transition from the state S to the state $\bar{S}$ and so may be used as a system of curvilinear coordinates in $\bar{S}$. We may therefore write

$$x_r = x_r(q_1, q_2, q_3), \quad \bar{x}_r = \bar{x}_r(q_1, q_2, q_3, t) \quad (r = 1, 2, 3) \tag{2.7}$$

and it follows from Eq. (2.1) that we may write the displacement vector in the form

$$\boldsymbol{u} = \boldsymbol{u}(q_1, q_2, q_3, t). \tag{2.8}$$

3. The strain tensor. Corresponding to the vectors (2.7) we obtain the expressions

$$ds^2 = g_{rs} dq^r dq^s, \quad d\bar{s}^2 = \bar{g}_{rs} d\bar{q}^r d\bar{q}^s \tag{3.1}$$

for the squares of the line elements in S and $\bar{S}$ respectively. In these equations $g_{ij}, \bar{g}_{ij}$ are the covariant metric tensors in S and $\bar{S}$ respectively, the quantities $\bar{g}_{ij}$ being calculated for a given time t and therefore being functions of t as well as of q_1, q_2, q_3. The state of strain of the elastic body may obviously be characterized by the $d\bar{s}^2 - ds^2$ and it follows from the Eqs. (3.1) that

$$d\bar{s}^2 - ds^2 = 2\gamma_{rs} dq^r dq^s \tag{3.2}$$

where

$$\gamma_{rs} = \tfrac{1}{2}(\bar{g}_{rs} - g_{rs}). \tag{3.3}$$

The quantities γ_{rs} are obviously the symmetrical components of a covariant tensor of the second rank; they are said to form the *strain tensor*. It will be observed that when the strain tensor vanishes at all points, $d\bar{s}=ds$ and the body is rigid.

To find the geometrical interpretation of the strain tensor we consider first the extension e_r of a line element ds_r along the q_r coordinate curves. Since $ds_r = \sqrt{g_{rr}}\, dq^r$, $d\bar{s}_r = \sqrt{\bar{g}_{rr}}\, dq^r$, ($r$ not summed), we find that

$$e_r = \sqrt{\frac{\bar{g}_{rr}}{g_{rr}}} - 1 = \sqrt{\left(1 + \frac{2\gamma_{rr}}{g_{rr}}\right)} - 1,$$

from which it follows that

$$\gamma_{rr} = g_{rr}[e_r + \tfrac{1}{2} e_r^2] \quad (r \text{ not summed}). \tag{3.4}$$

Further suppose that ϑ_{rs} is the angle between two line elements $d\bar{s}_r$, $d\bar{s}_s$ in the strained body then it is readily shown that

$$\cos\vartheta_{rs} = \frac{\bar{g}_{rs}}{\sqrt{\bar{g}_{rr}\bar{g}_{ss}}} \quad (r, s \text{ not summed})$$

and hence that

$$\cos\vartheta_{rs} = \frac{g_{rs} + 2\gamma_{rs}}{\sqrt{(g_{rr} + 2\gamma_{rr})(g_{ss} + 2\gamma_{ss})}} \tag{3.5}$$

so that the angles between the line elements are determined by γ_{rs}.

We shall now calculate the components of the strain tensor in terms of the components of the displacement vector $\boldsymbol{u}$. We note first of all that if $\boldsymbol{e}_r$, $\bar{\boldsymbol{e}}_r$ are base vectors in the states S, $\bar{S}$ respectively then

$$\bar{\boldsymbol{e}}_r = \partial_r \bar{\boldsymbol{x}}, \quad \boldsymbol{e}_r = \partial_r \boldsymbol{x}$$

where, it will be recalled, $\partial_r f$ denotes $\partial f/\partial q_r$. Hence it follows from Eq. (2.1) that

$$\bar{\boldsymbol{e}}_r = \partial_r \boldsymbol{u} + \boldsymbol{e}_r. \tag{3.6}$$

Now $\bar{g}_{rs} = \bar{\boldsymbol{e}}_r \cdot \bar{\boldsymbol{e}}_s$ so that

$$\bar{g}_{rs} = (\boldsymbol{e}_r + \partial_r \boldsymbol{u}) \cdot (\boldsymbol{e}_s + \partial_s \boldsymbol{u})$$
$$= g_{rs} + \boldsymbol{e}_r \cdot \partial_s \boldsymbol{u} + \boldsymbol{e}_s \cdot \partial_r \boldsymbol{u} + \partial_r \boldsymbol{u} \cdot \partial_s \boldsymbol{u}$$

and therefore

$$\gamma_{rs} = \tfrac{1}{2}(\boldsymbol{e}_r \cdot \partial_s \boldsymbol{u} + \boldsymbol{e}_s \cdot \partial_r \boldsymbol{u} + \partial_r \boldsymbol{u} \cdot \partial_s \boldsymbol{u}). \tag{3.7}$$

Now we may write

$$\boldsymbol{u} = u_m \boldsymbol{e}^m = u^m \boldsymbol{e}_m \tag{3.8}$$

and hence

$$\partial_r \boldsymbol{u} = \nabla_r u_m \boldsymbol{e}^m \tag{3.9}$$

where $\nabla_r u_m$ denotes the covariant derivative

$$\nabla_r u_m = \partial_r u_m - \begin{Bmatrix} s \\ m\, r \end{Bmatrix} u_s, \tag{3.10}$$

$\begin{Bmatrix} s \\ m\, r \end{Bmatrix}$ denoting the Christoffel symbol of the second kind calculated for the configuration S from the metric tensors g_{rs}, g^{rs}. Substituting from Eq. (3.9) into Eq. (3.7) and making use of the relations

$$\boldsymbol{e}_r \cdot \boldsymbol{e}^m = g_r^m = \delta_r^m, \quad \boldsymbol{e}^m \cdot \boldsymbol{e}^n = g^{mn}$$

we find that
$$\gamma_{rs} = \tfrac{1}{2}(\nabla_s u_r + \nabla_r u_s + \nabla_r u^n \nabla_s u_n), \qquad (3.11)$$
which gives the components of the strain tensor in terms of the components of the displacement vector with respect to S.

Alternatively we could write Eq. (3.6) in the form
$$e_r = \bar{e}_r - \overline{\nabla}_r \bar{u}_m \bar{e}^m$$
where we have expressed the displacement vector in terms of the contravariant base vectors $\bar{e}^r$ of the state $\bar{S}$ and $\overline{\nabla}$ denotes covariant differentiation with respect to $\bar{S}$ i.e. with respect to q_r and $\bar{g}_{rs}$. Hence
$$g_{rs} = e_r \cdot e_s = \bar{g}_{rs} - \overline{\nabla}_s \bar{u}_r - \overline{\nabla}_r \bar{u}_s + \overline{\nabla}_r \bar{u}^n \overline{\nabla}_s \bar{u}_n$$
so that in terms of the components of u with respect to $\bar{S}$
$$\gamma_{rs} = \tfrac{1}{2}(\overline{\nabla}_s \bar{u}_r + \overline{\nabla}_r \bar{u}_s - \overline{\nabla}_r \bar{u}^n \overline{\nabla}_s \bar{u}_n). \qquad (3.12)$$
For example, referred to the cartesian system (x_1, x_2, x_3) in the state S we find that
$$\gamma_{11} = \frac{\partial u_1}{\partial x_1} + \frac{1}{2}\left\{\left(\frac{\partial u_1}{\partial x_1}\right)^2 + \left(\frac{\partial u_2}{\partial x_1}\right)^2 + \left(\frac{\partial u_3}{\partial x_1}\right)^2\right\},$$
$$\gamma_{23} = \frac{1}{2}\left(\frac{\partial u_2}{\partial x_3} + \frac{\partial u_3}{\partial x_2}\right) + \frac{1}{2}\left(\frac{\partial u_1}{\partial x_2} \cdot \frac{\partial u_1}{\partial x_3} + \frac{\partial u_2}{\partial x_2} \cdot \frac{\partial u_2}{\partial x_3} + \frac{\partial u_3}{\partial x_2} \cdot \frac{\partial u_3}{\partial x_3}\right),$$
whereas referred to the cartesian system $(\bar{x}_1, \bar{x}_2, \bar{x}_3)$ in the state $\bar{S}$ we find that
$$\gamma_{11} = \frac{\partial \bar{u}_1}{\partial \bar{x}_1} - \frac{1}{2}\left\{\left(\frac{\partial \bar{u}_1}{\partial \bar{x}_1}\right)^2 + \left(\frac{\partial \bar{u}_2}{\partial \bar{x}_1}\right)^2 + \left(\frac{\partial \bar{u}_3}{\partial \bar{x}_1}\right)^2\right\},$$
$$\gamma_{23} = \frac{1}{2}\left(\frac{\partial \bar{u}_2}{\partial \bar{x}_3} + \frac{\partial \bar{u}_3}{\partial \bar{x}_2}\right) - \frac{1}{2}\left(\frac{\partial \bar{u}_1}{\partial \bar{x}_2} \cdot \frac{\partial \bar{u}_1}{\partial \bar{x}_3} + \frac{\partial \bar{u}_2}{\partial \bar{x}_2} \cdot \frac{\partial \bar{u}_2}{\partial \bar{x}_3} + \frac{\partial \bar{u}_3}{\partial \bar{x}_2} \cdot \frac{\partial \bar{u}_3}{\partial \bar{x}_3}\right).$$

From the covariant strain tensor γ_{mn} of the second rank we can form a mixed tensor in either of two ways according as whether we use the metric tensor g^{mn} of the initial unstrained state or the metric tensor $\bar{g}^{mn}$ of the final strained state. Making use of the metric tensor g^{mn} we can construct the mixed tensor
$$\gamma_n^m = g^{mp} \gamma_{pn} = \tfrac{1}{2}(g^{mp} \bar{g}_{pn} - \delta_n^m).$$
In a precisely similar way we can set up a contravariant tensor
$$\varepsilon^{mn} = g^{mp} \varepsilon_p^n = \bar{g}^{nq} \gamma_q^m = g^{mp} \bar{g}^{nq} \gamma_{pq}.$$

It should also be noted that if g is the determinant of the array g_{mn} then $\sqrt{g}\, dq^1 dq^2 dq^3$ is the volume of an element in the unstrained state. If therefore we denote by ϱ the density in the unstrained state and by $\bar{\varrho}$ the density in the strained state we have
$$\bar{\varrho}\sqrt{\bar{g}} = \varrho\sqrt{g}. \qquad (3.13)$$

4. The physical components of the displacement vector. If $\boldsymbol{d}_m$ is the unit vector in the direction of $\boldsymbol{e}_m$, then
$$\boldsymbol{d}_m = \frac{\boldsymbol{e}_m}{\sqrt{(\boldsymbol{e}_m \cdot \boldsymbol{e}_m)}} = \frac{\boldsymbol{e}_m}{\sqrt{g_{mm}}}$$
so that we may write the second of the Eqs. (3.8) in the form
$$\boldsymbol{u} = v^m \boldsymbol{d}_m \qquad (4.1)$$
where
$$v^m = u^m \sqrt{g_{mm}}. \qquad (4.2)$$

The quantities v^m are simply the components of the vector $\mathbf{u}$ referred to an oblique cartesian system of axes coinciding with the directions $\mathbf{e}_m$. They are called the *physical components of displacement* along the q_m—coordinate curves in S.

5. Strain invariants. It is a well known result of the tensor calculus that three invariants may be formed from a symmetrical tensor of the second order, so that strain invariants may be obtained from the mixed strain tensor γ_n^m. These invariants are the coefficients of powers of λ in the expansions of the determinant

$$|\lambda \delta_n^m + \delta_n^m + 2\gamma_n^m| = |\lambda \delta_n^m + g^{mp}\bar{g}_{pn}|$$

so that, if I_1, I_2, I_3 denote the strain invariants, we have

$$\lambda^3 + I_1\lambda^2 + I_2\lambda + I_3 = \begin{vmatrix} \lambda + g^{1p}\bar{g}_{p1} & g^{1p}\bar{g}_{p2} & g^{1p}\bar{g}_{p3} \\ g^{2p}\bar{g}_{p1} & \lambda + g^{2p}\bar{g}_{p2} & g^{2p}\bar{g}_{p3} \\ g^{3p}\bar{g}_{p1} & g^{3p}\bar{g}_{p2} & \lambda + g^{3p}\bar{g}_{p3} \end{vmatrix}.$$

Expanding the determinant and equating coefficients of powers of λ we find after a little reduction that

$$I_1 = g^{mn}\bar{g}_{mn}, \quad I_2 = \bar{g}^{mn}g_{mn}I_3, \quad I_3 = \frac{\bar{g}}{g} \tag{5.1}$$

where $\bar{g}$ and g denote the determinants $|\bar{g}_{mn}|$, $|g_{mn}|$ respectively.

If the body is *incompressible*, elements of volume remain unaltered so that $\bar{g} = g$ and hence for an incompressible solid

$$I_1 = g^{mn}\bar{g}_{mn}, \quad I_2 = \bar{g}^{mn}g_{mn}, \quad I_3 = 1. \tag{5.2}$$

Another set of strain invariants J_1, J_2, J_3 may be defined as the coefficients of powers of λ in the expansion of the determinant

$$|\lambda \delta_n^m + \delta_n^m - 2\varepsilon_n^m| = |\lambda \delta_n^m + \bar{g}^{mp}g_{pn}|.$$

It is found that in the general case

$$J_1 = g_{mn}\bar{g}^{mn}, \quad J_2 = \frac{1}{2}(J_1^2 - \bar{g}^{mp}\bar{g}^{nq}g_{mn}g_{pq}), \quad J_3 = \frac{g}{\bar{g}}. \tag{5.3}$$

It is known from the general theory that the two sets of strain invariants are not independent and it is indeed readily shown that

$$J_1 = \frac{I_2}{I_3}, \quad J_2 = \frac{I_1}{I_3}, \quad J_3 = \frac{1}{I_3}. \tag{5.4}$$

6. Infinitesimal strains. In the classical or infinitesimal theory of elasticity it is assumed that the components of the displacement vector and their derivatives with respect to q_m are infinitesimals of the first order so that we neglect products and squares of these quantities in comparison with their first powers. If we make this approximation in Eq. (3.11) we find that the covariant strain tensor becomes

$$\gamma_{mn} = \tfrac{1}{2}(\nabla_m u_n + \nabla_n u_m). \tag{6.1}$$

The differential of the displacement vector is

$$d\mathbf{u} = \nabla_n u_m \, dq^n \mathbf{g}^m$$

and this may be written in the form

$$d\mathbf{u} = (\gamma_{mn} + \omega_{mn}) \, dq^n \mathbf{g}^m \tag{6.2}$$

where ω_{mn} is the skew-symmetric tensor defined by the equation

$$\omega_{mn} = \tfrac{1}{2}(\nabla_n u_m - \nabla_m u_n). \tag{6.3}$$

If the body is not deformed, i.e. if $\gamma_{mn} = 0$, then

$$d\boldsymbol{u} = \omega_{mn} dq^n \boldsymbol{g}^m, \tag{6.4}$$

showing that the skew-symmetric tensor ω_{mn} represents an infinitesimal rigid-body displacement. If we find the covariant derivative of both sides Eq. (6.3) we obtain the relation

$$\nabla_p \omega_{mn} = \tfrac{1}{2}(\nabla_{np} u_m - \nabla_{mp} u_n)$$
$$= \tfrac{1}{2}\nabla_n(\nabla_p u_m + \nabla_m u_p) - \tfrac{1}{2}\nabla_m(\nabla_p u_n + \nabla_n u_p)$$

which is equivalent to

$$\nabla_p \omega_{mn} = \nabla_n \gamma_{pm} - \nabla_m \gamma_{pn}. \tag{6.5}$$

7. The compatibility relations. So far we have not made use of our assumption that both the initial unstrained state S and the strained state $\bar{S}$ are situated in three-dimensional Euclidean space. It is a well known result of Riemannian geometry[1] that for such a space the Riemann-Christoffel tensor must vanish. Thus if we denote the Riemann-Christoffel tensor of the unstrained state by R_{mnpq} and that of the strained state by $\bar{R}_{mnpq}$, then

$$R_{mnpq} = 0, \quad \bar{R}_{mnpq} = 0. \tag{7.1}$$

Now from the definition of the Riemann-Christoffel tensor

$$\bar{R}_{mnpq} = \tfrac{1}{2}(\partial_{np}\bar{g}_{mq} + \partial_{mq}\bar{g}_{np} - \partial_{nq}\bar{g}_{mp} - \partial_{mp}\bar{g}_{nq}) +$$
$$+ \bar{g}^{ij}\{\overline{[m,qi]} \cdot \overline{[n,pj]} - \overline{[m,pi]} \cdot \overline{[n,qi]}\} \tag{7.2}$$

where $\overline{[m,qi]}$ denotes the Christoffel symbol of the first kind calculated for the body in the state $\bar{S}$. Making use of the relation

$$\bar{g}_{mn} = g_{mn} + 2\gamma_{mn}$$

in the definition of the Christoffel symbol for the state $\bar{S}$, we find that it is related to that for the state S through the equation

$$\overline{[m,np]} = [m,np] + 2\gamma_{mnp} \tag{7.3}$$

where the symbol γ_{mnp} denotes the expression

$$\tfrac{1}{2}(\partial_n \gamma_{pm} + \partial_m \gamma_{pn} - \partial_p \gamma_{mn}).$$

Substituting from (7.3) into (7.2) we find that the difference between the Riemann-Christoffel tensor in the strained state and that in the unstrained state is

$$\begin{aligned}
\bar{R}_{mnpq} &- R_{mnpq} \\
&= \partial_{np}\gamma_{mq} + \partial_{mq}\gamma_{np} - \partial_{mp}\gamma_{nq} - \partial_{nq}\gamma_{mp} - \\
&\quad - 2\varepsilon^{ij}\{[m,qi][n,pj] - [m,pi][n,qi]\} + \\
&\quad + 2g^{ij}\{[n,pj]\gamma_{mqi} + [m,qi]\gamma_{npj} - [n,qj]\gamma_{mpi} - [m,pi]\gamma_{nqj}\} - \\
&\quad - 4\varepsilon^{ij}\{[n,pj]\gamma_{mqi} + [m,qi]\gamma_{nqj} - [n,qj]\gamma_{mpi} - \\
&\quad - [m,pi]\gamma_{nqj} + 2\gamma_{mqi}\gamma_{npj} - 2\gamma_{mpi}\gamma_{nqj}\} + \\
&\quad + 4g^{ij}(\gamma_{mqi}\gamma_{npj} - \gamma_{mpi}\gamma_{nqj}).
\end{aligned} \tag{7.4}$$

---
[1] J. A. Schouten: Tensor Analysis for Physicists, p. 99. Oxford 1951.

If we insert the conditions (7.1) into Eq. (7.4) we see that Eq. (7.4) reduces to one in which the right hand side of (7.4) is equated to zero. Since there are only six independent components of the Riemann-Christoffel tensor it follows that the tensor Eq. (7.4) reduces to six independent equations. These six equations, which express the fact that the components γ_{mn} of the strain tensor are not independent but have to be such that the space of S and $\overline{S}$ are Euclidean, are known as the *compatibility relations*.

If the strains are infinitesimal we can neglect second order terms in the compatibility relations and they then reduce to

$$\partial_{np}\gamma_{mq} + \partial_{mq}\gamma_{np} - \partial_{mp}\gamma_{nq} - \partial_{nq}\gamma_{mp} -$$
$$- 2\varepsilon^{ij}\{[m,qi][n,pj] - [m,pi][n,qj]\} +$$
$$+ 2g^{ij}\{[n,pj]\gamma_{mqi} + [m,qi]\gamma_{npj} - [n,qj]\gamma_{mpi} - [m,pi]\gamma_{nqj}\} = 0.$$

Making use of the relations

$$[m,ns] = g_{rs}\left\{{r \atop mn}\right\}, \qquad \varepsilon^{ij} = g^{ir}g^{js}\gamma_{rs},$$

we see that these equations may be put in the form

$$\left. \begin{aligned} &\partial_{np}\gamma_{mq} + \partial_{mq}\gamma_{np} - \partial_{mp}\gamma_{nq} - \partial_{nq}\gamma_{mp} - \\ &- 2\gamma_{rs}\left(\left\{{r \atop qm}\right\}\left\{{s \atop pn}\right\} - \left\{{r \atop mp}\right\}\left\{{s \atop qn}\right\}\right) + \\ &+ 2\left\{{r \atop np}\right\}\gamma_{mqr} + 2\left\{{r \atop qm}\right\}\gamma_{npr} - 2\left\{{r \atop nq}\right\}\gamma_{mpr} - 2\left\{{r \atop mp}\right\}\gamma_{nqr} = 0. \end{aligned} \right\} \quad (7.5)$$

From the definition of covariant differentiation we see that these equations may be written quite simply as

$$\nabla_{np}\gamma_{mq} + \nabla_{mq}\gamma_{np} - \nabla_{mp}\gamma_{nq} - \nabla_{nq}\gamma_{mp} = 0. \qquad (7.6)$$

It should be observed that the compatibility relations (7.6), which are valid for infinitesimal strains, can be obtained directly from the definition of the skew-symmetric tensor ω_{mn}. From the relation (6.5) in the form

$$\nabla_n\gamma_{mq} - \nabla_m\gamma_{nq} = \omega_{mn}$$

we see that

$$\nabla_{np}\gamma_{mq} - \nabla_{mp}\gamma_{nq} = \nabla_{qp}\omega_{mn}$$
$$= \nabla_{pq}\omega_{mn}$$
$$= \nabla_{nq}\gamma_{mp} - \nabla_{mq}\gamma_{np}$$

which is identical with Eq. (7.6).

II. The analysis of stress.

8. The stress tensor. To specify the state of stress at a point P in the strained body S we consider an element ΔA of a surface which is situated in S and which contains P. We may assume that the force exerted by the part of the body which is on one side of the surface on the remainder is statically equivalent to a force $\Delta \boldsymbol{T}$ at P and a couple $\Delta \boldsymbol{G}$. It may further be shown that as $\Delta A \to 0$ $(\Delta \boldsymbol{G}/\Delta A) \to 0$ and $\Delta \boldsymbol{T}/\Delta A$ tends to a vector $\boldsymbol{t}$, called the stress-vector, belonging to the element of area and representing the force per unit area of the surface in the deformed state. It should be observed that the stress on a surface element in a solid body does not, in general, act normally to that surface, but has components both normal to the plane and tangential to it. For instance, if we take ΔA to be a square through P perpendicular to the x-axis we get the situation in Fig. 2. The stress-vector, belonging to this surface, will not in general be parallel to the axis Ox so that it can be resolved into three components t_{xx}, t_{yx}, t_{zx} along the (x, y, z)-axes respectively. In the general case, at any point P of the strained state $\bar{S}$, we can construct a tetrahedron $PABC$ whose edges are formed by the coordinate curves PA, PB, PC through P of lengths ds_1, ds_2, ds_3 respectively. Considering the infinitesimal triangle PAB we see that

$$\overrightarrow{AB} = \overrightarrow{PB} - \overrightarrow{PA} = dq^1\, e_1 - dq^2\, e_2.$$

The surface $q_1 = const$ of this tetrahedron has area

$$dS_1 = \tfrac{1}{2}\sqrt{\bar{g}\,\bar{g}^{11}}\, dq^2\, dq^3$$

and may be represented vectorially by

$$\frac{e^1\, dS_1}{\sqrt{\bar{g}^{11}}}.$$

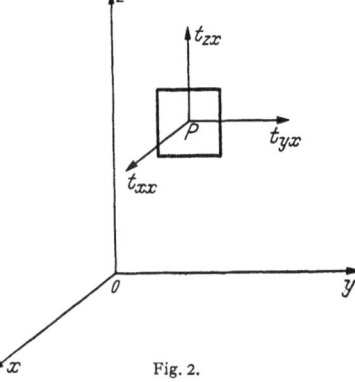

Fig. 2.

The surfaces $q_2 = const$, $q_3 = const$ have areas given by similar expressions. The area ABC may be represented by the vector

$$\boldsymbol{n}\, dS.$$

Since the area ABC is the vector sum of the areas of the three other faces of the tetrahedron, we find that

$$\boldsymbol{n}\, dS = \sum_{i=1}^{3} \frac{\bar{e}^i\, dS_i}{\sqrt{\bar{g}^{ii}}} \qquad (8.1)$$

so that, if n_i are the covariant components of $\boldsymbol{n}$ i.e. if.

$$\boldsymbol{n} = n_i\, \bar{e}^i,$$

we find that

$$n_i \sqrt{\bar{g}_{ii}}\, dS = dS_i. \qquad (8.2)$$

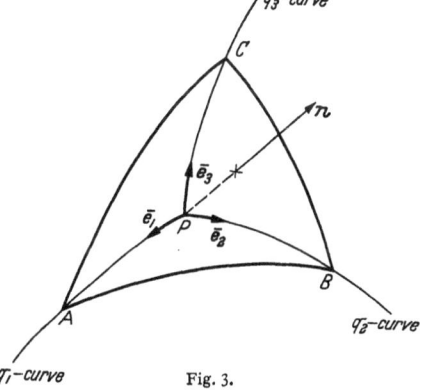

Fig. 3.

By considering the equilibrium of the tetrahedron $PABC$ we find that

$$\boldsymbol{t}\, dS = \sum_{i=1}^{3} \boldsymbol{t}_i\, dS_i. \qquad (8.3)$$

Substituting in Eq. (8.3) the value of dS_i given by Eq. (8.2) and dividing both sides of the resulting equation by dS we find that

$$\boldsymbol{t} = \sum_{i=1}^{3} n_i\, \boldsymbol{t}_i\, \sqrt{\bar{g}^{ii}}. \qquad (8.4)$$

Now $\boldsymbol{t}$ is an invariant under general transformations of the coordinate system and the quantities n_1, n_2, n_3 form a covariant vector, so that it follows from Eq. (8.4) that we may write

$$\boldsymbol{t}_i\, \sqrt{\bar{g}^{ii}} = \tau^{ij}\, \bar{e}_j \qquad (8.5)$$

where τ^{ij} is a contravariant tensor of the second order, called the *contravariant stress tensor*. Mixed and covariant stress tensors may be defined by the equations[1]

$$\tau^i_j = \bar{g}_{rj}\tau^{ir}, \qquad \tau_{ij} = \bar{g}_{ir}\tau^r_j. \tag{8.6}$$

If we introduce t^i, t_i the contravariant and covariant components respectively of the stress vector $\boldsymbol{t}$, so that

$$\boldsymbol{t} = t^i \bar{\boldsymbol{e}}_i = t_i \bar{\boldsymbol{e}}^i \tag{8.7}$$

then it follows from Eqs. (8.4) to (8.7) that

$$t^i = \tau^{ij} n_j, \qquad t_i = \tau^j_i n_j. \tag{8.8}$$

We may introduce *physical components of the stress tensor* in precisely the same way as we did the physical components of the displacement (cf. Sect. 4 above). We define the physical components of the stress tensor by the equations

$$\sigma_{ij} = \left(\frac{\bar{g}_{ij}}{\bar{g}^{ii}}\right)^{\frac{1}{2}} \tau^{ij}. \tag{8.9}$$

Substituting from Eq. (8.9) into Eq. (8.5) we find that

$$t_i = \sum_{j=1}^{3} \sigma_{ij} \frac{\bar{\boldsymbol{e}}_j}{\sqrt{\bar{g}_{jj}}}. \tag{8.10}$$

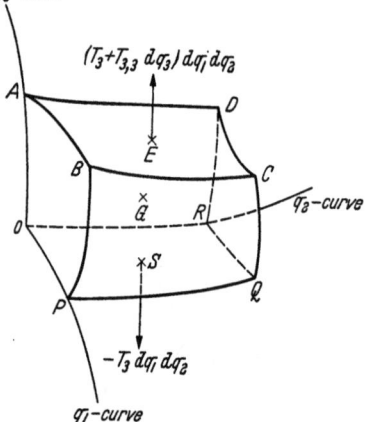

Fig. 4.

9. The equations of motion and of equilibrium. In addition to the stresses discussed above, there may be external forces acting upon the body in its strained state $\bar{S}$. These forces are of two kinds: (i) surface tractions; (ii) body forces. The surface tractions may be denoted by a vector $\boldsymbol{\Sigma}$ such that $\boldsymbol{\Sigma}\,dS$ is the surface force acting on an element of surface of area dS. The body forces, of which the force due to gravity is a typical example, are proportional to the masses of the particles upon which they act; body forces may be denoted by a vector $\boldsymbol{P}$ with the interpretation that if ϱ is the density of the solid, the body force may be represented by the vector field $\boldsymbol{P}_\varrho\,d\tau$, $d\tau$ denoting element of volume.

Let us consider now the equation of motion of a small parallelepiped $OPQRABCD$ of the strained body $\bar{S}$, bounded by the faces $q_i = $ const, $q_i + dq^i = $ const (cf. Fig. 4). The force on the face $OPQR$ is

$$-t_3 \sqrt{\bar{g}\bar{g}^{33}}\,dq^1\,dq^2$$

since the area of $OPQR$ is $\sqrt{\bar{g}\bar{g}^{33}}\,dq^1\,dq^2$, and we may write this force as $-\boldsymbol{T}_3\,dq^1\,dq^2$ where

$$\boldsymbol{T}_3 = \sqrt{\bar{g}\bar{g}^{33}}\,\boldsymbol{t}_3. \tag{9.1}$$

The force on the face $ABCD$ will be given by

$$(\boldsymbol{T}_3 + \partial_3\boldsymbol{T}_3\,dq^3)\,dq^1\,dq^2 \tag{9.2}$$

where $\partial_3\boldsymbol{T}_3$ denotes $\partial\boldsymbol{T}_3/\partial q^3$. It follows from (9.1) and (9.2) that the net force due to the stresses on the opposite faces $OPQR$, $ABCD$ is

$$\partial_3\boldsymbol{T}_3\,dq^1\,dq^2\,dq^3. \tag{9.3}$$

---

[1] We shall see later (Sect. 9 below) that τ^{ij} is a symmetric tensor.

Similar results will hold for the other faces. Summing these forces we find that the net force on the parallelepiped due to the stresses is

$$\partial_i T_i \, dq^1 \, dq^2 \, dq^3. \tag{9.4}$$

Now the volume of the parallelepiped is $\sqrt{\bar{g}} \, dq^1 \, dq^2 \, dq^3$ so that the body force acting upon it will be

$$\bar{\varrho} \, \boldsymbol{P} \sqrt{\bar{g}} \, dq^1 \, dq^2 \, dq^3, \tag{9.5}$$

where $\bar{\varrho}$ is the density in the strained state and the rate of change of momentum of the parallelepiped will be

$$\bar{\varrho} \, \boldsymbol{f} \sqrt{\bar{g}} \, dq^1 \, dq^2 \, dq^3 \tag{9.6}$$

where $\boldsymbol{f}$ is its acceleration. Making use of Eqs. (9.4) to (9.6) in Newton's second law of motion, and dividing both sides of the resulting relation by $dq^1 \, dq^2 \, dq^3$ we find that

$$\partial_i \boldsymbol{T}_i + \bar{\varrho} \sqrt{\bar{g}} \, \boldsymbol{P} = \bar{\varrho} \sqrt{\bar{g}} \, \boldsymbol{f} \tag{9.7}$$

is the equation of motion of the solid. If the body is in equilibrium $\boldsymbol{f} = 0$ and Eq. (9.7) reduces to

$$\partial_i \boldsymbol{T}_i + \bar{\varrho} \sqrt{\bar{g}} \, \boldsymbol{P} = 0. \tag{9.8}$$

The conditions at the boundary of the solid are that the surface stresses are equal to the prescribed surface tractions, i.e. we must have

$$\boldsymbol{t} = \boldsymbol{\Sigma} \tag{9.9}$$

on the boundary.

Now $\boldsymbol{T}_i = \sqrt{\bar{g}} \, \tau^{ij} \bar{\boldsymbol{e}}_j$ so that

$$\partial_i \boldsymbol{T}_i = \tau^{ij} \bar{\boldsymbol{e}}_j \, \partial_i \sqrt{\bar{g}} + \sqrt{\bar{g}} \, \bar{\boldsymbol{e}}_j \, \partial_i \tau^{ij} + \sqrt{\bar{g}} \, \tau^{ij} \, \partial_i \bar{\boldsymbol{e}}_j.$$

From the definition of the Christoffel symbol of the second kind[1] it is easily shown that

$$\partial_i \sqrt{\bar{g}} = \overline{\left\{{r \atop r\,i}\right\}} \sqrt{\bar{g}},$$

and it is also a simple matter to show that

$$\partial_i \bar{\boldsymbol{e}}_j = \overline{\left\{{r \atop j\,i}\right\}} \boldsymbol{e}_r$$

so that

$$\tau^{ij} \partial_i \bar{\boldsymbol{e}}_j = \overline{\left\{{r \atop j\,i}\right\}} \tau^{ij} \boldsymbol{e}_r \equiv \overline{\left\{{j \atop r\,i}\right\}} \tau^{ir} \boldsymbol{e}_j.$$

We therefore find that

$$\partial_i \boldsymbol{T}_i = \sqrt{\bar{g}} \, \bar{\boldsymbol{e}}_j \left(\partial_i \tau^{ij} + \overline{\left\{{r \atop r\,i}\right\}} \tau^{ij} + \overline{\left\{{j \atop r\,i}\right\}} \tau^{ir} \right) = \sqrt{\bar{g}} \, \bar{\boldsymbol{e}}_j \bar{\nabla}_i \tau^{ij} \tag{9.10}$$

where $\bar{\nabla}_i \tau^{ij}$ denotes the covariant derivative of τ^{ij} with respect to the strained state S.

If we write

$$\boldsymbol{P} = P^j \boldsymbol{e}_j, \quad \boldsymbol{f} = f^j \boldsymbol{e}_j \tag{9.11}$$

---

[1] H. Tietz: Geometrie. This Encyclopedia, Vol. II, p. 179. 1955.

and substitute from Eqs. (9.11) and (9.10) into Eqs. (9.7) and (9.8) we find that the equation of motion can be written

$$\bar{V}_i \tau^{ij} + \bar{\varrho} P^j = \bar{\varrho} f^j, \qquad (9.12)$$

and the equation of equilibrium can be written

$$\bar{V}_i \tau^{ij} + \bar{\varrho} P^j = 0. \qquad (9.13)$$

Similarly, if we write

$$\Sigma = \Sigma^j \bar{e}_j \qquad (9.14)$$

for the surface tractions, we find that the boundary conditions (9.9) can be written in the form

$$\tau^{ij} n_i = \Sigma^j \qquad (9.15)$$

where n_i are the covariant components of $\mathbf{n}$, the unit vector which is normal to the surface.

If the strains are infinitesimal we may refer to axes in the *unstrained* state S when we are discussing the state of stress. We may write

$$\mathbf{t} = \tau^{ij} n_i \mathbf{e}_j \qquad (9.16)$$

and write

$$\sigma_{ij} = \tau^{ij} \sqrt{\frac{g_{jj}}{g^{ii}}} \qquad (9.17)$$

for the physical components of the stress. Since $\bar{g} = g$, $\bar{\varrho} = \varrho$ and the equations of motion become

$$V_i \tau^{ij} + \varrho P^j = \varrho f^j \qquad (9.18)$$

while the equations of equilibrium become

$$V_i \tau^{ij} + \varrho P^j = 0. \qquad (9.19)$$

We obtained the equations of motion of a continuum by considering the translational motion of an element of a body. We shall now consider the equilibrium of such an element in order to show that the stress tensor τ^{ij} is a symmetric tensor. If we denote the centre of the parallelepiped by G and the centres of the faces $ABCD$, $OPQR$ by E and S respectively then it is easily show $\overrightarrow{GE} = \frac{1}{2} dq^3 \bar{e}_3$, $\overrightarrow{GS} = -\frac{1}{2} dq^3 \bar{e}_3$ so that, to the first order, the moment of the forces on the faces $ABCD$, $OPQR$ is $\bar{e}_3 \times \mathbf{T}_3 \, dq^1 \, dq^2 \, dq^3$. There are similar terms for the other faces; adding these and making use of the fact that the net moment of the forces must be zero we find (on dividing by $dq^1 \, dq^2 \, dq^3$) that

$$\bar{e}_i \times \mathbf{T}_i = 0.$$

Substituting from the equation

$$\mathbf{T}_i = \sqrt{\bar{g}} \, \tau^{ij} \bar{e}_j$$

for $\mathbf{T}_i$ we find that

$$\tau^{ij} (\bar{e}_i \times \bar{e}_j) = 0,$$

which when written out in full becomes

$$(\tau^{23} - \tau^{32}) \bar{e}_1 + (\tau^{31} - \tau^{13}) \bar{e}_2 + (\tau^{12} - \tau^{21}) \bar{e}_3 = 0$$

showing that

$$\tau^{ij} = \tau^{ji}, \qquad (9.20)$$

so that the stress tensor is symmetric.

Now it is a well known theorem in tensor calculus that, if a tensor τ^{ji} is symmetric, it is always possible to find a coordinate system (in $\bar{S}$) in which the non-diagonal components of the tensor (i.e. τ^{ij}, $i \neq j$) all vanish. In such a system the directions of the coordinate curves are called the *principal axes of stress* and the diagonal components of the tensor are called the *principal components of stress*.

10. The elastic potential. Suppose that when a body is strained the point P of the unstrained state S is transformed into the point $\bar{P}$ of the strained state in such a way that $\overrightarrow{P\bar{P}} = \boldsymbol{v}(q_1, q_2, q_3, t)$ is the displacement vector. Suppose further that S' is any other geometrically possible state which is reached from S in the same time and in which the point P moves not to the point $\bar{P}$ but to a neighbouring point P' where $\overrightarrow{P P'} = \boldsymbol{v}'(q_1, q_2, q_3, t)$. The difference

$$\delta \boldsymbol{v} = \boldsymbol{v}'(q_1, q_2, q_3, t) - \boldsymbol{v}(q_1, q_2, q_3, t) \tag{10.1}$$

is called a *variation* of $\boldsymbol{v}$. The variation of $\partial_i \boldsymbol{v}$ is

$$\delta(\partial_i \boldsymbol{v}) = \partial_i(\delta \boldsymbol{v}). \tag{10.2}$$

Since

$$\bar{\boldsymbol{e}}_i = \boldsymbol{e}_i + \partial_i \boldsymbol{v}$$

and $\boldsymbol{e}_i$ does not depend on $\boldsymbol{v}$ it follows that

$$\delta \bar{\boldsymbol{e}}_i = \partial_i(\delta \boldsymbol{v}). \tag{10.3}$$

It follows immediately from this result that

$$\delta(\bar{g}_{ij}) = \bar{\boldsymbol{e}}_i \cdot \partial_j(\delta \boldsymbol{v}) + \bar{\boldsymbol{e}}_j \cdot \partial_i(\delta \boldsymbol{v}) \tag{10.4}$$

and, since $\delta(g_{ij}) = 0$, it is obvious that

$$2\delta(\gamma_{ij}) = \bar{\boldsymbol{e}}_i \cdot \partial_j(\delta \boldsymbol{v}) + \bar{\boldsymbol{e}}_j \cdot \partial_i(\delta \boldsymbol{v}). \tag{10.5}$$

If we multiply both sides of Eq. (9.7) by $dq^1 dq^2 dq^3$ and integrate throughout the whole of $\bar{S}$ we have

$$\int_{\bar{S}} \left\{ \frac{1}{\sqrt{\bar{g}}} \partial_i \boldsymbol{T}_i \cdot \delta \boldsymbol{v} + \bar{\varrho}(\boldsymbol{P} - \boldsymbol{f}) \cdot \delta \boldsymbol{v} \right\} d\tau = 0 \tag{10.6}$$

where $d\tau = \sqrt{\bar{g}} \, dq^1 dq^2 dq^3$.

Now the first term on the left hand side of Eq. (10.6) may be written in the form

$$\int_{\bar{S}} \frac{d\tau}{\sqrt{\bar{g}}} \partial_i(\boldsymbol{T}_i \cdot \delta \boldsymbol{v}) - \int_{\bar{S}} \frac{d\tau}{\sqrt{\bar{g}}} \boldsymbol{T}_i \cdot \partial_i(\delta \boldsymbol{v}).$$

Using GREEN's theorem we find that the first term of this expression can be written in the form

$$\int_{\bar{B}} (\boldsymbol{T}_i \cdot \delta \boldsymbol{v}) \frac{n_i \, dS}{\sqrt{\bar{g}}}$$

where $\bar{B}$ is the boundary of the region $\bar{S}$. Furthermore

$$\frac{n_i \boldsymbol{T}_i}{\sqrt{\bar{g}}} = \boldsymbol{t} = \boldsymbol{\Sigma}, \quad \text{on } \bar{B}$$

so that Eq. (10.6) may be written in the form

$$\Delta A - \int_{\bar{S}} \bar{\varrho} \boldsymbol{f} \cdot \delta \boldsymbol{v} \, d\tau = \Delta U \tag{10.7}$$

where

$$\Delta A = \int_{\bar{B}} (\boldsymbol{\Sigma} \cdot \delta \boldsymbol{v}) \, dS + \int_{\bar{S}} \bar{\varrho} (\boldsymbol{P} \cdot \delta \boldsymbol{v}) \, d\tau \tag{10.8}$$

and

$$\Delta U = \int_{\bar{S}} \boldsymbol{T}_i \cdot \partial_i (\delta \boldsymbol{v}) \, \frac{d\tau}{\sqrt{\bar{g}}}. \tag{10.9}$$

Now since τ^{ij} is a symmetric tensor

$$\begin{aligned}\boldsymbol{T}_i \cdot \partial_i (\delta \boldsymbol{v}) &= \sqrt{\bar{g}} \cdot \tau^{ij} (\bar{\boldsymbol{e}}_j \cdot \partial_i (\delta \boldsymbol{v})) \\ &= \tfrac{1}{2} \tau^{ij} \sqrt{\bar{g}} \{\bar{\boldsymbol{e}}_j \cdot \partial_i (\delta \boldsymbol{v}) + \bar{\boldsymbol{e}}_i \cdot \partial_j (\delta \boldsymbol{v})\} \\ &= \tau^{ij} (\bar{g}) \, \delta(\gamma_{ij})\end{aligned}$$

so that we have

$$\Delta U = \int_{\bar{S}} \Delta E \, \bar{\varrho} \, d\tau \tag{10.10}$$

where

$$\Delta E = \tau^{ij} \, \delta(\gamma_{ij})/\bar{\varrho}. \tag{10.11}$$

We now *define* an *elastic body* $\bar{S}$ to be such that, for all virtual displacements $\delta \boldsymbol{v}$, ΔE is a complete differential of a continuous single-valued function which depends only on the state of strain in $\bar{S}$ at time t and tensors which represent the state S.

In such instances we may write

$$E = E(\gamma_{ij}). \tag{10.12}$$

We have defined an elastic body as one characterized by a function E with the above property. It may be questioned as to whether such a function does in fact exist. This question can be answered easily (in the affirmative) in two simple physical cases—when the change of state from the unstrained state S to the strained state $\bar{S}$ is reversible and is either adiabatic or isothermal[1]. Comparing Eqs. (10.11) and (10.12) and making use of the symmetry of the tensor τ^{ij} we find that

$$\Delta E = \delta E = \frac{\partial E}{\partial \gamma_{ij}} \delta \gamma_{ij} \tag{10.13}$$

and that

$$\tau^{ij} = \frac{1}{2} \bar{\varrho} \left(\frac{\partial E}{\partial \gamma_{ij}} + \frac{\partial E}{\partial \gamma_{ji}} \right). \tag{10.14}$$

Since mass elements are conserved in a virtual displacement, $\delta(\bar{\varrho} \, d\tau) = 0$,

$$\delta U = \int_{\bar{S}} \bar{\varrho} \, \delta E \, d\tau,$$

so that, apart from an additive constant,

$$U = \int_{\bar{S}} \bar{\varrho} E \, d\tau. \tag{10.15}$$

The function U is called the *total elastic potential energy* of the strained state $\bar{S}$, and E is called the *elastic potential of the body per unit mass*.

---

[1] For a simple proof of this see I. S. SOKOLNIKOFF: Mathematical Theory of Elasticity, pp. 83—87. New York 1946.

11. Variational principles. We return now to Eq. (10.7) and assume that the variation $\delta \boldsymbol{v}$ is such that it vanishes at times t_0 and t_1 so that

$$\int_{t_0}^{t_1}\!\!\int_{\overline{S}} \varrho \boldsymbol{f} \cdot \delta \boldsymbol{v}\, d\tau\, dt = -\int_{t_0}^{t_1}\!\!\int_{\overline{S}} \varrho\, \frac{\partial \boldsymbol{v}}{\partial t}\, \frac{\partial (\delta \boldsymbol{v})}{\partial t}\, d\tau\, dt = -\delta \int_{t_0}^{t_1} K\, dt$$

where K is the kinetic energy of the strained body $\overline{S}$. Eq. (10.7) is then seen to be equivalent to

$$\int_{t_0}^{t_1} \Delta A\, dt = \int_{t_0}^{t_1} \delta(U - K)\, dt.$$

Now ΔA denotes the virtual work done by the body forces and the surface tractions acting on $\overline{S}$ during a virtual displacement $\delta \boldsymbol{v}$. If these external forces are derivable from a potential energy function V then $\Delta A = -\delta V$ and we see that Eq. (10.7) is equivalent to

$$\delta \int_{t_0}^{t_1} L\, dt = 0 \tag{11.1}$$

where L denotes the Lagrangian function

$$L = K - U - V. \tag{11.2}$$

In the equilibrium case $\boldsymbol{f} = 0$ so that Eq. (10.7) leads to the equation

$$\delta(U + V) = 0 \tag{11.3}$$

which states that the total potential energy has a stationary value in the equilibrium state.

III. The relation between stress and strain.

12. Hooke's law. Till now we have been considering independently the state of strain and the state of stress in a solid body. We shall now consider the relation between stress and strain in a certain class of bodies which we shall call *elastic*. To derive such a relation we should have discuss the structure of the solid and then, by applying the techniques of statistical mechanics, derive the mechanical properties of the bulk solid from the nature of the atoms (or other constituent units such as molecular chains comprising it). Attempts to carry out such a programme[1] have been made only within the last hundred years; before then the theory was based on empirical relations such as Hooke's law which stated that if a thin rod or wire of length l_0 in the unstrained state is stretched the force necessary to stretch it to a length l is directly proportional to the elongation $l - l_0$. Before proceeding to the discussion of the general theory of elasticity we shall show how thermodynamical theory applied to a very simple system establishes a stress-strain relation of the form of Hooke's law.

We shall consider only *reversible* phenomena, that is we shall confine our attention to process in which the state of the system is defined unambiguously by the thermodynamic variables. From the first and second laws of thermodynamics we know that when any change takes place in a *unit mass* of the solid, the increase de in the internal energy is related to the work performed on the solid through the equation

$$de = T\, ds + \Delta w, \tag{12.1}$$

---

[1] Cf. Leibfried's report on elastic properties of crystals in Vol. VII, part 1, and the articles of Vol. X, of this Encyclopedia.

where T denotes the absolute temperature and ds is the change in s, the entropy per unit mass[1]. In discussing the equilibrium of a simple system of this kind it is often convenient to make use of the *Helmholtz free energy*, f defined by the equation

$$f = e - Ts. \tag{12.2}$$

For isothermal changes it is obvious that $\Delta f = \Delta w$, i.e. the change in the Helmholtz free energy is equal to the work done on the system by the external forces.

If we consider a specimen of length l acted upon by a tensile force, σ, and a hydrostatic pressure p, the total work done is

$$\Delta w = \sigma\, dl - p\, dv \tag{12.3}$$

where dv denotes the change in the specific volume v. Combining Eqs. (12.1) to (12.3) we find that

$$df = \sigma\, dl - p\, dv - s\, dT. \tag{12.4}$$

If we consider the deformation of rubber at atmospheric pressure we find that, in that case, $p\,dv$ is less than $\sigma\,dl$ by a factor of the order of 10^{-3}. In that case we may therefore write[2]

$$df = \sigma\, dl - s\, dT \tag{12.5}$$

from which it follows that

$$\sigma = \left(\frac{\partial f}{\partial l}\right)_T \tag{12.6}$$

and that

$$s = -\left(\frac{\partial f}{\partial T}\right)_l. \tag{12.7}$$

From Eqs. (12.6) and (12.7) it follows that

$$\left(\frac{\partial s}{\partial l}\right)_T = -\left(\frac{\partial \sigma}{\partial T}\right)_l. \tag{12.8}$$

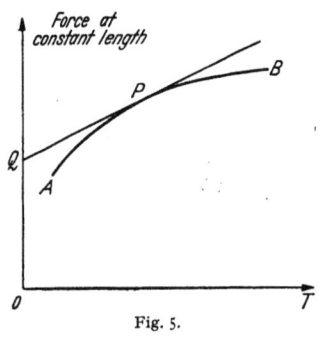

Fig. 5.

Eliminating f and s from Eqs. (12.2) and (12.6) we obtain the further relation

$$\left(\frac{\partial e}{\partial l}\right)_T = \sigma - T\left(\frac{\partial \sigma}{\partial T}\right)_l. \tag{12.9}$$

We shall consider the physical significance of these relations. Fig. 5 shows the variation with the absolute temperature of the force necessary to maintain a specimen at constant length. If P is the point on the curve corresponding to absolute temperature T and if PQ is the tangent to the curve at P, then

$$\text{slope of } PQ = \left(\frac{\partial \sigma}{\partial T}\right)_l = -\left(\frac{\partial \sigma}{\partial l}\right)_T, \tag{12.10}$$

while, by virtue of Eq. (12.9),

$$OQ = \sigma - T\left(\frac{\partial \sigma}{\partial T}\right)_l = \left(\frac{\partial e}{\partial l}\right)_T. \tag{12.11}$$

Eqs. (12.10) and (12.11) enable us to use experimental curves of the type shown in Fig. 5 to calculate the changes in the internal energy and entropy of the specimen during deformation. In some cases these curves are linear[3] showing

---
[1] The notation used here is that recommended by a joint committee of the Chemical Society, The Faraday Society and the Physical Society of London. Cf. R. A. SMITH: The Physical Principles of Thermodynamics. London 1952.

[2] For a more refined thermodynamic analysis in which volume changes are taken into account the reader is referred to: G. GEE, Trans. Faraday Soc. **42**, 585 (1946).

[3] MEYER and FERRI: Helv. chim. Acta **18**, 570 (1935). — ANTHOONE, GASTON and GUTH: J. Phys. Chem. **46**, 826 (1942).

that for the substances concerned $(\partial s/\partial l)_T$ and $(\partial e/\partial l)_T$ are both independent of the absolute temperature T. If, in addition to being linear, the curve passes through the origin of coordinates so that $OQ=0$, then $(\partial e/\partial l)_T=0$ and it follows from Eq. (12.6) that $\sigma=-T(\partial s/\partial l)_T$, showing that, in this case, the elastic force arises solely from the change in the entropy of the solid.

We return to the interpretation of Eq. (12.6). For isothermal changes involving small changes $l-l_0$ from the equilibrium value l_0 we may represent the Helmholtz free energy by the first three terms of its Taylor series

$$f = f_0 + \left(\frac{\partial f}{\partial l_0}\right)_T (l - l_0) + \tfrac{1}{2} \left(\frac{\partial^2 f}{\partial l_0^2}\right)_T (l - l_0)^2 \qquad (12.12)$$

where f_0, $(\partial f/\partial l_0)_T$, $(\partial^2 f/\partial l_0^2)_T$ denote the values of f, $(\partial f/\partial l)_T$ $(\partial^2 f/\partial l^2)_T$ for the value $l=l_0$. Now, in the equilibrium position l_0, the Helmholtz free energy is a minimum so that

$$\left(\frac{\partial f}{\partial l_0}\right)_T = 0, \quad \left(\frac{\partial^2 f}{\partial l_0^2}\right)_T > 0$$

and (12.12) reduces to

$$f = f_0 + \frac{1}{2}\left(\frac{\partial^2 f}{\partial l_0^2}\right)(l - l_0)^2.$$

Inserting this form into Eq. (12.6) we find that

$$\sigma = \left(\frac{\partial^2 f}{\partial l_0^2}\right)_T (l - l_0), \qquad (12.13)$$

showing that for *small* elongations $l-l_0$ the tensile force is directly proportional to the elongation in agreement with Hooke's law. If the elongations are large it is not permissible to represent the Taylor series for f by its first three terms, as we have done in Eq. (12.12). Retaining higher order terms we see that the dependence of σ on $l-l_0$ is no longer linear.

13. Homogeneous isotropic bodies. We shall return now to the discussion of the general theory. We saw in Eq. (10.14) that the stresses which occur in an elastic body can be calculated from a knowledge of E, the elastic potential of the body unit mass. Since elements of mass are conserved during deformation we have $\varrho\sqrt{g}=\bar\varrho\sqrt{\bar g}$. Hence, if we introduce $W=\varrho E$ the elastic potential per unit volume of the unstrained body, Eq. (10.14) may be written in the form

$$\sqrt{\bar g}\cdot\tau^{ij} = \frac{1}{2}\sqrt{g}\left(\frac{\partial W}{\partial \gamma_{ij}} + \frac{\partial W}{\partial \gamma_{ji}}\right). \qquad (13.1)$$

If, therefore, we wish to obtain an explicit relation between stress and strain, we must calculate the function W and this in turn implies that we must make some physical assumptions about the nature of the bodies being deformed. We shall restrict our attention to bodies which are of constant density in the unstrained state and whose elastic potential depends only on the three strain invariants I_1, I_2, I_3 defined by the Eqs. (5.1) and on scalar functions of the coordinates. A solid which has this last property is said to be *isotropic*; if, further, these scalar functions are constants, we say that the solid is *homogeneous*. Hence for a *homogeneous isotropic* solid W is a function of I_1, I_2, I_3 alone. It is with homogeneous isotropic solids which we shall deal here.

If we substitute a function $W(I_1, I_2, I_3)$ into Eq. (13.1) we find that

$$\tau^{ij} = \frac{1}{2}(g/\bar g)^{\frac{1}{2}}\left\{\frac{\partial W}{\partial I_1}\left(\frac{\partial I_1}{\partial \gamma_{ij}} + \frac{\partial I_1}{\partial \gamma_{ji}}\right) + \frac{\partial W}{\partial I_2}\left(\frac{\partial I_2}{\partial \gamma_{ij}} + \frac{\partial I_2}{\partial \gamma_{ji}}\right) + \frac{\partial W}{\partial I_3}\left(\frac{\partial I_3}{\partial \gamma_{ij}} + \frac{\partial I_3}{\partial \gamma_{ji}}\right)\right\}. \quad (13.2)$$

Now it can be shown from the Eqs. (5.1) that

$$\frac{\partial I_1}{\partial \gamma_{ij}} + \frac{\partial I_1}{\partial \gamma_{ji}} = 4g^{ij},$$

$$\frac{\partial I_2}{\partial \gamma_{ij}} + \frac{\partial I_2}{\partial \gamma_{ji}} = 4(g^{ij}g^{rs} - g^{ir}g^{js})\bar{g}_{rs},$$

$$\frac{\partial I_3}{\partial \gamma_{ij}} + \frac{\partial I_3}{\partial \gamma_{ji}} = \frac{4\bar{g}\,\bar{g}^{ij}}{g} = 4I_3\bar{g}^{ij}.$$

Substituting these expressions into Eq. (13.2) we obtain the stress-strain equations

$$\tau^{ij} = \Phi g^{ij} + \Psi B^{ij} + p\bar{g}^{ij} \tag{13.3}$$

where

$$B^{ij} = I_1 g^{ij} - g^{ir}g^{js}\bar{g}_{rs} \tag{13.4}$$

and

$$\Phi = \frac{2}{\sqrt{I_3}} \frac{\partial W}{\partial I_1}, \quad \Psi = \frac{2}{\sqrt{I_3}} \frac{\partial W}{\partial I_2}, \quad p = 2\sqrt{I_3}\frac{\partial W}{\partial I_3}. \tag{13.5}$$

In the special case in which the unstrained body is not only homogeneous and isotropic but incompressible, $I_3 = 1$ and W is now a function of I_1 and I_2 only. The stress-strain Eqs. (13.3) are of the same form but, in the case of incompressible bodies,

$$\Phi = 2\frac{\partial W}{\partial I_1}, \quad \Psi = 2\frac{\partial W}{\partial I_2}. \tag{13.6}$$

The function p remaining in the stress-strain equations for an incompressible body cannot be calculated from the third of the Eqs. (13.5); it is an unknown invariant representing a uniform hydrostatic pressure which can be determined from the equations of equilibrium and the appropriate boundary conditions.

The problem remains of determining the actual form of the function W for an elastic solid. The calculation of this function in terms of the structure of the solid lies outwith the scope of this article. For example, for such calculations for perfect crystal lattices the reader is referred to Chap. III of "Dynamical Theory of Crystal Lattices" by M. BORN and K. HUANG, (Oxford, 1954) and for those for rubber-like bodies the reader is referred to Chap. III to VII of "The Physics of Rubber Elasticity" by L. R. G. TRELOAR, (Oxford, 1949). We shall merely mention here some aspects of the work on rubber-like solids.

KUHN[1] has calculated the form of the function W for vulcanized rubber by considering the solid as an assembly of long-chain molecules, linked at a relatively small number of points so as to form an irregular three-dimensional network. The basic assumptions of the KUHN-TRELOAR theory are:

(a) The chains forming the network all have the same total length.

(b) The distribution of chain displacement lengths (i.e. rectilinear distance between the ends) is Gaussian.

(c) The solid is incompressible.

(d) The deformation changes the components of the displacement length of each chain in the same ratio as it changes the corresponding dimensions of the bulk rubber.

TRELOAR applied these assumptions in the calculation (by KUHN's method) of the deformation of a cube of bulk rubber, initially of unit edge, to a rectangular

---
[1] W. KUHN: Kolloid-Z. **76**, 258 (1936). See also L. R. G. Treloar: Trans. Faraday Soc. **39**, 36, 241 (1943).

parallelepiped with edges λ_1, λ_2 and λ_3. If N is the number of molecular chains per unit volume, it is found that

$$W = \tfrac{1}{2} G (\lambda_1^2 + \lambda_2^2 + \lambda_3^2 - 3) \tag{13.7}$$

where

$$G = \tfrac{1}{2} N k T, \tag{13.8}$$

k being BOLTZMANN's constant. For this simple deformation the strain invariants are (if we assume incompressibility)

$$I_1 = \lambda_1^2 + \lambda_2^2 + \lambda_3^2, \quad I_2 = \frac{1}{\lambda_1^2} + \frac{1}{\lambda_2^2} + \frac{1}{\lambda_3^2}, \quad I_3 = \lambda_1^2 \lambda_2^2 \lambda_3^2 = 1 \tag{13.9}$$

so that Eq. (13.7) can be written in the form

$$W = \tfrac{1}{2} G (I_1 - 3). \tag{13.10}$$

The experimental evidence established without doubt that the form (13.10) does not adequately express the behaviour of bulk rubbers. The discrepancy between theory and experiment has to a certain extent been bridged by MOONEY[1] who, on the basis of very simple assumptions which are independent of any structural model of the rubber, has shown that if the stress-strain relation for one kind of strain (e.g. simple shear) is given, that for another kind of strain can be derived. Considering the case in which the rubber is incompressible and the stress-strain relation in simpler shear is assumed to be linear, MOONEY showed that for the deformation of the unit cube the elastic potential W has the form

$$W = C_1 (\lambda_1^2 + \lambda_2^2 + \lambda_3^2 - 3) + C_2 \left(\frac{1}{\lambda_1^2} + \frac{1}{\lambda_2^2} + \frac{1}{\lambda_3^2} - 3 \right) \tag{13.11}$$

where C_1 and C_2 are fundamental constants for the given rubberlike material. In terms of the strain invariants MOONEY's equation may be written in the form

$$W = C_1 (I_1 - 3) + C_2 (I_2 - 3) \tag{13.12}$$

so that for such a solid the stress-strain equations are of the form (13.3) with

$$\Psi = 2C_1, \quad \Psi = 2C_2 \tag{13.13}$$

i.e. with Φ and Ψ constants.

MOONEY has developed a more general type of relation based on a non-linear shear relation. For the deformation of the cube he finds a relation of the type

$$W = \sum_{r=1}^{n} \{ A_{2r} (\lambda_1^{2r} + \lambda_2^{2r} + \lambda_3^{2r} - 3) + B_{2r} (\lambda_1^{-2r} + \lambda_2^{-2r} + \lambda_3^{-2r} - 3) \}$$

involving $2n$ independent parameters $A_2, A_4, \ldots, A_{2n}, B_2, B_4, \ldots, B_{2n}$. In terms of the strain invariants this becomes

$$W = A_2 (I_1 - 3) + B_2 (I_2 - 3) + A_4 (I_1^2 - 2I_2 - 3) + B_4 (I_2^2 - 2I_1 - 3) +$$
$$+ A_6 (I_1^3 - 3I_1 I_2) + B_6 (I_2^3 - 3I_1 I_2) + \cdots,$$

so that for a solid of this kind the Φ and Ψ occurring in the stress-strain Eqs. (13.3) are given by

$$\Phi = 2(A_2 - 2B_4) + 4A_4 I_1 - 6(A_6 + B_6) I_2 + 6A_6 I_1^2 + \cdots$$
$$\Psi = 2(B_2 - 2A_4) - 6(A_6 + B_6) I_1 + 4B_4 I_2 + 6B_6 I_2^2 + \cdots$$

---

[1] M. MOONEY: J. Appl. Phys. **11**, 582 (1940).

14. The classical theory.

When the strains are infinitesimal we have $\bar{g} = g$ so that Eq. (13.1) reduces to

$$\tau^{ij} = \frac{1}{2}\left(\frac{\partial W}{\partial \gamma_{ij}} + \frac{\partial W}{\partial \gamma_{ji}}\right). \tag{14.1}$$

In this section we shall derive the general stress-strain relation in the classical theory (i.e. the theory corresponding to infinitesimal strains), and then discuss the form it takes under certain special circumstances. Since the elastic potential per unit volume is an invariant, the form

$$W = W_0 + C^{ij}\gamma_{ij} + \tfrac{1}{2} C^{ijrs}\gamma_{ij}\gamma_{rs}, \tag{14.2}$$

with W_0 a constant and the C's satisfying the symmetry relations

$$C^{ij} = C^{ji}, \quad C^{ijrs} = C^{jirs} = C^{ijsr} = C^{jisr} \tag{14.3}$$

is consistent with the approximations we have already admitted in the theory of infinitesimal strains. In the Eq. (14.2) C^{ij} is a tensor of the second order and C^{ijrs} is a tensor of the fourth order. Inserting the expression (14.2) for W into Eq. (14.1) we find that

$$\tau^{ij} = C^{ij} + C^{ijrs}\gamma_{rs}.$$

If the body is unstrained and unstressed in its initial state then when $\gamma_{ij} = 0$, for all i and j, we must have $W = 0$, $\tau^{ij} = 0$. It follows immediately that $W_0 = 0$, $C^{ij} = 0$ for all i and j and hence that

$$W = \tfrac{1}{2} C^{ijrs}\gamma_{ij}\gamma_{rs} \tag{14.4}$$

and

$$\tau^{ij} = C^{ijrs}\gamma_{rs}. \tag{14.5}$$

From Eqs. (14.4) and (14.5) we have

$$W = \tfrac{1}{2} \tau^{ij}\gamma_{ij}. \tag{14.6}$$

If we introduce the mixed tensor

$$C^{ij}_{mn} = C^{ijrs} g_{rm} g_{sn} \tag{14.7}$$

then it follows from Eq. (14.6) that

$$\tau^{ij} = C^{ij}_{mn}\gamma^{mn} \tag{14.8}$$

where, because of the relations (14.3), we have

$$C^{ij}_{mn} = C^{ji}_{mn} = C^{ij}_{nm} = C^{ji}_{nm}. \tag{14.9}$$

If the density ϱ of the unstrained state S is constant and if all the components of the mixed tensor C^{ij}_{rs} are constant throughout S we say that the body is *homogeneous*.

Corresponding to the relation (14.8) we have

$$\tau_{ij} = C^{mn}_{ij}\gamma_{mn}. \tag{14.10}$$

In a system of rectangular coordinates there is no distinction between covariant, contravariant and mixed tensors, so that in this system of coordinates

$$C^{ij}_{mn} = C^{mn}_{ij}. \tag{14.11}$$

In general the constants C_{mn}^{ij} depend on the orientation of the coordinate axes. In the special case in which the form of W is independent of the direction of these axes we say that the body is *isotropic*. If W does not have this property the body is said to be *anisotropic* or *aeolotropic*.

If we take into account the symmetry properties (14.9) and (14.11) we find that there are 21 independent elastic constants C_{mn}^{ij} which may be represented by the matrix

$$\begin{pmatrix} C_{11}^{11} & C_{22}^{11} & C_{33}^{11} & C_{23}^{11} & C_{13}^{11} & C_{12}^{11} \\ C_{22}^{11} & C_{22}^{22} & C_{33}^{22} & C_{23}^{22} & C_{13}^{22} & C_{12}^{22} \\ C_{33}^{11} & C_{33}^{22} & C_{33}^{33} & C_{23}^{33} & C_{13}^{33} & C_{12}^{33} \\ C_{23}^{11} & C_{23}^{22} & C_{23}^{33} & C_{23}^{23} & C_{13}^{23} & C_{12}^{23} \\ C_{13}^{11} & C_{13}^{22} & C_{13}^{33} & C_{13}^{23} & C_{13}^{13} & C_{12}^{13} \\ C_{12}^{11} & C_{12}^{22} & C_{12}^{33} & C_{12}^{23} & C_{12}^{13} & C_{12}^{12} \end{pmatrix}. \tag{14.12}$$

α) *Symmetry with respect to the $x_1 x_2$-plane.* If we make the change of axes

$$x_1' = x_1, \quad x_2' = x_2, \quad x_3' = -x_3$$

then

$$\gamma_{11}' = \gamma_{11}, \quad \gamma_{12}' = \gamma_{12}, \quad \gamma_{22}' = \gamma_{22}, \quad \gamma_{33}' = \gamma_{33},$$

$$\gamma_{23}' = -\gamma_{23}, \quad \gamma_{13}' = -\gamma_{13}$$

so that

$$\tau_{11}' = C_{11}^{11}\gamma_{11}' + C_{11}^{22}\gamma_{22}' + C_{11}^{33}\gamma_{33}' + 2(C_{11}^{12}\gamma_{12}' + C_{11}^{13}\gamma_{13}' + C_{11}^{23}\gamma_{23}')$$
$$= C_{11}^{11}\gamma_{11} + C_{11}^{22}\gamma_{22} + C_{11}^{33}\gamma_{33} + 2(C_{11}^{12}\gamma_{12} - C_{11}^{13}\gamma_{13} - C_{11}^{23}\gamma_{23}).$$

If there is symmetry about the $x_1 x_2$-plane then

$$\tau_{11}' = \tau_{11} = C_{11}^{11}\gamma_{11} + C_{11}^{22}\gamma_{22} + C_{11}^{33}\gamma_{33} + 2(C_{11}^{12}\gamma_{12} + C_{11}^{13}\gamma_{13} + C_{11}^{23}\gamma_{23}).$$

Comparing this with the previous equation we see that

$$C_{11}^{13} = C_{11}^{23} = 0.$$

By precisely similar arguments we can show that

$$C_{13}^{22} = C_{23}^{22} = C_{13}^{33} = C_{22}^{33} = C_{12}^{23} = C_{12}^{13} = 0$$

so that, in this instance we have the 13 constants

$$\begin{pmatrix} C_{11}^{11} & C_{22}^{11} & C_{33}^{11} & 0 & 0 & C_{12}^{11} \\ C_{22}^{11} & C_{22}^{22} & C_{33}^{22} & 0 & 0 & C_{12}^{22} \\ C_{22}^{11} & C_{33}^{22} & C_{33}^{33} & 0 & 0 & C_{12}^{33} \\ 0 & 0 & 0 & C_{23}^{23} & C_{13}^{23} & 0 \\ 0 & 0 & 0 & C_{13}^{23} & C_{13}^{13} & 0 \\ C_{12}^{11} & C_{12}^{22} & C_{12}^{33} & 0 & 0 & C_{12}^{12} \end{pmatrix}. \tag{14.13}$$

β) *Orthotropy.* If a body is symmetric with respect to the $x_2 x_3$-plane as well as to the $x_1 x_2$-plane then it follows by arguments similar to those used in (a) above that of the 13 constants listed above $C_{12}^{11}, C_{12}^{22}, C_{12}^{33}$ and C_{13}^{23} also vanish showing that there are 9 non-vanishing elastic constants

$$\begin{pmatrix} C_{11}^{11} & C_{22}^{11} & C_{33}^{11} & 0 & 0 & 0 \\ C_{22}^{11} & C_{22}^{22} & C_{33}^{22} & 0 & 0 & 0 \\ C_{33}^{11} & C_{33}^{22} & C_{33}^{33} & 0 & 0 & 0 \\ 0 & 0 & 0 & C_{23}^{23} & 0 & 0 \\ 0 & 0 & 0 & 0 & C_{13}^{13} & 0 \\ 0 & 0 & 0 & 0 & 0 & C_{12}^{12} \end{pmatrix}. \qquad (14.14)$$

It is obvious from this result that such a body is also symmetric with respect to the $x_1 x_3$-plane. Such a body is called *orthotropic*.

γ) *Hexagonal system.* A system is said to possess hexagonal symmetry if it is unaltered by a transformation of the type

$$x_1' = x_1 \cos\vartheta + x_2 \sin\vartheta, \qquad x_2' = - x_1 \sin\vartheta + x_2 \cos\vartheta, \qquad x_3' = x_3$$

where ϑ is any angle.
For this change of axes

$$\gamma_{11}' = \cos^2\vartheta\, \gamma_{11} + 2\cos\vartheta \sin\vartheta\, \gamma_{12} + \sin^2\vartheta\, \gamma_{22},$$
$$\gamma_{22}' = \sin^2\vartheta\, \gamma_{11} - 2\cos\vartheta \sin\vartheta\, \gamma_{12} + \cos^2\vartheta\, \gamma_{22},$$
$$\gamma_{33}' = \gamma_{33},$$
$$\gamma_{23}' = - \sin\vartheta\, \gamma_{13} + \cos\vartheta\, \gamma_{23},$$
$$\gamma_{31}' = \cos\vartheta\, \gamma_{13} + \sin\vartheta\, \gamma_{23},$$
$$\gamma_{12}' = (\gamma_{22} - \gamma_{11}) \cos\vartheta \sin\vartheta + \gamma_{12}(\cos^2\vartheta - \sin^2\vartheta).$$

The components of the stress tensor τ_{ij} will transform in exactly the same way. For instance

$$\tau_{33}' = \tau_{33},$$

a relation which may be written in the form

$$C_{33}^{11}\gamma_{11}' + C_{33}^{22}\gamma_{22}' + C_{33}^{33}\gamma_{33}' + C_{33}^{12}\gamma_{12}' + C_{33}^{13}\gamma_{13}' + C_{33}^{23}\gamma_{23}'$$
$$= C_{33}^{11}\gamma_{11} + C_{33}^{22}\gamma_{22} + C_{33}^{33}\gamma_{33} + C_{33}^{12}\gamma_{12} + C_{33}^{13}\gamma_{13} + C_{33}^{23}\gamma_{23}.$$

Inserting the above values for the γ_{ij}' we find that

$$C_{33}^{11}(\cos^2\vartheta\, \gamma_{11} + 2\cos\vartheta \sin\vartheta\, \gamma_{12} + \sin^2\vartheta\, \gamma_{22}) +$$
$$+ C_{33}^{22}(\sin^2\vartheta\, \gamma_{11} - 2\cos\vartheta \sin\vartheta\, \gamma_{12} + \cos^2\vartheta\, \gamma_{22}) +$$
$$+ 2 C_{33}^{12}\{(\gamma_{22} - \gamma_{11})\cos\vartheta \sin\vartheta + \gamma_{12}(\cos^2\vartheta - \sin^2\vartheta)\} +$$
$$+ 2 C_{33}^{13}(\cos\vartheta\, \gamma_{13} + \sin\vartheta\, \gamma_{23}) - 2 C_{33}^{23}(\sin\vartheta\, \gamma_{13} - \cos\vartheta\, \gamma_{23})$$
$$= C_{33}^{11}\gamma_{11} + C_{33}^{22}\gamma_{22} + 2(C_{33}^{12}\gamma_{12} + C_{33}^{13}\gamma_{13} + C_{33}^{23}\gamma_{23}).$$

Equating to zero the coefficient of γ_{11} in this equation we find that, for all values of ϑ,

$$(C_{33}^{11} - C_{33}^{22})\sin^2\vartheta + 2\sin\vartheta \cos\vartheta\, C_{33}^{12} = 0$$

from which it follows that

$$C_{33}^{22} = C_{33}^{11}, \qquad C_{33}^{12} = 0. \qquad (14.15)$$

If we equate the coefficients of γ_{22}, γ_{12} to zero we obtain exactly these relations, and if we equate the coefficients of γ_{13}, γ_{23} to zero we find that

$$C_{33}^{13} = C_{33}^{23} = 0. \qquad (14.16)$$

If we carry out a similar calculation for τ_{13} and τ'_{13} we find, on equating coefficients of γ_{11} to zero, that

$$C^{11}_{13} = C^{11}_{23} = C^{22}_{13} = C^{12}_{13} = 0. \tag{14.17}$$

The equating to zero of the coefficient of γ_{22} leads to these relations again, and the coefficient of γ_{33} leads to the pair (14.16). The coefficient of γ_{12} leads to

$$C^{11}_{13} = C^{22}_{13}, \quad C^{12}_{13} = C^{12}_{23} = 0 \tag{14.18}$$

while the coefficient of γ_{13} yields the relation

$$C^{13}_{23} = 0. \tag{14.19}$$

Repeating these calculations for τ_{11}, τ'_{11} we find, on equating to zero the coefficients of γ_{11}, γ_{22}, and γ_{33}, that

$$C^{11}_{12} = C^{33}_{12} = 0, \quad C^{33}_{11} = C^{33}_{22}, \quad C^{11}_{11} = C^{22}_{22}. \tag{14.20}$$

The coefficient of γ_{12} yields

$$C^{12}_{12} = \tfrac{1}{2}(C^{11}_{11} - C^{21}_{11}), \quad C^{12}_{22} = 0, \tag{14.21}$$

but the coefficients of γ_{23}, γ_{31} lead to nothing new.

Finally if we consider the coefficient of γ_{13} in the equation obtained from τ_{23} and τ'_{23} we have

$$C^{23}_{23} = C^{13}_{13}. \tag{14.22}$$

Substituting from Eqs. (14.15) to (14.22) into the scheme (14.12) we find that, for a system possessing hexagonal symmetry, there are five independent elastic constants given by the scheme

$$\begin{pmatrix} C^{11}_{11} & C^{11}_{22} & C^{11}_{33} & 0 & 0 & 0 \\ C^{11}_{22} & C^{11}_{11} & C^{11}_{33} & 0 & 0 & 0 \\ C^{11}_{33} & C^{11}_{33} & C^{33}_{33} & 0 & 0 & 0 \\ 0 & 0 & 0 & C^{13}_{13} & 0 & 0 \\ 0 & 0 & 0 & 0 & C^{13}_{13} & 0 \\ 0 & 0 & 0 & 0 & 0 & \tfrac{1}{2}(C^{11}_{11}-C^{11}_{22}) \end{pmatrix}. \tag{14.23}$$

δ) *Isotropic bodies.* An isotropic body is one which has hexagonal symmetry about any two perpendicular axes. By repeating the arguments of subsection γ above we see that the elastic constants of an isotropic body are given by a scheme of the type (14.23) but with

$$C^{13}_{13} = \tfrac{1}{2}(C^{11}_{11} - C^{11}_{22}), \quad C^{33}_{33} = C^{11}_{11}, \quad C^{11}_{33} = C^{11}_{22}$$

so that there are in fact only two independent constants. If we write

$$C^{11}_{22} = \lambda, \quad \tfrac{1}{2}(C^{11}_{11} - C^{11}_{22}) = \mu$$

we find that the scheme of elastic constants for an isotropic body is

$$\begin{pmatrix} \lambda + 2\mu & \lambda & \lambda & 0 & 0 & 0 \\ \lambda & \lambda + 2\mu & \lambda & 0 & 0 & 0 \\ \lambda & \lambda & \lambda + 2\mu & 0 & 0 & 0 \\ 0 & 0 & 0 & \mu & 0 & 0 \\ 0 & 0 & 0 & 0 & \mu & 0 \\ 0 & 0 & 0 & 0 & 0 & \mu \end{pmatrix}. \tag{14.24}$$

It will be recalled that, in the derivation of Eq. (14.11) and the subsequent discussion, we have been working in a system of rectangular coordinates. It is easily seen from the scheme (14.24) that, in this system of coordinates, the elastic constants of a homogeneous isotropic elastic solid can be written in the form

$$C_{rs}^{ij} = \lambda \delta^{ij} \delta_{rs} + \mu (\delta_r^i \delta_s^j + \delta_s^i \delta_r^j). \tag{14.25}$$

Since the quantities C_{rs}^{ij} are the components of mixed tensors of order 4, it follows that, in a general system of coordinates the elastic constants of a homogeneous isotropic body are given by

$$C_{rs}^{ij} = \lambda g^{ij} g_{rs} + \mu (\delta_r^i \delta_s^j + \delta_s^i \delta_r^j) \tag{14.26}$$

where λ and μ are constants.

Using the relations

$$\tau^{ij} = C_{rs}^{ij} \gamma^{rs}, \qquad \gamma^{rs} = g^{rm} g^{ns} \gamma_{mn}$$

we find that the components of stress of homogeneous isotropic body are given by the equation

$$\tau^{ij} = \{\lambda g^{ij} g^{mn} + \mu (g^{im} g^{jn} + g^{in} g^{jm})\} \gamma_{mn}. \tag{14.27}$$

If we define the *dilatation* by the equation

$$\Delta = g^{mn} \gamma_{mn} \tag{14.28}$$

we find that

$$\tau^{ij} = \lambda \Delta g^{ij} + \mu (g^{im} g^{jn} + g^{in} g^{jm}) \gamma_{mn}. \tag{14.29}$$

In rectangular cartesian coordinates

$$\tau^{ij} = \lambda \Delta \delta^{ij} + 2\mu \gamma_{ij} \tag{14.30}$$

where the dilatation

$$\Delta = \gamma_{11} + \gamma_{22} + \gamma_{33} \tag{14.31}$$

denotes the change in volume per unit volume in the deformation.

15. Relations among elastic constants of homogeneous isotropic bodies. The constants λ, μ occurring in the stress-strain relation (14.29) are known as LAME'S elastic constants. The Eqs. (14.29) are the generalization of Eq. (12.13) which expresses HOOKE's law.

In addition to these constants four other elastic constants are in common use. These are:

(i) YOUNG's modulus (E);
(ii) POISSON's ratio (ν);
(iii) The bulk modulus (k);
(iv) The rigidity modulus (n).

We shall now indicate how these constants are related to LAME'S elastic constants.

α) *Young's modulus.* When a long cylinder, whose axis is parallel to the x-axis, is subjected to a uniform load over its plane ends, the ratio τ_{xx}/γ_{xx} defines the YOUNG's modulus E of the material of which the cylinder is made. Putting $\tau_{xx} = E\gamma_{xx}$, $\tau_{yy} = 0$, $\tau_{zz} = 0$ in Eqs. (14.30) we have

$$\left. \begin{aligned} E\gamma_{xx} &= (\lambda + 2\mu) \gamma_{xx} + \lambda (\gamma_{yy} + \gamma_{zz}), \\ 0 &= (\lambda + 2\mu) \gamma_{yy} + \lambda (\gamma_{zz} + \gamma_{xx}), \\ 0 &= (\lambda + 2\mu) \gamma_{zz} + \lambda (\gamma_{xx} + \gamma_{yy}) \end{aligned} \right\} \tag{15.1}$$

Sect. 15. Relations among elastic constants of homogeneous isotropic bodies.

from which it follows that
$$E = \frac{(3\lambda + 2\mu)\mu}{\lambda + \mu} \tag{15.2}$$
which expresses Young's modulus in terms of λ and μ.

β) *Poisson's ratio* (ν). In the extension of a cylinder described above, Poisson's ratio, ν, is defined as the ratio of the lateral contraction to the longitudinal extension of the cylinder, i.e.
$$\nu = -\gamma_{yy}/\gamma_{xx}. \tag{15.3}$$
If we solve Eqs. (15.1) for the ratio γ_{yy}/γ_{xx} we find that
$$\nu = \frac{\lambda}{2(\lambda + \mu)}. \tag{15.4}$$
Since both λ and μ are positive it follows that Poisson's ratio is always less than (or equal to) $\tfrac{1}{2}$. It is found experimentally that Poisson's ratio always lies somewhere in the range $\tfrac{1}{4}$ to $\tfrac{1}{2}$.

γ) *The bulk modulus* (k). When a uniform hydrostatic pressure δp is applied to a volume v of an elastic solid causing its volume to decrease by an amount δv the bulk modulus of the material is defined by the equation

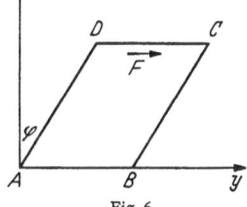

Fig. 6.

$$k = v\frac{\delta p}{\delta v}. \tag{15.5}$$

As we noted previously the dilatation Δ is the measure of the volume changes so that $\Delta = -\delta v/v$ and Eq. (15.5) becomes
$$k = -\frac{\delta p}{\Delta}. \tag{15.6}$$
If we put $\tau_{xx} = \tau_{yy} = \tau_{zz} = -\delta p$ in Eqs. (14.30) we have
$$-\delta p = \lambda\Delta + 2\mu\gamma_{xx},$$
$$-\delta p = \lambda\Delta + 2\mu\gamma_{yy},$$
$$-\delta p = \lambda\Delta + 2\mu\gamma_{zz}.$$
Adding these three equations and making use of Eqs. (14.30) and (15.6) we find that
$$k = \lambda + \frac{2\mu}{3}. \tag{15.7}$$

δ) *The rigidity modulus* (n). When a long rectangular parallelepiped of square cross-section whose long axis is parallel to Ox is deformed by a shearing force F per unit area acting on the face CD (cf. Fig. 6) then the rigidity modulus n is defined by the relation
$$n = F/\varphi \tag{15.8}$$
where φ is the angle of shear. Now in a deformation of this kind the point (x, y, z) goes overinto the point $(x, y, +\varphi z, z)$ so that all the components of strain are zero except $\gamma_{yz} = \tfrac{1}{2}\varphi$. Further, all the components of stress are zero except $\tau_{yz} = F$ so that Eq. (15.8) becomes $n = \tau_{yz}/2\gamma_{yz}$. It follows from Eqs. (14.30) that
$$n = \mu \tag{15.9}$$
so that the rigidity modulus n is identical with the second of the Lamé constants.

It is obvious that other relations exist between the elastic constants. We can take any two of these as independent and express the others in terms of them. The results of calculations of this kind are shown in Table 1.

Table 1. *Relations between the elastic constants.*

Constant	Basic Pair				
	(λ, μ)	(k, n)	(μ, ν)	(E, ν)	(E, n)
λ	λ	$k - \dfrac{2}{3}n$	$\dfrac{2\mu\nu}{1-2\nu}$	$\dfrac{\nu E}{(1+\nu)(1-2\nu)}$	$\dfrac{n(E-2n)}{3n-E}$
$\mu(n)$	μ	n	μ	$\dfrac{E}{2+2\nu}$	n
k	$\lambda + \dfrac{2}{3}\mu$	k	$\dfrac{2\mu(1+\nu)}{3(1-2\nu)}$	$\dfrac{E}{3(1-2\nu)}$	$\dfrac{En}{3(3n-E)}$
E	$\dfrac{(3\lambda+2\mu)\mu}{\lambda+\mu}$	$\dfrac{9kn}{3k+n}$	$2(1+\nu)\mu$	E	E
ν	$\dfrac{\lambda}{2(\lambda+\mu)}$	$\dfrac{3k-2n}{6k+2n}$	ν	ν	$\dfrac{1}{2}\dfrac{E}{n} - 1$

16. von Kármán's notation. In the discussion of special problems in three dimensional rectangular cartesian coordinates (x, y, z) we shall adopt v. Kármán's notation and write

$$\sigma_x = \tau_{xx}, \quad \sigma_y = \tau_{yy}, \quad \sigma_z = \tau_{zz}$$

so that the stress tensor becomes

$$\begin{pmatrix} \sigma_x & \tau_{xy} & \tau_{xz} \\ \tau_{xy} & \sigma_y & \tau_{yz} \\ \tau_{xz} & \tau_{yz} & \sigma_z \end{pmatrix}.$$

We shall denote the components of the displacement vector $\boldsymbol{v}$ by (u, v, w).

With this notation the equations of motion (9.19) assume the forms

$$\frac{\partial \sigma_x}{\partial x} + \frac{\partial \tau_{xy}}{\partial y} + \frac{\partial \tau_{xz}}{\partial z} + \varrho X = \varrho \frac{\partial^2 u}{\partial t^2}, \tag{16.1}$$

$$\frac{\partial \tau_{xy}}{\partial x} + \frac{\partial \sigma_y}{\partial y} + \frac{\partial \tau_{yz}}{\partial z} + \varrho Y = \varrho \frac{\partial^2 v}{\partial t^2}, \tag{16.2}$$

$$\frac{\partial \tau_{xz}}{\partial x} + \frac{\partial \tau_{yz}}{\partial y} + \frac{\partial \sigma_z}{\partial z} + \varrho Z = \varrho \frac{\partial^2 w}{\partial t^2}, \tag{16.3}$$

while the stress strain relations (14.30) may be put in the forms

$$(\sigma_x, \sigma_y, \sigma_z) = \lambda \Delta + 2\mu \left(\frac{\partial u}{\partial x}, \frac{\partial v}{\partial y}, \frac{\partial w}{\partial z} \right), \tag{16.4}$$

$$(\tau_{yz}, \tau_{zx}, \tau_{xy}) = \mu \left(\frac{\partial v}{\partial z} + \frac{\partial w}{\partial y}, \frac{\partial w}{\partial x} + \frac{\partial u}{\partial z}, \frac{\partial u}{\partial y} + \frac{\partial v}{\partial x} \right) \tag{16.5}$$

and

$$\Delta = \frac{\partial u}{\partial x} + \frac{\partial v}{\partial y} + \frac{\partial w}{\partial z}. \tag{16.6}$$

A similar notation is employed in the case of orthogonal curvilinear coordinates (ξ, η, ζ), say. We write

$$\sigma_\xi = \tau_{\xi\xi}, \quad \sigma_\eta = \tau_{\eta\eta}, \quad \sigma_\zeta = \tau_{\zeta\zeta}.$$

Sect. 17.   Torsion of a circular cylinder.

B. Torsion and flexure.

Before proceeding to the discussion of general solutions of the equations of elasticity we shall consider two very simple problems which have great technical importance—the torsion of cylinders and the flexure of beams. This section is meant to serve only as the briefest of introductions to these problems. No attempt will be made to discuss the great number of special problems which have been treated by many authors. For a much fuller discussion the reader is referred to pp. 91 to 248 of the 2nd edition of SOKOLNIKOFF's "*Mathematical Theory of Elasticity*". Here we shall consider only a few typical problems.

I. The torsion problem.

17. Torsion of a circular cylinder. We consider first the deformation of a right circular cylinder, of radius a, due to the application of a couple whose moment lies along the axis of the cylinder. We may suppose that one of the circular ends of the cylinder lies in the xy-plane and that the other lies in the plane $z = l$, the z-axis being chosen to coincide with the axis of the cylinder. The effect of the couple will be to deform the generators of the cylinder into helical curves. If the free end of the cylinder is rotated through an angle τl relative to the fixed end

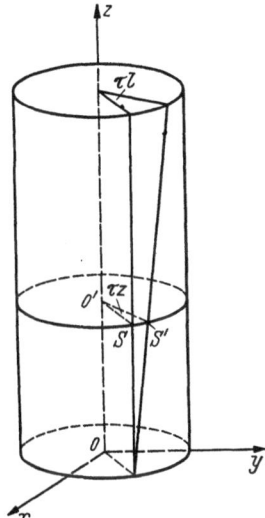

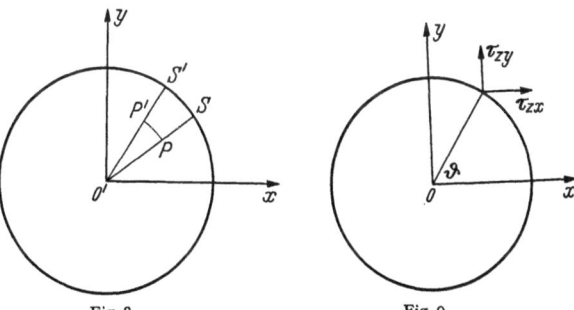

Fig. 7.   Fig. 8.   Fig. 9.

then, assuming that the amount of rotation of a section of the cylinder is proportional to its distance from the fixed end, we see that a section, centre O', distance z from the free end will have a rotation τz relative to the fixed section. If we now consider the displacement of a typical point P of this section (cf. Fig. 8) we see that the point P moves to P' where $O'P = O'P'$ and $< PO'P' = \tau z$. It is reasonable to assume that sections of the cylinder remain plane after the deformation so that we may take w, the z-component of the displacement of P to be zero. If $OP = r$ and $< xO'P = \vartheta$ then $PP' = r\tau z$ and the displacement of P has components $(u, v, 0)$ where

$$u = r\tau z \cos(\vartheta + \tfrac{1}{2}\pi) = -\tau z y,$$
$$v = r\tau z \sin(\vartheta + \tfrac{1}{2}\pi) = \tau z x.$$

Hence a typical point (x, y, z) in the cylinder has a displacement vector with components

$$u = -\tau z y, \quad v = \tau z x, \quad w = 0 \qquad (17.1)$$

referred to the fixed set of axes $O(x, y, z)$.

It follows from Eq. (17.1) that the components of stress at the point (x, y, z) are given by the equations

$$\sigma_x = \sigma_y = \sigma_z = \tau_{xy} = 0, \tag{17.2}$$

$$\tau_{zx} = -\mu\tau y, \qquad \tau_{zy} = \mu\tau x \tag{17.3}$$

from which it is immediately obvious that the equations of motion are satisfied identically provided, of course, that there are no body forces present. On the free surface $r = a$ we find that the normal component of the stress is

$$\tau_{zx} \cos\vartheta + \tau_{zy} \sin\vartheta \tag{17.4}$$

and this vanishes at every point since for this circle $\cos\vartheta = x/a$, $\sin\vartheta = y/a$.

The quantity in which we are most interested is the resultant couple $\boldsymbol{M}$ acting over the end of the cylinder. It is readily seen that $\boldsymbol{M}$ is of the form $(0, 0, M)$ where

$$M = \iint (x\,\tau_{zy} - y\,\tau_{zx})\,dx\,dy, \tag{17.5}$$

the integral being taken over the circle $x^2 + y^2 = a^2$. Substituting from Eqs. (17.3) into this last equation we see that

$$M = D\tau \tag{17.6}$$

where

$$D = \mu \iint (x^2 + y^2)\,dx\,dy = \tfrac{1}{2}\pi\mu a^4. \tag{17.7}$$

The resultant couple is therefore proportional to τ, the angle of twist per unit length. The constant of proportionality, D, provides a measure of the resistance offered by the beam to a twist of this kind; it is known as the *torsional rigidity* of the cylinder.

18. Torsion of non-circular cylinders. We consider now the twisting of a cylinder, of any cross-section, by the application of a couple whose moment lies along the axis of the cylinder. The cylinder is assumed to be free from body forces and its lateral surface to be free from external forces. As in the case of the circular cylinder we assume that one end of the cylinder is fixed in the plane $z = 0$ while the other end, in the plane $z = l$ is twisted by a couple $\boldsymbol{M} = (0, 0, M)$.

Since the section of the cylinder is no longer circular we can no longer assume that the sections will remain plane during the torsion. If, however, we assume that each section is deformed in precisely the same way we may take the components of the displacement vector of a typical point (x, y, z) to be

$$u = -\tau z y, \qquad v = \tau z x, \qquad w = \tau \varphi(x, y) \tag{18.1}$$

where $\varphi(x, y)$ is a function of x and y only known, as the *torsion function*. These equations are an obvious generalization of the Eqs. (17.1).

Substituting the components (18.1) in the stress-strain relations we find that the corresponding components of the stress tensor are

$$\tau_{yz} = \mu\tau\left(\frac{\partial\varphi}{\partial y} + x\right), \qquad \tau_{zx} = \mu\tau\left(\frac{\partial\varphi}{\partial x} - y\right), \tag{18.2}$$

$$\sigma_x = \sigma_y = \sigma_z = \tau_{xy} = 0. \tag{18.3}$$

The equation of equilibrium

$$\frac{\partial\tau_{xz}}{\partial x} + \frac{\partial\tau_{yz}}{\partial y} + \frac{\partial\sigma_z}{\partial z} = 0$$

will be satisfied by these expressions if φ satisfies the two-dimensional Laplace equation

$$\nabla_1^2 \varphi \equiv \frac{\partial^2 \varphi}{\partial x^2} + \frac{\partial^2 \varphi}{\partial y^2} = 0. \tag{18.4}$$

The two other equations of equilibrium are satisfied identically. In order that the lateral surface of the cylinder should be free from stress we must ensure that the torsion function $\varphi(x, y)$ is such that the equation

$$\tau_{xz} \cos \vartheta + \tau_{zy} \sin \vartheta = 0 \tag{18.5}$$

in which ϑ is the angle the normal n to the section makes with the x-axis, is satisfied at each point of the boundary of the section. In terms of the torsion function this equation becomes

$$\cos \vartheta \left(\frac{\partial \varphi}{\partial x} - y \right) + \sin \vartheta \left(\frac{\partial \varphi}{\partial y} + x \right) = 0 \quad (18.6)$$

on the bounding curve Γ of the cylinder (cf. Fig. 10), an equation which may be written in the form

$$\frac{\partial \varphi}{\partial n} = y \cos \vartheta - x \sin \vartheta \tag{18.7}$$

where $\partial/\partial n$ denotes differentiation along the outward drawn normal to Γ.

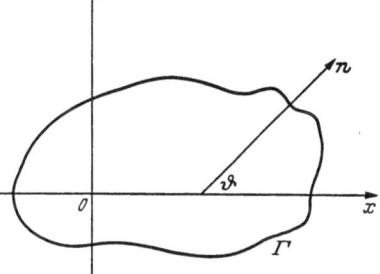

Fig. 10.

The solution of the torsion problem for a non-circular cylinder thus reduces to the solution of the Laplace equation (18.4) subject to the boundary condition (18.6) or (18.7), i.e. it reduces to determining a function which is harmonic in a certain region R and whose normal derivative takes a prescribed value on the boundary Γ of the region. This problem is known in Potential Theory as *Neumann's Problem* and it is known that if the integral of the normal derivative of the function, calculated over the boundary Γ, vanishes then the function φ is determined uniquely to within an arbitrary constant. It is readily verified from (18.7) that $\int_\Gamma (\partial \varphi / \partial n) \, ds$ vanishes so that Eqs. (18.4) and (18.7) determine a torsion function to within an arbitrary constant. It can also be shown that a change of origin of coordinates results in a new function φ_1 which differs from the original φ by terms which do not contribute to the expressions for the components of stress. It follows that the two solutions φ, φ_1 lead to displacement vectors which only differ by a rigid body displacement so that the choice of origin of coordinates is immaterial in the torsion problem.

We must now find the system of forces over the free end of the cylinder to which the system of stresses (18.2) is statically equivalent. Since φ is a harmonic function it follows that

$$\frac{\partial \varphi}{\partial x} - y = \frac{\partial}{\partial x} \left\{ x \left(\frac{\partial \varphi}{\partial x} - y \right) \right\} + \frac{\partial}{\partial y} \left\{ x \left(\frac{\partial \varphi}{\partial x} + x \right) \right\}$$

so that by GREEN's theorem

$$\iint_R \tau_{zx} \, dx \, dy = \mu \tau \int_\Gamma x \left\{ \frac{\partial \varphi}{\partial n} - y \cos \vartheta + x \sin \vartheta \right\} ds = 0.$$

Similarly

$$\iint_R \tau_{zy} \, dx \, dy = 0$$

so that there is zero resultant force acting on the end of the cylinder. The resultant moment of the external forces applied to the end of the cylinder is $(0, 0, M)$ where M is given by the equation

$$M = \iint (x\tau_{zy} - y\tau_{zx})\, dx\, dy. \tag{18.8}$$

Thus

$$M = D\tau \tag{18.9}$$

where D, the torsional rigidity, is given by the equation

$$D = \mu \iint_R \left(x^2 + y^2 + x\frac{\partial \varphi}{\partial y} - y\frac{\partial \varphi}{\partial x} \right) dx\, dy. \tag{18.10}$$

The expression (18.10) may be put into a form which shows explicitly that D is positive. From GREEN's theorem,

$$\iint_R \left(x\frac{\partial \varphi}{\partial y} - y\frac{\partial \varphi}{\partial x} \right) dx\, dy = \iint_R \left\{ \frac{\partial (x\varphi)}{\partial y} - \frac{\partial (y\varphi)}{\partial x} \right\} dx\, dy$$

$$= -\int_\Gamma \varphi (y\cos\vartheta - x\sin\vartheta)\, ds$$

$$= -\int_\Gamma \varphi \frac{\partial \varphi}{\partial n}\, ds$$

$$= -\iint_R \left\{ \left(\frac{\partial \varphi}{\partial x}\right)^2 + \left(\frac{\partial \varphi}{\partial y}\right)^2 \right\} dx\, dy,$$

so that by Eq. (18.10) we find for the torsional rigidity

$$D = \mu \iint_R \left\{ \left(x + \frac{\partial \varphi}{\partial y} \right)^2 + \left(y - \frac{\partial \varphi}{\partial x} \right)^2 \right\} dx\, dy \tag{18.11}$$

from which it is obvious that $D > 0$[1].

Torsion function for an elliptical cylinder. At the point (x, y) on the boundary of the ellipse

$$\frac{x^2}{a^2} + \frac{y^2}{b^2} = 1 \tag{18.12}$$

we have, in the notation of Eq. (18.6),

$$\tan \vartheta = \frac{a^2 y}{b^2 x}$$

so that we have to find a function $\varphi(x, y)$ which satisfies Eq. (18.4) and is such that

$$b^2 x \left(\frac{\partial \varphi}{\partial x} - y \right) + a^2 y \left(\frac{\partial \varphi}{\partial y} + x \right) = 0 \tag{18.13}$$

at every point on the ellipse (18.12). Now $\varphi = kxy$ satisfies Eq. (18.4) for all values of k and satisfies (18.13) if k is chosen so that

$$b^2(k-1) + a^2(k+1) = 0.$$

Solving this simple equation for k we obtain the torsion function

$$\varphi = -\frac{a^2 - b^2}{a^2 + b^2}\, xy. \tag{18.14}$$

---

[1] $D = 0$ only if $\dfrac{\partial \varphi}{\partial x} = y$, $\dfrac{\partial \varphi}{\partial y} = -x$ and this is not possible since $y\, dx - x\, dy$ is not a perfect differential.

The function $\varphi(x, y)$ given by this equation determines the z-component of the displacement vector at the point (x, y) of the cross-section, so that the contour lines $\varphi = $ const illustrate graphically the warping cross-sections of the elliptical cylinder. These contours are shown in Fig. 11 for the case in which the cylinder is twisted by a couple whose direction is shown by the arrows. The full lines indicate the parts of the cross-section which become convex during the twisting; the dotted lines indicate the parts which become concave.

The torsional rigidity of an elliptical cylinder is therefore given by the equation

$$D = \frac{2\mu}{a^2 + b^2} \iint_R (b^2 x^2 + a^2 y^2)\, dx\, dy = \frac{\pi \mu a^3 b^3}{a^2 + b^2}. \tag{18.15}$$

Substituting the expression (18.14) into Eqs. (18.2) we find that the non-vanishing components of the stress tensor are given by the equations

$$\left.\begin{array}{l} \tau_{yz} = \dfrac{2\mu \tau b^2 x}{a^2 + b^2}, \\[4pt] \tau_{zx} = -\dfrac{2\mu \tau a^2 y}{a^2 + b^2}. \end{array}\right\} \tag{18.16}$$

The magnitude $\bar{\tau}$ of the tangential stress at the point $P(x, y)$ on the boundary is given by

$$\bar{\tau} = (\tau_{xz}^2 + \tau_{yz}^2)^{\frac{1}{2}} = \frac{2\mu \tau a b}{a^2 + b^2} \bar{r} \tag{18.17}$$

where $\bar{r} = (b^2 x^2/a^2 + a^2 y^2/b^2)^{\frac{1}{2}}$ is the length of the radius of the ellipse conjugate to the radius OP. Eq. (18.17) shows that the maximum stress occurs at the extremities of the minor axis of the ellipse.

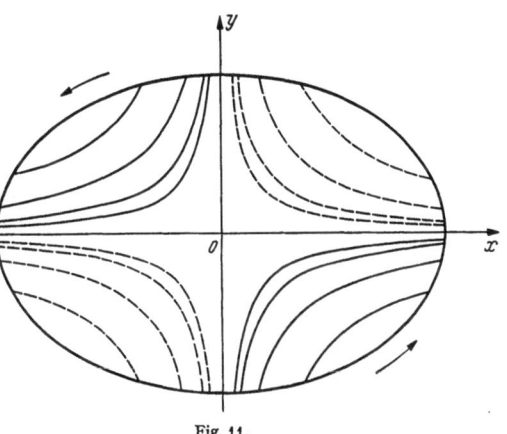

Fig. 11.

19. Stress functions. We saw in the last section that the torsion function $\varphi(x, y)$ is a plane harmonic function in the region R, the cross-section of the cylinder. Now from the theory of functions of a complex variable we know that there exists another plane harmonic function $\psi(x, y)$ such that the function $\varphi(x, y) + i\psi(x, y)$ is an analytic function of the complex variable $x + iy$. The functions φ and ψ are related to each other through the Cauchy-Riemann equations

$$\frac{\partial \varphi}{\partial x} = \frac{\partial \psi}{\partial y}, \qquad \frac{\partial \varphi}{\partial y} = -\frac{\partial \psi}{\partial x}. \tag{19.1}$$

The analytic function $\varphi(x, y) + i\psi(x, y)$ is known as the *complex torsion function*.

It is a simple matter to formulate the torsion problem in terms of the function $\psi(x, y)$ conjugate to the torsion function. Substituting from Eq. (19.1) into Eq. (18.2) we find that the non-vanishing components of stress are given by

$$\tau_{yz} = -\mu \tau \left(\frac{\partial \psi}{\partial x} - x\right), \qquad \tau_{zx} = \mu \tau \left(\frac{\partial \psi}{\partial y} - y\right) \tag{19.2}$$

and the boundary condition (18.6) becomes

$$\cos \vartheta \left(\frac{\partial \psi}{\partial y} - y\right) - \sin \vartheta \left(\frac{\partial \psi}{\partial x} - x\right) = 0. \tag{19.3}$$

If d/ds denotes differentiation along the boundary Γ,
$$\cos \vartheta = \frac{dy}{ds}, \quad \sin \vartheta = -\frac{dx}{ds}$$
so that Eq. (19.3) is equivalent to the relation
$$\frac{d}{ds}\left\{\psi - \frac{1}{2}(x^2+y^2)\right\} = 0.$$
The torsion problem is therefore solved if we can find a function which satisfies

$$\left. \begin{array}{ll} \text{(i)} & \nabla_1^2 \psi = 0 \quad \text{in } R; \\ \text{(ii)} & \psi - \frac{1}{2}(x^2+y^2) = \text{const} \quad \text{on } \Gamma. \end{array} \right\} \quad (19.4)$$

In terms of this function the torsional rigidity of the cylinder assumes the forms
$$D = \mu \iint_R \left(x^2+y^2 - x\frac{\partial \psi}{\partial x} - y\frac{\partial \psi}{\partial y}\right) dx\, dy \quad (19.5)$$
$$= \mu \iint_R \left\{\left(x - \frac{\partial \psi}{\partial x}\right)^2 + \left(y - \frac{\partial \psi}{\partial y}\right)^2\right\} dx\, dy. \quad (19.6)$$

The boundary value problem (19.4) is a Dirichlet problem so that its solution is unique.

It is convenient to introduce a *stress-function* $\Psi(x, y)$ in terms of which the boundary condition (ii) of Eq. (19.4) assumes a particularly simple form. If we write
$$\Psi(x, y) = \psi(x, y) - \tfrac{1}{2}(x^2+y^2),$$
it follows that the torsion problem is equivalent to the boundary-value problem

$$\left. \begin{array}{ll} \text{(i)} & \nabla_1^2 \Psi = -2 \quad \text{in } R; \\ \text{(ii)} & \Psi = \text{const} \quad \text{on } \Gamma. \end{array} \right\} \quad (19.7)$$

The components of stress then take the forms
$$\tau_{yz} = -\mu\tau\frac{\partial \Psi}{\partial x}, \quad \tau_{xz} = \mu\tau\frac{\partial \Psi}{\partial y}, \quad (19.8)$$
and the torsional rigidity is equal to
$$\left. \begin{array}{l} D = -\mu \iint_R \left\{x\frac{\partial \Psi}{\partial x} + y\frac{\partial \Psi}{\partial y}\right\} dx\, dy \\ = 2\mu \iint_R \Psi\, dx\, dy + \mu \int_\Gamma \Psi(y\, dx - x\, dy) \end{array} \right\} \quad (19.9)$$
or to
$$D = \mu \iint_R \left\{\left(\frac{\partial \Psi}{\partial x}\right)^2 + \left(\frac{\partial \Psi}{\partial y}\right)^2\right\} dx\, dy. \quad (19.10)$$

The function $\Psi(x, y)$ is capable of a simple geometrical interpretation. If $P(x, y)$ is any point on one member of the one-parameter family of curves
$$\Psi(x, y) = \text{const} \quad (19.11)$$
then the tangent at P to this curve has slope dy/dx where
$$\frac{\partial \Psi}{\partial x} + \frac{\partial \Psi}{\partial y} \cdot \frac{dy}{dx} = 0. \quad (19.12)$$

Eliminating $\partial \Psi/\partial x$ and $\partial \Psi/\partial y$ from Eqs. (19.8) and (19.12) we find that

$$\frac{dy}{dx} = \frac{\tau_{zy}}{\tau_{zx}}.$$

In other words, at each point of the curve (19.11) the stress vector is directed along the tangent to the curve. For this reason the curves (19.11) are called the *lines of shearing stress*.

20. The torsion of hollow cylinders. The discussion of the torsion problem given above assumes that the cylinder which is being twisted is of solid cross-section. The case in which the cylinder has one or more hollows is of great practical significance and may be treated by similar methods. A cylinder of this kind is shown in Fig. 12; it is made up of a region R bounded by an external contour Γ and a series of internal ones $C_1, C_2, \ldots, C_n$. By reasoning similar to that employed in the derivation of Eqs. (19.4) and (18.7) we can show that the torsion function φ must satisfy $\nabla_1^2 \varphi = 0$ in R and that on each of the contours $\Gamma, C_1, C_2, \ldots, C_n$

$$\frac{d\varphi}{dn} = y \cos \vartheta - x \sin \vartheta. \qquad (20.1)$$

It follows that the conjugate function $\psi(x, y)$ will satisfy $\nabla_1^2 \psi = 0$ and that

$$\psi - \tfrac{1}{2}(x^2 + y^2)$$

will take values $\gamma, c_1, c_2, \ldots, c_n$ on the contours $\Gamma, C_1, C_2, \ldots, C_n$. In the ordinary way the function $\psi(x, y)$ would be many-valued but if this last condition is to be satisfied ψ must return to each original value after the point $x+iy$ has gone round the contour C_k. Furthermore, since φ is determined to within an arbitrary constant by Eq. (20.1) its complex conjugate ψ must similarly be determined. Therefore only one of the constants $\gamma, c_1, c_2, \ldots, c_n$ can be chosen arbitrarily. The remaining n constants must be such that $\psi(x, y)$ is single-valued throughout the region R.

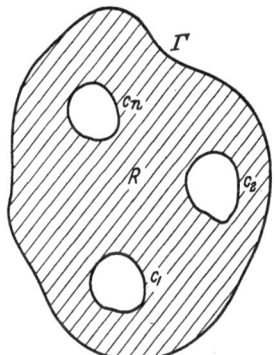

Fig. 12.

The stress function $\Psi(x, y)$ defining the problem will then satisfy the conditions

$$\begin{aligned}\text{(i)} \quad & \nabla_1^2 \Psi = -2 \quad \text{in} \quad R; \\ \text{(ii)} \quad & \Psi = \gamma \quad \text{on} \quad \Gamma; \\ \text{(iii)} \quad & \Psi = c_j \quad \text{on} \quad C_j \end{aligned} \qquad (20.2)$$

where only one of the constants $\gamma, c_1, c_2, \ldots, c_n$ may be chosen arbitrarily. If we chose $\gamma = 0$ then the torsional rigidity is given by the equation

$$D = 2\mu \iint_R \Psi \, dx \, dy + \sum_{i=1}^{n} \mu c_i \int_{C_i} (y \, dx - x \, dy) \qquad (20.3)$$

[cf. Eq. (19.9) above]. Now

$$\int_{C_i} (y \, dx - x \, dy) = 2A_i,$$

where A_i is the area enclosed by the curve C_i. Hence

$$D = 2\mu \iint_R \Psi \, dx \, dy + 2\mu \sum_{i=1}^{n} c_i A_i. \qquad (20.4)$$

21. Solution of the torsion problem for certain particular cases. α) *Elliptic cylinder.* The stress function $\Psi(x, y)$ is to be constant on

$$\frac{x^2}{a^2} + \frac{y^2}{b^2} = 1.$$

This suggests that we try

$$\Psi = A\left(\frac{x^2}{a^2} + \frac{y^2}{b^2}\right).$$

This function satisfies the equation $\nabla_1^2 \Psi = -2$ if

$$-2 = 2A\left(\frac{1}{a^2} + \frac{1}{b^2}\right).$$

Hence the conditions (19.7) are satisfied by the function

$$\Psi = -\frac{b^2 x^2 + a^2 y^2}{a^2 + b^2},$$

from which we have

$$\psi = \Psi + \frac{1}{2}(x^2 + y^2) = \frac{(a^2 - b^2)(x^2 - y^2)}{2(a^2 + b^2)}.$$

The conjugate (torsion) function is

$$\varphi = -\frac{(a^2 - b^2) xy}{(a^2 + b^2)}$$

as we found previously [Eq. (18.14) above]. The complex torsion function is

$$\varphi + i\psi = \frac{(a^2 - b^2) i z^2}{2(a^2 + b^2)}.$$

β) *Equilateral triangle.* Suppose the triangle is bounded by the lines $y = c \pm mx$, $y = 0$. Then the function

$$\Psi = Ay\{(y-c)^2 - m^2 x^2\}$$

vanishes along the boundary. For this function

$$\nabla_1^2 \Psi = \{(6 - 2m^2) y - 4c\} A$$

so that in order that $\nabla_1^2 \Psi = -2$ we must choose $m^2 = 3$, $A = 1/(2c)$. We have therefore found that the stress function for the equilateral triangle bounded by the lines $y = c \pm \sqrt{3} x$, $y = 0$ is

$$\Psi = \frac{y}{2c}\{(y-c)^2 - 3x^2\}.$$

From Eq. (19.9) the torsional rigidity is

$$D = \frac{\mu}{c} \iint \{y(y-c)^2 - 3x^2 y\}\, dx\, dy = \frac{\sqrt{3}}{45}\mu c^4.$$

Corresponding to this value of Ψ we have the functions

$$\varphi(x, y) = (3xy^2 - x^3 - 2cxy + c^2 x)/2c,$$
$$\psi(x, y) = (y^3 - 3x^2 y + c^2 y + cx^2 - cy^2)/2c.$$

The nature of the warping of the cross-sections calculated from this value of φ is shown in Fig. 13.

γ) *Rectangle.* Consider the rectangle bounded by the lines $x = \pm a$, $y = \pm b$. The torsion problem will be solved if we can find a harmonic function $\psi(x, y)$

which takes the form $\frac{1}{2}(x^2+y^2)$ on the boundary of the square, i.e. we must have
$$\psi(\pm a, y) = \tfrac{1}{2}(a^2 + y^2), \quad \psi(x, \pm b) = \tfrac{1}{2}(x^2 + b^2). \tag{21.1}$$
If we write
$$\psi(x, y) = a^2 + \tfrac{1}{2}(y^2 - x^2) + f(x, y) \tag{21.2}$$
then it follows that f must be a harmonic function satisfying the boundary conditions
$$f(\pm a, y) = 0 \tag{21.3}$$
$$f(x, \pm b) = x^2 - a^2. \tag{21.4}$$

The harmonic function
$$f(x, y) = \sum_n c_n \cosh(k_n y) \cos(k_n x) \tag{21.5}$$
will satisfy the condition (21.3) provided that
$$k_n = \left(n + \tfrac{1}{2}\right)\frac{\pi}{a} \tag{21.6}$$
and the condition (21.4) provided that the constants c_n are chosen so that
$$x^2 - a^2 = \sum_n c_n \cosh(k_n b) \cos k_n x.$$

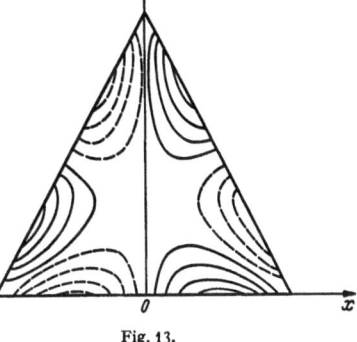

Fig. 13.

We must therefore have
$$c_n = \frac{1}{a \cdot \cosh(k_n b)} \int_{-a}^{a} (x^2 - a^2) \cos(k_n x)\, dx = -\frac{4}{a k_n^3} \cdot \frac{\sin(k_n a)}{\cosh(k_n b)}$$

giving finally for the function $\psi(x, y)$ the series
$$\psi(x, y) = a^2 + \frac{1}{2}(y^2 - x^2) - \frac{32 a^2}{\pi^3} \sum_{n=0}^{\infty} \frac{(-1)^n}{(2n+1)^3} \cdot \frac{\cosh(k_n y)}{\cosh(k_n b)} \cos(k_n x)$$
where k_n is defined by Eq. (21.6)

δ) *Circular shaft with circular groove.* To illustrate the use of the complex torsion function we consider an example due to C. WEBER[1]. Consider the analytical function
$$\varphi + i\psi = -ia\left(z - \frac{b^2}{z}\right) - \frac{1}{2} i b^2. \tag{21.7}$$

Putting $z = re^{i\vartheta}$ and separating real and imaginary parts we find that
$$\varphi = a\left(r + \frac{b^2}{r}\right) \sin \vartheta \tag{21.8}$$
and that
$$\psi = a x \left(1 - \frac{b^2}{r^2}\right) + \frac{1}{2} b^2. \tag{21.9}$$

Now when $r = b$, $\psi = \tfrac{1}{2} b^2$ i.e. $\psi - \tfrac{1}{2}(x^2 + y^2) = 0$. Also when $r = 2a \cos \vartheta$, $\psi = 2a^2 \cos^2 \vartheta$, i.e. $\psi - \tfrac{1}{2}(x^2 + y^2) = 0$. Thus (21.7) is the complex torsion function for the area bounded by the circles $r = b$, $r = 2a \cos \vartheta$ (cf. Fig. 14). Differentiating (21.9) we find that
$$\frac{\partial \psi}{\partial x} = a\left(1 - \frac{b^2}{r^2}\right) + \frac{2 a b^2 x^2}{r^4}, \quad \frac{\partial \psi}{\partial y} = \frac{2 a b^2 x y}{r^4}$$

---
[1] C. WEBER: Forschungsarbeiten, VDI No. 249 (1921).

so that from the Eqs. (19.2) we have
$$\tau_{xz}=\mu\tau y\left(\frac{2ab^2 x}{r^4}-1\right), \quad \tau_{yz}=-\mu\tau\left\{a\left(1-\frac{b^2}{r^2}\right)+\frac{2ab^2 x^2}{r^4}-x\right\}.$$

If we put $r=b$ in these equations we find that
$$\tau_{xz}=\mu\tau\sin\vartheta\,(2a\cos\vartheta-b), \quad \tau_{yz}=-\mu\tau\cos\vartheta\,(2a\cos\vartheta-b)$$

so that the resultant shearing stress in the grove is given by the equation
$$\tau=\sqrt{\tau_{xz}^2+\tau_{yz}^2}=\mu\tau\,(2a\cos\vartheta-b)$$

and takes the maximum value $2\mu\tau a(1-b/2a)$. If the ratio b/a is small this maximum value is approximately $2a\mu\tau$ i.e. it is twice the corresponding maximum for a circular shaft with no groove (Sect. 17 above).

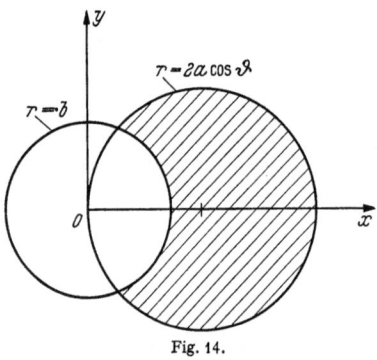

Fig. 14.

22. Finite torsion of a circular cylinder.

To illustrate the procedure to be followed in problems in which the deformations are finite, we consider a pure torsional deformation of a cylinder, i.e. a deformation in which planes normal to the axis in the unstrained state remain plane and suffer only a rotation proportional to their distance from the free end of the cylinder. If we suppose the cylinder has length l and radius a in the unstressed state and if we assume the material of the cylinder to be incompressible, then the body will retain cylindrical form in the strained state and will have the same length and radius.

If we use cylindrical polar coordinates (ϱ, φ, z) to denote the position of a typical point in the strained state then
$$\bar{x}_1=\varrho\cos\varphi, \quad \bar{x}_2=\varrho\sin\varphi, \quad \bar{x}_3=z \qquad (22.1)$$

and according to the above assumption the coordinates of this point in the unstrained state would be
$$x_1=\varrho\cos(\varphi-\psi z), \quad x_2=\varrho\sin(\varphi-\psi z), \quad x_3=z. \qquad (22.2)$$

It is easily shown that
$$\bar{g}_{ij}=\begin{pmatrix}1 & 0 & 0\\ 0 & \varrho^2 & 0\\ 0 & 0 & 1\end{pmatrix}, \quad g_{ij}=\begin{pmatrix}1 & 0 & 0\\ 0 & \varrho^2 & -\varrho^2\psi\\ 0 & -\varrho^2\psi & 1+\varrho^2\psi^2\end{pmatrix} \qquad (22.3)$$

from which it follows that $\bar{g}=g=\varrho^2$ so that $I_3=1$ showing that the incompressibility condition is satisfied. From Eq. (5.1) we have also
$$I_1=I_2=3+\varrho^2\psi^2. \qquad (22.4)$$

The quantities Φ and Ψ (Sect. 13) are therefore functions of the single variable ϱ. From Eqs. (22.3), (22.4) and the definition (13.4) we have
$$\bar{g}^{ij}=\begin{pmatrix}1 & 0 & 0\\ 0 & \varrho^{-2} & 0\\ 0 & 0 & 1\end{pmatrix}, \quad g^{ij}=\begin{pmatrix}1 & 0 & 0\\ 0 & \psi^2+\varrho^{-2} & \psi\\ 0 & \psi & 1\end{pmatrix}, \quad B^{ij}=\begin{pmatrix}2+\psi^2\varrho^2 & 0 & 0\\ 0 & \psi^2+2\varrho^{-2} & \psi\\ 0 & \psi & 2\end{pmatrix}$$

substituting these expressions into the stress-strain relations (13.3) we obtain the equations

$$\begin{aligned}
\tau^{11} &= \Phi + (2 + \varrho^2 \psi^2) \Psi + p, \\
\tau^{22} &= (\varrho^{-2} + \psi^2) \Phi + (2\varrho^{-2} + \psi^2) \Psi + p \varrho^{-2}, \\
\tau^{33} &= \Phi + 2\Psi + p, \\
\tau^{23} &= (\Phi + \Psi) \psi, \\
\tau^{31} &= \tau^{12} = 0,
\end{aligned} \qquad (22.5)$$

where we have adopted the convention $q_1 = \varrho$, $q_2 = \varphi$, $q_3 = z$ for the curvilinear coordinates (cf. Sect. 3 above).

Now the only non-zero Riemann-Christoffel symbols derived from the metric of the strained body are

$$\left\{\begin{matrix}1\\2\,2\end{matrix}\right\} = -\varrho, \quad \left\{\begin{matrix}2\\1\,2\end{matrix}\right\} = \left\{\begin{matrix}2\\2\,1\end{matrix}\right\} = \varrho^{-1} \qquad (22.6)$$

so that the equations of equilibrium (9.19),

$$\partial_i \tau^{ij} + \left\{\begin{matrix}i\\i\,k\end{matrix}\right\} \tau^{kj} + \left\{\begin{matrix}j\\i\,k\end{matrix}\right\} \tau^{ik} = 0 \qquad (22.7)$$

with the forms (22.5) reduce to the simple set of equations

$$\frac{\partial}{\partial \varrho} \{\Phi + (2 + \varrho^2 \psi^2) \Psi + p\} - \psi^2 \varrho \Phi = 0, \qquad (22.8)$$

$$\frac{\partial p}{\partial \varphi} = 0, \quad \frac{\partial p}{\partial z} = 0. \qquad (22.9)$$

It will be recalled that Φ, Ψ are functions of ϱ only and it follows from (22.9) that p is a function of ϱ only so that from Eq. (22.8)

$$p = -\Phi - (2 + \varrho^2 \psi^2) \Psi + \psi^2 \int_c^\varrho \varrho \Phi \, d\varrho \qquad (22.10)$$

where c is an arbitrary constant. If we insert this expression in the first Eq. (22.5) we find that

$$\tau^{11} = \psi^2 \int_c^\varrho \varrho \Phi \, d\varrho. \qquad (22.11)$$

If the curved surface $\varrho = a$ of the cylinder is free from stress then we must have $\tau^{11} = \tau^{12} = \tau^{13} = 0$ at $\varrho = a$. It follows from (22.11) that $\tau^{11} = 0$ when $\varrho = a$ if the constant c is taken to be a; τ^{12} and τ^{13} are identically zero. Putting $c = a$ in Eq. (22.10) we have

$$p = -\Phi - (2 + \varrho^2 \psi^2) \Psi + \psi^2 \int_a^\varrho \varrho \Phi \, d\varrho. \qquad (22.12)$$

The physical components of stress are (cf. Sect. 9 above)

$$\begin{aligned}
\sigma_\varrho &= \tau^{11} = \psi^2 \int_a^\varrho \varrho \Phi \, d\varrho, \\
\sigma_\varphi &= \varrho^2 \tau^{22} = \psi^2 \left\{ \int_a^\varrho \varrho \Phi \, d\varrho + \varrho^2 \Phi \right\}, \\
\sigma_z &= \tau^{33} = \psi^2 \left\{ \int_a^\varrho \varrho \Phi \, d\varrho - \varrho^2 \Psi \right\}, \\
\tau_{\varphi z} &= \varrho \tau^{23} = \psi \varrho (\Phi + \Psi), \\
\tau_{\varrho \varphi} &= \tau_{\varrho z} = 0.
\end{aligned} \qquad (22.13)$$

At the end $z=l$ of the cylinder the unit normal is parallel to $\bar{g}^3$ so that $n = (0, 0, 1)$ and the physical components of the surface tractions are $(0, \tau_{\varphi z}, \sigma_z)$. These distributed surface forces are statically equivalent to a couple

$$M = 2\pi \int_0^a \varrho^2 \tau_{\varphi z} d\varrho = 2\pi \psi \int_0^a \varrho^3 (\Phi + \Psi) d\varrho \tag{22.14}$$

about the axis of the cylinder and a force

$$F = 2\pi \int_0^a \varrho \sigma_z d\varrho = 2\pi \psi^2 \int_0^a \varrho d\varrho \left\{ \int_a^\varrho \varrho' \Phi(\varrho') d\varrho' - \varrho^2 \Psi \right\} \tag{22.15}$$

parallel to the axis.

For example, for a Mooney solid characterized by Eq. (13.12) we have

$$\Phi = 2C_1, \quad \Psi = 2C_2 \tag{22.16}$$

where C_1 and C_2 are constants. Substituting these values into Eqs. (22.14) and (22.15) we find, that for a cylinder made from such a substance, the equivalent couple and force are given by the equations

$$M = (C_1 + C_2) \pi a^4 \psi, \quad F = -\tfrac{1}{2}(C_1 + 2C_2) \pi a^4 \psi^2. \tag{22.17}$$

II. Bending of beams.

23. Elementary theory of bending. We consider a beam bounded by a cylindrical surface of any cross-section and two plane ends normal to it. The z-axis is taken to be the *central line* passing through the centroids of the cross-sections of the beam, while the xy-plane coincides with one of the ends. The beam is bent by a pair of couples of magnitude M applied to the ends and acting in the xz-plane, the cylindrical surface being free from external forces. The elementary theory of beam bending assumes that the central line remains unaltered in length and that plane sections remain plane and normal to the deformed central line. It follows easily that the longitudinal strain in the beam is given by

$$\frac{\partial w}{\partial z} = -\frac{x}{R} \tag{23.1}$$

where R is the radius of curvature of the central line. If now it is assumed that only non-zero component of stress is σ_z, we have from Eq. (23.1) and the stress-strain relation

$$\sigma_z = -\frac{E}{R} x. \tag{23.2}$$

It is obvious that this stress distribution satisfies the equations of equilibrium if there are no body forces and that the boundary conditions on the cylindrical surface are satisfied. The bending moment M is given by

$$M = -\int_A x \sigma_z dA = \int_A \frac{E x^2}{R} dA = \frac{EI}{R} \tag{23.3}$$

where A is the cross-sectional area of the beam and I is its second moment about the line $y=0$. Then from Eqs. (23.2) and (23.3) we have

$$\sigma_z = -\frac{M}{I} x. \tag{23.4}$$

24. Validity of the elementary theory.

A more rigorous treatment of the problem omits the initial assumptions concerning the configuration of the bent beam, assuming that the only non-zero stress component is given by Eq. (23.4). By substitution in the stress-strain relations we find that the components of strain are given by

$$\frac{\partial u}{\partial x} = \frac{\nu M}{EI} x, \qquad \frac{\partial v}{\partial y} = \frac{\nu M}{EI} x, \qquad \frac{\partial w}{\partial z} = -\frac{M}{EI} x,$$

$$\frac{\partial v}{\partial x} + \frac{\partial u}{\partial y} = 0, \qquad \frac{\partial w}{\partial y} + \frac{\partial v}{\partial z} = 0, \qquad \frac{\partial u}{\partial z} + \frac{\partial w}{\partial x} = 0.$$

If we ensure that there is no rigid body motion at or about the origin, so that

$$u = v = w = \frac{\partial u}{\partial z} = \frac{\partial v}{\partial z} = \frac{\partial v}{\partial x} = 0 \quad \text{at } (0,0,0),$$

these equations yield, upon integration, the components of displacement

$$\left.\begin{array}{l} u = \dfrac{M}{2EI}(z^2 + \nu x^2 - \nu y^2), \\[6pt] v = \dfrac{\nu M}{EI} x y, \qquad w = -\dfrac{M}{EI} x z. \end{array}\right\} \tag{24.1}$$

Using the notation $x' = x + u$, $y' = y + v$, $z' = z + w$ we have, on the central line $(0, 0, z)$

$$x' = \frac{M}{2EI} z^2, \qquad y' = 0, \qquad z' = z$$

and at once we see that the first assumption in the elementary theory, that the central line undergoes no change in length, is still valid. The slope of the central line is given by

$$\frac{dx'}{dz'} = \frac{dx'}{dz} = \frac{Mz}{EI}, \qquad x = y = 0 \tag{24.2}$$

while a point on a plane $z = $ const. is displaced to the point

$$z' = z\left(1 - \frac{Mx}{EI}\right). \tag{24.3}$$

Thus the second assumption, that plane sections remain plane and normal to the central line, only holds if u is so small that we can substitute x' for x in Eq. (24.3), which would then define a plane in the deformed beam with a normal given by Eq. (24.2). The radius of curvature of the central line is given by the formula

$$R = \frac{\left[1 + \left(\dfrac{dx'}{dz'}\right)^2\right]^{\frac{3}{2}}}{\dfrac{d^2 x'}{dz'^2}}, \qquad x = y = 0$$

which agrees with the elementary formula of Eq. (23.3) if dx'/dz' is sufficiently small.

The transition from pure bending, caused solely by couples at the ends of the beam, to the cantilever beam, where one end is fixed and the other is loaded by forces statically equivalent to a force perpendicular to the length of the beam, introduces considerably greater difficulty. The longitudinal stress, σ_z, is still found to be proportional to the bending moment, but the shear stresses τ_{xz}, τ_{yz} can no longer be regarded as zero, and because they must be zero on the lateral surface the shape of the cross-section becomes important. Accounts of the theory of flexure of beams may be found in various textbooks[1].

---
[1] I. S. SOKOLNIKOFF: Mathematical Theory of Elasticity, 2nd ed. New York: McGraw-Hill 1956. — S. TIMOSHENKO and J. N. GOODIER: Theory of Elasticity. New York: McGraw-Hill 1951. — CHI-TEH WANG: Applied Elasticity. New York: McGraw-Hill 1953.

C. Two-dimensional problems in elasto-statics.

It will be recalled that, in the theory of electrostatics, the discussion of "two-dimensional" problems is most fruitful since it enables the powerful theory of functions of a complex variable to be employed. A similar situation holds in the theory of elastostatics though, no doubt on account of the more complicated nature of the basic equations, it was much later before the theory of functions was brought to bear on elasto-static problems. This development was due almost entirely to the work of Kolosov and Muskhelishvili. Two kinds of two-dimensional problems arise in elasticity, but it emerges that the mathematical formulation of the two types is identical so that they can both be treated by the same methods.

I. General theory.

Two-dimensional problems in elasto-statics arise naturally in two distinct ways. In the first case the body being deformed is a long right cylinder acted upon by external forces which are so arranged that the component of the displacement in the direction of the axis of the cylinder vanishes, and the remaining components remain constant along the length of the cylinder; such a body is said to be in a state of *plane strain*. In the second case the body being deformed is a thin plate acted upon by external forces so distributed that the normal component of stress across the plate vanishes; the plate is said to be in a state of *plane stress*.

25. Plane strain. In a state of plane strain, we can choose the axis of the cylinder to be the z-axis so that $w=0$ and u and v are functions of x and y alone. That is, in plane strain, we can write

$$\frac{\partial}{\partial z} \equiv 0, \quad w = 0 \tag{25.1}$$

so that $\tau_{yz} = \tau_{zx} = 0$, $\gamma_{xz} = \gamma_{yz} = \gamma_{zz} = 0$. If we substitute these conditions into Eqs. (9.19) we find that the equations of equilibrium reduce to

$$\frac{\partial \sigma_x}{\partial x} + \frac{\partial \tau_{xy}}{\partial y} + \varrho X = 0, \tag{25.2}$$

$$\frac{\partial \tau_{xy}}{\partial x} + \frac{\partial \sigma_y}{\partial y} + \varrho Y = 0. \tag{25.3}$$

Similarly the stress-strain relations (14.30) yield the equations

$$\sigma_x = \lambda \left(\frac{\partial u}{\partial x} + \frac{\partial v}{\partial y} \right) + 2\mu \frac{\partial u}{\partial x}, \tag{25.4}$$

$$\sigma_y = \lambda \left(\frac{\partial u}{\partial x} + \frac{\partial v}{\partial y} \right) + 2\mu \frac{\partial v}{\partial y}, \tag{25.5}$$

$$\sigma_z = \lambda \left(\frac{\partial u}{\partial x} + \frac{\partial v}{\partial y} \right), \tag{25.6}$$

$$\tau_{xy} = \mu \left(\frac{\partial u}{\partial y} + \frac{\partial v}{\partial x} \right), \tag{25.7}$$

$$\tau_{yz} = \tau_{zx} = 0. \tag{25.8}$$

It follows immediately from Eqs. (25.4), (25.5) and (25.6) that, in a state of plane strain,

$$\sigma_z = \frac{\lambda}{\lambda + 2\mu} (\sigma_x + \sigma_y) = \frac{\nu}{1-\nu} (\sigma_x + \sigma_y). \tag{25.9}$$

In certain problems of plane strain it is advantageous to use polar coordinates r and ϑ instead of plane cartesian coordinates x and y. In these coordinates the equations of equilibrium reduce to

$$\frac{\partial \sigma_r}{\partial r} + \frac{1}{r}\frac{\partial \tau_{r\vartheta}}{\partial \vartheta} + \frac{\sigma_r - \sigma_\vartheta}{r} = 0, \tag{25.10}$$

$$\frac{1}{r}\cdot\frac{\partial \sigma_\vartheta}{\partial \vartheta} + \frac{\partial \tau_{r\vartheta}}{\partial r} + \frac{2\tau_{r\vartheta}}{r} = 0 \tag{25.11}$$

and the stress strain relations to

$$\sigma_r = \lambda\left(\frac{\partial u_r}{\partial r} + \frac{u_r}{r} + \frac{1}{r}\frac{\partial u_\vartheta}{\partial \vartheta}\right) + 2\mu\frac{\partial u_r}{\partial r}, \tag{25.12}$$

$$\sigma_\vartheta = \lambda\left(\frac{\partial u_r}{\partial r} + \frac{u_r}{r} + \frac{1}{r}\frac{\partial u_\vartheta}{\partial \vartheta}\right) + 2\mu\left(\frac{u_r}{r} + \frac{1}{r}\frac{\partial u_\vartheta}{\partial \vartheta}\right), \tag{25.13}$$

$$\sigma_z = \frac{\nu}{1-\nu}(\sigma_r + \sigma_\vartheta), \tag{25.14}$$

$$\tau_{r\vartheta} = \mu\left(\frac{\partial u_\vartheta}{\partial r} - \frac{u_\vartheta}{r} + \frac{1}{r}\frac{\partial u_r}{\partial \vartheta}\right), \tag{25.15}$$

$$\tau_{zr} = \tau_{z\vartheta} = 0. \tag{25.16}$$

26. Plane stress. In a state of plane stress parallel to the xy-plane, the stress components τ_{xz}, τ_{yz}, σ_z all vanish but the components u, v, w of the displacement vector are not in general independent of z. We find, therefore, that the equations of equilibrium assume the forms (25.2) and (25.3) and that the stress-strain equations yield the set

$$\sigma_x = \lambda\left(\frac{\partial u}{\partial x} + \frac{\partial v}{\partial y} + \frac{\partial w}{\partial z}\right) + 2\mu\frac{\partial u}{\partial x}, \tag{26.1}$$

$$\sigma_y = \lambda\left(\frac{\partial u}{\partial x} + \frac{\partial v}{\partial y} + \frac{\partial w}{\partial z}\right) + 2\mu\frac{\partial v}{\partial y}, \tag{26.2}$$

$$0 = \lambda\left(\frac{\partial u}{\partial x} + \frac{\partial v}{\partial y} + \frac{\partial w}{\partial z}\right) + 2\mu\frac{\partial w}{\partial z}, \tag{26.3}$$

$$0 = \mu\left(\frac{\partial v}{\partial z} + \frac{\partial w}{\partial y}\right), \tag{26.4}$$

$$0 = \mu\left(\frac{\partial u}{\partial z} + \frac{\partial w}{\partial x}\right), \tag{26.5}$$

$$\tau_{xy} = \mu\left(\frac{\partial u}{\partial y} + \frac{\partial v}{\partial x}\right). \tag{26.6}$$

If we eliminate $\partial w/\partial z$ from Eqs. (26.1), (26.2) and (26.3) and if we write $\lambda' = (1-2\nu)\lambda/(1-\nu)$, we obtain the equations

$$\left.\begin{aligned}\sigma_x &= \lambda'\left(\frac{\partial u}{\partial x} + \frac{\partial v}{\partial y}\right) + 2\mu\frac{\partial u}{\partial x}, \\ \sigma_y &= \lambda'\left(\frac{\partial u}{\partial x} + \frac{\partial v}{\partial y}\right) + 2\mu\frac{\partial v}{\partial y}, \\ \tau_{xy} &= \mu\left(\frac{\partial u}{\partial y} + \frac{\partial v}{\partial x}\right)\end{aligned}\right\} \tag{26.7}$$

for the non-vanishing components of the stress tensor. Similarly Eqs. (26.3), (26.4) and (26.5) give the relations

$$\frac{\partial w}{\partial x} = -\frac{\partial u}{\partial z}, \quad \frac{\partial w}{\partial y} = -\frac{\partial v}{\partial z}, \quad \frac{\partial w}{\partial z} = -\frac{\nu}{1-\nu}\left(\frac{\partial u}{\partial x} + \frac{\partial v}{\partial y}\right) \quad (26.8)$$

which enable w to be calculated from u und v apart from a constant term.

27. Generalized plane stress. The plane stress problem is not a truly two-dimensional problem since the variable z may appear as a parameter in each of the equations above. However, it was shown by Filon[1] that the system of equations may be modified by supposing that the stress component σ_z vanishes throughout the plate but that the shearing stresses τ_{xz}, τ_{yz} vanish only on the faces $z = \pm h$ of the plate. Filon's idea was that, if the plate is thin, a knowledge of the *average* values of the components of the displacement vector and of the stress tensor is as useful as that of the exact values at each point. For this reason we replace each physical quantity f by its average value $\bar{f}$ defined by the equation

$$\bar{f} = \frac{1}{2h}\int_{-h}^{h} f \, dz. \quad (27.1)$$

If we integrate both sides of the equations

$$\frac{\partial \sigma_x}{\partial x} + \frac{\partial \tau_{xy}}{\partial y} + \frac{\partial \tau_{xz}}{\partial z} + \varrho X = 0, \quad \frac{\partial \tau_{xy}}{\partial x} + \frac{\partial \sigma_y}{\partial y} + \frac{\partial \tau_{yz}}{\partial z} + \varrho Y = 0$$

over the thickness of the plate and use the fact that τ_{xz} and τ_{yz} both vanish on the faces of the plate we find that the average stress components $\bar{\sigma}_x, \bar{\sigma}_y, \bar{\sigma}_z$ satisfy the equations

$$\frac{\partial \bar{\sigma}_x}{\partial x} + \frac{\partial \bar{\tau}_{xy}}{\partial y} + \varrho \bar{X} = 0, \quad \frac{\partial \bar{\tau}_{xy}}{\partial x} + \frac{\partial \bar{\sigma}_y}{\partial y} + \varrho \bar{Y} = 0. \quad (27.2)$$

Since $\bar{\sigma}_z$ vanishes identically we find that

$$\overline{\left(\frac{\partial w}{\partial z}\right)} = -\frac{\nu}{1-\nu}\left(\frac{\partial \bar{u}}{\partial x} + \frac{\partial \bar{v}}{\partial y}\right) \quad (27.3)$$

so that

$$\bar{\sigma}_x = \lambda'\left(\frac{\partial \bar{u}}{\partial x} + \frac{\partial \bar{v}}{\partial y}\right) + 2\mu \frac{\partial \bar{u}}{\partial x}, \quad (27.4)$$

$$\bar{\sigma}_y = \lambda'\left(\frac{\partial \bar{u}}{\partial x} + \frac{\partial \bar{v}}{\partial y}\right) + 2\mu \frac{\partial \bar{v}}{\partial y}, \quad (27.5)$$

$$\bar{\tau}_{xy} = \mu\left(\frac{\partial \bar{u}}{\partial y} + \frac{\partial \bar{v}}{\partial x}\right) \quad (27.6)$$

where

$$\lambda' = \frac{2\lambda\mu}{\lambda + 2\mu} = \frac{1-2\nu}{1-\nu}\lambda \quad (27.7)$$

ν being Poisson's ratio.

II. Airy stress function.

28. The biharmonic equation. A fruitful method for solving the two-dimensional problem was originated by Airy[2]. If, in the equations of equilibrium (25.2) and (25.3) it is supposed that the body force components X, Y are derived

---
[1] L. N. G. Filon: Phil. Trans. Roy. Soc. Lond., Ser. A **201**, 63 (1903).
[2] G. B. Airy: Brit. Assoc. Rep. 1862. Phil. Trans. Roy. Soc. Lond. **153**, 49 (1863).

from a potential function $V(x, y)$ such that

$$X = -\frac{\partial V}{\partial x}, \quad Y = -\frac{\partial V}{\partial y}, \tag{28.1}$$

and that there exists a function χ such that

$$\tau_{xy} = -\frac{\partial^2 \chi}{\partial x \partial y},$$

then they reduce to the forms

$$\frac{\partial}{\partial x}\left(\sigma_x - \frac{\partial^2 \chi}{\partial y^2} - \varrho V\right) = 0, \quad \frac{\partial}{\partial y}\left(\sigma_y - \frac{\partial^2 \chi}{\partial x^2} - \varrho V\right) = 0.$$

It follows at once that the equations of equilibrium are satisfied by the expressions

$$\sigma_x = \frac{\partial^2 \chi}{\partial y^2} + \varrho V, \quad \sigma_y = \frac{\partial^2 \chi}{\partial x^2} + \varrho V, \quad \tau_{xy} = -\frac{\partial^2 \chi}{\partial x \partial y}, \tag{28.2}$$

where the function χ is known as the *Airy stress function*.

In order that the expressions (28.2) should provide valid solutions of elastic problems they must not only satisfy the equations of equilibrium but also the compatibility relation for the stresses. This takes two forms, depending on whether the case of plane strain or that of plane stress is considered. For plane strain the relation reduces to

$$\frac{\partial^2}{\partial y^2}[\sigma_x - \nu(\sigma_x + \sigma_y)] + \frac{\partial^2}{\partial x^2}[\sigma_y - \nu(\sigma_x + \sigma_y)] = 2\frac{\partial^2 \tau_{xy}}{\partial x \partial y} \tag{28.3}$$

and for plane stress to

$$\frac{\partial^2}{\partial y^2}(\sigma_x - \nu \sigma_y) + \frac{\partial^2}{\partial x^2}(\sigma_y - \nu \sigma_x) = 2(1 + \nu)\frac{\partial^2 \tau_{xy}}{\partial x \partial y}. \tag{28.4}$$

By substituting the expressions (28.2) into these equations we find that the stress function χ must satisfy, in the case of plane strain, the partial differential equation

$$\nabla_1^4 \chi + \frac{1 - 2\nu}{1 - \nu}\varrho \nabla_1^2 V = 0 \tag{28.5}$$

or, in the case of plane stress, the equation

$$\nabla_1^4 \chi + (1 - \nu)\varrho \nabla_1^2 V = 0, \tag{28.6}$$

where ∇_1^2 denotes the two-dimensional Laplacian operator $\partial^2/\partial x^2 + \partial^2/\partial y^2$. When the body force components (X, Y) are constant throughout the material,

$$\nabla_1^2 V = 0,$$

and Eqs. (28.5) and (28.6) both reduce to the two-dimensional biharmonic equation

$$\nabla_1^4 \chi = 0. \tag{28.7}$$

A number of interesting solutions of Eq. (28.7) can be obtained by postulating functions in the form of polynomials of various degrees, and adjusting their coefficients to satisfy different conditions[1]. The simplest case arises when $V(x, y) = 0$ and we assume that

$$\chi = \tfrac{1}{2} a_2 x^2 + b_2 x y + \tfrac{1}{2} c_2 y^2. \tag{28.8}$$

---

[1] A. MESNAGER: C. R. Acad. Sci., Paris **132**, 1475 (1901). See also A. TIMPE: Z. Math. Phys. **52**, 348 (1905).

Then, from Eqs. (28.2)
$$\sigma_x = c_2, \quad \sigma_y = a_2, \quad \tau_{xy} = -b_2. \tag{28.9}$$

so that all three stress components are constant, and we have a combination of uniform normal stresses and uniform shear with no body force. We consider two other cases of polynomial stress functions and show how they may be superposed to obtain further solutions. Eq. (28.7) is satisfied by the cubic polynomial

$$\chi = \tfrac{1}{6} a_3 x^3 + \tfrac{1}{2} b_3 x^2 y + \tfrac{1}{2} c_3 x y^2 + \tfrac{1}{6} d_3 y^3, \tag{28.10}$$

and, from Eq. (28.2) we see that in the absence of body forces, the corresponding stresses are

$$\sigma_x = c_3 x + d_3 y, \quad \sigma_y = a_3 x + b_3 y, \quad \tau_{xy} = -b_3 x - c_3 y. \tag{28.11}$$

Various cases are obtained from this set of stresses by making some of the coefficients zero. If, for example, only a_3 is non-zero, σ_y is the only non-zero stress component; but if one of b_3, c_3 is the only non-zero coefficient, a shearing stress is present in addition to a normal stress. When polynomials of higher degree than the third are used, the bi-harmonic Eq. (28.7) is satisfied only if certain conditions are imposed on the coefficients. Proceeding to the quintic polynomial, we put

$$\chi = \tfrac{1}{20} a_5 x^5 + \tfrac{1}{12} b_5 x^4 y + \tfrac{1}{6} c_5 x^3 y^2 + \tfrac{1}{6} d_5 x^2 y^3 + \tfrac{1}{12} e_5 x y^4 + \tfrac{1}{20} f_5 y^5, \tag{28.12}$$

which satisfies Eq. (28.7) if

$$e_5 = -(2c_5 + 3a_5), \quad f_5 = -\tfrac{1}{3}(b_5 + 2d_5).$$

From Eqs. (28.2) we find that in the absence of body forces the stress components are given by the equations

$$\left.\begin{aligned}
\sigma_x &= \tfrac{1}{3} c_5 x^3 + d_5 x^2 y - (2c_5 + 3a_5) x y^2 - \tfrac{1}{3}(b_5 + 2d_5) y^3, \\
\sigma_y &= a_5 x^3 + b_5 x^2 y + c_5 x y^2 + \tfrac{1}{3} d_5 y^3, \\
\tau_{xy} &= -\tfrac{1}{3} b_5 x^3 - c_5 x^2 y - d_5 x y^2 + \tfrac{1}{3}(2c_5 + 3a_5) y^3.
\end{aligned}\right\} \tag{28.13}$$

29. The loading of a light beam. To demonstrate the utility of the method we use the stress solutions (28.13) as the basis in discussing the stress in a light beam supported at its ends and bearing a uniform load. We suppose that the beam $-l \leq x \leq l, -c \leq y \leq c$ can be considered as a thin plate of unit thickness and bears a uniformly distributed load of intensity q on the surface $y = -c$ acting in the direction of y increasing. Then

$$\begin{aligned}
\tau_{xy} &= 0 \quad \text{at} \quad y = \pm c, \\
\sigma_y &= -q \quad \text{at} \quad y = -c, \\
\sigma_y &= 0 \quad \text{at} \quad y = +c,
\end{aligned}$$

while at $x = \pm l$

$$\int_{-c}^{c} \tau_{xy}\, dy = \mp ql, \quad \int_{-c}^{c} \sigma_x\, dy = 0, \quad \int_{-c}^{c} \sigma_x y\, dy = 0.$$

These boundary conditions can be satisfied by a combination of the stress solutions (28.9), (28.11) and (28.13). Since, in the usual elementary theory of bending, the stress component σ_x is given by an expression of the third degree in x and y, we take Eqs. (28.13) as the basic expressions, putting $a_5 = b_5 = c_5 = 0$ in order that σ_y should be uniform along the surface $y = -c$. By superposing solutions (28.9) with $b_2=c_2=0$ and solutions (28.11) with $a_3=c_3=d_3=0$ the remaining conditions at $y=\pm c$ can be satisfied. For, then,

$$\begin{aligned}\sigma_x &= d_5(x^2 y - \tfrac{3}{2} y^3), \\ \sigma_y &= \tfrac{1}{3} d_5 y^3 + b_3 y + a_2, \\ \tau_{xy} &= - d_5 x y^2 - b_3 x,\end{aligned} \quad (29.1)$$

and these conditions are satisfied if

$$\begin{aligned}&- d_5 c^2 - b_3 = 0, \\ &\tfrac{1}{3} d_5 c^3 + b_3 c + a_2 = 0, \\ &- \tfrac{1}{3} d_5 c^3 - b_3 c + a_2 = -q,\end{aligned}$$

that is, if

$$a_2 = -\tfrac{1}{2} q, \quad b_3 = \frac{3q}{4c}, \quad d_5 = -\frac{3q}{4c^3}.$$

Using these values and substituting I for $\tfrac{2}{3} c^3$, the second moment of the cross-sectional area of the beam about $y=0$, the expressions (29.1) become

$$\begin{aligned}\sigma_x &= -\frac{q}{2I}\left(x^2 y - \frac{2}{3} y^3\right), \\ \sigma_y &= -\frac{q}{2I}\left(\frac{1}{3} y^3 - c^2 y + \frac{2}{3} c^3\right), \\ \tau_{xy} &= -\frac{q}{2I}(c^2 - y^2) x.\end{aligned} \quad (29.2)$$

These solutions satisfy the first two conditions at $x=\pm l$; in order that the couples at the ends should vanish we now superpose a pure bending, $\sigma_x=d_3 y$, represented by solutions (28.11) when only d_3 is non-zero. Then we must have, at $x=\pm l$

$$\int_{-c}^{c} \sigma_x y\, dy = \int_{-c}^{c}\left\{-\frac{q}{2I}\left(l^2 y - \frac{2}{3} y^3\right) + d_3 y\right\} y\, dy = 0,$$

from which

$$d_3 = \frac{q}{2I}\left(l^2 - \frac{2}{5} c^2\right)$$

so that

$$\sigma_x = \frac{q}{2I}\left\{(l^2 - x^2) y + \frac{2}{3} y^3 - \frac{2}{5} c^2 y\right\} \quad (29.3)$$

in place of the first of expressions (29.2). The first term in this expression represents the solution given by the elementary theory of bending. It will be seen that the normal stress σ_x is non-zero at $x=\pm l$, but since the resultant force and resultant couple are zero, the effect at considerable distances from the ends will be practically the same as if σ_x were zero at $x=\pm l$.

30. Airy's function in polar coordinates. The stress function may be used in a polar coordinate system (r, ϑ). When the body force is zero the equations of equilibrium (25.10) and (25.11) are satisfied by putting

$$\begin{aligned}\sigma_r &= \frac{1}{r}\frac{\partial \chi}{\partial r} + \frac{1}{r^2}\frac{\partial^2 \chi}{\partial \vartheta^2}, \\ \sigma_\vartheta &= \frac{\partial^2 \chi}{\partial r^2}, \\ \tau_{r\vartheta} &= \frac{1}{r^2}\frac{\partial \chi}{\partial \vartheta} - \frac{1}{r}\frac{\partial^2 \chi}{\partial r\, \partial \vartheta}.\end{aligned} \quad (30.1)$$

These are valid solutions for plane strain or plane stress if χ, considered as a function r and ϑ, satisfies the bi-harmonic equation (28.7), which becomes, in polar coordinates,

$$\left(\frac{\partial^2}{\partial r^2} + \frac{1}{r}\frac{\partial}{\partial r} + \frac{1}{r^2}\frac{\partial^2}{\partial \vartheta^2}\right)\left(\frac{\partial^2 \chi}{\partial r^2} + \frac{1}{r}\frac{\partial \chi}{\partial r} + \frac{1}{r^2}\frac{\partial^2 \chi}{\partial \vartheta^2}\right) = 0. \tag{30.2}$$

We consider the case of stress distribution symmetrical about an axis $r=0$. Since the distribution is now independent of ϑ, Eq. (30.2) reduces to the ordinary differential equation

$$\frac{d^4\chi}{dr^4} + \frac{2}{r}\frac{d^3\chi}{dr^3} - \frac{1}{r^2}\frac{d^2\chi}{dr^2} + \frac{1}{r^2}\frac{d\chi}{dr} = 0,$$

which has the general solution

$$\chi = A \log r + B r^2 \log r + C r^2 + D. \tag{30.3}$$

Substitution of this function in the expressions (30.1) gives, for axial symmetry, the stress components

$$\left.\begin{array}{l} \sigma_r = \dfrac{A}{r^2} + B(1 + 2\log r) + 2C, \\[4pt] \sigma_\vartheta = -\dfrac{A}{r^2} + B(3 + 2\log r) + 2C, \\[4pt] \tau_{r\vartheta} = 0. \end{array}\right\} \tag{30.4}$$

If there is no hole at $r=0$ the constants A and B must be zero, and it follows that then there is only one type of axially symmetrical stress distribution, given by $\sigma_r = \sigma_\vartheta = $ const.

31. The deformation of a hollow cylinder. A problem which is readily solved by means of the results (30.4) is that of a hollow cylinder subjected to uniform pressures on the inner and outer surfaces. The original solution is due to LAMÉ[1]. The boundary conditions are

$$\left.\begin{array}{ll} \sigma_r = -p_a, & \text{at } r = a, \\ \sigma_r = -p_b, & \text{at } r = b, \end{array}\right\} \tag{31.1}$$

$a<b$. Since there are three arbitrary constants in expressions (30.4) but only two boundary conditions, it appears that the stresses are indeterminate. This is a situation normally found in the case of a multiply-connected body, and to determine the stress distribution completely in such an instance equations representing the conditions that the displacements should be single valued must be considered. Here, if we suppose that there is a cut at $\vartheta = \alpha$ say, so that an axially symmetrical stress distribution does not necessarily produce symmetrical displacement, the condition is effectively that the displacement should be continuous at $\vartheta = \alpha$. If u, v are the radial and tangential components of the displacement, respectively, then, for plane strain, the radial component of strain is given by

$$\frac{\partial u}{\partial r} = \frac{1+\nu}{E}\{(1-\nu)\sigma_r - \nu\sigma_\vartheta\}$$

$$= \frac{1+\nu}{E}\left\{\frac{A}{r^2} + B[(1-4\nu) + (2-4\nu)\log r] + c(2-4\nu)\right\}$$

---
[1] Leçons sur la théorie ... de l'élasticité. Paris 1852.

and hence
$$u = \frac{1+\nu}{E}\left\{-\frac{A}{r} - Br + B(2-4\nu)r\log r + c(2-4\nu)r\right\} + f(\vartheta) \qquad (31.2)$$

where $f(\vartheta)$ is a function of ϑ only; while the tangential component of strain is

$$\frac{u}{r} + \frac{1}{r}\frac{\partial v}{\partial \vartheta} = \frac{1+\nu}{E}\{-\nu\sigma_r + (1-\nu)\sigma_\vartheta\}$$
$$= \frac{1+\nu}{E}\left\{-\frac{A}{r^2} + B[(3-4\nu)\log r] + C(2-4\nu)\right\}$$

and, using Eq. (31.2), we find that

$$v = \frac{4B(1-\nu^2)}{E}r\vartheta - \int f(\vartheta)\,d\vartheta + g(r) \qquad (31.3)$$

where $g(r)$ is a function of r only. Since

$$\tau_{r\vartheta} = 0, \qquad \gamma_{r\vartheta} = \frac{1}{r}\frac{\partial u}{\partial \vartheta} + \frac{\partial v}{\partial r} - \frac{v}{r} = 0$$

and by substitution from Eqs. (31.2) and (31.3)

$$\frac{1}{r}f'(\vartheta) + g'(r) = \frac{1}{r}\int f(\vartheta)\,d\vartheta + \frac{1}{r}g(r)$$

so that
$$g(r) = Fr, \qquad f(\vartheta) = H\sin\vartheta + K\cos\vartheta$$

where F, H and K are constants to be determined. Substituting these functions in Eq. (31.3) we find that the tangential displacement is

$$v = \frac{4B(1-\nu^2)r\vartheta}{E} + Fr + H\cos\vartheta - K\sin\vartheta.$$

Clearly, the displacement is not single valued unless $B=0$, and this is the required condition, enabling the stress distribution to be determined unambiguously.

Putting $B=0$ in the expressions (30.4) and using the boundary conditions (31.1), we find that

$$\left.\begin{aligned}A &= \frac{(p_b - p_a)a^2 b^2}{b^2 - a^2}, \\ C &= \frac{a^2 p_a - b^2 p_b}{2(b^2 - a^2)}\end{aligned}\right\} \qquad (31.4)$$

and, by substitution, the stress components are

$$\left.\begin{aligned}\sigma_r &= \frac{a^2 b^2 (p_b - p_a)}{(b^2 - a^2)r^2} + \frac{a^2 p_a - b^2 p_b}{b^2 - a^2}, \\ \sigma_\vartheta &= -\frac{a^2 b^2 (p_b - p_a)}{(b^2 - a^2)r^2} + \frac{a^2 p_a - b^2 p_b}{b^2 - a^2}.\end{aligned}\right\} \qquad (31.5)$$

In the present case, the displacement must be axially symmetrical, so that $f(\vartheta) \equiv 0$. Then, substituting from Eqs. (31.4) in expressions (31.2) and (31.3), the components of displacement are found to be

$$\left.\begin{aligned}u &= \frac{1+\nu}{E(b^2-a^2)}\left\{\frac{a^2 b^2(p_a - p_b)}{r} + (1-2\nu)(a^2 p_a - b^2 p_b)r\right\}, \\ v &= Fr\end{aligned}\right\} \qquad (31.6)$$

where the tangential component, v, is merely a pure rotation.

III. Complex potentials.

32. Derivation of potentials. Two-dimensional elasticity seems an obvious field in which to use complex variable methods, but the early work of KOLOSOV[1] and, later, MUSKHELISHVILI[2] remained unknown outside Russia until recent years. The fundamental equations were derived independently by STEVENSON[3], whose approach is adopted here since it seems more direct and, unlike the work of KOLOSOV and MUSKHELISHVILI, does not neglect body forces.

The variables (x, y) are replaced by the complex conjugate variables

$$z = x + iy, \quad \bar{z} = x - iy$$

from which we have

$$\frac{\partial}{\partial x} = \frac{\partial}{\partial z} + \frac{\partial}{\partial \bar{z}}, \quad \frac{\partial}{\partial y} = i\left(\frac{\partial}{\partial z} - \frac{\partial}{\partial \bar{z}}\right). \tag{32.1}$$

The body forces are represented in terms of a potential function $V(x, y) = U(z, \bar{z})$ such that

$$X = -\frac{\partial V}{\partial x}, \quad Y = -\frac{\partial V}{\partial y}$$

so that the equations of equilibrium can be written

$$\frac{\partial}{\partial x}(\sigma_x - \varrho V) + \frac{\partial \tau_{xy}}{\partial y} = 0, \tag{32.2}$$

$$\frac{\partial}{\partial y}(\sigma_y - \varrho V) + \frac{\partial \tau_{xy}}{\partial x} = 0. \tag{32.3}$$

Then, multiplying Eq. (32.3) by i and adding to Eq. (32.2), we have by changing the variables to $(z, \bar{z})$ the complex equation

$$\frac{\partial}{\partial z}(\sigma_x - \sigma_y + 2i\tau_{xy}) + \frac{\partial}{\partial \bar{z}}(\sigma_x + \sigma_y - 2\varrho U) = 0 \tag{32.4}$$

which is obviously satisfied by a function $F(z, \bar{z})$ where

$$\frac{\partial F}{\partial z} = \sigma_x + \sigma_y - 2\varrho U, \tag{32.5}$$

$$-\frac{\partial F}{\partial \bar{z}} = \sigma_x - \sigma_y + 2i\tau_{xy}. \tag{32.6}$$

The complex displacement D is defined by the equation

$$D = u + iv.$$

Using the relations (32.1) we have the equation

$$2\frac{\partial D}{\partial \bar{z}} = \frac{\partial u}{\partial x} - \frac{\partial v}{\partial y} + i\left(\frac{\partial v}{\partial x} + \frac{\partial u}{\partial y}\right)$$

and then, by the stress-strain relations (25.4), (25.5) and (25.7)

$$4\mu \frac{\partial D}{\partial \bar{z}} = \sigma_x - \sigma_y + 2i\tau_{xy}. \tag{32.7}$$

---

[1] G. V. KOLOSOV: C. R. Acad. Sci., Paris **146**, 522 (1908); **148**, 1242 (1909). — Z. Math. Phys. **62**, 384 (1914). Application of the Complex Variable to the Theory of Elasticity. Moscow-Leningrad 1935.

[2] N. I. MUSKHELISHVILI: Math. Ann. **107**, 282 (1932). — C. R. Acad. Sci. USSR. **3**, 7, 73, 141 (1934). Some Basic Problems of the Mathematical Theory of Elasticity (Translated by J. R. M. RADOK). Groningen: Noordhoff 1953.

[3] A. C. STEVENSON: Proc. Roy. Soc. Lond., Ser. A **184**, 129 (1945). — Phil. Mag., Ser. VII **34**, 766 (1943).

Sect. 32.   Derivation of potentials.

From Eqs. (32.6) and (32.7)
$$4\mu \frac{\partial D}{\partial \bar{z}} = -\frac{\partial F}{\partial \bar{z}}$$
and hence
$$4\mu D = f(z) - F(z, \bar{z}) \tag{32.8}$$
where $f(z)$ is a function of z alone. Differentiating D with respect to z and using the relations (32.1) we find that
$$2\frac{\partial D}{\partial z} = \frac{\partial u}{\partial x} + \frac{\partial v}{\partial y} + i\left(\frac{\partial v}{\partial x} - \frac{\partial u}{\partial y}\right)$$
and hence, from the Eqs. (25.4) and (25.5), in the case of plane strain,
$$4(\lambda + \mu) \frac{\partial D}{\partial z} = \sigma_x + \sigma_y + 2i(\lambda + \mu)\left(\frac{\partial v}{\partial x} - \frac{\partial u}{\partial y}\right). \tag{32.9}$$
Differentiating Eq. (32.8) with respect to z and substituting for $\partial F/\partial z$ from Eq. (32.5) we obtain the equation
$$4\mu \frac{\partial D}{\partial z} = f'(z) - (\sigma_x + \sigma_y) + 2\varrho U. \tag{32.10}$$
Elimination of $\partial D/\partial z$ between Eqs. (32.9) and (32.10) yields the equation
$$\sigma_x + \sigma_y - \frac{\varrho U}{1-\nu} - \frac{i\mu}{1-\nu}\left(\frac{\partial v}{\partial x} - \frac{\partial u}{\partial y}\right) = 4\varphi'(z) \tag{32.11}$$
where
$$\varphi(z) = \frac{f(z)}{8(1-\nu)} \tag{32.12}$$
and ν is Poisson's ratio. The real part of Eq. (32.11) is
$$\sigma_x + \sigma_y - \frac{\varrho U}{1-\nu} = 2[\varphi'(z) + \overline{\varphi}'(\bar{z})]$$
so that, from Eq. (32.5),
$$\frac{\partial F}{\partial z} = 2[\varphi'(z) + \overline{\varphi}'(\bar{z})] - \left(\frac{1-2\nu}{1-\nu}\right)\varrho \frac{\partial W}{\partial z} \tag{32.13}$$
where
$$\frac{\partial}{\partial z} W(z, \bar{z}) = U.$$
Then, integrating with respect to z
$$F(z, \bar{z}) = 2[\varphi(z) + z\overline{\varphi}'(\bar{z})] - \left(\frac{1-2\nu}{1-\nu}\right)\varrho W(z, \bar{z}) + 2\overline{\psi}(\bar{z}) \tag{32.14}$$
where $\overline{\psi}(\bar{z})$ is a function of $\bar{z}$ alone. From Eqs. (32.8), (32.12) and (32.14) the complex displacement is given by
$$2\mu(u + iv) = \varkappa\varphi(z) - z\overline{\varphi}'(\bar{z}) - \overline{\psi}(\bar{z}) + \tfrac{1}{2}c\varrho W(z, \bar{z}) \tag{32.15}$$
where
$$\varkappa = 3 - 4\nu, \qquad c = \frac{1-2\nu}{1-\nu}$$
while, from Eqs. (32.5), (32.6) and (32.14),
$$\sigma_x + \sigma_y = 2[\varphi'(z) + \overline{\varphi}'(\bar{z})] + (2-c)\varrho \frac{\partial W}{\partial z}, \tag{32.16}$$
$$\sigma_x - \sigma_y + 2i\tau_{xy} = -2[z\overline{\varphi}''(\bar{z}) + \overline{\psi}'(\bar{z})] + c\varrho \frac{\partial W}{\partial \bar{z}}. \tag{32.17}$$

Handbuch der Physik, Bd. VI.

Thus, for a given body force function $W(z, \bar{z})$ the displacement and stress configurations are given completely in terms of the complex potentials $\varphi(z)$, $\psi(z)$ through Eqs. (32.15), (32.16) and (32.17). In Sect. 27 it was shown that solutions true for plane strain hold also for generalised plane stress if, instead of ν, a "modified Poisson's ratio", $\sigma = \nu/(1+\nu)$ is substituted. Here, as Stevenson[1] shows a further condition must be imposed, namely that the body force potential $V(x, y)$ should satisfy the biharmonic equation

$$\nabla_1^4 V = 0.$$

33. Relation to Airy's stress function.
Muskhelishvili derived the above equations, with zero body forces, from a consideration of the Airy stress function. The relation between the stress function and the complex potentials is readily demonstrated. When $V(x, y) \equiv 0$ we have, from Eq. (28.2) and the relations (32.1),

$$\sigma_x + \sigma_y = 4 \frac{\partial^2 \chi}{\partial z \, \partial \bar{z}},$$

$$\sigma_x - \sigma_y + 2i\tau_{xy} = -4 \frac{\partial^2 \chi}{\partial \bar{z}^2}$$

where χ is the stress function. Then, from Eqs. (32.16) and (32.17) we have

$$\left. \begin{array}{l} 2 \dfrac{\partial^2 \chi}{\partial z \, \partial \bar{z}} = \varphi'(z) + \overline{\varphi}'(\bar{z}), \\[2mm] 2 \dfrac{\partial^2 \chi}{\partial \bar{z}^2} = z \overline{\varphi}''(\bar{z}) + \overline{\psi}'(\bar{z}). \end{array} \right\} \quad (33.1)$$

It is clear at once that χ satisfies the biharmonic equation

$$\nabla_1^4 \chi = \frac{\partial^4 \chi}{\partial z^2 \, \partial \bar{z}^2} = 0.$$

Integrating Eqs. (33.1) we have

$$2 \frac{\partial \chi}{\partial \bar{z}} = \varphi(z) + z \overline{\varphi}'(\bar{z}) + f_1(\bar{z}),$$

$$2 \frac{\partial \chi}{\partial \bar{z}} = z \overline{\varphi}'(z) + \overline{\psi}(\bar{z}) + f_2(z)$$

from which

$$f_1(\bar{z}) = \overline{\psi}(\bar{z}),$$
$$f_2(z) = \varphi(z),$$

and hence

$$2 \frac{\partial \chi}{\partial \bar{z}} = \varphi(z) + z \overline{\varphi}'(z) + \overline{\psi}(\bar{z})$$

and

$$2\chi = \bar{z}\,\varphi(z) + z\overline{\varphi}(\bar{z}) + \int \overline{\psi}(\bar{z})\, d\bar{z} + g(z) \quad (33.2)$$

where $g(z)$ is a function of z alone. Since all its second derivatives are real χ must be real apart from a possible imaginary constant. We must have then, in Eq. (33.2)

$$\bar{g}(\bar{z}) = \int \overline{\psi}(\bar{z})\, d\bar{z} + c$$

where c is an arbitrary complex constant, and the relation between the stress function and the complex potentials is determined except for this constant. Putting $c = 0$ we have

$$2\chi = x\,\varphi_1(x, y) + y\,\varphi_2(x, y) + g_1(x, y) \quad (33.3)$$

---
[1] Phil. Mag., Ser. VII **34**, 766 (1943).

where
$$\varphi = \varphi_1 + i\varphi_2,$$
$$g = g_1 + ig_2$$

and, since φ and g are analytic functions $\varphi_1, \varphi_2, g_1, g_2$ are harmonic functions. MUSKHELISHVILI arrived at the complex potentials by observing that every biharmonic function may be represented in terms of harmonic functions in the manner of Eq. (33.3).

34. Transformation to orthogonal curvilinear systems. In many applications of this theory it will be necessary to have expressions for the stress and displacement components in an orthogonal curvilinear coordinate system (ξ, η) which we take to be defined by the conformal transformation $z = w(\zeta)$ where $\zeta = \xi + i\eta$. First consider the rectangular coordinate axes Pn, Ps formed by the normal, n, to the curve $\eta = \text{const}$ at the point $P(x, y)$ and the normal, s, to the curve $\xi = \text{const}$ at the same

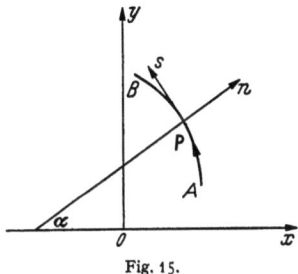

Fig. 15.

point, taken in the directions ξ increasing, η increasing, respectively. Since the stress components form a tensor it follows that if n makes an angle α with the x-axis (cf. Fig. 15), the stress components $\sigma_n, \sigma_s, \tau_{ns}$ in the (n, s) coordinate system are obtained by the transformation

$$\sigma_n = \sigma_x \cos^2\alpha + \sigma_y \sin^2\alpha + 2\tau_{xy}\sin\alpha\cos\alpha,$$
$$\sigma_s = \sigma_x \sin^2\alpha + \sigma_y \cos^2\alpha - 2\tau_{xy}\sin\alpha\cos\alpha,$$
$$\tau_{ns} = -\sigma_x \sin\alpha\cos\alpha + \sigma_y \sin\alpha\cos\alpha + \tau_{xy}(\cos^2\alpha - \sin^2\alpha).$$

It follows easily that

$$\left.\begin{aligned}\sigma_n + \sigma_s &= \sigma_x + \sigma_y, \\ \sigma_n - \sigma_s + 2i\tau_{ns} &= e^{-2i\alpha}(\sigma_x - \sigma_y + 2i\tau_{xy}), \\ \sigma_n - \sigma_s - 2i\tau_{ns} &= e^{2i\alpha}(\sigma_x - \sigma_y - 2i\tau_{xy}).\end{aligned}\right\} \quad (34.1)$$

Now if an increment dz_1 is taken along the curve $\eta = \text{const.}$ at the point P

$$dz_1 = J e^{i\alpha} d\xi$$

where J is real, and since $z = w(\zeta)$ is analytic at P

$$J e^{i\alpha} = w'(\zeta).$$

Hence
$$J^2 = |w'(\zeta)|^2 = w'(\zeta)\overline{w}'(\bar\zeta) \qquad (34.2)$$

and
$$e^{-2i\alpha} = \left[\frac{\overline{w}'(\bar\zeta)}{|w'(\zeta)|}\right]^2 = \frac{\overline{w}'(\bar\zeta)}{w'(\zeta)}. \qquad (34.3)$$

Then, writing $\sigma_\xi, \sigma_\eta, \tau_{\xi\eta}$ for the stress components in the coordinate system (ξ, η) and using Eq. (34.3) the relations (34.1) become, for any point $P(x, y)$

$$\left.\begin{aligned}\sigma_\xi + \sigma_\eta &= \sigma_x + \sigma_y, \\ \sigma_\xi - \sigma_\eta + 2i\tau_{\xi\eta} &= \frac{\overline{w}'(\bar\zeta)}{w'(\zeta)}(\sigma_x - \sigma_y + 2i\tau_{xy}), \\ \sigma_\xi - \sigma_\eta - 2i\tau_{\xi\eta} &= \frac{w'(\zeta)}{\overline{w}'(\bar\zeta)}(\sigma_x - \sigma_y - 2i\tau_{xy}).\end{aligned}\right\} \quad (34.4)$$

It may be noted here that the reason for the importance of the quantities

$$\sigma_x + \sigma_y, \quad \sigma_x - \sigma_y \pm 2i\tau_{xy}$$

lies in the fact, shown by GREEN and ZERNA[1], that they are the contravariant components, T^{12}, T^{11}, T^{22} of the stress tensor in the coordinate system $(z, \bar{z})$.

If u_ξ, u_η denote the components of displacement at $P(x, y)$ resolved along the normals n, s respectively then it is clear that

$$u_\xi + i u_\eta = e^{-i\alpha}(u + iv) \tag{34.5}$$

where $e^{-i\alpha}$ is given by Eq. (34.3). Putting

$$\varphi(z) = \varphi_0(\zeta), \quad \psi(z) = \psi_0(\zeta)$$

we have from Eqs. (32.15), (34.3) and (34.5)

$$2\mu(u_\xi + i u_\eta) = \frac{\overline{w'(\bar{\zeta})}}{|w'(\zeta)|}\left[\varkappa \varphi_0(\zeta) - \frac{w(\zeta)}{\overline{w'(\bar{\zeta})}}\overline{\varphi_0'(\bar{\zeta})} - \overline{\psi_0(\bar{\zeta})} + \frac{1}{2}c\varrho W(z, \bar{z})\right]. \tag{34.6}$$

If we put

$$\varphi'(z) = \Phi(z) = \Phi_0(\zeta), \quad \psi'(z) = \Psi(z) = \Psi_0(\zeta),$$

then from Eqs. (32.16), (32.17) and (34.1)

$$\sigma_\xi + \sigma_\eta = 2[\Phi_0(\zeta) + \overline{\Phi_0(\bar{\zeta})}] + (2 - c)\varrho \frac{\partial W}{\partial z}, \tag{34.7}$$

$$\sigma_\xi - \sigma_\eta + 2i\tau_{\xi\eta} = -\frac{2}{w'(\zeta)}[w(\zeta)\overline{\Phi_0'(\bar{\zeta})} + \overline{w'(\bar{\zeta})}\Psi_0(\zeta)] + \frac{\overline{w'(\bar{\zeta})}}{w'(\zeta)}c\varrho \frac{\partial W}{\partial z}, \tag{34.8}$$

$$\sigma_\eta - \sigma_\xi + 2i\tau_{\xi\eta} = \frac{2}{\overline{w'(\bar{\zeta})}}[\overline{w(\bar{\zeta})}\Phi_0'(\zeta) + w'(\zeta)\overline{\Psi_0(\bar{\zeta})}] - \frac{w'(\zeta)}{\overline{w'(\bar{\zeta})}}c\varrho \frac{\partial \overline{W}}{\partial \bar{z}}. \tag{34.9}$$

When mapping into circular regions, MUSKHELISHVILI puts $\zeta = \varrho e^{i\vartheta}$ and considers the orthogonal curvilinear coordinate system (ϱ, ϑ). If, at P, an increment dz_1 is taken along the curve $\vartheta = const.$, in the direction of ϱ increasing, the corresponding increment $d\zeta$ will be in the radial direction. Then, if β is the angle between the tangent to the curve $\vartheta = const$ and Ox,

$$dz_1 = e^{i\beta}|dz_1|$$

at P, while

$$d\zeta = e^{i\vartheta}|d\zeta|.$$

It follows that

$$\left.\begin{aligned} e^{i\beta} &= \frac{dz_1}{|dz_1|} = \frac{w'(\zeta)\,d\zeta}{|w'(\zeta)\,d\zeta|} = e^{i\vartheta}\frac{w'(\zeta)}{|w'(\zeta)|} = \frac{\zeta w'(\zeta)}{\varrho|w'(\zeta)|} \\ e^{-i\beta} &= \frac{\bar{\zeta}}{\varrho}\cdot\frac{\overline{w'(\bar{\zeta})}}{|w'(\zeta)|}. \end{aligned}\right\} \tag{34.10}$$

and

Comparison of Eqs. (34.3) and (34.10) demonstrates the relation

$$\beta = \alpha + \vartheta.$$

Relations involving the stress and displacement components in the ϱ, ϑ coordinate system, analogous to Eqs. (34.6) to (34.9) for the (ξ, η) system, follow at once by substituting the expressions (34.10) for $e^{i\beta}$ instead of the expressions (34.3)

---

[1] A. E. GREEN and W. ZERNA: Theoretical Elasticity, p. 193. London: Oxford University Press 1954.

for $e^{i\alpha}$ in Eqs. (34.1) and (34.5). When the body force is zero we have MUSKHELI-SHVILI's results (but with a slightly different notation):

$$2\mu(u_\varrho + i u_\vartheta) = \frac{\bar\zeta\, \overline{w'(\zeta)}}{\varrho |w'(\zeta)|}\left[\varkappa\varphi_0(\zeta) - \frac{w(\zeta)}{\overline{w'(\zeta)}}\overline{\varphi'_0(\zeta)} - \overline{\psi_0(\zeta)}\right], \qquad (34.11)$$

$$\sigma_\varrho + \sigma_\vartheta = 2[\Phi_0(\zeta) + \overline{\Phi_0(\zeta)}], \qquad (34.12)$$

$$\sigma_\vartheta - \sigma_\varrho + 2i\tau_{\varrho\vartheta} = \frac{2\zeta^2}{\varrho^2\, \overline{w'(\zeta)}}[\overline{w(\zeta)}\,\Phi'_0(\zeta) + w'(\zeta)\,\Psi_0(\zeta)]. \qquad (34.13)$$

When $z = w(\zeta) = \zeta$ these equations reduce simply to those relevant to a polar coordinate system:

$$2\mu(u_\varrho + i u_\vartheta) = e^{-i\vartheta}[\varkappa\varphi(z) - z\overline{\varphi'(z)} - \overline{\psi(z)}], \qquad (34.14)$$

$$\sigma_\varrho + \sigma_\vartheta = 2[\Phi(z) + \overline{\Phi(z)}], \qquad (34.15)$$

$$\sigma_\vartheta - \sigma_\varrho + 2i\tau_{\varrho\vartheta} = 2e^{2i\vartheta}[\bar z\,\Phi'(z) + \Psi(z)] \qquad (34.16)$$

where $z = \varrho\, e^{i\vartheta}$, and, by previous definition, $\Psi(z) = \psi'(z)$, $\Phi(z) = \varphi'(z)$.

35. Restrictions on potentials. If the elastic body occupies a simple connected region in the plane, the functions $\varphi(z)$, $\psi(z)$ will be single-valued in that region because of their analytic character. But if the region is multiply connected, representing, say, a plate with holes, then the functions may be no longer single valued. Physically, it is clear that the stress components must be single-valued, and we shall also impose the same condition on the displacement, although multi-valued displacements may be interpreted physically as *dislocations*[1]. The notation $[f]_A^B$ is used to denote the increase in the function f when it is taken along a path AB in the material. If B coincides with A so that the path becomes a closed contour, it can be shown that the operation of taking the increase $[f]_A^A$ is commutative with differentiation with respect to z or $\bar z$. When there are no body forces the conditions that the stresses should be single-valued are, from Eqs. (32.16) and (32.17),

$$[\varphi'(z) + \overline{\varphi'(\bar z)}]_A^A = 0, \qquad (35.1)$$

$$[z\overline{\varphi''(\bar z)} + \overline{\psi'(\bar z)}]_A^A = 0 \qquad (35.2)$$

where AA is any closed contour in the region. From these equations it follows that

$$[\varphi''(z)]_A^A = [\overline{\varphi''(\bar z)}]_A^A = [\overline{\psi'(\bar z)}]_A^A = 0. \qquad (35.3)$$

If the displacements are single-valued we have from Eq. (32.15)

$$[\varkappa\varphi(z) - z\overline{\varphi'(\bar z)} - \overline{\psi(\bar z)}]_A^A = 0.$$

Differentiating this with respect to z we have

$$[\varkappa\varphi'(z) - \overline{\varphi'(\bar z)}]_A^A = 0,$$

and comparison with Eq. (35.1) shows that

$$[\varphi'(z)]_A^A = [\overline{\varphi'(\bar z)}]_A^A = 0. \qquad (35.4)$$

We now find expressions for the components X, Y of the resultant stress and for the resultant couple, M, acting on an arc AB in the material. We have

$$X + iY = \int_{AB}(p_{xn} + i p_{yn})\, ds \qquad (35.5)$$

---

[1] COKER and FILON: Photoelasticity. Cambridge 1931.

where p_{xn}, p_{yn} are the components of stress exerted on the element of arc ds in the Ox and Oy directions respectively. Now,

$$p_{xn} = \sigma_n \cos\alpha - \tau_{ns} \sin\alpha,$$
$$p_{yn} = \sigma_n \sin\alpha + \tau_{ns} \cos\alpha,$$

and hence

$$p_{xn} + i p_{yn} = e^{i\alpha}(\sigma_n + i\tau_{ns}).$$

From Eqs. (34.1)

$$2(\sigma_n + i\tau_{ns}) = \sigma_x + \sigma_y + e^{-2i\alpha}(\sigma_x - \sigma_y + 2i\tau_{xy}),$$

and it follows that

$$2(p_{xn} + i p_{yn}) = e^{i\alpha}(\sigma_x + \sigma_y) + e^{-i\alpha}(\sigma_x - \sigma_y + 2i\tau_{xy}) \tag{35.6}$$

while

$$\frac{dz}{ds} = i e^{i\alpha}, \qquad \frac{d\bar{z}}{ds} = -i e^{-i\alpha}. \tag{35.7}$$

Then, from Eqs. (32.16), (32.17), (35.5), (35.6) and (35.7)

$$i(X+iY) = \int_{AB} [\varphi'(z) + \overline{\varphi}'(\bar{z})] \frac{dz}{ds} ds + \int_{AB} [z\overline{\varphi}''(\bar{z}) + \overline{\psi}'(\bar{z})] \frac{d\bar{z}}{ds} ds$$
$$= \int_{AB} \frac{d}{ds}[\varphi(z) + \overline{\psi}(\bar{z}) + z\overline{\varphi}'(\bar{z})] ds,$$

and hence

$$X + iY = -i[\varphi(z) + \overline{\psi}(\bar{z}) + z\overline{\varphi}'(\bar{z})]_A^B. \tag{35.8}$$

The moment M is given by

$$M = \int_{AB} (x p_{yn} - y p_{xn}) ds$$
$$= \operatorname{Re}\left\{-i\int_{AB} \bar{z}(p_{xn} + i p_{yn}) ds\right\}$$
$$= -\operatorname{Re}\left\{\int_{AB} \bar{z}[\varphi'(z) + \overline{\varphi}'(\bar{z})] \frac{dz}{ds} ds + \int_{AB} \bar{z}[z\overline{\varphi}''(\bar{z}) + \overline{\psi}'(\bar{z})] \frac{d\bar{z}}{ds} ds\right\}$$
$$= -\operatorname{Re}\left\{\int_{AB} \frac{d}{ds}[z\bar{z}\overline{\varphi}'(\bar{z}) + \bar{z}\overline{\psi}(\bar{z}) - \chi(z)] ds\right\}$$

in a similar fashion, where

$$\chi'(z) = \psi(z),$$

and we have

$$M = \operatorname{Re}[\chi(z) - z\psi(z) - z\bar{z}\varphi'(z)]_A^B. \tag{35.9}$$

If the arc is a closed contour AA, we have, using Eq. (35.4)

$$X + iY = -i[\varphi(z) + \overline{\psi}(\bar{z})]_A^A, \tag{35.10}$$
$$M = \operatorname{Re}[\chi(z) - z\psi(z)]_A^A. \tag{35.11}$$

Suppose the region S occupied by the body is bounded by the closed contours $L_1, L_2, \ldots, L_m, L_{m+1}$ where L_{m+1} contains the remainder, and $z_1, z_2, \ldots, z_m$ are any fixed points within the respective contours, while X_k, Y_k are the components of the stress resultant over L_k. Then it can be shown that

$$\varphi(z) = -\frac{1}{2\pi(1+\varkappa)} \sum_{k=1}^{m} (X_k + iY_k) \log(z - z_k) + \varphi^*(z),$$
$$\psi(z) = \frac{\varkappa}{2\pi(1+\varkappa)} \sum_{k=1}^{m} (X_k - iY_k) \log(z - z_k) + \psi^*(z)$$
(35.12)

where $\varphi^*(z)$ and $\psi^*(z)$ are functions holomorphic in S. When L_{m+1} recedes to infinity so that S becomes an infinite region these results still hold. Denote by L_R a circle $|z| = R$ containing all the finite L_k. Then for all $|z| \geq R$ we can write for the infinite region in place of expressions (35.12)

$$\varphi(z) = -\frac{X + iY}{2\pi(1+\varkappa)} \log z + \varphi^{**}(z),$$
$$\psi(z) = \frac{\varkappa(X - iY)}{2\pi(1+\varkappa)} \log z + \psi^{**}(z)$$
(35.13)

where

$$X = \sum_{k=1}^{m} X_k, \qquad Y = \sum_{k=1}^{m} Y_k$$

and $\varphi^{**}(z), \psi^{**}(z)$ are functions holomorphic in $|z| \geq R$ except possibly at infinity. Introducing these expressions into Eqs. (32.16) and (32.17) and imposing the condition that the stress components should be bounded throughout the infinite region S it is found that

$$\varphi^{**}(z) = \Gamma z + \varphi_1(z),$$
$$\psi^{**}(z) = \Gamma' z + \psi_1(z)$$
(35.14)

where Γ, Γ' are complex constants, and $\varphi_1(z)$, $\psi_1(z)$ are functions holomorphic in $|z| \geq R$ including the point at infinity, thus having expansions of the form

$$a_0 + \frac{a_1}{z} + \frac{a_2}{z^2} + \cdots$$

for sufficiently large $|z|$. The constants, $a_0, a_1, \ldots$ in these series, and also $\operatorname{Im}(\Gamma)$, may be assumed zero without affecting the state of stress. Also, using Eqs. (32.16) and (32.17) with $W = 0$, it is readily shown that

$$\operatorname{Re}(\Gamma) = \tfrac{1}{4}(N_1 + N_2), \qquad \Gamma' = -\tfrac{1}{2}(N_1 - N_2) e^{-2i\alpha} \tag{35.15}$$

where N_1, N_2 are the principal stresses at infinity and α the angle between the direction of N_1 and Ox. Using Eq. (35.13) and (35.14) with Eq. (32.15), the complex displacement is given for large $|z|$, in the absence of body forces, by

$$2\mu(u + iv) = -\frac{\varkappa(X + iY)}{2\pi(1+\varkappa)} \log(z\bar{z}) + (\varkappa\Gamma - \bar{\Gamma})z - \Gamma'\bar{z} + \cdots \tag{35.16}$$

where the further terms are bounded as $|z| \to \infty$. Clearly, the displacements are not bounded at infinity unless

$$X = Y = 0, \qquad \Gamma = \Gamma' = 0.$$

Such conditions imply that the resultant of the stresses acting on all the L_k is zero and that the stresses at infinity are zero, while φ and ψ would have expansions, for sufficiently large $|z|$,

$$\varphi(z) = a_0 + \frac{a_1}{z} + \frac{a_2}{z^2} + \cdots,$$

$$\psi(z) = b_0 + \frac{b_1}{z} + \frac{b_2}{z^2} + \cdots.$$

36. Case of an annulus.

A number of special cases have been solved by STEVENSON and others, building up the results by tentative methods similar to those adopted when using the Airy stress function. The working, however, is usually much simpler. We shall give some examples of this type before proceeding to an account of the more sophisticated developments of MUSKHELISHVILI.

First we consider an annulus $a \leq |z| \leq b$ in equilibrium under

(i) uniform shears applied to the boundaries,

(ii) constant pressures on the boundaries.

Substracting Eq. (34.16) from Eq. (34.15) and making the substitution $e^{2i\vartheta} = z/\bar{z}$ gives the equation

$$\sigma_\varrho - i\tau_{\varrho\vartheta} = \varphi'(z) + \overline{\varphi'(z)} - z\varphi''(z) - \frac{z}{\bar{z}}\psi'(z). \tag{36.1}$$

From Eqs. (35.12), if there are no stress resultants, X, Y, over $|z| = a$,

$$\varphi(z) = \varphi^*(z), \quad \psi(z) = \psi^*(z),$$

holomorphic functions in the annulus. But, under conditions (i) there is a resultant moment M which, from Eq. (35.11) is given by

$$M = \operatorname{Re}[\chi(z)]_\gamma$$

where γ is either of the contours $|z| = a$, $|z| = b$. This equation is satisfied by

$$\chi(z) = C \log z, \quad C = C_1 + iC_2$$

giving, in fact,

$$M = -2\pi C_2 \tag{36.2}$$

for all closed contours, γ, in the annulus containing the origin, while the function

$$\psi(z) = \chi'(z) = \frac{C}{z} \tag{36.3}$$

is holomorphic in the annulus. By putting

$$\varphi(z) = 0, \quad C_1 = 0$$

the conditions (i) are completely satisfied, and, from Eqs. (36.2) and (36.3),

$$\psi(z) = -\frac{iM}{2\pi z}.$$

Substituting these functions in Eq. (35.15) we have

$$\sigma_\varrho - i\tau_{\varrho\vartheta} = -\frac{iM}{2\pi z\bar{z}}.$$

Then, using Eq. (34.15) in addition, the components of stress are

$$\sigma_\varrho = \sigma_\vartheta = 0, \quad \tau_{\varrho\vartheta} = \frac{M}{2\pi\varrho^2}. \tag{36.4}$$

From Eq. (34.14)

$$2\mu(u_\varrho + iu_\vartheta) = -\frac{\varrho}{z}\overline{\psi(z)} = \frac{iM}{2\pi\varrho},$$

and hence

$$u_\varrho = 0, \quad u_\vartheta = \frac{M}{4\pi\mu\varrho}. \tag{36.5}$$

To satisfy conditions (ii), M must be zero, as well as X and Y. The function $\psi(z)$ of Eq. (36.3) can be retained, but with C real, so that

$$\psi(z) = \frac{C_1}{z}. \tag{36.6}$$

The introduction of a function

$$\varphi(z) = A z, \quad (A \text{ real}), \tag{36.7}$$

does not violate the conditions, and then there are two constants, A and C_1, to be evaluated in terms of the pressures p_a, p_b at $|z|=a$, $|z|=b$. Substituting the functions (36.6) and (36.7) in Eqs. (34.15) and (36.1) we find that

$$\sigma_\varrho = 2A + \frac{C_1}{\varrho^2}, \quad \sigma_\vartheta = 2A - \frac{C_1}{\varrho^2}, \quad \tau_{\varrho\vartheta} = 0. \tag{36.8}$$

Determination of A and C_1 by means of the conditions

$$\sigma_\varrho = -p_a \text{ on } \varrho = a; \quad \sigma_\varrho = -p_b \text{ on } \varrho = b,$$

and substitution in Eqs. (36.8) yield the expressions

$$\left.\begin{aligned}\sigma_\varrho &= \frac{p_a a^2 - p_b b^2}{b^2 - a^2} + \frac{a^2 b^2 (p_a - p_b)}{(b^2 - a^2) \varrho^2}, \\ \sigma_\vartheta &= \frac{p_a a^2 - p_b b^2}{b^2 - a^2} - \frac{a^2 b^2 (p_b - p_a)}{(b^2 - a^2) \varrho^2}.\end{aligned}\right\} \tag{36.9}$$

The displacement components corresponding to the complex potentials (36.6) and (36.7) are easily determined from Eq. (34.14), and are, in terms of the pressures

$$\left.\begin{aligned}u_\varrho &= \frac{(1 - 2\nu)(p_a a^2 - p_b b^2)}{2\mu (b^2 - a^2)} \varrho + \frac{a^2 b^2 (p_a - p_b)}{2\mu(b^2 - a^2)\varrho}, \\ u_\vartheta &= 0.\end{aligned}\right\} \tag{36.10}$$

It is readily seen that were A complex the stress components and the radial component of displacement would be unaltered but a pure rotation would take place. These results are the same as those of Eqs. (31.5) and (31.6), obtained by the stress function method.

37. Case of elliptic hole. As an example in the use of curvilinear coordinates we consider the case of an unstressed elliptic hole in an infinite plate. Elliptic coordinates (ξ, η) are defined by the transformation

$$z = w(\zeta) = c \cosh \zeta, \quad \zeta = \xi + i\eta, \tag{37.1}$$

and in these coordinates the elliptic boundary is defined by $\xi = \alpha$, a constant. It is assumed that at infinity the plate is in a condition of "all-round" tension T, so that, from Eqs. (34.7) and (34.9), in the absence of body forces

$$\Phi_0(\zeta) + \overline{\Phi}_0(\bar{\zeta}) \to T, \quad \overline{w}(\bar{\zeta}) \Phi_0'(\zeta) + w'(\zeta) \Psi_0(\zeta) \to 0 \tag{37.2}$$

as $|z| \to \infty$. STEVENSON found that complex potentials of the forms

$$\varphi(z) = A c \sinh \zeta, \quad \psi(z) = B c \operatorname{cosech} \zeta, \tag{37.3}$$

A, B real, were sufficient to describe the stress distribution. It is easily verified that these functions satisfy the conditions (35.3) and (35.4), while from Eqs. (35.10) and (35.11) the resultant stresses and moment over a circuit containing the hole are zero. We have from Eq. (37.1)

$$\frac{\partial z}{\partial \zeta} = w'(\zeta) = c \sinh \zeta \tag{37.4}$$

and then, with Eqs. (37.3)

$$\left.\begin{aligned}\Phi_0(\zeta) &= \varphi'(z) = A \coth \zeta \\ \Psi_0(\zeta) &= \psi'(z) = -B \operatorname{cosech}^2 \zeta \coth \zeta.\end{aligned}\right\} \tag{37.5}$$

By substituting from Eqs. (37.2) and (37.3) into the conditions (37.2) it is easily seen that they are satisfied if $T = 2A$ while substitution in Eqs. (34.7) and (34.9) yields, after subtraction, the equation

$$\sigma_\xi - i\tau_{\xi\eta} = \coth \zeta \cdot (A + A \coth \zeta \coth \bar{\zeta} + B \operatorname{cosech} \zeta \operatorname{cosech} \bar{\zeta}). \tag{37.6}$$

This expression vanishes on the boundary $\zeta + \bar{\zeta} = 2\alpha$ if

$$A [\sinh \zeta \sinh (2\alpha - \zeta) + \cosh \zeta \cosh (2\alpha - \zeta)] = -B,$$

that is, if

$$B = -A \cosh 2\alpha = -\tfrac{1}{2} T \cosh 2\alpha,$$

and then, from Eqs. (37.3) the complex potentials are

$$\varphi(z) = \tfrac{1}{2} T c \sinh \zeta, \quad \psi(z) = -\tfrac{1}{2} T c \cosh 2\alpha \operatorname{cosech} \zeta. \tag{37.7}$$

Various problems have been solved by STEVENSON[1] by using these methods. Among them are problems on ring spaces bounded by both concentric and eccentric circles, finite and infinite plates bounded by ellipses and regular curvilinear polygons, and semi-infinite plates with a trochoidal boundary or one given in polar coordinates by $r = c (\sec \vartheta - \lambda \cos \vartheta)$, with $\lambda \leq 1$. In some of the problems the body forces are non-zero and cases of strain "nuclei" are also treated. Similar techniques have been used by ROTHMAN[2] to solve problems where isolated forces act on the boundaries of holes in infinite plates or of discs of certain shapes.

IV. Cauchy integral methods.

It was seen in the last section that the tentative approach usually associated with the stress function method is facilitated by use of the complex potential together with a suitable conformal transformation, but the great advantage of the complex potential representation is due to the methods developed by MUSKHELISHVILI[3] for deriving the potentials directly from the boundary conditions. These methods are readily applicable to a body occupying in the z-plane a simply connected region, finite or infinite, which can be mapped by a conformal transformation on to a circle or half-plane: the treatment of multiply connected regions is considerably more complicated and will not be discussed here. Regions mapped on to a circle and on to a half-plane may both be tackled by two methods, the first involving manipulation of ordinary Cauchy integrals, the second using more sophisticated properties of Cauchy integrals. The second method is the more suitable in the case of the *mixed* boundary problem, that is, where stress is prescribed on part of the boundary and displacement on another part.

---

[1] A. C. STEVENSON: Phil. Mag., Ser. VII **33**, 639 (1942); **34**, 105, 766 (1943); **36**, 178 (1945). — Proc. Roy. Soc. Lond., Ser. A **184**, 129, 218 (1945).
[2] M. ROTHMAN: Quart. J. Mech. Appl. Math. **3**, 279, 469 (1950).
[3] N. I. MUSKHELISHVILI: Some Basic Problems of the Mathematical Theory of Elasticity. Groningen: Noordhoff 1953.

38. Cauchy integrals. We shall state, without proof, a number of results on Cauchy integrals which will be required subsequently. By a Cauchy integral is meant an integral of the form

$$F(z) = \frac{1}{2\pi i} \int_L \frac{f(t)\, dt}{t - z}$$

where L is a simple smooth contour, a simple smooth arc, or the union of such contours and arcs, and $f(t)$ is a complex function given on and assumed to be finite and integrable. Initially we suppose that z does not lie on L, and then $F(z)$ defines a function holomorphic in each part into which the plane is divided by L.

Suppose that L is a simple smooth contour described in the anticlockwise direction, so that it divides the plane into a finite region S^+ on the left and an infinite region S^- on the right. Let $f(t)$ be a function given for all points t of L, where it is continuous, then we have the following theorems:

I. *A necessary and sufficient condition for $f(t)$ to be the boundary value of some function holomorphic in S^+ is*

$$\frac{1}{2\pi i} \int_L \frac{f(t)\, dt}{t - z} = 0, \quad \text{all } z \text{ in } S^-.$$

II. *A necessary and sufficient condition for $f(t)$ to be the boundary value of some function holomorphic in S^-, including infinity, is*

$$\frac{1}{2\pi i} \int_L \frac{f(t)\, dt}{t - z} = a, \quad \text{all } z \text{ in } S^+,$$

where a is a constant equal to $f(\infty)$.

When L is the unit circle, denoted by γ, the properties of complex conjugate functions can be used with these theorems to obtain the following results:

I a. *A necessary and sufficient condition for the function $f(\sigma)$ continuous on γ, to be the boundary value of some function holomorphic outside γ is*

$$\frac{1}{2\pi i} \int_\gamma \frac{\overline{f(\sigma)}\, d\sigma}{\sigma - z} = 0, \quad \text{all } |z| > 1.$$

II a. *A necessary and sufficient condition for the function $f(\sigma)$, continuous on γ, to be the boundary value of some function holomorphic inside γ is*

$$\frac{1}{2\pi i} \int_\gamma \frac{\overline{f(\sigma)}\, d\sigma}{\sigma - z} = \bar{a}, \quad \text{all } |z| < 1$$

where $\bar{a}$ is a constant equal to $\overline{f}(0)$.

Corresponding propositions hold when L is an infinite straight line, taken here to be the real axis, so long as the function $f(t)$ in addition to being continuous on L, satisfies the condition

$$f(t) = a + O(|t|^{-\mu}), \quad \mu > 0,$$

for sufficiently large $|t|$. Then, taking S^+ to be the upper halfplane, the following theorems hold:

III. *A necessary and sufficient condition for $f(t)$ to be the boundary value of a function holomorphic in S^+ and continuous in $S^+ + L$, including infinity, is*

$$\frac{1}{2\pi i} \int_L \frac{f(t)\, dt}{t - z} = -\frac{1}{2} a, \quad \text{all } z \text{ in } S^-.$$

IV. *A necessary and sufficient condition for $f(t)$ to be the boundary value of a function holomorphic in S^- and continuous in S^-+L including infinity, is*

$$\frac{1}{2\pi i} \int_L \frac{f(t)\,dt}{t-z} = \frac{1}{2} a, \quad \text{all } z \text{ in } S^+.$$

Under the same conditions we have also:

IIIa. *A necessary and sufficient condition for $f(t)$ to be the boundary value of a function holomorphic in S^- is*

$$\frac{1}{2\pi i} \int_L \frac{\overline{f(t)}\,dt}{t-z} = -\frac{1}{2}\bar{a}, \quad \text{all } z \text{ in } S^-.$$

IVa. *A necessary and sufficient condition for $f(t)$ to be the boundary value of a function holomorphic in S^+ is*

$$\frac{1}{2\pi i} \int_L \frac{\overline{f(t)}\,dt}{t-z} = \frac{1}{2}\bar{a}, \quad \text{all } z \text{ in } S^+.$$

When z tends to a point, t_0, say, of L it is necessary to distinguish between the limits (if they exist) of the Cauchy integral $F(z)$ as z tends to t_0 from the left or from the right of L. If these limits exist $F(z)$ is said to be continuous at L from the left and from the right, and the limits are denoted by $F^+(t_0)$ and $F^-(t_0)$. We assume that the function $f(t)$ satisfies the Holder conditions, that is, that for every pair of points t_1, t_2 of L inequalities of the following form hold:

$$|f(t_1) - f(t_2)| \leq A|t_1 - t_2|^\mu, \quad 0 \leq \mu \leq 1,$$

A a positive constant. If this condition is satisfied in the neighbourhood of a point t_0 other than an end of L, and if $F^+(t_0)$ and $F^-(t_0)$ exist, then they are given by the Plemelj formulae

$$\left. \begin{array}{l} F^+(t_0) = \dfrac{1}{2} f(t_0) + \dfrac{1}{2\pi i} \displaystyle\int_L \dfrac{f(t)\,dt}{t-t_0}, \\[2ex] F^-(t_0) = -\dfrac{1}{2} f(t_0) + \dfrac{1}{2\pi i} \displaystyle\int_L \dfrac{f(t)\,dt}{t-t_0} \end{array} \right\} \tag{38.1}$$

where the integrals are principal values.

We turn now to solutions of *the problem of linear relationship of the boundary values* of a function $F(z)$. Let L be the union of a finite number of simple non-intersecting arcs and contours in the z-plane. The function $F(z)$ is said to be *sectionally holomorphic* in the entire plane if it satisfies the following conditions:

(1) $F(z)$ is holomorphic in the plane (excluding L) cut along L.

(2) It is continuous from the left and from the right at all points of L other than the ends of arcs.

(3) Near the ends of arcs c_i

$$|F(z)| < A|z - c_i|^{-\mu}, \quad 0 \leq \mu \leq 1,$$

where A is a positive constant.

The problem is to find the sectionally holomorphic function $F(z)$ with the line of discontinuity L on which the boundary values satisfy the condition

$$F^+(t) - G(t) F^-(t) = f(t) \tag{38.2}$$

(except at the ends), where $G(t)$ and $f(t)$ are functions given on L and $G(t) \neq 0$ everywhere on L. The functions $G(t)$ and $f(t)$ are also assumed to satisfy the Hölder condition on L.

We are concerned chiefly with the case where $G(t) = g$, a constant, in Eq. (38.2) and L consists of n simple arcs L_k with ends a_k, b_k. Then the general solution is of the form

$$F(z) = \frac{X_0(z)}{2\pi i} \int_L \frac{f(t)\,dt}{X_0^+(t)(t-z)} + X_0(z)\,P(z) \tag{38.3}$$

where, when g is not real and positive

$$X_0(z) = \prod_{k=1}^{n} (z-a_k)^{-\gamma}(z-b_k)^{\gamma-1}, \tag{38.4}$$

$$\gamma = \frac{\log g}{2\pi i}, \quad 0 < \arg g < 2\pi \tag{38.5}$$

and $P(z)$ is an arbitrary polynomial. This solution admits a pole at $z = \infty$ unless the degree of $P(z)$ is not greater than n. In general the solution (38.3) is not bounded at the ends a_k, b_k but a solution bounded at any p ends, denoted by $c_1, c_2, \ldots, c_p$ is obtained by substituting $X_p(z)$ for $X_0(z)$ throughout, where

$$X_p = X_0(z)(z-c_1)(z-c_2)\cdots(z-c_p). \tag{38.6}$$

This solution is holomorphic at $z = \infty$ if $p \leq n+1$ and the degree of $P(z)$ is not greater than $n-p$ or, in the case $p = n+1$, $P(z) = 0$. If $P(z) \equiv 0$ and $f(t)$ satisfies the conditions

$$\frac{1}{2\pi i} \int_L \frac{t^{k-1} f(t)\,dt}{X_p^+(t)} = 0, \quad k = 1, 2, \ldots, p-n-1 \tag{38.7}$$

the solution is holomorphic at $z = \infty$ for $p > n+1$ while $F(\infty) = 0$ if condition (38.7) is satisfied for $k = p-n$ also.

When, in Eq. (38.2), $G(t) = g$ a *real, positive* constant, the solution is similar to (38.3), but $X_*(z)$ must be substituted for $X_0(z)$ throughout, where

$$X_*(z) = X_0(z) \prod_{k=1}^{n}(z-b_k) = \prod_{k=1}^{n}\left(\frac{z-b_k}{z-a_k}\right)^{\gamma}. \tag{38.8}$$

This solution is bounded at all ends, and is holomorphic at $z = \infty$ if $P(z)$ is a constant. In the case $g = 1$ we have $\gamma = 0$, $X_*(z) = 1$ and the solution reduces to

$$F(z) = \frac{1}{2\pi i} \int_L \frac{f(t)\,dt}{t-z} + P(z) \tag{38.9}$$

where L may now include not only arcs but contours. The solution (38.9) may be used to obtain more general solutions (in which L consists of both arcs and contours) of the equation

$$F^+(t) - g\,F^-(t) = f(t),$$

where g is some complex constant. If L is one contour, consider the function $F_*(z)$ where

$$F_*(z) = F(z), \quad z \text{ in } S^+,$$
$$F_*(z) = g\,F(z), \quad z \text{ in } S^-.$$

Then the problem becomes

$$F_*^+(t) - F_*^-(t) = f(t),$$

and the solution is, by Eq. (38.9),
$$F_*(z) = \frac{1}{2\pi i} \int_L \frac{f(t)\,dt}{t-z} + P(z).$$

The extension to the general case is not difficult.

Finally we consider a case in which the function $G(t)$ of Eq. (38.2) is not constant everywhere on L, but changes discontinuously at certain points. Let L be a simple contour, part of which, L', is the union of n arcs $L_k = a_k b_k$ without common ends, while L'' denotes the remainder of L, and assume that

$$G(t) = g \neq 1 \quad \text{on} \quad L', \qquad G(t) = 1 \quad \text{on} \quad L''.$$

Then the boundary condition (38.2) takes the form

$$\left.\begin{array}{l} F^+(t) - g F^-(t) = f(t) \quad \text{on} \quad L', \\ F^+(t) - F^-(t) = f(t) \quad \text{on} \quad L''. \end{array}\right\} \tag{38.10}$$

So long as $f(t)$ satisfies the Hölder condition on L' and L'' separately the general solution, when g is not real and positive, is

$$F(z) = \frac{X_0(z)}{2\pi i} \int_L \frac{f(t)\,dt}{X_0(t)(t-z)} + X_0(z) P(z) \tag{38.11}$$

where $X_0(z)$ is given by Eq. (38.4) and $P(z)$ is an arbitrary polynomial. The properties of the solution are similar to those of (38.3). When g is real and positive the function $X_*(z)$ defined by Eq. (38.8) must be substituted for $X_0(z)$.

39. First method: regions mapped on to a circle. Suppose that S is a finite or infinite region of the z-plane, bounded by the simple contour L and mapped on to the unit circle, γ, in the ζ-plane by the function

$$z = w(\zeta). \tag{39.1}$$

It is assumed that for S finite $\zeta = 0$ corresponds to $z = 0$ and for S infinite $\zeta = 0$ corresponds to $z = \infty$. Initially, we assume that the displacements are bounded at infinity when S is infinite. Then X, Y, Γ, Γ' are all zero in Eq. (35.16). It follows that not only are the stress resultants over L zero, but, from Eqs. (32.16), (32.17), (35.13) and (35.14) (in the absence of body forces), the stress components vanish at infinity. Under these conditions the functions $\varphi(z), \psi(z)$ are holomorphic in S (including $z = \infty$ when S is infinite). Then, using the notation $\varphi(z) = \varphi_0(\zeta)$, $\psi(z) = \psi_0(\zeta)$ the functions $\varphi_0(\zeta), \psi_0(\zeta)$ are holomorphic in $|\zeta| < 1$. We assume that $\varphi_0(\zeta), \varphi_0'(\zeta), \psi_0(\zeta)$ are continuous up to the circle γ, and, without loss of generality, that $\varphi_0(0) = 0$.

In the *first boundary value problem* the stress components p_{xn}, p_{yn} are given everywhere on L, and from them we define the function of the arc coordinate, s,

$$f(s) = f_1(s) + i f_2(s) = i \int_0^s (p_{xn} + i p_{yn})\,ds + \text{const}, \tag{39.2}$$

where the constant and the choice of the point $s = 0$ on L are arbitrary. Then, from Eqs. (35.5) and (35.8), with a suitable choice of constant

$$\varphi(z) + \overline{\psi(\bar{z})} + z\overline{\varphi'(\bar{z})} = f \quad \text{on} \quad L. \tag{39.3}$$

Sect. 39. First method: regions mapped on to a circle.

Under the transformation (39.1), Eq. (39.3) becomes

$$\varphi_0(\zeta) + \overline{\psi}_0(\bar{\zeta}) + \frac{w(\zeta)}{\overline{w'(\zeta)}} \overline{\varphi'_0(\zeta)} = f \quad \text{on} \quad |\zeta| = 1 \tag{39.4}$$

which may be written

$$\varphi_0(\sigma) + \overline{\psi}_0(\bar{\sigma}) + \frac{w(\sigma)}{\overline{w'(\sigma)}} \overline{\varphi'_0(\sigma)} = f \tag{39.5}$$

where $\sigma = e^{i\vartheta}$ is any point on γ, and f is a known function of σ or ϑ. We assumt that f is not only single-valued and continuous but that its derivative with respece to ϑ is continuous and satisfies the Hölder condition. This is true if p_{xn} and p_{yn} satisfy the Hölder condition.

Now, by CAUCHY's Theorem,

$$\varphi_0(\zeta) = \frac{1}{2\pi i} \int_\gamma \frac{\varphi_0(\sigma) \, d\sigma}{\sigma - \zeta}$$

while, from Theorem IIa,

$$\frac{1}{2\pi i} \int_\gamma \frac{\overline{\psi}_0(\bar{\sigma}) \, d\sigma}{\sigma - \zeta} = \overline{\psi}_0(0) = 0.$$

Then, from the complex conjugate of Eq. (39.5)

$$\psi_0(\zeta) = \frac{1}{2\pi i} \int_\gamma \frac{\bar{f} \, d\sigma}{\sigma - \zeta} - \frac{1}{2\pi i} \int_\gamma \frac{\overline{w}(\bar{\sigma}) \, \varphi'_0(\sigma) \, d\sigma}{w'(\sigma) (\sigma - \zeta)}. \tag{39.6}$$

A functional equation for $\varphi_0(\zeta)$ is now constructed from Eq. (39.5), which we write

$$\overline{\psi}_0(\bar{\sigma}) = f - \varphi_0(\sigma) - \frac{w(\sigma)}{\overline{w'(\sigma)}} \overline{\varphi'_0(\sigma)}.$$

Now, from Theorem IIa,

$$\frac{1}{2\pi i} \int_\gamma \frac{\overline{\psi}_0(\bar{\sigma})}{\sigma - \zeta} d\sigma = \overline{\psi}_0(0) = \bar{a},$$

say, and it follows that

$$\frac{1}{2\pi i} \int_\gamma \frac{f \, d\sigma}{\sigma - \zeta} - \varphi_0(\zeta) - \frac{1}{2\pi i} \int_\gamma \frac{w(\sigma) \, \overline{\varphi'_0(\bar{\sigma})} \, d\sigma}{\overline{w'(\bar{\sigma})} (\sigma - \zeta)} = \bar{a} \tag{39.7}$$

for all $|\zeta| < 1$. In the first place the constant a remains arbitrary, but it can be fixed later by the condition $\varphi_0(0) = 0$. In principle, Eqs. (39.6) and (39.7) are sufficient to determine $\varphi_0(\zeta)$ and $\psi_0(\zeta)$ and thus to solve the problem.

If, when S is infinite, the stress resultant (X, Y) over L and the stresses at infinity are not all zero, the functions $\varphi(z)$ and $\psi(z)$ have, from Eqs. (35.13) and (35.14), the forms

$$\left.\begin{aligned}\varphi(z) &= -\frac{X + iY}{2\pi(1 + \varkappa)} \log z + \Gamma z + \varphi_1(z), \\ \psi(z) &= \frac{\varkappa(X - iY)}{2\pi(1 + \varkappa)} \log z + \Gamma' z + \psi_1(z),\end{aligned}\right\} \tag{39.8}$$

where $\varphi_1(z)$ and $\psi_1(z)$ are holomorphic in S, including infinity. From its definition $w(\zeta)$ must be of the form

$$w(\zeta) = \frac{c}{\zeta} + \text{a holomorphic function,}$$

and it follows that

$$\varphi_0(\zeta) = \frac{X+iY}{2\pi(1+\varkappa)} \log \zeta + \frac{\Gamma c}{\zeta} + \varphi_{1,0}(\zeta),$$
$$\psi_0(\zeta) = -\frac{\varkappa(X-iY)}{2\pi(1+\varkappa)} \log \zeta + \frac{\Gamma' c}{\zeta} + \psi_{1,0}(\zeta)$$ (39.9)

where $\varphi_{1,0}(\zeta)$ and $\psi_{1,0}(\zeta)$ are functions holomorphic inside and continuous up to γ. Substituting the functions (39.9) in Eq. (39.5) it is easily seen that $\varphi_{1,0}(\zeta)$ and $\psi_{1,0}(\zeta)$ must satisfy the same equations as $\varphi_0(\zeta)$ and $\psi_0(\zeta)$ did previously if f is replaced everywhere by f_0 where

$$f_0 = f - \frac{X+iY}{2\pi} \log \sigma - \frac{\Gamma c}{\sigma} - \frac{w(\sigma)}{\overline{w'(\overline{\sigma})}} \left\{ \frac{X-iY}{2\pi(1+\varkappa)} \sigma - \overline{\Gamma} \overline{c} \, \sigma^2 \right\} - \overline{\Gamma}' \overline{c} \, \sigma. \quad (39.10)$$

Since $\log \sigma = i\vartheta$ and, from Eqs. (35.5) and (39.2), f increases by $i(X+iY)$ for a complete circuit of L, f_0 is single-valued and continuous on γ.

For the *second boundary value problem* the displacements

$$u = g_1, \quad v = g_2 \qquad (39.11)$$

are given on L, and then from Eq. (32.15), when $W = 0$,

$$\varkappa \varphi_0(\sigma) - \frac{w(\sigma)}{\overline{w'(\overline{\sigma})}} \overline{\varphi}_0'(\overline{\sigma}) - \overline{\psi}(\overline{\sigma}) = 2\mu (g_1 + i g_2) = 2\mu g, \qquad (39.12)$$

say. By methods similar to those used for the first problem the following equations for $\varphi_0(\zeta)$ and $\psi_0(\zeta)$ are obtained on the assumption that $X = Y = \Gamma = \Gamma' = 0$:

$$\varkappa \varphi(\zeta) - \frac{1}{2\pi i} \int_\gamma \frac{w(\sigma) \overline{\varphi}'(\overline{\sigma})}{\overline{w'(\overline{\sigma})}(\sigma - \zeta)} d\sigma - \frac{\mu}{\pi i} \int_\gamma \frac{g \, d\sigma}{\sigma - \zeta} = \bar{a}, \qquad (39.13)$$

$$\psi(\zeta) = -\frac{\mu}{\pi i} \int_\gamma \frac{\bar{g} \, d\sigma}{\sigma - \zeta} - \frac{1}{2\pi i} \int_\gamma \frac{\overline{w}(\overline{\sigma}) \varphi'(\sigma) \, d\sigma}{w'(\sigma)(\sigma - \zeta)} \qquad (39.14)$$

for all $|\zeta| < 1$. When S is infinite and X, Y, Γ, Γ' are not all zero the problem may be reduced to the preceding case in the same way as before, so that, in place of Eq. (39.12) we have an equation in $\varphi_{1,0}(\sigma), \overline{\varphi}'_{1,0}(\overline{\sigma}), \overline{\psi}_{1,0}(\overline{\sigma})$ and g_0 appropriately defined.

40. Elliptic hole in an infinite plate. As an example of the application of this method we consider once again the case of an unstressed elliptic hole in an infinite plate subjected to an "all-round" tension T at infinity. That is to say, the principal stresses at infinity are given by

$$N_1 = N_2 = T$$

so that, from Eqs. (35.15),

$$\Gamma = \tfrac{1}{2} T, \quad \Gamma' = 0, \qquad (40.1)$$

assuming that $\mathrm{Im}(\Gamma) = 0$ since the stress distribution is not affected thereby. The outside of the ellipse, centre $z = 0$, with semiaxes $R(1+m)$, $R(1-m)$ is mapped on to the unit circle $|\zeta| < 1$ by the transformation

$$z = w(\zeta) = R\left(m\zeta + \frac{1}{\zeta}\right), \quad R > 0, \quad 0 \leq m < 1, \qquad (40.2)$$

which is single-valued if we take the branch for which z lies outside the ellipse when $|\zeta| < 1$. Since the hole is unstressed $X = Y = f = 0$ and putting $c = R$, we have from Eqs. (39.9), (39.10) and (40.1)

$$\varphi_0(\zeta) = \frac{TR}{2\zeta} + \varphi_{1,0}(\zeta), \tag{40.3}$$

$$\psi_0(\zeta) = \psi_{1,0}(\zeta), \tag{40.4}$$

$$f_0 = -\frac{1}{2} TR \left[\frac{1}{\sigma} + \frac{\sigma(1+m\sigma^2)}{\sigma^2 - m} \right], \tag{40.5}$$

using the relation $\sigma\bar{\sigma}=1$ in the last equation. The functions $\varphi_{1,0}(\zeta)$, $\psi_{1,0}(\zeta)$, f_0 now take the place of $\varphi_0(\zeta)$, $\psi_0(\zeta)$, f in Eqs. (39.6) and (39.7), which we now proceed to solve, assuming that $\varphi_{1,0}(0)=0$. Substituting in Eq. (39.7), we have

$$\varphi_{1,0}(\zeta) + \frac{1}{2\pi i} \int_\gamma \frac{(1+m\sigma^2)\,\varphi'_{1,0}(\sigma)\,d\sigma}{\sigma(m-\sigma^2)(\sigma-\zeta)} = -\bar{a} - \frac{1}{2\pi i} \int_\gamma \frac{TR}{2(\sigma-\zeta)} \left\{ \frac{1}{\sigma} + \frac{\sigma(1+m\sigma^2)}{\sigma^2 - m} \right\} d\sigma. \tag{40.6}$$

Now

$$\frac{1+m\sigma^2}{\sigma(m-\sigma^2)}\,\varphi'_{1,0}(\sigma)$$

is the boundary value of

$$F(\zeta) = \frac{1+m\zeta^2}{\zeta(m-\zeta^2)}\,\varphi'_{1,0}\!\left(\frac{1}{\zeta}\right)$$

which is easily shown to be holomorphic in $|\zeta|>1$ so that, by Theorem II,

$$\frac{1}{2\pi i} \int_\gamma \frac{F(\sigma)}{\sigma-\zeta}\,d\sigma = F(\infty) = 0, \quad |\zeta|<1.$$

Then, substituting this in Eq. (40.6) and calculating the remaining integral by Cauchy's theorem on residues, we find that

$$\varphi_{1,0}(\zeta) = -\bar{a} - \tfrac{1}{2} TR m\zeta. \tag{40.7}$$

Substituting from Eqs. (40.2), (40.5) and (40.7) into Eq. (39.6) we have

$$\psi_{1,0}(\zeta) = -\frac{1}{2\pi i} \int_\gamma \frac{1}{2} TR \left\{ \sigma + \frac{\sigma^2 + m}{\sigma(1 - m\sigma^2)} \right\} \frac{d\sigma}{\sigma - \zeta} \\
- \frac{1}{2\pi i} \int_\gamma \frac{\sigma(\sigma^2 + m)\,TRm\,d\sigma}{2(1 - m\sigma^2)(\sigma - \zeta)}. \tag{40.8}$$

Evaluating these integrals by the residue theorem we find that

$$\psi_0(\zeta) = \psi_{1,0}(\zeta) = -\frac{TR(1+m^2)\zeta}{1 - m\zeta^2}, \quad |\zeta|<1. \tag{40.9}$$

Hence, $a = \psi_{1,0}(0) = 0$, and then from Eqs. (40.3) and (40.7)

$$\varphi_0(\zeta) = \tfrac{1}{2} TR \left(\tfrac{1}{\zeta} - m\zeta\right), \quad |\zeta|<1. \tag{40.10}$$

If now $\varphi_0(\zeta)$ and $\psi_0(\zeta)$ are expressed as the functions of z, $\varphi(z)$, $\psi(z)$ by means of Eq. (40.2), the displacement components u, v and the stress components $\sigma_x, \sigma_y, \tau_{xy}$ can be found from Eqs. (32.15), (32.16) and (32.17). Using Eqs. (34.11) to (34.13) the stress and displacement components in directions tangential and normal to an ellipse defined by $|\zeta|=\text{const}<1$ can be found.

The relation of the potentials given by Eqs. (40.9) and (40.10) to those of Eq. (37.7) deduced by STEVENSON, is easily demonstrated. The transformation

$$\zeta = m^{-\frac{1}{2}} e^{-t} \qquad (40.11)$$

maps the circle $|\zeta| \leq 1$ on to $\mathrm{Re}(t) > \log m^{-\frac{1}{2}}$. A single-valued mapping requires a restriction of the kind $0 \leq \mathrm{Im}(t) \leq 2\pi$. If $m = e^{-2\alpha}$ the combination of the transformations (40.2) and (40.11) maps the outside of the original ellipse on to a region $\mathrm{Re}(t) > \alpha$, as was done by the single transformation (37.1) in STEVENSON's work, where ζ took the place of the present variable t. Substituting from Eq. (40.11) into Eq. (40.2) we have

$$z = 2R m^{\frac{1}{2}} \cosh t, \qquad (40.12)$$

and the correspondence with Eq. (37.1) is complete if $2R m^{\frac{1}{2}} = c$, while the functions (40.10) and (40.9) then become

$$\varphi_0(\zeta) = \tfrac{1}{2} T c \sinh t, \qquad \psi_0(\zeta) = - \tfrac{1}{2} T c \cosh 2\alpha \operatorname{cosech} t,$$

agreeing with Eqs. (37.7).

MUSKHELISHVILI (loc. cit.) solves several other problems concerned with the elliptic hole by this method and also the problem of concentrated forces and couples acting on a circular disc. A general theory for regions mapped on to the unit circle by rational functions is worked out, together with the application of approximate methods to regions such as an equilateral triangle or square.

41. First method: Half-plane. The essential ideas of the method employed with a region mapped on to the unit circle may be applied if the region is mapped on to the half-plane. However the general method will not be described but merely the application to a semi-infinite region in the z-plane itself.

We consider a body occupying the lower half-plane, $\mathrm{Im}(z) \leq 0$, denoted by S, and assume that in that region, for large z,

$$\left. \begin{array}{l} \varphi(z) = \gamma \log z + o(1) + \mathrm{const.}, \\ \psi(z) = \gamma' \log z + o(1) + \mathrm{const.} \end{array} \right\} \qquad (41.1)$$

where γ, γ' are constants. Under these conditions the stress components are zero at infinity. If (X', Y') is the resultant vector of the external forces acting on a segment, AB, of $\mathrm{Im}(z) = 0$ we have from Eq. (35.8)

$$X' + i Y' = i \left[\varphi(z) + \overline{\psi}(\bar z) + z \overline{\varphi'}(\bar z) \right]_A^B \qquad (41.2)$$

where the sign has been changed since S lies on the right of the boundary. Substituting the functions (41.1) into Eq. (40.2) we find that

$$X' + i Y' = \gamma \log \frac{r''}{r'} + \gamma \pi i + \overline{\gamma'} \log \frac{r''}{r'} - \overline{\gamma'} \pi i + \varepsilon \qquad (41.3)$$

as $A \to -\infty$, $B \to -\infty$ independently, where r', r'' are the distances of A, B from the origin and ε is a quantity tending to zero as $r', r'' \to \infty$. In order that X', Y' should remain finite as $r', r'' \to \infty$ independently, it is necessary and sufficient that

$$\gamma + \overline{\gamma'} = 0.$$

It follows that the resultant forces over the whole boundary are given by

$$X + i Y = \pi (\overline{\gamma'} - \gamma)$$

and that
$$\gamma = -\frac{1}{2\pi}(X+iY), \quad \gamma' = \frac{X-iY}{2\pi}$$

from Eqs. (41.1), which can now be written

$$\left.\begin{array}{l}\varphi(z) = -\dfrac{X+iY}{2\pi}\log z + o(1) + \text{const.},\\[6pt] \psi(z) = \dfrac{X-iY}{2\pi}\log z + o(1) + \text{const.}\end{array}\right\} \qquad (41.4)$$

In the first boundary value problem we suppose that on $\operatorname{Im}(z)=0$ the stress components are
$$\sigma_y = N(t), \quad \tau_{xy} = T(t),$$
given as functions of the abscissa, t. Then, from Eqs. (32.16) and (32.17) we have the equation
$$N - iT = \varphi'(t) + \overline{\varphi'(t)} + t\overline{\varphi''(t)} + \overline{\psi'(t)} \qquad (41.5)$$
and its conjugate,
$$N + iT = \varphi'(t) + \overline{\varphi'(t)} + t\varphi''(t) + \psi'(t). \qquad (41.6)$$

It is assumed that N and T are $o(t^{-1})$ for large t. Now $\psi'(t)$ is, by Eq. (41.4), the boundary value of a function $\psi'(z)$ holomorphic in $\operatorname{Im}(z) < 0$ and zero at infinity, so that by Theorem IV a

$$\frac{1}{2\pi i}\int_{-\infty}^{\infty}\frac{\overline{\psi'(t)}\,dt}{t-z} = 0, \quad \operatorname{Im}(z) < 0.$$

Then, from Eq. (41.5)

$$\int_{-\infty}^{\infty}\frac{N-iT}{t-z}dt - \int_{-\infty}^{\infty}\frac{\varphi'(t)}{t-z}dt - \int_{-\infty}^{\infty}\frac{\overline{\varphi'(t)}\,dt}{t-z} - \int_{-\infty}^{\infty}\frac{t\,\overline{\varphi''(t)}\,dt}{t-z} = 0. \qquad (41.7)$$

Now the last two integrals in this equation are zero by Theorem IV a, and the second integral, since $\varphi'(\infty)=0$, is easily evaluated by a contour integration to give the result

$$\Phi(z) = \varphi'(z) = -\frac{1}{2\pi i}\int_{-\infty}^{\infty}\frac{N-iT}{t-z}dt. \qquad (41.8)$$

Similarly,
$$\Psi(z) = \psi'(z) = -\frac{1}{2\pi i}\int_{-\infty}^{\infty}\frac{\psi'(t)\,dt}{t-z}$$

and, using Eqs. (41.6) and (41.8), we find eventually that

$$\Psi(z) = \frac{1}{2\pi i}\int_{-\infty}^{\infty}\frac{N-iT}{(t-z)^2}t\,dt - \frac{1}{2\pi i}\int_{-\infty}^{\infty}\frac{N+iT}{(t-z)}dt. \qquad (41.9)$$

Sufficient conditions that the expressions (41.8) and (41.9) should define functions satisfying the required conditions are that $N(t)$, $T(t)$, $N'(t)$, $T'(t)$ should satisfy the Hölder condition at all finite points and $tN(t)$, $tT(t)$, $t^2N'(t)$, $t^2T'(t)$ near infinity. The following example is one in which these conditions are not all satisfied but where $\Phi(z)$ and $\Psi(z)$ nevertheless behave satisfactorily. Consider the boundary conditions $T=0$, $N=-p$ for $-a \leq t \leq a$, zero otherwise. Then, from Eqs. (41.8) and (41.9)

$$\left. \begin{array}{l} \Phi(z) = \dfrac{p}{2\pi i} \displaystyle\int_{-a}^{a} \dfrac{dt}{t-z} = \dfrac{p}{2\pi i} \log \dfrac{z-a}{z+a}, \\[2ex] \Psi(z) = -\dfrac{zp}{2\pi i} \displaystyle\int_{-a}^{a} \dfrac{dt}{(t-z)^2} = -\dfrac{p\,a\,z}{\pi i (z^2-a^2)}. \end{array} \right\} \quad (41.10)$$

For the second boundary value problem we have, from Eq. (32.15), the condition

$$\varkappa \varphi(t) - t\overline{\varphi'}(t) - \overline{\psi}(t) = 2\mu(g_1 + i g_2) = 2\mu g \quad (41.11)$$

where g_1, g_2 are the components of displacement on the boundary $\mathrm{Im}(z)=0$. The problem is solved by methods similar to those used in the first boundary value problem, but an additional condition is imposed on the complex potentials, namely, that $X=Y=0$. This condition, however, may be relaxed, as before, by considering Eq. (41.11) as an equation in the boundary values of the *holomorphic* parts of $\varphi(z), \psi(z)$ and modifying the function g appropriately. The solution (for $X=Y=0$) is given by the equations

$$\varkappa \varphi(z) = -\dfrac{\mu}{\pi i} \int_{-\infty}^{\infty} \dfrac{g\,dt}{t-z} - \mu G, \quad (41.12)$$

$$\psi(z) = \dfrac{\mu}{\pi i} \int_{-\infty}^{\infty} \dfrac{g\,dt}{t-z} - z\,\varphi'(z) - \mu \overline{G}, \quad (41.13)$$

where G is given by the equation

$$g = G + o(1), \quad |t| \text{ large}.$$

42. Second method: Half-plane. Again we suppose that the body occupies the half-plane $\mathrm{Im}(z) \leq 0$ denoted now by S^- in which the complex potentials are defined. The first step is to provide an analytic continuation of the function $\Phi(z) = \varphi'(z)$ into the upper half-plane, S^+, through the unloaded parts of the boundary, L. From Eqs. (32.16) and (32.17) we have

$$\sigma_y - i\tau_{xy} = \Phi(z) + \overline{\Phi(z)} + z\overline{\Phi'(z)} + \overline{\Psi(z)}. \quad (42.1)$$

Now the equation

$$\Phi(z) = -\overline{\Phi}(\bar z) - z\overline{\Phi}'(\bar z) - \overline{\Psi}(\bar z), \quad z \text{ in } S^+ \quad (42.2)$$

defines $\Phi(z)$ as a function of z which is holomorphic in S^+ since the same function of $\bar z$ is holomorphic in S^-. On L the relation (42.2) becomes

$$\Phi^+(t) = -\overline{\Phi}^+(t) - t\overline{\Phi}'^+(t) - \overline{\Psi}^+(t) \quad (42.3)$$

and taking the boundary value of Eq. (42.1) and comparing with Eq. (42.3) we see that wherever σ_y, τ_{xy} are zero on L

$$\Phi^+(t) = \Phi^-(t).$$

Thus by means of the definition (42.2) $\Phi(z)$ becomes a function sectionally holomorphic in the whole plane cut along the loaded segments of L. It should be noted that the method of defining $\Phi(z)$ in S^+ is quite arbitrary, the present method being chosen for its practical utility.

Replacing z by $\bar{z}$ in Eq. (42.2), which then holds for z in S^-, and taking the complex conjugate, we find that

$$\Psi(z) = -\Phi(z) - \bar{\Phi}(z) - z\Phi'(z), \quad z \text{ in } S^-, \tag{42.4}$$

which is an expression for $\Psi(z)$ in terms of $\Phi(z)$ defined in the whole plane. Substituting from Eq. (42.4) into Eq. (42.1), we have

$$\sigma_y - i\tau_{xy} = \Phi(z) - \Phi(\bar{z}) + (z - \bar{z})\overline{\Phi'(z)}. \tag{42.5}$$

Extending $\varphi(z)$ into S^+ by assuming that $\varphi'(z) = \Phi(z)$ there also, we find that Eq. (32.15) becomes

$$2\mu(u + iv) = \varkappa\varphi(z) + \varphi(\bar{z}) - (z - \bar{z})\overline{\varphi'(z)} + \text{const.} \tag{42.6}$$

It is assumed that $\varphi(z)$ and $\psi(z)$ are of the forms (41.4) in S^-: it is not difficult to show that $\varphi(z)$ is then of the same form in S^+ as in S^-. We also suppose that, except possibly at a finite number of points t_j, $\Phi(z)$ is continuous from the left and from the right on L, and that

$$\lim_{y \to 0} y\,\Phi'(z) = 0 \tag{42.7}$$

while at t_j

$$|\Phi(z)| < A\,|z - t_j|^\alpha, \quad 0 \le \alpha < 1. \tag{42.8}$$

In the first boundary value problem the stresses on L are given in the form

$$\sigma_y^- = -P(t), \quad \tau_{xy}^- = T(t)$$

where P and T are assumed to satisfy the Hölder condition and to vanish at $|t| = \infty$. Then, by Eq. (42.5) the boundary condition is

$$\Phi^+(t) - \Phi^-(t) = P(t) + i\,T(t). \tag{42.9}$$

The solution of this follows immediately from result (38.9) and is

$$\Phi(z) = \frac{1}{2\pi i}\int_L \frac{P(t) + iT(t)}{t - z}\,dt, \tag{42.10}$$

the polynomial being omitted since $\Phi(z)$ is to vanish at infinity.

To solve the second boundary value problem we differentiate Eq. (42.6) with respect to x, so that the boundary condition takes the form

$$\Phi^+(t) + \varkappa\Phi^-(t) = 2\mu[g_1'(t) + i\,g_2'(t)] \tag{42.11}$$

where g_1 and g_2 are the components of displacement on L. Put

$$\Omega(z) = \Phi(z) \text{ in } S^-, \quad \Omega(z) = -\frac{1}{\varkappa}\Phi(z) \text{ in } S^+, \tag{42.12}$$

then Eq. (42.11) becomes

$$\Omega^+(t) - \Omega^-(t) = -\frac{2\mu}{\varkappa}(g_1' + i\,g_2'), \tag{42.13}$$

and the solution is

$$\Omega(z) = -\frac{\mu}{\varkappa\pi i}\int_L \frac{g_1'(t) + i\,g_2'(t)}{t - z}\,dt. \tag{42.14}$$

If $\Phi(z)$ is written in place of $\Omega(z)$ then Eqs. (42.12) may be considered as defining a new continuation of $\Phi(z)$ into S^+ and in place of Eq. (42.2) we have

$$\Phi(z) = \frac{1}{\varkappa}[\overline{\Phi}(z) + z\,\overline{\Phi'}(z) - \overline{\Psi}(z)], \quad z \text{ in } S^+. \tag{42.15}$$

By virtue of Eq. (42.13), this equation defines an analytic continuation of $\Phi(z)$ into S^+ through those parts of L where $\partial u/\partial x$ and $\partial v/\partial x$ are zero.

The mixed boundary value problem includes the so-called *punch problems*. Let L' be the union of n segments $L_k(a_k, b_k)$ of the real axis L, on which the components of displacement are given, while the external forces are given on the remainder, L''. Since the solution of the first boundary value problem is known, the effect of the external forces on L'' may be calculated separately and superposed to give the final result. Accordingly we suppose here that

$$\sigma_y^- = \tau_{xy}^- = 0, \quad \text{on } L''. \tag{42.16}$$

We consider first the boundary condition

$$u^- + iv^- = g(t) + c \quad \text{on } L', \tag{42.17}$$

where c is a constant which may be made zero without loss of generality since that would merely involve a rigid body motion not affecting the stress distribution. The resultant vector (X, Y) of the external forces on L' is supposed given. These conditions, corresponding to the case of n punches rigidly connected, define *Problem A*, in contrast to *Problem B*, which is a modification of the fundamental problem and allows independent vertical movement of the punches. In *Problem B* the resultant forces (X_k, Y_k) are given on each segment L_k and the boundary condition is written

$$u^- + iv^- = g(t) + c(t) \tag{42.18}$$

where $c(t) = c_k$ on L_k the constants c_k being unknown, although one constant may be assigned an arbitrary value so that we may put $c_1 = 0$, say, without affecting generality. The complex potentials are supposed to satisfy the conditions specified earlier with the points a_k, b_k in the place of the t_j in condition (42.8). In addition it is assumed that $g'(t)$ exists and satisfies the Hölder condition on L'.

In both problems A and B the boundary conditions (42.17) and (42.18) yield an equation corresponding to Eq. (42.11) (which, however, holds everywhere on L), namely,

$$\Phi^+(t) + \varkappa \Phi^-(t) = 2\mu g'(t) \quad \text{on } L'. \tag{42.19}$$

The boundary condition (42.16) implies that $\Phi^+(t) = \Phi^-(t)$ on L'' and it follows that the function $\Phi(z)$ is holomorphic in the entire plane cut along L'. The solution of the boundary value Eq. (42.19) is then of the form (38.3) with $X_0(z)$ defined by Eq. (38.4) where γ is given by Eq. (38.5) with $g = -\varkappa$; that is

$$\gamma = \frac{\log \varkappa}{2\pi i} + \frac{1}{2} = \frac{1}{2} - i\beta, \tag{42.20}$$

say. We have then

$$X_0(z) = \prod_{k=1}^{n} (z - a_k)^{-\frac{1}{2} + i\beta} (z - b_k)^{-\frac{1}{2} - i\beta}, \tag{42.21}$$

and the solution is

$$\Phi(z) = \frac{\mu X_0(z)}{\pi i} \int_{L'} \frac{g'(t)\,dt}{X_0^+(t)(t-z)} + X_0(z) P_{n-1}(z) \tag{42.22}$$

where, since $\Phi(z)$ is to be zero at infinity, the polynomial $P_{n-1}(z)$ is of the form

$$P_{n-1}(z) = C_0 z^{n-1} + C_1 z^{n-2} + \cdots + C_{n-1}.$$

The remaining conditions are now used to determine the coefficients $C_0, C_1, \ldots, C_{n-1}$.

In problem B we must use the given resultants (X_k, Y_k). If $\sigma_y^- = -P$, $\tau_{xy}^- = T$ on L' we have by Eq. (42.5)

$$P(t_0) + i T(t_0) = \Phi^+(t_0) - \Phi^-(t_0), \quad t_0 \text{ on } L', \tag{42.23}$$

or, by Eq. (42.18),

$$P(t_0) + i T(t_0) = \frac{1+\varkappa}{\varkappa} \Phi^+(t_0) - \frac{2\mu}{\varkappa} g'(t_0), \quad \text{on } L'. \tag{42.24}$$

Applying the first of the Plemelj formulae (38.1) to Eq. (42.22) we have

$$\Phi^+(t_0) = \mu g'(t_0) + \frac{\mu X_0^+(t_0)}{\pi i} \int_{L'} \frac{g'(t)\,dt}{X_0^+(t)(t-t_0)} + X_0^+(t_0) P_{n-1}(t_0). \tag{42.25}$$

Substitution of this expression for $\Phi^+(t_0)$ into Eq. (42.24) then gives a formula for the external forces at any point of L'. But

$$\int_{L_k} [P(t_0) + i T(t_0)]\,dt_0 = -Y_k + i X_k, \quad k = 1, 2, \ldots, n. \tag{42.26}$$

where (X_k, Y_k) are given, and we thus have n linear equations for the determination of the constants $C_0, C_1, \ldots, C_{n-1}$.

To solve problem A we must use the condition $c_k = c$, $(k=1, 2, \ldots, n)$. Now $\Phi(z)$ is continuous across the unloaded part of the boundary L'', so that, by Eqs. (42.16) and (42.22)

$$\left.\begin{aligned}
2\mu(u' + i v')^- &= (\varkappa + 1)\Phi(t_0) \\
&= \frac{(\varkappa+1)\mu X_0^+(t_0)}{\pi i} \int_{L'} \frac{g'(t)\,dt}{X_0^+(t)(t-t_0)} + (\varkappa + 1) X_0^+(t_0) P_{n-1}(t_0)
\end{aligned}\right\} \tag{42.27}$$

for t_0 on L'', where u', v' are derivatives of u, v with respect to x. Also, it is clear that

$$\int_{b_k}^{a_{k+1}} (u' + i v')^- \, dt_0 = g(a_{k+1}) - g(b_k), \quad k = 1, 2, \ldots, n-1, \tag{42.28}$$

and substitution from Eq. (42.27) into Eq. (42.28) yields $n-1$ linear equations for the determination of the C_j. An additional equation is obtained by using the condition that the resultant of the external forces (X, Y) is known, for from Eqs. (42.22) and (41.4)

$$C_0 = \lim_{z \to \infty} z \Phi(z) = -\frac{1}{2\pi}(X + i Y). \tag{42.29}$$

43. Single punch with straight base. The simplest example of the mixed boundary problem is the case of a single punch with a linear profile pressed horizontally on to the line $\mathrm{Im}(z) = 0$ under a vertical force. We assume that the surface of the body is constrained to remain in contact with the punch, which covers the interval $-l \leq t \leq l$ forming the segment L' of the real axis. Then we have

$$g'(t) = 0 \quad \text{on } L', \tag{43.1}$$

while for the resultant forces on L' we suppose that

$$X = 0, \quad Y = -P_0. \tag{43.2}$$

From Eq. (42.21)

$$X_0(z) = (z+l)^{-\frac{1}{2}+i\beta}(z-l)^{-\frac{1}{2}-i\beta} \tag{43.3}$$

and, by Eq. (42.22),

$$\Phi(z) = C_0 X_0(z). \tag{43.4}$$

Then, from Eqs. (42.29), (43.3) and (43.4)

$$\Phi(z) = \frac{iP_0}{2\pi}(z+l)^{-\frac{1}{2}+i\beta}(z-l)^{-\frac{1}{2}-i\beta}, \tag{43.5}$$

and, using Eqs. (42.24) and (42.25), it is found after some reduction that the pressure and tangential stress acting beneath the punch are

$$\left. \begin{aligned} P(t) &= \frac{P_0}{\pi(l^2-t^2)^{\frac{1}{2}}} \cdot \frac{1+\varkappa}{\sqrt{\varkappa}} \cos\left(\frac{\log \varkappa}{2\pi}\log\frac{l+t}{l-t}\right), \\ T(t) &= \frac{P_0}{\pi(l^2-t^2)^{\frac{1}{2}}} \cdot \frac{1+\varkappa}{\sqrt{\varkappa}} \sin\left(\frac{\log \varkappa}{2\pi}\log\frac{l+t}{l-t}\right). \end{aligned} \right\} \tag{43.6}$$

It is not difficult to show that $P(t)$ remains positive for all t such that

$$\left|\frac{t}{l}\right| < 0.9997.$$

For greater values of $|t|$ the stresses become so great that the classical theory of elasticity does not apply.

More difficult problems arise when the actual displacements on L are not given, but only the profiles of the punches, which may not make complete contact with the surface, or may be frictionless. Various problems of these kinds are dealt with by MUSKHELISHVILI[1], as well as the problem of contact of two elastic bodies. The "linear relationship method" is also applicable to cases in which the body occupies the whole plane but is cut along one or more arcs.

The corresponding methods for a circle can be developed in a similar way, and the extension to regions which can be mapped on to the half-plane or circle is clearly possible. The reader is referred to MUSKHELISHVILI's book for a detailed treatment, together with various examples and references to other work, chiefly Russian.

V. Fourier transform methods.

When the solid is infinite, semi-infinite, or is an infinite strip with parallel boundaries, the Fourier transform provides an effective method of solution. Problems on infinite wedges may be solved by somewhat similar methods using the Mellin transform[2].

44. Infinite solid with body forces. The two-dimensional Fourier transform of a function $f(x, y)$ is defined by

$$\bar{f}(\xi, \eta) = \frac{1}{2\pi} \int_{-\infty}^{\infty} \int_{-\infty}^{\infty} f(x, y) \, e^{i(\xi x + \eta y)} \, dx \, dy.$$

We suppose that the components of displacement and stress tend to zero as $|x|, |y| \to \infty$. Multiplying the equations of equilibrium (25.2) and (25.3) by $e^{i(\xi x + \eta y)}$ and integrating over the whole plane we find the relations.

$$\xi \bar{\sigma}_x + \eta \bar{\tau}_{xy} + i\varrho \bar{X} = 0, \quad \xi \bar{\tau}_{xy} + \eta \bar{\sigma}_y + i\varrho \bar{Y} = 0.$$

Similarly, from the compatibility Eq. (28.3) for plane strain we have

$$\bar{\sigma}_x[(1-\nu)\eta^2 - \nu\xi^2] + \bar{\sigma}_y[(1-\nu)\xi^2 - \nu\eta^2] = 2\xi\eta\bar{\tau}_{xy}.$$

---
[1] N. I. MUSKHELISHVILI: Some Basic Problems of the Mathematical Theory of Elasticity. Groningen: Noordhoff 1953.
[2] I. N. SNEDDON: Fourier Transforms, p. 439. New York: McGraw-Hill 1951.

Sect. 44.   Infinite solid with body forces.

Solving these three algebraic equations we find that

$$\bar{\sigma}_x + \bar{\sigma}_y = -\frac{\varrho}{1-\nu} \cdot \frac{i\xi \bar{X} + i\eta \bar{Y}}{\xi^2 + \eta^2},$$

$$\bar{\sigma}_x - \bar{\sigma}_y = -\frac{\varrho(1-2\nu)}{(1-\nu)} \cdot \frac{(\xi^2 - \eta^2)}{(\xi^2+\eta^2)^2}(i\xi \bar{X} + i\eta \bar{Y}) - 4\varrho \frac{\xi\eta(i\xi\bar{X} - i\eta\bar{Y})}{(\xi^2+\eta^2)^2},$$

$$\bar{\tau}_{xy} = -\frac{\varrho(\xi^2 - \eta^2)}{(\xi^2+\eta^2)^2}(i\xi \bar{Y} - i\eta \bar{X}) - \varrho\frac{(1-2\nu)}{(1-\nu)} \cdot \frac{\xi\eta(i\eta\bar{X} + i\xi\bar{Y})}{(\xi^2+\eta^2)^2},$$

where the functions $\bar{X}(\xi, \eta)$, $\bar{Y}(\xi, \eta)$ are known from the body force distribution. Using the inversion formula

$$f(x, y) = \frac{1}{2\pi} \int_{-\infty}^{\infty} \int_{-\infty}^{\infty} \bar{f}(\xi, \eta)\, e^{-i(\xi x + \eta y)}\, d\xi\, d\eta$$

and the Faltung theorem for Fourier integrals[1], we obtain the results

$$\left.\begin{aligned}
\sigma_x + \sigma_y &= -\frac{\varrho}{2\pi(1-\nu)} \int_{-\infty}^{\infty}\int_{-\infty}^{\infty} \frac{(x-\alpha) X(\alpha, \beta) + (y-\beta) Y(\alpha, \beta)}{(x-\alpha)^2 + (y-\beta)^2}\, d\alpha\, d\beta, \\
\sigma_x - \sigma_y &= -\frac{\varrho(1-2\nu)}{\pi(1-\nu)} \int_{-\infty}^{\infty}\int_{-\infty}^{\infty} \frac{(x-\alpha)(y-\beta)[(y-\beta) X(\alpha,\beta) - (x-\alpha) Y(\alpha,\beta)]}{[(x-\alpha)^2+(y-\beta)^2]^2}\, d\alpha\, d\beta - \\
&\quad -\frac{\varrho}{\pi} \int_{-\infty}^{\infty}\int_{-\infty}^{\infty} \frac{[(x-\alpha)^2 - (y-\beta)^2]}{[(x-\alpha)^2+(y-\beta)^2]^2}[(x-\alpha) X(\alpha,\beta)+(y-\beta) Y(\alpha,\beta)]\, d\alpha\, d\beta, \\
\tau_{xy} &= -\frac{\varrho(1-2\nu)}{4\pi(1-\nu)} \int_{-\infty}^{\infty}\int_{-\infty}^{\infty} \frac{[(y-\beta)^2 - (x-\alpha)^2]}{[(x-\alpha)^2+(y-\beta)^2]^2}[(y-\beta) X(\alpha,\beta) - (x-\alpha) Y(\alpha,\beta)]\, d\alpha\, d\beta - \\
&\quad -\frac{\varrho}{\pi} \int_{-\infty}^{\infty}\int_{-\infty}^{\infty} \frac{(x-\alpha)(y-\beta)[(x-\alpha) X(\alpha,\beta)+(y-\beta) Y(\alpha,\beta)]}{[(x-\alpha)^2+(y-\beta)^2]^2}\, d\alpha\, d\beta.
\end{aligned}\right\} \quad (44.1)$$

A particularly simple case is that of a point force of magnitude F acting at $x = y = 0$ in the direction of x decreasing. The force is represented by

$$X(x, y) = -\frac{F}{\varrho}\delta(x)\delta(y), \qquad Y(x, y) = 0,$$

where $\delta(x)$ is the Dirac delta function, with the properties

$$\delta(x) = 0, \quad x \neq 0, \qquad \int_{-\infty}^{\infty} f(x)\delta(x)\,dx = f(0).$$

Then, from Eqs. (44.1), we have for the stress components

$$\left.\begin{aligned}
\sigma_x &= \frac{Fx}{4\pi(1-\nu)r^2}\left\{(1-2\nu) + \frac{2x^2}{r^2}\right\}, \\
\sigma_y &= \frac{Fx}{4\pi(1-\nu)r^2}\left\{(1+2\nu) - \frac{2x^2}{r^2}\right\}, \\
\tau_{xy} &= \frac{Fy}{4\pi(1-\nu)r^2}\left\{(1-2\nu) + \frac{2x^2}{r^2}\right\},
\end{aligned}\right\} \quad (44.2)$$

where $r^2 = x^2 + y^2$.

---
[1] I. N. SNEDDON: loc. cit. p. 23.

45. Semi-infinite solid with surface loading.
We consider a solid not subject to body forces occupying the half-plane $x \geq 0$ and assume that the components of stress and displacement tend to zero as $x \to \infty$. On the boundary $x = 0$ it is supposed that the tangential stress τ_{xy} is zero, and either the pressure or displacement is known. Here we use the "bar" notation to represent the *one-dimensional transform* with respect to y;

$$\bar{f}(x, \xi) = \frac{1}{\sqrt{2\pi}} \int_{-\infty}^{\infty} f(x, y) e^{i\xi y} dy$$

to which corresponds the inversion formula

$$f(x, y) = \frac{1}{\sqrt{2\pi}} \int_{-\infty}^{\infty} \bar{f}(x, \xi) e^{-i\xi y} d\xi.$$

Multiplying the equations of equilibrium (25.2) and (25.3) and the stress-strain relations (25.4) to (25.7) for plane strain by $e^{i\xi y}$ and integrating along the whole y-axis, we have the following set of five ordinary differential equations:

$$\left.\begin{array}{l} D\bar{\sigma}_x - i\xi \bar{\tau}_{xy} = 0, \\ D\bar{\tau}_{xy} - i\xi \bar{\sigma}_y = 0, \\ \bar{\sigma}_x = \lambda(D\bar{u} - i\xi \bar{v}) + 2\mu D\bar{u}, \\ \bar{\sigma}_y = \lambda(D\bar{u} - i\xi \bar{v}) - 2i\mu \xi \bar{v}, \\ \bar{\tau}_{xy} = \mu(D\bar{v} - i\xi \bar{u}) \end{array}\right\} \quad (45.1)$$

where D denotes the operator d/dx. In order that the Eqs. (45.1) should be consistent, their eliminant, operating on any of the dependent variables, must be zero. We find that this condition reduces to

$$(D^2 - \xi^2)^2 \bar{g}(x, \xi) = 0 \quad (45.2)$$

where $\bar{g}(x, \xi)$ is any of the quantities $\bar{\sigma}_x, \bar{\sigma}_y, \bar{\tau}_{xy}, \bar{u}, \bar{v}$. The solution of Eq. (45.2) which tends to zero as $x \to \infty$ is

$$\bar{g} = (A + Bx) e^{-|\xi|x} \quad (45.3)$$

where A, B are functions of ξ depending on the boundary conditions and different, in general, for each solution.

We take the case in which the external pressure, p, is prescribed, that is,

$$\sigma_x = -p(y), \quad x = 0$$

or, in terms of the Fourier transforms,

$$\bar{\sigma}_x = -\bar{p}(\xi), \quad x = 0.$$

In addition, since $\tau_{xy} = 0$ on $x = 0$ the first of Eqs. (45.1) gives the condition

$$D\bar{\sigma}_x = 0, \quad x = 0.$$

Taking $\bar{g}$ in Eq. (45.3) to be the solution for $\bar{\sigma}_x$ we find from these conditions that

$$A = -\bar{p}(\xi), \quad B = -|\xi|\bar{p}(\xi).$$

Sect. 45. Semi-infinite solid with surface loading.

The remaining solutions are easily found by substitution in equations (45.1), and we have, after using the inversion formula,

$$\left.\begin{aligned}
\sigma_x &= -\frac{1}{\sqrt{2\pi}} \int_{-\infty}^{\infty} \bar{p}(\xi)\{1+|\xi|x\} e^{-|\xi|x - i\xi y} d\xi, \\
\sigma_y &= -\frac{1}{\sqrt{2\pi}} \int_{-\infty}^{\infty} \bar{p}(\xi)\{1-|\xi|x\} e^{-|\xi|x - i\xi y} d\xi, \\
\tau_{xy} &= -\frac{ix}{\sqrt{2\pi}} \int_{-\infty}^{\infty} \bar{p}(\xi)\,\xi\, e^{-|\xi|x - i\xi y} d\xi, \\
u &= \frac{1+\nu}{E\sqrt{2\pi}} \int_{-\infty}^{\infty} \frac{\bar{p}(\xi)}{|\xi|}\{2(1-\nu)+|\xi|x\} e^{-|\xi|x - i\xi y} d\xi, \\
v &= -\frac{i(1+\nu)}{E\sqrt{2\pi}} \int_{-\infty}^{\infty} \frac{\bar{p}(\xi)}{|\xi|}\{(1-2\nu)-|\xi|x\} e^{-|\xi|x - i\xi y} d\xi.
\end{aligned}\right\} \quad (45.4)$$

As an example we suppose that a uniform pressure p_0 acts over the segment $-a \leq y \leq a$ of the boundary $x=0$. Then

$$\bar{p}(\xi) = \frac{p_0}{\sqrt{2\pi}} \int_{-a}^{a} e^{i\xi y}\, dy = \sqrt{\frac{2}{\pi}}\, p_0\, \frac{\sin(\xi a)}{\xi}$$

and, since this is an even function of ξ, we find after substituting in Eqs. (45.4) the stress components

$$\sigma_x = -\frac{2p_0}{\pi} \int_0^{\infty} \frac{1+\xi x}{\xi} e^{-\xi x} \sin(\xi a) \cos(\xi y)\, d\xi,$$

$$\sigma_y = -\frac{2p_0}{\pi} \int_0^{\infty} \frac{1-\xi x}{\xi} e^{-\xi x} \sin(\xi a) \cos(\xi y)\, d\xi,$$

$$\tau_{xy} = -\frac{2p_0 x}{\pi} \int_0^{\infty} e^{-\xi x} \sin(\xi a) \sin(\xi y)\, d\xi.$$

Evaluating these integrals we have

$$\left.\begin{aligned}
\sigma_x &= \frac{p_0}{2\pi} \{2(\vartheta_1 - \vartheta_2) + \sin 2\vartheta_1 - \sin 2\vartheta_2\}, \\
\sigma_y &= \frac{p_0}{2\pi} \{2(\vartheta_1 - \vartheta_2) - \sin 2\vartheta_1 + \sin 2\vartheta_2\}, \\
\tau_{xy} &= \frac{p_0}{2\pi} (\cos 2\vartheta_2 - \cos 2\vartheta_1),
\end{aligned}\right\} \quad (45.5)$$

where

$$\vartheta_1 = \arctan \frac{y-a}{x}, \qquad \vartheta_2 = \arctan \frac{y+a}{x}.$$

Similar methods may be used for the case of an infinite strip with parallel boundaries[1].

---
[1] See I. N. SNEDDON: Fourier Transforms. New York: McGraw-Hill 1951.

46. Point force in a semi-infinite solid. The results for a point force in an infinite solid may be used with the general results for a semi-infinite solid to find the stresses produced in a semi-infinite solid by a point force acting at the point $(h, 0)$ in the direction of x decreasing. The boundary $x=0$ is assumed to be free of stress, so that

$$\sigma_x = \tau_{xy} = 0, \quad x = 0.$$

First, we write down from the results (44.2) the stress components in an infinite solid when, in addition to the point force F at $(h, 0)$ there is an equal and opposite, "image" force at $(-h, 0)$

$$\left. \begin{aligned} \sigma_x &= \frac{F}{4\pi(1-\nu)} \left\{ (1-2\nu)\left(\frac{x-h}{\varrho_1^2} - \frac{x+h}{\varrho_2^2}\right) + \frac{2(x-h)^3}{\varrho_1^4} - \frac{2(x+h)^3}{\varrho_2^4} \right\}, \\ \sigma_y &= \frac{F}{4\pi(1-\nu)} \left\{ (1+2\nu)\left(\frac{x-h}{\varrho_1^2} - \frac{x+h}{\varrho_2^2}\right) - \frac{2(x-h)^3}{\varrho_1^4} + \frac{2(x+h)^3}{\varrho_2^4} \right\}, \\ \tau_{xy} &= \frac{Fy}{4\pi(1-\nu)} \left\{ (1-2\nu)\left(\frac{1}{\varrho_1^2} - \frac{1}{\varrho_2^2}\right) + \frac{2(x-h)^2}{\varrho_1^4} - \frac{2(x+h)^2}{\varrho_2^4} \right\} \end{aligned} \right\} \quad (46.1)$$

where $\varrho_1^2 = (x-h)^2 + y^2$, $\varrho_2^2 = (x+h)^2 + y^2$. The effect of adding the "image" force is to make the tangential stress component τ_{xy} zero on $x=0$ and the problem is solved by superposing on these solutions those for a semi-infinite solid which will reduce the normal stress component σ_x to zero on $x=0$. In other words, we must add to the expressions (46.1) solutions obtained by putting

$$-p(y) = \frac{Fh}{2\pi(1-\nu)} \left\{ \frac{1-2\nu}{h^2+y^2} + \frac{2h^2}{(h^2+y^2)^2} \right\}$$

in the semi-infinite case, or

$$\bar{p}(\xi) = \sqrt{\frac{2}{\pi}} \int_0^\infty p(y) \cos(\xi y)\, dy = -\frac{F}{\sqrt{2\pi}} \left\{ 1 + \frac{h\xi}{2(1-\nu)} \right\} e^{-h\xi}$$

in Eqs. (45.4).

47. Crack and punch problems. When the boundary conditions in the semi-infinite case are of the mixed type, the Fourier integral formulation[1] leads to dual integral equations. We consider the Griffith crack[2] as being opened under the action of a variable internal pressure. If the crack occupies the interval $-c \leq y \leq c$, $x=0$, in an infinite two-dimensional solid it is clear from the symmetry about the y-axis that the problem is equivalent to that of a semi-infinite solid where the boundary conditions on $x=0$ are

$$\begin{aligned} u &= 0 & |y| &> c; \\ \sigma_x &= -p_0(y), & |y| &< c; \\ \tau_{xy} &= 0, & &\text{for all values of } y; \end{aligned}$$

the pressure $p_0(y)$ being an even function of y which is assumed known. Supposing that the components of stress and displacement vanish at infinity the theory of Sect. 45 is applicable, and using the above conditions we have from

---

[1] I. N. SNEDDON and H. A. ELLIOTT: Quart. Appl. Math. **4**, 262 (1946). — I. N. SNEDDON: Fourier Transforms. New York: McGraw-Hill 1951.
[2] A. A. GRIFFITH: Phil. Trans. Roy. Soc. Lond., Ser. A **221**, 180 (1921). — Proc. Internat. Congr. Applied Mech. (Delft), 1924, p. 55.

the first and fourth of Eqs. (45.4), when $x=0$,

$$\sqrt{\frac{2}{\pi}} \int_0^\infty \bar{p}(\xi) \cos(\xi y) d\xi = p_0(y), \quad 0 \le y \le c,$$
$$\int_0^\infty \bar{p}(\xi) \frac{\cos(\xi y)}{\xi} d\xi = 0, \quad y \ge c,$$
(47.1)

where

$$\bar{p}(\xi) = \sqrt{\frac{2}{\pi}} \int_0^\infty p(y) \cos(\xi y) dy$$

and $-p(y)$ is the value of σ_z on $x=0$. By making the substitutions

$$\xi = \varrho/c, \quad y = c\eta,$$
$$\bar{p}\left(\frac{\varrho}{c}\right) = \varrho^{\frac{1}{2}} F(\varrho), \quad g(\eta) = c p_0(c\eta),$$
(47.2)

and remembering that

$$\cos z = (\tfrac{1}{2}\pi z)^{\frac{1}{2}} J_{-\frac{1}{2}}(z)$$

we see that the Eqs. (47.1) can be put in the standard form

$$\int_0^\infty \varrho F(\varrho) J_{-\frac{1}{2}}(\varrho\eta) d\varrho = g(\eta), \quad 0 \le \eta \le 1,$$
$$\int_0^\infty F(\varrho) J_{-\frac{1}{2}}(\varrho\eta) d\varrho = 0, \quad \eta > 1.$$
(47.3)

This pair of equations is the case $\alpha=1$, $\nu=-\tfrac{1}{2}$ of the equations

$$\int_0^\infty y^\alpha f(y) J_\nu(xy) dy = g(x), \quad 0 \le x \le 1,$$
$$\int_0^\infty f(y) J_\nu(xy) dy = 0, \quad x > 1$$

which has the solution, due to BUSBRIDGE[1],

$$f(x) = \frac{2^{-\frac{1}{2}\alpha} x^{-\alpha}}{\Gamma(1+\tfrac{1}{2}\alpha)} \left\{ x^{1+\frac{1}{2}\alpha} J_{\nu+\frac{1}{2}\alpha}(x) \int_0^1 y^{\nu+1} (1-y^2)^{\frac{1}{2}\alpha} g(y) dy + \right.$$
$$\left. + \int_0^1 u^{\nu+1}(1-u^2)^{\frac{1}{2}\alpha} du \int_0^1 g(yu)(xy)^{2+\frac{1}{2}\alpha} J_{\nu+1+\frac{1}{2}\alpha}(xy) dy \right\}$$

so long as $g(y)$ is integrable over the range $(0, 1)$ and α satisfies the conditions $\alpha > -2$, $-\nu-1 < \alpha-\tfrac{1}{2} < \nu+1$. Using this result we find the solution of Eqs. (47.3) is

$$F(\varrho) = \left(\frac{2\varrho}{\pi}\right)^{\frac{1}{2}} \left\{ J_0(\varrho) \int_0^1 y^{\frac{1}{2}} (1-y^2)^{\frac{1}{2}} g(y) dy + \right.$$
$$\left. + \varrho \int_0^1 u^{\frac{1}{2}}(1-u^2)^{\frac{1}{2}} du \int_0^1 g(yu) y^{\frac{3}{2}} J_1(\varrho y) dy \right\}.$$
(47.4)

When $p_0(y)$ is a uniform pressure p_0 this expression can be reduced to obtain, using the relations (47.2), the equation

$$\bar{p}(\xi) = \sqrt{\frac{\pi}{2}} p_0 c J_1(c\xi)$$

---
[1] I. W. BUSBRIDGE: Proc. Lond. Math. Soc. **44**, 115 (1938).

and explicit expressions for the components of stress and displacement can then be found by means of Eqs. (45.4). In particular we find that

$$u = \frac{2(1-v^2)p_0}{E}(c^2-y^2)^{\frac{1}{2}}$$

showing that the crack is deformed into an ellipse of semi-axes $2(1-v^2)p_0c/E$ and c.

If a frictionless punch causes a known indentation of a part of the surface of a semi-infinite solid, the remainder being free of external stress, the problem takes a similar form. The boundary conditions on $x=0$ are:

$$\tau_{xy} = 0, \quad \text{for all values of } y;$$

$$u = u_0(y), \quad |y| \leq c; \quad \sigma_x = 0, \quad |y| > c.$$

If the punch is symmetrical, so that $u_0(y)$ is an even function of y, we have from the first and fourth of Eqs. (45.4)

$$\frac{2(1-v^2)}{E}\sqrt{\frac{2}{\pi}}\int_0^\infty \frac{\bar{p}(\xi)}{\xi}\cos(\xi y)\,d\xi = u_0(y), \quad 0 \leq y \leq c$$

$$\int_0^\infty \bar{p}(\xi)\cos(\xi y)\,d\xi = 0, \quad y > c.$$

These equations may be solved for $\bar{p}(\xi)$ by means of the result quoted above. A full account together with calculations of special cases is given by SNEDDON[1].

VI. Real potential methods.

48. Payne's potential functions. PAYNE[2] has used a decomposition of the Airy stress function into potential functions of real variables. The method is readily applied to problems involving semi-infinite bodies or infinite strips and can also be adapted to some anisotropic solids and to three-dimensional problems where there is axial symmetry. The discussion here is chiefly concerned with semi-infinite isotropic two dimensional bodies, occupying the half-plane $x \geq 0$, $-\infty < y < \infty$.

It is easily shown that the Airy stress function $\chi(x,y)$ satisfying the bi-harmonic Eq. (28.7) can be represented by

$$\chi = \varphi_1' - x\frac{\partial \varphi_2'}{\partial x} \tag{48.1}$$

where φ_1' and φ_2' are any harmonic functions. Substituting for χ in Eqs. (28.2) we find that in the absence of body forces the stress components are given by

$$\begin{aligned}
\sigma_x &= \frac{\partial^2 \varphi_1'}{\partial y^2} - x\frac{\partial^3 \varphi_2'}{\partial x\,\partial y^2} = -\frac{\partial^2 \varphi_1'}{\partial x^2} + x\frac{\partial^3 \varphi_2'}{\partial x^3}, \\
\sigma_y &= \frac{\partial^2 \varphi_1'}{\partial x^2} - 2\frac{\partial^2 \varphi_2'}{\partial x^2} - x\frac{\partial^3 \varphi_2'}{\partial x^3}, \\
\tau_{xy} &= -\frac{\partial}{\partial y}\left\{\frac{\partial \varphi_1'}{\partial x} - \frac{\partial \varphi_2'}{\partial x} - x\frac{\partial^2 \varphi_2'}{\partial x^2}\right\}.
\end{aligned} \tag{48.2}$$

---

[1] Loc. cit. pp. 431—439.
[2] L. E. PAYNE: University of Maryland, Technical Note BN-66. 1955.

Then, using these expressions in the stress-strain relations for plane strain and integrating, we have for the components of displacement

$$\left.\begin{aligned}2\mu u &= -\frac{\partial \varphi_1'}{\partial x} + x\frac{\partial^2 \varphi_2'}{\partial x^2} - (1-2\nu)\frac{\partial \varphi_2'}{\partial x}, \\ 2\mu v &= -\frac{\partial \varphi_1'}{\partial y} + 2(1-\nu)\frac{\partial \varphi_2'}{\partial y} + x\frac{\partial^2 \varphi_2'}{\partial x \partial y}.\end{aligned}\right\} \quad (48.3)$$

We now introduce the harmonic conjugate functions, ψ_1', and ψ_2' of φ_1' and φ_2' respectively, so that

$$\frac{\partial \varphi_i'}{\partial x} = \frac{\partial \psi_i'}{\partial y}, \qquad \frac{\partial \varphi_i'}{\partial y} = -\frac{\partial \psi_i'}{\partial x}, \qquad i=1,2$$

and put

$$\varphi_i = \frac{\partial \varphi_i'}{\partial x}, \qquad \psi_i = \frac{\partial \psi_i'}{\partial x}, \qquad i=1,2$$

where, of course, the φ_i, ψ_i are again harmonic functions. Eqs. (48.2) and (48.3) then reduce to

$$\left.\begin{aligned}\sigma_x &= -\frac{\partial \varphi_1}{\partial x} + x\frac{\partial^2 \varphi_2}{\partial x^2}, \\ \sigma_y &= \frac{\partial \varphi_1}{\partial x} - 2\frac{\partial \varphi_2}{\partial x} - x\frac{\partial^2 \varphi_2}{\partial x^2}, \\ \tau_{xy} &= -\frac{\partial}{\partial y}\left(\varphi_1 - \varphi_2 - x\frac{\partial \varphi_2}{\partial x}\right), \\ 2\mu u &= -\varphi_1 + x\frac{\partial \varphi_2}{\partial x} - (1-2\nu)\varphi_2, \\ 2\mu v &= \psi_1 - 2(1-\nu)\psi_2 + x\frac{\partial \varphi_2}{\partial y}.\end{aligned}\right\} \quad (48.4)$$

Any one of the components of stress or displacement can now be made to vanish everywhere on $x=0$ by specifying an appropriate linear relationship between φ_1 and φ_2. In fact, on $x=0$,

$$\begin{aligned}\sigma_x &= 0, \quad \text{if} \quad \varphi_1 \equiv 0, \\ \sigma_y &= 0, \quad \text{if} \quad \varphi_1 \equiv 2\varphi_2, \\ \tau_{xy} &= 0, \quad \text{if} \quad \varphi_1 \equiv \varphi_2, \\ u &= 0, \quad \text{if} \quad \varphi_1 \equiv -(1-2\nu)\varphi_2, \\ v &= 0, \quad \text{if} \quad \psi_1 \equiv 2(1-\nu)\psi_2.\end{aligned}$$

The development of the method is essentially the same whichever of these cases we take, but since that of a shear-free surface is commonly encountered we confine the remainder of the discussion to that case. If $\varphi_1 = \varphi_2 = \varphi$ then $\psi_1 = \psi_2 = \psi_1$ say, and Eqs. (48.4) become

$$\left.\begin{aligned}\sigma_x &= -\frac{\partial \varphi}{\partial x} + x\frac{\partial^2 \varphi}{\partial x^2}, \\ \sigma_y &= -\frac{\partial \varphi}{\partial x} - x\frac{\partial^2 \varphi}{\partial x^2}, \\ \tau_{xy} &= x\frac{\partial^2 \varphi}{\partial x \partial y}, \\ 2\mu u &= -2(1-\nu)\varphi + x\frac{\partial \varphi}{\partial x}, \\ 2\mu v &= -(1-2\nu)\psi + x\frac{\partial \varphi}{\partial y}.\end{aligned}\right\} \quad (48.5)$$

The stress components are assumed to be continuous in the half-plane and to vanish at infinity, but the displacements are not necessarily bounded at infinity, as may be seen by reference to the forms of the complex potentials (41.4) for the same problem.

49. Homogeneous boundary value problems. Fourier transform methods are used to solve the first and second boundary value problems. In the second boundary value problem we must assume that the displacement vanishes at infinity, while the displacement on the boundary is given by

$$\mu u = -(1-\nu) f(y), \quad x = 0.$$

We put

$$\bar{\varphi} = \frac{1}{\sqrt{2\pi}} \int_{-\infty}^{\infty} \varphi \, e^{i\alpha y} \, dy, \qquad (49.1)$$

and then $\bar{\varphi}$ satisfies the transformed LAPLACE's equation

$$\frac{d^2 \bar{\varphi}}{dx^2} - \alpha^2 \bar{\varphi} = 0.$$

Using the fourth of Eqs. (48.5) with the boundary conditions we find the solution

$$\bar{\varphi} = \bar{f}(\alpha) \, e^{-|\alpha| x}. \qquad (49.2)$$

The two potential functions can then be written in the forms

(a) $f(y)$ even:

$$\left. \begin{array}{l} \displaystyle \varphi = \frac{2}{\pi} \int_0^\infty e^{-\alpha x} \cos \alpha y \, d\alpha \int_0^\infty f(y') \cos \alpha y' \, dy', \\[1em] \displaystyle \psi = -\frac{2}{\pi} \int_0^\infty e^{-\alpha x} \sin \alpha y \, d\alpha \int_0^\infty f(y') \cos \alpha y' \, dy', \end{array} \right\} \qquad (49.3)$$

(b) $f(y)$ odd:

$$\left. \begin{array}{l} \displaystyle \varphi = \frac{2}{\pi} \int_0^\infty e^{-\alpha x} \sin \alpha y \, d\alpha \int_0^\infty f(y') \sin \alpha y' \, dy', \\[1em] \displaystyle \psi = \frac{2}{\pi} \int_0^\infty e^{-\alpha x} \cos \alpha y \, d\alpha \int_0^\infty f(y') \sin \alpha y' \, dy'. \end{array} \right\} \qquad (49.4)$$

To solve the first boundary value problem we first note that $\partial \varphi / \partial x$ also satisfies LAPLACE's equation. In addition to σ_x vanishing at infinity we have the condition

$$\sigma_x = -g(y), \quad x = 0.$$

These conditions are then used with the first of Eqs. (48.5) by a method similar to that above to obtain the solutions

(a) $g(y)$ even:

$$\left. \begin{array}{l} \displaystyle \frac{\partial \varphi}{\partial x} = \frac{2}{\pi} \int_0^\infty e^{-\alpha x} \cos \alpha y \, d\alpha \int_0^\infty g(y') \cos \alpha y' \, dy', \\[1em] \displaystyle \frac{\partial \psi}{\partial x} = -\frac{2}{\pi} \int_0^\infty e^{-\alpha x} \sin \alpha y \, d\alpha \int_0^\infty g(y') \cos \alpha y' \, dy', \end{array} \right\} \qquad (49.5)$$

(b) $g(y)$ odd:

$$\frac{\partial \varphi}{\partial x} = \frac{2}{\pi} \int_0^\infty e^{-\alpha x} \sin \alpha y \, d\alpha \int_0^\infty g(y') \sin \alpha y' \, dy',$$

$$\frac{\partial \psi}{\partial x} = \frac{2}{\pi} \int_0^\infty e^{-\alpha x} \cos \alpha y \, d\alpha \int_0^\infty g(y') \sin \alpha y' \, dy'.$$

(49.6)

50. Mixed boundary value problems. We now consider the single punch problem, the component of displacement, u, being prescribed everywhere on $x=0$, $|y|<b$, the remainder of the boundary being stress-free. Two types of conformal transformation are introduced to deal with this problem. The transformation

$$x + iy = b \sinh(\xi + i\eta)$$

defines the elliptic coordinate system (ξ, η) in which

$$\left. \begin{array}{l} x = b \sinh \xi \cos \eta, \\ y = b \cosh \xi \sin \eta. \end{array} \right\}$$

(50.1)

In this case the half-plane $x>0$ is mapped onto the *semi-infinite* strip $\xi>0$, $|\eta|<\frac{\pi}{2}$, the segment $x=0$, $|y|<b$ corresponding to $\xi=0$, $|\eta|<\frac{\pi}{2}$ and $x=0$, $y>b$, $y<-b$ corresponding to $\xi>0$, $\eta=\frac{\pi}{2}$, $\eta=-\frac{\pi}{2}$ respectively.

Bipolar coordinates (ξ, η) are defined by

$$x + iy = -b \cot \tfrac{1}{2}(\xi + i\eta)$$

or

$$\left. \begin{array}{l} x = -\dfrac{b \sin \xi}{\cosh \eta - \cos \xi}, \\ y = \dfrac{b \sinh \eta}{\cosh \eta - \cos \xi}. \end{array} \right\}$$

(50.2)

This transformation maps the half-plane $x>0$ on to the infinite strip $-\pi<\xi<0$, $-\infty<\eta<\infty$. The interval $x=0$, $|y|<b$ corresponds to the line $\xi=-\pi$, and the intervals $x=0$, $y>b$, $y<-b$ to $\xi=0$, $\eta>0$, $\eta<0$. It should be noted that under any conformal transformation the harmonic functions φ, ψ become harmonic functions of the new variables ξ and η.

The boundary conditions on $x=0$ are, from Eqs. (49.1),

$$\left. \begin{array}{l} -\dfrac{\mu}{1-\nu} u = \varphi = f(y), \quad -b<y<b, \\ -\sigma_x = \dfrac{\partial \varphi}{\partial x} = 0, \quad y<-b, \quad y>b. \end{array} \right\}$$

(50.3)

In elliptic coordinates ξ and η defined by Eqs. (50.1) these conditions become

$$\varphi = f_1(\eta), \quad \xi = 0, \quad -\frac{\pi}{2} < \eta < \frac{\pi}{2},$$

$$\frac{\partial \varphi}{\partial \eta} = 0 \quad \xi > 0, \quad \eta = \pm \frac{\pi}{2}.$$

If $f(y)$ is an even function of either of the forms

$$f(y) = \sum_{n=0}^\infty a_n b^{-2n}(b^2 - y^2)^n,$$

(50.4)

$$f(y) = \sum_{n=0}^\infty c_{2n} y^{2n},$$

(50.5)

it may be represented by the series

$$f_1(\eta) = \sum_{n=0}^{\infty} A_{2n} \cos(2n\eta)$$

and then

$$\varphi = \sum_{n=0}^{\infty} A_{2n} e^{-2n\xi} \cos(2n\eta) \tag{50.6}$$

since this is an harmonic function satisfying the boundary conditions. Similarly, if $f(y)$ is an odd function represented by one of the series

$$f(y) = y \sum_{n=0}^{\infty} \alpha_n b^{-2n} (b^2 - y^2)^n, \tag{50.7}$$

$$f(y) = \sum_{n=0}^{\infty} \beta_{2n+1} y^{2n+1}, \tag{50.8}$$

φ can be obtained in the form

$$\varphi = \sum_{n=0}^{\infty} B_{2n+1} e^{-(2n+1)\xi} \sin\{(2n+1)\eta\}. \tag{50.9}$$

The stress components become infinite at the ends $y = \pm b$ of the punch profile unless the conditions are such that there is no cusp of the deformed boundary at the ends. When the displacement on $x=0$ is symmetrical, and $f(y)$ is of the form (50.4) or (50.5) the infinities at $y = \pm b$ can be avoided by adding to the potential function (50.6) a quantity $A\xi$ where A is a constant to be determined. It is clear that ξ is an harmonic function and its addition does not violate the boundary conditions: however the displacement component u will no longer be bounded but increase logarithmically as x, $|y| \to \infty$. When $f(y)$ is of the form (50.4) we have from the transformation (50.1)

$$\sum_{n=0}^{\infty} A_{2n} \cos(2n\eta) = \sum_{m=0}^{\infty} a_m \cos^{2m} \eta$$

where, by Fourier analysis,

$$\left. \begin{array}{l} A_0 = \dfrac{1}{2} \displaystyle\sum_{m=0}^{\infty} a_m \dfrac{\Gamma(2m+1)}{2^{2m-1}\{\Gamma(m+1)\}^2}, \\[2mm] A_{2n} = \displaystyle\sum_{m=0}^{\infty} a_m \dfrac{\Gamma(2m+1)}{2^{2m-1}\Gamma(m+n+1)\Gamma(m-n+1)}. \end{array} \right\} \tag{50.10}$$

Substituting the coefficients (50.10) in Eq. (50.6) and reversing the order of summation leads to the expression

$$\varphi = \sum_{m=0}^{\infty} \sum_{n=0}^{m}{}' a_m \frac{\Gamma(2m+1) e^{-2n\xi} \cos(2n\eta)}{2^{2m-1}\Gamma(m+n+1)\Gamma(m-n+1)} \tag{50.11}$$

where the prime indicates that the term $n=0$ is to be halved. On $\xi=0$

$$\frac{\partial \varphi}{\partial x} = \frac{1}{b \cos \eta} \cdot \frac{\partial \varphi}{\partial \xi}$$

which becomes infinite (and hence σ_x, σ_y also) as $\eta \to \pm \tfrac{1}{2}\pi$, the end points of the punch. Adding a multiple of ξ to φ we form the new function

$$\varphi' = \sum_{m=0}^{\infty} \sum_{n=0}^{m}{}' a_m \frac{\Gamma(2m+1)\{e^{-2n\xi} \cos(2n\eta) + 2(-1)^n n\xi\}}{2^{2m-1}\Gamma(m+n+1)\Gamma(m-n+1)}, \tag{50.12}$$

thus removing the infinite stresses and also the discontinuities of $(\partial u/\partial y)_{x=0}$ at the end points $y=\pm b$. It should be noted that an analogous method proposed in the original paper for removing the infinities when $f(y)$ is odd violates the boundary conditions.

As a simple example of the method of solution in elliptic coordinates consider the case

$$f(y) = k(b^2 - y^2), \quad -b < y < b.$$

which is of the form (50.4), the only non-zero coefficient being $a_1 = kb^2$. Substitution in Eq. (50.12) gives the potential function

$$\varphi' = \tfrac{1}{2} k b^2 \{1 + e^{-2\xi} \cos(2\eta) - 2\xi\} \tag{50.13}$$

where the final term serves to eliminate stress infinities at $x=0$, $y=\pm b$. Apart from an arbitrary constant corresponding to a rigid body displacement, the conjugate function is easily seen to be

$$\psi' = -\tfrac{1}{2} k b^2 \{e^{-2\xi} \sin(2\eta) + 2\eta\}. \tag{50.14}$$

In the bipolar coordinate system (ξ, η) given by Eq. (50.2), the boundary conditions (50.3) become

$$\varphi = f_2(\eta), \quad \xi = -\pi;$$

$$\frac{\partial \varphi}{\partial \xi} = 0, \quad \xi = 0.$$

So long as $f(y)$ vanishes at $y=\pm b$ so that $f_2(\eta) \to 0$ as $\eta \to \pm \infty$, the harmonic function φ can be found by solving LAPLACE's equation,

$$\frac{\partial^2 \varphi}{\partial \xi^2} + \frac{\partial^2 \varphi}{\partial \eta^2} = 0$$

by means of a Fourier transform with respect to η. When $f(y)$ is an *even* function the result is

$$\varphi = \frac{2}{\pi} \int_0^\infty \frac{\cosh \alpha \xi \cos \alpha \eta \, d\alpha}{\cosh \alpha \pi} \int_0^\infty f_2(\eta') \cos \alpha \eta' \, d\eta' \tag{50.15}$$

and for an *odd* function

$$\varphi = \frac{2}{\pi} \int_0^\infty \frac{\cosh \alpha \xi \sin \alpha \eta \, d\alpha}{\cosh \alpha \pi} \int_0^\infty f_2(\eta') \sin \alpha \eta' \, d\eta'. \tag{50.16}$$

The term to be added to the right hand side of Eq. (50.15) in order to avoid infinite stresses at the corners of the punch becomes in these coordinates

$$A \log \left(\frac{\cosh \tfrac{1}{2} \eta + \cos \tfrac{1}{2} \xi}{\cosh \tfrac{1}{2} \eta - \cos \tfrac{1}{2} \xi} \right).$$

This procedure is effective when $f(y)$ is of the form (50.4) or (50.5) but has not the general application implied in the original paper, the example below being one of the exceptions. The term proposed by PAYNE[1] for the same purpose when $f(y)$ is odd again violates the boundary conditions.

An elementary example is provided by the case of an elliptic deformation given by

$$f(y) = k(b^2 - y^2)^{\frac{1}{2}}, \quad -b < y < b.$$

---

[1] Loc. cit.

We have, by the transformations (50.2),
$$f_2(\eta) = kb \operatorname{sech}(\tfrac{1}{2}\eta), \quad \xi = -\pi$$
and, substituting in Eq. (50.15) we find that
$$\varphi = \frac{2kb}{\pi} \left\{ \frac{\xi \sin\tfrac{1}{2}\xi \cosh\tfrac{1}{2}\eta + \eta \cos\tfrac{1}{2}\xi \sinh\tfrac{1}{2}\eta}{\cosh\eta - \cos\xi} \right\}.$$

On $x=0$, $|y|<b$,
$$\sigma_x = \sigma_y = -\frac{1+\cosh\eta}{b}\left(\frac{\partial\varphi}{\partial\xi}\right)_{\xi=-\pi}$$
$$= \frac{2k}{\pi}\left\{\cosh\tfrac{1}{2}\eta - \tfrac{1}{2}\eta \sinh\tfrac{1}{2}\eta\right\} = \frac{2k}{\pi\sqrt{b^2-y^2}}\left\{b - y\operatorname{Ar}\tanh\frac{y}{b}\right\}.$$

Other mixed boundary value problems are easily reduced to formulations similar to those above.

51. Infinite strip problems. When the body occupies an infinite strip $0<x<d$, $-\infty<y<\infty$ the problem may be solved by a Fourier integral method so long as the boundary conditions are homogeneous on each boundary, $x=0$, $x=d$. We suppose that the potential functions φ_1, φ_2 in Eqs. (48.4) are given by
$$\varphi_1 = k_1 \varphi + \varphi_1',$$
$$\varphi_2 = k_2 \varphi + \varphi_2'$$
where φ_1' and φ_2' are harmonic functions making no contribution to the boundary conditions on $x=0$ and k_1, k_2 are chosen as in the semi-infinite case according to the boundary condition on $x=0$. The function φ is then the same as that for the semi-infinite case having the same boundary conditions on $x=0$. It is possible to find φ_1' and φ_2' such that the boundary conditions on $x=d$ are satisfied and thus to solve the problem.

D. Three-dimensional problems in elasto-statics.

We now consider the solution of elasto-static problems in three dimensions. Here there do not exist the same powerful analytic tools as in the theory of two-dimensional problems, but the method of Betti provides a general theory and the theory of integral transforms and the use of curvilinear coordinates provide useful methods for tackling a restricted range of special problems.

We begin by giving the formulae by means of which the equations of elastic equilibrium may be cast into curvilinear coordinates.

I. General theory.

52. Orthogonal curvilinear coordinates. When the coordinate curves are orthogonal we obtain a formula of the type
$$ds^2 = h_1^2 dq_1^2 + h_2^2 dq_2^2 + h_3^2 dq_3^2 \tag{52.1}$$
for the fundamental quadratic form, so that $g_{mn} = h_m^2 \delta_{mn}$ and $g^{mn} = \delta_{mn}/h_m^2$, where δ_{mn} denotes the Kronecker delta. From the definition
$$[m, np] = \frac{1}{2}\left(\frac{\partial g_{mn}}{\partial q_p} + \frac{\partial g_{mp}}{\partial q_n} - \frac{\partial g_{np}}{\partial q_m}\right)$$

of the Christoffel symbol of the first kind we find that, for the metric (52.1)

$$[m, np] = h_m \frac{\partial h_m}{\partial q_p} \delta_{mn} + h_m \frac{\partial h_m}{\partial q_n} \delta_{mp} - h_n \frac{\partial h_n}{\partial q_m} \delta_{np}.$$

If, therefore, we assume that m, n, p are all different, we find that (repeated indices *not* summed)

$$\left.\begin{array}{c} [m, np] = 0, \\ [m, nm] = [m, mn] = -[n, mm] = h_m \dfrac{\partial h_m}{\partial q_n}, \\ [m, mm] = h_m \dfrac{\partial h_m}{\partial q_m}. \end{array}\right\} \quad (52.2)$$

With the same convention, we find from the definition

$$\left\{\begin{matrix} m \\ np \end{matrix}\right\} = g^{im}[i, np]$$

that

$$\left.\begin{array}{c} \left\{\begin{matrix} m \\ np \end{matrix}\right\} = 0, \quad \left\{\begin{matrix} m \\ mm \end{matrix}\right\} = \dfrac{1}{h_m} \dfrac{\partial h_m}{\partial q_m}, \\ \left\{\begin{matrix} m \\ mn \end{matrix}\right\} = \left\{\begin{matrix} m \\ nm \end{matrix}\right\} = \dfrac{1}{h_m} \dfrac{\partial h_m}{\partial q_n}, \quad \left\{\begin{matrix} n \\ mm \end{matrix}\right\} = -\dfrac{h_m}{h_n^2} \dfrac{\partial h_m}{\partial q_n}. \end{array}\right\} \quad (52.3)$$

If we denote the physical components of displacement along the q_m-coordinate curves by $(u_{q_1}, u_{q_2}, u_{q_3})$ then from Eq. (4.2)

$$u_{q_1} = h_1 u^1, \quad u_{q_2} = h_2 u^2, \quad u_{q_3} = h_3 u^3. \quad (52.4)$$

Similarly from Eq. (9.17) we see that the physical components of the stress tensor referred to rectangular axes along the q_m-coordinate curves are given by the formula

$$\sigma_{mn} = h_m h_n \tau^{mn} \quad (52.5)$$

From the definition

$$\Delta = \nabla_s u^s$$

of the dilatation, we find that

$$\Delta = \sum_s \frac{\partial}{\partial q_s}\left(\frac{u_{q_s}}{h_s}\right) + \sum_{i,s} \left\{\begin{matrix} s \\ i \, s \end{matrix}\right\} \frac{u_{q_i}}{h_i}$$

so that, by Eq. (52.3) we have

$$\Delta = \sum_s \frac{\partial}{\partial q_s}\left(\frac{u_{q_s}}{h_s}\right) + \sum_{i,s} \frac{u_{q_i}}{h_i h_s} \cdot \frac{\partial h_i}{\partial q_i}. \quad (52.6)$$

α) *Cylindrical polar coordinates.* If we take $q_1 = \varrho$, $q_2 = \varphi$, $q_3 = z$ where ϱ, φ, z are the cylindrical polar coordinates of a point then, since

$$ds^2 = d\varrho^2 + \varrho^2 d\varphi^2 + dz^2,$$

we have

$$h_1 = 1, \quad h_2 = \varrho, \quad h_3 = 1$$

and it follows that all the Christoffel symbols vanish except

$$[2, 12] = [2, 21] = -[1, 22] = \varrho, \quad (52.7)$$

$$\left\{\begin{matrix} 2 \\ 21 \end{matrix}\right\} = \left\{\begin{matrix} 2 \\ 12 \end{matrix}\right\} = \frac{1}{\varrho}, \quad \left\{\begin{matrix} 1 \\ 22 \end{matrix}\right\} = -\varrho. \quad (52.8)$$

Denoting the physical components of displacement by $(u_\varrho, u_\varphi, u_z)$ we see from Eq. (52.4) that

$$u_\varrho = u^1, \qquad u_\varphi = \varrho u^2, \qquad u_z = u^3. \tag{52.9}$$

Similarly, from Eqs. (52.5) we have the formulae

$$\left.\begin{array}{lll} \sigma_\varrho = \tau^{11}, & \sigma_\varphi = \varrho^2 \tau^{22}, & \sigma_z = \tau^{33}, \\ \tau_{\varphi z} = \varrho \tau^{23}, & \tau_{\varrho z} = \tau^{13}, & \tau_{\varrho\varphi} = \varrho \tau^{12} \end{array}\right\} \tag{52.10}$$

for the physical components of the stress.

All the quantities $\partial h_s/\partial q_i$ are zero except when $s=2$, $i=1$ so that Eq. (52.6) reduces to

$$\Delta = \sum_s \frac{1}{h_s} \frac{\partial u_{q_s}}{\partial q_s} + \frac{u_\varrho}{h_1 h_2}$$

that is, to

$$\Delta = \frac{\partial u_\varrho}{\partial \varrho} + \frac{u_\varrho}{\varrho} + \frac{1}{\varrho} \frac{\partial u_\varphi}{\partial \varphi} + \frac{\partial u_z}{\partial z}. \tag{52.11}$$

β) *Spherical polar coordinates.* If we take $q_1 = r$, $q_2 = \vartheta$, $q_3 = \varphi$ where r, ϑ, φ are the spherical polar coordinates of a point then, since

$$ds^2 = dr^2 + r^2 d\vartheta^2 + r^2 \sin^2 \vartheta \, d\varphi^2$$

we have

$$h_1 = 1, \quad h_2 = r, \quad h_3 = r \sin \vartheta,$$

and it follows that all the Christoffel symbols vanish except

$$\left.\begin{array}{l}[2,12] = [2,21] = -[1,22] = r, \\ [3,13] = [3,31] = -[1,33] = r \sin^2 \vartheta, \\ [3,23] = [3,32] = -[2,33] = r^2 \sin \vartheta \cos \vartheta \end{array}\right\} \tag{52.12}$$

and

$$\left.\begin{array}{ll} \left\{\begin{array}{c}2\\12\end{array}\right\} = \left\{\begin{array}{c}2\\21\end{array}\right\} = \dfrac{1}{r}, & \left\{\begin{array}{c}1\\22\end{array}\right\} = -r, \\[2mm] \left\{\begin{array}{c}3\\13\end{array}\right\} = \left\{\begin{array}{c}3\\31\end{array}\right\} = \dfrac{1}{r}, & \left\{\begin{array}{c}1\\33\end{array}\right\} = -r \sin^2 \vartheta, \\[2mm] \left\{\begin{array}{c}3\\23\end{array}\right\} = \left\{\begin{array}{c}3\\32\end{array}\right\} = \cot \vartheta, & \left\{\begin{array}{c}2\\33\end{array}\right\} = -\sin \vartheta \cos \vartheta. \end{array}\right\} \tag{52.13}$$

The physical components of the displacement are given by

$$u_r = u^1, \qquad u_\vartheta = r u^2, \qquad u_\varphi = r \sin \vartheta \, u^3, \tag{52.14}$$

while the physical components of the stress are given by

$$\left.\begin{array}{lll} \sigma_r = \tau^{11}, & \sigma_\vartheta = r^2 \tau^{22}, & \sigma_\varphi = r^2 \sin^2 \vartheta \, \tau^{33}, \\ \tau_{\vartheta\varphi} = r^2 \sin \vartheta \, \tau^{23}, & \tau_{r\varphi} = r \sin \vartheta \, \tau^{13}, & \sigma_{r\vartheta} = r \tau^{12}. \end{array}\right\} \tag{52.15}$$

The quantities $\partial h_s/\partial q_i$ are all zero except

$$\frac{\partial h_2}{\partial q_1} = 1, \qquad \frac{\partial h_3}{\partial q_1} = \sin \vartheta, \qquad \frac{\partial h_3}{\partial q_2} = r \cos \vartheta$$

so that from Eq. (52.6)

$$\Delta = \frac{\partial u_r}{\partial r} + \frac{1}{r} \frac{\partial u_\vartheta}{\partial \vartheta} + \frac{1}{r \sin \vartheta} \frac{\partial u_\varphi}{\partial \varphi} + \frac{2 u_r}{r} + \frac{\cot \vartheta}{r} u_\vartheta. \tag{52.16}$$

53. The vector form of the equilibrium equations.
In order to facilitate the derivation of general solutions of the equations of equilibrium we shall put these equations into vector form.

If we write Eqs. (9.19) and (14.29) in rectangular Cartesian coordinates x_j ($j=1, 2, 3$) we find that the equations of equilibrium of an elastic solid can be put in the form

$$\partial_i \tau^{ij} + \varrho P^j = 0 \tag{53.1}$$

where

$$\tau^{ij} = \lambda \Delta \, \delta^{ij} + \mu (\partial_i u_j + \partial_j u_i). \tag{53.2}$$

Substituting from Eqs. (53.2) into Eqs. (53.1), we obtain the set

$$\lambda \partial_j \Delta + \mu \partial_{ii} u_j + \mu \partial_{ij} u_i + \varrho P^j = 0.$$

Noting that

$$\partial_{ij} u_i = \partial_j(\partial_i u_i) = \partial_j \Delta,$$

we see that this is equivalent to

$$(\lambda + \mu) \partial_j \Delta + \mu \partial_{ii} u_j + \varrho P^j = 0$$

which may be written in the vectorial form

$$\mu \nabla^2 \boldsymbol{u} + (\lambda + \mu) \operatorname{grad} \Delta + \varrho \boldsymbol{P} = \boldsymbol{0}. \tag{53.3}$$

In the absence of body forces $\boldsymbol{P}=\boldsymbol{0}$ the zero vector, and this equation can be written as

$$(1 - 2\nu) \nabla^2 \boldsymbol{u} + \operatorname{grad} \operatorname{div} \boldsymbol{u} = \boldsymbol{0} \tag{53.4}$$

where ν is Poisson's ratio.

Alternatively we can make use of the vector identity

$$\operatorname{curl} \operatorname{curl} = \operatorname{grad} \operatorname{div} - \nabla^2 \tag{53.5}$$

to substitute the value

$$\operatorname{grad} \Delta - \operatorname{curl} \operatorname{curl} \boldsymbol{u}$$

for $\nabla^2 \boldsymbol{u}$ in Eq. (53.3) and so obtain the equation

$$(\lambda + 2\mu) \operatorname{grad} \operatorname{div} \boldsymbol{u} - \mu \operatorname{curl} \operatorname{curl} \boldsymbol{u} + \varrho \boldsymbol{P} = \boldsymbol{0}. \tag{53.6}$$

54. Kelvin's solution of the equilibrium equations.
In this section we shall derive a general form of particular solution of the vector equation of elastic equilibrium. To do this we decompose the displacement vector in a manner due to Kelvin[1]. We express the displacement by means of a scalar potential φ and a vector potential $\boldsymbol{f}$ by means of a formula of the type

$$\boldsymbol{u} = \operatorname{grad} \varphi + \operatorname{curl} \boldsymbol{f}, \tag{54.1}$$

and we suppose that the body forces are of such a kind that they can by expressed in a like manner by means of formulae of the type

$$\boldsymbol{P} = \operatorname{grad} \boldsymbol{\Phi} + \operatorname{curl} \boldsymbol{F}. \tag{54.2}$$

Since

$$\operatorname{grad} \operatorname{div} \boldsymbol{u} = \operatorname{grad} \nabla^2 \varphi$$

and

$$\operatorname{curl} \operatorname{curl} \boldsymbol{u} = - \operatorname{curl} \nabla^2 \boldsymbol{f},$$

---
[1] Sir W. Thomson: Camb. and Dublin Math. J., 1848 reprinted in Math. and Phys. Papers 1, 97.

it follows from Eq. (53.6) that the equations of equilibrium can be written in the vector form

$$\operatorname{grad}\left[(\lambda+2\mu)\nabla^2\varphi+\varrho\,\Phi\right]+\operatorname{curl}\left[\mu\,\nabla^2\boldsymbol{f}+\varrho\,\boldsymbol{F}\right]=0.$$

We see immediately that particular solutions of the equilibrium equations can be obtained from particular solutions of the scalar equation

$$(\lambda+2\mu)\,\nabla^2\varphi+\varrho\,\Phi=0 \tag{54.3}$$

and of the vector equation

$$\mu\,\nabla^2\boldsymbol{f}+\varrho\,\boldsymbol{F}=0. \tag{54.4}$$

Now the body force $\boldsymbol{P}$ can be expressed by a formula of the type (54.2) by putting

$$\Phi(\boldsymbol{r})=-\frac{1}{4\pi}\int_V \boldsymbol{P}(\boldsymbol{r}')\cdot\operatorname{grad}\frac{1}{|\boldsymbol{r}-\boldsymbol{r}'|}\,d^3\tau',$$

$$\boldsymbol{F}(\boldsymbol{r})=-\frac{1}{4\pi}\int_V \boldsymbol{P}(\boldsymbol{r}')\times\operatorname{grad}\frac{1}{|\boldsymbol{r}-\boldsymbol{r}'|}\,d^3\tau'$$

where the integration is taken over the volume V outside of which the body forces vanish. Using the relation

$$\operatorname{grad}\frac{1}{|\boldsymbol{r}-\boldsymbol{r}'|}=-\frac{\boldsymbol{r}-\boldsymbol{r}'}{|\boldsymbol{r}-\boldsymbol{r}'|^3}$$

we have

$$\Phi(\boldsymbol{r})=\frac{1}{4\pi}\int_V \frac{\boldsymbol{P}(\boldsymbol{r}')\cdot(\boldsymbol{r}-\boldsymbol{r}')}{|\boldsymbol{r}-\boldsymbol{r}'|^3}\,d\tau', \tag{54.5}$$

$$\boldsymbol{F}(\boldsymbol{r})=\frac{1}{4\pi}\int_V \frac{\boldsymbol{P}(\boldsymbol{r}')\times(\boldsymbol{r}-\boldsymbol{r}')}{|\boldsymbol{r}-\boldsymbol{r}'|^3}\,d\tau'. \tag{54.6}$$

Once the scalar quantity Φ and the vector $\boldsymbol{F}$ have been found by means of these equations it is a simple matter to obtain particular solutions of Eqs. (54.3) and (54.4). We know from potential theory that particular solutions of these equations are

$$\varphi(\boldsymbol{r})=\frac{\varrho}{4\pi(\lambda+2\mu)}\int\frac{\Phi(\boldsymbol{r}')\,d\tau'}{|\boldsymbol{r}-\boldsymbol{r}'|}, \tag{54.7}$$

$$\boldsymbol{f}(\boldsymbol{r})=\frac{\varrho}{4\pi\mu}\int\frac{\boldsymbol{F}(\boldsymbol{r}')\,d\tau'}{|\boldsymbol{r}-\boldsymbol{r}'|}. \tag{54.8}$$

55. Point force in an infinite solid. As a special case of KELVIN's solution we shall consider the case of a point force of magnitude Z acting at the origin parallel to the z-axis in an infinite elastic solid. In the notation of the last section, we may take V to be the cubical volume bounded by the planes $x=\pm\tfrac{1}{2}\varepsilon$, $y=\pm\tfrac{1}{2}\varepsilon$, $z=\pm\tfrac{1}{2}\varepsilon$ and put

$$\boldsymbol{P}=\left(0,0,\frac{Z}{\varrho\,\varepsilon^3}\right)$$

where ε is infinitesimal. Substituting these results in Eqs. (54.5) and (54.6) we obtain the expressions

$$\Phi=\frac{Z}{4\pi\varrho}\left(\frac{z}{r^3}\right),\quad \boldsymbol{F}=\frac{Z}{4\pi\varrho}\left(-\frac{y}{r^3},\frac{x}{r^3},0\right) \tag{55.1}$$

for the functions corresponding to a point force. We must therefore find particular solutions of the equations

$$\nabla^2 \varphi + \frac{Z}{4\pi(\lambda + 2\mu)} \left(\frac{z}{r^3}\right) = 0 \tag{55.2}$$

and

$$\nabla^2 \boldsymbol{f} + \frac{Z}{4\pi\mu} \left(-\frac{y}{r^3}, \frac{x}{r^3}, 0\right) = 0. \tag{55.3}$$

Now

$$\nabla^2 \left(\frac{x}{r}\right) = -\frac{2x}{r^3}$$

etc. so that we can write these equations in the forms

$$\nabla^2 \left\{\varphi - \frac{zZ}{8\pi(\lambda + 2\mu)r}\right\} = 0,$$

$$\nabla^2 \left\{\boldsymbol{f} + \frac{Z}{8\pi\mu} \left(\frac{y}{r}, \frac{-x}{r}, 0\right)\right\} = 0$$

from which it follows immediately that

$$\varphi = \frac{Z}{8\pi(\lambda + 2\mu)} \left(\frac{z}{r}\right), \tag{55.4}$$

$$\boldsymbol{f} = \frac{Z}{8\pi\mu} \left(-\frac{y}{r}, \frac{x}{r}, 0\right). \tag{55.5}$$

Now from (55.4) we have

$$\operatorname{grad} \varphi = \frac{Z}{8\pi(\lambda + 2\mu)} \left(-\frac{xz}{r^3}, \frac{-yz}{r^3}, \frac{1}{r} - \frac{z^2}{r^3}\right)$$

and from (55.5)

$$\operatorname{curl} \boldsymbol{f} = \frac{Z}{8\pi\mu} \left(\frac{xz}{r^3}, \frac{yz}{r^3}, \frac{1}{r} + \frac{z^2}{r^3}\right)$$

so that substituting in Eq. (54.1) we find that the displacement vector is given by the equation

$$\boldsymbol{u} = \frac{(\lambda + \mu) Z}{8\pi\mu(\lambda + 2\mu)} \left(\frac{xz}{r^3}, \frac{yz}{r^3}, \frac{z^2}{r^3} + \frac{\lambda + 3\mu}{\lambda + \mu} \cdot \frac{1}{r}\right). \tag{55.6}$$

More generally, the displacement due to a force $\boldsymbol{P}$ acting at the point with position vector $\boldsymbol{a}$, is given by the equation

$$\boldsymbol{u} = \frac{\lambda + \mu}{8\pi\mu(\lambda + 2\mu)} \left\{(\boldsymbol{r} - \boldsymbol{a}) \frac{(\boldsymbol{r} - \boldsymbol{a}) \cdot \boldsymbol{P}}{|\boldsymbol{r} - \boldsymbol{a}|^3} + \frac{\lambda + 3\mu}{\lambda + \mu} \cdot \frac{\boldsymbol{P}}{|\boldsymbol{r} - \boldsymbol{a}|}\right\}. \tag{55.7}$$

Applying the principle of superposition to this equation we see that when forces $\boldsymbol{P}(\boldsymbol{r})$ act through a volume V of finite size, particular integrals of the equations of equilibrium can be written in the form

$$\boldsymbol{u} = \frac{\lambda + \mu}{8\pi\mu(\lambda + 2\mu)} \int_V \left\{(\boldsymbol{r} - \boldsymbol{r}') \frac{(\boldsymbol{r} - \boldsymbol{r}') \cdot \boldsymbol{P}(\boldsymbol{r}')}{|\boldsymbol{r} - \boldsymbol{r}'|^3} + \frac{\lambda + 3\mu}{\lambda + \mu} \frac{\boldsymbol{P}(\boldsymbol{r}')}{|\boldsymbol{r} - \boldsymbol{r}'|}\right\} d\tau'. \tag{55.8}$$

This solution, which is equivalent to KELVIN's solution of the last section, was first derived by BOUSSINESQ.

56. The Boussinesq-Papkovich solution. In this section we shall derive a general solution of the equations of equilibrium in terms of a scalar function and a vector function, and then, in the following Sects. 57, 58, illustrate its use in the discussion of special problems.

Let

$$\boldsymbol{u} = A \operatorname{grad}(\varphi + \boldsymbol{r} \cdot \boldsymbol{\psi}) + B \boldsymbol{\psi}, \tag{56.1}$$

where r denotes the position vector of a field point and A and B are constants, whose values are as yet undetermined. Making use of the vector identity (53.5) and the fact that the operator curl grad is a null operator we find that

$$\operatorname{curl} \operatorname{curl} \boldsymbol{u} = B(\operatorname{grad} \operatorname{div} \boldsymbol{\psi} - \nabla^2 \boldsymbol{\psi}). \tag{56.2}$$

On the other hand, since

$$\operatorname{div} \operatorname{grad} = \nabla^2$$

and

$$\nabla^2 (\boldsymbol{r} \cdot \boldsymbol{\psi}) = \boldsymbol{r} \cdot \nabla^2 \boldsymbol{\psi} + 2 \operatorname{div} \boldsymbol{\psi},$$

we find that

$$\operatorname{div} \boldsymbol{u} = A \nabla^2 \varphi + A \boldsymbol{r} \cdot \nabla^2 \boldsymbol{\psi} + (2A + B) \operatorname{div} \boldsymbol{\psi}. \tag{56.3}$$

Substituting from Eqs. (56.2) and (56.3) into Eq. (53.6) we obtain the vector relation

$$(\lambda + 2\mu) A \operatorname{grad}(\nabla^2 \varphi + \boldsymbol{r} \cdot \nabla^2 \boldsymbol{\psi}) + $$
$$+ [(\lambda + 2\mu)(B + 2A) - \mu B] \operatorname{grad} \operatorname{div} \boldsymbol{\psi} + B \nabla^2 \boldsymbol{\psi} + \frac{\varrho}{\mu} \boldsymbol{P} = 0$$

which is obviously satisfied if we can choose A, B, φ, $\boldsymbol{\psi}$ so that

$$\nabla^2 \varphi + \boldsymbol{r} \cdot \nabla^2 \boldsymbol{\psi} = 0,$$
$$(\lambda + 2\mu)(B + 2A) - \mu B = 0,$$
$$B \nabla^2 \boldsymbol{\psi} + \frac{\varrho}{\mu} \boldsymbol{P} = 0.$$

The second of these equations is satisfied if we take $A = 1$ and

$$B = -\frac{2(\lambda + 2\mu)}{\lambda + \mu} = -4(1 - \nu).$$

Putting these values in Eq. (56.1) we see that

$$\boldsymbol{u} = \operatorname{grad}(\varphi + \boldsymbol{r} \cdot \boldsymbol{\psi}) - 4(1 - \nu) \boldsymbol{\psi} \tag{56.4}$$

is a solution of the vector equation of equilibrium (53.6) provided that φ and $\boldsymbol{\psi}$ are solutions of the equations

$$4(1 - \nu) \nabla^2 \varphi + \frac{\varrho}{\mu} (\boldsymbol{r} \cdot \boldsymbol{P}) = 0, \quad 4(1 - \nu) \nabla^2 \boldsymbol{\psi} = \frac{\varrho}{\mu} \boldsymbol{P}. \tag{56.5}$$

The general solution (56.4) was established by PAPKOVICH[1] and discovered independently later by NEUBER[2]; elements of this solution were known to BOUSSINESQ[3].

In the absence of body forces the Eqs. (56.5) reduce to

$$\nabla^2 \varphi = 0, \quad \nabla^2 \boldsymbol{\psi} = 0 \tag{56.6}$$

showing that, in this case, Eq. (56.4) yields a solution provided that φ and each of the components of $\boldsymbol{\psi}$ is a harmonic function.

The case in which there are no body forces present is rather a special one and yields interesting particular results. If we form the divergence of both sides of Eq. (56.4) we find that

$$\nabla^2 (\operatorname{div} \boldsymbol{u}) = 0$$

---

[1] P. F. PAPKOVICH: Izv. Akad. Nauk SSSR., Phys.-Math. Ser. **10**, 1425 (1932).
[2] H. NEUBER: Z. angew. Math. Mech. **14**, 203 (1934).
[3] M. J. BOUSSINESQ: Application des Potentiels a l'Étude de l'Équilibre et du Movement des Solides Élastiques. Paris: Gauthier-Villars 1885.

showing that $\Delta = \operatorname{div} \boldsymbol{u}$ is a harmonic function. Similarly if we form the curl of both sides of Eq. (56.4) and note that curl grad is a null operator we see that

$$\nabla^2 (\operatorname{curl} \boldsymbol{u}) = 0$$

showing that each of the components of the rotation

$$\boldsymbol{\omega} = \tfrac{1}{2} \operatorname{curl} \boldsymbol{u}$$

is a harmonic function.

Another special case arises when the body force is a constant, $\boldsymbol{F}$ say. It is readily shown that

$$\nabla^2 (r^2 \boldsymbol{F}) = 6\boldsymbol{F}, \qquad \nabla^2 (\boldsymbol{R} \cdot \boldsymbol{F}) = 6 (\boldsymbol{r} \cdot \boldsymbol{F})$$

where $\boldsymbol{R}$ denotes the vector with components (x^3, y^3, z^3) so that, if the body force is constant, the equations of equilibrium (56.5) have a solution

$$\varphi = \varphi_1 - \frac{\varrho}{24(1-\nu)\mu} (\boldsymbol{R} \cdot \boldsymbol{F}), \qquad \boldsymbol{\psi} = \boldsymbol{\psi}_1 + \frac{\varrho r^2 \boldsymbol{F}}{24(1-\nu)\mu} \tag{56.7}$$

where φ_1 and each of the components of $\boldsymbol{\psi}_1$ is a harmonic function.

57. Simple solutions. Certain very simple solutions of the elastic equations are contained in the Boussinesq-Papkovich solution.

For instance, if in Eq. (56.1) we put $\boldsymbol{\psi} = 0$ we obtain the simple solution

$$\boldsymbol{u} = A \operatorname{grad} \varphi \tag{57.1}$$

where A is a constant and φ is a harmonic function. For such a displacement field

$$\operatorname{div} \boldsymbol{u} = A \nabla^2 \varphi = 0 \tag{57.2}$$

so that the dilatation is zero everywhere. The components of stress are given by the simple equations

$$\left.\begin{array}{lll}
\sigma_x = 2\mu A \dfrac{\partial^2 \varphi}{\partial x^2}, & \sigma_y = 2\mu A \dfrac{\partial^2 \varphi}{\partial y^2}, & \sigma_z = 2\mu A \dfrac{\partial^2 \varphi}{\partial z^2}, \\[4pt]
\tau_{yz} = 2\mu A \dfrac{\partial^2 \varphi}{\partial y \partial z}, & \tau_{zx} = 2\mu A \dfrac{\partial^2 \varphi}{\partial z \partial x}, & \tau_{xy} = 2\mu A \dfrac{\partial^2 \varphi}{\partial x \partial y}.
\end{array}\right\} \tag{57.3}$$

As an example, let

$$\varphi = \log(z + r), \tag{57.4}$$

then

$$\boldsymbol{u} = A \left\{ \frac{x}{r(z+r)}, \frac{y}{r(z+r)}, \frac{1}{r} \right\} \tag{57.5}$$

and

$$\sigma_x = 2\mu A \left\{ \frac{y^2 + z^2}{r^3(z+r)} - \frac{x^2}{r^2(z+r)^2} \right\}, \qquad \tau_{yz} = -2\mu A \frac{y}{r^3},$$

$$\sigma_y = 2\mu A \left\{ \frac{z^2 + x^2}{r^3(z+r)} - \frac{y^2}{r^2(z+r)^2} \right\}, \qquad \tau_{zx} = -2\mu A \frac{x}{r^3},$$

$$\sigma_z = -2\mu A \frac{z}{r^3}, \qquad \tau_{xy} = -2\mu A \frac{xy(z+2r)}{r^3(z+r)^2}.$$

This is known as *Boussinesq's second type of simple solution*.

58. The use of potential functions in elasticity. It is a simple matter to deduce from the Boussinesq-Papkovich solution a simpler solution involving only one harmonic function which is such that the shearing stress vanishes across a plane which may be chosen to be $z = 0$.

We shall employ rectangular Cartesian coordinates (x, y, z) in terms of which system the components of $\mathbf{u}$ will be denoted by (u, v, w). If, in Eq. (56.4), we let

$$\varphi = (1 - 2\nu)\chi, \qquad \psi = \left(0, 0, \frac{\partial \chi}{\partial z}\right)$$

where χ is a harmonic function, we obtain the solution

$$u = z \frac{\partial^2 \chi}{\partial x \partial z} + (1 - 2\nu) \frac{\partial \chi}{\partial x}, \tag{58.1}$$

$$v = z \frac{\partial^2 \chi}{\partial y \partial z} + (1 - 2\nu) \frac{\partial \chi}{\partial y}, \tag{58.2}$$

$$w = z \frac{\partial^2 \chi}{\partial z^2} - 2(1 - \nu) \frac{\partial \chi}{\partial z}, \tag{58.3}$$

which equations in turn lead to the expressions

$$\sigma_x = 2\mu \left\{ z \frac{\partial^3 \chi}{\partial x^2 \partial z} + (1 - 2\nu) \frac{\partial^2 \chi}{\partial x^2} - 2\nu \frac{\partial^2 \chi}{\partial z^2} \right\}, \tag{58.4}$$

$$\sigma_y = 2\mu \left\{ z \frac{\partial^3 \chi}{\partial y^2 \partial z} + (1 - 2\nu) \frac{\partial^2 \chi}{\partial y^2} - 2\nu \frac{\partial^2 \chi}{\partial z^2} \right\}, \tag{58.5}$$

$$\sigma_z = 2\mu \left(z \frac{\partial^3 \chi}{\partial z^3} - \frac{\partial^2 \chi}{\partial z^2} \right), \tag{58.6}$$

$$\tau_{yz} = 2\mu z \frac{\partial^3 \chi}{\partial y \partial z^2}, \tag{58.7}$$

$$\tau_{zx} = 2\mu z \frac{\partial^3 \chi}{\partial x \partial z^2}, \tag{58.8}$$

$$\tau_{xy} = 2\mu \left\{ z \frac{\partial^3 \chi}{\partial x \partial y \partial z} + (1 - 2\nu) \frac{\partial^2 \chi}{\partial x \partial y} \right\} \tag{58.9}$$

for the components of the stress tensor.

From Eqs. (58.7) and (58.8) it follows that, if $\partial^3 \chi / \partial x \partial z^2$ and $\partial^3 \chi / \partial y \partial z^2$ both remain finite as $z \to 0$,

$$\tau_{xz} = \tau_{yz} = 0 \quad \text{on} \quad z = 0. \tag{58.10}$$

Similarly, if $\partial^2 \chi / \partial z^2$ and $\partial^3 \chi / \partial z^3$ do not become infinite as $z \to 0$, the values of the normal stress and displacement on the boundary $z = 0$ are given by the equations

$$\sigma_z = -2\mu \left(\frac{\partial^2 \chi}{\partial z^2} \right)_{z=0} \tag{58.11}$$

and

$$w = -2(1 - \nu) \left(\frac{\partial \chi}{\partial z} \right)_{z=0}. \tag{58.12}$$

In cylindrical polar coordinates (ϱ, φ, z) we obtain the following solution

$$u_\varrho = z \frac{\partial^2 \chi}{\partial \varrho \partial z} + (1 - 2\nu) \frac{\partial \chi}{\partial \varrho}, \tag{58.13}$$

$$u_\varphi = \frac{z}{\varrho} \frac{\partial^2 \chi}{\partial \varphi \partial z} + (1 - 2\nu) \frac{1}{\varrho} \frac{\partial \chi}{\partial \varphi}, \tag{58.14}$$

$$u_z = z \frac{\partial^2 \chi}{\partial z^2} - 2(1 - \nu) \frac{\partial \chi}{\partial z}, \tag{58.15}$$

for which the components of stress are given by the equations

$$\sigma_\varrho = 2\mu \left\{ z \frac{\partial^3 \chi}{\partial \varrho^2 \partial z} + \frac{\partial^2 \chi}{\partial \varrho^2} - 2\nu \left(\frac{\partial^2 \chi}{\partial \varrho^2} + \frac{\partial^2 \chi}{\partial z^2} \right) \right\}, \tag{58.16}$$

$$\sigma_\varphi = -2\mu \left\{ z \frac{\partial^3 \chi}{\partial \varrho^2 \partial z} + z \frac{\partial^3 \chi}{\partial z^3} + \frac{\partial^2 \chi}{\partial z^2} - (1-2\nu) \frac{\partial^2 \chi}{\partial \varrho^2} \right\}, \tag{58.17}$$

$$\sigma_z = 2\mu \left\{ z \frac{\partial^3 \chi}{\partial z^3} - \frac{\partial^2 \chi}{\partial z^2} \right\}, \tag{58.18}$$

$$\tau_{\varrho\varphi} = \mu \left\{ \frac{2z}{\varrho} \frac{\partial^3 \chi}{\partial \varrho \partial \varphi \partial z} - \frac{z}{\varrho^2} \cdot \frac{\partial^2 \chi}{\partial \varphi \partial z} - \frac{(1-2\nu)}{\varrho^2} \cdot \frac{\partial \chi}{\partial \varphi} \right\}, \tag{58.19}$$

$$\tau_{\varrho z} = 2\mu z \frac{\partial^3 \chi}{\partial \varrho \partial z^2}, \tag{58.20}$$

$$\tau_{\varphi z} = \frac{2\mu z}{\varrho} \frac{\partial^3 \chi}{\partial \varphi \partial z^2}, \tag{58.21}$$

where

$$\frac{\partial^2 \chi}{\partial \varrho^2} + \frac{1}{\varrho} \frac{\partial \chi}{\partial \varrho} + \frac{1}{\varrho^2} \frac{\partial^2 \chi}{\partial \varphi^2} + \frac{\partial^2 \chi}{\partial z^2} = 0. \tag{58.22}$$

In problems in which there is axial symmetry $\partial \chi / \partial \varphi \equiv 0$.

59. Axially symmetrical pressure on the surface of a semi-infinite solid. To illustrate the use of the results derived in the last section we shall consider the distribution of stress in the semi-infinite solid $z \geq 0$ when a radially symmetrical pressure acts normally on its surface. Because of the symmetry we may put

$$\frac{\partial}{\partial \varphi} \equiv 0 \tag{59.1}$$

in the equations of the last section.

If, in Eqs. (58.13) and (58.15), we substitute

$$\chi = -\frac{F}{4\pi\mu} \log(r_1 + z + a) \tag{59.2}$$

where

$$r_1^2 = \varrho^2 + (z+a)^2 \tag{59.3}$$

we find that

$$u_\varrho = \frac{F}{4\pi\mu} \left\{ \frac{z\varrho}{r_1^3} - (1-2\nu) \left[\frac{1}{\varrho} - \frac{z+a}{r_1 \varrho} \right] \right\}, \tag{59.4}$$

$$u_z = \frac{F}{4\pi\mu} \left\{ \frac{z(z+a)}{r_1^3} + \frac{2(1-\nu)}{r_1} \right\}. \tag{59.5}$$

Similarly substituting from Eq. (59.2) into Eq. (58.18) we find that

$$\sigma_z = -\frac{F}{2\pi} \left[\frac{z+a}{r_1^3} - z \left\{ \frac{1}{r_1^3} - \frac{3(z+a)^2}{r_1^5} \right\} \right]. \tag{59.6}$$

It follows immediately that the displacement field specified by Eqs. (59.4) and (59.5) arises from a normal surface pressure,

$$p(\varrho) = -(\sigma_z)_{z=0} = \frac{Fa}{2\pi(a^2 + \varrho^2)^{\frac{3}{2}}}. \tag{59.7}$$

This surface pressure has the property that

$$2\pi \int_0^\infty \varrho \, p(\varrho) \, d\varrho = F \int_0^\infty \frac{\varrho a \, d\varrho}{(a^2 + \varrho^2)^{\frac{3}{2}}} = F \tag{59.8}$$

showing that the total force applied normally to the surface is of magnitude F. The variation of $p(\varrho)$ with ϱ is shown in Fig. 16. It is obvious from this curve and from Eq. (59.8) that, if we let $a \to 0$ in this solution, we obtain the solution corresponding to a point force of magnitude F acting normally to the surface at the origin. The components of the displacement vector in this case are

$$u_\varrho = \frac{F}{4\pi\mu}\left\{\frac{z\varrho}{r^3} - (1-2\nu)\left[\frac{1}{\varrho} - \frac{z}{r\varrho}\right]\right\}, \qquad (59.9)$$

$$u_z = \frac{F}{4\pi\mu}\left\{\frac{z^2}{r^3} + \frac{2(1-\nu)}{r}\right\}, \qquad (59.10)$$

where $r^2 = \varrho^2 + z^2$.

60. The completeness of the Boussinesq-Papkovich solution. The Boussinesq-Papkovich solution (56.4) involves *four* scalar harmonic functions, i.e. φ and the three Cartesian scalar components of the vector ψ. We have seen that some problems may be solved by using less than the full number. The question then arises as to whether all four functions are needed for a complete solution. PAPKOVICH claimed, as a result of an erroneous argument, that φ may be taken to be zero without any loss of generality. Later SLOBODYANSKY[1] claimed to show that φ may be taken to be zero without loss of completeness, if the region occupied by the elastic material is bounded and simply connected or is the exterior of an arbitrary closed surface. In addition, since SLOBODYANSKY's argument fails when 4ν is an integer, he concludes (without justification) that, in this case, the reduction fails. A second theorem of SLOBODYANSKY's states that any one of the components of ψ may be taken to be zero. As pointed out by SOKOLNIKOFF, the proof of the first of these theorems is valid only if the region concerned is interior or exterior to a sphere, and the second is restricted to a region interior to a sphere.

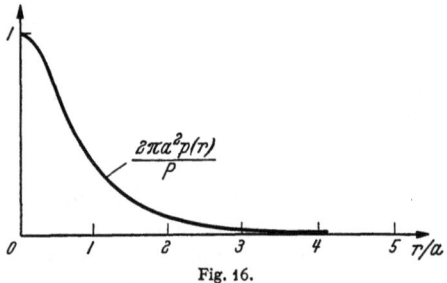

Fig. 16.

Recently EUBANKS and STERNBERG[2] have examined the precise conditions under which the completeness of the Boussinesq-Papkovich representation, in terms of four harmonic functions, has been established. This then enables them to prove that any one of these four functions may be taken to be zero, without loss in completeness, provided that the elastic solid occupies an open region of space which is "star-shaped" with respect to one of its points[3], that the coordinate system is chosen appropriately, and that 4ν is not an integer. In addition, EUBANKS and STERNBERG show with the aid of a counter-example that SLOBODYANSKY's conjecture is correct and that the Boussinesq-Papkovich representation with $\varphi = 0$ is *incomplete* if 4ν is an integer.

II. BETTI's method.

The fundamental problem in the mathematical theory of elasticity is that of finding the solution of the equations of equilibrium of an elastic body of given shape when either the surface displacements or the surface forces are prescribed.

---

[1] M. G. SLOBODYANSKY: Prikl. Mat. Mekh. Akad. Nauk. SSSR. **18**, 55 (1954).
[2] R. A. EUBANKS and E. STERNBERG: J. Rat. Mech. and Analysis **5**, 735 (1956).
[3] A region D is said to be "star-shaped" with respect to one of its points P if every straight line segment joining P to another point of D lies wholly in D. A region is star-shaped with respect to all of its points if it is convex.

The corresponding problems in the theory of electrostatics (or classical hydrodynamics) are familiar, and we know that there are two main methods of solution: for simple boundaries the main method is the use of curvilinear coordinates, and for more general types of boundary there has been developed a theory of Green's functions. Methods of both these types are available in the theory of elasticity though the resulting formulae are a good deal more complex. We shall illustrate the use of curvilinear coordinates in subdivision IV below. In the present section we shall outline the method of Betti which is the analogue of methods in potential theory based on Green's functions.

61. Betti's method. Betti's method[1] rests on the fact, proved in Sect. 56 above, that the dilatation Δ is a harmonic function. From this it follows that

$$\nabla^2 (r\,\Delta) = 2\,\mathrm{grad}\,\Delta$$

so that the equation of motion (53.4) can be written in the form

$$(1 - 2\nu)\,\nabla^2 u + \tfrac{1}{2}\,\nabla^2 (r\,\Delta) = 0$$

which is equivalent to

$$\nabla^2 \left(u + \frac{1}{2(1-2\nu)}\, r\,\Delta \right) = 0. \tag{61.1}$$

Hence, if Δ is known throughout the solid and the surface value of the vector u is prescribed, then the determination of the displacement field reduces to a problem in Potential Theory.

The basic problem therefore consists in determining Δ when the surface values of either the displacement or the applied forces are known. This is the problem which we shall consider in detail in the next section. Just as in Potential Theory the general theory of Dirichlet's and Neumann's problems rest upon Green's theorem, so in the theory of elasticity the basic tool is Betti's reciprocal theorem[2] which states that, if there are no body forces acting in the region bounded by a surface S, then

$$\int_S (F \cdot u')\, dS = \int_S (F' \cdot u)\, dS \tag{61.2}$$

where u is the displacement satisfying the equilibrium equations and F is the corresponding surface force, and also u' is a second displacement and F' is the corresponding surface force.

62. Formulae for the dilatation. The problem considered here is that of determining the value of the dilatation at a point in a solid bounded by a surface S when the surface values of either the components of displacement or of the surface forces are prescribed. For convenience we may take the point in which we are interested as the origin of coordinates. We surround that point by a sphere Σ of infinitesimal radius, and we pass to a limit by letting this radius tend to zero. Eq. (61.2) then takes the form

$$\int_\Sigma (X_i u'_i - X'_i u_i)\, dS = \int_S (X'_i u_i - X_i u'_i)\, dS \tag{62.1}$$

where we have written $F = (X_1, X_2, X_3)$, $u = (u_1, u_2, u_3)$ etc.

If we take

$$u'_i = \frac{\partial}{\partial x_i}\left(\frac{1}{r}\right) \tag{62.2}$$

---
[1] E. Betti: Nuovo Cim., Ser. II 6—10 (1872 et seq.).
[2] E. Betti: Nuovo Cim., Ser. II 7; 8 (1872). — Lord Rayleigh: Proc. Lond. Math. Soc. **4** (1873).

then it is readily shown that
$$X'_i = 2\mu n_j \frac{\partial^2}{\partial x_i \partial x_j}\left(\frac{1}{r}\right) \tag{62.3}$$
where $\mathbf{n} = (n_1, n_2, n_3)$ is the normal to the surface. On the sphere Σ
$$n_j = -\frac{x_j}{r} \tag{62.4}$$
so that
$$\int_\Sigma X_i u'_i\, dS = -\int_\Sigma \frac{x_j}{r}(\lambda \Delta \delta_{ij} + \mu \partial_j u_i + \mu \partial_i u_j)\frac{\partial}{\partial x_i}\left(\frac{1}{r}\right) dS$$
$$= \int_\Sigma \frac{x_i x_j}{r^4}(\lambda \Delta \delta_{ij} + \mu \partial_j u_i + \mu \partial_i u_j)\, dS.$$

Now each integral of type $\int_\Sigma x_1 x_2\, dS$ vanishes and each of the type $\int_\Sigma x_1^2\, dS$ has the value $\tfrac{4}{3}\pi r^4$ so that in the limit as the radius of Σ tends to zero
$$\int_\Sigma X_i u'_i\, dS = 4\pi(\lambda + \tfrac{2}{3}\mu)\Delta_0$$
where Δ_0 denotes the value of Δ at the origin.

From (62.3) we have similarly
$$\int_\Sigma u_i X'_i\, dS = -4\mu \int_\Sigma \frac{x_i u_i}{r^4}\, dS.$$

Expanding u_i by Maclaurin's theorem
$$u_i = (u_i)_0 + x_j(\partial_j u_i)_0 + \cdots$$
we find that
$$\int_\Sigma u_i X'_i\, dS = -4\mu \int_\Sigma \frac{x_i(u_i)_0 + x_i x_j(\partial_j u_i)_0}{r^4}\, dS$$
and in the limit as the radius of Σ tends to zero we find that the integral on the right has the value
$$-\tfrac{16}{3}\pi\mu(\partial_i u_i)_0 = -\tfrac{16}{3}\pi\mu\Delta_0.$$

Substituting these values into Eq. (62.1) we have the formula
$$\Delta_0 = \frac{1}{4\pi(\lambda + 2\mu)}\int_S (X'_i u_i - X_i u'_i)\, dS. \tag{62.5}$$

To evaluate the integral on the right hand side of Eq. (62.5) as it stands we should need to know the surface values of both the displacement and the applied force. Further analysis is needed to convert the result into one containing the surface values of one or the other. We consider the two cases separately:

α) *Surface displacements prescribed.* Suppose we can find a displacement vector $\mathbf{u}''$ which satisfies the vector equation of equilibrium at all points within S and which is identical with the vector $\mathbf{u}'$ at all points on S. Let the corresponding surface forces be $X''_i\,(i = 1, 2, 3)$. By the reciprocal theorem (61.2) we have
$$\int X''_i u_i\, dS = \int X_i u''_i\, dS = \int X_i u'_i\, dS. \tag{62.6}$$
Combining Eqs. (62.5) and (62.6) we obtain the solution
$$\Delta_0 = \frac{1}{4\pi(\lambda + 2\mu)}\int_S (X'_i - X''_i) u_i\, dS. \tag{62.7}$$

From their definition $X_i' - X_i''$ are the components of the surface force necessary to hold the surface fixed when there is a "centre of compression" at the origin. Thus, to determine the dilatation at any point we must therefore find a displacement which (a) satisfies the usual conditions of continuity and the equations of equilibrium everywhere except at the point, (b) in the neighbourhood of the point tends to become infinite, as if there were a centre of compression at the point, (c) vanishes at the surface. This displacement is analogous to a GREEN's function in Potential Theory.

β) *Surface forces prescribed.* The components X_i' form a system of surface forces which satisfy the conditions of rigid body equilibrium. Suppose that u_i'' are the components of the displacement produced in the body by the application of these surface tractions. We may then apply the reciprocal theorem to the displacements $\boldsymbol{u}, \boldsymbol{u}''$ which have no singularities within S, and obtain

$$\int X_i' u_i \, dS = \int X_i u_i'' \, dS. \tag{62.8}$$

Combining Eqs. (62.8) and (62.5) we obtain the solution

$$\Delta_0 = \frac{1}{4\pi(\lambda + 2\mu)} \int_S X_i (u_i'' - u_i') \, dS. \tag{62.9}$$

To find the dilatation at any point we need therefore only find the displacement field produced in the body when the surface is free from traction and there is a centre of dilatation at the point. This displacement is an analogue of a GREEN's function.

63. CERRUTI's solution of the problem of the plane.

To illustrate the use of BETTI's method we shall consider the case of a semi-infinite solid $z > 0$ bounded by the plane $z = 0$ when it is assumed that the surface displacements are prescribed[1]. Let (x', y', z') be any point of the body and $(x', y', -z')$ its mirror image in the boundary and let

$$r^2 = (x - x')^2 + (y - y')^2 + (z - z')^2, \quad R^2 = (x - x')^2 + (y - y')^2 + (z + z')^2 \tag{63.1}$$

so that r and R are respectively the distances of any point (x, y, z) from the point and its image. To calculate the dilatation in the case where the surface displacements are prescribed we require a displacement $\boldsymbol{u}''$ which on the plane $z = 0$ reduces to $\boldsymbol{u}' = \mathrm{grad}(r^{-1}) \equiv (\partial R^{-1}/\partial x, \partial R^{-1}/\partial y, -\partial R^{-1}/\partial z)$. It is readily shown that

$$\boldsymbol{u}'' = \left(\frac{\partial}{\partial x} + \frac{2(\lambda+\mu) z}{\lambda + 3\mu} \cdot \frac{\partial^2}{\partial x \, \partial z}, \; \frac{\partial}{\partial y} + \frac{2(\lambda+\mu) z}{(\lambda + 3\mu)} \cdot \frac{\partial^2}{\partial y \, \partial z}, \\ -\frac{\partial}{\partial z} + \frac{2(\lambda+\mu) z}{(\lambda + 3\mu)} \cdot \frac{\partial^2}{\partial z^2} \right) \frac{1}{R}. \tag{63.2}$$

It follows immediately from this expression and that for $\boldsymbol{u}'$ that on the plane $z = 0$

$$\boldsymbol{X}' = 2\mu \, \mathrm{grad}\left(\frac{\partial r^{-1}}{\partial z} \right), \quad \boldsymbol{X}'' = -\frac{2\mu(\lambda + \mu)}{\lambda + 3\mu} \, \mathrm{grad}\left(\frac{\partial r^{-1}}{\partial z} \right)$$

so that

$$\boldsymbol{X}' - \boldsymbol{X}'' = \frac{4\mu(\lambda + 2\mu)}{\lambda + 3\mu} \, \mathrm{grad}\left(\frac{\partial r^{-1}}{\partial z} \right). \tag{63.3}$$

Substituting this expression into Eq. (62.7) we find that

$$\Delta = -\frac{\mu}{\pi(\lambda + 3\mu)} \cdot \frac{\partial \varphi}{\partial z'} \tag{63.4}$$

---

[1] V. CERRUTI: Mem. fis. mat., Accad. Lincei (Roma) 1882.

where
$$\varphi = \operatorname{div}' \boldsymbol{L} = \frac{\partial L_1}{\partial x'} + \frac{\partial L_2}{\partial y'} + \frac{\partial L_3}{\partial z'}$$
with
$$\boldsymbol{L} = \int \boldsymbol{u}\, r^{-1}\, dx\, dy.$$

The functions L_i are harmonic on either side of the plane $z' = 0$ and at this plane

$$\boldsymbol{u} = -\frac{1}{2\pi} \lim_{z' \to +0} \frac{\partial \boldsymbol{L}}{\partial z}. \tag{63.5}$$

We must therefore solve the Eq. (61.1) which can be put in the form

$$\nabla'^2 \left(\boldsymbol{u} - \frac{\lambda + \mu}{2\pi(\lambda + 3\mu)} z'\, \operatorname{grad}' \varphi \right) = 0. \tag{63.6}$$

Now the three functions $\partial \boldsymbol{L}/\partial z'$ are harmonic in the region considered and the value of $\boldsymbol{u}$ as $z' \to +0$ is given by Eq. (63.5) so that the required solution of (63.6) is

$$\boldsymbol{u} = -\frac{1}{2\pi} \frac{\partial \boldsymbol{L}}{\partial z'} + \frac{(\lambda + \mu)\, z'}{2\pi(\lambda + 3\mu)}\, \operatorname{grad}' \varphi. \tag{63.7}$$

The corresponding solution for the case in which the surface forces are prescribed can be derived in a similar manner. For details the reader is referred to pp. 239 to 241 of Love's "A Treatise on the Mathematical Theory of Elasticity" 4th edit. (Cambridge 1934).

III. The method of integral transforms.

In this subdivision we shall consider how the theory of integral transforms may be applied to the solution of special three-dimensional problems.

64. The stresses in an infinite medium due to body forces. We shall first apply the theory of multiple Fourier transforms to the solution of the equations of equilibrium in an infinite elastic medium in the interior of which body forces of known magnitude are operating. If we choose rectangular Cartesian coordinates then the equations we have to solve are

$$\left. \begin{array}{l} \dfrac{\partial \sigma_x}{\partial x} + \dfrac{\partial \tau_{xy}}{\partial y} + \dfrac{\partial \tau_{yz}}{\partial z} + \varrho X = 0, \\[4pt] \dfrac{\partial \tau_{xy}}{\partial x} + \dfrac{\partial \sigma_y}{\partial y} + \dfrac{\partial \tau_{yz}}{\partial z} + \varrho Y = 0, \\[4pt] \dfrac{\partial \tau_{xz}}{\partial x} + \dfrac{\partial \tau_{yz}}{\partial y} + \dfrac{\partial \sigma_z}{\partial z} + \varrho Z = 0, \end{array} \right\} \tag{64.1}$$

where the components of the displacement vector are connected with the stress components through the equations

$$(\sigma_x, \sigma_y, \sigma_z) = \lambda \left(\frac{\partial u}{\partial x} + \frac{\partial v}{\partial y} + \frac{\partial w}{\partial z} \right) + 2\mu \left(\frac{\partial u}{\partial x}, \frac{\partial v}{\partial y}, \frac{\partial w}{\partial z} \right), \tag{64.2}$$

$$(\tau_{yz}, \tau_{zx}, \tau_{xy}) = \mu \left(\frac{\partial v}{\partial z} + \frac{\partial w}{\partial y},\; \frac{\partial u}{\partial z} + \frac{\partial w}{\partial x},\; \frac{\partial u}{\partial y} + \frac{\partial v}{\partial x} \right). \tag{64.3}$$

To solve these equations we introduce the Fourier transform $\bar{f}$ of every quantity f occurring in them, defined by the equation

$$\bar{f} = \frac{1}{(2\pi)^{\frac{3}{2}}} \int_{-\infty}^{\infty} \int_{-\infty}^{\infty} \int_{-\infty}^{\infty} f\, e^{i(\xi x + \eta y + \zeta z)}\, dx\, dy\, dz. \tag{64.4}$$

If we multiply both sides of each equation of the sets (64.1), (64.2), (64.3) by $e^{i(\xi x+\eta y+\zeta z)}$, integrate throughout the entire space, and make use of the results

$$\frac{1}{(2\pi)^{\frac{3}{2}}} \int_{-\infty}^{\infty}\int_{-\infty}^{\infty}\int_{-\infty}^{\infty} \frac{\partial f}{\partial x} e^{i(\xi x+\eta y+\zeta z)} dx\, dy\, dz = -i\xi \bar{f}, \quad \text{etc.,}$$

we see that these partial differential equations are equivalent to the set of algebraic equations

$$\begin{aligned}\xi\bar{\sigma}_x + \eta\bar{\tau}_{xy} + \zeta\bar{\tau}_{xz} &= -i\varrho\bar{X}, \\ \xi\bar{\tau}_{xy} + \eta\bar{\sigma}_y + \zeta\bar{\tau}_{yz} &= -i\varrho\bar{Y}, \\ \xi\bar{\tau}_{xz} + \eta\bar{\tau}_{yz} + \zeta\bar{\sigma}_z &= -i\varrho\bar{Z},\end{aligned} \quad (64.5)$$

$$i(\bar{\sigma}_x, \bar{\sigma}_y, \bar{\sigma}_z) = \lambda(\xi\bar{u} + \eta\bar{v} + \zeta\bar{w}) + 2\mu(\xi\bar{u}, \eta\bar{v}, \zeta\bar{w}), \quad (64.6)$$

$$i(\bar{\tau}_{yz}, \bar{\tau}_{zx}, \bar{\tau}_{xy}) = \mu(\zeta\bar{v} + \eta\bar{w}, \xi\bar{w} + \zeta\bar{u}, \eta\bar{u} + \xi\bar{v}). \quad (64.7)$$

If we substitute from Eqs. (64.6), (64.7) into Eqs. (64.5) we obtain the set

$$\begin{aligned}[\xi^2 + (1-2\nu)\bar{\xi}^2]\bar{u} + \xi\eta\bar{v} + \xi\zeta\bar{w} - (1-2\nu)(\varrho\bar{X}/\mu) &= 0, \\ \xi\eta\bar{u} + [\eta^2 + (1-2\nu)\bar{\xi}^2]\bar{v} + \eta\zeta\bar{w} - (1-2\nu)(\varrho\bar{Y}/\mu) &= 0, \\ \xi\zeta\bar{u} + \eta\zeta\bar{v} + [\zeta^2 + (1-2\nu)\bar{\xi}^2]\bar{w} - (1-2\nu)(\varrho\bar{Z}/\mu) &= 0\end{aligned} \quad (64.8)$$

where

$$\bar{\xi}^2 = \xi^2 + \eta^2 + \zeta^2. \quad (64.9)$$

Solving these equations we find the following expressions

$$\bar{u} = \frac{\varrho}{2(1-\nu)\mu\bar{\xi}^4}[2(1-\nu)\bar{\xi}^2\bar{X} - \xi(\xi\bar{X} + \eta\bar{Y} + \zeta\bar{Z})], \quad (64.10)$$

$$\bar{v} = \frac{\varrho}{2(1-\nu)\mu\bar{\xi}^4}[2(1-\nu)\bar{\xi}^2\bar{Y} - \eta(\xi\bar{X} + \eta\bar{Y} + \zeta\bar{Z})], \quad (64.11)$$

$$\bar{w} = \frac{\varrho}{2(1-\nu)\mu\bar{\xi}^4}[2(1-\nu)\bar{\xi}^2\bar{Z} - \zeta(\xi\bar{X} + \eta\bar{Y} + \zeta\bar{Z})], \quad (64.12)$$

for the Fourier transforms of the components of the displacement vector.

65. Terezawa's solution for a semi-infinite medium. If we let

$$\chi = J_0(\xi\varrho) e^{-\xi z} \quad (65.1)$$

in Eqs. (58.13) and (58.15) we find that the equations

$$u_\varrho = -\xi[(1-2\nu) - \xi z] J_1(\xi\varrho) e^{-\xi z}, \quad (65.2)$$
$$u_z = \xi[2(1-\nu) + \xi z] J_0(\xi\varrho) e^{-\xi z} \quad (65.3)$$

constitute a solution of the equilibrium equations in the case of axial symmetry. It follows from Eq. (58.18) that, when $z = 0$,

$$\sigma_z = -2\mu\xi^2 J_0(\xi\varrho). \quad (65.4)$$

It will be recalled that for this type of solution the shearing stress $\tau_{\varrho z} = 0$ on $z = 0$. It follows immediately from these equations that the solution of the equilibrium equations satisfying the boundary conditions

$$\sigma_z = -p(\varrho), \quad \tau_{z\varrho} = 0, \quad \text{on } z = 0 \quad (65.5)$$

is given by the pair of equations

$$u_\varrho = -\frac{1}{2\mu}\int_0^\infty (1-2\nu-\xi z)f(\xi)e^{-\xi z}J_1(\xi\varrho)\,d\xi, \tag{65.6}$$

$$u_z = \frac{1}{2\mu}\int_0^\infty [2(1-\nu)+\xi z]f(\xi)e^{-\xi z}J_0(\xi\varrho)\,d\xi, \tag{65.7}$$

provided $f(\xi)$ is chosen so that

$$p(\varrho) = \int_0^\infty \xi f(\xi) J_0(\varrho\xi)\,d\xi. \tag{65.8}$$

The problem of determining the distribution of stress and displacement in the semi-infinite solid is therefore solved if the integral equation (65.8) can be solved for $f(\xi)$. The solution of this equation is well-known[1]; it is

$$f(\xi) = \int_0^\infty \varrho\, p(\varrho) J_0(\varrho\xi)\,d\varrho. \tag{65.9}$$

If we calculate the stress-components from Eqs. (58.16) to (58.20) we find that

$$\sigma_z = -\int_0^\infty \xi(1+\xi z)f(\xi)J_0(\xi\varrho)e^{-\xi z}\,d\xi, \tag{65.10}$$

$$\left.\begin{aligned}\sigma_\varrho = &-\int_0^\infty \xi(1-\xi z)f(\xi)J_0(\xi\varrho)e^{-\xi z}\,d\xi-\\&-\frac{1}{\varrho}\int_0^\infty f(\xi)[\xi z-(1-2\nu)]J_1(\xi\varrho)e^{-\xi z}\,\frac{d\xi}{\xi},\end{aligned}\right\} \tag{65.11}$$

$$\sigma_\varrho+\sigma_\varphi+\sigma_z = -2(1+\nu)\int_0^\infty \xi f(\xi)J_0(\xi\varrho)e^{-\xi z}\,d\xi, \tag{65.12}$$

$$\tau_{\varrho z} = -z\int_0^\infty \xi^2 f(\xi)J_1(\xi\varrho)e^{-\xi z}\,d\xi. \tag{65.13}$$

The solution given by Eqs. (65.10) to (65.13) is equivalent to the solution derived by TEREZAWA[2].

As a special example, we consider the case of a point force of magnitude F acting normally to the surface. In this case we may take

$$p(\varrho) = \frac{F\,\delta(\varrho)}{2\pi\varrho}$$

where $\delta(\varrho)$ denotes the Dirac delta function[3], so that

$$f(\xi) = \frac{F}{2\pi}.$$

---
[1] This is merely HANKEL's inversion theorem; see I. N. SNEDDON: Fourier Transforms, p. 48. New York 1951.
[2] K. TEREZAWA: J. Coll. Sci. Imp. Univ. Tokyo 37, Art. 7 (1916).
[3] I. N. SNEDDON: Functional Analysis, Vol. II of this Encyclopedia, p. 229. 1955.

Substituting this value in Eq. (65.6) and (65.7) we find that the corresponding expressions for the components of the displacement vector are

$$u_\varrho = \frac{F}{4\pi\mu} \int_0^\infty [1 - (1 - 2\nu)\xi z] J_1(\xi\varrho) e^{-\xi z} d\xi,$$

$$u_z = \frac{F}{4\pi\mu} \int_0^\infty [2(1 - \nu) + \xi z] J_0(\xi\varrho) e^{-\xi z} d\xi.$$

The integrals are elementary[1] and give

$$u_\varrho = \frac{F}{4\pi\mu} \left\{ \frac{\varrho z}{(\varrho^2 + z^2)^{\frac{3}{2}}} - (1 - 2\nu) \left[\frac{1}{\varrho} - \frac{z}{\varrho(\varrho^2 + z^2)^{\frac{1}{2}}} \right] \right\},$$

$$u_z = \frac{F}{4\pi\mu} \left\{ \frac{2(1-\nu)}{(\varrho^2 + z^2)^{\frac{1}{2}}} + \frac{z^2}{(\varrho^2 + z^2)^{\frac{3}{2}}} \right\},$$

in agreement with the expressions (59.9), (59.10) derived above.

If, instead of the simple form (65.1) we take

$$\chi = J_0(\xi\varrho)(A \cosh \xi z + B \sinh \xi z)$$

we can, by a similar process[2], determine the distribution of stress in a thick elastic plate when axially symmetrical distributions of pressure are applied to its faces.

66. Mixed boundary value problems. The method of Hankel transforms, used in the last section, can be extended to the solution of mixed boundary value problems.

α) *Indentation problems.* We shall consider the axially symmetrical problem in which a perfectly rigid solid of revolution, whose shape is prescribed and whose axis of revolution coincides with the z-axis, is pressed normally into the boundary $z = 0$ of the semi-infinite elastic medium $z \geq 0$. The surface of the solid will fit the rigid punch between its apex and a certain circular section of radius a. If we assume that over the bounding plane the shearing stress vanishes at every point, the normal component of the displacement is prescribed over the region $\varrho \leq a$ and the normal stress vanishes for $\varrho > a$, i.e. we must have

$$\left. \begin{array}{ll} u_z = g(\varrho) & \varrho \leq a \\ \sigma_z = 0 & \varrho > a \\ \tau_{z\varrho} = 0 & \text{all values of } \varrho \end{array} \right\} \text{ on } z = 0 \qquad (66.1)$$

The solution expressed by Eqs. (65.6), (65.7) and Eqs. (65.10) to (65.13) will therefore be a solution of the problem, provided we can find a function $f(\xi)$ such that

$$\left. \begin{array}{ll} \int_0^\infty f(\xi) J_0(\xi\varrho) d\xi = G(\varrho), & \varrho \leq a \\ \int_0^\infty \xi f(\xi) J_0(\xi\varrho) d\xi = 0, & \varrho > a \end{array} \right\} \qquad (66.2)$$

where

$$G(\varrho) = \frac{\mu}{1 - \nu} g(\varrho).$$

---

[1] I. N. SNEDDON: The Special Functions of Mathematical Physics and Chemistry, p. 112. Edinburgh: Oliver & Boyd 1956.

[2] I. N. SNEDDON: Proc. Cambridge Phil. Soc. **42**, 260 (1946). See also I. N. SNEDDON: Fourier Transforms, pp. 468—480. New York: McGraw-Hill 1951.

In this way the problem is reduced to that of solving the pair of dual integral equations (66.2). The solution of these equations and a discussion of the stress distribution to which they lead in certain special cases is given in detail in pp. 455 to 468 of SNEDDON's *"Fourier Transforms"*, (McGraw-Hill, New York, 1951) to which the reader is referred.

β) *The circular crack.* The theory of Hankel transforms can also be used to determine the distribution of stress in the neighbourhood of a circular crack. We assume that the crack is created in the interior of an infinite medium and that it is "penny-shaped", occupying the circle $\varrho^2 = x^2 + y^2 \leq c^2$ in the plane $z = 0$. If we suppose that the crack is deformed by the application of an axially symmetrical pressure, then we see that, when $z = 0$,

$$\sigma_z = -p(\varrho), \quad \varrho < c.$$

Just as in the two-dimensional case (Sect. 47) we see that we may assume that the distribution of stress in the neighbourhood of the crack is the same as that produced in a semi-infinite elastic medium $z \geq 0$ when its free surface $z = 0$ is subjected to the boundary conditions

$$\left.\begin{array}{l} \tau_{\varrho z} = 0 \quad \text{for all values of } \varrho, \\ \sigma_z = -p(\varrho), \quad \varrho < c, \\ u_z = 0, \quad \varrho > c. \end{array}\right\} \quad (66.3)$$

The solution expressed by Eqs. (65.6), (65.7), and Eqs. (65.10) to (65.13) will therefore be a solution if it is possible to find a function $f(\xi)$ which satisfies the dual integral equations

$$\left.\begin{array}{l} \int_0^\infty \xi f(\xi) J_0(\xi \varrho) d\xi = p(\varrho), \quad \varrho \leq c, \\ \int_0^\infty f(\xi) J_0(\xi \varrho) d\xi = 0, \quad \varrho > c. \end{array}\right\} \quad (66.4)$$

This pair of dual integral equations can be solved and the resulting stress distribution determined. For details the reader is referred to pp. 486 to 500 of SNEDDON's *"Fourier Transforms"* (New York: McGraw-Hill 1951).

IV. Applications of curvilinear coordinates.

In this section we shall consider solutions of the equations of elastic equilibrium of an isotropic elastic solid in terms of systems of curvilinear coordinates. No attempt is made to develop a general theory but rather to select one or two typical problems by means of which to illustrate the method.

67. The problem of the sphere. To illustrate the use of spherical polar coordinates we shall consider "the problem of the sphere", i.e. the determination of the distribution of stress in the interior of a solid sphere when normal pressure is applied to the surface of the sphere. A general solution of this problem, expressed in terms of spherical harmonics but regarding these functions as functions of cartesian coordinates, has been derived by KELVIN[1] in the course of his discussion of the rigidity of the earth. To simplify the details of the solution we shall assume that the distribution of applied pressure is axially symmetrical so that we assume that the normal pressure on the surface $r = a$ of the sphere is $\mu f(\vartheta)$ and that the surface is free from shear.

---

[1] W. THOMSON (Lord KELVIN): Phil. Trans. Roy. Soc. Lond. **153** (1863).

Sect. 67. The problem of the sphere.

Referred to spherical polar coordinates r, ϑ, φ the physical components of the displacement may be denoted by

$$\mathbf{u} = (u_r, u_\vartheta, u_\varphi)$$

and the physical components of the stress by

$$\begin{pmatrix} \sigma_r & \tau_{r\vartheta} & \tau_{r\varphi} \\ \tau_{r\vartheta} & \sigma_\vartheta & \tau_{\vartheta\varphi} \\ \tau_{r\varphi} & \tau_{\vartheta\varphi} & \sigma_\varphi \end{pmatrix}.$$

Thus the boundary conditions are:

$$\sigma_r = -\mu f(\vartheta), \quad \tau_{r\vartheta} = 0 \quad \text{on} \quad r = a.$$

If it is assumed that the displacement vector is symmetrical about the z-axis then

$$u_\varphi = 0, \quad \tau_{\vartheta\varphi} = \tau_{r\varphi} = 0,$$

and it is easily shown that [cf. Eq. (52.16)]

$$\Delta = \frac{\partial u_r}{\partial r} + \frac{2 u_r}{r} + \frac{1}{r} \cdot \frac{\partial u_\vartheta}{\partial \vartheta} + \frac{\cot \vartheta}{r} u_\vartheta \tag{67.1}$$

and that

$$\operatorname{curl} \mathbf{u} = (0, 0, \omega) \tag{67.2}$$

where

$$\omega = \frac{1}{r} \frac{\partial}{\partial r}(r u_\vartheta) - \frac{1}{r} \frac{\partial u_r}{\partial \vartheta}. \tag{67.3}$$

Furthermore

$$\operatorname{curl} \operatorname{curl} \mathbf{u} = \left\{ \frac{1}{r \sin \vartheta} \frac{\partial}{\partial \vartheta}(\omega \sin \vartheta), -\frac{1}{r} \frac{\partial}{\partial r}(\omega r), 0 \right\}. \tag{67.4}$$

For the displacement components

$$u_r = -\frac{a f_0}{3\beta^2 - 4}\left(\frac{r}{a}\right) - \frac{a f_1 \cos \vartheta}{2(3\beta^2 - 4)}\left(\frac{r}{a}\right)^2 + \\ + \frac{1}{2} a \sum_{n=2}^{\infty} C_n \left\{ \frac{\beta^2 n^2 (n+2) - n(n+1)^2}{n^2 - 1} - [(\beta^2 - 1)n - 2]\left(\frac{r}{a}\right)^2 \right\} \left(\frac{r}{a}\right)^{n-1} P_n[(\cos \vartheta)], \tag{67.5}$$

$$u_\vartheta = -\frac{a f_1 \sin \vartheta}{2(3\beta^2 - 4)}\left(\frac{r}{a}\right)^2 + \\ + \frac{1}{2} a \sum_{n=2}^{\infty} C_n \left\{ \frac{(n+1)^2 - \beta^2 n(n+2)}{n^2 - 1} + \left[\frac{\beta^2(n+3)}{n+1} - 1\right]\left(\frac{r}{a}\right)^2 \right\} \left(\frac{r}{a}\right)^{n-1} \sin \vartheta\, P_n(\cos \vartheta) \tag{67.6}$$

in which $\beta^2 = (\lambda + 2\mu)/\mu$, $f_0, f_1, C_2, \ldots, C_n, \ldots$ are constants, we find from Eqs. (67.1) and (67.3) that

$$\Delta = -\frac{3 f_0}{3\beta^2 - 4} - \frac{3 f_1 \cos \vartheta}{3\beta^2 - 4}\left(\frac{r}{a}\right) + \sum_{n=2}^{\infty}(2n+3) c_n \left(\frac{r}{a}\right)^n P_n(\cos \vartheta) \tag{67.7}$$

and

$$\omega = -\frac{2 f_1 \sin \vartheta}{3\beta^2 - 4}\left(\frac{r}{a}\right) + \beta^2 \sum_{n=2}^{\infty} \frac{2n+3}{n+1} c_n \left(\frac{r}{a}\right)^n \sin \vartheta\, P_n'(\cos \vartheta). \tag{67.8}$$

From these expressions we find that

$$(\lambda + 2\mu) \operatorname{grad} \Delta - \mu \operatorname{curl} \operatorname{curl} \mathbf{u} = \frac{\mu f_1}{a}(-\cos \vartheta, \sin \vartheta, 0). \tag{67.9}$$

It follows from the vector equation of motion (53.6) that the displacement vector given by Eqs. (67.5) and (67.6) must be maintained by a force $\boldsymbol{P}$ such that in this coordinate system

$$\boldsymbol{P} = \frac{\mu f_1}{\varrho a}(\cos\vartheta, -\sin\vartheta, 0). \qquad (67.10)$$

This body force is of magnitude $\mu f_1/\varrho a$ and is in the positive z-direction.

The radial stress is given by the equation

$$\sigma_r = \lambda \Delta + 2\mu \frac{\partial u_r}{\partial r}$$

so that substituting from Eqs. (67.5) and (67.7) we find that

$$\sigma_r = -\mu f_0 - \mu f_1 \left(\frac{r}{a}\right)\cos\vartheta +$$

$$+ \mu \sum_{n=2}^{\infty} c_n \left[\frac{n^2(n+2)\beta^2 - n(n+1)^2}{n+1}\left(\frac{r}{a}\right)^{n-2} - \{n(n-1)(\beta^2-1) - (3\beta^2-4)\}\left(\frac{r}{a}\right)^n\right] P_n(\cos\vartheta).$$

On the surface of the sphere $r = a$ we have

$$(\sigma_r)_{r=a} = -\mu f_0 - \mu f_1 \cos\vartheta + \mu \sum_{n=2}^{\infty} c_n \left[\frac{(2n^2+4n+3)\beta^2}{n+1} - 2(n+2)\right] P_n(\cos\vartheta).$$

If the Fourier-Legendre series of $f(\vartheta)$ is

$$f_0 + f_1 \cos\vartheta + \sum_{n=2}^{\infty} f_n P_n(\cos\vartheta),$$

it follows that if $(\sigma_r)_{r=a} = -\mu f(\vartheta)$ then

$$c_n = -\frac{(n+1)f_n}{2(\beta^2-1)n^2 + 2(2\beta^2-3)n + (3\beta^2-4)}. \qquad (67.11)$$

The non-vanishing shear stress $\tau_{r\vartheta}$ is given by the equation

$$\tau_{r\vartheta} = \mu\left(\frac{\partial u_\vartheta}{\partial r} - \frac{u_\vartheta}{r} + \frac{1}{r}\frac{\partial u_r}{\partial \vartheta}\right)$$

where u_r and u_ϑ are given by Eqs. (67.5) and (67.6). We find that

$$\tau_{r\vartheta} = \mu\left(1 - \frac{a^2}{r^2}\right)\sum_{n=2}^{\infty} c_n \left\{\frac{n(n+2)(\beta^2-1)-1}{n+1}\right\}\left(\frac{r}{a}\right)^n \sin\vartheta\, P_n'(\cos\vartheta) \qquad (67.12)$$

from which it is obvious that $\tau_{r\vartheta} = 0$ on the surface of the sphere $r = a$.

The system of tractions on the spherical surface is not in statical equilibrium, its resultant being a force in the negative z-direction of magnitude

$$2\pi^2 \mu \int_0^\pi \sin\vartheta \cos\vartheta\, f(\vartheta)\, d\vartheta = \tfrac{4}{3}\pi a^2 \mu f_1$$

so that an acceleration $\mu f_1/\varrho a$ would be produced in a sphere of density ϱ.

The calculations for a constant pressure, applied to the part $0 \leq \vartheta \leq \alpha$ of the surface of the sphere have been carried out by DEAN, SNEDDON and PARSONS[1] to whose paper the reader is referred.

---

[1] W. R. DEAN, I. N. SNEDDON and H. W. PARSONS: Selected Government Research Reports, Vol. 6; Strength and Testing of Materials, Report No. 29. London: H.M.S.O. 1952.

68. Symmetrical strain in a solid of revolution.

When a solid of revolution is deformed symmetrically by surface tractions it is convenient to use cylindrical coordinates ϱ, φ, z. If we denote the displacement vector by $\boldsymbol{u} = (u_\varrho, u_\varphi, u_z)$ then if the strain is symmetrical $u_\varphi = 0$ and we have

$$\boldsymbol{u} = (u_\varrho, 0, u_z). \tag{68.1}$$

The components of strain are

$$\gamma_{\varrho\varrho} = \frac{\partial u_\varrho}{\partial \varrho}, \quad \gamma_{\varphi\varphi} = \frac{u_\varrho}{\varrho}, \quad \gamma_{zz} = \frac{\partial u_z}{\partial z}, \quad \gamma_{\varrho z} = \frac{1}{2}\left(\frac{\partial u_\varrho}{\partial z} + \frac{\partial u_z}{\partial \varrho}\right) \tag{68.2}$$

the remaining components $\gamma_{\varrho\varphi}, \gamma_{z\varphi}$ being zero. It follows immediately that, if E, ν denote Young's modulus and Poisson's ratio respectively then

$$\sigma_\varrho = \frac{\nu E}{(1+\nu)(1-2\nu)}\left[\frac{2(1-\nu)}{\nu}\frac{\partial u_\varrho}{\partial \varrho} + \frac{u_\varrho}{\varrho} + \frac{\partial u_z}{\partial z}\right], \tag{68.3}$$

$$\sigma_\varphi = \frac{\nu E}{(1+\nu)(1-2\nu)}\left[\frac{\partial u_\varrho}{\partial \varrho} + \frac{2(1-\nu)}{\nu}\frac{u_\varrho}{\varrho} + \frac{\partial u_z}{\partial z}\right], \tag{68.4}$$

$$\sigma_z = \frac{\nu E}{(1+\nu)(1-2\nu)}\left[\frac{\partial u_\varrho}{\partial \varrho} + \frac{u_\varrho}{\varrho} + \frac{2(1-\nu)}{\nu}\frac{\partial u_z}{\partial z}\right], \tag{68.5}$$

$$\tau_{\varrho z} = \frac{E}{2(1+\nu)}\left[\frac{\partial u_\varrho}{\partial z} + \frac{\partial u_z}{\partial \varrho}\right], \tag{68.6}$$

and the equations of equilibrium are (in the absence of body forces)

$$\frac{\partial \sigma_\varrho}{\partial \varrho} + \frac{\partial \tau_{\varrho z}}{\partial z} + \frac{\sigma_\varrho - \sigma_\varphi}{\varrho} = 0, \tag{68.7}$$

$$\frac{\partial \tau_{\varrho z}}{\partial \varrho} + \frac{\partial \sigma_z}{\partial z} + \frac{\tau_{\varrho z}}{\varrho} = 0. \tag{68.8}$$

Now if we put

$$u_\varrho = -\frac{1+\nu}{E}\frac{\partial^2 \chi}{\partial \varrho \, \partial z}, \tag{68.9}$$

$$u_z = \frac{1+\nu}{E}\left\{(1-2\nu)\nabla^2 \chi + \frac{\partial^2 \chi}{\partial \varrho^2} + \frac{1}{\varrho}\frac{\partial \chi}{\partial \varrho}\right\} \tag{68.10}$$

in Eqs. (68.3) to (68.6) we find that

$$\sigma_\varrho = \frac{\partial}{\partial z}\left\{\nu \nabla^2 \chi - \frac{\partial^2 \chi}{\partial \varrho^2}\right\}, \tag{68.11}$$

$$\sigma_\varphi = \frac{\partial}{\partial z}\left\{\nu \nabla^2 \chi - \frac{1}{\varrho}\frac{\partial \chi}{\partial \varrho}\right\}, \tag{68.12}$$

$$\sigma_z = \frac{\partial}{\partial z}\left\{(2-\nu)\nabla^2 \chi - \frac{\partial^2 \chi}{\partial z^2}\right\}, \tag{68.13}$$

$$\tau_{\varrho z} = \frac{\partial}{\partial \varrho}\left\{(1-\nu)\nabla^2 \chi - \frac{\partial^2 \chi}{\partial z^2}\right\}. \tag{68.14}$$

If we substitute these expressions into Eq. (68.7) we find that it is satisfied identically, while if we substitute them into Eq. (68.8) we find that

$$\nabla^4 \chi = 0. \tag{68.15}$$

In this way the problem of determining the distribution of stress in a solid of revolution reduces to that of finding solutions of the biharmonic equation (68.15) satisfying appropriate boundary conditions. For a discussion of a series of special problems by this method the reader is referred to Chap. 13 of "Theory of Elasticity" by S. Timoshenko and J. N. Goodier.

69. The use of oblate spheroidal coordinates.
To illustrate the use of oblate spheroidal coordinates we consider Sack's solution[1] of the problem of determining the distribution of stress in the vicinity of a circular crack in a homogeneous isotropic solid.

For problems relating to a plane circular disc or crack of infinitely small thickness, the appropriate coordinates are oblate spheroidal coordinates ψ, ϑ, φ, where

$$\left.\begin{aligned} x &= c \sinh \psi \cos \vartheta, \\ y &= c \cosh \psi \sin \vartheta \cos \varphi, \\ z &= c \cosh \psi \sin \vartheta \sin \varphi \end{aligned}\right\} \quad (69.1)$$

so that the surfaces

$\psi = $ const. represent oblate spheroids,
$\vartheta = $ const. represent hyperboloids of revolution of one sheet,
$\varphi = $ const. represent planes through the x-axis.

The surface $\psi = 0$ corresponds to the circular disc

$$x = 0, \quad y^2 + z^2 \leq c^2; \quad (69.2)$$

each point of the circle is represented twice, corresponding to the two sides of the disc. The fundamental differential form is

$$ds^2 = h_\psi^2 \, d\psi^2 + h_\vartheta^2 \, d\vartheta^2 + h_\varphi^2 \, d\varphi^2 \quad (69.3)$$

where

$$h_\psi^2 = h_\vartheta^2 = c^2 (\sinh^2 \psi + \cos^2 \vartheta) = c^2 h^2, \quad (69.4)$$

say, and

$$h_\varphi = c \cosh \psi \sin \vartheta \quad (69.5)$$

so that the Laplacian operator transforms to

$$\nabla^2 = \frac{1}{c^2 h^2 \cosh \psi \sin \vartheta} \left[\frac{\partial}{\partial \psi} \cosh \psi \sin \vartheta \frac{\partial}{\partial \psi} + \frac{\partial}{\partial \vartheta} \cosh \psi \sin \vartheta \frac{\partial}{\partial \vartheta} \right] + \frac{\partial^2}{c^2 \cosh^2 \psi \sin^2 \vartheta \, \partial \varphi^2}. \quad (69.6)$$

The normal solutions of Laplace's equation

$$\nabla^2 V = 0 \quad (69.7)$$

are

and

$$\left.\begin{aligned} \Pi_n^m (\sinh \psi) \, P_n^m (\cos \vartheta) \, e^{i m \varphi} \\ T_n^m (\sinh \psi) \, P_n^m (\cos \vartheta) \, e^{i m \varphi} \end{aligned}\right\} \quad (69.8)$$

where the functions P_n^m are the associated Legendre polynomials and Π_n^m, T_n^m represent the Lengendre functions of the first and second kind respectively of the argument ix:

$$P_n^m (x) = \text{const} \cdot (1 - x^2)^{\frac{1}{2}m} \frac{d^{n+m}}{dx^{n+m}} (1 - x^2)^n,$$

$$\Pi_n^m (x) = \text{const} \cdot (1 + x^2)^{\frac{1}{2}m} \frac{d^{n+m}}{dx^{n+m}} (1 + x^2)^n,$$

$$T_n^m (x) = \text{const} \cdot \Pi_n^m (x) \int_x^\infty \frac{d\xi}{[\Pi_n^m(\xi)]^2 (1 + \xi^2)}.$$

If one of the principal stresses of the external system is normal to the crack, there will be no shear stresses acting across its faces, neither will the other two

---
[1] R. A. Sack: Proc. Phys. Soc. Lond. **58**, 729 (1946).

principal stresses be affected by the presence of the crack but the normal component σ_ψ must be compensated for as there should be no tractions exerted on the inner surface. This disturbance will exhibit axial symmetry and fall off with increasing distance. Thus the only terms which need be considered are those containing T_n^0; terms with $m \neq 0$ or Π_n^0 need not be considered.

The first three functions T_n^0 are:

$$T_0^0(\sinh \psi) = \cot^{-1}(\sinh \psi),$$
$$T_1^0(\sinh \psi) = T_0^0(\sinh \psi) \sinh \psi - 1, \qquad (69.9)$$
$$T_2^0(\sinh \psi) = T_0^0(\sinh \psi)(3 \sinh^2 \psi + 1) - 3 \sinh \psi.$$

It is readily shown that if U and V are solutions of Laplace's equation the displacements u_ψ, u_ϑ are given by the equations

$$u_\psi = \frac{1}{h}\left[\frac{1}{c}\frac{\partial U}{\partial \psi} + (3 - 4\nu)\cosh \psi \cos \vartheta \cdot V - \sinh \psi \cos \vartheta \frac{\partial V}{\partial \psi}\right], \qquad (69.10)$$

$$u_\vartheta = \frac{1}{h}\left[\frac{1}{c}\frac{\partial U}{\partial \psi} - (3 - 4\nu)\sinh \psi \sin \vartheta \cdot V - \sinh \psi \cos \vartheta \frac{\partial V}{\partial \vartheta}\right], \qquad (69.11)$$

where ν is Poisson's ratio. From these it follows that

$$(\sigma_\psi)_{\psi=0} = \frac{2\mu}{c^2 \cos^2 \vartheta}\left[\frac{\partial^2 U}{\partial \psi^2} - \tan \vartheta \frac{\partial U}{\partial \vartheta} + 2(1-\nu)c \cos \vartheta \frac{\partial V}{\partial \psi}\right]_{\psi=0},$$

$$(\tau_{\psi\vartheta})_{\psi=0} = \frac{2\mu}{c^2 \cos^2 \vartheta}\left[\frac{\partial^2 U}{\partial \psi \partial \vartheta} + \tan \vartheta \frac{\partial U}{\partial \vartheta} + (1-2\nu)c \cos \vartheta \frac{\partial V}{\partial \vartheta}\right]_{\psi=0}.$$

If we take

$$U = -\frac{(1-2\nu)c^2 A}{3\pi\mu}\left[T_0^0(\sinh \psi) + \frac{1}{2}\left(\frac{3}{2}\cos^2 \vartheta - \frac{1}{2}\right)T_2^0(\sinh \psi)\right], \qquad (69.12)$$

$$V = \frac{cA}{\pi\mu} T_1^0(\sinh \psi) \cos \vartheta \qquad (69.13)$$

in these expressions, we find that for $\psi = 0$

$$\sigma_\psi = A, \qquad \tau_{\psi\vartheta} = 0 \qquad (69.14)$$

so that Eqs. (69.10) to (69.13) yield the solution of the equations of elastic equilibrium corresponding to the conditions (69.14) on the surfaces of the circular crack.

E. Dynamical problems.

In the discussion of special problems in the mathematical theory of elasticity we have so far assumed that the forces acting upon the elastic body are static. We shall now consider briefly some problems in which the applied forces vary with time so that dynamic stresses are set up in the body. By far the greatest body of work in this field has been done on the theory of the propagation of elastic waves. We shall only discuss such problems briefly since nowadays these are considered as belonging more properly to theoretical geophysics. For example, they are discussed in some detail in the articles by K. E. Bullen, W. M. Ewing and J. A. Jacobs in Vol. XLVII of this Encyclopedia.

I. Elastic waves.

70. P-waves and S-waves. It follows from Eq. (53.3) that we may write the equations of motion of an elastic solid in the vector form

$$\mu \nabla^2 \boldsymbol{u} + (\lambda + \mu)\,\mathrm{grad}\,(\mathrm{div}\,\boldsymbol{u}) + \varrho \boldsymbol{P} = \varrho \frac{\partial^2 \boldsymbol{u}}{\partial t^2} \qquad (70.1)$$

where $\boldsymbol{u}$ denotes the displacement vector, $\boldsymbol{P}$ the body force, ϱ the density and λ and μ are LAMÉ's elastic constants. If we write the body force $\boldsymbol{P}$ in the form

$$\boldsymbol{P} = \operatorname{grad} \boldsymbol{\Phi} + \operatorname{curl} \boldsymbol{\Psi}. \tag{70.2}$$

then introducing a scalar function φ and a vector $\boldsymbol{\psi}$ we write

$$\boldsymbol{u} = \operatorname{grad} \varphi + \operatorname{curl} \boldsymbol{\psi}. \tag{70.3}$$

For this representation of the displacement vector we have

$$\operatorname{div} \boldsymbol{u} = \nabla^2 \varphi. \tag{70.4}$$

Substituting from Eqs. (70.2), (70.3) and (70.4) into Eq. (70.1) we find the relation

$$\mu \nabla^2 (\operatorname{grad} \varphi + \operatorname{curl} \boldsymbol{\psi}) + (\lambda + \mu) \operatorname{grad} (\nabla^2 \varphi) + \varrho \operatorname{grad} \boldsymbol{\Phi} + \varrho \operatorname{curl} \boldsymbol{\Psi}$$
$$= \varrho \frac{\partial^2}{\partial t^2} (\operatorname{grad} \varphi) + \varrho \frac{\partial^2}{\partial t^2} (\operatorname{curl} \boldsymbol{\psi})$$

which may be written in the form

$$\operatorname{grad} \left[(\lambda + 2\mu) \nabla^2 \varphi + \varrho \boldsymbol{\Phi} - \varrho \frac{\partial^2 \varphi}{\partial t^2} \right] + \operatorname{curl} \left[\mu \nabla^2 \boldsymbol{\psi} + \varrho \boldsymbol{\Psi} - \varrho \frac{\partial^2 \boldsymbol{\psi}}{\partial t^2} \right] = 0$$

and therefore shows that the vector equation (70.3) yields a solution of the equations of motion provided that φ and ψ are chosen to be solutions of the equations

$$c_1^2 \nabla^2 \varphi - \frac{\partial^2 \varphi}{\partial t^2} = - \boldsymbol{\Phi} \tag{70.5}$$

and

$$c_2^2 \nabla^2 \boldsymbol{\psi} - \frac{\partial^2 \boldsymbol{\psi}}{\partial t^2} = - \boldsymbol{\Psi} \tag{70.6}$$

respectively, where

$$c_1^2 = \frac{\lambda + 2\mu}{\varrho}, \quad c_2^2 = \frac{\mu}{\varrho}. \tag{70.7}$$

Now Eq. (70.5) is an inhomogeneous wave equation with wave velocity c_1 showing that the φ part of the displacement is transmitted with speed c_1. From Eq. (70.4) it follows that the dilatation $\Delta = \operatorname{div} \boldsymbol{u}$ satisfies a wave equation with the same velocity. In seismology this wave is called a *primary wave* or simply a *P-wave*; it is a condensation-rarefaction wave involving change of volume. On the other hand, Eq. (70.6) shows that the $\boldsymbol{\psi}$ part of the displacement is transmitted at the lower speed c_2. From the fact that $\operatorname{curl} \boldsymbol{u} = \operatorname{curl} \operatorname{curl} \boldsymbol{\psi}$ it follows that the rotation $\boldsymbol{\omega} = \frac{1}{2} \operatorname{curl} \boldsymbol{u}$ satisfies a wave equation with velocity c_2. In seismology this wave is called a *secondary wave* or *S-wave*; it is a shear wave in which there is distortion without change of volume. In the case in which the rigidity μ is zero, the velocity c_2 is zero, showing that shear waves cannot be propagated in a medium with zero rigidity. In some investigations it is useful to introduce a constant β which is the ratio of the two wave-velocities, i.e.

$$\beta = \frac{c_1}{c_2} = \sqrt{\frac{\lambda + 2\mu}{\mu}} \tag{70.8}$$

or, in terms of POISSON's ratio ν,

$$\beta = \sqrt{\frac{2 - 2\nu}{1 - 2\nu}} \tag{70.9}$$

showing that $\beta = \sqrt{3}$ when $\nu = \frac{1}{4}$. It should further be noted that the ratio β is independent of the density ϱ and the YOUNG's modulus E of the material.

The reflection and refraction of plane seismic waves can be studied in exactly the same way as in the electromagnetic theory of light, though the situation in elasticity is, in general, more complicated since a wave of one type incident upon a plane surface may give rise to reflected and refracted waves of both types. For a brief amount of the theory the reader is referred to BULLEN's article in Vol. XLVII of this Encyclopedia; for a fuller account of the calculations and their application to seismological problems the reader should consult L. CAGNIARD, "*Réflexion et Réfraction des Ondes Séismiques Progressives*", (Paris, 1939).

In a homogeneous isotropic solid of infinite extent any disturbance can be represented as a superposition of P-waves and S-waves, but if the medium is inhomogeneous or bounded other types of waves occur, the most important of which are the surface waves which can be propagated in the neighbourhood of the boundary of an elastic solid. We shall discuss these surface waves in the next two sections.

71. Rayleigh waves. Consider now a semi-infinite solid $z \geq 0$ whose only boundary, the plane $z=0$, is free from stress in which waves whose amplitude diminishes exponentially with z are being propagated along the x-axis. In other words we seek solutions of Eq. (70.1), which may be written

$$c_2^2 \nabla^2 \boldsymbol{u} + (c_1^2 - c_2^2) \operatorname{grad} \operatorname{div} \boldsymbol{u} = \frac{\partial^2 \boldsymbol{u}}{\partial t^2} \tag{71.1}$$

of the form

$$\boldsymbol{u} = \boldsymbol{a}\, e^{-\alpha z + i q(x - ct)}. \tag{71.2}$$

For the solution (71.2),

$$\nabla^2 \boldsymbol{u} = (\alpha^2 - q^2)(a_1, a_2, a_3)\, e^{-\alpha z + i q(x - ct)}, \tag{71.3}$$

$$\operatorname{grad} \operatorname{div} \boldsymbol{u} = -(q^2 a_1 + i\alpha q a_3,\, 0,\, i\alpha q a_1 - \alpha^2 a_3)\, e^{-\alpha z + i q(x - ct)}, \tag{71.4}$$

$$\frac{\partial^2 \boldsymbol{u}}{\partial t^2} = -c^2 q^2 (a_1, a_2, a_3)\, e^{-\alpha z + i q(x - ct)}. \tag{71.5}$$

Substituting these forms in the vector equation of motion (71.1) and writing it in component form we obtain the relations

$$[c_2^2 \alpha^2 - (c_1^2 - c_2^2) q^2] a_1 - i(c_1^2 - c_2^2)\alpha q a_3 = 0, \tag{71.6}$$

$$[c_2^2 \alpha^2 + (c^2 - c_2^2) q^2] a_2 = 0, \tag{71.7}$$

$$-i(c^2 - c_2^2)\alpha q a_1 + [c_1^2 \alpha^2 + (c^2 - c_2^2) q^2] a_3 = 0. \tag{71.8}$$

From the second of these equations we have

$$a_2 = 0,$$

and eliminating a_1 and a_3 from the first and third of them we find that

$$\alpha^4 - \left(2 - \frac{c^2}{c_1^2} - \frac{c^2}{c_2^2}\right)\alpha^2 q^2 + \left(1 - \frac{c^2}{c_1^2}\right)\left(1 - \frac{c^2}{c_2^2}\right) q^4 = 0$$

so that for a given value of q there are two values of α, α_1 and α_2, where

$$\alpha_1^2 = \left(1 - \frac{c^2}{c_1^2}\right) q^2, \qquad \alpha_2^2 = \left(1 - \frac{c^2}{c_2^2}\right) q^2. \tag{71.9}$$

Putting $\alpha = \alpha_1$ in Eq. (71.8) we find that $q a_3 = i\alpha_1 a_1$ so that we have the solution

$$\boldsymbol{u}_1 = (-iq C_1,\, 0,\, \alpha_1 C_1)\, e^{-\alpha_1 z + i q(x - ct)} \tag{71.10}$$

and putting $\alpha=\alpha_2$ in Eq. (71.6) we find that $qa_1=-i\alpha_2a_3$ and hence

$$u_2 = (-i\alpha_2 C_2, 0, q C_2) e^{-\alpha_2 z + iq(x-ct)}. \tag{71.11}$$

Combining Eqs. (71.10) and (71.11) we obtain the general solution $u = (u, 0, w)$ where

$$u = -i(q C_1 e^{-\alpha_1 z} + \alpha_2 C_2 e^{-\alpha_2 z}) e^{iq(x-ct)}, \tag{71.12}$$

$$w = (\alpha_1 C_1 e^{-\alpha_1 z} + q C_2 e^{-\alpha_2 z}) e^{iq(x-ct)}. \tag{71.13}$$

For this displacement vector $\tau_{yz} = 0$ and

$$\tau_{xz} = i\mu q \left\{ 2\alpha_1 C_1 e^{-\alpha_1 z} - \left(\frac{c^2}{c_2^2} - 2\right) q C_2 e^{-\alpha_2 z} \right\} e^{iq(x-ct)}, \tag{71.14}$$

$$\sigma_z = \mu q \left\{ \left(\frac{c^2}{c_2^2} - 2\right) q C_1 e^{-\alpha_1 z} - 2\alpha_2 C_2 e^{-\alpha_2 z} \right\} e^{iq(x-ct)}. \tag{71.15}$$

If the surface $z=0$ is free from stress then $\tau_{xz} = \sigma_z = 0$ when $z=0$ so that

$$2\alpha_1 C_1 - \left(\frac{c^2}{c_2^2} - 2\right) q C_2 = 0,$$

$$\left(\frac{c^2}{c_2^2} - 2\right) q C_1 - 2\alpha_2 C_2 = 0.$$

Eliminating C_1 and C_2 from these equations we obtain the relation

$$\left(\frac{c^2}{c_2^2} - 2\right)^2 q^2 = 4\alpha_1 \alpha_2$$

which, by virtue of the Eqs. (71.9), can be expressed in the form

$$\left(2 - \frac{c^2}{c_2^2}\right)^2 = 4\left(1 - \frac{c^2}{c_1^2}\right)^{\frac{1}{2}} \left(1 - \frac{c^2}{c_2^2}\right)^{\frac{1}{2}}$$

which can be written as

$$\left(\frac{c}{c_2}\right)^6 - 8\left(\frac{c}{c_2}\right)^4 + \frac{8(3\beta^2 - 2)}{\beta^2} \left(\frac{c}{c_2}\right)^2 - \frac{16(\beta^2 - 1)}{\beta^2} = 0 \tag{71.16}$$

where β is defined by Eq. (70.8). The Eq. (71.16) is the equation for the determination of the velocity of the waves. If we denote the left-hand side of this equation by $f(c/c_2)$ we see that $f(0) = -16(\beta^2-1)/\beta^2$ and $f(1) = 1$ showing that Eq. (71.16) has a root between 0 and c_2. For example if the Poisson ratio is $\frac{1}{4}$, $\beta^2 = 3$ and Eq. (71.16) becomes

$$3\xi^3 - 24\xi^2 + 56\xi - 32 = 0, \qquad \xi = c^2/c_2^2$$

which is easily seen to have roots $4, 2 \pm 2/\sqrt{3}$. The value of c less than c_2 is therefore

$$c_2 \sqrt{2 - 2/\sqrt{3}} = 0.9194 c_2$$

and the corresponding values of the α's are

$$\alpha_1 = 0.8475 q, \qquad \alpha_2 = 0.3933 q$$

and we obtain the solution

$$u = A(e^{-\alpha_1 z} - 0.5773 e^{-\alpha_2 z}) \sin q(x - ct),$$

$$w = A(0.8475 e^{-\alpha_1 z} - 1.4679 e^{-\alpha_2 z}) \cos q(x - ct).$$

The other roots both make α_1 and α_2 purely imaginary and therefore do not lead to surface waves.

Surface waves of this type were first discovered by Lord RAYLEIGH[1] and for that reason are usually called *Rayleigh waves*.

72. Love waves. Other simple types of surface wave have been predicted and identified. If the solid is stratified so that the regions $z>0$ and $-h<z<0$ are of different materials, another type, called *Love Waves*[2], appears. LOVE assumed displacements of the type

$$\boldsymbol{u} = (0, V, 0)\, e^{iq(x-ct)}, \quad (-h<z<0) \tag{72.1}$$

and

$$\boldsymbol{u} = (0, V', 0)\, e^{iq(x-ct)}, \quad (z>0) \tag{72.2}$$

where V and V' are functions of z alone. We shall use unaccented letters to refer to the upper layer, accented ones to the lower medium.

For the displacement vector (72.1) we have

$$\operatorname{div} \boldsymbol{u} = 0,$$

$$\nabla^2 \boldsymbol{u} = \left(0, \frac{d^2 V}{dz^2} - q^2 V, 0\right) e^{iq(x-ct)}$$

so that substituting in the vector Eq. (71.1) we find that

$$c_2^2 \left(\frac{d^2 V}{dz^2} - q^2 V\right) = -c^2 q^2 V,$$

i.e.

$$\frac{d^2 V}{dz^2} + \sigma^2 q^2 V = 0$$

where

$$\sigma^2 = \frac{c^2}{c_2^2} - 1. \tag{72.3}$$

Hence we have in the region $-h \leq z \leq 0$

$$V = A \sin \sigma q z + B \cos \sigma q z. \tag{72.4}$$

Similarly in the region $z>0$ we have

$$V = C\, e^{-\sigma' q z} \tag{72.5}$$

where

$$\sigma'^2 = 1 - \frac{c^2}{c_2'^2}. \tag{72.6}$$

Now the physical conditions are that V and τ_{zy} must be continuous across $z=0$ and that $\tau_{zy}=0$ at $z=-h$. Hence V and $\mu(dV/dz)$ must be continuous across $z=0$ and $dV/dz=0$ at $z=-h$, so that we obtain the relations

$$B = C,$$
$$\mu A \sigma = -\mu' C \sigma',$$
$$A \cos \sigma q h + B \sin \sigma q h = 0.$$

Eliminating A, B, C from these equations we find that c must be such that

$$\begin{vmatrix} \cos \sigma q h & \sin \sigma q h & 0 \\ \mu \sigma & 0 & -\mu' \sigma' \\ 0 & 1 & -1 \end{vmatrix} = 0$$

---
[1] Lord RAYLEIGH (J. W. STRUTT): Proc. Lond. Math. Soc. **17**, 4 (1885).
[2] A. E. H. LOVE: Some Problems in Geodynamics. Cambridge 1911.

which reduces to
$$\mu' \sigma' = \mu \sigma \tan (q \sigma h).$$

The equation for c is therefore

$$\mu' \left(1 - \frac{c^2}{c_2'^2}\right)^{\frac{1}{2}} = \mu \left(-1 + \frac{c^2}{c_2^2}\right)^{\frac{1}{2}} \tan \left\{\left(\frac{c^2}{c_2^2} - 1\right)^{\frac{1}{2}} q h\right\} \tag{72.7}$$

where $c_2^2 = \mu/\varrho$, $c_2'^2 = \mu'/\varrho'$. It will be seen from this equation that this kind of wave exists only if the velocity of S-waves in the lower region is greater than the velocity of S-waves in the top layer. If c is a real root of this equation the solution of the problem is given by the equations

$$\left. \begin{array}{l} V = V_0 \cos \{\sigma q (z + h)\} \quad (-h < z < 0), \\ V = V_0 \cos (\sigma q h) \, e^{-\sigma' q z} \quad (z > 0) \end{array} \right\} \tag{72.8}$$

where V_0 is an arbitrary constant and σ, σ' are defined by Eqs. (72.3) and (72.6) respectively.

The study of such waves in stratified solids is of considerable importance in geophysical applications of elasticity since there appears to be a fairly abrupt change in physical properties of the Earth at the base of the crust (about 33 km below the surface).

It should be observed from Eq. (72.7) that (unlike Rayleigh waves) the speed of propagation of Love waves depends on the wavelength $(2\pi/q)$ of the waves so that dispersion will occur.

73. Propagation of elastic waves in cylinders. In cylindrical coordinates (r, φ, z) we may write for the physical components of displacement

$$\boldsymbol{u} = (u_r, u_\varphi, u_z),$$

and it is easily shown that

$$2\omega = \operatorname{curl} \boldsymbol{u} = \left(\frac{1}{r} \frac{\partial u_z}{\partial \varphi} - \frac{\partial u_\varphi}{\partial z}, \frac{\partial u_r}{\partial z} - \frac{\partial u_z}{\partial r}, \frac{1}{r} \frac{\partial (r u_\varphi)}{\partial r} - \frac{1}{r} \frac{\partial u_r}{\partial \varphi}\right) \tag{73.1}$$

and that

$$\Delta = \frac{1}{r} \frac{\partial (r u_r)}{\partial r} + \frac{1}{r} \frac{\partial u_\varphi}{\partial \varphi} + \frac{\partial u_z}{\partial z}. \tag{73.2}$$

Writing the equations of motion in the form [cf. Eq. (53.6)]

$$(\lambda + 2\mu) \operatorname{grad} \Delta - \mu \operatorname{curl} \operatorname{curl} \boldsymbol{u} = \varrho \frac{\partial^2 \boldsymbol{u}}{\partial t^2}, \tag{73.3}$$

we see that in these coordinates they become

$$(\lambda + 2\mu) \frac{\partial \Delta}{\partial r} - \frac{2\mu}{r} \frac{\partial \omega_z}{\partial \varphi} + 2\mu \frac{\partial \omega_\varphi}{\partial z} = \varrho \frac{\partial^2 u_r}{\partial t^2}, \tag{73.4}$$

$$(\lambda + 2\mu) \frac{1}{r} \frac{\partial \Delta}{\partial \varphi} - 2\mu \frac{\partial \omega_r}{\partial z} + 2\mu \frac{\partial \omega_z}{\partial r} = \varrho \frac{\partial^2 u_\varphi}{\partial t^2}, \tag{73.5}$$

$$(\lambda + 2\mu) \frac{\partial \Delta}{\partial z} - \frac{2\mu}{r} \frac{\partial}{\partial r} (r \omega_\varphi) + \frac{2\mu}{r} \frac{\partial \omega_r}{\partial \varphi} = \varrho \frac{\partial^2 u_z}{\partial t^2}. \tag{73.6}$$

We shall now discuss briefly the three types of elastic wave which may be propagated in cylinders, namely, torsional, longitudinal and flexural.

α) *Torsional waves.* If we assume that $u_r = u_z = 0$ and that u_φ is independent of φ it follows from (73.1) that

$$\omega_r = -\frac{1}{2} \frac{\partial u_\varphi}{\partial z}, \quad \omega_\varphi = 0, \quad \omega_z = \frac{1}{2} \left(\frac{\partial u_\varphi}{\partial r} + \frac{u_\varphi}{r}\right) \tag{73.7}$$

and from Eq. (73.2) that $\varDelta = 0$. Substituting these expressions into Eqs. (73.4) and (73.6) we find that they are satisfied identically while Eq. (73.5) becomes

$$\frac{1}{c_2^2}\frac{\partial^2 u_\varphi}{\partial t^2} = \frac{\partial^2 u_\varphi}{\partial r^2} + \frac{1}{r}\frac{\partial u_\varphi}{\partial r} - \frac{u_\varphi}{r^2} + \frac{\partial^2 u_\varphi}{\partial z^2} \tag{73.8}$$

where $c_2^2 = \mu/\varrho$. If we now consider harmonic waves of the form

$$u_\varphi = \varPhi(r)\, e^{i(qz+pt)} \tag{73.9}$$

we find that $\varPhi(r)$ satisfies the ordinary differential equation

$$\frac{d^2 \varPhi}{dr^2} + \frac{1}{r}\frac{d\varPhi}{dr} + \left(\varkappa_1^2 - \frac{1}{r^2}\right)\varPhi = 0 \tag{73.10}$$

in which

$$\varkappa_1^2 = \frac{p^2}{c_2^2} - q^2. \tag{73.11}$$

Since $\varPhi$ does not become infinite at $r=0$ it follows that the relevant solution of Eq. (73.10) for $\varkappa_1 \neq 0$ is

$$\varPhi(r) = A\, J_1(\varkappa_1 r) \tag{73.12}$$

where A is a constant, so that

$$u_\varphi = A\, J_1(\varkappa_1 r)\, e^{i(qz+pt)}. \tag{73.13}$$

For such a displacement σ_r and τ_{rz} are zero everywhere and

$$\tau_{r\varphi} = \mu\left\{\frac{1}{r}\frac{\partial u_\varphi}{\partial r} + r\frac{\partial}{\partial r}\left(\frac{u_\varphi}{r}\right)\right\}$$

$$= \mu A\, r\, \frac{\partial}{\partial r}\left(\frac{J_1(\varkappa_1 r)}{r}\right)\, e^{i(qz+pt)}.$$

Hence if $\tau_{r\varphi}$ is to vanish on the bounding surface $r=a$ of the cylinder then $\varkappa_1$ must be such that

$$\frac{\partial}{\partial a}\left[\frac{J_1(\varkappa_1 a)}{a}\right] = 0$$

which is easily shown to be equivalent to $\varkappa_1 a = \xi$ where ξ is a root of the equation

$$\xi J_0(\xi) = 2 J_1(\xi). \tag{73.14}$$

Introducing the wave-length $\varLambda = 2\pi/q$ and the phase velocity $c = p/q$ of the torsion waves we find that Eqs. (73.11) and (73.14) are equivalent to the relation

$$c = c_2 \left[\xi^2 \left(\frac{\varLambda}{2\pi a}\right)^2 + 1\right]^{\frac{1}{2}} \tag{73.15}$$

from which it follows that the group velocity of the torsion waves is c_2^2/c.

β) *Longitudinal waves.* If we assume that $u_\varphi = 0$ and that u_r and u_z are independent of φ it follows from (73.1) that $\omega_r = \omega_z = 0$ so that Eq. (73.5) is satisfied identically and Eqs. (73.4) and (73.6) reduce to

$$c_1^2 \frac{\partial \varDelta}{\partial r} + 2 c_2^2 \frac{\partial \omega_\varphi}{\partial z} = \frac{\partial^2 u_r}{\partial t^2} \tag{73.16}$$

and

$$c_1^2 \frac{\partial \varDelta}{\partial z} - \frac{2 c_2^2}{r}\frac{\partial}{\partial r}(r\omega_\varphi) = \frac{\partial^2 u_z}{\partial t^2}. \tag{73.17}$$

Now if we put

$$u_r = R(r)\, e^{i(qz+pt)}, \qquad u_z = Z(r)\, e^{i(qz+pt)} \tag{73.18}$$

where R and Z are functions of r alone, then, from Eq. (73.2), we obtain the equations

$$\Delta = \left(R' + \frac{1}{r} R + i q Z\right) e^{i(qz+pt)}, \tag{73.19}$$

$$2\omega_\varphi = (i q R - Z') e^{i(qz+pt)}. \tag{73.20}$$

If we substitute these expressions into Eqs. (73.16) and (73.17) we find that these latter equations are satisfied by expressions of the form

$$R = A \varkappa_2 J_0'(\varkappa_2 r) + B q J_1(\varkappa_3 r), \tag{73.21}$$

$$Z = A i q J_0(\varkappa_2 r) + \frac{Bi}{r} [J_1(\varkappa_3 r) + \varkappa_3 r J_1'(\varkappa_3 r)] \tag{73.22}$$

where A and B are constants and $\varkappa_2$, $\varkappa_3$ are defined by the equations

$$\varkappa_2^2 = \frac{p^2}{c_1^2} - q^2, \quad \varkappa_3^2 = \frac{p^2}{c_2^2} - q^2. \tag{73.23}$$

To satisfy the conditions $\sigma_r = \tau_{rz} = 0$ on $r = a$ we must have

$$\lambda \Delta + 2\mu \frac{\partial u_r}{\partial r} = 0, \quad \frac{\partial u_r}{\partial z} + \frac{\partial u_z}{\partial r} = 0$$

when $r = a$; it is readily shown that these equations lead to the relations

$$A \left[2c_2^2 \varkappa_2^2 J_0''(\varkappa_2 a) - p^2 c_1^2 J_0(\varkappa_2 a) \right] + 2B c_2^2 q \varkappa_3 J_1'(\varkappa_3 a) = 0,$$

$$2A \varkappa_2 q J_0'(\varkappa_2 a) + B \left(2q^2 - \frac{p^2}{c_2^2}\right) J_1(\varkappa_3 a) = 0.$$

Eliminating A and B from these equations we obtain the relation

$$\left. \begin{array}{r} \left(2q^2 - \dfrac{p^2}{c_2^2}\right) J_1(\varkappa_3 a) \left[2c_2^2 \varkappa_2^2 J_0''(\varkappa_2 a) - c_1^2 p^2 J_0(\varkappa_2 a)\right] \\ = 4 c_2^2 q^2 \varkappa_2 \varkappa_3 J_0'(\varkappa_2 a) J_1'(\varkappa_3 a) \end{array} \right\} \tag{73.24}$$

involving the frequency and wavelength of the vibrations, the radius a of the cylinder and the elastic wave velocities of the solid.

In the case where $\varkappa_2 a$, $\varkappa_3 a$ are small compared with unity the Bessel functions may be represented by the first two terms of the appropriate series expansions, and Eq. (73.24) reduces to

$$c = c_0 \left(1 - \tfrac{1}{4} \nu^2 q^2 a^2\right) \tag{73.25}$$

where $c_0^2 = E/\varrho$, and ν is Poisson's ratio.

The Eq. (73.24) was first derived by POCHHAMMER and CHREE[1]. The approximate form (73.25) was also derived by RAYLEIGH[2] by energy considerations. It is only recently however that numerical results have been computed from the frequency equation (73.24). For a discussion of these numerical results the reader is referred to the original papers[3] or to pp. 59 to 65 of H. KOLSKY's "Stress Waves in Solids" (Oxford 1953).

---

[1] L. POCHHAMMER: J. reine angew. Math. **81**, 324 (1876). — C. CHREE: Quart. J. Pure and Appl. Math. **23**, 335 (1889).

[2] Lord RAYLEIGH (J. W. STRUTT): Theory of Sound, 2nd edit., Vol. I, p. 252. London 1894.

[3] G. S. FIELD: Canad. J. Res. **5**, 619 (1931). — D. BANCROFT: Phys. Rev. **59**, 588 (1941). — E. v. CZERLINSKY: Akust. Z. **7**, 12 (1942). — J. A. MINDLIN: C. R. Doklady URSS. **1**, 11 (1946). — R. M. DAVIES: Phil. Trans. Roy. Soc. Lond., Ser. A **240**, 375 (1948).

γ) *Flexural waves.* In the case of flexural waves all three components of the displacement vector are non-zero and each of them depends on the angular coordinate φ. If we take

$$\boldsymbol{u} = (U \cos \varphi, V \sin \varphi, W \cos \varphi)\, e^{i(qz+pt)} \tag{73.26}$$

where U, V, W are functions of r alone, we find that

$$\varDelta = \left(U' + \frac{1}{r} U + \frac{1}{r} V + i q W\right) \cos \varphi\, e^{i(qz+pt)} \tag{73.27}$$

and that

$$2\omega = \left\{\sin \varphi \left(-\frac{W}{r} + iqV\right),\ \cos \varphi\,(iqU - W'),\ \sin \varphi \left(V' + \frac{1}{r} V + \frac{1}{r} U\right)\right\} e^{i(qz+pt)}. \tag{73.28}$$

Now from Eqs. (73.4) to (73.6) we have

$$\frac{\partial^2 \varDelta}{\partial r^2} + \frac{1}{r} \frac{\partial \varDelta}{\partial r} - \frac{\varDelta}{r^2} + \varkappa_2^2 \varDelta = 0 \tag{73.29}$$

where $\varkappa_2$ is given by the first of Eqs. (73.23), so that we may take

$$\varDelta = -\frac{p^2}{c_1^2} A\, J_1(\varkappa_2 r) \cos \varphi\, e^{i(qz+pt)} \tag{73.30}$$

where A is a constant of integration. In a similar way it is readily shown that

$$\omega_r = \frac{1}{2}\left\{iq\varkappa_3 C\, J_1'(\varkappa_3 r) + \frac{ip^2 B}{c_2^2}\, \frac{J_1(\varkappa_3 r)}{r}\right\} \sin \varphi\, e^{i(qz+pt)}, \tag{73.31}$$

$$\omega_z = \tfrac{1}{2} \varkappa_3^2 C\, J_1(\varkappa_3 r) \sin \varphi\, e^{i(qz+pt)} \tag{73.32}$$

where B and C are constants of integration and $\varkappa_3$ is defined by the second of Eqs. (73.23). Eqs. (73.27) and (73.28) can then be satisfied by putting

$$U = A \varkappa_2 J_1'(\varkappa_2 r) + B q \varkappa_3 J_1'(\varkappa_3 r) + C\, \frac{J_1(\varkappa_3 r)}{r}, \tag{73.33}$$

$$V = -A\, \frac{J_1(\varkappa_2 r)}{r} - Bq\, \frac{J_1(\varkappa_3 r)}{r} - C \varkappa_3 J_1'(\varkappa_3 r), \tag{73.34}$$

$$W = iq A\, J_1(\varkappa_2 r) - i \varkappa_3^2 B\, J_1(\varkappa_3 r). \tag{73.35}$$

Now if the cylindrical surface is free from tractions we must have $\sigma_r = \tau_{r\varphi} = \tau_{rz} = 0$ when $r = a$, and if we substitute the displacement vector (73.26) into the equations

$$\sigma_r = \lambda \varDelta + 2\mu\, \frac{\partial u_r}{\partial r},\quad \tau_{r\varphi} = \mu\left(\frac{1}{r}\frac{\partial u_r}{\partial \varphi} + \frac{\partial u_\varphi}{\partial r} - \frac{u_\varphi}{r}\right),\quad \tau_{rz} = \mu\left(\frac{\partial u_r}{\partial z} + \frac{\partial u_z}{\partial r}\right)$$

we find that the boundary conditions are equivalent to the statement that when $r = a$,

$$\beta^2 U' + (\beta^2 - 2)\left(\frac{U}{r} + \frac{V}{r} + iqW\right) = 0, \tag{73.36}$$

$$V' - \frac{1}{r}(U + V) = 0, \tag{73.37}$$

$$iqU + W' = 0, \tag{73.38}$$

with $\beta^2 = (\lambda + 2\mu)/\mu = c_1^2/c_2^2$. If we substitute from Eqs. (73.33) to (73.35) into Eqs. (73.36) to (73.38) we obtain three simultaneous homogeneous equations involving A, B and C. Eliminating A, B, C from these equations we obtain

$$\begin{vmatrix} x-1 & 2x-1 & 2x-1 \\ \varphi(\varkappa_2 a) - 2 & \varphi(\varkappa_3 a) - 2 & -2[\varphi(\varkappa_3 a) - 2] - q^2 a^2 (2x-1) \\ \varphi(\varkappa_2 a) - 1 & -(x-1)[\varphi(\varkappa_3 a) - 2] & 1 \end{vmatrix} = 0 \qquad (73.39)$$

where

$$\varphi(y) = y J_0(y)/J_1(y), \qquad x = \frac{p^2 \varrho (1+\nu)}{q^2 E}.$$

It should be noted that here, as in the case of longitudinal waves, the solutions will not be exact for the free vibrations of a cylinder of finite length, since we have not imposed the additional condition that there are no surface tractions across the plane ends of the cylinder. However we should expect the above solution to be sufficiently accurate for cylinders whose length greatly exceeds their radius.

The frequency equation (73.39) was first derived by BANCROFT[1] although the approximate equation

$$p^2 = \tfrac{1}{4} a^2 q^2 c_0^2 \qquad (73.40)$$

($c_0^2 = E/\varrho$), obtained by expanding the first few terms of each of the Bessel functions involved, was derived earlier by RAYLEIGH. A series of numerical computations based on BANCROFT's analysis have been published by HUDSON[2].

74. The generation of elastic waves by body forces.
We shall consider the problem of calculating the dynamic stresses produced in an elastic medium of infinite extent by time-dependent body forces[3]. If we introduce a space-like coordinate

$$\tau = c_1 t \qquad (74.1)$$

associated with the time t, c_1 being the velocity of propagation of P-waves in the solid defined by the first of the Eqs. (70.7), then we may write the equations of motion (9.12) in the form

$$\partial_k \tau^{kj} + \varrho P^j = \varrho c^2 \frac{\partial^2 u^j}{\partial \tau^2} \qquad (i, j = 1, 2, 3) \qquad (74.2)$$

where we have chosen rectangular Cartesian coordinates (x_1, x_2, x_3) to specify the position of a typical point of the solid and ∂_i denotes $\partial/\partial x_i$. It will be recalled from Eq. (14.30) that, for a homogeneous isotropic solid,

$$\tau^{kj} = \lambda \Delta \, \delta^{kj} + \mu(\partial_k u_j + \partial_j u_k) \qquad (74.3)$$

with

$$\Delta = \partial_m u_m. \qquad (74.4)$$

The problem we shall consider here is that of solving the set of Eqs. (74.2) to (74.4) when the mode of variation of the components P^j of the body force is prescribed throughout the solid. To solve these equations we introduce the four-dimensional Fourier transform of each of the components of stress and displacement. We shall denote the Fourier transform of a function, f, by placing a bar over it, thus, $\bar{f}$; in other words

$$\bar{f}(\xi_1, \xi_2, \xi_3, \omega) = \frac{1}{4\pi^2} \int_{E_4} f(x_1, x_2, x_3, \tau) \exp\{i(\xi_1 x_1 + \omega \tau)\} \, dV \qquad (74.5)$$

---

[1] D. BANCROFT: Phys. Rev. **59**, 588 (1941).
[2] G. E. HUDSON: Phys. Rev. **63**, 46 (1943).
[3] G. EASON, J. FULTON and I. N. SNEDDON: Phil. Trans. Roy. Soc. Lond. A **248**, 575 (1956).

where $dV = dx_1\, dx_2\, dx_3\, d\tau$ and E_4 denotes the entire $x_1 x_2 x_3 \tau$-space. If we multiply both sides of Eqs. (74.2) to (74.4) by $\exp\{i(x_p \xi_p + \omega \tau)\}$ and integrate over E_4 then, making use of the results[1]

$$\frac{1}{4\pi^2} \int_{E_4} \left(\partial_q f, \frac{\partial^2 f}{\partial \tau^2}\right) \exp\{i(x_p \xi_p + \omega \tau)\}\, dV = -(i\xi_q, \omega^2)\,\bar{f}$$

we find that Eqs. (74.2) to (74.4) are equivalent to the algebraic equations

$$-i\xi_k \bar{\tau}^{ki} + \varrho\, \bar{P}^i = -\varrho\, c^2 \omega^2 \bar{u}_j, \tag{74.6}$$

$$\bar{\tau}^{ki} = \lambda \bar{\Delta}\, \delta^{ki} - i\mu\,(\xi_k \bar{u}_j + \xi_j \bar{u}_k), \tag{74.7}$$

$$\bar{\Delta} = -i\xi_m \bar{u}_m \tag{74.8}$$

by means of which the Fourier transforms $\bar{\tau}^{ki}, \bar{v}^j$ of the components of stress and displacement can be determined in terms of $\bar{P}^i$, the Fourier transforms of the components of the body force.

Solving these equations we find that the Fourier transforms of the components of the displacement vector are given by the equations

$$\bar{u}_j = \frac{\beta^2(\gamma^2 - \omega^2)\bar{P}^j - (\beta^2 - 1)\,\xi^j(\xi_k \bar{P}^k)}{c_1^2(\gamma^2 - \omega^2)(\gamma^2 - \beta^2 \omega^2)} \tag{74.9}$$

where β is defined by Eq. (70.8) and

$$\gamma^2 = \xi_m \xi_m. \tag{74.10}$$

Similarly the Fourier transforms of the components of stress are given by the equation

$$\bar{\tau}^{kj} = -\frac{\lambda i \xi_r \bar{P}^r\, \delta^{kj}}{c_1^2(\gamma^2 - \omega^2)} - \frac{\mu i}{c_1^2}\left\{\frac{\beta^2(\gamma^2 - \omega^2)(\xi_k \bar{P}^j + \xi_j \bar{P}^k) - 2(\beta^2 - 1)\,\xi_k \xi_j(\xi_r \bar{P}^r)}{(\gamma^2 - \omega^2)(\gamma^2 - \beta^2 \omega^2)}\right\}. \tag{74.11}$$

To obtain the corresponding expressions for the components of the displacement vector we make use of FOURIER's integral theorem for four-dimensional transforms which states that if $\bar{\varphi}(\xi_1, \xi_2, \xi_3, \omega)$ is defined in terms of $\varphi(x_1, x_2, x_3, \tau)$ by Eq. (74.5) then

$$\varphi(x_1, x_2, x_3, \tau) = \frac{1}{4\pi^2} \int_{W_4} \bar{\varphi}(\xi_1, \xi_2, \xi_3, \omega) \exp\{-i(\xi_r x_r + \omega \tau)\}\, dW \tag{74.12}$$

where $dW = d\xi_1\, d\xi_2\, d\xi_3\, d\omega$ and W_4 is the entire $\xi_1 \xi_2 \xi_3 \omega$-space.

Inverting Eq. (74.9) by this rule we find that the components of the displacement vector are given by

$$u_j = \frac{1}{4\pi^4} \int_{W_4} \frac{\beta^2(\gamma^2 - \omega^2)\bar{P}^j - (\beta^2 - 1)\,\xi_j(\xi_k \bar{P}^k)}{c_1^2(\gamma^2 - \omega^2)(\gamma^2 - \beta^2 \omega^2)} \exp\{-i(\xi_r x_r + \omega \tau)\}\, dW. \tag{74.13}$$

Inverting the Eq. (74.11) by means of the Fourier integral theorem, we find that the components of the stress tensor are given by the integral formula

$$\tau^{kj} = -\frac{\varrho(\beta^2 - 2)}{4\pi^2 \beta^2}\, \delta^{kj} \int_{W_4} \frac{i\xi_r \bar{P}^r}{\gamma^2 - \omega^2} \exp\{-i(x_s \xi_s + \omega \tau)\}\, dW -$$

$$-\frac{\varrho}{4\pi^2 \beta^2} \int_{W_4} \frac{\beta^2(\gamma^2 - \omega^2)(\xi_k \bar{P}^j + \xi_j \bar{P}^k) - 2(\beta^2 - 1)\,\xi_j \xi_k(\xi_r \bar{P}^r)}{(\gamma^2 - \omega^2)(\gamma^2 - \beta^2 \omega^2)} \times \tag{74.14}$$

$$\times \exp\{-i(x_s \xi_s + \omega \tau)\}\, dW.$$

---

[1] I. N. SNEDDON: Functional Analysis, in Vol. II of this Encyclopedia, p. 279. 1955.

From the general solution given by Eqs. (74.13) and (74.14) it is possible to derive the general solutions for the statical problem and for the two-dimensional problem. Similarly the solution of the equations of motion in the case in which the distribution of body forces possesses axial symmetry may be obtained by means of multiple integral transforms with kernels of the form

$$\frac{r}{2\pi} J_\nu(\xi r) e^{i(\zeta z + \omega \tau)}.$$

For details of these solutions and of the particular forms assumed by them when the body forces P^j assume certain simple forms, the reader is referred to the paper by EASON, FULTON and SNEDDON cited above.

II. Boundary value problems of dynamic elasticity.

75. Introduction. We shall now consider the solution of boundary value problems in dynamic elasticity. It is only recently that solutions of such problems have been derived. The main methods of attack are based respectively on the theory of the complex variable and the theory of integral transforms. The complex variable method is applicable only to two-dimensional problems and the integral transform method, though applicable to three-dimensional problems, leads to simpler results in the two-dimensional case. For those reasons we shall restrict our attention to the case of plane strain in which the equations of motion are

$$\frac{\partial \sigma_x}{\partial x} + \frac{\partial \tau_{xy}}{\partial y} = \varrho \frac{\partial^2 u}{\partial t^2}, \qquad \frac{\partial \tau_{xy}}{\partial x} + \frac{\partial \sigma_y}{\partial y} = \varrho \frac{\partial^2 v}{\partial t^2} \qquad (75.1)$$

where

$$\sigma_x = (\lambda + 2\mu) \frac{\partial u}{\partial x} + \lambda \frac{\partial v}{\partial y}, \qquad \sigma_y = \lambda \frac{\partial u}{\partial x} + (\lambda + 2\mu) \frac{\partial v}{\partial y}, \qquad \tau_{xy} = \mu \left(\frac{\partial u}{\partial y} + \frac{\partial v}{\partial x} \right).$$

Furthermore the equation of compatibility takes the form

$$\nabla_1^2 (\sigma_x + \sigma_y) - \frac{2(\lambda + \mu)}{\lambda + 2\mu} \varrho \frac{\partial^2}{\partial t^2} \left(\frac{\partial u}{\partial x} + \frac{\partial v}{\partial y} \right) = 0. \qquad (75.3)$$

Using the fact that

$$2(\lambda + \mu) \left(\frac{\partial u}{\partial x} + \frac{\partial v}{\partial y} \right) = \sigma_x + \sigma_y$$

we see that (75.3) can be written in the form

$$\left(\nabla_1^2 - \frac{1}{c_1^2} \frac{\partial^2}{\partial t^2} \right) (\sigma_x + \sigma_y) = 0 \qquad (75.4)$$

where c_1^2 is the velocity of P-waves [defined by Eq. (70.7) above].

Differentiating the first of Eqs. (75.1) with respect to x, the second with respect to y and subtracting we find that

$$\frac{\partial^2 \sigma_x}{\partial x^2} - \frac{\partial^2 \sigma_y}{\partial y^2} = \frac{1}{2c_2^2} \frac{\partial^2}{\partial t^2} (\sigma_x - \sigma_y). \qquad (75.5)$$

Adding, instead of subtracting, we find that

$$\frac{\partial^2}{\partial x \partial y} (\sigma_x + \sigma_y) + \left(\nabla_1^2 - \frac{1}{c_2^2} \frac{\partial^2}{\partial t^2} \right) \tau_{xy} = 0. \qquad (75.6)$$

76. The complex variable method[1]. In this section we consider the application of complex variable methods to the solution of problems of dynamic plane elasti-

---
[1] J. R. M. RADOK: Quart. Appl. Math. **14**, 289 (1956).

The complex variable method.

city. If we rewrite Eq. (75.5) in the form

$$\left(\frac{\partial^2}{\partial x^2} - \frac{\varrho}{2\mu}\frac{\partial^2}{\partial t^2}\right)\sigma_x = \left(\frac{\partial^2}{\partial y^2} - \frac{\varrho}{2\mu}\frac{\partial^2}{\partial t^2}\right)\sigma_y \qquad (76.1)$$

we see that it will be satisfied identically if we express σ_x and σ_y in terms of a function U through the equations

$$\sigma_x = \frac{\partial^2 U}{\partial y^2} - \frac{\varrho}{2\mu}\cdot\frac{\partial^2 U}{\partial t^2}, \qquad \sigma_y = \frac{\partial^2 U}{\partial x^2} - \frac{\varrho}{2\mu}\cdot\frac{\partial^2 U}{\partial t^2}. \qquad (76.2)$$

From these equations we have that

$$\sigma_x + \sigma_y = V_1^2 U - \frac{1}{c_2^2}\cdot\frac{\partial^2 U}{\partial t^2}. \qquad (76.3)$$

Substituting this expression into Eq. (75.6) we find that

$$\left(V_1^2 - \frac{1}{c_2^2}\frac{\partial^2}{\partial t^2}\right)\left(\tau_{xy} + \frac{\partial^2 U}{\partial x\,\partial y}\right) = 0. \qquad (76.4)$$

We find the equation satisfied by U by substituting from Eq. (76.3) into (75.4):

$$\left(V_1^2 - \frac{1}{c_1^2}\frac{\partial^2}{\partial t^2}\right)\left(V_1^2 - \frac{1}{c_2^2}\frac{\partial^2}{\partial t^2}\right) U = 0, \qquad (76.5)$$

from all of which it is obvious that U is the dynamic counterpart of the Airy stress function.

If we confine our attention to problems in which the disturbing influences are moving with a velocity c parallel to the x-axis we may make use of the transformation

$$\xi = x - ct, \qquad \eta = y. \qquad (76.6)$$

Transforming the Eq. (76.5) to the new variables ξ and η we find that it can be written in the form

$$\left(\frac{\partial^2}{\partial \xi^2} - \frac{1}{s_1^2}\frac{\partial^2}{\partial \eta^2}\right)\left(\frac{\partial^2}{\partial \xi^2} - \frac{1}{s_2^2}\frac{\partial^2}{\partial \eta^2}\right) U = 0 \qquad (76.7)$$

where

$$s_1 = i\beta_1 = i\left(1 - \frac{c^2}{c_1^2}\right)^{\frac{1}{2}}, \qquad s_2 = i\beta_2 = i\left(1 - \frac{c^2}{c_2^2}\right)^{\frac{1}{2}}. \qquad (76.8)$$

In terms of these variables we have also from Eqs. (78.2) and (76.3) that

$$\sigma_x = \frac{\partial^2 U}{\partial \eta^2} - \frac{c^2}{2c_2^2}\frac{\partial^2 U}{\partial \xi^2}, \qquad \sigma_y = \frac{1}{2}(1 + \beta_2^2)\frac{\partial^2 U}{\partial \xi^2} \qquad (76.9)$$

and

$$\sigma_x + \sigma_y = \frac{\partial^2 U}{\partial \eta^2} + \beta_2^2\frac{\partial^2 U}{\partial \xi^2}. \qquad (76.10)$$

If we now let

$$z_1 = \xi + s_1\eta, \qquad z_2 = \xi + s_2\eta \qquad (76.11)$$

we find that Eq. (76.7) possesses the real solution

$$U = F_1(z_1) + \bar{F}_1(\bar{z}_1) + F_2(z_2) + \bar{F}_2(\bar{z}_2) = 2\,\mathrm{Re}\,[F_1(z_1) + F_2(z_2)] \qquad (76.12)$$

where $F(z_1)$, $F_2(z_2)$ are analytic functions of the complex variables z_1, z_2 respectively and $\bar{F}(\bar{z}_1)$, $\bar{F}_2(\bar{z}_2)$ are the corresponding conjugate functions. If we substitute the form (76.12) for U into Eqs. (76.9) and (76.10) we find that

$$\sigma_x = -2\,\mathrm{Re}\,[(\tfrac{1}{2} + \beta_1^2 - \tfrac{1}{2}\beta_2^2)\,F_1''(z_1) + \tfrac{1}{2}(1 + \beta_2^2)\,F_2''(z_2)], \qquad (76.13)$$

$$\sigma_y = (1 + \beta_2^2)\,\mathrm{Re}\,[F_1''(z_1) + F_2''(z_2)], \qquad (76.14)$$

$$\sigma_x + \sigma_y = -2(\beta_1^2 - \beta_2^2)\,\mathrm{Re}\,[F_1''(z_1)]. \qquad (76.15)$$

Substituting these expressions into Eqs. (75.2) and integrating, we obtain the expressions

$$\mu u = - \operatorname{Re}\left[F_1'(z_1) + \tfrac{1}{2}(1+\beta_2^2) F_2'(z_2)\right], \tag{76.16}$$

$$\mu v = \operatorname{Im}\left[\beta_1 F_1'(z_1) + \frac{1+\beta_2^2}{2\beta_2} F_2'(z_2)\right] \tag{76.17}$$

for the non-vanishing components of the displacement vector. Substituting these in turn into the third of Eqs. (75.2), we find that

$$\tau_{xy} = 2 \operatorname{Im}\left\{\beta_1 F_1''(z_1) + \frac{(1+\beta_2^2)^2}{4\beta_2} F_2''(z_2)\right\}. \tag{76.18}$$

Eqs. (76.13) to (76.18) comprise RADOK's solution, first published in 1956. It should be observed however that apart from a simple difference of notation this solution is identical with one previously derived by SNEDDON[1]. If we let

$$F_1'(z_1) = -\mu f'(z_1)$$

$$F_2'(z_2) = -\frac{2i\mu\beta_2}{1+\beta_2^2} g'(z_2)$$

in RADOK's solution we obtain SNEDDON's solution.

To illustrate the method we shall consider two very simple problems:

α) *Pulse of pressure moving along the surface of a semi-infinite solid.* We consider the distribution of stress in the semi-infinite two-dimensional elastic medium $y \geq 0$ when a moving pulse of pressure is applied to the boundary $y=0$. Taking the x-axis along the boundary we may assume boundary conditions of the form

$$\sigma_y = -\tfrac{1}{2}\{P''(x-vt) + \overline{P}''(x-vt)\}, \quad y=0 \tag{76.19}$$

and

$$\tau_{xy} = 0, \quad y = 0. \tag{76.20}$$

Putting $\eta = 0$ in Eq. (76.18) we find that Eq. (76.20) is satisfied if we take

$$\beta_1 F_1''(\xi) + \frac{(1+\beta_2^2)^2}{4\beta_2} F_2''(\xi) = 0$$

so that

$$F_2''(z_2) = -\frac{4\beta_1\beta_2}{(1+\beta_2^2)^2} F_1''(z_2).$$

Substituting this expression into Eq. (76.14) we obtain the expression

$$\sigma_y = (1+\beta_2^2)^{-1} \operatorname{Re}\left[(1+\beta_2^2)^2 F_1''(z_1) - 4\beta_1\beta_2 F_1''(z_2)\right]$$

for the normal stress σ_y, so that the boundary condition (76.19) is satisfied if we take

$$F_1(\xi) = -\frac{(1+\beta_2^2) P(\xi)}{(1+\beta_2^2)^2 - 4\beta_1\beta_2}.$$

Hence from Eqs. (76.16) and (76.17) we obtain the solution

$$u = \frac{1+\beta_2^2}{\mu[(1+\beta_2^2)^2 - 4\beta_1\beta_2]} \operatorname{Re}\left[P'(z_1) - \frac{2\beta_1\beta_2}{(1+\beta_2^2)^2} P'(z_2)\right],$$

$$v = \frac{-\beta_1}{\mu[(1+\beta_2^2)^2 - 4\beta_1\beta_2]} \operatorname{Im}\left[(1+\beta_2^2) P'(z_1) - 2P'(z_2)\right]$$

which apart from differences of notation is SNEDDON's solution[2].

---

[1] I. N. SNEDDON: Rend. Circ. Mat. Palermo (ii) **1**, 57 (1952).
[2] I. N. SNEDDON: Op. cit., Eqs. (11) and (12).

β) *Moving dislocation.* In the problem of the moving dislocation we seek a solution of the equations of motion satisfying

$$\sigma_y = 0, \quad [u]_L = \text{const}, \quad \text{on} \quad y = 0 \tag{76.21}$$

where $[u]_L$ denotes the change undergone by u for one circuit of L. It follows immediately from the Eq. (76.14) that the first of these conditions is satisfied if we take

$$F_2''(z_2) = -F_1''(z_2) \tag{76.22}$$

and from (76.16), (76.22) we see that the second is satisfied if we take

$$F_1'(z_1) = b \log z_1, \quad F_2'(z_2) = -b \log z_2. \tag{76.23}$$

Substituting from Eqs. (76.23) into Eqs. (76.16) and (76.17) we obtain the solution

$$u = -\frac{1}{\mu} \operatorname{Re}\left\{ b \log z_1 - \frac{1}{2} b (1+\beta_2^2) \log z_2 \right\}, \tag{76.24}$$

$$v = \frac{1}{\mu} \operatorname{Im}\left\{ b \log z_1 - \frac{1}{2} b \frac{(1+\beta_2^2)^2}{\beta_2} \log z_2 \right\}, \tag{76.25}$$

which is equivalent to Eshelby's solution[1].

In the paper cited above Radok has also derived by this method Yoffe's solution[2] of the problem of the moving Griffith crack, and Galin's solution[3] of the problem of the moving punch; in Sneddon's paper the method is applied to the solution of the boundary value problem:

$$\sigma_y = 0, \quad \tau_{xy} = -\tfrac{1}{2}\{G''(x-vt) + \overline{G}''(x-vt)\} \quad \text{on} \quad y = 0.$$

77. The integral transform method.
To illustrate the integral transform method we shall consider the distribution of stress in the semi-infinite elastic medium $y \geq 0$ when a variable pressure $p(x, t)$ is applied to the boundary $y = 0$[4]. If we replace the time variable t by a space-like variable τ defined by $\tau = c_1 t$, then we may write the boundary conditions of the problem in the form

$$\sigma_y = -p(x, \tau), \quad \tau_{xy} = 0, \quad \text{on} \quad y = 0. \tag{77.1}$$

To solve this dynamic boundary value problem we introduce the two-dimensional Fourier transform defined by the relation

$$\bar{f}(\xi, y, \omega) = \frac{1}{2\pi} \int_{-\infty}^{\infty} \int_{-\infty}^{\infty} f(x, y, \tau)\, e^{i(\xi x + \omega \tau)}\, dx\, d\tau \tag{77.2}$$

of all the quantities occurring in the Eqs. (75.1) and (75.2). If we multiply both sides of each of the equations in the sets (75.1) and (75.2) by $\exp\{i(\xi x + \omega \tau)\}$ and integrate from $-\infty$ to $+\infty$ with respect to each of the variables x and τ we obtain the set of simultaneous ordinary differential equations

$$i\xi \bar{\sigma}_x - D\bar{\tau}_{xy} = (\lambda + 2\mu)\omega^2 \bar{u}, \tag{77.3}$$

$$i\xi \bar{\tau}_{xy} - D\bar{\sigma}_y = (\lambda + 2\mu)\omega^2 \bar{v}, \tag{77.4}$$

$$\bar{\sigma}_x = -i\xi(\lambda + 2\mu)\bar{u} + \lambda D\bar{v}, \tag{77.5}$$

$$\bar{\sigma}_y = -i\xi \lambda \bar{u} + (\lambda + 2\mu)D\bar{v}, \tag{77.6}$$

$$\bar{\tau}_{xy} = \mu(D\bar{u} - i\xi \bar{v}), \tag{77.7}$$

---
[1] J. D. Eshelby: Proc. Phys. Soc. Lond. **62**, 307 (1949).
[2] E. Yoffe: Phil. Mag. **42**, 739 (1951).
[3] L. A. Galin: The Contact Problems of the Theory of Elasticity. Gastechizdat 1953.
[4] A solution of this problem by an integral transform method different from that presented here (in that it makes use of "potential" functions) is given on pp. 445—447 of I. N. Sneddon's "Fourier Transforms," (New York, 1951).

where we have written
$$D = \frac{d}{dy}.$$

If we substitute from Eqs. (77.5) to (77.7) into Eqs. (77.3) and (77.4) we obtain the pair of simultaneous ordinary differential equations

$$[\beta^2(\xi^2 - \omega^2) - D^2]\bar{u} + (\beta^2 - 1)i\xi D\bar{v} = 0, \tag{77.8}$$

$$(\beta^2 - 1)i\xi D\bar{u} + [\xi^2 - \beta^2(\omega^2 + D^2)]\bar{v} = 0 \tag{77.9}$$

for the determination of the Fourier transforms of the components of the displacement vector. In these equations we have written, as before, $\beta^2 = (\lambda + 2\mu)/\mu$. By a simple elimination these equations may be put into the simple form.

$$(D^2 - n_1^2)(D^2 - n_2^2)(\bar{u}, \bar{v}) = 0 \tag{77.10}$$

where

$$n_1^2 = \xi^2 - \omega^2, \quad n_2^2 = \xi^2 - \beta^2\omega^2. \tag{77.11}$$

The solutions of the Eqs. (77.10) are elementary:

$$\bar{u} = A_1 e^{-n_1 y} + A_2 e^{-n_2 y}, \quad \bar{v} = B_1 e^{-n_1 y} + B_2 e^{-n_2 y} \tag{77.12}$$

where the constants of integration A_1, A_2, B_1, B_2, are independent of y but may, of course, depend on ξ and ω. In deriving this solution we have assumed that the components of the displacement vector and those of the stress tensor all tend to zero as $\sqrt{x^2 + y^2}$ tends to infinity. When we substitute from Eqs. (77.12) into Eqs. (77.8) and (77.9) we obtain the following relations between the constants of integration:

$$n_1 A_1 = i\xi B_1, \quad \xi A_2 = i n_2 B_2. \tag{77.13}$$

Further relations for these constants are provided by the boundary conditions (77.1) which, in transformed form, are equivalent to

$$\mu[(\beta^2 - 2)D\bar{v} - i\xi\beta^2\bar{u}]_{y=0} = -\bar{p}(\xi, \omega) \tag{77.14}$$

$$[D\bar{u} - i\xi\bar{v}]_{y=0} = 0. \tag{77.15}$$

The set of Eqs. (77.13) to (77.15) provide us with four algebraic equations for the four constants of integration. Solving these equations we find then that

$$A_1 = \frac{i\xi\bar{p}(\xi^2 - \tfrac{1}{2}\beta^2\omega^2)}{2\mu G}, \quad A_2 = -\frac{i\xi n_1 n_2 \bar{p}}{2\mu G},$$

$$B_1 = \frac{n_1 \bar{p}(\xi^2 - \tfrac{1}{2}\beta^2\omega^2)}{2\mu G}, \quad B_2 = -\frac{\xi^2 n_1 \bar{p}}{2\mu G},$$

in which

$$G(\xi, \omega) = (\xi^2 - \tfrac{1}{2}\beta^2\omega^2)^2 - n_1 n_2 \xi^2. \tag{77.16}$$

Substituting these constants into the Eqs. (77.12) and inverting the expressions so obtained by means of the two-dimensional form of the Fourier integral theorem we obtain finally the expressions

$$u = \frac{1}{4\pi\mu} \int_{-\infty}^{\infty}\int_{-\infty}^{\infty} \frac{\bar{p}}{G}\left\{i\xi\left(\xi^2 - \tfrac{1}{2}\beta^2\omega^2\right)e^{-n_1 y} - i\xi n_1 n_2 e^{-n_2 y}\right\} e^{-i(\xi x + \omega\tau)} d\xi\, d\omega,$$

$$v = \frac{1}{4\pi\mu} \int_{-\infty}^{\infty}\int_{-\infty}^{\infty} \frac{\bar{p}}{G}\left\{n_1\left(\xi^2 - \tfrac{1}{2}\beta^2\omega^2\right)e^{-n_1 y} - \xi^2 n_1 e^{-n_2 y}\right\} e^{-i(\xi x + \omega\tau)} d\xi\, d\omega$$

for the components of the displacement vector.

F. Thermoelasticity.

In the preceding sections of this article it has been assumed that the elastic bodies being deformed are kept at a constant temperature. In this section we shall consider the stresses set up in an elastic body undergoing thermal changes.

78. The basic equations of thermoelasticity. Under free thermal expansion an isotropic body experiences a strain whose components $\gamma_{ij}^{(1)}$ referred to a set of orthogonal Cartesian axes $o(x_1, x_2, x_3)$ are specified by the equation

$$\gamma_{ij}^{(1)} = \alpha \vartheta \delta_{ij} \tag{78.1}$$

in which ϑ denotes the temperature change from T, the temperature of the solid in a state of zero stress and strain, and α denotes the coefficient of linear expansion of the solid. It is assumed that ϑ is sufficiently small for the thermal properties of the solid to remain constant throughout the times in which we are interested. In terms of the components, u_i, of the displacement vector, the total strain in the solid is given by the equation

$$\gamma_{ij} = \tfrac{1}{2}(\partial_j u_i + \partial_i u_j) \tag{78.2}$$

where $\partial_j u_i$ denotes the partial derivative $\partial u_i/\partial x_j$. This total strain is made up of the thermal strain and the elastic strain whose components $\gamma_{ij}^{(2)}$ are specified by the equation

$$\gamma_{ij}^{(2)} = \frac{1}{2\mu} \tau_{ij} - \frac{\lambda \Theta \delta_{ij}}{2\mu(3\lambda + 2\mu)} \tag{78.3}$$

where τ_{ij} are the components of the stress tensor,

$$\Theta = \tau_{ii} \tag{78.4}$$

is the sum of the principal stresses, λ and μ LAMÉ's elastic constants for the body. Substituting from Eqs. (78.1) to (78.3) into the equation

$$\gamma_{ij} = \gamma_{ij}^{(1)} + \gamma_{ij}^{(2)}$$

we obtain the tensor equation

$$\gamma_{ij} = \frac{1}{2\mu} \tau_{ij} - \left\{ \frac{\lambda \Theta}{2\mu(3\lambda + 2\mu)} - \alpha \vartheta \right\} \delta_{ij}. \tag{78.5}$$

Solving this tensor equation for the components of the stress tensor we find that

$$\tau_{ij} = (\lambda \Delta - \gamma \vartheta) \delta_{ij} + 2\mu \gamma_{ij} \tag{78.6}$$

where

$$\Delta = \gamma_{ii} = \partial_i u_i \tag{78.7}$$

denotes the dilatation in the solid and

$$\gamma = \alpha(3\lambda + 2\mu). \tag{78.8}$$

The physical relationship expressed by the tensor equation (78.6) is called the *Duhamel-Neumann law*.

The thermodynamic variables describing the state of the elastic solid are the strain components (78.2) and the absolute temperature $T + \vartheta$. By making use of the methods of reversible thermodynamics, BIOT[1] has shown that the entropy s per unit volume of the solid is given by the equation

$$s = c\varrho \log\left(1 + \frac{\vartheta}{T}\right) + \gamma \Delta, \tag{78.9}$$

---

[1] M. A. BIOT: J. Appl. Phys. **27**, 240 (1956).

where the additive constant, involved in the definition of the entropy, has been chosen so that the entropy is zero in the reference state. In this equation ϱ is the density of the solid, c is the specific heat per unit mass (assumed independent of temperature in the vicinity of the equilibrium temperature T), and γ is defined by Eq. (78.8). If ϑ is small in comparison with T we find that Eq. (78.9) gives the simple equation

$$s = \frac{\varrho c \vartheta}{T} + \gamma \varDelta \qquad (78.10)$$

for the entropy per unit volume, so that the quantity of heat absorbed by unit volume of the solid in the course of small deformations and small variations in temperature is given by the formula

$$h = Ts = \varrho c \vartheta + \gamma T \varDelta. \qquad (78.11)$$

Now it is known from the theory of the conduction of heat in solids[1] that the variation of temperature within an isotropic solid is governed by the equation

$$\frac{\partial h}{\partial t} = k \varDelta^2 \vartheta + q \qquad (78.12)$$

where k is the heat conductivity of the solid, and q is the quantity of heat per unit volume generated in the solid. Substituting from Eq. (78.10) into Eq. (78.11) we find that

$$\varrho c \frac{\partial \vartheta}{\partial t} + \gamma T \frac{\partial \varDelta}{\partial t} = k \nabla^2 \vartheta + q. \qquad (78.13)$$

If we introduce the diffusity

$$\varkappa = \frac{k}{\varrho c}$$

we can write this equation in the form

$$\frac{\partial \vartheta}{\partial t} = \varkappa \nabla^2 \vartheta + Q - \gamma' \frac{\partial \varDelta}{\partial t} \qquad (78.14)$$

where $Q = q/(\varrho c)$, $\gamma' = \gamma T/(\varrho c)$.

To complete the set of basic equations we have the equations of motion in the form

$$\tau_{ij,j} + \varrho F_i = \varrho \ddot{u}_i \qquad (78.15)$$

where (F_1, F_2, F_3) denotes the body force at the point (x_1, x_2, x_3) and $\ddot{u}_i$ denotes the i-th component, $\partial^2 u_i/\partial t^2$, of the acceleration of an infinitesimal element centred at the same point.

The set of sixteen equations symbolised by Eqs. (78.2), (78.6), (78.14) and (78.15) is sufficient, when taken with the appropriate boundary conditions, to determine the temperature variation and the components of stress and displacement when the heat sources and the body forces are prescribed.

79. Dimensionless form of the equations. It is convenient to write the basic set of thermoelastic equations in dimensionless form. If we take a typical length l as our unit of length, a time τ as our unit of time, the reference temperature T as the unit of temperature, and the rigidity modulus μ as unit of stress we find that the Eqs. (78.6), (78.14) and (78.15) respectively take the dimensionless forms

$$\sigma_{ij,j} + X_i = a \ddot{u}_i, \qquad (79.1)$$

$$\sigma_{ij} = [(\beta^2 - 2) \varDelta - b \vartheta] \delta_{ij} + 2\gamma_{ij}, \qquad (79.2)$$

$$\nabla^2 \vartheta + \bar{\Theta} = f \frac{\partial \vartheta}{\partial t} + g \frac{\partial \varDelta}{\partial t} \qquad (79.3)$$

---

[1] H. S. Carslaw and J. C. Jaeger: The Conduction of Heat in Solids. Oxford 1947.

where
$$X_i = \frac{l\varrho}{\mu} F_i, \quad \bar{\Theta} = \frac{Q\,l^2}{kT} \tag{79.4}$$
define the new source functions and
$$a = \left(\frac{l}{c_2\tau}\right)^2, \quad b = \frac{\gamma T}{\mu}, \quad f = \frac{\varrho c l^2}{k\tau}, \quad g = \frac{\gamma l^2}{k\tau}. \tag{79.5}$$

In the definition of a we have written c_2 for $(\mu/\varrho)^{\frac{1}{2}}$, the velocity of S-waves in the solid. $\beta^2 = (\lambda + 2\mu)/\mu$ is the square of the ratio of the velocity of P-waves to that of S-waves. In terms of Poisson's ratio $\beta^2 = 2(1+\nu)/(1-2\nu)$.

80. Steady-state problems.
If the body forces and the heat sources are time-independent and if the surface tractions are also statical then the basic set of equations—Eqs. (78.6), (78.14) and (78.15)—reduces to the forms
$$\tau_{ij} = (\lambda \Delta - \gamma \vartheta)\,\delta_{ij} + 2\mu\,\gamma_{ij}, \tag{80.1}$$
$$\varkappa\,\nabla^2\vartheta + Q = 0, \tag{80.2}$$
$$\tau_{ij,j} + \varrho F_i = 0. \tag{80.3}$$

Writing Eq. (80.1) in terms of the Young's modulus E and the Poisson ratio ν we obtain the equation
$$\tau_{ij} = \frac{E}{1+\nu}\left[\gamma_{ij} + \frac{1}{1-2\nu}\{\nu\Delta - (1+\nu)\,\alpha\vartheta\}\,\delta_{ij}\right]. \tag{80.4}$$

For an elastic solid free from body forces we find by inserting (80.4) in (80.3), putting $F_i = 0$ and using
$$\gamma_{ij,j} = \frac{1}{2}\nabla^2 u_i + \frac{1}{2}\frac{\partial \Delta}{\partial x_i}$$
that
$$\frac{\partial \Delta}{\partial x_i} + (1-2\nu)\nabla^2 u_i = 2(1+\nu)\,\alpha\,\frac{\partial \vartheta}{\partial x_i}. \tag{80.5}$$

To solve this equation Goodier[1] introduced a thermoelastic potential φ in terms of which the displacement vector (u_1, u_2, u_3) is defined by the relation
$$u_i = \frac{\partial \varphi}{\partial x_i}. \tag{80.6}$$

Inserting the expression (80.6) into Eq. (80.5) we see that the condition on φ is
$$\frac{\partial}{\partial x_i}[2(1-\nu)\nabla^2\varphi - 2(1+\nu)\,\alpha\vartheta] = 0$$
so that if we choose φ so that
$$\nabla^2 \varphi = m\,\vartheta(\mathbf{r}) \tag{80.7}$$
where $m = \alpha(1+\nu)/(1-\nu)$, the displacement vector defined by Eq. (80.6) is a solution of the steady-state equations of thermoelasticity. Eq. (80.7) is just Poisson's equation, and it is well known that a particular integral of this equation is
$$\varphi(\mathbf{r}) = \frac{m}{-4\pi}\int \frac{\vartheta(\mathbf{r}')}{|\mathbf{r}-\mathbf{r}'|}, \tag{80.8}$$
the integration extending throughout the solid.

The state of stress and strain represented by the particular integral (80.8) will require for its maintenance not only the prescribed distribution of temperature, but also certain boundary tractions which can be calculated from it by means of Eq. (80.4) and the conditions of equilibrium at the boundary. To solve the problem completely it only remains to determine a complementary stress distribution due to equal and opposite boundary tractions, which is merely a problem of given boundary tractions in the theory of elasticity, the fact that the body is heated being immaterial so long as the elastic constants are unaffected.

---
[1] J. N. Goodier: Phil. Mag. (vii) **23**, 1017 (1937).

Integrals of the type (80.8) were employed by BORCHARDT[1] in a general discussion of the theory of thermoelasticity and also to solve certain special problems involving asymmetric distributions of temperature in solids with spherical or circular boundaries. The distributions of stress due to special temperature distribution in infinite and semi-infinite solids have been discussed by a variety of authors[2]. There are very few *exact* solutions even of these steady state equations and such as there are are limited to spheres and cylinders[3] but several *approximate* solutions of the engineering problems concerned with thermal stresses in plates and rods are discussed in Chap. 14 of TIMOSHENKO and GOODIER's "Theory of Elasticity" (New York 1951).

Bibliography.

AUERBACH, F., and W. HORT, eds.: Handbuch der Physikalischen und Technischen Mechanik. Leipzig Bd. 3, 1927; Bd. 4, 1931.
BARTELS, J. ed.: Handbuch der Physik, Bd. XLVII, Geophysik I. Berlin 1956.
BIEZANO, C. B., and R. GRAMMEL: Technische Dynamik.
BRILLOUIN, L.: Les Tenseurs en Mécanique et en Élasticité. Paris 1938.
BULLEN, K. E.: An Introduction to the Theory of Seismology. Cambridge 1947.
BULLEN, K. E.: Seismology. London 1954.
BUTTY, E.: Tratado de elasticidad teoricotecnica (Elastotecnia), Vol. 1. Buenos Aires 1946.
CAGNIARD, L.: Réflexion et Réfraction des Ondes Séismiques. Paris 1939.
DINNIK, A. N.: Torsion: Theory and Application. Moscow 1938.
FÖPPL, A. u. L.: Drang und Zwang, 2. Aufl. Berlin Bd. 1, 1924; Bd. 2, 1928.
GOLDENVEISER, A. L.: Theory of Thin Elastic Shells. Moscow 1953.
GREEN, A. E., and W. ZERNA: Theoretical Elasticity. Oxford 1954.
JAEGER, J. C.: Elasticity, Fracture and Flow. London 1956.
JEFFREYS, H.: Cartesian Tensors. Cambridge 1931.
— The Earth, 3rd ed. Cambridge 1952.
KANTOROVICH, L. V.: Functional Analysis and Applied Mathematics. Nat. Bur. Stand. (Translation.) Washington 1952.
KLITCHIEFF, J.: Poglavja iz Teorije Elastičnosti sa Primenama. Belgrade 1950.
KOLOSSOV, G. V.: Application of the Complex Variable to the Theory of Elasticity. Moscow 1935.
KOLSKY, H.: Stress Waves in Solids. Oxford 1953.
LECORNU, L.: Théorie Mathématique de l'Élasticité. Paris 1938.
LEIBENSON, L. S.: Variational Methods of Solution of Problems in the Theory of Elasticity. Moscow 1943.
LORENZ, H.: Technische Elastizitätslehre. Berlin 1913.
LOVE, A. E. H.: Some Problems of Geodynamics. Cambridge 1911.
— The Mathematical Theory of Elasticity, 4th ed. Cambridge 1927.
MACELWANE, J. B.: An Introduction to Theoretical Seismology, Part I, Geodynamics. New York 1936.
MURNAGHAN, F. D.: Finite Deformations of an Elastic Solid. New York 1951.
MUSKHELISHVILI, N. I.: Some Basic Problems of the Mathematical Theory of Elasticity, 4th ed. Leningrad 1954.
NOVOZHILOV, V. V.: Foundations of Non-Linear Theory of Elasticity. Moscow 1953.
PIPPARD, A. J. S.: Strain Energy Methods of Stress Analysis. London 1928.
PRESCOTT, J.: Applied Elasticity. London 1924.
SEARLE, G. F. C.: Experimental Elasticity, 2nd ed. Cambridge 1933.
SOKOLNIKOFF, I. S.: Mathematical Theory of Elasticity, 2nd ed. New York 1956.
SOUTHWELL, R. V.: An Introduction to the Theory of Elasticity for Engineers and Physicists. Oxford 1936.
SYNGE, J. L.: The Hypercircle in Mathematical Physics. Cambridge 1957.
TIMOSHENKO, S. P.: Theory of Elastic Stability. New York 1934.
— Theory of Plates and Shells. New York 1940.
TIMOSHENKO, S. P., and J. N. GOODIER: Theory of Elasticity, 2nd ed. New York 1951.
TRELOAR, L. R. G.: The Physics of Rubber Elasticity. Oxford 1949.
WESTERGAARD, H. M.: Theory of Elasticity and Plasticity. Harvard 1952.

---

[1] C. W. BORCHARDT: Monogr. Akad. Wiss. Berlin 1873, 9.

[2] R. D. MINDLIN and D. H. CHENG: J. Appl. Phys. **21**, 926 (1950). — N. O. MYKLESTAD: J. Appl. Mech. A **1942**, 131.

[3] See, for instance, pp. 362—367 of the second edition of I. S. SOKOLNIKOFF's Mathematical Theory of Elasticity (New York 1956) and the original papers cited there.

Photoelasticity.

By

H. T. JESSOP.

With 114 Figures.

Introduction.

When loads are applied to any solid body the stresses induced in the body vary, in general, from point to point, and there will be regions of high stress which will constitute the weak points of the body at which fracture will occur if the loads are sufficiently large. A knowledge of the stresses at these points and of the way in which they may be modified by alterations in the shape of the body is of great importance to the designer of engineering components.

The engineer gains such knowledge mainly from three sources; practical experience, calculations based upon the mathematical theory of elasticity, and experimental methods of stress-analysis. His experience gives him a large amount of qualitative information, but this frequently fails to be applicable especially when radically new designs are contemplated. The mathematical theory yields much information about stress-distribution, but this can rarely be applied directly to practical problems without employing drastic approximations which leave the accuracy of the results very much in question. Experimental methods of stress-analysis provide the most reliable means of obtaining quantitative measurements, and the photoelastic method frequently gives the fullest and most accurate information, expecially in the regions of highest stress-concentration which are the most vital to the strength of the component.

The photoelastic method depends upon the property of certain transparent solids by which they become doubly refractive under the action of stress, the magnitude of the optical effect bearing a definite relation to that of the stress. In applying the method to the exploration of the stresses in any body subjected to given loads, a scale model of the body is first made in one of the stress-optically sensitive materials. The model is subjected to loads similar to those applied to the prototype, and the optical effects produced are measured in some form of polariscope. The material of the model having been calibrated, these observations lead directly to numerical values of certain of the stresses and stress-differences, and application of the results of the mathematical theory to the optical observations leads to the complete determination of the stresses at all points of the model.

Thus for its proper understanding, photoelasticity needs a knowledge of both crystal optics, and the theory of elasticity. These subjects are fully dealt with elsewhere[1], and the reader is advised to study the full treatment of both unless he has already a good knowledge of them. In this article, short summaries are given of the main facts and principles, both optical and mechanical, which are involved in the study and practice of photoelasticity, but these are intended to serve as reminders and quick references to basic data and not as texts.

---

[1] Crystal optics in Vol. XXV, theory of elasticity in this volume.

The theory of photoelasticity itself has also of necessity been treated in a somewhat brief way, and the reader is referred to Coker and Filon's book [1] for a full and detailed exposition. The practical methods of photoelasticity have been treated rather more fully, since there is no text-book available which contains all the developments which have taken place in recent years.

The text-books to which reference is made contain full bibliographies of the earlier work on the subject, and these have not been duplicated in this article. Individual references have been confined to the more important recent publications.

A. Theory.

I. History.

1. The establishment of the physical basis of photoelasticity. In 1816 David Brewster[1] discovered that a plate of glass subjected to a simple pressure or tension exhibited the double refraction characteristic of a uniaxial crystal the optic axis of which was parallel to the direction of the applied load. Brewster, however, did not succeed in obtaining a uniform stress in his model and was not able to make any quantitative estimate of the relation between the stress and the optical effect produced. Four years later Biot demonstrated that a strip of glass became doubly refracting when put into a state of longitudinal vibration. At about the same time Fresnel repeated Brewster's experiment and attempted to measure the changes in the velocities of the two oppositely polarised rays. That his results were indecisive is perhaps not surprising in view of the fact that methods of measurement were at that time very crude, and that the mathematical theory of elasticity was only in its early stages of development.

The first theory of the effect was presented by Franz Neumann in 1841. He expressed the velocities of the two waves in terms of the three principal strains in the medium. Some twelve years later Maxwell produced, independently, a theory in which the velocities were related to the principal stresses. The two theories produced relations of precisely similar form, and they were obviously equally applicable to an isotropic material strained within the linear elastic range. Both theories were moreover applicable to bodies under any system of combined stress, and they formed the foundation for all subsequent work on the theory of photoelasticity.

The first numerical determinations of the stress-optical coefficients of several materials were made by Wertheim in 1854, and he was followed by Mach in 1872 and Kerr in 1888. Both of the latter attempted also measurements of the separate retardations of the two oppositely polarised waves in glass using interferometer methods. Their results however, were not reliable owing to difficulties in measuring the strains produced in their materials by the applied loads. More accurate determinations of the separate coefficients for a number of glasses of different compositions were made by Pockels in 1902, while between 1902 and 1912 Filon carried out a number of investigations of the effect in a more extended range of glasses. Filon's work in this period included some very careful determinations of the dispersion of the effect as well as of the variation in the separate coefficients with the composition of the glasses and with temperature.

This work completed what may be called the first phase in the development of photoelasticity—the phase in which attention was confined almost exclusively to the investigation of the physical basis of the phenomenon of artificial double refraction. The possibility that the effect might be employed to measure the

---

[1] D. Brewster: Phil. Trans. Roy. Soc. Lond. **156** (1816).

stresses in a body subjected to load had been visualised by some investigators. It had, in fact, been mentioned by BREWSTER in his original memoir, and from about 1900 there were attempts, notably by MESNAGER, to explore the stress-distribution in very simple glass models under load. It was MESNAGER who eventually in 1912 carried out the most successful of the attempts to determine the stresses in an actual engineering structure by using the photoelastic effect in a glass model of a bridge [1]. The difficulty of fabricating models in glass however, was a serious obstacle and very little progress was made with this medium.

2. The exploration of two-dimensional stresses. In 1911, COKER, who had already done some work on the stress-optical effect in glass, discovered that xylonite exhibited the effect in a very marked degree, its sensitivity being about three times that of the most sensitive glass. The material was, moreover, obtainable in large plates from which it was extremely easy to machine two-dimensional models of any required shape. COKER then began an extensive series of investigations into the stress-distribution in various engineering components, in the course of which he developed techniques and methods of measurement which were much simpler than those hitherto employed. These investigations extended over a period of twentyfive years, during which time he lectured on the subject in many countries, while through his laboratory at University College, London, there passed a constant stream of research students from all parts of the world.

Very early in this period, COKER and FILON met and began a collaboration which was to result in the publication in 1931 of their treatise on photoelasticity [*1*] FILON, who already had an interest in the physics of the photoelastic effect as well as in the mathematical theory of elasticity, set out to establish a firm foundation for the new method of stress-exploration. His *Theory of Generalised Plane Stress* established the validity of the method and laid down the conditions in which photoelastic observations on a plate of finite thickness would yield reliable information on the stresses in the plate. He produced, also, mathematical analyses of many of the problems which COKER had investigated practically. Comparisons of the theoretical and experimental results confirmed the reliability of the latter in general, but also gave valuable indications of the way in which stress-distributions were affected by differences between the ideal conditions assumed in the theory and the actual conditions found in any practical problem.

During this period various investigations on similar lines were being carried out by other workers chief among whom were BAUD and HEYMANS in America, MESNAGER in France, KÖNIG, SCHULTZ and WÄCHTLER in Germany and TUZI in Japan, but the work of COKER and FILON constituted by far the greatest contribution to the advancement of the subject. Though techniques have been improved, the principles and methods established by them remain in force, and their book is still the most authoritative work on photoelasticity.

It is surprising, in view of the many important practical results published by investigators during this time, that engineers on the whole distrusted the method and did not take advantage of the means it offered of obtaining information by which the design of components could be improved. They certainly made use of such results as were published, but except in very few cases they did not initiate investigations of their own problems.

Up to 1935 also, no means had been found of applying the photoelastic method to three-dimensional stress systems, and the usefulness of the method was confined to those problems which were of two-dimensional character. A certain amount of qualitative information on some three-dimensional problems was certainly

---

[1] M. MESNAGER: Ann. Ponts Chauss. **16**, 133 (1913).

obtainable by the use of two-dimensional analogies, but there remained a very large number of practical problems in which the engineer possessed very little knowledge of the stress-distributions and in which the photoelastic method offered little or no help.

3. The exploration of three-dimensional stresses. The first approach to a practical solution of three-dimensional stress problems was made possible by the development of what is now called the "stress freezing" method. Residual optical patterns had been observed by several investigators in various materials which has been subjected to temperature changes under load. MAXWELL in 1853 and FILON and HARRIS in 1923 had both obtained the effect accidentally, while TUZI in 1927 and SOLAKIAN in 1935 both attempted to calculate the residual stresses and to relate them to the applied loads. None of these, however, fully appreciated the significance of the phenomenon, and it was OPPEL[1] who in 1936 first produced definite evidence that residual stress patterns could be obtained which were similar to the patterns produced under direct load, and that in certain materials the residual stresses were not released when the model was cut into small pieces. He was followed by HETENYI, who in 1938 first established the strict proportionality of the stresses in the two types of loading in one of the newly developed synthetic resins, and investigated the technique of the method. This discovery opened the way to the complete solution of three-dimensional problems, and resulted in a very sharp increase in activity in photoelastic research.

It was now possible to freeze the stresses in a loaded model of any shape, and, by cutting slices from it, to explore the stresses at internal points. The development of the method, however, required the devising of new techniques and the re-examination of the theory of the photoelastic effect and its applications to the conditions existing in the three-dimensional models. In this work many investigators took part. HETENYI, DRUCKER, FROCHT and MINDLIN developed methods of measurement of the frozen-stress, and JESSOP and FROCHT devised integration methods for determining the separate stresses. HILTSCHER, KUSKE and JESSOP adapted methods used in crystallography to determine the principal axes of stress at any point of a slice from a frozen model. WELLER, MENGES and JESSOP developed a method of measuring the stress-optical effect in a three-dimensional model by observation of the interference fringes in light scattered from a polarised beam traversing the model.

During this period also progress was greatly assisted by the discovery of many new photoelastic materials among the synthetic resins and "plastics" which were being produced for industrial purposes. These had stress-optical coefficients many times larger than that of xylonite, and were suitable for casting or machining to any desired shape. Thus by 1951 a variety of means were available for the complete determination of the stresses in any model under prescribed loads, and the limitations to the application of the method to practical stress-distribution problems were mainly the technical difficulties involved in manufacturing the photoelastic model and in reproducing in it the exact loading conditions of the prototype. Meanwhile the method was being increasingly adopted in industry, where it is now firmly established as one of the most valuable aids to sound engineering design.

II. The optical basis of photoelasticity.

4. The nature and characteristics of light. For all the phenomena of photoelasticity the electromagnetic theory of light offers an adequate explanation. We

---

[1] G. OPPEL: Forsch. Ing.-Wes. **7**, 5, 240 (1936).

shall accordingly adopt this theory and regard light as a wave of transverse electromagnetic disturbance propagated in space. There will then be associated with any light wave two varying fields, the electric and the magnetic, each of which may be defined by a vector. In an isotropic medium the two vectors are perpendicular to each other, and each is perpendicular to the direction of propagation (Fig. 1).

Evidence from the scattering of light and from the double refraction in crystals indicates that it is the electric field which is the primary factor in the propagation of waves. We shall accordingly take the electric vector as the vector defining the disturbance and refer to it as "the light vector".

Any beam of "ordinary" light which is of sufficient intensity to be detectable by optical methods will contain a vast number of elementary waves the directions of whose light vectors are distributed at random in the plane normal to the direction of propagation. A beam of light in which some system is introduced into the orientations of the vectors is said to be polarised. If all the waves have their light vectors in the same direction, the light is said to be plane polarised, the plane of polarisation being defined as the plane containing the magnetic vectors OH and the direction of propagation OX (Fig. 1). The plane containing the electric vectors OE and the direction OX is fre-

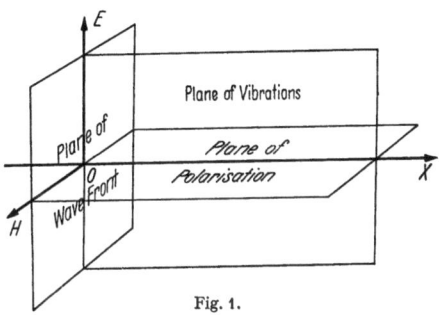

Fig. 1.

quently referred to as the plane of the vibrations. The direction of propagation OX is the direction of the rays of light which may be defined as the paths of flow of energy.

α) *Frequency and colour.* The frequency of a light vibration determines the colour of the light, and this frequency is not affected by transmission of the light through any medium. Changes in colour of light in passing through a medium are due to absorption by the medium, not to changes of frequency. What we call "white" light generally contains a mixture of frequencies corresponding to the range of colours of the visible spectrum from violet to dark red. The frequency is usually denoted by f (in cycles per sec) or p (in radians per sec) and $p = 2\pi f$.

β) *Velocity.* The velocity of light waves of any frequency in a vacuum is 3×10^{10} cm/sec, and it is sensibly the same in air. We shall denote this velocity by V. The velocity in all transparent liquids and solids is less than this value and varies from one medium to another. The ratio of the velocity in air to the velocity in the medium is called the *refractive index* of the medium and is denoted by n.

γ) The *wave-length* of light of any given frequency, being the distance travelled during one complete vibration, will vary from one medium to another. The wave-length in air is taken as defining the colour or frequency, and is measured in angströms, ($1 \text{ Å} = 10^{-8}$ cm). The range of wave-lengths in the visible spectrum is from about 3900 Å for the extreme violet to 7700 Å for the extreme red.

Monochromatic light contains vibrations of one wave-length only. Sodium light (5890 Å) or mercury green (5460 Å), obtained by using filters to exclude the violet and yellow light from a mercury lamp, are the most useful sources of approximately monochromatic light.

δ) *Amplitude and intensity.* The energy transmitted in any light wave is some function of both the amplitude, which measures the magnitude of the disturbance, and the frequency of the vibration. The intrinsic intensity of a light wave may be taken to be proportional to the energy transmitted. In photoelastic work we shall have no need to consider the intrinsic energy or to compare the intensities of light waves of different frequencies. Our only concern will be to estimate the variations in intensity of light of a single frequency, and for this purpose the intensity may be taken as directly proportional to the square of the amplitude of the vibration.

ε) *Phase.* The phase of a vibration at any instant defines the stage of the cycle reached at that instant, and so far as any isolated light wave is concerned, it cannot be measured, and would in any case have no practical significance. When, however, we are considering the effect of two light waves vibrating in the same plane and travelling along the same path, the difference in phase of the waves becomes all important, for on it depends the whole phenomenon of interference by which photoelastic effects are measured.

ζ) *The wave-front.* A wave-front is defined as a locus in space of points at which the wave has reached the same phase. If we consider a "point" source in an isotropic medium, such loci must obviously be spheres having the source as centre. At a large distance from the source the wave-front for a very small area of field may be considered as sensibly plane. The propagation of a beam of finite intensity from point to point of any medium may be considered as due to the total effect of the light vectors in the wave-front.

5. Propagation of light in a crystal. The results of the electromagnetic theory show that in an isotropic medium any disturbance, whatever the direction of the light vector, gives rise to a wave which is propagated through the medium with its form unchanged, the direction of propagation being normal to the wave-front.

In a crystal, however, it is shown that for any particular wave-front there are in general only two directions in which the light vector can lie in order to propagate a wave through the medium. These directions are perpendicular to each other, and any light vibration entering the crystal may be resolved into its component vibrations in these directions, each of which may be considered to give rise to its own wave. The light is thus polarised into two plane polarised waves. These two waves will travel with different velocities, and, in general, their directions of propagation will be different, neither direction being normal to the wave-front.

In the application of crystal optics to photoelasticity, however, some simplifications may be introduced. When a transparent "photoelastic" material is subjected to stress it acquires quasi crystalline properties as regards the transmission of light, and the results of the electromagnetic theory for crystals may be applied to it. But the changes in the velocity of light produced in a medium by stress are very small—never exceeding in practice 0.1%. While differences of this order can produce relative retardations which amount to a number of wave-lengths and may be measured to a high degree of accuracy, their effect upon the direction of propagation of the waves is very small, the divergences being of the order of 0.001 radian. Such divergences are less than those present in the collimated beam of any ordinary polariscope, and may accordingly be neglected. Thus we have the simplification that we may assume that the two polarised waves associated with any wave-front are both propagated along the same path, this path being normal to the wave-front.

FRESNEL's *ellipsoid*. The directions of polarisation in any wave-front and the relation between these directions and the velocities of the two oppositely polarised waves are very simply illustrated by FRESNEL's ellipsoid. If a, b and c are the three principal wave-velocities in a crystal—that is the velocities of the waves whose vibrations occur respectively in the directions of the three crystalline axes ox, oy, oz (Fig. 2), then the ellipsoid

$$a^2 x^2 + b^2 y^2 + c^2 z^2 = 1,$$

is the FRESNEL's ellipsoid for that crystal, the semi-axes OA, OB, OC being the reciprocals of the principal wave-velocities.

If OPQ be any central plane section of the ellipsoid, OP and OQ being the principal axes of the section, then a comparison of the geometry of the ellipsoid with the equations derived from the electromagnetic theory shows that if OPQ corresponds to the plane of a wave-front in the crystal, (1) OP and OQ are the directions of polarisation for that wave-front, and (2) the lengths of the semi-axes OP and OQ are the reciprocals of the velocities v_1 and v_2 of the two waves whose vibrations are respectively in those directions.

Any plane-polarised wave of specific wave-length λ may be represented by

$$u = a \cos \frac{2\pi}{\lambda} (vt - x)$$

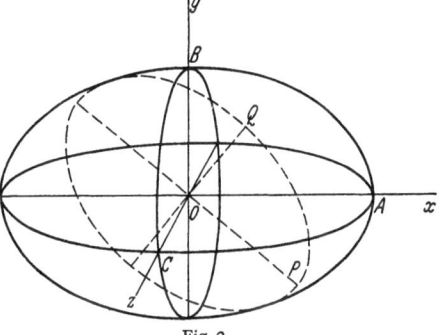

Fig. 2.

where u is the displacement in the vibration direction, λ is the wave-length in air, v the velocity in the medium and x the direction of propagation.

Alternatively, this may be written

$$u = a \cos \left(pt - \frac{2\pi x}{\lambda} \right)$$

where p is the angular frequency in radians per second, and $2\pi x/\lambda$ is the phase retardation suffered by the wave in travelling a distance x in the medium.

Or, since $\lambda = v/f = 2\pi v/p$, we may write

$$u = a \cos p \left(t - \frac{x}{v} \right) = a \cos p \left(t - \frac{nx}{V} \right)$$

expressing the retardation respectively in terms of the velocity of the wave in the medium and the refractive index of the medium.

6. Passage of polarised light through a crystal. If a beam of plane-polarised light is passed through a plate of crystal it will, in general, be repolarised in the crystal into two beams whose vibration directions are parallel to the secondary principal axes of the crystal in the plane of the wave-front.

Let ox and oy (Fig. 3) be the directions of these axes, and op the direction of the vibrations in the incident plane-polarised beam, and let any wave of this beam at the surface of the plate be represented by

$$u_p = a \cos pt. \tag{6.1}$$

In the crystal this wave will be repolarised into

and
$$\left. \begin{array}{l} u_x = a \cos \alpha \cos pt \\ u_y = a \sin \alpha \cos pt. \end{array} \right\} \tag{6.2}$$

In general these waves will travel with different velocities v_1 and v_2, and after passing through the thickness d of the plate, will emerge as

$$u'_x = a \cos\alpha \cos\left(pt - \frac{pd}{v_1}\right)$$
and
$$u'_y = a \sin\alpha \cos\left(pt - \frac{pd}{v_2}\right). \quad (6.3)$$

If the two waves now travel through any distance in air, their velocities being then equal, their added retardations will be equal, and Eqs. (6.3) become

$$u'_x = a \cos\alpha \cos\left(pt - \frac{pd}{v_1} - \varepsilon\right)$$
and
$$u'_y = a \sin\alpha \cos\left(pt - \frac{pd}{v_2} - \varepsilon\right). \quad (6.4)$$

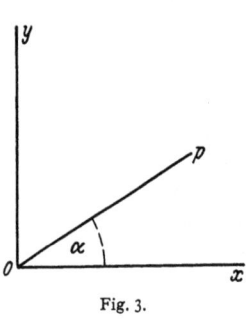

Fig. 3.

If they now pass through another polarising filter, called the analyser, whose vibration direction is normal to that of the polariser, only the resolute of each wave in that direction will be transmitted.

The light passing through the analyser will then be

$$u_a = u'_x \sin\alpha - u'_y \cos\alpha,$$

or

$$u_a = a \sin\alpha \cos\alpha \left\{\cos\left(pt - \frac{pd}{v_1} - \varepsilon\right) - \cos\left(pt - \frac{pd}{v_2} - \varepsilon\right)\right\}$$
$$= a \sin 2\alpha \sin\left\{pd\left(\frac{1}{2v_1} - \frac{1}{2v_2}\right)\right\} \sin\left\{pt - pd\left(\frac{1}{2v_1} + \frac{1}{2v_2}\right) - \varepsilon\right\}. \quad (6.5)$$

The amplitude of this vibration becomes zero (1) if $\alpha = 0°$ or $90°$; i.e. if the vibrations of the incident light are parallel to one of the vibration axes the crystal. In this case the beam is transmitted by the crystal without repolarisation, and is completely stopped by the analyser, (2) if $pd\left(\frac{1}{v_1} - \frac{1}{v_2}\right)$ is zero or an integral multiple of 2π. Writing this in terms of the wave-length the condition becomes

$$\frac{2\pi V}{\lambda} d\left(\frac{1}{v_1} - \frac{1}{v_2}\right) = 0 \quad \text{or} \quad 2i\pi$$

or

$$d(n_1 - n_2) = 0 \quad \text{or} \quad i\lambda \quad (6.6)$$

where i is an integer, and n_1 and n_2 are the refractive indices for the two waves.

Thus light of any particular wave-length is extinguished when the difference of the "optical paths" of the two waves is zero or an integral number of wavelengths. If white light is used, the proportions of light of different colours transmitted will thus be different, and the colour of the plate as seen in the polariscope will vary with the thickniss of the plate and with the difference of the two refractive indices.

α) *Circularly polarised light.* A plate of crystal of such a thickness that $d(n_1 - n_2) = \frac{1}{4}\lambda$ is called a quarter-wave plate. If such a plate is placed in front of a polariser with its polarising axes inclined at 45° to the vibration axis of the polariser, the amplitudes of the two waves in the plate will be equal, and, after transmission they will emerge with a phase difference of $\pi/2$.

They may thus be written as

$$u'_x = \tfrac{1}{2}\sqrt{2}\,a \cos p\,t$$

and

$$u'_y = \tfrac{1}{2}\sqrt{2}\,a \sin p\,t$$

using Eqs. (6.3) and omitting phase changes common to the two waves.

The light is then said to be circularly polarised, and has the property that the sum of the resolutes of the two vibrations in any direction will have the same amplitude

$$u'_x \sin \alpha + u'_y \cos \alpha = \tfrac{1}{2}\sqrt{2}\,a \sin(p\,t + \alpha).$$

Such quarter-wave plates are usually made of sheets of mica.

β) *Optical polarisers and compensators.* In order to measure the double refraction effect in any material it is necessary to start with a beam of plane-polarised light. Such a beam is generally obtained by making use of the property of double-refraction in a crystal to produce two oppositely polarised beams, and then eliminating one of them.

In a NICOL's prism the two beams are produced in a crystal of calcite. The difference of the velocities of the two oppositely polarised waves in this material is large enough to cause an appreciable divergence of the two paths, and it is possible to get rid of one of the beams by reflection while allowing the other to pass through the prism. This produces a perfect plane-polarised beam and such polarisers are generally used in the Physics laboratory. It is, however, impossible to obtain single crystals of calcite large enough for most photoelastic purposes.

A polarising material which can be obtained in large sheets is "polaroid". This consists of a thin layer of plastic material in which are embedded very small rod-like crystals of iodoquinine sulphate, oriented so as to have their axes parallel to one direction. These crystals are doubly refractive, and are also highly dichroic:—that is, they absorb one of the polarised waves much more strongly than the other, with the result that even a very thin layer of the crystals transmits almost completely plane-polarised light. Polarisation is not quite complete in the violet end of the spectrum, and there is a certain loss of light due to partial absorption of the other wave, but these disadvantages are more than compensated by the large field obtainable.

Another method of obtaining plane-polarised light is by reflection. If light is incident obliquely on the polished surface of a transparent medium, it is found that the reflected light contains a proportion of light which is polarised in the plane of incidence;—that is, whose vibrations are perpendicular to this plane, while the refracted light has a proportion polarised in the opposite direction. At a certain angle of incidence called the polarising angle for the medium the reflected light is practically completely polarised. The polarising angle varies with the refractive index of the material, its theoretical value being given by $\tan i = n$, where i is the angle of incidence.

For most glasses the polarising angle is between 56 and 60°.

This method is rarely used in practice as it is very cumbersome, but the fact that such polarisation by reflection and refraction does occur makes it desirable to avoid oblique incidence of any kind when observing the effects of double refraction in a polariscope.

Optical compensators are used to measure the relative retardation between two oppositely polarised waves. They are made from plates of crystal, usually quartz, which are so arranged as to provide a range of known relative retardations.

In the BABINET compensator there are two quartz wedges arranged with their optic axes at right angles, so that the wave which is relatively retarded in one wedge is relatively accelerated in the other (Fig. 4). At the mid-point O where the edges are of equal thickness, light passing through will have zero relative retardation, at other points there will be a relative retardation varying linearly with distance from O.

The compensator is placed in the polariscope in front of the plate to be measured, with its polarising axes parallel to those in the plate. Zero total relative retardation will then be observed at the point of the wedge at which the retardation in the plate is equal and opposite to that in the compensator.

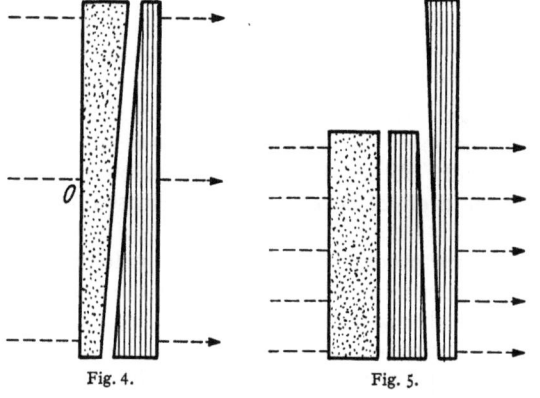

Fig. 4.   Fig. 5.

Alternatively, one wedge may be moved relative to the other, the relative retardation at O being then proportional to the distance moved.

The SOLEIL-BABINET compensator consists of three quartz plates, one of uniform thickness and two wedges, the optic axes being as indicated in Fig. 5. This gives a uniform relative retardation over the field whose magnitude may be altered by moving one of the wedges. This type of compensator is most frequently used as a means of providing a known relative retardation in a beam of light.

III. The mechanical basis of photoelasticity.

7. Stresses at a point. A stress in a continuous medium may be defined as a measure of the intensity of the forces operating between adjacent particles.

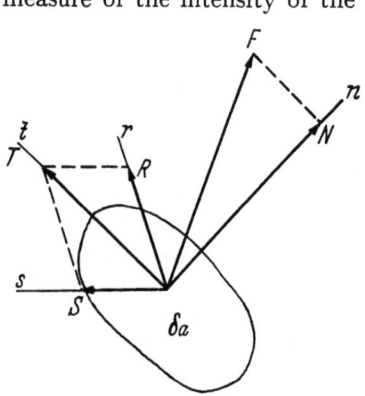

Fig. 6.

If we consider a small element of area, δa, in a stressed medium, the particles on one side of the area will, in general, exert forces upon the particles on the other side. The orientation of the element of area may be defined by the direction of the normal, n (Fig. 6), the side from which the normal is drawn being the positive side of the element. If the element be sufficiently small the resultant of the forces exerted by the material on the positive side upon the material on the other side reduces to a single force F. The mean total stress, or the mean force intensity *across this element of area* is denoted by $F/\delta a$.

F may be resolved into components N and T respectively normal and tangential to the element, and T may be resolved into its components R and S in arbitrarily chosen perpendicular directions r and s in the plane of the element. Then $N/\delta a$ is the mean normal stress (tension or compression according to the sign of N), across the element, while $R/\delta a$ and $S/\delta a$ are the mean tangential or shear stresses across the area in their respective directions.

If we now imagine the area to shrink to zero at a point P, then the limiting values to which the expressions for the mean stresses tend will be the stresses at the point P over an infinitesimal area perpendicular to n. Thus $\lim_{\delta a \to 0} N/\delta a$ is the normal stress in direction n at the point P and is denoted by σ_n. In the same way $\lim_{\delta a \to 0} R/\delta a$ and $\lim_{\delta a \to 0} S/\delta a$ are the tangential or shear stresses at P in their respective directions, across an area normal to n and are denoted by the symbols τ_{nr} and τ_{ns} respectively.

The stresses over such a single element of area as has been considered above do not completely specify the state of stress at the point P, for they take no account of the forces between pairs of particles lying on the same side of the area. If however, we consider the stresses acting across three such elements which are mutually perpendicular, and formulate the equations of equilibrium for a small element of the solid we find that the stresses in any direction are completely defined by the stresses across the three elements. Thus taking interfaces normal to the rectangular axes x, y, z, we shall have three *normal stresses*, σ_x, σ_y, σ_z and six *shear stresses*

$$\tau_{yz}, \tau_{zy}, \tau_{zx}, \tau_{xz}, \tau_{xy}, \text{ and } \tau_{yx}.$$

It can be shown, however, that for equilibrium

$$\tau_{yz} = \tau_{zy}, \quad \tau_{zx} = \tau_{xz}, \quad \text{and} \quad \tau_{xy} = \tau_{yx}.$$

The state of stress at the point is thus specified in terms of the six quantities $\sigma_x, \sigma_y, \sigma_z, \tau_{yz}, \tau_{zx}, \tau_{xy}$. The normal stress in a direction r whose direction cosines referred to x, y, z are l, m, n is then shown to be

$$\sigma_r = l^2 \sigma_x + m^2 \sigma_y + n^2 \sigma_z + 2mn\,\tau_{yz} + 2nl\,\tau_{zx} + 2lm\,\tau_{xy}. \tag{7.1}$$

This expression for σ_r leads to an elegant geometrical representation of the normal stresses in any direction at a point. If on a radius vector in direction r from the point we mark off a point (X, Y, Z) at distance R such that $\sigma_r = 1/R^2$ then the locus of this point is the quadric surface

$$1 = X^2 \sigma_x + Y^2 \sigma_y + Z^2 \sigma_z + 2YZ\,\tau_{yz} + 2ZX\,\tau_{zx} + 2XY\,\tau_{xy} \tag{7.2}$$

known as the *Stress Quadric* at the point.

This is obviously a central quadric, and will take the form of an ellipsoid if the normal stresses are all of the same sign, or a hyperboloid if one normal stress is of different sign from the others. In either case the figure will have three principal axes, and its equation referred to these will be

$$X^2 \sigma_x + Y^2 \sigma_y + Z^2 \sigma_z = 1. \tag{7.3}$$

The directions of these three axes will thus be directions in which no shear stresses act. They are known as *Axes of Principal Stress*, and will be referred to as directions p, q, r, while $\sigma_p, \sigma_q, \sigma_r$ will be used to denote *Principal Stresses* at a point. Referred to the principal axes, the normal stress in direction r then becomes,

$$\sigma_r = l^2 \sigma_p + m^2 \sigma_q + n^2 \sigma_r. \tag{7.4}$$

Thus the state of stress at a point may be completely defined by a knowledge of the directions and magnitudes of the three principal stresses at the point.

If we consider the central section of the quadric (7.2) by the plane $Z = 0$, we have the central conic

$$X^2 \sigma_x + Y^2 \sigma_y + 2XY\,\tau_{xy} = 1 \tag{7.5}$$

called the *Stress conic* in that plane, and the normal stress in any direction l, m in the plane is then given by

$$\sigma_r = l^2 \sigma_x + m^2 \sigma_y + 2lm\, \tau_{xy}. \qquad (7.6)$$

This figure again has two principal axes, referred to which its equation is

$$X^2 \sigma_x + Y^2 \sigma_y = 1. \qquad (7.7)$$

These will be axes along which there is no shear stress in that plane. They are known as *Secondary Principal Axes* in the plane and their directions are denoted by $p_1, q_1 (p_2, q_2$ etc. if more than one plane is in question). $\sigma_{p_1}, \sigma_{q_1}$ etc. then denote the magnitudes of the secondary principal stresses, and the normal stress in direction r in that plane is

$$\sigma_r = l^2 \sigma_{p_1} + m^2 \sigma_{q_1}. \qquad (7.8)$$

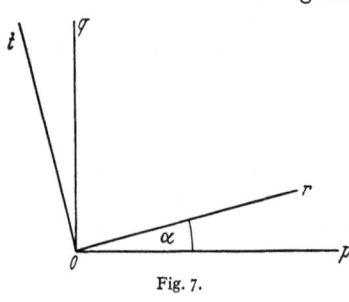

Fig. 7.

When dealing with stresses in one plane it is frequently more convenient to express the stresses in terms of a single angle. Thus the normal stress in a direction r making an angle α with an arbitrarily chosen axis ox in the plane is given by

$$\sigma_r = \sigma_x \cos^2\alpha + \sigma_y \sin^2\alpha + \tau_{xy} \sin 2\alpha. \qquad (7.9)$$

The normal stress perpendicular to this,

$$\sigma_t = \sigma_x \sin^2\alpha + \sigma_y \cos^2\alpha - \tau_{xy} \sin 2\alpha, \qquad (7.10)$$

and the shear stress

$$\tau_{rt} = \tfrac{1}{2}(\sigma_y - \sigma_x) \sin 2\alpha + \tau_{xy} \cos 2\alpha. \qquad (7.11)$$

The shear stress τ_{rt} is zero when

$$\tan 2\alpha = \frac{2\tau_{xy}}{\sigma_x - \sigma_y} \qquad (7.12)$$

and this gives the inclination α of the secondary principal axes to ox and oy.

If we refer to the principal axes of stress in the plane, the stresses in any two perpendicular directions r, t (Fig. 7) are

$$\sigma_r = \sigma_p \cos^2\alpha + \sigma_q \sin^2\alpha, \qquad (7.13)$$

$$\sigma_t = \sigma_p \sin^2\alpha + \sigma_q \cos^2\alpha, \qquad (7.14)$$

$$\tau_{rt} = \tfrac{1}{2}(\sigma_q - \sigma_p) \sin 2\alpha. \qquad (7.15)$$

An alternative form for the normal stresses is

$$\sigma_r = \tfrac{1}{2}(\sigma_p + \sigma_q) + \tfrac{1}{2}(\sigma_p - \sigma_q) \cos 2\alpha, \qquad (7.16)$$

$$\sigma_t = \tfrac{1}{2}(\sigma_p + \sigma_q) - \tfrac{1}{2}(\sigma_p - \sigma_q) \cos 2\alpha. \qquad (7.17)$$

Adding these two equations we get

$$\sigma_r + \sigma_t = \sigma_p + \sigma_q, \qquad (7.18)$$

i.e. the sum of the normal stresses in any two perpendicular directions in the plane is constant. Subtracting

$$\sigma_r - \sigma_t = (\sigma_p - \sigma_q) \cos 2\alpha. \qquad (7.19)$$

Also from (7.15) we see that the shear-stress has its maximum when $\alpha = 45°$, and that this maximum value is given by

$$\tau_{\max} = \tfrac{1}{2}(\sigma_p - \sigma_q). \qquad (7.20)$$

8. Distribution of stresses.

From considerations of continuity it can be seen that, except at points of application of finite loads, both the magnitudes and the directions of the stresses in any plane must change continuously as we move from point to point of the plane. Thus if we start from any point A, moving always in the direction of one principal stress in the plane σ_p (Fig. 8) we shall trace out a curve ABC the tangent to which at any point is in the direction of the principal stress σ_p at that point. This curve is called a *line of principal stress* or a *stress-trajectory*. A similar line can be drawn through A following the direction of σ_q at all points. Such curves could be drawn from any point, and the plane could be thus covered by two infinite systems of orthogonal curves which may be called the p and q systems. In practice a number of these curves are traced, more or less uniformly spaced, and the resulting network gives a picture of the distribution of stress, so far as the directions of the principal stresses are concerned.

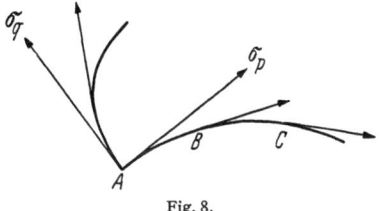

Fig. 8.

In a three-dimensional system of forces we should have of course three such orthogonal trajectories passing through each point. In general none of these would be a plane curve, and it would be virtually impossible to represent the three systems graphically. There is, however, one important case in which some simplification occurs. If the stresses in a solid body are symmetrical obout any plane, then at any point in the plane two of the principal stresses must lie in that plane. The two corresponding sets of trajectories will thus lie in the plane of symmetry, and may be represented graphically. It should be noted, however, that while lines of the third set must meet the plane orthogonally, their curvatures at the point of intersection are in general unknown, both in respect of magnitude and of the plane of curvature.

The variations of the stresses from point to point of a stressed body are determined from the Body Force Equations, derived from consideration of the equilibrium of a small rectangular element of the material. In Cartesian coordinates these take the form

$$\left.\begin{array}{l}\dfrac{\partial \sigma_x}{\partial x}+\dfrac{\partial \tau_{xy}}{\partial y}+\dfrac{\partial \tau_{xz}}{\partial z}+\varrho\, X=0,\\[6pt] \dfrac{\partial \tau_{xy}}{\partial x}+\dfrac{\partial \sigma_y}{\partial y}+\dfrac{\partial \tau_{yz}}{\partial z}+\varrho\, Y=0,\\[6pt] \dfrac{\partial \tau_{zx}}{\partial x}+\dfrac{\partial \tau_{yz}}{\partial y}+\dfrac{\partial \sigma_z}{\partial z}+\varrho\, Z=0,\end{array}\right\} \quad (8.1)$$

where ϱ is the density of the material and X, Y, Z are the components of body force operating.

In very many practical applications the body forces are negligible in comparison with the applied loads, and the last term on the left hand side of these equations may be omitted. We shall omit them in what follows.

For distribution of the stresses in one plane only we have, neglecting body forces,

$$\left.\begin{array}{l}\dfrac{\partial \sigma_x}{\partial x}+\dfrac{\partial \tau_{xy}}{\partial y}=0,\\[6pt] \dfrac{\partial \tau_{xy}}{\partial x}+\dfrac{\partial \sigma_y}{\partial y}=0.\end{array}\right\} \quad (8.2)$$

If instead of using Cartesian coordinates we consider the equilibrium in one plane of an element bounded by four neighbouring lines of principal stress (Fig. 9), we get the LAMÉ-MAXWELL form of the equations of equilibrium in a plane in terms of the principal stresses. These are

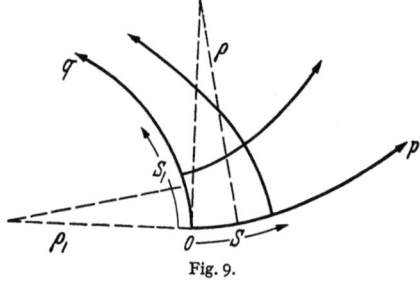

Fig. 9.

$$\left. \begin{array}{l} \dfrac{\partial \sigma_p}{\partial s} + \dfrac{\sigma_p - \sigma_q}{\varrho_1} = 0, \\[6pt] \dfrac{\partial \sigma_q}{\partial s_1} + \dfrac{\sigma_p - \sigma_q}{\varrho} = 0 \end{array} \right\} \quad (8.3)$$

where s and s_1 are measured along lines of principal stress p and q respectively, and ϱ and ϱ_1 are the radii of curvature of these respective lines.

In the particular case of a body in which the stresses are symmetrical about an axis (e.g. a solid of revolution loaded symmetrically about its axis), these equations may usefully be extended to three dimensions. If σ_p and σ_q be principal stresses in a plane of symmetry of such a body, the stress trajectories corresponding to σ_r will in this case all be circles whose centres lie on the axis of symmetry (Fig. 10). The equations in three dimensions then become

$$\left. \begin{array}{l} \dfrac{\partial \sigma_p}{\partial s} + \dfrac{\sigma_p - \sigma_q}{\varrho_1} + \dfrac{\sigma_r - \sigma_p}{\varrho_2} \cos \alpha = 0, \\[6pt] \dfrac{\partial \sigma_q}{\partial s_1} + \dfrac{\sigma_p - \sigma_q}{\varrho} + \dfrac{\sigma_r - \sigma_q}{\varrho_2} \sin \alpha = 0 \end{array} \right\} \quad (8.4)$$

where ϱ_2 is the radius of curvature of the r stress trajectory, i.e. the perpendicular AN to the axis of symmetry, and α is the angle made by this radius with the direction of the p trajectory at the point.

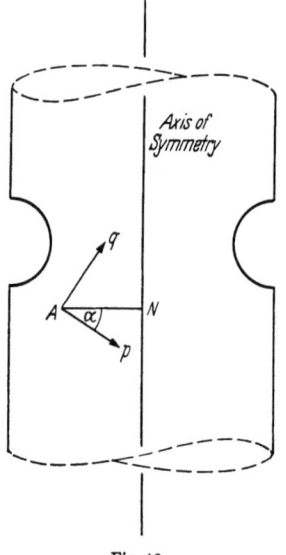

Fig. 10.

9. The strains at a point. The strains in a solid material may be of two kinds—normal strains which are either extensions or contractions and which correspond to normal stresses, and tangential or shear strains which correspond to shear stresses and may be regarded as produced by the slide of one layer of material over the other.

In general both the direction and the length of any line in the material will be altered by strain, and this introduces complications if the strains are large. If, however, we restrict the strains to such magnitudes that terms of the second order are negligible in comparison with first order terms, then we find that the changes in direction produced have only a second order effect upon the expressions for the normal strains, while the changes in length have only a second order effect upon the expressions for the shear-strains.

If l be the length before strain of a line in the material whose direction is r, and l_1 be its length after strain, then the normal strain in direction r, denoted by ε_r, is defined by the relation

$$l_1 = l(1 + \varepsilon_r). \quad (9.1)$$

Thus a normal strain will be positive, if it is an extension, and negative, if a contraction.

A shear strain, denoted by ε_{rs} deforms a rectangular element whose edges were initially parallel to the axes or, os into a parallelogram. Thus the shear strain ε_{rs} in Fig. 11a, has transformed the rectangle $ABCD$ into the Parallelogram AB_1C_1D and the magnitude of the strain is measured by the amount of the slide BB_1 divided by the distance AB. To the first order of small quantities this is equal to $\sin(BAB_1)$ or to $\cos(ADC_1)$. The rectangle would have suffered precisely the same deformation if it had been subjected to an equal slide ε_{sr} as shown in Fig. 11b. We have accordingly the result that a shear strain parallel to axes r and s may be represented as either ε_{rs} or ε_{sr}. In particular, when referring to a set of Cartesian axes we find that ε_{xy} and ε_{yx} are alternative expressions for the same shear strain, and that the same applies of course to ε_{yz} and ε_{zy}, and to ε_{zx} and ε_{xz}.

If, as was done in the case of stresses, we analyse the strains at a point referring them to three arbitrary orthogonal axes we find that the state of strain at

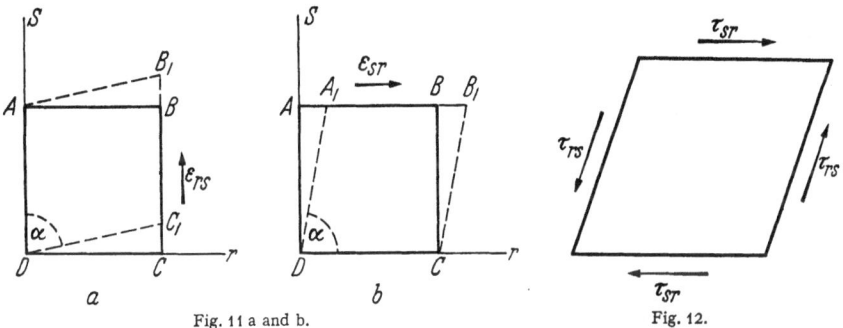

Fig. 11 a and b.   Fig. 12.

the point is completely defined by the three normal strains ε_x, ε_y, ε_z, and the three shear strains ε_{yz}, ε_{zx}, ε_{xy}, the normal strain in any direction r being given by

$$\varepsilon_r = l^2 \varepsilon_x + m^2 \varepsilon_y + n^2 \varepsilon_z + mn\, \varepsilon_{yz} + nl\, \varepsilon_{zx} + lm\, \varepsilon_{xy}. \qquad (9.2)$$

As in the case of stresses this leads to the existence of a *strain quadric* at the point,

$$\varepsilon_x X^2 + \varepsilon_y Y^2 + \varepsilon_z Z^2 + \varepsilon_{yz} YZ + \varepsilon_{zx} ZX + \varepsilon_{xy} XY = 1, \qquad (9.3)$$

a surface such that the normal strain in any direction at the point is equal to the reciprocal of the square of the radius vector to the quadric in that direction.

Also considerations of the geometry of this surface lead to the deduction of three principal directions in which the strains are purely normal strains. These are the principal axes of strain at the point and the corresponding strains are the three principal strains.

Consideration of any plane central section of the strain quadric leads to expressions for the strains in one plane which are precisely similar in form to those already quoted for plane stresses, except for the fact that terms involving the shear strain appear without the factor 2. The explanation of this, of course, is that in order to maintain equilibrium in an element under shear, the single shear strain ε_{rs} must be associated with the two equal shear stresses τ_{rs} and τ_{sr} which exert equal and opposite couples on the element (Fig. 12).

From the expressions for the strains in one plane we also deduce the result that two perpendicular normal strains of equal magnitude and opposite sign produce the same deformation as a shear strain in a direction bisecting the angle between the normal strains.

Thus the shear strain illustrated (very much exaggerated) in Fig. 11, whose magnitude is $\cos\alpha$, might equally well have been defined as a normal strain of magnitude $\frac{1}{2}\cos\alpha$ in the direction of the diagonal DB together with another, $-\frac{1}{2}\cos\alpha$, in the direction of the diagonal AC.

10. Stress-strain relations in an isotropic elastic material. The basis of all the stress-strain relations in an elastic medium is Hook's law of the proportionality of stress and strain.

In the mathematical theory of elasticity it is usual to start with the assumption of a generalised form of Hook's law and to derive from this an expression for the work done in straining an element of the medium, from which the stress-strain relations are derived.

A somewhat simpler method, which leads to the same relations, is to start with four well established experimental results from which the relations follow directly. The four results may be stated:

(1) If an isotropic elastic solid is subjected to a simple tension in any direction it undergoes a stretch in that direction proportional to the applied tensile stress. E.g. if σ_x is the only stress acting, $\sigma_x = E\varepsilon_x$.

The constant of proportionality, E, is Young's *modulus* for the material.

(2) Under a simple tension in any direction the solid also exhibits a stretch in all directions perpendicular to the applied tension, the transverse stretch being of opposite sign to the direct stretch, and proportional to it in magnitude.

Thus if σ_x be the only stress applied and ε_x be the direct stretch produced by it then

$$\varepsilon_y = \varepsilon_z = -\gamma\,\varepsilon_x$$

for all directions y and z perpendicular to x.

The constant γ is Poisson's *ratio* for the material.

(3) If an element of an isotropic elastic solid be subjected to a pair of equal shear stresses, τ_{xy} and τ_{yx}, it undergoes a shear strain ε_{xy} proportional to the shear stress, i.e.

$$\tau_{xy} = G\,\varepsilon_{xy}.$$

The constant G is the *modulus of rigidity* of the material.

(4) If a body be subjected to a number of stresses acting simultaneously, then each stress produces its own strain or strains as specified above, and these strains may be superposed to give the complete state of strain of the body.

Applying these results we find that if $\sigma_x, \sigma_y, \sigma_z, \tau_{yz}, \tau_{zx}, \tau_{xy}$ are the stress components at any point of an isotropic elastic solid, the six strain components are given by

$$\left.\begin{aligned}\varepsilon_x &= \frac{1}{E}\{\sigma_x - \gamma(\sigma_y + \sigma_z)\},\\ \varepsilon_y &= \frac{1}{E}\{\sigma_y - \gamma(\sigma_z + \sigma_x)\},\\ \varepsilon_z &= \frac{1}{E}\{\sigma_z - \gamma(\sigma_x + \sigma_y)\},\end{aligned}\right\} \quad (10.1)$$

$$\left.\begin{aligned}\varepsilon_{yz} &= \frac{1}{G}\tau_{yz},\\ \varepsilon_{zx} &= \frac{1}{G}\tau_{zx},\\ \varepsilon_{xy} &= \frac{1}{G}\tau_{xy}.\end{aligned}\right\} \quad (10.2)$$

If we eliminate from Eqs. (10.1) we get the expressions for the normal strains in terms of the normal stresses

$$\left.\begin{array}{l}\sigma_x = \lambda(\varepsilon_x + \varepsilon_y + \varepsilon_z) + 2\mu\varepsilon_x, \\ \sigma_y = \lambda(\varepsilon_x + \varepsilon_y + \varepsilon_z) + 2\mu\varepsilon_y, \\ \sigma_z = \lambda(\varepsilon_x + \varepsilon_y + \varepsilon_z) + 2\mu\varepsilon_z, \end{array}\right\} \quad (10.3)$$

where

$$\lambda = \frac{\gamma E}{(1+\gamma)(1-2\gamma)} \quad \text{and} \quad \mu = \frac{E}{2(1+\gamma)}.$$

The quantity $(\varepsilon_x + \varepsilon_y + \varepsilon_z)$ is called the *dilatation* and is denoted by δ. If in Eqs. (10.3) we put $\sigma_x = \sigma_y = \sigma_z = T$, we have the stress-strain relations for a body under a hydrostatic tension T, and adding the equations we get

$$3T = (3\lambda + 2\mu)\delta$$

or

$$T = k\delta.$$

k is the *bulk modulus* of the material.

Also by equating the expressions for the work done by a shear stress τ_{xy} and by the pair of equal and opposite perpendicular normal stresses which produce the same deformation, it can be shown that $\mu = G$. There are thus only two independent elastic constants involved in the six stress-strain relations in (10.2) and (10.3), and there must therefore be two independent relations between the four constants, E, G, γ and k used in practice.

11. Generalised plane stress. In photoelastic investigations of stress-distributions, observations have to be made on light which has passed through a plate of stressed material of finite thickness. We obtain then effects which depend upon the mean values of the stresses throughout the thickness of the plate. It is therefore necessary to know the conditions in which the relations between the mean stresses and strains conform to the relations derived for stresses in one plane.

If we consider a flat plate of uniform thickness, $2c$, the mid-plane of the plate being taken as the x, y plane, then integration of the first two Eqs. (8.1) through the thickness of the plate gives, omitting body forces,

$$\frac{\partial X}{\partial x} + \frac{\partial S}{\partial y} + (\tau_{xz})_{+c} - (\tau_{xz})_{-c} = 0,$$

$$\frac{\partial S}{\partial x} + \frac{\partial Y}{\partial y} + (\tau_{yz})_{+c} - (\tau_{yz})_{-c} = 0,$$

where X, Y and S are the mean values through the plate of the stresses σ_x, σ_y, and τ_{xy} respectively.

If we now specify that the plane faces of the plate are free from shear stress, these reduce to

$$\left.\begin{array}{l}\dfrac{\partial X}{\partial x} + \dfrac{\partial S}{\partial y} = 0, \\ \dfrac{\partial S}{\partial x} + \dfrac{\partial Y}{\partial y} = 0.\end{array}\right\} \quad (11.1)$$

Equations of precisely the same form as those given in (8.2) for the forces in one plane.

Similarly we can integrate the two-dimensional LAMÉ-MAXWELL equations (8.3) through the plate, though this necessitates the assumption that the directions of the principal stresses do not change from point to point through the thickness of the plate. These integrations lead to

$$\left. \begin{array}{l} \dfrac{\partial P}{\partial s} + \dfrac{P-Q}{\varrho_1} = 0, \\[2mm] \dfrac{\partial Q}{\partial s_1} + \dfrac{P-Q}{\varrho} = 0, \end{array} \right\} \quad (11.2)$$

where P and Q are the mean values through the plate of the principal stresses σ_p and σ_q.

In order to obtain the stress-strain relations in a form independent of z, we have to make the further assumption that σ_z is equal to zero throughout the plate.

Putting $\sigma_z = 0$ into the third equation of (10.3) we get

$$\varepsilon_z = -\frac{\lambda}{\lambda + 2\mu}(\varepsilon_x + \varepsilon_y)$$

and substituting this into the other two equations

$$\left. \begin{array}{l} \sigma_x = \lambda'(\varepsilon_x + \varepsilon_y) + 2\mu\,\varepsilon_x, \\ \sigma_y = \lambda'(\varepsilon_x + \varepsilon_y) + 2\mu\,\varepsilon_y. \end{array} \right\} \quad (11.3)$$

These may now be integrated through the plate giving

$$\left. \begin{array}{l} X = \lambda'(E_x + E_y) + 2\mu\,E_x, \\ Y = \lambda'(E_x + E_y) + 2\mu\,E_y \end{array} \right\} \quad (11.4)$$

where E_x, E_y are mean values of the corresponding strains through the thickness of the plate.

The state of stress in which these conditions apply was called by FILON, who first investigated the problem "*Generalised Plane Stress*". If a flat plate satisfies these conditions, then all the results of the analysis of two-dimensional stresses and strains may be applied to the mean values of these quantities through the plate. These include Eqs. (7.6) to (7.20) inclusive, (8.2) and (8.3).

There are also two results of considerable importance which will be merely quoted here, for the derivation of which the reader is referred to the standard text-books on the mathematical theory of elasticity.

The first is that in the absence of body forces the sum of the principal stresses in a two-dimensional distribution satisfies LAPLACE's equation. In terms of mean stresses this is

$$\left(\frac{\partial^2}{\partial x^2} + \frac{\partial^2}{\partial y^2}\right)(P+Q) = 0. \quad (11.5)$$

The second result relates to the effect of the elastic constants of the material upon the stress-distribution.

If a plate is multiply connected, that is if it has a hole, or holes, in it the boundaries of which are not continuous with the boundary of the plate, then, under certain conditions of loading, the stress-distribution is a function of the elastic constants. The result, for practical purposes may be stated: — The stresses in a plate of any shape, pierced by any number of holes, and subjected to given forces over its boundaries, are independent of the elastic constants if, and only if, the force resultants of the forces applied to the boundary of each hole separately vanish.

IV. Theory of photoelasticity.

12. The photoelastic laws. In 1841, F. E. NEUMANN[1] published the first tentative theory of the stress-optical effect. He assumed that the effect was due to changes in the molecular arrangement of the molecules produced by the loads, and accordingly based his theory upon the strains in the medium. His experiments with a rectangular beam in flexure led to the conclusion that the difference of velocities of the two oppositely polarised waves in light passing through the beam in a direction normal to the plane of bending was directly proportional to the difference of the two principal strains in the plane of the wave front.

He therefore started by assuming that the three principal wave velocities, i.e. the velocities of waves whose vibrations were parallel to the principal axes of strain, were represented by expressions containing terms proportional to the three principal strains. These were

$$\left.\begin{array}{l} a = v_0 + \alpha\,\varepsilon_x + \beta\,\varepsilon_y + \gamma\,\varepsilon_z, \\ b = v_0 + \alpha\,\varepsilon_y + \beta\,\varepsilon_z + \gamma\,\varepsilon_x, \\ c = v_0 + \alpha\,\varepsilon_z + \beta\,\varepsilon_x + \gamma\,\varepsilon_y, \end{array}\right\} \quad (12.1)$$

where v_0 was the velocity in the unstrained medium.

Hence

$$a - b = (\alpha - \gamma)\,\varepsilon_x + (\beta - \alpha)\,\varepsilon_y + (\gamma - \beta)\,\varepsilon_z.$$

But his experimental results showed $(a-b)$ to be independent of ε_z and therefore led to $\gamma = \beta$ and

$$a - b = (\alpha - \beta)(\varepsilon_x - \varepsilon_y), \quad (12.2)$$

his three equations then reducing to

$$\left.\begin{array}{l} a = v_0 + \alpha\,\varepsilon_x + \beta\,(\varepsilon_y + \varepsilon_z), \\ b = v_0 + \alpha\,\varepsilon_y + \beta\,(\varepsilon_z + \varepsilon_x), \\ c = v_0 + \alpha\,\varepsilon_z + \beta\,(\varepsilon_x + \varepsilon_y). \end{array}\right\} \quad (12.3)$$

MAXWELL[2] approached the problem from the point-of-view of stress in the medium. The experimental results showed the difference of wave velocities to be proportional also to the difference of principal stresses, and he expressed the principal wave velocities in terms of stresses in equations precisely similar to those of NEUMANN, and in the same way arrived at expressions containing only two coefficients:

$$\left.\begin{array}{l} a = v_0 + C_1\sigma_x + C_2(\sigma_y + \sigma_z), \\ b = v_0 + C_1\sigma_y + C_2(\sigma_z + \sigma_x), \\ c = v_0 + C_1\sigma_z + C_2(\sigma_x + \sigma_y) \end{array}\right\} \quad (12.4)$$

with the difference of principal velocities expressed by

$$a - b = (C_1 - C_2)(\sigma_x - \sigma_y). \quad (12.5)$$

---

[1] F. E. NEUMANN: Abh. Akad. Berlin, Part II **1841**.
[2] J. C. MAXWELL: Roy. Soc. Edinburgh. Trans. **20**, 87 (1853).

Whether the optical effect is, in fact, directly due to stress or to strain is not known, but this question will be discussed later in connection with the properties of photoelastic materials. For all applications of the photoelastic effect at present employed in stress-explorations, observations are confined to materials and conditions in which stress and strain are proportional, and either may be taken as producing the effect. Since, however, loads and mean stresses are the usual quantities employed to specify the conditions of stress in a body, it is generally more convenient to work in terms of the stress-optical coefficient. We shall accordingly continue by examining MAXWELL's stress-optical relations, although precisely similar reasoning may be applied to the strain-optical relations.

If we compare the properties of FRESNEL's ellipsoid and the stress-quadric, we can deduce from considerations of symmetry that the two surfaces must have the same principal axes. Their equations referred to these axes are, for the FRESNEL ellipsoid

$$a^2 x^2 + b^2 y^2 + c^2 z^2 = 1$$

and for the stress quadric

$$\sigma_x x^2 + \sigma_y y^2 + \sigma_z z^2 = 1.$$

In these equations we find the squares of the principal wave velocities related to the principal stresses, and it would seem logical to assume that it may be the squares of the velocities rather than the velocities themselves which should be linear functions of the stresses.

Thus we might start with expressions

$$\left. \begin{aligned} a^2 &= v_0^2 + C_1' \sigma_x + C_2'(\sigma_y + \sigma_z), \\ b^2 &= v_0^2 + C_1' \sigma_y + C_2'(\sigma_z + \sigma_x), \\ c^2 &= v_0^2 + C_1' \sigma_z + C_2'(\sigma_x + \sigma_y) \end{aligned} \right\} \quad (12.6)$$

instead of MAXWELL's expressions of Eqs. (12.4).

If MAXWELL's expressions are squared they give

$$a^2 = v_0^2 + 2v_0 C_1 \sigma_x + 2v_0 C_2(\sigma_y + \sigma_z) + \{C_1 \sigma_x + C_2(\sigma_y + \sigma_z)\}^2$$

and two similar equations, and if we are to neglect terms of the second order in the small coefficients C_1 and C_2, these reduce to the form (12.6) where

$$C_1' = 2v_0 C_1 \quad \text{and} \quad C_2' = 2v_0 C_2. \quad (12.7)$$

Eqs. (12.6) may therefore be taken either as exact relations or as approximate forms of MAXWELL's equations when second order terms are neglected.

They may conveniently be written

$$\left. \begin{aligned} a^2 &= v_0^2 + C_2'(\sigma_x + \sigma_y + \sigma_z) + (C_1' - C_2') \sigma_x, \\ b^2 &= v_0^2 + C_2'(\sigma_x + \sigma_y + \sigma_z) + (C_1' - C_2') \sigma_y, \\ c^2 &= v_0^2 + C_2'(\sigma_x + \sigma_y + \sigma_z) + (C_1' - C_2') \sigma_z. \end{aligned} \right\} \quad (12.8)$$

If we now substitute these values for a^2, b^2, c^2 into the equation for FRESNEL's ellipsoid we get

$$\{v_0^2 + C_2'(\sigma_x + \sigma_y + \sigma_z)\}(x^2 + y^2 + z^2) + (C_1' - C_2')(x^2 \sigma_x + y^2 \sigma_y + z^2 \sigma_z) = 1$$

or

$$\frac{1}{R^2} = v_0^2 + C_2'(\sigma_x + \sigma_y + \sigma_z) + (C_1' - C_2')(l^2 \sigma_x + m^2 \sigma_y + n^2 \sigma_z) \quad (12.9)$$

where R is the radius vector from the centre to the ellipsoid in the direction l, m, n.

The maximum and minimum values of R in (12.9) correspond to minimum and maximum values of the expression $l^2\sigma_x + m^2\sigma_y + n^2\sigma_z$ and this is the expression for the normal stress in direction l, m, n. Therefore the maximum and minimum values of R in any central section of FRESNEL's ellipsoid correspond to minimum and maximum values of the normal stress in that section, and we are thus able to deduce that *the directions of polarisation in any wave-front are the directions of the secondary principal stresses in the plane of the wave-front*. This is the first photoelastic law as derived by MAXWELL.

If we call these secondary principal stresses in a given wave-front σ_p and σ_q, and if R_p and R_q are the lengths of the radii-vectors in their directions, then

and
$$\left. \begin{aligned} \frac{1}{R_p^2} &= v_0^2 + C_2'(\sigma_x + \sigma_y + \sigma_z) + (C_1' - C_2')\sigma_p \\ \frac{1}{R_q^2} &= v_0^2 + C_2'(\sigma_x + \sigma_y + \sigma_z) + (C_1' - C_2')\sigma_q \end{aligned} \right\} \quad (12.10)$$

and we have

$$\frac{1}{R_p^2} - \frac{1}{R_q^2} = (C_1' - C_2')(\sigma_p - \sigma_q). \quad (12.11)$$

But in FRESNEL's ellipsoid

$$\frac{1}{R_p} = v_p \quad \text{and} \quad \frac{1}{R_q} = v_q,$$

v_p and v_q being the velocities of the waves whose vibration directions are parallel to p and q respectively. Therefore,

$$\frac{1}{R_p^2} - \frac{1}{R_q^2} = v_p^2 - v_q^2 = (v_p + v_q)(v_p - v_q) \approx 2v_0(v_p - v_q)$$

so that, neglecting second order terms, we get

$$v_p - v_q = C(\sigma_p - \sigma_q). \quad (12.12)$$

This equation expresses the second photoelastic law as derived by MAXWELL: *The difference of the velocities of the two oppositely polarised waves in any wave-front is proportional to the difference of the (secondary) principal stresses in the plane of the wave-front.*

From the preceding analysis, however, it would appear that if the first law is exact the exact form of the second law should be

The difference of the squares of the velocities of the two oppositely polarised waves is proportional to the difference of the (secondary) principal stresses in the plane of the wave-front.

Since no material has yet been found which exhibits the photoelastic effect sufficiently strongly to enable differentiation to be made between the two forms of the law, MAXWELL's form, being the simpler, is accepted as the basis for all photo-elastic measurements.

13. Passage of polarised light through a stressed plate. Consider a beam of plane polarised light whose vibrations are in direction ox (Fig. 13) incident upon a flat parallel plate to the edges of which are applied forces acting parallel to the faces of the plate, the faces themselves being free from stress.

Let op and oq be the directions of the principal stresses at a point 0 of the plate.

Then, if the original plane-polarised wave be

$$u_x = a \cos pt, \quad (13.1)$$

it will be polarised on entering the plate at 0 into the two waves

$$u_p = a \cos \alpha \cos p t, \\ u_q = - a \sin \alpha \cos p t. \quad (13.2)$$

If these waves travel with velocities v_1 and v_2 respectively, they will emerge, after travelling through the thickness d of the plate, as

and
$$u'_p = a \cos \alpha \cos p \left(t - \frac{d}{v_1}\right) \\ u'_q = - a \sin \alpha \cos p \left(t - \frac{d}{v_2}\right). \quad (13.3)$$

If the light now passes through an analyser "crossed" with the polariser, only the components of these vibrations parallel to oy will emerge, giving the vibration

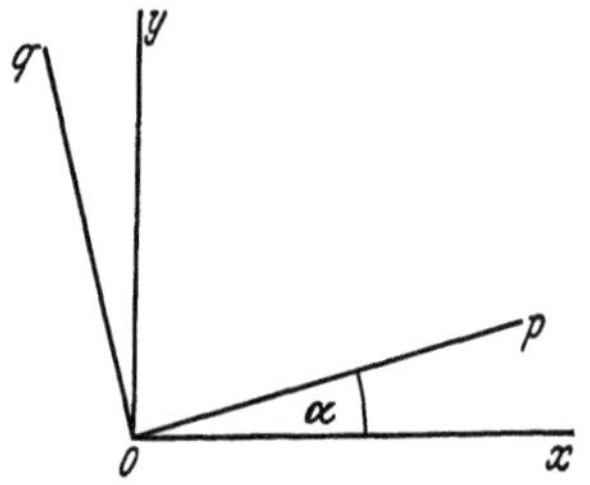

$$u_y = u'_p \sin \alpha + u'_q \cos \alpha, \\ = a \sin \alpha \cos \alpha \left\{\cos p\left(t - \frac{d}{v_1}\right) - \cos p \left(t - \frac{d}{v_2}\right)\right\},$$

i.e.
$$u_y = a \sin 2\alpha \sin p \left(\frac{d}{2v_1} - \frac{d}{2v_2}\right) \times \\ \times \sin p \left(t - \frac{d}{2v_1} - \frac{d}{2v_2}\right). \quad (13.4)$$

Fig. 13.

Let us suppose that the light is monochromatic. The intensity of the transmitted light will then be proportional to the square of the amplitude of this vibration, i.e. to

$$\sin^2 2\alpha \sin^2 p \left(\frac{d}{2v_1} - \frac{d}{2v_2}\right). \quad (13.5)$$

Considering the first factor of this expression, we see that the intensity will in general vary from point to point of the plate with variation in the directions of the principal stresses. If there are regions of the plate at which $\alpha = 0$ or $90°$, the intensity will be zero in these regions, and will increase gradually as we move toward regions in which α has a different value, becoming a maximum where $\alpha = 45°$. Since the rate of change of $\sin^2 2\alpha$ with α is very small in the neighbourhood of $\alpha = 0$ and $\alpha = 90°$, the intensity increases very slowly in regions where the rate of change of direction of the stresses is small, and the dark areas have the appearance of fringes whose edges are more or less diffuse and ill-defined.

These are *isoclinic fringes*. The regions of zero intensity, (generally lines which are approximately the central lines of the fringes), are the locus of points in the plate at which the principal stresses are parallel to the vibration directions of polariser and analyser. If the inclination of polariser and analyser axes to the chosen standard reference direction be φ, the lines observed are referred to as *isoclinic lines of parameter φ*.

If we alter the value of φ by rotating the polariser and analyser, still keeping them "crossed", the isoclinic fringes move continuously across the plate, enabling the observer to determine the directions of the principal stresses at all points. The positions of these fringes are independent of the magnitudes of the stresses and so are not affected by changes in the magnitudes of the applied loads.

Considering the second factor of expression (13.5), we see that the intensity of the transmitted light varies also with $\sin^2 p \left(\dfrac{d}{2v_1} - \dfrac{d}{2v_2} \right)$, becoming zero when

$$p \left(\frac{d}{2v_1} - \frac{d}{2v_2} \right) = 0 \quad \text{or} \quad i\pi,$$

i.e.

$$2\pi \frac{V}{\lambda} \left(\frac{d}{2v_1} - \frac{d}{2v_2} \right) = 0 \quad \text{or} \quad i\pi,$$

where i is an integer, or

$$d(n_1 - n_2) = 0 \quad \text{or} \quad i\lambda \tag{13.6}$$

where n_1 and n_2 are the indices of refraction for the two oppositely polarised waves.

The expressions $d(n_1 - n_2)$ and $d\left(\dfrac{V}{v_1} - \dfrac{V}{v_2} \right)$ are alternative forms for denoting the difference in "optical paths" of the two waves, generally referred to as the *relative retardation* and having the dimensions of a length. We see then that light is extinguished when the relative retardation is zero or an integral number of wave-lengths.

Now

$$d\left(\frac{V}{v_1} - \frac{V}{v_2} \right) = \frac{V d}{v_1 v_2} (v_2 - v_1) \approx \frac{V d}{v_0^2} (v_2 - v_1),$$

where v_0 is the velocity in the unstressed medium, and second order small terms are neglected.

But we have shown that $(v_2 - v_1)$ is proportional to the difference of the principal stresses, P and Q in the plane of the wave-front, and we see therefore that as the stress-difference increases from one point of the plate to another, there will be successive extinctions of the monochromatic light corresponding to equal increments in $P - Q$. These give a second series of dark fringes, the *isochromatic fringes*, which provide a scale of measurement of the magnitude of the principal stress-differences at different points of the plate. If the stresses are sufficiently great there will be a large number of them forming what is usually called the *stress fringe-pattern*, or the fringe-pattern. These fringes are unaffected by rotation of the crossed polariser and analyser.

The total relative retardation of the two waves in passing through a thickness d of the plate at a point where the difference of the mean principal stresses in the plate is $P - Q$ may be expressed as

$$\text{Relative retardation} = C \cdot d \cdot (P - Q), \tag{13.7}$$

where C is a *stress-optical coefficient* for the material of which the plate is made. The dimensions of C are those of the reciprocal of a stress, and FILON introduced a unit which he called the BREWSTER for this quantity. Using this unit, the relation becomes

Relative retardation measured in Å = stress-optical coefficient in BREWSTERS × × thickness in mm × stress difference in megadynes/cm².

This unit is still sometimes used in theoretical work, but for all practical applications it has been replaced by a coefficient called *the fringe-value* of the material. This is denoted by f and represents the stress-difference in lb. per sq. in. which produces a relative retardation of one wave-length in light passing through one inch thickness of the stressed material.

Thus the relation between stress and relative retardation is expressed by $(P-Q)$ in lb. per sq. in.

$= f \times$ relative retardation in wave-lengths $\div$ thickness in inches.

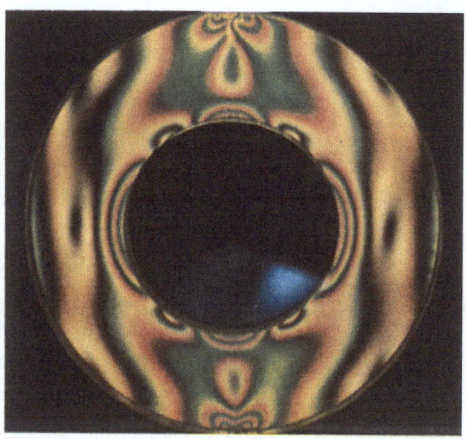

Fig. 14a. A photograph of the isochromatic fringes in a ring subjected to pressure along its vertical diameter. The material is C.R. 39, and the photograph was taken on Kodachrome film in circularly polarised white light.

This unit has the disadvantage that its value for any material varies with the wave-length of light used, so that when quoting the fringe-value of a material the wave-length for which it applies must be specified.

If white light be used instead of monochromatic light the isoclinic fringes will still be black, for their formation depends only upon the directions of the stresses. The isochromatic fringes, however, will now be coloured, except for the zero-relative retardation fringe, for a given stress-difference will give a relative retardation which amounts to an integral number of wave-lengths for certain colours only. These colours will be extinguished, but all other colours of the spectrum will be transmitted with greater or less intensity, giving the light a certain tint.

Fig. 14b. A photograph of the "frozen" stress-pattern in the upper part of one of the araldite plates used in the investigation of Sect. 45. The small holes near the top edge of the plate are those through which the loading pins passed. The fringe patterns around these indicate that the load has been equally distributed among the pins.

Thus as the stress-difference increases from zero, light of all wave-lengths is at first partially transmitted, giving sensibly white light, but with further increase the relative retardation approaches one wave-length of violet light and the transmitted light becomes yellow. This changes through orange and red as the blue and green wave-lengths are successively extinguished, then through a neutral tint corresponding fairly closely to extinction of the yellow of the spectrum, to a deep blue and blue-green as the orange and red wave-lengths are in turn extinguished. This sequence is called the first order colour band, and it is followed by other colour bands each corresponding roughly to one of the black fringes seen in monochromatic light. The colours, however, change from band to band, and it is not possible to identify any particular colour with a definite amount of relative retardation. For example, in the second band a relative retardation of 12000 Å which extin-

14. Use of circularly polarised light.

guishes orange light of 6000 Å for the second time also extinguishes violet light of 4000 Å for the third time, so that the tint of the transmitted light is quite different from that produced at the first extinction of the same orange colour.

An illustration of isochromatic fringe-patterns in white light is shown in the coloured figures. The blue tint in the fringes is absent owing to the yellow colour of the plastic used.

14. Use of circularly polarised light. If in the examination of the stressed plate of the previous section we insert between the polariser and the plate a quarter-wave plate whose vibration axes are inclined at 45° to the vibration axis of the polariser, the light incident on the stressed plate will be circularly polarised.

Then, if the initial plane polarised wave be denoted by

$$u_x = \sqrt{2}\, a \cos pt, \qquad (14.1)$$

it will be resolved on entering the quarter-wave plate into (Fig. 15)

$$\left.\begin{array}{l} u_a = a \cos pt, \\ u_b = - a \cos pt. \end{array}\right\} \qquad (14.2)$$

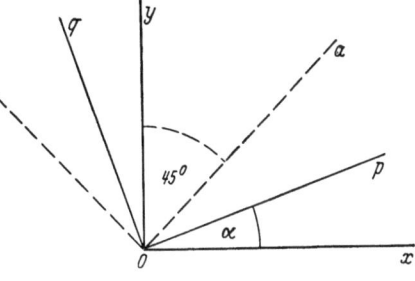

Fig. 15.

In travelling through the quarter-wave plate, one wave will be retarded on the other by a quarter wave-length, or will suffer a phase retardation of $\pi/2$. If then we take the axis oa as the "fast" axis, the waves will emerge as

$$\left.\begin{array}{l} u'_a = a \cos pt, \\ u'_b = - a \sin pt \end{array}\right\} \qquad (14.3)$$

neglecting, as we shall do throughout, changes in phase which are common to the two waves.

If the principal stresses P and Q at a point O of the plate are inclined at an angle α to the polariser and analyser axes, these two waves will be re-resolved on entering the stressed plate at O into

$$\left.\begin{array}{l} u_p = u'_a \cos\left(\dfrac{\pi}{4} - \alpha\right) - u'_b \sin\left(\dfrac{\pi}{4} - \alpha\right), \\ u_q = u'_a \sin\left(\dfrac{\pi}{4} - \alpha\right) + u'_b \cos\left(\dfrac{\pi}{4} - \alpha\right). \end{array}\right\} \qquad (14.4)$$

Substituting the expressions for u'_a and u'_b these reduce to

$$\left.\begin{array}{l} u_p = a \cos\left(pt - \dfrac{\pi}{4} + \alpha\right), \\ u_q = - a \sin\left(pt - \dfrac{\pi}{4} + \alpha\right) \end{array}\right\} \qquad (14.5)$$

which, since we may neglect phase changes common to the two, may be written

$$u_p = a \cos pt, \quad u_q = - a \sin pt. \qquad (14.6)$$

After passing through the plate these emerge as before as

$$u'_p = a \cos p\left(t - \dfrac{d}{v_1}\right), \quad u'_q = - a \sin p\left(t - \dfrac{d}{v_2}\right). \qquad (14.7)$$

Let the light now be made to pass through a second quarter-wave plate "crossed" with the first one, that is with its "slow" axis parallel to oa and its "fast" axis parallel to ob. We then shall have entering the quarter-wave plate

$$\left.\begin{aligned}u_a &= a\cos\left(\frac{\pi}{4}-\alpha\right)\cos p\left(t-\frac{d}{v_1}\right)-a\sin\left(\frac{\pi}{4}-\alpha\right)\sin p\left(t-\frac{d}{v_2}\right),\\ u_b &= -a\cos\left(\frac{\pi}{4}-\alpha\right)\sin p\left(t-\frac{d}{v_2}\right)-a\sin\left(\frac{\pi}{4}-\alpha\right)\cos p\left(t-\frac{d}{v_1}\right).\end{aligned}\right\} \quad (14.8)$$

In traversing this plate, u_a will be retarded relative to u_b by a quarter period, and will emerge as

$$u_a' = a\cos\left(\frac{\pi}{4}-\alpha\right)\sin p\left(t-\frac{d}{v_1}\right)+a\sin\left(\frac{\pi}{4}-\alpha\right)\cos p\left(t-\frac{d}{v_2}\right) \quad (14.9)$$

while $u_b' = u_b$ as in (14.8)

We can analyse the emergent light in either of two ways: First, by setting the analyser "crossed" with the polariser, in which case the light transmitted will be given by

$$u_y = \tfrac{1}{2}\sqrt{2}(u_a' + u_b')$$

which reduces to

$$u_y = \frac{1}{2}\sqrt{2}\,a\sin p\left(\frac{d}{2v_2}-\frac{d}{2v_1}\right)\cos\left\{p\left(t-\frac{d}{2v_1}-\frac{d}{2v_2}\right)-\left(\frac{\pi}{4}-\alpha\right)\right\}.$$

That is

$$u_y = \frac{1}{2}\sqrt{2}\,a\sin p\left(\frac{d}{2v_2}-\frac{d}{2v_1}\right)\cos(pt-\beta). \quad (14.10)$$

The amplitude of this wave is zero when

$$p\left(\frac{d}{2v_2}-\frac{d}{2v_1}\right)=0 \quad\text{or}\quad i\pi,$$

where i is an integer.

In monochromatic light, the field of the polariscope will be dark, with dark fringes occurring at points where the relative retardation is zero or an integral number of wave-lengths, as in the case of the plane-polariscope, but now there are no isoclinic fringes.

The second way of analysing the emergent light is to set the analyser "parallel" with the polariser, that is with its vibration axis parallel to ox. In this case the light transmitted will be given by

$$u_x' = \tfrac{1}{2}\sqrt{2}(u_a' - u_b')$$

which reduces to

$$u_x' = \frac{1}{2}\sqrt{2}\,a\cos p\left(\frac{d}{2v_2}-\frac{d}{2v_1}\right)\sin(pt-\beta). \quad (14.11)$$

Here the amplitude is zero when

$$p\left(\frac{d}{2v_2}-\frac{d}{2v_1}\right)=(2i-1)\frac{\pi}{2}$$

and a maximum when

$$p\left(\frac{d}{2v_2}-\frac{d}{2v_1}\right)=0 \quad\text{or}\quad i\pi,$$

with i again an integer.

Thus the field is bright, and the dark fringes occur at points where the relative retardation is an odd number of half wave-lengths.

Sect. 15. The TARDY and SÉNARMONT methods. 153

There are two other possible arrangements of the circular polariscope in which the same two alternative positions of the analyser occur, but with corresponding axes of the quater-wave plates parallel to each other instead of crossed. One of these gives a light field and half wave-length fringes and the other a dark field and whole wave-length fringes exactly as in the two cases analysed above. The cases may be summarised:

1. Dark field and whole wave-length fringes given by (a) polariser and analyser crossed and quarter-wave plates crossed or (b) polariser and analyser parallel and quarter-wave plates parallel.

2. Light field and odd half wave-length fringes given by (c) polariser and analyser crossed and quarter-wave plates parallel or (d) polariser and analyser parallel and quarter-wave plates crossed.

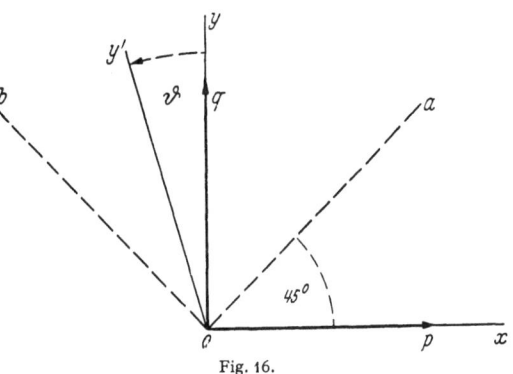

Fig. 16.

15. **The TARDY and SÉNARMONT methods of measuring fractional relative retardations.** If in the "crossed" circular polariscope of Sect. 14 we were to make $\alpha = 0$, by setting the polariser axis parallel to the direction of the principal stress P at the point under examination in the stressed plate, the analysis of Sect. 14 would be unaltered until we reach Eqs. (14.8). These will become, for the waves entering the second quarter-wave plate

$$u_a = \tfrac{1}{2}\sqrt{2}\,a\left\{\cos p\left(t - \tfrac{d}{v_1}\right) - \sin p\left(t - \tfrac{d}{v_2}\right)\right\}, \quad (15.1)$$
$$u_b = -\tfrac{1}{2}\sqrt{2}\,a\left\{\cos p\left(t - \tfrac{d}{v_1}\right) + \sin p\left(t - \tfrac{d}{v_2}\right)\right\}$$

and after traversing the plate, u_a being retarded relative to u_b we shall have

$$u'_a = \tfrac{1}{2}\sqrt{2}\,a\left\{\sin p\left(t - \tfrac{d}{v_1}\right) + \cos p\left(t - \tfrac{d}{v_2}\right)\right\}, \quad (15.2)$$
$$u'_b = u_b \quad \text{as in (15.1)}$$

If now we rotate the analyser axis through an angle ϑ from its crossed position to oy' (Fig. 16), and resolve the waves u'_a, u'_b parallel to oy' we shall have transmitted by the analyser

$$u_{y'} = u'_a \cos\left(\tfrac{\pi}{4} + \vartheta\right) + u'_b \cos\left(\tfrac{\pi}{4} - \vartheta\right)$$
$$= \tfrac{1}{2}\sqrt{2}\,a\left[\cos\left(\tfrac{\pi}{4} + \vartheta\right)\left\{\sin p\left(t - \tfrac{d}{v_1}\right) + \cos p\left(t - \tfrac{d}{v_2}\right)\right\} - \cos\left(\tfrac{\pi}{4} - \vartheta\right)\left\{\cos p\left(t - \tfrac{d}{v_1}\right) + \sin p\left(t - \tfrac{d}{v_2}\right)\right\}\right]. \quad (15.3)$$

Expanding the $\cos\left(\tfrac{\pi}{4} + \vartheta\right)$ and $\cos\left(\tfrac{\pi}{4} - \vartheta\right)$ we get

$$u'_y = \tfrac{1}{2}a\left[\sin\left\{p\left(t - \tfrac{d}{v_1}\right) - \vartheta\right\} - \sin\left\{p\left(t - \tfrac{d}{v_2}\right) + \vartheta\right\} - \cos\left\{p\left(t - \tfrac{d}{v_1}\right) - \vartheta\right\} + \cos\left\{p\left(t - \tfrac{d}{v_2}\right) + \vartheta\right\}\right] \quad (15.4)$$

and combining the sine and cosine terms this becomes

$$u_{y'} = a \sin\left\{p\left(\frac{d}{2v_2} - \frac{d}{2v_1}\right) - \vartheta\right\}\left[\cos p\left(t - \frac{d}{2v_1} - \frac{d}{2v_2}\right) + \sin p\left(t - \frac{d}{2v_1} - \frac{d}{2v_2}\right)\right]$$

or

$$\sqrt{2}\, a \sin\left\{p\left(\frac{d}{2v_2} - \frac{d}{2v_1}\right) - \vartheta\right\} \cos(pt - \gamma), \qquad (15.5)$$

where

$$\gamma = \frac{d}{2v_1} + \frac{d}{2v_2} + \frac{\pi}{4}.$$

The amplitude of this vibration is zero when

$$p\left(\frac{d}{2v_1} - \frac{d}{2v_2}\right) = \vartheta \quad\text{or}\quad \vartheta + i\pi \qquad (15.6)$$

which, as shown in Eqs. (13.6) is equivalent to

$$\text{Relative retardation} = d(n_2 - n_1) = \frac{\vartheta}{\pi}\cdot\lambda \quad\text{or}\quad \left(i + \frac{\vartheta}{\pi}\right)\lambda. \qquad (15.7)$$

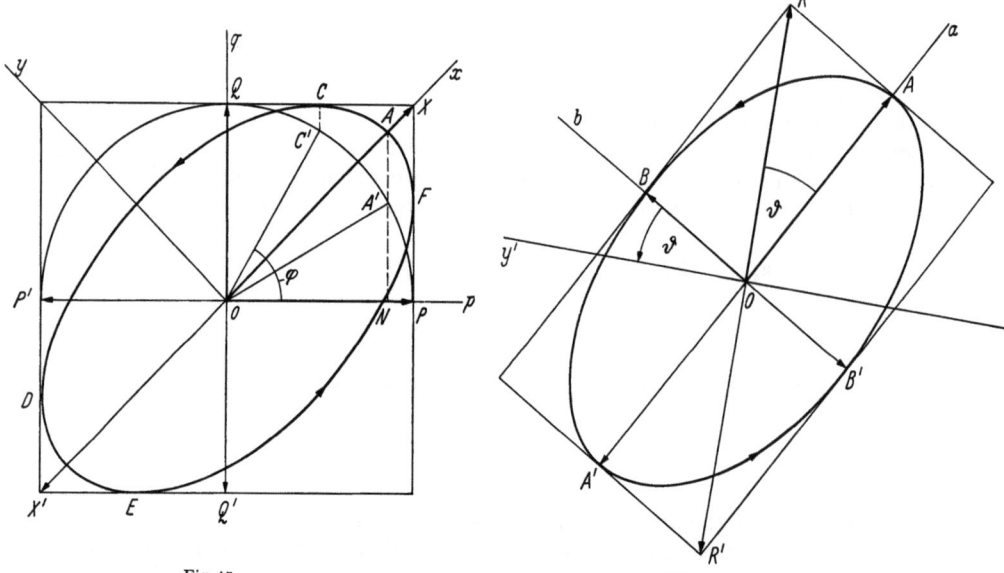

Fig. 17a.   Fig. 17b.

There will thus always be an angle ϑ, between 0 and π for which extinction of the emergent light will be produced at any point of the plate. The relative retardation at the point will then be given by (15.7), in which the value of the integer i can be determined by examination of the fringe-pattern in the plate.

This is the TARDY method of measurement.

A modification of this method, due to SÉNARMONT, involves the use of only one quarter-wave plate. The graphical analysis of this which follows is of interest as it presents a clear picture of the manner in which the quarter-wave plate operates.

In this method plane polarised light is passed directly into the stressed plate, the vibration axis of the polariser being set at 45° to the axes of principal stress at the point of the plate under examination.

If we represent the plane polarised wave entering the plate by a linear harmonic vibration of amplitude OX in direction x (Fig. 17a), this will be resolved in the plate into vibrations OP and OQ, initially in phase with each other.

If on emergence from the plate the wave OQ has been retarded relative to OP by an angular phase difference φ or $(2i\pi + \varphi)$, the two vibrations will combine to give the elliptic vibration $CDEF$ where the angle POC' is equal to φ.

The light is now passed through a quarter-wave plate, whose axes oa and ob are parallel to ox and oy (Fig. 17b). The elliptic vibration is resolved in these directions on entering the quarter-wave plate, producing vibrations OA and OB the phase of OA being $\pi/2$ or a quarter of a period in advance of that of OB.

If oa is the "slow" axis of the quarter-wave plate, the two waves will be in phase again on emergence, producing the linear vibration OR. This will be extinguished if the analyser vibration axis is in direction oy', inclined at an angle ϑ to its original "crossed" position oy (Fig. 17a).

From the symmetry of the elliptic vibration the angle POA' in Fig. 17a is seen to be equal to $\tfrac{1}{2}\varphi$ and the triangles NOA' and AOR in Figs. 17a and b respectively can be shown to be similar. Hence $\vartheta = \tfrac{1}{2}\varphi$ and the angle through which the analyser must be turned for extinction is, as in the TARDY method, one half the fractional angular retardation occurring in the stressed plate.

The operation of the second quater-wave plate in the TARDY method converts the elliptic vibration emerging from the stressed plate into a linear vibration in exactly the same way as is illustrated above, and a similar geometrical analysis could, of course, be made of that method.

B. Exploration of two-dimensional stress systems.

I. Photoelastic equipment.

16. The lens system. The photoelastic effect may be observed qualitatively with the aid of very simple equipment. Thus a parallel plate model may be held between two sheets of polaroid in front of any diffused light source and viewed by eye, when the stress-patterns produced by the application of loads may be observed. If, however, accurate observations and measurements are to be made, care must be taken to ensure that the paths of the light rays through the model approach very nearly the ideal condition assumed in the theory, i.e. that they are approximately normal to the plane of the plate. Also, if extended measurements are to be made, it will be necessary to include a means of measurement, and mechanical devices to facilitate the adjustment of the polaroids and other optical elements.

There are three possible methods of observation in general use, and each has advantages for the observation of different types of stress-distribution. They are: (1) direct observation by eye, or though a microscope or telescope; (2) observation of an enlarged image of the model projected on to a screen; (3) photography. In each of these methods we are dependent upon a lens to give us the image observed, and we must consider the way in which the lens forms an image of the fringe-pattern[1].

Ignoring errors due to lens aberrations, we may assume that with a single lens the cone of convergent rays which produce an image at a point I of the screen all diverge from some point O in the model (Fig. 18).

The image at I will therefore represent some sort of mean-value of the relative retardations which have taken place in all these rays, and therefore will give a

---

[1] H. T. JESSOP: J. Sci. Instrum. **25**, 124 (1948).

measure of the mean-value of the stresses operating in the region of the model through which the rays of the cone pass.

If then the image at I is to record with reasonable accuracy the stresses on a line through O normal to the plane of the model, two conditions must be satisfied. The angle α subtended by the lens at the model must be small, and, if the whole model is to be projected, the angle β subtended by the model at the lens must also be small.

In direct observation by eye, the first condition will usually be satisfied unless the model is a very thick one. The second condition need not be considered if observations are confined to a very small region of the model at a time, the eye being moved so that the rays from the point observed are always sensibly normal to the model. In photographing the whole model the angle β can be kept small only by placing the camera lens at a distance from the model. It will usually be found that this also makes the angle α sufficiently small.

To obtain an image in this way with a single lens necessitates the use of a source of light large enough to cover the area of the model, and such a source generally gives insufficient intensity to produce a good image on a screen. The standard pattern of photoelastic bench, which can be adapted to any of the methods of observation, employs a compact source of light of high intensity, with a lens or system of lenses by means of which the light beam is collimated so that the axes of all the small cones of light passing through the model are parallel. The angle of these cones will be defined by the size of the source and its distance from the collimating lens.

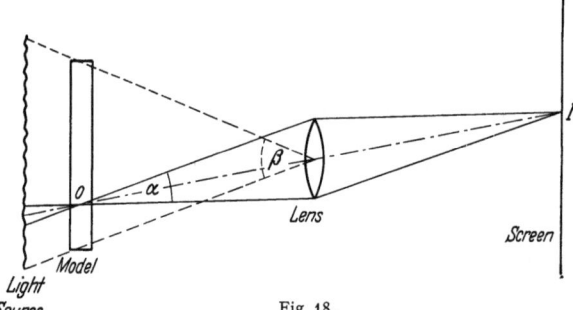

Fig. 18.

If the source S is of diameter d, (Fig. 19), and is situated at the principal focus of the lens L, then the collimated beam will consist of cones of light of angle $\alpha \approx d/f$ whose axes are all parallel to the axis of the lens. The only rays passing through any point O of the model will be those comprising one of these cones, and these will form the image of O on the screen. Thus if the ratio d/f is kept small the conditions for accurate localisation of the rays are ensured at the source, and it is immaterial what method of observation is employed.

Fig. 20 shows diagrammatically the optical system of a standard projecting bench. The short focus lens L_1 produces an image of the source S at S_1, and an iris diaphragm D at that point allows of adjustment of the size of the effective source supplying light to the collimating lens L_2. The polariser P, analyser A, and quarter-wave plates Q_1 and Q_2 are placed with the model M as shown, in the collimated beam. A field lens L_3 brings the parallel cones of light into a projecting lens L_4 which forms the image on a screen or on a photographic plate. If distortion of the image is to be avoided, the field lens L_3 must be a well corrected compound lens.

Light sources. The most efficient source of monochromatic light for a projecting bench is a high pressure mercury arc lamp, with colour filters to remove the yellow and violet light. This gives sensibly monochromatic green light of wave-length 5460 Å. The source is small and very intense, the arc of a 250 watt lamp being only about ¼ inch long and giving ample intensity for a field of 4 to 6 inches diameter.

For a white source a compact source filament projection lamp of 250 watt is quite adequate, but such a source is not really needed if a mercury lamp is employed, as the latter, used without colour filters, gives sufficient colour effects for a qualitative exploration of the isochromatic fringes.

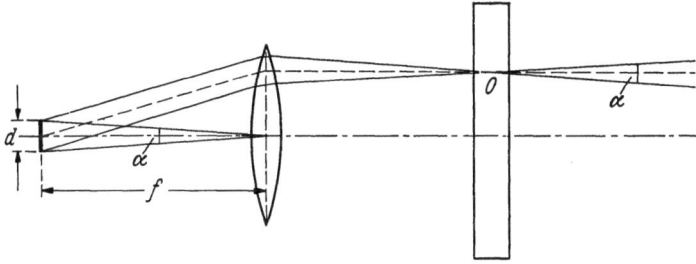

Fig. 19.

17. Quarter-wave plates and polarisers. Quarter-wave plates are made by splitting sheets of mica by hand to the required thickness. Since this thickness is only from 0.001 to 0.002 inch, and since the optical quality and the birefringence of mica vary greatly from one sheet to another, the accuracy of such plates varies considerably and it is rarely that one finds a plate which has not a measurable error.

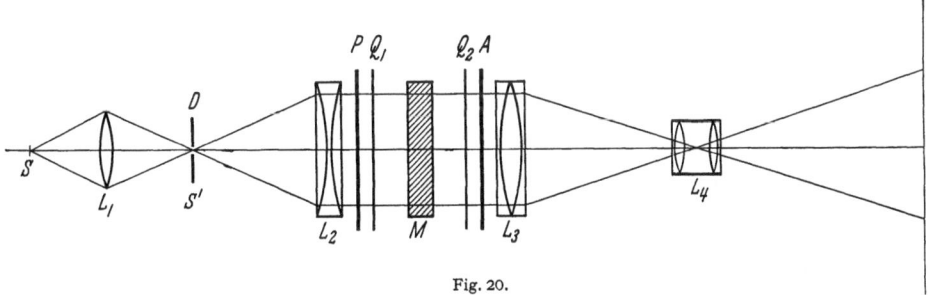

Fig. 20.

In the measurement of fractional relative retardations by the TARDY or SÉNARMONT method, the quarter-wave plate errors lead only to errors of the second order in the measurements, their greatest effect being to give merely a minimum in light intensity instead of total extinction on the final setting of the analyser[1]. The greater the wave plate error, however, the more difficult it is to identify this minimum exactly, and the greater is the probable error of the observation.

The errors in a fringe-pattern observed in a "circular" polariscope vary from point to point according to the relative directions of the principal stresses in the model and the quarterwave plate axes. At points where the stresses are inclined at 45° to these axes, the errors produced will be of the second order, and generally negligible, but where the stresses are parallel to the axes the error produced will be a maximum and will be equal to the algebraic sum or difference of the quarter-wave plate errors according as the quarter-wave plates have their corresponding axes parallel or crossed. If therefore the plates have errors of the same sign they should always be used crossed [arrangements (a) and (d) of

---

[1] H. T. JESSOP: Brit. J. Appl. Phys. **4**, 138, (1953).

Sect. 4], and if of opposite sign they should be used parallel [arrangements (b) and (c)].

Polarisers are almost invariably made of sheets of polaroid with celluloid mounting. There are various grades of polaroid and it is advisable to use the best; i.e. that which gives most complete extinction of light when "crossed".

Both polaroids and quarter-wave plates should be placed between stress-free glass plates for protection, and some fluid such as liquid paraffin or glycerin should be introduced in these assemblies to reduce loss of light by reflection at the intermediate faces. Each element is then mounted in a graduated circle which can be rotated about the optical axis of the polariscope.

Since the quarter-wave plates are not needed for some observations, it is desirable to have some convenient way of moving them into and out of position. A suitable method is to have these elements pivoted so that they swing out of the field when not required, and can be brought back into position without any adjustment.

For exploration of the isoclinic fringes it is practically a necessity to have the polariser and analyser "coupled" so that they can be rotated together. This coupling, however, must be of such a type that it can be disconnected when the analyser alone has to be rotated, as in TARDY or SÉNARMONT measurements. It is also very desirable that it shall not restrict movement of the polariser and analyser along the axis of the polariscope.

Fig. 21. Spring loaded straining frame. (a) Loading nut; (b) Connecting link for variable lever-arm ratios.

18. Straining frames. For general photoelastic work it is convenient to have incorporated in the bench some kind of straining frame which can be adapted to apply accurately measured loads of varying types and magnitudes to models of varying shapes and sizes. The frame should also be provided with a means of traversing it vertically and horizontally in a plane normal to the light beam so that different parts of the model under load may be brought into the field of observation.

There are several different methods of providing the necessary loads:

1. Dead weight loading by standard weights, either directly or through a lever. For models requiring a very small load this is sometimes the most accurate

method, but if the loads required are large, it is inconvenient. It has also the disadvantage that loads can only be varied by finite steps, and that it is not possible to observe continuously the changes in the optical pattern as the load is increased.

2. *Dead weight loading by water-tank and lever.* In this method the flow of water into and out of a tank is regulated by valves, and the loading can thus be carried out smoothly and from a distance, permitting of continuous observation. The tank is provided with gauges which enable accurate measurement of the applied load. The method has, however, the disadvantage of any deadweight loading, that the whole load has to be moved when the straining frame is traversed vertically or horizontally, and this gives rise to a certain amount of oscillation

3. *Spring loading through a lever.* In this method the spring, calibrated to record load accurately in terms of extension, is attached to the frame itself. The spring is extended by means of a nut on a threaded rod, and this can be operated smoothly, by means of a remote control if desired. Such a loading device is simple to construct and free from trouble in operation. Fig. 21 shows a spring loading frame arranged to apply load to a tension bar. In this straining frame two bell-crank levers are mounted on ball bearings and connected by a link having four possible positions (numbered 1 to 4 in the figure), so giving four different lever-arm ratios and ensuring accuracy of measurement over a wide range of loads.

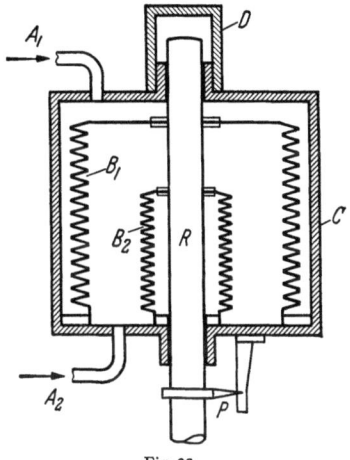

Fig. 22.

4. *Loading by hydraulic or pneumatic pressure.* Hydraulic rams are sometimes used when very large loads are required. They are generally inefficient for small loads owing to the amount of friction inevitably present. Air pressure may be employed through the medium of a "metal bellows" as illustrated diagrammatically in Fig. 22. The loading rod, R, moves vertically in guides at the top and bottom of the brass cylinder C, and is secured by airtight seals to the tops of two corrugated cylindrical bellows B_1, B_2 made of phosphor bronze. The lower ends of these are sealed to the base of the outer cylinder, and the top of the cylinder is sealed by the cap D. Two alternative air inlets A_1, A_2 allow the entry of air at high pressure to exert downward or upward force on the top of the bellows B_1. The only purpose of the second bellows is to act as a seal to the lower guide. By attaching suitable end fittings to the bottom of the loading rod, vertical pressure or tension may be applied to a model. The device is calibrated to give the applied load in terms of the air pressure recorded on a gauge. Errors due to the resistance offered by the bellows to extension or compression are eliminated by adjusting the height of the whole cylinder in the straining frame so that, when readings are taken, the bellows are in their unstrained position as indicated by the pointer P registering upon an index mark.

The chief advantage which this method of loading gives is that it enables a direct pressure to be easily applied to a model, a type of loading very difficult to achieve with accuracy by any method of loading through a lever.

19. Construction of the photoelastic bench. The photograph of Fig. 23 shows a photoelastic bench designed to give all the facilities needed for systematic stress-explorations. The components of the optical system are not shown in the positions they occupy when the bench is in use but have been separated for clearness.

On the left are the two lamp housings for white and mercury lamps, either of which may be brought into operation by a rotation of the platform on which they are mounted. The small condensers bring the light source to a focus at the iris diaphragm which will be at the principal focus of the collimating lens. The polaroids and quarter-wave plates are all mounted on ball-bearings, the quarter-wave plate mountings being pivoted to allow of their being swung out of the field when not in use.

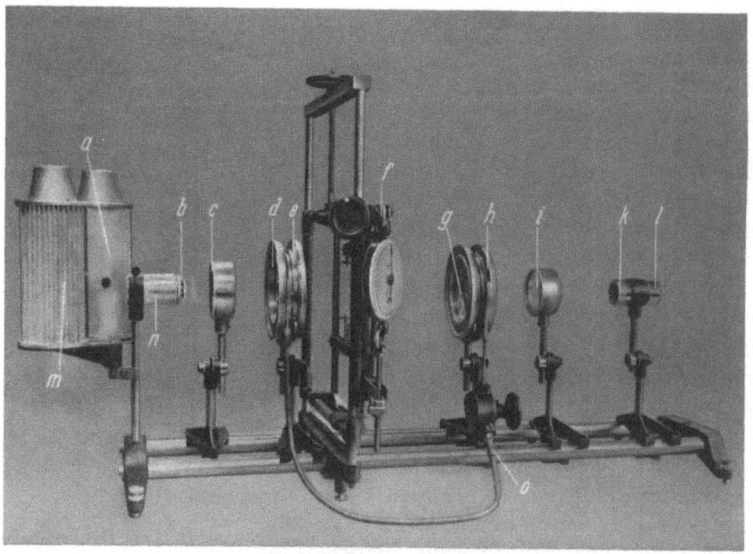

Fig. 23. The JESSOP-LEECH standard projecting photoelastic bench. (a) Lamp-housing for white light source; (b) Iris diaphram; (c) 5″ dia. collimating lens; (d) Polariser; (e), (g) Quarter-wave plates; (f) Straining frame; (h) Analyser; (i) Triple achromatic collimating lens; (k) Green filter; (l) Projecting lens; (m) Lamphousing for high-pressure mercury arc lamp; (n) Condensing lens; (o) Flexible drive connecting polaroids. (The optical elements are shown spaced out for clearness.)

The polariser and analyser are rotated by worm-wheel drives operated through a bevel-gear by a hand-wheel on the analyser stand. A flexible shaft is used to operate the polariser and this may be disconnected by turning a cam on the polariser mount which throws the worm drive out of mesh. The flexible drive allows good freedom of movement of the polariser and analyser assemblies, thus permitting easy access to the model for adjustment, while allowing the analyser to be brought very close to the model as it usually needs to be for projection purposes.

Two useful features are not shown in the photograph. The first is a remote control for operation of the polaroids. A telescopic shaft may be connected to the hand-wheel on the analyser stand enabling adjustments to be made from the end of the bench while the operator is observing the image on a camera-screen. The second is a loose circle on the analyser mounting. Graduated with 200 divisions round the circle, this enables the operator to read fractional relative retardations directly in hundredths of a wave-length when using the TARDY or SÉNARMONT method.

The bench shown has a field of 4 inch diameter, and this is probably the smallest convenient size for general purposes. Although the whole area of a large model cannot be viewed at one time, the traversing movements on the straining frame allow of observation of the model in sections, and this very rarely gives rise to any serious inconvenience. Benches of larger field can of course be constructed, but the cost of quarter-wave plates and fully corrected lenses rises very steeply as the size of the field becomes larger.

No provision is usually made on the bench itself for projection screen or other method of observation, since the conditions which may be required are very varied. Also, in most cases in which an enlarged projected image is used, the screen needs to be at such a distance that provision for it on the bench would necessitate a very long base.

In addition to the usual methods of photography and projection of an image on to an opaque screen there are three methods of observation and measurement which are useful for special purposes.

For tracing isoclinic lines it is very convenient to have the image projected on to a translucent horizontal screen. This is readily done by use of a reflecting box as shown in Fig. 24. The image is focussed on a sheet of tracing material which rests on a sheet of plate glass. In conjunction with remote control of the polaroid rotations this leads to easy and accurate sketching of the fringes.

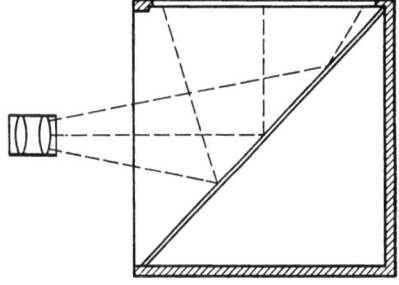

Fig. 24.

For the detailed examination of both isoclinics and isochromatics in regions where the stresses are changing rapidly from point to point, one may use a travelling microscope to examine an enlarged image. No screen is employed, but a low powered microscope provided with vertical and horizontal movements is focused upon the real image formed by the projecting lens.

For very accurate highly localised observations when using the TARDY or SÉNARMONT method, the condition of minimum light intensity at a given point may be determined by use of some type of photo-electric cell. The most convenient method of using this is to have the cell[1] mounted behind an opaque projection screen which is provided with vertical and horizontal traversing screws. A small pin hole in the screen allows light to fall upon the grid of the cell. An enlarged image of the model is projected on to the screen, and the latter is traversed until the pin hole is at the point of the image at which measurements are to be made. A galvanometer records the intensity of light falling on the cell, and the polariscope is adjusted to give a minimum deflection on the instrument. For satisfactory use of this method the voltage of the circuit supplying the light source must be stabilised. The method is capable of giving readings of relative retardation to 0.002 wave-length, but such a degree of accuracy is of course only justified if all the other factors involved (e.g. loads, optical properties and dimensions of model, etc), are known to an equal degree of accuracy.

II. Photoelastic materials.

20. Properties of materials. The ideal photoelastic material should possess a number of special properties. First and most important of these is that it must show a linear relation between stress and birefringence over a reasonably large range of stress. Secondly its stress-optical sensitivity must be such that

[1] A. F. C. BROWN and V. M. MICKSON: Brit. J. Appl. Phys. **1**, 39 (1950).

the optical effects observed can be measured to a good degree of accuracy. Its YOUNG's modulus should be high, so that distortions under load are small, and its ultimate strength should be great enough for it to bear loads up to the limit of the linear stress-optical range. The material must be sensibly homogeneous, both optically and mechanically, it must be obtainable in reasonably large sheets, and it must be readily machinable. It is also desirable, of course, that it should not be too expensive.

There are two defects from which all photoelastic materials except glass appear to suffer to a greater ot less extent. These are known as *time-creep* and *time-edge-effect*.

If a bar photoelastic material be subjected to pure tension under a constant load, both its extension and the optical relative retardation produced in it are found to change with time. The rate and amount of this time-creep in any material increases with temperature, and some materials with a high creep rate cannot be used satisfactorily in hot weather.

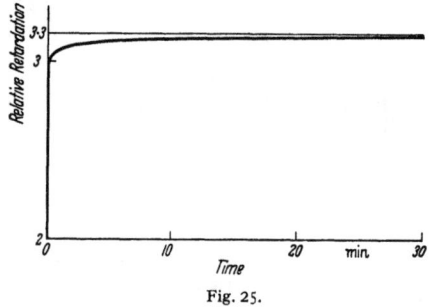

Fig. 25.

For most materials both instantaneous effect and rate of time-creep are proportional to the applied load, so that at any instant after loading the optical effect at any point will be proportional to the stress at the point and the fringe-pattern will still indicate the correct stress-distribution. The effect of the creep will then be to cause the effective fringe-value of the material to change with time. Fig. 25 shows a typical creep curve for a material under constant load. It will be seen that the greater part of the creep takes place in the first few minutes after loading, and that after a short time the rate of creep becomes so small that errors due to it will be negligible if observations are delayed until that time.

Time-edge-stress is produced by either absorption or evaporation of moisture at the free edges of a model, especially at freshly machined edges. In some materials moisture is absorbed causing the material at the egdes to swell and thus producing a compression parallel to the edges, which creates a compensating tension in the adjacent material farther from the edge. Other materials lose moisture, producing the opposite effect. The development of time-edge-effect may be considerably reduced, if not prevented, by keeping models in a suitably controlled atmosphere, but that is generally impracticable while a model is being examined.

It should be noted that the properties of all the synthetic resins are liable to vary with variations in the manufacturing processes, especially with the amount of "maturing" the materials have undergone. Both creep and edge-effect are more pronounced in insufficiently matured materials, and they may frequently both be reduced by a number of hours of further "curing" at a temperature of about 50° C. Such curing will usually alter both the mechanical properties and the fringe value.

There are very many photoelastic resins at present in use, and new ones are continually being developed. They appear often under different names in different countries and there are frequently many different grades of material known by the same name. It is therefore impossible give the properties of these resins exactly, but we may note the general characteristics of the various groups of materials, and the approximate values of their mechanical and optical properties.

α) Glass. Optically and mechanically an ideal material, but has very poor stress-optical sensitivity, is difficult to machine and very liable to brittle fracture [1].

β) Nitro-cellulose (Xylonite, celluloid, etc.). Moderate stress-optical sensitivity and easy to machine. Has moderate time-creep which is not strictly proportional to stress. Develops edge-stress through loss of volatile constituents. Moderate strength [1].

γ) Phenol-formaldehyde resins[1,2] (Catalin, Dekorit, Trolon, etc). Have high stress-optical sensitivity and are easy to machine. Have large time-creep, especially in hot weather. Develop large edge-stress due to loss of volatile constituents. Moderate strength.

δ) Ethoxyline resins[2,3,4] (Araldite, Epoxy, etc.). Have a high stress-optical sensitivity, combined with good mechanical properties and strength. Creep and edge effects are both small. Optical homogeneity is not always good.

ε) Bakelite BT/61/893 [3]. A resin produced especially for photoelastic work. Has moderate stress-optical sensitivity combined with high strength and good optical and mechanical properties. Subject to moderate time-edge stress.

ζ) Columbia resin—C.R. 39[5]. A special type of alkyd resin. Has moderate stress-optical sensitivity and moderate mechanical strength. Shows moderate amount of creep and time-edge effect.

η) Acrylic resins (Perspex, Plexiglass, Lucite, etc.). Generally have poor stress-optical sensitivity, but good optical and mechanical properties and good strength. Show small amounts of creep and time-edge stress.

With the exception of the ethoxyline resins all these materials are supplied in flat sheets ready for use. The ethoxyline materials are supplied in the form of a resin, either liquid or solid and a hardener. These are mixed and heated in order to cast sheets of the material. The mechanical and optical properties may be varied to some extent by varying the temperature conditions during the casting and maturing of the sheets.

Table 1 gives very approximate values of the mechanical and stress-optical properties of the most frequently used photo-elastic materials. The figures given may be taken as indicating the relative orders of the various quantities.

Table 1. *Properties of materials for loading at room temperature.*

Material	Young's modulus E lb/sq. in.	Tensile strength lb/sq. in.	Fringe value f lb/sq in./fr/ in. $\lambda = 5460$ Å (Hg green)	Fringe-strain ratio E/f (Figure of merit)
Glass	9×10^6	12×10^3	800 upward	10000
Xylonite	3.5×10^5	7×10^3	300	1100
Catalin 800	1.9×10^5	5×10^3	50	4000
Araldite B	4.5×10^5	15×10^3	56	8000
Bakelite (BT/61/893)	6.3×10^5	17×10^3	84	7500
Columbia Resin (C.R. 39)	3×10^5	6×10^3	90	3300
Perspex	4.5×10^5	7×10^3	500	900

21. Preparation of models. *α) Annealing.* Most photoelastic materials are liable to have initial internal stresses due to processes of manufacture, and

---

[1] C. MYLONAS: Proc. 7-th Int. Congr. Appl. Mech. 1948.
[2] E. MÖNCH: Ing.-Arch. **16** (1948).
[3] M. BALLET and G. MALLET: C. R. Acad. Sci., Paris **233**, 16 (1951).
[4] H. SPOONER and L. D. McCONNEL: Brit. J. Appl. Phys. **4**, 181 (1953).
[5] D. J. COOLIDGE: Proc. Soc. Exp. Str. An. **6**, No. 1 (1948).

where these are thermal gradient stresses they may be removed by annealing. This involves heating the material to a temperature at which it undergoes a partial softening (the temperature varying from one material to another) and then allowing it to cool very slowly. All the plastics have very low thermal conductivities, so rates of cooling of from one to five degrees centigrade per hour are needed to free the material from thermal gradient stresses. This annealing should be done on the plates of material before the models are cut, since it is liable to produce edge-effects.

β) *Shaping the model.* A large part of the success of a photoelastic investigation depends upon the care and accuracy with which the model is made. It is especially important that the edges of the model shall be free from stresses due to machining, for it is at the edges that accurate measurement of the stresses is most important.

With the exception of glass, all of the photoelastic materials mentioned in the previous section may be shaped by the processes normally used in engineering practice. The material may be sawn, drilled, filed, turned or milled, provided certain precautions are taken which are generally not necessary when machining metals. The most important considerations are to avoid overheating and strain of the material at the worked surface. To this end the action of the tools used should always be a definite cutting action. The cutting edges must be sharp and have a good rake, and ample clearance should be left so that there is no "rubbing" effect and so that the material cut off may escape freely and not clog the tool. The speed of cutting may be quite high, but the depth of cut taken must be kept very small and the tool must not operate for more than a very short time at any one point of the model. The use of an air blast helps to cool the material and to prevent clogging of the tool.

For the shaping of two-dimensional models from flat plates the most useful tool is a pantograph engraving machine using a small end-milling cutter. A pattern is first made by hand in wood, metal or xylonite, and the model is cut from this on a reduced scale, the cutter being taken repeatedly round the boundary of the model making a succession of shallows cuts. The plate from which the model is being cut should be firmly supported on a sheet of hardboard.

For the drilling of holes a special technique is necessary since ordinary twist-drills invariably produce machining stresses around the hole. There are two methods of avoiding this. One is to drill first a pilot hole with a twist drill, a little smaller than the hole required and then to use a centre-pin cutter with a milling action to enlarge the hole. The other is to secure the model to a face-plate in a lathe and to turn the hole with a cutting tool.

It will usually be found that immediately after cutting the model shows optical effects indicative of stress near its edges. These are thermal effects, and they may persist for a short time, but if the machining has been carefully carried out they should completely disappear.

The development of time-edge stresses may be considerably reduced by smoothing and polishing the machined surfaces, but this processes must be carried out with great care if permanent stresses due to friction are to be avoided. Fine emery cloth and polishing pads may be used, but with very light pressures and only for very short periods of time.

22. Calibration of the material. Any new photoelastic material must be calibrated, not only to find its fringe-value but also to establish the range over which the stress-optical relation is linear, and to obtain an estimate of the magnitude of any time-creep which may occur.

Sect. 22. Calibration of the material.

The usual way of doing this is by means of tests on a bar of uniform cross section under simple tension. Fig. 26 shows a suitable type of tension test-bar.

A flat bar is cut from a plate of thickness t. The central part of the bar is of uniform with b, and if the stress at the mid-point is to be uniform, the length l of this parallel part must be at least equal to $2b$. Circular fillets run smoothly out from this part to widened ends, which are bolted between pairs of metal plates, paper gaskets being inserted between the model and the plates. The tensile load is applied through pins passing through holes in the plates. The bar is subjected to a series of loads, first increasing and then decreasing, and the relative retardation at its mid-point is measured for each load. If, when relative retardation is plotted against load, a straight line is obtained, then creep is sensibly absent, and the fringe-value may be deduced from the slope of the line.

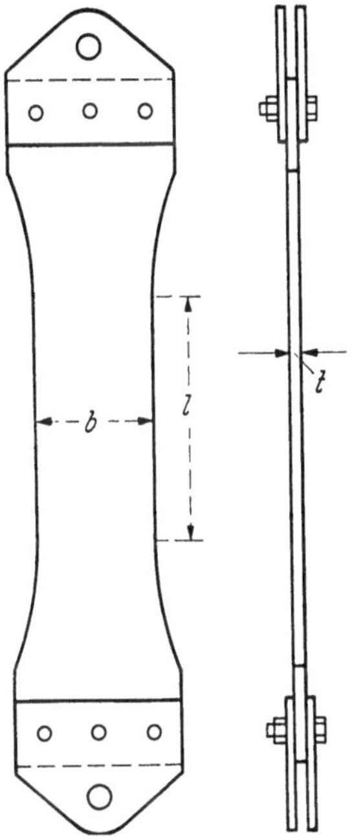

Thus if the load is W lbs., the vertical stress in the bar is $\frac{W}{tb}$ lbs. per sq. in., and in the uniform bar the transverse stress is zero, so this is the value of the stress-difference. If this produces a relative retardation r wave-lengths in light passing through a thickness t of the material, the relative retardation per inch is $\frac{r}{t}$. The fringe-value is then $\frac{W}{rb}$ or $\frac{1}{b}$ times the slope of the straight line graph. If the points do not lie on a straight line then measurements of creep with time under constant load must be carried out to estimate the possible errors which might arise through this factor.

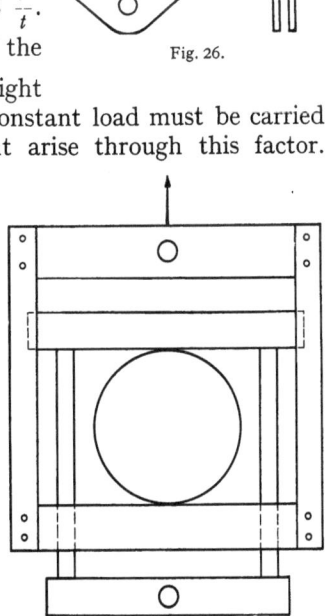

Fig. 26.

Even when using a type of material whose general characteristics are known it is frequently necessary, for accurate work, to measure the fringe-value of each individual plate of material. A convenient type of test-model for this purpose is a circular disc under a compressive load applied along a diameter. Such a disc is quickly cut and may be loaded in a frame as shown in Fig. 27. If a standard sized disc is used the centering in the frame is easily effected by the use of gaugeblocks cut to fit the spaces on either side of the disc.

For this case of a disc under diametral compression there is an exact mathematical solution which shows that the stress-difference at the centre of the disc is equal to $\frac{8W}{\pi t d}$ where W is the applied load, t the thickness and d the diameter of the disc. Then if r is the relative retardation in wavelengths at the centre produced by a load W lb., the fringe-value of the material will be $\frac{8W}{\pi r d}$. Although this solution is strictly exact only for point loads, it has been found by experiment that the stress-difference at the centre of the disc is not sensibly altered by the flattening

Fig. 27.

which occurs when the loads are applied by flat bars unless the length of the contact area approaches one half the radius of the disc. Such an amount of deformation is not needed for any practical calibration.

III. Exploration of two-dimensional stresses.

23. The plotting of isoclinic lines. For observation of the isoclinic fringes the model must be placed in the straining frame between crossed polaroids and subjected to the load under which it shows the fringes most clearly. The crossed polaroids are then rotated, and the fringes recorded for a range of inclinations of their axes from 0 to 90°.

The isoclinic lines may be plotted, either by direct tracing upon a projected image on a screen or by taking a number of photographs, and using an enlarging projector to enable all the lines to be traced on to a single drawing. The first method is usually the more satisfactory, and is also quicker and cheaper. In many cases, especially when a material with low fringe-value is used for the stress-measurements, it may be desirable to use a model made of a less sensitive material, (e.g. perspex) for tracing the isoclinics. This avoids the occurrence of numbers of isochromatic fringes which may render observation of the isoclinics difficult.

There are many factors also which tend to make the isoclinic fringes indistinct, and which may sometimes cause them to disappear entirely. Lack of homogeneity in the material or incorrect loading conditions may cause variations in the directions of the stresses from point to point through the thickness of the model. In such cases the fringes may become very faint, but their positions and the way in which they move as the coupled polaroids are rotated may be detected by close observation of the projected image, although a series of photographs might fail to reveal them.

Care must be taken, however, to distinguish between cases in which "fading" is due to such factors, and those in which the fringe fades because the inclination of the stresses in that region has reached a maximum value.

In general two isoclinic fringes of different parameters cannot cross each other, for this would mean two different directions for the principal stresses at the point of intersection. If, however, there is a point at which the principal stress-difference $(P-Q)$, is zero, the stress-conic at that point will be a circle, and any pair of perpendicular diameters may be considered as principal axes. In this case any number of isoclinic lines may pass through the point. Such a point is called an *isotropic point*.

In most cases tracing the isoclinics at intervals of 10° in the setting of the polaroids will be enough to enable the distribution of the stresses to be plotted with sufficient accuracy for qualitative purposes. If, however, the isoclinic lines in certain regions are to be used in computation of the separate stresses, as will be described later, they will have to be traced in those regions at much closer intervals.

When drawing an isoclinic line, the observer has to estimate the line of least intensity in a more or less diffuse fringe, and this again is better done on the actual projected image than on a photograph, for this line is not necessarily the apparent geometrical centre of the fringe. A certain amount of error is to be expected in regions where the fringes are diffuse, but in such regions errors in locating the position of an isoclinic line are less important, and it is justifiable to do some "smoothing" of the curves when making the final drawing.

The course of an isoclinic fringe very near to a free boundary may often be disturbed by the presence of time edge stress. In such regions, however, there is a guide to the drawing of the correct line, for at a free boundary there can be only one principal stress, and that will be parallel to the boundary. If an iso-

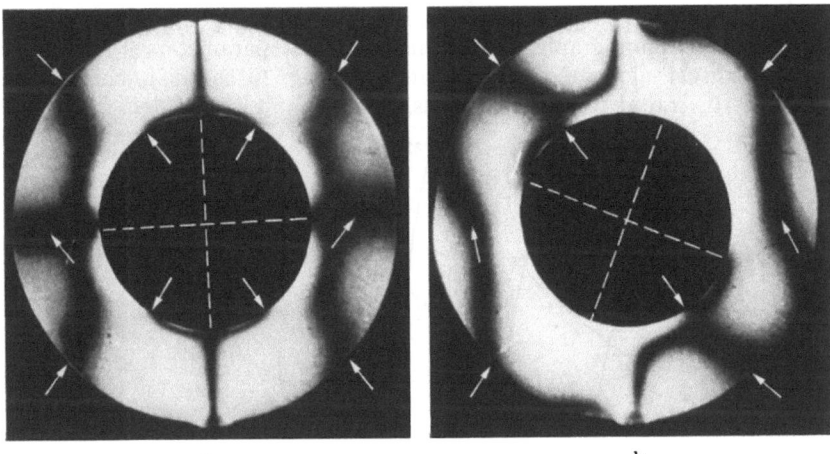

a b

Fig. 28 a and b. Isoclinic fringes in a perspex ring model under vertical diametral compression. Loci of points where the principal stresses are (a) vertical and horizontal, (b) inclined at 70° to vertical and horizontal. Isotropic points are indicated by arrows.

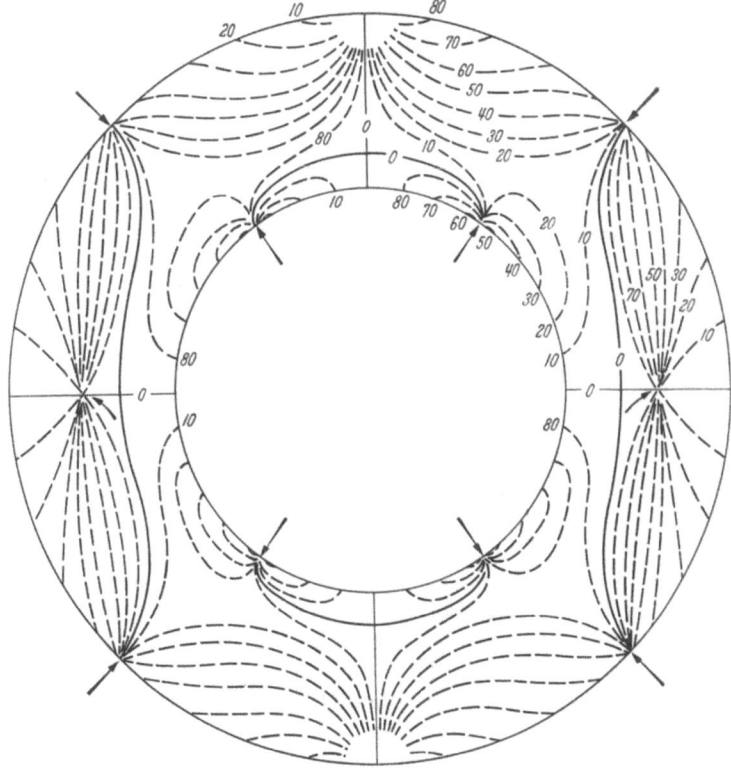

Fig. 29.

clinic of parameter angle φ meets a free-boundary, therefore, it must in general do so at a point where the tangent to the boundary is inclined at φ or $(90° - \varphi)$

to the reference direction. The exception to this rule occurs when there is an isotropic point on the boundary.

A further guide to the checking of the distribution of the isoclinics is that the parameter must change continuously from point to point of the model.

Fig. 28 shows photographs of isoclinic fringes of parameters 0 and 70° in a ring under thrust along a vertical diameter. Fig. 29 shows isoclinic lines at intervals of 10° traced from such fringes.

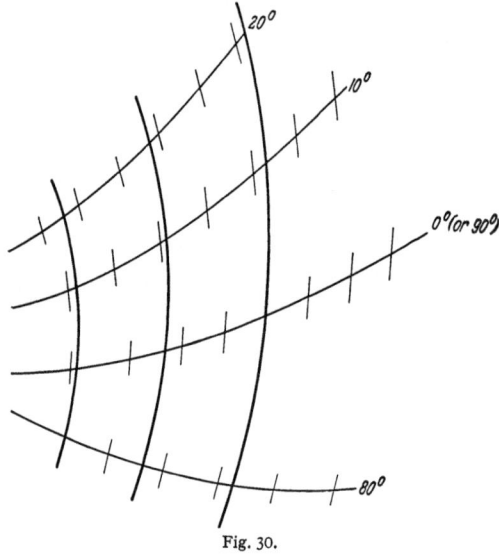

Fig. 30.

24. Stress trajectories. The only object of drawing stress-trajectories is to show in a clear manner the distribution of directions of principal stress. No great accuracy is needed, and the lines may be sketched free-hand at arbitrarily

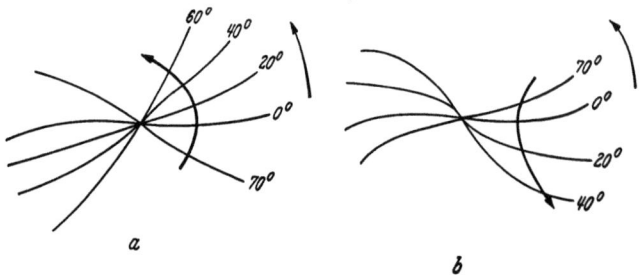

Fig. 31 a and b.

chosen intervals. A convenient way of doing this is to mark the inclination φ at a number of points on each isoclinic line as shown in Fig. 30. These serve as a guide for drawing one set of trajectories, and the other set may generally be sketched in normal to them without any further guide. At a free-boundary the stress-trajectories will always be parallel to and normal to the boundary. A number of trajectories will radiate from a point only when that point is one at which a finite load is being applied.

Two types of isotropic point frequently occur in two-dimensional models, and the distributions of the stress-trajectories around them follow very well defined patterns. In one type, called a positive isotropic point, the isoclinic line through the point moves round in the same direction as that in which the polaroids

Sect. 24. Stress trajectories. 169

are rotated. In the other, the negative point, the movement is in the opposite direction. Fig. 31 illustrates the distribution of the isoclinics in these two types.

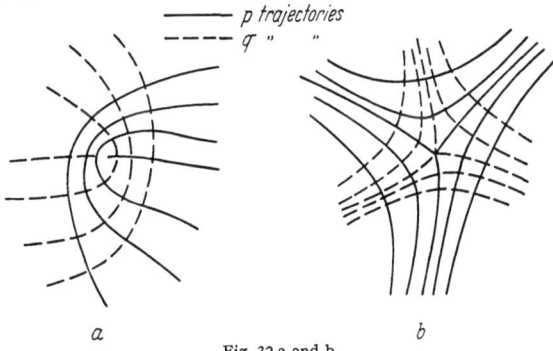

Fig. 32 a and b.

It will be seen that the stress-trajectories in the positive case (a) turn around the isotropic point, while in the negative case (b) they turn away from the point. Typical sets of stress-trajectories for the two types are shown in Fig. 32. The

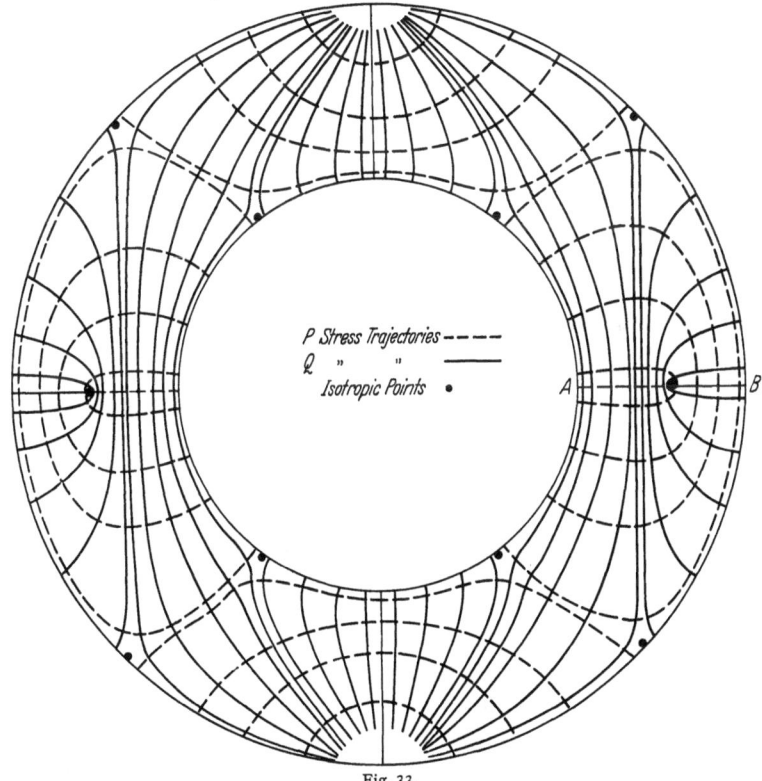

Fig. 33.

positive type is sometimes known as the interlocking type. Fig. 33 shows stress trajectories constructed from the isoclinics of Fig. 29.

Note: The formulae and equations which are used in the calculation of the stresses from the photoelastic data are all based upon certain sign conventions, and these must be consistently adhered to when recording observations.

The convention adopted for rotations in one plane is that rotations in a counter-clockwise direction are positive. This convention is applied not only to ordinary Cartesian coordinates but to the relation between the directions of the p and q lines of principal stress. Thus if at any point the positive direction of the p line has been specified, that of the q line will always be at 90° to it in the counter-clockwise direction. The sign of the curvature of these lines, and of the radius of curvature, is governed by the same rule. Thus in Fig. 34 the angle ψ between the positive direction of the tangent to the p line and the reference direction is increasing as S increases, and the curvature of the line is therefore positive. In the case of the q line as drawn, ψ_1 decreases as S_1 increases and the curvature is negative.

Some care needs to be taken in recording rotation directions when dealing with a projected image in a polariscope. The presence of lenses and/or mirrors may invert the image so that the direction of rotation of the polaroids may not correspond to the direction of rotation recorded on the image. These directions should always be checked, and the safest rule of procedure is always to record directions as they would appear on the drawing of the model which is used for the calculations.

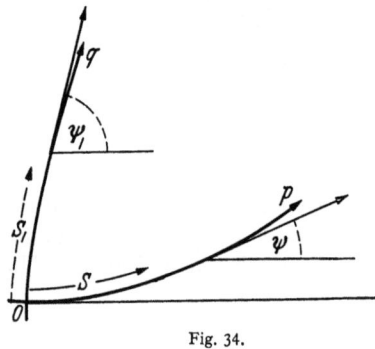

Fig. 34.

25. Determination of stress-differences from the fringe-pattern. For observations of the fringe-pattern it is usually necessary to eliminate the isoclinic fringes by using circularly polarised light. The model is placed in the straining frame and using one of the four arrangements of the polariscope given in Sect. 14 it is loaded to a point at which the number of fringes observed is adequate for the measurements which are required. Tracings or photographs of the fringe-pattern in both light and dark fields are obtained, using, of course, monochromatic light.

The orders of the fringes have first to be identified, and this may be done either by observation of the development of the pattern as the load is gradually applied, or by a study of the final fringe-pattern in white light. If the former method is used it will be found that fringes appear first at points of highest stress-difference and move across the model in the direction of decreasing stress-difference. There will frequently be found points of zero stress-difference where no change occurs, and starting from these the fringes may be numbered in the order in which they have appeared, as representing 1, 2, 3, etc. wave-lengths relative-retardation in the dark field, and $\frac{1}{2}, \frac{3}{2}$ etc. wave-lengths in the light field. In the second method, fringes of zero relative retardation will remain black while all others will be coloured, and directions of increasing relative retardation will be directions in which the colour of any fringe changes from red to blue (or pink to green for the fringes of high order). If the fringe pattern is close enough, the relative retardation at any point of the model may then be determined to a sufficient degree of accuracy by interpolation, using graphs plotted from the patterns in light and dark fields.

Fig. 35 shows photographs of these patterns in the loaded ring. The graph of $P-Q$ in Fig. 40 is plotted directly from these.

One of the most important problems is that of determining the boundary stress at a point of the model where a high stress-concentration occurs. The measurement of the relative retardation at such points is rendered difficult by two factors. First the possibility of an edge-stress effect, and second the impossibility of obtaining a reading at the extreme boundary. Fig. 36 shows a graph from which the stress at the boundary of a hole in a bar under tension has been found by extrapolation. In this model the hole was of 0.25 inch diameter and the nearest reliable reading of relative retardation was at about 0.01 inch from the boundary. In view of the high rate of increase in stress as the boundary is approached, a very considerable error would have been made if the relative retardation recorded in the fringe-pattern had been taken as the boundary value.

When carrying out stress-explorations in a model it is generally convenient to leave all observations in terms of wave-lengths of relative retardation (or "fringes"), until the final results are obtained, when the "fringe-value" of the model (that is the stress-difference in lb. per sq. inch which produces a relative retardation of one wave-length in the model) is used to convert the results to standard engineering units. Thus, the stress-difference $P-Q$ may be expressed in "fringe" or wave-length units throughout the calculations.

The signs of the quantities recorded will always be the signs of the stress-differences which will not necessarily be those of the relative retardations actually produced in the model, for some photoelastic materials have a positive and some a negative stress-optical coefficient, and the fringe-pattern does not differentiate between positive and negative relative retardations. There is, however, no difficulty in determining the signs of the stress-differences in practice, for since both P and Q must vary continuously from point to point of the model, $P-Q$ must be continuous, and cannot change sign without passing through a zero value [5]. It is always possible to pass from any one point to any other without passing through an isotropic point, and therefore the sign of $P-Q$ must be constant all over the model. In practice, there is always some point of a model at which the sign of the stress-difference is obvious from general considerations, so that, provided the p and q lines of principal stress have been correctly plotted, the sign of the stress-difference is known at all points.

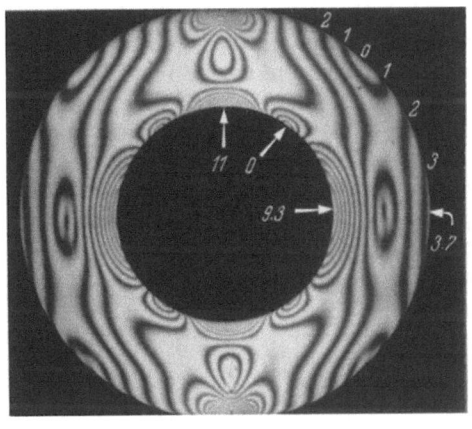

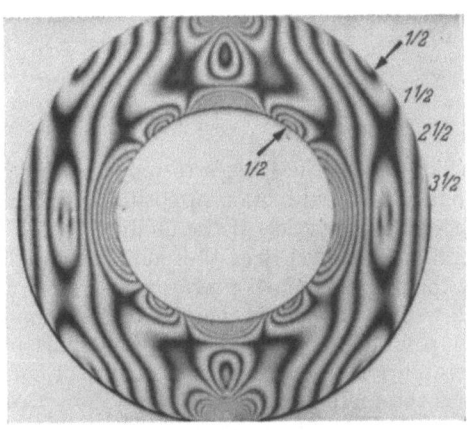

Fig. 35 a and b. Isochromatic fringes in dark and light fields in C.R. 39 ring model. Mercury green light.

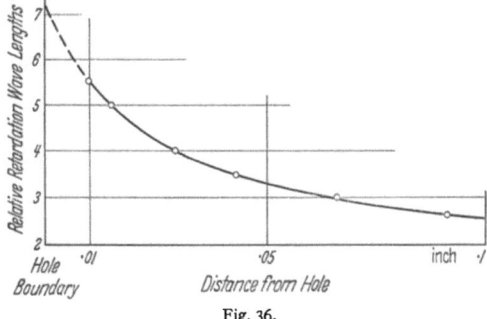

Fig. 36.

26. Measurement of fractional relative retardations. When the fringe-pattern is not close enough to enable relative-retardations to be interpolated with sufficient accuracy, some method of measurement must be adopted which will record fractions of a wave-length of relative-retardation. The TARDY and SÉNARMONT methods, the principles of which were given in Sect. 15, are usually the most convenient and most accurate to use. The theory of these methods shows that the direction in which the final rotation of the analyser must be made in order to measure the relative-retardation in the model depends not only on the sign of this retardation, but also on the relative orientations of the "fast" and "slow" axes of the quarter-wave plate or plates. It is not necessary, however, to work out these relations for any practical measurement. It will be seen that for both TARDY and SÉNARMONT methods the arrangement of the polariscope immediately before the final rotation of the analyser is one which shows the normal stress-pattern in the model, and as the analyser is rotated, the fringes move across the field. The direction in which the rotation must be made in order that the angle of rotation shall be one half the fractional angular relative-retardation, is that

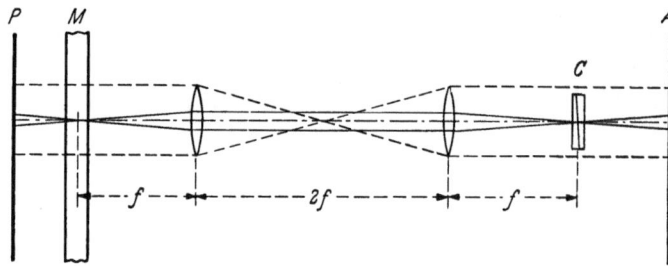

Fig. 37.

which causes the fringes to move in the direction of increasing stress-difference; that is in the direction opposite to that in which they move as the load on the model is increased. If the analyser is rotated in the other direction until extinction of light occurs at the point, the angle of rotation will be $(\pi - \tfrac{1}{2}\varphi)$ where φ is the angular relative-retardation. It is, in fact, desirable to obtain this reading as well as the direct one and to take the mean of the two values of φ obtained.

Another method of measurement is by direct compensation of the relative-retardation by the addition of one of equal magnitude and opposite sign. This was the method originally devised by COKER [1], who used as a compensator a strip of material, cut from the same sheet as the model, to which known tensions could be applied. This tension strip was mounted in a frame which could be placed in any part of the field and rotated so that the tension axis of the strip was set in any desired direction. It was placed close to the model with its axis parallel to one of the principal axes of stress at the point of the model under examination, so that the effect of applying tension to the strip was to reduce the total relative-retardation observed through model and compensator. Tension was then applied until the total relative-retardation was zero. This method possessed the advantage that the stress-difference in the model was measured directly in terms of the stress applied to the compensator.

The same compensation method may be used but with an optical compensator of the SOLEIL-BABINET type which has been calibrated so that its scale records fractions of a wave-length of the monochromatic light used.

For some purposes, for example in the measurement of edge-stress in a model, a compensator of the BABINET type is preferable, but since the relative-retardation in such a compensator is not uniform over the field, a special arrangement of the optical system of the polariscope is needed in order to bring into focus simultaneously a point in the model and one in the compensator. Such an arrangement is shown in Fig. 37. Between the model M and the compensator C are placed two lenses each of focal length f as indicated in the diagram. Both model and compensator then lie in a collimated beam, and the projecting system will focus the mid-planes of the two simultaneously. Since measurements with this apparatus will always be made at one point of the model at a time, the apertures of the lenses need not be large, but as the lenses lie between the polariser and analyser they must be free from stress.

27. Separation of the principal stresses. So far the photoelastic observations have yielded (a) the directions of the principal stresses, and (b) the magnitudes of the principal stress-differences, $P-Q$, at all points of the model. Our information gives us also the magnitudes of the actual stresses at all points on a "free" boundary, for at such points the stress normal to the boundary must be zero, and the magnitude of the tangential stress will be that of the stress-difference, while its sign will depend upon whether it belongs to the p or the q group as indicated by the stress-trajectories. For many practical problems this information will be sufficient, for the critical tensile stresses usually lie on a free boundary, while the maximum shear-stress, given by $\frac{1}{2}(P-Q)$, is the most important factor at all internal points. For a complete solution, however, we need to separate the stresses, and there are several ways of doing this, the one most suitable for any particular problem depending upon the type of stress-distribution, and upon the amount of information required. Of the four methods given below the third and fourth are the most generally useful.

α) *Measurement of the lateral extension.* If P and Q are the principal stresses in its plane at any point of the model, the strain in the direction normal to the plane will be $-\frac{r}{E}(P+Q)$, and the change in thickness of the model at that point will be $-\frac{r}{E}(P+Q)t$ where t is the thickness of the model. If then we measure the change of thickness of the model at any point when load is applied we shall be able to deduce the value of $P+Q$ at the point and hence obtain the values of P and Q. The measurement of such changes, however, is a matter of some difficulty, as they are usually of the same order of magnitude as the variation in thickness of a model from point to point, and an accuracy of the order of 10^{-6} inch in the measurements is required. COKER [1] designed a mechanical lateral extensometer which gave an accuracy of this order, but it presented considerable difficulty in operation, and the accuracy of the final evaluation of the stresses was not very high.

A more accurate method, in principle, of measuring the lateral extension is by obtaining interference fringes in light reflected from the two surfaces of the model and counting the number of fringes which pass across any given point as the load is applied. There are many practical difficulties in this method, chief of which is that of avoiding movement of the model as a whole during application of the load. Modifications of the method are however in use in some photoelastic laboratories[1,2], and new optical methods are in course of development[3].

β) *Numerical solution of* LAPLACE'S *equation*[4]. As already mentioned in Sect. 11 the sum of the principal stresses, in a model which conforms to the conditions for generalised plane stress and in which body forces are absent, will satisfy LAPLACE'S equation. This affords another means of determining the value of $P+Q$ at all points of the model. It may be shown [9] that in such a case the value

---
[1] H. FAVRE: Rev. Opt. **11**, 1 (1932).
[2] V. TESAR: Rev. Opt. **11**, 97 (1932).
[3] D. POST: Proc. Soc. Exp. Str. An. **12**, No. 1, 99, 191 (1954).
[4] M. M. FROCHT and M. M. LEVEN: J. Appl. Phys. **12**, 596 (1941).

of $P+Q$ at the centre of any small square area of the model is approximately equal to the mean of the values at the four corners of the square. Using this condition and the known values of $P+Q$ at the free boundary it is possible by applying the "relaxation" technique to obtain the values to any required degree of accuracy over most of the model [5]. Regions of difficulty occur near loaded boundaries, where the values of $P+Q$ are unknown, but these regions present difficulty by all methods.

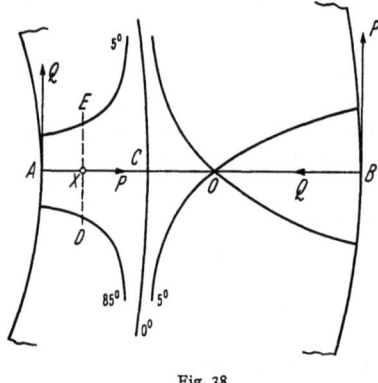

Fig. 38.

γ) *Integration of the* LAMÉ-MAXWELL *equations*. This method is most useful when it is required to separate the stresses at points on an axis of symmetry of the model. Such an axis will, of course, be also an axis of principal stress at all points.

Consider for example the horizontal axis of symmetry AB (Fig. 38) across the ring under diametral load illustrated in Fig. 35. The diagram of Fig. 38 shows the isoclinic lines of parameters 0° (or 90°), 5° and 85°. If we denote the principal stresses at A by P and Q as shown, then we may start by using the LAMÉ-MAXWELL equation

$$\frac{\partial P}{\partial s} + \frac{P-Q}{\varrho_1} = 0 \tag{27.1}$$

where s is measured along AB and ϱ_1 is the radius of curvature of a transverse line of principal stress such as DXE. Then at the point X of AB the value of ϱ_1 will be given by

$$\varrho_1 = \frac{\partial s_1}{\partial \psi} = \frac{DE \times 180}{10 \times \pi}.$$

Values of $\partial P/\partial s$ may therefore be calculated for a number of points on AO, and plotted against s. As we approach the point C where the zero isoclinics cross, the value of $\partial P/\partial s$ becomes very small, and it vanishes at C where ϱ_1 is infinite and $P-Q$ is still finite.

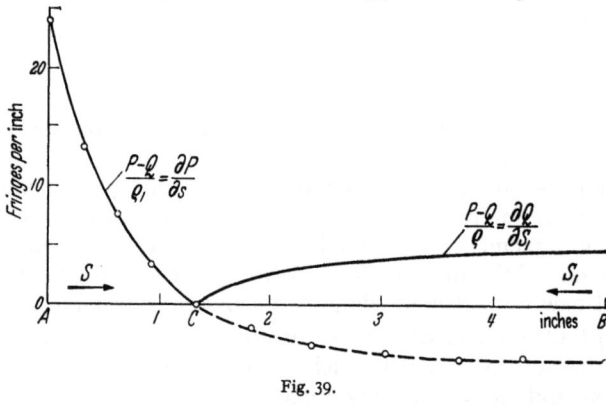

Fig. 39.

An inspection of the stress-trajectory pattern of Fig. 33 shows that at the isotropic point O, the horizontal line ceases to belong to the p set of trajectories, and becomes one of the q set, so that Eq. (27.1) is no longer applicable. The best procedure then is to start afresh from point B where the P and Q stresses will be related as shown in Fig. 38, and to determine the values of $\partial Q/\partial s_1$ at points along BO using the equation

$$\frac{\partial Q}{\partial s_1} + \frac{P-Q}{\varrho} = 0. \tag{27.2}$$

These values may be plotted on the same graph as those of $\partial P/\partial s$ as shown in Fig. 39.

Graphical integration of these graphs gives the values of the horizontal stress at all points of AB. The vertical stress may be deduced from these and the known values of $P-Q$ shown in the graph of Fig. 40. Fig. 41 shows these final values.

δ) *Integration of the Cartesian equations of equilibrium*[1]. When it is required to separate the stresses at points which do not lie on an axis of symmetry, we may use one of the Cartesian equations of equilibrium

$$\left. \begin{array}{l} \dfrac{\partial X}{\partial x} + \dfrac{\partial S}{\partial y} = 0; \\[2mm] \dfrac{\partial S}{\partial x} + \dfrac{\partial Y}{\partial y} = 0, \end{array} \right\} \quad (27.3)$$

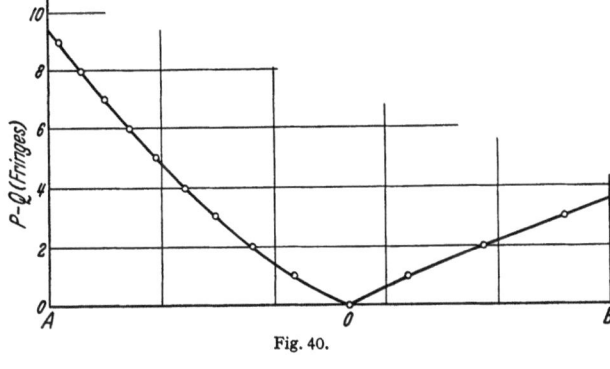

Fig. 40.

where X and Y are the normal stresses and S the shear stress referred to arbitrarily chosen rectangular axes x and y.

If it is required to separate the stresses at a point A of any model, a line OA which meets a free boundary at O is taken as the x axis (Fig. 42). Then if P_0 be the boundary stress at O, in a direction inclined at angle ϑ to OA, the normal stress X_0 at O will be $P_0 \cos^2 \vartheta$. If the shear-stresses at two points C and D a short distance on either side of OA be S_1 and S_2, then at the point B,

$$\dfrac{\partial S}{\partial y} \approx \dfrac{S_2 - S_1}{CD}$$

and the rate of change of X at B is given by

$$\dfrac{\partial X}{\partial x} = \dfrac{S_1 - S_2}{CD}.$$

The shear-stress S at any point is given by

$$S = \tfrac{1}{2}(P - Q) \sin 2\alpha,$$

where α is the angle between the direction x and the direction of the principal stress P.

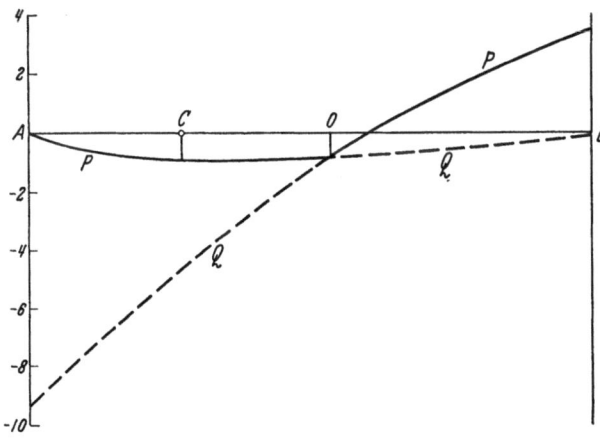

Fig. 41.

If then we measure the isoclinic angle φ and the stress-difference $P-Q$ at a number of points such as C and D, (say a number of points on two lines parallel to OA), we can find the value of $\partial X/\partial x$ at a series of points on OA.

Then

$$X_A = X_0 + \int_0^A \dfrac{\partial X}{\partial x} dx,$$

---

[1] L. N. G. FILON and E. G. COKER: Brit. Assoc. Rep. 201. 1914.

and using the relation
$$X = \tfrac{1}{2}(P+Q) + \tfrac{1}{2}(P-Q)\cos 2\alpha$$
we can find the value of $P+Q$ at A, and hence the separate values of P and Q.

28. Transference of results to prototype. In an investigation of the stresses in a component of a structure or machine the photoelastic tests are usually carried out on a scale model of the component. When the investigation is a purely two-dimensional one, the linear dimensions of the model in the plane of the observations must be strictly proportional to those of the prototype, but the thickness of the model need not be to the same scale. The stresses at any point of the prototype will then be expressed in terms of those in the model by the relation

$$\frac{P}{p} = \frac{W}{w}\frac{l}{L}\frac{t}{T}$$

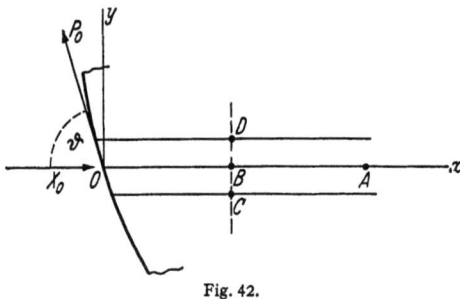

Fig. 42.

where P is the stress in the prototype, p that in the model, W/w the ratio of the loads applied, L/l the ratio of linear dimensions in the plane of the plate, and T/t the ratio of thicknesses.

When applying the results of photoelastic tests to an actual component, however, there are certain factors which may lead to error. The first of these is the discrepancy which may occur in load distribution owing to the difference in elastic constants of the materials of model and prototype.

Under the loads applied in photoelastic tests the model will generally undergo deformations near the load-points which are much larger than those occurring in, say, a steel component under its normal working loads. These deformations may alter the distribution of the load, causing an appreciable difference in the stresses in their neighbourhood. The stress-distribution in the immediate vicinity of load points may therefore not be the same in model and prototype. If such regions are to be investigated the loads employed should be such as to make the strains in the model as nearly as possible comparable with those which would occur in the prototype. The use of a thicker model in such a case enables a sufficiently large optical effect to be obtained with smaller stresses, though it introduces more difficulty in ensuring uniform distribution of load across the thickness.

In general the errors due to such differences in yield under loads will be confined to the immediate neighbourhood of the loads and will be negligible at a distance from the loaded boundary approximately equal to the length of boundary over which the load is spread. An illustration of this is found in Sect. 22 in the case of the diametrally loaded disc, where the stresses at the centre of the disc are virtually unaffected by a considerable flattening of the edge under load.

Another factor is the difference in elastic behaviour of the model material and of most structural materials. Within the range of loads used in photoelastic work the behaviour of the model material is perfectly elastic. That is to say, both strain and optical effect are proportional to the stress. In most structural materials there occurs a more or less well defined "yield point"—a stress beyond which the material exhibits plastic deformation, the result of which is to transfer stress from the most highly stressed regions to neighbouring regions. Such local plastic yield is an important factor in engineering design. In some structures which are subjected mainly to static loads it affords a considerable safety factor by removing the danger of fracture at isolated points of stress-concentration. In other cases, especially where high rates of alternation or repeated application of loads are involved, the occurrence of plastic yield may be closely associated with fatigue failure, and it may be essential to design so that the yield point is not reached. A photoelastic investigation indicates only the stresses which would be obtained if no plastic yield occurred, and from these the engineer, using his knowledge of the behaviour of the structural material must deduce how much yield will occur and how this will alter the distribution of stress.

A third factor is the effect of the difference in Poisson's ratio of the model and prototype. In a two-dimensional model this factor occurs only when an unbalanced load is applied to the boundary of a hole in the model. In problems of this type for which a mathematical solution exists it is found that the effect of a change in Poisson's ratio from 0.25 to 0.4, (the approximate values for steel and photoelastic plastics respectively), is to diminish the shear-stresses by amounts varying from point to point of the model, but not exceeding about 7% of the maximum values [1]. These analytical solutions, however, are only obtainable for problems in which the loads and boundary conditions can be expressed in terms of comparatively simple mathematical functions, and the conclusions from such investigations will not necessarily apply to the stress-distributions which occur in practice.

Probably the most important practical problem in which this factor will occur is that of the stresses in a plate produced by a load applied to a bolt or rivet passing through a hole in the plate. In this problem both the yield of the material and its Poisson's ratio have an effect upon the stresses directly under the load. The greater yield of the photoelastic material may lead to a more uniform distribution of load on the hole boundary, giving smaller maximum direct compressive stresses, while its larger Poisson's ratio will give larger transverse compressive stresses. The total result is that the stress-differences, and therefore the shear-stresses, may have considerably lower values in the photoelastic model than they would have in the metal prototype. This factor is more fully considered in a later section.

C. Exploration of three-dimensional stress systems.

I. Frozen stress materials and techniques.

29. The phenomenon of frozen stress. The fact that residual optical effects were frequently found in photoelastic materials after the removal of the loads had been noted by many observers, but it was not until 1936 that the technique was found for producing residual effects which were proportional to the stresses which had existed under load. It was then discovered that a number of the newly developed synthetic resins possessed a remarkable property.

If a bar of the material were loaded under simple tension at room temperature, its behaviour was sensibly elastic, though it showed a certain amount of plastic creep with time. As the temperature was raised the initial extension under load became greater, and the time-creep became much more rapid until at some critical temperature the whole of the extension of the bar took place immediately on applying the load and there was no time-creep. The same sequence of changes was shown, qualitatively, by the optical effect in the bar, and both strain and optical effect at the critical temperature were very many times greater than they were under the same load at room temperature. If the temperature were raised still further there was no significant change in either the mechanical or the stress-optical behaviour of the material.

If now the bar were allowed to cool to room temperature with the load still applied, it was found that *on removal of the load* the greater part of the strain and of the optical effect still remained. If a two-dimensional model which showed a distinct fringe-pattern under load were treated in this way, the residual fringe-pattern was precisely similar to that obtained under a much larger load at room temperature. The stresses and the stress-pattern were said to be "frozen" into the model.

The further discovery was made that the fringe-pattern was not disturbed by cutting the model, and that in most of the materials there was no appreciable change in the pattern with time. It was thus possible to freeze the stresses into any three-dimensional model, and by cutting thin slices from it to explore the stresses in any region.

There have been various theories to account for the stress-freezing effect, but the one which appears to be the most satisfactory is as follows. It is known that the resins which exhibit the effect most strongly are all compounds which polymerise into long-chain molecules, cross-linked by chains of different molecular groups, and it is assumed that when fully matured any mass of such a resin virtually consists of a number of very large, highly complex molecules, interlinked with one another and forming, in effect, one large deformable molecule. It is further assumed that at low temperatures physical cohesion bonds operate between the molecular groups of the different chains forming a rigid mass which deforms under load like any ordinary elastic solid. The critical temperature corresponds to the melting point of the ordinary solid, the temperature at which the cohesive bonds break down, and at this temperature the molecular framework is free to deform in a different way, while remaining perfectly elastic. When the model cools under load the cohesive bonds reform, but now they hold the molecular framework in its deformed position.

That there are two distinct types of deformation involved is indicated by two observed facts. The first is that the strain-optical coefficient of any material in the frozen state is different from that under cold loading. In some materials, in fact, the two coefficients have opposite signs, and in such cases any time-creep which occurs under cold loading is of the same sign as the frozen stress, i.e. the optical effect decreases with time. The second fact is that the POISSON's ratio in the frozen state is very nearly 0.5, while that in the cold state lies between 0.35 and 0.4. The high value of the ratio in the frozen state is compatible with the occurrence of a strain of the nature of a lattice deformation.

If this theory of the effect is adopted, the time creep under constant load at temperatures below the critical may be attributed to the successive breaking down of some of the cohesive bonds. The elastic state of the material at any instant may then be expressed in terms of different proportions of the two types of strain, in each of which the optical effect is proportional to the strain, but with a different strain-optical coefficient. For all practical purposes, however, the loads used will be such that the strains produced will be proportional to the stresses, both in cold loading and in the frozen stress method, so that the proportionality of relative-retardation to stress may be accepted in both cases.

30. Stress freezing materials and techniques. α) *Materials*. Some of the materials which show the photoelastic effect under direct load at room temperature exhibit also the frozen stress properties. Of these, bakelite, the phenolic resins and the ethoxyline resins are the most notable. Another group, the polyester resins, are unsuitable for direct loading on account of a large time-creep, but possess very good properties for stress-freezing. Fosterite[1] and the Marco resins[2] are examples of this type. Bakelite BT/61/893, Fosterite, and the phenolic resins are all factory produced, and are obtainable in blocks or rods from which the photoelastic models may be machined. The Marco and ethoxyline resins are supplied in the form of constituents which are combined in the laboratory to form the finished product. With these latter, one may either cast blocks from which to machine models, or, in suitable cases, one may cast models in their final shape. Very careful control of the casting conditions is necessary if the models are to be homogeneous and stress-free.

Some of the properties desirable in a material for frozen-stress-work are slightly different from those needed for direct loading. The most important is that the material shall have a large ratio of relative-retardation to strain. One of the factors which may introduce difficulty in any photoelastic work is that the strains in the model are generally larger than those occurring in the prototype, and this discrepancy becomes much greater in the case of models under frozen-stress conditions. It is desirable therefore to use materials in which the strains for a given optical effect are as small as possible. The quantity generally used to define this relation, the optical-strain coefficient, is the ratio of the YOUNG's modulus in the hot loaded state to the fringe value of the material in terms of the frozen optical effect. It denotes the number of wave-lengths of relative-retardation per inch thickness which would be produced by unit frozen strain, and is sometimes called the *figure of merit* of the material. Other desirable properties are freedom from time edge-stresses, freedom from recovery creep, and a reasonably high mechanical strength.

Of the materials mentioned above some of the ethoxyline resins, (e.g. araldite B), have by far the highest optical-strain coefficients and the highest mechanical strength. They show no appreciable recovery creep, and very little time-edge stress. They are about one tenth the price of Fosterite, the next best material for stress-freezing, and have the advantage that models may be directly cast with them. In common with the other materials they invariably show initial stresses when freshly cast, but these may generally be removed by careful annealing. The chief defect of araldite is a certain lack of optical homogeneity, the effect of which sometimes causes the isoclinic fringes to become faint and indistinct. The approximate values of mechanical and stress-optical properties are given in Table 2.

β) *Freezing procedure.* For freezing the models, as well as for annealing, an electric oven with a range up to 150° C and a controlled thermo-static device capable of raising or lowering the temperature at rates of from 1° C to 5° C per hour is required. Since the loads needed for stress-freezing are generally quite small they may be applied by dead weight loading, and unless very large models are used, the whole assembly of model and loads may be placed in the oven. Alternatively a frame for supporting the model may be fixed inside the oven and the loads applied by wires or rods passing through openings in the oven wall. The loads may be attached at room temperature, and the oven then switched on and brought up to the critical temperature which is maintained for a period varying from a few minutes to two hours according to the mass of material in the model. The rate at which the oven is allowed to cool must also be adjusted according to the dimensions of the model. When cold, the model is removed and is then ready for examination. Since the fringe-value of the material may vary

---

[1] M. M. LEVEN: Proc. Soc. Exp. Str. An. **6**, No. 1 u. No. 2 (1948).
[2] B. SUGARMAN, G. O. MOXLEY and I. A. MARSHALL: Brit. J. Appl. Phys. **3**, 233 (1952).

from one casting to another it is desirable to freeze with each model a calibration model, made from the same batch of resin. The disc under diametral compression is probably the simplest to machine and to load.

γ) Slicing the model. Slices are cut from the model either according to some pre-arranged scheme of exploration, or as a result of a preliminary inspection of the stress-pattern seen when the whole model is viewed in the polariscope. For such an inspection the model is immersed in a liquid with a refractive index equal to its own. A suitable immersion liquid is a mixture of α-bromonaphthalene (refractive index 1.66) and liquid paraffin (1.44). The stress-patterns seen will, of course, represent only the mean effect of stresses which vary from point to point of the model, but they will frequently give useful indications of the places at which slices should be cut to obtain measurements of the peak stresses. For such purposes as this the indices of refraction of liquid and model need not be accurately matched, though greater accuracy will be essential when measurements with oblique incidence are to be made.

Slices of the model may be cut by hand, using a sharp hack-saw, or by a milling cutter, the surfaces afterwards being finished by filing, and if desired, polishing. Great care must be taken in all these processes to avoid overheating which may produce partial annealing of the material. Polishing is the most dangerous in this respect, and this need not be done if the slices are placed in an immersion liquid for examination. Slices as thin as 0.03 inch may be prepared quite satisfactorily in this way.

Table 2. *Properties of materials for stress freezing.*

Material	"Freezing" temperature °C	Young's modulus E lb/sq .in.	Fringe value f lb/sq. in/fr/in. $\lambda = 5460$ Å	Fringe-strain ratio E/f (Figure of merit)
Bakelite (B.T/61/893)	110	1.5×10^3	3.5	400
Fosterite	85	2×10^3	4	500
Catalin 800	85	1.5×10^3	1.6	1000
Columbia Resin (C.R. 39)[1]	110	40×10^3	25	1600
Marco Resin (Sb. 26. C.)	90	3×10^3	5	600
Araldite B	135	2×10^3	1.2	1700

II. The determination of the stresses.

31. The stresses in a plane of symmetry. In very many practical stress-distribution problems there exists a plane of symmetry, i.e. a plane with respect to which both the distribution of loads and the geometrical shape of the model are symmetrical. The exploration of the stresses in such a region is simpler than in any other, for we know that at any point in a plane of symmetry two of the principal stresses will be in that plane. To examine such a region of a photoelastic model, however, we have to cut a slice of finite thickness whose mid-plane is the plane of symmetry, and the above condition does not apply throughout the thickness of the slice. In general, both the magnitudes and the directions of the principal stresses will change from point to point of the thickness, so that the effect upon light passing normally through the slice will give only an approximation to the values of the principal stresses in the mid-plane. This approximation will, of course, become closer as the thickness of the slice is decreased, but how thin the slice will need to be in any particular problem will depend upon the size of the model and the nature of the stress-distribution. The magnitudes of the

---

[1] C.R. 39 is unsuitable for three-dimensional investigations, but may be used when it is desired to freeze a two-dimensional stress-system.

stresses must, by symmetry, have either maximum or minimum values at the midplane and their rates of change at a small distance from this plane will therefore generally be small. Rates of change in direction may frequently be estimated from the geometry of the model with sufficient accuracy to give an indication of the thickness of the slice which will be permissible. In general it may be taken that if the isoclinic fringes seen in a slice are everywhere sharp and clear, then the changes in direction of the stresses are very small, and the observations through the slice may be taken as representing the effects of the plane of symmetry stresses to a good degree of accuracy. If a check on this accuracy is required one may begin with observations on a comparatively thick slice, and then reduce the thickness by filing and repeat the measurements.

Observation of such a slice by normal transmission in the polariscope yields the directions, and the magnitudes of the differences of the principal stresses in the plane exactly as in the two-dimensional case. At any point on a free boundary, also, the stress normal to the boundary must be zero, and the stress difference gives the principal stress in the plane of symmetry at that point.

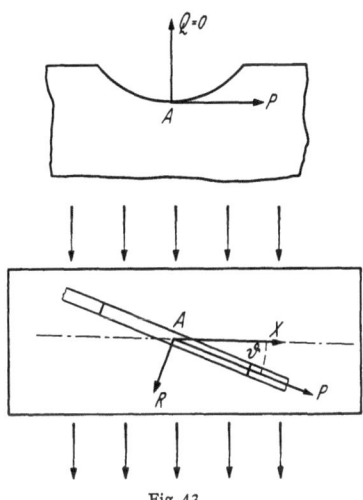

Fig. 43.

Thus if A (Fig. 43), be any point on the free boundary of a slice cut in a plane of symmetry, the observations with normal transmission give the value of the stress P at A. If we now rotate the slice through an angle ϑ about the normal to the boundary at A, using an immersion fluid to eliminate refraction, the principal stresses in the plane of the wave-front will be the zero stress Q and the stress X where $X = P \cos^2\vartheta + R \sin^2\vartheta$ while the length of the light path through the slice will be $t \sec \vartheta$ where t is the thickness of the slice. We can therefore find the values of both principal stresses at all points which lie on the free boundary of a plane of symmetry.

At any other point in the slice, we know the directions of P and Q, but not the magnitude of either, and if we carry out the same procedure of rotating the slice about the line of action of the Q stress and measuring the relative-retardation, our observation gives the value of the stress-difference $P \cos^2\vartheta + R \sin^2\vartheta - Q$. Subtracting this from the value of $P - Q$ obtained by normal transmission we get the value of $(P - R) \sin^2\vartheta$. Hence at points other than those on the free boundary, direct measurements of relative-retardation yield only the values of the principal stress-differences. From these, of course, we deduce the magnitudes of the three principal shear-stresses, $\frac{1}{2}(P - Q)$, $\frac{1}{2}(Q - R)$, $\frac{1}{2}(R - P)$, and for internal points these are frequently more important than the values of the separate stresses.

There is one special case in which the separation of the stresses in a plane of symmetry may be accomplished by a method simpler than the general method to be described in Sect. 33[1]. This is the case in which a body has symmetry about an axis, and also about a plane which is normal to that axis. An example of this is the notched cylindrical bar used frequently for tensile strength tests. If it is required to find the distribution of stresses on the radius OB in the horizontal plane of symmetry, the stress differences $P - Q$ and $R - P$ are measured at

---
[1] H. T. JESSOP: J. Sci. Instrum. **26**, 27 (1949).

points along this line, and two neighbouring isoclinic lines are traced as shown in Fig. 44. We then apply the three-dimensional LAMÉ-MAXWELL equation [Eq. (8.4)] in which, for this case, $\alpha=0$, giving

$$\frac{\partial P}{\partial s}+\frac{P-Q}{\varrho_1}+\frac{R-P}{\varrho_2}=0.$$

The radius of curvature of the Q trajectory, ϱ_1 is found from the isoclinics as in Sect. 27, while ϱ_2 at any point A is the radius AB. A graphical integration then gives the value of P at all points on the cross-section.

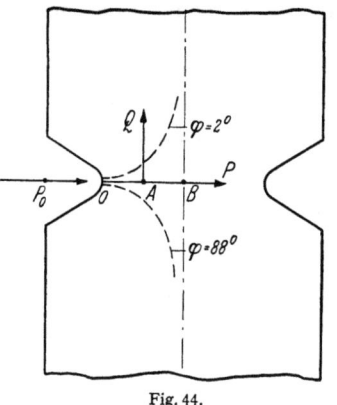

Fig. 44.

32. The stresses on a free boundary.
At a point on the free boundary which does not lie in a plane of symmetry we have in general no knowledge of the directions of the principal stresses. We do know, however, that the tangent plane to the boundary is a principal plane, and if we cut a thin slice from the model in this tangent plane, an inspection of the isoclinic fringes in the slice will determine the directions of the principal stresses at the point. In theory, too, the relative retardation through such a slice will give the magnitude of the principal stress-difference at the point, provided that the slice is sufficiently thin. In practice, however, at any point of high stress on the boundary, the magnitudes of the stresses decrease very rapidly as we proceed along the normal to the surface, and the mean-value of the stress-difference through even a thin slice may be much lower than the actual value at the surface. A tangential slice, therefore, will generally give only the directions of the stresses at the point, and we shall have to make a fresh model in order to determine their magnitudes.

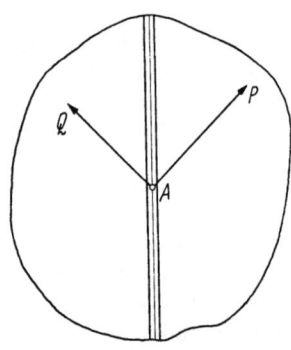

Fig. 45.

If from the second model we cut a slice whose plane contains the normal to the surface at the point under investigation, and is at 45° to the directions of the principal stresses P and Q as determined from the tangential slice (Fig. 45), then examination of the slice in light incident at $\pm 45°$ to its plane will yield the values of the separate stresses P and Q.

An alternative method[1] which does not involve the use of two models is to cut a slice whose plane contains the normal at the point but whose direction is arbitrary, and to examine it with three different directions of transmission.

Thus if P and Q be the stresses, and α be the angle the slice makes with the direction of P, the slice may be rotated about the normal to the edge at A so that the incident light is successively normal, and inclined at $+\vartheta°$ and $-\vartheta°$ (Fig. 46).

The stresses in the plane of the wave-front in the three cases will be, using Eqs. (7.16).

$$\tfrac{1}{2}(P+Q)+\tfrac{1}{2}(P-Q)\cos 2\alpha,$$
$$\tfrac{1}{2}(P+Q)+\tfrac{1}{2}(P-Q)\cos 2(\alpha+\vartheta),$$
$$\tfrac{1}{2}(P+Q)+\tfrac{1}{2}(P-Q)\cos 2(\alpha-\vartheta).$$

---
[1] V. M. HICKSON: Brit. J. Appl. Phys. **2**, 261 (1951).

Determination of the principal stresses at any point.

If r_1, r_2, r_3 are the measured relative retardations in wave lengths, and if ϑ be made $45°$ for convenience, we shall then have

$$\frac{1}{2}(P+Q) + \frac{1}{2}(P-Q)\cos 2\alpha = \frac{fr_1}{t},$$

$$\frac{1}{2}(P+Q) - \frac{1}{2}(P-Q)\sin 2\alpha = \frac{fr_2}{\sqrt{2}t},$$

$$\frac{1}{2}(P+Q) + \frac{1}{2}(P-Q)\sin 2\alpha = \frac{fr_3}{\sqrt{2}t},$$

where f is the fringe-value of the material and t the thickness of the slice.

From these three equations we can obtain the values of P, Q and α but it should be noted that in the course of solving the equations there will in general be additions of any errors in the observed quantities r_1, r_2, r_3 so that it is essential for these to be measured to a high degree of accuracy.

Another method of examining such a slice is to rotate about the normal, and to measure the relative retardation r for a series of values of ϑ and plot the values of $r\cos\vartheta$. These will have a maximum or minimum when one of the principal stresses is in the plane of the wavefront. The slice may then be set at the angles corresponding to maximum and minimum, and the values of the two stresses are then measured directly.

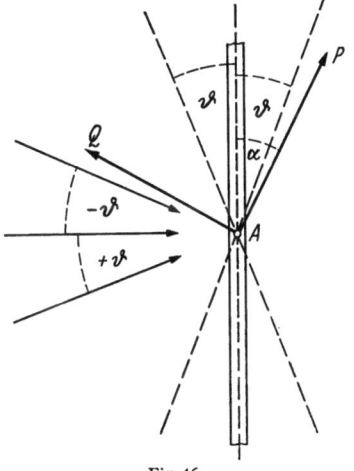

Fig. 46.

33. Determination of the principal stresses at any point of a three-dimensional model. In regions of a model which are neither free surfaces nor planes of symmetry, we do not known the direction of any principal stress and so have no guidance as to the directions in which slices should be cut to enable the stresses to be measured directly. In theory, it is possible to deduce the principal stress-differences at a point from a number of observations at different directions of transmission through a slice cut in any direction through the point.

Thus if we cut any slice through a chosen point O of the model and select arbitrary rectangular axes ox, oy, oz, fixed relative to the slice, we can tilt the slice in an immersion liquid so that light passes through it in a direction which is defined relative to x, y, z. Observation of the isoclinic fringe passing through O then gives the directions of the secondary principal stresses P_1, Q_1 in the plane of the wave-front, and we can calculate the direction cosines, l_1, m_1, n_1 and λ_1, μ_1, ν_1, of these relative to x, y, z. We can also measure the secondary stress-difference $(P_1 - Q_1)$.

Then if $X, Y, Z, S_{yz}, S_{zx}, S_{xy}$ are the stress-components at O referred to x, y, z, we have

$$\left.\begin{array}{l} P_1 = l_1^2 X + m_1^2 Y + n_1^2 Z + 2m_1 n_1 S_{yz} + 2n_1 l_1 S_{zx} + 2l_1 m_1 S_{xy}, \\ Q_1 = \lambda_1^2 X + \mu_1^2 Y + \nu_1^2 Z + 2\mu_1 \nu_1 S_{yz} + 2\nu_1 \lambda_1 S_{zx} + 2\lambda_1 \mu_1 S_{xy} \end{array}\right\} \quad (33.1)$$

and hence

$$\left.\begin{array}{l} (P_1 - Q_1) = (l_1^2 - \lambda_1^2) X + (m_1^2 - \mu_1^2) Y + (n_1^2 - \nu_1^2) Z + \\ \quad + 2(m_1 n_1 - \mu_1 \nu_1) S_{yz} + 2(n_1 l_1 - \nu_1 \lambda_1) S_{zx} + 2(l_1 m_1 - \lambda_1 \mu_1) S_{xy}. \end{array}\right\} \quad (33.2)$$

But
$$l_1^2 + m_1^2 + n_1^2 = \lambda_1^2 + \mu_1^2 + \nu_1^2 = 1, \quad (33.3)$$

and Eq. (33.2) becomes

$$\begin{aligned}(P_1 - Q_1) = (l_1^2 - \lambda_1^2)(X - Z) + (m_1^2 - \mu_1^2)(Y - Z) + \\ + 2(m_1 n_1 - \mu_1 \nu_1) S_{yz} + 2(n_1 l_1 - \nu_1 \lambda_1) S_{zx} + 2(l_1 m_1 - \lambda_1 \mu_1) S_{xy},\end{aligned} \quad (33.4)$$

a linear equation in which all the coefficients are known.

Five such observations giving five distinct equations would enable us to determine the differences of the normal stresses and the shear-stresses.

It should be noted that no set of observations of stress-differences alone can lead to the determination of the magnitudes of the separate stresses, for the application of the relations (33.3) between the direction cosines will always reduce our equations to the form of relations between stress-differences as shown above.

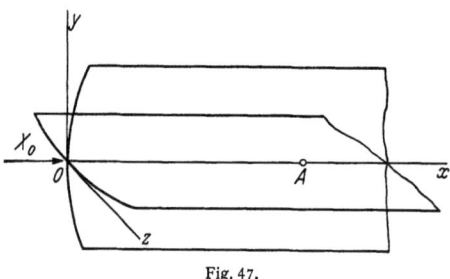

Fig. 47.

In practice this method is rarely satisfactory except in regions where the stresses vary very little from point to point of the model. In the five observations the light passes along five different paths through the slice, and though the error in any single observation might be small, the combination of the five different errors which is involved in the solution of the equations is liable to lead to serious errors in the final results.

A practicable method, introduced by FROCHT[1], depends upon the extension to three dimensions of the integration of the Cartesian equation of equilibrium as described for the two-dimensional case in Sect. 27. The method involves the use of two identical models, but it results in the complete solution for the stresses at a point.

The equation of equilibrium in three dimensions is

$$\frac{\partial X}{\partial x} + \frac{\partial S_{xy}}{\partial y} + \frac{\partial S_{xz}}{\partial z} = 0, \quad (33.5)$$

where X is the normal stress and S_{xy} and S_{xz} are the shear-stresses, referred to arbitrarily chosen axes.

If A is the point in the model at which the stresses are required, a line OA is taken as axis of x which meets a free boundary at O (Fig. 47). The stress X_0 at the boundary can be deduced from a knowledge of the principal stresses at O, determined as described in Sect. 32. In order to obtain the value of X at A, we have to integrate the rates of change of the shear-stresses in both the xy and the xz planes. We accordingly choose any convenient pair of perpendicular directions for the y and z axes, and cut slices in the xy and xz planes, one from each of two identical frozen models. Measurements in the two slices are made and integrations carried out exactly as in the two-dimensional case, the value of X at A being obtained from the sum of the two integrals.

$$X_A = X_0 - \int_0^A \frac{\partial S_{xy}}{\partial y} dx - \int_0^A \frac{\partial S_{xz}}{\partial z} dx. \quad (33.6)$$

---

[1] M. M. FROCHT and R. GUERNSEY jr.: Nat. Congr. Appl. Mech., June 1951.

Sect. 33. Determination of the principal stresses at any point.

We then measure the isoclinic angle and the secondary principal stress difference at A in each slice. Then if these are denoted by α_1 and (P_1-Q_1) in the slice in the xy plane, we have two relations

$$X - Y = (P_1 - Q_1) \cos 2\alpha_1, \qquad (33.7)$$

and

$$S_{xy} = \tfrac{1}{2}(P_1 - Q_1) \sin 2\alpha_1. \qquad (33.8)$$

Similarly the observations in the xz plane give

$$X - Z = (P_2 - Q_2) \cos 2\alpha_2, \qquad (33.9)$$

and

$$S_{xz} = \tfrac{1}{2}(P_2 - Q_2) \sin 2\alpha_2. \qquad (33.10)$$

Thus the values of X, Y, Z, S_{xy} and S_{xz} are determined.

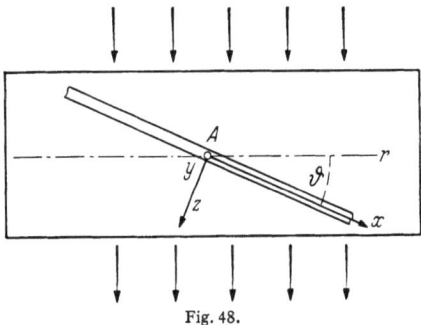

Fig. 48.

The value of the third shear-stress is obtained by an oblique transmission measurement through one of the slices.

If we rotate the xy slice through an angle ϑ about the y axis (Fig. 48) and observe the isoclinic angle α_3 and the secondary principal stress-difference (P_3-Q_3) we get the shear-stress in the plane of the wave front as

$$S_{yr} = \tfrac{1}{2}(P_3 - Q_3) \sin 2\alpha_3. \qquad (33.11)$$

But it can be seen from the diagram that

$$S_{yr} = S_{xy} \cos \vartheta - S_{yz} \sin \vartheta,$$

and hence

$$S_{yz} = S_{xy} \cos\vartheta - \tfrac{1}{2}(P_3 - Q_3) \sin 2\alpha_3. \qquad (33.12)$$

We thus have the six stress-components at A and from them we must derive the values of the principal stresses.

If we consider the equilibrium of a small element $ABCD$ (Fig. 49) at the point A, we find

$$\left.\begin{array}{l} R_x = lX + m S_{xy} + n S_{xz}, \\ R_y = l S_{xy} + m Y + n S_{yz}, \\ R_z = l S_{xz} + m S_{yz} + nZ, \end{array}\right\} \qquad (33.13)$$

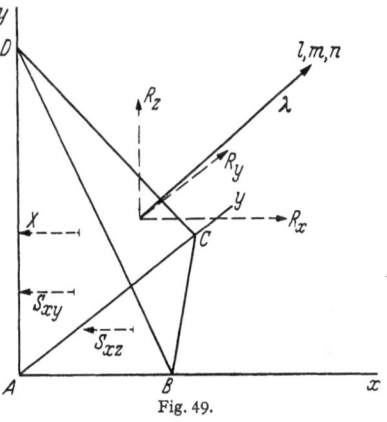

Fig. 49.

where R_x, R_y, R_z are the resolutes of the stresses across the face BCD, and l, m, n are the direction cosines of the normal to this face.

If l, m, n is a principal axis of stress at A, then the total stress over BCD is a normal stress λ, and we then have

$$R_x = l\lambda, \quad R_y = m\lambda, \quad R_z = n\lambda, \qquad (33.14)$$

and Eqs. (33.13) become

$$\left.\begin{array}{l} l(X-\lambda) + m S_{xy} + n S_{xz} = 0, \\ l S_{xy} + m(Y-\lambda) + n S_{yz} = 0, \\ l S_{xz} + m S_{yz} + n(Z-\lambda) = 0, \end{array}\right\} \qquad (33.15)$$

giving

$$\begin{vmatrix} X-\lambda, & S_{xy}, & S_{xz} \\ S_{xy}, & Y-\lambda, & S_{yz} \\ S_{xz}, & S_{yz}, & Z-\lambda \end{vmatrix} = 0. \qquad (33.16)$$

The three solutions to this equation, $\lambda_1, \lambda_2, \lambda_3$ will then be the three principal stresses P, Q and R at A, and their directions may be found by substituting into Eqs. (33.15).

Eq. (33.16) is used also for finding the principal stress differences from the two stress-differences and the shear-stresses obtained by solving five equations of the type of (33.4). In this case the equation is written

$$\begin{vmatrix} (X-Z)+(Z-\lambda), & S_{xy}, & S_{xz} \\ S_{xy}, & (Y-Z)+(Z-\lambda), & S_{yz} \\ S_{xz}, & S_{yz}, & (Z-\lambda) \end{vmatrix} = 0. \qquad (33.17)$$

It is solved for $(Z-\lambda)$, and the differences of the solutions give the principal stress-differences.

III. The tilting stage method of exploration of three-dimensional stresses[1].

34. Outline of the method. As shown in the previous section the determination of the principal stresses, or even of the principal stress-differences, by taking

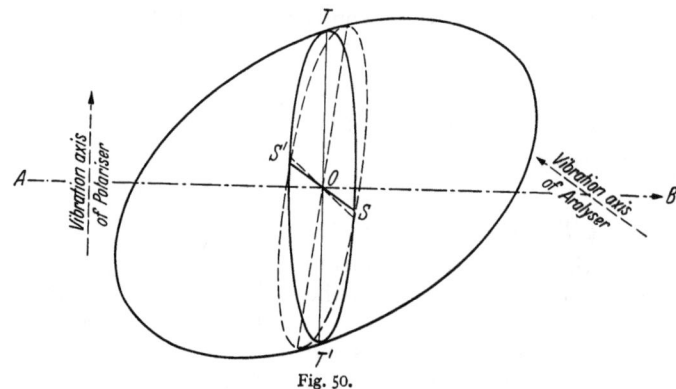

Fig. 50.

measurements of the relative retardations in arbitrary directions involves algebraic calculations in the course of which errors of observation are liable to be multiplied to an undesirable extent. There is, however, a method by which the directions of the principal axes at any point of a frozen slice may be determined directly, after which the slice may be suitably oriented for measurement of the principal stress-differences. This method depends upon the properties of FRESNEL's ellipsoid, and its operation involves only observation of the isoclinic fringe in the slice as the latter is rotated about different axes passing through the point.

As in all observations on frozen-stress models the optical effects are due to the mean of the stresses along the light path through the model, and for satisfactory results to be obtained it is essential to use as thin a slice as possible, since in some cases the light path may be in a direction inclined at a considerable angle to the normal. In what follows, we shall ignore this thickness effect and assume that the directions of the stresses are constant along any light path through the slice. In practice, the fact that clear isoclinic fringes can be observed at the angles of incidence used will be sufficient evidence that this is substantially the case.

If light is passed in any direction AB through a point O of a thin slice of a frozen model (Fig. 50), the directions of polarisation in the material will be the

---
[1] H. T. Jessop and M. K. Wells: Brit. J. Appl. Phys. **1**, 84 (1950).

directions of the principal axes of the section $STS'T'$ of the FRESNEL's ellipsoid at the point by the plane of the wave-front. If the slice is in a plane polariscope with polariser and analyser crossed parallel to the principal axes OS and OT of this section, an isoclinic fringe will be observed passing through the point O.

If now the slice is given a small rotation about an axis coinciding with one principal axis OS, the ellipsoid is also rotated, the new section by the plane of the wave-front being indicated by the broken curve in the diagram. In general the principal axes of the new section will not

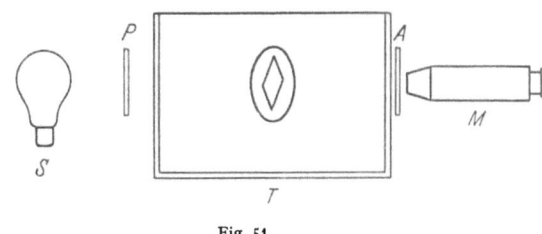

Fig. 51.

be parallel to the polariser and analyser and the isoclinic fringe will be seen to move away from the point O. The only cases in which this displacement of the fringe will not occur, will be those in which the axis OS about which the rotation is made is one of the three principal axes of the ellipsoid.

What is required therefore is some scheme for systematically varying the orientation of the slice in the polariscope in such a way as to arrive at positions in which rotation about one of the secondary principal axes in the plane of the wave-front shall not displace the isoclinic fringe from the point under investigation. The direction of this axis relative to the slice when this condition is fulfilled will define one principal axis of stress at the point.

The general principle of such a scheme was first introduced by VON FEDOROV who designed a universal tilting stage for use with a petro-

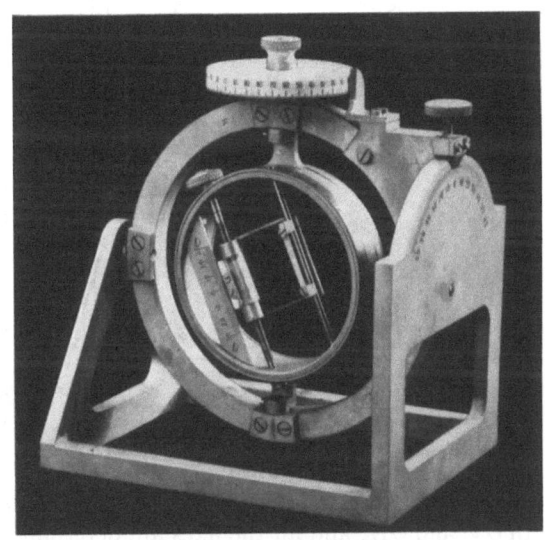

Fig. 52. Universal tilting stage.

graphic microscope for the examination of small crystals in geological work. His stage, and some form of the method, have been used by geologists for many years, but it is only recently that a complete theory has been produced which leads to the systematic operation of the method. It was found also that the large magnifications which were needed in geological work were not necessary when examining photoelastic models and a simpler form of tilting stage has been designed which can be used with a much less elaborate microscope.

The arrangement for photoelastic work is shown in Fig. 51. The tilting stage stands in a tank of immersion fluid, T, and the model is viewed through a low powered microscope M. A source of diffused white light, S, and a small polariser, P, and analyser, A, complete the apparatus. Fig. 52 is a photograph of the stage, which is shown diagrammatically in Fig. 53.

The slice of the model is held by spring clips on a glass platform mounted in a ring R_1, the point under examination being at the centre O of the ring. R_1 can turn in a second ring R_2 so as to rotate the slice about an axis ON normal to its plane. R_2 can turn about a diametral axis OV the bearings of which are on a third ring R_3 which can itself turn about a horizontal axis OH fixed in the stand S. Scales are provided to measure the three rotations. If desired a means may be provided of adjusting the plane of the platform relative to the plane of the axes so as to cater for slices of different thicknesses.

35. Procedure.
The sequence of operations is then as follows:

(i) all scales are set at zero reading and the slice is placed in position with some reference line in it, oriented with regard to the scale on R_1. The stage is then placed in the tank, the point O of the model is focussed on the cross-wires of the microscope, and the polaroids are crossed with their vibration axes vertical and horizontal.

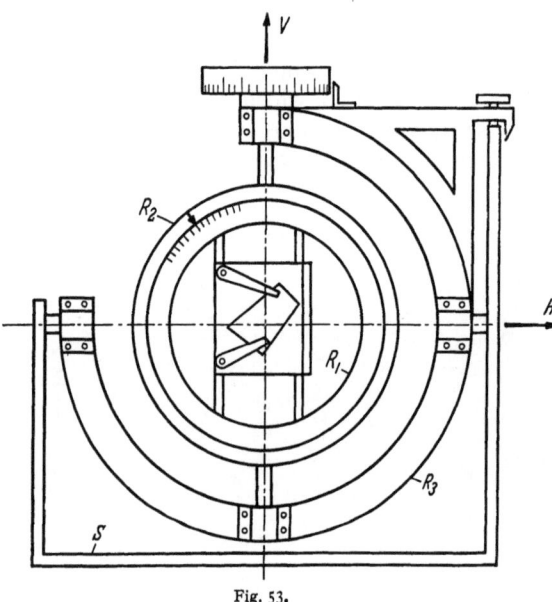

Fig. 53.

(ii) The ring R_1 is rotated until an isoclinic fringe passes through the point O. The secondary principal axes in the plane of the slice will now be vertical and horizontal.

(iii) A preliminary test is carried out by observing the effect of rotating the model about first the horizontal axis and then the vertical axis. If in either case no displacement of the fringe at O is observed, then that axis is one of the principal axes at O and the first part of the problem is solved.

(iv) The principal axis so found is set in the vertical direction, and the slice is then rotated about OV until a position is reached at which the fringe is no longer displaced by rotation about OH. In this position two principal axes coincide with OV and OH, and all the axes are determined.

(v) If, in the preliminary test, rotation about both axes displaces the fringe the following procedure is adopted.

(a) A small rotation $\delta\alpha$ is given about OH. This displaces the fringe from O.

(b) The fringe is brought back to O by a rotation $\delta\vartheta$ of ring R_1 about the normal axis ON.

(c) The fringe is again displaced, but in the opposite direction, by a small rotation $-\delta\alpha$ about OH.

(d) The fringe is brought back to O by a rotation $\delta\varphi$ about the axis OV.

(vi) The sequence of operations in (v) is repeated until a position is reached where rotation about OH produces no displacement of the fringe. OH is then the direction of one principal axis at O.

(vii) Returning to the setting of paragraph (ii) the operations of (v) are repeated, but in the order a, d, c, b, to find the direction of a second principal axis.

(viii) The stage is returned to its original zero setting and the ring R_1 rotated to an inclination differing from that of (ii) by 90°. An isoclinic fringe will again pass through O, but the secondary principal axes will have interchanged their positions.

(ix) The sequences given in (v) and (vii) are now repeated.

There are thus four ways of approaching a principal axis, and it will be shown that these will lead, in theory, to the determination of the three axes, one of the axes being determined twice.

In practice there is, of course, a limit to the angle to which the slice may be tilted while still showing a clear isoclinic fringe, and in some cases the principal axis may not be reached within that limit of tilt. It will be shown, however, that two of the axes can always be found without the tilt exeeding 45°, and the third can then be obtained by a graphical construction.

The most convenient way of presenting the results is to plot the coordinates of the three principal axes on a stereographic projection. This is illustrated in Fig. 54. The circle represents an imaginary sphere whose centre O is the point under examination, and the diametral plane $a\,d\,a'\,d'$ is the plane of the slice. Any axis through O, such as Op, meets the sphere in a point, while the plane perpendicular to it meets the sphere in a great circle lmn. Projected on to the diametral plane through the pole P these give the point p' and the circular arc $l'm'n'$. If Op represents the direction of a principal axis at O, then in the final setting of the tilting stage Op will be coincident with OH, dd' will be coincident with OV, and the angle φ will be the angle through which the slice has been rotated about OV.

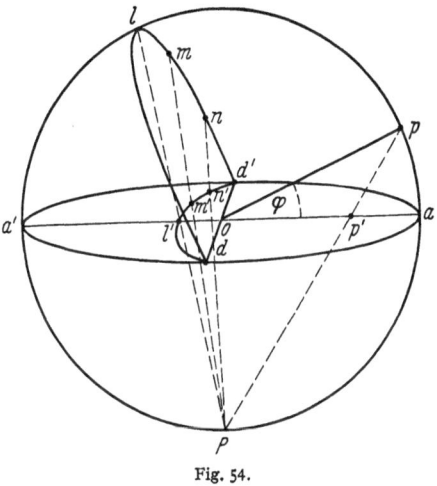

Fig. 54.

In order to facilitate the plotting, a stereographic net is used as a protractor. Fig. 55 shows such a net with two sets of curves, one the projections of great circles having a common diameter dd', the other the projections of small circles whose centres lie on dd'. The intersections of these latter with the perimeter of the net provide a plane protractor for measuring rotations in the plane of the slice. The net may be considered to represent the ring R_2, the line dd' corresponding to the axis OV about which it is tilted. The inner ring carrying the slice may be represented by a sheet of tracing paper, on which an outline of the slice is drawn, and which may be rotated relative to the net, using a drawing pin at the centre as pivot, to reproduce the rotation ϑ of R_1 about the normal axis ON.

If the final setting of the stage when a principal axis is aligned with OH is as shown in Fig. 56, with the slice turned through an angle ϑ about ON, and an angle φ about OV, then the direction of that principal axis relative to the slice is shown on the projection as indicated by the point p' in Fig. 57. The curve $dq'r'd'$ then represents the principal plane normal to Op and the projections of the other two principal axes will lie upon it.

When a second axis is found and plotted on the same diagram we get therefore a check on the accuracy of the observations, and when the second principal plane is plotted its intersection with the first one gives the projection of the third axis.

Fig. 58 shows the stereogram of the principal axes and planes, when the axes are all equally inclined to the plane of the slice. The angle α is the tilt of the slice needed to bring each axis into the plane of the wave-front, which in this

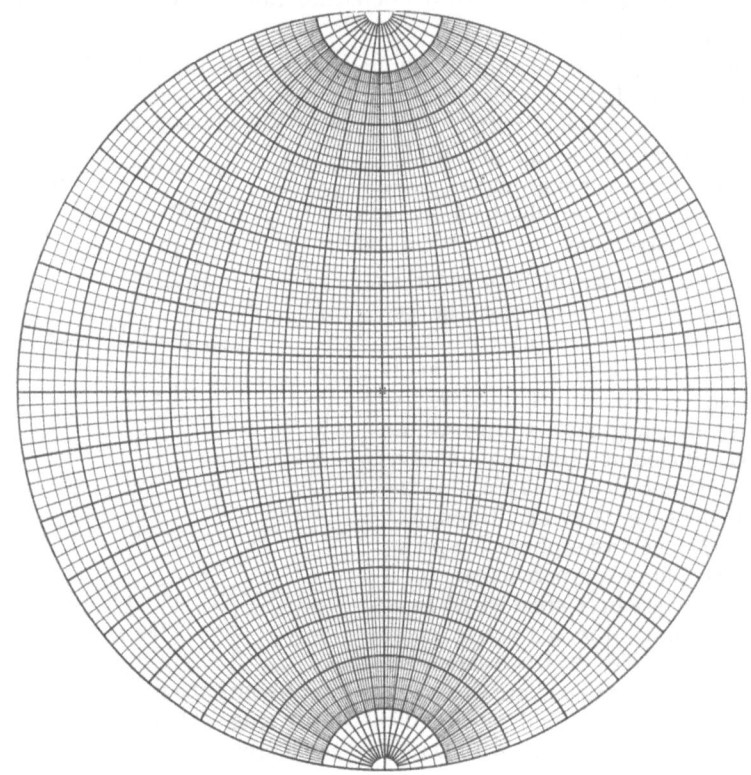

Fig. 55.

case is arc $\sin 1/\sqrt{3}$, or about $35°$. If one axis, say OP, were in the plane of the slice, then one of the other axes at least would be inclined to the plane at not more than $45°$. Two axes therefore will always lie within the permissible angle of tilt.

Fig. 58 represents the most unfavourable case for measurement of the principal stress-differences, for the tilt in this case to bring any principal plane into the wave front is arc $\cos 1/\sqrt{3}$ or about $55°$, and this may be too great for accurate measurement. We can, however, in such a case, rotate the slice so that one principal stress, say P, is brought into the wavefront while the other two are equally inclined to it. The principal stresses in the plane of the wave-front are then P, and $Q\cos^2 45° + R\sin^2 45°$, and the measured relative-retardation will give the value of
$$P - \tfrac{1}{2}(Q + R).$$

If we repeat this process bringing another principal stress, say Q, into the plane of the wave-front, we can find the value of
$$Q - \tfrac{1}{2}(P + R),$$
and from these two quantities the values of $P - Q$ and $P - R$ are found.

Sect. 36. Theory of the method. 191

For the determination of the separate stresses we still have to use some integration method, but a knowledge of the directions of the principal stresses greatly simplifies the calculations. The integration method of Sect. 33 leads to

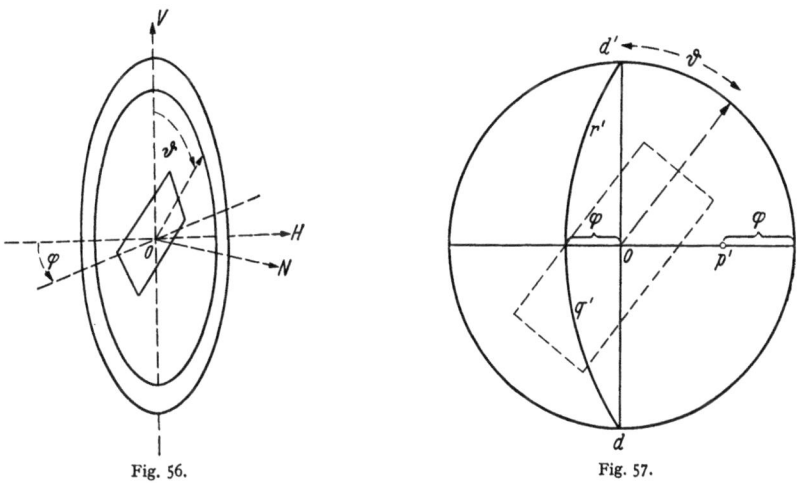

Fig. 56. Fig. 57.

the value of the normal stress X, and this is expressed in terms of the principal stresses by

$$X = l^2 P + m^2 Q + n^2 R,$$

or

$$X = l^2 (P - R) + m^2 (Q - R) + R$$

where l, m, and n are the direction cosines of x relative to the principal axes. With these directions and the principal stress-differences known, the values of the separate stresses follow immediately.

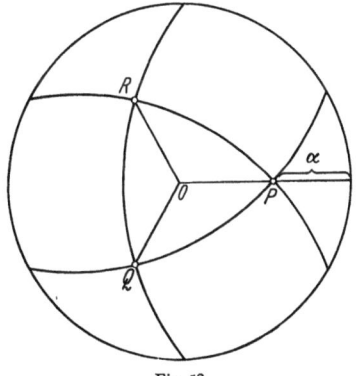

Fig. 58.

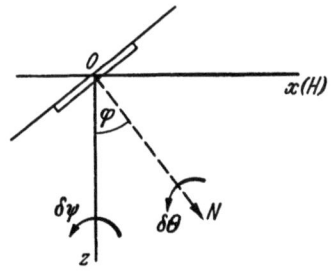

Fig. 59.

36. Theory of the method. We shall consider a sequence of operations as detailed in Sect. 35, starting from an arbitrary setting in which a rotation φ has already been made about the axis OV, so that the plane of the slice does not coincide with the plane of the wave-front. This setting is shown in plan in Fig. 59.

Take rectangular axes x, y, z such that ox coincides with the rotation axis OH and oy is vertical, and let the equation of the FRESNEL ellipsoid at O be

$$a^2 x^2 + b^2 y^2 + c^2 z^2 + 2f y z + 2g z x + 2h x y = 1. \qquad (36.1)$$

Then the xy plane will be the plane of the wave-front, and when the slice is set so that ox and oy are principal axes in their plane, the equation becomes

$$a^2 x^2 + b^2 y^2 + c^2 z^2 + 2f yz + 2g zx = 1, \tag{36.2}$$

a and b being the reciprocals of the principal semi-axes of the section by the plane of the wave-front.

If now we give the ellipsoid a small positive rotation $\delta\alpha$ about the x (or H) axis, the new section in the xy plane will be, (neglecting squares of $\delta\alpha$),

$$\left. \begin{array}{c} a^2 x^2 + (b^2 - 2f \delta\alpha) y^2 - \\ - 2g \delta\alpha \, xy = 1. \end{array} \right\} \tag{36.3}$$

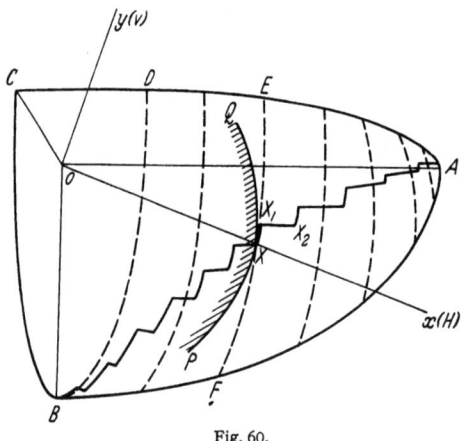

Fig. 60.

The principal axes of this section will then be inclined to ox and oy at an angle $\delta\psi$ where

$$\delta\psi \approx \frac{-g \delta\alpha}{a^2 - b^2} \tag{36.4}$$

and the isoclinic fringe will be displaced from O unless g is zero, in which case ox must be a principal axis of the ellipsoid.

The fringe may now be brought back to O by a rotation of the ellipsoid in the plane of the section, and this is effected by giving a rotation $\delta\vartheta$ about ON, where

$$\delta\vartheta = - \delta\psi \sec \varphi,$$

φ being the inclination of the plane of the slice to the wave-front.

It may be noted that this involves giving an additional rotation $\delta\vartheta \sin \varphi$ about ox, (Fig. 59), and if the angle φ becomes large this may in certain cases result in a further rotation of the principal axes so large as to neutralise the corrective rotation we wish to apply. In such a case the method might fail to determine an axis which lay within the permissible limit of tilt.

The ellipsoid is now given a rotation $-\delta\alpha$ about ox which results in a rotation of the principal axes in the wave-front of

$$- \delta\psi = \frac{g \delta\alpha}{a^2 - b^2}.$$

The rotation about OV by which we restore the axes to the horizontal and vertical is a rotation $\delta\varphi$ of the ellipsoid (36.2) about the y axis. This gives a new section by the wave-front

$$(a^2 + 2g \delta\varphi) x^2 + b^2 y^2 + 2f \delta\varphi \, xy = 1 \tag{36.5}$$

in which the principal axes have been rotated by an angle $\dfrac{f \delta\varphi}{a^2 - b^2}$ and therefore

$$\delta\varphi = \frac{a^2 - b^2}{f} \delta\psi \quad \text{or} \quad \frac{g}{f} \delta\alpha. \tag{36.6}$$

The operation of the method may best be seen by considering a particular case. Fig. 60 represents one octant of the FRESNEL ellipsoid, OA, OB, and OC being the principal semi-diameters, and $OA > OB > OC$. PXQ is part of the section by the xy plane, OX being a principal axis of the section and coinciding with the x axis. The broken lines are loci on the ellipsoid of points equidistant from O, BD being a circular section. Then since OX is a principal axis of the section PXQ, the length of the radius vector of the section has a stationary value at X, and PXQ and the locus EXF will have a common tangent at that point. Also if X lies in the segment ABD, OX will be a major axis of the section, so that (a^2-b^2) will be negative. We can deduce also from the geometry of the ellipsoid that in Eq. (36.2), the coefficient g will always be positive, while f will be positive and negative in alternate octants, being positive in the octant ABC as shown.

Operation (a) of Sect. 35 moves the principal axis of the section through an angle $\delta\psi$ to OX_1, and operation (b) rotates the plane of the section so that relative to the ellipsoid X moves to X_1. Operation (c) moves the principal axis in the reverse direction, and operation (d), the rotation about oy, moves X relative to the ellipsoid from X_1 to X_2, where it again becomes a principal axis of the section.

Thus the sequence of operations causes X to move in a series of steps alternately tangential to and normal to the curves EF until it reaches the extremity A of the major axis. The alternative sequence starting from the same point X leads to the extremity of the medium axis B. The third and fourth sequences start with the other principal axis of the section, which will be a minor axis and will therefore lie in a segment such as BCD. In this case it can be shown that the two alternative paths lead to C and B respectively.

It should be noted that there is automatic adjustment of the magnitudes of the successive steps in any path, for as X approaches either of the principal sections AB or AD, f tends to zero and the displacement $\delta\varphi$, $\left(=\frac{g}{f}\delta\alpha\right)$, which moves X parallel to the boundary, tends to become infinite. On the other hand, as X approaches the circular section BD, (a^2-b^2) tends to zero, and $\delta\varphi$ becomes very small due to Eq. (36.6) while $\delta\psi$ increases, Eq. (36.4). Thus if the steps are kept sufficiently small none of these boundaries can be crossed, and the path must terminate at a principal axis.

IV. The scattered light method of observation.

37. Theory of the method. When a beam of light passes through any medium, there is always a certain amount of "scattering" of light from all points on the path of the beam. In some very transparent materials such as clear glass this scattering is extremely small, and the path of the beam through the material is practically invisible. In others, which contain fine particles in suspension, for example a dust-laden atmosphere, there is a large amount of scattering by reason of which the path of the beam is seen as a bright band traversing the medium. There are, however, involved in this effect two distinct types of scattering. Where the scattering is due to particles whose dimensions are large in comparison with the wave-length of the light, the effect is very largely one of diffused reflections from the particles similar to those occurring when any opaque body is illuminated by the incidence of light upon it. The scattered light in this case is unpolarised, whatever be the characteristics of the initial beam, and the scattering has sensibly the same intensity in all directions. Where, however, the scattering is due to fine particles or to very large molecular groups whose dimensions are of the order of a wave-length of light, each point of the path acts as a secondary source of light, and if the light in the original beam is polarised, the scattered waves have different intensities in different directions. It is the second type of scattering which exhibits the photoelastic effect[1], and which we shall consider. If, as is frequently the case, a proportion of the first type is also present, its effect will be merely to render the interference phenomena observed less distinct.

If the plane of the paper in Fig. 61 represents the plane of the wave-front of a plane polarised beam passing through an isotropic scattering medium, OA being the amplitude of the vibrations in the beam, then the scattered wave propagated in any direction OP inclined at an angle α to OA will have an amplitude $\sigma OA \sin \alpha$, where σ is a coefficient of scattering for the medium. If the beam of light, initially plane polarised, is passing through a stressed photo-elastic material, then at any point of its path it may be considered to consist of two plane polarised waves whose vibrations are in the directions of the secondary principal stresses in the plane of the wave-front. The velocities of the two waves will in general be different, and consequently their phase-difference will be continually changing from point to point of the path of the beam.

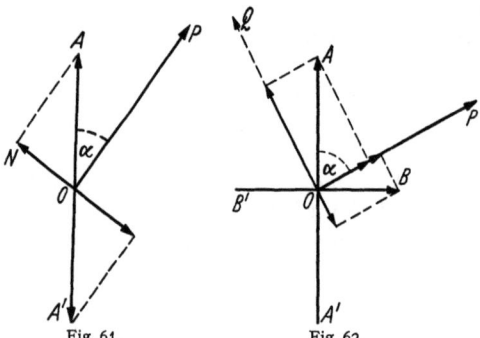

Fig. 61. Fig. 62.

Let Fig. 62 represent the plane of the wave-front at any point of the path in such a case, OA and OB being the amplitudes a and b of the two waves. Each of the waves may be considered to scatter its own quota of light according to the law stated above, and there will therefore be propagated in any direction OP two waves whose amplitudes are respectively $\sigma a \sin \alpha$ and $\sigma b \cos \alpha$. The vibrations of these secondary waves will be parallel, since both lie in the plane of the wave-front and normal to the direction of propagation OP. The displacements in the two waves may thus be added algebraically to give the displacement in the resultant scattered wave.

Let the displacements in the two primary waves at the point O of the path be represented by

$$u = a \cos p t \quad \text{in direction } OA$$

and

$$v = b \cos (p t + \psi) \quad \text{in direction } OB.$$

The light scattered in direction OP will then be the sum of the two waves

$$\left. \begin{array}{l} q_1 = \sigma a \sin \alpha \cos p t \\ q_2 = - \sigma b \cos \alpha \cos (p t + \psi). \end{array} \right\} \quad (37.1)$$

These add to give

$$q = \sigma (a^2 \sin^2 \alpha + b^2 \cos^2 \alpha - 2 a b \sin \alpha \cos \alpha \cos \psi)^{\frac{1}{2}} \cos (p t + \beta). \quad (37.2)$$

As the value of the phase-difference ψ changes from point to point of the beam, the amplitude of this scattered wave varies from

$$\sigma (a \sin \alpha - b \cos \alpha) \quad \text{when } \psi = 0 \text{ or } 2i\pi$$

to

$$\sigma (a \sin \alpha + b \cos \alpha) \quad \text{when } \psi = (2i - 1) \pi,$$

---

[1] R. Weller: J. Appl. Phys. **10**, 266 (1939); **12**, 610 (1941).

and if these amplitudes differ sufficiently the beam will appear to an observer at P to be crossed by a series of dark fringes. This effect will be most marked when $a=b$ and $\alpha=45$ or $135°$, for then the fringes will be black, (if the light is monochromatic) and the intensity of light between them will be greatest. If white light be used the initial zero fringe will be black, but all the others will be coloured as in the two-dimensional fringe-pattern.

It should be noted that the positions of maximum and minimum amplitudes of the scattered wave are interchanged when the direction of viewing is changed

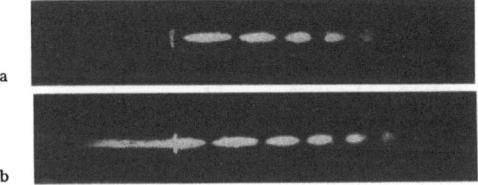

Fig. 63 a and b. Photographs of fringes in light scattered from a "needle" ray traversing a stressed model. (a) Zero and whole wave-length fringes, (b) half wave-length fringes. Model material, Marco resin with added silica.

from 45 to $135°$, and that while for one direction the fringes are at points where the relative-retardation of the two waves is zero or a whole number of wavelengths, for the other direction they indicate odd numbers of half wave-lengths relative-retardation. In practice these two conditions are fulfilled by keeping the direction of observation constant, at $45°$ to the directions of the principal stresses in the plane of the wave-front, and setting the polariser so that its vi-

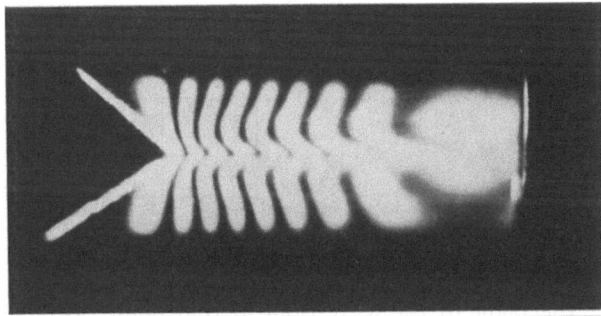

Fig. 64. Fringes in a „ribbon" beam traversing a notched plate model under vertical compression. Model of Marco resin with added silica.

bration axis is (a) parallel to and (b) normal to the direction of observation. Fig. 63 shows photographs of the fringes observed in these two conditions in a narrow pencil of light passing through a stressed model. Fig. 64 shows a "ribbon" beam passing through a model of a notched plate under vertical compression. By measurement of the positions of these we are able to derive the rate of change of relative-retardation at all points on the path of the beam, and thus to obtain some information about the stress-distribution in a three-dimensional model without having to isolate elements of it by cutting slices. The method can therefore be used with a model under direct load when the conditions of a problem would render the stress-freezing technique undesirable.

It should be noted that, since the stresses in general change from point to point in all directions in the material, observation must be restricted to the scattered rays emanating from a very small area of the wave-front. This is done by using as the primary beam a very narrow pencil of rays. In practice it is frequently of advantage to use a thin "ribbon" beam and to take observations only at points lying on one line parallel to the path of the beam. One then obtains a picture of fringes over a considerable area which renders their identification and measurement much easier.

38. The interpretation of the fringe observations. *Case 1.* In attempting to establish the relations between the interference fringes observed in the scattered light and the stresses in the medium, there are two cases to be considered. The first, and simplest, is that in which the secondary principal stresses in the plane of the wave-front of the primary beam maintain constant directions for all points on the path of the beam. In this case the two oppositely polarised waves maintain constant directions of vibration, and the rate of change of phase between them at any point as they traverse the stressed model is a measure of the difference of the secondary prin-

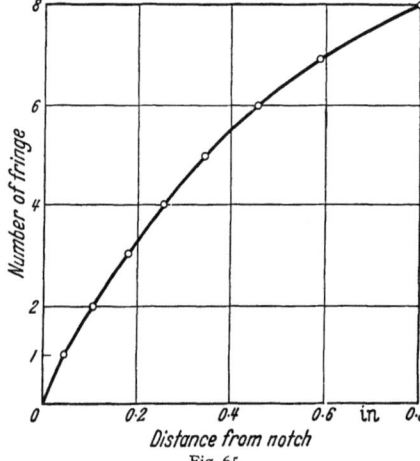

Fig. 65.

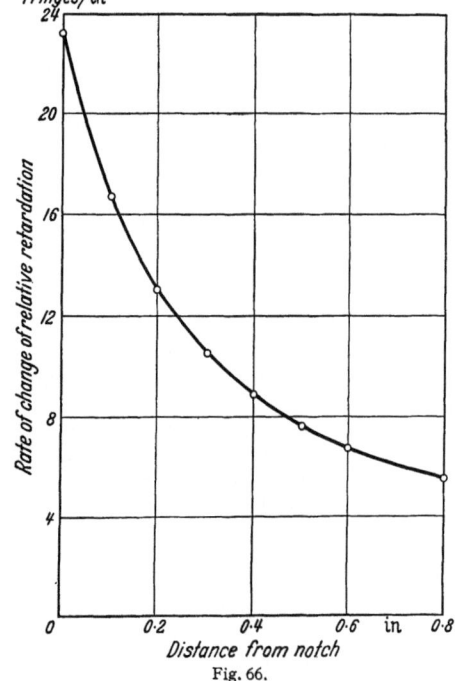

Fig. 66.

cipal stresses in the plane of the wave-front at that point. If in such a case we measure the positions of the fringes on the path of a narrow beam of monochromatic light, and plot a graph of phase-difference against distance, the rate of change of relative retardation at any point of the path is given by the slope of the graph at that point, and the product of this quantity and the fringe-value of the material gives the difference of the secondary principal stresses.

Fig. 65 is the graph of the fringes shown in Fig. 64. If the phase-difference is in wave-lengths of the monochromatic light used and the distance is in inches, the stress-difference, $f \times$ slope of graph, will be in lb. per sq. inch. Fig. 66 shows the slope plotted against distance across section.

The sign of the relative-retardation in the primary beam is not, of course, indicated by the fringes. In many cases we shall know, from general considerations of the loading of the model, what the sign is. If not, we may determine it by introducing, by means of a compensator, a small relative retardation of known sign into the beam before it enters the model. Observation of the direction of movement of the fringes produced by this additional retardation will then give the required information.

An important case to note is that in which the stress-difference in the plane of the wave-front changes sign at some point in the path of the beam. In such a case the order of the fringes will begin to go down again at the point of reversal. Such a condition will be clearly shown by use of the compensator method, for an additional relative-retardation will cause the fringes on opposite sides of the reversal point to move in opposite directions.

Case 2. The second case is that in which the secondary principal stresses in the plane of the wave-front are changing in direction from point to point of the path of the beam. In this case the two oppositely polarised waves do not retain their identities and their independence of one another, but are continually being repolarised in new directions. Hence the two waves whose phase difference is observed at any one point are not the same waves as are observed at another point, and the change of phase-difference observed is not, in general, equal to the relative-retardation due to stress, but contains also some factor due to the rotation of the axes.

We can obtain some information as to the magnitude of the rotation effect by considering the changes in relative-retardation due to stress and the changes in orientation of the axes to take place in successive small steps[1].

In the most general case we may take the displacements in the two polarised waves at time t at a point x of the path of the beam to be given by

$$\left. \begin{array}{l} u = a \cos p t \\ v = b \cos (p t + \psi). \end{array} \right\} \quad (38.1)$$

Suppose the stresses to maintain their directions for a small distance δx of the path, and let the relative phase-retardation produced in this distance by the stresses be $\delta\varphi$. Then at the point $x + \delta x$ the two waves will be

$$\left. \begin{array}{l} u = a \cos p t, \\ v = b \cos (p t + \psi + \delta\varphi) \end{array} \right\} \quad (38.2)$$

(neglecting phase changes common to the two waves).

Now suppose the directions of the axes of stress in the wave-front to turn through a small angle $\delta\alpha$. The two waves, re-resolved in these directions will be

$$\left. \begin{array}{l} u' = a \cos \delta\alpha \cos p t + b \sin \delta\alpha \cos (p t + \psi + \delta\varphi) \\ v' = b \cos \delta\alpha \cos (p t + \psi + \delta\varphi) - a \sin \delta\alpha \cos p t, \end{array} \right\} \quad (38.3)$$

or

$$\left. \begin{array}{l} u' = [a \cos \delta\alpha + b \sin \delta\alpha \cos (\psi + \delta\varphi)] \cos p t - \\ \qquad - b \sin \delta\alpha \sin (\psi + \delta\varphi) \sin p t, \\ v' = [b \cos \delta\alpha \cos \delta\varphi - a \sin \delta\alpha \cos \psi] \cos (p t + \psi) - \\ \qquad - [b \cos \delta\alpha \sin \delta\varphi + a \sin \delta\alpha \sin \psi] \sin (p t + \psi). \end{array} \right\} \quad (38.4)$$

If we neglect terms of the second order in the small quantities $\delta\alpha, \delta\varphi$, these reduce to

$$\left. \begin{array}{l} u' = (a + b\, \delta\alpha \cos \psi) \cos p t - b\, \delta\alpha \sin \psi \sin p t, \\ v' = (b - a\, \delta\alpha \cos \psi) \cos (p t + \psi) - (b\, \delta\varphi + a\, \delta\alpha \sin \psi) \sin (p t + \psi) \end{array} \right\} \quad (38.5)$$

---

[1] H. T. JESSOP: Brit. J. Appl. Phys. **2**, 259 (1951).

which may be written

$$u' = (a + \delta a) \cos(pt + \delta\psi_1),$$
$$v' = (b + \delta b) \cos(pt + \psi + \delta\psi_2)$$
(38.6)

where

$$a + \delta a = (a^2 + 2ab\,\delta\alpha\cos\psi)^{\frac{1}{2}} \approx a + b\,\delta\alpha\cos\psi,$$
$$b + \delta b = (b^2 - 2ab\,\delta\alpha\cos\psi)^{\frac{1}{2}} \approx b - a\,\delta\alpha\cos\psi,$$
(38.7)

$$\tan\delta\psi_1 \approx \delta\psi_1 = \frac{b\,\delta\alpha\sin\psi}{a + b\,\delta\alpha\cos\psi} \approx \frac{b}{a}\delta\alpha\sin\psi,$$
$$\tan\delta\psi_2 \approx \delta\psi_2 = \frac{b\,\delta\varphi + a\,\delta\alpha\sin\psi}{b - a\,\delta\alpha\cos\psi} \approx \delta\varphi + \frac{a}{b}\delta\alpha\sin\psi.$$
(38.8)

The increase in phase-difference between the two waves will be

$$\delta\psi = (\delta\psi_2 - \delta\psi_1) \approx \delta\varphi + \left(\frac{a}{b} - \frac{b}{a}\right)\delta\alpha\sin\psi.$$
(38.9)

Proceeding to the limit we may write

$$\frac{d\psi}{dx} = \frac{d\varphi}{dx} + \left(\frac{a}{b} - \frac{b}{a}\right)\sin\psi\,\frac{d\alpha}{dx}$$
(38.10)

or

$$\frac{d\psi}{dx} = \frac{d\varphi}{dx}\left\{1 + \left(\frac{a}{b} - \frac{b}{a}\right)\sin\psi\,\frac{d\alpha}{d\varphi}\right\}.$$
(38.11)

This equation cannot be integrated without a knowledge of the value of $d\alpha/dx$ which is not available in any practical problem. From it we can deduce, however, that while in general the space rate of change of the observed phase-difference ψ differs from that of the relative-retardation φ the two rates will be equal when either $a=b$ or $\sin\psi=0$.

In practice the method can only be usefully applied in cases where the direction of one principal stress is known, and in most such cases the rotation of axes in the plane of the wave-front does not occur.

39. Applications of the method. There are four types of problem in which the scattered light method of observation may be employed with advantage. These are:

(a) The separation of the principal stresses in a two-dimensional model.

(b) The determination of the second principal stress-difference at points in a plane of symmetry of a three-dimensional model.

(c) The complete determination of the stresses in a prismatic bar of any section under pure torsion.

(d) The determination of the principal stress-difference, and of the approximate directions of the principal stresses, at any point on the "free" surface of a three-dimensional model.

Case (a). In a two-dimensional model examination by normal methods in the polariscope yields the magnitude of the stress-difference $(P-Q)$ and the directions of the principal stresses at any point O (Fig. 67).

If now we pass through the model a narrow polarised beam in any direction AB parallel to the plane faces, the principal stresses in the plane of the wave-front will be the stress X, $(=P\sin^2\alpha + Q\cos^2\alpha)$, and the stress normal to the plate, which will be zero. The slope of the graph of the scattered light fringes at O will therefore give the value of X, and this, combined with the value already obtained of $P-Q$ yields at once the values of the separate stresses.

A particularly simple case, and the case most frequently met in practice, occurs when the stresses on an axis of symmetry of the model such as CD are required. Here the slope of the graph at any point gives directly the value of the stress Q.

It should be noted, however, that before applying this method to a two-dimensional problem, care must be taken to ensure that the plate of material used is free from transverse stress effects. Some two-dimensional materials, e.g. C.R.39, have initial stresses which have no effect upon the ordinary two-dimensional fringe-pattern, but would give the same optical effect in scattered light observations as a stress normal to the plane of the plate.

Case (b). The operation of the method in the case of a plane of symmetry in a three-dimensional model is precisely similar to that in case (a), except that

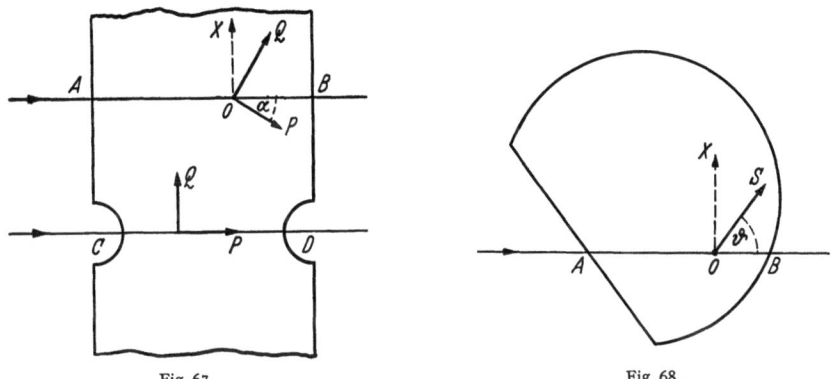

Fig. 67. Fig. 68.

we have now to take into account the third stress R normal to the plane of symmetry. The scattered light observations might be made, if desired, by passing the beam, in the plane of symmetry, through the uncut model, but since in general a slice containing the plane of symmetry must be cut in order to measure $P - Q$, it is usually more convenient to use this slice for the scattered light observations also. If we take Fig. 67 to represent the slice, then for the beam AB the stresses in the plane of the wave-front are X and R, so that the slope of the graph gives $P \sin^2\alpha + Q \cos^2\alpha - R$, and adding this quantity to $(P-Q)\cos^2\alpha$ we get $P - R$.

In both the two and the three-dimensional cases this method is particularly valuable in regions close to notches or load-points where the magnitudes of the principal stresses are changing very rapidly from point to point and large errors might be introduced by either oblique transmission or integration methods.

Case (c). The case of a prismatic bar in torsion is a unique one in that the stresses acting across any normal section of the bar are everywhere pure shear-stresses, and that the stresses over all normal sections are identical. It should be remembered, however, that these conditions apply only to regions of the bar sufficiently far from points of application of loads, or from points where the cross-section changes, for the effects of the local stresses produced by these to have become negligible.

Let Fig. 68 represent a normal cross-section in the "uniform" portion of the bar, and consider a beam of polarised light passing along any line AB in the plane of the section. At any point O the stress across the section is a shear-stress S acting in some direction ϑ to AB. The only stress in the plane of the wave-front at O will then be a shear-stress $S \sin \vartheta$ in direction OX, and this is equivalent to principal stresses $S \sin \vartheta$ and $-S \sin \vartheta$ in directions inclined at 45° to the plane of the section. The principal stresses in the plane of the wave-front will thus be in the same directions at all points on the path AB, and the slope of the relative-retardation graph at any point will be

$$\frac{\partial \varphi}{\partial x} = 2S \sin \vartheta / f.$$

Thus the shear-stress at any point will be given by

$$S = \frac{1}{2} f \operatorname{cosec} \vartheta \frac{\partial \varphi}{\partial x} = \frac{1}{2} f \frac{\partial \varphi}{\partial n}$$

where n is the direction normal to S, and also the direction in the section in which the relative retardation is changing most rapidly, i.e. in the direction normal to the fringes observed.

This same condition applies for any beam of light parallel to AB, and if therefore we use a broad ribbon beam of light which covers the whole section we shall get a fringe pattern such that (a) the direction of the shear-stress at any point is tangential to the fringe passing through the point, and (b) the magnitude of the shear-stress at the point is $\frac{1}{2} f$ times the gradient of relative-retardation in the direction normal to the fringe.

The conditions for observing in this case are that the vibrations of the incident light are either parallel to or normal to the plane of the section, and the direction of viewing is normal to the section. Then since the stresses are the same at all points on a line parallel to the axis of the bar, the same pattern will be observed in any section, and we can use quite a thick beam of light without affecting the accuracy of the observations.

Case (d). The determination of the surface stresses at a point where no conditions of symmetry exist is the only practical problem in which the question of rotation of the axes occurs. At a point on an unloaded boundary we know that two principal stresses lie in the tangent plane, and therefore if the incident beam is normal to the surface at the point of entry, the stresses in the wave-front at that point are principal stresses. As the beam penetrates into the model, however, the stresses in the wave-front will not, in general, maintain their directions, neither will they, in general, remain principal stresses. Thus, while fringes will be observed in the beam, the graph of these will give us useful information at one point only—the point of entry of the beam—for here we can ensure that the phase-difference ψ is zero, and hence that the second term in Eq. (38.10) vanishes. At this point therefore the slope $\partial \psi/\partial x$ of the graph obtained from the fringes is equal to $\partial \varphi/\partial x$ the rate of change of relative-retardation due to stress, thus enabling us to find the difference of principal stresses on the boundary.

The directions of these stresses are obtained by rotating the model about an axis coincident with the axis of the light beam, while viewing from a direction parallel to the polariser vibration axis. The fringes appear with greatest distinctness when the principal stresses are at 45° to the direction of viewing, and while this direction cannot be accurately determined it can usually be found to within about 5°, which is accurate enough for most purposes.

40. Method and apparatus. *α) Materials.* The scattering property of any piece of unstressed photoelastic material may easily be tested by passing a beam of plane-polarised light through it and observing the variation of intensity of the scattered light as the polariser is rotated. For a material to give satisfactory results the path of the beam should appear brightly illuminated when the vibration direction in the polarised beam is normal to the direction of viewing, and should be practically invisible when the polariser is turned through 90° from this position. Some photoelastic materials possess good natural scattering properties, but the degree of these appears to vary greatly from one specimen to another. In the case of materials which are cast in the laboratory it has been found that good scattering properties can be produced by introducing into the material during casting very small quantities of finely divided silica or alumina, sufficient only to give a slight opalescence to the material.

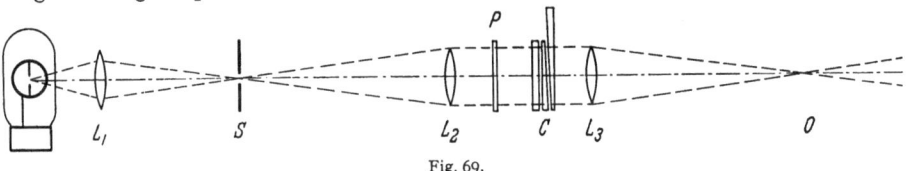

Fig. 69.

β) *The illuminating system.* The three requirements in the illuminating system are that the beam of light through the model shall (a) be very narrow, (b) have a high intensity and (c) be sufficiently "parallel" to avoid errors due to variation in the directions of the rays of the beam. These requirements are all satisfied by the optical system shown in Fig. 69. The light from a high pressure mercury arc lamp of 250 or 500 watts is focused on to an adjustable parallel slit, S by a lens L_1. The slit is at the principal focus of a lens L_2 which gives a well collimated beam in which are placed the polariser P and a uniform field compensator C of the SOLEIL-BABINET type. A third lens L_3 brings the beam of light to a focus at O. If L_2 and L_3 are of the same focal length we have at O a ribbon of light whose breadth and thickness are respectively approximately equal to the length and width of the slit. If also the ratio of the focal length of L_3 to its aperture is fairly large, say 10; 1, then the thickness of the beam will remain small over a few millimetres of its length, allowing for accurate localisation of observations in this region. Also with this ratio of 10: 1, no ray of the beam is inclined to the normal axis at more than 3°, and errors ude to obliquity will be negligible. The model to be observed is therefore placed, in a glass tank of liquid of its own refractive index, so that the beam passes along the required path, with the focal point O at the point to be observed. If measurements over a long distance on the path are required, the focal point may be moved forward in the model by moving the lens L_3. The settings of polariser, compensator and model axes are shown in Fig. 70, the plane of which is the plane of the wave front. If OV is the direction of viewing, the vibration axis of the polariser will be either OP_1 or OP_2, the axes of principal stress in the model will be in directions Ox and Oy at 45° to these, and the compensator axes C_1, C_2 will also be parallel to Ox and Oy since it will be required to add directly the relative retardations in compensator and model. Observation will normally be by a travelling microscope with quite a small magnfication.

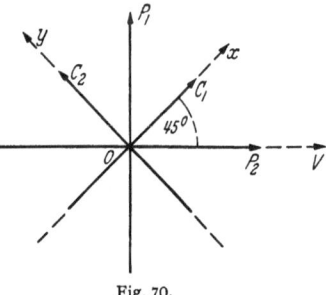

Fig. 70.

γ) *Apparatus.* In cases where the directions of the stresses to be measured are known, that is in all the cases mentioned in Sect. 39 except that of exploring the surface stresses, the only apparatus needed other than the optical equipment is some form of holder by means of which the model may be held in the tank of liquid at the correct orientation. In the case of surface stresses in unknown directions we need to have also a means of rotating the model about an axis which coincides with the axis of the incident beam. A model-holder which will allow of this, and which also provides convenient means of adjusting the position of the model in any case is illustrated in Fig. 71.

The model is held between jaws in a screw clamp which can be rotated about a horizontal axis A_1, perpendicular to the plane of the diagram, and a vertical axis A_2. The clamp is carried by an arm R movable vertically and horizontally on slides S_1, S_2, the whole assembly being mounted to turn about the axis A_3, fixed to coincide with the axis of the light-beam. When the model has been adjusted so that the light passes along the desired path, the holder is turned

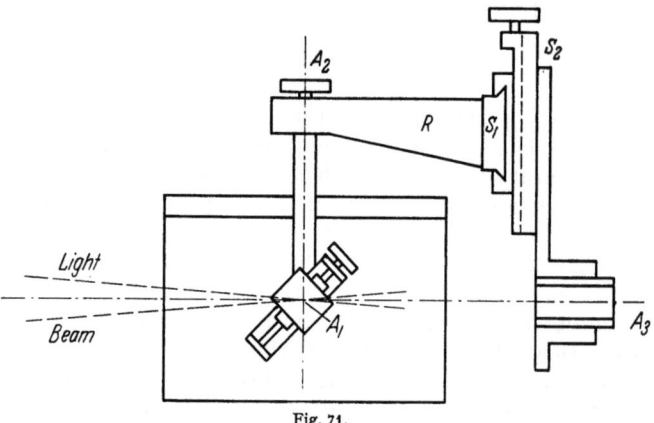

Fig. 71.

about A_3 to the position in which the fringes are most clearly defined. The principal axes in the model are then inclined at 45° to the direction of the polariser vibration axis.

δ) *Method of observation.* In cases in which the stresses are high and a large number of fringes are visible it may be sufficient to plot the graph of the whole and half wave-length fringes to get the stresses. To do this one has to use monochromatic light, and a green filter must be inserted in the primary beam. This reduces the intensity of the light considerably, and may render the fringes too indistinct for accurate measurement. The alternative method is to use the compensator to introduce into the beam, before it enters the model, a relative-retardation of opposite sign to that produced in the model by the stress-difference. This causes the black zero fringe to move along the beam from the boundary to a point where the magnitude of the retardation in the model is equal to that in the compensator. By measuring the displacements of the fringe for successive small increments of retardation imposed by the compensator we can get a very accurately plotted graph, even when the stresses in the model are small. It should be noted, however, that small stress-differences mean small rates of change of relative retardation, and therefore broader and more diffuse fringes the centres of which are more difficult to determine.

D. Practical applications.

I. Two-dimensional examples.

41. Stress concentrations. The danger point in any part of a structure or machine is usually associated with some feature in the shape of the part, called by engineers a "stress-raiser". A stress-raiser may be an isolated feature in a part whose shape is otherwise a simple one; for example a hole or notch in a straight bar. In such a case it is usually possible to calculate the value the stress in the bar would have if the stress raiser were not present, and the effect of its presence is then expressed in terms of a *stress-concentration factor* (S.C.F.). This factor is the ratio of the greatest stress in the neighbourhood of the stress-raising feature to the value of the corresponding stress in the absence of that feature.

In other cases the shape of the whole component is a complex one, and the problem becomes one of finding the distribution of stress everywhere in it. There will still be regions of high stress-concentration which will constitute the danger points, and these may still be associated with features such as holes, notches or fillets, but it is not possible to express their effects in terms of stress-concentration factors. In such cases the stress-distribution characteristics are represented by graphs or tables showing the stresses produced in various critical regions by the application of specified loads to the component.

The photoelastic method of stress-exploration may be employed in many of these problems, and in the cases in which it is applicable it will usually be found to be quicker and more accurate than any other method. In some cases the photoelastic investigation will be required merely as a check upon a design which the engineer has evolved on the basis of approximate theoretical calculations, and many industrial laboratories are concerned only with this type of problem. There is, however, wide scope for the method in the carrying out of systematic investigations into the more general aspects of stress-distribution in order to obtain reliable data on which to base future designs.

The principles and techniques described in previous sections will form the basis of any such investigations, but the particular technique to be used in any case has to be selected with due consideration to the mechanical conditions of the problem and to the amount of information which is required.

The examples which follow give some idea of the range of problems to which the method may be applied and illustrate some of the special techniques used.

Acknowledgments. The examples are all taken from work carried out by staff and postgraduate students in the Photoelastic Laboratories of the Department of Civil and Municipal Engineering at University College, London.

The writer wishes to acknowledge the work of staff, students and technicians of the Department in the course of these investigations, and to thank the Head of the Department, the journals in which parts of the work have been published, and the firms for whom some of the investigations were carried out, for permission to reproduce photographs and other details.

42. Loading the model. In many practical problems interest may be confined to one special region of a component, and it may be impracticable to use a model of the whole component. In such cases we frequently cannot reproduce the exact loading conditions of the prototype. For example, in an investigation of the stresses in the root of an aeroplane wing-spar we do not need a model of the whole spar, and we cannot reproduce on a model of a part of it the exact forces which would be acting on that part in the prototype. In such cases we must ensure that the load system which we apply to the model shall produce *in the region to be examined* the same stress-distribution as would occur in that region as part of the whole component.

In designing the loading system to be applied to any model we rely upon a principle first propounded by DE ST. VENANT, which states in effect that the stress-distribution in any region of an elastic body distant from a region of applied forces depends only upon the static resultant of the forces and is independent of their actual distribution. This statement of the principle is not exact, but both experimental and analytical results confirm that the effect upon the stresses at any point of local variations in the manner of applying loads decreases rapidly as the distance from the loaded region increases, and becomes negligible at a comparatively small distance. Just what that distance may be in any particular case is a question which can only be decided by experience. In the case of a long bar loaded at its ends a very rough estimate is that it is of the same order as the width of the bar.

An illustration of this is shown in Fig. 72. Equal and opposite couples have been applied to the ends of a rectangular bar by means of four concentrated loads. The local effects of these loads have sensibly disappeared at a short distance from the inner load points, leaving a stress-pattern of parallel fringes indicating a uniform bending moment across all sections in the central part of the bar, — the simplest distribution whose resultant is a pure couple.

Fig. 72. Fringe-pattern in beam under flexure, showing rapid disappearance of effects due to local stresses produced by end loads.

In the case of a wide plate to which equal tensile loads are applied at points equally spaced along one edge, the stress in the plate becomes sensibly uniform at a distance from the load points equal to the spaces between them. This result is employed in loading the plate models of Sects. 44 and 45.

It should be noted that DE ST. VENANT's principle applies not only in the case of external forces but also in the case of internal forces across any section due to the stresses in the model. Thus in the example of Sect. 43 the upper half of the bar is in equilibrium under whatever forces are applied at its upper end plus the system of forces due to the stresses acting across the section through the hole. This latter system, while statically equivalent to a uniform load across the section, is by no means uniform, but the stress system in the bar tends to a uniform distribution at a sufficient distance from the hole.

Fig. 73 is an illustration of a tension test-bar in which it was desired to obtain a uniform stress across the bar in its central region. The length of the narrow parallel portion of the bar however is insufficient to produce this, and the fringe-pattern shows a considerable variation in stress across the mid-section.

The figure also illustrates very well the stress-concentrations which occur at a change of cross section. In this case the circular fillets by which the width of the bar is increased at each end are of comparatively small radius, and at each point of sudden change of curvature, i.e. where the straight edge of the central portion runs into the circular arc, there is a stress about 30% greater than the mean stress in the central part. The greater the radius of the fillet the smaller will be the stress-concentration, but some concentration will always occur with a circular fillet.

Fig. 74 shows the ideal type of transition fillet, cut in the shape of part of a sine curve whose point of inflexion occurs at the junction with the straight edge of the central part. Thus there is no sudden change of curvature, and if the fillet is properly designed the stress at the edge of the bar decreases smoothly as the width of the bar increases. Photoelastic tests provide by far the best method of exploring the design of such fillets.

43. Rectangular bar with circular hole under tension[1] (I. JONES and P. A. HOLLISTER). This problem has been the subject of a large number of investigations, both theoretical and experimental, but it is of considerable importance in many engineering applications, and is typical of the kind of problem which may be systematically explored by photoelastic tests.

A uniform rectangular bar of photoelastic material of width D and thickness t has a circular hole

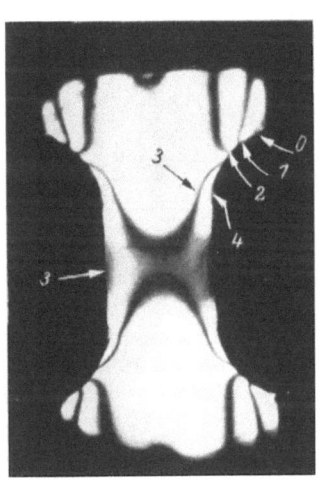

Fig. 73.

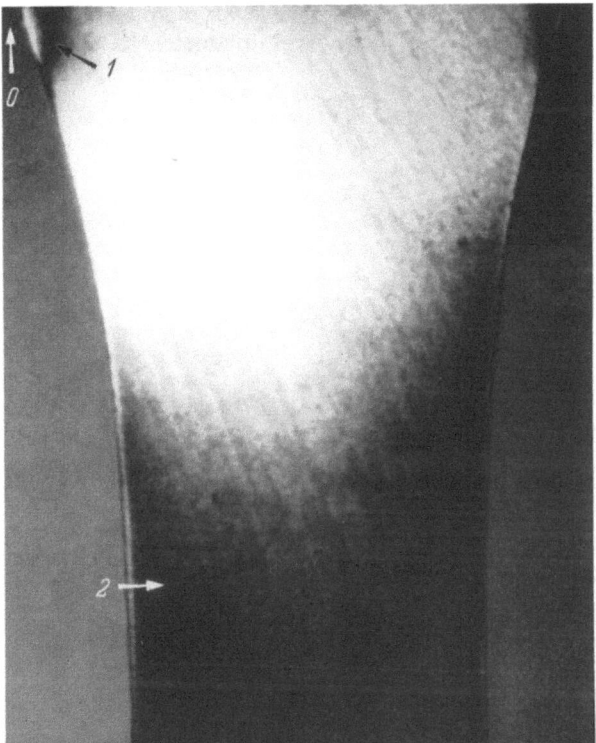

Fig. 74.

Fig. 73. Illustrating non-uniform distribution of stress in a short tension bar with circular fillets. There is a stress-concentration of 1.3 at the beginning of the fillets.

Fig. 74. Illustrating uniform distribution of stress in a long tension bar with "sine" fillets. The stress decreases steadily with no concentration in the fillets.

of diameter d drilled at the mid-point of its axis, and is then subjected to a uniform tension T lb. per sq. in. in the direction of its length. The mean stress across the transverse section through the centre of the whole will then be $\frac{D}{D-d}T$, which we shall call T'. The stress, however, will no longer be uniform, but will be found to have a maximum value T_m on the hole boundary. In this case the stress-concentration factor is usually measured as T_m/T'.

Figs. 75 and 76 show typical fringe patterns in such bars. As in the case of the ring in Sect. 23 there are four isotropic points on the hole boundary, and these are seen as points of zero stress-difference in the fringe-pattern.

The graphs of Fig. 77 show the distribution of the principal stress-difference, $(P-Q)$, and of the principal stresses, P and Q, on the vertical and horizontal

[1] By courtesy of the J. Roy. Aeron. Soc. **59**, 64 (1955). — Aero Quart. **6**, 230 (1955).

axes of symmetry. In each case the graph of $(P-Q)$ was first plotted from the fringe-pattern, (supplemented in the case of the vertical axis by SÉNARMONT readings of fractional relative retardations). The stress in the direction of each axis was then obtained by graphical integration of the LAMÉ-MAXWELL equations as shown in Sect. 27. The greatest tensile stress occurs on the horizontal axis, and this is therefore the important one to consider from the point of view of the strength of the bar.

Tests on a number of similar bars with holes of varying diameter showed that as the ratio of hole-diameter to width of bar increases, the stress concentration factor T_m/T' decreases from a value of 3 in the case of a very small hole to a value ap-

Fig. 75.

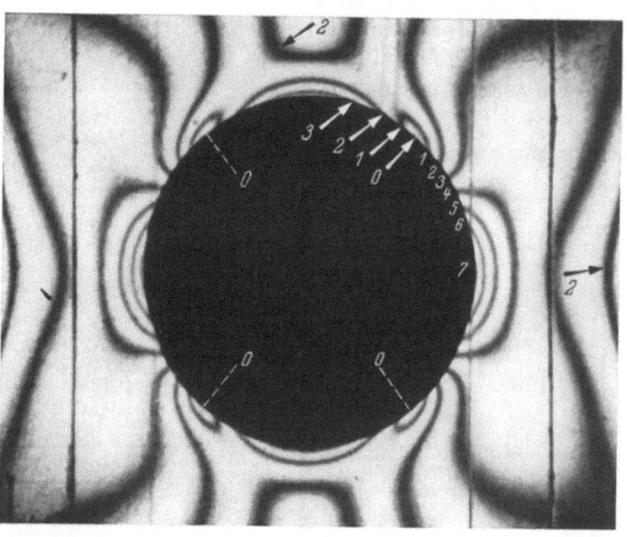

Fig. 76.

Fig. 75. Fringe-pattern in uniform rectangular bar with central circular hole under simple tension.

Fig. 76. Fringe-pattern showing distribution of stress around the boundary of the hole. The edges at top and bottom are in compression, the sides are in tension.

proaching 2 as the diameter of the hole approaches the width of the bar. It is important to note, however, that in the case of the higher stress concentrations, the high stresses occur in relatively smaller areas of the plate, and in some circumstances this will mitigate the weakening effect of the concentration. The graphs of Fig. 78 show the distribution of the stress-difference across the horizontal axis of symmetry for holes of different sizes under loads which give the same mean stress across the section.

44. Flat plate with non-circular hole under tension. When a hole in a bar or plate is not circular, any kind of theoretical evaluation of the stress-distribution is usually very difficult and gives only approximate results. A photoelastic investigation of a hole of any shape under any system of forces in the plane of the plate is very easily carried out, and the method has been widely used in such problems as those occurring in rectangular openings in the decks or sides of ships, and in cut-outs in the webs of wing-spars of aircraft.

The two examples which follow are special cases of this type of problem.

Sect. 44. Flat plate with non-circular hole under tension. 207

α) *The stresses around a "drilled" crack.* It is well known that if a plate has a crack in it, and tension is applied in a direction normal to the crack, the stress-

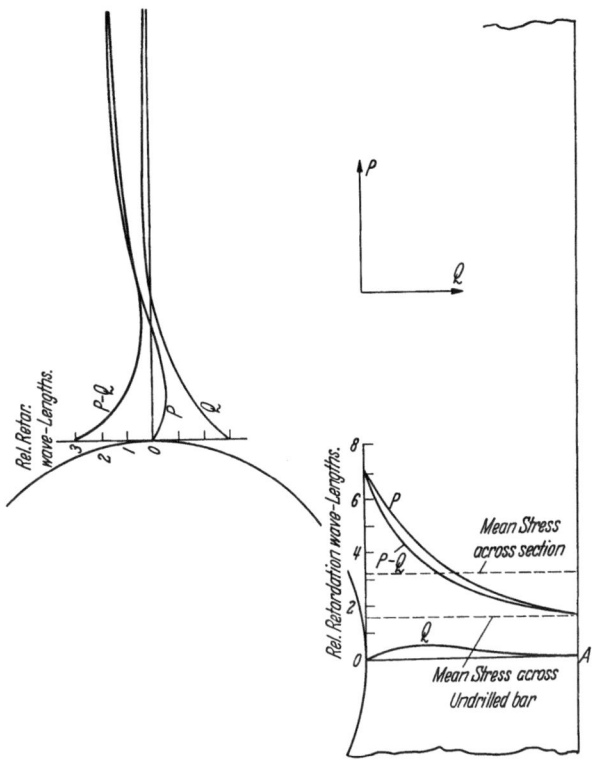

Fig. 77. Distribution of stresses on vertical and horizontal axes of symmetry in a tension bar with central circular hole Ratio of diameter of hole width of bar = $\frac{1}{4}$. The magnitudes of the stresses have been left in terms of relative retardations The stress in lb. per sq.in. = rel. retar. in wave-lengths × f/t, where f = fringe value of material and t = thickness of model.

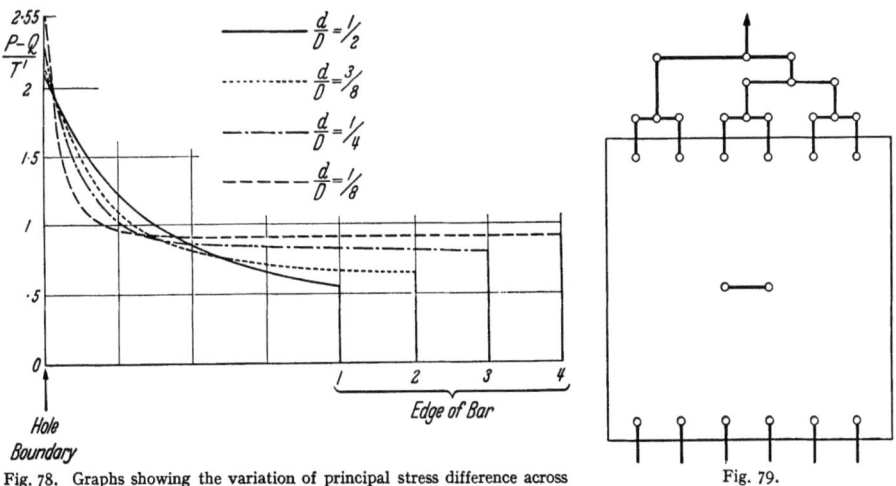

Fig. 78. Graphs showing the variation of principal stress difference across horizontal section with different sized holes.

Fig. 79.

concentrations at the ends of the crack reach extremely large values with the result that the crack tends to spread. It is also well known that if circular holes are drilled at the ends of the crack the tendency to spread is reduced. This method has long been used by engineers as an emergency "first-aid" treatment when a crack develops in a component and no immediate replacement can be made.

The stresses at the end of an actual crack, although large, are confined to such a small region that it is impracticable to obtain any measurement of them, but

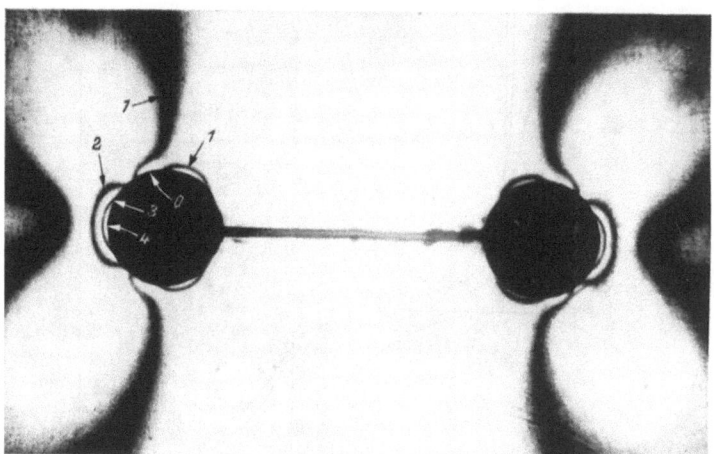

Fig. 80. Fringe-pattern showing stresses around a "drilled" crack in a plate under tension.

we can explore the stress-concentrations which remain when the crack has been treated by drilling holes.

Fig. 79 shows the arrangement used for this investigation. An araldite plate $6^{1}/_{2}$ in. square and $^{1}/_{8}$ in. thick was loaded through linkages as shown so that equal tensions were applied at six points on the boundary. Examination of the undrilled plate showed that this produced a sensibly uniform stress over the central region. A drilled crack was simulated by cutting a slit between two circular holes of 0.15 in. diameter whose centres were 1 inch apart. The diameters of the holes were subsequently increased by stages to 0.45 in. the fringe-pattern around the holes under a given load being observed at each stage.

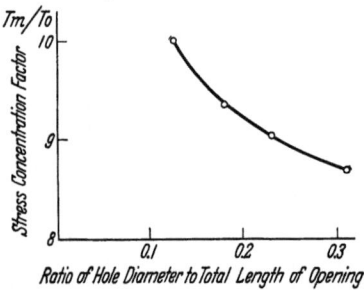

Fig. 81. Graph showing relation of stress concentration factor to diameter of hole in a "drilled" crack.

Fig. 80 shows the fringe-pattern in the case of 0.3 in. diameter holes, the stress-concentration factor in this case being 9. Fig. 81 shows the relation between the S.C.F. and the ratio of hole-diameter to total length of opening. The stress-concentration factor is still dangerously high, but decreases as the hole-diameter increases.

In such a case as this, where the area of material around the crack is comparatively large it would obviously be better, if practicable, to drill a single hole to cut out the whole crack. The S.C.F. would then be of the order of 3.

It should be noted, however, that in attempting to obtain the S.C.F. in any individual practical case, it is necessary to reproduce in the model the exact loading conditions which operate in the prototype. In the case illustrated, the use of multiple load-points with a linkage ensured that the distribution of the load across the plate near to the loading pins remained uniform even though the yield of the plate along the lines of tension which crossed the crack was greater than that in the other parts of the plate. The results obtained therefore would be applicable to any large plate to which uniform tension had been applied at a distance from the crack.

β) *Rectangular opening with rounded corners.* Fig. 82 shows the fringe-pattern around a rectangular opening in a flat plate under a uniform tension applied in a direction parallel to one edge of the opening. Here again a multiple point loading has been used to ensure a uniform applied stress, but in this example the loads have been applied at a distance from the opening at one end (the top of the photograph), and close to the opening at the other. The load applied was sufficient to produce one "fringe" in the uncut plate, so that the S.C.F.'s at the top and bottom corners were respectively 3.5 and 5.0.

The explanation of this difference is, of course, that in the case of the load at a distance the stress has been able to diffuse gradually, leaving a region immediately above the opening comparatively free from stress. The stress-distribution along the vertical axis above the opening is, in fact, very similar to that above the circular hole in the example of Sect. 43. Below the opening, however, where the load is applied close to the free edge, there is no room for gradual diffusion, and the resulting stresses are more concentrated around the corners.

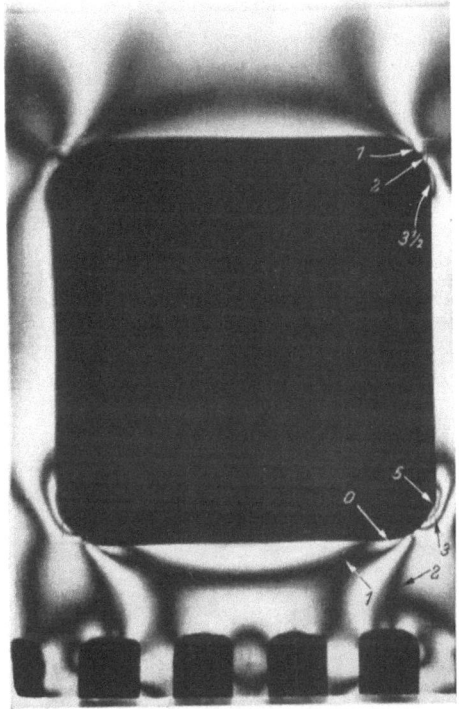

Fig. 82. Stresses around a rectangular opening with rounded corners in a plate under tension. A distributed load is applied at a distance from the opening at the top, and close to the opening at the bottom. (The links through which the loading pins pass can be seen at the bottom of the photograph.)

This variation of stress-distribution with loading conditions has an important application in the problem of assessing the stress-concentrations around openings in the cylindrical wall of a vessel subjected to internal pressure, for example in the wall of the pressurised cabin of an aircraft. In a complete cylindrical shell under internal pressure P lb. per sq. in., there will be a uniform circumferential tension $T = RP/t$, where R is the radius of the shell and t its thickness. In a shell where R is very large compared with t, the curvature may be neglected, and it is the accepted practice to estimate the stress-concentrations due to features such as window-openings by carrying out tests on flat plates under a tension corresponding to T. In the pressurised cylinder, however, the internal pressure acts right up the edge of the opening, thus producing a condition which approaches that of a uniform tension T applied very close to the edge. While therefore the actual stress-distribution in the three-dimensional prototype cannot be reproduced exactly in a flat plate model, the actual stress-concentrations will be much more nearly given by applying a uniform load close to the opening than by applying it at a distance.

45. The stresses around steam-pipe holes in a boiler drum[1] **(P. L. COUTTS).** This is another example of a pressurised cylinder, but in this case the walls of the vessel are comparatively thick and the openings are small. The loading conditions in the boiler-drum also are complex, for in addition to the internal

Fig. 83. Loading device for exerting simultaneous vertical and horizontal uniform tensions on a plate. D, D are ring dynamometers for measuring the loads which are applied by screws. The frame F which carries the horizontal loading mechanism is suspended on springs. L, L are bell-crank levers through which the horizontal load is applied.

pressure there are bending loads which introduce different longitudinal stresses in different regions of the drum.

The complete investigation of this problem involves the photoelastic examination of a large variety of hole arrangements under a number of different ratios of circumferential to longitudinal stress. The investigation may however be divided into two parts: a complete series of tests on flat plates to give the characteristic distributions of the mean stresses, supplemented by a few frozen-stress tests on three-dimensional models by which the variations of stress through the thickness of the wall of the drum may be examined.

The method of investigation of the flat plates is illustrated by the photograph of Fig. 83. Vertical and horizontal loads are applied to the plate by linkages, the frame carrying the horizontal loading system being suspended on long springs

---

[1] By courtesy of the Water Tube Boiler Makers Association.

so as to "float" freely and allow small movements of the model under load. Fig. 84 shows photographs of the fringes in one such model under two different loading conditions.

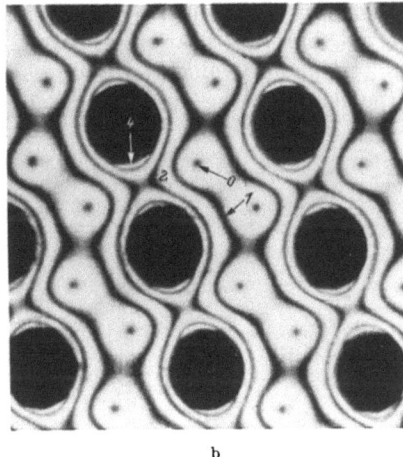

Fig. 84 a and b. Fringe-patterns in a plate drilled with a "raked" array of holes, (a) under vertical tension only, (b) under equal vertical and horizontal tensions.

Fig. 85 shows a typical set of graphs recording the peak stresses on a hole-boundary for a given arrangement of hole centres. Each curve shows the variation for a hole of given diameter as the ratio of longitudinal to circumferential tension increases. The different curves apply to different hole diameters.

46. The stresses in a plate due to a loaded bolt or rivet (M. CLUTTERBUCK).

A bolted or riveted connection between plates or bars is a very common feature in most types of structure, and in many cases it constitutes a point of weakness. The problem is not a purely two-dimensional one, for the bending of the bolt causes a variation of the load applied at different points in the thickness of the plate. In practice, too,

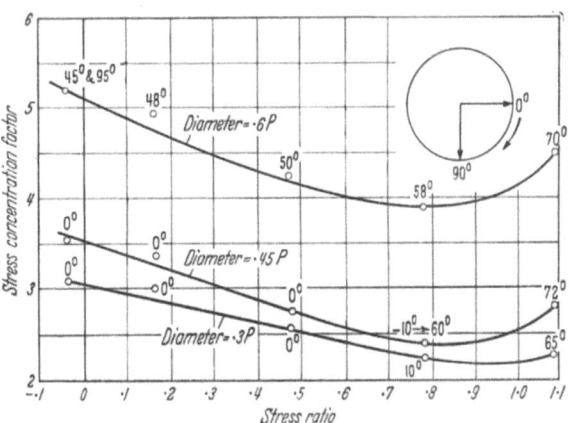

Fig. 85. Graphs showing how the S.C.F. in a plate with a given arrangement of hole centres varies with the size of hole and the ratio of horizontal to vertical tension applied. The photographs of Fig. 84 correspond to two points on the middle curve—hole diameter = 0.45 × horizontal pitch. The figures beside the curves indicate the point on the boundary at which the maximum stress occurs.

there is a compressive stress through the thickness of the plate due to the pressure of the bolthead and nut. The first approach is to find the two-dimensional distribution of stress, using a simple pin instead of a bolt, and employing loads which do not cause any appreciable bending. The three-dimensional factors are then investigated by frozen-stress methods.

Fig. 86 shows the fringe-pattern in a flat plate due to a load symmetrically applied to a pin passing through a circular hole in the plate. This photograph is of a model in which the stresses have been frozen, since in the case of a model under direct load the fringe-pattern would have been partially obscured by the loading wires. The characteristics of the fringe-patterns in the two cases are the same.

There is a very high concentration of compressive stress immediately under the pin, but the shear-stress, whose magnitude is given by the fringe-pattern, has a maximum value at a point slightly below the edge of the hole. This point is the centre of the oval fringes. High tensile stresses occur at the sides of the hole a little below the horizontal diameter.

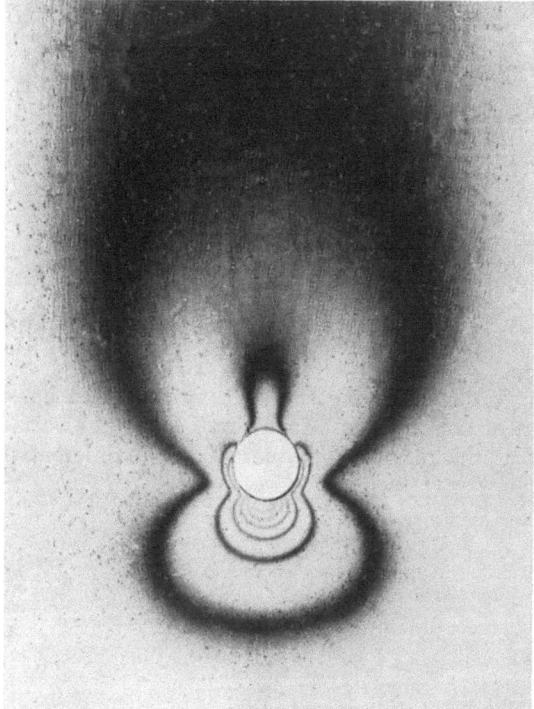

Fig. 86. Fringe-pattern in a flat plate due to a load applied to a pin passing through a hole.

This case of the loaded hole is the only case in which the value of the Poisson's ratio of the material affects the stress-distribution, and is therefore the only one in which the results of photoelastic tests should fail to give the correct distribution of elastic stress in a metal prototype. Tests were carried out upon three similar models,—araldite in frozen-stress conditions, araldite loaded cold, and glass,—the respective Poisson's ratios being 0.5, 0.38, 0.26. These tests indicated that the effect upon the stress-distribution of the different contact conditions in the three cases far outweighed any effect the different Poisson's ratios might have. Slight differences in the fit of the pins, in the yield of the surfaces under load, and in the friction between pin and hole were found to cause variations of up to 20% in the magnitudes of the stresses in the immediate neighbourhood of the hole, and this is the only region in which the Poisson's ratio would be expected to have any marked effect.

This is an illustration of the general principle that photoelastic tests cannot be expected to give accurate information about stresses in regions very close to contact loads unless it is possible to ensure that the contact conditions are the same in model and prototype.

Fig. 87 shows a plate in which equal loads have been applied to three pins, simulating the loads due to a row of bolts. The load on each pin was the same as that used in the previous example, and it can be seen that the distribution immediately around any hole is not sensibly affected by the neighbouring holes. Fig. 88 shows the distribution of tension across a section of the plate between two holes along a line just below the line of centres. As the spacing of such a row of bolts connecting two plates and carrying a fixed total load becomes closer, the load on each bolt decreases, tending to make the peak tension on the hole boundaries smaller. At the same time the mean tension across the plate between the bolts becomes greater, and this increase would eventually begin to increase the boundary tension again. There is, therefore, an optimum spacing for which the peak stresses are a minimum.

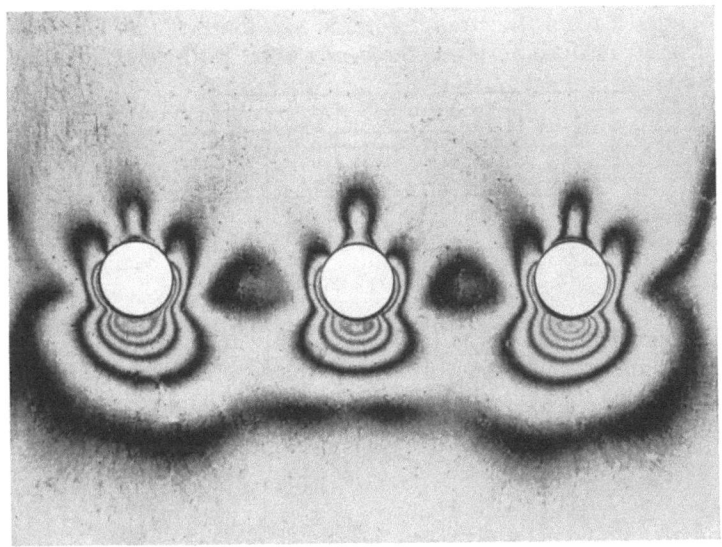

Fig. 87. The same plate as in Fig. 86 with equal loads applied to three pins.

47. Bolted spar joint[1]. Fig. 89 shows a type of bolted joint which is frequently used for the wing-spars of aircraft. From the point of view of photoelasticity it presents a problem which cannot be dealt with by standard methods unless the bolts are of the same material as the spar, which is not usually the case. If the materials are the same the whole model may be constructed in araldite, and the stresses frozen under load, after which the model is taken apart and the separate components examined in the ordinary way.

If in the prototype the bolts are of steel and the spars of light alloy, we have to reproduce in the model the corresponding ratio of YOUNG's moduli of the two materials since the relative yield of bolts and spars is an important factor in the distribution of load among the bolts. This neces-

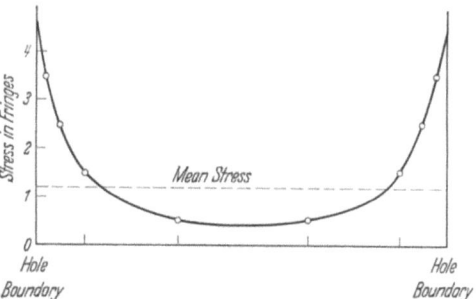

Fig. 88. Distribution of tension between holes of the plate shown in Fig. 87.

sitates the use of different materials, and if frozen stress method were used would introduce an unknown factor in the differential thermal expansions of the materials.

We can, however, examine the stresses in an outer component of such a joint by adapting the polariscope to the use of a "reflection" method. If the inner surface of one of the outer components be silvered, polarised light passing through

---

[1] By courtesy of the J. Roy. Aeron. Soc. **57**, 125 (1953).

the component will be reflected back and may be analysed in the usual way. Fig. 90 shows how a standard photoelastic bench may be adapted for this purpose. The parallel beam after passing through the polariser and first quarter-wave plate is reflected by two surface silvered mirrors F_1 and F_2 so that it passes, in a directional normal to the axis of the bench, on the other side of the model M which is under load in the straining frame. In front of the model is a half-silvered mirror H, inclined at $45°$ to the bench axis. Half the light of the polarised

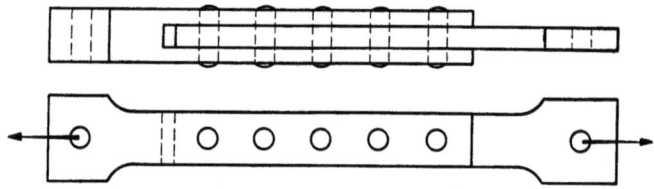

Fig. 89. Model of multi-bolt spar joint.

beam is reflected by this so as to pass normally through the outer component of the model, and is then reflected back by the silvered surface. One half of this reflected beam now passes through the half-silvered mirror, and then through the second quarter-wave plate and the analyser to the projecting lenses.

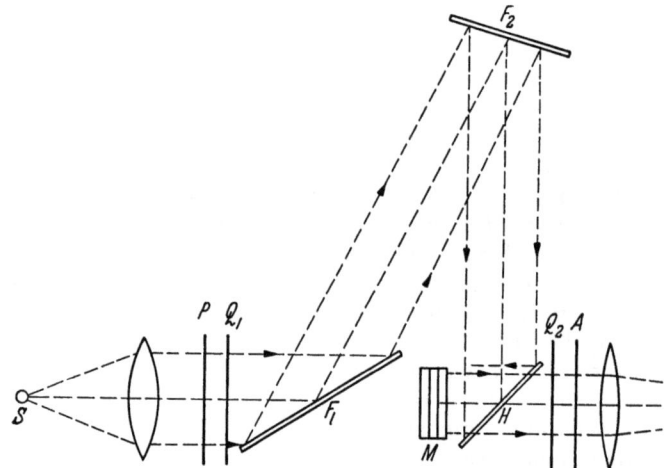

Fig. 90. Diagram of reflecting polariscope.

Three quarters of the light intensity of the beam is lost in this method, and there is also a certain amount of depolarisation produced by the reflections which may make the isoclinic fringes faint in some areas, but the isochromatic fringes are still quite clearly visible.

The mirrors F_1 and F_2 can, of course, be dispensed with if the bench is built in two sections so that the light source and polarising filters can be moved so as to direct the polarised beam directly on to the mirror H.

When such a joint consists of more than two superimposed components, as in the case illustrated, one method of investigating the central component is by the frozen-stress method and we must then use bolts of the same material as the spar. It is, however, possible in such a case to examine the effects of different bolt materials by a separate test, using the reflection method, first with bolts of the same material as the spar and then with bolts made of material whose

Young's modulus is, say, three times that of the spar. A comparison of the stress-distributions in the outer component in these two cases will give, at least qualitatively, an indication of the way in which this factor will affect the distribution among the bolts in the central component, and a correction can then be applied to the frozen-stress results.

An alternative method is to use a model in which the outer components are made of material with very low optical sensitivity, such as perspex, while the centre component is made of sensitive material such as araldite, which has the same elastic modulus as perspex. The model is then loaded in the standard polariscope, and the fringe-pattern observed will be substantially that due to the stresses in the araldite. It is however impossible to obtain isoclinic fringes by this method since at any point the directions of the stresses in the inner and outer components are different and the light is repolarised in each component.

Fig. 91 is a photograph of two such composite models under load, demonstrating that a closer pitching of the bolts leads to a reduction of the peak-stress in the model with a more uniform distribution of load among the bolts.

In all such multiple load-point problems it is extremely difficult to reproduce in a photoelastic model the same relative tolerances and strains as would occur in a metal prototype, and very little more than qualitative correspondence in the stress-distributions can be expected. The photoelastic tests however can show how changes in various features of the design, such as hole diameter and pitch or width of spar, affect the load distribution.

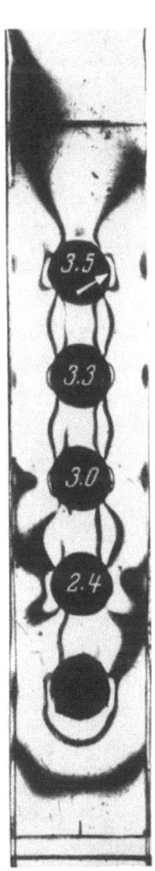

Fig. 91. Fringe patterns in two models of bolted joints under the same load. The figures indicate the relative magnitudes of the peak stresses on the hole boundaries. The stress-patterns in the lower holes are not reliable since the stresses in the perspex cleat cause interference in this region.

II. Three-dimensional examples.

48. Stresses in a boiler drum[1] (J. A. Coutts).

Fig. 92 is a photograph of one of the three-dimensional models used to investigate the effects of curvature of the drum upon the stress-distribution in the problem of which the two-dimensional aspects are given in Sect. 45.

This is a good example of a built-up model. The central cylindrical portion of the drum, the ends, and the "stub" tubes were separately machined from castings of araldite B (hot setting), and then cemented together with araldite D (cold setting).

The assembled model was given a final annealing, and was then loaded and frozen. The internal pressure was provided by a head of liquid paraffin of about 11 feet and the drum itself was immersed in a tank of the same liquid, a stirrer being employed to ensure uniform distribution of temperature over the model.

---

[1] By courtesy of the Water Tube Boilermakers' Association.

Under internal pressure alone the ratio of the longitudinal to circumferential stress in the undrilled drum would be one half. In order to take measurements corresponding to other ratios, ranging from 0 to 1, longitudinal stresses were superimposed on the pressure stresses by applying dead weight loads to the ends of the drum.

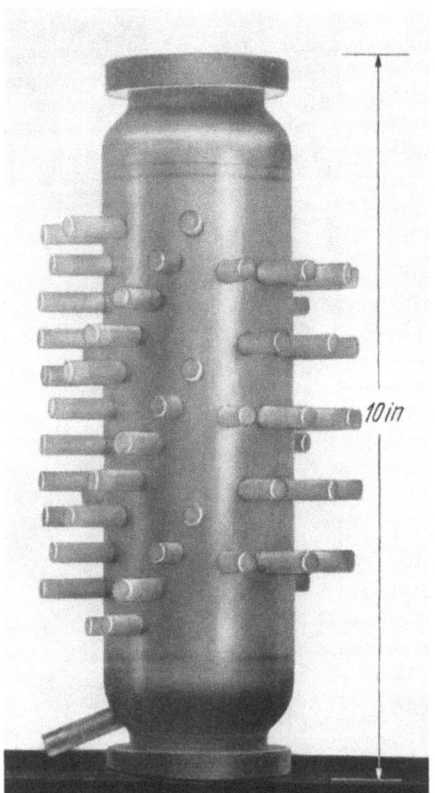

Fig. 92.

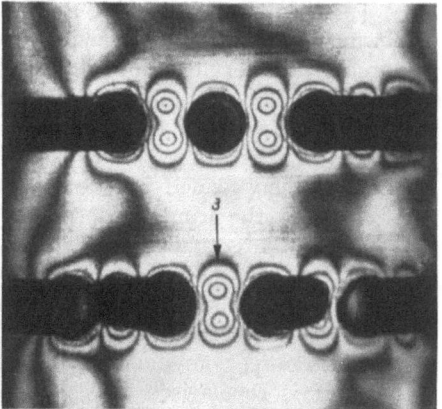

Fig. 93.

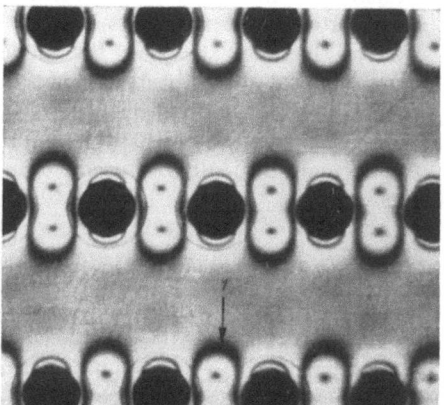

Fig. 94.

Fig. 92. Araldite model of a boiler-drum.

Fig. 93. Fringe-pattern through the wall of boiler-drum, frozen under an internal pressure of $4\frac{1}{2}$ lb. per sq. in.

Fig. 94. Fringe-pattern in a flat plate under loading similar to that of the drum in Fig. 93, but of one third the equivalent magnitude.

Fig. 93 is a photograph of the fringe-pattern seen through the thickness of the drum wall in a model frozen under internal pressure only. In the region where the light is transmitted normally the pattern indicates the distribution of the mean stresses through the drum. Fig. 94 shows the pattern in a flat plate under the same ratio of longitudinal to transverse loads. Taking into account the relative loads, the thicknesses of the two models and the cold and frozen fringe-values, the relative retardation observed in the drum should be almost exactly three times that in the plate. The photographs show that this is the case in the region surrounding the two isotropic points in the narrow ligaments. The regions in which the peak stresses occur on the hole boundaries are masked in Fig. 93 by the presence of the tubes.

Fig. 95 shows two photographs through a slice cut transversely to include the point of highest stress on a hole-boundary. The tube in this case has been cut off flush with the surface of the drum. Photograph (a) shows that some stress has been introduced by the shrinkage of the araldite cement between tube and drum, causing some discontinuity of the fringes at the join. It also shows that the stress due to loading continues to increase through the wall of the tube itself, reaching its maximum value on the inner boundary of the tube, where it is approximately equal to the peak stress on the hole boundary in the flat plate model of Fig. 94.

Photograph (b) shows that on the line AB, where the light passes normally through the hole boundary, the mean stress-difference at the inner surface of the drum exceeds that at the outer surface by about 35%. To obtain the actual boundary stresses in this case the slice must be progressively thinned by cutting off thin layers from the hole-boundary and measuring the relative retardations through the section AB at each stage. Extrapolation from a graph of these measurements yields the best value obtainable of the peak stress.

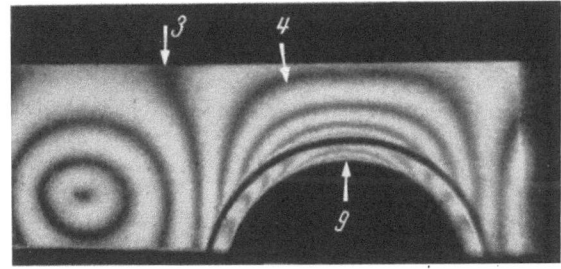

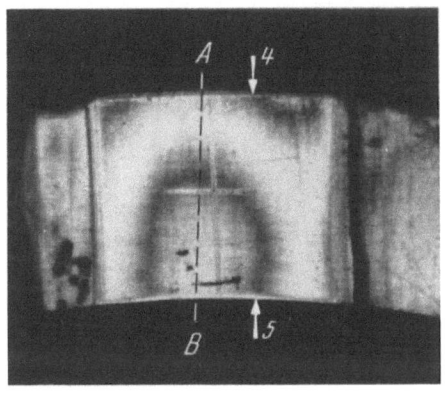

Fig. 95a and b. (a) Detail of fringe-pattern shown in Fig. 93, with tube cut-off. (b) Fringe-pattern in slice of Fig. (a) viewed in direction of arrow marked 9.

Fig. 96. Component parts of an araldite model of the end connection for a piston-rod.

49. Stresses in the end connection of a piston-rod[1] (G. D. GIMLETT). The piston-rod of a steam locomotive is connected to the cross-head by a slightly

---

[1] By courtesy of British Railways, Western Region.

tapered key fitting into slots in the rod and in a socket attached to the crosshead. In service these components are subjected to repeated heavy loads. Fig. 96 shows models of the socket, the key and the end of the piston-rod. These were used in an investigation of the stresses in the key and around the slots. The models were all machined from solid blocks of araldite B, and were then assembled and frozen in the oven under load.

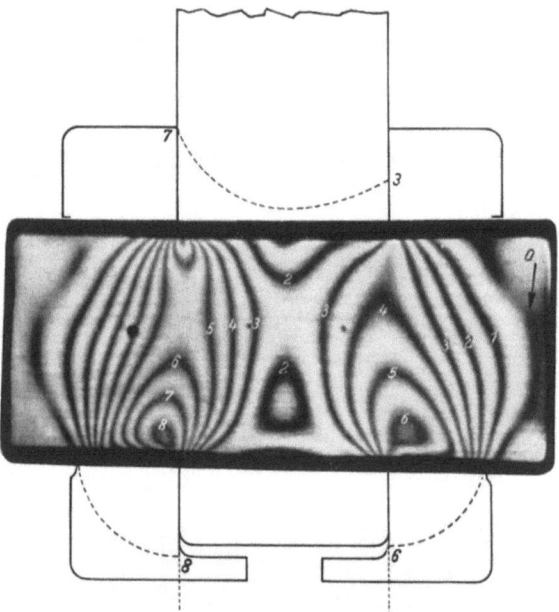

Fig. 97. Fringe-pattern through the key of the piston-rod connection.

Fig. 97 is a photograph of the fringe-pattern taken through the whole thickness of the key with the positions of the rod and socket drawn in. It illustrates well the effect of the bending of the key in producing concentrations of load at the outer ends of the slot in the piston-rod and at the inner ends of the slots in the socket. The photograph shows also that the load has not been shared equally by the two sides, an effect probably due to differences in the accuracy of fit of the two ends of the tapered key. This fringe-pattern, of course, shows only the magnitudes of the mean stress-differences in a stress-system resulting from a combination of contact pressures and bending moments. Also, owing to the curvature of the top and bottom edges of the key, it does not show the stresses up to these edges, so that it cannot indicate with any accuracy the distribution of the actual load. It does, however, indicate the relative magnitudes of the shear-stresses in different regions of the key. The graphs sketched above and below the photograph show the distribution of these as shown by the fringepattern.

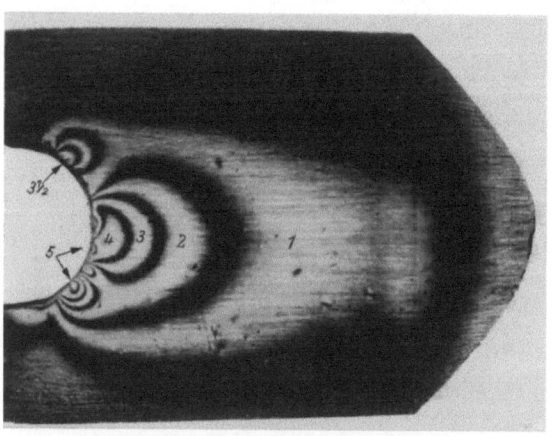

Fig. 98. Fringe-pattern in a surface slice from the piston-rod, showing uneven distribution of pressure on the key.

Fig. 98 shows the fringes in a slice cut from the piston-rod as shown by the shaded area in Fig. 99. It shows that the distribution of load round the curved edge of the key is also far from uniform. There appear to be three "high spots" which take an undue proportion of the load.

In such a problem as this, involving contact stresses, it is important to assess the probability that such irregularities in load-distribution as are found in the model will occur in the prototype. The relative degrees of smoothing out of irregularities will depend upon the relative strains which occur in the two cases.

If L and l are corresponding linear dimensions in prototype and model, W and w the loads applied, E_1 and E_2 the respective Young's moduli, then the ratio of the strains in the two cases will be

$$\frac{W}{w} \cdot \frac{l^2}{L^2} \cdot \frac{E_2}{E_1}.$$

In this case W, the maximum load on the piston was 37 tons, w was $11\frac{1}{4}$ lbs and l/L was $\frac{1}{2}$. E_1 was 30×10^6 and E_2 was 2×10^3.

Using these values we find that the strains in the prototype will be about $\frac{1}{8}$ those in the model. Then since the model is half size the relative displacements due to strain in the prototype will be about $\frac{1}{4}$ those in the model.

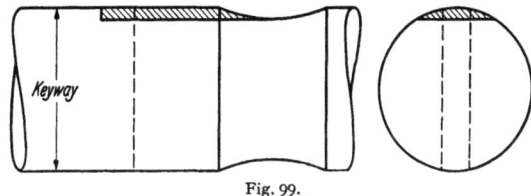

Fig. 99.

Unless then the accuracy of fit of the key in the prototype were more than four times as good as that in the model we should expect to find initial irregularities of distribution at least as marked as those found in the model. Some of the stress-concentrations resulting from these, however, would almost certainly lead to plastic yield, and this would result in a smoothing out of high spots and a more even distribution than is shown by the elastic stresses.

The bending effect observed in this example is similar to that which will occur in the case of a bolted or riveted joint between two plates or bars. The contact loads will be unevenly distributed through the thickness of the plates, making the maximum stresses greater than the mean values observed in a two-dimensional test. How great this effect will be will of course depend upon the thickness of the plates, the diameter of the bolt, and the relative Young's moduli of the two. The effect is unlikely to be negligible in any practical design, and will only be determined by a full three-dimensional test.

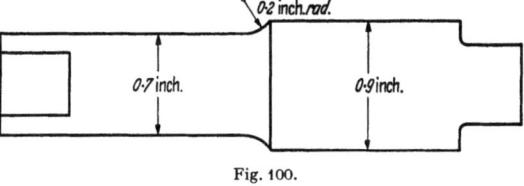

Fig. 100.

50. The stresses in a shouldered shaft[1]. When a cylindrical shaft is subjected to torsion the distribution of stress in any uniform portion is obtained quite simply from elementary theory. If, however, the section of the shaft changes at any point, local stress-concentrations are introduced and the simple theory no longer applies.

To reduce such concentrations the change of section is usually made through a circular fillet of as large a radius as the design of the shaft will conveniently allow. Fig. 100 shows the dimensions of a model of such a shaft, used for a photo-elastic investigation of the surface stress in the fillet. Flats were machined on the ends of the model to fit the clamps through which the torque was applied. The model in this case was machined from a rod of Marco resin containing a

---

[1] By courtesy of the Brit. J. Appl. Phys. **2**, 249 (1951).

small quantity of finely divided silica to improve its scattering properties, and the stresses were frozen into it.

The stresses in the surface of the shaft in the region of the fillet were of chief interest, and these were explored by the scattered-light method. The model, in a tank of immersion fluid, was adjusted so that the beam of polarised light was incident normally at a point of the fillet, and from measurements of the position of the zero fringe for a range of settings of the compensator, a graph was drawn of relative-retardation against distance along the normal. The slope of this graph at the point of entry of the beam was a measure of the shear-stress at that point. Graphs were obtained for a number of points along the fillet, and from these the distribution of the stress was plotted.

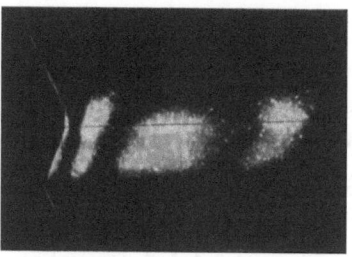

Fig. 101. Fringes seen by scattered light in a beam entering the fillet of a shouldered shaft in torsion.

Fig. 101 is a photograph of the fringes observed at one point. In making the observations, one cross-wire of the microscope was made to coincide with the normal to the fillet surface, and measurements were made along this line. Fig. 102 shows some of the graphs of observed relative retardations. When the slopes of the tangents are plotted for points along the fillet it is found that the peak stress occurs at a point where the tangent to the fillet makes an angle of about 15° with the axis of the shaft, and the stress concentration factor at this point is about 1.7.

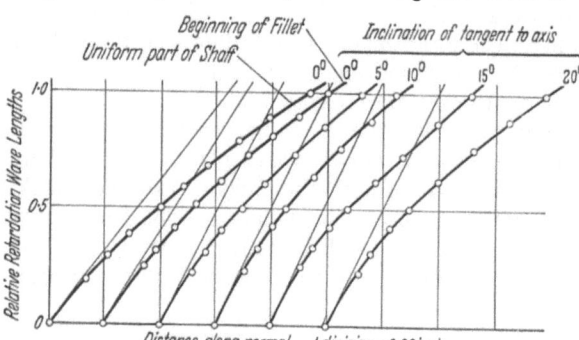

Fig. 102. Graphs of scattered light fringes in the shouldered shaft.

The graphs of Fig. 102 show that while there is some latitude in drawing the shape of the curve between the point of entry of the beam and the first measurement which can be obtained of a fringe position, the slope at the point of entry as obtained from such a graph will give a greater accuracy than would a mean value of the relative retardation through a slice even as thin as 0.01 inch.

51. Stresses in a hinge-bracket (JESSOP and SNELL[1]). Fig. 103 shows a photoelastic model of an elevator hinge-bracket from the tail-plane of an aircraft, with the loading rig attached[2]. This model was cast, full size, in Marco resin using a mould of "vinamold", and the actual bracket as pattern. In order to avoid the formation of rind-stress during freezing, the model had to be immersed in a bath of glycerin and water, and the special loading frame was designed to allow this.

Preliminary inspection and calculation indicated that the main web of the bracket was amply strong for the maximum load it had to carry, but that there might be serious stress-concentrations in the recesses which had been machined

---
[1] By courtes of the Quart. J. Roy. Aeron. Soc. **3**, 161 (1951).
[2] The pattern from which this model was taken is now obsolete.

in the base of the bracket to take the securing bolts. It was therefore necessary to ensure that the contact conditions at these bolts in the loaded model should simulate those in the prototype. Accordingly washers, made of Marco resin, were inserted under the bolt-heads, and springs, S, were placed under the nuts which would be subjected to tension. This ensured first that the bolt-heads

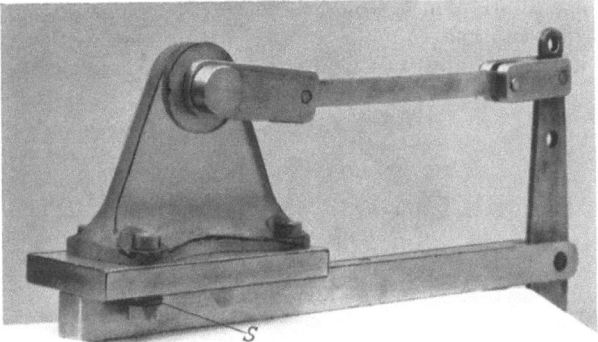

Fig. 103. Marco model of a hinge-bracket mounted on its loading rig.

would not cut into the resin, and second that any thermal expansion of the bolt would not allow the model to lift from the plate on which it was mounted.

After freezing, a slice $1/16$ inch thick was cut through the centres of the two bolt holes as shown in Fig. 104. Fig. 105 is a photograph, taken through a low-

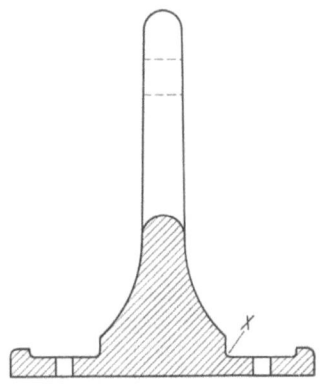

Fig. 104.

Fig. 105. Fringes in a slice cut through the bolt-holes of the hinge-bracket

powered microscope, of the fringes in the small fillet at X in this slice. The second fringe is clearly visible at the boundary, and calculations from this showed that the stress at that point in the prototype reached a value considerably above the nominal yield-stress of the material.

The region of very high stress is, of course, exceedingly small, and there is no doubt that local plastic yield would ease the concentration and prevent immediate failure. But such a point of high stress constitutes a potential source of fatigue cracks which might lead to failure after repeated loading, and is therefore an undesirable feature in such a component. Such recesses for securing bolts are now machined with fillets of much larger radius which reduce the stress-concentrations to safe values.

52. Stresses in an insulator cap[1] (JESSOP and SNELL).

The prototype in this case was a porcelain cap 16 inches in diameter which formed the lower end of a large insulated terminal. In order to save space while still retaining a sufficient length of insulating surface, the cap was designed with its lower end "tucked in" as shown in Fig. 106 which is a photograph of a sectioned quarter scale model made in Marco resin. (This investigation was carried out before the introduction of araldite, which would have been better suited to it.) In service the top rim of the cup had to make an airtight junction, with the upper part of the insulator assembly, and was held in position by tension applied to a bolt

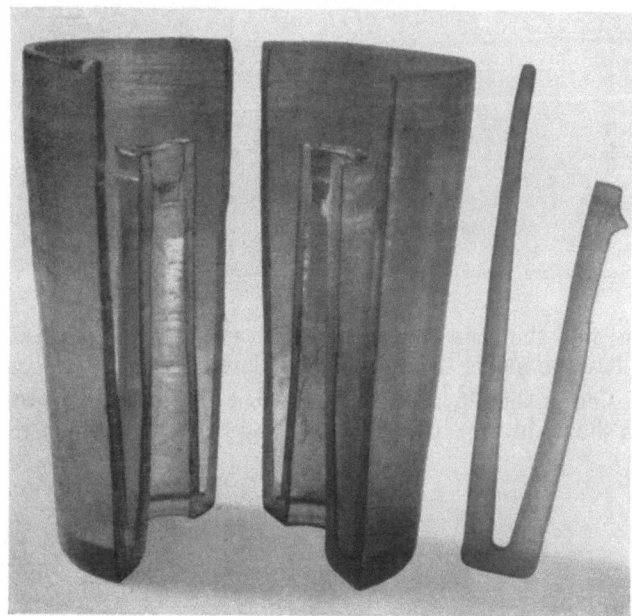

Fig. 106. Sectioned Marco model of a porcelain insulator-cap.

the head of which bore upon a collar in the upper end of the central tube. This is shown in the sectional drawing of Fig. 107.

Patterns were cut to the shape of the inner and outer surfaces of the cap, and from these a "vinamold" mould was made in which the model was cast. After freezing, the model was cut, and examination of a slice from the plane of symmetry showed that there were serious stress concentrations in the fillets at A and B (Fig. 107). Modifications in the shape of the central tube reduced the stress at B to a tolerable value, but no modification in the shape of the collar which was permissible in respect of other design requirements, would reduce the stress at A, and eventually the method of applying the tension had to be altered.

Owing to the time and difficulty involved in casting the three-dimensional models, a number of comparative tests were carried out using a two-dimensional analogy. A model of the central section of the cap was cut from a flat sheet of C.R.39, and this was stiffened by a number of pairs of celluloid strips cemented and pinned to the model as shown in Fig. 107. The cross-sections of these strips were calculated so that the tensions in them when the model tended to deform under load reproduced the equivalent of the mean hoop tensions acting in the three-dimensional model. (This case of a comparatively thin-walled cylinder

---

[1] By courtesy of Micanite and Insulators Ltd.

Sect. 53. The stresses in a pre-stressed concrete beam (P. Christodoulides). 223

under strains which were symmetrical about its axis was a particularly suitable one for this type of treatment.) It was, in fact, discovered later that the lateral displacements which occurred in the model when stiffened in this way were so small that no measurable difference in the fillet stresses was produced by substantial increases in the strength of the stiffening strips. Any constraint which sensibly prevented distortion of the model appeared to be adequate for the purpose so long as its local effects did not extend to the region under examination.

The agreement obtained between the results for similar two and three-dimensional models was of the order of 20% — as good as could be expected in view of the different conditions in the two cases.

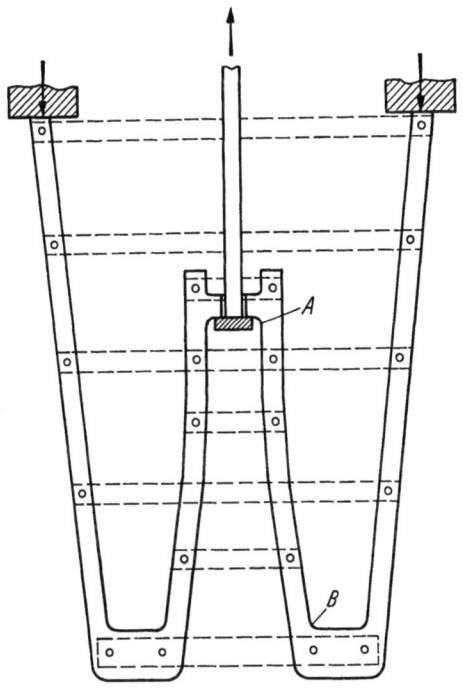

Fig. 107. Stiffened two-dimensional model representing the mid-section of the insulator-cap.

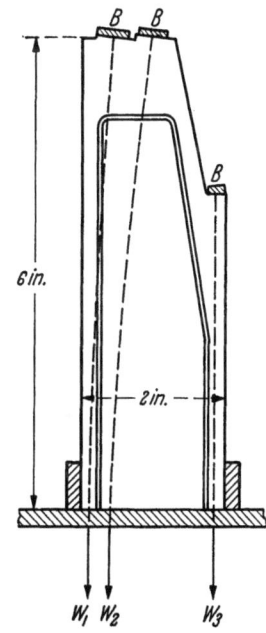

Fig. 108. Post-tensioning of a pre-stressed concrete beam.

Fig. 109. Method of loading a model of one end of a pre-stressed beam.

53. The stresses in a pre-stressed concrete beam[1] (P. Christodoulides).

In the type of pre-stressing most frequently used in large beams, the concrete beam is cast with ducts running through its length, and steel cables are afterwards passed through the ducts and placed under tension. The concrete itself is reinforced with steel rods, especially in regions around the end anchorages where the loading of the cables will introduce tensile stresses. For this reinforcement to be effective it is necessary to know where these tensions will occur and how large they will be. Exact calculations of these are impossible, and approximate calculations have frequently proved to be unreliable, but the problem is one which permits of photoelastic investigation.

Fig. 108 illustrates the general design of such a beam. The ducts are usually of parabolic form, and the pre-stressing cables, under tension, are secured to

---

[1] By courtesy of British Railways, Midland Region.

plates or blocks at the ends of the beam. The regions around these end anchorages are the only ones in which tensile stresses will be produced, so it is only necessary to use a model of one end of the beam, and to reproduce in the model the loads transmitted by the cables. Fig. 109 shows a suitable arrangement for this. The model of the end section of the beam stands on a platform, and wires passing

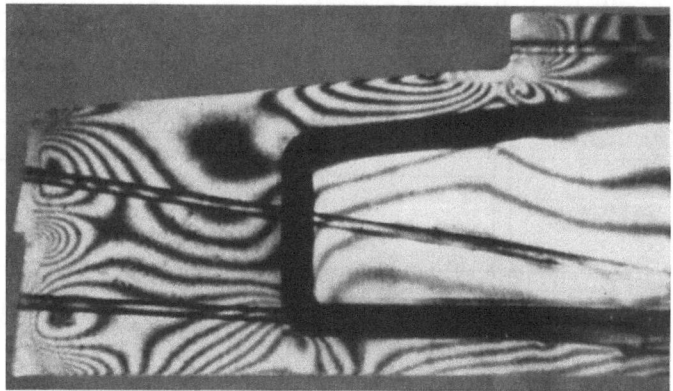

Fig 110. Fringe-pattern through the whole thickness of a frozen model of a pre-stressed beam (Araldite).

through the ducts are attached to anchor-plates B, B and loaded with weights. The stresses under load are frozen into the model.

In the example illustrated the model was made from araldite and its dimensions are shown in Fig. 109. Such a model may be cast direct in a suitable mould, greased steel wires being fixed in the mould to produce the ducts, or it may be machined from a block of the material and the ducts afterwards drilled in it. In the latter case the parabolic shape of the duct is replaced by two straight sections. The curvature of the duct is so small that this produces a negligible difference in the loading conditions. After freezing, slices are cut from the end of the model for examination.

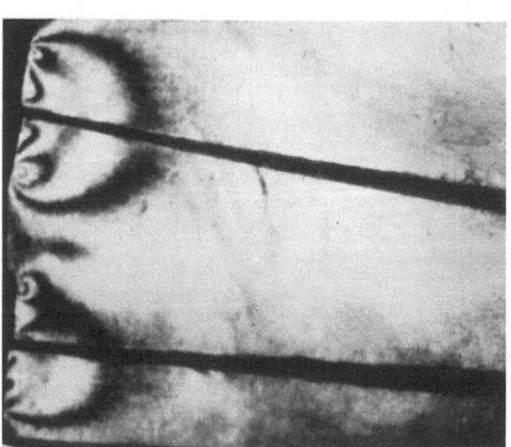

Fig. 111. Fringe-pattern in the central slice of the beam of Fig. 110.

For a complete exploration of the stresses it is necessary to use at least two identical models. In this case three were used. From the first were cut three parallel slices: one in the plane of symmetry in which all the ducts were situated, and one from each side face of the model. From the second and third were cut a number of slices in planes normal to the plane of symmetry. The surface stresses were obtained directly from the surface slices, while the stresses in the plane of symmetry were obtained by the three-dimensional integration method of Sect. 33. The slices were from 0.03 to 0.04 inch thick.

The investigation showed that in some regions the tensile stresses in the beam were considerably higher than those obtained by calculation.

Fig. 110 is a photograph of the fringe-pattern in a model, the light passing through the whole thickness of the model. This gives a qualitative picture of the distribution of stress in the vertical plane. Fig. 111 shows the fringes in the slice cut in the plane of symmetry, one of the slices used in the detailed exploration.

III. Present state and possible future developments.

54. Present state of photoelasticity. We have seen that methods are now available for the investigation of a wide range of stress-distribution problems, and that techniques have been evolved whereby such investigations can be carried out with a degree of accuracy comparable, in most cases, with the degree of accuracy of the engineer's knowledge of the conditions operating in his prototype.

Elastic stresses in two-dimensional components, and on an unloaded boundary or in a plane of symmetry in three-dimensional components, are determinable with comparative ease and rapidity, and to a high degree of accuracy. Internal stresses in regions of three-dimensional components devoid of symmetry require a more elaborate investigation, but are still determinable by the photoelastic method and by no other means.

But the above statements are limited by two considerations: we must be able to make a scale model of the prototype in a photoelastic material, and we must be able to reproduce in the model the loading conditions which operate in the prototype. In the case of three-dimensional models, moreover, it is generally necessary that these loading conditions shall be maintained for a number of hours at an elevated temperature. At the present time our inability to satisfy these conditions excludes from accurate photoelastic analysis certain types of stress, the most important of which are: stresses in the region of plastic yield, many kinds of dynamic stresses, and stresses due to thermal gradients.

It is to be expected that progress in the application of photoelasticity in the next few years will be largely concerned with methods of approach to these problems. Already a certain amount of work has been done in exploring new materials and new techniques and it appears to be probable that methods will soon be developed which, if they may not give complete solutions, will at least add considerably to our knowledge of the stresses occurring in these more difficult problems. Brief notes on some of this work are given in the following sections.

55. Stresses in the plastic region. As noted in a previous section the stress-strain and stress-optical relations in the materials generally used in stress explorations are sensibly linear within the range of loads employed. The materials are, in fact, chosen on account of these linear relations.

There are, however, many materials which exhibit the photoelastic effect but which are subject to a fairly large amount of "plastic" yield at room temperature. This yield is generally progressive with time, and increases in rate as the load increases, but in some cases there is something resembling a yield-point in that there is a fairly well defined load at which the plastic yield increases much more rapidly. This plastic yield is characterised not only by a rapid increase in strain, but also by a rapid change in the optical effect.

Qualitatively the local plastic yield occurring in overstressed regions of a model made of one of these materials appears to be similar in distribution to that occurring in a similar model made of steel or light alloy, but it is not yet possible to obtain quantitative comparison of the relative stress-distributions. Investigations are still proceeding into the stress-strain and stress-optical relations in the photoelastic materials themselves.

Among workers in this field, HILTSCHER[1] has done work on the elasto-plastic optical effect in polystyrene, one of the materials in which the optical effect of the "plastic" strain is of opposite sign to that of the "elastic" strain. Exploring a different property, MÖNCH[2] has carried out investigations with celluloid, in which material he found that the dispersion of the double-refraction was markedly greater in the plastic range than in the elastic.

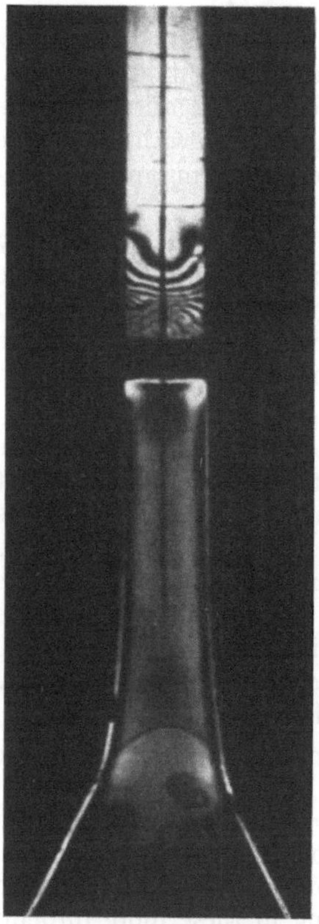

In both polystyrene and celluloid, however, the effects are complicated by the time-creep factor, and it is difficult to interpret the effects observed in a model in which the stress-distribution continually changes with time.

HETENYI found that a certain type of nylon exhibited a marked yield-point, and the writer has observed similar characteristics in a strip of transparent polythene. These materials appear to be sensibly elastic up to a critical strain at which plastic flow occurs with no apparent time-creep. Unfortunately the materials are at present only obtainable in the form of thin sheets which lack rigidity.

More recently COUTTS has observed in araldite loaded at room temperature something much more nearly approaching the yield-point characteristic of metals. Testing a thin uniform bar in tension he observed a sudden increase in the fringe order occurring near one point of the bar just before fracture occurred. On fracture there was an instantaneous almost complete disappearance of the optical effect in the greater part of the bar, but a marked residual pattern in the region close to the break. This is shown in the top part of the bar in the photograph of Fig. 112. It indicates that the "plastic" yield has taken place in a small region near some point of weakness, as usually happens in metal test-bars.

Fig. 112. Residual fringe-pattern, (top) in an araldite bar broken under tension at room temperature. The residual pattern disappears upon annealing (bottom).

The same pattern was observed on both sides of the break, but was found to disappear when the model was annealed at the „freezing" temperature. The lower part of the bar in the photograph has been so annealed. It may be noted that the fringe-pattern in the upper part did not show any appreciable change in a period of 9 months.

It would appear that the "plastic" strain in this instance, at least, is in fact the "frozen stress" type of elastic strain, occurring under very high load at room temperature. The characteristics of this phenomenon and the possibilities of using it in the exploration of stresses in the plastic region are being investigated.

---

[1] R. HILTSCHER: Z. VDI **95**, 777 (1953); **97**, 49 (1955).
[2] E. MÖNCH: Z. angew. Phys. **6**, 371 (1954).

56. Dynamic stresses. Methods of exploring certain types of dynamic stress have already been developed and used in practical problems. For example the stresses due to centrifugal forces in any body rotating at constant speed about a fixed axis may be determined completely by normal frozen-stress methods. All that is required is the means of rotating the model at a suitable speed in the oven. Fig. 113 is a photograph of a very simple model, a shaped bar, in which the stresses have been frozen while it was rotating at high speed. In this case the speed used was too high for a complete photoelastic analysis, for the fringe-pattern near the shaft is too close to be distinguished.

In the case of transient or variable forces, whether those due to accelerations or to contact pressures, the freezing technique is of course impracticable, and any investigation is confined to two-dimensional problems. Acceleration stresses in reciprocating components, impact stresses, and stresses due to rapidly varying contact pressures such as occur in the teeth of spur-wheels, transmitting loads at speed, have all been investigated, in two-dimensional components, by flash photography, stroboscopic methods and high speed cinematography. In practically all such cases the duration of the applied force is long in comparison with the time taken for a stress wave to travel through the particular component under observation, and the stresses build up to a maximum and die away again comparatively smoothly and uniformly, very much in the same way as they do when static loads are applied and removed gradually.

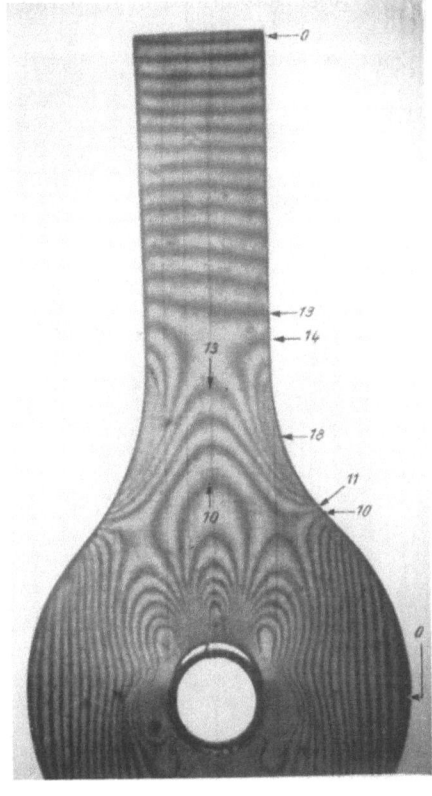

Fig. 113. Frozen fringe-pattern in an araldite bar showing the stresses due to rotation at high speed. (Photograph by courtesy of the G. E. C. Research Laboratories, Stanmore).

A closer analysis of this type of impact has been made by FROCHT, who used "streak" photography to investigate the stress waves in a circular disc subjected to a hammer blow in the direction of a diameter. His photographs recorded the build-up of the fringe-pattern along the load axis, and showed the small variations from the "static" build-up which were produced by successive reflected waves.

A second type of impact is that in which the duration of the impulse is very short, producing a shock wave. In this case the magnitudes of the peak stresses will be strongly affected by the timing and disposition of the various reflected waves.

D. G. CHRISTIE[1] has obtained a series of photographs showing the shock waves in a glass plate due to the detonation of a small explosive charge on one edge. Using a series of 9 flashes from different sources, the photographs from the different flashes being recorded through different lenses, he avoided the necessity for a moving film, while obtaining photographs of the whole plate.

These techniques await further development.

---
[1] D. G. CHRISTIE: Trans. Soc. Glass Techn. **36**, 74 (1952).

57. Thermal gradient stresses. The stresses due to differential thermal expansion in photoelastic materials constitute one of the difficulties of the investigator. The very low thermal conductivity of most of the materials leads to the formation of initial stresses in castings, and even undue handling of the material may produce appreciable thermal stresses which take some time to disappear.

There seems to be a possibility, however, that this sensitivity of the materials may be used to obtain at any rate some qualitative information regarding thermal stresses in structures and machines. Fig. 114 is a photograph of an araldite plate which has drilled in it, parallel to one edge, a hole through which steam is being passed, thus establishing a thermal gradient across the width of the plate. A few holes of different shapes have been cut in the plate, and the fringe-patterns indicate the distribution of the stresses induced by the heat flow around the openings.

Even in a flat plate, of course, the heat flow is three-dimensional, since radiation occurs from the faces of the plate as well as from edges, so that the actual stresses in the plate cannot be deduced from the fringe-pattern. There would also appear to be no line of approach to even a simple three-dimensional analysis except by observations in scattered light. The difficulties therefore of carrying out any kind of quantitative evaluation of stresses would probably be very great, but the rough experiment illustrated here suggests that such a method might well be used to give a qualitative comparison of the thermal gradient stresses which would be present in different designs, and possibly in some cases to yield the order of magnitude of the stresses. So far as the writer is aware, no work has yet been done on this effect.

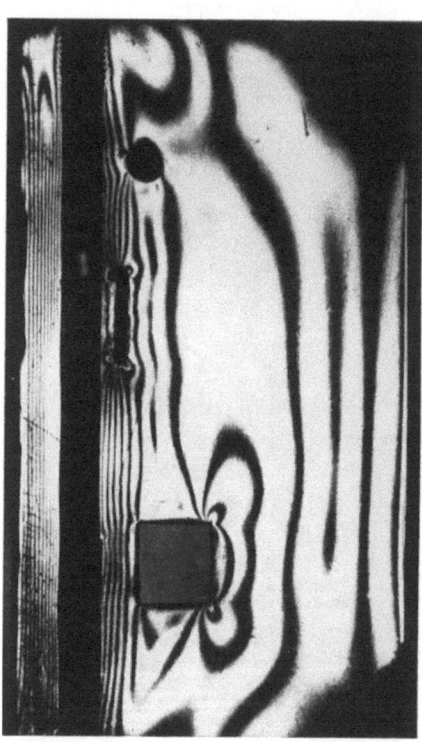

Fig. 114. Fringe-pattern around holes in an araldite plate, produced by thermal gradient.

Bibliography.
Text books on photoelasticity.
[1] Coker, E. G., and L. N. G. Filon: A Treatise on Photoelasticity. Cambridge 1931.
[2] Föppl, L., and H. Neuber: Festigkeitslehre mittels Spannungsoptik. Munich and Berlin 1935.
[3] Mesmer, G.: Spannungsoptik. Berlin 1939.
[4] Boiteux, H. le, and R. Boussard: Elasticité et Photoélasticimétrie.' Paris 1940.
[5] Frocht, M. M.: Photoelasticity. New York: Vol. I 1941; Vol. II 1948.
[6] Kammerer, A.: Recherches sur la photoélasticimtérie. Paris 1944.
[7] Pirard, A.: La Photoélasticité. Paris 1947.
[8] Hendry, A. W.: An Introduction to Photoelastic Analysis. Glasgow 1948.
[9] Jessop, H. T., and F. C. Harris: Photoelasticity, Principles and Methods. London 1949.
[10] Föppl, L., and E. Mönch: Praktische Spannungsoptik. Berlin 1950.
[11] Hetényi, M.: Handbook of Experimental Stress Analysis. New York 1950.
[12] Kuske, A.: Verfahren der Spannungsoptik. Düsseldorf 1951.
[13] Heywood, R. B.: Designing by Photoelasticity. London 1952.

The Mathematical Theories of the Inelastic Continuum.

By

ALFRED M. FREUDENTHAL and HILDA GEIRINGER.

With 60 Figures.

This article is an attempt to present the current stage of development of the theories of the inelastic continuum. In contrast to the classical theories of elasticity and hydrodynamics, the present status of which is the result of two centuries of extensive research, the theories dealt with in this article are still in a comparatively early stage of development notwithstanding the present steadily increasing volume of contributions.

While the equally recent theory of compressible flow has, for several decades, been explored by the foremost mathematicians and physicists, the perhaps even more difficult subject of the deformation of the inelastic solid or pseudo-solid continuum has not found a nearly comparable interest. It is therefore obvious that in various aspects a presentation of this subject will be imperfect, and open to objections both from the physical and the mathematical points of view.

If, when finding physical considerations for which the mathematical theory appears inadequate or mathematical theories the physical relevance of which is not clarified, the reader will attempt to close some of these gaps, one important purpose of this article will have been achieved.

The first part (Chap. A to D) has been prepared by A. M. FREUDENTHAL, the second part (Chap. E to H) by H. GEIRINGER. While the authors have attempted to coordinate their treatment of the various subjects, the two parts represent independent contributions and each author is responsible for his own work. It did not seem desirable to fully coordinate the notations, since the type of notation used is adapted to the treatment of the subject matter.

First part.
The inelastic continuum.

A. Mechanics and thermodynamics of the inelastic continuum.

I. The inelastic behavior of solids.

1. The physical basis of inelasticity. The deviation from perfect elasticity of the deformational response to applied forces in a continuous, homogeneous, isotropic or quasi-isotropic solid may conveniently be attributed to three principal sources which, individually or jointly, produce the wide variety of inelastic behavior that can be observed in solid and pseudo-solid materials under various conditions:

(a) dissipation of mechanical energy through its *interaction* with the *flux* of non-mechanical (thermal, electrical, magnetic) energy or with the *flux* of discrete particles of matter (diffusion), producing *anelastic* deformation and relaxation;

(b) dissipation of mechanical energy through viscous or quasi-viscous *flow* of fluid or quasi-fluid components each of which is considered as a continuous medium dispersed through the other components; these fluid "phases" interacting either with a continuous solid elastic matrix or with equally dispersed elastic "phases" produce the different types of *visco-elastic* response;

(c) dissipation of mechanical energy through *solid friction* and heat developed in *slip* in a multitude of randomly oriented but crystallographically well defined planes in the quasi-isotropic polycrystalline aggregate, or in the slippage of oriented long molecular chains in high polymeric substances, producing the general *plastic* response.

Although formally *anelasticity* might be considered as a type of linear *visco-elasticity* its non-mechanical origin not only justifies but requires separate consideration.

Both anelasticity and visco-elasticity are phenomena that depend on the *total* stress-intensity and are therefore associated with any stress-intensity larger than zero; plasticity, on the other hand depends on a limiting stress-intensity, *the yield stress* σ_0, which is associated with sudden structural breakdown or with sudden changes in the dissipation mechanism, being attained or exceeded. Structurally, the difference can be attributed to the fact that plastic deformation proceeds by motion and concentration of dislocations within the continuum.

Phenomenologically, the viscous and the plastic dissipation of the work of the applied forces differ by the character of the relation between the stress-intensity and the resulting rate of flow in the medium. On the basis of extensive observations of these relations in metallic[1] and non-metallic[2] substances over wide ranges of temperature it appears that the uni-axial quasi-viscous relation which, in general, represents the response at a given strain of materials to relatively low stress intensities has the form

$$\dot{\varepsilon} \exp \frac{\Delta H}{RT} = (a\sigma)^n \quad \text{or} \quad c_1\sigma + c_2\sigma^2 + c_3\sigma^3 + \cdots,$$

while the plastic relation, which represents the response to high stress-intensities, has the form

$$\dot{\varepsilon} \exp \frac{\Delta H}{RT} = \text{const.} \exp(c\sigma),$$

valid for $\sigma > \sigma_0$; the constants a and c depend on structure, ΔH denotes the *activation energy* for self-diffusion in cal/mole, $R \approx 2$ cal/mole is the gas constant and T the absolute temperature in °K. For metals the activation energy is closely related to the melting point T_m and its magnitude can be very roughly estimated from the relation[1]

$$\log \Delta H = 4 + \log(0.0036\, T_m);$$

for high polymers the scarce reported values of ΔH suggest an order of magnitude of ΔH between 10^3 and 10^5 cal/mole depending on molecular weight.

Secondary sources of inelasticity arise as a result of the deviation of the structure of *real* solids from the assumed continuity, homogeneity and isotropy of the *ideal* solid. A porous elastic matrix with fluid or soft plastic phases filling the pores will, under an applied stress-field, show a rather complex inelastic response resulting from the interaction between capillary fluid motion through the pore-space under over-pressure producing transient density changes, or between permanent yielding of the plastic filler, and permanent density increase by localized breakdown of the solid matrix and reduction of the pore-volume. The Inelastic response of porous metals, of stone, of concrete and of soils is the result of these density changes, superimposed on the effects produced by the principal

---

[1] J. E. DORN: J. Mech. Phys. of Solids **3**, 85 (1955).
[2] A. J. STAVERMAN and F. SCHWARZL: Chap. 2, 140, in H. A. STUART, Physik der Hochpolymeren, vol. 4, Berlin: Springer 1956.

sources of inelasticity; the larger the number of simultaneously operative dissipation mechanisms, the more difficult the idealization of the inelastic response to applied forces.

It is expedient to assume that deviation from a perfectly elastic response to hydrostatic stress fields is due either to anelastic effects associated with transport phenomena in the continuum (heat transfer, diffusion, etc.) or within a porous solid medium (pore-fluid motion), or to local structural breakdown or yielding within *polyphase*, inhomogeneous or locally anisotropic solid materials. Adiabatic as well as very slow isothermal volume changes in the isotropic continuum due to an imposed hydrostatic pressure are therefore perfectly elastic, apparent volume-viscosity under intermediate conditions being attributed to anelastic mechanisms[1].

An appearance of perfect elasticity in a real solid, implying instantaneous recovery of deformation upon removal of the forces that have caused it, can be the result either of methods of observation of insufficient sensitivity to detect the deviation from perfect elasticity or of imposed testing-conditions under which the rate of energy dissipation is too low to be perceptible within the time-scale of the experiment; the assumption of perfect elasticity thus implies the absence of any significant interaction between the fluxes of mechanical and non-mechanical energy.

The time-scale of a phenomenon or problem as implied in the velocity of the relative motions within the continuum is of considerable importance in determining the relative significance of the various energy dissipation mechanisms under the specific conditions imposed. Because the effectiveness of those mechanisms depends in various degrees on the stress, the rate of deformation and the absolute temperature level, the time-scale and the temperature-field will affect the mechanical response of the inelastic continuum not less than the stress-intensity itself.

Under conditions of extremely rapid inelastic deformation for instance, under which only a small portion of the energy dissipated in this deformation and transformed into heat is carried away (nearly adiabatic conditions), local temperature increases proportional to the locally dissipated energy of the deformation intensify, in turn, the subsequent dissipation process by increasing the (temperature-sensitive) rate of deformation and thus the apparent "ductility" of the material.

The momentary response of a real solid or pseudo-solid is therefore a mixture of the elastic and of different types of inelastic responses, strongly affected by and changing with the stress intensity and its rate of change, as well as with environment, particularly temperature. Only ideal materials are *elastic, viscous* or *plastic*, because this is how they are defined. Real materials have no definite mechanical "properties" independent of the momentarily imposed conditions. Simple idealizations can therefore only refer to particular conditions under which a specific type of inelasticity dominates the behavior to such an extent as to make the effect of the other types negligible. At elevated temperatures simple idealizations are not even approximately valid; the various types of inelasticity occur simultaneously.

2. Idealization of inelastic response. The simplest idealizations of inelastic response are the equations of the isotropic viscous, the isotropic visco-elastic and the isotropic plastic continuum formulated in terms of the components of the deviations of the stress and strain tensors and their time derivatives; such tensor relations may be formulated either on the basis of considerations of mathematical expediency alone or they may reproduce physical observations. The approach selected depends necessarily on the use to be made of the established relations.

---

[1] L. ROSENHEAD: Proc. Roy. Soc. Lond., Ser. A **226**, No. 1164, 1—64 (1954).

The representation of real solids by the homogeneous, isotropic continuum has the shortcomings of all idealizations of real phenomena: the relations defining the behavior of the continuum represent statistical averages of actual behavior over the locally inhomogeneous and anisotropic elements of the real materials. The discrepancies between the behavior of the idealized continuum and of the elements of the real material are of little significance if only mean values of the response to imposed conditions can actually be observed or measured, as in the case of visco-elastic and anelastic phenomena. In polycrystalline materials, however, the unit process of inelastic deformation is heterogeneous slip on crystallographic planes, which itself is easily observable under adequate magnification. The postulated plastic response of the isotropic homogeneous continuum assumed to represent the behavior of the polycrystalline metal aggregate is therefore the statistical average over a multitude of identifiable heterogeneous deformation processes within the individual crystals. Plastic deformation can be considered as isotropic, homogeneous and continuous only in the sense that the discontinuous, anisotropic slip processes are so numerous, their orientation so random, and the crystal sizes and orientations so randomly distributed that the discontinuity and anisotropy of the unit process can be neglected in considering the total plastic response[1].

The character of the transition from elastic to plastic response *(yield limit)* depends necessarily on the number and directional preferences of simultaneous slip processes occurring within the transition range. As long as a multitude of slip processes over short distances along randomly oriented planes are associated with every force-increment above the elastic limit, the transition is gradual and the plastic response may be considered to be quasi-isotropic. On the other hand, a sudden, sharp transition from elastic to plastic response is the result of elastically unrestrained avalanche-like propagation of a limited number of extensive slip-processes of preferential orientation, by which the heterogeneous character of the deformation by slip is promoted from the sub-microscopic to the macroscopic scale, as illustrated by the appearance of so-called LUEDER's or glide-bands which interfere with the assumed isotropy of the plastic deformation[2]. A certain incompatibility thus exists between the assumption of a sharp transition between the elastic and the plastic response, implicit in the idealized perfect elastic-plastic continuum, and the physical reality underlying such a transition.

The heterogeneity of the plastic deformation is necessarily the more pronounced the less severe the restrictions imposed on the propagation of such deformation by the surrounding elastic matrix. The significance of the discrepancy between the results of an analysis based on the assumption of an isotropic elastic-plastic response and the behavior of a real material with a sharp yield limit, such as mild carbon steel, will depend on the type of the imposed stress-field; the less homogeneous this field, and the more severely restrained therefore the propagation of slip, the better the approximation.

The two non-linear relations between stress and rate of strain characteristic of the observed general viscous and general plastic response (see Sect. 1) are idealized by postulating a linear viscous relation and a time-insensitive ideal plastic relation as mathematically expedient approximations of the two basic types of inelasticity, which, in conjunction with the linear elastic relation, form the basis of the theories of the visco-elastic and of the ideal elastic-plastic continuum. Physically the linear viscous approximation of inelastic behavior is justified only within the range of very low stresses; the time-insensitive ideal

---

[1] W. BOAS and M. E. HARGREAVES: Proc. Roy. Soc. Lond., Ser. A **193**, 89 (1948).
[2] D. KUHLMANN: Z. Metallkde. **41**, 129 (1950).

plastic response, on the other hand, is a fair approximation of the inelastic behavior at the yield limit, being based on the assumption that the observed exponential relation between strain-rate and stress, expressed in the form $\sigma = f(T) +$ const. $\log \dot{\varepsilon}$ is sufficiently closely approximated by $\sigma = \sigma_0(T)$ provided the range of variation of $\dot{\varepsilon}$ is moderate. In fact, for polycrystalline metals at room temperature, for instance mild carbon steel, a variation of $\dot{\varepsilon}$ between 10^{-5} and 10 in./in. sec. produces an increase of σ_0 by less than 60%.

3. Change of state of the continuum. The *state* of the continuum and its change are specified in terms of a continuous relation between certain functions of the mechanical variables (stresses, displacements and their derivatives) and the thermodynamical variables; this relation constitutes the *equation of state* of the considered medium. The existence of such an equation implies that the momentary state depends only on the momentary values of the variables and of their derivatives involved in this equation, but not on the path over which this state has been attained.

In real solid materials in which a change of state is accompanied by a permanent change of the internal structure as, for instance, in polycrystalline metals, an equation of state cannot, in principle, be expected to exist. Only under conditions under which identifiable structural changes are too slight to significantly affect the mechanical response of the continuum, or do not occur at all, can, in first approximation, the existence of an equation of state be assumed. Such conditions are, in general, associated either with *small* inelastic deformations of media with identifiable structure or with deformation of media that are practically without structure or *quasi-amorphous*.

Since the internal structure of permanently deformed real solids is usually not in stable thermodynamic equilibrium and is, in fact, the further removed from such equilibrium the more extensive the deformation, "spontaneous" changes of state, which are the changes that proceed independently of the changes of the mechanical variables, being produced by structural processes governed by the trend towards increasing thermodynamic stability or entropy, (diffusion, chemical reactions, transformation) must be expected to interfere with the existence of an equation of state, the more the less stable the undeformed structure or the larger the inelastic deformation.

Equations of state describe processes of energy flux and energy transformation within the considered medium, involving mechanical energy as well as heat- and other types of non-mechanical energy. In the isotropic continuum they are therefore scalar relations between invariants of the mechanical variables and the thermodynamic variables. In purely mechanical equations of state the thermodynamic variables are considered as parameters, delimiting *isothermal* conditions $(dT=0)$ or *adiabatic* conditions $(dQ=0)$. Such energy equations can be transformed into stress-strain relations under the simplifying assumption that all energy terms involving mechanical variables can be expressed as functions of *one* of these variables alone[1]. This is, therefore, the condition under which stress-strain relations may themselves be considered to have the character of equations of state. Their continuity, for a continuously changing independent variable, expressing continuity of change of the dependent variable, depends on changes of the structural pattern being the exclusive result of changes of the independent variable. For real materials the assumption of such continuity is therefore an approximation without which, however, the establishment of mathematical theories of inelasticity would not be possible.

---

[1] K. WEISSENBERG: Mitt. dtsch. Mat.-Prüf.-Anst. **19**, 54 (1932).

II. Mechanics of deformable media.

4. Basic equations. In the following the coordinates a are assigned to the particles of a material at time t_0, the coordinates x to the coordinates of the same particle at time $t > t_0$. Greek indices refer to a coordinate system attached to the material particles throughout their motion (*material* coordinates), Latin subscripts to a coordinate system attached to points fixed in space (*spatial* coordinates). Strictly speaking, therefore, Cartesian coordinates a_α are transformed by the motion into curvilinear coordinates and vice versa[1]. In first approximation this difference, which involves differentials of second and higher order, can be neglected.

The motion of particles of the continuum from a position $a_\alpha = a_i$ through various locations x_i, with displacement, velocity and acceleration components expressed, respectively, by

$$u_i = x_i - a_i, \quad v_i = \frac{dx_i}{dt} = \frac{du_i}{dt}, \quad \dot{v}_i = \frac{dv_i}{dt} \tag{4.1}$$

can be described by the single-valued, continuously differentiable transformations

$$x_i = x_i(a_\alpha, t) \quad \text{and} \quad a_\alpha = a_\alpha(x_i, t) \tag{4.2}$$

with the derivatives

$$x_{i,\alpha} = \frac{\partial x_i}{\partial a_\alpha}, \quad a_{\alpha,i} = \frac{\partial a_\alpha}{\partial x_i}, \quad x_{\alpha,i} = \frac{\partial x_\alpha}{\partial a_i}, \quad a_{i,\alpha} = \frac{\partial a_i}{\partial x_\alpha}.$$

Problems of mechanics of the deformable isotropic continuum can be reduced to the determination, as functions of space-coordinates and of time, of the tensor of stress over a closed surface surrounding a material point of the continuum, and of the displacement vector or velocity vector at this point, under conditions specified in terms of forces, displacements or velocities along boundaries delimiting the considered geometrical shape.

The first condition, that of *conservation of mass* in any motion of the continuum, implies that the mass flux across the parts of the surface dA of orientation (cosines of surface normal) n_i surrounding the volume V is compensated by the change of mass within this volume, or

$$\int_A \varrho v_i n_i\, dA + \frac{\partial}{\partial t} \int_V \varrho\, dV = 0. \tag{4.3}$$

Applying the divergence theorem to the first term, Eq. (4.3) is transformed into the *continuity equation*

$$(\varrho v_i)_{,i} + \frac{\partial \varrho}{\partial t} = \frac{d\varrho}{dt} + \varrho v_{i,i} = 0. \tag{4.4}$$

For stationary flow with $\partial \varrho / \partial t = 0$ therefore

$$(\varrho v_i)_{,i} = 0, \tag{4.4a}$$

while for incompressible flow with $\varrho = \text{const.}$

$$v_{i,i} = 0. \tag{4.4b}$$

In curvilinear coordinates v_i is replaced by the contravariant vector v^i.

The second condition, that of the *existence of a stress tensor*, implies that the nine components of the stress tensor σ_{ij} in a Cartesian coordinate system determining the state of stress at a point are related to the three components of the

---

[1] V. V. Novozhilov: Foundations of the Nonlinear Theory of Elasticity. Prikl. Math. Mech. **15** (2) (1951).

vector of traction t_i acting upon a surface through the point by the equation

$$t_i = \sigma_{ji} \cdot n_j \tag{4.5}$$

where n_j denotes the orientation of the surface normal. In the absence of external moments and associated rigid body rotations, symmetry conditions furnish the three equations

$$\sigma_{ij} = \sigma_{ji}. \tag{4.6}$$

The equation of motion can be derived from NEWTON's principle of *conservation of linear momentum* for the volume V enclosed by the surface A

$$\frac{d}{dt}\int_V \varrho v_i \, dV = \int_V \frac{\partial}{\partial t}(\varrho v_i) \, dV + \int_A \varrho v_i v_j n_j \, dA = \int_V X_i \, dV + \int_A t_i \, dA \tag{4.7}$$

where X_i denotes the components of the external force vector. According to Eq. (4.5) and the divergence theorem

$$\int_A t_i \, dA = \int_A \sigma_{ji} n_j \, dA = \int_V \sigma_{ji,j} \, dV.$$

Eq. (4.7), valid for a moving element for which ϱdV is conserved, can thus be written in the form

$$\varrho \frac{dv_i}{dt} = \varrho \frac{\partial v_i}{\partial t} + \varrho v_{i,j} v_j = X_i + \sigma_{ji,j} = X_i + \sigma_{ij,j} \tag{4.8}$$

which represents the *equation of motion* of the continuum. Introducing the components of the stress deviation

$$s_{ij} = \sigma_{ij} + p\delta_{ij} \tag{4.9}$$

where the hydrostatic pressure $p = -\tfrac{1}{3}\sigma_{kk}$, Eq. (4.8) takes the form

$$\varrho \frac{\partial v_i}{\partial t} + \varrho v_{i,j} v_j = X_i - p_{,j}\delta_{ij} + s_{ij,j} \tag{4.10}$$

which for $s_{ij}=0$ is transformed into EULER's equation of motion of the ideal fluid.

When $v_{i,j}\to 0$ or $v_i\to 0$ Eq. (4.8) represents the stress-equation of motion (with $X_i=0$)

$$\sigma_{ij,j} = \varrho \frac{\partial v_i}{\partial t} = s_{ij,j} - p_{,j}\delta_{ij} \tag{4.11a}$$

or the stress-equation of equilibrium or slow stationary flow (with $X_i\neq 0$)

$$\sigma_{ij,j} + X_i = s_{ij,j} - p_{,j}\delta_{ij} + X_i = 0. \tag{4.11b}$$

For curvilinear orthogonal coordinates Eq. (4.8) is usually written in the form

$$\sigma^{ij}{}_{,j} + X^i = \varrho \frac{dv^i}{dt} = \varrho\left(\frac{\partial v^i}{\partial t} + v^i{}_{,j} v^j\right) \tag{4.11}$$

where the rules of covariant differentiation have to be applied[1] to the contravariant stress tensor σ^{ij}.

For the determination of the 10 unknowns ϱ, σ_{ij} and v_i or u_i only the 4 equations (4.4) and (4.8) are, so far, available. The missing 6 equations must be provided by the relations between the 6 components of the stress tensor and the 6 components of a suitably specified symmetric tensor function of the velocity or of the displacement vector. Depending on whether this function is derived from the displacement or from the velocity vector it represents the *strain tensor*

---

[1] A. E. GREEN and W. ZERNA: Theoretical Elasticity, pp. 25, 67. Oxford: University Press 1954.

or the *rate of strain tensor*; its relation with the stress tensor describes the mechanical behavior or the *"constitution"* of the isotropic continuum, thus providing the six *constitutive equations* of the continuum.

The parameters of the "constitutive" tensor relation may either be true physical constants which depend only on temperature or entropy, as in the case of the elastic or of the linear relaxing or viscous continuum, or they may be functions of invariants of the stress and the strain tensor or the rate of strain tensor, as in the case of the plastic, visco-plastic or general viscous continuum. In the first case the 10 equations for the 10 unknowns are sufficient for the solution of (isothermal) boundary-value problems in which a uniform temperature $T = \text{const.}$ appears only as a parameter, determining the values of the physical constants of the constitutive equations. In the second case additional equations such as the *yield condition* have to be provided, on the basis of which the undetermined parameters of the constitutive equations can be evaluated.

The frequently made assumption of incompressibility of the continuum ($\varrho = \text{const.}$) eliminates the density as an unknown and imposes by the relation $v_{i,i} = 0$ [see Eq. (4.4b)] or $u_{i,i} = 0$ a condition on the constitutive relations by which the number of independent equations is reduced from six to five, providing thus 9 equations for the remaining 9 unknowns. In the limiting case of the incompressible ideal fluid for which the 5 independent unknowns $s_{ij} = 0$, the four Eqs. (4.4b) and (4.10) are sufficient for the evaluation of the remaining 4 unknowns p and v_i.

When the analysis is extended to include the thermodynamical aspect, two additional equations are required for the two additional variables, the (absolute) temperature T and entropy S. These two equations are furnished by the *equation of state*, describing the exchange between mechanical energy, thermal energy and entropy, and the *equation of heat transfer* by which the relation between heat-energy flux and the temperature field is established. Moreover, the constitutive equations must be expanded so as to include the density change due to change of temperature. Thus, in the general case of the heat-conducting isotropic elastic, linear viscous or visco-elastic continuum 12 equations exist for the 12 unknowns $\varrho, \sigma_{ij}, v_i, T$ and S; for the isotropic plastic continuum the yield condition provides an additional equation for the undetermined parameter of the constitutive relations. By the limiting assumption of isothermal ($dT = 0$) or adiabatic ($dQ = 0$) conditions one of the thermodynamical variables is eliminated as well as the coupling between the mechanical and thermodynamical variables.

5. Measures of strain. The deformation of the isotropic continuum is most expediently expressed by the change of the distance between two neighboring material points in the course of this deformation. Introducing the transformation Eqs. (4.2), their differentials

$$da_\alpha = a_{\alpha,i} dx_i \quad \text{and} \quad dx_i = x_{i,\alpha} da_\alpha \tag{5.1}$$

and the material and spatial metric tensors $g_{\alpha\beta}$ and g_{ij} respectively, the length of this distance in the reference configuration and in the terminal configuration can be expressed in the form:

$$ds_0^2 = g_{\alpha\beta} da_\alpha da_\beta = c_{ij} dx_i dx_j \tag{5.2}$$

and

$$ds^2 = g_{ij} dx_i dx_j = c_{\alpha\beta} da_\alpha da_\beta \tag{5.3}$$

where

$$c_{ij} = g_{\alpha\beta} a_{\alpha,i} a_{\beta,j} \quad \text{and} \quad c_{\alpha\beta} = g_{ij} x_{i,\alpha} x_{j,\beta} \tag{5.4}$$

are the Cauchy-Green *deformation tensors*[1]; they determine two quadric surfaces $c_{ij} dx_i dx_j = 1$ and $c_{\alpha\beta} da_\alpha da_\beta = 1$, with centers at x_i and a_α, into which the respective quadrics in the reference configuration and the deformed configuration $g_{\alpha\beta} da_\alpha da_\beta = 1$ and $g_{ij} dx_i dx_j = 1$ with centers at a_α and x_i are transformed.

In Cartesian coordinates with $g_{\alpha\beta} = \delta_{\alpha\beta}$ and $g_{ij} = \delta_{ij}$ these quadrics represent the *deformation ellipsoid* and the *reciprocal deformation ellipsoid*. Their axes define the directions of the principal extensions c_i and c_α with respect to which the matrices $[c_{ij}]$ and $[c_{\alpha\beta}]$ of the tensors are transformed into their diagonal forms $[c_i]$ and $[c_\alpha]$; their terms are the non-trivial roots of the characteristic matrix equations $[c_{ij} - c\delta_{ij}] = 0$ and $[c_{\alpha\beta} - c\delta_{\alpha\beta}] = 0$, respectively.

The difference $(ds^2 - ds_0^2)$ can therefore be expressed by

$$(ds^2 - ds_0^2) = (c_{\alpha\beta} - \delta_{\alpha\beta}) da_\alpha da_\beta = 2\varepsilon_{\alpha\beta} da_\alpha da_\beta \\ = (\delta_{ij} - c_{ij}) dx_i dx_j = 2\varepsilon_{ij} dx_i dx_j \quad (5.5)$$

where $\varepsilon_{\alpha\beta}$ and ε_{ij} represent the Lagrangian and Eulerian *strain tensors* defined, respectively, with reference to the undeformed and the deformed configuration. For a rigid body displacement for which $ds = ds_0$ it follows that $c_{\alpha\beta} = \delta_{\alpha\beta}$ and $c_{ij} = \delta_{ij}$. The principal strains ε_i and ε_α, being the eigenvalues of the tensor matrices $[\varepsilon_{ij}]$ and $[\varepsilon_{\alpha\beta}]$, are again obtained as the non-trivial roots of the characteristic matrix equations $[\varepsilon_{ij} - \varepsilon\delta_{ij}] = 0$ and $[\varepsilon_{\alpha\beta} - \varepsilon\delta_{\alpha\beta}] = 0$.

Since according to Eqs. (5.2) and (5.3) for the directions of principal extensions, $ds_0^2 = c_i ds^2$ and $ds^2 = c_\alpha ds_0^2$, and thus $c_i = c_\alpha^{-1}$, the principal strain components

$$\varepsilon_\alpha = \tfrac{1}{2}(c_\alpha - 1) = \tfrac{1}{2}[(1 + \Delta)^2 - 1]$$

and

$$\varepsilon_i = \tfrac{1}{2}(1 - c_i) = \tfrac{1}{2}[1 - (1 + \Delta)^{-2}],$$

where the specific principal extension in the specified direction

$$\Delta = (ds - ds_0)/ds_0.$$

For small extensions by binomial expansion

$$(1 + \Delta)^{-2} \approx 1 - 2\Delta \quad \text{and} \quad (1 + \Delta)^2 \approx 1 + 2\Delta$$

and therefore $\varepsilon_\alpha = \varepsilon_i$, while for finite extensions

$$\Delta = \sqrt{1 + 2\varepsilon_\alpha} - 1 \quad \text{and} \quad \Delta = (1/\sqrt{1 - 2\varepsilon_i} - 1),$$

and thus $\varepsilon_\alpha \neq \varepsilon_i$.

The two measures of principal (finite) strain become identical for the same specific extension only if principal strain components h_α and h_i are defined in the alternative forms

$$h_\alpha = \tfrac{1}{2}\log(1 + 2\varepsilon_\alpha) = \tfrac{1}{2}\log c_\alpha \quad \text{and} \quad h_i = -\tfrac{1}{2}\log(1 - 2\varepsilon_i) = -\tfrac{1}{2}\log c_i \quad (5.6)$$

so that $(1+\Delta) = \sqrt{1 + 2\varepsilon_\alpha} = 1/\sqrt{1 - 2\varepsilon_i}$ and $h_\alpha = h_i = \log(1 + \Delta)$. This so-called logarithmic definition of strain[2] has the advantage that the sum of the logarithmic strains in consecutive deformations is equal to the total logarithmic strain. For any other definition this simple superposition rule holds only for infinitesimal extensions. The disadvantage of the logarithmic strain measure, however, is the lack of a simple relation between the mixed components $h_{\alpha\beta}$ and h_{ij} and the change of the angles between the axes of the deformed coordinate system which

---
[1] A. L. CAUCHY: Oeuvres (2) **7**, 82 (1827). — G. GREEN: Papers, pp. 245, 293 (1839—1842).
[2] H. HENCKY: Z. techn. Phys. **9**, 214, 457 (1928).

exists for other definitions of strain[1]. The logarithmic definition is therefore useful only in problems of irrotational deformation.

In order to express the strains in terms of displacements the first of the Eqs. (4.1) is introduced into the Eqs. (5.4). Hence

and
$$c_{\alpha\beta} = \delta_{ij}(a_{i,\alpha} + u_{i,\alpha})(a_{j,\beta} + u_{j,\beta}) = \delta_{\alpha\beta} + u_{\alpha,\beta} + u_{\beta,\alpha} + u_{\gamma,\alpha} u_{\gamma,\beta}$$
$$c_{ij} = \delta_{\alpha\beta}(x_{\alpha,i} - u_{\alpha,i})(x_{\beta,j} - u_{\beta,j}) = \delta_{ij} - u_{i,j} - u_{j,i} + u_{k,i} u_{k,j}. \qquad (5.7)$$

Therefore, according to Eq. (5.5)

and
$$\varepsilon_{\alpha\beta} = \tfrac{1}{2}(u_{\alpha,\beta} + u_{\beta,\alpha} + u_{\gamma,\alpha} u_{\gamma,\beta})$$
$$\varepsilon_{ij} = \tfrac{1}{2}(u_{i,j} + u_{j,i} - u_{k,i} u_{k,j}). \qquad (5.8)$$

If differential terms of higher than first order are neglected, Eqs. (5.8) become identical, expressing the commonly used measure of infinitesimal strain

$$\varepsilon_{\alpha\beta} = \varepsilon_{ij} = \tfrac{1}{2}(u_{\alpha,\beta} + u_{\beta,\alpha}) = \tfrac{1}{2}(u_{i,j} + u_{j,i}). \qquad (5.9)$$

In orthogonal curvilinear coordinates an approximate expression for infinitesimal strains is obtained by a Taylor series expansion of the metric tensor $g_{ij}(x_i)$ at $x_i = a_i + u_i$. Limiting this expansion to linear terms for small values of u_i the expression is obtained

$$g_{ij}(x_i) = g_{ij}(a_i) + g_{ij,k} u_k. \qquad (5.10)$$

Hence, according to Eq. (5.4)

and
$$c_{\alpha\beta} = [g_{ij}(a_i) + g_{ij,k} u_k](a_i + u_i)_{,\alpha}(a_j + u_j)_{,\beta} \qquad (5.11)$$

$$2\varepsilon_{\alpha\beta} = \frac{c_{\alpha\beta} - g_{\alpha\beta}}{g_{\alpha\beta}}. \qquad (5.12)$$

Neglecting terms of higher than first order and considering that $g_{ij} = 0$ and $g_{\alpha\beta} = 0$ respectively for $i \neq j$ and $\alpha \neq \beta$, the following expressions are obtained for the infinitesimal strain-components (no summation of α and β)

$$\varepsilon_{\alpha\alpha} = u_{\alpha\alpha} + \frac{1}{2g_{\alpha\alpha}} g_{\alpha\alpha,\gamma} u^{\gamma} \qquad (5.13)$$

and for $\alpha \neq \beta$

$$\varepsilon_{\alpha\beta} = \frac{1}{2\sqrt{g_{\alpha\alpha} g_{\beta\beta}}} (g_{\alpha\alpha} u_{\alpha,\beta} + g_{\beta\beta} u_{\beta,\alpha}). \qquad (5.14)$$

Eqs. (5.9), considered as a system of partial differential equations to be solved for the displacement when the strain-components are prescribed functions of the coordinates, represent a set of conditions *(compatibility equations)* that must be imposed on the functions ε_{ij} if the existence of a unique continuous solution $u_i(x_i)$ is to be assured. Considering an infinitesimal displacement du_i, the linear transformation equations indicate that

$$du_i = u_{i,j} dx_j = \tfrac{1}{2}[(u_{i,j} + u_{j,i}) + \tfrac{1}{2}(u_{i,j} - u_{j,i})] dx_j = (\varepsilon_{ij} + \omega_{ij}) dx_j \qquad (5.15)$$

where the skew-symmetric (rotation) tensor $\omega_{ij} = -\omega_{ji}$ is a measure of the rigid body rotation. When the displacement gradients are infinitesimal the strain is infinitesimal; the converse, however, is not true. It is therefore obvious that specification of ε_{ij} does not uniquely determine u_i unless ω_{ij} is also specified.

---

[1] C. TRUESDELL: J. Rat. Mech. Analysis **1**, 144 (1952).

When this has been done, cross-differentiation of Eqs. (5.9) provides the equations of compatibility ensuring continuity of the displacement:

$$\varepsilon_{ij,kl} + \varepsilon_{kl,ij} - \varepsilon_{ik,jl} - \varepsilon_{jl,ik} = 0. \tag{5.16}$$

Of these 81 equations only six are independent, the others being either identically satisfied or repetitions because of symmetry in the indices. Explicitly

$$2\varepsilon_{xy,xy} - \varepsilon_{xx,yy} - \varepsilon_{yy,xx} = 0 \tag{5.16a}$$

and two similar equations obtained by cyclic permutation of the (unsummed) subscripts, as well as

$$\varepsilon_{xx,yz} + \varepsilon_{yz,xx} - \varepsilon_{zx,yx} - \varepsilon_{xy,zx} = 0 \tag{5.16b}$$

and two similar equations obtained by cyclic permutation. Eqs. (5.16a) and (5.16b) can be transformed into compatibility equations in terms of stresses by expressing the strains (or strain-increments) in terms of the stresses, using the stress-strain relations characterizing the medium. These equations for the elastic medium without body forces

$$\nabla^2 \sigma_{ij} + \frac{1}{1+\nu} \sigma_{kk,ij} = 0 \tag{5.17}$$

have been obtained by BELTRAMI[1].

In order to express the rate of deformation of the isotropic continuum, Eq. (5.3) is differentiated with respect to time

$$\frac{d}{dt}(ds^2) = g_{ij}\left[\frac{dx_i}{dt}dx_j + \frac{dx_j}{dt}dx_i\right] \tag{5.18}$$

since, for a stationary coordinate system $\dot{g}_{ij}=0$. Considering that

$$\frac{dx_i}{dt} = \dot{x}_{i,j}dx_j \quad \text{and} \quad \frac{dx_j}{dt} = \dot{x}_{j,i}dx_i \tag{5.18a}$$

the material derivative of ds^2 is

$$\frac{d}{dt}(ds^2) = (\dot{x}_{i,j} + \dot{x}_{j,i})dx_i dx_j = 2d_{ij}dx_i dx_j \tag{5.18b}$$

where d_{ij} represents the *rate of deformation tensor*[2]. In the case of infinitesimal strain the *strain-rate tensor* is identical with the rate of deformation tensor

$$\dot{\varepsilon}_{\alpha\beta} = \dot{\varepsilon}_{ij} = \tfrac{1}{2}(v_{i,j} + v_{j,i}) = d_{ij} \tag{5.19}$$

provided the *vorticity tensor*

$$w_{ij} = \tfrac{1}{2}(\dot{x}_{i,j} - \dot{x}_{j,i}) \tag{5.20}$$

is of an order of magnitude[3] not greater than d_{ij}.

6. Invariants. The principal invariants of a symmetric tensor of second rank are the 3 elementary functions of the "eigenvalues" of the tensor matrix that are obtained as the non-trivial roots of the characteristic matrix equation, the coefficients of which are the invariants themselves. For the stress tensor and the strain tensor with the characteristic equations

$$[\sigma_{ij} - \sigma\delta_{ij}] = (\sigma - \sigma_1)(\sigma - \sigma_2)(\sigma - \sigma_3) = \sigma^3 - I_\sigma \sigma^2 + II_\sigma \sigma - III_\sigma = 0 \tag{6.1}$$

and

$$[\varepsilon_{ij} - \varepsilon\delta_{ij}] = (\varepsilon - \varepsilon_1)(\varepsilon - \varepsilon_2)(\varepsilon - \varepsilon_3) = \varepsilon^3 - I_\varepsilon \varepsilon^2 + II_\varepsilon \varepsilon - III_\varepsilon = 0 \tag{6.2}$$

---

[1] E. BELTRAMI: Accad. Linc. Rend. (5) **1**, 141 (1892).
[2] E. BELTRAMI: Opere **2**, pp. 202—379 (1872—1874).
[3] See footnote 1 p. 238 (p. 153).

the principal invariants are:

$$\begin{aligned} &I_\sigma = \sigma_1 + \sigma_2 + \sigma_3, & &I_\varepsilon = \varepsilon_1 + \varepsilon_2 + \varepsilon_3, \\ &II_\sigma = \sigma_1\sigma_2 + \sigma_2\sigma_3 + \sigma_3\sigma_1, & &II_\varepsilon = \varepsilon_1\varepsilon_2 + \varepsilon_2\varepsilon_3 + \varepsilon_3\varepsilon_1, \\ &III_\sigma = \sigma_1\sigma_2\sigma_3 = \det[\sigma_{ij}], & &III_\varepsilon = \varepsilon_1\varepsilon_2\varepsilon_3 = \det[\sigma_{ij}]. \end{aligned} \quad (6.3)$$

In the theory of inelasticity another set of invariants is widely used. For the stress tensor they are

$$\begin{aligned} J_1 &= \sigma_{ii} = I_\sigma, \\ J_2 &= \tfrac{1}{2}\sigma_{ij}\sigma_{ji} = \tfrac{1}{2}\sum_i(\sigma_i)^2 = \tfrac{1}{2}I_\sigma^2 - II_\sigma, \\ J_3 &= \tfrac{1}{3}\sigma_{ij}\sigma_{jk}\sigma_{ki} = \tfrac{1}{3}\sum_i(\sigma_i)^3 = \tfrac{1}{3}I_\sigma^3 - I_\sigma II_\sigma + III_\sigma. \end{aligned} \quad (6.4)$$

Similar expressions I_1, I_2 and I_3 can be written for the strain tensor.

Resolving the stress tensor σ_{ij} into its hydrostatic part $p\delta_{ij}$ and into the stress deviation s_{ij} according to Eq. (4.9) and, similarly, the tensor of infinitesimal strain into its volumetric part $\varepsilon_v\delta_{ij}$, where $\varepsilon_v = -\tfrac{1}{3}\varepsilon_{kk}$, and into the component of the strain deviation $e_{ij} = \varepsilon_{ij} + \varepsilon_v\delta_{ij}$, invariants of the stress and the strain deviation are defined by the coefficients of the characteristic equations

$$s^3 + II_s s - III_s = 0 \quad \text{and} \quad e^3 + II_e e - III_e = 0. \quad (6.5)$$

They are of the same form as Eqs. (6.3) and (6.4) with $I_s = I_e = 0$. Hence

$$\begin{aligned} &J_1' = I_s = 0, & &I_1' = I_e = 0, \\ &J_2' = \tfrac{1}{2}s_{ij}s_{ji} = -II_s, & &I_2' = \tfrac{1}{2}e_{ij}e_{ji} = -II_e, \\ &J_3' = \tfrac{1}{3}s_{ij}s_{jk}s_{ki} = III_s, & &I_3' = \tfrac{1}{3}e_{ij}e_{jk}e_{ki} = III_e. \end{aligned} \quad (6.6)$$

Explicitly

$$\begin{aligned} J_2' &= \tfrac{1}{2}(s_{11}^2 + s_{22}^2 + s_{33}^2 + 2s_{12}^2 + 2s_{23}^2 + 2s_{31}^2) = \tfrac{1}{2}(s_1^2 + s_2^2 + s_3^2) \\ &= \tfrac{1}{6}[(\sigma_{11}-\sigma_{22})^2 + (\sigma_{22}-\sigma_{33})^2 + (\sigma_{33}-\sigma_{11})^2 + 6(\sigma_{12}^2 + \sigma_{23}^2 + \sigma_{31}^2)] \\ &= \tfrac{1}{6}[(\sigma_1-\sigma_2)^2 + (\sigma_2-\sigma_3)^2 + (\sigma_3-\sigma_1)^2], \\ J_3' &= \tfrac{1}{3}(s_1^3 + s_2^3 + s_3^3) = \tfrac{1}{27}(2\sigma_1-\sigma_2-\sigma_3)(2\sigma_2-\sigma_3-\sigma_1)(2\sigma_3-\sigma_1-\sigma_2). \end{aligned} \quad (6.7)$$

Moreover

$$J_2' = J_2 - \tfrac{1}{6}J_1^2 \quad \text{and} \quad J_3' = J_3 - \tfrac{2}{3}(J_1 J_2 - \tfrac{1}{9}J_1^3). \quad (6.6a)$$

Similar expressions can be written for the invariants of the strain deviation.

Other forms in which the second invariants of the stress deviation and strain deviation are frequently used are the shear stress in the plane $(1/\sqrt{3}, 1/\sqrt{3}, 1/\sqrt{3})$ (octahedral plane), the *octahedral shear stress*, and the associated *octahedral shear strain*[1] which are related to the respective invariants by the equations:

$$\tau_0^2 = \tfrac{2}{3}J_2' \quad \text{and} \quad \gamma_0^2 = \tfrac{8}{3}I_2' \quad (6.8a)$$

as well as the so-called stress-and strain-*"intensities"*[2]

$$\bar{\sigma}^2 = 3 J_2' \quad \text{and} \quad \bar{\varepsilon}^2 = \tfrac{4}{3}I_2'. \quad (6.8b)$$

---

[1] A. Nadai: Theory of Flow and Fracture of Solids, Vol. 1, 103. New York: McGraw-Hill 1950.

[2] H. Hencky: Z. angew. Math. Mech. **4**, 323 (1924).

The volume change is related to the strain invariants (6.3) by the expression

$$\left(\frac{dV_i}{dV_\alpha}\right)^2 = 1 + 2\mathrm{I}_\varepsilon + 4\mathrm{II}_\varepsilon + 8\mathrm{III}_\varepsilon \tag{6.9}$$

where the subscript ε refers to the strain $\varepsilon_{\alpha\beta}$. For infinitesimal strain

$$\frac{dV_i}{dV_\alpha} \approx \sqrt{1+2\mathrm{I}_\varepsilon} \approx 1 + \mathrm{I}_\varepsilon = 1 - 3\varepsilon_v. \tag{6.9a}$$

Finally, two dimensionless invariants can be defined in terms of the ratio between the third and second invariants of the stress and the strain deviations[1]

$$\bar{\mu} = \frac{J_3'}{(2J_2')^{\frac{3}{2}}} \quad \text{and} \quad \bar{\nu} = \frac{I_3'}{(2I_2')^{\frac{3}{2}}} \tag{6.10}$$

which, by determining the ratios between the components of the principal deviatoric stresses and the principal deviatoric strains specify the "*form*" of the tensors and establish the criterion $\bar{\mu} = \bar{\nu}$ for similarity of the tensors.

In geometric representation this criterion is equivalent to that of similarity of the *Mohr* circles of these tensors[2], but independent of the order in which the principal axes are numbered and by which the six alternative arrangements of the circles are defined. The Mohr circles are obtained by projecting the 3 principal circles on the unit sphere in the coordinate systems $(\sigma_1 \sigma_2 \sigma_3)$ or $(\varepsilon_1 \varepsilon_2 \varepsilon_3)$ on the (σ, τ)-plane or (ε, γ)-plane, where σ, τ and ε, γ denote, respectively, the normal and the tangential components of the traction t_i and of the specific extension ε_n acting on a plane with unit normal of orientation n_i. In order to obtain the components σ and τ for any plane the three equations

$$\sigma = t_i n_i = \sigma_{ij} n_i n_j, \quad |\tau|^2 = \sum |t_i|^2 - |\sigma|^2, \quad n_i n_i = 1 \tag{6.11}$$

must be solved for the three unknowns n_i. The solutions for $n_i = \text{const.}$ represent the projections of circles on the unit sphere in planes parallel to the j-k plane. For instance for n_3 explicitly

$$n_3^2 (\sigma_1 - \sigma_3)(\sigma_2 - \sigma_3) = (\sigma_1 - \sigma)(\sigma_2 - \sigma) + \tau^2.$$

For $n_3 = 0$ (principal circle) this equation is transformed into the equation of the stress circle in the (σ, τ)-plane representing the intersection of the (σ_1, σ_2)-plane with the unit-sphere:

$$\left(\sigma - \frac{\sigma_1+\sigma_2}{2}\right)^2 + \tau^2 = \left(\frac{\sigma_1 - \sigma_2}{2}\right)^2$$

with center at $\sigma = \frac{1}{2}(\sigma_1 + \sigma_2)$ and radius $\tau_3 = \frac{1}{2}(\sigma_1 - \sigma_2)$. By setting respectively $n_1 = 0$ and $n_2 = 0$ similar expressions are obtained for the other two principal stress circles with centers at $\sigma = \frac{1}{2}(\sigma_2 + \sigma_3)$, $\sigma = \frac{1}{2}(\sigma_3 + \sigma_1)$ and radii $\tau_1 = \frac{1}{2}(\sigma_2 - \sigma_3)$, $\tau_2 = \frac{1}{2}(\sigma_3 - \sigma_1)$. The three components τ_i are the principal shearing stresses; they fulfil the condition $\tau_1 + \tau_2 + \tau_3 = 0$. The octahedral normal and shear stress components are the components σ and τ in the octahedral planes $n_1 = n_2 = n_3 = 1/\sqrt{3}$.

A similar representation specifies the Mohr circles of infinitesimal strain with centers at $\varepsilon_i = \frac{1}{2}(\varepsilon_k + \varepsilon_j)$ and radii $\gamma_i = \frac{1}{2}(\varepsilon_k - \varepsilon_j)$.

For a particular order in which the principal axes of stress and of strain are numbered the similarity of two deviations $\sigma_{ij} = s_{ij}$ and $\varepsilon_{ij} = e_{ij}$ can be expressed

---

[1] H. FROMM: Ing.-Arch. **4**, 436 (1933).
[2] O. MOHR: Z. VDI **1900**, 1524. See also footnote 1, p. 240 (p. 96).

in terms of the associated ratios $\mu_i = \nu_i$ defined by

$$\mu_i = \frac{2\sigma_i - \sigma_j - \sigma_k}{\sigma_j - \sigma_k} \quad \text{and} \quad \nu_i = \frac{2\varepsilon_i - \varepsilon_j - \varepsilon_k}{\varepsilon_j - \varepsilon_k} \tag{6.12}$$

which determine the position of the intermediate principal stress on the σ-axis and intermediate principal strain on the ε-axis[1].

The values of these ratios vary between $\mu_i = -1$ for $\sigma_i = \sigma_k$ or $\nu_i = -1$ for $\varepsilon_i = \varepsilon_k$ and $\mu_i = 1$ for $\sigma_i = \sigma_j$ or $\nu_i = 1$ for $\varepsilon_i = \varepsilon_j$. The relations between the invariants μ_i and $\bar{\mu}$ as well as between ν_i and $\bar{\nu}$ are cubic equations in μ_i^2 and in ν_i^2, respectively, the solutions of which for given values of $\bar{\mu}$ or $\bar{\nu}$ furnish three pairs of roots $\pm \mu_i$ and $\pm \nu_i$.

Frequently, it is expedient to express the components of the stress tensor in terms of its invariants. The hydrostatic pressure being proportional to J_1, the components of the stress deviations s_{ij} are obtained by solving the first characteristic Eq. (6.5) for the principal stress deviations s_k and expressing s_{ij} in terms of s_k and the directions between the coordinate axes and the axes of the principal components

$$s_{ij} = \sum_k s_k \cos(i,k) \cos(j,k). \tag{6.13}$$

Replacing II_s and III_s by J_2' and J_3' according to Eqs. (6.6), Eq. (6.5) takes the form

$$s^3 - J_2' s - J_3' = 0. \tag{6.5a}$$

This equation is solved by the introduction of the auxiliary angle 3ω through the expression[2]

$$\cos 3\omega = \frac{J_3' \sqrt{2}}{\tau_0^3} = \frac{s_1 s_2 s_3}{\tau_0^3} \sqrt{2} \tag{6.14}$$

as a result of which the three roots s_k can be directly written in the form

$$\left.\begin{array}{l} s_1 = \dfrac{1}{\sqrt{2}} \tau_0 (\cos\omega - \sqrt{3} \sin\omega) = \tau_0 \sqrt{2} \cos\left(\omega + \dfrac{\pi}{3}\right), \\[4pt] s_2 = \dfrac{1}{\sqrt{2}} \tau_0 (\cos\omega + \sqrt{3} \sin\omega) = \tau_0 \sqrt{2} \cos\left(\omega - \dfrac{\pi}{3}\right), \\[4pt] s_3 = -\tau_0 \sqrt{2} \cos\omega \end{array}\right\} \tag{6.15}$$

where $0 \lesseqgtr 3\omega \leq \pi$ and therefore $0 \lesseqgtr \omega \leq \dfrac{\pi}{3}$.

Considering that for the particular order of the roots s_k in Eq. (6.15)

$$\tan\omega = \frac{1}{\sqrt{3}} \frac{s_1 - s_2}{s_3} = \frac{\sigma_1 - \sigma_2}{2\sigma_3 - \sigma_1 - \sigma_2} \sqrt{3} = \sqrt{3}/\mu_3$$

the angle ω can be interpreted as the angle between the direction of τ_0 and that of the (negative) σ_3-axis in the projection of the stress-space on the octahedral plane (see Fig. 3), provided that the direction of τ_0 falls within the sextant enclosed between the $(-\sigma_3)$ and $(+\sigma_2)$ axes[3]. For an arbitrary location of this direction the above equation can be generalized in the form

$$\tan\omega_i = \pm \frac{(\sigma_j - \sigma_k)\sqrt{3}}{2\sigma_i - \sigma_j - \sigma_k} = \pm \sqrt{3}/\mu_i \tag{6.16}$$

---

[1] W. Lode: Mitt. Forschungsarb. Ingenieurw. **303** (1928).
[2] H. M. Westergaard: Theory of Elasticity and Plasticity, p. 71. New York: J. Wiley & Sons 1952.
[3] V. V. Sokolovsky: Theory of Plasticity. Berlin: VEB Verlag 1955.

and therefore
$$\cos^2 \omega_i = (1 + 3/\mu_i^2)^{-1}. \tag{6.16a}$$

Solving Eq. (6.14) written in the alternative form
$$4 \cos^3 \omega - 3 \cos \omega = 3 \sqrt{6} \bar{\mu} \tag{6.14a}$$

for $\cos \omega$ and introducing the three roots into Eq. (6.16a) the three pairs of values $\pm \mu_i$ are obtained in terms of the form invariant $\bar{\mu}$.

III. Thermodynamic considerations.

7. Energy and entropy balance. Multiplying Eq. (4.8) by v_i and integrating over the volume

$$\int_V \varrho v_i \frac{dv_i}{dt} dV = \int_V v_i X_i dV + \int_V v_i \sigma_{ij,j} dV \tag{7.1}$$

the principle of conservation of mechanical energy can be derived from Newton's equation of conservation of linear momentum; the left side of Eq. (7.1) represents the rate of change of kinetic energy W_k in the volume V

$$\frac{dW_k}{dt} = \frac{d}{dt} \int_V \frac{1}{2} \varrho v_i v_i dV = \int_V \varrho v_i \dot{v}_i dV \tag{7.2}$$

and the second term on the right side

$$\int_V v_i \sigma_{ij,j} dV = \int_V (v_i \sigma_{ij})_{,j} dV - \int_V \sigma_{ij} v_{i,j} dV. \tag{7.3}$$

Applying the divergence theorem as well as Eqs. (5.19) and (5.20)

$$\int_V v_i \sigma_{ij,j} dV = \int_A v_i \sigma_{ij} n_j dA - \int_V \sigma_{ij} (\dot{\varepsilon}_{ij} + w_{ij}) dV = \int_A v_i t_i dA - \int_V \sigma_{ij} \dot{\varepsilon}_{ij} dV \tag{7.4}$$

since $\sigma_{ij} w_{ij} = 0$, and also $\sigma_{ij} n_j = t_i$. Eq. (7.1) is thus transformed into

$$\int_A v_i t_i dA + \int_V v_i X_i dV = \int_V \varrho v_i \dot{v}_i dV + \int_V \sigma_{ij} \dot{\varepsilon}_{ij} dV = \frac{dW_k}{dt} + \frac{dU}{dt} \tag{7.5}$$

which expresses the principle of *conservation of (mechanical) energy*, the left side representing the rate of total work done on the body, which is the sum of the rate of work of the surface forces and of the external forces, the right side representing the rate of change of the total energy of the system, consisting of the rate of change of the kinetic energy W_k and of the *internal energy* U of the medium. With $dW_k/dt = 0$ Eq. (7.5) states that the rate at which energy is transferred into the body is equal to the rate at which its internal energy U increases.

Eq. (7.5) can be expanded to include the thermal energy by adding the rate of heat input

$$\frac{dQ}{dt} = \int_A q \, dA = -\int_A q_i n_i dA = -\int_V q_{i,i} dV \tag{7.6}$$

where q_i denotes the heat flux vector. The internal and kinetic energy increase thus balances the external work done and the heat supplied[1]

$$\int_A v_i t_i dA + \int_V v_i (X_i - \varrho \dot{v}_i) dV - \int_V q_{i,i} dV = \frac{dU}{dt}. \tag{7.7}$$

---
[1] G. Jaumann: Sitzgsber. Akad. Wiss. Wien (IIa) **120**, 385 (1911). — Denkschr. Akad. Wiss. Wien **95**, 461 (1918).

or for $X_i = 0$

$$\int_A v_i t_i \, dA - \int_V q_{i,i} \, dV = \frac{dU}{dt} + \int_V \varrho v_i \dot{v}_i \, dV. \qquad (7.7a)$$

The *First Law of Thermodynamics* introduces the internal energy U of the system as a *state function* by which the change of state of the continuum is phenomenologically described independently of the path over which the changed state has been reached; dU is therefore a total differential, and U has the character of a potential of the state variables. The rate of change of internal energy of the continuum represents the sum of "*internal energy production*" resulting from the application of mechanical energy and heat to the body. One part of this energy represents the reversible increase of the average *potential* energy of the material particles of the continuum; it can be expressed in terms of a specific *elastic potential* or strain energy function φ. The remaining internal energy production represents the increase of the average *kinetic* energy of the material particles, and thus of the heat energy into which the applied mechanical energy is irreversibly transformed or "dissipated" as a result of irreversible deformation of the continuum. It can be phenomenologically expressed in terms of a sum of specific functions $\dot{\chi}_i$ of the mechanical and thermodynamical state variables by which the different processes of *internal energy dissipation* can be described.

Introducing the rate of change of the *specific* internal energy du/dt per unit volume into Eq. (7.7) and considering Eq. (7.4) the equation is obtained

$$-\int_V q_{i,i} \, dV + \int_V (\sigma_{ij,j} + X_i - \varrho \dot{v}_i) v_i \, dV = -\int_V \sigma_{ij} \dot{\varepsilon}_{ij} \, dV + \int_V \frac{du}{dt} \, dV \qquad (7.8)$$

which, considering the equilibrium condition (4.8), is transformed into the expression of the First Law for the volume element

$$\left.\begin{array}{l} \dfrac{du}{dt} = \sigma_{ij} \dot{\varepsilon}_{ij} - q_{i,i} = (-p \delta_{ij} + s_{ij})(-\dot{\varepsilon}_v \delta_{ij} + \dot{e}_{ij}) - q_{i,i} \\[4pt] = 3 p \dot{\varepsilon}_v + s_{ij} \dot{e}_{ij} - q_{i,i} = \dfrac{p}{\varrho} \dfrac{d\varrho}{dt} + s_{ij} \dot{e}_{ij} - q_{i,i} \end{array}\right\} \qquad (7.9)$$

since according to Eqs. (4.4) and (5.19)

$$\frac{d\varrho}{dt} = -\varrho v_{i,i} = -\varrho \dot{\varepsilon}_{ii} = 3 \varrho \dot{\varepsilon}_v.$$

Identifying the specific internal energy of the *elastic* medium with the strain-energy potential

$$\frac{du}{dt} = \varrho \dot{\varphi}, \qquad (7.10)$$

Eq. (7.9) is transformed into the energy equation of the elastic continuum

$$\varrho \dot{\varphi} = \sigma_{ij} \dot{\varepsilon}_{ij} - q_{i,i}. \qquad (7.11)$$

For the *dissipative* medium for which

$$s_{ij} \dot{e}_{ij} = \sum_i \varrho \dot{\chi}_i$$

expresses the condition of total dissipation of the power of the stress deviation, this stress power is converted into heat, so that the total heat input[1]

$$\frac{d\overline{Q}}{dt} = \frac{dQ}{dt} + \sum_i \int_V \varrho \dot{\chi}_i \, dV = -\int_V q_{i,i} \, dV + \sum_i \int_V \varrho \chi_i \, dV \qquad (7.12a)$$

---

[1] H. S. CARSLAW and J. C. JAEGER: Conduction of Heat in Solids, p. 9. New York: Oxford Univ. Press 1948.

and therefore for the volume element

$$-q_{i,i} + \sum_i \varrho \dot{\chi}_i = \varrho c \frac{dT}{dt} = \varrho c \left(\frac{\partial T}{\partial t} + T_{,i} v_i\right) \tag{7.12b}$$

where c is the heat capacity of the medium at constant volume and T the absolute temperature.

The validity of the First Law for real materials is subject to the limitations imposed by the *Second Law*, which introduces the entropy S as a second path-independent state function depending only on the internal state variables of the system, and specifies that the total rate of entropy change dS/dt in a materially closed but thermally open system which exchanges heat with its surroundings, but not matter, is the sum of the rate of entropy change dS_e/dt resulting from such heat exchange and of the rate of entropy increase dS_i/dt resulting from *internal* irreversible processes in the system itself. Hence for the volume V

$$\frac{dS}{dt} = \frac{dS_e}{dt} + \frac{dS_i}{dt} \tag{7.13}$$

where the first term on the right-hand side represents an *external entropy flux*, the second term the *internal entropy production*[1].

The entropy is related to the heat-input by the definitions

$$\frac{dS_e}{dt} = \frac{1}{T} \frac{dQ}{dt} = -\int_V \frac{1}{T} q_{i,i} dV \quad \text{and} \quad \frac{d\bar{S}}{dt} = \frac{1}{T} \frac{d\bar{Q}}{dt}. \tag{7.14}$$

While dQ is not an exact differential, the absolute temperature T represents the integrating denominator which makes dS_e an exact differential[2].

For the volume element, defining $\varrho \eta$ as the specific entropy,

$$T \frac{d}{dt}(\varrho \eta_e) = -q_{i,i} \quad \text{and} \quad T \frac{d}{dt}(\varrho \bar{\eta}) = -q_{i,i} + \sum_i \varrho \dot{\chi}_i. \tag{7.15}$$

Because the external entropy flux may be positive, zero or negative, depending on the direction of the heat flow, the total rate of entropy change can also be positive, zero or negative; it can therefore not serve as a criterion of irreversibility of a change of state. Thus, the Second Law must be formulated in terms of the *entropy production* alone, defining the irreversibility of a change of state by a *positive* rate of entropy production

$$\frac{dS_i}{dt} > 0. \tag{7.16}$$

The condition $dS_i/dt = 0$ therefore delimits a reversible change of state. This is, however, not its only meaning: it also defines the condition of thermodynamic equilibrium in a thermodynamically closed system for which the total entropy reaches its maximum (see Sect. 8).

Introducing the total entropy density per unit volume $s = \varrho \eta$, the external entropy flux vector (q_i/T) and the rate of non-mechanical internal entropy production per unit volume ϑ', the equation of entropy balance in the absence of mechanical energy dissipation according to Eq. (7.13) and the first Eq. (7.14) can be written in the form

$$\frac{dS}{dt} = \int_V \frac{ds}{dt} dV = -\int_V \frac{1}{T} q_{i,i} dV + \int_V \vartheta' dV. \tag{7.17}$$

---

[1] I. PRIGOGINE: Etude Thermodynamique des Phénomènes Irréversibles. Paris: Dunod 1947.
[2] C. CARATHEODORY: Math. Ann. **67**, 355 (1909).

Considering that

$$\int_V \frac{1}{T} q_{i,i} dV = \int_V \left(\frac{q_i}{T}\right)_{,i} dV + \int_V \frac{q_i T_{,i}}{T^2} dV = \int_A \left(\frac{q_i}{T}\right) n_i dA + \int_V \frac{q_i T_{,i}}{T^2} dV, \quad (7.18)$$

Eq. (7.17) is transformed into

$$\frac{dS}{dt} + \int_A \left(\frac{q_i}{T}\right) n_i dA = \frac{dS_i}{dt} = -\int_V \frac{q_i T_{,i}}{T^2} dV + \int_V \vartheta' dV > 0. \quad (7.19)$$

from which follows that the first term on the right-hand side represents the entropy production due to heat flow. Because according to Eq. (7.16) this term must be positive and therefore $q_i T_{,i} < 0$, the relation between the heat flux vector and the temperature gradient under the assumption of linearity (FOURIER's law of heat conduction) must be of the form

$$q_i = -k T_{,i} \quad (7.20)$$

where the coefficient of conductivity $k > 0$.

Using $d\bar{S}/dt$ in place of dS/dt the entropy balance becomes

$$\frac{dS}{dt} + \int_A \left(\frac{q_i}{T}\right) n_i dA = -\int_V \frac{q_i T_{,i}}{T^2} dV + \int_V \vartheta' dV + \sum_i \int_V \frac{1}{T} \varrho \dot{\chi}_i dV > 0 \quad (7.21)$$

The complete entropy balance for the volume element[1] is therefore

$$\frac{ds}{dt} + \left(\frac{q_i}{T}\right)_{,i} = -\frac{q_i T_{,i}}{T^2} + \frac{\sum_i \varrho \dot{\chi}_i}{T} + \vartheta'. \quad (7.22)$$

Combining Eqs. (7.22), (7.12) and (7.9), the relation is obtained

$$T \frac{ds}{dt} = \varrho c \frac{dT}{dt} = \frac{du}{dt} - \sigma_{ij} \dot{\varepsilon}_{ij} + \sum_i \varrho \dot{\chi}_i + T \vartheta' \quad (7.23\text{a})$$

or

$$T \frac{ds}{dt} = \frac{du}{dt} - \frac{p}{\varrho} \frac{d\varrho}{dt} - \left(s_{ij} \dot{e}_{ij} - \sum_i \varrho \dot{\chi}_i\right) + T \vartheta'. \quad (7.23\text{b})$$

Introducing $\vartheta' = 0$ and either $s_{ij} = 0$, $\dot{\chi}_i = 0$ (ideal fluid) or $s_{ij} \dot{e}_{ij} = \sum_i \varrho \dot{\chi}_i$ (dissipative medium), Eq. (7.23 b) degenerates into GIBBS' entropy equation

$$T \frac{ds}{dt} = \frac{du}{dt} - \frac{p}{\varrho} \frac{d\varrho}{dt}.$$

The last two terms on the right hand side of Eq. (7.23 a) can be combined into the *total* entropy production, ϑ, representing the dissipation of applied mechanical energy and the entropy production due to non-mechanical effects:

$$\vartheta = \frac{1}{T} \sum_i \varrho \dot{\chi}_i + \vartheta'. \quad (7.24)$$

---

[1] E. LOHR: Denkschr. Akad. Wiss. Wien **93**, 339 (1917). — C. ECKART: Phys. Rev. **58**, 267 (1940). — J. MEIXNER: Ann. Phys. (5) **39**, 333 (1941).

Introducing Eq. (7.10) and $\vartheta = 0$ into Eq. (7.23a), the entropy equation for the elastic continuum due to Voigt[1] is obtained

$$T \frac{ds}{dt} = T \frac{ds_e}{dt} = \varrho \dot{\varphi} - \sigma_{ij} \dot{\varepsilon}_{ij}. \tag{7.25}$$

Since for adiabatic conditions $ds_e/dt = 0$ Eq. (7.25) defines the *adiabatic* elastic potential $\varrho \dot{\varphi} = \sigma_{ij} \dot{\varepsilon}_{ij}$.

For the elastically compressible dissipative continuum Eq. (7.23b) is transformed into

$$T \frac{ds}{dt} = \frac{du}{dt} - \frac{p}{\varrho} \frac{d\varrho}{dt} + T\vartheta' = T\vartheta - q_{i,i} = \varrho c \frac{dT}{dt} \tag{7.26}$$

a form similar to that proposed by Eckart[2] for the anelastic medium

$$T \frac{ds}{dt} = -q_{i,i} + s_{ij} \dot{e}_{ij} + T\vartheta'. \tag{7.27}$$

Eq. (7.23) together with the continuity Eq. (4.4), the 3 equations of motion (4.8), the 6 stress-strain-temperature relations to be specified and, for the plastic or combined plastic medium, the plasticity condition, provide the 11 (12) equations for the 11 (12) unknowns ϱ, v_i, σ_{ij}, T, (and the undetermined coefficient of the plastic stress-strain relation). Eq. (7.20) provides the necessary relation for the conversion of heat-flow into temperature. This equation is, however, only an approximation of the more general relation connecting the heat-flux vector with temperature as well as with the mechanical state variables (see Sect. 8). If this relation is used instead of Eq. (7.20), three additional equations are added for the three additional unknowns q_i.

For adiabatic conditions $dQ/dt = 0$; therefore introducing $q_{i,i} = 0$ into Eq. (7.12b)

$$\varrho c \left(\frac{\partial T}{\partial t} + v_i T_{,i} \right) = \sum_i \varrho \dot{\chi}_i \tag{7.28}$$

from which the temperature-field T associated with adiabatic conditions in a dissipative medium can be determined.

Writing Eq. (7.23a) in the form

$$\sigma_{ij} \dot{\varepsilon}_{ij} = \frac{du}{dt} - T \frac{ds_e}{dt} \tag{7.29}$$

isothermal state functions can be defined. For the elastic medium with $u = \varrho \varphi$ therefore

$$\sigma_{ij} \dot{\varepsilon}_{ij} = \frac{d}{dt} (\varrho \varphi - T \varrho \eta_e) + \dot{T} \varrho \eta_e = \varrho \dot{\psi} + \varrho \eta_e \dot{T} \tag{7.29a}$$

where $\psi = (\varphi - T \eta_e)$ represents the "free energy" of the medium or the *isothermal* elastic potential, since for $\dot{T} = 0$

$$\sigma_{ij} \dot{\varepsilon}_{ij} = \varrho \dot{\psi} \tag{7.30}$$

For dissipative media with $\dot{T} = 0$, Eq. (7.12b) in the form

$$\sum \varrho \dot{\chi}_i = s_{ij} \dot{e}_{ij} = q_{i,i} \tag{7.31}$$

indicates that isothermal conditions for finite rates of deformation can only be sustained by the forced removal of the heat generated in such deformation.

---
[1] W. Voigt: Ann. Phys. 36, 743 (1889).
[2] C. Eckart: Phys. Rev. 73, 373 (1948).

8. State functions. The variables defining the state of an element of the *elastic* continuum are its (absolute) temperature T (or, rather, temperature increment above a reference temperature associated with the *natural* state) and the strain components ε_{ij} describing the deviation of the momentary state from this *natural* unstrained and unstressed state to which the element will immediately return when released from all external forces; it is assumed that the strained state is very close to the natural state.

A change of state is described by a change of the strain energy which is identical with the total internal energy according to Eq. (7.10); the state function is therefore a function $\varrho\,\varphi(\varepsilon_{\alpha\beta}, T)$ or $\varrho\,\varphi(\varepsilon_{ij}, T)$ which vanishes for $\varepsilon_{\alpha\beta}=0$ or $\varepsilon_{ij}=0$. If the medium is inherently heterogeneous the potential $\varrho\varphi(x_i, \varepsilon_{ij}, T)$ contains the coordinates themselves, in addition to their gradients. It should, however, be realized that even in the isotropic medium the appearance of $T(x_i)$ in the strain energy function implies a certain type of heterogeneity associated with any imposed inhomogeneous temperature field.

The variables defining the state of an element of a *dissipative* continuous medium are its temperature and the deviatoric stress or strain-rate components s_{ij} or $\dot{e}_{ij}$, respectively, and their time-derivatives. Any irreversible change of state proceeds in the direction of increasing entropy towards thermodynamic equilibrium at which the entropy attains its maximum, and at which therefore the rate of entropy-production is zero. It can be described by the change of entropy or rather by the rate of internal entropy production $\vartheta = (s_{ij}\,\dot{e}_{ij})/T$ associated with an irreversible change of state; the state functions $\vartheta = \frac{1}{T}\varrho\dot{\chi}$ can thus be expressed in the two alternative forms $T\vartheta_v = \varrho\dot{\chi}_v(\dot{e}_{ij}, \ddot{e}_{ij}, \ldots, T)$ and $T\vartheta_p = \varrho\dot{\chi}_p(s_{ij}, \dot{s}_{ij}, \ldots, T)$ depending on whether the state of the medium is primarily defined by the deformation rates or by the stresses. It is, moreover, assumed that any considered state is so close to the state of thermodynamic equilibrium for which $T\vartheta=0$ that, in first approximation, the relations between the mechanical state variables s_{ij} and $\dot{e}_{ij}$ can be linearized, whatever their actual form[1].

The above considerations are also extended to the irreversible non-mechanical phenomena which must be expected to affect the mechanical state of the continuum, and which, in analogy to the mechanical phenomena, can be described by state variables representing generalized "forces" F_i and generalized "fluxes" f_i. Because of the assumption of "near equilibrium" and the resulting linearization of the relations between the state variables, the state functions ϑ' defining the non-mechanical dissipation processes can be expressed as bi-linear forms in the state variables:

$$T\vartheta' = \sum_i F_i f_i. \tag{8.1}$$

The assumed linear relations between the state variables are

$$f_i = \sum_k c_{ik} F_k \quad \text{and} \quad F_k = \sum_i m_{ki} g_i \tag{8.2}$$

where c_{ik} and m_{ki} represent, respectively, general *compliances* and *moduli*. Thus, for instance, according to Eq. (7.22) a temperature gradient can be considered as a generalized "force", with the resulting heat flow as the associated "flux" in the same way in which the stress deviation s_{ij} may represent a "stress-flux" produced by the "velocity forces" associated with material flow and

---

[1] S. R. DE GROOT: Thermodynamics of Irreversible Processes, p. 9. New York: Interscience 1952. — R. HAASE: Ergebn. exakt. Naturw. **26**, 56 (1952). — R. O. DAVIES: Rep. Progr. Phys., Lond. **19**, 326 (1956).

expressed by $\dot{e}_{ij}$. The relations (8.2) are however, subject to the *Curie principle of tensorial symmetry*: any generalized "force" can give rise to a multitude of "fluxes", provided that only variables of the same tensorial rank are coupled. Coefficients c_{ik} and m_{ki} for $i \neq k$ will automatically vanish for all uncoupled processes[1].

It can be shown on the basis of classical as well as of quantum-mechanical considerations that the laws of motion of individual particles of a system are symmetric in time, and thus invariant against the transformation $t \rightarrow -t$, unless the forces are odd, for instance linear, functions of the particle velocities. ONSAGER[2] has shown that this principle leads, on the basis of considerations of statistical mechanics, to the *principle of symmetry of the matrices of the coefficients* in the Eqs. (8.2)

$$c_{ik} = c_{ki} \quad \text{and} \quad m_{ik} = m_{ki}. \tag{8.3}$$

Eqs. (8.1) to (8.3) in conjunction with Eq. (7.24) determine the state function of the volume element of the dissipative continuum

$$T \vartheta = \sum_i \varrho \dot{\chi}_i + \sum_i F_i f_i = \sum_{i,k} c_{ik} F_i F_k = \sum_{k,i} m_{ki} f_k f_i \tag{8.4}$$

where the product sum of generalized forces F_i and fluxes f_i includes the terms for the mechanical dissipation $\sum_i \varrho \dot{\chi}_i = s_{ij} \dot{e}_{ij}$ provided such dissipation depends on the deviations of the stress- and strain-rate tensors only. Since in these terms both "forces" and "fluxes" are second-order tensor deviations, they cannot give rise to vectorial fluxes or forces. There will, therefore, be no interaction of mechanical dissipation with non-mechanical effects unless the latter depend on second order tensor deviations F_{ij} or f_{ij}.

Considering linear relations $s_{ij} = f(\dot{e}_{ij}, f_{ij})$ and $\dot{e}_{ij} = f(s_{ij}, F_{ij})$, Eq. (8.4) is a quadratic positive form in the state variables. Therefore $c_{ii} > 0$, $m_{kk} > 0$, the determinants $|c_{ik}| > 0$ and $|m_{ik}| > 0$, as well as all principal minors $c_{ii} c_{kk} - c_{ik}^2 > 0$ and $m_{kk} m_{ii} - m_{ik}^2 > 0$. These inequalities permit the estimate of the upper limits of the values of the coefficients c_{ik} or m_{ki} of coupled processes $(i \neq k)$ when the coefficients governing the simple processes themselves $(i = k)$ are known.

The thermodynamic state functions of the isotropic homogeneous continuum $\varrho \varphi$, ϑ_v and ϑ_p are necessarily functions of the invariants of the mechanical state variables and of temperature. Introducing the relevant mechanical state variables ε_{ij} for $(\varrho \varphi)$ and $\dot{e}_{ij}$ and s_{ij} for ϑ_v and ϑ_p, respectively, thus implicitly neglecting energy dissipation by volume flow (volume viscosity or volume plasticity), the expressions for the potential and entropy production functions are:

$$\left. \begin{array}{l} \varrho \varphi = \varrho \varphi (I_\varepsilon, II_\varepsilon, III_\varepsilon, T), \quad \vartheta_v = \varrho \dot{\chi}_v (I_2', I_3', T)/T, \\ \vartheta_p = \varrho \dot{\chi}_p (J_2', J_3', T)/T \end{array} \right\} \tag{8.5}$$

where I_2' and I_3' are the invariants of the strain-rate deviation tensor.

To define the state of the medium independently of the path over which it has been reached, the state functions must be potentials from which the relations between the state variables in the thermally isolated elastic and dissipative continua $(q_{i,i} = 0)$ can be derived. Hence

$$\sigma_{ij} = \frac{\partial \Phi}{\partial \varepsilon_{ij}}, \quad \sigma_{ij} = \frac{\partial \dot{X}_v}{\partial \dot{e}_{ij}}, \quad \dot{e}_{ij} = \frac{\partial \dot{X}_p}{\partial \sigma_{ij}} \tag{8.6}$$

where Φ, $\dot{X}_v$ and $\dot{X}_p$ are the *potentials* of the *elastic* and of the *incompressible dissipative* medium per unit volume.

---

[1] See footnote 1, p. 248.
[2] L. ONSAGER: Phys. Rev. **37**, 405 (1931); **38**, 2265 (1932).

The *adiabatic* elastic potential $\Phi(\mathrm{I}_\varepsilon, \mathrm{II}_\varepsilon, \mathrm{III}_\varepsilon, T)$ is usually presented as a power series expansion in the invariants, with coefficients depending on T. For infinitesimal strains implied in the assumption of small deviations from the unstrained state, powers higher than squares of the strain-components are neglected, so that

$$\Phi(\mathrm{I}_\varepsilon, \mathrm{II}_\varepsilon, T) = \tfrac{1}{2}\sigma_{ij}\varepsilon_{ij} = \tfrac{1}{2}(3p\,\varepsilon_v + s_{ij}e_{ij}) \\ = \tfrac{1}{6}(3K - 2G)\,\mathrm{I}_\varepsilon^2 - 2G\,\mathrm{II}_\varepsilon = \tfrac{1}{2}K\,\mathrm{I}_1^2 + 2G\,\mathrm{I}_2' \qquad (8.7)$$

where K and G denote, respectively, the bulk modulus and the shear modulus of the elastic medium. Hence, according to Eq. (8.6)

$$\sigma_{ij} = (3K - 2G)\,\delta_{ij}\tfrac{1}{3}\varepsilon_{kk} + 2G\,\varepsilon_{ij} = K\,\varepsilon_{kk}\delta_{ij} + 2G\,e_{ij}. \qquad (8.8)$$

For the heat-conducting elastic continuum an isothermal potential $\psi = (\varphi - T\eta_e)$ or, per volume element, $\Psi = \Phi - Ts_e$ has been defined by Eqs. (7.25) and (7.30); both in adiabatic and isothermal processes the state functions are exact differentials. Comparison of

$$\frac{d\Phi}{dt} = \frac{\partial\Phi}{\partial\varepsilon_{ij}}\dot\varepsilon_{ij} + \frac{\partial\Phi}{\partial s_e}\dot s_e \quad \text{and} \quad \frac{d\Psi}{dt} = \frac{\partial\Psi}{\partial\varepsilon_{ij}}\dot\varepsilon_{ij} + \frac{\partial\Psi}{\partial T}\dot T \qquad (8.9)$$

with Eqs. (7.25) and (7.29a) written for the volume element under the assumption $\dot\varrho = 0$ produces the relations

$$T = \frac{\partial\Phi}{\partial s_e} \quad \text{and} \quad s_e = -\frac{\partial\Psi}{\partial T} \qquad (8.10)$$

between the specific potentials and the thermodynamical variables, provided σ_{ij} is independent of $\dot\varepsilon_{ij}$.

The bi-linear form of the elastic potentials and the associated Eq. (8.8) are consistent with the general thermodynamic approach which, by using bi-linear state functions and linear relations of the form (8.2) between the state variables simplifies the joint consideration of mechanical and non-mechanical phenomena in the deformable continuum.

For the incompressible dissipative medium *dissipation potentials* $\dot X_v$ and $\dot X_p$ have been defined by Eqs. (8.6). Such potentials will exist only if the deformations are slow enough to suppress the effect of the heat produced in the process. There are two alternative ways in which bi-linear state functions of the form (8.4) for the incompressible dissipative medium close to thermodynamic equilibrium (for which such functions necessarily vanish) can be interpreted:

The assumption that the state of the medium depends primarily on the velocity gradients of the motion, the shear stresses arising as the result of the internal resistance to the flow and vanishing as the velocity gradients vanish, leads to the formulation

$$T\dot\vartheta_v = \varrho\dot\chi_v = s_{ij}\dot e_{ij} = 2\eta\,\dot e_{ij}\dot e_{ij} = 4\eta\,\dot I_2' \qquad (8.11)$$

representing the *dissipation function* of the incompressible, irrotational *viscous fluid* defined by the relations

$$s_{ij} = 2\eta\,\dot e_{ij} \quad \text{and} \quad w_{ij} = 0 \qquad (8.12)$$

where η denotes the coefficient of shear viscosity and w_{ij} the vorticity (see Sect. 5). On the other hand, the assumption that the state of the medium depends primarily on the applied shear stresses, the irreversible deformation being the result of the intensity of those stresses above a certain critical limit and vanishing when this limit is not attained leads to the formulation

$$T\dot{\vartheta}_p = \varrho\dot{\chi}_p = s_{ij}\dot{e}_{ij} = 2\lambda s_{ij}s_{ij} = 4\lambda J_2' \tag{8.13}$$

representing the *dissipation function* of the incompressible *plastic solid* defined by the relation

$$\dot{e}_{ij} = 2\lambda s_{ij} \tag{8.14}$$

provided $\lambda(I_2')$, the "plastic compliance", is an isotropic function of the strain-velocity, which makes Eq. (8.14) homogeneous in time [see Eq. (23.4)].

The dissipation functions $T\dot{\vartheta}_v$ and $T\dot{\vartheta}_p$ are not themselves potentials. In order to fulfil Eqs. (8.6) in conjunction with Eqs. (8.12) and (8.14) dissipation potentials must be introduced of the form

$$\dot{X}_v(\dot{e}_{ij}) = \tfrac{1}{2} T\dot{\vartheta}_v = \eta\,\dot{e}_{ij}\dot{e}_{ij} \quad \text{and} \quad \dot{X}_p(s_{ij}) = \tfrac{1}{2} T\dot{\vartheta}_p = \lambda s_{ij}s_{ij} \tag{8.15}$$

representing one-half of the rate of energy dissipation. Obviously, by using the relations (8.12) and (8.14) alternative expressions $\tfrac{1}{2}T\dot{\vartheta}_v(s_{ij}) = \dot{X}_v(s_{ij})$ but $T\dot{\vartheta}_p(\dot{e}_{ij}) = \dot{X}_p(\dot{e}_{ij})$ can be obtained, in which $\dot{e}_{ij}$ and s_{ij} are replaced by s_{ij} and $\dot{e}_{ij}$ respectively. These alternative forms are very useful in the development of variational principles; however, they do not clearly reflect the physical character of the dissipation processes. When $\lambda = \lambda(\dot{I}_2', J_2')$, the plastic potential $\dot{X}_p(s_{ij}) = T\dot{\vartheta}_p$.

Dissipation potentials of the form $\dot{X}_v(x_i, \dot{e}_{ij})$ or $\dot{X}_p(x_i, s_{ij})$ express the fact that the coefficients $\eta(x_i)$ and $\lambda(x_i)$ vary with the coordinates as a result of the initial inhomogeneity of the medium. While the Eqs. (8.6) remain valid, the resulting relations between the state variables are no longer isotropic[1].

9. Variational principles.
On the basis of the elastic and dissipation potentials general variational principles for the continuum can be formulated, considering that the approach to thermodynamic equilibrium in stationary motion of the near-equilibrium dissipative continuum is governed by variational principles involving the rate of entropy production[2] and the related dissipation potentials in the same way in which mechanical equilibrium in the thermally isolated elastic continuum is governed by variational principles involving the elastic potential. The respective principles for isotropic continua of any type are easily derived from the energy balance Eqs. (7.5) and (7.8) by introducing into the stress-power (internal energy rate) term $\sigma_{ij}\dot{\varepsilon}_{ij}$ the relevant types of internal energy expressed with the aid of the associated potentials, and by comparing two closely adjacent states of energy balance.

For the elastic continuum for which this comparison involves the displacement field in the state of equilibrium and an adjacent state defined by the *virtual displacements* from the former $\delta u_i = v_i dt$ compatible with the continuity of the system but not necessarily with the conditions of equilibrium, the respective variational principle generally known as the principle of virtual work is obtained by establishing the energy balance for the displacements δu_i replacing in Eq. (7.5) v_i by δu_i as well as $\sigma_{ij}\dot{\varepsilon}_{ij}$ by $\sigma_{ij}\delta\varepsilon_{ij}$ and developing the potential $\Phi(\varepsilon_{ij} + \delta\varepsilon_{ij})$ of the adjacent (varied) state into a Taylor series, considering the first Eq. (8.6):

$$\begin{aligned}\Phi(\varepsilon_{ij} + \delta\varepsilon_{ij}) &= \Phi(\varepsilon_{ij}) + \frac{\partial\Phi}{\partial\varepsilon_{ij}}\delta\varepsilon_{ij} + \frac{1}{2}\frac{\partial^2\Phi}{\partial\varepsilon_{ij}\partial\varepsilon_{kl}}\delta\varepsilon_{ij}\delta\varepsilon_{kl} \\ &= \Phi(\varepsilon_{ij}) + \sigma_{ij}\delta\varepsilon_{ij} + \Phi(\delta\varepsilon_{ij})\end{aligned} \tag{9.1}$$

---
[1] W. Olszak: Arch. Math. Stos., Warsaw **6**, 493, 639 (1954). — W. Olszak and W. Urbanowski: Arch. Math. Stos **8**, 85, 671 (1956). — Bull. Polish Acad. Sci. (4) **5**, 29, 39 (1957).
[2] See footnote 1, p. 248.

from which the expression for $\sigma_{ij}\delta\varepsilon_{ij}$ is obtained

$$\begin{aligned}\int \sigma_{ij}\delta\varepsilon_{ij}\,dV &= \int \Phi(\varepsilon_{ij}+\delta\varepsilon_{ij})\,dV - \int \Phi(\varepsilon_{ij})\,dV - \int \Phi(\delta\varepsilon_{ij})\,dV \\ &= \delta\int \Phi_e\,dV - \int \Phi(\delta\varepsilon_{ij})\,dV\end{aligned} \quad (9.1\text{a})$$

in terms of the variation of the elastic potential $\delta\int \Phi_e\,dV$. Hence Eq. (7.5) for the displacements δu_i and equilibrium conditions $(v_i=0)$ is transformed into the variational principle

$$\delta\left[\int_V \Phi_e\,dV - \int_A t_i u_i\,dA - \int_V X_i u_i\,dV\right] = \delta\Pi = 0 + \delta^2\int_V \Phi_e\,dV \quad (9.2)$$

where the surface integral extends over those parts of the surface on which the forces are prescribed. The expression in brackets represents the total potential energy; a stationary value defines the deformed condition associated with stable equilibrium. Since the second variation $\delta^2\int \Phi_e\,dV$ is the strain energy $\Phi(\delta\varepsilon_{ij})$ due to $\delta\varepsilon_{ij}$, which is positive, $\Pi\to$minimum. For $X_i=0$ and given boundary displacements $\delta u_i=0$ the external forces do no work and therefore $\delta\Pi=\delta\int \Phi_e\,dV$ and $\int \Phi_e\,dV\to$minimum.

The equivalent principle for the incompressible dissipative medium is obtained by comparing the state of stationary flow with velocities v_i with an adjacent state with velocities $v_i+\delta v_i$, where the *virtual* velocities δv_i are compatible with the continuity equation, but not necessarily with the stress-equation of motion. Replacing in Eqs. (9.1) the elastic potential by the rate of energy dissipation $T\vartheta(\dot{e}_{ij}+\delta\dot{e}_{ij})$ and considering that according to Eqs. (8.6) two different expressions are obtained for the variations of the viscous and plastic dissipation functions in terms of $s_{ij}\delta\dot{e}_{ij}$

or

and

or

$$\left.\begin{aligned}\delta\int T\vartheta_v(\dot{e}_{ij})\,dV &= \int 2s_{ij}\delta\dot{e}_{ij}\,dV + \int T\vartheta_v(\delta\dot{e}_{ij})\,dV \\ \int s_{ij}\delta\dot{e}_{ij}\,dV &= \delta\int \dot{X}_v(\dot{e}_{ij})\,dV - \int \dot{X}_v(\delta\dot{e}_{ij})\,dV \\ \delta\int T\vartheta_p(\dot{e}_{ij})\,dV &= \int s_{ij}\delta\dot{e}_{ij}\,dV + \int T\vartheta_p(\delta\dot{e}_{ij})\,dV \\ s_{ij}\delta\dot{e}_{ij}\,dV &= \delta\int \dot{X}_p(\dot{e}_{ij})\,dV - \int \dot{X}_p(\delta\dot{e}_{ij})\,dV\end{aligned}\right\} \quad (9.3)$$

it follows that if the variations of the dissipation *functions* are replaced by the variations of the dissipation *potentials* the variational principle for the incompressible dissipative medium takes the form

$$\delta\left[\int_V \dot{X}(\dot{e}_{ij})\,dV - \int_A t_i v_i\,dA - \int_V X_i v_i\,dV\right] = 0 + \delta^2\int_V \dot{X}(\dot{e}_{ij})\,dV \quad (9.4)$$

which governs the velocity field in stationary flow; the surface integral extends over those parts of the surface on which the forces are prescribed. The expression in brackets represents the excess of one-half or of the total rate of dissipation of energy over the rate at which work is done by surface tractions and body forces; its second variation $\dot{X}(\delta\dot{e}_{ij})$ is positive, being one-half or the total internal energy dissipation $\dot{X}(\delta\dot{e}_{ij})$ due to $\delta\dot{e}_{ij}$. That the quantity in brackets is minimized by the actual velocity field in incompressible, slow stationary flow, which is thus unique, has been established for the linear viscous fluid by HELMHOLTZ[1]

---

[1] H. v. HELMHOLTZ: Wiss. Abh., Lpz. **1**, 223 (1882).

and Korteweg[1], and generalized by Rayleigh[2] for certain types of viscous flow that are neither slow nor stationary, but fulfil the condition that the velocity components can be derived from a single function $H(x_i)$ according to $v_{i,jj}=H_{,i}$. For ideal plastic flow the same principle has been developed first by Markov[3] and re-established in a somewhat more general form by Hill[4]. For $X_i=0$ and given boundary velocities for which $\delta v_i=0$, the external forces do work at constant rate; therefore $\delta \int \dot{X}(\dot{e}_{ij})dV = 0$ and $\int \dot{X}(\dot{e}_{ij})dV \to$ minimum. Eq. (9.4) expresses a principle of minimum rate of energy dissipation or of entropy production in terms of dissipation *potentials* which ensures uniqueness of the velocity fields. Alternatively Eq. (9.4) in the form

$$\delta\left[\int_A t_i v_i \, dA + \int_V X_i v_i \, dV - \int_V \dot{X}(\dot{e}_{ij}) \, dV\right] = 0 - \int_V \dot{X}(\delta \dot{e}_{ij}) \, dV = 0 + \\ + \text{negative second order term in } \delta \dot{e}_{ij} \quad (9.5)$$

expresses the principle that the actual velocity field maximizes the expression in brackets.

Fundamentally different variational principles are obtained when the comparisons of closely adjacent varied states of the system refer to the stress-field rather than to the displacement or velocity-field. Comparing, for the elastic medium, the strainwork of the actual state $\sigma_{ij}\varepsilon_{ij}=\Phi(\sigma_{ij})$ with that of an adjacent state $\Phi(\sigma_{ij}+\delta\sigma_{ij})$, where the stress-increments fulfil the equilibrium condition $\delta\sigma_{ij,j}=0$, by developing the Taylor series

$$\Phi(\sigma_{ij}+\delta\sigma_{ij}) = \Phi(\sigma_{ij}) + \frac{\partial \Phi}{\partial \sigma_{ij}} \delta\sigma_{ij} + \frac{1}{2}\frac{\partial^2 \Phi}{\partial \sigma_{ij}\partial \sigma_{kl}} \delta\sigma_{ij}\delta\sigma_{kl} \\ = \Phi(\sigma_{ij}) + \varepsilon_{ij}\delta\sigma_{ij} + \Phi(\delta\sigma_{ij}) \quad (9.6)$$

and considering that

$$\int_V \varepsilon_{ij}\delta\sigma_{ij}\,dV = \int_V (\varepsilon_{ij}+\omega_{ij})\delta\sigma_{ij}\,dV = \int_V \delta\sigma_{ij} u_{i,j}\,dV$$
$$= \int_V (\delta\sigma_{ij}u_i)_{,j}\,dV - \int_V u_i \delta\sigma_{ij,j}\,dV = \int_A \delta\sigma_{ij}n_j u_i\,dA = \int_A \delta t_i u_i\,dA.$$

Eq. (9.6) can be written in the form

$$\delta\int_V \Phi_\sigma\,dV = \int_V \Phi(\sigma_{ij}+\delta\sigma_{ij})\,dV - \int_V \Phi(\sigma_{ij})\,dV = \int_A \delta t_i u_i\,dA + \int_V \Phi(\delta\sigma_{ij})\,dV$$

or

$$\delta\left[\int_V \Phi_\sigma\,dV - \int_A u_i t_i\,dA\right] = \delta\Sigma = 0 + \text{positive second order terms in } \delta\sigma_{ij}. \quad (9.7)$$

The surface integral extends over those parts of the surface on which the displacements are prescribed. The expression in brackets represents the "*complementary energy*"; its second variation, the strain energy $\Phi(\delta\sigma_{ij})$ due to $\delta\sigma_{ij}$ is positive. A minimum value of the complementary energy Σ therefore distinguishes the equilibrium state of stress. For prescribed boundary forces $\delta t_i=0$ and $\delta\Sigma=\delta\int\Phi_\sigma\,dV=0$; therefore $\int \Phi_\sigma\,dV \to$ minimum[5].

The equivalent variational principle for the incompressible dissipative medium is obtained by comparing the rate of energy dissipation for the actual state of stationary flow $T\vartheta(s_{ij})$ with that of an adjacent state $T\vartheta(s_{ij}+\delta s_{ij})$. Expanding the latter into a Taylor series in analogy with Eq. (9.6) in which Φ is replaced

---

[1] D. J. Korteweg: Phil. Mag. (5) **16**, 112 (1883).
[2] Lord Rayleigh: Phil. Mag. (6) **26**, 776 (1913).
[3] A. M. Markov: Prikl. Math. Mech. **11**, 339 (1947).
[4] A. Nadai: Proc. First U.S. Congress Appl. Mech., p. 479, New York 1952. — R. Hill: J. Appl. Mech. **17**, 64 (1950).
[5] F. Engesser: Z. Arch. Ing. Ver. Hannover **35**, 738 (1889).

by $T\vartheta$, and considering that according to Eqs. (8.11) and (8.13) the variations of the dissipation *functions*

$$\left.\begin{aligned}\delta\int_V T\vartheta_v(s_{ij})dV &= 2\int_V \dot{e}_{ij}\delta s_{ij}dV + \int_V T\vartheta_v(\delta s_{ij})dV = 2\int_A \delta t_i u_i dA + \int_V T\vartheta_v(\delta s_{ij})dV\\ \text{and for }\lambda &= \lambda(\dot{I}_2')\\ \delta\int_V T\vartheta_p(s_{ij})dV &= 2\int_V \dot{e}_{ij}\delta s_{ij}dV + \int_V T\vartheta_p(\delta s_{ij})dV = \int_A \delta t_i u_i dA + \int_V T\vartheta_p(\delta s_{ij})dV\end{aligned}\right\} \quad (9.8)$$

it follows that in terms of dissipation *potentials* $\dot{X}(s_{ij})$

$$\delta\left[\int_V \dot{X}(s_{ij})dV - \int_A v_i t_i dA\right] = 0 + \text{positive second order terms in } \delta s_{ij} \quad (9.9)$$

where the surface integral extends over those parts of the surface on which the velocities are prescribed. The expression in brackets represents the excess of one-half or of the total rate of energy dissipation over the rate of work of the surface forces; since its second variation $\dot{X}(\delta s_{ij})$ is positive, a minimum value defines the actual stress field as unique. Alternatively, Eq. (9.9) in the form

$$\delta\left[\int_A v_i t_i dA - \int_V \dot{X}(s_{ij})dV\right] = 0 + \text{negative second order term in } \delta s_{ij} \quad (9.10)$$

expresses the condition that the term in brackets, which is the excess of the rate at which work is being done by the surface forces over, respectively, half or the total rate of energy dissipation, is maximized by the actual stress field.

For constant power of the surface forces $\int v_i t_i dA = \text{const.}$, Eq. (9.9) is transformed into $\delta\int \dot{X}(s_{ij})dV = 0$ and $\int \dot{X}(s_{ij})dV \to \text{minimum}$. The actual stress-field in stationary flow thus minimizes the rate of energy dissipation, a principle established by HELMHOLTZ for the linear viscous fluid[1].

Since for the incompressible medium according to the last Eq. (8.6)

$$\delta\dot{X}(s_{ij}) = \frac{\partial \dot{X}(s_{ij})}{\partial s_{ij}}\delta s_{ij} = \dot{e}_{ij}\delta s_{ij} = 0 \quad (9.11)$$

because the potential of the ideal plastic continuum, as introduced by v. MISES[2], is proportional to the yield condition (see Sect. 23), Eq. (9.11) indicates that when at the yield limit $\dot{X}(s_{ij}) = \text{const.}$ the velocity-field is such that a stress variation within this limit performs no work and therefore $\int \dot{X}(s_{ij})dV \to$ minimum. Alternatively, Eq. (9.10) with $\dot{X}(s_{ij}) = \text{const.}$ is transformed into $\delta\int v_i t_i dA = 0$ and therefore $\int v_i t_i dA \to$ maximum, indicating that the actual stress field in the plastic medium maximizes the rate of work of the surface forces. Eq. (9.11) which is, in fact, the basis of the theory of plasticity, is due to v. MISES[2], whose plastic potential is a continuous function. It has recently been extended by KOITER[3] to "*piece-wise continuous*" plastic potentials (see Sect. 23). The principle of maximum work of the surface forces, originally suggested by v. MISES, has been restated explicitly several times[4] as have been, in fact, all the Eqs. (9.4) to (9.9) for plastic media[5].

Since according to Eqs. (7.5) and (4.11)

$$\int_A v_i t_i dA = \int_V \sigma_{ij}\dot{\varepsilon}_{ij}dV + \int_V v_i(\varrho\dot{v}_i - X_i)dV$$

---

[1] See footnote 1, p. 252.
[2] R. v. MISES: Z. angew. Math. Mech. **8**, 184 (1928).
[3] W. T. KOITER: Quart. Appl. Math. **11**, 350 (1953).
[4] M. A. SADOWSKY: J. Appl. Mech. **10**, 65 (1943). — A. H. PHILLIPPIDES: J. Appl. Mech. **15**, 241 (1948). — R. HILL: Quart. J. Appl. Math. **1**, 18 (1948). — Phil. Mag. (7) **42**, 868 (1951).
[5] R. HILL: Mathematical Theory of Plasticity, Chap. 3, Oxford: University Press 1950.

the right hand side of this equation is also a maximum. Hence for incompressible stationary plastic flow for which $\sigma_{ij,j} = \varrho \dot{v}_i - X_i = 0$ the actual stress-field σ_{ij} maximizes the rate of energy dissipation $\sigma_{ij} \dot{\varepsilon}_{ij} = s_{ij} \dot{e}_{ij}$ for a given velocity-field v_i for which $v_{i,i} = 0$. Defining the state of the plastic continuum by "generalized stresses" $Q_1 \ldots Q_n$ and "generalized strain-rates" $\dot{q}_1 \ldots \dot{q}_n$, and expressing the yield condition in the form $f(Q_1 \ldots Q_n) \propto \dot{X}(Q_1 \ldots Q_n)$ PRAGER[1] has established Eq. (9.11) in the alternative form

$$\delta \dot{X}(Q_1 \ldots Q_n) \propto \delta f = \frac{\partial f}{\partial Q_1} \delta Q_1 + \cdots + \frac{\partial f}{\partial Q_n} \delta Q_n = \dot{q}_i \delta Q_i = 0 \quad (9.11\text{a})$$

and the associated extremum principle

$$Q_1 \dot{q}_1 + \cdots + Q_n \dot{q}_n \rightarrow \text{maximum} \quad (9.11\text{b})$$

as the basis of the "limit analysis" of rigid-plastic structures in which the elastic parts are considered rigid, so that the effect of the elastic potential in formulating the pertinent variational principle is disregarded and the uniqueness condition of the structure determined solely by the dissipative (plastic) potential (see Sect. 35).

It follows from the Taylor expansion (9.1) that the variational principles expressed by Eqs. (9.4), (9.5), (9.9) and (9.10) remain valid for stationary states of non-linear incompressible dissipative media with dissipation potentials specified as homogeneous functions in $\dot{I}'_2$ or J'_2 of degree n such that $T \vartheta(\dot{e}_{ij}) = 2n \dot{X}(\dot{I}'_2)$ and $T \vartheta(s_{ij}) = 2n \dot{X}(J'_2)$, while Eqs. (9.2) and (9.7) are valid for non-linear elastic media in equilibrium defined by potentials Φ that are homogeneous functions of stress or of strain of degree $2n$ such that the elastic strainwork $W = 2n\Phi$.

Variational principles for media with combined properties are derived in the same way as those for simple media by establishing the energy balance according to Eq. (7.5) using all relevant potentials. Thus, for instance, for the incompressible dissipative medium with several component viscous dissipation potentials differing by the their viscosity coefficients, or with a viscous and a plastic potential (viscoplastic medium), Eqs. (9.4) and (9.9) remain valid, with the sum of the dissipation potentials per volume element replacing the single potential $\dot{X}(\dot{e}_{ij})$ or $\dot{X}(s_{ij})$.

For the visco-plastic medium for instance, the extension of Eq. (9.4) provides the minimum principle for the strain-velocity field

$$\delta \left\{ \int_V [\dot{X}_v(\dot{e}_{ij}) + \dot{X}_p(\dot{e}_{ij})] \, dV - \int_A t_i v_i \, dA - \int_V X_i v_i \, dV \right\} = 0, \quad (9.12)$$

the extension of Eq. (9.9) the maximum principle for the stress-field

$$\delta \left\{ \int_A t_i v_i \, dA - \int_V [\dot{X}_v(s_{ij}) + \dot{X}_p(s_{ij})] \, dV \right\} = 0. \quad (9.13)$$

Eqs. (9.12) and (9.13) have been derived independently by ILYUSHIN, OLDROYD and PRAGER[2].

For media with combined elastic and incompressible dissipative response the stress power in Eq. (7.5) is replaced by the sum of the rate of change of the

---

[1] W. PRAGER: Probleme der Plastizitäts-Theorie. Basel: Birkhäuser 1955 and references there. See also P. S. SYMONDS and B. G. NEAL: J. Franklin Inst. **252**, 383, 469 (1951). — R. HILL: Phil. Mag. **42**, 868 (1951).

[2] A. A. ILYUSHIN: Ucheneye Zap. Mosk. Univ. Mekhanika **39**, 1 (1940). — J. G. OLDROYD: Proc. Cambridge Phil. Soc. **43**, 100 (1947). — W. PRAGER: von Mises Memorial Volume, 208, New York: Acad. Press 1954.

elastic potential $\dot{\Phi}$ and the relevant dissipation potential or sum of dissipation potentials. Thus, the extension of Eq. (9.4) to include the rate of change of the elastic potential $\dot{\Phi}_e$ and various dissipation potentials $\dot{X}_n(\dot{e}_{ij})$ produces the minimum condition for the strain-velocity field

$$\delta\left[\int_V \dot{\Phi}_e dV + \int_V \sum_n \dot{X}_n(\dot{e}_{ij}) dV - \int_A t_i v_i dA - \int_V X_i v_i dV\right] = 0, \qquad (9.14)$$

the extension of Eq. (9.9) to include the rate of change of the elastic potential $\dot{\Phi}_\sigma$ and various dissipation potentials $\dot{X}_m(s_{ij})$ produces the maximal condition for the stress field

$$\delta\left[\int_A t_i v_i dA - \int_V \dot{\Phi}_\sigma dV - \int_V \sum_m \dot{X}_m(s_{ij}) dV\right] = 0. \qquad (9.15)$$

For stationary conditions of the visco-elastic body with given boundary velocities or given surface stresses ($\delta v_2 = 0$ or $\delta t_i = 0$) or constant rate of work of the external forces the condition $\int[\dot{\Phi}_e + \sum \dot{X}_v(e_{ij})] dV \to$ minimum determines the actual strain-rate field, the condition $\int[\dot{\Phi}_\sigma + \sum \dot{X}_v(s_{ij})] dV \to$ minimum the actual stress field[1].

For the ideal elastic-plastic body for which $\dot{X}_p = $ const. and therefore $\delta \dot{X}_p = 0$, the respective conditions are

$$\int \Phi_e dV \to \text{minimum} \quad \text{and} \quad \int \Phi_\sigma dV \to \text{minimum}. \qquad (9.16)$$

Hence the elastic strain-energy of the elastic-plastic medium is stationary; the strain-increment-field as well as the stress-field of the elastic-plastic medium are therefore governed by minimum principles involving the elastic potential alone, as first suggested by HAAR and v. KARMAN and later proved formally by GREENBERG[2].

B. Stress-strain relations.

I. General formulation.

10. Stress in the isotropic continuum. The relations between stress and strain and their time-derivatives express the fact that deformation or flow of the continuum proceed against internal forces or, conversely, that stresses acting on the continuum produce deformations. Therefore, the stress-tensor can be specified by the relation

$$\sigma_{ij} = f_{ij}(a_\alpha, x_i, x_{i,\alpha}, x_{i,\alpha\beta} \ldots, \dot{x}_{i,\alpha} \ddot{x}_{i,\alpha} \ldots \dot{x}_{i,\alpha\beta} \ldots, t) \qquad (10.1)$$

where f_{ij} denotes a symmetric tensor function of second rank. The appearance of a_α in Eq. (10.1) is an indication of initial inhomogeneity of the continuum, that of x_i of effects depending on the momentary location of the particle; the gradients $x_{i,\alpha}$ represent the local strain, while the appearance of terms $x_{i,\alpha\beta}$ is an indication of the effect on the stress at a point of the strains at neighboring points in terms of the local strain-gradients, reflecting very rapid spatial changes of the deformational state of the medium. The influence of the time t can be either explicit, and thus indicating history-effects, or implicit through the appearance of time derivatives of the gradients $x_{i,\alpha}$, $x_{i,\alpha\beta}$, representing velocity effects.

---

[1] M. A. BIOT: IUTAM Colloquium Madrid, p. 251. Berlin: Springer 1956.
[2] A. HAAR and TH. v. KARMAN: Göttinger Nachr., Math.-phys. Kl. 204 (1909). — H. J. GREENBERG: Quart. Appl. Math. **7**, 85 (1949). — L. FINZI: Acta Pontif. Acad. Sci. **15**, 121 (1953).

If the stress at x_i at time t depends on the momentary conditions at this point, as well as on the conditions of neighboring points $\bar{x}_i$ and on conditions at different times $\bar{t} < t$ the stress tensor can be expressed by the integral

$$\sigma_{ij} = \int_{-\infty}^{t} \left[\int_V f_{ij}(a_\alpha, x_i, x_{i,\alpha} \ldots \bar{a}_\alpha, \bar{x}_i, \bar{x}_{i,\alpha}) \, d\bar{V} \right] d\bar{t} \tag{10.2}$$

over the total strain-or stress-field and the complete stress-history. If the influence on σ_{ij} of effects at times $\bar{t} < t$ decays rapidly towards zero with increasing elapsed time $(t-\bar{t})$, and if the influence at x_i of strains at $\bar{x}_i$ decays rapidly towards zero with increasing distance $(x_i - \bar{x}_i)$, the functions under the integral in Eq. (10.2) containing $(t - \bar{t})$ and $(x_i - \bar{x}_i)$ can be developed into Taylor series at t and x_i. The approximations obtained by breaking the series off after a few terms represent relations of the form (10.1) containing space and time-derivatives of x_i of higher than first order.

Stress-strain relations in the isotropic continuum are, in general, based on the assumption that the stress tensor at x_i depends only on x_i, the gradient of x_i and its time derivatives:

$$\sigma_{ij} = f_{ij}(x_i, x_{i,\alpha}, \dot{x}_{i,\alpha}, \ddot{x}_{i,\alpha} \ldots) = f(x_i, \varepsilon_{ij}, \dot{\varepsilon}_{ij} \ldots). \tag{10.3}$$

The effect of x_i on the stress-strain relation may reflect either an initial or strain-induced inhomogeneity of the mechanical response of the medium[1] or the interaction with a non-mechanical non-uniform flux within the medium producing a non-homogeneous field, such as a temperature-or a magnetic field, or a combination of both. In the perfectly homogeneous isotropic continuum without history effects the above relation has the form

$$\sigma_{ij} = f(\varepsilon_{ij}, \dot{\varepsilon}_{ij}, \ddot{\varepsilon}_{ij} \ldots). \tag{10.4}$$

Eq. (10.4) is obviously not the only way of postulating an isotropic, homogeneous stress-strain relation. A wider class of materials can be specified by relations involving the *rate of stress-tensor*

$$\frac{d}{dt} \sigma_{ij} = \frac{\partial}{\partial t} \sigma_{ij} + v_k \sigma_{ij,k} = f'_{ij}(x_{i,j}, \dot{x}_{i,j}, \ddot{x}_{i,j}, \ldots) = f'(\varepsilon_{ij}, \dot{\varepsilon}_{ij} \ldots) \tag{10.5}$$

where v_k denotes the relative velocity of x_i with respect to a reference system moving at uniform velocity, without rotation.

A still more general relation is obtained by combining Eqs. (10.4) and (10.5) specifying the sum of the stress tensor and the rate of stress tensor as a function of the strain and its time derivatives, while the most general relation would be of the form

$$F(\sigma_{ij}, \dot{\sigma}_{ij}, \ddot{\sigma}_{ij} \ldots \varepsilon_{ij}, \dot{\varepsilon}_{ij}, \ddot{\varepsilon}_{ij} \ldots) = 0. \tag{10.6}$$

Relations of the type (10.6) including derivatives of higher than second order are of little practical importance, as their use requires the specification of a larger number of initial conditions than are usually available on the basis of physical considerations.

11. Isotropic relations between two tensors. The postulated isotropy of the continuum imposes certain limitations on the form of the relations (10.1) to (10.6), arising from requirements of tensor invariance, which these relations must satisfy. The general analysis and solution of this problem for the dissi-

---
[1] See footnote 1, p. 251.

pative medium, using tensor analysis, is due to REINER[1]; his principal results have been restated for isotropic media in a more formal version by RIVLIN and RIVLIN and ERICKSEN[2]. Attempts have also been made to extend the considerations of tensorial invariance to relations between more than two tensors[3].

The analysis is based on the postulate that the stress tensor is isotropically related to either the strain tensor or the strain-rate tensor so that their principal axes coincide if the stress matrix $[\sigma_{ij}]$ can be expressed as a matrix polynomial in the strain or strain-rate matrix $[\varepsilon_{ij}]$ or $[\dot{\varepsilon}_{ij}]$ respectively, with coefficients that are themselves polynomials in the principal matrix invariants. Hence a simplified form of Eq. (10.4) in matrix notation for a general dissipative medium the state of which depends on $\dot{\varepsilon}_{ij}$ alone can be written in form:

$$[\sigma_{ij}] = f[\dot{\varepsilon}_{ij}] = \overline{M}_0[1] + \overline{M}_1[\dot{\varepsilon}_{ij}] + \overline{M}_2[\dot{\varepsilon}_{ij}]^2 + \overline{M}_3[\dot{\varepsilon}_{ij}]^3 + \cdots \quad (11.1)$$

where [1] denotes the unit matrix and $\overline{M}_i = \overline{f}_i(I_{\dot{\varepsilon}}, II_{\dot{\varepsilon}}, III_{\dot{\varepsilon}})$. A similar equation can be written for the relation $[\sigma_{ij}] = f[\varepsilon_{ij}]$ representing a general elastic solid.

Since according to the Cayley-Hamilton theorem of matrix analysis a symmetric matrix satisfies its own characteristic equation, a recursion formula for powers of $[\dot{\varepsilon}_{ij}]$ is obtained from Eq. (6.2) in the form

$$[\dot{\varepsilon}_{ij}]^3 = III_{\dot{\varepsilon}}[1] - II_{\dot{\varepsilon}}[\dot{\varepsilon}_{ij}] + I_{\dot{\varepsilon}}[\dot{\varepsilon}_{ij}]^2 \quad (11.2)$$

by the aid of which any power of $[\dot{\varepsilon}_{ij}]$ higher than the square can be reduced to a linear combination of $[1]$, $[\dot{\varepsilon}_{ij}]$ and $[\dot{\varepsilon}_{ij}]^2$. The third and all higher powers may thus be eliminated from Eq. (11.1) producing the relation

$$[\sigma_{ij}] = M_0[1] + M_1[\dot{\varepsilon}_{ij}] + M_2[\dot{\varepsilon}_{ij}]^2 \quad (11.3)$$

for the isotropic dissipative medium, and a similar relation with $[\varepsilon_{ij}]$ replacing $[\dot{\varepsilon}_{ij}]$ for the isotropic elastic medium. The *moduli* $M_i = f_i(I_{\dot{\varepsilon}}, II_{\dot{\varepsilon}}, III_{\dot{\varepsilon}})$ are polynomials in the principal invariants of $[\dot{\varepsilon}_{ij}]$ and thus scalar functions of the coordinates and of temperature, different from $\overline{M}_i$. Obviously, by starting from the expression $\dot{\varepsilon}_{ij} = f[\sigma_{ij}]$ a matrix relation can be obtained of the form

$$[\dot{\varepsilon}_{ij}] = C_0[1] + C_1[\sigma_{ij}] + C_2[\sigma_{ij}]^2 \quad (11.4)$$

where the *compliances* $C_i = F_i(I_\sigma, II_\sigma, III_\sigma)$ are polynomials in the principal invariants of stress. The equations for the elastic continuum are of a form identical with Eqs. (11.3) and (11.4) with $[\varepsilon_{ij}]$ replacing $[\dot{\varepsilon}_{ij}]$.

The above equations written in tensor notation are

$$\sigma_{ij} = M_0 \delta_{ij} + M_1 \dot{\varepsilon}_{ij} + M_2 \dot{\varepsilon}_{ik}\dot{\varepsilon}_{kj} \quad (11.3\text{a})$$

and

$$\dot{\varepsilon}_{ij} = C_0 \delta_{ij} + C_1 \sigma_{ij} + C_2 \sigma_{ik}\sigma_{kj}. \quad (11.4\text{a})$$

The coefficients M_0 and C_0 are evaluated from the condition $\dot{\varepsilon}_{ii} = 0$ for which the hydrostatic pressure $\sigma_{ii}/3 = -p$. Introducing this condition into Eqs. (11.3a) and (11.4a):

$$-3p = 3M_0 + M_2 \dot{\varepsilon}_{ik}\dot{\varepsilon}_{ik} \quad \text{and} \quad 3C_0 - 3C_1 p + C_2 \sigma_{ik}\sigma_{ik} = 0,$$

---

[1] M. REINER: Amer. J. Math. **67**, 350 (1945).
[2] R. S. RIVLIN: Phil. Trans. Roy. Soc. Lond., Ser. A **241**, 379 (1948). — Proc. Roy. Soc. Lond., Ser. A **193**, 260 (1948). — R. S. RIVLIN and J. L. ERICKSEN: J. Rat. Mech. Analysis **4**, 323 (1955).
[3] W. NOLL: J. Rat. Mech. Analysis **4**, 3 (1955). — R. S. RIVLIN: Quart. Appl. Math. **13**, 177 (1955). — R. S. RIVLIN and J. L. ERICKSEN: ibid., p. 420.

the coefficients are obtained:

$$M_0 = -p - \tfrac{1}{3} M_2 \dot\varepsilon_{ik}\dot\varepsilon_{ik} \quad \text{and} \quad C_0 = C_1 p - \tfrac{1}{3} C_2 \sigma_{ik}\sigma_{ik}. \tag{11.5}$$

Therefore the stress-strain-rate relations of the incompressible dissipative medium, considering that $\dot\varepsilon_{ij} = \dot e_{ij}$, $s_{ij} = \sigma_{ij} + p\delta_{ij}$ and $\sigma_{ik}\sigma_{ik} = 2J_2$, $\dot\varepsilon_{ik}\dot\varepsilon_{ik} = 2\dot I_2'$, as well as the relations between the second invariants of the tensor and of the tensor deviations (6.6)

$$s_{ij} = M_1 \dot e_{ij} + M_2(\dot e_{ik}\dot e_{kj} - \tfrac{2}{3}\dot I_2' \delta_{ij}) \tag{11.3b}$$

and

$$\dot e_{ij} = C_1' s_{ij} + C_2(s_{ik}s_{kj} - \tfrac{2}{3} J_2' \delta_{ij}) \tag{11.4b}$$

where the coefficients $C_1' = C_1 + \tfrac{2}{3} C_2 J_1$, and $M_i(\dot I_2', \dot I_3')$ and $C_i(J_2', J_3')$ are polynomials in the invariants of the strain-rate and stress deviations; the second terms on the right-hand side represent the deviations of the squares of the strain-rate and the stress deviations respectively. Eqs. (11.3b) and (11.4b) are due to REINER[1]; Eq. (11.4b) was proposed independently by PRAGER[2] for the general plastic medium.

That both Eqs. (11.3b) and (11.4b) admit dissipation potentials $\dot X_v(\dot I_2', \dot I_3')$ and $\dot X_p(J_2', J_3')$ respectively is easily shown by comparing the relations

$$\frac{\partial \dot X_v}{\partial \dot\varepsilon_{ij}} = \frac{\partial \dot X_v}{\partial \dot I_2'} \frac{\partial \dot I_2'}{\partial \dot\varepsilon_{ij}} + \frac{\partial \dot X_v}{\partial \dot I_3'} \frac{\partial \dot I_3'}{\partial \dot\varepsilon_{ij}} = s_{ij} \tag{11.6}$$

and

$$\frac{\partial \dot X_p}{\partial \sigma_{ij}} = \frac{\partial \dot X_p}{\partial J_2'} \frac{\partial J_2'}{\partial \sigma_{ij}} + \frac{\partial \dot X_p}{\partial J_3'} \frac{\partial J_3'}{\partial \sigma_{ij}} = \dot e_{ij} \tag{11.7}$$

respectively with Eqs. (11.3b) and (11.4b). Since

$$\frac{\partial \dot I_2'}{\partial \dot\varepsilon_{ij}} = \dot e_{ij}, \qquad \frac{\partial \dot I_3'}{\partial \dot\varepsilon_{ij}} = \dot e_{ik}\dot e_{kj} - \tfrac{2}{3} \dot I_2' \delta_{ij} \tag{11.8}$$

and

$$\frac{\partial J_2'}{\partial \sigma_{ij}} = s_{ij}, \qquad \frac{\partial J_3'}{\partial \sigma_{ij}} = s_{ik}s_{kj} - \tfrac{2}{3} J_2' \delta_{ij}, \tag{11.9}$$

the moduli and compliances are the derivatives of the dissipation potentials

$$M_1 = \frac{\partial \dot X_v}{\partial \dot I_2'}, \quad M_2 = \frac{\partial \dot X_v}{\partial \dot I_3'} \tag{11.10}$$

and

$$C_1' = \frac{\partial \dot X_p}{\partial J_2'}, \quad C_2 = \frac{\partial \dot X_p}{\partial J_3'}. \tag{11.11}$$

The necessary and sufficient relations between the moduli as well as between the compliances are obtained by cross-differentiation

$$\frac{\partial M_1}{\partial \dot I_3'} - \frac{\partial M_2}{\partial \dot I_2'} = 0 \quad \text{and} \quad \frac{\partial C_1'}{\partial J_3'} - \frac{\partial C_2}{\partial J_2'} = 0. \tag{11.12}$$

---

[1] See footnote 1, p. 258.
[2] W. PRAGER: J. Appl. Phys. **16**, 837 (1945); **19**, 540 (1948). — G. H. HANDELMAN, C. C. LIN and W. PRAGER: Quart. Appl. Math. **4**, 397 (1947).

By combining Eqs. (11.6) to (11.9) the rate of energy dissipation for the volume element therefore

$$s_{ij}\dot{e}_{ij} = 2\dot{I}_2' \frac{\partial \dot{X}_v}{\partial \dot{I}_2'} + 3\dot{I}_3' \frac{\partial \dot{X}_v}{\partial \dot{I}_3'} \qquad (11.13)$$

and

$$s_{ij}\dot{e}_{ij} = 2J_2' \frac{\partial \dot{X}_p}{\partial J_2'} + 3J_3' \frac{\partial \dot{X}_p}{\partial J_3'}. \qquad (11.14)$$

Quasi-linear relations for the general incompressible dissipative medium are obtained by specifying $M_2 = 0$ and $C_2 = 0$ and therefore, according to Eqs. (11.10) and (11.11) $\dot{X}_v = \dot{X}_v(\dot{I}_2')$ and $\dot{X}_p = \dot{X}_p(J_2')$, from which the classical relations of the theory of linear viscosity and of ideal plasticity are derived by assuming the simplest possible forms of the dissipation potentials $\dot{X}_v = 2\mu \dot{I}_2'$ and $\dot{X}_p = 2\lambda J_2'$ respectively (see Sect. 8). In the quasi-linear theories of dissipative media the dissipation potentials are polynomials in either $\dot{I}_2'$ or J_2' and so are the coefficients M_1 and C_1' according to Eqs. (11.10) and (11.11). Hence

$$\frac{s_{ij}}{\dot{e}_{ij}} = \frac{(\sigma_i - \sigma_j)}{(\dot{\varepsilon}_i - \dot{\varepsilon}_j)} = M_1 = \frac{1}{C_1'} \qquad (11.15)$$

is the condition for the existence of a quasi-linear or a truly linear tensor relation, depending on whether the coefficients are functions of the invariants and thus of the coordinates, or true physical constants.

The general tensor relation and the associated potentials for the elastic continuum can be obtained by considerations similar to those leading to Eqs. (11.1) to (11.15), except for the condition of incompressibility, which is replaced by the condition that $\sigma_{ij} = 0$ for $\varepsilon_{ij} = 0$.[1]

12. Second order effects. When the coefficients of the quadratic terms in the stress-strain-rate relations (11.3b) and (11.4b) $M_2 > 0$ and $C_2 > 0$, certain second order effects appear that do not exist in linear media. Such phenomena exist in dissipative as well as in elastic media[2]. They be can illustrated by considering the effects of a pure shearing strain or of a pure shear stress with, respectively, the matrices of strain-rate and of stress

$$[\dot{e}_{ij}] = \dot{\gamma} \begin{vmatrix} 0 & 1 & 0 \\ 1 & 0 & 0 \\ 0 & 0 & 0 \end{vmatrix}, \quad [s_{ij}] = \tau \begin{vmatrix} 0 & 1 & 0 \\ 1 & 0 & 0 \\ 0 & 0 & 0 \end{vmatrix} \qquad (12.1)$$

where $\dot{\gamma}$ is a measure of the rate of shear, τ of the shear stress. For $M_2 = 0$ and $C_2 = 0$, the linear relations hold $\tau = M_1 \dot{\gamma}$ and $\dot{\gamma} = C_1' \tau$.

The effect of the quadratic terms $M_2 > 0$ and $C_2 > 0$ is twofold: the appearance of the squares of the matrices $[\varepsilon_{ij}]$ and $[\sigma_{ij}]$ and of the diagonal matrices $\frac{2}{3}\dot{I}_2'[1] = \frac{2}{3}\dot{\gamma}^2[1]$ and $\frac{2}{3}J_2'[1] = \frac{2}{3}\tau^2[1]$. Hence with

$$[\dot{e}_{ij}]^2 = \dot{\gamma}^2 \begin{vmatrix} 1 & 0 & 0 \\ 0 & 1 & 0 \\ 0 & 0 & 0 \end{vmatrix} \quad \text{and} \quad [s_{ij}]^2 = \tau^2 \begin{vmatrix} 1 & 0 & 0 \\ 0 & 1 & 0 \\ 0 & 0 & 0 \end{vmatrix} \qquad (12.2)$$

---

[1] M. REINER: Amer. J. Math. **70**, 433 (1948).
[2] M. REINER: Quart. J. Mech. Appl. Math. **5**, 42 (1952). See also footnote 1 p. 258 and 260.

the matrix relations are obtained:

$$[s_{ij}] = -\tfrac{2}{3} M_2 \dot{\gamma}^2 \begin{vmatrix} 1 & 0 & 0 \\ 0 & 1 & 0 \\ 0 & 0 & 1 \end{vmatrix} + M_1 \dot{\gamma} \begin{vmatrix} 0 & 1 & 0 \\ 1 & 0 & 0 \\ 0 & 0 & 0 \end{vmatrix} + M_2 \dot{\gamma}^2 \begin{vmatrix} 1 & 0 & 0 \\ 0 & 1 & 0 \\ 0 & 0 & 0 \end{vmatrix} \quad (12.3\mathrm{a})$$

and

$$[\dot{e}_{ij}] = -\tfrac{2}{3} C_2 \tau^2 \begin{vmatrix} 1 & 0 & 0 \\ 0 & 1 & 0 \\ 0 & 0 & 1 \end{vmatrix} + C_1' \tau \begin{vmatrix} 0 & 1 & 0 \\ 1 & 0 & 0 \\ 0 & 0 & 0 \end{vmatrix} + C_2 \tau^2 \begin{vmatrix} 1 & 0 & 0 \\ 0 & 1 & 0 \\ 0 & 0 & 0 \end{vmatrix}. \quad (12.3\mathrm{b})$$

Comparison of Eqs. (12.3) with (12.1) shows the new terms: a hydrostatic pressure proportional to $\dot{\gamma}^2$ or a rate of volume change proportional to τ^2 (which becomes effective only if the medium is compressible) that have been termed the "*Kelvin effect*"[1] or *dilatancy*[2], and a normal stress in the plane of motion proportional to $\dot{\gamma}^2$ or a normal extension rate in this plane proportional to τ^2, referred to as the *Poynting effect*[3] in solids and as the *Weissenberg effect*[4] in fluids.

Both effects are characteristic of non-linear continuum mechanics and are governed by the material characteristics M_2 and C_2. As REINER has pointed out, the form of the general isotropic relations (11.3b) and (11.4b) is independent of the definition of strain introduced, which will only affect the numerical values of the coefficients M_1 and C_1. The two effects are also independent of the type of non-linearity that might implicitly exist in the quasi-linear terms of Eqs. (12.3), as a result of possible dependence of M_1 and C_1' on higher than first powers of I_2' and J_2', according to Eqs. (11.10) and (11.11). Such non-linearity can only appear in the form of simple odd-order correction terms to the linear terms of the classical theories, and is independent of the coefficients M_2 and C_2.

13. General isotropic tensor relations. For the more general case in which the stress matrix is isotropically related to both the strain matrix and the rate of strain matrix and the principal axes of the three tensors coincide, it has been shown[5] by the aid of an extended Cayley-Hamilton theorem that $[\sigma_{ij}]$ can be expressed in the following form:

$$\begin{aligned}
[\sigma_{ij}] = {}& M_0 [1] + M_1 [\varepsilon_{ij}] + M_2 [\varepsilon_{ij}]^2 + M_3 [\dot{\varepsilon}_{ij}] + \\
& + M_4 [\dot{\varepsilon}_{ij}]^2 + M_5 ([\varepsilon_{ij}][\dot{\varepsilon}_{ij}] + [\dot{\varepsilon}_{ij}][\varepsilon_{ij}]) + \\
& + M_6 ([\varepsilon_{ij}]^2 [\dot{\varepsilon}_{ij}] + [\dot{\varepsilon}_{ij}][\varepsilon_{ij}]^2) + M_7 ([\varepsilon_{ij}][\dot{\varepsilon}_{ij}]^2 + \\
& + [\dot{\varepsilon}_{ij}]^2 [\varepsilon_{ij}]) + M_8 ([\varepsilon_{ij}]^2 [\dot{\varepsilon}_{ij}]^2 + [\dot{\varepsilon}_{ij}]^2 [\varepsilon_{ij}]^2)
\end{aligned} \quad (13.1)$$

where the M_i are polynomials of the invariants of $[\varepsilon_{ij}]$ and $[\dot{\varepsilon}_{ij}]$ as well as of the various matrix products. Neglecting all second and higher order terms a quasi-linear approximation to Eq. (13.1) is obtained

$$[\sigma_{ij}] = M_0 [1] + M_1 [\varepsilon_{ij}] + M_3 [\dot{\varepsilon}_{ij}]. \quad (13.2)$$

A similar quasi-linear relation can be established by expressing the rate of strain matrix in terms of the stress-and rate of stress matrices

$$[\dot{\varepsilon}_{ij}] = C_0 [1] + C_1 [\sigma_{ij}] + C_3 [\dot{\sigma}_{ij}] \quad (13.3)$$

where $\dot{\sigma}_{ij}$ is specified according to Eq. (10.5).

---

[1] W. THOMSON (Lord KELVIN) and P. G. TAIT: Treatise on Natural Philosophy, Part I, 2nd edit., Vol. 2, No. 679. Cambridge 1883.
[2] See footnote 1, p. 258.
[3] J. H. POYNTING: Proc. Roy. Soc. Lond., Ser. A **82**, 546 (1909); **86**, 534 (1912).
[4] K. WEISSENBERG: Nature, Lond. **159**, 310 (1947).
[5] R. S. RIVLIN: J. Rat. Mech. Analysis **4**, 681 (1955).

The elimination of M_0 and C_0 can again be based on the assumptions of incompressible flow or incompressible deformation or both. In the latter case M_0 reduces to the hydrostatic pressure p and $C_0 = -C_1 p - C_3 \dot{p}$. Hence the simplest form of Eqs. (13.2) and (13.3) for the incompressible medium written in tensor form is

$$s_{ij} = M_1 e_{ij} + M_3 \dot{e}_{ij} \tag{13.4}$$

and

$$\dot{e}_{ij} = C_1 s_{ij} + C_3 \dot{s}_{ij} \tag{13.5}$$

where M_i and C_i are either polynomials in the second invariants $M_1(I_2')$, $M_3(\dot{I}_2')$, $C_1(J_2')$, $C_3(\dot{J}_2')$ and thus functions of the coordinates, or true physical constants.

14. "Hypo-elastic" relations. A generalization of the isotropic tensor relations has been proposed by TRUESDELL[1] following previous formulations[2], by postulating a relation between the rate of deformation d_{ij} and the rate of stress

$$\frac{d}{dt}\sigma_{ij} = f(d_{ij}) \quad \text{or} \quad \frac{d}{dt}\sigma_{ij} = f(d_{ij}, \sigma_{ij}) \tag{14.1}$$

depending on whether the response is independent of or dependent on the stress-field itself. For nearly irrotational motion the left side of Eq. (14.1) is identical with Eq. (10.5), containing the convective but no rotational terms. Eq. (14.1) for finite deformation defines the "hypo-elastic" medium, which has no preferred state of strain or state of stress and thus permits an arbitrarily stressed initial state; its stress-strain relations are obtained by integration of Eq. (14.1) over the loading path and the deformation and are therefore not independent of either.

Eq. (14.1) reduces to that of the classical linear elastic medium under the assumptions usually made in formulating that theory. For finite deformations, however, for which the hypo-elastic medium is pronouncedly non-linear, integration of Eq. (14.1) produces an amazing variety of non-linear responses, including "strain-hardening" and "strain-softening", depending on the loading and on assumed values of dimensionless "material constants". It has even been shown that at infinite strain a response suggesting a "yield limit" is asymptotically approached[3]. This fact appears to have been responsible for the claims that a physical condition of plastic yielding can be predicted on the basis of the theory implied in Eq. (14.1), and that plastic yielding itself is a phenomenon associated with very large strains.

These claims that have been made in the physical interpretation of the equations of the hypo-elastic body are, however, flatly contradicted by the fact, confirmed in innumerable experiments, that the start of plastic yielding is associated with strains of an order of magnitude justifiably considered small, and that such yielding is physically sharply distinct from elastic straining[4]. For large strains, however, the latter may produce conditions of elastic instability (buckling), with stress-strain relations resembling *in shape though not in scale of the deformation* asymptotic conditions of yielding. The hypo-elastic medium, being

---

[1] C. TRUESDELL: J. Rat. Mech. Analysis **4**, 83 (1955).

[2] S. ZAREMBA: Bull. Int. Acad. Sci., Cracovie **85**, 380, 594, 614, (1903). — Mém. Sci. Math. 1937, No. 82. See also footnote 1, p. 241.

[3] A. E. GREEN: Proc. Roy. Soc. Lond., Ser. A **234**, 46 (1956). — J. Rat. Mech. Analysis **5**, 725 (1956). — T. Y. THOMAS: Proc. Nat. Acad. Sci. U. S. A. **41**, 716, 720, 762, 908 (1955). — C. TRUESDELL: J. Appl. Phys. **27**, 441 (1956).

[4] A. SEEGER: IUTAM Colloquium Madrid, p. 90. Berlin: Springer 1956.

an interesting generalization of the classical elastic medium, is therefore unrelated to the plastic continuum in which the existence of the mechanism of energy dissipation through the motion of dislocations does not depend on the assumption that the over-all strains are large.

II. Anelastic relations.

15. Anelasticity due to pressure relaxation. The build-up and relaxation of isotropic pressure $-p = \frac{1}{3}\sigma_{ii}$ in the elastic continuum as a result of non-mechanical effects, such as temperature changes and density changes due to internal motion of particles (volume-diffusion), or of the effects of actual fluid pressure in the pores of a quasi-continuous porous elastic medium, produces the types of anelasticity known as *thermo-elasticity*, *pore-fluid anelasticity* or *consolidation*, *diffusion-relaxation*, etc. Their common feature is the (scalar) interaction of the assumedly linear phenomena with the elastic continuum through its compressibility. The general theory of interaction effects in solids is due to Voigt[1]; certain simple problems have been solved by Lord Kelvin[2] and others[3]. The rigorous mechanical-thermodynamic analysis of the 3-dimensional problem of pressure relaxation has been developed by Biot[4].

α) *Thermo-anelasticity.* The thermo-elastic equations are obtained[4] by combining the stress-strain relations of the elastic continuum Eq. (8.8) including the volume strain due to a temperature increment ΔT above the (absolute) reference temperature T

$$\sigma_{ij} = K \varepsilon_{kk} \delta_{ij} + 2G e_{ij} - 3K\alpha \Delta T \delta_{ij} \qquad (15.1)$$

where α is the coefficient of linear thermal expansion, with the entropy Eq. (7.29),

$$\frac{ds}{dt} = \frac{1}{T_1} \frac{du}{dt} - \frac{1}{T_1} \sigma_{ij} \dot{\varepsilon}_{ij} = \frac{1}{T_1} \frac{\partial u}{\partial T_1} \dot{T}_1 + \frac{1}{T_1}\left(\frac{\partial u}{\partial \varepsilon_{ij}} - \sigma_{ij}\right) \dot{\varepsilon}_{ij} \qquad (15.2)$$

where $T_1 = T + \Delta T$. Since (ds/dt) must be an exact differential in T_1 and ε_{ij} and therefore

$$\frac{\partial}{\partial \varepsilon_{ij}}\left(\frac{1}{T_1}\frac{\partial u}{\partial T_1}\right) = \frac{\partial}{\partial T_1}\left[\frac{1}{T_1}\left(\frac{\partial u}{\partial \varepsilon_{ij}} - \sigma_{ij}\right)\right] \qquad (15.3)$$

it follows that

$$\frac{\partial u}{\partial \varepsilon_{ij}} - \sigma_{ij} = -T_1 \frac{\partial \sigma_{ij}}{\partial T_1}. \qquad (15.4)$$

Hence from Eq. (15.1)

$$\frac{\partial u}{\partial \varepsilon_{ij}} - \sigma_{ij} = 3K\alpha T_1 \delta_{ij}. \qquad (15.5)$$

Eq. (15.2) therefore becomes

$$\frac{ds}{dt} = c \frac{\Delta \dot{T}}{T_1} + 3K\alpha \dot{\varepsilon}_{kk} \approx c\frac{\Delta \dot{T}}{T} + 3K\alpha \dot{\varepsilon}_{kk} \quad \text{or} \quad s = c\frac{\Delta T}{T} + 3K\alpha \varepsilon_{kk} \qquad (15.6)$$

where $c = (\partial u/\partial T_1)$ is the specific heat per unit volume of the unstrained continuum and ΔT is assumed to be small in relation to T, so that in first approximation $T_1 \approx T$.

---
[1] W. Voigt: Lehrbuch der Krystallphysik. Berlin: B. G. Teubner 1910.
[2] W. Thomson (Lord Kelvin): Phil. Mag. (5) **5** (1878).
[3] K. Bennewitz and H. Roetger: Phys. Z. **37**, 578 (1936). — C. Zener: Phys. Rev. **52**, 230 (1937); **53**, 90, 192, 582, 686, 1010 (1938). — M. Paesler: Z. Physik **122**, 357 (1944). — K. Terzaghi: Erdbaumechanik. Wien: Franz Deuticke 1925.
[4] M. A. Biot: J. Appl. Phys. **12**, 155, 426, 578 (1941); **25**, 1385 (1954); **26**, 182 (1955); **27**, 240 (1956). — Phys. Rev. **97**, 1463 (1955).

For adiabatic conditions for which

$$\Delta \dot{T} = -\frac{T}{c} 3 K \alpha \dot{\varepsilon}_{kk} \qquad (15.7)$$

the temperature increment can be eliminated from the stress-strain relations (15.1), which take the form

$$\sigma_{ij} = \left[K + 9K^2 \alpha^2 \frac{T}{c} \right] \varepsilon_{kk} \delta_{ij} + 2G e_{ij} = \overline{K} \varepsilon_{kk} \delta_{ij} + 2G e_{ij} \qquad (15.8)$$

where $\overline{K} > K$ is the *adiabatic* bulk modulus.

Combining Eqs. (15.1) with the equilibrium Eqs. (4.8) and the definition of infinitesimal strain Eq. (5.9) the three equations are obtained

$$G u_{i,kk} + \left(K + \frac{G}{3} \right) u_{k,ki} - 3 K \alpha \Delta T_{,i} = \varrho \ddot{u}_i - X_i \qquad (15.9)$$

for the four unknowns u_i and ΔT. The fourth equation is obtained by combining Eq. (15.6) with the linear equation of heat conduction (7.20) considering that in the elastic continuum $T \, ds/dt = -q_{i,i}$ [see Eq. (7.15)]. Hence

$$\frac{ds}{dt} = \frac{1}{T} k \Delta T_{,ii} = \frac{c}{T} \Delta \dot{T} + 3 K \alpha \dot{u}_{k,k} \qquad (15.10)$$

indicating that the displacement- and temperature-fields are thermodynamically coupled. Taking the divergence of Eq. (15.9) with $X_i = 0$

$$\varepsilon_{kk,ii} = a \Delta T_{,ii} + \frac{a \varrho}{3 K \alpha} \ddot{\varepsilon}_{kk} \qquad (15.9\text{a})$$

where $a = 9K\alpha/(3K + 4G)$, and combining it with Eq. (15.10), the differential equation for the entropy is obtained (for $\ddot{\varepsilon}_{kk} = 0$)

$$s_{,ii} = \frac{1}{h^2} \frac{\partial s}{\partial t} \qquad (15.10\text{a})$$

where

$$h^2 = \frac{k}{c} \frac{K + 4G/3}{\overline{K} + 4G/3}.$$

Eq. (15.10a) shows that the entropy satisfies the diffusion equation under the assumption $\ddot{\varepsilon}_{kk} = 0$.

The rate of entropy production per unit volume due to heat flow according to Eqs. (7.19) and (7.20)

$$\vartheta'_T = \frac{-q_i \Delta T_{,i}}{T^2} = \frac{1}{k} \left(\frac{q_i}{T} \right)^2 = \frac{1}{k} s_i s_i = k \left(\frac{\Delta T_{,i}}{T} \right) \left(\frac{\Delta T_{,i}}{T} \right) \qquad (15.11)$$

where s_i is the entropy flux vector; the rate of energy dissipation is $T \vartheta'_T$, the thermal dissipation potential $\dot{X}_T = \frac{1}{2} T \vartheta'_T$. Introducing a thermal energy potential of the form $\Phi_T = \frac{1}{2} c \, (\Delta T)^2 / T$, the variational principles Eqs. (9.14) and (9.15) can be extended to include temperature effects[1].

Because of Eqs. (15.10) and (15.9a) the temperature and the volume-expansion satisfy the equations

$$\nabla^2 \left(h^2 \nabla^2 - \frac{\partial}{\partial t} \right) \Delta T = 0 \quad \text{and} \quad \nabla^2 \left(h^2 \nabla^2 - \frac{\partial}{\partial t} \right) \varepsilon_{kk} = 0. \qquad (15.10\text{b})$$

---

[1] M. A Biot: J. Appl. Phys. **27**, 244 (1956). — J. Aero Sci. **24**, 857 (1957). — J. H. Weiner: Quart. Appl. Math. **15**, 102 (1957).

β) *Pore-pressure anelasticity.* A set of equations similar to Eqs. (15.9) and (15.10) is obtained[1] for the interaction of the deformation of a quasi-isotropic porous elastic solid of porosity f (ratio of pore-volume to bulk-volume or cross-sectional pore-area to bulk-section), assuming randomly distributed pores, with the flow of compressible pore-fluid under the hydrostatic over-pressure $\bar{p}$ producing the average stress $\bar{\sigma}$ on the fluid phase

$$\bar{\sigma} = -f\bar{p}. \tag{15.12}$$

The equilibrium conditions (4.8) must obviously be fulfilled by the total stress tensor

$$(\sigma_{ij} + \bar{\sigma}\delta_{ij})_{,j} = \varrho \ddot{u}_i - X_i \tag{15.13}$$

where the stress tensor σ_{ij} represents the average stress on the solid phase.

The stress-strain relations are obtained by assuming a combined elastic potential made up of the elastic strain energies of the solid and of the liquid phases

$$\bar{\Phi} = \Phi + Q\bar{\varepsilon}\varepsilon_{kk} + \tfrac{1}{2}R\bar{\varepsilon}^2 \tag{15.14}$$

where Φ is the elastic potential of the solid phase according to Eq. (8.7), $\bar{\varepsilon}$ denotes the volume compression of the fluid and Q and R are two additional material constants; the stress components are therefore

$$\sigma_{ij} = \frac{\partial \bar{\Phi}}{\partial \varepsilon_{ij}} \quad \text{and} \quad \bar{\sigma} = \frac{\partial \bar{\Phi}}{\partial \bar{\varepsilon}} \tag{15.15}$$

or

$$\sigma_{ij} = K\varepsilon_{kk}\delta_{ij} + 2Ge_{ij} + Q\bar{\varepsilon}\delta_{ij} \tag{15.16}$$

and

$$\bar{\sigma} = Q\varepsilon_{kk} + R\bar{\varepsilon}.$$

Eliminating $\bar{\varepsilon}$ and introducing Eq. (15.12) the combined stress-strain relations for the bulk medium

$$(\sigma_{ij} + \bar{\sigma}\delta_{ij}) = S\varepsilon_{kk}\delta_{ij} + 2Ge_{ij} - \bar{\alpha}\bar{p}\delta_{ij} \tag{15.17}$$

where the constants $S = K - Q^2/R$ and $\bar{\alpha} = (Q+R)f/R$. Eq. (15.17) corresponds to Eq. (15.1) if K is replaced by S, ΔT by $\bar{p}$ and $3K\alpha$ by $\bar{\alpha}$. Combining it with Eqs. (15.13) and (5.9) the three equations obtained

$$Gu_{i,kk} + (S+G)u_{k,ki} - \bar{\alpha}\bar{p}_{,i} = \varrho \ddot{u}_i - X_i \tag{15.18}$$

correspond to Eqs. (15.9).

The fourth equation for the evaluation of the four unknowns u_i and $\bar{p}$ is obtained by introducing DARCY's law of fluid motion through a porous medium, which corresponds to the heat-flow Eq. (7.20) in the thermo-elastic problem. It relates the pressure gradient in the fluid phase to the relative motion of this phase

$$\bar{\sigma}_{,k} = b(\bar{v}_k - \dot{u}_k) = -f\bar{p}_{,k} \tag{15.19}$$

where $\bar{v}_k$ denotes the pore-fluid velocity. By taking the divergence of Eq. (15.19) and eliminating $\dot{\bar{\varepsilon}} = \bar{v}_{k,k}$ with the aid of the second Eq. (15.16) the relation corresponding to Eq. (15.10) is obtained

$$k_1 \bar{p}_{,kk} = \frac{f^2}{R}\dot{\bar{p}} + \frac{(Q+R)f}{R}\dot{\varepsilon}_{kk} = \bar{c}\dot{\bar{p}} + \bar{\alpha}\dot{u}_{k,k} \tag{15.19a}$$

where $\bar{c} = f^2/R$ and $k_1 = f^2/b$ represents a coefficient of permeability. The left-hand side of Eq. (15.19a)

$$k_1 \bar{p}_{,kk} = -f(\dot{\bar{\varepsilon}} - \dot{\varepsilon}_{kk}) = \frac{\partial \xi}{\partial t} \tag{15.20}$$

---

[1] See footnote 4, p. 263.

is proportional to the rate of increment of fluid content ξ in the porous volume element, corresponding to the rate of entropy increase in the thermo-elastic problem.

Combining Eqs. (15.19a) and (15.20) the relation corresponding to Eq. (15.6) is obtained

$$\frac{\partial \xi}{\partial t} = \bar{c}\dot{\bar{p}} + \bar{\alpha}\dot{\varepsilon}_{kk} \quad \text{or} \quad \xi = \bar{c}\bar{p} + \bar{\alpha}\varepsilon_{kk}. \tag{15.21}$$

For $\partial \xi/\partial t = 0$ the relation holds

$$\dot{\bar{p}} = -\frac{\bar{\alpha}}{\bar{c}}\dot{\varepsilon}_{kk} \tag{15.21a}$$

so that Eq. (15.17) is transformed into the "unrelaxed" form

$$(\sigma_{ij} + \bar{\sigma}\delta_{ij}) = \left(S + \frac{\bar{\alpha}^2}{\bar{c}}\right)\varepsilon_{kk}\delta_{ij} + 2G e_{ij} = \overline{S}\,\varepsilon_{kk}\delta_{ij} + 2G e_{ij} \tag{15.17a}$$

where $\overline{S} = (K + 2Q + R) > S$ represents the "unrelaxed" bulk modulus corresponding to the adiabatic modulus $\overline{K}$.

Taking the divergence of Eq. (15.18)

$$\varepsilon_{kk,ii} = \bar{a}\,\bar{p}_{,ii} + \frac{\bar{a}\varrho}{\bar{\alpha}}\,\ddot{\varepsilon}_{kk} \tag{15.18a}$$

where $\bar{a} = \bar{\alpha}/(2G+S)$, combined with Eqs. (15.19a) and (15.20) the equation for the fluid content ξ is obtained in the form of the diffusion equation (for $\ddot{\varepsilon}_{kk} = 0$):

$$\xi_{,ii} = \frac{1}{\bar{h}^2}\frac{\partial \xi}{\partial t} \tag{15.20a}$$

where

$$\bar{h}^2 = \frac{k_1}{\bar{c}}\,\frac{2G+S}{2G+S+\bar{\alpha}^2/\bar{c}}$$

corresponding to Eq. (15.10a). The pressure $\bar{p}$ and the volume expansion ε_{kk} satisfy the Eqs. (15.10b) with $h = \bar{h}$.

The variables of the pressure relaxation phenomena are thus governed by diffusion equations of the general form

$$h^2 \varphi_{,ii} - \frac{\partial \varphi}{\partial t} = \psi(x_i). \tag{15.22}$$

By putting $\varphi_\nu = \bar{\varphi}_\nu(x_i) f_\nu(t)$ the homogeneous equation is separated into

$$\dot{f}_\nu + h^2 \nu^2 f_\nu = 0 \quad \text{and} \quad (\bar{\varphi}_\nu)_{,ii} + \nu^2 \bar{\varphi} = 0 \tag{15.23}$$

so that

$$\varphi_\nu = \bar{\varphi}_\nu(x_i)\, e^{-t/\tau_\nu} \quad \text{and} \quad \varphi = \sum_0^\infty \varphi_\nu$$

where $\bar{\varphi}_\nu$ are the orthogonal modes of the eigenvalue problem, φ_ν are the "*relaxation modes*", or solutions with exponential decay governed by the characteristic relaxation time of the mode $\tau_\nu = 1/\nu^2 h^2$, obtained from the solution of the characteristic equation of the second Eq. (15.23) for specified boundary conditions. The relaxation times τ_ν are thus parameters combining the effects of the relevant physical constants with the boundary conditions of the specific problem.

Solutions of Eqs. (15.22) for particular problems of pressure relaxation have been obtained by Biot using various methods, for instance Papkovitch-Boussinesq potentials[1].

---

[1] P. E. Papkovitch: C. R. Acad. Sci., Paris **195**, 513, 754 (1932). — See also pp. 89—91 in this volume.

Introducing the solutions $\varepsilon_{kk}(x_i, t)$, $\Delta T(x_i, t)$ or $\bar{p}(x_i, t)$ into the stress-strain relations (15.1) and (15.16) the resulting expressions for σ_{ij} indicate that the superposition of the number of processes of relaxation (relaxation modes) of the built-up hydrostatic pressure produces the volume anelasticity of the medium, while the elastic relations between the deviatoric components remain unaffected. The medium thus behaves as a continuum with apparent volume viscosity, defined by a spectrum of volume viscosity coefficients, but perfect elasticity of volume-constant distortion.

16. Homogeneous relaxation. Temperature or pore-fluid pressure are only two out of a number of phenomena the interaction of which with the elastic deformation gives rise to anelastic effects[1]. Obviously such interaction is not limited to the volumetric part of the strain but may involve the distortional strain as well as, for instance, in the case of the interaction with grain-boundary diffusion in metals (grain-boundary relaxation)[2], magnetic eddy currents[3], as well as various types of atomic or molecular diffusion in crystal lattices and polymers[4].

Introducing the simplifying assumption that the interaction involves only a single phenomenon and is homogeneous throughout the considered volume, it can be represented by the decay with time or *relaxation* of the momentary value of a non-mechanical homogeneous generalized force X_ν towards the thermodynamically stable zero value by a process governed by an equation of the form of the first Eq. (15.23)

$$\left(\frac{dX_\nu}{dt}\right)' = -\tau_\nu^{-1} X_\nu \tag{16.1}$$

which reflects the underlying assumption that the rate of relaxation is proportional to the momentary deviation from equilibrium, as well as by a linear interaction between this generalized force and the elastic strain ε_{ij} of the type expressed by Eqs. (15.7) and (15.21)

$$\left(\frac{dX_\nu}{dt}\right)'' = \left(\frac{\partial X_\nu}{\partial \varepsilon_{ij}}\right) \dot{\varepsilon}_{ij} = -\gamma_\nu \dot{\varepsilon}_{ij} \tag{16.2}$$

where γ_ν denotes the constant interaction coefficient in the un-relaxed state (unrelaxed modulus); Eq. (16.2) exists only when X_ν and ε_{ij} are of equal tensorial rank. Combining Eqs. (16.1) and (16.2)

$$\left(\frac{dX_\nu}{dt}\right) = \left(\frac{dX_\nu}{dt}\right)' + \left(\frac{dX_\nu}{dt}\right)'' = -\tau_\nu^{-1} X_\nu - \gamma_\nu \dot{\varepsilon}_{ij}. \tag{16.3}$$

Eq. (16.3) is the homogeneous equivalent of Eqs. (15.10) and (15.21). When the ν-th relaxation process depends on $\varkappa$ parameters $X_\varkappa$, Eq. (16.3) is generalized in the form

$$\left(\frac{dX_\nu}{dt}\right) = -\sum_\varkappa m_{\nu\varkappa} X_\varkappa \tag{16.3a}$$

reflecting interaction between different molecular processes.

---

[1] See footnote 1, p. 263. — C. ZENER: Elasticity and Anelasticity. Chicago: University Press 1951.

[2] C. ZENER: Phys. Rev. **58**, 87 (1940); **60**, 906 (1941). — T. S. KÊ: Phys. Rev. **71**, 533; **72**, 41 (1947). — J. Appl. Phys. **20**, 274 (1949).

[3] K. HONDA, S. SHIMIZU and S. KUSAKABE: Phys. Z. **3**, 380 (1901). — W. BROWN: Phys. Rev. **50**, 1156 (1936). — C. ZENER: Phys. Rev. **53**, 1010 (1938).

[4] W. GORSKI: Phys. Z. SSSR. **8**, 443, 562 (1935). — C. ZENER: Trans. Amer. Inst. Mining Metallurg. Eng. **167**, 155 (1946). — H. A. STUART: Physik der Hochpolymeren, Band 4. Berlin: Springer 1956. — F. H. MÜLLER: Relaxationsverhalten der Materie. Darmstadt: Steinkopff 1953.

Introducing the linear relation between stress, strain and the non-mechanical variable X_ν in the elastic medium in the simplified form

$$\sigma_{ij} = M_r \varepsilon_{ij} - \lambda_\nu X_\nu \tag{16.4}$$

where M_r is the "relaxed" elastic modulus for $X_\nu = 0$ and λ_ν represents a coefficient of interaction, and eliminating X_ν between Eqs. (16.3) and (16.4) the homogeneous anelastic relation is obtained[1]

$$M_r \varepsilon_{ij} + M_u \tau \dot{\varepsilon}_{ij} = \sigma_{ij} + \tau \dot{\sigma}_{ij} \tag{16.5}$$

where $\tau = \tau_\nu$ and $M_u = (1 + \lambda_\nu \gamma_\nu / M_r)$ is the elastic modulus in the unrelaxed state. Alternatively Eq. (16.5) can be written in the form

$$M_r (\varepsilon_{ij} + \tau_u \dot{\varepsilon}_{ij}) = \sigma_{ij} + \tau \dot{\sigma}_{ij} \tag{16.6}$$

where

$$\tau_u = M_u \tau / M_r = \tau_r (1 + \lambda_\nu \gamma_\nu / M_r).$$

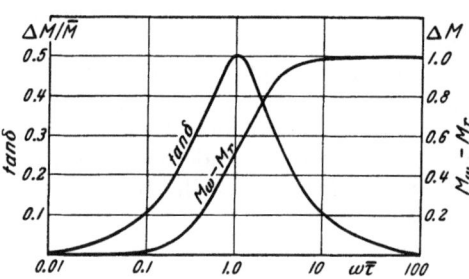

Fig. 1. Frequency dependence of dynamic modulus and damping loss angle for single relaxation process (standard solid)[2].

Eq. (16.6) defines the linear anelastic or "standard" solid into which the elastic solid is transformed whenever the relaxation of a non-mechanical generalized force X_ν is governed by Eq. (16.3). The constants τ_u and τ represent the relaxation times of strain at constant stress or of stress at constant strain respectively. With $\tau = \tau_u$ Eq. (16.6) is transformed into that of the linear elastic solid.

Eq. (16.4) includes only one elastic constant; the components σ_{ij} and ε_{ij} may therefore be considered to represent either the deviatoric stress and strain-components or the volumetric components alone. In the first case $\sigma_{ij} = s_{ij}$, $\varepsilon_{ij} = e_{ij}$, $M_r = 2G$ and X_ν is a tensor of second rank producing shear anelasticity; in the second case for which $i = j$, $M_r = K$ and X_ν is a scalar quantity, the equations are formally those of pressure relaxation.

The equations of homogeneous anelasticity permit a simple illustration of the response of anelastic media to periodic forces or deformations producing stresses $\sigma_{ij} = \sigma_{ij}^0 \exp(i\omega t)$ and strains $\varepsilon_{ij} = \varepsilon_{ij}^0 \exp(i\omega t)$. According to Eq. (16.6) the stress-strain relation is

$$\sigma_{ij}^0 = M_r \frac{1 + i\omega \tau_u}{1 + i\omega \tau} \varepsilon_{ij}^0 = M^* \varepsilon_{ij}^0 \tag{16.7}$$

where M^* is a complex modulus; the ratio of its imaginary and real parts is a measure of the energy dissipation by defining the angle δ by which the strain lags behind the stress[2]

$$\tan \delta = \omega \frac{\tau_u - \tau}{1 + \omega^2 \tau_u \tau} = \frac{\Delta \omega \tau}{1 + \omega^2 \tau^2 (1 + \Delta)} = Q^{-1} \tag{16.8}$$

where $\Delta = (\tau_u - \tau)/\tau$. Introducing the geometric means

$$\bar{\tau} = (\tau_u \tau)^{\frac{1}{2}} \quad \text{and} \quad \bar{M} = (M_u \cdot M_r)^{\frac{1}{2}}, \tag{16.9}$$

Eq. (16.8) is transformed into

$$\tan \delta = \frac{M_u - M_r}{\bar{M}} \frac{\omega \bar{\tau}}{1 + \omega^2 \bar{\tau}^2} = \frac{\Delta M}{\bar{M}} \frac{\omega \bar{\tau}}{1 + \omega^2 \bar{\tau}^2}. \tag{16.10}$$

---

[1] C. ZENER: Elasticity and Anelasticity, p. 70. Chicago: University Press 1948.
[2] Ibid. 44.

The second factor in Eq. (16.10), which gives the frequency variation of tan δ, has a maximum of 0.5 $\Delta M/\overline{M}$ for $\omega\bar{\tau}=1$; hence $(\tan\delta)_{max}=(M_u-M_r)/2\sqrt{M_u M_r}$. The frequency dependence of the dynamic modulus M_ω defined as the ratio of σ_{ij} to that part of ε_{ij} which is in phase with σ_{ij} is

$$M_\omega = M_r + \Delta M\,(\omega\bar{\tau})^2/(1+\omega^2\bar{\tau}^2). \tag{16.11}$$

The variation of (tan δ) and of M_ω with ($\omega\bar{\tau}$) is shown in Fig. 1. For the Maxwell body tan $\delta=(\omega\tau)^{-1}$, for the Kelvin body tan $\delta=\omega\tau$.

III. Visco-elastic relations.

17. Simple linear relations. When the relaxation processes in the continuum are due to the interaction between elastic (solid) and viscous (fluid) "phases" of the quasi-isotropic medium, the resulting stress-strain relations are referred to as "visco-elastic". Depending on whether in such interaction the over-all deformation is governed by the solid or by the fluid "phase", simple visco-elastic media are classified as *"retarded elastic"* and as *"relaxing"*. The former is obtained by superposition of the stress deviations, and its behavior in pure shear can be described by a mechanical model of a spring and of a linear viscous dashpot coupled *in parallel*, the latter is obtained by superposition of the strain deviations, and its behavior in pure shear can be described by a spring and a linear viscous dashpot coupled *in series*[1].

It is expedient to assume that the viscous flow responsible for both retarded elasticity and relaxation is incompressible; the visco-elastic relations therefore involve only the deviators of stress, strain and their time derivatives. Density changes in the absence of non-mechanical effects are assumed to be elastic, volume viscosity being attributed to anelastic effects alone.

The simplest linear retarded elastic medium, known as the Voigt[2] or Kelvin[3] body is essentially an elastic solid with linear velocity-dependent damping of shear-vibrations. It has been introduced by Meyer[4] and extensively studied by Duhem[5], Thompson[6] and others[7]. It is defined by the equations

$$\left.\begin{array}{l} s_{ij} = 2G e_{ij} + 2\eta_k \dot{e}_{ij} = 2G(e_{ij}+\tau_k \dot{e}_{ij}), \\ \sigma_{ii} = 3K(\varepsilon_{ii}-3\alpha T) \end{array}\right\} \tag{17.1}$$

or, combined

$$\sigma_{ij} = 2G\varepsilon_{ij} + 2G\tau_k\dot{\varepsilon}_{ij} + \delta_{ij}[(K-\tfrac{2}{3}G)\varepsilon_{kk}-\tfrac{2}{3}G\tau_k\dot{\varepsilon}_{kk}-3K\alpha T] \tag{17.2}$$

where η_k is the coefficient of shear viscosity and $\tau_k=\eta_k/G$ denotes the time of strain-relaxation or *retardation time*.

The simplest linear relaxing medium known as the Maxwell body[8], is an elastically compressible viscous fluid with linear velocity-dependent elastic resistance to shear. It has been introduced by Boltzmann on the basis of an extension

---
[1] J. M. Burgers: First Report on Viscosity and Plasticity, Academy of Sciences 5—67. Amsterdam: North-Holland Publ. Co. 1935.
[2] W. Voigt: Göttinger Abh. **36**, No. 1 (1889). — Ann. d. Physik (2) **47**, 671 (1892).
[3] Lord Kelvin: Math. Phys. Papers **3**, 27 (1875).
[4] O. E. Meyer: J. reine angew. Math. **78**, 130 (1874). — Ann. d. Physik (6) **1**, 108 (1874).
[5] P. Duhem: C. R. Acad. Sci., Paris **136**, 281, 343, 592, 733, 858, 1032, 1379 (1903).
[6] J. H. C. Thompson: Phil. Trans. Roy. Soc. Lond., Ser. A **231**, 339 (1933).
[7] M. J. O. Strutt: Ann. d. Physik (4) **87**, 153 (1928). — H. J. Jeffreys: The Earth. Cambridge 1929. — B. Gutenberg u. H. Schlechtweg: Phys. Z. **37**, 745 (1930). — A. N. Gerasimov: Prikl. Math. Mech. (2) **2**, 379, 467 (1938). — C. Torre: Z. angew. Math. Mech. **33**, 300 (1953).
[8] J. C. Maxwell: Phil. Mag. (4) **35**, 133 (1868). — Sci. Papers **2**, 623, Cambridge 1890.

of Maxwell's relaxation hypothesis and generalized by Natanson[1], Zaremba[2] and others[3]. It is defined by the equations

$$2G\dot{e}_{ij} = \dot{s}_{ij} + \tau_M^{-1} s_{ij},$$
$$3K(\dot{\varepsilon}_{ii} - 3\alpha\dot{T}) = \dot{\sigma}_{ii}$$
(17.3)

and combined

$$2G\dot{\varepsilon}_{ij} = \dot{\sigma}_{ij} + \tau_M^{-1}\sigma_{ij} - \frac{1}{3K}\delta_{ij}\left[(3K-2G)\frac{1}{3}\dot{\sigma}_{kk} + 3K\tau_M^{-1}\frac{1}{3}\sigma_{kk}\right] + 2G\alpha\dot{T}\delta_{ij} \quad (17.4\text{a})$$

or

$$\dot{\sigma}_{ij} + \tau_M^{-1}\sigma_{ij} = 2G\dot{\varepsilon}_{ij} + \delta_{ij}[K\varepsilon_{kk}\tau_M^{-1} + (K-\tfrac{2}{3}G)\dot{\varepsilon}_{kk} - 3K(\dot{T} + \tau_M^{-1}T)] \quad (17.4\text{b})$$

where η_M is the coefficient of shear viscosity and $\tau_M = \eta_M/G$ is the *stress-relaxation time*. Eqs. (17.4) are valid only as long as the approximation $d\sigma_{ij}/dt \approx \dot{\sigma}_{ij}$ is justified (see Sect. 10).

Eqs. (17.1) and (17.3) are linear approximations of the general isotropic tensor matrix relations defined by Eqs. (13.3) and (13.4), with parameters that are true physical constants. They admit therefore a quadratic elastic potential and a quadratic viscous dissipation function as specified by Eqs. (8.7) and (8.11).

The characteristic behavior of the simple linear visco-elastic bodies can be illustrated by integrating their stress-strain relations under the assumption of incompressibility: $K \to \infty$, $\varepsilon_{kk} \to 0$ and $K\varepsilon_{kk} \to p$, as well as $\alpha \to 0$ for various conditions. Thus for instance for constant strain rate $\dot{\varepsilon}_{ij} = \dot{e}_{ij} = c_{ij}$

$$s_{ij}(\text{VOIGT}) = 2G\varepsilon_{ij} + 2G\tau_k c_{ij}$$

and

$$s_{ij}(\text{MAXWELL}) = s_{ij}^0 e^{-\varepsilon_{ij}/c_{ij}\tau_M} + 2G\tau_M c_{ij}(1 - e^{-\varepsilon_{ij}/c_{ij}\tau_M}).$$

For $c_{ij} \to 0$ the Voigt body tends towards the elastic medium, the Maxwell body relaxes towards a stress-free condition at the decreasing rate $s_{ij}^0 \tau_M^{-1} e^{-t/\tau_M}$ while for $c_{ij} \to \infty$ the Voigt body becomes rigid, the Maxwell body elastic. A sustained stress σ_{ij} produces in the Voigt body a retarded strain that asymptotically approaches the elastic strain at the decreasing rate $\dot{\varepsilon}_{ij} = (s_{ij}/2G)\tau_k^{-1} e^{-t/\tau_k}$ while it produces in the Maxwell body *creep* at the constant rate $\dot{e}_{ij} = s_{ij}/2\eta_M$. If the stress is released the strain in the Voigt body relaxes at the same decreasing rate at which it approaches the elastic strain under load, while in the Maxwell body only the elastic strain is immediately recovered.

18. Linear operators. General linear visco-elastic media show various combinations of retarded elasticity and relaxation and can be constructed by suitably combining the responses of simple Kelvin and Maxwell bodies. The stress-strain relations for combined incompressible media in the form of differential equations containing s_{ij}, ε_{ij} and their time derivatives are obtained by solving the equations of superposition of strains or of stresses $s_{ij}^\nu = s_{ij}$ and $\sum_\nu \dot{\varepsilon}_{ij}^\nu = \dot{\varepsilon}_{ij}$ or $\dot{\varepsilon}_{ij}^\nu = \dot{\varepsilon}_{ij}$ and $\sum_\nu s_{ij}^\nu = s_{ij}$, where ν is the number of component simple responses, and $\dot{\varepsilon}_{ij}^\nu$ and s_{ij}^ν the individual component strain-rates and deviatoric stresses. The larger their number, the higher the order of the time-derivatives in the resulting relations between the resultant stresses and strains s_{ij} and ε_{ij}[4].

---

[1] L. Natanson: Z. phys. chem. **38**, 690 (1901). — Bull. Acad. Sci. Cracovie **1901**, 95, 161; **1902**, 488, 494; **1903**, 268, 283.

[2] S. Zaremba: See footnote 2, p. 262.

[3] C. Torre: Öst. Ing.-Arch. **8**, 55 (1954). — Kolloid-Z. **138**, 11 (1954). — R. S. Rivlin: J. Rat. Mech. Analysis **4**, 681 (1955); **5**, 179 (1956).

[4] T. Alfrey: Quart. Appl. Math. **3**, 143 (1945). — T. Alfrey, and P. Doty: J. Appl. Phys. **16**, 700 (1945).

The "standard solid" referred to in Sect. 16, the deviator equation of which is of the form

$$2G\left(1+\tau_u \frac{\partial}{\partial t}\right)e_{ij} = \left(1+\tau \frac{\partial}{\partial t}\right)s_{ij} \tag{18.1}$$

is made up of an elastic response with modulus G and a Maxwell response with modulus G_M and viscosity coefficient η_M coupled in parallel; the relaxation times $\tau = \eta_M/G_M$ and $\tau_u = \tau(1+G_M/G) > \tau$. Since the total deformation of the standard solid is governed by the elastic matrix, it is a retarded elastic medium. A deviator equation of similar form

$$2\eta\left(1+\tau \frac{\partial}{\partial t}\right)\dot{e}_{ij} = \left(1+\tau_r \frac{\partial}{\partial t}\right)s_{ij} \tag{18.2}$$

is made up of a Maxwell body with modulus G_M and viscosity coefficient η_M coupled in parallel with a viscous response of viscosity coefficient η_V; the relaxation times are $\tau_r = \eta_M/G_M$ and $\tau = \tau_r \eta_V/2\eta < \tau_r$, where $2\eta = \eta_M + \eta_V$. The medium is essentially a fluid[1] and degenerates into the viscous fluid as $\tau_r \to \tau$ for $\eta_M = \eta_V$.

The simplest linear visco-elastic medium showing both retarded elasticity and either stress-relaxation or creep has a deviator equation of the form[2]

$$2G\left(b_1 \frac{\partial}{\partial t} + b_2 \frac{\partial^2}{\partial t^2}\right)e_{ij} = \left(1 + a_1 \frac{\partial}{\partial t} + a_2 \frac{\partial^2}{\partial t^2}\right)s_{ij} \tag{18.3}$$

and can be made up either of a Maxwell and a Kelvin body coupled in series with $G = G_M$, $a_1 = \tau_M[1+(G_M/G_K)] + \tau_K$, $a_2 = \tau_M \tau_K$, $b_1 = \tau_M$ and $b_2 = \tau_M \tau_K G_K/G_M$, or of two Maxwell elements coupled in parallel, with $G = G_1 + G_2$, $a_1 = \tau_1 + \tau_2$, $a_2 = \tau_1 \tau_2$, $b_1 = \tau_1$ and $b_2 = a_2 = \tau_1 \tau_2$.

The deviator equation of a general linear visco-elastic medium can therefore be written in the form[3]

$$2G\boldsymbol{Q} e_{ij} = \boldsymbol{P} s_{ij} \tag{18.4}$$

where $\boldsymbol{P}$ and $\boldsymbol{Q}$ are linear operators

$$\boldsymbol{P} = a_0 + a_1 \frac{\partial}{\partial t} + \cdots + a_m \frac{\partial^m}{\partial t^m} \quad \text{and} \quad \boldsymbol{Q} = b_0 + b_1 \frac{\partial}{\partial t} + \cdots + b_n \frac{\partial^n}{\partial t^n},$$

with constants a_i and b_i that are combinations of relaxation times and of shear moduli.

Combining Eq. (18.4) with the second Eq. (17.1) the stress-strain-temperature relations of the elastically compressible general linear visco-elastic medium are obtained in the alternative forms

$$\begin{aligned}\boldsymbol{P}\sigma_{ij} &= 2G\boldsymbol{Q}\varepsilon_{ij} + (3K\boldsymbol{P}-2G\boldsymbol{Q})\tfrac{1}{3}\delta_{ij}\varepsilon_{kk} - 3\alpha K\boldsymbol{P}\delta_{ij}T \\ &= 2G\boldsymbol{Q}\varepsilon_{ij} + \lambda\delta_{ij}\varepsilon_{kk} - 3\alpha K\boldsymbol{P}\delta_{ij}T \end{aligned} \tag{18.5a}$$

and

$$\begin{aligned}2G\boldsymbol{Q}\varepsilon_{ij} &= \tfrac{1}{9K}(2G\boldsymbol{Q}-3K\boldsymbol{P})\delta_{ij}\sigma_{kk} + \boldsymbol{P}\sigma_{ij} + 2G\alpha\boldsymbol{Q}\delta_{ij}T \\ &= -\tfrac{\lambda}{3K}\delta_{ij}\sigma_{kk} + \boldsymbol{P}\sigma_{ij} + 2G\alpha\boldsymbol{Q}\delta_{ij}T \end{aligned} \tag{18.5b}$$

where a new operator λ is defined by $\lambda = \tfrac{1}{3}(3K\boldsymbol{P}-2G\boldsymbol{Q})$. Eq. (17.2) is a special case of Eq. (18.5a) with $\boldsymbol{P}=1$ and $\boldsymbol{Q}=(1+\tau_K \partial/\partial t)$, while Eq. (17.4) is a special case of Eq. (18.5b) with $\boldsymbol{P}=(\tau_M^{-1}+\partial/\partial t)$ and $\boldsymbol{Q}=\partial/\partial t$.

---
[1] J. G. OLDROYD: Proc. Roy. Soc. Lond., Ser. A 200, 523 (1950).
[2] R. D. MINDLIN: J. Appl. Phys. 20, 206 (1949).
[3] T. ALFREY: Quart. Appl. Math. 2, 113 (1944).

In order to obtain the visco-elastic stress-strain relations for one- or two-dimensional states of stress and strain the Eqs. (18.5) must actually be solved for the conditions considered. The operator forms for such particular conditions therefore differ from that of the general relations (18.5). Only for the completely incompressible medium are the operators for the one- and two-dimensional relations identical with those of the three-dimensional relation. Thus, for instance for plane stress ($\sigma_{33}=0$) Eq. (18.5a) is transformed into

$$P(3KP+2GQ)\sigma_{11} = 2GQ[(6KP+2GQ)\varepsilon_{11} + (3KP-2GQ)\varepsilon_{22} - 9\alpha KPT] \quad (18.6)$$

with a similar equation for σ_{22}; for the mixed components ($i \neq j$), however, the relation remains always $P\sigma_{12} = 2GQ\varepsilon_{12}$. For uniaxial stress ($\sigma_{11} \neq 0$, all other components vanish)

$$(6KP+2GQ)\sigma_{11} = 18KGQ(\varepsilon_{11} - \alpha T). \quad (18.7)$$

For plane strain, on the other hand, Eq. (18.5b) is transformed into

$$4GQ(3KP+GQ)\varepsilon_{11} = P(3KP+4GQ)\sigma_{11} - P(3KP-2GQ)\sigma_{22} + 2GQ(3KP+4GQ)\alpha T \quad (18.8)$$

and a similar equation for ε_{22}; again, for the mixed components, $P\sigma_{12} = 2GQ\varepsilon_{12}$.

19. Superposition theories. Considering Eqs. (18.5) as differential equations for σ_{ij} and ε_{ij} respectively, they can be solved by the use of Laplace transforms under the simplifying assumption of zero initial conditions. Introducing the operator transforms[1]

$$\bar{P}(p) = \sum_m a_k p^k, \quad \bar{Q}(p) = \sum_n b_k p^k, \quad \bar{\lambda}(p) = \left(K\bar{P} - \frac{2G\bar{Q}}{3}\right) \quad (19.1)$$

and the ratios

$$\bar{L}(p) = \frac{\bar{\lambda}(p)}{\bar{P}(p)}, \quad \bar{M}(p) = \frac{\bar{Q}(p)}{\bar{P}(p)}, \quad \text{and} \quad \bar{L}'(p) = \frac{\bar{\lambda}(p)}{\bar{Q}(p)}, \quad \bar{M}'(p) = \frac{\bar{P}(p)}{\bar{Q}(p)} \quad (19.2)$$

the transforms of Eqs. (18.5) are

$$\bar{\sigma}_{ij} = 2G\bar{M}(p)\bar{\varepsilon}_{ij} + \bar{L}(p)\delta_{ij}\bar{\varepsilon}_{kk} - 3\alpha K\delta_{ij}\bar{T} \quad (19.3\,\text{a})$$

and

$$2G\bar{\varepsilon}_{ij} = \bar{M}'(p)\bar{\sigma}_{ij} - \frac{1}{3K}\bar{L}'(p)\delta_{ij}\bar{\sigma}_{KK} + 2G\alpha\delta_{ij}\bar{T}. \quad (19.3\,\text{b})$$

The inverse transforms of Eqs. (19.3) are the solutions of Eqs. (18.5). They are obtained with the aid of the convolution theorem in the form

$$\sigma_{ij}(t) = 2G\int_0^t M(t-\vartheta)\varepsilon_{ij}(\vartheta)\,d\vartheta + \delta_{ij}\int_0^t L(t-\vartheta)\varepsilon_{kk}(\vartheta)\,d\vartheta - 3\alpha K\delta_{ij}T(t) \quad (19.4\,\text{a})$$

and

$$2G\varepsilon_{ij}(t) = \int_0^t M'(t-\vartheta)\sigma_{ij}(\vartheta)\,d\vartheta - \frac{1}{3K}\delta_{ij}\int_0^t L'(t-\vartheta) \times \\ \times \sigma_{kk}(\vartheta)\,d\vartheta + 2G\alpha\delta_{ij}T(t). \quad (19.4\,\text{b})$$

The stress or strain at time t due to transient effects at times ϑ is thus expressed as an integral over the momentary increments of strain $\varepsilon_{ij}(\vartheta) = \dot{\varepsilon}_{ij}$ and stress

---

[1] A. C. ERINGEN: J. Appl. Mech. **22**, 563 (1955).

$\sigma_{ij}(\vartheta) = \dot{\sigma}_{ij}$ applied during the interval between ϑ and $(\vartheta + d\vartheta)$ of past time $0 < \vartheta < t$. The functions $M(t-\vartheta)$ and $L(t-\vartheta)$ represent the response to a unit strain-impulse in terms of the stress relaxation rate, the functions $M'(t-\vartheta)$ and $L'(t-\vartheta)$ that to a unit stress-impulse in terms of the creep rate. These functions therefore represent "*memory-functions*" of stress *(relaxation-functions)* and of strain *(creep-functions)* respectively. Eqs. (19.4) thus state the principle of linear superposition of stress and strain residuals with the aid of the "memory functions" L, M and L', M' of elapsed time $(t-\vartheta)$. Such functions have been directly postulated as definitions of specific materials with "memory", or "heredity"[1]. Eqs. (19.1) to (19.4), however, indicate that they can be considered simply as alternative representations of the linear operators Eqs. (18.5)[2].

According to Eqs. (19.1) and (19.2) the memory-functions are obtained as inverse transforms of quotients of two polynomials. Since in the pair of quotients in one of the Eqs. (19.3) the denominator will be of higher degree than the numerator this pair of inverse transforms can be expressed in the form of series of negative exponentials $\sum_{k} C(\alpha_k) \exp[-\alpha_k(t-\vartheta)]$ where the number of terms is equal to the degree of the denominator. In this representation the memory-functions appear as superpositions of exponential decay (relaxation) functions, indicating the equivalence of the concept of memory-functions with that of multiple relaxation times.

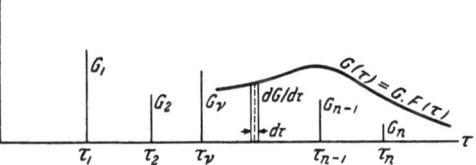

Fig. 2. Discrete and continuous spectrum of relaxation times.

20. Relaxation spectra. When the number of linear elements forming the total visco-elastic response of an incompressible medium is no longer small, it is expedient to represent this response by n Maxwell "elements" coupled in parallel. Hence for the ν-th element

$$2G_\nu \dot{e}_{ij}^\nu = \dot{s}_{ij}^\nu + \tau_\nu^{-1} s_{ij}^\nu \quad \text{and} \quad s_{ij} = \sum_\nu s_{ij}^\nu \quad \text{while} \quad \dot{e}_{ij} = \dot{e}_{ij}^\nu. \tag{20.1}$$

Plotting the values of G_ν as ordinates at the respective locations τ_ν on a horizontal τ-axis, a discrete spectrum of relaxation times $G_\nu[\tau_\nu]$ is obtained (Fig. 2). Since $\sum_n G_\nu = G$ the normalized spectrum represents the spectrum (distribution function) of relaxation times existing within the medium $F_\nu[\tau_\nu] = G_\nu[\tau_\nu]/G$.

With $\dot{e}_{ij}^\nu = \dot{e}_{ij}$ (superposition of stresses) Eq. (20.1) can be written in the form

$$2G \mathbf{Q} e_{ij} = 2G \dot{e}_{ij} = \sum_\nu (\dot{s}_{ij}^\nu + \tau_\nu^{-1} s_{ij}^\nu) = \mathbf{P} s_{ij} \tag{20.2}$$

where $\mathbf{P}^{-1} = \sum_\nu F_\nu/(\partial/\partial t + \tau_\nu^{-1})$.

Applying the Laplace transform to Eq. (20.2) and introducing $\overline{M}(p) = \overline{Q}(p)/\overline{P}(p)$

$$\bar{s}_{ij} = 2G \bar{e}_{ij} \overline{M}(p) = 2G \bar{e}_{ij} \sum_\nu \left[\frac{p F_\nu}{(p + \tau_\nu^{-1})} \right], \tag{20.3}$$

an equation is obtained which can be solved for s_{ij} if e_{ij} is given, or vice versa, by using the convolution theorem. The relaxation-function obtained for $e_{ij} = \text{const.}$

---

[1] L. BOLTZMANN: Ann. Phys. **7**, 624 (1876). — Ges. Abh. **1**, 616 (1909). — V. VOLTERRA: Drei Vorlesungen, p. 155. Leipzig und Berlin 1914.

[2] R. v. MISES: Proc. 3rd Internat. Congr. Appl. Mech., Stockholm, p. 7, 1930. — B. GROSS: Math. Struct. of the Theories of Visco-Elasticity. Paris: Hermann 1953.

and the creep-function, obtained for $s_{ij} = $ const. are the inverse transforms of $\overline{M}(p)/p$ and $1/\overline{M}(p)\,p$ respectively, which, being the inverse transforms of quotients of two polynomials can again be expressed in the form of series of negative exponentials.

For an infinite number of relaxation times the discrete spectrum of shear moduli $G_\nu[\tau_\nu]$ or of relaxation times $F_\nu[\tau_\nu]$ is replaced by a continuous distribution function defined by

$$G_\nu[\tau_\nu] = \frac{dG}{d\tau}\,d\tau = G F(\tau)\,d\tau \qquad (20.4)$$

subject to the normalizing conditions

$$\int_0^\infty F(\tau)\,d\tau = 1 \quad \text{or} \quad \int_0^\infty G(\tau)\,d\tau = G. \qquad (20.5)$$

The contribution to the total stress deviation s_{ij} of elements having relaxation times between τ and $(\tau + d\tau)$, according to Eq. (20.1),

$$2\frac{dG}{d\tau}\dot{e}_{ij}\,d\tau = \frac{\partial^2 s_{ij}(\tau,t)}{\partial t\,\partial\tau}\,d\tau + \tau^{-1}\frac{\partial s_{ij}(\tau,t)}{\partial\tau}\,d\tau \qquad (20.6)$$

where, for unlimited distributions

$$s_{ij} = \int_0^\infty \frac{\partial s_{ij}(\tau,t)}{\partial\tau}\,d\tau. \qquad (20.7)$$

Applying the Laplace transform to Eq. (20.6) with the initial conditions $e_{ij}(0) = 0$ and $\partial s_{ij}(\tau, 0)/\partial\tau = 0$ (no internal stresses), and integrating over the relaxation spectrum, the equation is obtained

$$2G\int_0^\infty p\bar{e}_{ij}(p)\frac{1}{p+\tau^{-1}}\frac{dF}{d\tau}\,d\tau = \int_0^\infty \frac{\partial \bar{s}_{ij}(\tau,p)}{\partial\tau}\,d\tau = \bar{s}_{ij}(p) \qquad (20.8)$$

which, for a specified function $F(\tau)$, can be solved directly for $\bar{s}_{ij}$ if the strain-sequence is given. Considering that $\bar{e}_{ij}(p)$ is not a function of τ it can be written in the form

$$\bar{e}_{ij}(p) = \frac{1}{2G\int_0^\infty \frac{1}{p+\tau^{-1}}\frac{dF}{d\tau}\,d\tau}\frac{\bar{s}_{ij}(p)}{p} \qquad (20.9)$$

in which it can be solved for $\bar{e}_{ij}$ if the stress sequence is given.

The *relaxation-function* $\psi(t)$ for a continuous spectrum of relaxation times $F(\tau)$ is obtained by inverse transform of Eq. (20.8) with $e_{ij}(t) = e_{ij}^0(0) = $ const. and therefore $\bar{e}_{ij}(p) = e_{ij}^0/p$

$$s_{ij}(t) = 2e_{ij}^0\,G\int_0^\infty F(\tau)\,e^{-t/\tau}\,d\tau = 2e_{ij}^0\,G\,\psi(t) \qquad (20.10\text{a})$$

where $\psi(t)$ represents the decay, with time, of the stress deviation s_{ij} from its initial value $s_{ij}^0 = 2Ge_{ij}^0$ induced by the strain e_{ij}^0 applied instantaneously at time $t = 0$. For various assumptions of the form of $F(\tau)$ various decay functions $\psi(t)$ are obtained.

Because of the wide range of the distribution of relaxation times expected in real materials it has been found expedient to use a logarithmic time-scale and to define the distribution $F^*(\log \tau)$ by

$$F^*(\log \tau) = \tau F(\tau). \qquad (20.11)$$

Eq. (20.10) is therefore transformed into

$$s_{ij} = 2e_{ij}^0 \, G \int_{-\infty}^{+\infty} F^*(\log \tau) \, e^{-t/\tau} \, d(\log \tau). \quad (20.10\,\text{b})$$

Solutions of Eqs. (20.8) and (20.9) for arbitrary strain- or stress-sequences can be obtained if the integrals in these equations can be inverted, which depends on the assumed form of $F(\tau)$. The simplest form that has been extensively used in high-polymer research[1] is the limited rectangular distribution within the range between τ_1 and τ_2 with the equation $F^*(\log \tau) = \tau F(\tau) = 1/\log(\tau_2/\tau_1)$. Introducing this function into Eq. (20.8) the equation is obtained:

$$\bar{s}_{ij} = \frac{2G}{\log(\tau_2/\tau_1)} \bar{e}_{ij} \log \frac{p\tau_2 + 1}{p\tau_1 + 1} \quad (20.12)$$

the inverse transforms of which, for simple strain sequences such as relaxation or constant strain-rate, can be expressed in terms of tabulated exponential integrals. For zero strain rate $e_{ij} = e_{ij}^0 = \text{const.}$

$$s_{ij}(t) = 2G \, e_{ii}^0 \, [\text{Ei}(-t/\tau_1) - \text{Ei}(-t/\tau_2)]/\log(\tau_2/\tau_1) = s_{ij}(0)\,\psi(t) \quad (20.13)$$

while for constant strain rate

$$s_{ij}(\dot{e}_{ij}) = 2G \, \dot{e}_{ij} \{[(1 - e^{-t/\tau_2})/(t/\tau_2)] - [(1 - e^{-t/\tau_1})/(t/\tau_1)] + \\ + [\text{Ei}(-t/\tau_1) - \text{Ei}(-t/\tau_2)]\}/\log(\tau_2/\tau_1) = 2G(t)\,\dot{e}_{ij}. \quad (20.14)$$

Assuming for $\log \tau$ a normal (Gaussian) distribution, WIECHERT and WAGNER[2] have derived the associated relaxation-functions.

The inverse transform of Eq. (20.9) is, however, not a standard form or a tabulated function even for the simplest stress sequence and a limited rectangular distribution. For certain stress sequences applied in tests it has been obtained by integration in the complex plane[3].

Eqs. (20.10) can also be used for the determination of the characteristic relaxation spectrum of a material from an observed relaxation-function $\psi(t)$. For this purpose $\psi(t)$ in Eq. (20.10a) is transformed by the substitution of a new variable $v = 1/\tau$, the relaxation frequency, into

$$\psi(t) = \int_0^\infty N(v) \, e^{-tv} \, dv \quad (20.15)$$

where

$$N(v) = F(1/v)/v^2. \quad (20.16)$$

Eqs. (20.15) is the Laplace transform of the function $N(v)$ which, for an analytically specified relaxation-function $\psi(t)$, can be obtained by inversion of the transform; the relaxation spectrum is then obtained from Eq. (20.16)[4].

Physically, Eq. (20.15) represents the response function to a unit force of an infinite number of damped uncoupled linear oscillators, $N(v)\,dv$ representing the

---

[1] R. BECKER: Z. Phys. **33**, 192 (1925). — R. SIMHA: J. Appl. Phys. **13**, 201 (1952). — W. KUHN, O. KUNZLE and H. PREISSMANN: Helv. chim. Acta **30**, 307, 464 (1947). — R. SIPS: J. Polymer Sci. **5**, 69 (1950). — A. V. TOBOLOKY, B. A. DUNELL and R. D. ANDREWS: J. Text. Res. **7**, 221 (1951).
[2] E. WIECHERT: Ann. Phys. **50**, 335, 546 (1893). — K. W. WAGNER: Ann. Phys. (4) **40** 817 (1913).
[3] A. M FREUDENTHAL and B. ALBRECHT: Trans. Soc. Rheology **2** (1958).
[4] B. GROSS: J. Appl. Phys. **18**, 212 (1947).

response of the oscillator with frequency v. It should be borne in mind, however, that this fact does not imply the physical existence of such a system of oscillators within the medium, since the function $\psi(t)$ can be represented by different, mathematically equivalent integrals.

The relaxation spectrum $F(\tau)$ of the linear visco-elastic medium as inversion of Eq. (20.15) can be obtained either with the aid of tables of Laplace transforms or by Bromwich integrals, provided $\psi(t)$ is or can be closely represented by an analytic function. However, in order to deal directly with experimentally observed relaxation-functions, approximations have been proposed[1], most of which are based on the representation of the relaxation-spectrum in terms of the derivatives of the relaxation-function with the aid of WIDDER's equation[2]. Spectra can be calculated from experimental relaxation-functions with increasing accuracy by using higher approximations[3]. In first approximation

$$F_1^*(\log t) \approx - d\psi(t)/d(\log t) = - t\, d\psi(t)/dt;$$

the spectrum is represented by the (negative) slope of the relaxation-function in logarithmic scale. The n-th approximation involves all logarithmic derivatives of the experimental function up to the order n. Thus, for instance, for $n = 2$

$$F_2^*(\log t/2) = - [d\psi/d\log t - d^2\psi/d(\log t)^2] = t^2 d^2\psi/dt^2.$$

The precision of the methods of approximation depends on the form of the relaxation-function; for a straight-line relaxation-function in logarithmic time-scale, for instance, implying a uniform spectrum, the first derivative is no longer an approximation but a rigorous representation. The precision is also, in general, limited by the fact that whereas Eq. (20.15) establishes the dependence of $\psi(t)$ at t on *all* values of the function $N(v)$, although the factor e^{-tv} reduces the contribution of distant ranges of t, the approximations imply *unique* relations between points of the relaxation-function and points of the associated spectrum. Moreover, the use of higher derivatives of $\psi(t)$ requires a sharply increasing accuracy of the observed function.

Alternatively, the visco-elastic response of specific linear materials can be derived from dynamic investigations establishing stationary conditions of forced vibrations at various frequencies ω and observing the variation with ω of the real and of the imaginary parts of the complex mechanical impedance $\mathfrak{G}(i\omega) = G'(\omega) + i G''(\omega)$ of the system. The solution of Eq. (20.8) with the imposed strain-sequence $e_{ij} = e_{ij}^0 \exp(i\omega t)$ furnishes the necessary relations between the observed functions and the unknown relaxation-spectrum of the material for the determination of the latter[4]

$$s_{ij} = 2\,\mathfrak{G}(i\omega)\, e_{ij} = 2[G'(\omega) + i\, G''(\omega)]\, e_{ij} \qquad (20.17)$$

where

$$\mathfrak{G}(i\omega) = i\omega \int_0^\infty e^{-i\omega\tau} \psi(\tau)\, d\tau \qquad (20.18)$$

---

[1] T. ALFREY and P. DOTY: J. Appl. Phys. **16**, 700 (1945). — D. TER HAAR: Physica, Haag **16**, 719, 738, 839 (1950).
[2] D. V. WIDDER: Trans. Amer. Math. Soc. **36**, 107 (1934).
[3] F. SCHWARZL: Physica, Haag **17**, 830, 923 (1951). — F. SCHWARZL and A. J. STAVERMAN: Physica, Haag **18**, 791 (1952). See also Chap. 1 in H. A. STUART, Physik der Hochpolymeren, vol. 4. Berlin: Springer 1956.
[4] B. GROSS: J. Appl. Phys. **19**, 257 (1948). — See also footnote 2, p. 273.

Sect. 20. Relaxation spectra.

or, separating into real and imaginary components

$$G'(\omega) = \omega \int_0^\infty \psi(\tau) \sin \omega \tau \, d\tau,$$
$$G''(\omega) = \omega \int_0^\infty \psi(\tau) \cos \omega \tau \, d\tau. \tag{20.19}$$

The inversion of these integrals, which are Fourier transforms of $\psi(\tau)$, furnishes the relation between the relaxation-function and the dynamic characteristics

$$\psi(t) = \frac{2}{\pi} \int_0^\infty \frac{G'(\omega)}{\omega} \sin(\omega t) \, d\omega = \frac{2}{\pi} \int_0^\infty \frac{G''(\omega)}{\omega} \cos(\omega t) \, d\omega \tag{20.20}$$

from which the spectrum $F(\tau)$ can, in turn, be derived using the inversion of Eq. (20.15).

The direct relation between $\mathfrak{G}(i\omega)$ and the spectrum $F(\tau)$ is obtained by introducing $\psi(t)$ from Eq. (20.15) into Eq. (20.18), inverting the order of integration and evaluating the integral over the exponential function. The resulting equation

$$\mathfrak{G}(i\omega) = i\omega \int_0^\infty \frac{N(v)}{v + i\omega} \, dv \tag{20.21}$$

is STIELTJES[1] integral. The real and imaginary parts

$$G'(\omega) = \int_0^\infty N(v) \frac{\omega^2}{\omega^2 + v^2} \, dv,$$
$$G''(\omega) = \int_0^\infty N(v) \frac{\omega v}{\omega^2 + v^2} \, dv \tag{20.22}$$

express the dynamic elastic modulus $G'(\omega)$ and the energy loss modulus $G''(\omega)$ in term of the relaxation frequency spectrum $N(v)$ or, using Eq. (20.16), in terms of the spectrum of relaxation-times:

$$G'(\omega) = \int_0^\infty F(\tau) \frac{\omega^2 \tau^2}{1 + \omega^2 \tau^2} \, d\tau = \int_{-\infty}^{+\infty} F^*(\log \tau) \frac{\omega^2 \tau^2}{1 + \omega^2 \tau^2} \, d(\log \tau),$$
$$G''(\omega) = \int_0^\infty F(\tau) \frac{\omega \tau}{1 + \omega^2 \tau^2} \, d\tau = \int_{-\infty}^{+\infty} F^*(\log \tau) \frac{\omega \tau}{1 + \omega^2 \tau^2} \, d(\log \tau). \tag{20.23}$$

Eqs. (20.23) are generalizations of Eqs. (16.11) and (16.10); the loss-angle $\tan \delta = G''(\omega)/G'(\omega)$, the "dynamic viscosity" $\eta(\omega) = G''(\omega)/\omega$.

Rigorous methods of inversion of Eqs. (20.22) have been developed[2]; they lead to the simple relations

$$N(\omega) = \pm \frac{1}{\pi \omega} \operatorname{Im} \mathfrak{G}(\omega e^{\pm i\pi}) = \pm \frac{2}{\pi \omega} \operatorname{Im} \mathfrak{G}'(\omega e^{\pm i\pi/2})$$
$$= \frac{2}{\pi \omega} \operatorname{Re} \mathfrak{G}''(\omega e^{\pm i\pi/2}) \tag{20.24}$$

---

[1] G. DOETSCH: Laplace Transformation, p. 163. Berlin: Springer 1937.
[2] R. FUOSS and J. G. KIRKWOOD: J. Amer. Chem. Soc. 63, 385 (1941).

where Im and Re designate the imaginary and the real parts of the complex functions, from which $N(\omega)$ can be easily derived if $\mathfrak{G}$, G' or G'' are given as or reproduced by analytic functions in which ω is subsequently explaced by $\omega e^{\pm i\pi}$ or $\omega e^{\pm i\pi/2}$. In order to deal directly with experimental functions, methods of approximation similar to those used for the relaxation function have also been proposed. Thus using derivatives up to the third order[1]

$$F^*(\log 1/\omega) = d\,G'(\omega)/d\,(\log \omega) - \tfrac{1}{4} d^3 G'/d\,(\log \omega)^3$$

or

$$F^*(\log 1/\omega) = \frac{2}{\pi}\,[G''(\omega) - d^2 G''/d\,(\log \omega)^2]$$

the relaxation spectrum can be derived either from observations of the dynamic modulus or of the energy loss (damping or loss-angle).

Numerical methods of inversion by solution of the convolution-type equations arising from Eq. (20.3) by successive approximation have also been developed[2].

The effective use of the methods that have been established for the study of relaxation spectra of real materials, especially polymers, is limited by the restricted time-scale of the observed functions. That relaxation-spectra of real materials, particularly polymers, are extremely wide and not uni-modal has been shown by the investigation of the response to applied load-sequences over a wide range of time-scales between a few milliseconds and several days of metals as well as of certain phenolics, by combining various testing procedures[3]. Tests within a rather narrow time-scale can therefore provide no more than the characteristic response of the material within the *same* scale; the interpretation of such tests with the aid of the methods outlined in this section will therefore provide only that segment of the relaxation-spectrum that is actually involved in the particular test. Conversely, prediction of the mechanical behavior of linear visco-elastic materials on the basis of equations embodying a discrete number of relaxation times or a limited continuous spectrum, such as the rectangular, will be valid only within the time-scale of the selected parameters and can not be extrapolated beyond their range.

IV. Plastic relations.

21. Basic assumptions. The state of the isotropic continuum is defined as perfectly plastic if all admissible states of stress fulfil an isotropic condition, the *yield condition* $f(\sigma_{ij}) = 0$, assumed to be expressible in terms of a function of the invariants of the stress tensor. Physically this definition implies that at the yield limit the rate of plastic flow associated with any stress-increment $d\sigma_{ij}$ would increase so rapidly that states of stress for which $f(\sigma_{ij} + d\sigma_{ij}) > 0$ can, in fact, not be sustained, while any stress-increment for which the stress-point remains on the *yield surface* produces no plastic strain work. The condition of perfect plasticity is therefore $df(\sigma_{ij}) = 0$ at $f(\sigma_{ij}) = 0$.

In the stress-space of the isotropic continuum, represented by a Cartesian coordinate system with the directions of the principle stresses as axes, the yield condition $f(\sigma_{ij}) = 0$ describes either a single, continuously differentiable surface enclosing the origin, or a number of such surfaces, intersecting along sharp edges and forming a polyhedron: the yield surface. States of stress defined by points inside the yield surface ($f < 0$) are "pre-plastic" and may be rigid,

---

[1] See footnote 3, p. 276.

[2] F. C. ROESLER and J. R. A. PEARSON: Proc. Phys. Soc. Lond. B **67**, 338 (1954). — F. C. ROESLER and W. A. TWYMAN: Proc. Phys. Soc. Lond. B **68**, 97 (1955).

[3] C. ZENER: See footnote 1, p. 268. See also footnote 3, p. 275.

elastic or visco-elastic; states of stress defined by points outside the yield surface cannot exist; points on the yield surface ($f=0$) represent the entity of admissible plastic states of stress. The yield condition and thus the actual form of the yield surface are selected either on the basis of considerations of mathematical expediency or are fitted to reproduce experimental results. The yield surface can obviously not be cut twice by a radius from the origin.

The assumed isotropy of the plastic continuum requires yield conditions of the form $f(J_1, J_2, J_3)=0$. In the case of a single continuous function admissible plastic states of stress are defined by $f=0$; for a yield condition consisting of several "*piece-wise continuous*" intersecting functions this condition is modified: it is sufficient if for a certain state of stress one of the ν functions f_ν vanishes, while all others assume negative values, to define this state of stress as plastic.

The experimental evidence appears to justify the assumption of incompressibility of plastic flow, at least in first approximation; the stress-invariants in the yield condition can therefore be replaced by the invariants of the stress-deviation. Under this assumption the yield condition reduces to $f(\sigma_{ij}) = f(J_2', J_3') = 0$ and therefore

$$df = \frac{\partial f}{\partial \sigma_{ij}} d\sigma_{ij}$$
$$= \frac{\partial f}{\partial J_2'} dJ_2' + \frac{\partial f}{\partial J_3'} dJ_3' = 0 \quad (21.1)$$

and

$$\frac{\partial f}{\partial J_1} = \frac{\partial f}{\partial \sigma_{ii}} = 0.$$

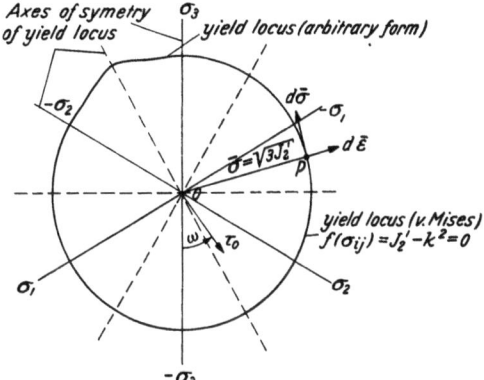

Fig. 3. Representation of yield locus in plane $J_1 = 0$ of stress-space.

Since, moreover, the *direction* of stressing is, in general, assumed to have no effect on the yield surface, the plastic states of stress σ_{ij} and $-\sigma_{ij}$ being symmetric, while J_3' changes its sign with that of the stress-components, it follows that the yield condition must be even in J_3'.

Because the function $f(s_{ij})$ is independent of J_1 it can be projected on the $J_1 = \sigma_1+\sigma_2+\sigma_3 = 0$-plane in the stress-space with the directional cosines of its normal equal to $(1/\sqrt{3}, 1/\sqrt{3}, 1/\sqrt{3})$, without loss of generality. Its curve of intersection with this plane represents the yield condition not only on this particular stress-plane of zero hydrostatic pressure but, because of the assumed incompressibility of the plastic flow, it is identical with the intersection of any parallel plane defined by an arbitrary value of J_1. In the coordinate system obtained by a projection of the axes of principal stress on the stress-plane $J_1=0$, this curve represents the general "locus" of incompressible yield (Fig. 3). Because of the assumed isotropy and symmetry of plastic states this locus is symmetrical with respect to the projected principal stress-axes, inclined to each other at 120°, as well with respect to three other axes perpendicular to the principal stress-axes. In this representation the admissible stress-increment path $d\sigma_{ij}$ lies within the yield locus, the position of the stress vector is defined by the angles of the radius to the stress-point[1].

Since a stress-increment $d\sigma_{ij}$ within the yield locus produces no strain work the increment of the rate of plastic strain work along any stress path $d\sigma_{ij}$ within the yield surface $d\sigma_{ij}\dot{\varepsilon}_{ij}''=0$, where double primes are used to distinguish the

---

[1] See footnote 5, p. 254 (p. 35).

purely *plastic* strain-components and dots are used to identify the strain-*rates* or strain-*increments*, although differentiation is not necessarily with respect to time; the components of the plastic strain rate $\dot{\varepsilon}''_{ij}$ or plastic strain-increment $d\varepsilon''_{ij}$ can be identified with the deviatoric components $\dot{e}''_{ij}$ or de''_{ij}. The direction of the vector representing the plastic strain-increment or strain rate must therefore be normal to the yield surface at the point of intersection with the stress-increment vector, provided that this direction $(\partial f/\partial \sigma_{ij})$ is uniquely defined. For yield conditions made up of several intersecting surfaces the differential along the edges of intersection is not unique. It has, however, been proposed that for states of stress within corners of the yield locus the total plastic strain-increment be obtained as the resultant of the contributions to the strain-increment of the individual surfaces of the discontinuous yield surface (see Sect. 23).

A work-hardening or strain-hardening continuum is defined by the assumption of a positive value of the increment of the rate of plastic strain-work $d\sigma_{ij}\dot{\varepsilon}''_{ij} > 0$ beyond the yield condition $f(\sigma_{ij}) = 0$. At the yield condition $\dot{\varepsilon}''_{ij} = 0$ and therefore $d\sigma_{ij}\dot{\varepsilon}''_{ij} = 0$, which here defines a state of no loading. The concept of the yield condition is therefore generalized by the introduction of a *"loading function"* which is designed to represent the entity of yield conditions associated with the consecutive stages of increasing plastic deformation[1]. This loading function is related either to the irrecoverable plastic strain work $W_p = \int \sigma_{ij}\dot{\varepsilon}''_{ij}dt$ that has been expended in reaching any particular condition *(work-hardening function)* or in terms of an invariant plastic strain $\bar{\varepsilon}'' = \int f(I'_2)dt$ associated with this condition *(strain-hardening function)*. If it is assumed that the history of plastic straining enters through a single parameter only, in the form $q = F(W_p)$ or $q = H(\bar{\varepsilon}'')$ the loading function is a generalization of the yield criterion[2]:

$$f(J'_2, J'_3) = q \qquad (21.2)$$

where f itself does not depend on the strain history. The isotropy of the plastic deformation is thus maintained by the specification of the loading function as an invariant function of stress alone; the consecutive yield surfaces are geometrically similar expansions in scale of the initial yield surface.

The stress-path which for the work-hardening continuum satisfies Eq. (21.1) is designated as *"neutral"*, since the associated plastic strain rate $\dot{\varepsilon}''_{ij} = 0$; this condition can be satisfied by specifying[3]

$$\dot{\varepsilon}''_{ij} = h_{ij}df \qquad (21.3)$$

where h_{ij} is a symmetric tensor depending on σ_{ij} and q, but not on $d\sigma_{ij}$; "loading" is associated with $df > 0$, "unloading" with $df < 0$. Since the total strain-rate $\dot{\varepsilon}_{ij} = \dot{\varepsilon}'_{ij} + \dot{\varepsilon}''_{ij}$, where $\dot{\varepsilon}'_{ij}$ denotes the elastic component, becomes $\dot{\varepsilon}_{ij} = \dot{\varepsilon}'_{ij}$ for $df \leq 0$ and therefore $\dot{\varepsilon}''_{ij} = 0$, the continuity of the elastic-plastic loading path and the elastic unloading path for a neutral change of stress $df = 0$ is assured by Eq. (21.3)[4].

Since for plastic straining $\sigma_{ij}\dot{\varepsilon}''_{ij} > 0$ in both the ideal plastic and the work- or strain-hardening continuum, the angle between the radius vector of the stress-point on the yield locus and the normal to the yield surface at this point will be smaller than $\pi/2$. Loading conditions for which the two directions remain parallel throughout the load history because the ratios between the principal stresses

---
[1] See footnote 1, p. 241.
[2] See footnote 5, p. 254 (p. 26).
[3] Ibid. (p. 34).
[4] G. H. HANDELMAN, C. C. LIN and W. PRAGER: Quart. Appl. Math. **4**, 397 (1947).

remain constant are designated as *"radial loading"*[1]. As a result of the associated straight-line path of the integration $\varepsilon''_{ij} = \int \dot{\varepsilon}''_{ij} dt$, the plastic stress-strain-rate relations can be replaced by stress-strain relations. A similar simplification results from the introduction of yield conditions or loading functions consisting of intersecting plane surfaces and producing a polygonal (*"piece-wise linear"* [2]) yield locus, provided the states of stress during loading remain on the same or on parallel planes. Since the directions of the strain-rate vectors remain normal to these planes and therefore parallel, the total plastic strain is independent of the loading path. Under conditions of radial loading or under the assumption of a piece-wise linear yield locus or loading surface the general stress-strain-rate or "incremental" relations of the theory of plasticity usually designated as "*flow*"-*theory*[3] can be replaced by the stress-strain relations of the "*deformation*"-*theory* which permit a simplified approach to plastic boundary value problems by eliminating the necessity of cumbersome integration over the path of straining.

Because of the assumption of incompressibility of plastic flow, the total rate of deviatoric strain

$$\dot{e}_{ij} = \dot{e}''_{ij} + \frac{1}{2G} \dot{s}_{ij} \quad \text{while} \quad \varepsilon_{ii} = \frac{1}{3K} \sigma_{ii}. \tag{21.4}$$

For the incompressible elastic-plastic continuum $K \to \infty$ and $\varepsilon_{ii} \to 0$ while $\sigma_{ii} \to 3p$; for the rigid-plastic continuum both $K \to \infty$ and $G \to \infty$ so that $\dot{e}_{ij} = \dot{e}''_{ij}$. The latter assumption is considered justified if the elastic strains are small in relation to the plastic strains; since this condition is never realized in the vicinity of the elastic-plastic boundary, *rigid-plastic* approximations to solutions of *elastic-plastic* boundary value problems are admissible only at a sufficient distance from the elastic-plastic boundary, the shape of which they cannot reproduce[4].

22. Yield conditions. The simplest continually differentiable isotropic condition of incompressible yielding is obtained by neglecting, in first approximation, the effect of J'_3 and assuming

$$f(\sigma_{ij}) = J'_2 - k^2 = 0 \tag{22.1}$$

where $\pm k$ represents the yield stress in shear. Eq. (22.1) has been proposed by von Mises[5]; in stress-space it defines a circular cylinder of radius $k\sqrt{2}$, the axis of which coincides with the space-diagonal, and which therefore intersects the plane $J_1 = 0$, which is normal to the space diagonal, along a circle of the same radius. Since for unaxial tension $\sigma_{11} = \sigma_0$ and therefore $J'_2 = \frac{1}{3}\sigma_0^2$, the ratio $k/\sigma_0 = 1/\sqrt{3} = 0.576$ determines the yield stress in tension σ_0 for a given value of k or vice versa. Eq. (22.1) is in excellent agreement with experiments on ductile metals[6]. Introducing Eq. (22.1) into Eqs. (6.8a) and (6.15) a parametric representation, in terms of k and ω, is obtained of the stress deviations that satisfy this yield condition.

For plane stress with $\sigma_3 = 0$ Eq. (22.1) takes the form of a family of ellipses with parameter k in the (σ_1, σ_2)-plane

$$\sigma_1^2 - \sigma_1 \sigma_2 + \sigma_2^2 = 3k^2 = \sigma_0^2 \tag{22.2}$$

---

[1] A. A. Ilyushin: Prikl. Math. Mech. 9, 207 (1945); 10, 347 (1946).
[2] See footnote 3, p. 254; also J. L. Sanders: Proc. 2nd U. S. Nat. Congr. Appl. Mech. p. 455, 1954. — W. Prager: J. Appl. Mech. 20, 317 (1953). — P. G. Hodge: IUTAM Colloquium Madrid 1955, p. 147. Berlin: Springer 1956.
[3] W. Prager: J. Appl. Mech. 15, 226 (1948).
[4] A. M. Freudenthal: Prelim. Rep. 2nd Congr. Bridge and Struct. Eng. 3. Berlin: W. Ernst 1936.
[5] R. v. Mises: Göttinger Nachr., Math.-phys. Kl. 582 (1913). — M. T. Huber: Czasopismo techn., Lwow 22, 81 (1904).
[6] W. Lode: Z. angew. Math. Mech. 5, 142 (1925). — Z. Physik 36, 913 (1926). — G. I. Taylor and H. Quinney: Phil. Trans. Roy. Soc. Lond., Ser. A 230, 323 (1931).

the major axes of which are inclined under 45° to the stress-axes. For elastic plane strain with $\varepsilon_3 = 0$ the yield condition depends on the additional parameter ν (Poisson's ratio)

$$(\sigma_1 + \sigma_2)^2 (1 - \nu + \nu^2) - 3\sigma_1\sigma_2 = 3k^2 = \sigma_0^2. \tag{22.3}$$

For $\nu = 0$ Eqs. (22.2) and (22.3) are identical; for $\nu = 0.5$ (incompressible elastic medium) Eq. (22.3) degenerates into the straight lines

$$\sigma_1 - \sigma_2 = \pm 2k = \pm 2\sigma_0/\sqrt{3}. \tag{22.3a}$$

The assumption of incompressible plastic flow is abandoned in the *generalized* von Mises yield condition; expressing the fact that yielding in a compressible medium depends on J_1, it has the form[1]

$$f(\sigma_{ij}) = J_2' - \varphi(J_1) = 0 \tag{22.4}$$

where various assumptions can be made with respect to the function $\varphi(J_1)$, either to fit experimental results[2] or for reasons of mathematical expediency.

Considerations of the first kind suggest functions of the form $\varphi = k(k - a J_1)$ or $\varphi = (k - a J_1)^2$ leading to the alternative yield conditions

$$J_2' + a k J_1 - k^2 = 0 \quad \text{or} \quad \sqrt{J_2'} + a J_1 - k = 0, \tag{22.5}$$

the first of which defines in stress-space a paraboloid of rotation, the second a right circular cone the axes of which coincide with the space diagonal and thus represents a generalization of Coulomb's criterion to three dimensions[3]. The constant $a = k/3 p_t$ can be related to a limiting hydrostatic tension p_t at which the yield surface intersects its axis, while k is the yield stress in shear for the plane $J_1 = 0$. The intersection with this plane remains a circle with radius $k/\sqrt{2}$ or $\sigma_0\sqrt{\tfrac{2}{3}}$.

Considerations of mathematical expediency, on the other hand, justify a yield condition of the "*parabola*" form proposed by von Mises[4]

$$J_2' + \frac{1}{12} J_1^2 - \left[\frac{(4a^2 k^2 - J_1^2)}{8ak}\right]^2 = 0 \tag{22.6}$$

where $|J_1| \leq 2ak$ and $a = (1 + \sqrt{2})$. For plane stress Eq. (22.6) defines two branches of parabolas

$$(\sigma_1 + \sigma_2)^2 - 4a^2 k^2 \pm 4a k (\sigma_1 - \sigma_2) = 0 \tag{22.6a}$$

which intersect on the diagonal $\sigma_1 = \sigma_2$ at the distance $\pm a$ from the origin.

The simplest expression of a general yield condition $f(J_2', J_3') = 0$ that satisfies Eq. (21.1) is of the form

$$f(\sigma_{ij}) = J_2'^3 - c J_3'^2 - k^6 = J_2'^3 (1 - c J_3'^2/J_2'^3) - k^6 = J_2'^3 (1 - 8c\bar{\mu}^2) - k^6 = 0 \tag{22.7}$$

where k is the yield stress in shear and $\bar{\mu}$ the "form"-invariant of the stress deviation defined by the first Eq. (6.10). Eq. (22.7) coincides with the von Mises yield condition for $J_3' = 0$ and therefore $\bar{\mu} = 0$, which, on the intersection with the plane $J_1 = 0$, are the corners of the St. Venant-Tresca hexagon [Eq. (22.8)]

---

[1] R. v. Mises: Z. angew. Math. Mech. **5**, 147 (1925). — F. Schleicher: Z. angew. Math. Mech. **6**, 199 (1926).
[2] A. Leon: Ing.-Arch. **4**, 421 (1933). — C. Torre: Öst. Ing.-Arch. **1**, 36, (1946); **4**, 174 (1950).
[3] D. C. Drucker and W. Prager: Quart. Appl. Math. **10**, 157 (1952).
[4] R. v. Mises: Reissner Anniversary Volume, p. 428. Ann Arbor: Edwards 1949.

inscribed into the circle of the von Mises yield condition (see Fig. 4). Between these points it is intermediate between this hexagon and the circle. It has been established[1] that with $c \sim 0.70$ Eq. (22.7) reproduces the experimentally observed slight deviations[2] from the behavior defined by Eq. (22.1) fairly well.

The simplest form of a piece-wise continuous yield condition in the form of three pairs of parallel planes enclosing a right hexagonal prism with the space diagonal as axis is represented by the St. Venant-Tresca yield condition[3]

$$4f_{1,4} = [(\sigma_2 - \sigma_3)^2 - 4k^2] = 0; \quad 4f_{2,5} = [(\sigma_3 - \sigma_1)^2 - 4k^2] = 0;$$
$$4f_{3,6} = [(\sigma_1 - \sigma_2)^2 - 4k^2] = 0 \quad (22.8)$$

or, combined, $f_{1,4} \cdot f_{2,5} \cdot f_{3,6} = 0$. With $k = \sigma_0/\sqrt{3}$ Eq. (22.8) describes the hexagonal prism circumscribed, with $k = 0.5\sigma_0$ the prism inscribed into the cylindrical von Mises yield surface (22.1), producing on intersection with the plane $J_1 = 0$ the circumscribed and inscribed hexagons of the circular yield locus. Eq. (22.3a) shows that for conditions of plane strain in the incompressible continuum the von Mises condition is identical with the St. Venant condition in which $k = \sigma_0/\sqrt{3}$ has been introduced.

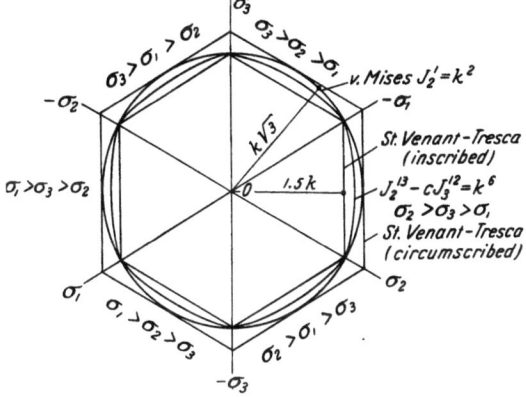

Fig. 4. Representation of yield conditions (22.1), (22.7) and (22.8) in plane $J_1 = 0$ of stress-space.

The isotropic yield conditions $f(\sigma_{ij})$ of the ideal plastic medium can be used as "loading functions" of the work-hardening medium according to Eqs. (21.2) and (21.3). Such use implies isotropic strain- or work-hardening, independent of strain-history, an assumption that is justified only for a simple loading path and small plastic strains. A better approximation of real behavior beyond this limitation would require the introduction of loading functions of the form $f(J_2', J_3', I_2', I_3')$ or $f(J_2', J_3', s_{ij} e_{ij}'')$. The simplest example of such functions, which can, however, no longer be represented in stress-space, is of the form [see Eq. (9.11)]

$$f(Q_i, q_i) = f(\sigma_{ij}, \varepsilon_{ij}'') = f_1(J_2', J_3') - a\, s_{ij} e_{ij}'' - k_i^2 = 0 \quad (22.9)$$

where a is an experimental constant and k_i are the consecutive values of the yield stress in shear, $k_i > k_{i-1}$, describing the work-hardening stages of the medium. These functions contain all features of the isotropic functions with the addition of a *Bauschinger effect* under conditions of reversed loading. The consecutive yield surfaces, while still geometrically similar, are no longer concentric[4].

23. Stress-strain-rate relations. The state of the incompressible plastic continuum is determined by the plastic dissipation potential[5] $\frac{1}{2}T\vartheta_p = \dot{X}_p(s_{ij})$ according to the last Eq. (8.15). Using this equation in the definition of the

---

[1] W. PRAGER: Mem. Sci. Math. **87** (1937). — J. Appl. Phys. **16**, 837 (1945). — Quart. Appl. Math **4**, 397 (1947).

[2] See footnote 6, p. 281.

[3] H. TRESCA: C. R. Acad. Sci., Paris **59**, 754 (1864); **64**, 809 (1867).

[4] See footnote 5, p. 254 (p. 26) and H. FROMM: footnote 1, p. 241 (p. 458). See also F. EDELMAN and D. C. DRUCKER: Franklin. Inst. **250**, 581 (1951). — W. PRAGER: Proc. Inst. Mech. Engrs. London **169**, 41 (1955).

[5] See footnote 2, p. 254.

incompressible ideal plastic medium $d\sigma_{ij}\dot{e}''_{ij}=0$ considering that $\dot{e}_{ij}=\dot{e}''_{ij}$, the relation is obtained

$$d\sigma_{ij}\dot{e}''_{ij} = d\sigma_{ij}\frac{\partial \dot{X}_p(s_{ij})}{\partial \sigma_{ij}} = 0 \tag{23.1}$$

which is compatible with Eq. (21.1) only if the plastic potential $\dot{X}_p(s_{ij})$ is identical to a constant factor 2λ with the yield condition $f(s_{ij})$, and therefore $\dot{X}_p(s_{ij})=2\lambda f(J'_2, J'_3)$. Because the existence of a plastic potential ensures the uniqueness of the stress-and velocity-fields in the ideal plastic continuum (see Sect. 9), this identity is the necessary condition for the application of the extremum principles, Eqs. (9.9) to (9.11), to the ideal plastic medium defined by Eq. (21.1).

Hence the most general stress-strain-rate relation of the incompressible isotropic ideal plastic medium is expressed by Eq. (11.4b) with $\dot{X}_p(s_{ij})=f(J'_2, J'_3)$[1]

$$\frac{1}{2\lambda}\dot{e}''_{ij} = \frac{\partial f}{\partial J'_2} s_{ij} + \frac{\partial f}{\partial J'_3}\left(s_{ik}\cdot s_{kj}-\frac{2}{3}J'_2 \delta_{ij}\right). \tag{23.2}$$

This relation can be considerably simplified by the two assumptions $\partial f/\partial J'_3=0$ and therefore $f(s_{ij})=f(J'_2)$ and $f(J'_2)=J'_2-k^2=0$, by which Eq. (23.2) degenerates into the linear relation (8.14)

$$\dot{e}''_{ij} = 2\lambda\frac{\partial f}{\partial J'_2} s_{ij} = 2\lambda s_{ij}. \tag{23.3}$$

The scalar invariant $\lambda>0$ for $f(s_{ij})=0$ and $df=0$; $\lambda=0$ for $f(s_{ij})<0$ or for $f(s_{ij})=0$ and $df<0$ (unloading). Its value is obtained by squaring Eq. (23.3)

$$\dot{e}''_{ij}\dot{e}''_{ij} = 4\lambda^2 s_{ij}s_{ij} \quad \text{or} \quad 2\lambda = \sqrt{\frac{\dot{I}'_2}{J'_2}} = \sqrt{\frac{\dot{I}'_2}{k^2}}. \tag{23.4}$$

Eq. (23.3) represents the Mises-Lévy stress-strain-rate relation[2]

$$\dot{e}''_{ij} = \sqrt{\frac{\dot{I}'_2}{k^2}} s_{ij}. \tag{23.5}$$

The rate of plastic strain work considering Eq. (23.4) is

$$\begin{aligned} s_{ij}\dot{e}''_{ij} &= 2\lambda\frac{\partial f}{\partial \sigma_{ij}} s_{ij} = 2\lambda s_{ij}s_{ij} = \sqrt{4\dot{I}'_2 J'_2} \\ &= \sqrt{2\dot{I}'_2}\sqrt{2J'_2} = \sqrt{3J'_2}\sqrt{\tfrac{4}{3}\dot{I}'_2} = \bar{\sigma}\dot{\bar{\varepsilon}} > 0 \end{aligned} \tag{23.6}$$

where $\sqrt{2J'_2}$ and $\sqrt{2\dot{I}'_2}$ might be considered as "effective" stress- and strain-rate, respectively. It is, however, more expedient to introduce the stress-and strain-"intensities" $\bar{\sigma}$ and $\bar{\varepsilon}$ defined by Eqs. (6.8b) because of their easy correlation with conditions of uniaxial stress. For the von Mises yield condition the rate of plastic work $s_{ij}\dot{e}''_{ij}=2k\sqrt{\dot{I}'_2}=4\lambda^2 k$. Since necessarily $2\lambda(\partial f/\partial \sigma_{ij})s_{ij}>0$ and $\lambda>0$, the yield surface must be convex[3].

---

[1] See footnote 1, p. 283 and 4, p. 280.

[2] M. Lévy: C. R. Acad. Sci., Paris **70**, 1323 (1870). — B. de St. Venant: C. R. Acad. Sci., Paris **70**, 473 (1870). — R. v. Mises: See footnote 2, p. 254 and 5, p. 281.

[3] D. C. Drucker: Proc. First U. S. Congr. Appl. Mech., Chicago p. 487, 1951. — Quart. Appl. Math. **14**, 35 (1956).

Combining the elastic and the plastic strain-components according to Eq. (21.4) stress-strain-rate relations are obtained

$$\dot{e}_{ij} = 2\bar{\lambda} s_{ij} + \frac{1}{2G} \dot{s}_{ij} \quad \text{and} \quad \dot{\varepsilon}_{ii} = \frac{\dot{\sigma}_{ii}}{3K} \qquad (23.7)$$

which have been introduced by PRANDTL[1] and REUSS[2]. The rate of strain work

$$s_{ij}\dot{e}_{ij} = 2\bar{\lambda} s_{ij} s_{ij} + \frac{1}{2G} s_{ij}\dot{s}_{ij} = 4\bar{\lambda} J_2' = 4\bar{\lambda} k^2 = s_{ij}\dot{e}_{ij}'' > 0 \qquad (23.8)$$

since $s_{ij}\dot{s}_{ij} = 0$ when $J_2' = k^2$. Hence $2\bar{\lambda} = s_{ij}\dot{e}_{ij}''/2k^2$ so that

$$2G\dot{e}_{ij} = \frac{G s_{mn} \dot{e}_{mn}''}{k^2} s_{ij} + \dot{s}_{ij} \quad \text{or} \quad \dot{s}_{ij} = 2G\left(\dot{e}_{ij} - \frac{s_{mn}\dot{e}_{mn}''}{2k^2} s_{ij}\right). \qquad (23.9)$$

Thus Eq. (23.9), which is the equivalent of the first Eq. (23.7) and is to be combined with the second equation to furnish the rate of stress $\dot{\sigma}_{ij}$, holds only for $s_{ij}\dot{e}_{ij}'' > 0$; when $J_2' < k^2$ or when $J_2' = k^2$ but $s_{ij}\dot{e}_{ij}'' < 0$ (unloading) it degenerates into the elastic relation; as $G \to \infty$ or $\dot{s}_{ij} = 0$ Eq. (23.7) is transformed into the Mises-Levy relation (23.3). It is easy to see that Eq. (23.9) can also be interpreted as the constitutive equation of a fictitious anisotropic elastic medium, the anisotropy proportional to $s_{ij} s_{mn}$ indicating the plastic flow[3].

Stress-strain-rate relations of the incompressible work-hardening continuum are obtained by specifying the tensor h_{ij} in Eq. (21.3) in terms of a plastic potential $g(J_2', J_3')$ and another isotropic function $h(J_2', J_3')$ that may depend on strain-history[4]

$$h_{ij} = h \frac{\partial g}{\partial \sigma_{ij}}. \qquad (23.10)$$

Hence considering Eqs. (11.9), Eq. (21.3) can be written in the form

$$\dot{e}_{ij}'' = h \frac{\partial g}{\partial \sigma_{ij}} df = h(J_2', J_3') \left[\frac{\partial g}{\partial J_2'} s_{ij} + \frac{\partial g}{\partial J_3'} \left(s_{ik}s_{kj} - \frac{2}{3} J_2' \delta_{ij}\right)\right] df \qquad (23.11)$$

which has been proposed by PRAGER[5]. This equation is considerably simplified by the identification $g = f$. The necessary relation between the functions g and f if they are not identical has been discussed by HILL[6]. Introducing a monotonically increasing loading function $f(J_2', J_3') = F(\int s_{ij}\dot{e}_{ij}'')$ or, less elaborately, $f(J_2') = F(\int \bar{\sigma} \dot{\bar{\varepsilon}})$ associated with a plastic potential $g = J_2'$, HILL has shown that the simplest workhardening relation of the incompressible isotropic continuum can be written in the form

$$\dot{e}_{ij}'' = \frac{1}{2F'J_2'} s_{ij} df \qquad (23.12)$$

where $F' = \frac{1}{\bar{\sigma}} H$ and $H = \frac{d\bar{\sigma}}{d\bar{\varepsilon}}$ denotes the slope (tangent modulus) of the observed $\bar{\sigma}$-$\bar{\varepsilon}$ relation[7].

In a compressible plastic medium in which the yield condition is a function of the first and second invariants $f(J_1, J_2) = 0$ and the plastic potential therefore

---

[1] L. PRANDTL: Proc. First Internat. Congr. Appl. Mech., Delft, p. 43, 1924.
[2] E. REUSS: Z. angew. Math. Mech. **10**, 266 (1930).
[3] L. FINZI: Rend. Sci. Ist. Lombardo **90**, 532 (1956).
[4] E. MELAN: Ing.-Arch. **9**, 116 (1938). — Sitzgsber. Akad. Wiss. Wien **145**, 195 (1936); **147**, 73 (1938).
[5] See footnote 4, p. 280.
[6] See footnote 5, p. 254 (p. 35).
[7] Ibid. (p. 39).

$\dot{X}_p(\sigma_{ij}) = 2\lambda f(J_1, J_2')$, the associated stress-strain-rate relation

$$\frac{1}{2\lambda}\dot{\varepsilon}_{ij}'' = \frac{\partial f}{\partial \sigma_{ij}} = \frac{\partial f}{\partial J_2'} s_{ij} + \frac{\partial f}{\partial J_1} \delta_{ij} = s_{ij} + \frac{\partial f}{\partial J_1} \delta_{ij} \qquad (23.13)$$

since $\partial J_1/\partial \sigma_{ij} = \delta_{ij}$. The rate of plastic cubic dilatation therefore is

$$\dot{\varepsilon}_{ii}'' = 6\lambda \frac{\partial f}{\partial J_1} > 0 \qquad (23.14)$$

if it is assumed that a hydrostatic pressure increases the yield limit in shear, according to either of the Eqs. (22.5). The stress-strain-rate relations associated with these yield conditions are obtained by introducing Eqs. (22.5) into Eq. (23.13)

$$\dot{\varepsilon}_{ij}'' = 2\lambda'(s_{ij} + a k \delta_{ij}) \quad \text{and} \quad \dot{\varepsilon}_{ij}'' = 2\lambda''\left(\frac{s_{ij}}{2\sqrt{J_2'}} + a \delta_{ij}\right) \qquad (23.15)$$

where the coefficients of proportionality λ' and λ'' can be determined by squaring the relations (23.15); hence

$$\dot{\varepsilon}_{ij}'' \dot{\varepsilon}_{ij}'' = 4\lambda'^2(2k^2 + 3 a^2 k^2) \quad \text{and} \quad \dot{\varepsilon}_{ij}'' \dot{\varepsilon}_{ij}'' = 4\lambda''^2(\tfrac{1}{2} + 3 a^2) \qquad (23.16)$$

from which $2\lambda' = \sqrt{2 I_2'}/k\sqrt{2 + 3 a^2}$ and $2\lambda'' = \sqrt{2 I_2'}/\sqrt{0.5 + 3 a^2}$. The plastic cubic dilatations are respectively $\dot{\varepsilon}_{ii}'' = 6\lambda' a k$ and $\dot{\varepsilon}_{ii}'' = 6\lambda'' a$.

The extension of the concept of the plastic potential to a discontinuous yield surface consisting of several functions was first proposed by von Mises[1] with respect to the Tresca-St. Venant yield condition (22.8); it has recently been applied to a general yield surface consisting of several functions $f_\nu(\sigma_{ij})$ by Koiter[2] specifying the relation

$$\dot{\varepsilon}_{ij}'' = \sum_\varkappa 2\lambda_\varkappa \frac{\partial f_\varkappa(\sigma_{ij})}{\partial \sigma_{ij}} \qquad (23.17)$$

as the "flow rule" associated with the yield condition $f_\varkappa(\sigma_{ij}) = 0$. The $\varkappa < \nu$ functions $f_\varkappa(\sigma_{ij}) = 0$ are those of the ν functions $f_\nu(\sigma_{ij})$ that vanish and thus satisfy the yield condition for the considered state of stress, while the remaining $(\nu - \varkappa)$ functions take negative values; $\lambda_\varkappa$ are arbitrary positive factors of proportionality.

For the Tresca-St. Venant yield surface, which represents a specific case of a general discontinuous yield surface, considering the portion $f_{2,5} = 0$ of the surface for which $\sigma_1 < \sigma_2 < \sigma_3$ while $f_{1,4} < 0$ and $f_{3,6} < 0$, von Mises has obtained the relation[1]

$$\dot{\varepsilon}_1'' : \dot{\varepsilon}_2'' : \dot{\varepsilon}_3'' = \frac{\partial f_{2,5}}{\partial \sigma_1} : \frac{\partial f_{2,5}}{\partial \sigma_2} : \frac{\partial f_{2,5}}{\partial \sigma_3} = 1 : 0 : -1 \qquad (23.18)$$

since the principal strain-rates

$$\dot{\varepsilon}_1'' = 2\lambda \frac{\partial f_{2,5}}{\partial \sigma_1} = \lambda(\sigma_1 - \sigma_3), \quad \dot{\varepsilon}_3'' = 2\lambda \frac{\partial f_{2,5}}{\partial \sigma_3} = \lambda(\sigma_3 - \sigma_1); \qquad (23.19)$$

therefore

$$\dot{\varepsilon}_3'' - \dot{\varepsilon}_1'' = 2\lambda(\sigma_3 - \sigma_1) = 4\lambda k \quad \text{and} \quad 2\lambda = \frac{\dot{\varepsilon}_3'' - \dot{\varepsilon}_1''}{2k}. \qquad (23.20)$$

Similarily, for the portion $f_{3,6} = 0$ for which $\sigma_1 < \sigma_3 < \sigma_2$

$$\dot{\varepsilon}_1'' : \dot{\varepsilon}_2'' : \dot{\varepsilon}_3'' = 1 : -1 : 0. \qquad (23.21)$$

---

[1] See footnote 2, p. 254.
[2] See footnote 3, p. 254.

The matrices corresponding to Eqs. (23.18) and (23.21) have vanishing determinants; they represent therefore not only incompressible but also plane deformations, consisting of slip in the direction of the maximum shear and defined by $v_1 = 2\dot\gamma x_3$, $v_2 = v_3 = 0$ and $v_1 = 2\dot\gamma x_2$, $v_2 = v_3 = 0$, respectively. This follows from the fact that the non-vanishing components of the strain-rate tensors according to Eq. (5.19) are, respectively, $\dot\varepsilon''_{13} = \dot\varepsilon''_{31} = \dot\gamma$ and $\dot\varepsilon''_{12} = \dot\varepsilon''_{21} = \dot\gamma$; therefore from Eq. (6.2) with $II_{\dot\varepsilon} = -\dot\gamma^2$ and $I_{\dot\varepsilon} = III_{\dot\varepsilon} = 0$ the principal strain-components $\dot\varepsilon''_1 = -\dot\varepsilon''_3 = \dot\gamma$ and $\dot\varepsilon''_1 = -\dot\varepsilon''_2 = \dot\gamma$, according to Eqs. (23.18) and (23.21).

Under the assumption that along the edges of the Tresca-St. Venant yield surface where the direction of the normal to this surface is not unique (uniaxial and bi-axial states of stress) the flow is characterized by a linear combination of Eqs. (23.18) and (23.21) in the form[1]

$$\dot\varepsilon''_1 : \dot\varepsilon''_2 : \dot\varepsilon''_3 = 1 : -\alpha : -(1-\alpha) \qquad (23.22)$$

where $0 \leq |\alpha| = |\dot\varepsilon''_2/\dot\varepsilon''_1| \leq 1$ is a constant ratio along the strain path; the plastic strain-rate intensity according to Eq. (6.8b)

$$\dot{\bar\varepsilon} = \sqrt{\frac{2}{3}}\sqrt{\dot\varepsilon''_{ij}\dot\varepsilon''_{ij}} = \sqrt{\frac{2}{3}}\sqrt{\dot\varepsilon''^2_1 + \dot\varepsilon''^2_2 + \dot\varepsilon''^2_3} = \frac{2}{\sqrt{3}}(1-\alpha+\alpha^2)^{\frac{1}{2}}\dot\varepsilon''_1. \qquad (23.23)$$

Integrating over the strain-path,

$$\int \dot{\bar\varepsilon}\, dt = \frac{2}{\sqrt{3}}(1-\alpha+\alpha^2)^{\frac{1}{2}}\varepsilon''_1 \qquad (23.24)$$

with similar expressions for ε''_2 and ε''_3. Eqs. (23.17) to (23.23) are therefore valid for the principal strains as well as for the strain-rates.

THOMAS[2] has shown that the linear isotropic stress-strain-rate relations of the incompressible plastic continuum associated with the portion of the Tresca yield condition $(\sigma_3 - \sigma_1) = 2k$ can be written in the form

$$\dot{e}''_{ij} = \frac{\dot{e}''_3 - \dot{e}''_1}{2k} s_{ij} = \sqrt{\frac{\dot{e}''_3 \cdot \dot{e}''_1}{s_3 \cdot s_1}}\, s_{ij} = 2\lambda s_{ij} \qquad (23.25)$$

with similar expressions for the other portions. Considering that $(\dot{e}''_3 - \dot{e}''_1)^2 = 4\dot{I}'_2$ and therefore $2\lambda = \sqrt{\dot{I}'_2}/k$, these expressions are identical with the Mises-Lévy equations (23.6), except for the difference in the numerical value of k for the two yield conditions.

For a material with work-hardening Eq. (23.17) is replaced by[3]

$$\dot\varepsilon''_{ij} = \sum_{\varkappa} h_\varkappa \frac{\partial f_\varkappa}{\partial \sigma_{ij}}\, df_\varkappa \qquad (23.26)$$

where $h_\varkappa$ are positive functions of stress- and strain history and the sum is taken over the $\varkappa$ of the ν functions $f_\nu(\sigma_{ij})$ for which $f_\varkappa(\sigma_{ij}) = 0$ and $df_\varkappa \geq 0$.

24. Stress-strain relations. The use of plastic stress-strain relations instead of the stress-strain-rate relations so far considered is theoretically admissible only if the former can be obtained from the latter by path-independent integration, as in the case of proportional or "radial" loading with fixed stress-axes and principal stress ratios, or for segment-wise plane yield surfaces. However, since

---
[1] W. PRAGER: J. Appl. Mech. **20**, 317 (1953).
[2] T. Y. THOMAS: Proc. Nat. Acad. Sci. U. S. A. **40**, 593 (1954).
[3] See footnote 3, p. 254.

the simplification of the solutions of boundary value problems through the use of stress-strain relations instead of stress-strain-rate relations is usually quite drastic, such relations have been extensively used even when the above conditions were not rigorously satisfied, and relatively simple solutions of technically important boundary value problems were obtained[1]. From a purely physical point of view it very frequently seems preferable to obtain a reasonably approximate general solution of a boundary value problem in closed form rather than a rigorous solution that can only be evaluated for specific conditions by elaborate numerical methods.

The stress-strain relations replacing, respectively, Eqs. (23.3) and (23.7) are usually known as the Hencky equations[2]

$$e''_{ij} = 2\varphi s_{ij} \tag{24.1}$$

and

$$e_{ij} = \left(\frac{1}{2G} + 2\varphi\right) s_{ij} = \frac{1}{2\overline{G}} s_{ij} \quad \text{with} \quad \varepsilon_{ii} = 3K\sigma_{ii} \tag{24.2}$$

where $\varphi(x_i)$ is a scalar proportionality factor. Within the plastic region $\varphi > 0$, for unloading $\varphi = 0$; the secant shear modulus $\overline{G} = G/(1 + 4G\varphi) < G$ within the plastic region. The equation $\varphi = 0$ associated with Eq. (24.2) and the yield condition $J'_2 = k^2$ delimits the elastic and the plastic domains. The factor φ can be evaluated by squaring Eq. (24.2)

$$2G^2 e_{ij} e_{ij} = k^2 (1 + 4G\varphi)^2 \tag{24.3}$$

and therefore $\varphi = [(2G I'_2/k) - 1]/4G \geq 0$.

Differentiation of Eq. (24.2) under the assumption $\overline{G} = \overline{G}(J'_2)$

$$\dot{s}_{ij} = 2\overline{G} \dot{e}_{ij} + 2 \frac{\partial \overline{G}}{\partial J'_2} dJ'_2 e_{ij} \tag{24.4}$$

shows the lack of continuity between the loading and unloading stress-strain-relation for a *neutral* change for which $dJ'_2 = 0$; the continuity provided by Eq. (23.7) would require $\overline{G} = G$, while within the plastic range necessarily $\overline{G} < G$. This lack of continuity is one of the objections against the use of plastic stress-strain relations[3].

Eq. (23.24) has shown that for a singular plane yield surface the associated stress-strain-rate relation can be replaced by the stress-strain relation. Hence for the respective portion of the St. Venant-Tresca yield condition in analogy to Eq. (23.25)

$$e''_{ij} = \frac{e''_3 - e''_1}{2k} s_{ij} = 2\varphi s_{ij} \tag{24.4}$$

as long as the stress path remains within the same portion of the yield surface. The use of stress-strain relations associated with the St. Venant-Tresca condition is therefore subject to less stringent limitations than those associated with the von Mises condition, providing thus an expedient approach to the rigorous solution of boundary value problems in the theory of plasticity[4]. It should, however, be remembered that the Tresca-St. Venant yield condition differs significantly from experimental results obtained with ductile metals[5].

---

[1] See footnotes 1, p. 240; 3, p. 242; and A. A. ILYUSHIN: Plasticity. Moscow 1948.
[2] H. HENCKY: Z. angew. Math. Mech. **4**, 323 (1924).
[3] See footnote 2. p. 259, also 5, p. 254 (p. 47).
[4] W. T. KOITER: Biezeno Anniversary Vol., p. 232; Amsterdam 1953. See also footnote 2, p. 281.
[5] See footnote 6, p. 281.

V. Combined quasi-linear relations.

25. Visco-plasctic relation. The linear visco-plastic continuum, also known as the Bingham body[1] is a linear viscous fluid at stresses exceeding the yield limit; below this limit it is usually assumed to be rigid. The stress-deviation above the yield limit is therefore the sum of the stress-deviation sustained at the yield limit k_{ij} and the deviation associated with the viscous flow. The general stress-strain-rate relations are due to HOHENEMSER-PRAGER[2], ILYUSHIN and OLDROYD[3]. Combining Eq. (23.3) with Eq. (8.12)

$$s_{ij} = k_{ij} + 2\eta \dot{e}''_{ij} = \left(\frac{1}{2\lambda} + 2\eta\right)\dot{e}''_{ij} = 2\eta_1 \dot{e}''_{ij} > k_{ij} \qquad (25.1)$$

where $\eta_1 = \eta\left(1 + \frac{1}{4\lambda\eta}\right)$ denotes a variable coefficient of shear viscosity of an equivalent *non-linear viscous fluid*, the expression $\frac{1}{4\lambda\eta}$ being a characteristic of visco-plastic behavior.

For the von Mises condition $k_{ij}k_{ij} = 2k^2$; with λ according to Eq. (23.5) and therefore $\dot{e}''_{ij}/k_{ij} = \sqrt{\dot{I}'_2}/k$, Eq. (25.1) is transformed into

$$s_{ij} = k_{ij}(1 + 2\eta\sqrt{\dot{I}'_2}/k) = k_{ij}\psi \qquad (25.2a)$$

where $\psi = s_{ij}/k_{ij} > 1$ is a scalar function that tends towards unity as J'_2 tends towards k^2 and $\dot{e}''_{ij}$ tends to zero; or explicitly

$$s_{ij} = (2\eta + k/\sqrt{\dot{I}'_2})\dot{e}''_{ij} = 2\eta_1 \dot{e}''_{ij} \qquad (25.2b)$$

with $2\eta_1 = 2\eta(1 + k/2\eta\sqrt{\dot{I}'_2})$. The variable coefficient of viscosity is also obtained by squaring Eq. (25.1)

$$\left(\frac{1}{2\lambda} + 2\eta\right) = 2\eta_1 = \sqrt{\frac{J'_2}{\dot{I}'_2}} \qquad (25.3)$$

where $J'_2 = k^2 \psi^2$.

Alternatively, from Eq. (25.1) with the von Mises yield condition

$$2\eta \dot{e}''_{ij} = s_{ij} - k_{ij} = s_{ij}(1 - k/\sqrt{J'_2}) = s_{ij}(1 - 1/\psi) \qquad (25.4)$$

valid for $J'_2 > k$ or $\psi > 1$; hence

$$2\eta_1 = 2\eta/(1 - k/\sqrt{J'_2}) . \qquad (25.5)$$

On the yield surface η_1 tends therefore to infinity as J'_2 tends towards k^2 from above and $\dot{e}''_{ij}$ tends to zero. Eqs. (25.2b) and (25.4) are the alternative explicit expressions of s_{ij} in terms of $\dot{e}''_{ij}$ and vice versa. The rate of visco-plastic strain work

$$\left.\begin{aligned}s_{ij}\dot{e}_{ij} &= s_{ij}\dot{e}''_{ij} = k_{ij}\dot{e}''_{ij} + 4\eta \dot{I}'_2 \\ &= 2k\sqrt{\dot{I}'_2} + 4\eta \dot{I}'_2 = (J'_2 - k^2\psi)/\eta = 2k\psi\sqrt{\dot{I}'_2} = 2\sqrt{J'_2}\sqrt{\dot{I}'_2} > 0\end{aligned}\right\} \qquad (25.6)$$

is the sum of the rates of plastic and viscous strain work. The minimum principles governing the velocity- and the stress-field have been discussed in Sect. 9.

---
[1] E. C. BINGHAM: J. Soc. Chem. Ind. **50**, 419 (1931). — J. Rheol. **2**, 10 (1931).
[2] K. HOHENEMSER and W. PRAGER: Z. angew. Math. Mech. **12**, 216 (1932).
[3] See footnote 2, p. 255.

26. Combination of elastic, viscous and plastic relations. A variety of combined inelastic media can be defined by superposition of the elastic and dissipative responses. Such media might be expected to reproduce the behavior of real materials under conditions under which the simple idealizations are inadequate, such as elevated temperature. It is expedient to discuss only the superposition of deviations, and either to assume incompressibility or to add to all resulting deviatoric relations the elastic compressibility relation $\dot{\sigma}_{ii} = 3K\dot{\varepsilon}_{ii}$.

Addition to Eq. (23.7) of a linear viscous component produces a visco-elastic stress-relaxing (MAXWELL) medium with a yield limit; it is defined by the equation

$$\dot{e}_{ij} = \frac{1}{2G}\dot{s}_{ij} + \left(\frac{1}{2\eta} + 2\bar{\lambda}\right)s_{ij} \quad \text{for} \quad J_2' = k^2; \tag{26.1}$$

for $J_2' < k^2$ Eq. (17.3) is valid since $\bar{\lambda} = 0$. The rate of strain work at the yield limit

$$\left.\begin{aligned} s_{ij}\dot{e}_{ij} &= \frac{1}{2G}s_{ij}\dot{s}_{ij} + \left(\frac{1}{2\eta} + 2\bar{\lambda}\right)s_{ij}s_{ij} \\ &= s_{ij}\dot{e}_{ij}'' = 2k^2\left(\frac{1}{2\eta} + 2\bar{\lambda}\right) = 4k^2\bar{\lambda}^* > 0 \end{aligned}\right\} \tag{26.2}$$

since $s_{ij}\dot{s}_{ij} = 0$. The combined plastic compliance is $2\bar{\lambda}^* = 2\bar{\lambda}\left(1 + \frac{1}{4\bar{\lambda}\eta}\right) = s_{ij}\dot{e}_{ij}/2k^2$, as in the Prandtl-Reuss relation (23.9). However, the condition $\bar{\lambda} > 0$ for the validity of Eq. (26.1) is satisfied only if $s_{ij}\dot{e}_{ij} - k^2/\eta > 0$. For applied rates of strain work, $s_{ij}\dot{e}_{ij} < k^2/\eta$, Eq. (26.2) degenerates into Eq. (17.3); the yield-limit can not be attained, the stresses relaxing faster than they are built-up. Thus "unloading" in the visco-elastic-plastic medium is defined by the condition $s_{ij}\dot{e}_{ij} < k^2/\eta$, by which it differs from the elastic-plastic medium.

Addition to Eq. (25.1) of an elastic component produces an elastic-visco-plastic medium defined by the equation

$$\dot{e}_{ij} = \frac{1}{2G}\dot{s}_{ij} + \frac{1}{2\eta_1}s_{ij} \quad \text{for} \quad J_2' > k^2 \tag{26.3}$$

where η_1 is defined by Eq. (25.5); for $J_2' \leq k^2$ the coefficient $\eta_1 = \infty$ and Eq. (26.3) degenerates into the plastic relation. Eq. (26.3) is formally identical with Eq. (17.3) with a variable relaxation time $\bar{\tau}_M = \eta_1/G = \tau_M/(1 - k/\sqrt{J_2'})$ that tends asymptotically towards $\tau_M = \eta/G$ as the ratio $\sqrt{J_2'}/k$ tends towards infinity. The addition of an elastic component to the visco-plastic relation thus produces a relaxing medium above the yield limit $J_2' > k^2$; however, the stress-intensity can only relax towards $\sqrt{J_2'} = k$, which is the stress-intensity that can be permanently sustained.

The rate of strain work

$$s_{ij}\dot{e}_{ij} = s_{ij}\dot{e}_{ij}'' + s_{ij}\dot{e}_{ij}''' = \frac{1}{2G}\dot{J}_2' + \frac{1}{\eta_1}J_2' = \frac{1}{2G}[\dot{J}_2' + 2\bar{\tau}_M^{-1}(1 - k/\sqrt{J_2'})J_2'] \tag{26.4}$$

is the sum of the rates of elastic and dissipated work. Loading, defined by $\dot{J}_2' > 0$, requires that $2Gs_{ij}\dot{e}_{ij} > 2\bar{\tau}_M^{-1}(1 - k/\sqrt{J_2'})J_2'$. For smaller rates of applied strain work $\dot{J}_2' < 0$, which indicates "unloading" (relaxation) in spite of a positive applied work-rate $s_{ij}\dot{e}_{ij} > 0$; for $\dot{J}_2' = 0$ the loading is "neutral", the applied rate of strain work just balancing the energy-dissipation at the yield limit. If no further strain work is applied after reaching a certain stress level characterized by $J_{20}' > k^2$, Eq. (26.4) is transformed into the relaxation equation

$$\dot{J}_2' + 2\bar{\tau}_M^{-1}(1 - k/\sqrt{J_2'})J_2' = 0 \tag{26.5}$$

the solution of which for $J_2' \geq k^2$ is

$$\left[\left(\frac{\sqrt{J_2'}}{k}\right) - 1\right] = \left[\left(\frac{\sqrt{J_{20}'}}{k}\right) - 1\right] e^{-t/\tau_M}. \tag{26.6}$$

Finally, Eqs. (17.3) and (25.1) can be combined to form a relaxing visco-elastic-visco-plastic medium with constant viscosity coefficient $\bar{\eta}$ for $J_2' > 0$ and an additional variable coefficient η_1 defined by Eq. (25.5) for $J_2' > k^2$; its equation is

$$\dot{e}_{ij} = \frac{1}{2G} \dot{s}_{ij} + \left(\frac{1}{2\bar{\eta}} + \frac{1}{2\eta_1}\right) s_{ij} \quad \text{for} \quad J_2' > k^2 \tag{26.7}$$

which is formally identical with Eq. (17.3) if a combined variable relaxation time $\bar{\tau} = \eta_1 \bar{\eta}/G(\eta_1 + \bar{\eta}) < \tau_M$ is introduced. For $J_2' \leq k^2$ Eq. (26.7) degenerates into Eq. (17.3) of the relaxing visco-elastic medium with relaxation time τ_M. Thus stress-relaxation takes place at any stress level, but is governed by different relaxation times above and below the yield condition $J_2' = k^2$.

The rate of strain work for $J_2' > k^2$ is

$$s_{ij} \dot{e}_{ij} = s_{ij} \dot{e}'_{ij} + s_{ij} \dot{e}''_{ij} = \frac{1}{2G} \dot{J}_2' + \left[\frac{1}{\bar{\eta}} + \frac{1}{\eta}(1 - k/\sqrt{J_2'})\right] J_2'. \tag{26.8}$$

As in the elastic-visco-plastic medium, loading defined by $\dot{J}_2 > 0$ requires that the applied rate of strain work $s_{ij} \dot{e}_{ij} > \frac{\eta + \bar{\eta}(1 - k/\sqrt{J_2'})}{\eta \bar{\eta}} J_2'$. "Unloading" by stress relaxation alone occurs under a positive rate of applied strain work between this limit and zero. The relaxation equation for $s_{ij} \dot{e}_{ij} = 0$ and $J_2' = J_{20}'$ at time $t = 0$ has the form

$$\dot{J}_2' + 2\left[\bar{\tau}_M^{-1} + \tau_M^{-1}\left(1 - \frac{k}{\sqrt{J_2'}}\right)\right] J_2' = 0 \tag{26.9}$$

where $\bar{\tau}_M$ denotes the relaxation time of the visco-elastic component; the solution

$$\left[\frac{\sqrt{J_2'}}{k} - 1\right] + \frac{\tau_M}{\bar{\tau}_M} \frac{\sqrt{J_2'}}{k} = \left\{\left[\frac{\sqrt{J_{20}'}}{k} - 1\right] + \frac{\tau_M}{\bar{\tau}_M} \frac{\sqrt{J_{20}'}}{k}\right\} e^{-t/\tau^*} \tag{26.10}$$

with $\tau^* = \bar{\tau}_M \tau_M/(\bar{\tau}_M + \tau_M)$ is valid for $J_2' > k^2$, while for $J_2' \leq k^2$ the simple relaxation equation holds $\sqrt{J_2'} = \sqrt{J_{20}'} \exp(-t/\bar{\tau}_M)$.

The Eqs. (26.1), (26.3) and (26.7) are special forms of the general quasi-linear relation (13.5) of the incompressible medium

$$\dot{e}_{ij} = \dot{e}'_{ij} + \dot{e}''_{ij} = C_1(J_2') \dot{s}_{ij} + C_3(J_2') s_{ij} \tag{26.11}$$

with isotropic functions C_1 and C_3 assumed to be independent of J_3'. A constant $C_1 = 1/2G$ implies the existence of an elastic distortional potential $\Phi(s_{ij}) = J_2'/2G$, while the second term admits a (plastic) dissipation potential $\dot{X}_p(s_{ij}) = \int C_3(J_2') dJ_2'$. Eq. (26.11) is the equation of a generalized relaxing medium in which plastic potentials of higher than first order in J_2' can be specified to fit observed stress-strain-rate relations[1]. Thus, for instance, with $\dot{X}_p(s_{ij}) = aJ_2'^n/2\eta$ and $C_1 = 1/2G$ Eq. (26.11) can be written in the form

$$\dot{e}_{ij} = \frac{1}{2G}[\dot{s}_{ij} + an\tau^{-1} J_2'^{(n-1)} s_{ij}] \tag{26.12}$$

---

[1] F. K. G. Odquist: Trans. Roy. Inst. Techn., Stockh. **66** (1953). — J. Appl. Mech. **21**, 295 (1954).

where $\tau = \eta/G$ and a is a constant of dimension $(1/\text{stress})^{2(n-1)}$. The constants of this relation can be evaluated by squaring both sides of Eq. (26.12) to obtain the creep equation $(\dot{s}_{ij}=0)$ in terms of stress-and strain-intensities

$$\dot{\bar{\varepsilon}} = \frac{n\bar{\sigma}}{3^n \eta}\left(\frac{\bar{\sigma}}{\bar{\sigma}^*}\right)^{2(n-1)} \tag{26.13}$$

where $\bar{\sigma}^* = a^{-\frac{1}{2(n-1)}}$ denotes a reference level of stress-intensity, and comparing Eq. (26.13) with an experimental uniaxial creep equation; the exponent $(n-1)$ should be an even integer. Creep experiments with aluminum and steel alloys at elevated temperatures[1] suggest values of $3 < n < 9$ as representative. Such equations have been used by Odquist[1] and by Hoff[2] and coworkers in the stress- and stability-analysis of engineering structures at elevated temperatures. For values $n > 5$, Eq. (26.3) with suitable selected values of k and η can be made to represent Eq. (26.12) fairly well. The rate of strain work

$$s_{ij}\dot{e}_{ij} = \frac{1}{2G}[\dot{J}_2' + 2an\tau^{-1}J_2'^n] \tag{26.14}$$

determines the critical rate of applied work that delimits "loading" and "unloading by relaxation" at which $\dot{J}_2' = 0$. True relaxation is governed by Eq. (26.14) with $s_{ij}\dot{e}_{ij}=0$ or

$$\frac{dJ_2'}{2an J_2'^n} = -\frac{dt}{\tau} \tag{26.15}$$

which for $n=1$ degenerates into the linear relaxation equation, the assumption being that Eq. (26.12) is valid even for $\dot{s}_{ij} < 0$. Because $s_{ij}\dot{e}_{ij}'' = 2n\dot{X}_p(s_{ij})$ Eq. (26.11) can be written in the form

$$\dot{e}_{ij} = C_1 \dot{s}_{ij} + \left(s_{mn}\frac{\partial \dot{X}_p}{\partial s_{mn}}\right)^{-1} 2n\dot{X}_p \frac{\partial \dot{X}_p}{\partial s_{ij}} \tag{26.11a}$$

by which the general relation between C_3 and $\dot{X}_p$ is established. Alternatively, a relation involving $\dot{X}_v(\dot{e}_{ij}'')$ in the form

$$s_{ij} = M_1 e_{ij}' + M_3(\dot{I}_2')\dot{e}_{ij}'' = M_1 e_{ij}' + \left(\dot{e}_{mn}''\frac{\partial \dot{X}_v}{\partial \dot{e}_{mn}''}\right)^{-1} 2n\dot{X}_v\frac{\partial \dot{X}_v}{\partial \dot{e}_{ij}''} \tag{26.11b}$$

represents a generalized anelastic medium.

The relaxation of J_2' according to Eqs. (26.5), (26.9) and (26.15) represents a homogeneous relaxation of the stresses under the assumption of continuing similarity of the Mohr stress-circles. In this case the components of the stress deviation are easily obtained from Eq. (6.15) since $\omega = \text{const.}$ and therefore $d\omega/dt = 0$. In the general case in which both J_2' and ω change with time the relation

$$\frac{ds_{ij}}{dt} = \frac{\partial s_{ij}}{\partial J_2'}\dot{J}_2' + \frac{\partial s_{ij}}{\partial \omega}\dot{\omega}$$

permits the evaluation of the components of the stress deviation provided $\omega(t)$ has been determined from the imposed initial conditions.

It should be kept in mind that non-linear equations of the type (26.12) are usually established on the basis of creep tests in which a constant stress is applied, so that $s_{ij}\dot{e}_{ij}'' > 0$ and $\dot{J}_2' = s_{ij}\dot{s}_{ij} = 0$ since $\dot{s}_{ij}=0$. Unless the flow is linear viscous, as in the case of the visco-elastic-plastic or visco-elastic-visco-plastic

---

[1] See footnote 1, p. 291.
[2] N. J. Hoff: Quart. Appl. Math. **12**, 49 (1954). — J. Aeron. Sci. **22**, 661 (1955). — Aeron. Quart. **7**, February (1956).

media, there is no reason to assume a priori that the relaxation process for which $\dot{s}_{ij} < 0$ is associated with the type of flow observed when $\dot{s}_{ij} \geq 0$[1]. This is particularly so for polycrystalline metals in which the latter condition usually involves considerable amounts of slip, the propagation of which requires a constant or increasing shear stress. Thus, the relaxation process of a medium defined by Eq. (26.12) is probably not governed by Eq. (26.15), but by a similar relation with a value of the exponent n characteristic for unloading, and therefore smaller than that obtained from observations under constant or increasing stress.

It should be noted that in stress-strain-rate relations designed to represent the mechanical response of real materials at elevated temperatures the rapid variation of the viscosity coefficients with temperature can not be neglected, while the variation of the elastic and plastic constants over moderate temperature ranges is not significant[2]. Hence the viscosity coefficients and relaxation times are rapidly varying functions of the temperature-difference $\Delta T = T - T_0$, usually introduced in the form[3]

$$G\tau(\Delta T) = \eta(\Delta T) = \eta(x_i, t) = \eta_0 e^{\frac{Q}{R}\left(\frac{1}{T} - \frac{1}{T_0}\right)} \approx \eta_0 e^{-\frac{Q}{RT_0}\frac{\Delta T}{T_0}}. \quad (26.16)$$

Hence for any medium with a viscous component a non-homogeneous stationary or transient temperature field $\Delta T(x_i)$ or $\Delta T(x_i, t)$ significantly affects the stress-strain-rate relations by transforming the constant viscosity coefficients into functions $\eta(x_i)$ or $\eta(x_i, t)$. Eq. (26.16) implies that the coefficient of viscosity changes with temperature instantaneously. In reality, however, a change of over-all viscosity in a quasi-homogeneous medium indicates a multitude of local changes in the internal structure, which necessarily *follow* the establishment of conditions of thermal equilibrium after the change of ambient temperature. It is therefore to be expected that the "equilibrium viscosity" $\eta(\Delta T) = \eta_\infty$ associated with the temperature change ΔT and expressed by Eq. (26.16) will not be attained instantaneously. For glass, for instance, the rate of approach to equilibrium viscosity has been found to be of the form[4]

$$\frac{d\eta}{dt} = k\frac{\eta_\infty - \eta}{\eta} \quad (26.17)$$

where $\eta(t)$ is the momentary value and k a parameter that varies only slightly with temperature. Similar delays should be expected to exist in all media of temperature-sensitive internal structure. However, since these delays are very short, they are significant only when the time-scale of the temperature change is comparable to the delay in reaching equilibrium viscosity as, for instance, in problems of quenching and thermal shock[5].

C. The visco-elastic and the visco-plastic medium.

I. The visco-elastic continuum.

27. Quasi-static problems. The quasi-static problems of the elastically compressible linear visco-elastic continuum, in which inertia forces due to the deformation are neglected, are governed by the continuity condition (4.4), the three equilibrium Eqs. (4.11b), the six constitutive Eqs. (18.5) and the equations of energy balance

---

[1] A. H. COTTRELL and V. AYETKIN: J. Inst. Met. **77**, 413 (1950). — R. P. CARREKER, J. G. LESCHEN and J. D. LUBAHN: Trans. Amer. Soc. Min. Met. Engrs. **180**, 139 (1949).
[2] A. E. JOHNSON: Metallurgia **40**, 125 (1949). — M. B. MILLENSON and S. S. MANSON: Nat. Adv. Comm. Aeron., Rep. No. 906, 1948.
[3] See footnote 1, p. 230.
[4] H. R. LILLIE: J. Amer. Ceram. Soc. **14**, 502 (1931); **16**, 619 (1933).
[5] A. M. FREUDENTHAL: J. Appl. Phys. **25**, 1115 (1954). — Von Mises Memorial Volume, p. 251. New York: Academic Press 1954.

(7.9) and entropy balance (7.22) which, in conjunction with the prescribed boundary conditions in terms of the surface tractions t_i according to Eq. (4.5) or surface displacements u_i or velocities v_i, determine the 12 unknowns ϱ, σ_{ij}, v_i, T and S, provided certain initial conditions are satisfied. These initial conditions are due to the existence of time-derivatives in the constitutive equations; their number depends on the orders of the linear operators $\boldsymbol{P}$ and $\boldsymbol{Q}$, and is zero in the case of the elastic continuum, for which the order of $\boldsymbol{P}$ and $\boldsymbol{Q}$ is zero.

The time-dependence appearing in the solution of the basic equations may be the result of the prescribed time-dependence of the boundary conditions or of the change with time of the boundary configuration, and of the time-dependent response. The solutions will be functions of time t and of the space-coordinates x_i with the material constants a_m and b_n [see Eq. (18.4)] as parameters.

Under the simplifying assumption of zero initial conditions, which makes the surface tractions and displacements vanish for $t<0$, the time-dependence of the basic equations can be removed by the application of Laplace transforms[1]. The transformed Eqs. (4.11b)

$$\bar{\sigma}_{ij,i} + \bar{X}_i = 0 \tag{27.1}$$

and (18.5)

$$\boldsymbol{\bar{P}}\bar{\sigma}_{ij} = 2G\boldsymbol{\bar{Q}}\,\bar{\varepsilon}_{ij} + \bar{\lambda}\,\delta_{ij}\bar{\varepsilon}_{kk} - 3\alpha K\boldsymbol{\bar{P}}\,\delta_{ij}\bar{T} \tag{27.2a}$$

or

$$2G\boldsymbol{\bar{Q}}\,\bar{\varepsilon}_{ij} = -\frac{\bar{\lambda}}{3K}\,\delta_{ij}\bar{\sigma}_{kk} + \boldsymbol{\bar{P}}\bar{\sigma}_{ij} + 2G\alpha\boldsymbol{\bar{Q}}\,\delta_{ij}\bar{T} \tag{27.2b}$$

where $\bar{\lambda} = (3K\boldsymbol{\bar{P}} - 2G\boldsymbol{\bar{Q}})/3$, $\boldsymbol{\bar{P}}$ and $\boldsymbol{\bar{Q}}$ are polynomials in the transform variable, represent an equivalent boundary value problem for an elastic domain of the same shape as the visco-elastic domain, but with "constants" that are functions of the transform parameters.

Thus, for instance, combining Eq. (27.1) with $\bar{X}_i = 0$ and Eq. (27.2a) the relations are obtained

$$\bar{\sigma}_{ij,j} = 2G\left[\frac{\boldsymbol{\bar{Q}}}{\boldsymbol{\bar{P}}}\bar{\varepsilon}_{ij}\right]_{,j} + \left[\frac{\bar{\lambda}}{\boldsymbol{\bar{P}}}\bar{\varepsilon}_{jj}\right]_{,i} - 3\alpha K\bar{T}_{,i} = 0 \tag{27.3}$$

which, under the assumption that the material constants a_m and b_n do not vary in space, can be written in the form

$$\frac{2G\boldsymbol{\bar{Q}}}{\boldsymbol{\bar{P}}}\bar{\varepsilon}_{ij,j} + \frac{\bar{\lambda}}{\boldsymbol{\bar{P}}}\bar{\varepsilon}_{jj,i} - 3\alpha K\bar{T}_{,i} = 0 \tag{27.4}$$

or, considering Eq. (5.9),

$$\frac{G\boldsymbol{\bar{Q}}}{\boldsymbol{\bar{P}}}\bar{u}_{i,jj} + \frac{(G\boldsymbol{\bar{Q}} + \bar{\lambda})}{\boldsymbol{\bar{P}}}\bar{u}_{j,ji} - 3\alpha K\bar{T}_{,i} = 0 \tag{27.5}$$

and, taking the divergence

$$\frac{(2G\boldsymbol{\bar{Q}} + \bar{\lambda})}{\boldsymbol{\bar{P}}}\bar{u}_{j,jii} = \left(K + \frac{4G\boldsymbol{\bar{Q}}}{3\boldsymbol{\bar{P}}}\right)\bar{u}_{j,jii} = 3\alpha K\bar{T}_{,ii}. \tag{27.6}$$

The similarity of Eqs. (27.5) and (27.6) with Eqs. (15.9) of the elastic problem is immediately apparent. Hence, if in the Laplace transform of the elastic solution, $\bar{\sigma}_{ij}(x_i, p)$ and $\bar{u}_i(x_i, p)$, the modulus G is replaced by $(G\boldsymbol{\bar{Q}}/\boldsymbol{\bar{P}})$ and $(\lambda + G) = (K + G/3)$ by $[(G\boldsymbol{\bar{Q}} + \bar{\lambda})/\boldsymbol{\bar{P}}]$, the Laplace transform of the solution for the elastically

---

[1] R. Sips: J. Polymer Sci. **5**, 69 (1950); **6**, 285 (1951); **7**, 191 (1952). — I. N. Zverev: Prikl. Math. Mech. **15**, 295 (1950). — W. T. Read: J. Appl. Phys. **21**, 671 (1950). — M. Bieniek: Arch. Mech. Stos., Warsaw **4**, 43 (1952). — E. H. Lee: Quart. Appl. Math. **13**, 183 (1955).

compressible visco-elastic medium $\bar{\sigma}_{ij}^v(x_i, p)$ and $\bar{u}_i^v(x_i, p)$ is obtained. The visco-elastic solution can be obtained by inverse transform of $\bar{\sigma}_{ij}^v(x_i, p)$ and $\bar{u}_i^v(x_i, p)$. Alternatively, it might be possible to replace the elastic constants in the untransformed elastic solution directly by the visco-elastic operators and integrate the resulting ordinary differential equations in time[1]. Thus the large number of existing solutions of elastic boundary-value problems can be directly utilized for the derivation of the associated visco-elastic solutions.

The visco-elastic solution satisfies the same boundary conditions as the elastic solution from which it has been derived. This statement does not imply, however, that the elastic and the associated visco-elastic solutions are identical along the boundaries, nor that the geometrical similarities of the associated problems ensure similarity of the space distribution of the boundary conditions. At points where displacements are prescribed as, for instance, in problems of plane strain the resulting boundary stresses are obtained from operator equations and will, therefore, differ from the stresses of the elastic solution; when the boundary conditions are prescribed by functions $t_i(x_i, t)$ and $u_i(x_i, t)$ their Laplace transforms $\bar{t}_i(x_i, p)$ and $\bar{u}_i(x_i, p)$ will have different space-distributions, unless the boundary conditions can be represented by product functions

$$t_i(x_i, t) = t_i(x_i) f(t), \qquad u_i(x_i, t) = u_i(x_i) f_1(t) \qquad (27.7\text{a})$$

in which case

$$\bar{t}_i(x_i, t) = t_i(x_i) \bar{f}(p), \qquad \bar{u}_i(x_i, t) = u_i(x_i) \bar{f}_1(p) \qquad (27.7\text{b})$$

so that the space-functions are unaffected by the transformation. These simplifying conditions, designated as "proportional loading", permit the superposition of loading conditions differing in their time functions.

The assumption of proportional loading is expedient, but not necessary. The physical inconvenience of this assumption can be avoided by representing the boundary and initial conditions with the aid of the Heaviside unit-step-function $H(\vartheta, t)$

$$t_i(x_i, t) = t_i(x_i) H(\vartheta, t) \qquad (27.8)$$

where $H(\vartheta, t) = 0$ for $t < \vartheta$ and $H(\vartheta, t) = 1$ for $t \geq \vartheta$. Since the transform of this function $\bar{H}(\vartheta, p) = e^{-\vartheta p}/p$, or for $\vartheta = 0$, $\bar{H}(p) = 1/p$, its use permits, in general, the inversion of the solution $\bar{\sigma}_{ij}^v(x_i, p)$ by partial fractions.

Considering, as an example, a spherical cavity of radius R under internal pressure $-q(t)$ in the infinite incompressible medium, the elastic radial displacement at $r \geq R$

$$u_r = q(t) \frac{R^3}{4Gr^2}.$$

The visco-elastic solution is obtained when G is replaced by GQ/P and the resulting differential equation

$$2GQ u_r = P\left[q(t) \frac{R^3}{2r^2}\right]$$

is solved either directly or by inverting its transform. Thus, for instance, for the Maxwell body with $Q = \partial/\partial t$, $P = (\partial/\partial t + \tau_M^{-1})$ and constant R the equation

$$v_r = (\dot{q} + \tau_M^{-1} q) \frac{R^3}{4Gr^2}$$

---

[1] See footnote 3, p. 271; also A. M. FREUDENTHAL and H. G. LORSCH: Proc. 2nd U.S. Congr. Appl. Mech., p. 539, New York 1954.

establishes the pressure q as a function of a prescribed radial velocity v_r and vice-versa. When $q(t)$ is prescribed the elastic and visco-elastic stresses are identical; when $v_R(t)$ is prescribed the stresses are functions of the momentary value of q and therefore different from the elastic stresses. If the increase of R as a result of the pressure is no longer infinitesimal, the time function $R(t)$ must be considered and the operator applied to the product $\boldsymbol{P}[q(t) \cdot R^3(t)]$, producing the differential relation

$$4 G r^2 \dot{R}/R^3 = \dot{q} + [2\dot{R}/R + \tau_M^{-1}]\, q$$

between q and R. The same method is applied when the elastic solution is non-linear as, for instance, in the case of the problem of contact solved by Hertz[1]. The visco-elastic solution for impact between a visco-elastic relatively light and soft moving sphere and a heavy, almost rigid mass of smooth spherical surface is by elastic analogy

$$\alpha^{\tfrac{3}{2}} = \frac{C}{4\pi G}\; \frac{\boldsymbol{P}(K + \tfrac{4}{3} G\, \boldsymbol{Q})}{\boldsymbol{Q}(K + \tfrac{1}{3} G\, \boldsymbol{Q})}\; F(t)$$

where α is the "compliance" (relative displacement of mass centers), $F(t)$ the force and C a coefficient incorporating the geometry. This equation can be written directly and solved simultaneously with the equations of motion $m\ddot{x} = -F(t)$ by the use of Laplace transforms[2].

28. The effect of inhomogeneity.

The very pronounced structure- and temperature-sensitivity of the coefficient of shear viscosity as distinct from the insensitivity with respect to structure and temperature of the elastic moduli produces a highly structure- and temperature-sensitive visco-elastic response which has no elastic parallel.

Consideration of the structure-sensitivity of the visco-elastic response usually takes the form of dissipation potentials of higher than first order in the second invariants of the stress or the strain-velocity deviation (see Sect. 26), leading to quasi-linear stress-strain-rate relations with parameters that are not only invariant functions of the coordinates but, through the time-dependence of the stress-field, also functions of time. The associated differential equations are non-linear with variable coefficients and can at best be solved by numerical approximation.

In the presence of a stationary or transient temperature-field all parameters of the stress-strain relations that contain viscosity-coefficients become functions of this field according to Eq. (26.16); they are sharply varying functions of the space-coordinates when the temperature-field is stationary, as well as functions of time when the temperature-field is transient. In the latter case the transforms of the visco-elastic operators are therefore not easily obtained, considering the usual forms of the solutions of transient heat-conduction problems.

For a stationary temperature-field the gradients of the operators differ from zero. Therefore Eq. (27.3) takes the form

$$\bar{\sigma}_{ij,j} = 2G\left[\left(\frac{\overline{\boldsymbol{Q}}}{\overline{\boldsymbol{P}}}\right)_{,j} \bar{\varepsilon}_{ij} + \left(\frac{\overline{\boldsymbol{Q}}}{\overline{\boldsymbol{P}}}\right) \bar{\varepsilon}_{ij,j}\right] + \left(\frac{\bar{\lambda}}{\overline{\boldsymbol{P}}}\right)_{,i} \bar{\varepsilon}_{jj} + \left(\frac{\bar{\lambda}}{\overline{\boldsymbol{P}}}\right) \bar{\varepsilon}_{jj,i} - 3K\alpha\,\overline{T}_{,i} = 0. \quad (28.1)$$

Hence, Eq. (27.5) becomes

$$G\left(\frac{\overline{\boldsymbol{Q}}}{\overline{\boldsymbol{P}}}\right)_{,j}(\bar{u}_{i,j} + \bar{u}_{j,i}) + \left(\frac{\bar{\lambda}}{\overline{\boldsymbol{P}}}\right)_{,i} \bar{u}_{j,j} + \left[\frac{G\,\overline{\boldsymbol{Q}} + \bar{\lambda}}{\overline{\boldsymbol{P}}}\right] \bar{u}_{j,ji} + G\left(\frac{\overline{\boldsymbol{Q}}}{\overline{\boldsymbol{P}}}\right) \bar{u}_{i,jj} - 3\alpha K\,\overline{T}_{,i} = 0. \quad (28.2)$$

---
[1] H. Hertz: Ges. Werke: **1**, 155 (1895).
[2] Y. Pao: J. Appl. Phys. **26**, 1083 (1955). — I. E. Prakopovic: Prikl. Math. Mech. **20**, 680 (1956).

Taking the divergence,

$$\left[\frac{2G\bar{Q}+\bar{\lambda}}{\bar{P}}\right]\bar{u}_{j,jii} + 2\left[\frac{G\bar{Q}+\bar{\lambda}}{\bar{P}}\right]_{,i}\bar{u}_{j,ji} + 2G\left(\frac{\bar{Q}}{\bar{P}}\right)_{,j}\bar{u}_{j,ii} +$$
$$+ G\left(\frac{\bar{Q}}{\bar{P}}\right)_{,ij}(\bar{u}_{i,j}+\bar{u}_{j,i}) + \left(\frac{\bar{\lambda}}{\bar{P}}\right)_{,ii}\bar{u}_{j,j} - 3\alpha K\bar{T}_{,ii} = 0.$$
(28.3)

Comparison of Eqs. (28.2) and (28.3) with Eqs. (27.5) and (27.6) illustrates the difficulty introduced by the consideration of the space-dependence of the

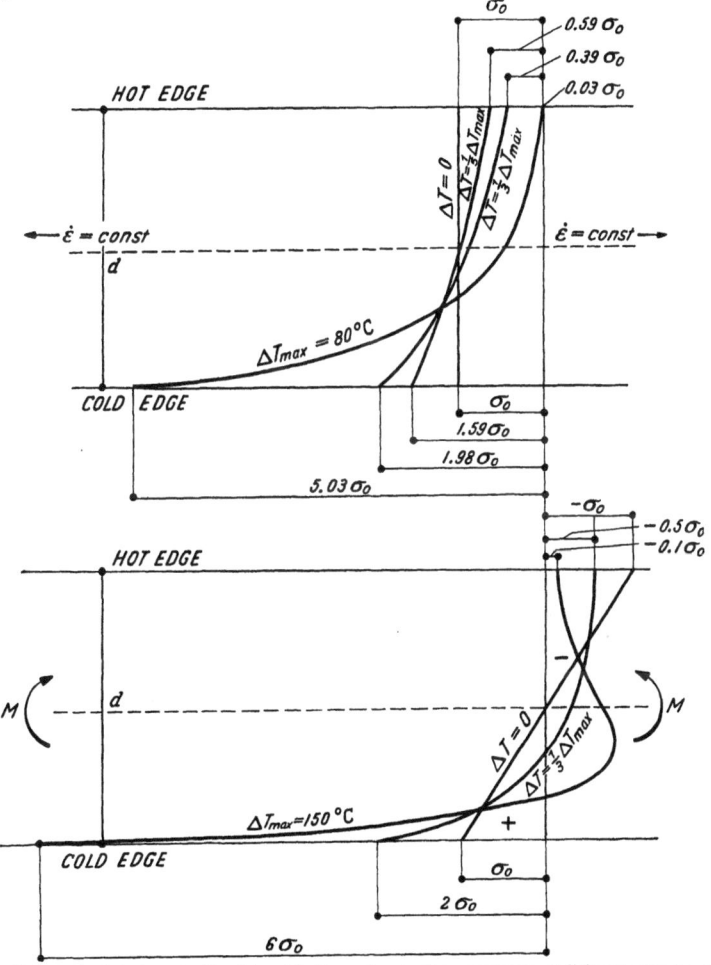

Fig. 5. Stress-distribution in flat visco-elastic strip with linear temperature gradients of different severity ΔT under conditions of (above) uniform extension ($\dot{\varepsilon}$ = const.) and (below) pure bending (M = const.) [Q/RT_0 = 50, T_0 = 800° K].

visco-elastic response in the form of space-derivatives of the operators. However, the assumption of temperature-insensitive viscosity coefficients in the presence of temperature-fields, while resulting in simpler equations, leads to unrealistic solutions. This is illustrated in Fig. 5 for the elementary problems of uniform extension and pure bending in the presence of stationary linear temperature gradients in a linear visco-elastic medium at time $t \to \infty$[1].

---

[1] A. M. FREUDENTHAL: Trans. N. Y. Acad. Sci. 19, 328 (1957). — Chap. 11 in C. BONILLA (ed.) Nuclear Engineering, p. 578. New York: McGraw-Hill 1957.

The effect on the distribution of stresses due to applied forces of the inhomogeneity of the visco-elastic response accompanying an inhomogeneous temperature-field may considerably exceed the effect of the thermal stresses themselves resulting from this temperature-field, particularly in the relaxing visco-elastic medium, such as the Maxwell body, in which thermal stresses are transient. The redistribution of the force-induced stresses persists under a stationary temperature-field long after the initial thermal stresses have been eliminated by relaxation; upon removal of the temperature-field and of the acting forces, complex fields of long persistent residual stresses remain, since their rate of relaxation is sharply decreased by the lowering of the temperature. The compatibility equations for the visco-elastic medium are obtained by introducing Eqs. (18.5b) into Eqs. (5.16):

$$\nabla^2 \sigma_{ij} + \frac{2}{3}\left(1 + \frac{G\boldsymbol{Q}}{3K\boldsymbol{P}}\right)\sigma_{kk,ij} = -2G\alpha \frac{\boldsymbol{Q}}{\boldsymbol{P}}\left[\frac{9\boldsymbol{P}}{3\boldsymbol{P} + 4G\boldsymbol{Q}/K}\nabla^2 T \delta_{ij} + T_{,ij}\right] \quad (28.4)$$

which for $T=0$ is the visco-elastic equivalent of BELTRAMI's Eqs. (5.17). Defining a stress-function F by $\sigma_{xx} = F_{,yy}$, $\sigma_{yy} = F_{,xx}$ and $\sigma_{xy} = -F_{,xy}$ so that the equilibrium equations are automatically fulfilled, the differential equation for the stress-function in plane stress has the form

$$\frac{3PK + G\boldsymbol{Q}}{K}\nabla^4 F = -9G\alpha \boldsymbol{Q}\nabla^2 T. \quad (28.5)$$

The equation for the stress-function in the inhomogeneous visco-elastic medium

$$\frac{1}{9K}\nabla^4 F + \frac{2}{3}\nabla^2\left[\left(\frac{\boldsymbol{P}}{2G\boldsymbol{Q}}\right)\nabla^2 F\right] - \left(\frac{\boldsymbol{P}}{2G\boldsymbol{Q}}\right)_{,xx} F_{,yy} - \left(\frac{\boldsymbol{P}}{2G\boldsymbol{Q}}\right)_{,yy} F_{,xx} + \\ + 2\left(\frac{\boldsymbol{P}}{2G\boldsymbol{Q}}\right)_{,xy} F_{,xy} = -\alpha \nabla^2 T \quad (28.6)$$

again illustrates the difficulty of actually solving boundary-value problems in such media. Certain simplifications in Eq. (28.6) are, however, obtained in the case of linear and exponential space-dependence of the coefficients a_i and b_i in the operators $\boldsymbol{P}$ and $\boldsymbol{Q}$ [see Eq. (18.4)].

In the case of spherical symmetry the differential equation for the radial component of the stress-deviation $s_r = \sigma_r + p = \frac{2}{3}(\sigma_r - \sigma_\vartheta)$ is obtained by combining the equilibrium condition

$$\frac{\partial \sigma_r}{\partial r} + \frac{2}{r}(\sigma_r - \sigma_\vartheta) = \frac{\partial p}{\partial r} + \frac{1}{r^3}\frac{\partial}{\partial r}(r^3 s_r) = 0$$

with the compatibility condition

$$\frac{\partial \varepsilon_\vartheta}{\partial r} - \frac{1}{r}(\varepsilon_r - \varepsilon_\vartheta) = \frac{\partial \varepsilon}{\partial r} + \frac{1}{r^3}\frac{\partial}{\partial r}(r^3 e_r) = 0$$

where $e_r = \varepsilon_r + \varepsilon = \frac{2}{3}(\varepsilon_r - \varepsilon_\vartheta)$, and with the stress-strain relations

$$\boldsymbol{P} s_r = 2G \boldsymbol{Q} e_r \quad \text{and} \quad p = 3K(\varepsilon - \alpha T).$$

The resulting equation[1]

$$\frac{\partial}{\partial r}\left[\left(\frac{\boldsymbol{P}}{\boldsymbol{Q}} + \frac{4G}{3K}\right) r^3 s_r\right] = 4G r^3 \frac{\partial}{\partial r}(\alpha T)$$

or, integrating with respect to r,

$$\left(\boldsymbol{P} + \frac{4G}{3K}\boldsymbol{Q}\right) r^3 s_r = 4G \boldsymbol{Q}\int_{r_0}^{r} \varrho^3 \frac{\partial}{\partial \varrho}(\alpha T)\, d\varrho + F(t),$$

---

[1] A. M. FREUDENTHAL: See footnote 5, p. 293.

where r_0 is the inner radius of the sphere and $F(t)$ a function of integration to be determined from the initial conditions, can be solved for relatively simple operators $\boldsymbol{P}$ and $\boldsymbol{Q}$ and specified solution $T(r, t)$ of the heat equation.

In the case of cylindrical symmetry two simultaneous differential equations for s_r and s_z are obtained in the same way.

29. Dynamic problems. The equations of motion of the elastically compressible linear visco-elastic continuum with space-dependent operators are obtained by introducing the transform of the first term of Eq. (4.8) into Eq. (28.2), setting $\bar{T}=0$. Hence

$$\bar{\sigma}_{ij,j} = G\left(\frac{\bar{Q}}{\bar{P}}\right)_{,j}(\bar{u}_{i,j}+\bar{u}_{j,i}) + \left(\frac{\bar{\lambda}}{\bar{P}}\right)_{,i}\bar{u}_{j,j} + \left[\frac{G\bar{Q}+\bar{\lambda}}{\bar{P}}\right]\bar{u}_{j,ji} + G\frac{\bar{Q}}{\bar{P}}\bar{u}_{i,jj} = \varrho p^2 \bar{u}_i. \quad (29.1)$$

The transform of the distortional wave-equation follows from $\bar{u}_{j,j}=0$

$$G\frac{\bar{Q}}{\bar{P}}\bar{u}_{i,jj} + G\left(\frac{\bar{Q}}{\bar{P}}\right)_{,j}(\bar{u}_{i,j}+\bar{u}_{j,i}) = \varrho p^2 \bar{u}_i, \quad (29.2)$$

that of the equation for irrotational waves from $\bar{u}_{i,j}=\bar{u}_{j,i}$

$$\left[\frac{2G\bar{Q}+\bar{\lambda}}{\bar{P}}\right]\bar{u}_{i,jj} + 2G\left(\frac{\bar{Q}}{\bar{P}}\right)_{,j}\bar{u}_{i,j} + \left(\frac{\bar{\lambda}}{\bar{P}}\right)_{,i}\bar{u}_{j,j} = \varrho p^2 \bar{u}_i; \quad (29.3)$$

by taking the divergence and neglecting the second derivatives of the viscoelastic operators, the first approximation to the dilatational wave-equation is obtained:

$$\left[\frac{2G\bar{Q}+\bar{\lambda}}{\bar{P}}\right]\bar{u}_{i,ijj} + 2\left[\frac{2G\bar{Q}+\bar{\lambda}}{\bar{P}}\right]_{,j} u_{i,ij} = \varrho p^2 \bar{u}_{i,i}. \quad (29.4)$$

It follows that the two types of waves are coupled only through the second space derivatives of the visco-elastic operators.

For homogeneous visco-elastic constants Eqs. (29.1) degenerate into

$$\bar{\sigma}_{ij,j} = \left[\frac{G\bar{Q}+\bar{\lambda}}{\bar{P}}\right]\bar{u}_{j,ji} + G\frac{\bar{Q}}{\bar{P}}\bar{u}_{i,jj} = \varrho p^2 \bar{u}_i, \quad (29.5)$$

Eqs. (29.2) into

$$G\frac{\bar{Q}}{\bar{P}}\bar{u}_{i,jj} = \varrho p^2 \bar{u}_i, \quad (29.6)$$

while Eqs. (29.3) and (29.4) can be written in the equivalent forms

$$\left[\frac{2G\bar{Q}+\bar{\lambda}}{\bar{P}}\right]\bar{u}_{i,jj} = \varrho p^2 \bar{u}_i \quad \text{or} \quad \left[\frac{2G\bar{Q}+\bar{\lambda}}{\bar{P}}\right]\bar{u}_{i,ijj} = \varrho p^2 \bar{u}_{i,i}. \quad (29.7)$$

Eqs. (29.5) to (29.7) are standard forms of transformed wave-equations in the elastic medium wich visco-elastic operators replacing the elastic constants. For instance for a plane wave with $u_i=(u_1,0,0)$ Eq. (29.5) has the form

$$\left[\frac{2G\bar{Q}+\bar{\lambda}}{\bar{P}}\right]\left(\frac{\partial^2 \bar{u}_1}{\partial x_1^2}\right) = \varrho p^2 \bar{u}_1. \quad (29.8)$$

Introducing the boundary conditions $u_1=f(t)$ at $x_1=0$ and $u_1\to 0$ as $x_1\to\infty$ so that $\bar{u}_1=\bar{f}(p)$ at $x_1=0$ and $\bar{u}_1\to 0$ as $x_1\to\infty$, the transform of the solution is obtained in the form

$$\left. \begin{aligned} \bar{u}_1 &= \bar{f}(p)\exp\left(-x_1 p\sqrt{\frac{\varrho}{(K+4G\bar{Q}/3\bar{P})}}\right) = \bar{f}(p)\exp\frac{-x_1 p}{\sqrt{a^2+4c^2\bar{Q}/3\bar{P}}} \\ &= \bar{f}(p)\bar{\psi}(x_1,p) \end{aligned} \right\} \quad (29.9)$$

where $[(2G\overline{Q}+\overline{\lambda})/\overline{P}] = [K+4G\overline{Q}/3\overline{P}]$, $a^2 = K/\varrho$ and $c^2 = G/\varrho$. If a velocity $v_1 = \varphi(t)$ at $x_1 = 0$ is prescribed instead of the displacement, $\bar{v}_1 = p\,\bar{f}(p) = \bar{\varphi}(p)$, the function $\bar{f}(p)$ is replaced by $\bar{\varphi}(p)/p$. For plane distortional waves the transformed visco-elastic operator $[(2G\overline{Q}+\overline{\lambda})/\overline{P}]$ is replaced by $G\overline{Q}/\overline{P}$ and therefore

$$\bar{u}_1 = \bar{f}(p)\exp\frac{-x_1 p}{c\sqrt{\overline{Q}/\overline{P}}} = \bar{f}(p)\,\bar{\psi}_1(x_1, p). \tag{29.10}$$

Thus the solution of the visco-elastic wave-equations for specified boundary and initial conditions becomes a problem of inversion of transform products by convolution and integration in the complex plane. Solutions have been obtained for relatively simple operators P and Q, and conditions of plane, spherical and cylindrical waves[1].

In the case of propagation of pulses the assumption of zero initial conditions must be abandoned and $\bar{\ddot{u}}_i = p^2 \bar{u}_i(p) - p u_i(0) - \dot{u}_i(0)$ introduced instead of $\bar{\ddot{u}}_i = p^2 \bar{u}_i(p)$, or the initial conditions be expressed with the aid of the Heaviside unit step function, for instance $v_i(0, t) = v_0 H(t)$ or $u_i(0, t) = v_0 t H(t)$ for impact at given velocity v_0 at $x_i = 0$.

For particularly simple forms of the operators the wave-equations can be solved directly. Thus, for instance, with $P = 1$, $Q = (1 + \tau_k\,\partial/\partial t)$ Eq. (29.8) is transformed into STOKES' equation[2]

$$\ddot{u} = c_1^2\,\frac{\partial^2}{\partial x_1^2}\left(u + \frac{4\eta}{3\varrho c_1^2}\dot{u}\right) = \frac{1}{\varrho}\,\frac{\partial\sigma}{\partial x_1} \tag{29.11}$$

where $c_1^2 = (\lambda+2G)/\varrho = (K+4G/3)/\varrho$ is the velocity of the elastic dilatational wave.

For a periodic disturbance of initial amplitude $u(0, t) = A\,e^{i\omega t}$ the solution is of the form

$$u = A\,e^{-\alpha x_1}\exp i\omega(t - x_1/c), \tag{29.12}$$

where α is the *attenuation coefficient* and c the real phase velocity of the travelling wave. Introducing this solution into Eq. (29.11) two simultaneous equations in α and c are obtained by equating real and imaginary parts[3]

and
$$\left.\begin{aligned}\left(\frac{c}{c_1}\right)^2 &= \frac{2}{k^2\omega^2}(1+k^2\omega^2)\left[\sqrt{(1+k^2\omega^2)} - 1\right] \\ \alpha &= \frac{c}{c_1}\,\frac{k^2\omega^2}{2k c_1(1+k^2\omega^2)} = \frac{\omega}{c_1\sqrt{2(1+k^2\omega^2)}}\left[\sqrt{(1+k^2\omega^2)} - 1\right]^{\frac{1}{2}}\end{aligned}\right\} \tag{29.13}$$

where $k = 4\eta/3\varrho c_1^2$. When $k\omega$ is small compared to unity the phase velocity $c = c_1$ is independent of frequency and $\alpha = \omega^2 k/2c_1$. When $k\omega$ is large $c = c_1\sqrt{2k\omega}$ and $\alpha = \sqrt{\omega/2k}/c_1$.

While the untransformed Eq. (29.8) is a standard form only for the Kelvin body, the untransformed Eq. (29.6) can be solved directly for both the Kelvin and Maxwell bodies. In the first case

$$u_{i,jj} + \tau_K \dot{u}_{i,jj} = \frac{1}{c_0^2}\ddot{u}_i, \tag{29.14}$$

---

[1] R. Sips, I.N. Zverev, M. Bieniek: See footnote 1, p. 294. — E.H. Lee and I. Kanter: J. Appl. Phys. **24**, 1115 (1953). — R.D. Glauz and E.H. Lee: J. Appl. Phys. **25**, 947 (1954). — D.S. Berry and S.C. Hunter: J. Mech. Phys. of Solids **4**, 72 (1956).
[2] G.G. Stokes: Cambridge Trans. **8**, 287 (1845).
[3] K.W. Hillier: Proc. Phys. Soc. Lond. B **62**, 701 (1949); **64**, 998 (1951). — H. Kolsky: Stress-Waves in Solids, p. 116. Oxford: University Press 1953.

in the second case

$$v_{i,jj} = \frac{1}{c_0^2}(\ddot{v}_i + \dot{v}_i/\tau_M) \tag{29.15}$$

where $c_0^2 = G/\varrho$. Eq. (29.14) is identical in form with Stokes equation, Eq. (29.15) is the "telegraph equation".

Satisfying one-dimensional visco-elastic wave-equations by solutions (29.12) written in the alternative form

$$u = A \exp\{i[\omega t - (\beta - i\alpha)x_1]\} \tag{29.16}$$

expressions for $\alpha^2(\omega)$ and $\beta^2(\omega)$ are obtained from which the phase velocity $c = \omega/\beta(\omega)$ and the specific damping $4\pi c \alpha(\omega)/\omega = 4\pi \alpha(\omega)/\beta(\omega)$ can be derived. Thus, for instance, for a progressing shear wave in a Maxwell body according to Eq. (29.15)[1],

$$\beta^2 = \frac{\omega^2}{2c_0^2}\left[\sqrt{1+\left(\frac{1}{\omega\tau}\right)^2}+1\right]; \quad \alpha^2 = \frac{\omega^2}{2c_0^2}\left[\sqrt{1+\left(\frac{1}{\omega\tau}\right)^2}-1\right] \tag{29.17}$$

with phase velocity $c^2 = \omega^2/\beta^2 = 2c_0^2/[\sqrt{1+\tau^{-2}\omega^{-2}}+1]$ and specific damping $4\pi\alpha/\beta = 4\pi[(\sqrt{1+\tau_M^{-2}\omega^{-2}}-1)/(\sqrt{1+\tau_M^{-2}\omega^{-2}}+1)]^{\frac{1}{2}}$. When $\omega \gg 1/\tau_M$ and therefore $\omega\tau_M$ is large, $c = c_0$ and $\alpha \to 0$. The specific damping $4\pi\alpha/\beta$ increases with decreasing frequency. While the double value of β is associated with the two directions of progressive motion with velocities $\pm c$ only the negative root of the second Eq. (29.17) has a physical meaning.

Stationary solutions of visco-elastic wave-equations are obtained by superimposing solutions of the form $\varphi_i(x_i) e^{\omega t}$. Thus, for instance, substituting this form into Eq. (29.14) the equation for the shear modes in a Kelvin body is obtained

$$\varphi_{i,jj} c_0^2 (1 + \tau_K \omega) = \omega^2 \varphi_i \tag{29.18}$$

The conditions that the solutions $\varphi_i(x_i)$ must satisfy the boundary conditions of the problem produce the frequency equation which, in the case of the Kelvin body, is quadratic with roots that are either conjugate complex with negative real part (damped oscillation) or real and negative (exponential decay). The damping of the individual model depends on geometry and is higher for high modes[2].

II. The visco-plastic medium.

30. Two-dimensional flow. The deviatoric stress-strain-rate relations of the visco-plastic continuum Eqs. (25.4) imply the coincidence of the principal directions of the stress deviation s_{ij}, the yield stress deviation k_{ij} and the strain-rate deviation $\dot{e}''_{ij}$, and thus the proportionality of the deviatoric components of the three tensors

$$s_{ij} = \psi k_{ij} \quad \text{and} \quad \dot{e}''_{ij} = (\psi - 1) k_{ij}/2\eta \tag{30.1}$$

where $\psi^2 = J_2'/k^2$. The von Mises yield criterion Eq. (22.1) is assumed to determine the transition from the rigid (or elastic) into the plastic state; at this limit therefore $\psi = 1$, $s_{ij} = k_{ij}$ and $\dot{e}''_{ij} = 0$.

The stress-strain-rate relation in invariant form is obtained by squaring Eqs. (30.1)

$$4\eta^2 \dot{I}_2' = \left[\frac{(\psi-1)}{\psi}\right]^2 J_2' \tag{30.2a}$$

and

$$\sqrt{J_2'} = \pm k + 2\eta \sqrt{\dot{I}_2'}. \tag{30.2b}$$

---

[1] N. Ricker: Bull. Seism. Soc. **33**, 197 (1943). — A. W. Nolle: J. Acoust. Soc. Amer. **19**, 184 (1947).
[2] See footnote 6, p. 269.

For incompressible plane flow $\ddot\varepsilon_{11}'' + \ddot\varepsilon_{22}'' + \ddot\varepsilon_{33}'' = \ddot\varepsilon_1'' + \ddot\varepsilon_2'' + \ddot\varepsilon_3'' = 0$ and $\ddot\varepsilon_3'' = 0$; therefore $\ddot\varepsilon_1'' = -\ddot\varepsilon_2''$ and $\ddot\varepsilon_1'' : \ddot\varepsilon_2'' : \ddot\varepsilon_3'' = 1 : -1 : 0$. As previously shown the corresponding motion is a plane shear defined by $v_1 = 2\dot\gamma x_2$, $v_2 = 0$, $v_3 = 0$ in the direction of the maximum shear stress $s_{12} = \sigma_{12} = \tau_{max}$, since the only non-vanishing components of the strain-rate tensor are $\ddot\varepsilon_{12}'' = \ddot\varepsilon_{21}'' = \dot\gamma$, and the principal strain components according to Eq. (6.2) are $\ddot\varepsilon_1'' = -\ddot\varepsilon_2'' = \dot\gamma$ (see Sect. 23). Hence Eq. (30.2b) with $\sqrt{J_2'} = \pm s_{12}$ and $\sqrt{I_2'} = \pm\dot\gamma$ is of the form

$$s_{12} = \tau_{max} = k + 2\eta\dot\gamma \qquad (30.3)$$

or, since $\tau_{max} = \tfrac{1}{2}(\sigma_1 - \sigma_2)$ and $\dot\gamma = \tfrac{1}{2}(\ddot\varepsilon_1'' - \ddot\varepsilon_2'')$,

$$(\sigma_1 - \sigma_2) = 2k + 2\eta(\ddot\varepsilon_1'' - \ddot\varepsilon_2''). \qquad (30.4)$$

Since the state of stress at any point can be considered as a superposition of three plane shears represented by the components $s_{ij}(i \neq j)$ upon an isotropic pressure which does not affect the incompressible flow and can therefore be neglected, the six Eq. (25.1) or (25.4) are reduced to three equations of the type (30.3)

$$s_{ij} = k_{ij} + 2\eta \ddot e_{ij}'' \quad (i \neq j) \qquad (30.5)$$

valid for $|s_{ij}| \geq k$.

Considering that for an arbitrary coordinate system (x, y) the relations between τ_{max} and $\dot\gamma$ and the components of the stress tensor and the strain-rate tensor respectively are

$$\tau_{max} = \tfrac{1}{2}\sqrt{(\sigma_y - \sigma_x)^2 + 4\sigma_{xy}^2}, \quad \dot\gamma = \tfrac{1}{2}\sqrt{(\ddot\varepsilon_y'' - \ddot\varepsilon_x'')^2 + 4\ddot\varepsilon_{xy}''^2} \qquad (30.6)$$

and that the angles between the x-axis and the directions of maximum shear and maximum stress, respectively, are defined by

$$\tan 2\vartheta = \frac{\sigma_y - \sigma_x}{2\sigma_{xy}} = \frac{\ddot\varepsilon_y'' - \ddot\varepsilon_x''}{2\ddot\varepsilon_{xy}''}; \quad \tan 2\vartheta' = \frac{2\sigma_{xy}}{\sigma_y - \sigma_x} = \frac{2\ddot\varepsilon_{xy}''}{\ddot\varepsilon_y'' - \ddot\varepsilon_x''}, \qquad (30.7)$$

the relations are obtained

$$(\sigma_y - \sigma_x) = 2\tau_{max}\sin 2\vartheta; \quad (\ddot\varepsilon_y'' - \ddot\varepsilon_x'') = 2\dot\gamma\sin 2\vartheta. \qquad (30.8)$$

Therefore, with $(\sigma_x + \sigma_y) = -2\sigma$ and $\ddot\varepsilon_y'' + \ddot\varepsilon_x'' = 0$,

$$\left.\begin{array}{ll} \sigma_x = -\sigma - \tau_{max}\sin 2\vartheta, & \ddot\varepsilon_x'' = -\dot\gamma\sin 2\vartheta, \\ \sigma_y = -\sigma + \tau_{max}\sin 2\vartheta, & \ddot\varepsilon_y'' = \dot\gamma\sin 2\vartheta, \\ \sigma_{xy} = \tau_{max}\cos 2\vartheta, & \ddot\varepsilon_{xy}'' = \dot\gamma\cos 2\vartheta \end{array}\right\} \qquad (30.9)$$

and, in conjunction with Eq. (30.3)

$$\left.\begin{array}{l} \sigma_x = -\sigma - 2\eta(\dot\gamma + K)\sin 2\vartheta, \\ \sigma_y = -\sigma + 2\eta(\dot\gamma + K)\sin 2\vartheta, \\ \sigma_{xy} = 2\eta(\dot\gamma + K)\cos 2\vartheta \end{array}\right\} \qquad (30.10)$$

where $K = k/2\eta$. Moreover, considering Eq. (5.19)

$$\left.\begin{array}{l} v_{x,x} = -\dot\gamma\sin 2\vartheta, \quad v_{y,y} = \dot\gamma\sin 2\vartheta, \\ v_{x,y} + v_{y,x} = 2\dot\gamma\cos 2\vartheta \end{array}\right\} \qquad (30.11)$$

the subscripts not to be summed. Eqs. (30.10) and (30.11) represent the basic equations of the theory of plane visco-plastic flow from which the differential

Two-dimensional flow.

equations for the unknown functions $\dot{\gamma}$ and ϑ can be obtained by combining Eq. (30.10) with the two equilibrium Eqs. (4.11 b) (with non-summed subscripts),

$$\left.\begin{array}{l} \sigma_{x,x} + \sigma_{xy,y} = 0 \\ \sigma_{xy,x} + \sigma_{y,y} = 0 \end{array}\right\} \qquad (30.12)$$

either directly or by specifying a stress-function F which automatically satisfies these equations, as well as by eliminating the velocity-components from Eqs. (30.11) directly or by introducing a stream-function W from which the velocity-components can be derived.

The direct method furnishes the system of two non-linear differential equations of second order for $\dot{\gamma}$ and ϑ, considering η to be a constant[1]

$$\left.\begin{array}{l} [(\dot{\gamma}+K)\cos 2\vartheta]_{,xx} - [(\dot{\gamma}+K)\cos 2\vartheta]_{,yy} + 2[(\dot{\gamma}+K)\sin 2\vartheta]_{,xy} = 0, \\ [\dot{\gamma}\sin 2\vartheta]_{,xx} - [\dot{\gamma}\sin 2\vartheta]_{,yy} - 2[\dot{\gamma}\cos 2\vartheta]_{,xy} = 0 \end{array}\right\} \qquad (30.13)$$

while the definitions of the stress- and stream-functions according to

$$\sigma_x = F_{,yy}, \qquad \sigma_y = F_{,xx}, \qquad \sigma_{xy} = -F_{,xy} \qquad (30.14)$$

and

$$v_x = -W_{,y}, \qquad v_y = W_{,x} \qquad (30.15\text{a})$$

and therefore

$$\dot{\varepsilon}''_x = -W_{,xy}, \qquad \dot{\varepsilon}''_y = W_{,xy}, \qquad 2\dot{\varepsilon}''_{xy} = W_{,xx} - W_{,yy} \qquad (30.15\text{b})$$

and their combination with Eqs. (30.10) and (30.11), respectively, provide the system of four differential equations

$$F_{,xx} - F_{,yy} = 4\eta(\dot{\gamma}+K)\sin 2\vartheta; \qquad F_{,xy} = -2\eta(\dot{\gamma}+K)\cos 2\vartheta \qquad (30.16)$$

and

$$W_{,xx} - W_{,yy} = 2\dot{\gamma}\cos 2\vartheta; \qquad W_{,xy} = \dot{\gamma}\sin 2\vartheta \qquad (30.17)$$

in the four inknown functions $F, W, \dot{\gamma}, \vartheta$. Eliminating $\dot{\gamma}$ and ϑ the two equations are obtained

$$\left.\begin{array}{l} (F_{,xx}-F_{,yy})(W_{,xx}-W_{,yy}) + 4F_{,xy} \cdot W_{,xy} = 0, \\ \sqrt{(F_{,xx}-F_{,yy})^2 + 4F_{,xy}^2} - 2\eta\sqrt{(W_{,xx}-W_{,yy})^2 + 4W_{,xy}^2} = 2k. \end{array}\right\} \qquad (30.18)$$

The simplest integrals of Eqs. (30.18) are

$$W = \mp\omega xy, \qquad F = \pm(k+2\eta\omega)y^2 \qquad (30.19)$$

where ω is an undetermined function of time, the upper signs referring to tension, the lower to compression, and represent the strip of uniform width subjected to uniform longitudinal stress $\sigma_x = \pm 2k + 4\eta\omega$.

Eqs. (30.13) or (30.18) can be reduced to one fourth-order non-linear differential equation for either F or W. For instance by eliminating $\dot{\gamma}$ and ϑ from Eq. (30.13) with the aid of Eqs. (30.17) the latter is obtained in the form

$$2\eta\nabla^4 W + k\left\{\left(\frac{\partial^2}{\partial x^2} - \frac{\partial^2}{\partial y^2}\right)\left[\frac{1}{\dot{\gamma}}(W_{,xx} - W_{,yy})\right] + 4\frac{\partial^2}{\partial x\partial y}\left[\frac{1}{\dot{\gamma}}W_{,xy}\right]\right\} = 0 \qquad (30.20)$$

where, according to the second Eq. (30.6)

$$\dot{\gamma} = \tfrac{1}{2}\sqrt{(W_{,xx}-W_{,yy})^2 + 4W_{,xy}^2} \qquad (30.20\text{a})$$

---

[1] A. A. ILYUSHIN: Uchenye Zapiski Mosk. Univ., Mekhanika 39, 1 (1940).

while according to Eq. (30.7)

$$\tan 2\vartheta = \frac{2W_{,xy}}{(W_{,xx}-W_{,yy})}. \tag{30.21}$$

For $k=0$, Eq. (30.20) reduces to the equation of slow motion of a viscous fluid, with $\eta=0$ and $\dot\gamma=\text{const.}$, to that of a perfectly plastic solid[1].

At a rigid boundary $W=\text{const.}$ and $\partial W/\partial n$ is given, as in the case of a viscous fluid. At the yield surface $W_{,xy}$ and $(W_{,xx}-W_{,yy})$ vanish together, while $W_{,x}$ and $W_{,y}$ are continuous.

An equation of form similar to Eq. (30.20) can be obtained for the stress-function F by eliminating $\dot\gamma$ and ϑ from Eqs. (30.13) with the aid of Eq. (30.16)

$$\nabla^4 F - k\left\{\left(\frac{\partial^2}{\partial x^2}-\frac{\partial^2}{\partial y^2}\right)\left[\frac{1}{\tau_{\max}}(F_{,xx}-F_{,yy})\right]+4\frac{\partial^2}{\partial x\partial y}\left[\frac{1}{\tau_{\max}}F_{,xy}\right]\right\}=0 \tag{30.22}$$

where, according to the first Eq. (30.6)

$$\tau_{\max}=\tfrac{1}{2}\sqrt{(F_{,xx}-F_{,yy})^2+4F_{,xy}^2} \tag{30.22a}$$

while

$$\tan 2\vartheta = -\frac{(F_{,xx}-F_{,yy})}{2F_{,xy}}. \tag{30.23}$$

For $k=0$ Eq. (30.22) reduces to the biharmonic equation governing the stress-field of the slowly moving viscous fluid, while for $\tau_{\max}=k$ Eq. (30.22) vanishes; the stress function for the ideal plastic solid is thus determined by Eq. (30.22a) with $\tau_{\max}=k$[1],

$$(F_{,xx}-F_{,yy})^2+4F_{,xy}^2=4k^2. \tag{30.22b}$$

Eqs. (30.20) and (30.22) were first obtained by ILYUSHIN[2]. Eq. (30.20) has also been derived independently by OLDROYD[3] directly from the equation of motion resulting from the combination of the stress-strain relations Eq. (25.1) with Eq. (4.10) and the incompressibility condition $v_{j,j}=0$

$$\varrho\frac{dv_i}{dt}+p_{,j}\delta_{ij}=s_{ij,j}=(2\eta_1\dot e''_{ij})_{,j}=(\eta_1 v_{i,j})_{,j}+\eta_{1,j}v_{j,i} \tag{30.24}$$

where $\eta_1=\eta+k/2\sqrt{I'_2}$ according to Eq. (25.2b) and therefore

$$\varrho\frac{\partial v_i}{\partial t}+\varrho v_{i,j}v_j+p_{,i}=\eta v_{i,jj}+k\left(\frac{v_{i,j}}{2\sqrt{I'_2}}\right)_{,j}+\left(\frac{k}{2\sqrt{I'_2}}\right)_{,j}v_{j,i}. \tag{30.25}$$

For slow rectilinear flow with $v_i(v,0,0)$ and the isotropic pressure-gradient $p_{,i}[\pi(t),0,0]$ Eq. (30.24) degenerates into

$$\varrho\frac{\partial v_i}{\partial t}+\pi(t)=(\eta_1 v_{i,j})_{,j}=\eta v_{i,jj}+k\left[\frac{v_{,j}}{|v_{,j}|}\right]_{,j} \tag{30.26}$$

since $2\sqrt{I'_2}=|v_{,j}|$ is the absolute value of the gradient of v; for stationary flow

$$\eta v_{i,jj}+k\left[\frac{v_{,j}}{|v_{,j}|}\right]_{,j}=\pi(t). \tag{30.26a}$$

Since $[v_{,j}/|v_{,j}|]$ is the unit vector normal to a curve of the family of curves $v=\text{const.}$, its divergence is the curvature of this line, positive if it is convex toward

---

[1] C. W. OSEEN: Ark. Mat. Astronom. Fysik, Ser. A **24**, No. 13 (1933).
[2] See footnote 1, p. 303.
[3] J. G. OLDROYD: Proc. Cambridge Phil. Soc. **44**, 214 (1948).

increasing v. Hence for straight line contours $v = \text{const.}$ and Eq. (30.26) reduces to

$$\varrho \frac{\partial v}{\partial t} + \pi(t) = \eta\, v_{,jj}. \tag{30.27}$$

Along a rigid boundary $v = v_0(t)$ with $v_0(t) = v_0(0)$ as the initial condition. Along the yield surface, across which the transition from the visco-plastic to the elastic or rigid state occurs, $v = \text{continuous}$ and $v_{,j} = 0$. Hence on the yield surface v is a function of time only. In stationary flow the elastic or rigid region within the stationary yield surface moves as a rigid body; in non-steady flow the yield surface will, in general, vary with time. Within the region of flow $v_{,j}$ must be continuous and $|v_{,j}| > 0$, while the incompressibility of the flow requires $v_{i,i} = v_{,i} = 0$. If $v_{,j}$ is continuous throughout the considered region, the flow is viscous; if the flow is plastic throughout so that $s_{ij} = k_{ij}$, a discontinuity of the velocity-field exists along a rigid boundary. If, however, a narrow layer exists between a fixed rigid boundary and a yield surface across which $v_{,j}$ varies continuously but rapidly from zero to the velocity of the elastic or rigid "plug" enclosed by this surface, conditions of a "boundary layer" are created which are typical of visco-plastic problems in which part of the considered region remains rigid or elastic because in this region $J'_2 < k^2$, while visco-plastic flow is concentrated within relatively narrow regions.

Eqs. (30.26) and (30.27) have been used by OLDROYD in the analysis of problems of rectilinear flow [1]; their similarity to the equation of heat-conduction for a material with variable diffusivity in which heat is generated at the rate $-\pi(t)$ per unit mass is obvious. When the non-vanishing components v_i of plane flow in Eq. (30.25) are replaced by the gradients of the stream-function W according to Eq. (30.15a) and accelerations and body forces are neglected, Eq. (30.20) is obtained. OLDROYD has used this equation as well as Eq. (30.26) directly for the derivation of visco-plastic boundary layer equations and the analysis of certain simple boundary layer problems[1,2].

31. Boundary value problems and the stability of visco-plastic flow. The difficulties of integrating Eqs. (30.13), 30.18), (30.20) or (30.22) are considerable, and only a very limited number of solutions of simple problems can be obtained.

The utilization of the variational principles, Eqs. (9.12) and (9.13), is limited by the fact that the integrand $\int [\dot{X}_v(\dot{e}_{ij}) + \dot{X}_p(\dot{e}_{ij})]\, dV$ is not a quadratic form, since according to Eqs. (8.15) Eq. (9.12) takes the explicit form

$$\delta\left[\int (2\eta\, \dot{I}'_2 + 2k\sqrt{\dot{I}'_2})\, dV - \int t_i v_i\, dA - \int X_i v_i\, dV\right] = 0 \tag{31.1}$$

or, for stationary plane flow [see Eq. (30.3)]

$$\delta \int (2\eta\, \dot{\gamma}^2 + 2k\, \dot{\gamma})\, dx\, dy = 0. \tag{31.2}$$

Comparison of the first integral of Eq. (31.1) with Eq. (25.6) shows that this integral is not the rate of energy dissipation in the visco-plastic medium, but the sum of one-half of the rate of the viscous and the total rate of plastic dissipation, in accordance with the definitions of the viscous and the plastic potentials as $2\dot{X}_v(\dot{e}_{ij}) = T\vartheta_v$ but $\dot{X}_p(\dot{e}_{ij}) = T\vartheta_p$ (see Sect. 8).

Transforming the viscous dissipation potential $\dot{X}_v(\dot{e}_{ij})$ into $\dot{X}_v(s_{ij})$ with the aid of Eq. (30.5) the explicit form of Eq. (9.13) is obtained

$$\delta\left[\int t_i v_i\, dA - \frac{1}{2\eta}\int (\sqrt{J'_2} - k)^2\, dV\right] = 0, \tag{31.1a}$$

---
[1] See footnote 3, p. 304.
[2] J. G. OLDROYD: Proc. Cambridge Phil. Soc. **43**, 383 (1947).

considering that $\delta \dot{X}_p(s_{ij}) = 0$ because of the yield condition. By introducing Eqs. (30.6) and (30.15b) into Eq. (31.2) ILYUSHIN[1] has shown that Eq. (30.19) can be derived from the extremum principle Eq. (31.1). However, the use of the Ritz method for finding solutions of specific boundary values by minimizing the function in Eq. (31.2) presumes that a small number of functions can be found the sum of which, with arbitrary multipliers, would closely approximate this function.

Another approach to the solution of visco-plastic boundary value problems due to ILYUSHIN is the method of "neighboring motions", in which a known stream function W of a certain problem A with boundary conditions $\Gamma(A)$ is used to obtain the stream function of the problem Z with boundary conditions $\Gamma(Z)$ by consecutively deriving the solutions of the "neighboring problems" $B, C, \ldots, Y$ with boundary conditions $\Gamma(B), \Gamma(C), \ldots, \Gamma(Y)$, each of which differs only slightly from the preceding one. Assuming that all quantities characterizing two neighboring motions differ by infinitesimal amounts, so that the stream functions W and $(W + \Delta W)$ of the two motions differ by ΔW, which is everywhere small in comparison to W, the differential equation and the boundary conditions of a neighboring motion with respect to the preceding one are linearized. The solution of problem Z is thus obtained as the last step in the consecutive integration of these equations for all intermediate problems.

While the computational difficulties of the outlined method are considerable, it permits the obtaining of solutions where more direct methods fail. The differential equations governing the neighboring motions are obtained by replacing in Eqs. (30.18) the (known) stress-function F by the neighboring function $(F + \Delta F)$ and the (known) stream-function W by the stream-function of the neighboring motion $(W + \Delta W)$, using the first two terms of a Taylor or binomial expansion and thus neglecting second and higher order derivatives of ΔF and ΔW in comparison to the corresponding derivatives of F and W. Considering, moreover, that the (known) functions F and W satisfy Eqs. (30.18), the resulting linear differential equations with variable coefficients for the functions ΔF and ΔW can be written in the form

$$\left. \begin{aligned} & (F_{,xx} - F_{,yy})(\Delta W_{,xx} - \Delta W_{,yy}) + (W_{,xx} - W_{,yy})(\Delta F_{,xx} - \Delta F_{,yy}) + \\ & \qquad\qquad + 4 F_{,xy}\,\Delta W_{,xy} + 4 W_{,xy}\,\Delta F_{,xy} = 0, \\ & \sqrt{(W_{,xx} - W_{,yy})^2 + 4 W_{,xy}^2}\,[(F_{,xx} - F_{,yy})(\Delta F_{,xx} - \Delta F_{,yy}) + 4 F_{,xy}\,\Delta F_{,xy}] - \\ & \qquad - 2\eta\sqrt{(F_{,xx} - F_{,yy})^2 + 4 F_{,xy}^2} \times \\ & \qquad \times [(W_{,xx} - W_{,yy})(\Delta W_{,xx} - \Delta W_{,yy}) + 4 W_{,xy}\,\Delta W_{,xy}] = 0. \end{aligned} \right\} \quad (31.4)$$

ILYUSHIN[1] has used these equations for the analysis of neighboring motions for the plane strip and the cylinder subject to an axial load, as well as for the hollow circular cylinder subject to internal pressure, particularly with reference to the problem of the stability of the visco-plastic flow, which he was the first to recognize and to define in terms of the stability of a "neighboring motion" introduced in the form of an arbitrary surface-perturbation; the motion is unstable or stable depending on whether the inital perturbation has a tendency to increase or to decrease with time; when the perturbation remains unaffected, the flow is "neutral". Problems of the stability of visco-plastic flow were also studied by ISHLINSKI[2].

---

[1] See footnote 1, p. 303.
[2] A. I. ISHLINSKI: Prikl. Math. Mech. **7**, 110 (1943).

Sect. 31.   Boundary value problems and the stability of visco-plastic flow.

For the simplest case of the plane strip of length $2l$ and width $2h$ with the stress- and stream-functions according to Eq. (30.19) and therefore $W_{,xx} - W_{,yy} = 0$, $W_{,xy} = \mp \omega$, $F_{,xx} - F_{,yy} = \mp 2(k + 2\eta\omega)$ and $F_{,xy} = 0$, Eqs. (31.4) take the form

$$\left. \begin{array}{l} (\Delta W_{,xx} - \Delta W_{,yy}) + \dfrac{2\omega}{k + 2\eta\omega} \Delta F_{,xy} = 0, \\ -4\eta \Delta W_{,xy} + (\Delta F_{,xx} - \Delta F_{,yy}) = 0. \end{array} \right\} \quad (31.5)$$

Under the assumption that the perturbed boundary remains symmetric with respect to the x-axis the solutions of Eqs. (31.5) can be assumed in the form

$$\Delta W = \sum_{n=1}^{n=\infty} w_n(y) \sin \frac{n\pi x}{l}, \quad \Delta F = \sum_{n=1}^{n=\infty} f_n(y) \cos \frac{n\pi x}{l} \quad (31.6)$$

which, introduced into the Eqs. (31.5), produce the system of linear differential equations for w_n and f_n

$$\left. \begin{array}{l} w_n'' + \left(\dfrac{n\pi}{l}\right)^2 w_n + \dfrac{2\omega}{k + 2\mu\omega} \left(\dfrac{n\pi}{l}\right) f_n' = 0, \\ f_n'' + \left(\dfrac{n\pi}{l}\right)^2 f_n + 4\eta \left(\dfrac{n\pi}{l}\right) w_n' = 0. \end{array} \right\} \quad (31.7)$$

The solutions are fitted to the boundary conditions including those along the surfaces with the total perturbation $\bar{\delta}$ represented by the Fourier series $\bar{\delta} = \sum \delta_n \cos(n\pi x/l)$. The stress-components of the disturbed flow are obtained from Eqs. (30.14) in which F is replaced by $(F + \Delta F)$, the velocity-components in a similar way from Eqs. (30.15a). The instability of the motion will depend on whether the resulting velocity-components v_y at the disturbed surface have the same sign as the perturbation or the opposite sign.

The analysis of the stability of the flow thus reduces to a study of the stability of the individual components $\delta_n \cos(n\pi x/l)$ of the perturbation $\bar{\delta}$. The m-th component will be stable, and the motion relative to this component unstable if the ratio $v_y(h)/\delta_m \cos(m\pi x/l) > 0$; if this is the only stable component, it will in time dominate all other components, and the boundary will finally be transformed into a cosine curve forming m waves within the length of the strip. The deformation of the strip will therefore be unstable if a single component of the perturbation is stable.

The analysis with the aid of Eq. (31.4) of the stability of two-dimensional visco-plastic flow in the vicinity of the perturbed boundary is thus rather elaborate; the characteristic instability of this flow can, however, be simply illustrated by the analysis of the volume-constant uniaxial visco-plastic extension under a force P of a thin cylindrical bar of initial length l_0 and uniform cross-section A_0, but with one slightly reduced ("perturbed") initial section $A_0' < A_0$ in which therefore the stress $\sigma' = P/A' > P/A = \sigma$. From Eq. (30.2b) with $J_2' = \sigma^2/3$, $\sigma_0 = k\sqrt{3}$, $\sqrt{I_2'} = \dot{\varepsilon}\sqrt{3}/2$,

$$\dot{\varepsilon} = \frac{(\sigma - \sigma_0)}{3\eta} = \frac{(P/A - \sigma_0)}{3\eta} \quad (31.8)$$

or, considering that $d(Al) = 0$ and therefore $dA/A = -dl/l = -\varepsilon$,

$$-A\dot{\varepsilon} = \frac{dA}{dt} = -\frac{(P - \sigma_0 A)}{3\eta} \quad (31.9)$$

the solutions of which, for the initial conditions $A = A_0$ and $A' = A_0'$ respectively at $t = 0$, are[1]

and

$$\left.\begin{array}{l}\sigma_0(A_0 - A) = (P - A_0\sigma_0)(e^{\frac{\sigma_0 t}{3\eta}} - 1) \\ \sigma_0(A_0' - A') = (P - A_0'\sigma_0)(e^{\frac{\sigma_0 t}{3\eta}} - 1)\end{array}\right\} \qquad (31.10)$$

and therefore,

$$\frac{(A - A')}{(A_0 - A_0')} = \frac{\Delta A}{\Delta A_0} = e^{\frac{\sigma_0 t}{3\eta}} \qquad (31.11)$$

where the increase with time of the ratio $\Delta A/\Delta A_0$ is a measure of the increase of the initial depth of the surface perturbation characterizing the instability of the motion. This instability vanishes as σ_0 tends towards zero (viscous flow), as for $\sigma_0 = 0$ Eq. (31.9) has the solutions

$$A_0 - A = A_0' - A' = \frac{Pt}{3\eta} \qquad (31.12)$$

and therefore $\Delta A/\Delta A_0 = 1$; the same result is obtained from Eq. (31.10) for stresses $\sigma \gg \sigma_0$, producing high velocities. Hence the instability of the visco-plastic extension tends to vanish as its character becomes increasingly viscous, either through high strain-velocities or very low yield limit.

D. Problems of structural mechanics.

I. Visco-elastic structures.

32. The elastic equations. The equations of structural mechanics for the analysis of struts, beams and plates the depth d of which is small in relation to their length or span l are derived on the basis of the Navier-Bernoulli assumptions[2] that in elastic bending the effect of shear can be neglected, that plane cross-sections remain plane and that, moreover, the deflections w in the direction y perpendicular to the axis x of the structure are small enough to replace the curvature of the deformed axis $1/R$ by $-(d^2w/dx^2)$. The Bernoulli-Euler differential equation for the straight or slightly curved elastic beam subject to a bending moment M resulting from the above assumptions is

$$\frac{d^2 w}{dx^2} = -\frac{M(x)}{EI} \qquad (32.1)$$

where E is the elastic (Hooke) modulus and I the moment of inertia of the cross-section. Combined with the relation $p(x) = -d^2M/dx^2$ between the distributed load $p(x)$ and the resulting moment, Eq. (32.1) becomes, considering that $p(x, t)$ may also be a function of time,

$$[EI w_{,xx}]_{,xx} = p(x, t). \qquad (32.2)$$

The equation for the uniform elastic plate obtained by Lagrange under the same assumptions[3] is

$$\nabla^2(D \nabla^2 w) = D \nabla^4 w = p(x, z, t) \qquad (32.3)$$

where $D = E d^3/12(1 - \nu^2)$.

---
[1] E. Orowan: Bedford Conference on Principles of Rheological Measurment, p. 156. London: Nelson 1949.
[2] J. Bernoulli: Oeuvres **1**, 976 (1744).
[3] C. L. M. H. Navier: Ann. Chim. Phys. **39**, 149, 207 (1828).

Considering that the total distributed load p in Eqs. (32.2) and (32.3) is the sum of the applied actual load intensity p_0, of the resistance of a supporting medium, usually assumed to be, in first approximation, proportional to the local deflection w and thus equal to βw, where β is a factor of proportionality (resistance modulus), of the "effective load" p^* due to the bending action of a longitudinal force $P(t)$ acting in the axis of the beam, $p^*(x) = -M^*_{xx} = -(Pw)_{,xx}$ or of longitudinal forces $N_x(t)$, $N_y(t)$ and shear forces $N_{xy}(t) = N_{yx}(t)$ acting in the central plane of the plate[1]

$$p^* = -(N_x w_{,xx} + 2 N_{xy} w_{,xy} + N_y w_{,yy})$$

and finally of the mass forces $\varrho \ddot{w}\, dV = m\ddot{w}$, Eqs. (32.2) and (32.3) can be written in the form

$$[E I w_{,xx}]_{,xx} = p_0(x, t) - \beta w - p^*(x, t) - m\ddot{w} \tag{32.4}$$

and

$$\nabla^2(D \nabla^2 w) = p_0(x, z, t) - \beta_1 w - p^*(x, z, t) - m\ddot{w}. \tag{32.5}$$

The negative sign of the term p^* indicates that with increasing deformation an increasing part of the load p_0 is carried in tension along the central axis or plane of the structure and thus reduces the bending moments. When P or N_x and N_y represent external compression forces applied at the supports, which increase the bending moments due to p_0, the term p^* in Eqs. (32.4) and (32.5) is added to p_0 as it increases the deflection of the structure due to p_0.

The solutions of problems of structural mechanics, except those encountered in the analysis of structures of large curvature (curved beams, shells), can be treated as solutions of boundary value problems of Eqs. (32.4) and (32.5) under various assumptions concerning the four terms of their right hand sides. Thus, for quasi-static problems of bending $p_0 \neq 0$, $\beta = 0$, $p^* = 0$, $m\ddot{w} = 0$, of buckling $p_0 = 0$, $\beta = 0$, $p^* > 0$, $m\ddot{w} = 0$, of bending with axial load $p_0 \neq 0$, $\beta = 0$, $p^* \gtrless 0$, $m\ddot{w} = 0$; the same conditions modified by the resistance of an elastic medium require $\beta > 0$ for $w > 0$ and $\beta = 0$ for $w < 0$ (one-sided resistance) or $\beta > 0$ (surrounding medium). Dynamic problems are characterized by $m\ddot{w} \neq 0$, while the three other terms may either all vanish, producing the simple wave equations, or be different from zero, producing the equations of forced dynamic motion.

33. Quasi-static visco-elastic problems. The general equations for linear visco-elastic beams and plates are obtained by replacing in the elastic Eqs. (32.4) and (32.5) the elastic coefficients by visco-elastic operators. According to Eq. (18.7) the operator equivalent of HOOKE's modulus E in the elastically compressible visco-elastic continuum is

$$E = \frac{9KGQ}{(3KP + GQ)} = \frac{9GQ}{3P + GQ/K} \tag{33.1}$$

and therefore Eq. (32.4) is transformed into[2]

$$9GQ[Iw_{,xx}]_{,xx} = \left(3P + \frac{GQ}{K}\right)(p_0 - \beta w - p^* - m\ddot{w}). \tag{33.2}$$

Comparing Eq. (18.6) with the elastic relation for plane stress

$$\sigma_{xx} = \frac{E}{1-\nu^2}(\varepsilon_{xx} + \nu\varepsilon_{zz}) = \frac{2G}{1-\nu}\left(\varepsilon_{xx} + \frac{1}{\nu}\varepsilon_{zz}\right)$$

---

[1] TH. v. KARMAN: Enzykl. math. Wiss. **4**, 349 (1910).
[2] A. M. FREUDENTHAL: 6th Internat. Congr. Appl. Mech., Paris 1946. — Proc. 5th Internat. Congr. Bridge and Struct. Eng., Lisbon 1956.

the operator equivalent of the plate rigidity D in the elastically compressible viscoelastic continuum is

$$D = \frac{d^3}{12} 4GQ(3KP + GQ)/P(3KP + 2GQ) \tag{33.3}$$

the equivalent of ν being

$$\nu = (3KP - 2GQ)/2(3KP + GQ).$$

Hence Eq. (32.5) is transformed into the visco-elastic plate equation

$$4G\frac{d^3}{12} Q\left(3P + \frac{GQ}{K}\right) \nabla^4 w = P\left(3P + 2\frac{GQ}{K}\right)(p_0 - \beta_1 w - p^* - m\ddot{w}). \tag{33.4}$$

The form of Eqs. (33.2) and (33.4) shows that the consideration of the elastic compressibility produces unmanageable differential equations assuming even a relatively simple visco-elastic response. A considerable simplification of these equations is achieved by introducing the assumption of incompressibility ($K \to \infty$ or $\nu \to \frac{1}{2}$), as a result of which Eq. (33.2) is transformed into

$$3GQ[Iw_{,xx}]_{,xx} = P(p_0 - \beta w - p^* - m\ddot{w}) \tag{33.5}$$

and Eq. (33.4) into

$$4G\frac{d^3}{12} Q \nabla^4 w = DQ \nabla^4 w = P(p_0 - \beta_1 w - p^* - m\ddot{w}) \tag{33.6}$$

in which form their solution becomes manageable either by direct integration or by inversion of the Laplace transform of the elastic solution.

The simplest idealization of the behavior of structures showing creep or relaxation is obtained by introducing the operators of the incompressible Maxwell body $P = (\partial/\partial t + \tau_M^{-1})$ and $Q = \partial/\partial t$ into Eqs. (33.5) and (33.6) and solving them for the various relevant conditions.

The principal difference in behaviour between elastic and relaxing or creeping visco-elastic structures is the *relaxation* of all stresses due to imposed deformation, such as the stresses due to external restraints, particularly temperature-stresses and stresses due to the motion of supports, as well as the *magnification* with time of initial excentricities of load action, leading to time-dependent instability of excentrically loaded structures, such as struts and arches. Expressing for the incompressible Maxwell body p_0 and p^* in Eq. (33.5) in terms of the bending moment $M(x, t)$ and $M^*(x, t)$ assuming $EI = \text{const}$, $\beta = 0$, and $\ddot{w} = 0$ the equation obtained

$$EI\dot{w}_{,xx} + (\dot{M} - \dot{M}^*) + \tau_M^{-1}(M - M^*) = 0 \tag{33.7}$$

where $M^* = \pm Pw$ and $\dot{M}^* = \pm(P\dot{w} + \dot{P}w)$, is the equation for quasi-static bending ($M^* = 0$) or buckling ($M = 0$, $M^* < 0$) or combined bending and axial load ($M \neq 0$, $M^* \neq 0$). The elastic moments $M_0 = -EIw_{0,xx}$ induced by an initial deformation w_0 applied instantaneously at $t=0$ and sustained ($\dot{w}_0 = 0$) for $t > 0$ relax according to $M = M_0 \exp(-t/\tau_M)$. On the other hand, the solution of the creep-buckling equation of a strut of length l with $P = \text{const}$ and $\dot{M}^* = -P\dot{w}$

$$\dot{w}_{,xx} + \frac{P}{EI}(\dot{w} + \tau_M^{-1} w) = 0 \tag{33.8}$$

which has the form[1]

$$w(x, t) = \sum_{n=1}^{n=\infty} \exp\frac{a_n t}{\tau_M} [A_n \sin(a_n x) + B_n \cos(a_n x)] \tag{33.9}$$

---

[1] See footnote 2, p. 309.

where $a_n = \sqrt{(1 + 1/\alpha_n) P/EI}$ is determined by the specific boundary conditions, indicates an increase with time of the deflection in all modes at a rapidly increasing rate proportional to $\exp(\alpha_n t \tau_M^{-1})$, and therefore rapidly increasing bending moments $M_n^* = P w_n(x, t)$. The strut can thus fail in any mode and under any load P, provided its time of application is long enough to produce a critical excentricity. "Creep-buckling" is therefore not a time-dependent instability of the configuration but one of gradual, no matter how rapid, failure in bending with axial load. The same phenomenon reduces the stability of flat arches constructed of materials showing creep[1], as well as of plates and shells. An extensive analysis of the effects of creep in the design of various types of engineering structures is due to ARUTUNIAN[2].

For space-dependent operators $\boldsymbol{P}$ and $\boldsymbol{Q}$ but uniform I and d Eqs. (33.5) and (33.6) take the form

$$3GI \left[\frac{\boldsymbol{Q}}{\boldsymbol{P}} w_{,xx}\right]_{,xx} = 3GI\left\{\left[\frac{\boldsymbol{Q}}{\boldsymbol{P}}\right]_{,xx} w_{,xx} + 2\left[\frac{\boldsymbol{Q}}{\boldsymbol{P}}\right]_{,x} w_{,xxx} + \left[\frac{\boldsymbol{Q}}{\boldsymbol{P}}\right] w_{,xxxx}\right\} \qquad (33.10)$$
$$= p_0 - \beta w - p^* - m\ddot{w}$$

and

$$D \nabla^2 \left[\frac{\boldsymbol{Q}}{\boldsymbol{P}} \nabla^2 w\right] = p_0 - \beta w - p^* - m\ddot{w}. \qquad (33.11)$$

34. Dynamic problems. The equation of longitudinal motion of a thin linear visco-elastic strut is obtained by combining Eq. (4.8) for one dimension with the visco-elastic stress-strain relation $\sigma = \sigma_x = \boldsymbol{E}\,\varepsilon_x = \boldsymbol{E} u_{,x}$; hence

$$\sigma_{,x} = [\boldsymbol{E} u_{,x}]_{,x} = \boldsymbol{E}_{,x} u_{,x} + \boldsymbol{E} u_{,xx} = \varrho \ddot{u}. \qquad (34.1)$$

The equations of transverse motion of technical beams and plates, neglecting the effect of shear and rotatory inertia, are obtained by eliminating on the right hand side of Eqs. (33.2) and (33.4) all but the acceleration terms. Hence for the isotropic beam

$$9G\boldsymbol{Q}[I w_{,xx}]_{,xx} + \left(3\boldsymbol{P} + \frac{G\boldsymbol{Q}}{K}\right) m\ddot{w} = 0 \qquad (34.2)$$

while for the plate

$$\boldsymbol{D} \nabla^4 w + m\ddot{w} = 0. \qquad (34.3)$$

As in the case of quasi-static problems, relatively simple solutions of these equations can be obtained only if the elastic compressibility is neglected. Hence according to Eqs. (34.1), (33.5) and (33.6) the following simplified equations of motion are obtained for the longitudinal vibrations of a strut

$$3G\boldsymbol{Q} u_{,xx} - \boldsymbol{P} \varrho \ddot{u} = 0, \qquad (34.4)$$

the transverse vibration of a beam, neglecting the effects of shear and rotatory inertia

$$3G\boldsymbol{Q}[I w_{,xx}]_{,xx} + \boldsymbol{P} m\ddot{w} = 0, \qquad (34.5)$$

and those of a plate of uniform thickness d

$$\frac{G d^3}{3} \boldsymbol{Q} \nabla^4 w + \boldsymbol{P} m\ddot{w} = 0. \qquad (34.6)$$

The correction for shear in Eq. (34.5) can be introduced by adding the elastic correction term multiplied by $\boldsymbol{P}$, the effect of an axial force by adding the term

---
[1] A. M. FREUDENTHAL: Publ. Internat. Assoc. Bridge and Struct. Eng. 4, 249 (1936).
[2] N. CH. ARUTUNIAN: Certain Problems in the Theory of Creep. Moscow 1952.

Pp^*, while the equations of forced vibration are obtained by setting the right-hand side of the equations of transverse motion equal to $Pp(x, t)$.

Introducing solutions of the standard form

$$u = \sum_n u_n(x)\, \varphi_n(t), \quad w = \sum_n w_n(x)\, \varphi_n(t) \quad \text{or} \quad \sum_n w_n(x, z)\, \varphi_n(t) \tag{34.7}$$

each of the Eqs. (34.4) to (34.6) is transformed into a system of equations of the type

and

$$\frac{3G}{\varrho}\frac{u_{n,xx}}{u_n} = \frac{P\ddot{\varphi}_n}{Q\varphi_n} = \text{const} \tag{34.8}$$

$$\frac{3GI}{m}\frac{w_{n,xxxx}}{w_n} = -\frac{P\ddot{\varphi}_n}{Q\varphi_n} = \text{const} \tag{34.9}$$

or

$$\frac{Gd^3}{3m}\frac{\nabla^4 w_n}{w_n} = -\frac{P\ddot{\varphi}_n}{Q\varphi_n} = \text{const}. \tag{34.10}$$

Hence for incompressible visco-elastic materials the shapes of the modes governed by the left-hand sides of Eqs. (34.8) to (34.10) are identical with the elastic modes; the time-dependence is obtained by solving the total differential equation for $\varphi_n(t)$ on the right hand side for specific operators. Solutions of specific problems have been obtained for the Maxwell and the Kelvin bodies for which the equations are of second order and therefore easily manageable[1].

For steady state oscillations at a certain imposed frequency ω_k the function $\varphi_k(t) = e^{i\omega_k t}$ and the right hand sides of Eqs. (34.8) to (34.10) $P\ddot{\varphi}/Q\varphi$ are transformed into $-\omega^2 P(i\omega_k)/Q(i\omega_k)$. Since the real and imaginary components of the complex modulus $GQ(i\omega_k)/P(i\omega_k) = G'(\omega_k) + iG''(\omega_k)$ can be determined experimentally at each test frequency of a simple structure, such as a shallow beam ("vibrating reed") by measurement of amplitude and phase difference or damping, the relaxation spectrum of the material can be derived, if the range of test frequencies is sufficiently wide[2] (see Sect. 20).

For space-dependent operators Eqs. (34.8) to (34.10) must be replaced respectively by Eq. (34.1) and Eqs. (33.10) and (33.11), on the right hand side of which only the term involving the acceleration remains. Thus, for instance Eq. (33.10) takes the form

$$\left[\frac{Q}{P}\right]_{,xx} w_{,xx} + 2\left[\frac{Q}{P}\right]_{,x} w_{,xxx} + \left[\frac{Q}{P}\right] w_{,xxxx} + \left(\frac{m}{3GI}\right)\ddot{w} = 0. \tag{34.11}$$

For media with $P = 1$, such as the Kelvin body, this equation is simplified, particularly if the ratios $Q_{,xx}/Q$ and $Q_{,x}/Q$ are constant, as in the case of exponential variability.

The analysis of forced vibrations proceeds in a similar manner as for elastic structures Thus, for instance, the problem of the longitudinal vibration of a visco-elastic rod the free end of which is subject to the axial force $P_0 \sin\omega t$ (or the formally identical problem of the torsional vibration with forcing sinusoidal torque) governed by the transformed Eq (34.1).

$$\bar{E}\bar{u}_{,xx} = \varrho\, p^2 \bar{u} \tag{34.12}$$

---

[1] K. Honda and S. Onno: Phil. Mag. 42, 115 (1921). — S. L. Quimby: Phys. Rev. 25, 558 (1925). — K. Sezawa: Bull. Earthq. Res. Inst. Toyko 3, 43 (1927). — H. Jeffreys: Geophys. Suppl. Roy. Astr. Soc. 2, 318 (1931). See also footnote 6, p. 269.

[2] A. J. Staverman and F. Schwarzl: Chap. 1, Sect. 6,7 of H. A. Stuart, Physik der Hochpolymeren, vol. 4. Berlin: Springer 1956.

with the solution

$$\bar{u} = C_1 \exp\left(x p \sqrt{\frac{\varrho}{E}}\right) + C_2 \exp\left(-x p \sqrt{\frac{\varrho}{E}}\right) \tag{34.13}$$

is solved by inversion of Eq. (34.13) the constants of which have been determined by the boundary conditions $u=0$ at $x=0$ and $E A u_{,x} = P_0 \sin \omega t$ or $A \bar{E} \bar{u}_{,x} = P_0 \frac{\omega}{p^2 + \omega^2}$. Problem sof forced transverse vibrations are solved by introducing on the right hand side of Eqs. (34.5) and (34.6) forcing functions $P p(x, t) = p(x) P \psi(t)$ or $P p(x, z, t) = p(x, z) P \psi(t)$ and expanding $p(x)$ or $p(x, z)$ in terms of the elastic modes. The time-functions $\varphi_n(t)$ are obtained by solving the equations

$$\frac{m}{3GI} P \ddot{\varphi}_n + \text{const.} \, Q \, \varphi_n = \frac{a_{0n}}{3GI} P \psi(t) \tag{34.14a}$$

and

$$\frac{3m}{G d^3} P \ddot{\varphi}_n + \text{const.} \, Q \, \varphi_n = \frac{3 a_{0n}}{G d^3} P \psi(t), \tag{34.14b}$$

where a_{0n} is a measure of the average load-intensity and the constant is defined by Eqs. (34.9) or (34.10). The differential equations governing the propagation of pulses in thin rods of visco-elastic materials with constant parameters are obtained by introducing the respective expressions for E according to Eq. (33.1) into Eq. (34.1). Thus for the Kelvin body with $P=1$ and $Q=1+\tau\frac{\partial}{\partial t}$ the differential equation is obtained

$$\ddot{u} = c^2 u_{,xx} + \left(\frac{c}{c_0}\right)^2 \frac{3\eta}{\varrho} \frac{\partial}{\partial t} \left[u_{,xx} - \frac{1}{3 c_1^2} \ddot{u}\right] \tag{34.15}$$

where $c_0^2 = 3 G/\varrho$, $c^2 = E/\varrho$ and $c_1^2 = 3K/\varrho$. For the incompressible medium $c_0 = c$ and $c_1 = \infty$. Hence

$$\ddot{u} = c^2 (u_{,xx} + \tau \dot{u}_{,xx}). \tag{34.15a}$$

Similarily for the Maxwell body with $P = \partial/\partial t + 1/\tau$, $Q = \partial/\partial t$ and $v = \dot{u}$

$$v_{,xx} = \frac{1}{c_0^2}\left[1 + \frac{1}{3}\left(\frac{c_0}{c_1}\right)^2\right]\ddot{v} + \frac{1}{c_0^2 \tau}\dot{v} \tag{34.16}$$

which, for the incompressible medium, degenerates into the one-dimensional "telegraph equation" (29.15)

$$c_0^2 v_{,xx} = \ddot{v} + \tau^{-1}\dot{v}. \tag{34.17}$$

Eqs. (34.15) and (34.17) are standard forms and can be conveniently treated by operational methods for various boundary conditions and the stress-waves obtained from the equation $\sigma_{,x} = \varrho \dot{v}$[1].

II. Elastic-plastic structures.

35. One-dimensional quasi-static problems. While the general theory of the plastic and elastic-plastic continuum, including boundary value problems in plane stress and plane strain, forms the subject of the Second Part of this article, the specific use of this theory in the analysis of the plastic carrying capacity of one-dimensional simple metallic structures represents an important part of the theory of inelastic structures and is therefore included here.

---

[1] See footnote 1, p. 300.

The basis of this analysis is the development of "*yield-hinges*" in one-dimensional elastic-plastic structural parts such as technical beams and struts subject to bending moments and axial loads; assumedly these "hinges" start to form when the extreme elastic fiber stress at a critical section attains the yield stress in uni-axial tension σ_0, and are completed when the triangular distribution of the elastic stresses has been transformed into the discontinuous fully plastic rectangular stress-distribution with vanishing elastic core (Fig. 6), the moment of which defines the maximum or "flow" moment M_0 which the "hinge" can carry in the absence of axial loads. Denoting by $\pm h$ the vertical distance from the neutral axis of the interface between the plastic region and the elastic core of a beam of depth d

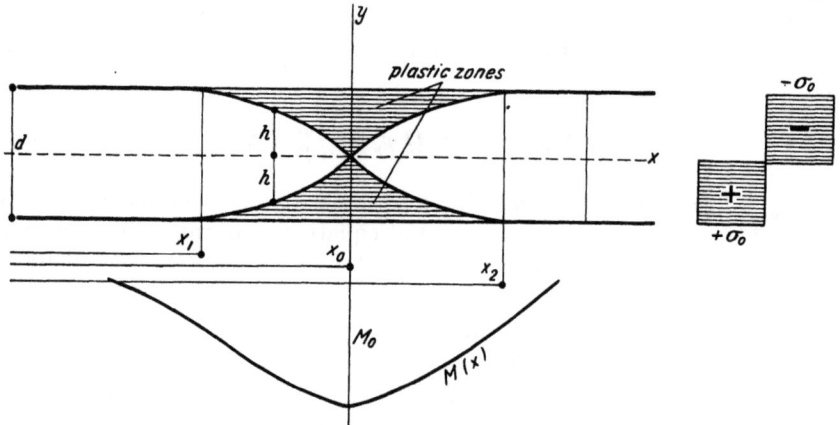

Fig. 6. Yield-hinge in beam of symmetric cross-section.

with symmetric cross section A of width $b(y)$ and by ξ the ratio $\xi = 2h/d$, the bending moment

$$M(x) = 2 \int_0^{d/2} \sigma(x, y)\, y\, dA \tag{35.1}$$

where $dA = b(y)\,dy$, is obtained by introducing for $-h \leq y \leq h$ the stress in the elastic core $\sigma = \pm \sigma_0 y/h$, while for $h < y \leq d/2$ and $-d/2 \leq y < -h$ the plastic stress $\sigma = \sigma_0$. Hence from Eq. (35.1) for a rectangular section $b(y) = b$

$$M(x) = \tfrac{1}{4}\sigma_0 b\, d^2 (1 - \tfrac{1}{3}\xi^2) = M_0 (1 - \tfrac{1}{3}\xi^2) \tag{35.2}$$

and therefore

$$\xi(x) = \sqrt{3\left[1 - \frac{M(x)}{M_0}\right]}, \qquad 0 \leq \xi \leq 1. \tag{35.3}$$

Thus the beam of span l is fully plastic ("yield-hinge") for vanishing thickness of the elastic core or $\xi(x_0) = 0$ and $M(x_0) = M_0 = \tfrac{1}{4}\sigma_0 b\, d^2$; at $x \leq x_1$ and $x \geq x_2$, where $\xi(x) = 1$ and therefore $|M(x)| \leq 2M_0/3$, the section is fully elastic with maximum fiber stress $\pm \sigma_0$ at x_1 and x_2. For $x_1 < x < x_2$ the beam is partly plastic; the shape of the elastic-plastic boundary and the length of the elastic-plastic zone are determined by the form of $M(x)$ according to Eq. (35.3).

The deflection of the elastic-plastic portion of the beam within the range $x_1 < x < x_2$ is governed by the deflection of the elastic core, as the plastic region provides no additional restraint. Therefore, according to the Navier-Bernoulli assumption, the relation between the elastic strain ε at the elastic-plastic interface and the radius of curvature R is

$$\varepsilon = \pm \frac{\sigma_0}{E} = \pm \frac{h}{R} \approx -h w_{,xx} = -\frac{1}{2}\xi\, d w_{,xx} \tag{35.4}$$

or, considering Eq. (35.3)

$$w_{,xx} = -2\sigma_0/E\,d\sqrt{3\left[1-\left|\frac{M(x)}{M_0}\right|\right]}. \tag{35.5}$$

At the "yield-hinge" $M(x)=M_0$, $w_{,xx}=\infty$ and therefore $R=0$; the slope of the deflected beam changes discontinuously, characteristic of the motion in a hinge.

For the joint action in a strut of an axial force $N(x)$ and a bending moment $M(x)$ under the assumption that plane sections remain plane and that (non-symmetric) plastic regions of depth $\left(\frac{d}{2}-h_1\right)$ and $\left(\frac{d}{2}-h_2\right)$ have developed from both fibers of the rectangular section, Eq. (35.3) is replaced by

$$\xi_{1,2} = \pm\frac{N}{N_0} + \sqrt{3\left[1-\left(\frac{N}{N_0}\right)^2-\left|\frac{M}{M_0}\right|\right]} \quad \text{for} \quad -1\leq\xi_{1,2}\leq 1 \tag{35.6}$$

where $\xi_1 = 2h_1/d$, $\xi_2 = 2h_2/d$, $N_0 = \pm\sigma_0 bd$, and Eq. (35.5) by

$$w_{,xx} = -2\sigma_0/E\,d\sqrt{3\left[1-\left(\frac{N}{N_0}\right)^2-\left|\frac{M}{M_0}\right|\right]}. \tag{35.7}$$

The "yield-hinge" defined by $-w_{,xx}=\infty$ and $R=0$ is excentric and develops with center at $\xi_{1,2} = \pm N/N_0$ when $M(x)$ and $N(x)$ satisfy the yield condition

$$\left(\frac{N}{N_l}\right)^2 + \left|\frac{M}{M_0}\right| = 1 \tag{35.8}$$

which for $N=0$ degenerates into the yield condition for the beam, $M=M_0$. The introduction of the absolute value of $|M/M_0|$ reflects the symmetry of the yield conditions.

Eqs. (35.3) and (35.5) are due to FRITSCHE[1] who, following some previous studies by engineering designers[2], was the first to develop systematic methods of analysis of the plastic carrying capacity ("*collapse*" or "*limit*" load) of statically indeterminate beam structures; Eqs. (35.6) to (35.8) are due to GIRKMANN[3] who extended these methods to the analysis of frames.

The procedures of analysis generally referred to as "limit design" (Traglastverfahren) utilize the fact that the formation of m hinges in n-fold statically indeterminate structures reduces the degree of indeterminacy to $(n-m)$. Hence, the development of $(n+1)$ yield-hinges in the course of the deformation produced by a monotonically increasing ("proportional") system of loads transforms the structure into an unstable kinematic mechanism of one degree of freedom, which thus constitutes a plastic failure (instability) condition of the structure; the load-intensity associated with this condition represents the plastic "collapse" or "limit" load for proportional loading.

While the actual disposition and sequence of formation of yield-hinges can, in most cases, be established *a priori* or easily determined by a step by step

---

[1] J. FRITSCHE: Bauingenieur **11**, 851, 873, 888 (1930). — Z. angew. Math. Mech. **11**, 176 (1931). — Prel. Rep. 2nd Internat. Congr. Bridge and Struct. Eng., Berlin, p. 15, 1936.

[2] G. v. KAZINCZY: Betonszemle, Budapest **2**, No. 5, 6, 7 (1914). — Final Rep., 2nd Internat. Congr. Bridge and Struct. Eng., Berlin, p. 56, 1936. — N. C. KIST: Eisenbau **11**, 425 (1920). — H. MAIER-LEIBNITZ: Bautechnik **6**, 11, 27 (1928); **7**, 313, 366 (1929). See also H. MAIER-LEIBNITZ: Prel. Rep. 2nd Internat. Congr. Bridge and Struct. Eng., Berlin, p. 103, 1936.

[3] K. GIRKMANN: Ber. Akad. Wiss. Wien, Math.-naturw. Kl. **140**, 679 (1931). — Stahlbau **5**, 121 (1932).

analysis, such an analysis becomes cumbersome for highly redundant frame-structures in which alternative dispositions of yield-hinges satisfying the equilibrium conditions may exist. For this case PRAGER and coworkers[1] have established the principle that the actual collapse load is bounded by a "safe (i.e. stable) statically admissible" stress-field at or below the yield stress, satisfying the equilibrium conditions, which represents the lower limit, and by an "unsafe (i.e. unstable) kinematically admissible" velocity field, expressed in terms of rates of deflection or of change of curvature satisfying the boundary conditions, as the upper limit, which, however, does not satisfy the uniqueness principle of maximum rate of plastic work, the total rate of plastic strain work being less than the work of the external forces [see Eqs. (9.11)]. Thus, instead of actually solving the problem, which may be difficult since it requires coincidence of the two limits, an estimate of the range of the collapse load is obtained.

By neglecting the deformation in the elastic and elastic-plastic regions and introducing the assumption that the rate of plastic strain work is the sum of the rates of work in the yield hinges, which is the sum of the products of $Q_i = M_0$ by the respective angular velocities $q_i = \dot{\varkappa}_i$ the elastic-plastic structure is transformed into a *rigid-plastic* structure in which angular motion in the hinges represents the only type of non-rigid motion. The rates of total plastic strain work $M_0 \sum_i |\dot{\varkappa}_i|$ in the i yield-hinges of the alternative, kinematically admissible velocity fields are thus easily computed and compared; the alternative with extremal sum $\sum_i |\dot{\varkappa}_i|$ represents the best approximation. Because of the time-insensitivity of the plastic deformation the values $\dot{\varkappa}_i$ can be identified, to an arbitrary scale factor, with the angular rotations in the i hinges of the kinematic chain[1].

However, the rigid-plastic idealization of elastic-plastic structures presupposes a sharply concentrated elastic-plastic region on both sides of the theoretical hinge-section, such as occurs under concentrated loads and at fixed ends or at supports of continuous structures. Moreover, in the course of the transformation through the successive formation of yield-hinges of the elastic structure into the final collapse configuration (kinematical mechanism) the initially formed hinges are not necessarily preserved. The location of these hinges, which depends on the moment-distribution in the elastic structure, does not necessarily coincide with the location of hinges determined by the conditions of plastic collapse. The transformation might therefore be accompanied by rather complex processes of local unloading in existing hinges, while new hinges are formed in neighboring locations, conditions which interfere with the basic assumptions of rigid-plastic analysis.

For moments slowly changing along the structure, and particularly for a combination of bending moments with axial forces, the elastic-plastic region extends to both sides of the yield-hinge over a substantial portion of the span of the structure or structural part. For a uniformly loaded beam of span l and rectangular cross-section with a fully developed yield-hinge at midspan the length of the elastic-plastic section $(l/\sqrt{3})$ exceeds half the span; if the beam is continuous with additional fully developed yield-hinges over the supports elastic-plastic regions also extend over $l/6$ from each support and thus almost reach the central zone. The addition of an axial force, by producing an excentric plastic hinge, further increases the length of the (non-symmetric) elastic-plastic zone which therefore may extend over the whole span. Under such conditions the effect on the deformation of the structure and, in the case of compressive forces, on its stability, of the reduced elastic rigidity over the elastic-plastic regions is therefore no longer

---

[1] See footnote 1, p. 255.

negligible and the rigid-plastic idealization definitely inadequate. Hence, the application of the approximate rigid-plastic methods, developed for the estimation of the collapse load of highly redundant metal frame structures with predominant bending, to redundant structures such as arches and heavy frames in which the effect of axial loads is significant is not justified, since an extensive elastic-plastic region can be considered neither rigid nor "safe"[1] even when the yield-hinge is not fully developed. Only a careful step by step analysis of such structures, considering the gradual development and effect of the elastic-plastic regions[2], leads to a physically significant, though very elaborate, estimate of the collapse load.

But even for the highly redundant, predominantly bent structures for which the rigid-plastic methods were originally developed they are of rather limited physical interest; the large deformations that necessarily occur in the first yield-hinges of a highly redundant real structure before the final plastic collapse load can be attained by the formation of the last hinge, will tend to cause premature failure by localized instability or fracture, as a result of which the theoretical plastic collapse load can, in fact, not be reached[3], unless strain-hardening is relied upon, an assumption which, however, is incompatible with the basis of the analysis.

A failure condition of considerably more physical interest than collapse under proportional loading is that of cyclic plastic deformation under a variety of repeated load configurations none of which causes immediate collapse. The criteria which these load combinations have to satisfy if failure by cyclic plastic deformation is to be avoided and the structure, after an initial period of plastic deformation, is to "shake-down" to an essentially elastic response by the formation of a system of counteracting residual stresses, are due to GRUENING[4], FRITSCHE[5], BLEICH[6], MELAN[7], PRAGER[8] and KOITER[9]. They are based on the consideration that if a system of self-equilibrating elastic residual stresses in the redundant structure can be found such that the sum of these stresses and the maximum positive or negative stresses produced by the loads in the assumedly perfectly elastic structure at every point and for all possible load combinations does nowhere exceed the yield stress, the structure will "shake-down" to such a distribution of residual stresses that all subsequent load applications of the same (or lower) intensity will produce a purely elastic response. If such a system cannot be found the structure will fail by gradually increasing[10] or repeated cyclic plastic deformation[11] after a finite number of load applications.

---

[1] E. T. ONAT and W. PRAGER: J. Mech. Phys. of Solids **1**, 77 (1953).

[2] See footnote 3, p. 315. — M. R. HORNE: J. Mech. Phys. Solids **4**, 104 (1956).

[3] F. STUESSI and C. F. KOLLBRUNNER: Bautechnik **13**, 264 (1935). — Final Rep. 2nd Internat. Congr. Bridge and Struct. Eng., Berlin, p. 74, 1936. — H. MEIER-LEIBNITZ: Final Rep. 2nd Internat. Congr. Bridge and Struct. Eng., Berlin, p. 71; also Stahlbau **9**, 153 (1936).

[4] M. GRUENING: Tragfähigkeit statisch unbest. Tragwerke etc. Berlin: Springer 1926.

[5] J. FRITSCHE: Bauingenieur **12**, 827 (1931).

[6] H. H. BLEICH: Bauingenieur **13**, 261 (1932). — F. BLEICH: Prel. Rep. 2nd Internat. Congr. Bridge and Struct. Eng., Berlin, p. 137, 1936.

[7] E. MELAN: Prel. Rep. 2nd Internat. Congr. Bridge and Struct. Eng., Berlin, p. 45, 1936. — See also footnote 4, p. 285.

[8] W. PRAGER and P. S. SYMONDS: J. Appl. Mech. **17**, 315 (1950).

[9] W. T. KOITER: Proc. Acad. Sci., Amst. B **59**, 24 (1956).

[10] B. G. NEAL and P. S. SYMONDS: J. Instn. Civ. Engrs. **35**, 186 (1950/51). See also footnotes 4, 5, 6 and B. G. NEAL: Plastic Methods of Struct. Analysis. New York: J. Wiley 1956.

[11] M. R. HORNE: Research, Eng. Struct. Suppl. (Colston Papers 2) 141, (1949). — J. Inst. Civ. Engrs. **34**, 174 (1950). See also footnotes 4, 5 and 6.

A simple illustration of the difference in the analysis of the carrying capacity of an elastic-plastic structure for proportional loading and for repeated variable load configurations is provided by a continuous beam of uniform cross-section and yield-hinge moment M_0 over two equal spans l, subject to the alternating load configurations of (a) one concentrated load P at midspan of one span producing a maximum positive moment, and (b) two concentrated loads at midspan of both spans producing the maximum negative moment at the support. The single redundant is the moment X over the support (Fig. 7).

The collapse load-intensity P_F is obtained by the conditions for the simultaneous formation of yield hinges over the support and at midspan: $X = -M_0$ and $M_m = P_F l/4 - M_0/2 = M_0$; hence $P_F = 6M_0/l$. The determination of the maximum "shake-down" load intensity P_F', however, presupposes the knowledge of the elastic moment-distribution for both load configurations; for configuration (a): $X_a = -3Pl/32$, $M_{ma} = 13Pl/64$, for configuration (b): $X_b = -3Pl/16$, $M_{mb} = 5Pl/32$. The establishment of a condition of "shake-down" through a residual support moment X_R requires that $M_{m\,max} + \tfrac{1}{2} X_R = \tfrac{13}{64} Pl - \tfrac{1}{2} X_R \leq M_0$ and $-X_{max} + X_R = -\tfrac{12}{64} Pl + X_R \geq -M_0$; this equation is satisfied by the residual support moment $X_R = -\tfrac{1}{96} Pl$ and the associated total support moment $X = -\tfrac{19}{96} Pl = -M_0$, from which the shake-down load

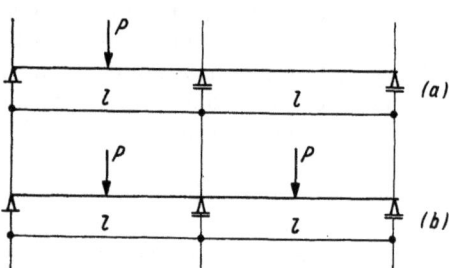

Fig. 7. Alternative load systems for "shake-down" analysis.

$P_F' \leq 5.05 M_0/l < P_F$. Considering that the load-intensity for which the first yield-hinge would develop under the critical of the two load configurations is $P_F'' = \tfrac{64}{13} M_0/l = 4.92 M_0/l$, it is obvious from the comparison of this value with P_F and P_F' that for real structures for which, with rare exceptions, a variety of load configurations must necessarily be considered, plastic collapse analysis under proportional loading is not relevant from a physical point of view for any but the simplest type of structure; the results of a "shake-down" analysis, on the other hand, will frequently not differ significantly enough from those of the elastic analysis to justify its use as a standard engineering design procedure, considering the order of magnitude of the uncertainty in the design assumptions that must be provided for in such design[1].

36. Thin plates. The elastic-plastic deformation of circular thin plates under circular symmetric loading conditions has first been analyzed by Sokolovsky[2], who established on the basis of the von Mises yield condition and elastic-plastic stress-strain relations the plastic limit load at which a purely plastic state is reached at the center of the plate, by investigating in detail the problem of the gradual spreading of the plastic zones. For the simple case of a uniformly loaded circular plate of radius R and thickness $2d$ the load intensity at which the yield limit is attained in the extreme fiber of the critical section at the center of the plate follows from the elastic solution to be

$$p_0 = \frac{32}{(3+\nu)\sqrt{3}} \frac{k d^2}{R^2} = \frac{32}{3(3+\nu)} \frac{\sigma_0 d^2}{R^2} \qquad (36.1)$$

---

[1] A. M. Freudenthal: Trans. Amer. Soc. Civ. Engrs. **112**, 125 (1947); **121**, 1337 (1956).
[2] V. V. Sokolovsky: Prikl. Mat. Mech. **8**, No. 2 (1944). — Theory of Plasticity, 2nd edit., Chap. 14. Moscow 1950. See also A. A. Ilyushin, footnote 1, p. 288.

where $k\sqrt{3}=\sigma_0$. Thus the characteristic ratio $5.26 \leq (p_0 R^2/k d^2) \leq 6.15$ depends on the value of ν. When $p > p_0$ two plastic zones gradually develop from the center, spreading towards the supports as well as towards the central plane, until a fully plastic region develops at the center. By an approximate numerical integration of the differential equations of the elastic-plastic problem Sokolovsky has obtained the shape of the remaining elastic core for various load-intensities $p > p_0$, as well as the relations between the characteristic ratio $(p R^2/k d^2)$ and the depth of this core as well as the deflection at midspan. These results, which are shown in Fig. 8, while not quite rigorous because of the use of the stress-strain

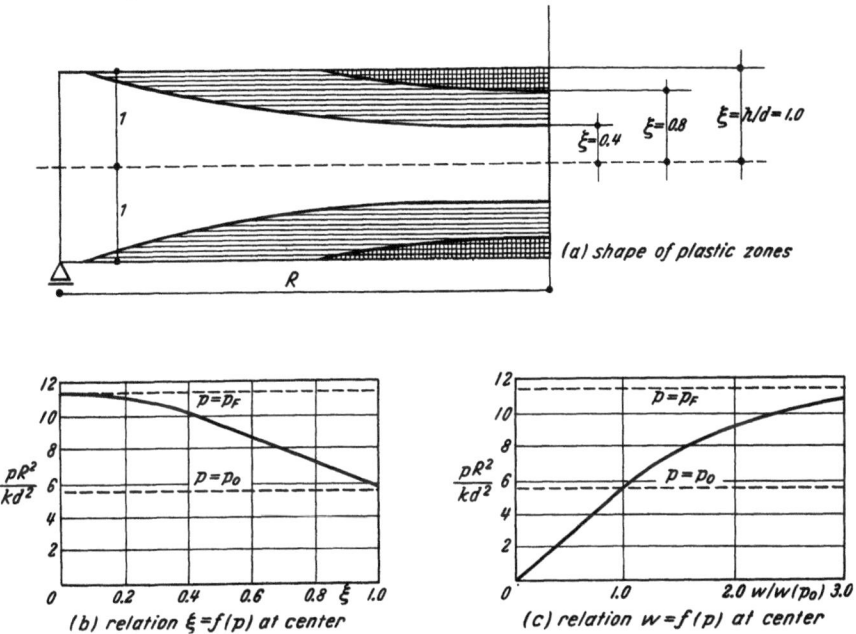

Fig. 8. Elastic-plastic circular plate (after Sokolovsky [1]).

relations instead of the stress-strain-rate relations (which, however, would make the problem much less manageable), illustrate the essential difference between the elastic-plastic deformation of two-dimensional plates and that of one-dimensional beams and struts: simultaneously with the development of a fully plastic condition at the center of the plate the entire plate is transformed into the fully plastic condition. Hence a concept similar to that of the yield-hinge in one-dimensional elastic-plastic structures in which the plastic deformation is more or less concentrated and which, therefore, under certain conditions justifies the introduction of the rigid-plastic idealization, does not exist in the uniformly loaded circular plate. Hence it should not, without evidence to the contrary, be assumed to exist in other types of plates or under different loads, unless the considerations are extended to reinforced concrete plates. Here the sharp concentration of the plastic deformation of the steel reinforcement along the crack pattern of the concrete forming a system of linear "yield-hinges", establishes almost perfect conditions for the application of methods of rigid-plastic analysis and the estimation of statically admissible lower limits and kinematically admissible upper limits of the ultimate carrying

---

[1] See footnote 2, p. 319.

capacity of plates of arbitrary shapes, as shown by Johansen[1], Olszak[2] and others[3]. Bending moments that satisfy the equilibrium and yield conditions represent a statically admissible system, angular rotations along yield hinges that satisfy the boundary restraints are kinematically admissible system.

The ultimate carrying capacity of the elastic-plastic plate can be determined directly by the analysis, on the basis of the equilibrium equations and the plasticity conditions alone, of the fully plastic plate, as originally proposed by Sokolovsky, who applied this analysis to freely supported circular and annular plates using both the von Mises and the Tresca yield conditions[4]. Combining the equilibrium conditions involving only the moments M_r and M_ϑ

$$r \frac{dM_r}{dr} + (M_r - M_\vartheta) = r\, Q_r \qquad (36.2)$$

where for uniformly distributed load p

$$Q_r = -\frac{1}{r} \int_0^r p\, r\, dr \qquad (36.3)$$

with $M_\vartheta = M_0 = \sigma_0 d^2$, which represents the Tresca yield condition for the range $M_r < M_\vartheta$ and fully developed plastic section in bending, and integrating the resulting differential equation with the boundary condition $M_r = M_\vartheta = M_0$ for $r=0$ the radial moment is

$$M_r = \sigma_0 d^2 - \tfrac{1}{6} p\, r^2. \qquad (36.4)$$

Since $M_r = 0$ for $r = R$ the limit load

$$p_F = 6 \frac{\sigma_0 d^2}{R^2} = \frac{6 M_0}{R^2}, \qquad (36.5)$$

while for the von Mises yield condition the factor 6 is replaced by 6.45, found by numerical integration; hence $0.56(3+\nu)p_0 < p_F < 0.6(3+\nu)p_0$, depending on the yield condition used. The same method is applied to other types of circular loading including a concentrated load at the center for which $P_F = 2\pi \sigma_0 d^2$ is found as the limiting load. Obviously, the use of the Tresca condition is more expedient then that of the von Mises condition, since it does not interfere with the linearity of Eq. (36.2).

Methods paralleling those of Sokolovsky have been used by Prager and co-workers for a more general analysis of plastic and rigid-plastic problems[5], producing essentially the same results. They have been extended to shells of revolution[6], including the combined efforts of bending moments and axial forces[7], considering admissible velocity fields in addition to stress fields.

It should be noted that the physical significance of this method of analysis is not yet clearly established. Only for the circular plate does the existing elastic-plastic analysis[8] permit a realistic evaluation of the physical relevance of the fully plastic solution in terms of the deflections associated with the load intensities

---

[1] K. W. Johansen: Brudlinieteorier. Kopenhagen: Gjellerup 1943. — Final Report, Third Congr. Int. Ass. Bridge Struct. Eng., p. 565, Liège, 1948.

[2] W. Olszak et al.: Arch. Mech. Stos. **5**, 329 (1953); **8**, 197 (1956); **9**, 467, 605 (1957). — Bull. Polish. Acad. Sci. **3**, (4), 195 (1955); **4** (4), 209 (1956).

[3] A. Sawczuk: Bauplanung, Bautechnik **11**, 315 (1957). — Arch. Mech. Stos. **8**, 549 (1956).

[4] See footnote 2, p. 318.

[5] H. G. Hopkins and W. Prager: J. Mech. Phys. of Solids **2**, 1 (1953). — H. G. Hopkins and A. J. Wang: J. Mech. Phys of Solids **3**, 117 (1954). See also footnote 1, p. 255.

[6] D. C. Drucker: Proc. First Midwestern Conf. Solid Mech., p. 158. Urbana 1953.

[7] E. T. Onat and W. Prager: Proc. Acad. Sci., Amst. B **57**, 534 (1954). — P. G. Hodge: J. Appl. Mech. **21**, 336 (1954); **23**, 73 (1956). See also footnote 2, p. 281.

[8] See footnote 2, p. 318.

$p_0 < p < p_F$. Fig. 8 indicates that in the case of a uniformly loaded circular plate the extremely rapid increase of the deflection for $p > 0.8 p_F$, at which the elastic core has been reduced to roughly one-half of the plate thickness, makes it rather doubtful whether the theoretical value of the plastic limit load can ever be attained before the plate is so severely deformed that the method of analysis used has become irrelevant.

37. Dynamic problems. Problems of plastic deformation of structures subject to dynamic forces are of considerable complexity for a variety of reasons, the most important of which are:

the complex relation between high strain-rate, stress and strain within the plastic region, resulting from the speed-effect on slip (slip-delay[1]) and on the quasi-viscous grain-boundary motion in conjunction with the nearly adiabatic temperature effect of the rapid plastic energy dissipation;

the existence of elastic-plastic domains, and

the non-linearity of the differential equations governing even the simplest problem.

The attempted theoretical treatment has been limited, so far, to problems of one-dimensional wave propagation and of transverse impact on simple structures, under the assumption of isothermal conditions and neglecting, in general, the effect of speed on the plastic response as well as the interaction of elastic and plastic regions.

Problems of one-dimensional plastic wave-propagation are governed by equations of motion of the form (4.11a)

$$\varrho \ddot{u} = \sigma_{,x} = \frac{d\sigma}{d\varepsilon} \varepsilon_{,x} = \frac{d\sigma}{d\varepsilon} u_{,xx} \qquad (37.1)$$

which is a non-linear wave equation of the form $\ddot{u} = c_p^2 u_{,xx}$ where the velocity of wave propagation

$$c(\varepsilon) = \sqrt{\left(\frac{d\sigma}{d\varepsilon}\right)/\varrho} < \sqrt{\frac{E}{\varrho}} = c_0 \qquad (37.2)$$

is a function of the tangent modulus $(d\sigma/d\varepsilon)$ of the stress-strain relation associated with the elastic-plastic deformation. Since $\dot{u} = \int_0^\varepsilon c(\varepsilon) d\varepsilon < c_0 \varepsilon$, the larger plastic strain propagates at a lower velocity than the elastic strain or the smaller plastic strains, so that the length of the pulse increases as it moves away from its origin, provided the $\sigma - \varepsilon$-diagram is concave towards the ε-axis. The phenomenon is, however, complicated by the interaction between loading and unloading waves as well as by wave-reflection[2].

Eqs. (37.1) and (37.2) are due to TAYLOR[3] and VON KARMAN[4], who discussed their solution, assuming identity of the dynamic and the quasi-static stress-strain relations. A substantially similar theory was developed independently by RAKHMATULIN[5] and SHAPIRO[6]. Modifications to include relatively simple strain-rate effects were proposed by SOKOLOVSKY[7] and others[8].

---

[1] D. S. CLARK: Trans. Amer. Soc. Met. **46**, 47 (1954) and references there. — J. M. KRAFFTE: Trans. Amer. Soc. Met. **48**, 249 (1956).

[2] R. M. DAVIS: Surveys in Mechanics, p. 64. Cambridge: University Press 1956.

[3] G. I. TAYLOR: J. Inst. Civil Eng. London **26**, 486 (1946). — Sci. Papers **1**, Cambridge: University Press 1957.

[4] TH. V. KARMAN and P. DUWEZ: J. Appl. Phys. **21**, 987 (1950).

[5] K. A. RHAKMATULIN: Prikl. Math. Mech. **9**, 91 (1945); **10**, 333 (1946); **12**, 39 (1948).

[6] G. S. SHAPIRO: Prikl. Math. Mech. **10**, 597 (1946).

[7] V. V. SOKOLOVSKY: Prikl. Math. Mech. **12**, 261 (1948).

[8] L. E. MALVERN: Quart. Appl. Math. **8**, 405 (1951). — J. Appl. Mech. **18**, 203 (1951). — R. J. RUBIN: J. Appl. Phys. **25**, 528 (1954).

Eq. (37.1) is valid only as long as the energy of the lateral motion is negligible compared with the energy of the longitudinal motion. Equations governing the propagation of plane, cylindrical and spherical waves in the infinite medium were established and discussed by various investigators[1].

In a number of experiments[2] the validity of the strain-rate-independent theory based on Eq. (37.1) was tested; the test results suggest the existence of a significant speed effect.

The numerous attempts to establish the resistance of elastic-plastic structures to transverse impulsive loading are, in general, based on the drastically simplifying assumption of arbitrary initial velocities and unlimited rotations in fully developed "plastic hinges" of linear structures, the parts of which are assumed to remain rigid between the "hinges"[3], or of fully plastic regions in otherwise rigid plates[4]. That the physical significance of such rigid-plastic analysis is rather problematic has been shown by the comparative analysis of linear structures in which the elastic deformation is not neglected[5], although the concept of the concentrated plastic hinge is retained (see Sect. 35); the fact that in this analysis speed-effects and nearly adiabatic temperature effects on the plastic response, as well as changes of geometry in the deformed structure are also neglected, although its validity depends on the assumption of a high ratio of the plastically dissipated to the elastic energy, necessarily associated with extensive plastic deformation, makes its physical relevance somewhat questionable.

Second part.

The ideal plastic body[6].

E. The basic equations.

I. The three-dimensional problem.

38. Quadratic yield condition[7]. We present in this section the basic equations of the three-dimensional perfectly plastic body under the assumption of the quadratic yield condition proposed by von Mises and of the associated stress-strain relations of Saint Venant and von Mises. We refrain from an attempt to sketch the early historical development, which leads to a complete statement of the concept of the ideal plastic body by von Mises[8], since this history has been often exhibited and may be found e.g. in the books by Prager and Hodge [23] (1951), Hill [10]

---

[1] K. A. Rhakmatulin and G. S. Shapiro: Prikl. Math. Mech. **12**, 369 (1948). — F. A. Bakhshian: Prikl. Math. Mech. **12**, 281 (1948). — Y. L. Lunc: Prikl. Math. Mech. **13**, 55 (1949). — D. S. Wood: J. Appl. Mech. **19**, 521 (1952).

[2] W. Ramberg and L. K. Irwin: Proc. Amer. Soc. Test. Mat. **55**, 1061 (1955). — E. J. Sternglass and D. A. Stuart: J. Appl. Mech. **20**, 247 (1953). — B. E. K. Alter and C. W. Curtis: J. Appl. Phys. **27**, 1079 (1956).

[3] E. H. Lee and P. S. Symonds: J. Appl. Mech. **19**, 308 (1952). — M. F. Conroy: J. Appl. Mech. **19**, 465 (1952). — P. S. Symonds: J. Appl. Mech. **20**, 475 (1953).

[4] H. G. Hopkins and W. Prager: Z. angew. Math. Phys. Zürich **5**, 317 (1954). — A. J. Wang and H. G. Hopkins: J. Mech. Phys. of Solids **3**, 27 (1954).

[5] H. H. Bleich and M. G. Salvadori: Trans. Amer. Soc. Civ. Engrs. **120**, 499 (1955. — J. A. Seiler, B. A. Cotter and P. S. Symonds: J. Appl. Mech. **23**, 515 (1956).

[6] Compare Sect. 21.

[7] Some of the contents of Chapter E, I. are already contained in Chapters A and B of the first part. Since too much duplication had to be avoided, the present Chapter cannot be completely self contained.

[8] R. v. Mises: Göttinger Nachr. Math. phys. Kl. 582—592 (1913).

(1950), NADAI [20] (1950) and [19] (1931), SOKOLOVSKY [26] (1950), GEIRINGER [6] (1937); see in this connection also a recent article by PRAGER[1].

The basic equations are, to begin with, the *continuity equation* [see Eq. (4.4)] and the *equation of motion*. In the first of these:

$$\varrho \operatorname{div} \boldsymbol{v} + \frac{d\varrho}{dt} = 0 \tag{38.1}$$

$\boldsymbol{v}$ denotes the vector of flow velocity, ϱ the density and d/dt the "material derivative". If $\varrho = $ constant, as assumed in general in the theory of perfectly plastic solids, $d\varrho/dt = 0$ and (38.1) reduces to

$$\operatorname{div} \boldsymbol{v} = 0 \quad \text{or} \quad \sum_{i=1}^{3} \frac{\partial v_i}{\partial x_i} = 0. \tag{38.1'}$$

In the equation of motion let $\boldsymbol{k}$ be the resultant of the external forces per unit of mass and σ_{ij} the components of the stress tensor Σ. The mean pressure p is then defined as the invariant expression

$$p = -\tfrac{1}{3}(\sigma_{11} + \sigma_{22} + \sigma_{33}), \tag{38.2}$$

and the resultant pressure force per unit of volume is equal to $\operatorname{grad} p$. Finally, there is the stress deviation force $\boldsymbol{d}$ (corresponding to the viscous force in viscous flow theory) with components $d_i = \sum_{j=1}^{3} \frac{\partial s_{ij}}{\partial x_j}$ $(j=1, 2, 3)$. The s_{ij}, the components of the stress deviation tensor S, are as in (4.9),

$$s_{ij} = \sigma_{ij} + p\,\delta_{ij}, \tag{38.3}$$

and the equation of motion appears in the form

$$\varrho \frac{d\boldsymbol{v}}{dt} = \varrho\,\boldsymbol{k} - \operatorname{grad} p + \boldsymbol{d}, \tag{38.4}$$

or in component form, with X_i as components of $\varrho \boldsymbol{k}$, [see (4.10)]

$$\varrho \frac{dv_i}{dt} = X_i - \frac{\partial p}{\partial x_i} + \sum_{j=1}^{3} \frac{\partial s_{ij}}{\partial x_j}, \tag{38.4'}$$

or [see (4.11)]

$$\varrho \frac{dv_i}{dt} = X_i + \sum_{j=1}^{3} \frac{\partial \sigma_{ij}}{\partial x_j}. \tag{38.4''}$$

Next, we need a relation which specifies s_{ij} in terms of the other variables p, ϱ, v_i, and possibly also x_1, x_2, x_3, t. The Saint Venant-Lévy-Mises relations[2] do this in the form

$$s_{ij} = \mu \cdot \frac{1}{2}\left(\frac{\partial v_i}{\partial x_j} + \frac{\partial v_j}{\partial x_i}\right), \quad \mu > 0 \tag{38.5}$$

or

$$\frac{1}{2}\left(\frac{\partial v_i}{\partial x_j} + \frac{\partial v_j}{\partial x_i}\right) = \lambda s_{ij}, \quad \lambda \geq 0. \tag{38.5'}$$

---

[1] W. PRAGER: James Clayton Lecture. Proc. Inst. Mech. Engrs. **169**, 41—57 (1955).

[2] B. DE SAINT-VENANT: C. R. Acad. Sci., Paris **70**, 473—480 (1870). — M. LÉVY: C. R. Acad. Sci., Paris **70**, 1323—1325 (1870). — R. v. MISES, see Ref. 8 on p. 322 of the present article.

Note that in contrast to viscous flow theory μ is an unknown (positive) function of the coordinates and possibly of t. Since $s_{11}+s_{22}+s_{33}=0$, Eqs. (38.5) are compatible with (38.1') and (38.5) stands for five independent scalar equations.

In the following we shall often use

$$x_1 = x, \quad x_2 = y, \quad x_3 = z, \quad \sigma_{11} = \sigma_{xx} = \sigma_x, \quad \sigma_{12} = \sigma_{xy} = \tau_{xy} = \tau_z, \ldots$$

and similar abbreviations for strains and for strain rates[1].

In addition to the $1+3+5=9$ Eqs. (38.1'), (38.4), (38.5') for ten unknowns v_i, p, s_{ij}, μ a yield condition holds (Sect. 22) which limits the admissible stress tensors to ∞^5. We shall give it in this section in VON MISES' "quadratic" form (22.1).

If we consider problems of equilibrium, the acceleration force $\varrho \frac{d\boldsymbol{v}}{dt}$ vanishes; usually, the body force $\varrho \boldsymbol{k}$ is also disregarded. The basic $1+3+5+1=10$ equations of the theory are then:

$$\frac{\partial v_1}{\partial x_1} + \frac{\partial v_2}{\partial x_2} + \frac{\partial v_3}{\partial x_3} = 0, \tag{38.6}$$

$$-\frac{\partial p}{\partial x_i} + \sum_{j=1}^{3} \frac{\partial s_{ij}}{\partial x_j} = 0, \quad i=1,2,3, \tag{38.7}$$

$$\frac{1}{2}\left(\frac{\partial v_i}{\partial x_j} + \frac{\partial v_j}{\partial x_i}\right) = \lambda s_{ij}, \quad \lambda \geq 0, \quad i,j=1,2,3 \tag{38.8}$$

$$\sum_{i,j}^{1\ldots 3} s_{ij} s_{ij} = 2\tau_0^2 = 2k^2, \tag{38.9}$$

where τ_0 is the yield stress in simple shear.

Another form of (38.9) is [compare (6.7)]

$$(\sigma_x - \sigma_y)^2 + (\sigma_y - \sigma_z)^2 + (\sigma_z - \sigma_x)^2 + 6(\tau_x^2 + \tau_y^2 + \tau_z^2) = 6\tau_0^2 = 2\sigma_0^2, \tag{38.9'}$$

where σ_0 is the yield stress in simple tension. (Thus, in VON MISES' theory, $\sigma_0 = \sqrt{3}\,\tau_0$.) We may also easily see that

$$J_2 \equiv -(s_x s_y + s_y s_z + s_z s_x) + (\tau_x^2 + \tau_y^2 + \tau_z^2) = \tau_0^2 = k^2, \tag{38.9''}$$

where J_2 is the second invariant of the deviator S.

Denoting by subscripts 1, 2, 3 the (common) *principal directions* of the tensors Σ and S we obtain also:

$$s_1^2 + s_2^2 + s_3^2 = 2k^2, \tag{38.$\bar{9}$}$$

$$(\sigma_1 - \sigma_2)^2 + (\sigma_2 - \sigma_3)^2 + (\sigma_3 - \sigma_1)^2 = 6k^2, \tag{38.$\bar{9}$'}$$

$$J_2 \equiv -(s_1 s_2 + s_2 s_3 + s_3 s_1) = k^2, \tag{38.$\bar{9}$''}$$

where $\sigma_1, \sigma_2, \sigma_3$ are the principal stresses, etc. In general, it is assumed that $\sigma_1 \geq \sigma_2 \geq \sigma_3$. Also, introducing the principal shear stresses

$$\tau_1 = \tfrac{1}{2}(\sigma_2 - \sigma_3) = \tfrac{1}{2}(s_2 - s_3), \quad \text{etc.}, \tag{38.10}$$

$$\tau_1^2 + \tau_2^2 + \tau_3^2 = \tfrac{3}{2}k^2, \quad \text{where} \quad \tau_1 + \tau_2 + \tau_3 = 0 \tag{38.9'''}$$

---

[1] The explicit expressions of Eqs. (38.4') in both cylindrical and spherical coordinates are e.g. in [26], p. 59, and p. 60, while we find on pp. 63, 64 the expressions for the left sides of (38.5') in these coordinates.

results[1]. The criterion (38.9) has been generalized by VON MISES[2] and by F. SCHLEICHER[3] by replacing the constant right side in it by a function of p in accordance with experiments by W. LODE[4] (cf. also O. MOHR's form, see Ref. 1, p. 328).

The geometric presentation of the yield condition in a $\sigma_1, \sigma_2, \sigma_3$-space has been given in Sect. 22.

In (38.5') we have on the left the tensor of plastic flow velocity, or plastic deformation velocity, which we shall denote by $\dot{E}''$, where the dot stands for d/dt, since we use flow *velocities* and the double primes stand for "plastic" in contrast to "elastic" (see Sect. 43). Eq. (38.8) then reads

$$\dot{\varepsilon}''_{ij} = \lambda s_{ij}, \quad \lambda \geq 0. \tag{38.8'}$$

Here λ can be eliminated by squaring Eqs. (38.8'), summing over all subscripts and using (38.9). We obtain [see Eq. (23.4)]

$$\lambda^2 = \frac{1}{2k^2} \cdot [(\dot{\varepsilon}''_1)^2 + (\dot{\varepsilon}''_2)^2 + (\dot{\varepsilon}''_3)^2]. \tag{38.11}$$

Alternatively, we may multiply both sides of (38.8') by s_{ij} and sum. Then, with

$$\dot{W}'' = \sum_{i,j} \dot{\varepsilon}''_{ij} s_{ij} = \sum_i \dot{\varepsilon}''_i s_i,$$

we obtain

$$\lambda = \frac{\dot{W}''}{2k^2}. \tag{38.12}$$

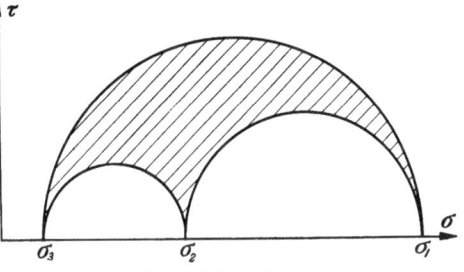

Fig. 9. Mohr circles.

Hence λ is proportional to the plastic work producing a change in shape—the change in volume being zero by (38.1'). Comparing (38.11) and (38.12) we find, using (38.9), the equality

$$\sum_i s_i^2 \sum_i (\dot{\varepsilon}''_i)^2 = \left(\sum_i s_i \dot{\varepsilon}''_i\right)^2. \tag{38.13}$$

39. Some basic formulas. Mohr circles. The stress tensor Σ, like any tensor, associates with a direction ν a vector t_ν, the stress vector or traction, which may be resolved into a component σ in the ν-direction, the normal stress, and into a component τ perpendicular to it, see Eqs. (6.11). From the formulas for σ and τ we deduce the well-known stress representation due to O. MOHR[5]. He has proved that *the points which correspond to all possible ∞^2 directions ν (for a fixed point in the material) lie in the shaded area between the three circles of Fig. 9, called circles of Mohr, which intercept the σ-axis in the three points with abscissae $\sigma_3, \sigma_2, \sigma_1$ respectively* (Fig. 9).

---

[1] If $\sigma_1 \geq \sigma_2 \geq \sigma_3$ then τ_1 and τ_3 are positive, and τ_2 negative and $-\tau_2 = |\tau|_{\max}$, where $|\tau|_{\max}$ means the greatest absolute value of τ. We also note, SOKOLOVSKY [26], p. 31, that (38.9''') is approximately equivalent to the condition $\left(\frac{1}{\sqrt{3}} + \frac{1}{2}\right) \cdot |\tau|_{\max} = k$. Cf. also HILL [10], p. 117.

[2] R. v. MISES: Z. angew. Math. Mech. **5**, 147—149 (1925).
[3] F. SCHLEICHER: Z. angew. Math. Mech. **6**, 199—216 (1926).
[4] W. LODE: Z. angew. Math. Mech. **5**, 142—144 (1925). — Z. Physik **36**, 913—939 (1926).
[5] OTTO MOHR: Abhandlungen aus dem Gebiete der technischen Mechanik, 2nd ed., pp. 192—235. Berlin 1914. See [20], p. 96, footnote, for more bibliographical details and p. 96 seq. for details of proof and illustrations. Cf. our Sect. 6.

If a_1, a_2, a_3 are the direction cosines $\cos(\nu 1)$, $\cos(\nu 2)$, $\cos(\nu 3)$, then for $a_3 = 0$, $a_1^2 + a_2^2 = 1$, the σ, τ satisfy the equation

$$\left(\sigma - \frac{\sigma_1 + \sigma_2}{2}\right)^2 + \tau^2 = \left(\frac{\sigma_1 - \sigma_2}{2}\right)^2, \tag{39.1}$$

i.e., the equation of the half-circle to the right, and similarly for $a_2 = 0$ and for $a_1 = 0$, respectively.

Nadai introduced and explained the octahedral shearing stress τ_{oct}^2 (see Sect. 6):

$$\left.\begin{array}{l} \tau_{oct}^2 = \dfrac{1}{9}[(\sigma_1 - \sigma_2)^2 + (\sigma_2 - \sigma_3)^2 + (\sigma_3 - \sigma_1)^2] = \dfrac{2}{3} k^2 = \dfrac{2}{3} J_2, \\ J_2 = \dfrac{3}{2} \tau_{oct}^2. \end{array}\right\} \tag{39.2}$$

This last equation has been used as a physical interpretation of the invariant J_2. Another physical interpretation in terms of the energy associated with a change of shape was given by Hencky[1]. Prager [23] very rightly remarks "Mises' yield condition derives its importance in the mathematical theory of plasticity not from the fact that the invariant J_2 appearing therein can be interpreted physically in this or that manner, but from the fact that it has the simplest form compatible with the general postulates which any yield condition must fulfill." Actually, it is also in very good agreement with experimental evidence, particularly for ductile metals (cf. e.g. [10], p. 22 and Sect. 22 of this article).

40. Plastic potential. Denote the left side of (38.9') by $6g(\sigma_x, \sigma_y, \ldots \tau_z)$. We see that

$$3 \frac{\partial g}{\partial \sigma_x} = 2\sigma_x - \sigma_y - \sigma_z = 3\left(\sigma_x - \frac{\sigma_x + \sigma_y + \sigma_z}{3}\right) = 3 s_x, \ldots$$

$$\frac{\partial g}{\partial \tau_{yz}} = \tau_{yz}, \quad \frac{\partial g}{\partial \tau_{zy}} = \tau_{zy}, \quad \text{etc.}$$

Accordingly we may replace Eqs. (38.8) by

$$\frac{\partial g}{\partial \sigma_{ij}} = \frac{\partial g}{\partial s_{ij}} = \mu \dot{\varepsilon}_{ij}'', \quad \text{or} \quad \text{Grad } g = \mu \dot{E}'', \tag{40.1}$$

where $\dot{E}''$ denotes the (plastic) flow velocity tensor, and "Grad" denotes a symbolic tensor, in analogy to the symbolic vector "grad". Or, if we use $\chi = \dfrac{1}{\mu} g$, we obtain

$$\frac{\partial \chi}{\partial \sigma_{ij}} = \dot{\varepsilon}_{ij}'', \tag{40.1'}$$

where $g(\sigma_{ij})$ may denote the left side of a yield condition which need not be that of Sect. 38. This rule which associates to a stress Σ a plastic strain $\dot{E}''$, to within an arbitrary factor, is called von Mises' *theory of plastic potential*[2] (compare Chapter B of this article). Mathematically, a proof is needed to assure that by the operation $\partial g/\partial \sigma_{ij}$ (or briefly by the operation "Grad") the components of a tensor are obtained, i.e., scalar quantities which obey the transformation laws. This is proved in tensor calculus. An elementary proof limited to rectangular Cartesian coordinates has been given by Geiringer[3]. In recent years the concept of the

---

[1] H. Hencky: Proc. 1st Internat. Congr. Appl. Mech. Delft 1924, pp. 312—317.
[2] R. v. Mises: Z. angew. Math. Mech. 8, 161—185 (1928), see p. 180 seq.
[3] H. Geiringer: Some recent results in the theory of an ideal plastic body. Advances in Applied Mechanics III, pp. 197—294. New York 1953. For a simplification we refer to this often quoted paper in the following as to Geiringer: Advances.

plastic potential has gained added importance in the theory of *limit analysis* which uses this relation between yield condition and flow rule[1].

We obtain a geometric interpretation of (40.1) by writing it in the form

$$\dot\varepsilon_1'' : \dot\varepsilon_2'' : \dot\varepsilon_3'' = \frac{\partial G}{\partial \sigma_1} : \frac{\partial G}{\partial \sigma_2} : \frac{\partial G}{\partial \sigma_3}. \tag{40.2}$$

The sign of the yield function $G(\sigma_1, \sigma_2, \sigma_3)$ may be chosen in such a way that the exterior normal points in the direction of increasing G. Then (40.2) associates to every point $(\sigma_1, \sigma_2, \sigma_3)$ on the yield surface the vector direction $\dot\varepsilon_1'' : \dot\varepsilon_2'' : \dot\varepsilon_3''$ as that of the exterior normal to the yield surface. This leads to a unique direction $\dot\varepsilon_1'' : \dot\varepsilon_2'' : \dot\varepsilon_3''$ only if there exists a uniquely determined exterior normal to the yield surface at each point (cf. Sect. 41).

An important extremum principle due to von Mises (p. 184 of the paper quoted above, Ref. 2 on p. 326) is explained in our Sect. 9, particularly Eqs. (9.11) seq.

Prager[2], using generalized stresses Q_i and "corresponding" generalized strains q_i, considers

$$L = Q_1 q_1 + Q_2 q_2 + \cdots + Q_n q_n$$

and shows that von Mises' extremum principle still applies to this generalization. For a generalization due to W. T. Koiter[3] see Sect. 9 and Sect. 22.

A function $g(\sigma_{ij})$ which is to serve as a yield function and as a plastic potential is subject to certain obvious restrictions[4]. In case of an isotropic material, g must not depend on the chosen coordinate system, but only as in (39.2) on the three invariants J_1, J_2, J_3 of the stress tensor, or, in other words, only on the values of the principal stresses, $\sigma_1, \sigma_2, \sigma_3$, and not on the principal directions; this dependence must be symmetric. Since g is to serve as the potential of an incompressible perfectly plastic medium we have to assume that it does not depend on the first invariant. It follows that g remains unchanged if Σ is replaced by $S = \Sigma + pI$, where I is the unit tensor, viz,

$$g(\sigma_{ij}) = g(s_{ij}) = G(\sigma_1, \sigma_2, \sigma_3) = G(s_1, s_2, s_3),$$

where G is a symmetric function of its arguments. Consequently G depends only on the differences $\sigma_1 - \sigma_2$, etc., hence only on the τ_1, τ_2, τ_3. Thus, we may write

$$g(\sigma_{ij}) = G(\sigma_1, \sigma_2, \sigma_3) = K(\tau_1, \tau_2, \tau_3) = 0 \tag{40.3}$$

as yield condition for an isotropic, incompressible, perfectly plastic body. It follows that

$$\frac{\partial G}{\partial \sigma_1} + \frac{\partial G}{\partial \sigma_2} + \frac{\partial G}{\partial \sigma_3} = \frac{\partial g}{\partial \sigma_x} + \frac{\partial g}{\partial \sigma_y} + \frac{\partial g}{\partial \sigma_z} = 0. \tag{40.4}$$

Thus, we see that in (40.1) or (40.1') or, as $\mu \neq 0$, in

$$\dot\varepsilon_{ij}'' = \lambda \frac{\partial g}{\partial \sigma_{ij}} \tag{40.5}$$

the incompressibility (38.6) of the plastic strain rate tensor and the condition (40.4) are interdependent. Since $\partial g/\partial \sigma_{ij} = \partial g/\partial s_{ij}$ we may also write (40.5) as

$$\dot e_{ij}'' = \lambda \frac{\partial g}{\partial s_{ij}}, \tag{40.5'}$$

---

[1] See for example, D. C. Drucker, W. Prager and H. J. Greenberg: Quart. Appl. Mech. **9**, 381—389 (1952). Regarding the plastic-potential rule see also R. Hill: Phil. Mag. (7) **40**, 971—983 (1949).

[2] W. Prager: Proc. 8th Internat. Congr. Appl. Mech., (1952), Istanbul, 1955, pp. 65—72.

[3] W. T. Koiter: Biezeno Anniversary Volume. Haarlem 1953, pp. 232—251; also W. T. Koiter: Quart. Appl. Math. **77**, 350—354 (1953).

[4] See also D. C. Drucker: Proc. 1st. Nat. Congr. Appl. Mech., Chicago 1950, pp. 487—491.

where $\dot{e}''_{ij} = \dot{\varepsilon}''_{ij} - \delta_{ij}(\dot{\varepsilon}''_{11} + \dot{\varepsilon}''_{22} + \dot{\varepsilon}''_{33}) = \dot{\varepsilon}''_{ij}$ is the strain rate deviator. The complete system of VON MISES' equations consists then of Eqs. (38.1), [or (38.7)], (40.3) — which has the property (40.4) — and (40.5). Eqs. (40.5) are often called associated flow conditions or associated stress-strain relations, where "associated" relates to the yield condition (40.3). [Compare Sect. 23, in particular (23.1), (23.3).]

41. Tresca's yield criterion. "Singular" yield conditions. We consider now the yield condition due to Tresca. It had been used in connection with the Lévy-Mises flow rule (38.8) by the early pioneers in the field and often this is still done. This is in contrast to the idea of the plastic potential explained in Sect. 40 which associates (38.8) with VON MISES' yield criterion. TRESCA's criterion is written as

$$|\tau|_{\max} = \text{constant} \tag{41.1}$$

and states that in perfectly plastic flow yielding occurs when the greatest shear stress reaches a certain limit value which is the same for all stress tensors throughout the material[1]. If the stresses are ordered so that

$$\sigma_1 \geq \sigma_2 \geq \sigma_3 \tag{41.2}$$

the condition is

$$2|\tau|_{\max} = \sigma_1 - \sigma_3 = 2\tau_0 = 2k. \tag{41.1'}$$

Actually we do not know which σ_i is the largest, which the smallest principal stress. All we say is that one of the three differences has the absolute value equal to $2k$ (see Sect. 22). It follows that the yield surface is a regular hexagonal prism, inscribed in the cylinder of the VON MISES yield condition. VON MISES, (see Ref. 2 on p. 326), applied to this yield function the theory of plastic potential, Eqs. (40.2) which gives here

$$\dot{\varepsilon}''_1 : \dot{\varepsilon}''_2 : \dot{\varepsilon}''_3 = 1 : 0 : -1, \tag{41.3}$$

and showed that the corresponding transformation is a plane slip in the directions of $\tau_{\max}$ (see Sect. 23).

We have seen how the theory of plastic potential applies to the faces of the Tresca prism, where there exists indeed a uniquely defined normal. KOITER[2] and PRAGER[3] complement the rule for the edges of the prism (see Sect. 23). E.g., for points on the edge formed by the adjacent faces $\sigma_1 - \sigma_2 = k$, and $\sigma_1 - \sigma_3 = k$, we obtain possible flow mechanisms by a linear combination of the flow mechanisms for these two faces in the form

$$\dot{\varepsilon}''_1 : \dot{\varepsilon}''_2 : \dot{\varepsilon}''_3 = -1 : r : 1 - r, \quad 0 \leq r \leq 1. \tag{41.4}$$

This rule may be adapted to yield surfaces with a corner, etc.[4]

42. "Compatibility" relations. Consider the system (38.6) — (38.9) of ten equations with ten unknowns v_i, σ_{ij}, λ. Following VON MISES (Ref. 8 on p. 322)

---

[1] H. TRESCA: Mém. prés. par divers savants **18**, 733—799 (1868). The condition had also been stated by B. DE SAINT VENANT, J. Math. pures appl. II **16**, 308 (1871). The condition (41.1') can be regarded as a particular case of COULOMB's (1773) condition basic in the theory of earth pressure, viz., $\sigma_1 - \sigma_3 = c_1 + c_2(\sigma_1 + \sigma_3)$. — J. J. GUEST: Phil. Mag., Ser. V **50**, 69 (1900) proposed to use it for ductile metals with c_2 a constant small compared to c_1. More general is O. MOHR's form: $\sigma_1 - \sigma_3 = f(\sigma_1 + \sigma_3)$ which states that in yielding the maximum shear stress $(\sigma_1 - \sigma_3)/2$ depends only on the corresponding normal stress $(\sigma_1 + \sigma_3)/2$; no influence is attributed to the median stress σ_2.

[2] W. T. KOITER: see Ref. 3 on p. 327.

[3] W. PRAGER: J. Appl. Mech. **20**, 317 (1953).

[4] For general discussion of edges, corners see also P. G. HODGE jr.: J. Rat. Mech. Analysis **5**, 917—938 (1956).

one may eliminate the σ_{ij} and obtain five equations for the five functions v_i, p and $\mu = 1/\lambda$. We simply replace in (38.7) the s_{ij} by means of (38.8) or (40.1), write (38.9') as

$$(\dot{\varepsilon}_x'' - \dot{\varepsilon}_y'')^2 + \cdots + 6(\gamma_x''^2 + \gamma_y''^2 + \gamma_z''^2) = 6\lambda^2 k^2, \tag{42.1}$$

and use as the fifth equation the incompressibility relation $\dot{\varepsilon}_x'' + \dot{\varepsilon}_y'' + \dot{\varepsilon}_z'' = 0$.

On the other hand, if we are only interested in the six stresses, we may consider the four Eqs. (38.7) and (38.9) (or a more general yield condition) and complement them by two *compatibility conditions*, (relations between the stress components) derived by W. JENNE[1]. These relations simplify considerably in the case of plane deformation and in the axial symmetric case.

43. The flow equations of PRANDTL[2] and REUSS[3]. The theory of VON MISES as explained so far is based on the following conception[4]. There exists a function g of the six stresses and a constant C such that at each point of the material $g \leq C$. Wherever $g < C$ the laws of elasticity hold; wherever $g = C$ the flow law (40.5) holds for the plastic flow velocities with a (non-negative) proportionality factor λ varying in space and time. In the framework of this theory it is permissible, but in no way necessary, to neglect the elastic stresses—if the problem warrants it—in the domain where $g < C$ (see also Sect. 95).

In the flow theory of PRANDTL and REUSS (PRANDTL dealt with the plane problem, REUSS with the general case), both plastic and elastic strains are considered simultaneously in the domain where $g = C$. Following REUSS we denote elastic strains by a prime, plastic strains by a double prime and total strains without a superscript, hence e.g., $E = E' + E''$ or, the dot denoting d/dt, $\dot{E} = \dot{E}' + \dot{E}''$ etc. Using as before

$$\sigma = -p = \tfrac{1}{3}(\sigma_x + \sigma_y + \sigma_z), \quad \varepsilon = \tfrac{1}{3}(\varepsilon_x + \varepsilon_y + \varepsilon_z),$$

$$\dot{\varepsilon}'' = \frac{1}{3}\left(\frac{\partial v_x}{\partial x} + \frac{\partial v_y}{\partial y} + \frac{\partial v_z}{\partial z}\right),$$

we need the deviator e_{ij} and likewise e_{ij}', e_{ij}'', $\dot{e}_{ij}$, etc. we put

$$\varepsilon_{ij} = e_{ij} + \varepsilon \delta_{ij}, \quad \dot{\varepsilon}_{ij} = \dot{e}_{ij} + \dot{\varepsilon} \delta_{ij}. \tag{43.1}$$

As always, plastic incompressibility, $\dot{\varepsilon}'' = 0$, is assumed, as well as $\varepsilon'' = 0$; hence $\varepsilon = \varepsilon'$, $\dot{\varepsilon} = \dot{\varepsilon}'$. HOOKE's law is written in the form

$$e_{ij}' = \frac{1}{2G} s_{ij}, \quad \varepsilon' = \varepsilon = \frac{1}{K}\sigma. \tag{43.2}$$

Here G and K are related to the elasticity modulus E and POISSON's ratio ν by

$$G = \frac{E}{2(1+\nu)}, \quad K = \frac{E}{1-2\nu}. \tag{43.2'}$$

From (43.2) we obtain

$$\dot{e}_{ij}' = \frac{1}{2G}\dot{s}_{ij}, \quad \dot{\varepsilon}' = \frac{1}{K}\dot{\sigma}. \tag{43.3}$$

For the plastic part of the deformation we assume, in addition to $\varepsilon'' = 0$, that (38.8') holds

$$\dot{\varepsilon}_{ij}'' = \lambda s_{ij}. \tag{43.4}$$

---
[1] W. JENNE: Z. angew. Math. Mech. 8, 18—44 (1928). Cf. his Eqs. (18).
[2] L. PRANDTL: Proc. 1st Internat. Congr. Appl. Mech., Delft 1924, pp. 43—54.
[3] E. REUSS: Z. angew. Math. Mech. 10, 266—274 (1930).
[4] See e.g. R. v. MISES: H. Reissner Anniversary Volume, Ann Arbor, Michigan 1949, pp. 415—429 (paper presented Feb. 1948 at the meeting on plasticity in Providence, R. I).

Adding the first Eqs. (43.3) and Eqs. (43.4), and observing that $\ddot{\varepsilon}_{ij}'' = \dot{e}_{ij}''$, we obtain

$$\dot{e}_{ij} = \frac{1}{2G} \dot{s}_{ij} + \lambda s_{ij}, \tag{43.5}$$

that is, using also (43.2), the system of six equations

$$\frac{d}{dt}(e_{ij}) = \frac{1}{2G}\frac{d}{dt}(s_{ij}) + \lambda s_{ij} \quad \text{and} \quad \frac{d\sigma}{dt} = K\frac{d\varepsilon}{dt}. \tag{43.5'}$$

These six equations together with three equilibrium equations and the yield condition are now again ten equations for the $10 = 5 + 1 + 3 + 1$ unknown quantities s_{ij}, σ, v_i, λ. Mathematically, Eqs. (43.5) are much more complicated than (38.8) since in (43.5) the s_{ij} and their time derivatives both appear. The Eqs. (43.5) apply only during plastic flow, i.e., they apply for $J_2 = k^2$. If in a problem $\dot{s}_{ij} = 0$, then from (43.3) $\dot{e}_{ij}' = 0$, hence $\dot{e}_{ij} = \dot{e}_{ij}'' = \ddot{\varepsilon}_{ij}''$ and the Reuss equations (43.5) reduce to the Mises equations (38.8'). Likewise, if the limit $G \to \infty$ is considered, $e_{ij}' \to 0$, in general, from (43.2) and therefore in this case again Eqs. (38.8') are obtained. It is, however, quite unjustified to take this—conversely—as a reason for identifying the Mises theory with a Reuss theory for $G \to \infty$. Fact is that in the plastic region of a body, D_{pl}, where $J_2 = k^2$, the Mises theory offers the theory of the *plastic* deformation, viz. (43.4), and it makes no specific statement concerning the elastic displacements in D_{pl}[1]. With respect to D_{el}, the region below the yield limit, VON MISES has stated repeatedly and used in much of his work, that there the material is elastic (see as one of many examples the treatment reproduced in our Sect. 92). Of course, in appropriate cases the elasticity in D_{el} may be disregarded (see Sect. 95 seq.). This is legitimate in certain problems of unrestricted plastic flow but, in general, not in problems of contained plastic deformation where this "neglecting" would blot out the difference between regions of contained plastic deformation and of elastic deformation since both regions would be considered rigid (see, however, footnote 2, p. 426).

To eliminate λ, in (43.5) we multiply each Eq. (43.5) by s_{ij} and add all six equations:

$$\sum_{i,j} s_{ij} \dot{e}_{ij} = \frac{1}{2G} \sum_{i,j} s_{ij} \dot{s}_{ij} + \lambda \sum_{i,j} s_{ij} s_{ij}.$$

Now according to Eq. (38.9') the second term to the right equals $2\tau_0^2 = 2k^2$ and the first term to the right is zero. Hence,

$$\sum_{i,j} s_{ij} \dot{e}_{ij} = 2\lambda k^2,$$

and, writing as an abbreviation

$$\sum_{i,j} s_{ij} \dot{e}_{ij} = \dot{W} \tag{43.6}$$

we have

$$\lambda = \frac{\dot{W}}{2k^2}. \tag{43.7}$$

---

[1] Considered in this sense the Reuss theory complements rather than contradicts the Mises theory. It is worth mentioning that in the discussion following v. MISES' General Lecture, given at the Third International Congress of Applied Mechanics, Stockholm, 1930 (see R. v. MISES: Proc. 3rd Internat. Congr. Appl. Mech. II, pp. 3—13) the new Reuss theory, just published 1930 in the Z. angew. Math. Mech., is denoted by the participants in the discussion as Mises-Reuss theory.

If in (43.6) we write $\dot{e}_{ij}=\dot{e}'_{ij}+\dot{e}''_{ij}$, we have $\dot{W}=\dot{W}'+\dot{W}''$ where $\dot{W}'$ and $\dot{W}''$ denote the elastic and plastic contribution respectively. Now, by (43.3), $\dot{W}'=(1/2G)\sum_{i,j}s_{ij}\dot{s}_{ij}=0$ in the plastic domain, where $\sum_{i,j}s_{ij}s_{ij}=2k^2$, hence there $\dot{W}=\dot{W}''$ and (43.7) is replaced by

$$\lambda = \frac{\dot{W}''}{2k^2}, \tag{43.7'}$$

as in (38.12). Hence $\dot{W}$ may be replaced by $\dot{W}''$, and vice versa. Eqs. (43.5) may then be written

$$\dot{e}_{ij} = \frac{1}{2G}\dot{s}_{ij} + \frac{\dot{W}''}{2k^2}s_{ij}. \tag{43.8}$$

Of these six equations only four are independent: First, the sum of the three equations for $\dot{e}_{11}$, $\dot{e}_{22}$ and $\dot{e}_{33}$ vanishes identically. In addition, the combination

$$\sum_{i,j}\dot{e}_{ij}s_{ij} = \frac{1}{2G}\sum_{i,j}\dot{s}_{ij}s_{ij} + \frac{\dot{W}''}{2k^2}\sum_{i,j}s_{ij}s_{ij}$$

vanishes identically since it stands for $\dot{W}=0+\frac{\dot{W}}{2k^2}\cdot 2k^2$. The balance of equations is therefore the same as before, since now there is one less unknown.

We note from (43.7), since λ is not negative, that the same holds for $\dot{W}$. Hence Eqs. (43.8) hold at the yield limit, $J_2=k^2$ and for $\dot{W}\geq 0$. If either the yield limit is not reached, $J_2<k^2$, or $J_2=k^2$ but $\dot{W}<0$, (elastic unloading from a plastic state), the flow is elastic and (43.3) valid.

Again, for $G\to\infty$ we obtain from (43.8) $\dot{e}_{ij}=\frac{\dot{W}''}{2k^2}s_{ij}$, i.e., Eqs. (38.8') with λ replaced by (38.12), if in (43.2) as $G\to\infty$ the ratio $s_{ij}/2G\to 0$, hence $\dot{e}'_{ij}\to 0$, so that the $\dot{e}_{ij}$ on the left side in (43.8) become $\dot{e}''_{ij}=\dot{\varepsilon}''_{ij}$.

It is certainly satisfactory that in the Prandtl-Reuss equations (43.5), or (43.8) the elastic strains in the plastic domain are incorporated. On the other hand the Reuss equations introduce the mathematical difficulty of containing both the s_{ij} and $\dot{s}_{ij}$. Even with MISES' relations the complete system of equations is difficult to handle. The replacement of (38.8) by (43.5) adds greatly to these mathematical difficulties.

44. Further stress strain laws. Many authors have proposed stress strain laws which are not derived from a plastic potential (see Sect. 23). We briefly mention the work of PRAGER.

Both the Mises-Lévy theory and the Prandtl-Reuss theory assume a sudden transition from the elastic to the plastic state. Hence two different sets of equations have to be used in the two domains, between which the boundary is in general not known, but has to be determined as part of the problem.

In an attempt to overcome this difficulty PRAGER[1] has proposed stress-strain relations which reflect a continuous transition from the elastic to the plastic state[2].

---

[1] W. PRAGER: Proc. 5th Internat. Congr. Appl. Mech., Cambridge 1938, pp. 234—237, W. PRAGER: Duke Math. J. 9, 228—233 (1942).

[2] We refer the reader to the original papers, in particular to the first one, Eqs. (5) and (6). Cf. also HILL [10], p. 49.

In a survey of proposed stress-strain laws Prager[1] introduces a useful terminology. He calls a stress-strain law of *flow type* or of *deformation type* depending on whether it links the stresses (stress deviations) and rates of stress (of stress deviation) *either to the rate of strain (and* maybe to the strain) *or* to the *strain* itself. We discussed here only flow type laws. The best known deformation-type law is due to Hencky[2]. It may be considered as a development of the theory of A. Haar and Th. von Kármán[3] (cf. discussion in Hill [10], p. 45). Regarding further stress-strain laws see Sect. 23, and for work up to 1930 the survey article on the mechanics of continua by von Mises[4].

45. Remarks on some three-dimensional problems. The problems discussed and solved in the theory of the perfectly plastic body are mainly problems of plane strain, of plane stress, problems with axial symmetry, with spherical symmetry, etc. Very few general three-dimensional problems have been considered.

The fundamentals reported in the preceding sections relate to the general problem. Results concerned with the characteristics of the three-dimensional problem will be discussed in the next article. Here we mention a few results of a more restricted importance.

W. Jenne[5] considers some generalizations of the problem of plane deformation. In plane deformation or plane strain the state of stress is the same in all planes perpendicular to one of the principal directions of stress, which is the same for all points of the body. Jenne studies the case of equal states of stress in planes perpendicular to an arbitrary space curve and further generalizations in this direction. The considerations are interesting but involved. We wish to report here an elegant auxiliary result, a particular case of which we shall use in Chap. G, Eqs. (71.4). Denote by u, v, w the principal axes of Σ. The change of the directions of this triad can be studied by means of a tensor of angular velocity. Consider, e.g., a point $P(x, y, z)$ and a neighboring point $P'(x+dx, y+dy, z+dz)$ at distance ds from P; then, in the transition from P to P' the triad of principal axes of Σ will rotate with the angular velocity $\bar{\omega}_x dx + \bar{\omega}_y dy + \bar{\omega}_z dz$. Denote by $\omega_{u u}, \omega_{u,v} \ldots$ the nine components of the tensor of angular velocity Ω with respect to the u, v, w-system. Then for the derivatives of the σ_i in the u, v, w-directions we have the formulas

$$\frac{\partial \sigma_1}{\partial u} = 2\tau_2 \omega_{wv} + 2\tau_3 \omega_{vw}, \quad \frac{\partial \sigma_2}{\partial v} = 2\tau_3 \omega_{uw} + 2\tau_1 \omega_{wu},$$
$$\frac{\partial \sigma_3}{\partial w} = 2\tau_1 \omega_{vu} + 2\tau_2 \omega_{uv}. \qquad (45.1)$$

In a paper of 1954, Prager[6] remarks that three-dimensional plastic flow is practically unexplored. He quotes a paper by Simoni[7] which he shows to reveal itself at closer inspection as a plane problem, and a genuinely three-dimensional flow field studied by Hill[8] where the incipient plastic flow in a prismatic bar of plastic-rigid material subject to combined tension, torsion and bending, is described. Prager then investigates completely the general three-dimensional plastic flow possible under a uniform state of stress[9].

---

[1] W. Prager: J. Appl. Mech. **14**, 226—233 (1948). It seems that the terms have already been used by A. A. Ilyushin, Prikl. Mat. Mekh. **9**, 207—218 (1945).

[2] W. Hencky: Proc. 1st Internat. Congr. Appl. Mech., Delft 1924, pp. 312—317.

[3] A. Haar and Th. von Kármán: Nachr. kgl. Ges. Wiss. Göttingen 1909, 204—218 (cf. also p. 427 of the present article).

[4] R. v. Mises: Proc. 3rd Internat. Congr. Appl. Mech. II, pp. 3—13 (pp. 9—13 contain an interesting discussion of the paper).

[5] W. Jenne: Z. angew. Math. Mech. **8**, 18—44 (1928).

[6] W. Prager: Rev. Fac. Sci. Univ. Istanbul **19**, 23—27 (1954).

[7] F. de Simoni: Ist. Lombardo Sci. Lett., Rendic. Cl. Sci. Mat. Nat. (3) **15**, 623—634 (1951).

[8] R. Hill: Quart. Appl. Math. **1**, 18—28 (1948).

[9] See also L. Finzi: Torino R. Accad. Sci. **76**, 1—19 (1941), and (received only at time of page-proof reading) L. Finzi: Ist. Lombarda Sci. Sett., Rendic. Cl. Sci. Mat. Nat. **90**, 528—535 (1956).

II. Discontinuous solutions[1].

a) Characteristics. Application to the three-dimensional problem of the perfectly plastic body.

46. Introduction. In many branches of mathematical physics the concepts of "discontinuity surface", "characteristic surface", "shock" play a great role. The mathematicians to whom we owe much of our present insight into these theories — here we only mention the names of B. RIEMANN, J. HADAMARD, T. LEVI-CIVITA, H. HUGONIOT — developed the problems and concepts with respect to the theory of fluids, in particular compressible fluid flow, and numerous investigations followed.

In recent years similar investigations have been attempted regarding perfectly plastic solids. This theory has not yet reached a clarity and completeness comparable to achievements in *compressible* fluid flow. We shall report on some of the results obtained so far; in doing so we thought it appropriate to explain some of the mathematical foundations, at least to a certain degree, which might prove helpful to one or the other research worker in a field where considerations of this type are comparatively new.

47. Examples. α) Denote by φ a function of x and y and consider one of the simplest partial differential equations of second order:

$$\frac{\partial^2 \varphi}{\partial x \partial y} = 0. \tag{47.1}$$

Along some *initial curve*, $y = f(x)$, e.g., the straight line $y = x$, the values of φ and of $\partial \varphi / \partial x$ may be prescribed, *initial data*, or *Cauchy data*, as arbitrarily given functions of one variable,

$$\varphi(x, x) = f(x), \quad \left(\frac{\partial \varphi}{\partial x}\right)_{y=x} = g(x). \tag{47.2}$$

We recognize immediately the truth of the well-known result that by these data a solution is uniquely given. For, introduce $G(x) = \int^x g(x)\, dx$, and put

$$\varphi(x, y) = G(x) + f(y) - G(y).$$

This function satisfies (47.1) and also (47.2), since

$$\varphi(x, x) = G(x) + f(x) - G(x) = f(x)$$

$$\left(\frac{\partial \varphi}{\partial x}\right)_{y=x} = G'(x) = g(x).$$

Next, we choose as initial curve the line $x = c$, and prescribe along this initial line the values of φ and of $\partial \varphi / \partial x$:

$$\varphi(c, y) = f(y), \quad \left(\frac{\partial \varphi}{\partial x}\right)_{x=c} = g(y). \tag{47.3}$$

The general form of the solution of (47.1) is $\varphi(x, y) = h(x) + k(y)$, where h and k are arbitrary functions of one variable; thus, from (47.3)

$$\varphi(c, y) = h(c) + k(y) = f(y), \quad h'(c) = g(y).$$

We thus see that *it is not possible to prescribe $g(y)$ arbitrarily*; this function must reduce to a constant, say k, since it is to be equal to $h'(c)$. On the other hand,

---

[1] The considerations of this division II apply to three dimensions except for Sect. 61.

if $g(y)$ is a constant, then there are clearly *infinitely many solutions* which all satisfy the given initial conditions. Take, e.g., $k=3$, $c=1$, and $h(x)=x^3$, $h'(x)=3x^2$, $h'(c)=3=k$. But for $h(x)=\frac{3}{10}x^{10}$, likewise $h'(1)=3$ and similarly for $h(x)=-\frac{6}{\pi}\cos\frac{\pi}{2}x$. We see that two completely different solutions

(1) $\varphi(x, y) = x^3 + f(y) - 1$, and (2) $\varphi(x, y) = f(y) - \frac{6}{\pi}\cos\frac{\pi}{2}x$

(and infinitely many others) both satisfy (47.1) and (47.3), viz.,

$$\varphi(1, y) = f(y), \quad \left(\frac{\partial \varphi}{\partial x}\right)_{x=1} = 3. \tag{47.3'}$$

Both solutions along $x=c$ have the same $\frac{\partial \varphi}{\partial x}$, $\frac{\partial \varphi}{\partial y}$, $\frac{\partial^2 \varphi}{\partial x \partial y}$, $\frac{\partial^2 \varphi}{\partial y^2}$, $\frac{\partial^3 \varphi}{\partial x^2 \partial y}$, ... etc. But the values of $\partial^2 \varphi/\partial x^2$, etc. are different. We may pronounce the result also in the following way: Along the *exceptional* or *characteristic* line $x=c$, the two different solutions (1) and (2) can be patched together without violating (47.1) or (47.3). Derivatives of φ along the characteristic line are determined by (47.1) and (47.3), but derivatives across this line remain undetermined.

β) Consider next the much more general partial differential equation[1]

$$A \frac{\partial^2 \varphi}{\partial x^2} + 2B \frac{\partial^2 \varphi}{\partial x \partial y} + C \frac{\partial^2 \varphi}{\partial y^2} = F, \tag{47.4}$$

where A, B, C, F may be functions of x, y, φ and of its first derivatives. Assume again, as before, initial conditions along the line $x=c$ namely:

$$\varphi(c, y) = f(y), \quad \left(\frac{\partial \varphi}{\partial x}\right)_{x=c} = g(y). \tag{47.5}$$

We now ask: To what extent does the differential equation (47.4) together with the initial conditions (47.5) determine a solution? Obviously we can compute from (47.5) for $x=c$: $\frac{\partial \varphi}{\partial y}$, $\frac{\partial^2 \varphi}{\partial x \partial y}$, $\frac{\partial^2 \varphi}{\partial y^2}$, etc. In order to compute $\partial^2 \varphi/\partial x^2$ we need the differential equation, (47.4) and find if $A \neq 0$:

$$\left[\frac{\partial^2 \varphi}{\partial x^2}\right]_{x=c} = \left[\frac{F}{A} - \frac{C}{A}\frac{\partial^2 \varphi}{\partial y^2} - \frac{2B}{A}\frac{\partial^2 \varphi}{\partial x \partial y}\right]_{x=c}.$$

By using both (47.5) and (47.4)—which we may differentiate—we can thus compute for $x=c$ as many derivatives as we wish and set up a Taylor expansion which determines $\varphi(x, y)$ in a neighborhood of the initial curve $x=c$.

Or, we may conclude as follows—always if $A \neq 0$. Consider the neighboring line $x=x_1=c+dc$. Then, approximately

$$\varphi(c+dc, y) = \varphi(c, y) + \frac{\partial \varphi}{\partial x}(c, y) dc,$$

$$\frac{\partial \varphi}{\partial x}(c+dc, y) = \frac{\partial \varphi}{\partial x}(c, y) + \frac{\partial^2 \varphi}{\partial x^2}(c, y) dc, \quad \text{etc.}$$

Thus, we know approximately φ and $\partial \varphi/\partial x$ along the line $x=x_1$. We then may proceed to a line $x=x_2$ adjacent to $x=x_1$ etc. Similarly we may proceed towards the left to $x=x_1'=c-dc$.

Neither of these simple considerations leads to an exact mathematical existence proof. They show, however, the role of the condition $A \neq 0$ for our present

---

[1] Cf. the presentation in R. v. MISES: Mathematical theory of compressible fluid flow. Completed by HILDA GEIRINGER and G. S. S. LUDFORD. New York: Academic Press 1958.

example. (In the previous example, the corresponding condition was that the initial curve must not be parallel to either of the two axes x or y.) Assume now that it is possible—as in the previous example—to find two different solutions of (47.4) and (47.5), say one, $\varphi=\varphi_1$, to the right of $x=c$, one, $\varphi=\varphi_2$, to the left, but such that along that line, $\varphi_1=\varphi_2$, $\dfrac{\partial \varphi_1}{\partial x} = \dfrac{\partial \varphi_2}{\partial x}$. We may then call this combination of φ_1 and φ_2 a "solution" in the combined domain to the right *and* to the left. In fact, this combined solution satisfies the differential equation in both domains, including the line $x=c$, and the boundary conditions along $x=c$. Also along this vertical line: $\dfrac{\partial^2 \varphi_1}{\partial x \partial y} = \dfrac{\partial^2 \varphi_2}{\partial x \partial y}$, $\dfrac{\partial^2 \varphi_1}{\partial y^2} = \dfrac{\partial^2 \varphi_2}{\partial y^2}$. It is true that, in general, $\dfrac{\partial^2 \varphi_1}{\partial x^2} \neq \dfrac{\partial^2 \varphi_2}{\partial x^2}$; this quantity, however has now the coefficient $A=0$. Hence we see again that *along $x=c$ two entirely different solutions can be patched together*, if $A=0$.

The preceding formulations are still dependent on the arbitrary coordinate system. We must obtain an *invariant interpretation* of a condition like $A \neq 0$. Such an interpretation can be achieved by mathematical considerations or by mechanical conclusions.

γ) *Examples of invariant interpretations*. Consider the Eqs. (38.7) and (38.9) for a "plane problem". We shall consider this theory in detail in Chap. F. For the present we write (with $\tau_{xy}=\tau$):

$$\frac{\partial \sigma_x}{\partial x} + \frac{\partial \tau}{\partial y} = 0, \quad \frac{\partial \tau}{\partial x} + \frac{\partial \sigma_y}{\partial y} = 0, \quad (\sigma_x - \sigma_y)^2 + 4\tau^2 = 4k^2.$$

To integrate these we introduce a function $\varphi(x, y)$ such that

$$\frac{\partial \varphi}{\partial x} = -\tau, \quad \frac{\partial \varphi}{\partial y} = \sigma_x.$$

The first equation is then identically satisfied, and the second combined with the yield condition gives

$$(\sigma_y - \sigma_x) \frac{\partial^2 \varphi}{\partial x^2} - 4\tau \frac{\partial^2 \varphi}{\partial x \partial y} + (\sigma_x - \sigma_y) \frac{\partial^2 \varphi}{\partial y^2} = 0,$$

an equation of type (47.4). Assume that along the y-axis φ and $\partial \varphi/\partial x$ are given. (If $\varphi(0, y)$ is given, this amounts to the same as if $\dfrac{\partial \varphi}{\partial y}(0, y)$ is given. Hence we may assume that along the y-axis τ and σ_x are given.) We know from the previous example that this will determine a solution uniquely unless $A=0$, viz. unless $\sigma_y - \sigma_x = 0$. If $A=0$, the y-direction is exceptional or characteristic. To obtain the condition $A=0$ in a meaningful form we consider the angle ϑ' of the first principal direction with the y-direction and find $\tan 2\vartheta' = 2\tau/(\sigma_y - \sigma_x)$. If $\sigma_y - \sigma_x = 0$, $2\vartheta' = 90°$, or $\vartheta' = 45°$. *Hence the characteristic bisects the angle of the principal directions*; and this is now a geometrical characterization, which is independent of any coordinate system (see Sect. 66).

As a second example we establish an important geometrical property of a characteristic. We consider the general yield condition $g(\sigma_x, \sigma_y, \tau) = 0$ and obtain, with φ defined as before, and using the abbreviation $\varphi_{xx} = \partial^2 \varphi / \partial x^2$, etc.

$$\frac{\partial g}{\partial \sigma_y} \varphi_{xx} - \frac{\partial g}{\partial \tau} \varphi_{xy} + \frac{\partial g}{\partial \sigma_x} \varphi_{yy} = 0.$$

The y-direction will be characteristic if $A = \partial g/\partial \sigma_y = 0$. But according to (40.5), this gives $\dot{\varepsilon}_y = 0$. Hence *the rate of extension in a characteristic direction vanishes*.

This is another invariant geometrical characterization valid for a general yield condition under the strain-stress law (40.5).

48. Systems of differential equations. Instead of the single partial differential equation (47.4) with two independent variables we now consider a system of m equations for m unknown functions $u_1, u_2, \ldots u_m$ of $n+1$ independent variables $x_0, x_1, \ldots x_n$. It is convenient for our purpose to consider a system of equations of order two rather than one of order one, although, theoretically, the former can be reduced to the latter.

We remember the condition, $A \neq 0$, which appeared in the discussion of (47.4), in connection with data given on the line $x = \text{const.} = c$. From the fact that $\varphi(c, y)$ and $\frac{\partial \varphi}{\partial x}(c, y)$ were given, we knew all derivatives with respect to y of $\varphi(c, y)$ or, as we may say, all *interior derivatives* along the line $x = c$. In addition our knowledge of the normal *exterior* derivative $\frac{\partial \varphi}{\partial x}(c, y)$ led us beyond the line $x = c$. The meaning of $A \neq 0$ was that exterior derivatives like $\frac{\partial^2 \varphi}{\partial x^2}(c, y)$ which cannot be found from the initial conditions could be computed from the differential equation since in this case, we could solve Eq. (47.4) with respect to $\partial^2 \varphi/\partial x^2$. If the equation is given in the explicit form $\frac{\partial^2 \varphi}{\partial x^2} = \Phi\left(x, y, \frac{\partial \varphi}{\partial x}, \frac{\partial \varphi}{\partial y}, \frac{\partial^2 \varphi}{\partial x \partial y}, \frac{\partial^2 \varphi}{\partial y^2}\right)$ then the condition $A \neq 0$ is clearly fulfilled and the line $x = c$ cannot be a characteristic. These ideas can be generalized.

Consider the m equations

$$\sum_{k=1}^{m} \sum_{i,j}^{0 \ldots n} a_{\mu k i j} \frac{\partial^2 u_k}{\partial x_i \partial x_j} + b_\mu = 0 \qquad \mu = 1, 2, \ldots m, \qquad (48.1)$$

where the $a_{\mu k i j}$ and the b_μ depend on the x_i, on the u_k, and the first partial derivatives $\partial u_k/\partial x_i$ where always $i, j = 0, 1, \ldots n; \mu, k = 1, \ldots m$. Instead of the line $x = c$ we consider now the hyperplane, $\omega: x_0 = c$, (an ordinary plane if $n + 1 = 3$) and assume the *initial data*: all u_k are given on ω, as well as all $\partial u_k/\partial x_0$; or, more explicitly

$$u_k(c, x_1, \ldots x_n) = f_k(x_1, \ldots x_n), \quad \frac{\partial u_k}{\partial x_0}(c, x_1, \ldots x_n) = g_k(x_1, \ldots x_n), \quad (48.2)$$

as generalization of (47.5). Note that from the knowledge of $u_k(c, x_1, x_2, \ldots x_n)$ we can compute *all interior derivatives* $\frac{\partial u_k}{\partial x_\nu}(c, x_1, \ldots x_n)$, $\nu = 1, 2, \ldots n$ along ω; in addition the *exterior derivatives* $\partial u_k/\partial x_0$ are given. By further differentiation in ω we can compute higher interior derivatives. But, in order to be able to compute the exterior derivatives $\partial^2 u_k/\partial x_0^2$, it is necessary that the system (48.1) can be solved with respect to $\frac{\partial^2 u_1}{\partial x_0^2}, \frac{\partial^2 u_2}{\partial x_0^2}, \ldots \frac{\partial^2 u_m}{\partial x_0^2}$. A system solved with respect to these derivatives is briefly called *normal*. One denotes as the *Cauchy problem* the determination in a neighborhood of ω of solutions $u_1, u_2, \ldots u_m$ of (48.1) which assume the values (48.2) on ω. This determination is uniquely possible for a normal system under appropriate assumptions for the $a_{\mu k i j}$, the b_μ the f_k and g_k which appear in (48.1) and (48.2)[1]. In order to be able to transform (48.1) to a normal system a certain determinant of coefficients must be different from zero.

---

[1] The proof is due to Cauchy and Sonja Kovalevska (see, e.g. E. Goursat: Cours d'Analyse, Vol. III, and many other sources).

Hence, with $a_{\mu k00}=a_{\mu k}$ and $\|a_{\mu k}\|$ denoting the $m\times m$ determinant of these coefficients,
$$\Omega = \|a_{\mu k}\| \neq 0 \tag{48.3}$$
is the generalization of the condition $A \neq 0$. In this case we can solve with respect to the $\partial^2 u_\mu/\partial x_0^2$, $(\mu = 1, 2, \ldots m)$, and by further differentiations of these equations, higher exterior derivatives of the u_μ can also be computed. It is then plausible (thinking in terms of Taylor expansions in the neighborhood of ω or of step by step numerical computation) that a solution of (48.1), which assumes the given values (48.2) on ω, exists in a certain neighborhood of ω.

These considerations are of a completely formal nature and we have not given a proof that—under appropriate additional assumptions—$\Omega \neq 0$ is sufficient for the existence of a solution.

On the other hand, we call ω, i.e. the plane $x_0 = c$, *exceptional* or *characteristic* if
$$\Omega = 0. \tag{48.4}$$
Clearly, the condition (48.3) depends on the coordinate system and must be *transformed into an invariant form* by means of physical or of purely mathematical considerations (see Sect. 51).

If the given system is of first order
$$\sum_{k=1}^{m}\sum_{i=0}^{n} a_{\mu k i}\frac{\partial u_k}{\partial x_i} + b_\mu = 0, \quad k = 1, 2, \ldots m, \tag{48.5}$$
the initial values or Cauchy data consist of values of the u_μ on ω. Analogous considerations as before lead to the condition
$$\Omega = \|a_{\mu k}\| \neq 0, \tag{48.6}$$
where $a_{\mu k} = a_{\mu k 0}$ is the coefficient of $\partial u_k/\partial x_0$ in the μ-th equation, and ω is called exceptional or characteristic if $\Omega = 0$. Again the system is called normal if it is solved for the $\partial u_k/\partial x_0$, $k = 1, 2, \ldots m$. Note that the $a_{\mu k}$ in (48.3) or (48.6) depend on the given Cauchy data. Hence the same geometrically specified plane may or may not be exceptional depending on these given data.

49. Characteristics of the von Mises plasticity equations. The following investigations are due to T. Y. Thomas[1]. Particular cases have been studied before[2].

From the basic equations (Sect. 38) we eliminate the five stress components s_{ij} and using (38.5) we obtain on the whole five equations:
$$\frac{\partial p}{\partial x} = \frac{\partial}{\partial x}(\mu\dot\varepsilon_{xx}) + \frac{\partial}{\partial y}(\mu\dot\varepsilon_{xy}) + \frac{\partial}{\partial z}(\mu\dot\varepsilon_{xz}), \tag{49.1}$$
and two similar equations[3]; then, as in (38.11),
$$\mu^2 = \frac{2\tau_0^2}{\dot\varepsilon_1^2 + \dot\varepsilon_2^2 + \dot\varepsilon_3^2}, \quad \text{or} \quad \mu = \frac{\sqrt{2}\,\tau_0}{A}, \quad \text{where } A^2 = \sum_{i,j} = \dot\varepsilon_{ij}^2, \tag{49.2}$$
and
$$\dot\varepsilon_1 + \dot\varepsilon_2 + \dot\varepsilon_3 = 0. \tag{49.3}$$

---
[1] T. Y. Thomas: J. Rat. Mech. a. Analysis 1, 343—357 (1952).
[2] Of course, the characteristics for plane deformation are well known. (See Chapters. F and G of this article.) P. S. Symond: Quart. Appl. Math. 6, 448—452 (1949) investigated the problem in the case of axial symmetry; cf. also R. Hill [10], p. 263. This problem is essentially elliptic, that means there are, in general, no real characteristics.
[3] Here plastic strains are denoted by ε_{ij}, $\dot\varepsilon_{ij}$, etc. since no confusion with elastic strains is possible. The A of (49.2) has nothing to do with the notation A in Sect. 47.

Handbuch der Physik, Bd. VI.

We now identify z with x_0 of the preceding section and try to solve our system with respect to the highest derivatives $\frac{\partial^2 v_x}{\partial z^2}, \frac{\partial^2 v_y}{\partial z^2}, \frac{\partial^2 v_z}{\partial z^2}, \frac{\partial p}{\partial z}$. From (49.1), it follows that for $i = 1, 2, 3$

$$\frac{\partial p}{\partial x_i} = \sum_j^{1\ldots 3} \frac{\partial}{\partial x_j}(\mu \dot{\varepsilon}_{ij}) = \sum_j \frac{\partial \mu}{\partial x_j} \dot{\varepsilon}_{ij} + \mu \frac{\partial \dot{\varepsilon}_{ij}}{\partial x_j}, \qquad (49.4)$$

and in particular

$$\frac{\partial \mu}{\partial z} = -\frac{\sqrt{2}\,\tau_0}{A^2} \frac{\partial A}{\partial z},$$

$$\frac{\partial A}{\partial z} = \frac{1}{A}\left(\dot{\varepsilon}_{xz}\frac{\partial^2 v_x}{\partial z^2} + \dot{\varepsilon}_{yz}\frac{\partial^2 v_y}{\partial z^2} + \dot{\varepsilon}_{zz}\frac{\partial^2 v_z}{\partial z^2} + \cdots\right),$$

$$\frac{\partial \mu}{\partial z} = -\frac{\sqrt{2}\,\tau_0}{A^3}\left(\dot{\varepsilon}_{xz}\frac{\partial^2 v_x}{\partial z^2} + \cdots\right).$$

Hence, if in the first Eq. (49.4) we write only the terms that matter, we have

$$\frac{\partial p}{\partial x} = \frac{\sqrt{2}\,\tau_0}{2A}\frac{\partial^2 v_x}{\partial z^2} + \cdots - \frac{\sqrt{2}}{A^3}\tau_0 \dot{\varepsilon}_{xz}\left(\dot{\varepsilon}_{xz}\frac{\partial^2 v_x}{\partial z^2} + \dot{\varepsilon}_{yz}\frac{\partial^2 v_y}{\partial z^2} + \dot{\varepsilon}_{zz}\frac{\partial^2 v_z}{\partial z^2} + \cdots\right)$$

$$\doteq \frac{\sqrt{2}\,\tau_0}{A}\left[\frac{\partial^2 v_x}{\partial z^2}\left(\frac{1}{2} - \frac{\dot{\varepsilon}_{xz}^2}{A^2}\right) - \frac{\partial^2 v_y}{\partial z^2}\frac{\dot{\varepsilon}_{xz}\dot{\varepsilon}_{yz}}{A^2} - \frac{\partial^2 v_z}{\partial z^2}\frac{\dot{\varepsilon}_{xz}\dot{\varepsilon}_{zz}}{A^2} + \cdots\right],$$

and two similar equations. Differentiation with respect to z of Eq. (49.3) gives

$$\frac{\partial^2 v_z}{\partial z^2} + \cdots = 0.$$

Hence for (48.3) we obtain the symmetric determinant

$$\begin{vmatrix} A^2 - 2\dot{\varepsilon}_{xz}^2 & -2\dot{\varepsilon}_{xz}\dot{\varepsilon}_{yz} & -2\dot{\varepsilon}_{xz}\dot{\varepsilon}_{zz} & 0 \\ -2\dot{\varepsilon}_{xz}\dot{\varepsilon}_{yz} & A^2 - 2\dot{\varepsilon}_{yz}^2 & -2\dot{\varepsilon}_{yz}\dot{\varepsilon}_{zz} & 0 \\ -2\dot{\varepsilon}_{xz}\dot{\varepsilon}_{zz} & -2\dot{\varepsilon}_{yz}\dot{\varepsilon}_{zz} & A^2 - 2\dot{\varepsilon}_{zz}^2 & 1 \\ 0 & 0 & 1 & 0 \end{vmatrix} = \begin{vmatrix} A^2 - 2\dot{\varepsilon}_{xz}^2 & -2\dot{\varepsilon}_{xz}\dot{\varepsilon}_{yz} \\ -2\dot{\varepsilon}_{xz}\dot{\varepsilon}_{yz} & A^2 - 2\dot{\varepsilon}_{yz}^2 \end{vmatrix}$$

$$= -A^2(A^2 - 2\dot{\varepsilon}_{xz}^2 - 2\dot{\varepsilon}_{yz}^2) = 0. \qquad (49.5)$$

It remains to find an invariant form of this last condition. For this purpose we write the last expression in (49.5) in terms of the stresses and replace $\sum_{i,j} s_{ij}^2$ by $2\tau_0^2$. We find simply

$$\tau_{xz}^2 + \tau_{yz}^2 = \tau_0^2 = k^2. \qquad (49.6)$$

If this relation holds, the plane $z = $ constant is tangent to a characteristic surface element. Hence we have so far the result: *The shear stress corresponding to a characteristic surface element equals $\pm k$.* It follows that this shear stress τ^2 must be equal to $\left(\frac{\sigma_1 - \sigma_3}{2}\right)^2 = \left(\frac{s_1 - s_3}{2}\right)^2$, with σ_1 the largest, σ_3 the smallest stress. Then, from

$$\left(\frac{s_1 - s_3}{2}\right)^2 = k^2 = \tfrac{1}{2}(s_1^2 + s_2^2 + s_3^2)$$

it follows that

$$(s_1 + s_3)^2 + 2s_2^2 = 0$$

or

$$s_2 = 0, \quad s_1 + s_3 = 0. \qquad (49.7)$$

From these we obtain

$$\sigma_2 = \frac{\sigma_1 + \sigma_3}{2} = -p, \quad (\sigma_1 - \sigma_3)^2 = 4\tau_0^2, \qquad (49.8)$$

well-known formulas of the theory of plane deformation.

Hence, if, and only if, $s_2 = 0$ there exist at a point two characteristic surface elements which are identical in direction with the planes of maximum and minimum shearing stress. The direction cosines of the normals to these characteristic elements with respect to a coordinate system which has the three principal directions are therefore

$$\left(\frac{1}{\sqrt{2}}, 0, \frac{1}{\sqrt{2}}\right) \text{ and } \left(\frac{1}{\sqrt{2}}, 0, -\frac{1}{\sqrt{2}}\right),$$

and with respect to an arbitrary x_1, x_2, x_3-system the direction cosines are[1]

$$\alpha_i = \frac{1}{\sqrt{2}} [\cos(x_i 1) + \cos(x_i 3)] \text{ and } \frac{1}{\sqrt{2}} [\cos(x_i 1) - \cos(x_i 3)]. \quad (49.9)$$

50. Further results and comments. We shall see (Sect. 51) that in a problem like that of the preceding sections there exists a whole cone of normals to characteristic surface elements which, however, may or may not be real. We see that in the present case this cone is either fully imaginary, or if $s_2 = 0$, viz. $\dot{\varepsilon}_2 = 0$, $s_1 = -s_3$, $\dot{\varepsilon}_1 + \dot{\varepsilon}_3 = 0$, it is imaginary with the exception of two real directions. The problem reduces to that of "plane strain" (or "plane deformation") if the directions 1, 2, 3 are the same at each point. We may thus say that the problem is either fully *elliptic* (i.e. there are no real characteristics) or it admits only two real characteristic directions at each point[2].

The result of Sect. 49 has been discussed by W. PRAGER[3]. As particular cases of THOMAS' result he mentions the case of plane strain, that of plastic torsion, and that of the plastic twisting of a circular ring studied by FREIBERGER[4] (Sect. 85). On the other hand, it is seen that in the vast majority of three-dimensional problems real characteristics will not arise if the quadratic yield condition with associated flow rule is used. The existence of real characteristic surfaces presents a great mathematical advantage and (as PRAGER remarks) inasmuch as characteristic surfaces are slip surfaces it seems almost a physical necessity. PRAGER[5] adds that it might perhaps be possible to apply some adjustment to the yield condition so as to make the problem hyperbolic. For a similar purpose VON MISES, 1948, introduced a modification of his quadratic yield condition in the case of plane stress (see Sect. 72).

THOMAS has also investigated the question of whether the characteristic surface elements of the VON Mises plasticity equations, if they exist, join so as to form characteristic surfaces. Starting at some point of the material we consider the space curve which has at every point the direction—say α—of one particular of the two characteristic normals and obtain a two-dimensional family of space curves, a congruence of curves. We ask whether these curves admit a family of surfaces normal to them—a family of orthogonal trajectories. This is not obvious, in space.

---

[1] In the same paper THOMAS has also discussed the VON MISES flow equations in connection with TRESCA's yield condition (thus abandoning the theory of plastic potential). The result is similar, except that the condition $s_2 = 0$ does not appear. There exist two and only two real characteristic directions at each point of the medium and the characteristic surface elements, whose normals they are, are the surface elements of maximum and minimum shearing stress.

[2] THOMAS' result was reestablished by him in 1953, T. Y. THOMAS: J. Rat. Mech. a. Analysis **2**, 339—381 (1953), and by J. L. ERICKSEN: J. Math. Phys. **34**, 74—79 (1955). W. PRAGER presented the same result in a seminar lecture (1954) at Brown University.

[3] W. PRAGER: Proc. 2nd U. S. Nat. Congr. Appl. Mech., Michigan 1954, pp. 21—32, see p. 25. This paper contains a wealth of material on discontinuous solutions in plasticity.

[4] W. FREIBERGER: Aeron. Res. Lab. Australia, Report SM 213, (1953).

[5] W. PRAGER: James Clayton Lecture. Proc. Inst. Mech. Engrs. **169**, 41—57 (1955), see particularly p. 52.

If a family of surfaces $f(x, y, z) = $ constant are the orthogonal trajectories of a given vector field $\boldsymbol{\alpha}$, then the gradient of f must have the $\boldsymbol{\alpha}$-direction. Hence, λ denoting a non-vanishing scalar function,

$$\operatorname{grad} f = \lambda \boldsymbol{\alpha}$$

must hold. Since curl grad $f = 0$, we obtain, using an easily verified formula of vector calculus:

$$0 = \operatorname{curl}(\lambda \boldsymbol{\alpha}) = \boldsymbol{\alpha} \times \operatorname{grad} \lambda - \lambda \operatorname{curl} \boldsymbol{\alpha},$$

or $\boldsymbol{\alpha} \times \operatorname{grad} \lambda = \lambda \operatorname{curl} \boldsymbol{\alpha}$. But $\boldsymbol{\alpha} \cdot (\boldsymbol{\alpha} \times \operatorname{grad} \lambda) = 0$. Hence $\lambda(\boldsymbol{\alpha} \cdot \operatorname{curl} \boldsymbol{\alpha}) = 0$. Since we may perform these steps also in reverse we see that

$$\boldsymbol{\alpha} \cdot \operatorname{curl} \boldsymbol{\alpha} = 0 \tag{50.1}$$

is the necessary and sufficient condition for a vectorfield in space to admit orthogonal trajectories. (In the plane, this is identically satisfied.) In our problem the $\boldsymbol{\alpha}$ are given by (49.9), varying as the directions 1, 2, 3 vary.

Recently THOMAS has also studied[1] the characteristics of the general three-dimensional problem, where instead of von Mises' flow equations the Prandtl-Reuss equations (Sect. 43) are used and total incompressibility (in contrast to plastic incompressibility) is assumed. Derivations and results are rather involved, and we refer the reader to the original paper and the references quoted there.

b) General consideration of discontinuous solutions.

51. Characteristic surfaces. Characteristic condition. We return to systems as considered in Sect. 48 and shall obtain the characteristic condition in a general and invariant form[2]. Consider the system of first order (48.5) of m equations for m unknown functions $u_k(x_0, x_1, \ldots x_n) = u_k(x)$, $k = 1, \ldots m$. Instead of the hyperplane $x_0 = c$ of Sect. 48, we consider a hypersurface S

$$z(x_0, x_1, \ldots x_n) = a_0, \quad \text{briefly} \quad z(x) = a_0, \tag{51.1}$$

assuming that by an appropriate change of the variables $x_0, x_1, \ldots x_n$ into new variables $z, z_1, \ldots z_n$ the plane $x_0 = c$ is transformed into S. Then, the families $z(x) = $ const., $z_1(x) = $ const., $\ldots z_n(x) = $ const. form the new coordinate surfaces. Cauchy data on S, i.e. values of the $u_1, \ldots u_m$ in terms of the $z_1, \ldots z_n$ will determine a solution in the neighborhood of S if the system (48.5) can be transformed into a normal system with respect to these variables (see Sect. 48).

Let us compute this condition. With $p_i = \partial z/\partial x_i$ we have for $k = 1, 2, \ldots m$:

$$\frac{\partial u_k}{\partial x_i} = \frac{\partial u_k}{\partial z} \frac{\partial z}{\partial x_i} + \frac{\partial u_k}{\partial z_1} \frac{\partial z_1}{\partial x_i} + \cdots + \frac{\partial u_k}{\partial z_n} \frac{\partial z_n}{\partial x_i} = \frac{\partial u_k}{\partial z} p_i + \cdots, \tag{51.2}$$

where the points indicate that we are only interested in the first term, $\frac{\partial u_k}{\partial z} p_i$. If we substitute the $\partial u_k/\partial x_i$ into Eqs. (48.5) we obtain

$$\sum_{k=1}^{m} \frac{\partial u_k}{\partial z} \sum_{i=0}^{n} a_{\mu k i} p_i + \cdots = 0.$$

The coefficient of $\partial u_k/\partial z$ in the μ-th equation is thus seen to be $\sum_{i=0}^{n} a_{\mu k i} p_i$. The p_i are proportional to the direction cosines α_i of the normal $\boldsymbol{\alpha}$ to S so that we may use these α_i instead of the p_i. The coefficients

$$\omega_{\mu k} = \sum_{i=0}^{n} a_{\mu k i} \alpha_i \tag{51.3}$$

---

[1] T. Y. THOMAS: J. Rat. Mech. a. Analysis **5**, 251—262 (1956).
[2] Attention is called to the beautiful presentation in T. LEVI-CIVITA. Caractéristiques des systèmes différentielles et propagation des ondes. Paris 1932.

generalize the $a_{\mu k}$ of Eq. (48.6). The transformed system can be solved for the $\frac{\partial u_1}{\partial z}, \ldots \frac{\partial u_m}{\partial z}$ if

$$\Omega = \|\omega_{\mu k}\| \neq 0. \tag{51.4}$$

The $\omega_{\mu k}$ are linear forms in the α_i and Ω is homogeneous of degree m in the α_i.

The equation

$$\Omega = \|\omega_{\mu k}\| = \|\Sigma a_{\mu k i} \alpha_i\| = 0, \tag{51.5}$$

with $\omega_{\mu k}$ defined by (51.3) singles out those *directions* $\boldsymbol{\alpha}^*$ *for which a surface* S^* *in connection with values* u_k^* *on* S^* *becomes exceptional or characteristic* (see also Sect. 52). We know from previous study that for such a characteristic surface the Cauchy problem cannot be solved in general.

In the *linear case* where the $a_{\mu k i}$ and the b_μ in Eqs. (48.5) depend only on the x_i but not on the u_k, the characteristic surfaces are determined once and for all for a given system. In the more general case which holds in plasticity—as well as in gas dynamics—the exceptional character depends on the surface S and on the given values on it. Eq. (51.5) is called *characteristic condition* or *direction condition*. Regarding the generalization of (48.3) (which refers to a system of second order) see the end of Sect. 57.

52. Compatibility conditions. In a similar way as before we denote at any point of S the derivatives of the u_k with respect to the $z_1, z_2, \ldots z_n$ as *interior derivatives* and $\partial u_k/\partial z$, $(k=1, 2, \ldots m)$ as *exterior derivatives*. We have seen that the m original Eqs. (48.5) fail to determine the m exterior derivatives if and only if Eq. (51.5) holds for some direction $\boldsymbol{\alpha}^*$. Equivalent to this is that at least one combination (of the original equations) exists—say $s \geq 1$ such combinations—which do not contain any derivative in the z-direction. This shows again that on an exceptional surface S^* the u_k^* are not "arbitrary", since there necessarily exist $s \geq 1$ relations between them. *These s relations between the interior derivatives on S^* are called compatibility relations* since they restrict the arbitrariness of the u_k^* on S^*. *We call the S^* with compatible values u_k^* on it exceptional or characteristic.* The analytic formulation of the compatibility relations follows in Sect. 58. In addition to these s compatibility relations *there remain $(m-s)$ equations each of which contains at least one exterior derivative*, i.e. derivative in the $\boldsymbol{\alpha}^*$-direction. These $(m-s)$ equations cannot determine m exterior derivatives. Hence in no neighborhood of S^* is a solution uniquely determined. Along S^* two different solutions can be patched together.

53. Discontinuous solutions. In the light of the preceding facts, following HADAMARD, LEVI-CIVITA and particularly VON MISES[1], we define *discontinuous solutions of a system (48.5) across a surface S^** as follows:

(1) On both sides of S^* all differential equations are satisfied.

(2) Across S^* at least one of the u_k or its derivatives has a jump.

A finite jump of u_k across S^* causes the derivative of u_k in the direction normal to S^* to become infinite. Then, if for example u_1 has such a jump, condition (1) can still hold if the normal derivative of u_1 does not appear in the $(m-s)$ "remaining" equations (see the end of Sect. 52)—or, if the value $+\infty$ or $-\infty$ for $\partial u_1/\partial z$ does not contradict these $(m-s)$ equations.

One may ask conversely whether such a "separation surface" S^* is necessarily characteristic. This can be answered affirmatively if we know that in a domain which includes S^* the given system is hyperbolic. It is impossible then that two different solutions meet along a surface S for which (51.4) holds. For,

---

[1] R. v. MISES: Proc. 1st Nat. Congr. Appl. Mech., Chicago 1950, pp. 667–671.

in this case, (under certain formal restrictions which are not severe in a hyperbolic region) the Cauchy problem would admit a unique solution in the neighborhood of S and neither a u_k nor a derivative of u_k could change abruptly across S.

The preceding definition of discontinuous solutions differs from usual ones, since we include the case that a u_k itself may jump across S*—*absolute discontinuity* in the terminology of HADAMARD. This is in contrast to occasional statements that for a system (48.5) only derivatives may jump across characteristics[1].

The above distinctions and considerations help to clarify the important question of *which variables may undergo abrupt changes across characteristics*. Both HADAMARD and LEVI-CIVITA use physical reasoning to show why in compressible flow, e.g., the pressure must not jump or why certain velocity components cannot change abruptly. From our present point of view the answer is simple: *A variable whose finite exterior derivative appears in the $(m-s)$ "remaining" equations cannot jump*. Other variables may undergo sudden changes subject to the conditions (1) and (2) of this section.

54. Preliminary comments on discontinuous solutions in plasticity. For the general space problem the discontinuities found by THOMAS present an important and general result. The investigation was based on the Eqs. (49.1) — (49.3), the first of which are equations of second order in the velocities. Following HADAMARD we call a surface S (a line S) *discontinuous of order n* with respect to a magnitude if all derivatives up to the order $(n-1)$ are continuous across S but the n-th derivative is discontinuous across S, $(n=1, 2, \ldots)$. A jump in the magnitude itself may be denoted as a discontinuity of order zero but, preferably, as an *absolute discontinuity*. With this notation the characteristic surface elements found by THOMAS are discontinuous of order two in the velocities and of order one with respect to the pressure.

Next, ERICSEN[2] has shown that *across the same surfaces also a discontinuity of order one is possible* for tangential components of the velocity[3]. *The stresses*, and consequently p, *are always continuous* across these characteristics.

We shall see that in the *plane problems* of "plane strain" as well as of "plane stress" an *absolute discontinuity of a tangential velocity component* is possible across a characteristic. This will follow from the principle pronounced at the end of Sect. 53. We shall see that stresses cannot jump across characteristics[4].

---

[1] Similarly, "weak" discontinuities (see Note 3) which for Eqs. (48.5) would be discontinuities of first order derivatives, are sometimes identified with discontinuities across characteristics. Actually, both "weak" and "strong" discontinuities can take place across characteristics. E.g. in a general compressible fluid flow, surfaces composed of streamlines are, in general, characteristic and may exhibit different values of the density ϱ and (or) of the tangential velocity-components on both sides.

The "shocks" of compressible fluid flow are, in general, not across characteristics. As pointed out repeatedly by v. MISES (see preceding footnote and footnote 1, p. 334, Sect. 15.2) for mathematical as well as physical reasons, these shocks cannot be considered as discontinuous solution (in the sense of the present section) of the ideal flow equations.

[2] I. L. ERICKSEN: J. Math. Phys. **34**, 74—79 (1955).

[3] Note that here a discontinuity of order one in the velocities is possible although Eqs. (49.1) contain second order derivatives in the velocities. A discontinuity has been called weak if its order is not lower than the order of the highest derivative of the respective quantity in the equations. Here we see that a discontinuity across a characteristic is not necessarily weak. (More details in Sect. 60.)

[4] In a recent paper, seen only at time of proofreading, T. Y. THOMAS, J. Math. Mech. **6**, 67—85 (1957) investigates discontinuities of order one for the problem of plane stress (see Sect. 64); "order one" means here that the stress components and velocity components are continuous across the discontinuity surface S while at least one of their first derivatives with respect to a space coordinate is discontinuous across S. He considers a moving discontinuity surface in the sense of HADAMARD (see Sect. 55).

A particular type of *absolute discontinuities of the stresses*, first systematically investigated by PRAGER, will be discussed in detail. Across such a discontinuity line the velocities cannot be discontinuous. See Sects. 61, 70, 84. We shall return to most of these questions.

c) HADAMARD's theory.

55. Waves. The theory of discontinuity surfaces can be approached from a different point of view, introduced mainly by HADAMARD[1]. These concepts and ideas, which are most suggestive, have recently been widely used by authors on plasticity, e.g. ERICKSEN, PRAGER, THOMAS and others[2]; they may indeed prove very useful in further studies; thus it may be justified to present them, very briefly. As one application we shall use them to derive the mathematical form of the compatibility conditions (Sect. 52, Sect. 58).

Let x stand for x_1, x_2, x_3. Consider a region $R(x, t)$ in ordinary space, varying in time, which is divided into two parts R_1 and R_2 by a *discontinuity surface* $S(x, t)$ of equation $z(x, t) = 0$. This discontinuity may be such that it always affects the same particles, and is then called a *stationary* or *material* discontinuity, or that it affects different particles in the course of time and is then called a *wave* ("onde") or *wave surface*.

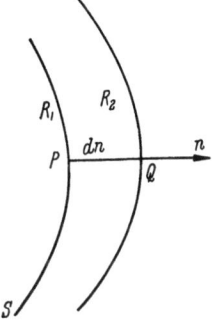

Fig. 10. Illustrating velocity of displacement.

Consider on the wave surface $S(x, t)$ a point $P = P(x)$ and the normal to S at P in direction from R_1 to R_2 (Fig. 10). At time $(t+dt)$ this normal will intersect the same surface (which has moved in time) at the point $Q = Q(x+dx)$. The quotient of PQ by dt is called the *velocity of displacement* of S at P at the instant t. The components of the vector $\overline{PQ}$ are the dx_i, and, if $p_i = \dfrac{\partial z}{\partial x_i}$, $g^2 = \sum_i^{1\ldots 3} p_i^2$, then $\alpha_i = p_i/g$ are the direction cosines of $\overline{PQ}$. Let $PQ = dn$; then $dx_i = \alpha_i\, dn$ and $dn = \sum_1^3 \alpha_i\, dx_i$. Next, with $p_0 = \dfrac{\partial z}{\partial t}$ from $z(x, t) = 0$, $z(x+dx, t+dt) = 0$, there follows

$$dz = p_0\, dt + \sum_i p_i\, dx_i = 0. \tag{55.1}$$

Thus

$$dn = \sum_{i=1}^{3} \alpha_i\, dx_i = \sum_1^3 \frac{p_i}{g}\, dx_i = -\frac{p_0}{g}\, dt,$$

and we obtain for the *velocity of displacement T of the discontinuity surface in ordinary space:*

$$T = \frac{dn}{dt} = -\frac{p_0}{g} = -\frac{\dfrac{\partial z}{\partial t}}{\left[\left(\dfrac{\partial z}{\partial x_1}\right)^2 + \left(\dfrac{\partial z}{\partial x_2}\right)^2 + \left(\dfrac{\partial z}{\partial x_3}\right)^2\right]^{\frac{1}{2}}}. \tag{55.2}$$

Next, we introduce ϑ, the *velocity of propagation*, or the *velocity of displacement of S in the medium*

$$\vartheta = T - v_n, \tag{55.3}$$

where v_n is the velocity of the medium in the n-direction. For a material discontinuity, ϑ vanishes[3].

---

[1] J. HADAMARD: Leçons sur la propagation des ondes, et les équations de l'hydrodynamique. Paris 1903.

[2] See the introduction to these ideas in T. Y. THOMAS: J. Rat. Mech. a. Analysis **2**, 339—381 (1953), Sects. 1—8.

[3] If the equation of S is normed so that $g = 1$, hence $p_i = \alpha_i$, we obtain from

$$\frac{dz}{dt} = \frac{\partial z}{\partial t} + \sum_{i=1}^{3} \frac{\partial z}{\partial x_i} v_i = \frac{\partial z}{\partial t} + v_n \quad \text{that} \quad T = -\frac{\partial z}{\partial t}, \quad \vartheta = -\frac{dz}{dt}.$$

56. Kinematical discontinuity conditions.

Consider a differentiable function $f(x, t)$ in $R(x, t)$. The partial derivatives of f are continuous in $R_1 + S$ and in $R_2 + S$, but across S jumps of derivatives may occur. Discontinuity of order n $(n = 1, 2, \ldots)$ of S, with respect to f has been defined in Sect. 54. An absolute discontinuity (discontinuity of f itself across S) is not considered in the present context. We shall use subscripts, or superscripts, 1 and 2, for values in the regions R_1, R_2 on both sides of S and denote in a customary manner $f_2 - f_1$ as $[f]$ etc., and occasionally we shall for brevity write $t = x_0$. A *discontinuity of order one* is then expressed by

$$[f] = 0, \quad \text{at least one} \left[\frac{\partial f}{\partial x_i}\right] \neq 0, \quad i = 0, 1, 2, 3.$$

Consider a point P on S. We have from the first of these equations: $f_P^1 = f_P^2$ and for a neighboring point Q on S: $f_Q^1 = f_Q^2$, hence $f_Q^1 - f_P^1 = f_Q^2 - f_P^2$. If we take Q very close to P and go to the limit, $df_P^1 = df_P^2$ obtains, or

$$\sum_{i=0}^{3} \frac{\partial f^1}{\partial x_i} dx_i = \sum_{i=0}^{3} \frac{\partial f^2}{\partial x_i} dx_i, \quad \text{or} \quad \sum_{i=0}^{3} \left[\frac{\partial f}{\partial x_i}\right] dx_i = 0. \tag{56.1}$$

Since the dx_i are in S we have also

$$dz = \sum_{i=0}^{3} p_i dx_i = 0, \quad \sum_{i=0}^{3} \frac{p_i}{g} dx_i = 0. \tag{56.2}$$

The coexistence of (56.1) and (56.2) is expressed by writing with an arbitrary factor λ:

$$\sum_{i=0}^{3} \left(\left[\frac{\partial f}{\partial x_i}\right] - \lambda \frac{p_i}{g}\right) dx_i = 0. \tag{56.3}$$

Assume $p_0 \neq 0$ and determine λ from $\left[\frac{\partial f}{\partial x_0}\right] - \lambda \frac{p_0}{g} = 0$. Since the dx_i are arbitrary this leads to $\left[\frac{\partial f}{\partial x_i}\right] - \lambda \frac{p_i}{g} = 0$, $i = 1, 2, 3$. Hence with $\frac{p_0}{g} = -T$, $\frac{p_i}{g} = \alpha_i$:

$$\left[\frac{\partial f}{\partial x_i}\right] = \lambda \alpha_i, \quad i = 1, 2, 3, \quad \left[\frac{\partial f}{\partial t}\right] = -\lambda T, \tag{56.4}$$

where [1] the α_i are the direction cosines of the normal to S.

For a discontinuity of order two we obtain:

$$\left[\frac{\partial^2 f}{\partial x_i \partial x_j}\right] = \lambda \alpha_i \alpha_j, \quad \left[\frac{\partial^2 f}{\partial x_i \partial t}\right] = -\lambda T \alpha_i, \quad \left[\frac{\partial^2 f}{\partial t^2}\right] = \lambda T^2, \quad i = 1, 2, 3. \tag{56.5}$$

Conditions (56.4), (56.5) etc. have been denoted by HADAMARD as kinematical conditions (of compatibility).

57. Application to a system of equations.

So far we have not considered any differential equations. Now we combine the conditions (56.4) or (56.5) with a system of differential equations. Consider Eqs. (48.5) where the $a_{\mu k i}$ and the b_μ are continuous across S. Therefore jumps of $\frac{\partial u_k}{\partial x_i}$ must satisfy the conditions

$$\sum_{k=1}^{m} \sum_{i=1}^{n} a_{\mu k i} \left[\frac{\partial u_k}{\partial x_i}\right] = 0, \quad \mu = 1, \ldots m,$$

and from (56.4), writing $p_0/g = \alpha_0$,

$$\left[\frac{\partial u_k}{\partial x_i}\right] = \lambda_k \alpha_i, \quad k = 1, \ldots m, \quad i = 0, \ldots n. \tag{57.1}$$

---

[1] Here λ is a scalar, just as f. If instead of f a tensor, say of order two with components t_{ij} is considered, then there exists a non zero-tensor a_{ij} defined on S such that $\left[\frac{\partial t_{ij}}{\partial x_k}\right] = a_{ij} \alpha_k$ and $\left[\frac{\partial t_{ij}}{\partial t}\right] = -a_{ij} T$. An analogous remark applies to Eqs. (56.5).

so that
$$\sum_{k=1}^{m}\sum_{i=0}^{n} a_{\mu k i}\lambda_k \alpha_i = 0, \quad \mu = 1,\ldots m. \tag{57.2}$$

Hence with the $\omega_{\mu k}$ of (51.3), since $g \neq 0$, we obtain

$$\sum_{k=1}^{m} \omega_{\mu k}\lambda_k = 0. \tag{57.3}$$

These are m homogeneous equations for the m unknowns $\lambda_1, \ldots \lambda_m$. The λ_k are parameters which characterize the discontinuities of the first derivatives of the m functions $u_1, \ldots u_m$. Eqs. (57.3) are termed by HADAMARD as *dynamical conditions* (of compatibility)[1]. This system has a solution not identically zero [i.e., a solution such that, according to (57.1), not all derivatives are continuous across S] *if and only if* $\|\omega_{\mu k}\| = 0$, i.e. if *condition* (51.5) *holds*. We see that here *the characteristic condition* (51.5) *is recovered by equating to zero the coefficient determinant of the dynamical conditions* (57.3). As mentioned before, Eq. (51.5) is that of a cone of order m in $(n+1)$-dimensional (here four-dimensional) space. Each plane normal to a solution $\boldsymbol{\alpha}^*$ of (51.5) is tangent to a characteristic surface element[2].

Analogous considerations may be applied to a system like (48.1) which has second order discontinuities; in correspondence to Eqs. (57.1), (51.3), (51.5) we obtain

$$\left[\frac{\partial^2 u_k}{\partial x_i \partial x_j}\right] = \varrho_k \alpha_i \alpha_j, \quad k = 1, \ldots m, \quad i,j = 0, \ldots n, \tag{57.4}$$

$$\omega_{\mu k} = \sum_{i,j=0}^{n} a_{\mu k i j}\alpha_i \alpha_j, \quad k,\mu = 1, \ldots m, \tag{57.5}$$

$$\sum_{k=1}^{m}\omega_{\mu k}\varrho_k = 0, \quad \mu = 1, \ldots m, \quad \text{dynamical conditions}, \tag{57.6}$$

$$\|\omega_{\mu k}\| = 0, \quad \text{characteristic condition}. \tag{57.7}$$

This characteristic condition which generalizes (48.3) has not been given before in our text. If $m = n = 1$ the last determinant reduces to a single element, and we obtain instead of Eq. (47.4)

$$A\alpha_1^2 + 2B\alpha_1\alpha_2 + C\alpha_2^2 = 0, \quad \tan\varphi = \frac{1}{A}\left(-B \pm \sqrt{B^2 - AC}\right), \tag{57.8}$$

where φ is the angle of a characteristic curve with the x-axis. For $\varphi = \pi/2$, this was our starting point in Sect. 47.

58. Compatibility conditions. We can now derive in a very simple way the analytic form of the compatibility conditions explained in Sect. 52. Let $\boldsymbol{\alpha}^*$ be a solution of Eq. (51.5) and put $\omega_{\mu k}^* = \sum_i a_{\mu k i}\alpha_i^*$ as in (51.3). Then according to (57.1) — (57.3)

$$\sum_{k=1}^{m}\omega_{\mu k}^*\lambda_k^* = 0, \quad \mu = 1, \ldots m, \tag{58.1}$$

defines the jumps of the derivatives of the u_k. Consider the transposed system (58.1),

$$\sum_{\mu=1}^{m}\omega_{\mu k}^*\gamma_\mu^* = 0. \tag{58.2}$$

We shall prove that these γ_μ^* determine *those linear combinations of the original equations*, explained in Sect. 52, *for which only interior derivatives*, i.e. derivatives perpendicular to $\boldsymbol{\alpha}^*$, *appear*. In fact, consider the linear combinations of Eqs. (48.5)

$$\sum_{\mu=1}^{m}\gamma_\mu^*\sum_{k=1}^{m}\sum_{i=0}^{n} a_{\mu k i}\frac{\partial u_k}{\partial x_i} = -\sum_\mu \gamma_\mu^* b_\mu = B^*. \tag{58.3}$$

---

[1] It is in my opinion confusing to speak of kinematical or dynamical conditions "*of compatibility*". The term "*compatibility conditions*" should in this context be reserved to the relations between the interior derivatives, as explained in Sects. 52 and 58.

[2] HADAMARD's theory has been applied by THOMAS and ERICKSEN to the v. Mises equations of plasticity, cf. T. Y. THOMAS: J. Rat. Mech. a. Analysis **2**, 339—381 (1953), and Ref. 4, p. 342; see ERICKSEN, Ref. 2, p. 342.

If vectors A_k^* with components

$$A_{ki}^* = \sum_{\mu=1}^m \gamma_\mu^* a_{\mu k i}, \quad k = 1, \ldots m, \quad i = 0, \ldots n, \tag{58.4}$$

are introduced, Eqs. (58.3) become

$$\sum_{k=1}^m \sum_{i=0}^n A_{ki}^* \frac{\partial u_k}{\partial x_i} = B^*. \tag{58.5}$$

Now, using (58.2) and (58.4), we have for all k:

$$\sum_\mu \left(\sum_i a_{\mu k i} \alpha_i^*\right) \gamma_\mu^* = \sum_i \left(\sum_\mu \gamma_\mu^* a_{\mu k i}\right) \alpha_i^* = \sum_i A_{ki}^* \alpha_i = 0, \tag{58.6}$$

and we see that the m vectors A_k^* defined by (58.4) are normal to α^*. Therefore the left side of (58.5) is seen to contain *only differentiations normal to* α^* and hence only "interior" derivatives (Sect. 52), and therefore (58.5) *is the compatibility condition corresponding to the direction* α^*, *the normal to* S^*.

d) Shock conditions. Stress discontinuities.

59. "Shock conditions". In our definition of discontinuous solutions it was not implied that discontinuities are necessarily across characteristics. We ask whether there is an analogue to the shocks of compressible fluid theory[1] and to the shock conditions which hold in this case.

We consider Eqs. (38.1) and (38.4). The method outlined in MISES' work (see preceding footnote) may be used in the derivation of shock conditions[2]. Since our final result will be very obvious we do not give complete derivations. Denote by S the surface which separates the two regions 1 and 2 under consideration, by T the normal velocity of the moving discontinuity surface S in the direction n leading from the side 1 to the side 2 of S. Then, the result corresponding to the continuity Eq. (38.1) is the condition

$$\varrho_1(v_{1n} - T) = \varrho_2(v_{2n} - T), \quad \text{or} \quad [\varrho(v_n - T)] = 0, \tag{59.1}$$

where $T - v_n$ is the velocity of propagation (Sect. 55).

Next, denote by $\boldsymbol{t}_n$ the stress vector corresponding to the n-direction [viz. $\boldsymbol{t}_n = \boldsymbol{t}_x \cos(nx) + \boldsymbol{t}_y \cos(ny) + \boldsymbol{t}_z \cos(nz)$, where $\sigma_x, \tau_{xy}, \tau_{xz}$ are the components of $\boldsymbol{t}_x$, etc.]. Then the result derived from (38.4) is

$$[\varrho(v_n - T)\boldsymbol{v}] = [\boldsymbol{t}_n], \tag{59.2}$$

or, in components,

$$[\varrho(v_n - T)v_i] = [\sigma_{ni}], \quad i = 1, 2, 3, \tag{59.3}$$

where σ_{ni} is the component of $\boldsymbol{t}_n$ in the i-direction.

However, in our present problem these conditions simplify very much. Since we assume $\varrho = $ constant, Eq. (59.1) reduces to $v_{1n} = v_{2n}$. If, next, we consider instead of Eqs. (38.4) merely the equilibrium conditions (38.7), the left side of (59.3) is seen to vanish and our results are simply

$$[v_n] = 0, \quad [\sigma_{ni}] = 0. \tag{59.4}$$

If, for example, a discontinuity surface has the normal in the x-direction at the point under consideration, then v_x must remain continuous, and likewise σ_x,

---

[1] We mentioned before (footnote 1 on p. 342) that these shocks cannot be considered as a phenomenon of ideal fluid flow. One has to assume a "transition zone" D_λ, where the fluid is viscous while outside viscosity is negligible. See R. v. MISES: Mathematical theory of compressible fluid flow. New York 1958, Arts. 11 and 14.

[2] See also more formal derivations in T. Y. THOMAS: J. Rat. Mech. a. Analysis **2**, 339—381 (1953) and Math. Magazine **22**, 169—189 (1949).

τ_{xy}, τ_{xz}, whereas $\sigma_y, \sigma_z, \tau_{yz}$ may change rapidly across the element. [This follows in this case also from direct consideration of our basic system. In fact in (38.7) the only stresses which are not differentiated in the x-direction are $\sigma_y, \sigma_z, \tau_{yz}$. A jump of a stress which *is* differentiated in the x-direction would imply an infinite value of the respective derivative.]

The general conditions (59.1), (59.2) may be used in "dynamic problems" where the above simplifying assumptions are not made. Of course, these equations, or (59.4) in the simplified case, are only necessary conditions. The possible jumps must be compatible with the basic equations of Sect. 38. (Compare also Sect. 37.)

60. On the classification of discontinuities. The definitions of Sect. 53 are general. They apply to discontinuities across any surface (line), characteristic or otherwise. HADAMARD's definitions and kinematical conditions (Sects. 55, 56) are even independent of any system of differential equations but they do not refer to absolute discontinuities.

PRAGER[1] classifies discontinuities as weak or strong, say with respect to a system of first order differential equations, like our Eqs. (48.5): The discontinuity of any u_k is weak if it is of order one, or higher, but not of order zero; in the latter case the discontinuity is called strong (see Note 3, p. 342). This distinction is perhaps not very essential since the same physical system can be written in various forms depending on whether any and which quantities are eliminated, etc. At any rate we must remember not to identify weak discontinuities with discontinuities across characteristics. In fact, in compressible flow as well as in our problem, weak and strong discontinuities are possible across characteristics. (See Sects. 53 and 54.)

Discontinuities for which $\vartheta = T - v_n = 0$ [see Eq. (55.3)], which we called material discontinuities, are called "contact discontinuities" by PRAGER.

In addition to the obvious distinction between the various physical quantities affected by a discontinuity we shall here simply distinguish between discontinuities across characteristics and discontinuities across non-characteristic surfaces, characteristics being defined by the vanishing of the respective characteristic determinants like (48.3), (51.5), (57.7). In each case the order of the discontinuity with respect to a chosen system of equations may also be investigated.

The characteristics of the three-dimensional problem of Sect. 38 have been studied (Sects. 49, 50), other discontinuities have been briefly indicated (Sect. 54), and details will follow. A survey of discontinuities in plasticity would, however, be very incomplete without the description of an important and frequent type of absolute stress discontinuity introduced and studied by PRAGER[2]. A general theory of the three-dimensional case is not available. Therefore we will give here briefly the description in the case of the *plane* problem and shall return to it when we know more details on the plane problem.

61. Stress discontinuities. In the problem of *plane strain* it is assumed that one of the three principal directions of the stress tensor Σ [and therefore, according to (38.8), also of the strain rate tensor $\dot{E}$] is, at any point, parallel to one and the same direction (see footnote 3 on p. 349). We take it as the z-direction. Then, $\dot{\gamma}_x = \dot{\gamma}_y = 0$, $\tau_x = \tau_y = 0$. In addition, it is assumed that all stresses and all

---

[1] W. PRAGER: Proc. 2nd U. S. Nat. Congr. Appl. Mech., Michigan 1954, pp. 21–32.
[2] W. PRAGER: R. Courant Anniversary Volume, pp. 289–299. New York 1948.

strains are independent of z and that $v_z =$ constant. Therefore

$$\left.\begin{array}{l} \dot{\varepsilon}_z = \dot{\varepsilon}_3 = 0, \quad \dot{\varepsilon}_x + \dot{\varepsilon}_y = 0 \\ \tau_x = \tau_y = 0, \quad s_z = \sigma_z + p = 0, \quad p = -\tfrac{1}{2}(\sigma_x + \sigma_y) = -\sigma_z, \end{array}\right\} \quad (61.1)$$

and the Eqs. (38.6) — (38.9) take on the simple form, with $\tau_{xy} = \tau$:

$$\frac{\partial \sigma_x}{\partial x} + \frac{\partial \tau}{\partial y} = 0, \qquad \frac{\partial \tau}{\partial x} + \frac{\partial \sigma_y}{\partial y} = 0, \qquad (61.2)$$

$$(\sigma_x - \sigma_y)^2 + 4\tau^2 = \tfrac{4}{3}\sigma_0^2 = 4\tau_0^2 = 4k^2, \qquad (61.3)$$

$$\dot{\varepsilon}_x + \dot{\varepsilon}_y = 0, \qquad \frac{\dot{\varepsilon}_x - \dot{\varepsilon}_y}{\dot{\gamma}_z} = \frac{\sigma_x - \sigma_y}{\tau}. \qquad (61.4)$$

Take the direction of a curve of discontinuity at a generic point P as the y-direction. Then, from (59.4), v_x as well as σ_x and τ must be continuous across that curve at P. Also [since $\partial v_y/\partial x$ appears in (61.4)] v_y must be continuous across the curve at P if infinite τ is excluded[1]. Consider Eqs. (56.4) with $f = v_y$. We find (since the n-direction is the x-direction) for $x_i = y$, $\alpha_i = \cos(ny) = \cos(xy) = 0$ that $[\partial v_y/\partial y] = 0$, or $[\dot{\varepsilon}_y] = 0$[2], and from the first Eq. (61.4) that $[\dot{\varepsilon}_x] = 0$.

We now write n and t instead of x and y and have from (61.3), since only σ_t may jump

$$\sigma_t = \sigma_n \pm 2\sqrt{k^2 - \tau^2}. \qquad (61.5)$$

Thus, the jump amounts to $4\sqrt{k^2 - \tau^2}$. The corresponding jump in pressure $p = \tfrac{1}{2}(\sigma_n + \sigma_t)$, equals $2\sqrt{k^2 - \tau^2}$. We think of the discontinuity line as of a narrow transition zone through which σ_n and τ remain unchanged while σ_t changes rapidly from $\sigma_t^{(1)}$ to $\sigma_t^{(2)}$, and $\sigma_t^{(2)} \leq \sigma_t \leq \sigma_t^{(1)}$ (Fig. 11). In the transition zone neither of the equality signs holds, otherwise the respective part of the transition zone would simply add to the continuous stress region. Hence, there

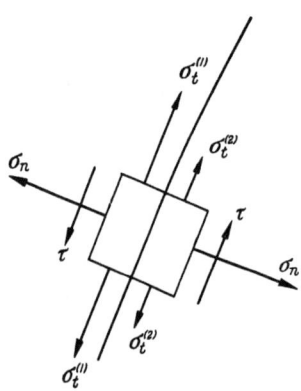

Fig. 11. Stress discontinuity.

$$(\sigma_t - \sigma_n)^2 + 4\tau^2 < 4k^2.$$

Now consider the velocities and strain rates. From (61.5) we conclude (using for simplicity again x, y for n, t)

$$\sigma_y^{(1)} - \sigma_x = -(\sigma_y^{(2)} - \sigma_x). \qquad (61.6)$$

On the other hand, according to (61.6), with $\dot{E} = \lambda S$

$$\dot{\varepsilon}_y^{(1)} = \lambda^{(1)} \frac{\sigma_y^{(1)} - \sigma_x}{2}, \qquad \dot{\varepsilon}_y^{(2)} = \lambda^{(2)} \frac{\sigma_y^{(2)} - \sigma_x}{2} = -\lambda^{(2)} \frac{\sigma_y^{(1)} - \sigma_x}{2}.$$

We have proved before that $\dot{\varepsilon}_y^{(1)} = \dot{\varepsilon}_y^{(2)}$ (which is physically obvious), also $\sigma_y^{(1)} - \sigma_x$ cannot vanish since otherwise $\sigma_y^{(1)} = \sigma_x = \sigma_y^{(2)}$ and no discontinuity. Hence

---

[1] If $\tau \to \infty$ the y-direction is characteristic, since it will be seen, (66.5), that $\tan 2\varphi = (\sigma_y - \sigma_x)/2\tau$ where φ is the angle of a characteristic with the x-axis. We shall, however, see that across a characteristic the stresses must be continuous (Sect. 70). That the velocity is continuous across a stress discontinuity has been shown for a more difficult problem by R. HILL: J. Mech. Phys. of Solids 1, 19—30 (1952).

[2] This had been assumed, for physical reasons, in H. GEIRINGER: Advances, p. 276.

$\lambda^{(1)} = -\lambda^{(2)}$; since a negative λ-value is excluded, it follows that at *every discontinuity point*

$$\lambda = 0, \quad \dot{\varepsilon}_x = \dot{\varepsilon}_y = \dot{\gamma} = 0, \quad \text{or} \quad \dot{E} = 0, \tag{61.7}$$

where $\dot{E}$ is the strain-rate tensor[1].

The last result, together with the inequality, which follows Eq. (61.5) suggests the previously mentioned consideration of the transition region as a thin, inextensible, elastic membrane between two plastic regions. This has been proposed by PRAGER, LEE, HILL for physical reasons[2]. We shall consider stress discontinuities repeatedly in the following. So far the main general results are: (a) The stress vector t_n associated to the direction normal to the discontinuity line is continuous across this line. (b) The tangential component σ_t may jump subject to the condition $\sigma_t^{(1)} + \sigma_t^{(2)} = 2\sigma_n$. (c) The velocity vector v is continuous across the line. (d) The strain-rate tensor, $\dot{E}$, vanishes for each point of the line. These properties describe, in our opinion, the discontinuity in a sufficient way. Other properties and interpretations follow from them.

Most of the results and consideration in Chap. E. II were concerned with the general three-dimensional problem. This last section, exceptionally, has dealt with a plane problem. We shall now give the theory of the plane problem of a perfectly plastic body.

F. The problem of plane strain[3].

I. Plane strain, plane stress, and generalizations.

62. Plane strain with quadratic yield condition derived from three-dimensional problem. The assumptions given at the beginning of Sect. 61 lead to the conditions

$$\dot{\gamma}_x = 0, \quad \dot{\gamma}_y = 0, \quad v_z = \text{constant} \tag{62.1}$$

Since consequently $\dfrac{\partial v_z}{\partial x} = \dfrac{\partial v_z}{\partial y} = \dfrac{\partial v_z}{\partial z} = 0$, we have

$$\frac{\partial v_x}{\partial z} = 0, \quad \frac{\partial v_y}{\partial z} = 0, \quad v_x = v_x(x,y), \quad v_y = v_y(x,y).$$

On account of (62.1) three of the Eqs. (38.8) reduce to

$$\tau_x = 0, \quad \tau_y = 0, \quad \sigma_z = \tfrac{1}{2}(\sigma_x + \sigma_y). \tag{62.2}$$

Of the seven remaining equations (38.6) — (38.9) the third of the equilibrium conditions is then identically satisfied since $\tau_x = 0$, $\tau_y = 0$, $\partial \sigma_z / \partial z = 0$. Hence, there remain six equations, two equilibrium conditions, the yield condition and three Eqs. (38.8), for the six unknown functions $\sigma_x, \sigma_y, \tau_z \equiv \tau, v_x, v_y, \lambda$; all these magnitudes depend only on x, y. These six equations form *the complete problem of plane strain*. An easy computation shows that after elimination of λ they take on the form (61.2) — (61.4).

---

[1] See also R. HILL [*10*], p. 160, and his paper in J. Mech. Phys. of Solids 1, 19—30 (1952), p. 26; of E. H. LEE: Proc. 3rd Symp. Appl. Math., New York 1950, pp. 213—228.

[2] See W. PRAGER: R. Courant Anniversary Volume, New York 1948, pp. 289—299, and particularly, E. H. LEE: Proc. 3rd Symp. Appl. Math., New York 1950, pp. 213—228; R. HILL [*10*], pp. 157—160. See also J. L. ERICKSEN: J. Math. Phys. **34**, 74—79 (1955).

[3] In Chapters F and G the strains and strain rates are plastic and a distinction between elastic and plastic strains does not arise. Hence the strains and strain rates are denoted by ε_{ij} or $\dot{\varepsilon}_{ij}$ etc. rather than by ε''_{ij} etc. This applies also to Sect. 61.

Next, we consider Tresca's condition: Since $\sigma_3 = \frac{1}{2}(\sigma_1 + \sigma_2)$, the three principal τ-values (38.10) are $\tau_1 = \frac{1}{4}(\sigma_2 - \sigma_1)$, $\tau_2 = \frac{1}{4}(\sigma_2 - \sigma_1)$, $\tau_3 = \frac{1}{2}(\sigma_1 - \sigma_2)$. Hence τ_3 is the greatest, and Tresca's condition takes the form:

$$(\sigma_1 - \sigma_2)^2 = (\sigma_x - \sigma_y)^2 + 4\tau^2 = 3\tau_0^2 = 3k^2. \tag{62.3}$$

It is seen that the conditions (61.3) and (62.3) are now of the same form.

63. Plane strain under general yield condition. We now consider a general yield function as in Eq. (40.3). We take the plastic potential equal to the yield function. For such a function g of the stresses *it can be shown that the tensor $\partial g/\partial \sigma_{ij}$ is co-axial with the stress tensor σ_{ij}*[1]. From Eqs. (40.5), together with (62.1), it follows that Eqs. (62.2) are now replaced by

$$\frac{\partial g}{\partial \tau_x} = 0, \quad \frac{\partial g}{\partial \tau_y} = 0, \quad \frac{\partial g}{\partial \sigma_z} = 0, \tag{63.1}$$

or, with previous notation (40.3)

$$\frac{\partial G}{\partial \sigma_3} = 0. \tag{63.2}$$

Using the above property of $\partial g/\partial \sigma_{ij}$ we conclude that $\tau_x = \tau_y = 0$ and $\sigma_z = \sigma_3$; from (63.2) we then compute σ_3 in terms of σ_1 and σ_2 and obtain

$$\tau_x = \tau_y = 0, \quad \sigma_3 = m(\sigma_1, \sigma_2) = l(\sigma_x, \sigma_y, \tau) = \sigma_z, \tag{63.3}$$

where m is some symmetric function of σ_1, σ_2. Hence with $G(\sigma_1, \sigma_2, m(\sigma_1, \sigma_2)) = F(\sigma_1, \sigma_2)$ the plane yield condition (with a symmetric F)

$$f(\sigma_x, \sigma_y, \tau) = F(\sigma_1, \sigma_2) = 0 \tag{63.4}$$

obtains. Next, we show that

$$\frac{\partial f}{\partial \sigma_x} + \frac{\partial f}{\partial \sigma_y} = \frac{\partial F}{\partial \sigma_1} + \frac{\partial F}{\partial \sigma_2} = 0. \tag{63.5}$$

Denote by the subscript m that, after differentiation, we put $\sigma_3 = m(\sigma_1, \sigma_2)$. Then:

$$\frac{\partial F}{\partial \sigma_1} + \frac{\partial F}{\partial \sigma_2} = \left[\frac{\partial G}{\partial \sigma_1} + \frac{\partial G}{\partial \sigma_2} + \frac{\partial G}{\partial \sigma_3}\left(\frac{\partial \sigma_3}{\partial \sigma_1} + \frac{\partial \sigma_3}{\partial \sigma_2}\right)\right]_m = \left[\frac{\partial G}{\partial \sigma_1} + \frac{\partial G}{\partial \sigma_2}\right]_m = 0,$$

on account of (63.2) and (40.4).

The three remaining relations (40.5) are, with $\tau_{xy} = \tau_{yx} = \tau$,

$$\dot{\varepsilon}_x = \lambda \frac{\partial f}{\partial \sigma_x}, \quad \dot{\varepsilon}_y = \lambda \frac{\partial f}{\partial \sigma_y}, \quad \dot{\gamma}_{xy} = \lambda \frac{\partial f}{\partial \tau_{xy}} \quad \left(\text{or } 2\dot{\gamma} = \lambda \frac{\partial f}{\partial \tau}\right), \tag{63.6}$$

of which only two are independent, on account of the first of the Eqs. (61.4) and (63.5), namely:

$$\dot{\varepsilon}_x + \dot{\varepsilon}_y = 0, \quad \frac{\partial f}{\partial \sigma_x} + \frac{\partial f}{\partial \sigma_y} = 0. \tag{63.7}$$

The present set of equations then consists of the two Eqs. (61.2), of the yield condition (63.4) with the restriction (63.5), and two independent Eqs. (63.6), hence of six equations for $\sigma_x, \sigma_y, \tau, v_x, v_y, \lambda$.

We see that, on account of $\frac{\partial F}{\partial \sigma_1} + \frac{\partial F}{\partial \sigma_2} = 0$, the yield function $F(\sigma_1, \sigma_2)$ only depends on the difference $\sigma_1 - \sigma_2$; hence $F(\sigma_1, \sigma_2) = A[(\sigma_1 - \sigma_2)]$, where A is a function of one variable. Finally, eliminating λ from (63.6), we obtain for plane

---

[1] For a formal proof see H. Geiringer: Advances, p. 206.

strain under a general yield condition the system:

$$\frac{\partial \sigma_x}{\partial x} + \frac{\partial \tau}{\partial y} = 0, \quad \frac{\partial \tau}{\partial x} + \frac{\partial \sigma_y}{\partial y} = 0,$$
$$f(\sigma_x, \sigma_y, \tau) = A[(\sigma_1 - \sigma_2)] = 0, \quad (63.8)$$
$$\dot\varepsilon_x + \dot\varepsilon_y = 0, \quad (\dot\varepsilon_x - \dot\varepsilon_y)/2\dot\gamma = \left(\frac{\partial f}{\partial \sigma_x} - \frac{\partial f}{\partial \sigma_y}\right)\Big/\frac{\partial f}{\partial \tau}.$$

In addition: $\tau_{yz} = \tau_{zx} = 0$, $\tau_{xz} = \tau_y = 0$, $\sigma_3 = m(\sigma_1, \sigma_2)$ and $v_z =$ constant. Here we have, indeed, a particular solution of the general problem of Sect. 40. We see, however, that the "general" yield condition in plane strain is not much more general than the quadratic condition of the preceding section.

64. Plane stress with quadratic yield condition. The main assumption is again that one of the principal directions of Σ be the same everywhere. This direction being chosen as the z-direction, the essential additional assumption is now that $\sigma_z = 0$ everywhere, hence

$$\tau_x = \tau_y = \sigma_z = 0 \quad (64.1)$$

throughout the body. In addition we assume that the remaining stresses are independent of z.

The third equilibrium condition is identically satisfied, as before. In plane strain, however, the relations (62.1) lead only to the restriction that v_x and v_y do not depend on z, whereas here (64.1) eliminates three unknowns. Thus, there remain $2+1+6=9$ equations (two equilibrium, one yield, six strain rate equations) for $3+3+1=7$ unknowns $\sigma_x, \sigma_y, \tau, v_x, v_y, v_z, \lambda$. As a consequence of (64.1) we obtain from corresponding stress-strain relations

$$\frac{\partial v_y}{\partial z} + \frac{\partial v_z}{\partial y} = 0, \quad \frac{\partial v_x}{\partial z} + \frac{\partial v_z}{\partial x} = 0, \quad 3\frac{\partial v_z}{\partial z} + \lambda(\sigma_x + \sigma_y) = 0. \quad (64.2)$$

It is not impossible that velocity distributions $\boldsymbol{v}(x, y, z)$ can be found which satisfy all conditions. It can, however, not be asserted.

In *the usual problem of plane stress* we wish to determine the stresses and the velocities v_x and v_y (and consequently λ) *as functions of x, y*, from the six Eqs. (64.3), which are analogous to the six Eqs. (61.2), (61.3), (63.6) of plane strain. This problem, as we shall see later cannot be considered as a particular case of the three-dimensional problem[1]. With von Mises' yield function as the plastic potential it is as follows:

$$\frac{\partial \sigma_x}{\partial x} + \frac{\partial \tau}{\partial y} = 0, \quad \frac{\partial \tau}{\partial x} + \frac{\partial \sigma_y}{\partial y} = 0,$$
$$\sigma_x^2 + \sigma_y^2 - \sigma_x\sigma_y + 3\tau^2 = \sigma_1^2 + \sigma_2^2 - \sigma_1\sigma_2 = 3\tau_0^2 = \sigma_0^2, \quad (64.3)$$
$$\dot\varepsilon_x = \lambda(2\sigma_x - \sigma_y), \quad \dot\varepsilon_y = \lambda(2\sigma_y - \sigma_x), \quad \dot\gamma = \lambda \cdot 3\tau.$$

Here the yield condition follows from (38.9') by putting $\sigma_z = 0$[2]. The last three equations are derived from the corresponding ones in (38.8), using $\sigma_z = 0$,

---

[1] The same is true for the plane stress problem of elasticity theory and for various other important problems in mechanics. In my paper quoted above (footnote 1 on p. 350) I tried, —erroneously—to derive the problem of plane stress as a particular case of the three-dimensional problem (see pp. 202, 203, and 209). That this is indeed not possible will be seen later (Sect. 73) since the characteristics formed there cannot be recovered as a special case of the general result of Sect. 49. Here we note that some difficulties arise immediately if Eqs. (64.2) in addition to (64.3), are taken into consideration.

[2] We mention also the form $(\sigma_1 + \sigma_2)^2 + 3(\sigma_1 - \sigma_2)^2 = 12\tau_0^2$, which shows the dependence on $(\sigma_1 + \sigma_2)$ in contrast to (62.3). See R. v. Mises: Z. angew. Math. Mech. **5**, 147—149 (1925). — F. Schleicher: Z. angew. Math. Mech. **6**, 199—216 (1926), footnote 1 on p. 328.

$p = -\frac{1}{3}(\sigma_x + \sigma_y)$. We note that in (64.3) $\dot{\varepsilon}_x + \dot{\varepsilon}_y$ is not zero since $\dot{\varepsilon}_z \neq 0$. The problem (64.3) must be considered on its own merit as a two-dimensional problem.

Generalizing both problems, (63.8) and (64.3), we now define a *general complete plane problem* with a general yield function $F(\sigma_1, \sigma_2) = f(\sigma_x, \sigma_y, \tau)$, F being symmetric in σ_1 and σ_2 (compare the "restrictions" discussed p. 327):

$$\left.\begin{aligned} \frac{\partial \sigma_x}{\partial x} + \frac{\partial \tau}{\partial y} &= 0, \quad \frac{\partial \tau}{\partial x} + \frac{\partial \sigma_y}{\partial y} = 0, \\ f(\sigma_x, \sigma_y, \tau) &= F(\sigma_1, \sigma_2) = 0, \\ \lambda \frac{\partial f}{\partial \sigma_x} &= \dot{\varepsilon}_x, \quad \lambda \frac{\partial f}{\partial \sigma_y} = \dot{\varepsilon}_y, \quad \lambda \frac{\partial f}{\partial \tau} = 2\dot{\gamma}. \end{aligned}\right\} \quad (64.4)$$

These reduce to (63.8) if $\frac{\partial f}{\partial \sigma_x} + \frac{\partial f}{\partial \sigma_y} = 0$, and to (64.3) if $f(\sigma_x, \sigma_y, \tau)$ is chosen as in the second line of (64.3); other "plane stress" yield functions will be considered[1] in Sect. 72 ff.

65. Generalized plane stress. A thin plate loaded only in its plane is approximately in a state of plane stress. In a more general approach the thickness h of the plate is not neglected. In formulating this problem R. HILL ([10] p. 300) uses the yield criterion as in (64.3). His considerations are, however, not restricted to this assumption.

Consider a plate of small thickness h under conditions of plane stress (viz. $\sigma_z = \tau_x = \tau_y = 0$). Forces are applied along the edge of the plate. The thickness h need not remain uniform during the deformation; but if $\partial h/\partial s$ (ds parallel to the surface of the plate) is small compared to unity, the state may still be considered as approximately "plane". We assume that at a certain instant t_0 the thickness $h(x, y)$ is a given function of x, y. Denote by σ_x, σ_y, τ the stress components averaged over the thickness of the plate and use these in the yield criterion. By v_x, v_y we denote the averaged velocity components, and the strain rates are defined as usual in terms of these components. Then, the equations of this problem are

$$\left.\begin{aligned} \frac{\partial}{\partial x}(h\sigma_x) + \frac{\partial}{\partial y}(h\tau) &= 0, \quad \frac{\partial}{\partial x}(h\tau) + \frac{\partial}{\partial y}(h\sigma_y) = 0, \quad f(\sigma_x, \sigma_y, \tau) = 0 \\ \dot{\varepsilon}_x : \dot{\varepsilon}_y : 2\dot{\gamma} &= \frac{\partial f}{\partial \sigma} : \frac{\partial f}{\partial \sigma_y} : \frac{\partial f}{\partial \tau}. \end{aligned}\right\} \quad (65.1)$$

The unknown functions are $v_x, v_y, \sigma_x, \sigma_y, \tau$.

After v_x and v_y have been found the equation of continuity may serve to compute *the change of h* at the instant t_0. If d/dt denotes the material deriv-

---

[1] The classification adopted in this paper differs from that used by several authors: We consider in the present Chap. F the problem of plane strain with dependence of the yield function on $(\sigma_1 - \sigma_2)$ only, and (63.7) satisfied. All other yield conditions which depend also on $(\sigma_1 + \sigma_2)$ are treated in Chap. G, among them those of "plane stress" by v. MISES, and by TRESCA [see Eqs. (72.1), (72.5), etc.]. Other authors, including SOKOLOVSKY, HILL, MANDEL consider separately the problem of plane strain (our Chap. F), that of plane stress, and finally other yield conditions which depend also on $(\sigma_1 + \sigma_2)$, considered as generalizations in the sense of O. MOHR, SCHLEICHER and MISES (see quotations of preceding footnote). In this connection the concept of the envelope of Mohr circles is then taken as starting point, and, for example, SOKOLOVSKY and MANDEL (not HILL) limit the problem to the case of real contact with the envelope (hyperbolic problem). — From a physical point of view the separate consideration of the problem of plane stress is certainly justified, particularly if this is considered in the general form of Sect. 65; from a mathematical point of view the first classification seems more logical.

ative, the rate of strain $\dot\varepsilon_z$ averaged through the thickness equals $(1/h)(dh/dt)$, or

$$\dot\varepsilon_z = \frac{1}{h}\frac{dh}{dt} = \frac{1}{h}\left(\frac{\partial h}{\partial t} + v_x\frac{\partial h}{\partial x} + v_y\frac{\partial h}{\partial y}\right),$$

and the continuity equation (38.6) takes the form

$$\frac{1}{h}\frac{dh}{dt} + \dot\varepsilon_x + \dot\varepsilon_y = 0,$$

or

$$\frac{\partial h}{\partial t} + \frac{\partial}{\partial x}(hv_x) + \frac{\partial}{\partial y}(hv_y) = 0. \tag{65.2}$$

If $h(x, y)$ is known at $t=t_0$ and v_x, v_y have been determined, Eq. (65.2) serves to determine $[\partial h/\partial t]_{t=t_0}$. Mathematically, this problem (65.1), together with (65.2), does not differ much from (64.4).

On the other hand HODGE[1] considers the system (65.1), (65.2) with $(\partial h/\partial t)_{t=t_0}=0$ and $h(x, y)$ as an unknown function. This is then a system of six equations with the additional unknown function h. Mathematically, this problem is much more difficult than (64.4) since we have to consider the six equations simultaneously. In this set-up a uniform h is excluded since from (65.2) it would amount to $\dot\varepsilon_x + \dot\varepsilon_y = 0$ — plane strain rather than plane stress.

II. The theory of plane strain.

This is the best developed branch of the mathematical theory of plasticity and has found a wide field of applications; probably the two facts are interdependent. We give here a direct presentation of the main facts of this theory rather than obtaining the case of plane strain as a particular case of the general problem (64.4)[2].

a) Differential relations.

66. Basic equations. Our system consists of the Eqs. (61.2) — (61.4) for the functions $\sigma_x(x, y)$, $\sigma_y(x, y)$, $\tau(x, y)$, $v_x(x, y)$, $v_y(x, y)$. As long as we do not consider boundary conditions, Eqs. (61.2) and (61.3) may be considered as three (nonlinear) equations for three unknown functions. The three equations may be reduced to one single equation by means of the *function of* AIRY[3], $F=F(x,y)$, where

$$\sigma_x = \frac{\partial^2 F}{\partial y^2}, \qquad \tau = -\frac{\partial^2 F}{\partial x \partial y}, \qquad \sigma_y = \frac{\partial^2 F}{\partial x^2}. \tag{66.1}$$

The equilibrium Eqs. (61.2) are thus satisfied, and (61.3) gives the second-order equation

$$\left(\frac{\partial^2 F}{\partial x^2} - \frac{\partial^2 F}{\partial y^2}\right)^2 + \left(2\frac{\partial^2 F}{\partial x \partial y}\right)^2 = 4k^2. \tag{66.2}$$

This equation—which has not been much used in modern presentations—can be reduced to a Monge-Ampère equation[4]. First introduce the new independent variables u, v by:

$$u = xi + y, \qquad v = x + yi,$$

and obtain by a simple computation

$$\frac{\partial^2 F}{\partial u^2}\frac{\partial^2 F}{\partial v^2} + h^2 = 0, \qquad h = \frac{k}{2}, \tag{66.2'}$$

---
[1] G. P. HODGE: Quart. Appl. Math. **8**, 381—386 (1951).
[2] This last approach was used by H. GEIRINGER: Advances, pp. 216—270.
[3] G. B. AIRY: British Ass. Rqs. 1862.
[4] Cf. E. STORCHI: Istituto Lombardo di Scienze e Lettere. Rendiconti Classe di Scienze Matematiche e Naturali **68**, 694—713 (1953).

which, with usual notation for second derivatives, is $rt+h^2=0$, or $r=-h^2/t$. Differentiating this last equation with respect to v we obtain

$$\left(\frac{\partial z}{\partial v}\right)^2 \frac{\partial^2 z}{\partial u^2} - h^2 \frac{\partial^2 z}{\partial v^2} = 0, \quad \text{where} \quad z = \frac{\partial F}{\partial v}. \tag{66.2''}$$

This equation, linear in the second derivatives, is a Monge-Ampère equation, and may be used as the basis for further discussion. Many of the results on our problem can be reached by this approach which, however, we shall not follow up here.

Denote now by ψ the angle which an arbitrary direction v makes with the positive x-direction, by $\sigma_v = \sigma$ the component in the v-direction of the stress vector t_v associated with that direction, by $\tau_v = \tau$ the component normal to v (see Sect. 39). Then, with $\tau_{yz} = \tau_z$:

$$\begin{aligned}
\sigma &= \sigma_x \cos^2\psi + 2\tau_z \cos\psi \sin\psi + \sigma_y \sin^2\psi \\
&= \frac{\sigma_x + \sigma_y}{2} + \frac{\sigma_x - \sigma_y}{2} \cos 2\psi + \tau_z \sin 2\psi, \\
\tau &= \qquad -\frac{\sigma_x - \sigma_y}{2} \sin 2\psi + \tau_z \cos 2\psi.
\end{aligned} \tag{66.3}$$

The *normal stress* σ reaches extremum values, $(d\sigma/d\psi = 0)$, for $\psi = \vartheta$ and $\psi = \vartheta + 90°$, where

$$\tan 2\vartheta = \frac{2\tau_z}{\sigma_x - \sigma_y}. \tag{66.4}$$

For these *principal directions*, we see from (66.3) that τ vanishes and the extreme values σ_1 and σ_2 of σ are equal to

$$\sigma_{1,2} = \frac{1}{2}(\sigma_x + \sigma_y) \pm \left[\left(\frac{\sigma_x - \sigma_y}{2}\right)^2 + \tau_z^2\right]^{\frac{1}{2}}.$$

On the other hand, $d\tau/d\psi = 0$ defines two directions $\psi = \varphi$ and $\psi = \varphi + 90°$ where

$$\tan 2\varphi = \frac{\sigma_y - \sigma_x}{2\tau_z} \tag{66.5}$$

for which the *shear stress* τ takes on extremum values. They are seen to bisect the angles of the principal directions and are called directions of *maximum shear stress*. The corresponding values are: $\tau = \pm \frac{1}{2}(\sigma_1 - \sigma_2)$; for each of the two directions $\sigma = \frac{1}{2}(\sigma_1 + \sigma_2)$ holds. All this holds formally for any symmetric tensor Σ.

Next, in the theory of plastic potential, the strain-rate tensor $\dot{E}$ is coaxial to grad Σ, and hence to Σ (see proof in Advances, p. 206). Hence the *shear strain* $\dot{\gamma}$ — derived from $\dot{E}$, as τ was from Σ—has its absolute maximum value $\frac{1}{2}|\dot{\varepsilon}_1 - \dot{\varepsilon}_2|$ for the same directions (66.5). *The directions of extremum shear strain are called slip directions or directions of the slip lines*. The corresponding *normal strain rate* is $\dot{\varepsilon} = \frac{1}{2}(\dot{\varepsilon}_1 + \dot{\varepsilon}_2)$, which vanishes in plane strain theory.

Now, we compute the *characteristic directions*. We take from Sect. 47 γ) the result that the *stress characteristics*, i.e. the characteristics of the system (61.2), (61.3) make the angles $\pm 45°$ with the first principal direction. It remains to compute from (61.4) the directions of the *velocity characteristics*. Eqs. (61.4), viz.

$$\frac{\partial v_x}{\partial x} + \frac{\partial v_y}{\partial y} = 0,$$

$$\frac{\partial v_x}{\partial x} - \left(\frac{\partial v_x}{\partial y} + \frac{\partial v_y}{\partial x}\right) \cot 2\vartheta - \frac{\partial v_y}{\partial y} = 0$$

are two linear first-order equations for v_x, v_y. Denote by m the slope of a characteristic direction and apply (51.5) for $k=2$. We obtain

$$m^2 - 2m\tan\vartheta - 1 = 0,$$

and

$$m = \frac{dy}{dx} = \frac{\sin 2\vartheta \pm 1}{\cos 2\vartheta} = \tan(\vartheta \pm 45°).$$

Hence, we have the following results (Fig. 12): *The directions of maximum and minimum shear stress, which coincide with the directions of maximum and minimum shear strain, make angles $\vartheta \pm 45°$ with the positive x-axis*[1], *ϑ being the angle of the first principal direction with this axis. In plane strain, these directions are also the directions of the stress and of the velocity characteristics and along each of them the normal strain rate is zero.*

We denote the curves which have at each point these distinguished directions, as slip lines, or also as characteristics C^- and C^+. They form a curvilinear net $x = x(\alpha, \beta)$, $y = y(\alpha, \beta)$ where α, β are the parameters of the net. A curve $\beta = $ constant, a C^-, is called an α-curve, a curve $\alpha = $ constant, a C^+, is called a β-curve.

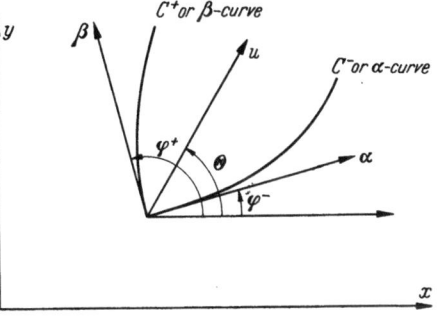

Fig. 12. Characteristic directions in plain strain.

Next, consider the well-known formulas, valid for any symmetric tensor,

$$\left.\begin{aligned}\sigma_x &= \sigma_1\cos^2\vartheta + \sigma_2\sin^2\vartheta = \frac{\sigma_1+\sigma_2}{2} + \frac{\sigma_1-\sigma_2}{2}\cos 2\vartheta,\\ \sigma_y &= \sigma_1\sin^2\vartheta + \sigma_2\cos^2\vartheta = \frac{\sigma_1+\sigma_2}{2} - \frac{\sigma_1-\sigma_2}{2}\cos 2\vartheta,\\ \sigma_z &= \sigma_1\cos\vartheta\sin\vartheta - \sigma_2\sin\vartheta\cos\vartheta = \frac{\sigma_1-\sigma_2}{2}\sin 2\vartheta.\end{aligned}\right\} \quad (66.6)$$

In our problem, where $(\sigma_1+\sigma_2)/2 = (\sigma_x+\sigma_y)/2 = -p$, $(\sigma_1-\sigma_2)/2 = k$, they take on the form[2]

$$\left.\begin{aligned}\sigma_x &= -p+k\cos 2\vartheta, & \sigma_y &= -p-k\cos 2\vartheta, & \tau &= k\sin 2\vartheta,\\ s_x &= -s_y = k\cos 2\vartheta, & \tau_z &= k\sin 2\vartheta, & s_x^2 + \tau_z^2 &= k^2.\end{aligned}\right\} \quad (66.6')$$

These equations form a representation of the stresses, in terms of p, k, ϑ, for which the yield condition holds.

Now compute $\dfrac{\partial\sigma_x}{\partial x} + \dfrac{\partial\tau_z}{\partial y}$ and likewise $\dfrac{\partial\tau_z}{\partial x} + \dfrac{\partial\sigma_y}{\partial y}$ by means of (66.6'). Since both are zero, we obtain:

$$\frac{\partial}{\partial x}\left(\frac{p}{2k}\right) = -\frac{\partial\vartheta}{\partial x}\sin 2\vartheta + \frac{\partial\vartheta}{\partial y}\cos 2\vartheta, \qquad \frac{\partial}{\partial y}\left(\frac{p}{2k}\right) = \frac{\partial\vartheta}{\partial x}\cos 2\vartheta + \frac{\partial\vartheta}{\partial y}\sin 2\vartheta.$$

The following may be added. The integration theory of the present Chap. F, which we are going to exhibit, as well as that of the more general problem of Chap. G, is based on the exchange of dependent and independent variables, used for "plane strain" as early as 1923 by several authors (see note 1, p. 356). In G, Sect. 71 we obtain, following VON MISES[3], by an exchange of variables the Eqs. (71.10). SOKOLOVSKY ([26], p. 176 seq.) uses the same variables, (71.9), in the particular case of plane strain and obtains the equations to which (71.10) reduce

---

[1] This holds also for the general problem (64.4).
[2] The magnitude $p/2k$, which plays an important role in all that follows, is often also denoted by χ.
[3] R. v. MISES: 415—429 Reissner Anniversary Volume. Ann Arbor, Michigan 1949.

in this particular case. However, in this case, where the angle of the characteristics with the first principal direction is constant, one gets more elegant results by using differentiation with respect to characteristic variables ξ, η. These are subsequently introduced by Sokolovsky in connection with an approximate method due to Christianowitsch [see his Eqs. (5.47), (5.48)]. Eqs. (67.12) or (68.1) are, however, simpler, without being approximate (cf. also footnote 1, p. 362).

Denote by $\partial/\partial u$, $\partial/\partial v$ directional derivatives in the first and second principal directions, viz. $\dfrac{\partial}{\partial u} = \cos\vartheta \dfrac{\partial}{\partial x} + \sin\vartheta \dfrac{\partial}{\partial y}$, etc. We obtain the intermediate result (which will become of interest in the case of Sect. 71):

$$\frac{\partial}{\partial u}\left(\frac{p}{2k}\right) = \frac{\partial\vartheta}{\partial v}, \qquad \frac{\partial}{\partial v}\left(\frac{p}{2k}\right) = \frac{\partial\vartheta}{\partial u}.$$

Using these formulas, we compute the directional derivative $\dfrac{\partial}{\partial l^+}$ of $\dfrac{p}{2k}$ in the β-direction, viz. $\dfrac{\partial}{\partial l^+} = \dfrac{1}{\sqrt{2}}\left(\dfrac{\partial}{\partial u} + \dfrac{\partial}{\partial v}\right)$, and similarly $\dfrac{\partial}{\partial l^-}$ for the α-direction. We obtain:

$$\frac{\partial}{\partial l^-}\left(\frac{p}{2k}\right) = -\frac{\partial\vartheta}{\partial l^-}, \qquad \frac{\partial}{\partial l^+}\left(\frac{p}{2k}\right) = \frac{\partial\vartheta}{\partial l^+}. \tag{66.7}$$

These are equivalent to

$$\frac{\partial}{\partial\alpha}\left(\frac{p}{2k}\right) = -\frac{\partial\vartheta}{\partial\alpha}, \qquad \frac{\partial}{\partial\beta}\left(\frac{p}{2k}\right) = \frac{\partial\vartheta}{\partial\beta}. \tag{66.7'}$$

Eqs. (66.7) or (66.7') are *compatibility equations*, along C^- or C^+, respectively, since each contains only differentiations along one characteristic direction. They can be integrated and we obtain[1], using for convenience the angle $\vartheta - 45° = \varphi$ rather than ϑ:

$$\left.\begin{array}{l} \dfrac{p}{2k} + \varphi = \text{constant along each } C^-, \\[4pt] \dfrac{p}{2k} - \varphi = \text{constant along each } C^+. \end{array}\right\} \tag{66.8}$$

The first constant must be a function of β alone (since $\beta=$ constant along each C^-) and the second constant a function of α alone. Thus we have the equations equivalent to (66.8):

$$\frac{p}{2k} = g(\beta) + f(\alpha), \qquad \varphi = g(\beta) - f(\alpha). \tag{66.9}$$

Also, as a consequence of (66.7') or (66.9),

$$\frac{\partial^2}{\partial\alpha\,\partial\beta}\left(\frac{p}{2k}\right) = 0, \qquad \frac{\partial^2\varphi}{\partial\alpha\,\partial\beta} = 0 \tag{66.10}$$

result ("integrability conditions").

We finally want the *velocity compatibility relations*. They must consist of formulas expressing the previous result that the rate of extension along a slip line is zero: $\dot{\varepsilon}_\alpha = 0$, $\dot{\varepsilon}_\beta = 0$. The rate of extension in the α-direction is, with $\varphi^- = \vartheta - 45° = \varphi$:

$$\dot{\varepsilon}_\alpha = \dot{\varepsilon}_x \cos^2\varphi + 2\dot{\gamma}_z \sin\varphi\cos\varphi + \dot{\varepsilon}_y \sin^2\varphi$$

$$= \cos\varphi\left(\frac{\partial v_x}{\partial x}\cos\varphi + \frac{\partial v_x}{\partial y}\sin\varphi\right) + \sin\varphi\left(\frac{\partial v_y}{\partial x}\cos\varphi + \frac{\partial v_y}{\partial y}\sin\varphi\right)$$

$$= \frac{\partial v_x}{\partial l^-}\cos\varphi + \frac{\partial v_y}{\partial l^-}\sin\varphi = 0,$$

$$\dot{\varepsilon}_\beta = \frac{\partial v_x}{\partial l^+}\cos\varphi^+ + \frac{\partial v_y}{\partial l^+}\sin\varphi^+ = 0.$$

---

[1] Cf. H. Hencky: Z. angew. Math. Mech. **3**, 241—251 (1923). — C. Carathéodory and E. Schmidt: Z. angew. Math. Mech. **3**, 468—475 (1923).

A simple geometric interpretation of this result will be discussed presently. We now express $\dot\varepsilon_\alpha$ and $\dot\varepsilon_\beta$ in terms of the components v_1, v_2 of $\boldsymbol{v}$ in the α- and β-directions[1]. Using $v_1 = v_x \cos \varphi^- + v_y \sin \varphi^-$, $v_2 = v_x \cos \varphi^+ + v_y \sin \varphi^+$ and the definitions of $\partial/\partial l^-$, $\partial/\partial l^+$, we find by an elementary computation

$$\frac{\partial v_1}{\partial l^-} = \left(\frac{\partial v_x}{\partial l^-} \cos \varphi^- + \frac{\partial v_y}{\partial l^-} \sin \varphi^-\right) + v_2 \frac{\partial \varphi^-}{\partial l^-},$$

$$\frac{\partial v_2}{\partial l^+} = \left(\frac{\partial v_x}{\partial l^+} \cos \varphi^+ + \frac{\partial v_y}{\partial l^+} \sin \varphi^+\right) - v_1 \frac{\partial \varphi^+}{\partial l^+}.$$

The expressions to the right, in parentheses, are zero, and we have the result:

$$\frac{\partial v_1}{\partial l^-} - v_2 \frac{\partial \varphi^-}{\partial l^-} = 0, \qquad \frac{\partial v_2}{\partial l^+} + v_1 \frac{\partial \varphi^+}{\partial l^+} = 0 \qquad (66.11)$$

or, since $d\varphi^- = d\varphi^+ = d\vartheta$:

$$dv_1 - v_2 \, d\vartheta = 0, \quad \text{along a } C^-; \quad dv_2 + v_1 \, d\vartheta = 0 \quad \text{along a } C^+. \qquad (66.11')$$

These are the compatibility relations in terms of v_1 and v_2 [2]. In terms of v_x, v_y they read

$$\frac{\partial v_y}{\partial l^-} = -\frac{\partial v_x}{\partial l^-} \cot \varphi^-, \qquad \frac{\partial v_y}{\partial l^+} = -\frac{\partial v_x}{\partial l^+} \cot \varphi^+. \qquad (66.12)$$

These are less useful since $\cot \varphi^-$ and $\cot \varphi^+$ depend on the stresses.

Consider a *velocity plane*, i.e. a plane with rectangular coordinates v_x, v_y, so that to a point P with coordinates x, y in the *physical plane* corresponds the point $\bar P$ with coordinates v_x, v_y. The images of the characteristics C^+, C^- at P are curves $\bar C^+, \bar C^-$ at $\bar P$; these are at the same time the *characteristics in the velocity plane*. Eqs. (66.12) state *that the velocity characteristic $\bar C^+$ at $\bar P$ is normal to the C^+ at P, and similarly for the C^- and $\bar C^-$* [3] (cf. Fig. 23).

More important than the velocity plane is the *stress plane*, or stress graph—introduced by von Mises, Neuber and Sauer[4]—since the stress equations are non-linear while the velocity equations are anyway linear. The stress plane is particularly useful in the study of the general problem (64.4) (see Chap. G). In the stress plane the independent variables are two stresses, e.g. σ_x and σ_y, or p and ϑ. With any of these as independent variables (and some physical variables as dependent variables) thus *by means of an interchange of independent and dependent variables*, the stress equations become linear. The images of the C^+, C^- in the stress plane with coordinates p, ϑ, say, may be called Γ^+, Γ^- and Eqs. (66.8) are the equations of the Γ^+ and Γ^-, respectively.

67. Continuation. Slip line field. The equations of the orthogonal slip lines, the characteristics, have been written (p. 355) as

$$x = x(\alpha, \beta), \quad y = y(\alpha, \beta).$$

---

[1] In previous publications I denoted by u, v the components of $\boldsymbol{v}$ in the α- and β-direction. Now the notation is changed since we use $\partial/\partial u, \partial/\partial v$ as directional derivatives in the two principal directions.

[2] H. Geiringer: Proc. 3rd Internat. Congr. Appl. Mech. II, 1930, pp. 185—190.

[3] H. Geiringer: Proc. Nat. Acad. Sci. U.S.A. **37**, 214—220 (1951); there the theorem is proved for the general problem (64.4). Cf. also W. Prager: Trans. Roy. Inst. Techn., Sweden **65**, 1—26 (1953), where the orthogonality for the present problem, viz. plane strain is proved.

[4] R. v. Mises: Reissner Anniversary Volume. Ann Arbor, Michigan 1949, pp. 415—429 (see footnote 4, p. 329). R. Sauer: Z. angew. Math. Mech. **29**, 274—279 (1949). H. Neuber: Z. angew. Math. Mech. **28**, 253—257 (1948).

Then, with $A>0$, $B>0$:
$$(dr)^2 = dx^2 + dy^2 = A^2 d\alpha^2 + B^2 d\beta^2, \quad A = \frac{dl^-}{d\alpha}, \quad B = \frac{dl^+}{d\beta}. \quad (67.1)$$

Writing
$$\varphi^- = \varphi, \quad \varphi^+ = 90° + \varphi,$$
we have
$$\left.\begin{aligned} \frac{\partial x}{\partial \alpha} &= A \cos \varphi, & \frac{\partial x}{\partial \beta} &= -B \sin \varphi, \\ \frac{\partial y}{\partial \alpha} &= A \sin \varphi, & \frac{\partial y}{\partial \beta} &= B \cos \varphi. \end{aligned}\right\} \quad (67.2)$$

Using the identities
$$\frac{\partial^2 x}{\partial \alpha \, \partial \beta} = \frac{\partial^2 x}{\partial \beta \, \partial \alpha} \quad \text{and} \quad \frac{\partial^2 y}{\partial \alpha \, \partial \beta} = \frac{\partial^2 y}{\partial \beta \, \partial \alpha},$$

we obtain
$$\cos \varphi \left(\frac{\partial A}{\partial \beta} + B \frac{\partial \varphi}{\partial \alpha} \right) + \sin \varphi \left(\frac{\partial B}{\partial \alpha} - A \frac{\partial \varphi}{\partial \beta} \right) = 0,$$
$$\sin \varphi \left(\frac{\partial A}{\partial \beta} + B \frac{\partial \varphi}{\partial \alpha} \right) - \cos \varphi \left(\frac{\partial B}{\partial \alpha} - A \frac{\partial \varphi}{\partial \beta} \right) = 0.$$

Therefore
$$\frac{\partial A}{\partial \beta} + B \frac{\partial \varphi}{\partial \alpha} = 0, \quad \frac{\partial B}{\partial \alpha} - A \frac{\partial \varphi}{\partial \beta} = 0. \quad (67.3)$$

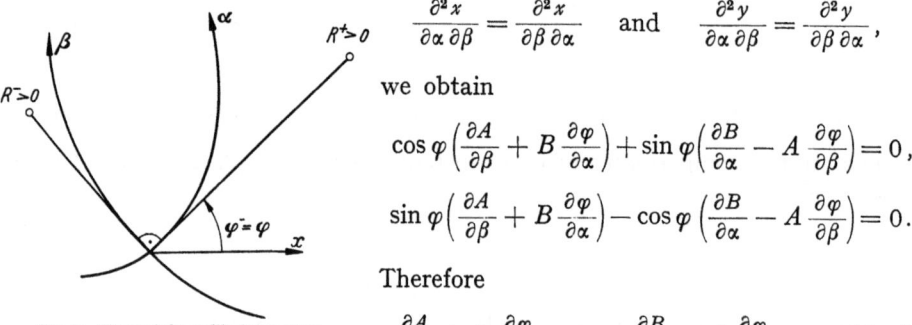

Fig. 13. Signs of the radii of curvature in slip-line field.

Denote by R^+ and R^- the radii of curvature of a C^+, or C^- chosing the sign positive if the center of curvature of an α-line (β-line) is in direction of increasing β (of increasing α). Put, accordingly,
$$\frac{1}{R^-} = \frac{d\varphi^-}{dl^-} = \frac{\frac{\partial \varphi}{\partial \alpha}}{\frac{dl^-}{d\alpha}} = \frac{1}{A} \frac{\partial \varphi}{\partial \alpha}, \quad \frac{1}{R^+} = -\frac{d\varphi^+}{dl^+} = -\frac{\frac{\partial \varphi}{\partial \beta}}{\frac{dl^+}{d\beta}} = -\frac{1}{B} \frac{\partial \varphi}{\partial \beta}. \quad (67.4)$$

[The negative sign in the second Eq. (67.4) must be chosen: Indeed (see Fig. 13), while α is increasing with increasing x, β is increasing with decreasing x. Hence, if as in the figure $R^- > 0$, $R^+ > 0$, by our definition, the $\varphi^- = \varphi$ is indeed increasing with increasing α along the α-curve, but φ is decreasing along the β-curve with increasing β.] Now we compute $\partial R^+/\partial \alpha$, using (67.3) and (67.1), and in the same way we find $\partial R^-/\partial \beta$. The result is
$$\frac{\partial R^+}{\partial \alpha} + \frac{dl^-}{d\alpha} = \frac{B}{\left(\frac{\partial \varphi}{\partial \beta}\right)^2} \frac{\partial^2 \varphi}{\partial \alpha \, \partial \beta}, \quad \frac{\partial R^-}{\partial \beta} + \frac{dl^+}{d\beta} = -\frac{A}{\left(\frac{\partial \varphi}{\partial \alpha}\right)^2} \frac{\partial^2 \varphi}{\partial \alpha \, \partial \beta}. \quad (67.5)$$

All these formulas hold *for any orthogonal net*.

Now, however, consider the second Eqs. (66.9) and (66.10), viz.:
$$\varphi = g(\beta) - f(\alpha), \quad \frac{\partial^2 \varphi}{\partial \alpha \, \partial \beta} = 0. \quad (67.6)$$

We then obtain from (67.5)
$$\frac{\partial R^+}{\partial \alpha} + \frac{dl^-}{d\alpha} = 0, \quad \frac{\partial R^-}{\partial \beta} + \frac{dl^+}{d\beta} = 0. \quad (67.7)$$

We actually see that from each of the four Eqs. (67.6), (67.7) the other three follow.

Eqs. (67.6), (67.7) form the analytic expression of two well-known geometric properties of the slip lines. For formulations and figures see GEIRINGER ([6], pp. 35 and 39), HILL ([10], p. 139), HODGE and PRAGER ([23], pp. 130—134). The equivalent of the first slip line property is Eq. (67.6), that of the second property are the two Eqs. (67.7), each for one family of curves, and we have shown, following CARATHÉODORY and E. SCHMIDT[1], that it is sufficient to *assume one of these properties for one family in order to obtain both properties for both families*[2].

We consider now a region where $f'(\alpha) \neq 0$, $g'(\beta) \neq 0$. Since, from (67.4) and (67.6),

$$\frac{1}{R^-} = -\frac{f'(\alpha)}{A}, \qquad \frac{1}{R^+} = -\frac{g'(\beta)}{B}, \tag{67.4'}$$

the last assumption implies that in the considered region the curvatures of the curves of the net are different from zero. (The opposite case which arises for solutions denoted as "degenerate solution", "simple wave", "fan", will be discussed in connection with the general plane problem[3].) In order to take care of the signs of the R^+, R^- we use two constants a, b, where $|a| = |b| = 1$, and an additive constant c for convenience and according to (66.9) we introduce the *characteristic coordinates* ξ, η by

$$g(\beta) = b\eta + c, \qquad f(\alpha) = -a\xi, \tag{67.8}$$

$$\varphi = a\xi + b\eta + c, \qquad \frac{p}{2k} = -a\xi + b\eta + c. \tag{67.9}$$

Then [see (67.4)] e.g., $a = -1$, $b = +1$ corresponds to $R^- < 0$, $R^+ < 0$. Denote by h_1 and h_2 the values of A and B for these new characteristic parameters. The simplified Eqs. (67.4), (67.7), (67.3) are then

$$h_1 = aR^-, \qquad h_2 = -bR^+. \tag{67.10}$$

$$\left. \begin{array}{ll} \dfrac{\partial R^+}{\partial \xi} + aR^- = 0, & \dfrac{\partial R^-}{\partial \eta} - bR^+ = 0, \\[6pt] \dfrac{\partial h_2}{\partial \xi} - bh_1 = 0, & \dfrac{\partial h_1}{\partial \eta} + ah_2 = 0. \end{array} \right\} \tag{67.11}$$

Since also $\partial\vartheta/\partial\xi = a$, $\partial\vartheta/\partial\eta = b$, Eqs. (67.11) may also be written

$$dR^+ + R^- d\vartheta = 0, \qquad dR^- - R^+ d\vartheta = 0, \tag{67.11'}$$

along a C^- or C^+, respectively, in obvious analogy to (66.11')[4]. From (67.11) we conclude that

$$\frac{\partial^2 R^+}{\partial \xi \partial \eta} + abR^+ = 0, \qquad \frac{\partial^2 R^-}{\partial \xi \partial \eta} + abR^- = 0. \tag{67.12}$$

Finally, Eqs. (67.2) take the form[5]

$$\left. \begin{array}{ll} \dfrac{\partial x}{\partial \xi} = aR^- \cos\varphi, & \dfrac{\partial x}{\partial \eta} = bR^+ \sin\varphi, \\[6pt] \dfrac{\partial y}{\partial \xi} = aR^- \sin\varphi, & \dfrac{\partial y}{\partial \eta} = -bR^+ \cos\varphi. \end{array} \right\} \tag{67.13}$$

---

[1] C. CARATHÉODORY and E. SCHMIDT: Z. angew. Math. Mech. 3, 468—475 (1923).
[2] A further geometric property has been found by I. KAPUANO: Rev. Fac. Sci. Univ. Istanbul A, 36—39 (1941).
[3] A very misleading name is used in the translation [26]. The "fans" are denoted as "integrals of the plasticity equations".
[4] More specifically, R^+ is replaced by v_1, and R^- by $-v_2$.
[5] From these equations follows $x = a \int R^- \cos\varphi \, d\varphi$, $y = a \int R^- \sin\varphi \, d\varphi$ along a ξ-line, with similar relations along an η-line.

If ξ, η are introduced into the velocity equations (66.11) they become:

$$\frac{\partial v_1}{\partial \xi} - a v_2 = 0, \qquad \frac{\partial v_2}{\partial \eta} + b v_1 = 0, \qquad (67.14)$$

and from these

$$\frac{\partial^2 v_1}{\partial \xi \partial \eta} + a b v_1 = 0, \qquad \frac{\partial^2 v_2}{\partial \xi \partial \eta} + a b v_2 = 0, \qquad (67.15)$$

equations of the same form as (67.12).

b) Integration. Particular solutions.

68. Integration. The actual solution of a concrete mechanical problem often meets with great difficulties as we shall discuss at the end of this section. However, the theory contained in the last few sections presents in the ξ, η-plane considerable advantages: (1) The equations are hyperbolic for the stresses as well as for the velocities. (2) The equations also for both, the stresses and the velocities, are simple, and their *Riemann function* is known.

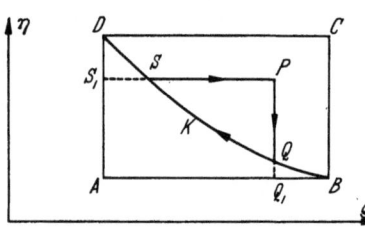

Fig. 14. Illustrating RIEMANN's integration method.

The Riemann method applies to the "linearized" equations (67.11), (67.12) and provides explicit solutions for certain fundamental initial-value problems, if in the ξ, η-plane appropriate initial values are given. Thus, two additional questions arise: (a) How do we find initial values in the ξ, η-plane from meaningful data in the physical plane? (b) If, by means of the Riemann method and of appropriate initial data, $R^+(\xi, \eta)$ and $R^-(\xi, \eta)$ have been found in a certain domain, how do we find the corresponding stress distribution in the physical plane? For the velocities these two questions do not arise since, after the stress problem has been solved, the velocity problem is linear and can be solved in the x, y-plane.

The exact explicit solutions are often cumbersome and therefore numerical as well as graphical approximation methods have been devised of which we shall also speak.

Cauchy data for the linear equation[1] $\frac{\partial^2 z}{\partial \xi \partial \eta} = z$ consist in giving z and $\frac{\partial z}{\partial \xi}$ along a curve K in the ξ, η-plane whose tangent is nowhere parallel to either axis. If then the Riemann function is known, the solution is determined in the characteristic rectangle whose horizontal and vertical sides pass through the two end-points of K (Fig. 14):

$$z(P) = z(S) + \int_{QS} \left(\omega \frac{\partial z}{\partial \xi} d\xi + z \frac{\partial \omega}{\partial \eta} d\eta \right),$$

where $\int_{QS}$ is the line integral along K. The Riemann function $\omega(\xi, \eta; \xi_0 \eta_0)$ for our equation is the Bessel function

$$\omega = \sum_{n=0}^{\infty} \frac{(\xi - \xi_0)^n (\eta - \eta_0)^n}{n! \, n!} = I[(\xi - \xi_0)(\eta - \eta_0)].$$

We do not enter into details regarding this well-known method[2]. The main point is that R^+ (and R^-) are explicitly determined in terms of ξ, η if R^+ and $\partial R^+/\partial \xi$

---
[1] To fix the ideas let us take here $ab = -1$.
[2] See e.g. GEIRINGER [6], pp. 44 seq., where the method is explained in relation to the present problem.

are known along K. [A similar formula for $z(P)$ holds if R^+ and $\partial R^+/\partial \eta$ are given along K.]

Now consider question (a). An arc $\mathscr{K}$ is given in the x,y-plane by $x = x(t)$, $y = y(t)$. Along this arc we know e.g. the stresses: $p/k = -s$ and ϑ [equivalent to giving σ_x, σ_y, τ, on account of Eq. (66.6')]. The ϑ given along $\mathscr{K}$ must be such that the actual inclination of $\mathscr{K}$ is at no point equal to $\vartheta + 45°$ or $\vartheta - 45°$, that means that $\mathscr{K}$ has nowhere a characteristic direction. Then, from (67.9) (since $\varphi = \vartheta - 45°$) we find ξ and η at every point of $\mathscr{K}$, say $\xi = \xi(t), \eta = \eta(t)$, and by (67.13), R^- and R^+ follow along $\mathscr{K}$.

To the arc $\mathscr{K}$ in the x,y-plane there corresponds K in the ξ, η-plane and we know R^-, R^+ along it. The preceding Riemann formula then gives R^- or R^+ in $ABCD$; the other of these functions follows by (67.11).

Finally (question b), if $R^-(\xi, \eta), R^+(\xi, \eta)$ have been found in $ABCD$, Eqs. (67.13) give $\partial x/\partial \xi, \partial x/\partial \eta, \partial y/\partial \xi, \partial y/\partial \eta$ in $ABCD$, and $x = x(\xi, \eta), y = y(\xi, \eta)$ are found by quadratures[1]. Thus the slip-line field in the physical plane is known. The slip-line field will have singularities at points where $J = \dfrac{\partial x}{\partial \xi} \dfrac{\partial y}{\partial \eta} - \dfrac{\partial x}{\partial \eta} \dfrac{\partial y}{\partial \xi} = 0$. (Since, by (67.13), $J = h_1 h_2 = R^- R^+$, we see that along such lines, say along $h_1(\xi \eta) = 0$, the ξ-curves will have cusps[2].) In a region D where $J \neq 0$ we can invert the above formulas and find $\xi = \xi(x, y), \eta = \eta(x, y)$ and therefore the stresses at every point (x, y) in D.

In the second or *characteristic initial-value problem* of $\dfrac{\partial^2 z}{\partial \xi \partial \eta} = z$ we need the value of z along AB and AD. Denote by Q_1 and S_1 the points of intersection of the vertical and horizontal straight line through P with AB and AD (see Fig. 14). The Riemann formula is

$$z(P) = z(S_1) + \int_{S_1}^{A} z \frac{\partial \omega}{\partial \eta} d\eta + \int_{A}^{Q_1} \omega \frac{\partial z}{\partial \xi} d\xi,$$

where ω is the same Bessel function as before. The initial data in the x,y-plane now simply consist in giving two curves intersecting at a right angle, and designated as C^-, C^+, respectively. In other words, we wish to determine a system of slip lines which contains these two curves as a ξ-curve and as an η-curve[3]. We take the x- and y-axis tangent to the C^- and C^+ at their point of intersection, O. Then, denoting by φ_0 and s_0 the values of φ and $s = -p/k$ at O and taking $b = 1$, $a = -1$, $c = 0$ in (67.9), we have: $\varphi_0 = 0 = \eta_0 - \xi_0$, $\tfrac{1}{2} s_0 = -\eta_0 - \xi_0 = -2\eta_0$, where η_0 must be given. Then, along the ξ-curve: $\xi = \eta_0 - \varphi$; along the other: $\eta = \varphi^+ + \xi_0 - 90° = \varphi^+ + \eta_0 - 90°$. On the other hand, we know by inspection the radii of curvature, R^- along the ξ-curve, R^+ along the η-curve. To find $R^+(\xi, \eta)$ in the rectangle we use the Riemann formula just given where we need R^+ along $S_1 A$, and $\partial R^+/\partial \xi$ along the horizontal piece. We find $\partial R^+/\partial \xi$ by (67.11) since it equals R^-. (Similarly we proceed if we wish to find R^-, using a second formula similar to the above, for $z(P)$.) Having found, say, R^+ in the rectangle $ABCD$, the problem continues as before.

We mention that in the integration theory other variables than R^-, R^+ can be used[4], viz. $X = x \cos \varphi + y \sin \varphi$, $Y = -x \sin \varphi + y \cos \varphi$. These X, Y are

---
[1] Compare the formulae in footnote 5, p. 359.
[2] More on these singularities in Sect. 77.
[3] This problem has been carried out by C. CARATHÉODORY and E. SCHMIDT: Z. angew. Math. Mech. **3**, 468—475 (1923).
[4] H. GEIRINGER: Mémoires sur la mécanique des fluides, offerts à M. D. RIABOUCHINSKY, Paris 1955, pp. 1—3.

the components of the radius vector r in the characteristic directions. Then

$$\frac{\partial Y}{\partial \xi} = \left(-\frac{\partial x}{\partial \xi} \sin\varphi + \frac{\partial y}{\partial \xi} \cos\varphi\right) + (x\cos\varphi + y\sin\varphi).$$

Since the expression in the first parenthesis vanishes along a ξ-line, we have with a similar computation for $\partial X/\partial \eta$:

$$\frac{\partial Y}{\partial \xi} = X, \quad \frac{\partial X}{\partial \eta} = Y, \quad \frac{\partial^2 X}{\partial \xi \partial \eta} = X, \quad \frac{\partial^2 Y}{\partial \xi \partial \eta} = Y. \tag{68.1}$$

Then, e.g. in the characteristic problem, both X and Y are readily computed along the two given C^- and C^+. The ξ and η along the curves follow as before and the Riemann method gives $X(\xi,\eta)$ and $Y(\xi,\eta)$ in a rectangle. Then:

$$\left.\begin{array}{l} x(\xi,\eta) = X(\xi,\eta)\cos(\eta-\xi) - Y(\xi,\eta)\sin(\eta-\xi), \\ y(\xi,\eta) = X(\xi,\eta)\sin(\eta-\xi) + Y(\xi,\eta)\cos(\eta-\xi), \end{array}\right\} \tag{68.2}$$

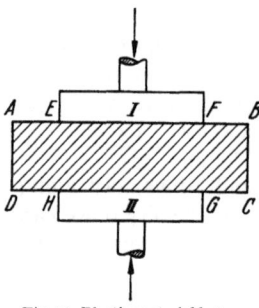

Fig. 15. Plastic material between two plates.

and the slip-line net may be drawn.

Consider, finally, for the *velocities*, say, the Cauchy problem. If stresses are given along $\mathscr{K}$ in the x,y-plane we know ξ,η there; if then the velocity components v_1, v_2 are given along $\mathscr{K}$, we know there $v_1(\xi,\eta), v_2(\xi,\eta)$, and the Riemann method gives these functions in $ABCD$. If, finally $\xi = \xi(x,y)$, $\eta = \eta(x,y)$ have been found (stress problem solved) in $ABCD$, we know there $v_1(\xi(x,y),\eta(x,y)) = v_1(x,y)$ and, similarly, $v_2(x,y)$.

We therefore have the remarkable result that we possess *explicit solutions* of important initial-value problems[1]. The Riemann method can be adapted to the "third" or "mixed" initial-value problem (see e.g. GEIRINGER [6], p. 46; see also the solutions in Sects. 96 to 98 of the present article[2]).

All these are exact analytic solutions. In addition, rapidly converging approximate solutions are available. A geometric method based on Eqs. (66.7) and (66.11') has been worked out by PRAGER[3], and other approximation methods are described by HILL [10], pp. 141—151 (see the original publication[4]). Some more will be added in connection with the general case.

Unfortunately, the concrete problems of plasticity theory do not reduce often to a succession of initial-value problems as described in this section. Fig. 15 indicates the following concrete problem. A plane plate of cross-section $ABCD$ is pressed between two pistons moved toward each other by the vertical forces P. Boundary conditions are:

1. Along BC and AD: $\quad \sigma_x = \tau = 0$.
    Along FB, CG, HD, AE: $\sigma_y = \tau = 0$.
2. Along EF and HG:

    a) The vertical velocity component equals $+c$ or $-c$,
    b) if there is no friction: $\quad\quad\quad \tau = 0$;
        if the coefficient of friction equals ψ: $\tau/\sigma = \psi$.

---

[1] A similar situation exists in the problem of unsteady parallel flow (x, t-problem) of a compressible fluid. This problem is also everywhere hyperbolic, and the Riemann function is known, as well as explicit solutions of the two initial-value problems. Similar results do not hold for the steady plane problem of a compressible fluid and for the general problem of plasticity, (64.4).

[2] See papers by W. HAACK and G. HELLWIG: Math. Z. **53**, 244—266 (1950), and 340—356 (1950), and H. BECKERT: Ber. Verh. Sächs. Akad. Wiss. Leipzig **97**, 68 (1950).

[3] W. PRAGER: Trans. Roy. Inst. Techn., Sweden **65**, 1—26 (1953).

[4] R. HILL, E. H. LEE and S. J. TUPPER: J. Appl. Mech. **8**, 46—52 (1951).

Hence along each piece of the boundary we know two scalar functions. The mathematical question is whether for this physically meaningful problem a solution exists[1].

69. Examples of exact particular solutions. The theory of the equation $\frac{\partial^2 u}{\partial \xi \partial \eta} = u$ (telegraph equation) is well known. Corresponding to any solution $u(\xi, \eta)$ of this equation, there is a slip-line field. An especially simple particular solution is the following:

1. Put here and in the following example $b=1$, $a=-1$ and consider

$$R^+ = A\, e^{c\xi + \frac{1}{c}\eta}, \qquad R^- = A\, c\, e^{c\xi + \frac{1}{c}\eta}, \tag{69.1}$$

$$\left. \begin{array}{l} x = -A\,\dfrac{c}{1+c^2}\, e^{c\xi+\frac{1}{c}\eta}(c\cos\varphi - \sin\varphi) = -\dfrac{Ac}{\sqrt{1+c^2}}\, e^{c\xi+\frac{1}{c}\eta}\cos(\varphi+\gamma), \\[2mm] y = -A\,\dfrac{c}{1+c^2}\, e^{c\xi+\frac{1}{c}\eta}(\cos\varphi + c\sin\varphi) = -\dfrac{Ac}{\sqrt{1+c^2}}\, e^{c\xi+\frac{1}{c}\eta}\sin(\varphi+\gamma), \end{array} \right\} \tag{69.2}$$

where $\sin\gamma = (1+c^2)^{-\frac{1}{2}}$, $\cot\gamma = c$. The slip lines form two families of logarithmic spirals of opposite sense and making angles γ and $90-\gamma$ with the radius vector. The origin is a singular point. Through each other point there passes one spiral of each family.

Obviously, we can write in (69.1), (69.2) c_r and γ_r for c and γ and consider the respective sums (finite or infinite) since the principle of superposition holds.

2. Next, consider with α a constant,

$$\left. \begin{array}{l} R^+ = A\cos\left(d\xi - \dfrac{\eta}{d}\right) - B\sin\left(d\xi - \dfrac{\eta}{d}\right), \\[2mm] R^- = -Bd\cos\left(d\xi - \dfrac{\eta}{d}\right) - A\,d\sin\left(d\xi - \dfrac{\eta}{d}\right), \end{array} \right\} \tag{69.3}$$

and, with $\varphi = \eta - \xi$, $\psi = d\xi - \dfrac{\eta}{d}$,

$$\left. \begin{array}{l} x = \dfrac{2Ad}{1-d^2}(d\cos\psi\cos\varphi - \sin\psi\sin\varphi) + \dfrac{2Bd}{1-d^2}(\cos\psi\sin\varphi + d\sin\psi\cos\varphi), \\[2mm] y = \dfrac{2Ad}{1-d^2}(d\cos\psi\sin\varphi - \sin\psi\cos\varphi) - \dfrac{2Bd}{1-d^2}(\cos\psi\cos\varphi - d\sin\psi\sin\varphi). \end{array} \right\} \tag{69.4}$$

The slip lines are epicycloids and hypocycloids. For $d=1$, $R^- = A\sin\varphi - B\cos\varphi$, $R^+ = A\cos\varphi + B\sin\varphi$, the slip lines are cycloids. Again, superposition gives new solutions.

3. Next, consider any solution $u = h_1$ of $\dfrac{\partial^2 h_1}{\partial \xi \partial \eta} = h_1$ and form, see (67.11), $\dfrac{\partial u}{\partial \eta} = h_2$. A slip-line field is found by forming, as in (68.2),

$$\left. \begin{array}{l} x(\xi, \eta) = h_1 \cos(\eta - \xi) - h_2 \sin(\eta - \xi), \\ y(\xi, \lambda) = h_1 \sin(\eta - \xi) + h_2 \cos(\eta - \xi). \end{array} \right\} \tag{69.5}$$

We compute $\partial x/\partial \xi$, $\partial x/\partial \eta$, etc. and find, with $\varphi = \eta - \xi$,

$$\left. \begin{array}{ll} \dfrac{\partial x}{\partial \xi} = \left(\dfrac{\partial h_1}{\partial \xi} + h_2\right)\cos\varphi, & \dfrac{\partial x}{\partial \eta} = -\left(h_1 + \dfrac{\partial h_2}{\partial \eta}\right)\sin\varphi, \\[2mm] \dfrac{\partial y}{\partial \xi} = \left(\dfrac{\partial h_1}{\partial \xi} + h_2\right)\sin\varphi, & \dfrac{\partial y}{\partial \eta} = \left(h_1 + \dfrac{\partial h_2}{\partial \eta}\right)\cos\varphi. \end{array} \right\} \tag{69.6}$$

---

[1] As early as 1925, R. v. Mises: Z. angew. Math. Mech. 5, 147—149 (1925), called attention to this situation and illustrated it. See also Geiringer-Prager: Ergebn. exakt. Naturw. 13, 310—363 (1934), especially p. 343. More recent contributions in this direction are due to E. H. Lee: J. Appl. Mech. 19, 97—103 (1952).

Thus we see that the radii of curvature R^-, R^+ of the slip-line field (69.5) are $R^- = -\left(\dfrac{\partial h_1}{\partial \xi} + h_2\right)$ and $R^+ = -\left(h_1 + \dfrac{\partial h_2}{\partial \eta}\right)$. The field (69.5) is the sum of two fields, one with $k_1 = \dfrac{\partial h_1}{\partial \xi}$, $k_2 = h_1$, the other with $l_1 = h_2$, $l_2 = \dfrac{\partial h_2}{\partial \eta}$.

4. The Bessel function $I(\xi \cdot \eta)$, where

$$I(z) = 1 + \frac{z}{1!} + \frac{z^2}{2!\,2!} + \frac{z^3}{3!\,3!} + \cdots$$

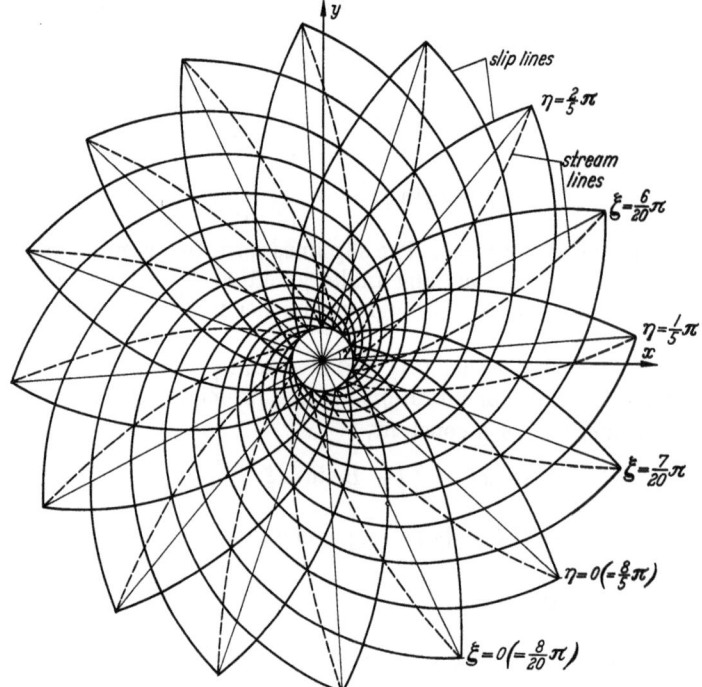

Fig. 16. Logarithmic spirals as slip-lines (full) and as streamlines (dashed).

satisfies our equation. Since this is also true for the partial derivatives $\partial^n I/\partial \xi^n$, $\partial^m I/\partial \eta^m$, we may set up a general solution in the form

$$u(\xi, \eta) = a\,I(\xi \cdot \eta) + a_1 \frac{\partial I}{\partial \xi} + a_2 \frac{\partial^2 I}{\partial \xi^2} + \cdots + b_1 \frac{\partial I}{\partial \eta} + b_2 \frac{\partial^2 I}{\partial \eta^2} + \cdots,$$

We do not go into considerations concerning convergence, completeness, etc., which would be of a rather mathematical character.

We now consider examples of velocities. Obviously, for the velocity components v_1, v_2 we can use the same particular solutions just considered. Thus, for example:

$$\left.\begin{aligned}
v_1 &= \sum_r A_r\, e^{c_r \xi + \frac{1}{c_r}\eta}, & v_2 &= -\sum_r A_r\, c_r\, e^{c_r \xi + \frac{1}{c_r}\eta}, \\
v_x &= \sum_r A_r \sqrt{1 + c_r^2}\, e^{c_r \xi + \frac{1}{c_r}\eta} \cos(\gamma_r - \varphi) & \sin \gamma_r &= \frac{c_r}{\sqrt{1 + c_r^2}}, \\
v_y &= \sum_r A_r \sqrt{1 + c_r^2}\, e^{c_r \xi + \frac{1}{c_r}\eta} \sin(\gamma_r - \varphi) & \cot \gamma_r &= c_r.
\end{aligned}\right\} \quad (69.7)$$

If in a flow $v_1 = v_2$, the flow direction at every point bisects the angle of the characteristics and has therefore a principal direction. An example of such a

diagonal flow is obtained from (69.7) by putting $c=-1$:
$$v_1 = v_2 = e^{-(\xi+\eta)}. \tag{69.8}$$

We may combine it with the slip-line field (69.2). The differential equations of the streamlines being $dy/dx = \tan \vartheta$, $\vartheta = \varphi + 45°$, we easily find

$$\tan \varphi = \frac{cy-x}{cx+y}, \quad \tan \vartheta = \frac{1+\tan \varphi}{1-\tan \varphi} = \frac{y(1+c)-x(1-c)}{x(1+c)+y(1-c)}. \tag{69.9}$$

These streamlines are again logarithmic spirals; they intersect the radius vector under the constant angle $\gamma = \arctan \frac{1-c}{1+c}$. In Fig. 16 the full lines are the two families of slip lines, the dashed lines are the streamlines.

For $c=1$, the slip-line spirals are symmetric to the radii and the streamlines are the radii (Fig. 17). This pattern is approximately realized if a punch of circular cross-section is forced into a plastic material. It has been observed in wrought iron plates and in the cross-section of thick-walled tubes (see NADAI [*19*], p. 228). Uniform normal pressure p acts along the circle of radius $r=a$. Hence at $r=a$, $\sigma_r = -p$, $\tau_{r\vartheta} = 0$. Thus, by the yield condition, $\sigma_\vartheta = -p+2k$. This inner circle plays the role of the curve $\mathscr{K}$ of a Cauchy problem. The solution of the equations

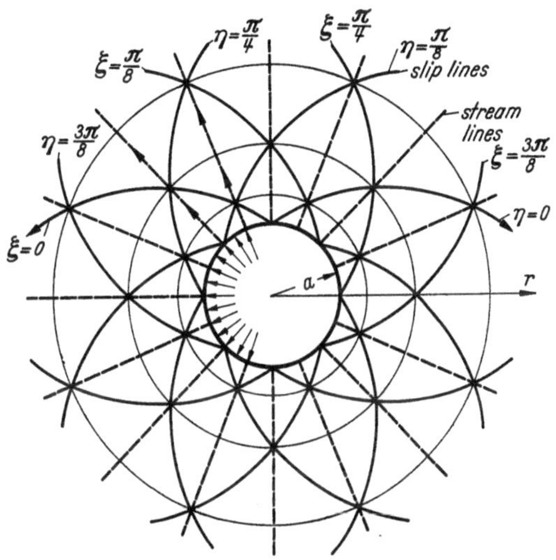

$$\frac{d}{dr}(r\sigma_r) - \sigma_\vartheta = 0, \quad \sigma_\vartheta - \sigma_r = 2k,$$

Fig. 17. Logarithmic spirals as slip-lines and radial streamlines for thick-walled tube under internal pressure.

with the above boundary conditions, is due to PRANDTL:

$$\sigma_r = -p + 2k \log \frac{r}{a}, \quad \sigma_\vartheta = -k + 2k\left(1+\log \frac{r}{a}\right), \quad \tau_{r\vartheta} = 0.$$

The characteristics are the two families of orthogonal spirals

$$\vartheta \pm \log \frac{r}{a} = \text{constant}.$$

The problem where along $r=a$ the more general stress distribution holds:

$$\sigma_r = -p, \quad \tau_{r\vartheta} = -t, \quad p \geq 0$$

has been considered by SOKOLOVSKY and by MICHLIN[1]. It has likewise a simple explicit solution for the stresses. The characteristics are two families of spirals which approach logarithmic spirals as the distance from the circular hole increases.

CHRISTIANOWITSCH, in a well-known and somewhat controversial paper[2], considered instead of the circular hole one of arbitrary continuous smooth boundary (cf. [*26*], pp. 184—192) for the case of a normal pressure at the hole; (see also pp. 192—195).

---

[1] S. G. MICHLIN: The mathematical theory of plasticity. Publ. Acad. Sci. USSR. 1938.
[2] S. A. CHRISTIANOWITSCH: Mat. Sbornik **1**, 511—543 (1936).

As another type of example consider a particular case of (69.3). But here (on account of the sign of the curvatures of the slip lines) we take $a=1$, $b=-1$, $\varphi^-=\xi-\eta-45°$, $\varphi^+=\xi-\eta+45°$. Then:

$$R^-=4A\cos(\xi-\eta-45°), \quad R^+=-4A\sin(\xi-\eta-45°),$$
$$x=2A(\xi+\eta)-A\sin 2(\xi-\eta+45°),$$
$$y= \quad A\cos 2(\xi-\eta+45°). \tag{69.12}$$

The slip lines form two families of orthogonal cycloids. PRANDTL, who first considered this slip-line field, found that it originates if a plastic mass is compressed between two rigid, rough plates. However, the streamlines which correspond to this problem cannot be found by guessing, as in the case of the thick-walled tube—a boundary problem must be solved (see Sect. 97 for the case of infinitely long plates).

70. Discontinuities. We return to our discussion of discontinuities (Sects. 54, 59, 61). Consider Eqs. (66.7'). For the characteristic C^- the first of Eqs. (66.7') is the compatibility equation, while the second one is the "remaining equation" (see Sects. 52, 53). Since in the "remaining equation" both variables ϑ and $p/2k$ are differentiated across C^-, none of these variables can jump. *The stresses must be continuous across a* C^-. In the same way we conclude for the C^+. Consider, next, the velocities and Eqs. (67.14) or (66.11). For a C^- the first Eq. (67.14) is the compatibilitiy equation and the second the "remaining equation". In this second equation v_2 is differentiated across the characteristic, but v_1 is not. Therefore *the tangential velocity of the C^- may jump, while the normal component must remain continuous*, and, of course, the same holds for the C^+.

Hence, stress discontinuities can only arise across non-characteristic lines. We have seen, Sect. 61, that across such a line σ_n, τ, and v must be continuous, and only σ_t may jump as in (61.5). Also, $\dot{E}=0$ at the discontinuity line.

We can now add some more information. Let us express the jump conditions in terms of $s=-p/k$ and ϑ. Denote by λ the angle of the normal to the discontinuity line with the first principal direction. Then Eqs. (66.6) hold with σ_n, σ_t, τ on the left, and λ (rather then ϑ) on the right-hand side, or, with

$$k=\frac{1}{2}(\sigma_1-\sigma_2), \quad s=\frac{1}{2k}(\sigma_1+\sigma_2),$$
$$\sigma_n=ks+k\cos 2\lambda, \quad \sigma_t=ks-k\cos 2\lambda, \quad \tau=-k\sin 2\lambda. \tag{70.1}$$

Since σ_n and τ remain continuous we obtain

$$\begin{rcases} s_2-s_1+\cos 2\lambda_2-\cos 2\lambda_1=0, \\ \sin 2\lambda_2-\sin 2\lambda_1=0, \end{rcases} \tag{70.2}$$

or

$$\lambda_1+\lambda_2=90°, \quad s_2=s_1+2\cos 2\lambda_1. \tag{70.3}$$

Or, with $\varphi^+=\vartheta+45°$ and γ the angle of the normal and the x-axis[1]:

$$\begin{rcases} \varphi_2^++\varphi_1^+=2\gamma, \\ s_2-s_1=2\sin 2(\varphi_1^+-\gamma). \end{rcases} \tag{70.3'}$$

At a point of the discontinuity line there now are four slip-line directions C_1^+, C_2^+ and C_1^-, C_2^- where $C_1^+ \perp C_1^-$, $C_2^+ \perp C_2^-$. The above equations show that the discontinuity line bisects both the angles of $C_1^- C_2^-$ and of $C_1^+ C_2^+$ (Fig. 18).

---

[1] This equation shows again that across a curve of characteristic direction there can be no discontinuity of stresses.

Fig. 19 shows the Mohr circles of two stresses $\Sigma^{(1)}$, $\Sigma^{(2)}$ on both sides of the discontinuity. The σ_n and τ are the same for both tensors. The figure shows the directions u_1, v_1 and u_2, v_2 of two pairs of principal directions compatible with a common stress vector t_n, which in both tensors corresponds to the direction n, normal to the discontinuity line. From the figure we read, with $\sigma_n = \sigma$:

$$4\sigma = (\sigma_1^{(1)} + \sigma_2^{(1)}) + (\sigma_1^{(2)} + \sigma_2^{(2)}),$$
$$2\sigma = \sigma_2^{(1)} + \sigma_1^{(2)} = \sigma_1^{(1)} + \sigma_2^{(2)};$$
(70.4)

$$\sigma_1^{(1)} + \sigma_1^{(2)} = 2\sigma - 2k,$$
$$\sigma_2^{(1)} + \sigma_2^{(2)} = 2\sigma + 2k.$$
(70.5)

It can be seen that the curvatures $1/R^+$, $1/R^-$ may change across a discontinuity line (see PRAGER[1], HILL [10], p. 159). Of the contributions concerned with

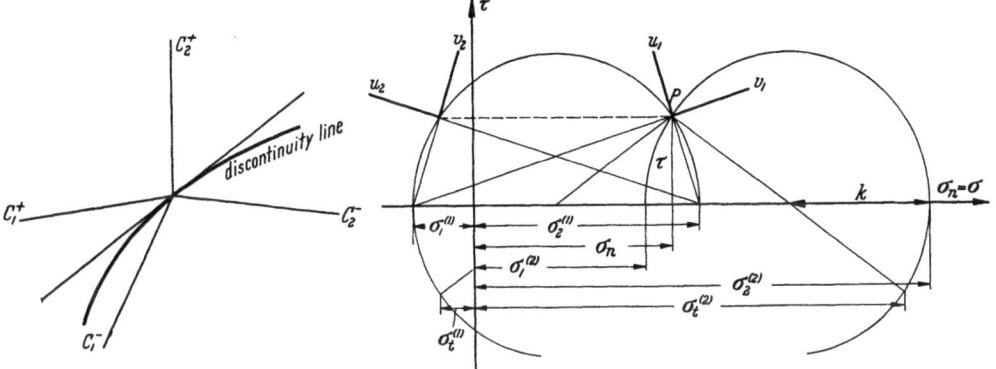

Fig. 18. Stress discontinuity lines with characteristics.

Fig. 19. Mohr's circles of two stress tensors with one stress vector in common.

stress discontinuities we mention a few in addition to those discussed in the preceding chapter. LEE[2] has given the velocity distribution for a problem with a discontinuity line. CARRIER and WINZER[3] have studied the interaction of discontinuities meeting at a point. HILL[4] has carried out the investigation of discontinuities for the problem of plane stress.

G. The general plane problem.

I. Basic theory.

a) The equations.

71. Linearization. In this chapter we shall consider the theory of the problem (64.4), which contains the problems of plane strain and of plane stress as particular cases[5]. Consider the first three Eqs. (64.4)

$$\frac{\partial \sigma_x}{\partial x} + \frac{\partial \tau}{\partial y} = 0, \quad \frac{\partial \tau}{\partial x} + \frac{\partial \sigma_y}{\partial y} = 0,$$
(71.1)

$$f(\sigma_x, \sigma_y, \tau) = F(\sigma_1, \sigma_2) = 0.$$
(71.2)

---
[1] W. PRAGER: R. Courant Anniversary Volume, pp. 289—299. New York 1948.
[2] E. H. LEE: Proc. 3rd Symp. Appl. Math., pp. 213—228. New York 1950.
[3] A. WINZER and G. F. CARRIER: J. Appl. Mech. 261—264 (1948).
[4] R. HILL: J. Math. Phys. of Solids 1, 19—30 (1952).
[5] Cf. the article H. GEIRINGER: Advances. We mention particularly the valuable monograph, H. MANDEL: Sur les équilibres par tranches parallèles des milieux plastiques à la limite d'écoulement. Paris: Louis Jean 1942. This article is concerned only with the stress

They are nonlinear but can be converted into a linear system in many ways. Following von Mises[1] we use a parametric representation of the yield condition

$$F(\sigma_1, \sigma_2) = 0, \quad \sigma_1 = \sigma_1(s), \quad \sigma_2 = \sigma_2(s), \tag{71.3}$$

where s is an adequate parameter, e.g. $s = -p/k = (\sigma_1 + \sigma_2)/2k$ with k a constant. Denote by ϑ the angle between the x-axis and the first principal direction, which we denote as the u-direction, the second being the v-direction. We assume in the usual way $\sigma_1 \geq \sigma_2$[2]. Using Eqs. (66.6), we compute $\dfrac{\partial \sigma_x}{\partial x} + \dfrac{\partial \tau}{\partial y}$ as well as $\dfrac{\partial \tau}{\partial x} + \dfrac{\partial \sigma_y}{\partial y}$. With directional derivatives, as in Sect. 66, we obtain

$$\frac{\partial \sigma_1}{\partial u} \cos \vartheta + \frac{\partial \sigma_2}{\partial v} \sin \vartheta + (\sigma_2 - \sigma_1)\left(\frac{\partial \vartheta}{\partial u} \sin \vartheta - \frac{\partial \vartheta}{\partial v} \cos \vartheta\right) = 0,$$

$$\frac{\partial \sigma_1}{\partial u} \sin \vartheta + \frac{\partial \sigma_2}{\partial v} \cos \vartheta - (\sigma_2 - \sigma_1)\left(\frac{\partial \vartheta}{\partial u} \cos \vartheta + \frac{\partial \vartheta}{\partial v} \sin \vartheta\right) = 0.$$

If we multiply the first of these equations by $\cos \vartheta$, the second by $\sin \vartheta$, and add, we obtain Jenne's equations (see Sect. 42), a particular case of (45.1):

$$\frac{\partial \sigma_1}{\partial u} = (\sigma_2 - \sigma_1) \frac{\partial \vartheta}{\partial v}, \quad \frac{\partial \sigma_2}{\partial v} = (\sigma_2 - \sigma_1) \frac{\partial \vartheta}{\partial u}. \tag{71.4}$$

Using (71.3) and putting

$$\frac{d\sigma_1}{ds} = \sigma_1', \quad \frac{d\sigma_2}{ds} = \sigma_2', \quad \frac{\sigma_2 - \sigma_1}{\sigma_1'} = g(s), \quad \frac{\sigma_2 - \sigma_1}{\sigma_2'} = h(s) \tag{71.5}$$

we obtain the "reducible" equations[3]

$$\frac{\partial s}{\partial u} = g(s) \frac{\partial \vartheta}{\partial v}, \quad \frac{\partial s}{\partial v} = h(s) \frac{\partial \vartheta}{\partial u}, \tag{71.6}$$

which generalize the equations preceding (66.7).

Together with (71.3) these determine σ_1, σ_2 and ϑ, hence Σ. The plane where s and ϑ are polar coordinates has been termed by von Mises the stress graph (see end of Sect. 66).

The dependent and independent variables may now be interchanged if $d = \dfrac{\partial(s, \vartheta)}{\partial(u, v)} \neq 0$. With

$$\frac{\partial s}{\partial u} = d \frac{\partial v}{\partial \vartheta}, \quad \frac{\partial s}{\partial v} = -d \frac{\partial u}{\partial \vartheta}, \quad \frac{\partial \vartheta}{\partial u} = -d \frac{\partial v}{\partial s}, \quad \frac{\partial \vartheta}{\partial v} = d \frac{\partial u}{\partial s}, \tag{71.7}$$

we find von Mises' linear equations

$$\frac{\partial v}{\partial \vartheta} = g(s) \frac{\partial u}{\partial s}, \quad \frac{\partial u}{\partial \vartheta} = h(s) \frac{\partial v}{\partial s}. \tag{71.8}$$

Note that in these equations u and v are not coordinates.

---

distributions, not with velocities; hence in our terminology Mandel does not consider the complete problem. Mandel's starting point is the envelope of Mohr circles (see Sect. 74) — which determines a yield condition — and all points of contact with the circles are assumed to be real. This limits the study to hyperbolic states of stress. The same approach is that of Sokolovsky [26], Chap. X, (based on a paper of 1948). Cf. also footnote 1 on p. 353, and the end of our Sect. 75.

[1] R. v. Mises: Reissner Anniversary Volume. Ann Arbor, Michigan 1949.

[2] In Advances, I assumed $\sigma_1 \leq \sigma_2$, which simplifies some formulas in the present problem; it is, however, unusual, and with this assumptions the formulas would not directly reduce to those of Chap. F.

[3] By that we mean equations, linear and homogeneous in the first order derivatives with coefficients which are functions of the dependent variables only.

Sect. 71. Linearization.

The basic equations of compressible fluid flow

$$\frac{\partial q}{\partial s} = \frac{q}{M^2-1}\frac{\partial \vartheta}{\partial n}, \quad \frac{\partial q}{\partial n} = q\frac{\partial \vartheta}{\partial s},$$

with q and ϑ the polar coordinates of the velocity vector, and M the Mach number, are of the same mathematical form as Eqs. (71.6); if φ and ψ denote potential and stream functions, the basic linearized equations for these functions,

$$\frac{\partial \varphi}{\partial \vartheta} = \frac{q}{\varrho}\frac{\partial \psi}{\partial q}, \quad \frac{\partial \psi}{\partial \vartheta} = -\frac{\varrho q}{1-M^2}\frac{\partial \varphi}{\partial q},$$

are similar to (71.8)[1]. The stress graph corresponds to the hodograph, our s and ϑ to q and ϑ, the yield condition to the adiabatic condition, principal stress trajectories to stream lines and potential lines, and characteristics to characteristics. Mathematical insight and heuristic ideas regarding our subject can be gained from the more developed field of gas dynamics[2].

We may now derive in various ways equations for quantities in the physical plane. Let e.g.[3]

$$X = x\cos\vartheta + y\sin\vartheta, \quad Y = y\cos\vartheta - x\sin\vartheta, \tag{71.9}$$

and, according to the definition of u- and v-directions,

$$du = dx\cos\vartheta + dy\sin\vartheta, \quad dv = dy\cos\vartheta - dx\sin\vartheta. \tag{71.9'}$$

Hence from (71.9) and (71.9')

$$dX = du + Y\,d\vartheta, \quad dY = dv - X\,d\vartheta$$

whence

$$\frac{\partial X}{\partial \vartheta} = \frac{\partial u}{\partial \vartheta} + Y, \quad \frac{\partial X}{\partial s} = \frac{\partial u}{\partial s}, \quad \frac{\partial Y}{\partial \vartheta} = \frac{\partial v}{\partial \vartheta} - X, \quad \frac{\partial Y}{\partial s} = \frac{\partial v}{\partial s}. \tag{71.9''}$$

Then, using (71.8) we obtain

$$\frac{\partial X}{\partial \vartheta} = h(s)\frac{\partial Y}{\partial s} + Y, \quad \frac{\partial Y}{\partial \vartheta} = g(s)\frac{\partial X}{\partial s} - X \tag{71.10}$$

and

$$\left.\begin{array}{l}\dfrac{\partial^2 X}{\partial \vartheta^2} - hg\dfrac{\partial^2 X}{\partial s^2} = -X + \dfrac{\partial X}{\partial s}(g'h + g - h), \\[6pt] \dfrac{\partial^2 Y}{\partial \vartheta^2} - hg\dfrac{\partial^2 Y}{\partial s^2} = -Y + \dfrac{\partial Y}{\partial s}(h'g + g - h).\end{array}\right\} \tag{71.11}$$

We see that the problem is hyperbolic, elliptic, or parabolic dependent as to whether

$$hg > 0, \quad hg < 0, \quad hg = 0 \text{ or } \infty. \tag{71.12}$$

In plane strain, $\sigma_1 = k(s+1)$, $\sigma_2 = k(s-1)$ is a parametric presentation of $(\sigma_1 - \sigma_2)^2 - 4k^2 = 0$. Hence $g = h = -2$, $gh = 4$; in this case (71.11) reduces to $\partial^2 X/\partial \vartheta^2 - 4\partial^2 X/\partial s^2 + X = 0$, which again is the telegraph equation[4].

---

[1] If a scale factor is given M and ϱ depend on q only; in (71.6), (71.8), $g(s)$ and $h(s)$ depend on s only.

[2] Essentially, the same analogy has been applied by Nobuo Inoue to some exact solutions in plane strain and soil mechanics. Nobuo Inoue: J. Phys. Soc. Japan 7, 518—523, 604—609, 610—618 (1952), where also literature is given. Cf. also R. Hill: J. Mech. Phys. of Solids 2, 110—116 (1954) who discusses and in some instances develops Inoue's gas dynamical analogy.

[3] Cf. R. v. Mises: Reissner Anniversary Volume. Ann Arbor, Michigan 1949.

[4] Other essentially equivalent equations have been dealt with in Geiringer: Advances, pp. 224—227.

Another linear equation, valid also in the case of non-isotropy (cf. p. 327), is due to SAUER[1]. The first Eq. (71.1) is satisfied if we put

$$\sigma_x = \frac{\partial \varphi}{\partial y}, \qquad \tau = -\frac{\partial \varphi}{\partial x}.$$

Assume that the yield condition can be solved with respect to σ_y, say:

$$\sigma_y = k(\sigma_x, \tau), \qquad \frac{\partial k}{\partial \sigma_x} = -k_1, \qquad \frac{\partial k}{\partial \tau} = k_2. \tag{71.13}$$

Then the second Eq. (71.1) gives

$$\frac{\partial^2 \varphi}{\partial x^2} + k_2 \frac{\partial^2 \varphi}{\partial x \partial y} + k_1 \frac{\partial^2 \varphi}{\partial y^2} = 0, \tag{71.14}$$

a non-linear equation whose coefficients depend on $\partial \varphi/\partial x$ and $\partial \varphi/\partial y$. This transforms into a linear equation by means of the Legendre transformation

$$\Phi = -\tau x + \sigma_x y - \varphi, \qquad d\Phi = -x\, d\tau + y\, d\sigma_x - d\varphi,$$
$$x = -\frac{\partial \Phi}{\partial \tau}, \qquad y = \frac{\partial \Phi}{\partial \sigma_x}, \tag{71.15}$$

and, if $\dfrac{\partial(\sigma_x, \tau)}{\partial(x, y)} \neq 0$, we obtain the linear equation

$$\frac{\partial^2 \Phi}{\partial \sigma_x^2} + k_2 \frac{\partial^2 \Phi}{\partial \sigma_x \partial \tau} + k_1 \frac{\partial^2 \Phi}{\partial \tau^2} = 0. \tag{71.16}$$

In the case of isotropy (symmetry), Eq. (71.11) seems preferable to (71.16), since the coefficients depend on one variable only. Eq. (71.11) simplifies by the use of the new independent variable μ, or λ, defined by

$$\frac{d\mu}{ds} = (hg)^{-\frac{1}{2}} \qquad \text{or} \qquad \frac{d\lambda}{ds} = (-hg)^{-\frac{1}{2}}$$

in the hyperbolic and elliptic case, respectively. A more detailed study of these equations then depends on the particular yield condition.

Equations analogous to (66.6'), with ω equal to our s and $\chi = (\sigma_1 - \sigma_2)/2k$, are

$$\sigma_x = k(\omega + \chi \cos 2\vartheta), \qquad \sigma_y = k(\omega - \chi \cos 2\vartheta), \qquad \tau = k\chi \sin 2\vartheta. \tag{71.17}$$

72. Various yield conditions. α) Assume VON MISES' *quadratic condition* (Sect. 38) and $\sigma_3 = 0$. The yield condition of (64.3) obtains which is the equation of an ellipse in the σ_1, σ_2-plane, with minor axis $2k\sqrt{2}/\sqrt{3}$, major axis $2k\sqrt{2}$. We use the following parametric representation and formulas:

$$\sigma_1^2 + \sigma_2^2 - \sigma_1 \sigma_2 - 4k^2 = 0,$$
$$s = \frac{1}{2k}(\sigma_1 + \sigma_2), \qquad (-2 \leq s \leq 2),$$
$$\sigma_1 = k\left(s + \sqrt{\frac{4-s^2}{3}}\right), \qquad \sigma_2 = k\left(s - \sqrt{\frac{4-s^2}{3}}\right),$$
$$\sigma_1' = k\left(1 - \frac{s}{\sqrt{12-3s^2}}\right), \qquad \sigma_2' = k\left(1 + \frac{s}{\sqrt{12-3s^2}}\right), \tag{72.1}$$
$$g = \frac{-2(4-s^2)}{\sqrt{12-3s^2} - s}, \qquad h = \frac{-2(4-s^2)}{\sqrt{12-3s^2} + s},$$
$$gh = \frac{(\sigma_1 - \sigma_2)^2}{\sigma_1' \sigma_2'} = \frac{(4-s^2)^2}{3-s^2}.$$

Another useful parametric presentation is due to NADAI. Let an angle δ be defined by

$$\sin \delta = \frac{s}{2}, \qquad \tan \delta = \frac{s}{\sqrt{4-s^2}}, \qquad -90° \leq \delta \leq 90°. \tag{72.2}$$

---

[1] R. SAUER: Z. angew. Math. Mech. **29**, 274—279 (1949).

Then, we see that
$$\sigma_1 = \frac{4k}{\sqrt{3}} \sin(\delta + 30°), \qquad \sigma_2 = \frac{4k}{\sqrt{3}} \sin(\delta - 30°). \tag{72.3}$$

(Practically the same presentation has been used by Sokolovsky [26], p. 268 seq.; his parameter is $w = 90° - \delta$.) Using the abbreviation $F_i = \partial F/\partial \sigma_i$, $i = 1, 2$, and anticipating the formula of the next section, $(F_1 + F_2)/(F_1 - F_2) = -\cos 2\alpha$ [see Eq. (73.3) etc.], we see that in the hyperbolic part
$$\tan \delta = -\sqrt{3} \cos 2\alpha. \tag{72.4}$$

It is easy to express in terms of δ the various statements which we shall make in terms of s.

β) TRESCA'S *condition*, with $\sigma_3 = 0$, leads to the intersection of the spatial hexagonal prism (Sect. 41) with the σ_1, σ_2-plane[1]. Hence, with the same s:

$$\begin{array}{l}\sigma_1 - \sigma_2 = 2k, \qquad\qquad \text{if } \sigma_1 \sigma_2 \leq 0, \\ \qquad\quad = 4k - |\sigma_1 + \sigma_2|, \text{ if } \sigma_1 \sigma_2 \geq 0, \\ \sigma_1 = k(s+1), \quad \sigma_2 = k(s-1), g = h = -2, \quad gh = 4, \quad |s| \leq 1 \\ \sigma_1 = k[2 + s - |s|], \qquad \sigma_2 = k[-2 + s + |s|], \qquad 1 \leq |s| \leq 2\end{array} \tag{72.5}$$

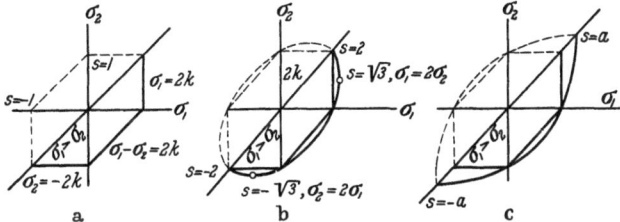

Fig. 20 a—c. Yield conditions in plane stress: (a) hexagonal condition; (b) quadratic condition; (c) parabola condition.

The first and the third line give the 45° line (to the right) of the hexagon (the third line in a parametric representation). The second and fourth line are the equations of the vertical and the horizontal line of the hexagon (the fourth line in a parametric representation). The 45°-lines correspond to a hyperbolic, the others to a parabolic problem[2].

γ) The quadratic limit (72.1) is hyperbolic for values $-\sqrt{3} < s < +\sqrt{3}$, while for $|s|$ between $\sqrt{3}$ and 2 the problem is elliptic. To avoid this considerable difficulty, von Mises proposed a comparatively small adjustment of his yield condition. He introduced a *parabola limit* for which the problem is hyperbolic throughout. In the σ_1, σ_2-plane this limit is represented by two branches of parabolas passing through four corners of the Tresca hexagon. Their equation is, with $1 + \sqrt{2} = a$, where $a^2 - 1 = 2a$

$$\begin{array}{l}\dfrac{\sigma_1 - \sigma_2}{k} = \pm \dfrac{1}{a}\left[a^2 - \left(\dfrac{\sigma_1 + \sigma_2}{2k}\right)^2\right], \quad \text{upper sign for right branch}, \\ \sigma_1 = ks + \dfrac{k}{2a}(a^2 - s^2), \quad \sigma_2 = ks - \dfrac{k}{2a}(a^2 - s^2), \quad |s| \leq a, \\ g = -(a+s), \quad h = -(a-s).\end{array} \tag{72.6}$$

The three conditions are shown in Fig. 20[3].

---

[1] A "generalized Tresca condition" was introduced by P. G. Hodge jr.: J. Math. Phys. **29**, 38—48 (1950).

[2] For both v. Mises' and Tresca's yield condition for plane stress, Sokolovsky was first to investigate the stress problem: V. V. Sokolovsky: C. R. Acad. Sci. USSR. **51**, 175—178 (1946) and same Journal, **51**, 421—424 (1946), cf. [26], p. 272 seq.

[3] C. Carathéodory and E. Schmidt, Z. angew. Math. Mech. **3**, 468—475 (1923), discussed the possibility of isometric nets of principal stress trajectories of the plane strain yield condition, $\sigma_1 - \sigma_2 = $ constant. P. F. Neményi and A. van Tuyl, Quart. J. Mech. a. Appl. Math. **5**, 1—11 (1952), consider a general yield condition and show that only for three distinct families of yield conditions do there exist isometric nets (beyond the trivial nets consisting of concentric circles and radial straight lines). Among these, one is the "parabola condition".

δ) Coulomb's condition used in soil mechanics (see Ref. 1, p. 328), written in the above variables, is as follows:

$$\pm \frac{\sigma_1 - \sigma_2}{2} = k \cos \Phi + \frac{\sigma_1 + \sigma_2}{2} \sin \Phi, \quad s \geq -\cot \Phi,$$

$$\sigma_1 = k(s + s \sin \Phi + \cos \Phi), \quad \sigma_2 = k(s - s \sin \Phi - \cos \Phi), \quad (72.7)$$

$$\sigma'_1 = k(1 + \sin \Phi), \quad \sigma'_2 = k(1 - \sin \Phi), \quad \sigma'_1 \sigma'_2 = k^2 \cos^2 \Phi.$$

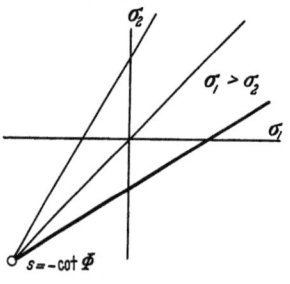

Fig. 21. Coulomb's yield condition.

Here k is the cohesion and Φ, a constant, the angle of internal friction (see Fig. 21). The problem is hyperbolic[1]. (More on these examples in Sect. 75.)

b) Characteristics of the complete plane problem.

73. Characteristic directions and compatibility relations. The first three Eqs. (64.4) have been replaced by Eqs. (71.6). Hence the complete problem is equivalent to Eqs. (71.6) and those in the last line of (64.4), which we write in the form

$$\frac{\dot{\varepsilon}_x - \dot{\varepsilon}_y}{2\dot{\gamma}} = \frac{\frac{\partial f}{\partial \sigma_x} - \frac{\partial f}{\partial \sigma_y}}{\frac{\partial f}{\partial \tau}}, \quad \frac{\dot{\varepsilon}_x + \dot{\varepsilon}_y}{2\dot{\gamma}} = \frac{\frac{\partial f}{\partial \sigma_x} + \frac{\partial f}{\partial \sigma_y}}{\frac{\partial f}{\partial \tau}}. \quad (73.1)$$

This is a system of four equations with the dependent variable s, ϑ, v_x, v_y and two independent variables x, y. The first two equations do not contain velocities, the last two contain no derivatives of the stresses. It is readily seen[2] that the characteristic determinant of order four resolves into the product of two determinants of order two which define separately the characteristics of the stress problem and those of the velocity problem.

Eqs. (71.6) relate to the principal directions as axes. Denote by α the angle of a characteristic direction with the first principal direction (the u-direction). The characteristic determinant equated to zero is:

$$\begin{vmatrix} \sin \alpha & g \cos \alpha \\ \cos \alpha & h \sin \alpha \end{vmatrix} = 0. \quad (73.2)$$

Hence, using (71.5) and writing with the notation of (64.4), and in analogy to (40.1), Grad $f = \mu \dot{E}$, we obtain, with $F_i = \frac{\partial F}{\partial \sigma_i}$

$$\tan^2 \alpha = \frac{g}{h} = \frac{\sigma'_2}{\sigma'_1} = -\frac{F_1}{F_2} = -\frac{\dot{\varepsilon}_1}{\dot{\varepsilon}_2},$$

$$\tan \alpha = \frac{dv}{du} = \pm \sqrt{\frac{d\sigma_2}{d\sigma_1}}. \quad (73.3)$$

We see again that hyperbolic points are those where g and h or σ'_1 and σ'_2 have the same sign and where $\dot{\varepsilon}_1$, $\dot{\varepsilon}_2$ or F_1, F_2 have opposite signs. The characteristics make angles $\vartheta \pm \alpha$ with the x-axis.

---

[1] Some more recent papers on the subject are: D. C. Drucker and W. Prager: Quart. J. Appl. Math. **10**, 157—165 (1952). — D. C. Drucker: J. Mech. Phys. of Solids **1**, 217—226 (1953). — R. T. Shield: Quart. Appl. Math. **11**, 61—75 (1953). — R. T. Shield: J. Math. Phys. **33**, 144—156 (1954). This paper contains also the previously quoted generalization of the theory of stress discontinuities.

[2] For details, cf. H. Geiringer: Advances, pp. 227—229.

In the first Eq. (73.1) the right side equals $\cot 2\vartheta$ since Grad f and Σ have the same axes and Eqs. (66.4) hold for any symmetric tensor. In the second Eq. (73.1) the right side equals $(F_1 + F_2)/(F_1 - F_2) \sin 2\vartheta = -\cos 2\alpha/\sin 2\vartheta$, and the velocity equations become

$$\frac{\dot{\varepsilon}_x - \dot{\varepsilon}_y}{2\dot{\gamma}} = \cot 2\vartheta, \quad \frac{\dot{\varepsilon}_x + \dot{\varepsilon}_y}{2\dot{\gamma}} = -\frac{\cos 2\alpha}{\sin 2\vartheta}, \tag{73.1'}$$

two linear differential equations for v_x, v_y. Let $m = dy/dx$ denote the slope of a velocity characteristic. The characteristic determinant equated to zero gives:

$$\left. \begin{aligned} m^2(\cos 2\alpha + \cos 2\vartheta) - 2m \sin 2\vartheta + (\cos 2\alpha - \cos 2\vartheta) &= 0, \\ m = \frac{dy}{dx} = \frac{\sin 2\vartheta \pm \sin 2\alpha}{\cos 2\vartheta + \cos 2\alpha} &= \tan(\vartheta \pm \alpha). \end{aligned} \right\} \tag{73.4}$$

Hence, the directions of stress characteristics and of velocity characteristics coincide under the theory of the plastic potential.

Consider finally the directions for which the normal strain vanishes. If such a direction makes the angle ψ with the first principal direction, then the corresponding strain rate is

$$\dot{\varepsilon} = \dot{\varepsilon}_1 \cos^2 \psi + \dot{\varepsilon}_2 \sin^2 \psi = 0, \quad \text{or} \quad \tan^2 \psi = -\dot{\varepsilon}_1/\dot{\varepsilon}_2 = \tan^2 \alpha.$$

Hence, we have the result (Fig. 22a):

At a hyperbolic point, i.e. at a point where $F_1 F_2 < 0$, there are two real double-counting characteristics C^+, C^- which represent the four characteristics of the system (71.6), (73.1). They form the angles $\vartheta + \alpha, \vartheta - \alpha$ with the x-axis, where α is given by (73.3). The normal strain in a characteristic direction vanishes. The directions of maximum shear stress, which coincide with the directions of maximum shear strain and bisect the angle of the principal directions, are in general not characteristic. This happens only if $F_1 + F_2 = 0$, $\dot{\varepsilon}_1 + \dot{\varepsilon}_2 = 0$ as in (generalized) plane strain.

The stress equations have been studied by many authors, the complete problem only more recently[1].

To derive one pair of *compatibility equations* we multiply the first Eq. (71.6) by $\sin \alpha$, the second by $\frac{\sin^2 \alpha}{\cos^2 \alpha} = \frac{g}{h} \cos \alpha$, and add. The right side then equals $g \frac{\partial \vartheta}{\partial l^+}$, the left side is $\frac{\partial s}{\partial u} \sin \alpha + \frac{\partial s}{\partial v} \frac{\sin^2 \alpha}{\cos \alpha} = \tan \alpha \left(\frac{\partial s}{\partial u} \cos \alpha + \frac{\partial s}{\partial v} \sin \alpha \right) = \tan \alpha \frac{\partial s}{\partial l^+}$, so that we obtain $\frac{\partial s}{\partial l^+} = g \cot \alpha \frac{\partial \vartheta}{\partial l^+}$, and, in a similar way, $\frac{\partial s}{\partial l^+} = h \tan \alpha \frac{\partial \vartheta}{\partial l^+}$. If, as before, $\tan \alpha = + \sqrt{\frac{g}{h}} = + \sqrt{\frac{\sigma_2'}{\sigma_1'}}$ then: $h \tan \alpha = \frac{\sigma_2 - \sigma_1}{\sigma_2'} \cdot \left(+ \sqrt{\frac{\sigma_2'}{\sigma_1'}} \right) = \frac{\sigma_2 - \sigma_1}{+\sqrt{\sigma_1' \sigma_2'}} < 0$ if $\sigma_1 > \sigma_2$.

Hence, $g \cot \alpha = h \tan \alpha = -\sqrt{gh}$, if $\sigma_1 > \sigma_2$, and[2]

$$\frac{\partial s}{\partial l^+} = -\sqrt{gh} \frac{\partial \vartheta}{\partial l^+}, \quad \frac{\partial s}{\partial l^-} = +\sqrt{gh} \frac{\partial \vartheta}{\partial l^-} \tag{73.5}$$

---

[1] For references see H. GEIRINGER: Advances, p. 241.
[2] This involved derivation was made to avoid arbitrariness of signs of roots, which would lead to materially wrong formulas (e.g. in Sects. 74 and 82). In my paper (preceding footnote) I used $\sigma_2 > \sigma_1$, hence $h \tan \alpha = +\sqrt{gh}$ and many formulas simplify due to the fact that then in (71.4), $\sigma_2 - \sigma_1 > 0$. It seems, however that in the present article the usual choice, $\sigma_1 > \sigma_2$, was to be followed up (see also Ref. 2, p. 368).

are the compatibility relations which generalize Eqs. (66.7)[1]. The C^-, C^+ which make the respective angles $\varphi^- = \vartheta - \alpha$, $\varphi^+ = \vartheta + \alpha$ with the x-axis form a net which is, in general, not orthogonal.

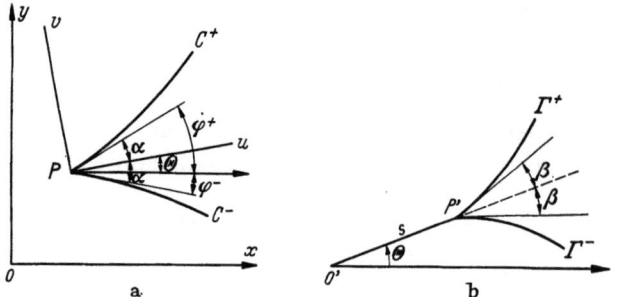

Fig. 22 a and b. Characteristics in physical plane and stress graph.

The C^-, C^+ are the images (Fig. 22) of the fixed characteristics Γ^-, Γ^+ in the stress graph which have the equations (upper sign for Γ^+):

$$\frac{d\vartheta}{ds} = \mp \frac{1}{\sqrt{gh}} \qquad (73.6)$$

and make equal angles β with the radius vector in the stress graph, where s and ϑ are polar coordinates and $\tan \beta = s\, d\vartheta/ds$. These Γ-curves can be found once

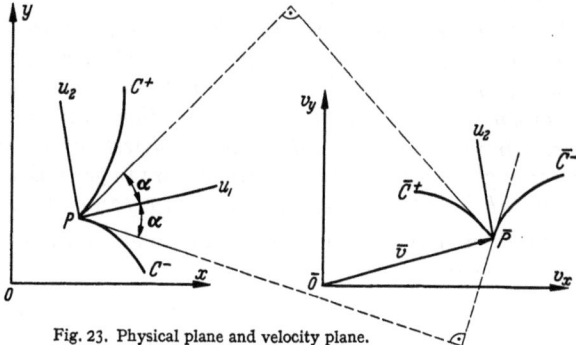

Fig. 23. Physical plane and velocity plane.

and for all for specific yield conditions. With

$$S(s) = -\int^s \frac{ds}{\sqrt{gh}} = \int^s \frac{\sqrt{\sigma_1' \sigma_2'}}{\sigma_2 - \sigma_1} ds, \qquad (73.7)$$

we obtain for Γ^-, Γ^+, respectively,

$$S(s) + \vartheta = \text{constant}, \qquad S(s) - \vartheta = \text{constant}, \qquad (73.8)$$

which generalize Eqs. (66.8). In plane strain $S = -\dfrac{s}{2} = \dfrac{p}{2k}$.

We need a *second pair of compatibility conditions* along the doubly counting characteristics. The fact that along a C^- and a C^+ the respective rate of extension is zero gives, exactly as in (66.12),

$$\frac{\partial v_x}{\partial l^-} \cos \varphi^- + \frac{\partial v_y}{\partial l^-} \sin \varphi^- = 0, \qquad \frac{\partial v_x}{\partial l^+} \cos \varphi^+ + \frac{\partial v_y}{\partial l^+} \sin \varphi^+ = 0. \qquad (73.9)$$

---

[1] Note that in "plane strain", from (72.5), $g = h = -2$, $+\sqrt{gh} = 2$; then Eqs. (73.5) agree with (66.7), since $p/k = -s$.

Sect. 74. Continuation. Relation to O. Mohr's theory.

The interpretation in terms of the velocity plane (Fig. 23) is the same as in Sect. 66 (cf. also footnote 3 on p. 357). (In the figure, u_1, u_2 are used instead of u, v in order to avoid confusion with velocity components.)

Next, we want equations analogous to (66.11). Denote by v_1, v_2, v_3, v_4 components of v in the direction of C^-, in the direction of C^+, in the direction normal to C^-, and in the direction normal to C^+ (Fig. 24a). Or: $v_1 = v_x \cos \varphi^- + v_y \sin \varphi^-$, $v_3 = v_y \cos \varphi^- - v_x \sin \varphi^-$, etc. By straightforward computation, analogous to that in Sect. 66, we obtain

$$\frac{\partial v_1}{\partial l^-} = \left(\frac{\partial v_x}{\partial l^-} \cos \varphi^- + \frac{\partial v_y}{\partial l^-} \sin \varphi^- \right) + (v_y \cos \varphi^- - v_x \sin \varphi^-) \frac{\partial \varphi^-}{\partial l^-},$$

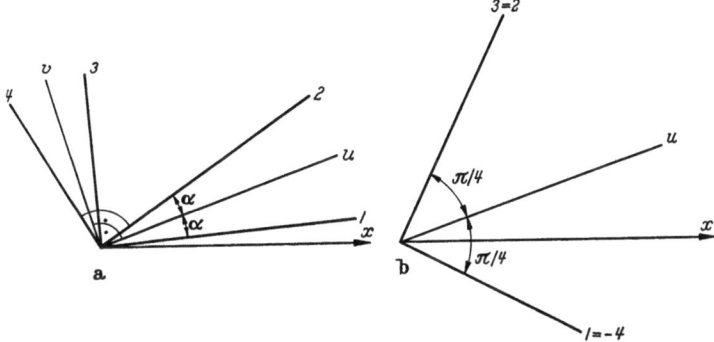

Fig. 24 a and b. Characteristic directions and their normals: (a) general case; (b) orthogonal case.

and, using (73.9), and computing likewise $\frac{\partial v_2}{\partial l^+}$ we obtain

$$\frac{\partial v_1}{\partial l^-} = v_3 \frac{\partial \varphi^-}{\partial l^-}, \qquad \frac{\partial v_2}{\partial l^+} = v_4 \frac{\partial \varphi^+}{\partial l^+}, \qquad (73.10)$$

or, briefly,

$$dv_1 = v_3 \, d\varphi^- \text{ along a } C^-, \qquad dv_2 = v_4 \, d\varphi^+ \text{ along a } C^+. \qquad (73.10')$$

In plane strain, Fig. 24b, $v_2 = v_3$, $v_4 = -v_1$, $d\varphi^- = d\varphi^+ = d\vartheta$ and we recover (66.11'). Finally, we express v_3 and v_4 in terms of v_1, v_2, the components in the characteristic directions, and obtain

$$\frac{\partial v_1}{\partial \varphi^-} = \frac{v_2 - v_1 \cos 2\alpha}{\sin 2\alpha}, \qquad \frac{\partial v_2}{\partial \varphi^+} = \frac{v_2 \cos 2\alpha - v_1}{\sin 2\alpha}, \qquad (73.11)$$

along a C^- and C^+, respectively.

74. Continuation. Relation to O. Mohr's theory. Differential equations in characteristic coordinates. In a σ, τ-coordinate system (see Sect. 39) the equation of a Mohr circle is (Fig. 25b)

$$\left(\sigma - \frac{\sigma_1 + \sigma_2}{2} \right)^2 + \tau^2 = \left(\frac{\sigma_1 - \sigma_2}{2} \right)^2. \qquad (74.1)$$

Here $OA = \sigma$, $AP = \tau$. From the figure we find:

$$\left. \begin{array}{l} \sigma = \dfrac{\sigma_1 + \sigma_2}{2} - \dfrac{\sigma_1 - \sigma_2}{2} \cos 2\varepsilon = \sigma_1 \sin^2 \varepsilon + \sigma_2 \cos^2 \varepsilon, \\[6pt] \tau = \dfrac{\sigma_1 - \sigma_2}{2} \sin 2\varepsilon = (\sigma_1 - \sigma_2) \sin \varepsilon \cos \varepsilon, \end{array} \right\} \qquad (74.2)$$

and from Eqs. (66.3) (Fig. 25a), taking for the x- and y-directions those of the first and second principal directions of Σ:

$$\sigma = \sigma_1 \cos^2 \psi + \sigma_2 \sin^2 \psi, \qquad \tau = (\sigma_2 - \sigma_1) \sin \psi \cos \psi. \tag{74.2'}$$

These coincide with (74.2) if $\psi = 90° + \varepsilon$. From Eqs. (74.2) or from Fig. 25b we obtain

$$\frac{\sigma_1 + \sigma_2}{2} = \sigma + \tau \cot 2\varepsilon, \qquad \frac{\sigma_1 - \sigma_2}{2} = \frac{\tau}{\sin 2\varepsilon}. \tag{74.3}$$

We now consider the one-dimensional family of Mohr circles, whose σ_1, σ_2 satisfy a given equation $F(\sigma_1, \sigma_2) = 0$, which we identify with the yield condition of a plasticity problem. These circles have an envelope whose point of

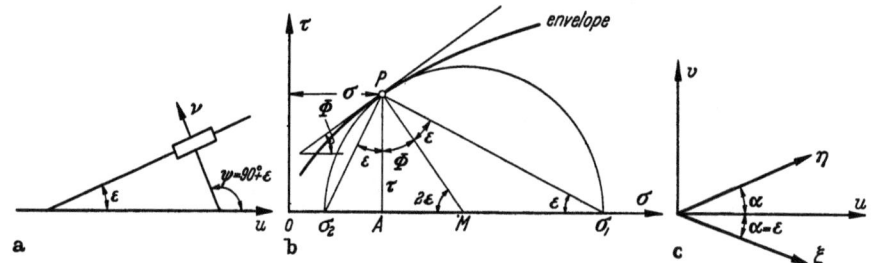

Fig. 25 a—c. Characteristic directions defined by means of envelope of Mohr circles.

contact with any circle may be real or imaginary. Consider a point P of real contact with the envelope. To each value of the parameter s corresponds a Mohr circle with $\sigma_1 = \sigma_1(s)$, $\sigma_2 = \sigma_2(s)$, which at the point of contact with the envelope has coordinates $\sigma = \sigma(s)$, $\tau = \tau(s)$, and an angle $\varepsilon = \varepsilon(s)$. Call Φ the angle which the tangent to the envelope makes with the positive σ-axis; then we have along the envelope

$$\frac{d\tau}{d\sigma} = \tan \Phi, \tag{74.4}$$

also $\Phi = 90° - 2\varepsilon$, and Eqs. (74.3) become

$$\frac{\sigma_1 + \sigma_2}{2} = \sigma + \tau \tan \Phi, \qquad \frac{\sigma_1 - \sigma_2}{2} = \frac{\tau}{\cos \Phi}. \tag{74.3'}$$

Now, differentiating and using (74.4) we obtain:

$$\left.\begin{aligned}\frac{\sigma_1' + \sigma_2'}{2} &= \sigma' + \tau' \tan \Phi + \frac{\tau}{\cos^2 \Phi} \Phi' = \frac{\sigma' + \tau \Phi'}{\cos^2 \Phi}, \\ \frac{\sigma_1' - \sigma_2'}{2} &= \frac{\sin \Phi}{\cos^2 \Phi} (\sigma' + \tau \Phi');\end{aligned}\right\} \tag{74.5}$$

dividing we find

$$\frac{\sigma_1' - \sigma_2'}{\sigma_1' + \sigma_2'} = \sin \Phi = \cos 2\varepsilon. \tag{74.5'}$$

Using (73.3) we find that $(\sigma_1' - \sigma_2')/(\sigma_1' + \sigma_2') = \cos 2\alpha$. Thus we see (cf. Figs. 25 a, c) that *the angle ε in (74.2) which corresponds on any circle of* Mohr *to a point of real contact with the envelope, is equal to the angle α which a characteristic makes with the first principal direction*[1]. The points with real contact with the envelope

---

[1] The first proof of this interesting relation is due to H. Mandel, Sur les équilibres par tranches parallèles des milieux plastiques à la limite d'écoulement. Paris: Louis Jean 1942, pp. 38, 39. The proof given here (which is even simpler than that of H. Geiringer in Advances (1953), p. 236, is independent of, and different from, a proof in Hill [10], p. 295. In several papers C. Torre has proved a closely related result (cf. Geiringer, Advances p. 237).

of Mohr circles are the hyperbolic points. Hence, *we have for σ and τ*:

$$\sigma = \sigma_1 \sin^2 \alpha + \sigma_2 \cos^2 \alpha, \quad \tau = (\sigma_1 - \sigma_2) \sin \alpha \cos \alpha. \tag{74.2''}$$

From these formulas we obtain a new expression for the S' of (73.7). From (74.5) and (74.3'):

$$\frac{\sigma' + \tau \Phi'}{2\tau} = \frac{\sigma'_1 + \sigma'_2}{2(\sigma_1 - \sigma_2)} \cos \Phi.$$

But $\cos \Phi = \sin 2\alpha = 2 \dfrac{\sqrt{\sigma'_1 \sigma'_2}}{\sigma'_1 + \sigma'_2}$, if $\dfrac{\sigma'_2}{\sigma'_1} = \tan^2 \alpha$ is used. Hence

$$\frac{\sigma' + \tau \Phi'}{2\tau} = \frac{\sqrt{\sigma'_1 \sigma'_2}}{\sigma_1 - \sigma_2} = -S'(s), \tag{74.6}$$

or, in terms of $\alpha = 45° - \dfrac{\Phi}{2}$:

$$S' = \alpha' - \frac{\sigma'}{2\tau}. \tag{74.6'}$$

We may also express S' in terms of the radius $R = \dfrac{\sigma_1 - \sigma_2}{2}$ of Mohr's circle and find easily

$$S' = -\frac{R'}{2R} \cot \Phi = -\frac{R'}{2R} \tan 2\alpha. \tag{74.7}$$

If we use S rather than s as parameter, the last formulas read with ` denoting differentiation with respect to S:

$$\alpha^` - \frac{\sigma^`}{2\tau} = 1, \quad -\frac{R^`}{2R} \tan 2\alpha = 1. \tag{74.8}$$

Characteristic coordinates. On account of Eqs. (73.8), and in analogy to Eqs. (67.9), we put[1]

$$S(s) = \eta + \xi, \quad \vartheta = \eta - \xi. \tag{74.9}$$

These characteristic coordinates[2],

$$S + \vartheta = 2\eta, \quad S - \vartheta = 2\xi, \tag{74.10}$$

do not apply in a region with rectilinear characteristics.

We then obtain from (73.3)

$$\frac{\partial y}{\partial \xi} = \frac{\partial x}{\partial \xi} \tan (\vartheta - \alpha), \quad \frac{\partial y}{\partial \eta} = \frac{\partial x}{\partial \eta} \tan (\vartheta + \alpha). \tag{74.11}$$

Now introduce, similarly as in Sect. 67,

$$h_1 = \frac{ds_1}{d\xi}, \quad h_2 = \frac{ds_2}{d\eta}, \tag{74.12}$$

where ds_1 and ds_2 denote line elements of a C^- and C^+, respectively. Then the following equations [corresponding to (67.13)] will hold:

$$\left.\begin{array}{ll} \dfrac{\partial x}{\partial \xi} = h_1 \cos \varphi^-, & \dfrac{\partial x}{\partial \eta} = h_2 \cos \varphi^+, \\[6pt] \dfrac{\partial y}{\partial \xi} = h_1 \sin \varphi^-, & \dfrac{\partial y}{\partial \eta} = h_2 \sin \varphi^+. \end{array}\right\} \tag{74.13}$$

Now,

$$\frac{\partial \varphi^+}{\partial \eta} = \frac{\partial \vartheta}{\partial \eta} + \frac{\partial \alpha}{\partial \eta} = \frac{\partial \vartheta}{\partial \eta} + \frac{\partial \alpha}{\partial S} \frac{\partial S}{\partial \eta} = 1 + \alpha^` = 1 + \frac{\alpha'}{S'},$$

---
[1] A theorem like Hencky's first theorem follows readily from Eqs. (74.9).
[2] The choice of (74.10) corresponds in (67.9) to $b = +1$, $a = -1$, $c = 0$.

where the two different accents denote differentiation with respect to s (') and with respect to S ('), respectively. There are three more such formulas and we obtain

$$\frac{\partial \varphi^+}{\partial \eta} = -\frac{\partial \varphi^-}{\partial \xi} = 1 + \alpha', \quad \frac{\partial \varphi^-}{\partial \eta} = -\frac{\partial \varphi^+}{\partial \xi} = 1 - \alpha', \tag{74.14}$$

and, introducing the radii of curvature with the same sign conventions as in (67.4), writing, however, R_1, R_2 rather than R^-, R^+ we have—corresponding to the relations (67.10):

$$h_1 = -R_1(1 + \alpha'), \quad h_2 = -R_2(1 + \alpha'). \tag{74.15}$$

Now using the conditions $\dfrac{\partial^2 x}{\partial \xi \partial \eta} = \dfrac{\partial^2 x}{\partial \eta \partial \xi}$ and $\dfrac{\partial^2 y}{\partial \xi \partial \eta} = \dfrac{\partial^2 y}{\partial \eta \partial \xi}$ we obtain from Eqs. (74.13) the relations[1]

$$\left.\begin{aligned}\frac{\partial h_1}{\partial \eta} \sin 2\alpha - \left(1 - \frac{\alpha'}{S'}\right)(h_1 \cos 2\alpha + h_2) &= 0, \\ \frac{\partial h_2}{\partial \xi} \sin 2\alpha - \left(1 - \frac{\alpha'}{S'}\right)(h_1 + h_2 \cos 2\alpha) &= 0,\end{aligned}\right\} \tag{74.16}$$

which reduce for $\alpha = 45°$ to (67.11) with $b = 1$, $a = -1$. With the abbreviation $A(S) = \sin 2\alpha / \left(1 - \dfrac{\alpha'}{S'}\right)$ we may write

$$A \frac{\partial h_1}{\partial \eta} = h_1 \cos 2\alpha + h_2, \quad A \frac{\partial h_2}{\partial \xi} = h_1 + h_2 \cos 2\alpha. \tag{74.16'}$$

These equations may be simplified further by introducing

$$h_1 \sqrt{\tau} = m_1, \quad h_2 \sqrt{\tau} = m_2 \tag{74.17}$$

Then

$$\frac{\partial h_1}{\partial \eta} = \frac{1}{\sqrt{\tau}} \frac{\partial m_1}{\partial \eta} - \frac{m_1}{2\tau \sqrt{\tau}} \frac{\partial \tau}{\partial \eta},$$

and, using (74.6') and $\tau'/\sigma' = \cot 2\alpha$, we have:

$$\left.\begin{aligned}A \frac{\partial h_1}{\partial \eta} &= \frac{A}{\sqrt{\tau}} \frac{\partial m_1}{\partial \eta} + \frac{m_1}{2\tau \sqrt{\tau}} \frac{\tau}{S'} \frac{\sin 2\alpha}{\sigma'} \cdot 2\tau S' \\ &= \frac{A}{\sqrt{\tau}} \frac{\partial m_1}{\partial \eta} + \frac{m_1}{\sqrt{\tau}} \cos 2\alpha.\end{aligned}\right\}$$

The second term to the right is $h_1 \cos 2\alpha$, hence, on account of (74.16'):

$$A \frac{\partial h_1}{\partial \eta} = \frac{A}{\sqrt{\tau}} \frac{\partial m_1}{\partial \eta} + h_1 \cos 2\alpha = h_1 \cos 2\alpha + h_2 = h_1 \cos 2\alpha + \frac{m_2}{\sqrt{\tau}}.$$

We obtain

$$A \frac{\partial m_1}{\partial \eta} = m_2, \quad A \frac{\partial m_2}{\partial \xi} = m_1. \tag{74.18}$$

These equations, due to MANDEL, are probably as simple as are obtainable. (For plane strain, $A = 1$ and $m_1 = h_1 \sqrt{k}$, $m_2 = h_2 \sqrt{k}$.) Cross differentiation of Eqs. (74.18)

---

[1] Just as in plane strain there is a parallelism between the velocity equations (73.11) and the Eqs. (74.16). If characteristic coordinates are used, the former become [use (74.14)]

$$\sin 2\alpha \frac{\partial v_1}{\partial \xi} = \left(1 + \frac{\alpha'}{S'}\right)(v_1 \cos 2\alpha - v_2), \quad \sin 2\alpha \frac{\partial v_2}{\partial \eta} = \left(1 + \frac{\alpha'}{S'}\right)(v_2 \cos 2\alpha - v_1).$$

which reduce to (67.14) for $\alpha = 45°$, $b = 1$, $a = -1$. These velocity equations are, however, not the same equations as Eqs. (74.16).

leads to

$$A^2 \frac{\partial^2 m_1}{\partial \xi \partial \eta} + A A' \frac{\partial m_1}{\partial \eta} - m_1 = 0,$$
$$A^2 \frac{\partial^2 m_2}{\partial \xi \partial \eta} + A A' \frac{\partial m_2}{\partial \xi} - m_2 = 0.$$
(74.19)

Here $A' = \dfrac{dA}{dS}$.

75. Examples to Sects. 73 and 74.

α) *Quadratic condition.* Using (72.1), we obtain from (73.6), (73.7)

$$\frac{d\vartheta}{ds} = \pm \frac{\sqrt{3-s^2}}{4-s^2}, \quad S(s) = -\int^s \frac{\sqrt{3-s^2}}{4-s^2} ds,$$
$$\pm \vartheta = \arctan \frac{s}{\sqrt{3-s^2}} - \frac{1}{2} \arctan \frac{s}{2\sqrt{3-s^2}} + \text{constant},$$
(75.1)

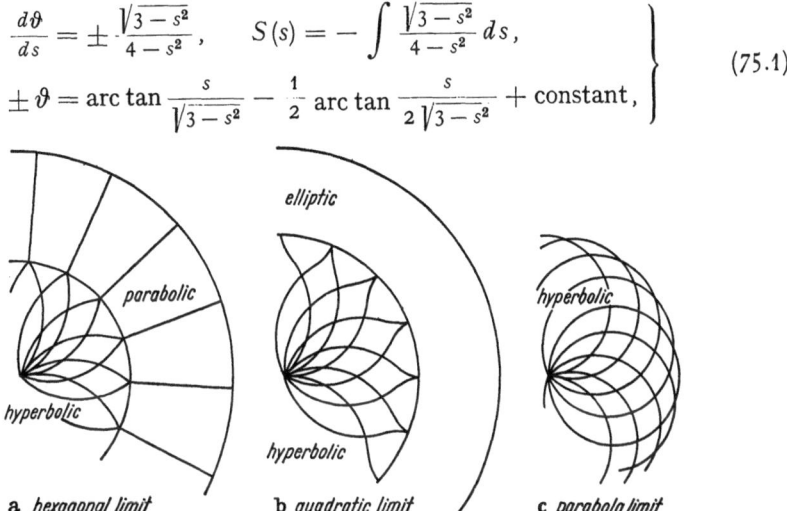

Fig. 26 a—c. Fixed characteristics for the three yield conditions of Fig. 20: (a) hexagonal limit; (b) quadratic limit; (c) parabola limit.

as the equations of the fixed characteristics in the stress plane and at the same time the compatibility conditions in the physical plane. The upper sign holds along the ξ-line. Also $\tan^2\alpha = \dfrac{\sigma_2'}{\sigma_1'} = (\sqrt{12-3s^2}+s)/(\sqrt{12-3s^2}-s)$, and

$$\tan\alpha = \pm \frac{\sqrt{12-3s^2}+s}{2\sqrt{3-s^2}}.$$
(75.2)

The points $s = \pm\sqrt{3}$ are parabolic: for $s = +\sqrt{3}$ we have $\sigma_1 = 2\sigma_2$ and $\alpha = 90°$; for $s = -\sqrt{3}$, $\sigma_2 = 2\sigma_1$ and $\alpha = 0°$; for both, $\beta = 0$, where β is the angle of Γ^+ or Γ^- with the radius vector in the stress graph. The hyperbolic part extends from $s = -\sqrt{3}$ to $s = +\sqrt{3}$, the elliptic part through $\sqrt{3} < |s| < 2$. Fig. 26b shows the Γ-curves in polar coordinates after von Mises[1].

Next we compute σ and τ using (74.2″) and (72.1):

$$\sigma = \sigma_1 \sin^2\alpha + \sigma_2 \cos^2\alpha = \frac{1}{2k}(\sigma_1 \sigma_2' + \sigma_2 \sigma_1')$$
$$= \frac{1}{2k}(\sigma_1\sigma_2)' = \frac{2k}{3}(s^2-1)' = \frac{4k}{3}s,$$
$$\tau = \frac{2k}{3}\sqrt{3-s^2}.$$

---

[1] R. v. Mises: Reissner Anniversary Volume. Ann. Arbor, Michigan 1949. In rectangular coordinates s, ϑ a Γ-line is shown in H. Geiringer, Advances, p. 238.

Hence
$$\sigma = \frac{4k}{3} s, \quad \tau = \frac{2k}{3}\sqrt{3 - s^2}, \quad \sigma^2 + 4\tau^2 = \frac{16k^2}{3}; \tag{75.3}$$

thus the yield condition appears as an ellipse in the σ, τ-plane[1].

β) *Hexagonal condition* (TRESCA, ST. VENANT). The problem is hyperbolic for $|s| < 1$; then $S(s) = -\frac{s}{2}$. The Γ-characteristics in the stress-graph are ordinary spirals $\pm \vartheta = \frac{s}{2} + $ constant, or straight lines, in polar or rectangular coordinates s, ϑ, respectively. In the parabolic region the unique set of characteristics consists of radii through O' (Fig. 26a). In the hyperbolic region $\alpha = 45°$, $\varphi^\pm = \vartheta \pm 45°$ and the characteristics are orthogonal[2].

γ) *Parabola condition* (Fig. 26c). The problem is hyperbolic everywhere, with the exception of the two points $s = \pm a$, where $hg = 0$ [see (76.2)]. We find

$$\begin{aligned}
\frac{d\vartheta}{ds} &= \pm(a^2 - s^2)^{-\frac{1}{2}}, \quad \vartheta = \pm \arcsin\frac{s}{a} + \text{constant}, \\
-S &= \arcsin\frac{s}{a}, \quad \sin S = -\frac{s}{a}, \\
\tan \alpha &= \sqrt{\frac{a+s}{a-s}}, \quad \cos 2\alpha = -\frac{s}{a}, \quad S = 90° - 2\alpha, \\
1 - \frac{\alpha'}{S'} &= \frac{3}{2}, \quad A(S) = \frac{\sin 2\alpha}{1 - \alpha'/S'} = \frac{2}{3a}\sqrt{a^2 - s^2} = \frac{2}{3}\cos S = \frac{2}{3}\cos(\xi + \eta).
\end{aligned} \tag{75.4}$$

We may see that Eqs. (74.18), (74.19) become here comparatively simple.

δ) *Coulomb condition*. The problem is hyperbolic since, as seen in (72.3), $\sigma'_1 \sigma'_2 = k^2 \cos^2 \Phi > 0$, if $\varphi < 90°$. From Sect. 74: $\alpha = 45° - \Phi/2$, and

$$\begin{aligned}
\varphi^- &= \vartheta - 45° + \Phi/2, \quad \varphi^+ = \vartheta + 45° - \Phi/2, \\
\sqrt{gh} &= 2(1 + s\tan\Phi), \\
S &= -\int \frac{ds}{\sqrt{gh}} = -\frac{1}{2}\cot\Phi \log(1 + s\tan\Phi).
\end{aligned} \tag{75.5}$$

Also from (74.4) we see that

$$\pm \tau = k + \sigma \tan \Phi \tag{75.6}$$

is the equation of the yield locus in the σ, τ-system. The characteristics, C^+, C^- are called *failure lines* and we see that $\vartheta \pm S$ are constant along these lines[3]. The S in (75.5) tends towards $-s/2$ as $\alpha \to 0$. The Γ-curves in the stress graph are logarithmic spirals.

Since $\alpha' = 0$, $A(S) = \cos \Phi = $ constant, we see from (74.15) that $h_1 = -R_1$, $h_2 = -R_2$. The two Eqs. (74.19) become

$$\cos^2 \Phi \frac{\partial^2 m_i}{\partial \xi \partial \eta} - m_i = 0, \quad i = 1, 2, \tag{75.7}$$

and the integration problem is similar to that in plane strain, if $\xi/\cos \Phi$ and $\eta/\cos \Phi$ are now considered as independent variables.

---

[1] A yield condition in the σ, τ-plane, considered as envelope of Mohr circles, is called by MANDEL "courbe intrinsèque".

[2] We have seen that the assumptions of isotropy and orthogonality lead to $\dot{\varepsilon}_1 + \dot{\varepsilon}_2 = 0$, and the yield condition is of the form as in (63.8). If isotropy is not required, the most general plane yield condition with orthogonal characteristics is of the form $\tau = f[(\sigma_x - \sigma_y)]$, where f is a function of one variable, as pointed out by R. SAUER: Z. angew. Math. Mech. 29, 274—279 (1949).

[3] References given by R. T. SHIELD: Ref. 1, p. 372.

Thus e.g.
$$m_1 = c\, e^{c\frac{\xi}{\cos \Phi} + \frac{1}{c}\frac{\eta}{\cos \Phi}}, \quad m_2 = e^{c\frac{\xi}{\cos \Phi} + \frac{1}{c}\frac{\eta}{\cos \Phi}}$$
is a particulur solution of (75.9) and new solutions can be derived by addition, integration with respect to a parameter, etc.

If, as proposed by DRUCKER and PRAGER[1], the theory of plastic potential is used in this problem we obtain from (73.1')

$$\begin{aligned}\dot\varepsilon_x : \dot\varepsilon_y : \dot\gamma &= (\cos 2\alpha - \cos 2\vartheta) : (\cos 2\alpha + \cos 2\vartheta) : -\sin 2\vartheta \\ &= (\sin \Phi - \cos 2\vartheta) : (\sin \Phi + \cos 2\vartheta) : -\sin 2\vartheta.\end{aligned} \quad (75.8)$$

The problem is almost as simple as that of plane strain, since Φ has a constant value. Consider the Eqs (73.11); now, $d\varphi^- = d\varphi^+ = d\vartheta$, $2\alpha = 90° - \Phi$, and

$$\left.\begin{aligned}dv_1 &= (v_2 \sec \Phi - v_1 \tan \Phi)\, d\vartheta, \\ dv_2 &= (-v_1 \sec \Phi + v_2 \tan \Phi)\, d\vartheta\end{aligned}\right\} \quad (75.9)$$

along a C^- and C^+, respectively.

ε) SOKOLOVSKY[2] introduces the yield condition

$$\frac{\sigma_1 - \sigma_2}{2K} = \sin\left(\frac{\sigma_0}{K} - \frac{\sigma_1 + \sigma_2}{2K}\right), \quad (75.10)$$

where σ_0 and K are constants, and

$$2\sigma_0 - \pi K \leq (\sigma_1 + \sigma_2) \leq 2\sigma_0. \quad (75.11)$$

It is easily seen that for this yield condition $F_1 F_2 \leq 0$: the stress problem is hyperbolic. In the σ, τ-plane this condition represents a cycloid; certain mathematical simplifications result which allow explicit solutions (see [26], p. 315 seq.). Such hyperbolic yield conditions, which depend on $(\sigma_1 + \sigma_2)$, may be considered as a generalization of the plane strain yield condition in the sense of SCHLEICHER[3] and VON MISES[4]; the parabola-condition considered in this and previous sections, which is also of this type, has the outspoken purpose to approximate the quadratic condition for plane stress $\sigma_1^2 + \sigma_2^2 - \sigma_1 \sigma_2 = \text{constant}$.

c) Remarks on integration. Examples.

76. On integration[5]. If for a specific yield condition a solution $m_1(\xi, \eta), m_2(\xi, \eta)$ of (74.18) has been found, then $h_1 = m_1/\sqrt{\tau}$, $h_2 = m_2/\sqrt{\tau}$, where $\tau = \tfrac{1}{2}(\sigma_1 - \sigma_2) \sin 2\alpha$ is a known function of $(\xi + \eta)$. Then, along a ξ-line

$$dx = h_1 \cos(\vartheta - \alpha)\, d\xi, \quad dy = h_1 \sin(\vartheta - \alpha)\, d\xi$$

with $\vartheta = \eta - \xi$, and α a known function of $(\eta + \xi)$; similar relations hold along an η-line. Hence

$$\left.\begin{aligned}dx &= \frac{1}{\sqrt{\tau}}[m_1 \cos(\vartheta - \alpha)\, d\xi + m_2 \cos(\vartheta + \alpha)\, d\eta], \\ dy &= \frac{1}{\sqrt{\tau}}[m_1 \sin(\vartheta - \alpha)\, d\xi + m_2 \sin(\vartheta + \alpha)\, d\eta],\end{aligned}\right\} \quad (76.1)$$

and $x = x(\xi, \eta), y = y(\xi, \eta)$ can be found by quadratures.

---
[1] D. C. DRUCKER and W. PRAGER: Quart. J. Appl. Math. **10**, 157—165 (1952).
[2] V.V. SOKOLOVSKY: J. Appl. Math. Mech. **13** (1949).
[3] F. SCHLEICHER: Z. angew. Math. Mech. **6**, 199—216 (1926).
[4] R. v. MISES: Z. angew. Math. Mech. **5**, 147—149 (1925).
[5] The very great mathematical difficulties which result from the combination of a *mixed problem* with *non-linearity* are well known in compressible flow theory. Here, in plasticity theory, these difficulties are exactly the same.

General integration procedures (as discussed for "plane strain") for finding solutions of (74.18) are available in the case of the Coulomb condition and to a certain degree in the case of the parabola yield condition. We add a few general remarks, due to MANDEL (Ref. 5 on p. 365) regarding the integration of (74.18).

Assume that we know two particular solutions of (74.18), say m_1, m_2 and γ_1, γ_2. Then $m_1 \gamma_1 d\xi + m_2 \gamma_2 d\eta$ is *a total differential*, as can be verified immediately. Now, conversely, consider the expressions

$$m_1 \delta_1 d\xi + m_2 \delta_2 d\eta, \qquad \gamma_1 \delta_1 d\xi + \gamma_2 \delta_2 d\eta,$$

each of which is assumed to be a total differential. If then m_1, m_2 and γ_1, γ_2 are independent solutions, it is easily seen then δ_1, δ_2 are likewise solutions. We know e.g. from (76.1) that $\frac{m_1}{\sqrt{\tau}} \cos(\vartheta - \alpha) d\xi + \frac{m_2}{\sqrt{\tau}} \cos(\vartheta + \alpha) d\eta$ is a complete differential for any m_1, m_2. It follows that

$$\frac{1}{\sqrt{\tau}} \cos(\vartheta - \alpha) = \delta_1, \qquad \frac{1}{\sqrt{\tau}} \cos(\vartheta + \alpha) = \delta_2 \tag{76.2}$$

is a solution of (74.18). We shall consider this example below.

It is useful to consider also *the net of principal trajectories*. Substituting in (71.9') for dx and dy the expressions (76.1), we obtain

$$-m_1 d\xi + m_2 d\eta = 0, \quad \text{differential equation of } u\text{-lines}, \tag{76.3}$$

$$m_1 d\xi + m_2 d\eta = 0, \quad \text{differential equation of } v\text{-lines}. \tag{76.4}$$

The expressions on the left are not exact differentials[1]; one can, however, find an integrating factor $\lambda(\xi, \eta)$ for the first, and $\mu(\xi, \eta)$ for the second expression. It may easily be verified that

$$\lambda = C e^{-\Sigma}, \quad \mu = C e^{+\Sigma}, \quad \Sigma = \int \frac{dS}{A(S)}, \tag{76.5}$$

where C is a constant and $A = \dfrac{\sin 2\alpha}{1 - (\alpha'/S')}$, as introduced in Sect. 74 are such integrating factors. Hence

$$du = e^{\Sigma}(m_1 d\xi + m_2 d\eta), \quad dv = e^{-\Sigma}(-m_1 d\xi + m_2 d\eta) \tag{76.6}$$

are now total differentials, and $dv = 0$ ($du = 0$) is the differential equation of the u-lines (the v-lines). By a brief computation we then obtain MANDEL's elegant result (l.c., p. 91):

$$u = \tfrac{1}{2} e^{\Sigma}(E_2 - E_1), \quad v = \tfrac{1}{2} e^{-\Sigma}(E_2 + E_1), \tag{76.7}$$

where E_1 and E_2 are solutions of (74.18), and $\dfrac{\partial E_1}{\partial \vartheta} = m_1$, $\dfrac{\partial E_2}{\partial \vartheta} = m_2$. The orthogonal trajectories have the equations $u(\xi, \eta) = $ constant, (v-lines), $v(\xi, \eta) = $ constant, (u-lines).

We note that $m_1 = m_2 = e^{\Sigma}$ as well as $m_1 = -m_2 = e^{-\Sigma}$ are particular solutions of (74.18).

As an example we now consider the particular solution of (74.18):

$$E_1 = \frac{1}{\sqrt{\tau}} \sin(\vartheta - \alpha), \qquad E_2 = \frac{1}{\sqrt{\tau}} \sin(\vartheta + \alpha). \tag{76.8}$$

Accordingly [cf. also Eqs. (76.2)] we investigate the particular solution

$$m_1 = \frac{1}{\sqrt{\tau}} \cos(\vartheta - \alpha), \qquad m_2 = \frac{1}{\sqrt{\tau}} \cos(\vartheta + \alpha). \tag{76.2'}$$

Carrying out the calculation, explained at the beginning of this section, we find

$$\left. \begin{array}{l} x = \displaystyle\int \frac{dS}{2\tau} + \frac{1}{4\tau} \sin 2\alpha \cos 2\vartheta, \\[4pt] y = \hphantom{\displaystyle\int \frac{dS}{2\tau} + {}} \dfrac{1}{4\tau} \sin 2\alpha \sin 2\vartheta, \end{array} \right\} \tag{76.9}$$

---

[1] With the notation of Sect. 71: $dy \cos\vartheta - dx \sin\vartheta = dX - Y\,d\vartheta$, $dx \cos\vartheta + dy \sin\vartheta = dY + X\,d\vartheta$, where dX and dY are exact differentials.

which gives x and y in terms of s and ϑ or rather S and ϑ. Let us verify: We expect to find e.g.

$$\frac{\partial x}{\partial \xi} = \frac{m_1}{\sqrt{\tau}} \cos(\vartheta - \alpha) = \frac{1}{\tau} \cos^2(\vartheta - \alpha) = \frac{\partial x}{\partial S} - \frac{\partial x}{\partial \vartheta};$$

now

$$\frac{\partial x}{\partial \vartheta} = -\frac{1}{2\tau} \sin 2\alpha \sin 2\vartheta,$$

and

$$\frac{\partial x}{\partial S} = \frac{1}{2\tau} - \frac{1}{4\tau^2} \frac{\partial \tau}{\partial S} \sin 2\alpha \cos 2\vartheta + \frac{1}{2\tau} \cos 2\alpha \cos 2\vartheta \frac{d\alpha}{dS}.$$

Using the formulas (74.6'), we obtain indeed the expected result.

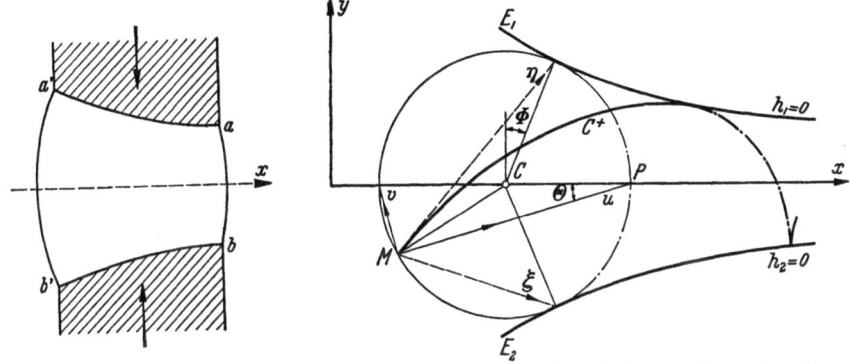

Fig. 27. Plastic mass between curved plates (MANDEL).

Fig. 28. Compression of a plastic mass between two plates (MANDEL).

The lengthy computation which leads to the result (76.9) can be omitted if we use the following simple remark, also due to MANDEL: If a total differential is of the form

$$d\Pi = A(S) B(\vartheta) dS + C(S) D(\vartheta) d\vartheta$$

then Π will be obtained, ω being a constant, as

$$\Pi = \omega C(S) B(\vartheta) + \text{constant},$$

as may be verified immediately. Now for the m_1, m_2 of (76.2) we have indeed, using (76.1):

$$dx = \frac{m_1}{\sqrt{\tau}} \cos \varphi^- d\xi + \frac{m_2}{\sqrt{\tau}} \cos \varphi^+ d\eta$$

$$= \frac{1}{\tau} \left[\cos^2(\vartheta - \alpha) \frac{dS - d\vartheta}{2} + \cos^2(\vartheta + \alpha) \frac{dS + d\vartheta}{2} \right]$$

$$= \frac{1}{2\tau} [dS + \cos 2\alpha \cos 2\vartheta \, dS - \sin 2\alpha \sin 2\vartheta \, d\vartheta],$$

$$dy = \frac{1}{2\tau} [\cos 2\alpha \sin 2\vartheta \, dS + \sin 2\alpha \cos 2\vartheta \, d\vartheta].$$

Here for dx, with $A = \cos 2\alpha$, $B = \cos 2\vartheta$, $C = \sin 2\alpha$, $D = -\sin 2\vartheta$, the above situation obtains, with $\omega = \frac{1}{2}$, and we find the first Eq. (76.9) and similarly, the second one.

In this example the lines of constant S (or s), the *isobars*, are the circles

$$(x-a)^2 + y^2 = \frac{1}{16\tau^2} \sin^2 \alpha, \quad \text{where} \quad a = \int \frac{dS}{2\tau}. \tag{76.10}$$

The radius equals $\sin 2\alpha/4\tau = 1/4R$, where $R = \tau/\sin 2\alpha$ is the radius of MOHR's circle.

The lines aa' and bb' (Fig. 27) which separate the plastic and the rigid material must have characteristic directions. They are (Fig. 28) actually envelopes of one family of characteristics and loci of cusps of the other family. Such lines, *limit lines* E_1 and E_2 (see general theory, Sect. 77 seq.), are the images in the physical plane of the lines $m_1=0$ and $m_2=0$. In our example $m_1=\tau^{-\frac{1}{2}}\cos(\vartheta-\alpha)=0$ for $\vartheta-\alpha=90°$ and $m_2=0$ for $\vartheta+\alpha=90°$. The first one is the envelope of characteristics C^+, and the C^- have cusps there forming the constant angle $90°$ with the envelope; the second is the envelope of the C^- and locus of the cusps of the C^+. These same curves E_1 and E_2 are also envelopes of the *isobars* (lines of constant s) and of the *isoclines* (lines of constant ϑ), while the principal stress lines have cusps at the limit lines. The solution corresponds to the physical problem of a plastic mass pressed between two completely rough plates, gliding in the direction of divergence; the traces which represent the walls must be similar to the envelopes. More details on this example may be found in MANDEL's book, p. 123 seq.

In the particular case $\alpha=$ constant the two envelopes E_1 and E_2 are straight lines which make the angle 2Φ with each other (Sect. 74). For $\alpha=0°$ we have parallel lines and the characteristics are cycloids.

II. Singular solutions and various remarks[1].

a) Limit line singularities and branch line singularities.

77. Limit line singularities. We have noticed before that the transition from a solution $x=x(s,\vartheta)$, $y=y(s,\vartheta)$ to the inverse $s=s(x,y)$, $\vartheta=\vartheta(x,y)$ is impossible at points or along lines where the Jacobian $\dfrac{\partial(x,y)}{\partial(s,\vartheta)}$ vanishes. Such a line will be called *limit line*. We found such lines in the preceding example.

We have to consider singularities of two kinds: *limit type singularities* and *branch type singularities*. The first type is characterized by the vanishing of a determinant like $D=\dfrac{\partial(x,y)}{\partial(s,\vartheta)}$ with stress graph coordinates in the denominator, physical plane coordinates in the numerator; the second by $d=0$, where $d=D^{-1}$. (Many other Jacobians are equivalent to D in this respect, e.g. $\dfrac{\partial(u,v)}{\partial(s,\vartheta)}$, or in a hyperbolic problem $\dfrac{\partial(x,y)}{\partial(\xi,\eta)}$ etc.; for instance

$$\frac{\partial(u,v)}{\partial(s,\vartheta)}=\frac{\partial(u,v)}{\partial(x,y)}\frac{\partial(x,y)}{\partial(s,\vartheta)}=\frac{\partial(x,y)}{\partial(s,\vartheta)}, \text{ or } \frac{\partial(x,y)}{\partial(\xi,\eta)}=\frac{2}{S'}\frac{\partial(x,y)}{\partial(s,\vartheta)}.$$

Hence these last two are equivalent to each other wherever S' is neither zero nor infinite.) Roughly speaking, the first type of singularity prevents the transition from the stress graph to the physical plane, the second that from the physical plane to the stress graph. Using Eqs. (71.8), (73.7) and (74.9) we find by a straight-forward computation that

$$D=\frac{\partial(u,v)}{\partial(s,\vartheta)}=\frac{1}{h}\frac{\partial u}{\partial\xi}\frac{\partial u}{\partial\eta}=-\frac{1}{g}\frac{\partial v}{\partial\xi}\frac{\partial v}{\partial\eta}. \tag{77.1}$$

We can deal briefly with the case, not important in the present connection, of an *elliptic region*[2], where $gh<0$. Using Eqs. (71.6) we obtain

$$d=\frac{\partial(s,\vartheta)}{\partial(x,y)}=\frac{\partial(s,\vartheta)}{\partial(u,v)}=g\left(\frac{\partial\vartheta}{\partial v}\right)^2-h\left(\frac{\partial\vartheta}{\partial u}\right)^2. \tag{77.2}$$

---

[1] Here, to a certain extent, the theory precedes the applications (in contrast to gas dynamics where the development of the theory was prompted by the knowledge of examples). At this moment, I cannot mention examples in our field which exhibit the various singularities. However, the presentation and classification of limit line and branch line singularities may help workers to clarify their problem. (E.g. an eminent author like SOKOLOVSKY fuses together limit lines and lines of stress discontinuity.)

[2] The parabolic case, where $gh=0$ or $gh\to\infty$ will be partly considered in Sect. 78. An s-value for which g or h or both vanish will in general designate an exceptional point of the yield limit. E.g., in the quadratic limit $g=h=0$ for $s=\pm 2$. This is the absolute maximum

Here both terms have the same signs, since $gh<0$, and therefore d can vanish only if $\frac{\partial \vartheta}{\partial u} = \frac{\partial \vartheta}{\partial v} = 0$. Then, also $\frac{\partial s}{\partial u} = \frac{\partial s}{\partial v} = 0$ from (71.6). Clearly, such a singularity must be isolated.

In a similar way consider $D = \frac{\partial(u,v)}{\partial(s,\vartheta)} = g\left(\frac{\partial u}{\partial s}\right)^2 - h\left(\frac{\partial v}{\partial s}\right)^2$. These two terms can vanish only [see (71.9'') and (71.8)] if $\frac{\partial X}{\partial s} = 0$, $\frac{\partial Y}{\partial s} = 0$, $\frac{\partial X}{\partial \vartheta} - Y = 0$, $\frac{\partial Y}{\partial \vartheta} + X = 0$. A brief computation shows that these imply that $\frac{\partial x}{\partial s} = 0$, $\frac{\partial y}{\partial s} = 0$, $\frac{\partial x}{\partial \vartheta} = 0$, $\frac{\partial y}{\partial \vartheta} = 0$, hence again an isolated singularity.

We now consider the main case, that of a *hyperbolic region*, and, using (74.13),

$$J = \frac{\partial(x,y)}{\partial(\xi,\eta)} = h_1 h_2 \sin 2\alpha. \qquad (77.3)$$

Assume that $\alpha \neq 0°$, $\alpha \neq 90°$; since $\tan \alpha = \sqrt{g/h}$ this can be excluded if neither g nor h is zero or infinite.

Consider now in the ξ, η-plane the locus $h_1(\xi,\eta) = 0$ for a given stress distribution, and in particular the mapping onto the x, y-plane in the neighborhood of this line[1]. Consider a point p with coordinates ξ_0, η_0, where $h_1(\xi_0, \eta_0) = 0$, $(\partial h_1/\partial \eta)_{\xi_0, \eta_0} \neq 0$. Then, there is a curve $\eta = g(\xi)$, with $\eta_0 = g(\xi_0)$, on which $h_1(\xi, \eta) = 0$. We call it the *critical curve* and denote it by l_1, and we call its image L_1 in the physical plane the *limit line*. A point P at which $h_1 = 0$ is called a *limit point*.

Consider a point m on l_1 where *both* $\partial h_1/\partial \xi$ and $\partial h_1/\partial \eta$ are *different from zero*. Hence l_1 does not have the ξ-direction at m, and, using (74.16) and $\alpha \neq 0°, \neq 90°$, we see that $h_2 \neq 0$ at m. Then (74.13) shows that at M (image of m) the L_1 has the η-direction, i.e. L_1 is tangent to the C^+ at M. The same conclusion holds for any curve C through M whose image c does not have the ξ-direction at m; all such curves C are tangent to the L_1 at M.

For example, the images in the ξ, η-plane of the lines of constant s, the isobars, or the lines of constant ϑ, the isoclines, do not have the *exceptional direction*, the ξ-direction, at m. In fact, e.g. for the latter, $d\vartheta = 0$ and $d\eta = \frac{1}{2}(dS + d\vartheta) = \frac{1}{2} dS \neq 0$ if $Sd = ds/\sqrt{hg} \neq 0$; similar conclusions hold for the isobars.

If, however, *the curve c has the ξ-direction at m*, that means $d\eta/d\xi = f'(\xi) = 0$, we can no longer conclude that C has the η-direction at M. *In this case C has a cusp at M*. In fact,

$$\frac{dx}{d\xi} = \frac{\partial x}{\partial \xi} + f'(\xi) \frac{\partial x}{\partial \eta}, \qquad \frac{dy}{d\xi} = \frac{\partial y}{\partial \xi} + f'(\xi) \frac{\partial y}{\partial \eta},$$

and since both $\partial x/\partial \xi$ and $\partial y/\partial \xi$ vanish because $h_1 = 0$, and since $f'(\xi) = 0$, both $dx/d\xi$ and $dy/d\xi$ are zero at M; it can also easily be shown that not both second derivatives $d^2x/d\xi^2$, $d^2y/d\xi^2$ vanish. It follows that the C^- has a cusp at M, its

---

of s, where $\sigma_1 = \sigma_2$. The circle $s = 2$ plays the role of the maximum circle $q = q_{max}$ in compressible flow; it delimits the region of possible plastic flow (for $|s| > 2$ the σ_i become imaginary while the yield condition is formally satisfied). This boundary $|s| = 2$ is not of great interest.

[1] Our presentation reproduces to a great extent the excellent study of MANDEL, l.c., Chap. V. We discuss, however, several features not considered by MANDEL. Cf. also R. v. MISES, Mathematical theory of compressible fluid flow. New York: Academic Press 1958, particularly Art. 19. SOKOLOVSKY considers envelopes of characteristics; he calls them "lines of rupture"; this same rather misleading notation is used by him for what we call stress-discontinuity lines.

tangent making the angle 2α with the direction of L_1. Since the principal trajectories bisect the angles of the C^-- and the C^+-directions, neither of them is tangent to the L_1 at M; hence both their images must have the ξ-direction at m (see Fig. 29) and therefore they both must have cusps at M. All these results have their counterparts for points of an L_2, along which $h_2 = 0$. We review: *Consider in the ξ, η-plane the locus $h_1(\xi, \eta) = 0$, the critical curve l_1, and points of l_1 at which $\partial h_1/\partial \xi \neq 0$, $\partial h_1/\partial \eta \neq 0$. Its image L_1, the limit line, is the envelope of the C^+, of the isobars, and of the isoclines; it is the locus of cusps of the C^- and of the principal stress trajectories* (Fig. 29).

We call *stream lines* such lines which have at every point the direction of $\boldsymbol{v}$. This direction will be known only after the whole problem has been solved. In general

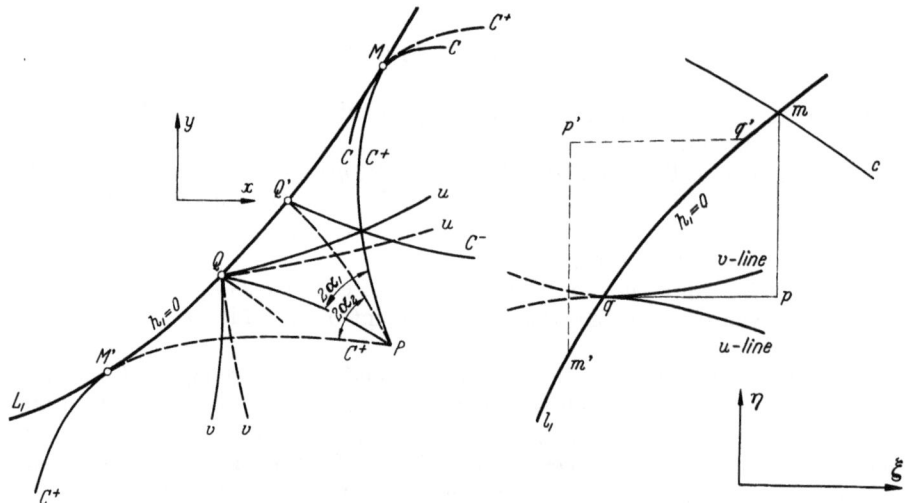

Fig. 29. Limit line and critical curve.

[see Eqs. (73.11)] a stream line will not coincide with (part of) a characteristic. At a point M of L_1, where a streamline does not have the C^+-direction, its image will have the ξ-direction and therefore the streamline has a cusp at L_1.

In general, two solutions meet at the limit line: A point P in the neighborhood of L_1 in the x, y-plane is the image of two points p, p' in the ξ, η-plane. In fact, through P pass two C^+-lines which touch the L_1 at M and M', respectively, with corresponding m and m' on l_1. On the η-line through m lies p; on that through m' lies p'. The points p, p' are on opposite sides of the l_1 since η varies in the opposite sense along PM and along PM'. Likewise two C^--lines PQ and PQ' pass through P making angles $2\alpha_1$ and $2\alpha_2$ with the C^+-characteristics PM and PM', respectively. The point q corresponding to Q lies on the ξ-line through p and on l_1, while q' lies on the ξ-line through p' on l_1. This correspondence may be discussed in more detail.

Let us compute $\partial s/\partial u$, $\partial s/\partial v$, viz. the changes of pressure in a principal direction. We now write ω [as in Eq. (71.17)] rather than s to avoid confusion with line elements and $\dfrac{\partial}{\partial s_u}$ for $\dfrac{\partial}{\partial u}$, $\dfrac{\partial}{\partial s_v}$ for $\dfrac{\partial}{\partial v}$. Then:

$$\frac{\partial \omega}{\partial s_1} = \frac{\partial \omega}{\partial s_u} \cos \alpha - \frac{\partial \omega}{\partial s_v} \sin \alpha, \qquad \frac{\partial \omega}{\partial s_2} = \frac{\partial \omega}{\partial s_u} \cos \alpha + \frac{\partial \omega}{\partial s_v} \sin \alpha.$$

There follow:
$$\frac{\partial \omega}{\partial s_u} = \frac{1}{2\cos\alpha}\left(\frac{\partial \omega}{\partial s_1} + \frac{\partial \omega}{\partial s_2}\right), \qquad \frac{\partial \omega}{\partial s_1} = \frac{\partial \omega}{\partial \xi}\frac{d\xi}{ds_1} = \frac{1}{h_1}\frac{d\omega}{dS} = \frac{1}{h_1 S'},$$

and therefore
$$\frac{\partial \omega}{\partial s_u} = \frac{1}{2\cos\alpha \cdot S'}\cdot\left(\frac{1}{h_1}+\frac{1}{h_2}\right), \qquad \frac{\partial \omega}{\partial s_v} = \frac{1}{2\sin\alpha \cdot S'}\left(-\frac{1}{h_1}+\frac{1}{h_2}\right). \tag{77.4}$$

Thus, grad ω = grad $s \to \infty$ at a limit line L_1 or L_2 unless $S' \to \infty$ and the same holds for grad ϑ, in agreement with the fact that the curvatures of the stress trajectories were seen to have cusps at L_1 and L_2.

78. Limit line singularities. Continuation. So far we assumed that at the point(s) m, where $h_1 = 0$, both $\partial h_1/\partial \xi \neq 0$, $\partial h_1/\partial \eta \neq 0$. If $h_1 = 0$ and $\partial h_1/\partial \eta = 0$, it follows from (74.16) that $h_2 = 0$ — a so-called *double limit point* $h_1 = h_2 = 0$ — which appears as a point of intersection of an L_1 and an L_2; this is an isolated point, which we do not discuss further.

Next, assume $h_1 = 0, \partial h_1/\partial \eta \neq 0$, $\partial h_1/\partial \xi = 0$, $\partial^2 h_1/\partial \xi^2 \neq 0$. Then, the l_1 has the ξ-direction and the Jacobian J changes sign there. Writing $h_1(\xi, \eta) = 0$ in the form $\eta = g(\xi)$, we find $g'(\xi) = -\dfrac{\partial h_1}{\partial \xi}\Big/\dfrac{\partial h_1}{\partial \eta} = 0$ at m while $g''(\xi) \neq 0$. At M, the image of m, $\dfrac{dx}{d\xi} = \dfrac{\partial x}{\partial \xi} + \dfrac{\partial x}{\partial \eta} g'(\xi) = 0$, and likewise $dy/d\xi = 0$, and it can be seen that $d^2x/d\xi^2$, $d^2y/d\xi^2$ can not both vanish if $h_2 \neq 0$. Hence it is seen that the L_1 has a cusp at M

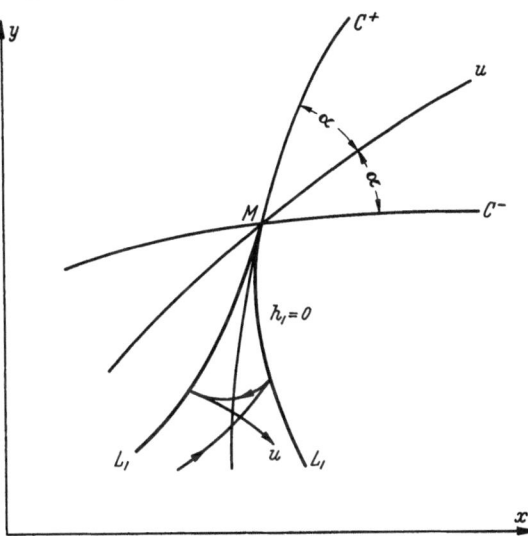

Fig. 30. Limit line with cusp.

(Fig. 30). Further study again shows that the C^+ touches the L_1 at M; this means it has the cusp tangent there, but it does not have itself a cusp; the same holds for any curve through M whose image does not have the ξ-direction at m, e.g. for the isoclines and isobars. On the other hand, the image of the C^- has, of course, the ξ-direction at m, and the same is true of the images of both principal trajectories, since neither has the L_1-direction at M. The image of both trajectories contacts the l_1, which has an extremum at this point m [1].

It remains briefly to describe the limit type singularities along a "sonic" line. (In Sect. 93 there is an example.) We mean by that a line s = constant along which gh changes sign [2]. *We consider the case* $gh \to \infty$, $S' \to 0$ and even more specifically only the two cases $g \to \infty$, h finite, and $h \to \infty$, g finite. (Cf. footnote 2, p. 384, for some remarks on $g = 0$, $h = 0$.) In our quadratic condition e.g., $S' = \sqrt{3-s^2}/(4-s^2) \to 0$ for $s = \pm\sqrt{3}$; for $s = +\sqrt{3}$, $g \to \infty$, h finite, $\alpha = 90°$ while for $s = -\sqrt{3}$, g finite, $h \to \infty$, $\alpha = 0°$. Of course, a sonic line, where *necessarily* $S' = 0$ since S = constant, is in general not a limit line [3].

We consider now the case $g \to \infty$, h finite, hence $S' = 0$, and $\alpha = 90°$, from (73.3). With a view to Eq. (77.1) we consider $\partial u/\partial \xi$ and $\partial u/\partial \eta$. Using (71.8)

---
[1] Unfortunately I do not know simple examples which would exhibit the various limit line singularities. Such examples would be very useful.
[2] We tend to avoid the word "parabolic" in order to escape confusion with terms derived from the "parabola condition".
[3] It seems to me that MANDEL assumes that his "isobare limite"—our sonic line—is always a limit line. This is not true.

we find, as before

$$\frac{\partial u}{\partial \xi} = \frac{\partial u}{\partial s}\frac{\partial s}{\partial \xi} + \frac{\partial u}{\partial \vartheta}\frac{\partial \vartheta}{\partial \xi} = \frac{1}{g}\frac{\partial v}{\partial \vartheta} \cdot \frac{1}{S'} - h\frac{\partial v}{\partial s},$$

$$\frac{\partial u}{\partial \eta} = \frac{\partial v}{\partial s}\frac{\partial s}{\partial \eta} + \frac{\partial v}{\partial \vartheta}\frac{\partial \vartheta}{\partial \eta} = \frac{1}{g}\frac{\partial v}{\partial \vartheta} \cdot \frac{1}{S'} + h\frac{\partial v}{\partial s}.$$

Now, $\frac{1}{gS'} = -\sqrt{\frac{h}{g}}$ tends to zero as $g \to \infty$ (and hence $\frac{1}{gS'}\frac{\partial v}{\partial \vartheta} \to 0$ unless $\frac{\partial v}{\partial \vartheta}$ would tend strongly towards infinity, which we exclude). Hence at a parabolic point as considered ($g \to \infty$, h finite)

$$\frac{\partial u}{\partial \xi} = -h\frac{\partial v}{\partial s}, \qquad \frac{\partial u}{\partial \eta} = h\frac{\partial v}{\partial s}.$$

Laying aside the case $D \to \infty$ which corresponds to branch type singularities we define as *ordinary sonic point* (in contrast to limit point and branch point) a point where $g \to \infty$, h finite, $\frac{\partial v}{\partial s} \neq 0$; hence, both $\frac{\partial u}{\partial \xi}$, $\frac{\partial u}{\partial \eta}$ are different from zero and $D \neq 0$. Likewise the sonic point where $h \to \infty$, g finite is *ordinary* if $\frac{\partial u}{\partial s} \neq 0$ and consequently $D \neq 0$ by (77.1). We return to the case $g \to \infty$, h finite.

A new type of limit point and limit line (different from a sonic point of at L_1 or of an L_2) which we shall call *sonic limit point* is characterized by $\alpha = 90°$ (viz. $g \to \infty$, h finite) and $\frac{\partial v}{\partial s} = 0$. At such a point either $\frac{\partial v}{\partial s} = 0$, $\frac{\partial v}{\partial \vartheta} \neq 0$, or $\frac{\partial v}{\partial s} = 0$, $\frac{\partial v}{\partial \vartheta} = 0$. The first case can clearly happen along a whole arc of curve which we then denote as *sonic limit line*, L_l. In the second case the point is isolated[1] (or it may present some more complicated singularity, like the intersection of an L_l and an L_1, etc.).

From $\partial u/\partial \xi = 0$, $\partial u/\partial \eta = 0$, the L_l under consideration is a v-line. Hence at every point of the L_l the u-direction is normal to the L_l, and since $\alpha = 90°$ it follows that both the C^+ and the C^- are enveloped by the L_l (see Fig. 31). Since at the L_l, $\partial v/\partial s = 0$, $\partial v/\partial \vartheta \neq 0$, $dv = \frac{\partial v}{\partial s}ds + \frac{\partial v}{\partial \vartheta}d\vartheta = \frac{\partial v}{\partial \vartheta}d\vartheta$, we see that if $dv = 0$ (u-line), also $d\vartheta = 0$, and vice versa, hence the line $\vartheta =$ constant, the isocline, is likewise normal to the L_l. More generally, and in analogy to our study of the L_1 and L_2 we conclude, since at the L_l: $du = 0$, $\partial v/\partial s = 0$, $\partial v/\partial \vartheta \neq 0$, that

$$dx = \frac{\partial x}{\partial v}dv = \frac{\partial x}{\partial v}\frac{\partial v}{\partial \vartheta}d\vartheta = -\sin\vartheta\frac{\partial v}{\partial \vartheta}d\vartheta,$$

$$dy = \frac{\partial y}{\partial v}dv = \frac{\partial y}{\partial v}\frac{\partial v}{\partial \vartheta}d\vartheta = \cos\vartheta\frac{\partial v}{\partial \vartheta}d\vartheta.$$

Thence

$$\frac{dy}{dx} = -\cot\vartheta, \qquad \text{if } d\vartheta \neq 0. \tag{78.1}$$

Therefore (Fig. 31) *an element of any curve in the stress graph on which $d\vartheta \neq 0$ maps onto an element in the x, y-plane which has the second principal direction.* (This is true for the sonic limit line itself *which therefore everywhere points in the v-direction and is itself a v-line*). The characteristics make the angle $\alpha = 90°$ with the u-direction, by which they are separated, hence are tangent to the v-direction and one is the continuation of the other.

---

[1] It can be shown that a limit line L_1 which is everywhere "sonic" is not possible. Hence our L_l is the only sonic limit line.

Since the sonic limit line is a v-line and since along it $s=$ constant, it follows that $\partial s/\partial v=0$ along the L_t and $\partial s/\partial u \to \infty$ across the L_t as may be seen from the first Eq. (77.4), with $h_1=h_2$, $S' \to 0$.

If, for an element in the stress graph, $d\vartheta=0$ (exceptional direction), its map in the physical plane does not have the v-direction, and it can be shown that it has a cusp at the L_t. This holds true for the u-line and for the isocline. For the latter, clearly, $d\vartheta=0$, and the former, being normal to the v-line in the x, y-plane must have the exceptional (radial) direction in the stress graph.

In the second case, where $h \to \infty$, g finite, $\alpha=0°$, Eq. (78.5) is replaced by $dy/dx = \tan\vartheta$, if $d\vartheta \neq 0$. The role of first and second principal directions is interchanged.

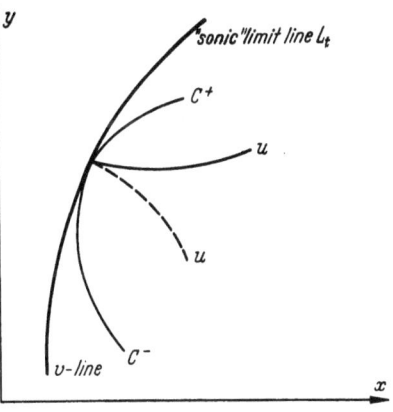

Fig. 31. "Sonic" limit line, L_t.

79. Branch line singularities. Here, $J = h_1 h_2 \sin 2\alpha$ cannot become infinite unless either h_1 or h_2 becomes infinite. The loci $h_1(x,y)=\infty$ and $h_2(x,y)=\infty$ are called branch lines B_1 and B_2; there is no analogue to the sonic limit line.

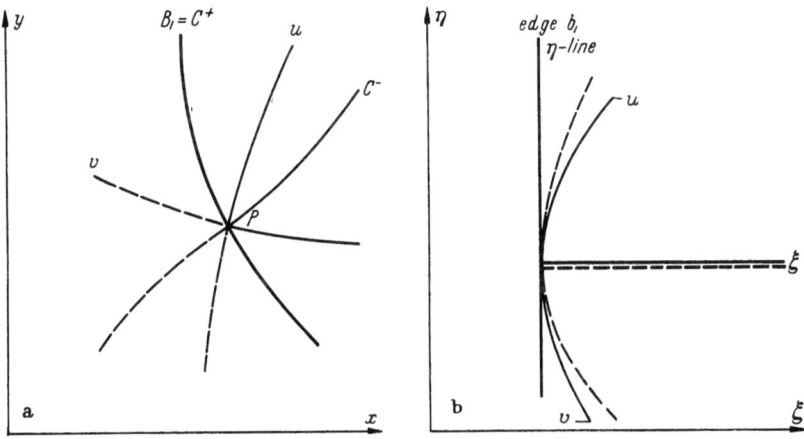

Fig. 32 a and b. Branch line in (a) physical plane and (b) characteristic plane.

We may achieve formal similarity with preceding considerations by putting

$$k_1 = (h_1 \sin 2\alpha)^{-1}, \quad k_2 = (h_2 \sin 2\alpha)^{-1}, \quad \alpha \neq 0°, 90°. \tag{79.1}$$

Interchanging (x, y) with (ξ, η) in Eqs. (74.13) we obtain

$$\begin{aligned}
\frac{\partial \xi}{\partial x} &= k_1 \sin \varphi^+, & \frac{\partial \eta}{\partial x} &= -k_2 \sin \varphi^-, \\
\frac{\partial \xi}{\partial y} &= -k_1 \cos \varphi^+, & \frac{\partial \eta}{\partial y} &= k_2 \cos \varphi^-, \\
j &= (h_1 h_2 \sin 2\alpha)^{-1} = k_1 k_2 \sin 2\alpha.
\end{aligned} \tag{79.2}$$

The image in the ξ, η-plane, or in the s, ϑ-plane, of the branch line B_1 of the x, y-plane is called *the edge*, b_1 (Fig. 32). If $k_1 = 0$, then $d\xi = 0$, hence $\xi =$ constant. Therefore b_1 is an η-line in the ξ, η-plane. It follows, just as before, that all lines in the x, y-plane, with the exception of such lines in the x, y-plane as have the C^--direction (the exceptional direction) at the point of intersection with the B_1, appear in the ξ, η-plane as tangents to the straight vertical line, b_1, the edge. Among these are *the principal trajectories, and b_1 is an envelope of them*. It can be concluded, as before, that a line which has ξ-direction at the intersection with the b_1, and in particular a ξ-line, must have a cusp there. Since, however, *the ξ-lines are straight horizontal lines, they return at the edge*. For a C^--line,

$$d(\vartheta - \alpha) = \frac{\partial(\vartheta - \alpha)}{\partial \xi} d\xi = -\left(1 + \frac{\alpha'}{S'}\right) d\xi,$$

and since $d\xi$ changes sign at an intersection with b_1, the same must be true for $d(\vartheta - \alpha)$ (unless[1], quite exceptionally, $1 + \frac{\alpha'}{S'} = 0$); hence *each C^- has an inflection point at its intersection with the B_1*. Since the isobars and isoclines are the lines $\eta \pm \xi =$ constant, they intersect the vertical edge at $\pm 45°$ and are not tangent to it. Therefore, in the flow plane *they have the exceptional, the C^--direction, at B_1*. Hence, at each point of B_1, *the C^-, the isobar and the isocline touch each other, while the trajectories bisect the angles between B_1 and the inflection tangent of the C^- and cross the B_1 without singularity*. The direction of the vector grad s is at B_1 perpendicular to the C^--direction, and the same holds for grad ϑ.

A branch line has physical reality. In fact, its characteristic property of dividing two plastic regions in which the same stress occurs at different points is nothing out of the ordinary. However, one and the same single-valued stress-graph solution cannot represent such a distribution. Hence a representation of such a distribution, e.g. in form of a series, must break down at the edge. To save space we do not review the above properties of branch lines.

We repeat: A limit line L_1 is the image of $h_1(\xi, \eta) = 0$, where a single-valued solution of the linearized stress-graph equations (or characteristic-plane equations) is assumed; $J(\xi, \eta) = 0$ along the L_1; a separate discussion is needed at "sonic" points. A branch line B_1, defined by $k_1(x, y) = 0$, corresponds to a single-valued solution of the original non-linear physical plane equations, and along it $j(x, y) = 0$, $j = J^{-1}$.

b) Simple waves.

In the next sections we shall study so-called *simple waves* where $j(x, y) = 0$, not along a line only, but in a two-dimensional region, and where the transition to the stress graph, which in Sect. 71 formed our starting point, is no longer possible.

80. Definition. Simple waves is the name of an important type of solution of the basic equations. They play a very useful role in building up solutions to boundary-value problems in plasticity[2]. They may be introduced in various essentially equivalent ways. We ask, e.g., for a stress distribution in which the lines $s =$ constant, the isobars, coincide with the isoclines; we call such lines w-lines. Each w-line is thus mapped, by definition, onto one point (s, ϑ) of the stress graph. We assume that these points do not all coincide, i.e. that the solution does not merely represent a region of constant state. Hence the whole set of w-lines, or the whole region R in the x, y-plane covered by w-lines, is mapped onto one line Λ of the stress graph.

---

[1] For the Coulomb condition $1 + \frac{\alpha'}{S'} = 1$; for the parabola condition $1 + \frac{\alpha'}{S'} = \frac{1}{2}$.

[2] In compressible fluid flow the term "Prandtl-Meyer solution" is used; the terms "fan" and "lost solution" are likewise in use.

The existence of the line Λ in the stress graph implies the existence of a relation between s and ϑ, and, as a consequence, the vanishing throughout R of the Jacobian d of (77.2).

We conclude, just as in Sect. 77, that if $gh<0$, the Jacobian d can only vanish at isolated points, and we consider the case where there are real characteristics.

Among the lines crossing the w-lines there must be at least one set of characteristics C, and the image of each of these must be on Λ. In fact, each point of such a line C must map onto a point of Λ. It follows that Λ is a Γ^+ or a Γ^- and that each characteristic of the other set —each C^- if Λ is a Γ^+—is mapped onto a single point of Λ. Thus, the w-lines form this second set of characteristics. Since on each of them both ϑ and s, and therefore α as a function of s, are constant, it follows that $\vartheta \mp \alpha$, that is, the slope of each w-line, is constant. Hence, the w-lines are straight.

Since the whole region R covered by the w-lines is mapped onto one characteristic, Γ^+ or Γ^-, the equation of this characteristic, $S \mp \vartheta =$ constant, is valid throughout R. Hence the definition: *A plane stress distribution is a simple wave solution if one set of characteristics consists of straight lines on each of which the stress tensor Σ is constant. The image of this region in the stress graph is an arc of a Γ-characteristic. If it is a Γ^+, the C^- are straight and $S - \vartheta$ has a constant value throughout R, and correspondingly in the case of a Γ^-.*

Throughout the simple wave region we have $d = \partial(s, \vartheta)/\partial(x, y) = 0$. Therefore, the interchange of variables (x, y) and (s, ϑ), so useful in general, is not possible here, and in this sense simple waves are "lost solutions" since they cannot be found as solutions of the linearized equations in the stress graph.

The simple wave pattern forms the transition between the general distribution, where a region of the x, y-plane is mapped onto an area of the stress graph, and the extremely degenerate "constant state", where an area of the x, y-plane maps onto a single point.

This is expressed in the theorem: *The region adjacent to a domain of constant state is either another domain of constant state or a simple wave region.* In other words, a region of constant state, which maps into a single stress graph point, cannot be directly adjacent to a "general" state of stress. A simple wave must form the link between them[1].

A simple wave can connect any uniform hyperbolic state Σ_1 with another uniform hyperbolic state Σ_2, provided either $S + \vartheta$ or $S - \vartheta$ has the same value in both states. By combining a Γ^+-wave and a Γ^--wave and inserting a uniform state between the two, a given final state Σ_2 can be reached, in general and in many cases in two ways.

Simple waves, denoted in our Chap. F as "degenerate solutions", have already been considered by the pioneers in our field. In the present, more general, case they have been studied by GEIRINGER[2] and MANDEL[3]. Cf. also remarks in [26], p. 323.

81. Simple waves. Continuation. An individual wave may be specified in several ways, e.g. by giving *a certain characteristic Γ_0^- as the image of the whole "minus" wave*; and, in addition, giving in the x, y-plane *a family of straight lines to represent the C^+*. If these C^+ have a point in common, we speak of a *centered wave*. The stress distribution for the wave follows immediately, if merely φ, the angle which a straight C^+ (a straight C^-) makes with the positive x-axis, is

---

[1] See proof in H. GEIRINGER: Advances, p. 258.
[2] GEIRINGER, Advances, pp. 257—270.
[3] H. MANDEL, reference 5 on p. 367 of this article.

given for each straight characteristic. Then for any specific yield condition the two relations hold

$$S(s) \mp \vartheta = \text{constant}, \quad \vartheta \mp \alpha = \varphi, \tag{81.1}$$

where, as in all similar formulas, the upper (lower) sign holds for a Γ^+ (a Γ^-) wave, and the "constant", as well as the φ, are given. Hence along each single C^- (C^+), s and ϑ are determined by the two equations (81.1).

Next, consider the computation of the *principal trajectories* and of the *cross characteristics*, i.e. the other set of characteristics. For that, we need also the equation of the particular family of lines which form the straight characteristics. An adequate way to specify this set of straight lines is by giving e.g. *one* cross characteristic or *one* principal trajectory. In fact, knowing one trajectory—say a u-line—in the x, y-plane we know ϑ at each point of this line; then, in case of a minus wave, $S(s)=$ constant $-\vartheta$ determines s and, therefore, α; hence, at each point of the given u-line, the direction $\vartheta+\alpha$ of the C^+ through this point is known. Similarly, if one C^- is given, $\vartheta-\alpha$ is known along it and $S(s)+\alpha(s)=$ constant $-(\vartheta-\alpha)$ provides s; then α, and finally $\alpha+\vartheta$ follow at all points of the C^-.

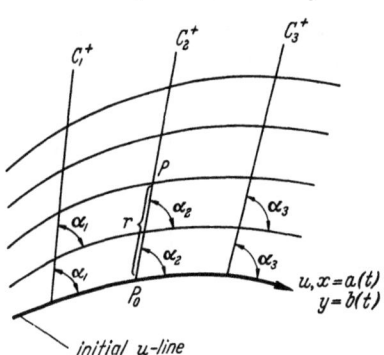

Fig. 33. Straight characteristics and cross characteristics in (backward) simple wave.

Consider now the first case, where one u-trajectory is given: $x=a(t)$, $y=b(t)$, $db/da = \tan\vartheta$; t is a parameter. Consider a minus wave with the C^+ as straight characteristics. For a point $P(x, y)$ on the C^+ passing through the point P_0 on the given initial trajectory: $x=a+r\cos\varphi^+$ and $y=b+r\sin\varphi^+$, where $r=P_0P$ (Fig. 33). If we consider, a, b, r and $\varphi^+ = \varphi$ as functions of t, the equation of the trajectory through P is

$$(db + r\cos\varphi\, d\varphi + dr\sin\varphi)\cos\vartheta - (da - r\sin\varphi\, d\varphi + dr\cos\varphi)\sin\vartheta = 0,$$

and since

$$db\cos\vartheta - da\sin\vartheta = 0,$$

it follows, with $\varphi-\vartheta=\alpha$, that

$$r\, d\varphi \cos\alpha + dr \sin\alpha = 0,$$

or

$$\frac{r\, d\varphi}{dr} = -\tan\alpha. \tag{81.2}$$

We note that this is the same equation which can be written down immediately for a centered wave, with r, φ as polar coordinates. For the cross characteristics, always assuming a Γ_0^--wave, we find in the same way

$$\frac{r\, d\varphi}{dr} = \tan 2\alpha. \tag{81.3}$$

Note that from (81.1), writing $2\eta_0$ for "constant", we have

$$\alpha = \varphi - \vartheta = \varphi - 2\eta_0 + S(s), \quad \text{or} \quad \alpha(s) - S(s) = \varphi - 2\eta_0.$$

Thus φ is determined in terms of s; likewise s and $\alpha(s)$ are determined in terms of φ. Examples follow in the next section.

We can proceed with the integration of (81.3) (cf. MANDEL, l. c., p. 57): In $dr/r = -\cot 2\alpha\, d\varphi$ we substitute $d\varphi = d\vartheta + d\alpha$ and $d\vartheta = -dS$. Then:

$$\frac{dr}{r} = \cot 2\alpha\,(dS - d\alpha).$$

Using (74.6') and (74.4) we obtain

$$\frac{dr}{r} = -\cot 2\alpha \frac{d\sigma}{2\tau} = -\frac{d\tau}{d\sigma}\frac{d\sigma}{2\tau} = -\frac{d\tau}{2\tau} \tag{81.4}$$

which gives

$$r\sqrt{\tau} = \text{constant}. \tag{81.5}$$

To obtain the above equation in r, φ we need τ in terms of φ for the specific yield condition. Exactly the same relation (81.5) results in case of a Γ^+-wave. The cross characteristics have cusps on the envelope of the rectilinear characteristics.

In the case of a centered wave, the cross characteristics are similar to each other and r in (81.5) is the radius vector. If $\alpha = \text{constant}$ they are logarithmic spirals (see Sect. 69 for $\alpha = 45°$).

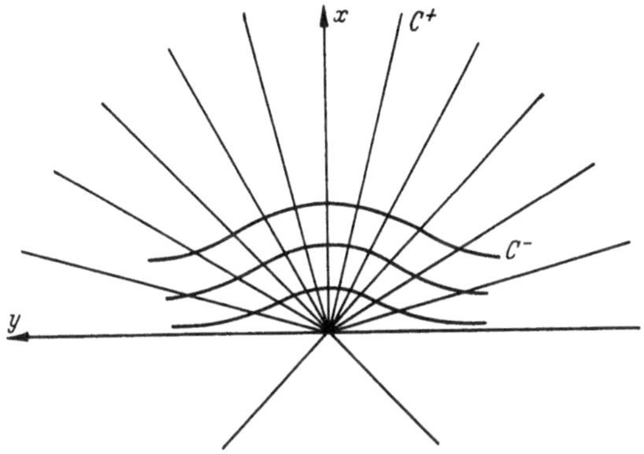

Fig. 34. Quadratic limit. Complete centered wave. Straight characteristics and cross characteristics.

82. Simple waves for particular yield conditions. α) *Quadratic yield condition.* Consider the plus wave

$$\vartheta = S(s) \tag{82.1}$$

where

$$S(s) = -\arctan\frac{s}{\sqrt{3-s^2}} + \frac{1}{2}\arctan\frac{s}{2\sqrt{3-s^2}}. \tag{82.1'}$$

The straight C^- makes the angle $\varphi^- = \varphi = \vartheta - \alpha$ with the x-axis.

Using (75.1), (75.2);

$$\varphi = \varphi^- = \vartheta - \alpha = -\arctan\frac{s}{\sqrt{3-s^2}} + \frac{1}{2}\arctan\frac{s}{2\sqrt{3-s^2}} - \arctan\frac{\sqrt{12-3s^2}+s}{2\sqrt{3-s^2}}.$$

To simplify we introduce

$$t = \frac{s}{\sqrt{3-s^2}}, \qquad s^2 = \frac{3t^2}{1+t^2}. \tag{82.2}$$

Then

$$\varphi^- = -\arctan t + \frac{1}{2}\arctan\frac{t}{2} - \arctan\left(\frac{1}{2}\sqrt{4+t^2} + \frac{t}{2}\right). \tag{82.3}$$

It may be verified that

$$\arctan\frac{\sqrt{4+t^2}+t}{2} - \frac{1}{2}\arctan\frac{t}{2} = 45°.$$

Therefore

$$\varphi = -\arctan t - 45°, \quad t = -\tan(\varphi + 45°), \tag{82.4}$$
$$s = -\sqrt{3}\sin(\varphi + 45°), \quad \vartheta = S(s).$$

As s goes from $-\sqrt{3}$ to zero, to $+\sqrt{3}$, t goes from $-\infty$, to zero, to $+\infty$, ϑ from 45°, to zero, to $-45°$, and φ from 45°, to $-45°$, to $-135°$. Hence, in a *complete wave*, φ rotates through 180° (Fig. 34).

The image of the Γ_0^-:

$$\vartheta = -S(s) - 45° = \arctan t - \frac{1}{2}\arctan\frac{t}{2} - 45°, \tag{82.5}$$

gives with $\varphi^+ = \varphi = \vartheta + \alpha$,

$$\varphi = -45° + \arctan t - \frac{1}{2}\arctan\frac{t}{2} + \arctan\left[\frac{1}{2}\sqrt{4+t^2} + \frac{t}{2}\right] = \arctan t. \tag{82.5'}$$

Thus

$$\tan\varphi = t, \quad \varphi = \arctan t, \quad s = \sqrt{3}\sin\varphi. \tag{82.6}$$

As s runs from $-\sqrt{3}$ to $+\sqrt{3}$, φ increases from $-90°$ to $+90°$, and ϑ from zero to $90°$. We compute the *cross characteristics* for this last wave using (81.5). To express τ in terms of φ we use (75.3) and (82.6) and find

$$\sigma = \frac{4k}{3}s = \frac{4k}{\sqrt{3}}\sin\varphi, \quad \tau = \frac{2k}{\sqrt{3}}\cos\varphi. \tag{82.7}$$

Hence from (81.5)

$$r^2\cos\varphi = \text{constant}. \tag{82.8}$$

β) *Parabola limit.* With $t = s/a$ we consider the Γ^- wave, using (75.5)

$$\left.\begin{array}{l}\vartheta = -S(s) = \arcsin t, \quad \text{thus}\quad t = \sin\vartheta, \\[4pt] \tan\alpha = \sqrt{\dfrac{1+t}{1-t}} = \tan\left(45° + \dfrac{\vartheta}{2}\right).\end{array}\right\} \tag{82.9}$$

$$\alpha = \frac{\vartheta}{2} + 45°, \quad \varphi^+ = \vartheta + \alpha = \frac{3\vartheta}{2} + 45°. \tag{82.10}$$

Hence, with $\varphi^+ = \varphi$:

$$\vartheta = \tfrac{2}{3}(\varphi - 45°), \quad s = a\sin\vartheta. \tag{82.11}$$

Thus we know s and ϑ along each straight C^+ which makes the angle φ with the x-axis. Here the range of ϑ is $180°$ (from $-90°$ to $+90°$), that for φ is $270°$, from $-90°$ to $+180°$.

For the *cross characteristics* (Fig. 35a) we use (81.3):

$$\frac{r\,d\varphi}{dr} = -\tan 2\alpha = -\tan(\vartheta + 90°) = -\tan\frac{2\varphi + 180°}{3}, \tag{82.12}$$

$$r = r_0\left[\sin\frac{2\varphi + 180°}{3}\right]^{-\tfrac{3}{2}}. \tag{82.13}$$

The *principal stress* lines may likewise be found (Fig. 35b).

Velocities. We first ask whether *velocity distributions* exist *such that* $v =$ constant *along each straight characteristic*.

In other words, is it possible to prescribe initial values v_x, v_y along a non-characteristic curve K such that the above holds? Obviously, then the initial distribution along K is subject to a condition. In fact, consider the image $\overline{K}$ of K in the velocity plane. Since to each C^+ there corresponds only one point on $\overline{K}$ (on account of $v_x = $ constant, $v_y = $ constant on C^+), the $\overline{K}$ is the image of the whole simple wave region, hence also of all C^-. Therefore, [see (73.9)] along $\overline{K}$ the relation

$$\frac{dv_y}{dv_x} = -\cot(\vartheta - \alpha) \tag{82.14}$$

must hold. More explicitly: Through every point of K there passes a straight C^+ to which there belong constant values s, ϑ and an angle $(\vartheta - \alpha)$ which is given in terms of s or of ϑ, say $\cot(\vartheta - \alpha) = H(s)$. Suppose K given as $x = x(s)$, $y = y(s)$ and along it $v_x = v_x(s)$, $v_y = v_y(s)$; then $v'_x/v'_y = -H(s)$ must hold along K. It can be shown that if this condition holds

Sect. 82. Simple waves for particular yield conditions.

along K, and consequently in the whole region, there will exist a corresponding velocity distribution.

Now, consider the general case. Let us have a coordinate system consisting of the characteristics with the straight C^+ as η-lines, the cross characteristics

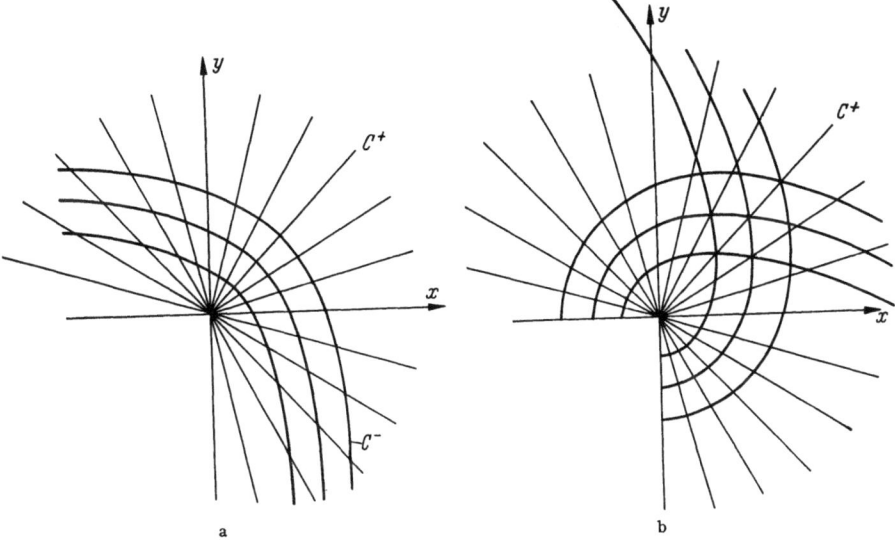

Fig. 35a and b. Parabola limit. Complete entered wave. (a) Straight characteristics and cross characteristics. (b) Lines of principal stress.

as ξ-lines. Since the angles $\varphi^-, \varphi^+, \alpha$ are all constant along the C^+, any of these may serve as ξ-coordinate. Choose $\xi = \varphi^+ = \varphi$. To define η we choose a certain C^+ with $\varphi = \varphi_0$, and on it a point O, and take for the η of an arbitrary point P the distance OP' where P' is the point of intersection of the C^- through P with the fixed C^+ (Fig. 36). Of course, the characteristic coordinates of the previous sections cannot be used here.

As before, v_1 and v_2 are the components of v in the directions of C^-, C^+ respectively. Consider Eqs. (73.10). The second gives $dv_2/dl^+ = 0$. The component of v in the direction of a straight C^+ is constant along it,

$$\frac{\partial v_2}{\partial \eta} = 0. \qquad (82.15)$$

The first Eq. (73.11) gives

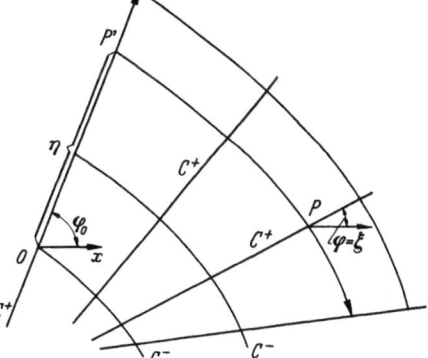

Fig. 36. Characteristic coordinates in simple wave.

$$\frac{\partial v_1}{\partial \varphi^-} = \frac{v_2 - v_1 \cos 2\alpha}{\sin 2\alpha}, \quad \text{or} \quad \frac{\partial v_1}{\partial \varphi^-} + v_1 \cot 2\alpha = \frac{v_2}{\sin 2\alpha}. \qquad (82.16)$$

For $\alpha = 45°$ this yields $\dfrac{\partial v_1}{\partial \xi} - v_2 = 0$. In general, since v_2 as well as α do not change along a C^+, we see: *Not only v_2, but also $\dfrac{\partial v_1}{\partial \varphi^-} + v_1 \cot \alpha$, remains constant along a C^+.*

To integrate (82.16) we observe that if v_2 is given along a curve which intersects the C^+, then v_2 is known everywhere. We use $\xi = \varphi^+ = \varphi$ as independent variable and have $\varphi^- = \varphi^-(\xi)$, $\alpha = \alpha(\xi)$, $v_2 = v_2(\xi)$. Putting

$$\cot 2\alpha \frac{\partial \varphi^-}{\partial \xi} = a(\xi), \qquad \frac{v_2}{\sin 2\alpha} \frac{\partial \varphi^-}{\partial \xi} = b(\xi),$$

Eq. (82.16) becomes the linear differential equation

$$\frac{\partial v_1}{\partial \xi} + a(\xi) v_1 - b(\xi) = 0$$

with the integral

$$v_1(\xi, \eta) e^{\int_{\xi_0}^{\xi} a \, dt} = \int_{\xi_0}^{\xi} b \, e^{\int_{\xi_0}^{\xi} a \, dt} d\xi + \psi(\eta), \qquad (82.17)$$

where

$$\psi(\eta) = v_1(\xi_0, \eta).$$

Consider a *Cauchy problem*. Along a non-characteristic curve K with equation $\eta = \eta(\xi)$ both v_1 and v_2 are given: $v_1 = g(\xi)$, $v_2 = h(\xi)$. We know v_2 along each C^+ intersecting K and we know v_1 from Eq. (82.17).

In a *characteristic initial-value problem* we e.g. assume that v_2 is given on an arc OC of the C^-: $\eta = \eta_0$, while v_1 is given on an arc OB of the straight C^+: $\xi = \xi_0$. We then know v_2 on each C^+ intersecting OC, we know v_1 from (82.17), and $\psi(\eta) = v_1(\xi_0, \eta)$. Further problems of this type are equally easy.

Thus, in a simple wave region we can determine explicitly both stresses and velocities.

c) Various remarks.

83. Remarks on the approximate solution of initial-value problems. In the Cauchy problem, along an arc K given by $x = x(t)$, $y = y(t)$, *with suitable regularity assumptions, values of s and ϑ are given in such a way that K nowhere has the characteristic direction*; that means that the slope of K nowhere is equal to either $\tan(\vartheta + \alpha(s))$ or to $\tan(\vartheta - \alpha(s))$. These data determine a solution on both sides in a neighborhood of K which is contained in the corresponding characteristic quadrangle. If, in addition, two components of $\boldsymbol{v}$ are given along K the velocity is uniquely determined in the characteristic quadrangle.

For an obvious (though crude) approximation procedure, consider on K the neighboring points 1, 2 (Fig. 37) and draw through them short rectilinear segments in the characteristic directions (knowing s and ϑ, we know $\vartheta \pm \alpha$). They have an intersection, 3, on the upper side of the arc K and another on the lower side. Considering all distances 12, 13, 23 as infinitesimal, and neglecting terms of higher order, this, in an obvious notation, amounts to

$$s_3 - s_1 = \pm (\sqrt{gh})_1 \cdot (\vartheta_3 - \vartheta_1), \qquad s_3 - s_2 = \mp (\sqrt{gh})_2 \cdot (\vartheta_3 - \vartheta_2). \qquad (83.1)$$

In both linear equations for s_3, ϑ_3 either the upper or the lower signs hold. The determinant $\pm[(\sqrt{gh})_1 + (\sqrt{gh})_2]$ is different from zero, since $\sqrt{gh}$ cannot change sign between 1 and 2, and both s_3, ϑ_3 can be evaluated. In this way, starting from a sequence of points on K, say between A and B, one can derive from the given values s, ϑ on AB the values along a second row of points, $A'B'$. Continuing in the same manner, s- and ϑ-values are eventually found for all lattice points within a curvilinear triangle ABC, where AC and BC are characteristics (of two different kinds)—provided the procedure does not break down earlier, which may happen if the direction of a cross line such as $A'B'$, $A''B''$, ... somewhere approaches a characteristic direction. However, due to the non-characteristic nature of AB and the continuity of all functions involved, this can happen only at a finite distance from AB. All these conclusions apply, of course, also to the triangle ABD on the lower side of AB.

Thus, while such a step procedure does not constitute an existence proof, it makes it clear that a solution exists in a neighborhood of AB. Besides, insofar as a solution exists in $ABCD$, it is uniquely determined there, and we can approximate it by this step procedure, unless the procedure breaks down in the manner explained above.

As a second case (Fig. 38), *consider data given along two intersecting lines AB, AC, one of them a characteristic.* Assume that AB is a minus characteristic and that the non-characteristic arc AC lies in the angular space between the minus characteristic AB and the positively directed plus characteristic through A. Values of s and ϑ along the C^- must be given in such a way that at each point its angle φ^- with the x-axis equals $\vartheta - \alpha(s)$ and such that the second Eq. (73.5) holds. It follows that, if the geometric shape of AB is given, we may prescribe only the value of either s or ϑ at *one* point of AB; then s, ϑ follow along AB.

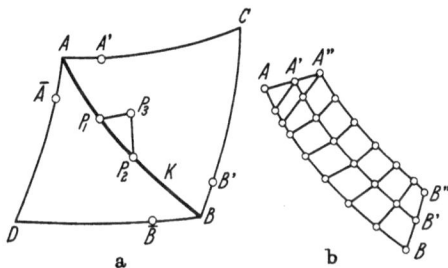
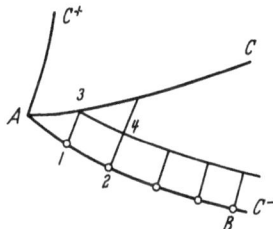

Fig. 37 a and b. Approximate solution of Cauchy problem. Fig. 38. Approximate solution of third or mixed problem.

We further suppose that either ϑ or s or a combination of both is given along the non-characteristic arc AC (*"third"* or *"mixed"* problem). From the data along AB, the initial elements of the plus characteristics at all points of AB can be derived, and we assume that they are plotted in the direction towards AC. If the point 1 is adjacent to A, the characteristic element through 1 will intersect the line AC in some point 3. From a given value at 3 and from $s_3 - s_1 = (\sqrt{gh})_1 (\vartheta_3 - \vartheta_1)$, the quantities ϑ_3, s_3 can be derived. This, then, enables us to find the beginning of the minus characteristic 34 through 3, and the above used compatibility relations, applied to the segments 34 and 24, give the values s_4, ϑ_4. In this way, step-by-step, the whole quadrangle $ABDC$, where CD and BD are characteristics, can be filled up by a net of points at which s and ϑ are known.

A slight modification of the procedure has to take place if AC is likewise a characteristic, here a C^+ (*second*, or *characteristic problem*). If we know that of two geometrically given intersecting arcs one is a C^-, the other a C^+, then neither ϑ nor s can be arbitrarily prescribed anywhere along these curves. Their values follow from the two characteristic conditions and the two compatibility conditions, with both s and ϑ at A being determined by the two characteristic conditions. The stepwise construction of the net inside the characteristic quadrangle $ABDC$ is analogous to the preceding one. The existence of a solution, however, is guaranteed only in a neighborhood of point A and, similarly, for the mixed problem.

In each case, the complete problem also calls for the determination of the velocities. Knowing $s = s(x, y)$, $\vartheta = \vartheta(x, y)$ at a point, we know $\alpha(x, y)$. We then may use the finite difference equivalent of the velocity equations (73.11), or perhaps (73.9), the choice dependent on what exactly are the initial data.

Returning to the stress problem, we recall that the knowledge of the net of characteristics in a region is equivalent to knowing the stress tensor. To construct this net we used the compatibility relations (73.5) in the form of finite difference equations and the characteristic conditions (73.3). This most direct procedure, essentially due to Massau[1], becomes a practical approximation procedure if the compatibility relations are used in the form (73.8), together with a tabulation of $S(s)$ and $\alpha(s)$ for a chosen yield condition. A first approximation thus found can then be improved by iteration.

We may arrange the approximation procedure also in the following not essentially different way. Consider, e.g. the characteristic initial-value problem where we know that two arcs OA, OB in the x, y-plane are parts of a ξ-line and an η-line, respectively. We assume that neither of them is straight. (If e.g. OB is straight, all C^+ characteristics in the simple wave region must be straight.) Consider a subdivision $\xi_0, \xi_1, \ldots \xi_m, \ldots$ on OA and $\eta_0, \eta_1, \ldots \eta_n, \ldots$ on OB and a corresponding lattice where a general nodal point has the coordinates (ξ_m, η_n) or, briefly, (m, n). We saw that, if we know that OA and OB are characteristics, we know both ϑ and s along them, as well as the constants ξ_0, η_0, where $\xi = \xi_0$ on OC, $\eta = \eta_0$ on OB. Then from Eqs. (74.9):

$$\vartheta_{mn} - \vartheta_{m0} - \vartheta_{0n} + \vartheta_{00} = 0, \tag{83.2}$$

and similarly,

$$S_{mn} - S_{m0} - S_{0n} + S_{00} = 0. \tag{83.3}$$

Hence, we know ϑ and S, i.e. ϑ and s, at all nodal points.

We still need the coordinates x_{mn}, y_{mn} in the x, y-plane to which the s_{mn}, ϑ_{mn} belong. To find these coordinates, our original independent variables, we need a step-by-step procedure. In the method explained at the beginning of this section, this procedure consisted in the actual drawing of characteristic segments. In the present method, it consists in setting down the equations

$$\left. \begin{array}{l} y_{mn} - y_{m-1,n} = \tan \dfrac{\varphi^+_{m-1,n} + \varphi^+_{m,n}}{2} \cdot (x_{mn} - x_{m-1,n}), \\[6pt] y_{mn} - y_{m,n-1} = \tan \dfrac{\varphi^-_{m,n-1} + \varphi^-_{m,n}}{2} \cdot (x_{mn} - x_{m,n-1}). \end{array} \right\} \tag{83.4}$$

These are two simultaneous linear equations for x_{mn}, y_{mn}, if $x_{m,n-1}, \ldots y_{m-1,n}$ are known. The $x_{m,0}$, $y_{m,0}$, $x_{0,n}$, $y_{0,n}$ are the known coordinates of the original subdivision points along OA and OB. The φ^+, φ^- are known at all nodal points since we know $s_{m,n}$, $\vartheta_{m,n}$. Again the procedure can be improved by iteration.

This and other more refined procedures, have been described by Hill ([10], p. 140) for plane strain. They can be generalized, to a certain extent, to the case of other yield conditions. A more geometric arrangement, using stress graph and velocity plane is also useful[2].

84. Summary remarks on some further problems. α) *Discontinuities.* The same principles as used in Sect. 70 may be applied to the general problem of this Chapter. Consider Eqs. (73.5). For a characteristic C^- the second Eq. (73.5) is the compatibility relation and the first the "remaining equation" (see Sect. 53). Since in this first equation both variables s and ϑ are differentiated across the C^- none of these variables can jump. *The stresses must be continuous across a characteristic.* Consider next the velocities and Eqs. (73.10), (73.11). For a C^- the first Eq. (73.11) is the compatibility equation and the second the "remaining equation". In this second equation v_2 is differentiated across the C^-, but v_1

---

[1] I. Massau: Mémoire sur l'intégration des équations aux dérivées partielles, Gand, Meya van Loos, 1899; reprinted as Edition du Centénaire, Mons, 1952.

[2] H. Geiringer: Advances, p. 250 seq.

is not. *Hence the tangential velocity v_1 may jump across C^- while the normal component must remain continuous,* and, of course, similarly for the C^+.

Hence stress discontinuities can arise only across non-characteristic lines. As in plane strain, it follows that *across such a discontinuity line d, which is assumed to have the y-direction at the point under consideration, σ_x, τ and v remain continuous, while σ_y changes abruptly*. It also follows, as in Sect. 61, that $\dot{E} = 0$ at every point of d. In fact

$$\dot{\varepsilon}_y^{(1)} = \lambda^{(1)} \left(\frac{\partial f}{\partial \sigma_y}\right)_1, \quad \dot{\varepsilon}_y^{(2)} = \lambda^{(2)} \left(\frac{\partial f}{\partial \sigma_y}\right)_2,$$

where f is the yield function, used as plastic potential, and the subscript of $\partial f/\partial \sigma_y$ means that $\sigma_y^{(1)}$ or $\sigma_y^{(2)}$ are used in $\partial f(\sigma_x, \sigma_y, \tau)/\partial \sigma_y$. These two values must be opposite in sign, just as in Sect. 61, and the conclusion remains the same.

In the case of COULOMB's condition [Eqs. (72.5) and (75.7)] the stress discontinuities have been studied by SHIELD[1]. He applies his results to the wedge which was also PRAGER's[2] first example. The discontinuities for plane stress, for VON MISES' yield condition, have been investigated by HILL[3].

β) *Plastic rigid boundary.* It has been shown by HILL ([*10*], p. 150) and by LEE[4] that in plane strain the boundary between plastic and rigid material must consist of slip lines. (Of course, it can also be an envelope of slip lines, a limit line.) The general problem, where slip lines and characteristics are not identical, is taken up by GEIRINGER[5]. The essential point in HILL's argument ([*3*], p. 150) which is indeed quite general is the following one. Consider a non-characteristic curve AB along which plastic material is adjacent to rigid material. The velocity v across AB must be continuous, since in a hyperbolic region a jump of the velocities v_x, v_y, which are subject to two first-order partial differential equations, cannot occur across a non-characteristic curve. Hence, if the rigid material is at rest, the same must be true for the adjacent plastic material in the characteristic quadrangle defined by AB (the velocity problem is linear). Such a rigid block can then be separated from adjacent plastically deforming material only by characteristics. Hence in the general case, where slip lines and characteristics are not identical, the separation line is characteristic.

H. Boundary-value problems.

We consider in this last chapter a small number of problems in some detail with the aim of pointing out the approach, the methods, and the difficulties. More space has been given to elastic-plastic than to rigid-plastic problems since they are physically more significant.

I. Some elastic-plastic problems[6].

a) The torsion problem.

85. Fully elastic and fully plastic torsion. We consider the equilibrium of a cylindrical bar under axial moments, in other words, a bar which is twisted

---
[1] R. T. SHIELD: J. Math. Phys. **33**, 144—156 (1954).
[2] W. PRAGER: Courant Anniversary Volume, 289—299. New York 1948.
[3] R. HILL: J. Mech. Phys. of Solids **1**, 19—30 (1952).
[4] E. H. LEE: J. Appl. Mech. **19**, 97—103 (1952).
[5] H. GEIRINGER: Advances p. 253 seq. In a letter of Fall 1953, Professor HILL informed me that he had reached the result already in J. Mech. Appl. Math. **2** (1949); and he pointed out that the discussion in his work, [*10*], pp. 150/151 applies to other cases, apart from plane strain.
[6] For this chapter cf. also HOFFMAN and SACHS [*11*].

about an axis parallel to the generators by equal and opposite couples. We take the z-axis in the direction of the generators.

If, then, we treat the stress distribution for a bar with concave corners as an elastic problem, infinite stresses are obtained at the corners. We conclude that at these corners the yield limit has been reached and Hooke's law is no longer valid. Outside of a vicinity of the corners the material will be elastic (Fig. 39). The general case is that, for any cross section, the material will be plastic in some region and elastic in the remainder of the cross section. A main problem is to determine in a given case the boundary between the plastic and elastic regions; in addition, we wish to evaluate the stresses, displacements, and velocities in the respective regions. In our problem of plane shear, the only non-vanishing stresses are $\tau_{xz} = \tau_x$, $\tau_{yz} = \tau_y$, the components of the shear vector τ (the notation is different from that in Chap. E). Thus, $\sigma_x = \sigma_y = \sigma_z = \tau_{xy} = 0$, and

$$\frac{\partial \tau_x}{\partial x} + \frac{\partial \tau_y}{\partial y} = 0 \tag{85.1}$$

is the only equilibrium equation which does not vanish identically. It is satisfied by a *stress function* $\psi(x, y)$, where

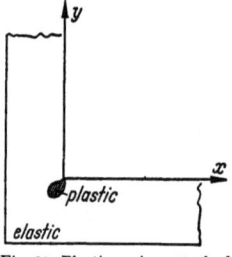

Fig. 39. Plastic region attached to concave corner of L-beam (Trefftz).

$$\frac{\partial \psi}{\partial y} = \tau_x, \qquad \frac{\partial \psi}{\partial x} = -\tau_y. \tag{85.2}$$

It follows that the *contour lines*, $\psi = $ constant (the "streamlines" if we think in terms of an incompressible flow) *have everywhere the direction of* τ, and hence we call them *stress lines*, while the normals to the contour lines (which correspond to the potential lines) *have the direction of the vector* $\mathrm{grad}\,\psi$. Thus, if we denote by $\partial/\partial s$ and $\partial/\partial n$ differentiation in direction of and normal to the contour lines respectively:

$$\frac{\partial \psi}{\partial s} = 0, \quad \frac{\partial \psi}{\partial n} = \tau, \quad \left(\frac{\partial \psi}{\partial x}\right)^2 + \left(\frac{\partial \psi}{\partial y}\right)^2 = \tau^2. \tag{85.3}$$

Since the lateral surface of the bar is free of stress, the stress vector τ must be tangent to the boundary of the cross section. Hence this boundary is a stress line, $\psi = $ constant. This last condition is general no matter whether the problem is completely elastic or completely plastic or partly elastic, partly plastic. Thus

$$\psi = 0 \quad \text{on the boundary}. \tag{85.4}$$

Denote by A the area of the cross section. The twisting moment or *torque* required to produce the *twist per unit length* α is

$$T = \iint_A (x\,\tau_y - y\,\tau_x)\,dx\,dy = -\iint_A \left(x\frac{\partial\psi}{\partial x} + y\frac{\partial\psi}{\partial y}\right)dx\,dy = 2\iint_A \psi\,dx\,dy. \tag{85.5}$$

The last equality follows by partially integrating and using $\psi = 0$ on the boundary. All this holds true for elastic as well as plastic torsion.

In *the elastic region*[1] Hooke's law shows that $\varepsilon_x = \varepsilon_y = \varepsilon_z = \gamma_{xy} = 0$ and $\gamma_{xz} = \gamma_x$, $\gamma_{yz} = \gamma_y$ are the only non-vanishing strains. If u_x, u_y, u_z are the small displacements of a point (x, y, z), the deformation is described by

$$u_x = -\alpha y z, \quad u_y = \alpha x z, \quad u_z = \alpha w(x, y), \tag{85.6}$$

---

[1] In this Chapter the distinction between ε', ε'', and ε (see Sect. 43) is made only where it seems desirable for clarity.

where α is the twist per unit length and $w(x, y)$ is the *warping function* (see Saint Venant[1] and Nadai[2], as well as text books on elasticity theory). Hence by Hooke's law

$$\tau_x = G\alpha\left(-y + \frac{\partial w}{\partial x}\right), \qquad \tau_y = G\alpha\left(x + \frac{\partial w}{\partial y}\right), \tag{85.7}$$

and eliminating w the compatibility relation

$$\operatorname{curl} \boldsymbol{\tau} = \frac{\partial \tau_y}{\partial x} - \frac{\partial \tau_x}{\partial y} = 2G\alpha \tag{85.8}$$

results. Using (85.2) this becomes

$$\Delta \psi = -2\alpha G. \tag{85.9}$$

The boundary-value problem (85.9), (85.4) determines ψ uniquely if the boundary is known.

Prandtl[3] has given a well-known intuitive interpretation of this problem. Suppose a thin membrane (soap film) is fastened along the contour C of the fully elastic cross section and loaded by a uniform surface pressure, p, proportional to α. If S is the surface tension of the membrane, its vertical displacement, u, satisfies the equation $\frac{\partial^2 u}{\partial x^2} + \frac{\partial^2 u}{\partial y^2} = -\frac{p}{S}$, and $u = 0$ along C. For $p = 2\alpha GS$ the right side of the above equation is $-2\alpha G$ and comparison with (85.9) shows that we may put $\psi = u$: the membrane surface reproduces the elastic stress surface. According to (85.3) the resultant shearing stress $|\boldsymbol{\tau}|$ is equal, at any point (x, y), to the greatest slope of the membrane, and its direction is that of the contour line through this point.

Next assume that all over the cross section *the material is plastic*. We then obtain for both von Mises' and Tresca's conditions

$$\tau_x^2 + \tau_y^2 = \tau_0^2 = k^2, \quad \text{or} \quad \tau = \sqrt{\tau_x^2 + \tau_y^2} = k, \tag{85.10}$$

and at the boundary (85.4) is valid[4]. By (85.2)

$$|\operatorname{grad} \psi| = \left[\left(\frac{\partial \psi}{\partial x}\right)^2 + \left(\frac{\partial \psi}{\partial y}\right)^2\right]^{\frac{1}{2}} = \frac{\partial \psi}{\partial n} = k. \tag{85.11}$$

If ϑ denotes the angle of $\boldsymbol{\tau}$ with the x-axis, Eq. (85.10) is satisfied by

$$\tau_x = k \cos \vartheta, \qquad \tau_y = k \sin \vartheta; \tag{85.12}$$

if this is substituted into (85.1)

$$\sin \vartheta \frac{\partial \vartheta}{\partial x} - \cos \vartheta \frac{\partial \vartheta}{\partial y} = 0 \tag{85.13}$$

results. Since by (85.11) *the gradient of the plastic stress surface is constant* this surface is a *surface of constant slope* (Böschungsfläche). Such surfaces have well-known interesting properties (see e.g. Nadai, l. c., p. 445). Denote by y' the slope of a stress line and by $\partial/\partial s$, $\partial/\partial n$ the same as in (85.3). Then

$$\frac{\partial \psi}{\partial s} = 0, \quad y' = -\frac{\partial \psi}{\partial x}\bigg/\frac{\partial \psi}{\partial y} = \tan \vartheta, \quad \frac{\partial \psi}{\partial n} = \tau = k \tag{85.14}$$

results, which is the same as in (85.3), except that now $\tau = k$. The contour lines have the direction of $\boldsymbol{\tau}$ everywhere, and, since $\partial \psi/\partial n = k$, they are *equidistant curves*.

---
[1] B. de Saint Venant: Mém. prés. par div. sav. Acad. Sci., math. et phys. **14**, 233—560 (1856).
[2] A. Nadai: Z. angew. Math. Mech. **3**, 442—454 (1923), see p. 448.
[3] L. Prandtl: Phys. Z. **4**, 758—759 (1903).
[4] For multiply-connected regions the analysis need be modified.

Next, using Eq. (85.13), we see that *the orthogonal trajectories of the stress lines are straight lines* of slope $-\cot\vartheta$ where ϑ, corresponds to the point of intersection with C. We call those straight trajectories *the normals*; along each of them τ remains the same; they are the lines of steepest descent, having the direction of the vector $\operatorname{grad}\psi$; the stress lines have the direction of $\operatorname{grad}\vartheta$.

From the first-order partial differential equation (85.13) we see that *the characteristics* have the slope $-\cot\vartheta$, hence they *coincide with the orthogonal trajectories*.

A physical construction of the surface of constant slope Σ belonging to a given cross section has been indicated by NADAI[1]. If the contour of the cross-section is cut out (physically), laid down horizontally and covered with a fine powder, the natural surface of constant slope will form above the cross-section. *The shape of Σ is independent of α.*

Fig. 40 shows the construction of the projection of Σ having a polygon as contour[2]. In case C consists of straight segments and arcs of circles, Σ consists of planes and conic surfaces.

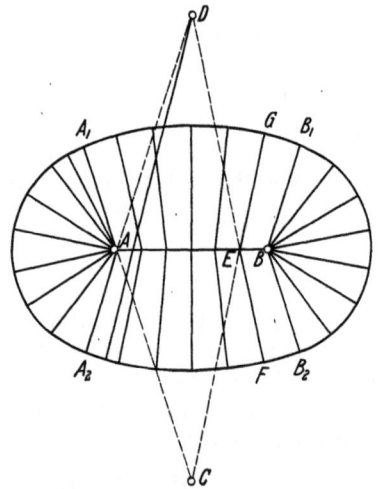

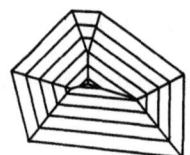

Fig. 40. Sandhill over polygon (NADAI).

Fig. 41. Fully plastic oval with characteristics and stress discontinuity line (PRAGER, HODGE).

Even if the contour C of the cross section has no corners, edges like those in Fig. 40 will in general appear. An instructive example is due to PRAGER and HODGE[3]. Fig. 41 shows an oval contour consisting of four circular arcs with centers A, B, C and D. The orthogonal trajectories, the characteristics, are therefore straight lines through these centers. All characteristics through points of the arc $\widehat{A_1 A_2}$ intersect in A, all through $\widehat{B_1 B_2}$ in B; each other point of AB is a point of intersection of two characteristics. Since at any point of any characteristic the stress vector τ is perpendicular to it, it follows that the direction of τ changes abruptly at a generic point E of AB. Thus, AB is a *stress discontinuity line*, as studied in Sects. 61, 70, etc. In general, the characteristics intersect on a line of stress discontinuity. Since the normal component of τ cannot jump, the discontinuity line must bisect the angle formed by the characteristics meeting at that line.

The strain rate $\dot{E}''$ in the fully plastic problem is given by VON MISES' relations

$$\dot{\varepsilon}_x'' = \dot{\varepsilon}_y'' = \dot{\varepsilon}_z'' = \dot{\gamma}_{xy}'' = 0, \quad \dot{\gamma}_{xz}'' = \lambda \tau_{xz}, \quad \dot{\gamma}_{yz}'' = \lambda \tau_{yz}. \tag{85.15}$$

---

[1] A. NADAI: Z. angew. Math. Mech. **3**, 442—454 (1923).
[2] See photos of "sandhills" in NADAI [*19*], p. 134 seq. For the case of concave corners, cf. also PRAGER and HODGE [*23*], p. 66.
[3] W. PRAGER and P. G. HODGE [*23*], p. 64 and W. PRAGER, Proc. 2nd U. S. Nat. Congr. Appl. Mech., Ann Arbor, Michigan 1954, pp. 21—32. Cf. also the fully plastic distribution in the oval cross-section similar to an ellipse in [*26*], pp. 126/127 and photographs p. 129.

It is seen, as in Sect. 61, that *the strain-rate tensor $\dot{E}''$ vanishes* at the discontinuity line. The characteristics drawn through an arc of the contour C will in general have an envelope and beyond this limit line they cannot be continued. In Prager and Hodge's example the limit line shrinks to the four points A, B, C, D. It may, however, be a curve inside the contour C.

W. Freiberger[1] has studied the fully plastic torsion of circular ring sectors. This problem is more general since the body is part of a torus rather than of a cylinder (Fig. 42). The characteristics are still the orthogonal trajectories of the stress lines in the x, y-plane and they intersect on a line of stress discontinuity which at every point bisects the angle of the characteristics. In this problem the characteristics as well as the discontinuity line are curved; $\dot{E}''$ vanishes at the discontinuity line.

86. Elastic-plastic torsion. It is natural to assume that in general a solution of the torsion problem will consist of an elastic core and one or several plastic

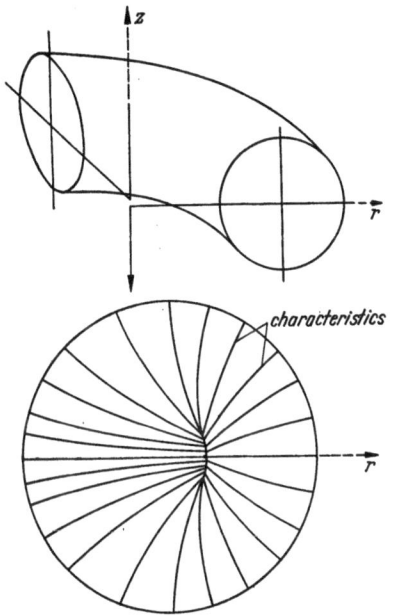

Fig. 42. Fully plastic torsion of circular ring sector with characteristics and stress discontinuity line (Freiberger).

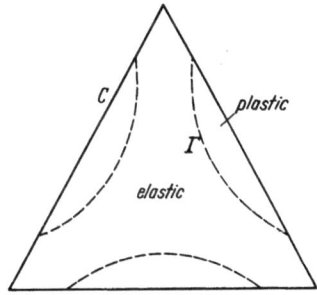

Fig. 43. Scheme of elastic-plastic triangle in torsion.

pieces, attached to the contour. For mathematical and physical reasons [see Eq. (85.9) and membrane analogy] yielding will start at the contour C. As the twisting moment is increased plastic zones spread out. The plastic-elastic boundary Γ is unknown. If α is small enough so that changes in the contour C can be neglected, we obtain the following model:

Above the given contour C the surface of constant slope, the "roof", has been erected—materially, in cardboard—corresponding to the given contour and the given slope k; above the base of the roof a stretched membrane is loaded by a pressure p, proportional to α. If this pressure reaches a certain value parts of the membrane will touch the surface of the roof from within; that means that along the lines of contact the slope of the membrane (which can never exceed k) has reached the slope k of the roof. The stress surface ψ is formed by the free parts of the membrane and by those parts which rest on the roof. The horizontal projection of those parts of the membrane which touch the roof gives the plastic region of the cross-section; below the free parts of the membrane the material

---

[1] W. Freiberger: Quart. Appl. Math. **14**, 259—264 (1956) and Commonwealth of Australia, Dept. of Supply, Aer. Res. Labor., Rept. SM 213 (1953). The fully plastic twisting of a circular ring sector of arbitrary cross-section is also considered by A. J. Wang and W. Prager: J. Mech. Phys. of Solids **3**, 169—175 (1955). With respect to the velocity problem the approach differs from Freiberger's.

remains elastic. If the pressure on the membrane is more and more increased, an increasing part will rest on the roof, but the part under the edges of the roof will still remain free, corresponding to the elastic neighborhood of the lines of discontinuity. This visualization of our problem is due to Nadai[1] and has been used by Trefftz[2] and by later authors.

This interpretation suggests that the mathematical problem has a solution. The mathematical problem is the following: A closed contour C is given (Fig. 43), enclosing a region A. We ask for a vector field $\boldsymbol{\tau}(x,y)$, having a given constant curl $\left(\text{viz.,} \dfrac{\partial \tau_y}{\partial x} - \dfrac{\partial \tau_x}{\partial y} = 2\alpha G\right)$ in some inner region, a given magnitude τ ($\tau = k$) in the remaining "outer" part, including the boundary Γ between the two parts, and given direction along C. Also, $\boldsymbol{\tau}$ should be continuous across Γ; this appears physically plausible (see Nadai, l. c.) and has been proved by Prager ([23], p. 70). The boundary Γ separating "inner" and "outer" regions is unknown. In terms of ψ the problem is as follows: We want, in A, a continuous function ψ, with continuous $\partial \psi / \partial x$, $\partial \psi / \partial y$, with given constant $\Delta \psi$ in some unknown inner part of A, given $|\text{grad } \psi|$ in the remaining outer part, and given constant value on the closed contour C. As pointed out by Trefftz and von Mises the problem is of the type of the well-known free-boundary problems, or jet problems of hydromechanics, where the separation line between the regions in which ψ is subject to different conditions is likewise unknown. Von Mises[3] remarks that this analogy, as well as the membrane analogy, leads to the assumption that the problem is determinate.

The *strains in the elastic region* are determined by Hooke's law

$$\gamma_x / \tau_x = \gamma_y / \tau_y. \tag{86.1}$$

In the *plastic region* the elastic strains are assumed as being so small that time-changes in the contour C and displacements of elements can be neglected—as in elasticity[4]. Once the element has become plastic the stress $\boldsymbol{\tau}$, at a fixed element, does not change, since its magnitude, τ, equals k and its direction is tangent to the contour line. Hence with the notation of Sect. 43 $\dot{\tau}_x = 0$, $\dot{\tau}_y = 0$, and from (43.3), $\dot{e}'_{ij} = 0$, $\dot{e}''_{ij} = \dot{e}_{ij}$, the Reuss equations reduce to von Mises' stress strain equations and $\dot{\gamma}_x / \tau_x = \dot{\gamma}_y / \tau_y$. On account of the time constancy of τ_x, τ_y from an instant t_0 on, we can even integrate these last equations from t_0 to t, and since $\gamma_x(t_0)/\tau_x - \gamma_y(t_0)/\tau_y = 0$, by (86.1), we see that (86.1) holds also for the strains in the plastic region.

We can no longer assume that the warping is proportional to the twist α; hence we replace the last Eq. (85.6) by $u_z = w(x, y, \alpha)$ and obtain in the elastic region, instead of (85.7),

$$\frac{\partial w}{\partial x} - \alpha y = \frac{\tau_x}{G}, \qquad \frac{\partial w}{\partial y} + \alpha x = \frac{\tau_y}{G}. \tag{86.2}$$

Introducing the strains from (86.1) we have

$$\frac{\gamma_x}{\gamma_y} = \left(\frac{\partial w}{\partial x} - \alpha y\right) \Big/ \left(\frac{\partial w}{\partial y} + \alpha x\right) = \frac{\tau_x}{\tau_y} \tag{86.3}$$

in both the plastic and elastic regions. Once the stresses have been evaluated this equation determines the only unknown function, $w(x, y, \alpha)$. Using $\tau_x/\tau_y =$

---

[1] A. Nadai: Z. angew. Math. Mech. 3, 442—454 (1923).
[2] E. Trefftz: Z. angew. Math. Mech. 5, 64—73 (1925).
[3] R. v. Mises: Reissner Anniversary Volume 415—429. Ann Arbor, Michigan 1949.
[4] See also Hill [10], p. 88.

cot ϑ, (86.3) becomes

$$\frac{\partial w}{\partial x}\sin\vartheta - \frac{\partial w}{\partial y}\cos\vartheta = \alpha(x\cos\vartheta + y\sin\vartheta). \tag{86.4}$$

The characteristics of this linear first-order partial differential equation, defined by
$$dx:dy:dw = \sin\vartheta : -\cos\vartheta : \alpha(x\cos\vartheta + y\sin\vartheta), \tag{86.5}$$

are again the normals to the contour. To determine w we write (86.4) as

$$\frac{\partial w}{\partial n} = -\alpha d, \tag{86.6}$$

where d is the distance from the normal through the point (x,y) to the origin. This can be integrated[1] and gives (Fig. 44),

$$w(x,y) - w(\xi,\eta) = -\alpha l d = \alpha(x\eta - y\xi), \tag{86.7}$$

where (ξ,η) are the coordinates of the point of intersection of the normal through (x,y) and the boundary Γ, and l is the distance between (x,y) and (ξ,η). Hence w is known in the plastic region after it has been determined in the elastic region. There, however, it follows from Eqs. (85.7), which are of the form $\partial w/\partial x = f(x,y)$, $\partial w/\partial y = g(x,y)$, with f and g known. Thus the strains can be found[2].

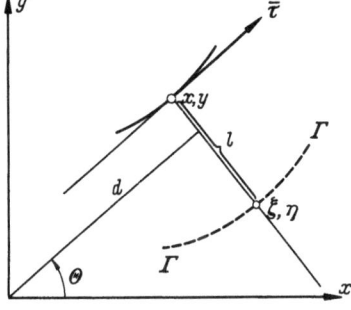

Fig. 44. Illustrating computation of warping in torsion.

87. Examples, further problems and concluding remarks. Returning to the stress problem we repeat that the boundary-value problem formulated at the beginning of this section has not been solved except when the shape of Γ can be estimated from the beginning. This is possible for the *bar of circular section*, radius a, where Γ is obviously a concentric circle of radius $\varrho < a$. The elastic stress function valid inside this circle is

$$\psi = \tfrac{1}{2}G\alpha(\varrho^2 - x^2 - y^2), \quad \tau_x = -G\alpha y, \quad \tau_y = G\alpha x, \quad \tau = G\alpha r. \tag{87.1}$$

The value $\tau = k$ is reached for $\alpha = k/G\varrho$ and, correspondingly, $\varrho = k/G\alpha$. Hence, for $\varrho \leq r \leq a$: $\tau_x = -ky/r$, $\tau_y = kx/r$, $\tau_y/\tau_x = -\dfrac{x}{y} = \tan\vartheta$, and the torque is therefore

$$T = 4\frac{\pi}{2}\int_0^a r^2\tau\,dr = 2\pi k\left[\int_0^\varrho \frac{r^3}{\varrho}dr + \int_\varrho^a r^2\,dr\right] = 2\pi k\left(\frac{a^3}{3} - \frac{\varrho^3}{12}\right), \tag{87.2}$$

where $\varrho = k/G\alpha$ may be substituted. The axial displacement $u_z = w$ is zero in the elastic as well as in the plastic zone. A "fully plastic solution" corresponds

---

[1] Cf. P. G. HODGE jr.: J. Appl. Mech. **16**, 399—406 (1949), p. 400.
[2] Arguing against MISES' flow equations, PRAGER [23], pp. 81—84, who insists on the identity of MISES' theory with that of a "rigid-plastic" material (see our Sect. 95), states that in MISES' theory necessarily $w = 0$, in both the elastic and (contained) plastic region, as a consequence of $\alpha \to 0$ in both regions which in turn follows from $G \to \infty$. It would indeed be strange to study torsion with $\alpha = 0$. But this is in no way implied by or connected with MISES' theory (see also p. 330, line 9 seq. of our article). Actually, in the torsion problem the Reuss equations reduce to MISES' equations on account of $\dot{s}_{ij} = 0$ (and not on account of $G \to \infty$); the wrong result $\alpha = 0$, $w = 0$, follows if torsion is considered as a problem of a rigid plastic material which, indeed, would be quite inadequate. (Cf. MISES' treatment of elastic plastic torsion problems in Reissner Anniversary Volume, l. c. and our Sects. 91, 92, and 95.)

to $\alpha \to \infty$; then $T = \frac{2\pi}{3} k a^3$. The above example was first treated by SAINT VENANT.

A direct problem has been solved by TREFFTZ[1], who computed the stress distribution for an L-shaped beam (the two legs of the L having equal widths), in particular in the vicinity of the re-entrant corner. The method is based in a natural way on complex function theory, and is approximate, but the errors are small.

Practically all known examples use inverse methods, where the boundary C is derived from an assumed solution and an assumed separation line Γ. Essentially the same indirect method has been described by SOKOLOVSKY[2] and by VON MISES[3]. We illustrate the method for an oval cross-section, as considered by SOKOLOVSKY.

We start with the solution of the elastic problem (85.9):

$$\frac{\psi}{\alpha G} = -\frac{b^2 x^2 + a^2 y^2}{a^2 + b^2} \tag{87.3}$$

and with the assumption of the elliptic boundary Γ

$$\frac{\xi^2}{a^4} + \frac{\eta^2}{b^4} = 1. \tag{87.4}$$

From (87.3)

$$\frac{\partial \psi}{\partial x} = -2\alpha G \frac{b^2}{a^2+b^2} x = -\tau_y, \quad \frac{\partial \psi}{\partial y} = -2\alpha G \frac{a^2}{a^2+b^2} y = \tau_x. \tag{87.5}$$

Hence, on Γ,

$$k^2 = \frac{4\alpha^2 G^2}{(a^2+b^2)^2}(b^4 \xi^2 + a^4 \eta^2) = \frac{4\alpha^2 G^2}{(a^2+b^2)^2} a^4 b^4, \tag{87.6}$$

and we obtain

$$2\alpha G = k \frac{a^2+b^2}{a^2 b^2} \tag{87.7}$$

as the relation valid for the twist at which yielding occurs along (87.4). If the equation of Γ is written in parametric form, using a', b' for a^2, b^2 viz. $\xi = -a'\sin\vartheta$, $\eta = b'\cos\vartheta$, the equation of a normal through (ξ, η) is

$$y - \eta = -\cot\vartheta(x - \xi), \quad \text{or} \quad y + x \cot\vartheta = (b' - a')\cos\vartheta. \tag{87.8}$$

The differential equation of the orthogonal trajectories to these normals, i.e. of the stress lines, is

$$\frac{dy}{dx} = \tan\vartheta = -\frac{x-\xi}{y-\eta} = -\frac{x + a'\sin\vartheta}{y - b'\cos\vartheta}. \tag{87.9}$$

and by integration, with c as an integration constant, we obtain

$$\begin{aligned} x &= -\sin\vartheta\left[\frac{a'}{2} + c + \frac{a'-b'}{2}\cos^2\vartheta\right], \\ y &= \cos\vartheta\left[\frac{b'}{2} + c - \frac{a'-b'}{2}\sin^2\vartheta\right]. \end{aligned} \tag{87.10}$$

These curves are ovals of double symmetry and differ very little from ellipses with semi-axes $\frac{a'}{2} + c$ and $\frac{b'}{2} + c$. We have thus a solution of the elastic-plastic problem for an approximately elliptic contour C (see Fig. 45).

---

[1] E. TREFFTZ: Z. angew. Math. Mech. **5**, 64—73 (1925).
[2] V. V. SOKOLOVSKY: J. Appl. Math. Mech. **6**, 241—246 (1942); see [26], p. 132.
[3] R. v. MISES: Reissner Anniversary Volume, l. c.

Sect. 87. Examples, further problems and concluding remarks. 407

On the other hand, if we assume the oval contour as given by the "axes" $A = \frac{a'}{2}+c$, $B=\frac{b'}{2}+c$, we can evaluate a' and b', the axes of the ellipse Γ, from the two equations

$$\frac{2\alpha G}{k} = \frac{a'+b'}{a'b'}, \qquad 2(A-B) = a'-b,$$

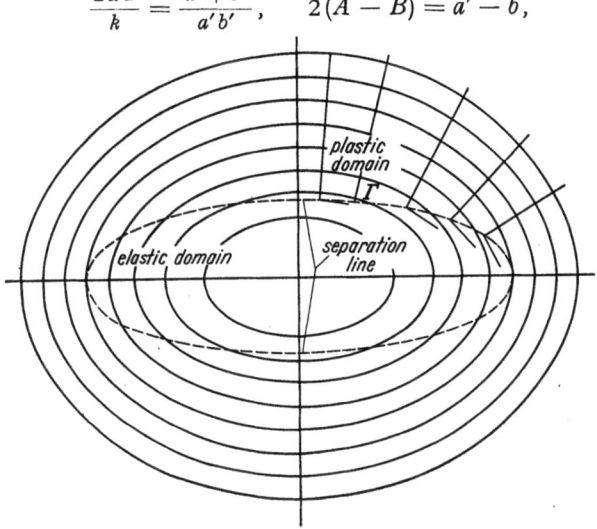

Fig. 45. Elastic-plastic torsion of approximately elliptical oval (SOKOLOVSKY).

where the first equation is (87.7) again, and we find, with $d = A - B$:

$$a' = \frac{k}{2G\alpha} + d + \left[d^2 + \left(\frac{k}{2G\alpha}\right)^2\right]^{\frac{1}{2}}, \quad b' = \frac{k}{2G\alpha} - d + \left[d^2 + \left(\frac{k}{2G\alpha}\right)^2\right]^{\frac{1}{2}}. \quad (87.11)$$

The expressions for a' and b' depend only on the difference $d = A - B$. Our solution makes sense only if Γ lies entirely in C, i.e. if $a' < A$, $b' < B$, which leads to the condition

$$\alpha > \frac{k}{G} \frac{B}{A(2B-A)}. \qquad (87.12)$$

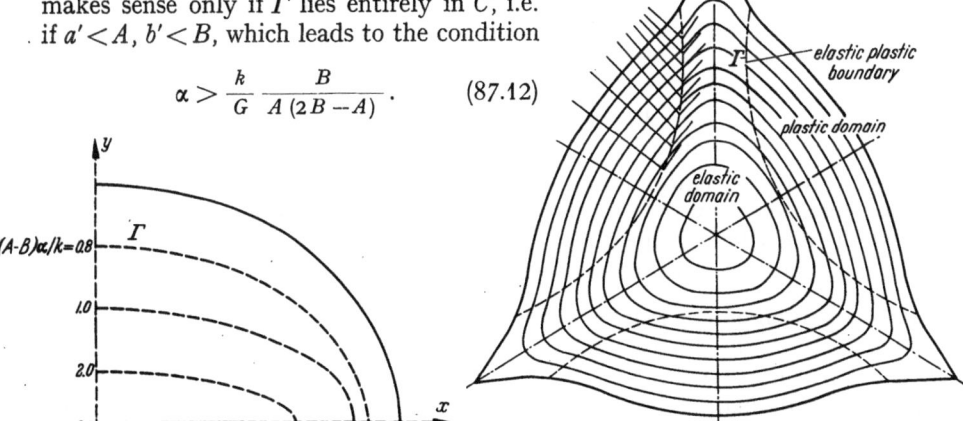

Fig. 46. Spreading of separation line Γ for increasing twist (PRAGER).

Fig. 47. Triangular solution for elastic-plastic torsion (VON MISES).

Figure 46 (cf. [23], p. 80) shows the spreading of the separation line Γ with increasing twist α.

The warping in the elastic region is found from (85.7) and (87.5) as

$$w = -xy\frac{a'-b'}{a'+b'};$$

[in this combined elastic plastic problem $w(x,y,\alpha)$ is not even in the elastic section proportional to α]; on Γ we have

$$w = \frac{a'-b'}{a'+b'}\, a'\, b'\, \sin\vartheta \cos\vartheta, \qquad (87.13)$$

and the warping in the plastic region is determined by (86.5) (cf. [26], p. 133). The moment T can likewise be found ([26], p. 134).

Von Mises (l. c.) has applied (Fig. 47) the same indirect method (combined with graphical integration) to find an approximate solution for an equilateral triangular section. Elastic-plastic torsion for particular cross sections has also been studied by Galin[1].

We have so far no general method to solve the direct problem, i.e. the boundary-value problem explained in Sect. 86. As stated above (p. 406), the particular problem of the L-shaped cross section with infinitely long legs and equal width of the legs has been solved by Trefftz by means of complex function theory. His procedure provides an approximate solution of the direct problem.

An approximate formula for warping under large twist has been given by Hodge[2]. In most actual solutions numerical or graphical methods are needed to carry out the steps. As a typical example we mention the numerical solution of the elastic-plastic torsion problem of a shaft of rotational symmetry and varying diameter[3]. We do not develop here this interesting problem. Note, however, that in this problem[4] the characteristics are no longer straight lines (although still normal to the contour lines); they may have an envelope, i.e. the solution may have a limit line, which may, however, be outside the contour). Non-rectilinear characteristics also appear in the elastic-plastic solution of the twisted ring sector[5]. Here, the envelope of the characteristics which, for the fully plastic solution is a limit line inside the contour, does not materialize in the plastic part of the elastic-plastic solution.

The technically important one-dimensional problem of the elastic and plastic behavior of *beams*, as well as the related technical theory of the thin plate, are considered in Sects. 35, 36 of this article, in connection with an exposition of the basic ideas of the fruitful method known as *limit analysis*.

b) The thick walled tube.

In Sect. 69 we briefly commented on the fully plastic radial yielding in a thick tube subjected to high internal pressure considered as a problem of plane strain. Here, we study the typical problem, where the tube yields only in part so that there results an inner plastic region surrounded by an elastic region; also $\dot\varepsilon_z = 0$ is no longer a necessary assumption.

88. Expansion of a cylindrical tube. Let the inner radius of the circular cylindrical ring be a, the outer b and the generators in the z-direction; the tube is expanded by an internal pressure p and loaded at both ends by some equal and opposite longitudinal forces of magnitude L. The tube is long enough to justify

---

[1] L. A. Galin: J. Appl. Math. Mech. **13** (1949).
[2] P. G. Hodge jr.: J. Appl. Mech. **16**, 399—406 (1949). He also considers in this paper the problem of repeated loading.
[3] R. P. Eddy and F. S. Shaw: J. Appl. Mech. **16**, 139—148 (1949), and further quotations of numerical methods in Hill [10], p. 94. See also the presentation in [26], p. 140 seq.
[4] See the presentation in Hill [10], pp. 94—96, in P. G. Hodge jr.: An introduction to the mathematical theory of perfectly plastic solids. Providence: Brown University 1950, pp. 107—118, also Sokolovsky. J. Appl. Math. Mech. **1**, 343—346 (1945).
[5] See footnote 1, p. 403.

the assumption that the stress and strain distribution is the same in normal cross-sections sufficiently far from the end. It follows by symmetry that any originally plane normal cross-section remains plane and that ε_z is the same all over a representative cross-section and remains the same through the expansion[1].

Three main *end conditions* are usually considered. The tube may be closed at both ends by plugs firmly attached to the ends: *closed-end* conditions; or it is closed by floating pistons: *open-end* conditions; a third condition is that of *zero extension* or *plane strain*.

The principal directions are radial, axial and circumferential and, with respect to such a coordinate system, $\tau_{r\vartheta} = \tau_{z\vartheta} = \tau_{rz} = 0$. Our problem is independent of ϑ and is also assumed independent of z. Thus, two of the three equilibrium conditions are identically satisfied, and the remaining one reduces to

$$\frac{\partial \sigma_r}{\partial r} + \frac{\sigma_r - \sigma_\vartheta}{r} = 0. \tag{88.1}$$

The non-vanishing strains are

$$\varepsilon_r = \frac{\partial u_r}{\partial r}, \quad \varepsilon_\vartheta = \frac{u_r}{r}, \quad \varepsilon_z = \frac{\partial u_z}{\partial z} = \text{constant}, \tag{88.2}$$

where $u_r = u$ is the radial, u_z the axial displacement. There therefore remain altogether four unknowns: σ_r, σ_ϑ, σ_z, and u.

We first consider *elastic expansion*. The radial displacement u is of the form

$$u = A r + \frac{B}{r}. \tag{88.3}$$

If E is YOUNG's modulus, ν POISSON's ratio, we have from HOOKE's law

$$\left. \begin{array}{l} E\,\varepsilon_r = E\dfrac{\partial u}{\partial r} = E\left(A - \dfrac{B}{r^2}\right) = \sigma_r - \nu(\sigma_\vartheta + \sigma_z), \\[4pt] E\,\varepsilon_\vartheta = E\dfrac{u}{r} = E\left(A + \dfrac{B}{r^2}\right) = \sigma_\vartheta - \nu(\sigma_z + \sigma_r), \\[4pt] E\,\varepsilon_z = \phantom{E\dfrac{u}{r} = E\left(A + \dfrac{B}{r^2}\right)} = \sigma_z - \nu(\sigma_r + \sigma_\vartheta). \end{array} \right\} \tag{88.4}$$

Solving these equations and using the abbreviation

$$\alpha = (1+\nu)(1-2\nu)/E$$

we obtain

$$\left. \begin{array}{l} \alpha\,\sigma_r = A - (1-2\nu)\dfrac{B}{r^2} + \nu\,\varepsilon_z \\[4pt] \alpha\,\sigma_\vartheta = A + (1-2\nu)\dfrac{B}{r^2} + \nu\,\varepsilon_z \\[4pt] \alpha\,\sigma_z = 2\nu A + (1-\nu)\varepsilon_z. \end{array} \right\} \tag{88.5}$$

It is seen that σ_z as well as $\sigma_r + \sigma_\vartheta$, and therefore $\sigma_r + \sigma_\vartheta + \sigma_z = 3\sigma$, are independent of r.

If the flow is elastic throughout, the boundary conditions are

$$\left. \begin{array}{rll} \sigma_r = -p & \text{for} & r = a \\ = 0 & \text{for} & r = b. \end{array} \right\} \tag{88.6}$$

---

[1] The presentation in this section follows, HILL [10], p. 106 seq. cf. also SOKOLOVSKY [26], Chap. 3. He takes strain hardening into consideration; his investigation uses HENCKY's stress-strain law.

Substituting these into (88.5), we find the values of A and B, and, using the abbreviation $p' = a^2 p/(b^2 - a^2)$, we obtain

$$\sigma_r = -p'\left(\frac{b^2}{r^2} - 1\right), \quad \sigma_\vartheta = p'\left(\frac{b^2}{r^2} + 1\right),$$
$$\sigma_z = E\varepsilon_z + 2\nu p', \quad \sigma = \frac{1}{3}[2(1+\nu)p' + E\varepsilon_z]$$
(88.7)

and

$$u = -\nu\varepsilon_z r + \frac{1+\nu}{E} p'\left[(1-2\nu)r + \frac{b^2}{r}\right].$$
(88.8)

In general, not ε_z but the axial force L is prescribed, and, from this, ε_z may be computed for the different end conditions. The axial force L is here

$$L = \pi(b^2 - a^2)\sigma_z.$$

Using the third Eq. (88.7) we find the axial strain ε_z, corresponding to given load L and pressure p,

$$\varepsilon_z = \frac{\sigma_z}{E} - \frac{2\nu}{E} p' = \frac{a^2}{(b^2 - a^2)E}\left(\frac{L}{\pi a^2} - 2\nu p\right).$$
(88.9)

In case of closed end conditions, L is a tension equal to the force $\pi a^2 p$ exerted by the internal pressure on each plug. Hence from (88.9')

$$\varepsilon_z = \frac{1-2\nu}{E} p', \quad \sigma_z = p' = \frac{1}{2}(\sigma_r + \sigma_\vartheta) \quad \text{(closed end)}.$$
(88.10)

When both ends are open, $L = 0$, and

$$\varepsilon_z = -\frac{2\nu}{E} p', \quad \sigma_z = 0 \quad \text{(open end)};$$
(88.11)

and under plane strain

$$\varepsilon_z = 0, \quad \sigma_z = 2\nu p' = \nu(\sigma_r + \sigma_\vartheta) \quad \text{(plane strain)}.$$
(88.12)

This case reduces to the first for $\nu = \frac{1}{2}$, which means (elastic) incompressibility, $\varepsilon_r + \varepsilon_\vartheta + \varepsilon_z = 0$, as may be verified from (88.4). We note that σ_r and σ_ϑ are not influenced by the end conditions while σ_z is different in the three cases and each time intermediate to σ_r and σ_ϑ.

The above formulas are valid throughout the tube for values of p such that a certain function of the stresses, the yield function, remains below the yield limit. We consider VON MISES' limit, that means the region where

$$J_M \equiv (\sigma_r - \sigma_\vartheta)^2 + (\sigma_\vartheta - \sigma_z)^2 + (\sigma_z - \sigma_r)^2 < 6k^2,$$
(88.13)

and from (88.7)

$$J_M = 6p'^2 \frac{b^4}{r^4} + 2(\sigma_z - p')^2.$$
(88.13')

Yielding occurs first at the inner surface $r = a$; hence the pressure p^* at which $J_M = 6k^2$ is determined by

$$3\left(\frac{b^2}{b^2 - a^2}\right)^2 p^{*2} + \left[\sigma_z - \frac{a^2}{b^2 - a^2} p^*\right]^2 = 3k^2.$$
(88.14)

Inserting for σ_z the values from Eqs. (88.10) — (88.12), we obtain

$$p^* = k\left(1 - \frac{a^2}{b^2}\right) \quad \text{(closed end)},$$
$$= k\left(1 - \frac{a^2}{b^2}\right)\bigg/\sqrt{1 + \frac{a^4}{3b^4}} \quad \text{(open end)},$$
$$= k\left(1 - \frac{a^2}{b^2}\right)\bigg/\sqrt{1 + (1-2\nu)^2 \frac{a^4}{3b^4}} \quad \text{(plane strain)}.$$
(88.15)

We see that of the three yield pressures that for "open end" is the smallest, that for "closed end" the largest.

With TRESCA's yield condition we have to consider

$$\sigma_\vartheta - \sigma_r = 2p'\frac{b^2}{r^2} = \frac{2p}{r^2}\frac{a^2 b^2}{b^2 - a^2}.$$

Putting $r = a$, we obtain the yield pressure p_0:

$$p_0 = \frac{k\sqrt{3}}{2}\left(1 - \frac{a^2}{b^2}\right), \tag{88.16}$$

the same for all end conditions.

89. Partly plastic tube. α) *Determination of σ_ϑ and σ_r for Tresca's yield.* For a pressure somewhat exceeding p^* or p_0, respectively, the tube will become partly plastic. On account of symmetry the elastic-plastic boundary Γ is a circle with radius ϱ. Consider first TRESCA's yield, where there is independence of the end conditions. We assume that σ_z remains the intermediate principal stress. Then, using (88.7)

$$\sigma_\vartheta - \sigma_r = \left[\frac{2p}{r^2}\frac{a^2 b^2}{b^2 - a^2}\right]_{r=\varrho} = k\sqrt{3}, \tag{89.1}$$

and substituting this value of p into Eqs. (88.7) and (88.8) we obtain the *solution for the partly plastic tube in the elastic part, in the case of Tresca's yield conditions*; with $p'' = \sqrt{3}\,k\varrho^2/2b^2$:

$$\left.\begin{array}{l}\sigma_r = -p''\left(\dfrac{b^2}{r^2} - 1\right), \quad \sigma_\vartheta = p''\left(\dfrac{b^2}{r^2} + 1\right), \\[6pt] \sigma_z = E\,\varepsilon_z + 2\nu p'', \quad u = -\nu\,\varepsilon_z r + \dfrac{1+\nu}{E} p''\left[(1-2\nu)r + \dfrac{b^2}{r}\right],\end{array}\right\} \tag{89.2}$$

where the p' of (88.7), (88.8) is now replaced by p''.

In the *plastic region* $a \leq r \leq \varrho$, there hold the equilibrium condition (88.1) and the yield condition

$$\sigma_\vartheta - \sigma_r = k\sqrt{3}, \tag{89.3}$$

by means of these two equations the two stresses $\sigma_\vartheta, \sigma_r$ can thus be found, and for all end conditions, without solving the complete problem. We obtain the simple differential equation

$$\frac{\partial \sigma_r}{\partial r} = k\sqrt{3}\cdot\frac{1}{r}.$$

Observing that σ_r and σ_ϑ at $r=\varrho$ must coincide with the values from (89.2), we find for $a \leq r \leq \varrho$:

$$\sigma_r = \sqrt{3}\,k\left[\frac{1}{2}\left(\frac{\varrho^2}{b^2} - 1\right) - \log\frac{\varrho}{r}\right], \quad \sigma_\vartheta = \sqrt{3}\,k\left[\frac{1}{2}\left(\frac{\varrho^2}{b^2} + 1\right) - \log\frac{\varrho}{r}\right], \tag{89.4}$$

$$p = -(\sigma_r)_{r=a} = \sqrt{3}\,k\left[\frac{1}{2}\left(1 - \frac{\varrho^2}{b^2}\right) + \log\frac{\varrho}{a}\right]. \tag{89.5}[1]$$

We may eliminate the unknown ϱ from (89.4) by means of (89.5) and obtain

$$\sigma_r = k\sqrt{3}\log\frac{r}{a} - p, \quad \sigma_\vartheta = k\sqrt{3}\left(1 + \log\frac{r}{a}\right) - p. \tag{89.6}$$

---

[1] According to HILL this formula for p was given by L. B. TURNER: Trans. Cambridge Phil. Soc. **21**, 377 (1909), and Engineering **92**, 115 (1911). The solution for $\sigma_\vartheta, \sigma_r, p$ agrees with NADAI's; A. NADAI: Trans. Amer. Soc. Mech. Engrs. **52**, 193 (1930), and [*19*], p. 196.

Thus, if a, b, k, p are given, ϱ follows from (89.5) and σ_r, σ_ϑ from (89.6). Hence the problem is partly statically determined, and this for all end conditions.

β) *Complete problem under assumption of plane strain. Tresca's yield.* In order to obtain σ_z and u in the plastic region we must use stress-strain relations[1]. To simplify the work we restrict ourselves to plane strain, $\varepsilon_z = 0$. From (88.4) we obtain by addition, with $\sigma_r + \sigma_\vartheta + \sigma_z = 3\sigma$:

$$\frac{\partial u}{\partial r} + \frac{u}{r} = 3\frac{1-2\nu}{E}\sigma. \tag{89.7}$$

We wish to consider both plastic and elastic strains in the region of plastic flow which is restricted by the surrounding elastic ring. We use the Reuss equations. With $s_z = \sigma_z - \sigma$ and $\sigma_r + \sigma_\vartheta = A$, we have $2\sigma = A + s_z$, $2\sigma_z = A + 3s_z$ where, from (89.4), $A = 2k\sqrt{3}\left(\frac{\varrho^2}{2b^2} - \log\frac{\varrho}{r}\right)$. Following HILL, we take $q = \frac{\sqrt{3}}{2k}s_z$ as unknown. Then

$$\sigma_z = k\sqrt{3}\left(q + \frac{\varrho^2}{2b^2} - \log\frac{\varrho}{r}\right), \quad \sigma = k\sqrt{3}\left(\frac{q}{3} + \frac{\varrho^2}{2b^2} - \log\frac{\varrho}{r}\right),$$

and (89.7) takes the form

$$\frac{\partial u}{\partial r} + \frac{u}{r} = \frac{1-2\nu}{E}k\sqrt{3}\left(q - 3\log\frac{\varrho}{r} + \frac{3\varrho^2}{2b^2}\right). \tag{89.8}$$

We write the Reuss equations in the form

$$\dot{\varepsilon}_r = \frac{1}{E}[\dot{\sigma}_r(1+\nu) - 3\nu\dot{\sigma}] + \lambda(\sigma_r - \sigma),$$

and two similar ones, and eliminate λ from the equations for $\dot{\varepsilon}_\vartheta$ and $\dot{\varepsilon}_z$. In these equations the dot (which means differentiation with respect to time) may be interpreted to mean differentiation with respect to some magnitude which increases monotonically with time. We choose ϱ as such a parameter, as long as $\varrho < b$ and obtain:

$$s_z[E\dot{\varepsilon}_\vartheta - (1+\nu)\dot{\sigma}_\vartheta + 3\nu\dot{\sigma}] + s_\vartheta[(1+\nu)\dot{\sigma}_z - 3\nu\dot{\sigma}] = 0, \tag{89.9}$$

where $\dot{\varepsilon}_\vartheta = \frac{1}{r}\frac{\partial u}{\partial \varrho}$, $s_\vartheta = \sigma_\vartheta - \sigma$, with σ_ϑ from (89.6), $\sigma = \frac{1}{2}(A + s_z)$ where $A = \sigma_r + \sigma_\vartheta$, $s_z = \frac{2k}{\sqrt{3}}q$. Carrying out differentiations and collecting terms we obtain by a laborious computation

$$\frac{E}{k\sqrt{3}}\frac{4q}{3r}\frac{\partial u}{\partial \varrho} + \left[1 - \frac{2}{3}(1-2\nu)q\right]\frac{\partial q}{\partial \varrho} + (1-2\nu)\frac{2q-1}{\varrho}\left(1 - \frac{\varrho^2}{b^2}\right) = 0. \tag{89.10}$$

The boundary conditions are supplied by the continuity of both q and u across Γ, hence, from (89.2), at $r = \varrho$, with $\varepsilon_z = 0$:

$$\left.\begin{array}{l} u = k\sqrt{3}\,\dfrac{1+\nu}{2E}\,\dfrac{\varrho^2}{b^2}\left[(1-2\nu)\varrho + \dfrac{b^2}{\varrho}\right], \\[6pt] q = (2\nu - 1)\dfrac{\varrho^2}{2b^2}. \end{array}\right\} \tag{89.11}$$

Eqs. (89.8), (89.10) are two non-linear partial differential equations of first order for the unknowns u, q and the independent variables r, ϱ; they are of the type $\frac{\partial u}{\partial x} = c$, $a\frac{\partial u}{\partial y} + b\frac{\partial v}{\partial y} = d$, where a, b, c, d may depend on all four variables.

---

[1] R. HILL, E. H. LEE and S. J. TUPPER: Proc. Roy. Soc. Lond., Ser. A **191**, 278—303 (1947). (Cf. also HILL [*10*], p. 114.) A numerical solution is included. See also R. HILL E. H. LEE, S. J. TUPPER: Proc. 1st U. S. Nat. Congr. Mech., Chicago, p. 561, 1951.

The equations are hyperbolic with fixed characteristics, parallel to the axes in the r, ϱ-plane. The boundary conditions are Cauchy data, u and q given along the non-characteristic line $\varrho = r$ (Fig. 48). A step-by-step method, as previously explained, can be used to find u and q in the shaded characteristic triangle of Fig. 48. It is obvious from the figure that if the problem has been solved for a certain ratio b/a, its solution is known for the smaller ratio b'/a. After q has been found, the longitudinal force L can be evaluated

$$L = 2\pi \left[\int_a^\varrho \sigma_z r\, dr + \int_\varrho^b \sigma_z r\, dr \right], \qquad (89.12)$$

with the plastic value of σ_z in the first, the elastic in the second integral. A graph of σ_z as a function of r/a for various values of ϱ/a may be found in [10], p. 116. It is not typically different from that in Fig. 42a.

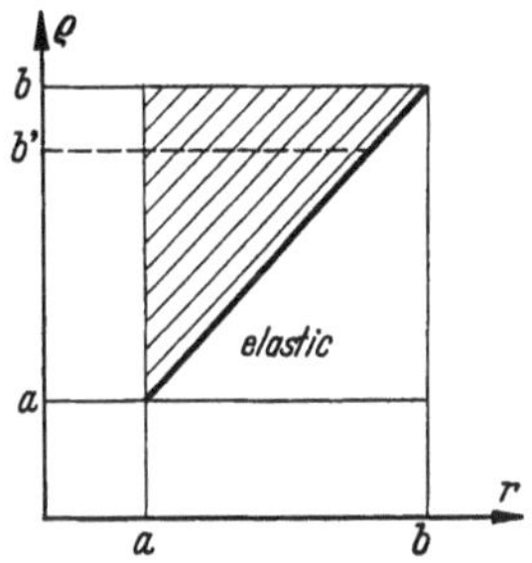

γ) *Complete problem for plane strain. Von Mises yield*[1]. Instead of (89.1) we now consider $J_M = 6k^2$, for $r = \varrho$. Using (88.13') with $\sigma_z = 2\nu p'$ from (88.7), we obtain

$$p'^2 \left[\frac{b^4}{\varrho^4} + \frac{1}{3}(2\nu - 1)^2 \right] = k^2, \qquad (89.13)$$

which is satisfied for a pressure

$$p'' = k \left[\frac{b^4}{\varrho^4} + \frac{1}{3}(2\nu - 1)^2 \right]^{-\frac{1}{2}}. \qquad (89.13')$$

Fig. 48. r, ϱ-plane in problem of thick-walled tube.

Thus, we obtain the solution for *the elastic part of the partly plastic tube under von Mises' yield*, if the p' of (88.7), (88.8) is replaced by the p'' of (89.13') [similar to the result (89.2)].

In the *plastic region*, $a \leq r \leq \varrho$, we now have no (partially) statically determinate problem. The unknown functions are three stresses and two strains, since $\varepsilon_z = 0$. The equations are: equilibrium condition, yield condition, two independent Prandtl-Reuss equations, and a "compatibility condition" for the strains ε_r and ε_ϑ. This problem has been solved by HODGE and WHITE[1], see also PRAGER and HODGE ([23], pp. 95—100). They use the unknowns $s_r, s_\vartheta, \varepsilon_r, \varepsilon_\vartheta$, or rather, for symmetry, $\varepsilon = \frac{1}{3}(\varepsilon_r + \varepsilon_\vartheta)$, $\varphi = \frac{1}{3}(\varepsilon_r - \varepsilon_\vartheta)$, s_r, s_ϑ. The yield condition is written

$$s_r^2 + s_\vartheta^2 + s_r s_\vartheta = k^2. \qquad (89.14)$$

In the equilibrium equation $\sigma_z = s_r + \sigma$ is used, with $\sigma = \dfrac{E}{1-2\nu}\varepsilon$, and we obtain:

$$\frac{\partial s_r}{\partial r} + \frac{E}{1-2\nu} \frac{\partial \varepsilon}{\partial r} = \frac{s_\vartheta - s_r}{r}. \qquad (89.15)$$

Since $\varepsilon_r = \partial u/\partial r$, $\varepsilon_\vartheta = u/r$ a relation holds between them, or between ε and φ, namely

$$\frac{\partial \varepsilon}{\partial r} - \frac{\partial \varphi}{\partial r} = 2 \frac{\varphi}{r}. \qquad (89.16)$$

Two Reuss equations are written in the form containing s_r, e_r and s_ϑ, e_ϑ, respectively, λ is eliminated, and $e_r = \frac{1}{2}(\varepsilon + 3\varphi)$, $e_\vartheta = \frac{1}{2}(\varepsilon - 3\varphi)$ is introduced. Thus:

$$\frac{E}{2(1+\nu)}\left[(s_\vartheta - s_r)\dot\varepsilon + 3(s_\vartheta + s_r)\dot\varphi\right] = s_\vartheta \dot s_r - s_r \dot s_\vartheta, \qquad (89.17)$$

---

[1] P. G. HODGE jr. and G. N. WHITE jr.: J. Appl. Mech. **17**, 180—184 (1950).

where "dot" means again differentiation with respect to ϱ. Eqs. (89.14) — (89.17) are four equations for s_r, s_ϑ, ε, φ. We may still eliminate s_ϑ, say, and obtain the three equations

$$\left.\begin{aligned}\frac{\partial \varepsilon}{\partial r}-\frac{\partial \varphi}{\partial r}&=2\frac{\varphi}{r},\\ \frac{\partial s_r}{\partial r}+\frac{E}{1-2\nu}\frac{\partial \varepsilon}{\partial r}&=\frac{1}{2r}\left(-3s_r+\sqrt{4k^2-3s_r^2}\right),\\ \dot{s}_r-\frac{E}{8(1+\nu)k^2}\Big[\big(4k^2-3s_r^2-3s_r\sqrt{4k^2-3s_r^2}\big)\dot{\varepsilon}&-\\ -3\big(4k^2-3s_r^2+s_r\sqrt{4k^2-3s_r^2}\big)\dot{\varphi}\Big]&=0.\end{aligned}\right\} \quad (89.18)$$

The boundary data are again the given values of all quantities for $\varrho = r$. These non-linear equations are much more complicated than (89.8), (89.10) but still hyperbolic with only two independent variables, and with known characteristics $r = $ constant, $\varrho = $ constant independent of the solution. The numerical procedure is therefore still straightforward. Again, the solution for a ratio b/a provides solutions under otherwise similar circumstances for smaller ratios. After s_r, s_ϑ, ε, φ have been found, there follow $s_z = -(s_r + s_\vartheta)$ and $\sigma_z = s_z + \frac{E}{1-2\nu}\varepsilon$. Hodge and White[1] have determined the stress distribution and the radial displacement for $b = 2a$ and various values of ϱ/a (Figs. 49, 50). The numerical results for the complete solutions of the problem, Sect. 89α), β) on one hand and γ) on the other hand, are not very different from each other; see Hodge and White, l.c. We note that the former is not in accordance with the theory of the plastic potential. It affords, however, a great simplification compared to the latter.

90. Further solutions. Comments. (a) If *the assumption $\varepsilon_z = 0$ is dropped*, then in (89.8), on the left side, ε_z must be added, and in (89.9) there appears $-E\dot{\varepsilon}_z$ in the second bracket. The equation for u is the same as in the second line of (89.2). The modified Eq. (89.10) then contains a term with $\dot{\varepsilon}_z$. A third finite relation between the three unknowns u, q, ε_z is supplied by the end condition (see [10], p. 114 for details).

(b) The above problem has been studied by Hill ([10], pp. 106—114) for *finite strains*[2]. The unknown u is then replaced by $v = dr/d\varrho$; one family of characteristics is still straight, $\varrho = $ constant, but the second consists of curves $dr - v\,d\varrho = 0$, which are not known in advance.

(c) Hill, Lee and Tupper have also obtained a solution (including numerical results) of the problem of Sect. 89α), β) for *closed-end conditions*[3].

(d) The equations in Sect. 89α), β) may be written in *non-dimensional form* with the variables σ_r/k, σ_ϑ/k, q and Eu/k. It can be seen ([10], p. 115) that these are functions of r/a, b/a, ν but not of E. Thus, scaling of a solution obtained for a given b/a and ν supplies solutions for the same b/a and ν, but varying k and E.

(e) Various assumptions have been made in order to simplify the procedures of Sect. 89, which necessitate numerical treatment. The best known is that of total incompressibility in the plastic region[4]. (We always assume

---

[1] See also P. G. Hodge, footnote 4 on p. 408, and [23].
[2] Cf. also C. W. MacGregor, L. F. Coffin jr. and J. C. Fisher: J. Appl. Phys. **19**, 291—297 (1948). — The same authors, in I. Franklin Inst. **245**, 135—158, (1948,) computed the closed-end problem for v. Mises' yield condition and Hencky's stress-strain law. For comparison see, Hodge and White, l. c. p. 183.
[3] R. Hill, E. H. Lee and S. J. Tupper: Ministry of Supply Armament Res. Dept. Theoret. Res. Report 11/46.
[4] A. Nadai: Trans. Amer. Soc. Mech. Engrs. **52** 193 (1930).

$\varepsilon'' = 0$, but the elastic part of the total strain is not *a priori* incompressible.) In order to avoid an unrealistic

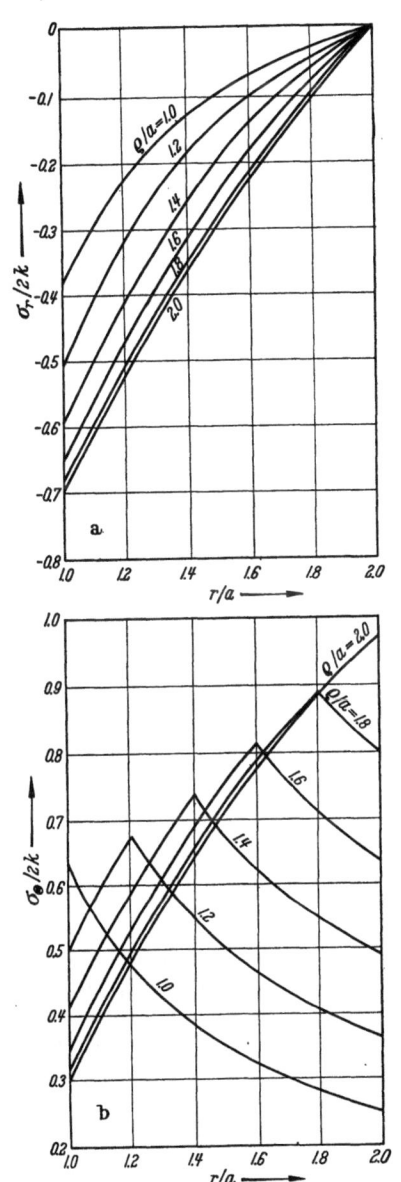

Fig. 49 a and b. Distribution of (a) radial stress and (b) circumferential stress in thick tubes under internal pressure (HODGE).

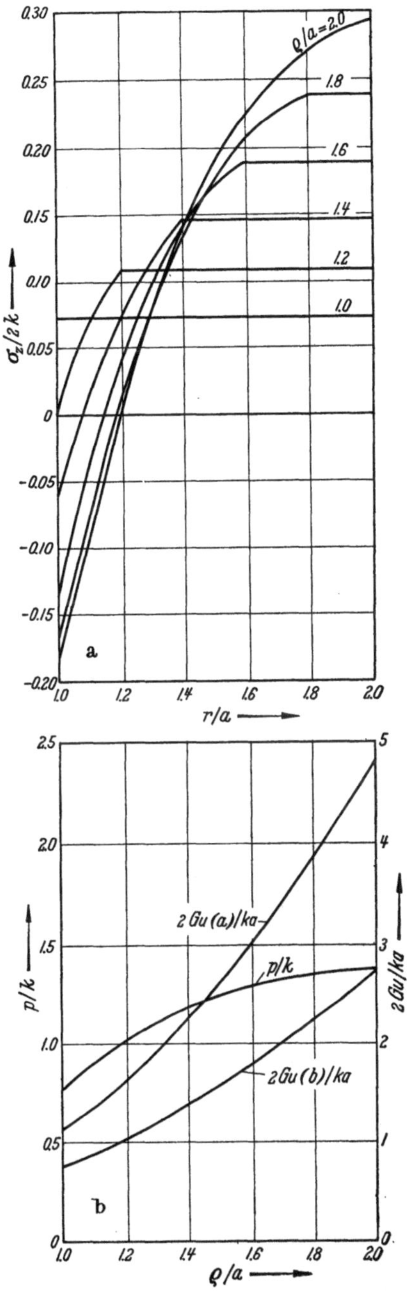

Fig. 50 a and b. (a) Axial stress and (b) radial displacement in thick-walled tube under internal pressure (HODGE).

discontinuity at $r = \varrho$, some authors also assume incompressibility in the elastic region[1]. This makes, however, no essential difference.

---
[1] W. PRAGER: Theory of Plasticity (mimeographed lecture notes) Brown University Providence, R. I. 1942. Cf. also the presentation in HODGE, Ref. 4, p. 408 and in [23], pp. 100—110.

We make again the assumption of plane strain, $\varepsilon_z = 0$; if now, in addition, $\varepsilon = 0$ we obtain

$$\varepsilon_r + \varepsilon_\theta = \frac{\partial u}{\partial r} + \frac{u}{r} = 0, \qquad (90.1)$$

and this equation is satisfied by

$$u(r, \varrho) = \frac{D(\varrho)}{r}, \qquad a \leq r \leq b. \qquad (90.1')$$

Hence, independent of the stresses and valid in the interval $a \leq r \leq b$,

$$\varepsilon_r = -\frac{D}{r^2}, \qquad \varepsilon_\theta = \frac{D}{r^2}, \qquad \varepsilon_z = 0, \qquad \varepsilon = 0, \qquad a \leq r \leq b. \qquad (90.2)$$

From (88.4) we see that elastic incompressibility, $\varepsilon' = \varepsilon = 0$, leads to $\nu = \tfrac{1}{2}$, unless $\sigma = 0$. To obtain the stresses in the elastic region we may thus use (89.13'), with $2\nu - 1 = 0$, in connection with the two lines that follow this equation. Then, with $p'' = k\varrho^2/b^2$, we obtain

$$\left. \begin{array}{ll} \sigma_r = p''\left(1 - \dfrac{b^2}{r^2}\right), & \sigma_\theta = p''\left(1 + \dfrac{b^2}{r^2}\right), \\ \sigma_z = p'' = \sigma, & u = k\varrho^2/2Gr. \end{array} \right\} \quad r \geq \varrho. \qquad (90.3)$$

The Reuss equations are:

$$\dot{s}_r : \dot{s}_\theta : \dot{s}_z = (2G\dot{e}_r - \lambda s_r) : (2G\dot{e}_\theta - \lambda s_\theta) : (2G\dot{e}_z - \lambda s_z). \qquad (90.4)$$

From Eqs. (90.2), holding for all stages of the deformation, we see that $\dot{e}_r : \dot{e}_\theta : \dot{e}_z = 1 : -1 : 0$, from (90.3), that $s_r : s_\theta : s_z = 1 : -1 : 0$ in the elastic region; using this and the continuity across Γ we see that also *in the plastic region*

$$s_r : s_\theta : s_z = \dot{s}_r : \dot{s}_\theta : \dot{s}_z = 1 : -1 : 0. \qquad (90.5)$$

Substituting these relations into the yield condition (89.14), we find for $r \leq \varrho$

$$\sigma_r = \sigma - k, \qquad \sigma_\theta = \sigma + k, \qquad \sigma_z = \sigma, \qquad (90.6)$$

and, using the equation of equilibrium, we obtain

$$\sigma = 2k\left(\frac{\varrho^2}{2b^2} + \log\frac{r}{\varrho}\right). \qquad (90.7)$$

Thus, in the present approach in the elastic as well as plastic domain, $\sigma_r + \sigma_\theta = 2\sigma_z$. This relation obtains in the plastic region in NADAI's original work, but in the elastic region he has, as from (89.2), $(\sigma_r + \sigma_\theta)\nu = \sigma_z$, and these coincide only for $\nu = \tfrac{1}{2}$; in general this amounts at $r = \varrho$ for σ_z to a percentage discontinuity of $(1 - 2\nu)/2\nu$, viz. 67% for $\nu = 0.3$[1]. However, there is no real advantage in avoiding this discontinuity. HODGE and WHITE, for $b/a = 2$, $\varrho/a = 0.5$, $\nu = 0.3$, find hardly a difference between the present and the exact theory [Sect. 89γ] as far as σ_r, σ_θ are concerned; however, for σ_z and u there is a similar discrepancy as in NADAI's original work[2] (Fig. 51). The authors suggest that by using correction factors for σ_z and u the simplified solution gives a good approximation to the exact one the in case of plane strain[3]. This "adjusted" solution is not recommended

---

[1] Cf. NADAI [19], p. 197, Eqs. (46), (47) and the discussion in the paper of HODGE and WHITE, quoted on p. 413.

[2] SOKOLOVSKY's procedure is similar to NADAI's, see, e.g. [26], p. 95.

[3] It should be remembered that for $\sigma_z = \tfrac{1}{2}(\sigma_r + \sigma_\theta)$, as in (90.6), the form of v. MISES' and TRESCA's criteria become the same viz. $\sigma_\theta - \sigma_r = 2k$ (MISES), and $\sigma_\theta - \sigma_r = \sqrt{3}k$ (TRESCA), so that there is only *one* incompressibility-approximation theory.

for closed-end tubes. Careful discussion and comparison of results found under various assumptions is given by HODGE and WHITE[1].

(f) HODGE (pp. 147—155 of the book quoted in footnote 4 on p. 408) and PRAGER and HODGE ([23], p. 110ff.) consider also the important problem of the variation of stresses and displacements for *unloading* and for *repeated loading*. For simplification, incompressibility is assumed[2]. The stress field after unloading is obtained by superposition of an elastic stress field, producing the stress-free inner surface by the applied pressure $-p$.

(g) We show finally, following a noteworthy idea of KOITER[3], that considerable simplification results if TRESCA's yield condition is used together with its associated flow rule (see Sect. 41). We *assume* that σ_z is the intermediate stress, $\sigma_r \leq \sigma_z \leq \sigma_\vartheta$, in the elastic as well as the plastic part. The stress distribution in the elastic part is the same as in (89.2). In the plastic part σ_r and σ_ϑ are given by (89.4) and the pressure by (89.5). So far, this is the same as above.

Now, in the plastic region, we write the components of total strain as sums of elastic and plastic strains:

$$\varepsilon_r = \varepsilon_r' + \varepsilon_r'', \quad \varepsilon_\vartheta = \varepsilon_\vartheta' + \varepsilon_\vartheta'', \quad \varepsilon_z = \varepsilon_z' + \varepsilon_z'', \quad (90.8)$$

where HOOKE's law (88.4) holds for the elastic strains, and VON MISES' law (41.3) for the plastic strain rates[4]. Thus, here

$$\dot\varepsilon_\vartheta'' : \dot\varepsilon_z'' : \dot\varepsilon_r'' = 1 : 0 : -1. \quad (90.9)$$

It follows—this is the decisive simplification—that here

$$\dot\varepsilon_z'' = 0, \quad \varepsilon_z'' = 0, \quad \varepsilon_z = \varepsilon_z'. \quad (90.10)$$

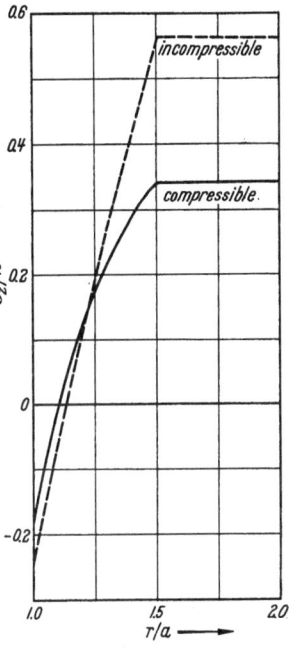

Fig. 51. Distribution of axial stress for compressible (full line) and incompressible (dashed line) material (HODGE, WHITE).

Therefore, the axial stress σ_z can be computed from the third Eq. (88.4) (which holds for the elastic strain ε_z', which here equals ε_z) together with (89.4):

$$\sigma_z = E\varepsilon_z + \nu(\sigma_r + \sigma_\vartheta) = E\varepsilon_z + \sqrt{3}\,k\nu\left(2\log\frac{r}{\varrho} + \frac{\varrho^2}{b^2}\right). \quad (90.11)$$

The radial displacement u follows, as before, from the dilatation, $\varepsilon = \dfrac{1-2\nu}{E}\cdot 3\sigma$, where ε is purely elastic, since $\varepsilon''=0$. Using this and (88.2), (89.4), we obtain easily the differential equation for u

$$\frac{\partial u}{\partial r} + \frac{u}{r} = \frac{1}{E}(1-2\nu)(1+\nu)\,k\sqrt{3}\left[2\log\frac{r}{\varrho} + \frac{\varrho^2}{b^2}\right] - 2\nu\,\varepsilon_z. \quad (90.12)$$

---

[1] D. N. DE G. ALLEN and D. G. SOPWITH: Proc. Roy. Soc. Lond., Ser. A **205**, 69 (1951) obtain a solution in closed form by using Hencky's stress-strain relations (Sect. 44). For the present problem the results based on these "deformation type" relations are acceptable. Cf. HODGE and WHITE, l.c.

[2] Cf. also RUTH MOUFANG: Z. angew. Math. Mech. **28**, 33—42 (1948).
We mention a study of our problem for non-homogeneous material, more specifically, for a material with "cylindrical" inhomogeneity: W. OLSZAK and W. URBANOWSKI: Bull. Acad. Polonaise Sci. **4**, 153—163 (1956).

[3] W. T. KOITER: Note 3, p. 327.

[4] There in some confusion in terminology: KOITER denotes v. MISES' (1913) Eqs. (38.8): $\dot\varepsilon_{ij}'' = \lambda s_{ij}$ as the "well-known Prandtl-Reuss equations" and with respect to (90.9) he quotes a passage of [10], 1950.

This equation can be integrated. If the continuity of u for $r=\varrho$ is used we obtain

$$u = \frac{1-\nu^2}{E} k \sqrt{3}\, \frac{\varrho^2}{r} + \frac{(1+\nu)(1-2\nu)}{E} k \sqrt{3}\, r \left[\log \frac{r}{\varrho} + \frac{1}{2}\left(\frac{\varrho^2}{b^2} - 1\right)\right] - \nu r \varepsilon_z. \qquad (90.13)$$

For plane strain, $\varepsilon_z = 0$, the solution is thus completed. For tubes with open ends or closed ends, ε_z must be expressed in terms of the axial load L[1].

KOITER then investigates his assumption that σ_z is the intermediate stress: At the boundary, for $r=\varrho$, this holds true. For $r<\varrho$ the results are: In plane strain, the assumption is always satisfied if $b/a \leq 5.75$. For open-end tubes and closed-end tubes the inequality depends on ν. For example, for $\nu=0.3$ (for which some of the previously reported investigations were made) the results are

$$b/a \leq 5.75 \text{ (plane strain)}, \quad \leq 6.19 \text{ (open end)}, \quad \leq 5.43 \text{ (closed end)}. \qquad (90.14)$$

We mention, finally, that the elastic-plastic problems considered here may, of course, be combined in various ways. B. CROSSLAND and R. HILL study e.g. the plastic behaviour of thick tubes under combined torsion and internal pressure[2].

c) Flat ring and flat sheet in plane stress. Further elastic-plastic problems.

91. Flat ring radially stressed as a problem of plastic-elastic equilibrium. Consider a flat ring with inner radius $r=a$ and outer radius $r=b$ stressed by a pressure p, uniformly distributed over the circumference of the inner circle and acting in the plane of the sheet, while the outer circumference is free of stress. The conditions of "plane stress" are assumed satisfied, and in the plastic domain there hold the radial equation of equilibrium

$$\sigma_\vartheta = \frac{d}{dr}(r\sigma_r) \qquad (91.1)$$

and the MISES' yield condition [see (72.1)]

$$\sigma_r^2 + \sigma_\vartheta^2 - \sigma_r \sigma_\vartheta = 4k^2. \qquad (91.2)$$

In general, in a plasticity problem the flow equations determine plastic flow velocities v. In certain problems, however, there exists a complete solution with $v=0$ for a certain range of stresses (as a consequence of $\lambda=0$ in (38.8)), i.e. a solution of plastic equilibrium. In the present problem, which we study as representative for this circumstance, this equilibrium is characterized by a one-to-one relation between the dimension of the ring, defined by b/a, and the applied pressure p (or rather $p/2k$). The combinations of plastic and elastic equilibrium and the characterization of the situation for plastic flow $(v\neq 0)$ have been fully investigated by VON MISES[3].

Plastic equilibrium. With the dimensionless variables $t=r/b$, $u=\sigma_r/2k$, $v=\sigma_\vartheta/2k$, Eqs. (91.1), (91.2) read (the prime denoting d/dr),

$$v^2 - uv + u^2 = 1, \quad \text{or} \quad v = \frac{u}{2} \pm \frac{1}{2}\sqrt{4-3u^2} = r u' + u.$$

Hence we obtain the ordinary differential equation

$$\frac{dr}{r} = \frac{2\,du}{-u \pm \sqrt{4-3u^2}}. \qquad (91.3)$$

---

[1] See KOITER: l. c., p. 240.

[2] B. CROSSLAND and R. HILL: J. Mech. Phys. of Solids **2**, 27—39 (1953).

[3] R. v. MISES: Reissner Anniversary Volume, 415—429. Ann Arbor, Michigan 1949. — See also NADAI [20], p. 472.

Putting, for abbreviation,
$$z = +\sqrt{4-3u^2}, \quad \text{hence: } v = \tfrac{1}{2}(u+z) \tag{91.4}$$
we obtain
$$\log r = 2\int \frac{du}{z-u} = -\frac{1}{2}\log(z-u) - \sqrt{3}\arctan\frac{z+2}{u\sqrt{3}} + \text{const.} \tag{91.5}$$
or
$$\frac{r^2}{b^2} = \frac{2}{z-u} e^{2\vartheta\sqrt{3}}, \quad \text{where } \tan\vartheta = \frac{u\sqrt{3}}{z+2}. \tag{91.6}$$

The last relation between u, z and ϑ can be written in terms of the parameter ϑ and gives $u = \frac{2}{\sqrt{3}}\sin 2\vartheta$, $z = 2\cos 2\vartheta$, and, using (91.4) we obtain:

$$\left.\begin{array}{ll} u = \dfrac{2}{\sqrt{3}}\sin 2\vartheta, & v = \dfrac{2}{\sqrt{3}}\sin\left(2\vartheta + \dfrac{\pi}{3}\right), \\[2mm] \dfrac{p}{2k} = -\dfrac{2}{\sqrt{3}}\sin 2\vartheta, & \dfrac{b^2}{r^2} = \dfrac{2}{\sqrt{3}}\sin\left(\dfrac{\pi}{3} - 2\vartheta\right)e^{-2\sqrt{3}\vartheta}. \end{array}\right\} \tag{91.7}$$

It is seen that for $\vartheta = 0$, $p/2k = 0$, $r = b$, while $p/2k$ reaches its maximum, viz. $\frac{2}{\sqrt{3}}$ for $\vartheta = -\frac{\pi}{4}$. Thus, collecting:

$$\vartheta = 0, \quad u = 0, \quad v = 1, \quad \frac{b^2}{r^2} = 1,$$

$$\vartheta = -\frac{\pi}{4}, \quad u = -\frac{2}{\sqrt{3}}, \quad v = -\frac{1}{\sqrt{3}}, \quad \frac{b^2}{r^2} = \frac{1}{\sqrt{3}} e^{\sqrt{3}\frac{\pi}{2}}.$$

By differentiation we see that, for $\vartheta = -\frac{\pi}{4}$, the right side of the last Eq. (91.7) reaches its maximum; hence

$$\left(\frac{b}{r}\right)_{\max} = 2.964, \quad \text{or} \quad \left(\frac{r}{b}\right)_{\min} = 0.338. \tag{91.7'}[1]$$

Thus, if for a flat ring $b/a > 2.964$, overall yielding cannot arise merely by the application of an internal pressure. On the other hand, if $b/a \leq 2.964$, then, corresponding to the specific b/a-value of this ring, there exists a unique pressure $p = -\frac{4k}{\sqrt{3}}\sin 2\vartheta_1$, where ϑ_1 follows from the third Eq. (91.7), with $r = a$ such that the whole ring is in plastic equilibrium; and vice versa, to any pressure between 0 and $4k/\sqrt{3}$ there follows the dimension b/a of a corresponding ring in fully plastic equilibrium. We now wish to determine the state of the body if this specific relation between b/a and $p/2k$ does not hold.

92. Continuation: Plastic-elastic equilibrium. *If the whole ring is elastic, we obtain, as in* (88.7)

$$\sigma_r = -p'\left(\frac{b^2}{r^2} - 1\right), \quad \sigma_\vartheta = p'\left(\frac{b^2}{r^2} + 1\right), \quad p' = \frac{pa^2}{b^2 - a^2}, \tag{92.1}$$

or, in terms of u, v and using $t = r/b$, $t_1 = a/b$, $u_1 = u_{r=a}$, etc.

$$u = \varkappa\left(1 - \frac{1}{t^2}\right), \quad v = \varkappa\left(1 + \frac{1}{t^2}\right), \quad \varkappa = \frac{p}{2k}\frac{t_1^2}{1 - t_1^2}; \tag{92.2}$$

or

$$u = u_1\left(1 - \frac{1}{t^2}\right)\Big/\left(1 - \frac{1}{t_1^2}\right), \quad v = u_1\left(1 + \frac{1}{t^2}\right)\Big/\left(1 + \frac{1}{t_1^2}\right). \tag{92.2'}$$

---

[1] The relation between Nadai's parameter δ, Sect. 72, and v. Mises' ϑ is: $2\vartheta = \delta - \frac{\pi}{6}$. In the present problem δ varies from $\frac{\pi}{6}$ to $-\frac{\pi}{3}$, ϑ from 0 to $-\frac{\pi}{3}$ and the parameter s, used by us in Chap. G, from 1 to $-\sqrt{3}$, where the hyperbolic problem becomes parabolic, and $\alpha = 0°$.

The condition for this fully elastic equilibrium is that $u_1^2 + v_1^2 - u_1 v_1 \leq 1$, even for $r=a$. Using (92.2) we obtain this condition in the form:

$$\frac{p}{2k} \leq \frac{1-t_1^2}{\sqrt{3+t_1^4}}, \quad \text{and} \quad \left(\frac{p}{2k}\right)_{max} = \frac{1}{\sqrt{3}}. \tag{92.3}$$

But if $\frac{p}{2k} > \frac{1-t_1^2}{\sqrt{3+t_1^4}}$, smaller, however, than the $p/2k$ value which corresponds in (91.7) to the particular t_1 (>0.338) under consideration, then the ring will be neither fully elastic nor fully plastic. For example, for a ring with $b=2a$, $t_1 = \frac{1}{2}$, the right side of (92.3) equals $\frac{3}{7} \sim 0.43$. If the applied $p/2k$ is less than 0.43 the ring is fully elastic; on the other hand, the value of $p/2k$ which in (91.7) corresponds to $b/a = 2$ is approximately 0.77 and for this $p/2k$ the whole ring is in plastic equilibrium. For each value of $p/2k$ between 0.43 and 0.77 the ring will be

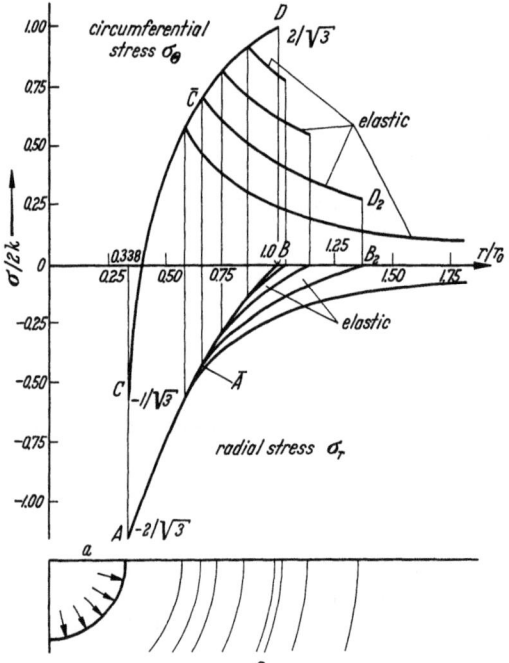
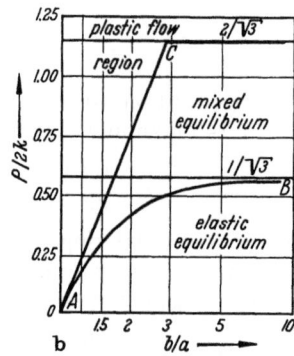

Fig. 52 a and b. Flat ring under internal pressure as elastic-plastic problem: (a) radial stress and circumferential stress; (b) elastic, plastic, and mixed equilibrium (VON MISES).

plastic in an inner annulus between $r=a$ and a specific $r=\varrho$, the plastic-elastic boundary Γ, and elastic between $r=\varrho$ and $r=b$; and, conversely, to each ϱ between $r=a$ and $r=2a$ there corresponds a $p/2k$ which generates this particular boundary. (We shall see that, for example, to $p/2k=0.5$ there corresponds $\varrho=1.58a$.)

We denote the values corresponding to ϱ by $\bar{t}, \bar{u}, \bar{v}$, etc. The formula, corresponding to (92.2'), then, is

$$u = \bar{u}\left(1 - \frac{1}{t^2}\right) \Big/ \left(1 - \frac{1}{\bar{t}^2}\right), \quad v = \bar{v}\left(1 + \frac{1}{t^2}\right) \Big/ \left(1 + \frac{1}{\bar{t}^2}\right), \tag{92.4}$$

We may eliminate $\bar{t}$ and, using (91.4), we obtain (92.4) in the form

$$u = \frac{1}{4}(\bar{z} + 3\bar{u}) - \frac{1}{4t^2}(\bar{z} - \bar{u}), \tag{92.5}$$

as the relation between u and $\bar{u}, \bar{v}$ at Γ [remember that $v = \frac{1}{2}(u+z)$]. Since from (92.2) $u+v = 2\varkappa = \bar{u}+\bar{v}$, we have for v in terms of u:

$$v = \bar{u} + \bar{v} - u = \frac{1}{2}(\bar{z} + 3\bar{u}) - u. \tag{92.6}$$

In Fig. 52a the line AB on the left gives the relation between u and $t=r/r_0$ for fully plastic equilibrium [Eqs. (91.6) or (91.7)], r_0 being the abscissa of the point B; the curve CD gives the corresponding v,t-relation. These relations only hold for t-values between 0.338 $(=1/2.964)$ and one. On the u-line we choose now an arbitrary point $\bar{A}$ with corresponding $\bar{u}$; and the point $\bar{C}$ above has the corresponding $\bar{v}$ as ordinate. The line $\bar{A}B_2$ is then plotted from (92.5) until its intersection with the t-axis, where $u=0$; and $\bar{C}D_2$ is the corresponding elastic v-line. [We easily see that the two u-lines, the elastic and the plastic one, are tangent at $u=\bar{u}$. In fact, with $x=\log r$ we have from (92.2)

$$u = \varkappa - \varkappa b^2 e^{-2\varkappa}, \quad \left(\frac{du}{dx}\right)_{\text{elast}} = 2\varkappa \frac{b^2}{r^2} = 2(\varkappa - u),$$

while for the u in the plastic region, from (91.5),

$$\frac{du}{dx} = \frac{z-u}{2};$$

at the transition

$$\left(\frac{du}{dx}\right)_{\text{elast}} = 2\varkappa - 2\bar{u} = \bar{u} + \bar{v} - 2\bar{u} = \bar{v} - \bar{u} = \left(\frac{du}{dx}\right)_{\text{pl}}.]$$

We thus obtain the "mixed" lines $A\bar{A}B_2$ for u, and $C\bar{C}D_2$ for v. This has the following meaning: B_2 (with abscissa ~ 1.5 in the figure) corresponds to a ring with $\frac{b}{a} \sim 1.35/0.338 \sim 4$ (>2.964). The point $\bar{A}$ has an abscissa $\varrho \sim 2a$; hence between $r=a$ and $r=2a$ the ring is in plastic, and between $r=2a$ and $r=4a$ in elastic equilibrium. The $p/2k$ which leads to this solution follows from (91.7): corresponding to the ratio 2 it is approximately 0.77; hence in a ring with $b=4a$ the above particular plastic-elastic distribution results for $p/2k=0.77$. For a smaller (larger) pressure the plastic annulus will be smaller (larger) than in the example, but its maximum width (for $p/2k=2/\sqrt{3}$) ranges from a to $2.964a$.

The limits of possible states of equilibrium are indicated in Fig. 52b. The line AC shows Eq. (92.3), with equality sign, and independent variable $1/t_1$. If, for *any* ring ratio, the pressure is below or on this line, the ring is fully elastic. If $p \leq 4k/\sqrt{3}$, any point in the "mixed region" defines a ring in mixed equilibrium; e.g. the point with abscissa 4 and ordinate 0.77 corresponds to the above discussed example. To all points with the same $p/2k$ there belongs one and the same intersection of this horizontal with AC. The abscissa of this point of intersection defines the outer boundary $\varGamma$ of the fully plastic annulus and the remainder is elastic. If, finally, to a ring of $b/a < 2.964$ we apply a pressure higher than the "corresponding" pressure on AC, or to a ring with $b/a \geq 2.964$ a pressure $>4k/\sqrt{3}$, then *free plastic flow* sets in if, for example, to a ring with $b/a=1.5$ the pressure $p/2k=1/\sqrt{3}=0.575$ is applied, the ring starts to flow plastically, while under a slightly smaller over-pressure (compare Fig. 52b) it would be entirely in *contained plastic equilibrium*. Such flow will cause a thickening around the edge of the hole. This thickening has been studied by G. I. Taylor and R. Hill[1], in the slightly simpler situation where $b \to \infty$, under the assumption of a plastic rigid material (see Sect. 95).

---

[1] Cf. [20], p. 477, and [26], p. 287 seq., with generalization for a non-circular hole. In the presentation of Hill [10], p. 307, where work of G. I. Taylor, Quart. J.Mech. a. Appl. Math. 1, 103—124 (1948), is incorporated, the deformation is studied which arises around the edge of the hole as soon as *free plastic flow* sets in. The material is considered as "rigid-plastic" (see Sect. 95).

93. Expansion of a circular hole in an infinite sheet.

This problem may be treated as a particular case of the preceding one, as $b \to \infty$, $t \to 0$. A fully plastic state, as studied in Sect. 91 for the ring, is now, of course, impossible.

The sheet is overall elastic if

$$\frac{p}{2k} \leq \frac{1}{\sqrt{3}}, \tag{93.1}$$

as seen from (92.3) and Fig. 52b. The corresponding stresses, from (92.1), are

$$\sigma_r = -p \cdot \frac{a^2}{r^2}, \quad \sigma_\vartheta = p \frac{a^2}{r^2}. \tag{93.2}$$

If $2k/\sqrt{3} \leq p \leq 4k/\sqrt{3}$, or, from (91.7), $-\frac{\pi}{4} \leq \vartheta_1 \leq -\frac{\pi}{12}$, there is an annulus in plastic equilibrium, which reaches its maximum width if $p = 4k/\sqrt{3}$, $\vartheta_1 = -\frac{\pi}{4}$. The value of b^2/r^2 from (91.7), for $\vartheta_1 = -\frac{\pi}{4}$, is $\frac{1}{\sqrt{3}} e^{\sqrt{3}\frac{\pi}{2}}$, that for $\vartheta_1 = -\frac{\pi}{12}$ is $\frac{2}{\sqrt{3}} e^{\sqrt{3}\frac{\pi}{6}}$; the ratio is $\frac{1}{2} e^{\sqrt{3}\frac{\pi}{3}} = (1.75)^2$. Thus the maximum plastic annulus has the width of $1.75 a$.

In this simpler case we easily compute directly: If $p/2k > 1/\sqrt{3}$, the elastic stresses in the region $r \geq \varrho$ are, from (93.2),

$$u = -\frac{1}{\sqrt{3}} \frac{\varrho^2}{r^2}, \quad v = \frac{1}{\sqrt{3}} \frac{\varrho^2}{r^2}. \tag{93.3}$$

We find the first three Eqs. (91.7) as before; the last takes on the form

$$\frac{C^2}{r^2} = \frac{2}{\sqrt{3}} \sin\left(\frac{\pi}{3} - 2\vartheta\right) e^{-2\sqrt{3}\vartheta}.$$

At $r = \varrho$, from (93.3), $u = -\frac{1}{\sqrt{3}}$, which arises in (91.7) for $2\vartheta = -\frac{\pi}{6}$, and this gives $\frac{C^2}{\varrho^2} = e^{\sqrt{3}\frac{\pi}{6}}$. Hence the last Eq. (91.7) is to be replaced by

$$\left.\begin{array}{c} \dfrac{\varrho^2}{r^2} = \dfrac{2}{\sqrt{3}} e^{-\sqrt{3}(2\vartheta + \frac{\pi}{6})} \sin\left(\dfrac{\pi}{3} - 2\vartheta\right) = e^{-\sqrt{3}\delta} \cos\delta, \\ -\dfrac{\pi}{3} \leq \delta \leq 0, \quad a \leq r \leq \varrho, \end{array}\right\} \tag{93.4}$$

where δ is NADAI's parameter (see (72.2)) which simplifies the formula. The discussion is simple: If, at $r = a$, a pressure $p/2k$, between $1/\sqrt{3}$ and $2/\sqrt{3}$ is applied, the corresponding δ_1 is between 0 and $-\frac{\pi}{3}$, and $\varrho^2 = a^2 e^{-\sqrt{3}\delta_1} \cos\delta_1$; the δ for the stresses in this ring is between 0 and δ_1, and beyond this annulus the sheet is elastic. This plastic equilibrium is hyperbolic corresponding to the range $0 \geq s \geq -\sqrt{3}$, in terms of the parameter s of Chap. G; in the problem of Sect. 91, s was between $+1$ and $-\sqrt{3}$ [remember that the "hyperbolic range" is $(-\sqrt{3}, +\sqrt{3})$]. For $\delta = \pi/3$, $s = -\sqrt{3}$ the problem becomes parabolic (see Sect. 72), the angle α between the first principal direction—that of the circumferential stress σ_ϑ—and the characteristics is zero and the coinciding characteristics envelop the edge of the hole. If p becomes greater than $4k/\sqrt{3}$, the radius ϱ may increase further to $\varrho' > \varrho = 1.75 a$ (at least at the beginning), but the ratio between the outer and the inner radius defined by (93.4) remains the same so that the inner radius of this annulus now equals $\varrho'/1.75 = 0.57\varrho' = a' > a$. At this circle, $r = a'$, the equations become parabolic; if we consider the constrained material in the ring between $r = a'$ and $r = \varrho'$ as rigid (see Sect. 95), the circle $r = a'$ forms the "plastic-rigid" boundary—enveloped by characteristics—between the "rigid" material and that which flows freely (see Sect. 84β and the sketch Fig. 53). This flow will cause a deformation in the domain from $r=a$ to $r=a'$ where the material near the hole piles up into a thickened crater. The mechanics of this deformation has been clarified by TAYLOR and by HILL (l. c.), based on the theory of Sect. 65, and under TRESCA's yield condition.

In the above problem, Eqs. (93.3) show that $u \to 0$, $v \to 0$ as $r \to \infty$. NADAI ([20], p. 481 and [19], p. 193) and SOKOLOVSKY ([26], p. 283) have pointed out that the plastic problem of an infinite disk with a circular hole (with or without pressure at $r=a$), stressed uniformly in its plane by a stress at infinity, can be treated in a similar way[1]. The whole sheet may then become plastic. In a domain corresponding to $\pi/3 \leq \delta \leq \pi/2$ (or s between $\sqrt{3}$ and 2) the stress distribution will be elliptic.

94. A few further elastic-plastic problems. (a) In problems of cylindrical or spherical symmetry the determination of the plastic-elastic interface is facilitated by the symmetry of the problem, which requires the boundary to be spherical or cylindrical. The expansion of a *spherical shell* by internal pressure is the problem originally studied by REUSS[2], which led him to propose his stress-strain relations.

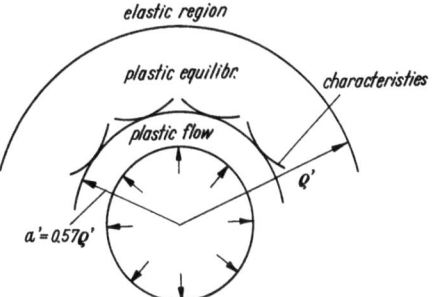

Fig. 53. Schematic distribution of states of stress in expansion of circular hole (HILL).

The problem may be found in detail in HILL [10], p. 97, for finite deformation[3]. The thorough analysis is straightforward.

(b) The problems of *rotating cylinders* and *rotating disks* are somewhat similar to those of thick-walled tubes and flat rings [see NADAI [20], p. 482, HODGE (Note 4, p. 408) p. 158 seq., SOKOLOVSKY [26], p. 100 seq.]. A related problem is that of a rotating disc with a hole[4].

(c) A problem different from that of a thick-walled tube is that of a thin-walled (circular) *cylindrical shell*. By means of a rigid plastic analysis the "collapse load" may be determined; however, in order to find the displacements, an elastic-plastic analysis is needed. We mention a recent paper by HODGE[5] where also references are given.

(d) A problem which, in contrast to the preceding ones, does not depend on r but only on the polar angle ϑ is that of a wedge under a uniform pressure acting on one face; it has been considered by many authors, cf. extensive study in SOKOLOVSKY [26], Chap. XI, particularly p. 342 seq. (see our Sect. 96).

(e) NADAI[6] indicated a problem, later solved rigorously by TREFFTZ[7], viz. that of stresses and displacements in pure shear in the neighborhood of a semi-circular grove. Around the hole a plastic domain will form, limited by an unknown elastic-plastic boundary, $\varGamma$. The problem is a plane one; $\tau_{xz}=\tau_x$, $\tau_{yz}=\tau_y$ are the only non-zero stresses, γ_{xz}, γ_{yz} the only non-zero strains. TREFFTZ determined

---

[1] SOKOLOVSKY considers these problems under both MISES' and TRESCA's yield conditions for plane stress. He also considers a non-circular hole. Cf. also the comments Sect. 94, (f).

[2] E. REUSS: Z. angew. Math. Mech. **10**, 266—274 (1930).

[3] R. HILL: J. Appl. Mech. **16**, 295 (1949). SOKOLOVSKY [26b], 95—100, presenting his paper in J. Appl. Math. Mech. **8** (1944), uses HENCKY's stress-strain relations.

[4] Cf. H. I. WEISS and W. PRAGER: J. Aer. Sci. **21**, 196 (1954).

[5] P. G. HODGE jr.: J. Appl. Mech. **77**, 22 (1955).

[6] A. NADAI: Z. angew. Math. Mech. **3**, 442—454 (1923), see p. 449.

[7] E. TREFFTZ: Z. angew. Math. Mech. **5**, 64—73 (1925).

the elastic-plastic boundary Γ and the complete solution by means of conformal mapping.

(f) In the *elastic-plastic problems of plane strain* a mechanical analogy plays a similar role to the sand hill, soap film analogies in elastic-plastic torsion: the so-called *plate analogy*. To fix the ideas, we consider an infinite plane region with a circular hole of radius a, free from loads along the circumference C of the hole. At infinity where the directions x, y are chosen as principal directions there acts a general uniform stress:

$$\sigma_x = A, \quad \sigma_y = B, \quad \tau_{x,y} = 0.$$

We use AIRY's stress function F, Eq. (66.1), by means of which we satisfy the two plane equilibrium equations identically. At infinity:

$$F_\infty = \tfrac{1}{2}(Bx^2 + Ay^2) + Cx + Dy + E, \tag{94.1}$$

with C, D, E arbitrary constants. Since the hole is free from surface traction, $\sigma_n = \dfrac{\partial^2 F}{\partial s^2} = 0$, $\tau_{sn} = -\dfrac{\partial^2 F}{\partial s \partial n} = 0$, we can assume that

$$F = 0, \quad \frac{\partial F}{\partial n} = 0 \quad \text{on} \quad C. \tag{94.2}$$

As long as the applied forces at infinity are such that the whole domain is elastic, the stress function F satisfies the biharmonic equation

$$\Delta_2 F = \frac{\partial^4 F}{\partial x^4} + 2\frac{\partial^4 F}{\partial x^2 \partial y^2} + \frac{\partial^4 F}{\partial y^4} = 0, \tag{94.3}$$

and the boundary conditions (94.1), (94.2). This stress function can be visualized by means of an infinite elastic plate exterior to C, clamped along C and bent by appropriate moments at infinity. The deflection w of this plate satisfies the biharmonic equation, and the moments at infinity may be such that (94.1) holds for w at infinity, while along the clamped contour conditione (94.2) hold for w [1]. If, however, the loads are increased such that local yielding occurs around the hole up to a boundary Γ, then, outside Γ elastic, inside, plastic, conditions prevail. In the plastic domain F satisfies the yield condition (66.2), say, and the boundary conditions (94.2), while outside Γ, (94.3) and the conditions at infinity must hold. Also, F is continuous across Γ, together with its first and second derivatives. The generalization of the plate analogy, found by PRAGER[2] and by GALIN[3], works under certain assumptions regarding the elastic plastic boundary Γ. (See [23], p. 202, cf. [26], p. 200 seq.)

L. A. GALIN[3] solved analytically the more general problem where a normal internal pressure p is applied to the hole in addition to the biaxial tension $\sigma_x = A, \sigma_y = B$ at infinity. The stresses in the plastic neighborhood of the cavity are given by (89.6) if TRESCA's yield condition is used. Since the present F is independent of ϑ, the definition of F in polar coordinates,

$$\sigma_r = \frac{1}{r}\frac{\partial F}{\partial r} + \frac{1}{r^2}\frac{\partial^2 F}{\partial \vartheta^2}, \quad \sigma_\vartheta = \frac{\partial^2 F}{\partial r^2}, \quad \tau_{r\vartheta} = \frac{1}{r^2}\frac{\partial F}{\partial \vartheta} - \frac{1}{r}\frac{\partial^2 F}{\partial r \partial \vartheta},$$

here becomes

$$\sigma_r = \frac{1}{r}\frac{\partial F}{\partial r}, \quad \sigma_\vartheta = \frac{\partial^2 F}{\partial r^2}, \quad \tau_{r\vartheta} = 0, \tag{94.4}$$

---

[1] K. WIEGHARDT: Mitt. Geb. Ingenieurwes. **49**, 15—30 (1908).

[2] W. PRAGER: Theory of Plasticity. Mimeographed lecture notes. Brown University, Providence, R. I. 1942.

[3] L. A. GALIN: J. Appl. Math. Mech. **10**, 365—386 (1946), Translation: GDAM, Brown University 1947, and L. A. GALIN: J. Appl. Math. Mech. **12**, 757—760 (1948).

and F can be found explicitly, using (89.6):

$$F = \frac{k\sqrt{3}}{2}\left[r^2 \log\frac{r}{a} - \frac{r^2}{2}\left(1 + \frac{2p}{k\sqrt{3}}\right)\right]. \tag{94.5}$$

Since this is a (degenerate) solution of the biharmonic equation, F in this problem satisfies the same equation as the stress function in the elastic region. GALIN took this as the starting point for his solution of this particular problem by means of complex function theory. (See HODGE (footnote 4, p. 408), p. 348, PRAGER and HODGE [23], p. 203, HILL [10], p. 253.)

II. Some plastic rigid problems.

We only present a few typical plastic-rigid problems, and these not in great detail, referring the reader to the excellent presentations in the literature quoted. Some attention is given to the *concept* of the plastic-rigid body.

a) Introductory remarks.

95. The plastic-rigid body. α) The perfectly plastic body as defined by VON MISES contains, in general, an elastic and a plastic part, dependent as to whether $F < C$ or $F = C$, where F is an appropriate yield function and C a constant. If $F = C$, the plastic strain rate $\dot{E}''$ is proportional to a tensor of stresses, in general derived from F according to the theory of the plastic potential (Sect. 40); we obtain $\dot{E}'' = \lambda S$, where S is the stress deviator, if F is Mises' yield function, and we find Tresca's "associated flow rule" (Sect. 41) if F is Tresca's yield function. The positive proportionality factor λ varies in space and time and can assume any value between zero and infinity. In a particular problem, the elastic displacements in either part may or may not be considered as negligible. It is not implied in the theory, as such, that the elastic displacements in regions below the yield limit or in regions of contained plastic deformation are to be neglected.

In the problem of the thick-walled tube (Sects. 88—90) the stresses in the plastic part have reached the yield limit, but the deformation is restrained by the surrounding elastic material. It is a natural approach to use Hooke's equations in the elastic, the Prandtl-Reuss equations in the plastic, part. In the problem of elastic-plastic torsion (Sects. 85—87) the elastic strains in the elastic part are relevant; in the plastic part, $\dot{S} = 0$ holds, thus the Reuss and the Mises equations become identical (see Sect. 43), and the elastic strains in the plastic region are automatically "neglected". In the elastic-plastic analysis of the problem of a flat ring, under internal pressure (Sects. 91 and 92) let it be assumed for the moment that $b/a > 2.964$ (so that fully plastic equilibrium is excluded) and $p/2k > 1/\sqrt{3}$ (so that the ring is not fully elastic). Then, as long as $p/2k < 2/\sqrt{3}$, there is an annulus of definite width in "plastic equilibrium", surrounded by an elastic region. This plastic annulus is in a state of contained plastic deformation. This appears mathematically as follows: If a complete solution of a problem obtains which satisfies all differential equations (of plasticity) and all boundary conditions, with identically vanishing proportionality factor λ (in the flow equations), then the plastic deformation velocities *are zero* (no matter whether we think of von Mises' or Reuss' relations), and we have a case of plastic equilibrium. If then a pressure $p > 4k/\sqrt{3}$ is applied, *free plastic flow* sets in.

In the problem of the thick-walled tube, as well as in that of elastic-plastic torsion, the regions of contained deformation show the same basic character. In each of these problems a complete solution is found, satisfying all equations and boundary conditions, with plastic deformation velocities zero. The warping function w in the torsion problem and the radial displacement u in the case of the tube are *strains*; at a fixed time t_0—for a fixed radius $\varrho = \varrho_0$ of the elastic-plastic boundary—the corresponding strain *rates* are zero. The problem of the ring as well as that of torsion show the situation particularly clearly because there we can compute, to begin with, all stresses in the domain of plastic equilibrium, whereas in the problem of the tube (see the treatment of Sect. 89β) the stress σ_z and the displacement u appeared as simultaneous "unknowns". The contained deformation in the three cases is, however, of the same type.

It seems instructive, in this connection, to look at Koiter's skillful treatment of the thick-walled tube problem (end of Sect. 90). He does not "neglect" elastic strains in the plastic region but he sets up separately Hooke's law for the elastic strains and Mises' for the plastic strain rates. A similar approach might be helpful in other problems[1], though probably not as simple as in the case of the Tresca yield condition and associated flow rule.

In the so-called *rigid-plastic* analysis, strains which are of the order of elastic strains are neglected throughout: the material below the yield limit—which otherwise would be considered as elastic—is treated as rigid; in a domain of unrestricted plastic flow the total strain (or strain rate) is identified with the plastic one since this part is overwhelmingly large; regions of contained plastic flow are also considered as rigid, since all strains there are of elastic order of magnitude. There is obviously no reason to use the Reuss equations in this case[2].

We shall consider, in the following, a few examples of this type of analysis.

β) The state of a material is completely described if the stress Σ and the velocity $\boldsymbol{v}$ are known as functions of the radius vector $\boldsymbol{r}$ and the time t: $\boldsymbol{v} = \boldsymbol{v}(\boldsymbol{r}, t)$, $\Sigma = \Sigma(\boldsymbol{r}, t)$. Another way of description is the "material" one, where stress and velocity of each individual particle are given functions of time. It is known that, correspondingly, two types of differentiation are distinguished: the local differentiation $\partial/\partial t$, at fixed $\boldsymbol{r}$ and the material differentiation, d/dt, for fixed particle, where

$$\frac{d}{dt} = \frac{\partial}{\partial t} + v_x \frac{\partial}{\partial x} + v_y \frac{\partial}{\partial y} + v_z \frac{\partial}{\partial z} = \frac{\partial}{\partial t} + v \frac{\partial}{\partial s}, \qquad (95.1)$$

and $\partial/\partial s$ means differentiation in the direction of $\boldsymbol{v}$. Streamlines are lines which, at a fixed time t_0, have at each point the direction of $\boldsymbol{v}$, principal trajectories are lines which at $t = t_0$ have everywhere the first (second) principal stress direction etc. In general, the one-dimensional infinity of streamlines (in the plane) or of trajectories *changes in time*; so do the characteristics, the slip-line pattern, etc.

We call a state *steady* or *stationary* if $\partial/\partial t = 0$ for all magnitudes under consideration, i.e. at a fixed place $\boldsymbol{r}$ the same $\boldsymbol{v}$ and Σ hold for all time. In this case there is *one* family of streamlines, of trajectories, of slip lines; the pattern does not

---

[1] It may be that the addition of the plastic and elastic strain rates, (Eqs. (43.3)—(43.5)) in the derivation of the Reuss equations, while leading to elegant equations, contributes in some cases to the mathematical complication. This addition is not essential for the consideration of elastic strains in the plastic region.

[2] Historically, the "plastic rigid analysis" is not a conception due to v. Mises, who emphasizes the basically elastic plastic character of the perfectly plastic body. This analysis is however accepted by him whenever the problem warrants neglecting the elastic contributions. (See, as one example, the problem of Sect. 97.) Rigid-plastic analysis is applied to the flat ring or sheet by G. I. Taylor and by Hill because they are mainly interested in the mechanics of the distortion, which appears as a thickening near the hole for $p > 4k/\sqrt{3}$ when free plastic flow has set in; on the other hand, v. Mises, who is interested in the elastic-plastic state for $p < 4k/\sqrt{3}$, and in particular in the contained plastic deformation, uses an elastic-plastic analysis (Sects. 91, 92).

change in time. Of course, the same spatial region is not filled at different moments by the same particles; but the new particle has again the same v and Σ at a fixed place as had the particle which left the place.

In an unsteady problem the theory gives the solution for a particular instant t_0. If necessary, we may then use this solution to formulate boundary conditions for the next time interval, and so on. This is extremely laborious and rarely done. The study for the fixed moment t_0 is termed the problem of *incipient* plastic flow. In addition to steady problems, so-called problems of *pseudosteady state* have been identified, where the slip-line pattern, while not remaining fixed, changes only in scale.

Numerous interesting and important problems of steady flow in plane strain have been worked out in the last ten years by HILL, LEE, PRAGER, HODGE and other authors; these are problems of sheet-drawing, of extrusion, piercing, strip-rolling, etc.[1].

Problems exhibiting *axial symmetry* have also been considered under a plastic-rigid analysis. (For a survey see [*10*], p. 262 seq., particularly pp. 280—281). The independent variables are r and z (and not r alone, as in most problems considered in Chap. HI) while it is assumed that stresses and strains are independent of ϑ. The non-zero stresses and strains are σ_r, σ_ϑ, σ_z, τ_{rz} and ε_r, ε_ϑ, ε_z, γ_{rz}. In a more recent paper R. T. SHIELD[2] surveys the problem; he assumes plastic-rigid material and TRESCA's yield condition. The governing equations become statically determinate if a well-known hypothesis of HAAR and v. KÁRMÁN[3] is assumed satisfied which, in this case, takes the form that the circumferential stress σ_ϑ is equal to one of the two principal stresses in the r, z-plane. Then, four equations for four unknowns are obtained but the legitimacy of the assumption remains questionable. SHIELD studies several problems under these conditions.

We shall present in the following only: the problem of the unilaterally loaded *wedge* as a problem of incipient plastic flow, which also exemplifies a *stress discontinuity*, and the pseudosteady problem of a *plastic mass pressed between two rough rigid plates*.

b) Wedge with pressure on one face.

96. General discussion and velocity distribution. Consider a wedge of perfectly plastic material with vertex angle $2\beta_0 \geq 90°$. There exists a continuous stress solution due to PRANDTL[4]. In Fig. 54a the triangles ABC and BDE are regions of constant state. In ABC the principal stresses are $-p_0$, normal to AB, and $2k - p_0$, parallel to AB. In BDE, they are zero normal to BE and $-2k$ parallel to it. The region BCD is filled by a centered "fan" where the slip lines are radii and arcs of circles. The value of p_0 that produces the plastic flow is

$$p_0 = 2k(1 + 2\beta_0 - 90°). \tag{96.1}$$

On the other hand, we may consider the discontinuous solution of Fig. 54b[5]; the corresponding pressure value is easily found to be

$$p_0' = 2k(1 - \cos 2\beta_0). \tag{96.2}$$

---

[1] See explanations and literature in W. PRAGER: James Clayton Lecture, Proc. Instn. Mech. Engrs. **169**, 41—57 (1955), particularly p. 51, compare HILL [*10*], Chap. VII.

[2] R. T. SHIELD: Proc. Roy. Soc. Lond., Ser. A **233**, 267—287 (1955).

[3] A. HAAR and TH. v. KÁRMÁN: Nachr. Ges. Wiss. Göttingen, math.-phys. Kl. 204—218 (1909).

[4] L. PRANDTL: Göttinger Nachr. 74—89 (1920), and L. PRANDTL: Z. angew. Math. Mech. **1**, 15—20 (1921).

[5] For details and additional references cf., for example, H. GEIRINGER, Advances, p. 278 seq.

Hence $p'_0 < p_0$ for $2\beta_0 > 90°$, while $p_0 = p'_0 = 2k$ for $2\beta_0 = 90°$. Hencky[1] and Prandtl[2] have suggested that, in the case of two alternate solutions, the correct one is that requiring the smaller force. It is, however, seen that in the present case only the continuous solution gives a consistent velocity distribution. Indeed, in Fig. 54b the rigid region below $A'AHCKEE'$ is at rest, hence the particles in the plastic region of constant state, ACB and BCE, can move only parallel to AC and CE, respectively. The discontinuity line BC is inextensible (see Sect. 61); its particles can only move perpendicular to this line or not at all. These requirements are not compatible (unless all velocities are zero) and therefore no complete solution of this type exists. On the other hand, one easily sees that a uniquely determined velocity field exists in the case of Fig. 54a if the velocities are e.g. given along AB.

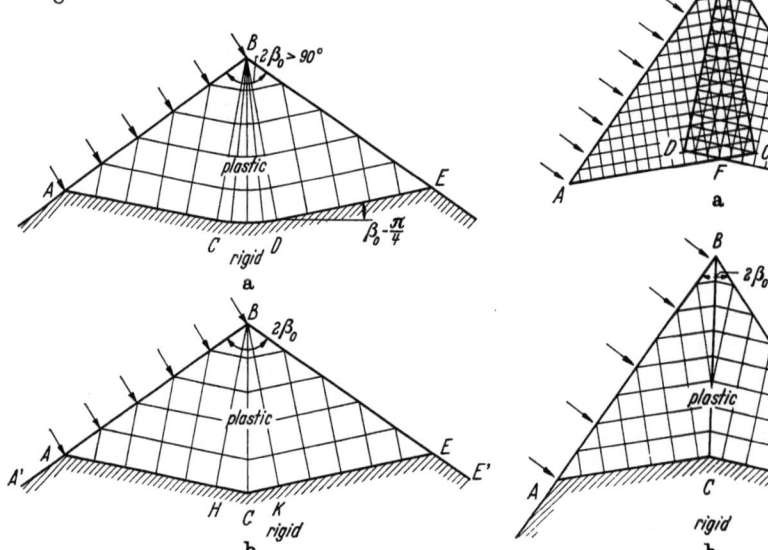

Fig. 54a and b. Wedge, $2\beta_0 > 90°$, under unilateral pressure: (a) continuous and (b) discontinuous slip-line field, (Prandtl).

Fig. 55a and b. Wedge, $2\beta_0 \leq 90°$, under unilateral pressure: (a) continuous and (b) discontinuous slip-line field (Prager).

We now turn to the more interesting case, $2\beta_0 < 90°$. Fig. 55a shows clearly that, without discontinuity line, an (impossible) multivalued stress field would result. Prager[3] and, in particular, Lee[4] have, however, determined a complete solution with a stress discontinuity. The slip lines are shown in the figure. We shall now determine the velocities.

The plastic-rigid boundary (Fig. 56) is a velocity characteristic; stress and velocity characteristics coincide, hence the characteristics ACE form this boundary. Along this boundary the normal velocity is zero. Along the discontinuity line BC, $\dot{\varepsilon}_y = 0$ (Sect. 70), $v_y = \text{constant} = 0$, since $v_y = 0$ at C. Along AB we prescribe one component of velocity, say the normal component; then along each of the lines BA, BC, AC, CE one velocity component is known. The two velocity equations are two linear differential equations of first order. In ACD we have a "mixed" boundary-value problem (Sect. 83) since we know one unknown

---

[1] H. Hencky: Z. angew. Math. Mech. **3**, 241—251 (1923).
[2] L. Prandtl: Z. angew. Math. Mech. **3**, 401—406 (1923).
[3] W. Prager: R. Courant Anniversary Volume, pp. 289—299. New York 1948.
[4] E. H. Lee: Proc. 3rd Symp. Appl. Math., pp. 213—228. New York 1950.

along the characteristic AC, one along AD; in CDF we again have a mixed problem, with data on CD and CF, and so on through smaller and smaller triangles. If we carry this out analytically, denoting by v_1 and v_2 the components of $\boldsymbol{v}$ in the characteristic directions, the expressions for v_1 and v_2 are seen to be sums of an increasing number of terms; as we approach the point B these infinite sums approach a limit. Actually, the velocity at B turns out to be normal to BC and of magnitude of $f_B/\cos\beta$ (as we could expect, since the component of $\boldsymbol{v}$ in direction of BC is zero). We can verify that the condition $\dot{E}=0$ at the discontinuity line holds true.

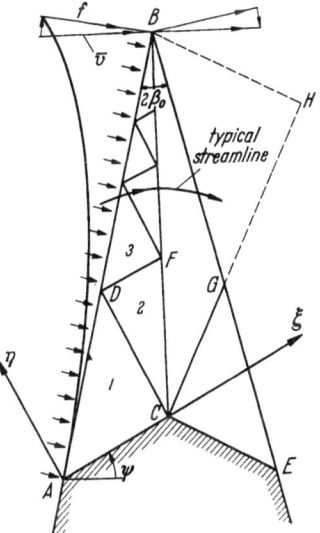

On the right side of the wedge the computation is easy, since now we know both components of velocity on BC. This determines the velocities in BCG; finally, from data along the characteristics CG and CE, the velocities in GCE follow. It can be seen that the velocities on the right are symmetric to those on the left—see the "typical streamline" in Fig. 56. (This is not obvious since the original data are not symmetric.) For details and comments, particularly regarding the given function f, which must have a positive second derivative, see LEE's original paper[1].

Fig. 56. Computation of velocities for wedge under unilateral pressure (LEE).

c) Plastic mass between rough rigid plates.

97. Infinite slab. An infinitely long slab of height $2h$ is pressed between two rough rigid plates which are forced to remain parallel, while they move at a uniform downward (upward) velocity c. The origin is at the center of the bar and the x-axis is parallel to the trace of the plates. While the mass becomes plastic, below each plate, symmetric to the y-axis, a rigid kernel will form,

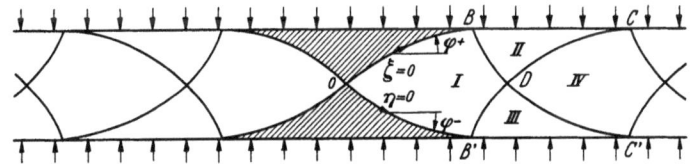

Fig. 57. Plastic mass between infinitely long rough plates.

bounded by characteristics (Fig. 57). Since material on the right side of this axis flows to the right, that on the left to the left, a discontinuity would arise if the slab were in plastic flow throughout. The stress field is that of (69.12) with $2A = h$, already given by PRANDTL[2]. The characteristics are two families of orthogonal cycloids with the horizontal lines $y = \pm h$ as envelopes.

To find the velocities, a boundary-value problem must be solved[3]. The boundary conditions are that everywhere where rigid material is adjacent to plastic material the normal velocities coincide. This boundary consists of parts of the slip lines $\xi = 0$ and $\eta = 0$ adjacent to the rigid kernel and of the horizontal envelope of the cycloids adjacent to the plates, beginning from the points of contact

---

[1] E. H. LEE, preceding footnote. See also H. GEIRINGER: Advances, pp. 278—281.
[2] L. PRANDTL: Proc. 1st Internat. Congr. Appl. Mech., Delft 1924, p. 43—54.
[3] H. GEIRINGER: Proc. 3rd Internat. Congr. Appl. Mech. II, 1930, pp. 185—190. See [6] for details.

with these slip lines. The domain I is limited by the four cycloids $\xi=0$, $\eta=0$, $\xi=\pi/4$, $\eta=\pi/4$. Along $\xi=0$ the normal component $c\cos\varphi'$ of the pressing rigid material equals the normal component v_1 of the plastic mass (v_1, v_2 are the components of $\boldsymbol{v}$ in the ξ- and η-directions) and similarly for $\eta=0$; hence, writing φ' and φ for the φ^+ and φ^- of Sect. 69:

$$\text{along } \xi=0: v_1 = c\cos\varphi' = c\cos\left(\frac{\pi}{4}-\eta\right) = \frac{c}{\sqrt{2}}(\sin\eta+\cos\eta) = g(\eta),$$
$$\text{along } \eta=0: v_2 = c\cos\varphi = c\cos\left(-\frac{\pi}{4}+\xi\right) = \frac{c}{\sqrt{2}}(\cos\xi+\sin\xi) = g(\xi).$$
(97.1)

By integration of (67.14), v_2 and v_1 are found along $\xi=0$ and $\eta=0$, respectively, where the integration constants are such that the same value of v_1 and v_2 results

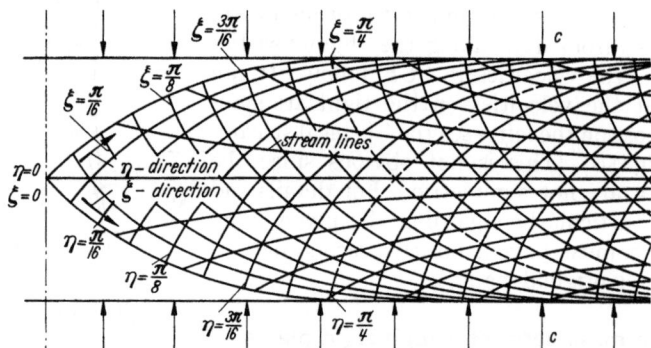

Fig. 58. Streamlines for plastic mass between infinitely long rough plates (GEIRINGER).

at $\xi=\eta=0$. Thus, with $f(x) = \frac{c}{\sqrt{2}}(\sin x - \cos x + 2)$, $g(x) = \frac{c}{\sqrt{2}}(\sin x + \cos x)$

$$\text{along } \xi = 0: v_1 = g(\eta),\ v_2 = f(\eta),$$
$$\text{along } \eta = 0: v_1 = f(\xi),\ v_2 = g(\xi).$$
(97.2)

We want a function $v_1(\xi,\eta)$ which satisfies Eq. (67.15) in I and takes on the values $g(\eta)$ and $f(\xi)$ on $\xi=0$ and $\eta=0$, respectively. Next there follows, by symmetry,

$$v_2(\xi,\eta) = v_1(\eta,\xi). \qquad (97.3)$$

The above boundary-value problem for v_1 can be solved explicitly by means of Riemann's method (Sect. 68) and it reduces to quadratures. Then, in the domain II, there is a mixed boundary-value problem since, in addition to knowing v_1 on the slip line between I and II, we know $v_1 = c$ on the horizontal straight line. To solve this problem, instead of v_1 we consider the function

$$w = v_1 - c\cos\left(\frac{\pi}{4}+\xi-\eta\right).$$

For w the same differential equation holds but with $w=0$ on BC[1]. This problem has been treated by substituting for it a characteristic-value problem with such an asymmetry in the data that $w=0$ on BC. Thus v_1, and consequently v_2 are found in II, and similarly in III. In IV we then have again a characteristic-value problem, etc. It should be noted that we obtain non-vanishing tangential velocities along $\xi=0$ and $\eta=0$ which are not equal to those of the pressing material. This tangential discontinuity across a characteristic is, however, admissible. Eventually, after v_1 and v_2 are found, the stream lines can be plotted (Fig. 58). Remember also the problem of MANDEL in Sect. 76.

---

[1] The use of w simplifies the formulas also in region I.

Thus, a complete solution has been found. It is, however, based on the assumption of plastic flow throughout (except for the rigid kernel at the center), of an infinite mass between infinite plates; these assumptions simplify the problem but may not be quite realistic. We now shall consider the same problem in a set-up which is physically more realistic.

98. Slab of material with overhanging ends between rough plates. We consider the compression of a slab of plastic material between parallel rigid rough plates of initial distance $2h$ and such that the slab overhangs them[1]. It is then clear that the material to the right of some curve ABA', where A is the right upper corner of the upper plate, will not reach the yield limit; it is considered rigid and AB and $A'B$ must be shear lines (Fig. 59). If ABA' is known, the stress field can in principle, be determined (Fig. 60) step-by-step in I, II, III, IV, ... by the second

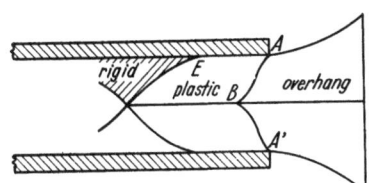
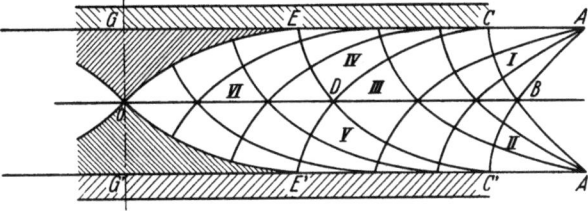

Fig. 59. Slab with overhanging ends between rough plates (HILL, LEE, TUPPER).

Fig. 60. Slip-line field for slab with overhanging ends between rough plates (HILL, LEE, TUPPER).

differential equation (66.10) and the fact that we know φ along ABA', along AE, and symmetrically. In fact, the ξ-lines (η-lines) meet the upper (lower) plate orthogonally on account of its complete roughness. The horizontal line AE is the envelope of η-lines, a limit line, just as in the preceding problem.

A suitable choice of AB is[2] to take it as a straight line of slope one, and $A'B$ symmetrically (Fig. 60). The singular points A and A' are centers of centered simple waves; the upper wave is between AB and AC, with concentric circular arcs as cross characteristics. For this choice of ABA', the material on the right of it is able to move as a rigid body. In fact, along AB the velocity v_2 in the direction of AB is constant by the second Eq. (66.11'), and since along this radius $v_2\,d\vartheta$ is constant, the first Eq. (66.11') shows that dv_1 is constant along it. Therefore, if in this fan the component v_1 normal to one radius is constant, the same holds for each radius; but along AC the normal component *is* constant. Hence the velocity along AB is constant.

The slip-line field in I and II is thus known. For the further computation Eqs. (67.11), (67.12) are used. The arrangement from here on is exactly the same as in the problem of the preceding section since the Eqs. (67.11) and (67.14) are of the same type. First, in III we have in $CBC'D$ a characteristic-value problem for either R^- or R^+ (which is explicitly solved by Bessel functions), and $R^+(\xi, \eta) = R^-(\eta, \xi)$, as in (97.3). In IV there is a mixed problem with R^- given along CD and $R^- = 0$ along the limit line EC. By the same trick as in the preceding section, we solve instead a characteristic problem with such asymmetric data that $R^- = 0$ results on CE. Then V and VI follow. The length-to-height ratio is now, of course, not arbitrary but depends on the stage of compression. In the work of HILL, LEE and TUPPER[1] (see Fig. 60) it is 6.72. The

---

[1] R. HILL, E. H. LEE and S. J. TUPPER: J. Appl. Mech. **8**, 46—52 (1951).
[2] See L. PRANDTL, Proc. 1st Internat. Congr. Appl. Mech., Delft 1924, pp. 43—54, who considers the problem with the overhanging plates.

curves in Fig. 60 towards the center are similar in form to cycloids and approach this form for increasing length-to-height ratio; as the characteristics approach the cycloidal form the stresses approach PRANDTL's stress distribution of the problem of Sect. 97. The value of p follows at each point from (66.8). The initial pressure at B equals k from equilibrium considerations, hence $p=k$ all along AB; the pressure along AC is easily found as $k\left(1+\frac{\pi}{2}\right)$, and so on. The pressure distribution approaches that of PRANDTL.

From the solution of the figure we can obtain solutions for smaller l/h ratios, the smallest being 3.64, corresponding to a solution progressing to point D only (from the right). Cutting off the field at various points between D and F corresponds to ratios between 3.64 and 6.72. By continuing the computations beyond field VI of the figure we obtain larger ratios; i.e., as the compression continues we deal with a sequence of blocks of increasing l/h ratio. HILL (see [10], p. 230) has also considered the case where $l/h < 3.64$, but greater than unity. There exists a solution where the central rigid kernel extends over the whole pressing plate.

After the slip-line field has been found (the problem is statically determinate like the preceding problems), the velocity distribution can be determined in exactly the same way as in the preceding section, of course based on the present slip-line field: we start in the middle and finish with the distribution across ABA', which must turn out constant. Again there is a tangential discontinuity along EF (and symmetrically)[1].

Other non-steady motion problems include those of *indentation*, started in 1920 by PRANDTL (see [10], p. 254 seq., [26], p. 206 seq.). See also the interesting critical discussion by LEE[2].

Reference Books.

[1] ALFREY, T.: Mechanical behavior of high polymers. New York: Interscience 1948.
[2] ARUTIUNIAN, N. CH.: Problems in the theory of creep. Moscow 1952.
[3] COLONETTI, G.: L'équilibre des corps déformables. Paris: Dunod 1955.
[4] FRENKEL, J.: Kinetic theory of liquids. Moscow 1943. Engl. transl. New York: Dover 1955.
[5] FREUDENTHAL, A. M.: Inelastic behavior of engineering materials and structures. New York: Wiley 1950. German transl. Berlin: VEB 1955.
[6] GEIRINGER, H.: Fondements mathématiques de la théorie des corps plastiques isotropes. Paris 1937.
[7] GOLDENBLATT, I. I.: Problems of the mechanics of deformable bodies. Moscow 1955.
[8] GRAMMEL, R. (editor): Deformation and flow of solids. Berlin: Springer 1956.
[9] GROSS, B.: Mathematical theories of viscoelasticity. Paris: Hermann 1953.
[10] HILL, R.: Mathematical theory of plasticity. Oxford: University Press 1950.
[11] HOFFMAN, O., and G. SACHS: Theory of plasticity for engineers. New York 1953.
[12] ILYUSHIN, A. A.: Plasticity. Moscow 1948. French transl. Paris: Eyrolles 1956.
[13] KACHANOV, L. M.: Theory of Plasticity. Moscow 1956.
[14] KACHANOV, L. M.: Problems of the theory of creep. Moscow 1949.
[15] KALISKI, S.: Dynamic boundary value problems in the theory of elastic and inelastic bodies. Warsaw 1957.

---

[1] The problem of the block pressed between overhanging plates, considered already by PRANDTL (Proc. 1st Internat. Congr. Appl. Mech., Delft 1924, pp. 43—54), has been studied by R. HILL: Dissertation, Cambridge 1948, Ministry of Supply Survey 1/48. Cf. also SOKOLOVSKY [26], p. 243 seq.

[2] E. H. LEE: J. Appl. Mech. **19**, 97—103 (1952). Discussed in H. GEIRINGER: Advances, p. 289 seq.

[16] LEBENSON, L. S.: Elements of the mathematical theory of plasticity. Moscow 1943.
[17] LOHR, E.: Mechanik der Festkörper. Berlin: W. de Gruyter & Co. 1952.
[18] MUELLER, F. H. (editor): Relaxationsverhalten der Materie. Darmstadt: Steinkopff 1953.
[19] NADAI, A.: Plasticity. New York: McGraw-Hill 1931.
[20] NADAI, A.: Theory of flow and fracture of solids, Vol. 1. New York: McGraw-Hill 1950.
[21] PHILLIPS, A.: Introduction to Plasticity. New York: Ronald 1956.
[22] NEAL, B. G.: Plastic Methods of Structural Analysis. New York: Wiley 1956.
[23] PRAGER, W., and P. G. HODGE: Theory of perfectly plastic solids. New York: Wiley 1951. German transl. Wien: Springer 1954.
[24] PRAGER, W.: Probleme der Plastizitätstheorie. Basel: Birkhäuser 1955.
[25] RZHANITSIN, A. R.: Problems of mechanics of time-deformable systems. Moscow 1949.
[26] SOKOLOVSKY, V. V.: Theory of plasticity. Moscow 1946, 1950. Revised German edition, Berlin: VEB 1955.
[27] WESTERGAARD, H. M.: Theory of Elasticity and Plasticity. Cambridge: Harvard University Press 1952.
[28] ZENER, C.: Elasticity and anelasticity of metals. Chicago: Chicago University Press 1948.

Rheology.

By
M. REINER.

With 45 Figures.

A. Preliminaries.

I. Introduction.

1. Some definitions[1]. Although the word "rheology" is derived from the Greek "ῥέω", which means "flow", it is used in the wider sense to mean that branch of physics which is concerned with the deformation of materials.

Deformation (D) is a movement of parts or particles of a material body with regard to one another such that the continuity of the body is not destroyed. If under the action of finite forces the deformation of the body increases with time continuously and irreversibly, the material is said to *flow* (f). *Plastic flow* occurs only when the forces exceed a certain limit, indicated by the *yield point* of the material. *Viscous flow* occurs under the action of any forces however small, albeit at a reduced rate of deformation, which rate vanishes with the forces. The deformation of a body is produced by *loads* (P), which together with the reactions from supports and the body-forces constitute the *external forces*. When the forces are removed, part of the deformation will always be recovered. This part is elastic, and is called *strain*[2] (e). From this definition it follows that we cannot determine the strain from a loading diagram in which the total deformation is recorded, but can upon unloading, when we see how much deformation is recovered. The recovery from deformation is a manifestation of the property of *elasticity*; accordingly all *strain is elastic*. In general all strain vanishes when the load is removed. However, as a result of certain manufacturing processes (e.g., the rolling of mild steel, the setting of concrete or hardening of cheese[3]) a body may become *self-strained*, when internal stresses are present in the absence of external forces. In every case there is *no strain without stress*. In contradistinction, while nascent viscous and plastic deformations are accompanied by stresses, these deformations do not disappear with the removal of the load and accordingly there will be *stressless deformations*. The investigation of these different kinds of deformations in relation to the stresses by which they are accompanied is the subject of *rheology*. It should be noted that there are *stressless volume changes* resulting from variations of temperature or of moisture content of such materials as wood, concrete and earth which, when not resisted, are not accompanied by stresses, even in the nascent state. While they do not form part of the subject-matter of rheology, they must be taken into account in rheological observations.

---

[1] When terms defined in this and other sections are applied upon real phenomena, logical difficulties often arise. For instance, it may not be easy to decide whether a deformation is irreversible or not. Sometimes a slight agitation may make it reversible.

[2] The reader must be warned that this definition is not in general use. For instance, the tensile-test diagram of mild steel is termed "stress-strain" diagram, even if the recoverable part of the elongation is negligibly small.

[3] M. REINER, G. W. SCOTT BLAIR and G. MOCQUOT: Lait **29**, 351 (1949).

Phenomenological rheology, which is the subject of the present article, deals with homogeneous or quasi-homogeneous and isotropic or quasi-isotropic materials on a phenomenological level, i.e. considering them as *continuous media*. Both the aeolotropic crystal and the ultimate discrete entities, such as molecules and atoms which constitute the medium, are outside its range. Rheology is separated from thermodynamics by the provision that rheological processes take place under *isothermal* conditions (e.g., placing viscometers in thermostatic baths).

Macro-rheology regards all materials as they may appear to superficial inspection by the naked eye; i.e. as if they were homogeneous and devoid of structure. Only pure liquids and perfect single crystals are phenomenologically homogeneous; most materials with which rheology deals are *dispersed systems* consisting of two or more phases of which one will usually be air contained in pores which may be visible or invisible. For instance such an apparently homogeneous material as gold has its density raised from 19.258 to 19.367 "by compression between dies used in coining" [Lord KELVIN (1875)][1].

Rheologically speaking, dispersed systems can be distinguished as *sols* which act in the manner of liquids and *jellies* which act in the manner of solids, with *gels* forming a continuous transition from the first to the second. In sols the continuous phase or the *dispersion medium* is a liquid; the *dispersed phase* can be a liquid, when one speaks of an *emulsion*, or a solid in the case of a *suspension*. In a gel the solid phase prevails, but does not necessarily form one continuous structure throughout the body. The gel may consist of "sponges" separated by thin layers of an amorphous phase. Using such pictures, a polycrystalline metal may be considered as a gel where the thin layers of atomic disorder between the crystals form the "liquid" phase. Even water has been considered from this view point with "water crystals" (cybomas) forming a structure, but distinguished from the polycrystalline metal-structure by the ease with which thermal vibrations may cause a water-molecule to leave its regular lattice point, thus creating a "Frenkel defect"[2]. As an extreme example of what the rheologist may consider as a gel, we may mention *concrete*. In good concrete a solid skeleton is formed by the coarse aggregates (broken stone or gravel), the interstices being filled with mortar. The mortar itself, consisting of fine aggregates (sand) and cement, can be considered as the liquid phase. The mortar will, in general, not completely fill the interstices of the coarse-aggregate skeleton, which in turn will not form one solid structure throughout the body of the concrete.

At the other extreme, sols are included in phenomenological rheology if the dispersed phase consists of molecules which can be considered as small bodies differing from large masses only in scale.

A material is *quasi-homogeneous* when the size of the largest dispersed element is smaller than the smallest volume element the deformation of which is under consideration. Concrete, as described above, can be treated as quasi-homogeneous when the dimensions of the structural elements made of it are much greater than the dimensions of the largest stones of which it is composed.

---

[1] When the year only of the publication is quoted, the complete reference is given in the Bibliography.

[2] Compare Chap. VII by FORSLIND in M. REINER (1954). The structure of liquids is discussed in detail in the article by H. S. GREEN in volume X in accordance with which "liquids have a molecular structure devoid of long-range order, but sufficiently closely packed to ensure that any molecule is in continual interaction with its neighbours". Even in gases clusters of two or more molecules may exist, and the incidence of such will increase with increase of density. This may be the cause of the elasticoviscosity of air as described in Sect. 35 below.

Materials are considered to be *quasi-isotropic* when the smallest volume element contains anisotropic dispersed elements *of all orientations*. Such is the case of a commercial polycrystalline metal, where the crystals which constitute it are oriented at random. A quasi-isotropic material can become anisotropic through deformation, e.g., a metal sheet after rolling or a metal rod through drawing.

Microrheology derives the rheological behaviour of two- and multiphase systems from the known rheological behaviour of their constituents. The first and the most famous microrheological investigation was EINSTEIN's (1906, 1911) derivation of the viscosity of a suspension, about which more is said in Sect. 40 below. Here the solid phase was assumed to consist of rigid spheres and the continuous phase of a simple viscous liquid. The actual composition of dispersed systems makes such a mathematical approach unworkable beyond the simplest cases. In these circumstances one substitutes for the unknown structure a *mechanical model* which is supposed to behave analogously to the real structure. Such models consist of different elements such as elastic springs, viscous dashpots, and sliding blocks, which will in general have no exact counterpart in the real material. More about this method is said in Sect. 13 below.

Where phenomenological rheology borders on other branches of physics, chemistry, or psycho-physiology, a belt of subjects is formed which may be collected under the term *metarheology*. They are dealt with in other contributions to the present volume. The last mentioned branch is not represented in the present Encyclopaedia but is treated in the publications of SCOTT BLAIR and his school (1949).

2. Mechanical foundations. Inasmuch as every particle of a body can be considered as a body subjected in its movements to the laws of mechanics, rheology is founded on mechanics. Considering a particle in the form of a *volume element* we may apply the six *dynamical equations* of mechanics

$$\left. \begin{array}{l} \Sigma P_l = 0, \\ \Sigma M_l = 0 \end{array} \right\} \qquad (2.1)[1]$$

which have here been written as equations of equilibrium by applying D'ALEMBERT's principle. The forces acting upon the volume-element are *body-forces* (B), such as gravity, proportional to the mass of the volume-element, and *surface forces*. The latter are exerted either by loads upon that part of the surface of the element which may be part of the boundary of the body under consideration, or by the rest of the body acting through the interface separating it from the element, in accordance with NEWTON's law of equal action and reaction. The surface force per unit area is called *traction*. The aggregate of all tractions acting upon surface elements of all orientations contained by a point in the body is called the *stress*[2] (s_{lm}) *at the point*. It is a tensor of rank two. Applying the second Eq. (2.1) makes the tensor symmetrical[3], so that

$$s_{lm} = s_{ml} \qquad (2.2)$$

with six components. The three first dynamical equations (2.1) are therefore insufficient to determine the tensor and the problem is threefold *mechanically indeterminate*.

---

[1] When no special mention is made of the reference system it is asssumed to be orthogonal Cartesian. Tensor indices are mainly l and m.

[2] In engineering science "stress" is used synonymously with "traction".

[3] This does not apply if the material is magnetic. The symmetry of the stress-tensor is therefore a postulate. It has been named "BOLTZMANN's postulate".

3. Classical kinematics. The kinematic state of a body is determined by the "displaced" position x_l of all its particles at all times t, the particles[1] being identified by their "initial" position $_l x$ at the time $t=0$. The particles make up the continuum of a single body, if particles adjacent at any time t remain so during the time-interval dt following t or, in other words, if the infinitesimal displacements du by which the particles change their positions during dt are continuous functions of the coordinates. Therefore

$$du_l = u_{l,\alpha} dx_\alpha. \qquad (3.1)[2]$$

Proceeding from the particle x_l to another particle $x_l + dx_l$ in a certain direction $\vec{ds}$, we have

$$dx = ds\,[ds \cdot x_l], \qquad (3.2)$$

where $[ds \cdot x_l]$ indicates the direction cosines of the angles (ds, x_l), and (3.1) becomes

$$\frac{du_l}{ds} = u_{l,\alpha}[ds \cdot x_\alpha]. \qquad (3.3)$$

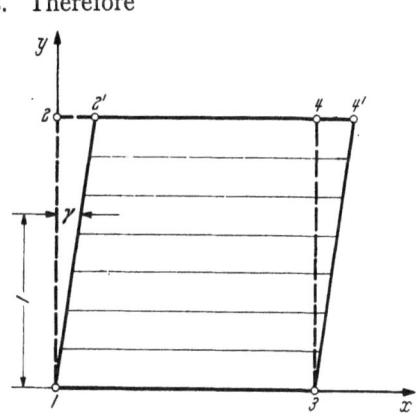

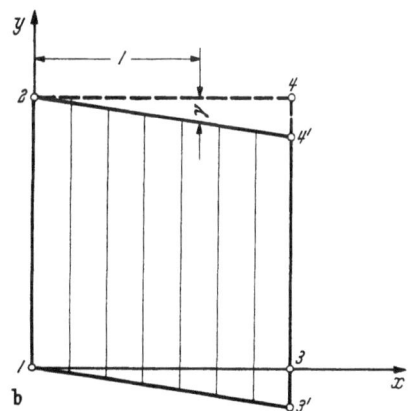
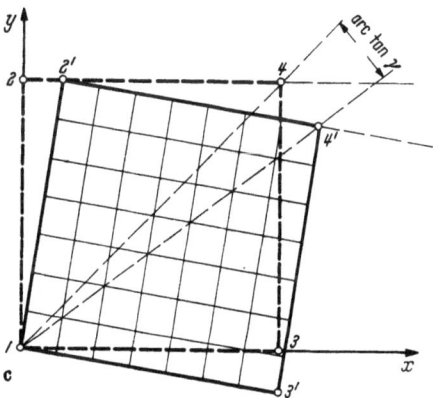

Fig. 1 a—c. Superposition of two simple shears resulting in a rotation. γ Amount of shear = tangent of angle of rotation. (a) Simple shear parallel to the xz plane. (b) Simple shear parallel to the yz plane. (c) Superposition of both shears.

This defines the *displacement-gradient tensor*

$$\gamma_{lm} = u_{l,m}. \qquad (3.4)$$

It should be noted that this tensor is not symmetrical.

A special case of such displacement is *simple shear* given by

$$u_x = \gamma y; \quad u_y = u_z = 0. \qquad (3.5)$$

If another simple shear

$$u_y = -\gamma x; \quad u_z = u_x = 0 \qquad (3.6)$$

---

[1] Mathematically, a particle is to be considered as a material point, i.e. a point which possesses mass and can be identified.

[2] We denote summation indices by Greek letters, omitting the summation sign Σ, and we use $u_{l,m}$ for $\partial u_l/\partial x_m$.

is superimposed and γ is infinitesimal, the result is a rotation as shown in Fig. 1, and not a deformation. This shows that the gradient-tensor is not suitable for the description of a deformation. The antisymmetrical tensor

$$\omega_{lm} = \tfrac{1}{2}(u_{l,m} - u_{m,l}) \tag{3.7}$$

is a measure of the *infinitesimal rotation* of the medium[1]. Therefore the *infinitesimal relative deformation*[2] during the time dt may be measured by the symmetrical tensor

$$d_{lm} = \gamma_{lm} - \omega_{lm} = \tfrac{1}{2}(u_{l,m} + u_{m,l}) \tag{3.8}$$

which defines its *Cauchy measure*. With

$$v_l = \dot{u}_l \tag{3.9}$$

we get analogously the *flow-tensor*

$$f_{lm} = \tfrac{1}{2}(v_{l,m} + v_{m,l}) = \dot{d}_{lm}. \tag{3.10}$$

If the displacement u_l is such that it is recovered upon the release of the load, the tensor given by the expression on the right of (3.8) defines the *infinitesimal strain tensor* ε_{lm}.

This is the procedure of the classical mechanics of continua. Like every tensor, the deformation tensor can be resolved into an isotropic component and a *deviator* indicated by the subscript (0) in accordance with

$$d_{lm} = \tfrac{1}{3} d_{\alpha\alpha}\,\delta_{lm} + d_{lm(0)}. \tag{3.11}$$

In this case

$$d_{\alpha\alpha} = I_d, \tag{3.12}$$

the first invariant of the infinitesimal deformation tensor, is identical with the *cubical dilatation* d_v, and the second term then determines a *distortion* or change of shape without change of volume. Similarly

$$\varepsilon_{lm} = \tfrac{1}{3}\varepsilon_{\alpha\alpha}\,\delta_{lm} + \varepsilon_{lm(0)} \tag{3.13}$$

and

$$\varepsilon_{\alpha\alpha} = I_\varepsilon = \varepsilon_v, \tag{3.14}$$

where ε_v is the *volumetric strain*. When the flow tensor f_{lm} is resolved on the model of (3.11), it is usually assumed that there is *no volume flow* i.e. no irreversible cubical dilatation, or

$$f_{\alpha\alpha} = I_f = f_v = 0, \tag{3.15}$$

and the flow tensor is therefore identical with its deviator. It will be shown in Sect. 24 that this view is too narrow.

No attempt is made in classical kinematics to define the tensor D_{lm} of *finite deformation*. This is not a straight-forward problem. It will be dealt with in Sect. 29 below.

---

[1] The rotation, being of the kind which a rigid body may suffer, is of no interest in rheology. However, phenomena such as described by the author [Physics Today **9**, 16 (1956)] form border cases.

[2] If the length l_0 of a bar is changed to l through elongating it by Δl, this constitutes a *deformation*; if Δl is related to l_0 this constitutes a measure of *relative deformation*. However, "deformation" is generally used for "relative deformation", and we shall follow this practice, except when the terms conflict.

4. Dynamics. The traction $\vec{s}_n$, where n indicates the direction of the external normal, can be resolved into a *normal component (tension or compression)*[1] in the direction $\vec{n}$

$$\sigma_n = s_{nn}, \tag{4.1}$$

and the *tangential component (shearing traction)*[1] in the direction $\vec{t}$, which is normal to $\vec{n}$ in the plane through $\vec{s}_n$ and $\vec{n}$,

$$\tau_n = s_{nt}. \tag{4.2}$$

The symbol n does not indicate a running index. This shows that $\vec{s}_n$ is a pseudo-vector, and really a degenerated tensor. For this reason SOKOLNIKOFF[2] writes $\overset{n}{s}$. The traction $\vec{s}_n$ is connected with the stress components by the three equations

$$s_{nl} = s_{\alpha l}[n \cdot x_\alpha]. \tag{4.3}$$

The application of the first of Eqs. (2.1) upon a prismatic volume-element yields

$$s_{\alpha l, \alpha} + \varrho (B_l - \ddot{x}_l) = 0. \tag{4.4}$$

These are called the *dynamical equations* (or *stress-equations* when $\ddot{x}$ vanishes). For their solution the tractions $\bar{s}_{lm}$ produced by the loads upon the boundary of the body or the *dynamical boundary conditions must be known*.

When resolving the stress-tensor in accordance with Eq. (3.11)

$$s_{lm} = \tfrac{1}{3} s_{\alpha\alpha} \cdot \delta_{lm} + s_{lm(0)}, \tag{4.5}$$

the negative isotropic component

$$-p_m = \tfrac{1}{3} s_{\alpha\alpha} \tag{4.6}$$

defines the *mean pressure*, while the deviator $s_{lm(0)}$ is sometimes spoken of as "stress-differences".

5. Energetics. When acting upon a body, the external forces P perform work w_P per unit volume, the *specific power input* being $\dot{w}_P$. The power (per unit volume) due to the surface forces is $(s_{\alpha\beta} \dot{x}_\beta)_{,\alpha}$, which is equal to $s_{\alpha\beta} \dot{x}_{\beta,\alpha} + s_{\alpha\beta,\alpha} \dot{x}_\beta$, while the power (again per unit volume) due to the body forces is $\varrho B_\alpha \dot{x}_\alpha$. Therefore (changing dummy indices)

$$\dot{w}_P = s_{\alpha\beta} \dot{x}_{\beta,\alpha} + (s_{\alpha\beta,\alpha} + \varrho B_\beta) \dot{x}_\beta. \tag{5.1}$$

However, from (4.4)

$$s_{\alpha l, \alpha} + \varrho B_l = \varrho \ddot{x}_l, \tag{5.2}$$

and therefore

$$\dot{w}_P = s_{\alpha\beta} \dot{x}_{\beta,\alpha} + \varrho \dot{x}_\alpha \ddot{x}_\alpha \tag{5.3}$$

or, considering (3.10)

$$\dot{w}_P = s_{\alpha\beta} \dot{d}_{\alpha\beta} + \varrho \dot{x}_\alpha \ddot{x}_\alpha. \tag{5.4}$$

But $\dot{e}_k$, the *kinetic energy* imparted to the unit element in unit time, is

$$\dot{e}_k = \tfrac{1}{2} \varrho \overline{(\dot{x}_\alpha)^2} = \varrho \dot{x}_\alpha \ddot{x}_\alpha. \tag{5.5}$$

Therefore the *specific stress power* $\dot{w}_s$ is

$$\dot{w}_s = \dot{w}_P - \dot{e}_k = s_{\alpha\beta} \overline{\dot{d}_{\alpha\beta}}. \tag{5.6}$$

---

[1] The σ und τ notation is used in engineering literature.
[2] I. S. SOKOLNIKOFF: Elasticity. New York and London 1946.

By applying the resolutions (3.11) and (4.5) we find

$$\begin{aligned}\dot{w}_s &= \dot{w}_v + \dot{w}_{(0)}, \\ \dot{w}_v &= s_{\alpha\alpha}/3 \cdot \dot{d}_{\beta\beta}, \\ \dot{w}_{(0)} &= s_{\alpha\beta(0)} \dot{d}_{\alpha\beta(0)}.\end{aligned} \right\} \qquad (5.7)$$

The stress power is therefore the sum of the power used for changing the volume and the power used for changing the shape.

The stress work must comply with the two thermodynamical laws. For isothermal processes both are combined in the *Gibbs-Helmholtz equation*[1]

$$\left.\begin{aligned}\delta w &= \delta \Phi + \delta \psi, \\ \delta \psi &\geq 0,\end{aligned} \right\} \qquad (5.8)$$

where Φ is the *intrinsic*[2] *free energy density* and ψ the *bound energy density*. The second Eq. (5.8) expresses the condition that while in the case of thermodynamic equilibrium ψ vanishes, in other cases it can only increase. Generally, in rheological processes of every *real* material both functions Φ and ψ will be involved, but in classical elasticity ψ vanishes identically, and the sign = in the second Eq. (5.8) refers to this ideal case. Taking into consideration our definition of strain as the recoverable part of deformation we can thus identify Φ with the *strain-work* w_e, which is the conserved part of the stress-work.

6. The rheological method and plan of this article. Rheology as the physics of deformation deals with the coordination of the forces acting upon a material body and the deformation produced. Proceeding from the force-vector to the stress-tensor, mechanics makes available Eq. (4.4) which, as has been said in Sect. 2, is threefold indeterminate.

Rheology solves the problem by *postulating* for every kind of material a rheological equation of state or, in short, a *rheological equation* between the stress tensor and the deformation tensor and their time derivatives[3], expressing the deformation tensor in terms of the three components of the *displacement*-vector u_l, or the *velocity*-vector v_l, thereby reducing the number of unknowns from six to three. The form of the rheological equation is suggested by rheological experiments; but being an *exact* relation between mathematical entities, it describes the behaviour of an *ideal* material which thus serves as a standard of comparison in the analysis of the behaviour of an actual substance. The ideal materials are each named after scientists who first postulated their respective equations[4].

The rheological equation is generally of the form

$$R(s_{lm}, D_{no}) = 0, \qquad (6.1)$$

where D and s stand also for their time derivatives, and D may denote finite or infinitesimal deformation and/or strain, while l, m and n, o are running indices. In an isotropic material, this can usually be split into two equations, one R_v related to volume changes, the other $R_{(0)}$ referring to distortion[5]. The scalar parameters appearing in the rheological equation are rheological "coefficients"

---

[1] K. Weissenberg: Abh. preuß. Akad. Wiss. **1931**, H. 2.

[2] i.e. after deduction of the kinetic energy produced. This is effected in viscometry through the "kinetic energy correction".

[3] Their space derivatives do not appear in the rheological equation.

[4] This terminology was introduced by v. Mises, Proc. 3rd Internat. Congr. Mech. **2**, 3 (1930).

[5] This cannot be done if the second order effect of dilatancy is taken into account, compare Sect. 29 below.

or "moduli", characterizing the rheological properties of the material[1]. Using this terminology, both the "Euclid solid" = rigid body (whose rheological equation degenerates to $D_{lm} \equiv 0$) and the "Pascalian liquid" = "ideal liquid" (whose rheological equation degenerates to $s_{lm(0)} \equiv 0$) are outside the field of rheology. These two "bodies" in which no rheological coefficients appear, belong to Mechanics proper. Rheology enters from the side of solids with the *Hooke-body*; from the side of liquids with the *Newtonian liquid*. The former, which is a perfectly elastic solid, is treated in the classical theory of elasticity. The latter, which is a "simple" viscous liquid, is treated in classical hydrodynamics. If to this is added the *St. Venant body*, a solid which possesses a yield point below which it is strained elastically, but above which it flows plastically at constant stress, and which is the subject of ideal plasticity, it will be shown that all other "rheological bodies" so far investigated can be considered as built up of a combination of these three fundamental bodies. In this rheological systematisation account must be taken of the basic fact that *under hydrostatic pressure all materials, whether solids or liquids, behave in a far reaching approximation essentially in the same manner:* they are perfectly elastic. Rheological differences are strikingly revealed under shear (or more generally, in distortion) only. Every material therefore possesses two rheological equations, one volumetric, common to all, and the other distortional, specific for itself.

In representing the rheological method, this article is planned on the following lines. After introducing some mathematical tools which we need in the development, we shall mention briefly in Part A, II, the Classical Bodies as far as is required for an exposition of the rheological bodies. We shall then treat in Part B Macrorheology, limiting ourselves in the field of elasticity to *infinitesimal strains*. Finite elasticity is treated in this Encyclopaedia in Vol. X. However, contrary to prevailing belief, infinitesimal elasticity is not restricted to first order phenomena, as will be shown later. Part B is accordingly divided into two Subparts, B, I treating first order and B, II higher order phenomena of deformation and flow; to these is added B, III, which deals with the limits imposed on deformation by the strength of the material. In Part C, we treat rather briefly Microrheology. Because of the great mathematical difficulties, the theory of this branch of rheology is still in a primitive stage[2]. This is followed in Part D by a sketch of rheometry, a subject in which physical considerations take second place behind purely technological ones. Finally, in Part E, we add miscellaneous subject matter which, if included in the body of exposition, would have formed inconvenient interruptions in a smooth representation.

We have not included a treatment of the rheological properties of different materials, as this would be passing from the field of physics into the field of technology. The reader will find such information in the books quoted in the Bibliography.

7. Some mathematical tools. *α) Physical tensors.* The physical quantities appearing in rheology are represented by tensors of zero rank (scalars), first rank (vectors) and second rank (tensors in the narrower sense) in a three-dimensional Euclidean space. These tensors are expressed by their physical components[3],

---

[1] In the classical approach they are assumed to be "constants".
[2] For a more detailed review see J. J. HERMANS (Editor), 1953.
[3] Physical tensor components were introduced by A. J. McCONNEL, Applications of the absolute differential calculus, London and Glasgow 1931—1947. Compare also OLLENDORF, Die Welt der Vektoren, Wien 1950, and C. TRUESDELL, Z. angew. Math. Mech. **33**, 345 (1953); **34**, 69 (1954). The concepts were further developed by BRAUN and KARNI in papers awaiting publication.

which are related to what may be named the "geometrical" components, as used in geometry and other branches of physics, by

$$\left.\begin{aligned} v(l) &= v^l (g_{ll})^{\frac{1}{2}} \\ t(lm) &= t^{lm} (g_{ll} g_{mm})^{\frac{1}{2}}. \end{aligned}\right\} \quad (7.1)$$

The physical components are indicated by placing the indices half way up, and in brackets. *Physical vector components act by the parallelogram rule to form the vector, and in contradistinction to the geometrical components they are all of the same physical dimension, namely, the same as the vector field to which they belong.* In orthogonal Cartesian coordinates, when $g_{lm} = \delta_{lm}$, the physical components are identical with the geometrical components. In a curvilinear orthogonal coordinate system the physical components at a given point are the components in a local Cartesian system whose axes are parallel to the coordinate curves at the point. To convert a tensor equation written in general geometrical components into one in physical components one replaces the *mixed* geometrical components by physical components[1], and the covariant derivative by the physical derivative, as explained below. To convert a tensor equation written in Cartesian components into one in orthogonal physical components one proceeds as follows[2]:

$$\left.\begin{array}{ll} \text{replace} & \text{by} \\ v_m & v(m) \\ t_{lm} & t(lm) \\ \varphi_{,l} & \varphi_{,(l)} = (g^{ll})^{\frac{1}{2}} \dfrac{\partial \varphi}{\partial x(l)} \\ v_{l,m} & v(l,m) = (g^{mm})^{\frac{1}{2}} \dfrac{\partial v(l)}{\partial x(m)} + G(l, \alpha\, m)\, v(\alpha) \\ t_{lm,n} & t(l m, n) = (g^{nn})^{\frac{1}{2}} \dfrac{\partial t(l\, m)}{\partial x(n)} + G(l, n \alpha)\, t(\alpha\, m) + \\ & \quad + G(m, n \alpha)\, t(l \alpha), \end{array}\right\} \quad (7.2)$$

where the physical Christoffel symbol $G_{l,mn}$ is defined by

$$G(l, m\, n) = \frac{1}{2} (g^{ll} g^{mm} g^{nn})^{\frac{1}{2}} \left(\frac{\partial g_{lm}}{\partial x^n} - \frac{\partial g_{mn}}{\partial x^l} \right). \quad (7.3)\,[3]$$

The physical Christoffel symbol is equal to the curvature of the l-coordinate curve at the point. To determine the physical Christoffel symbols in curvilinear coordinates the radii of curvatures of the coordinate curves may therefore be measured or calculated, and their reciprocals taken.

In rheological investigations the reference systems used are mostly orthogonal Cartesian and often orthogonal curvilinear, (especially cylindrical and spherical) occasionally rectilinear skew, and in exceptional cases general curvilinear. In all these cases physical components are used and indices are therefore subscripts and not half-way-up.

---

[1] I. BRAUN: Bull. Res. Council Israel 1, 127 (1951).
[2] The method has been developed by BRAUN, l. c.
[3] Note that the physical Christoffel symbol is antisymmetric, in contradistinction to the geometrical Christoffel symbol and the $l\, n$ indices therefore cannot be exchanged.

Because of the wide use made in rheology of *cylindrical coordinates* we shall write out the relevant expressions explicitly. We have

$$x^1 = r, \quad x^2 = \theta, \quad x^3 = z \qquad (7.4)$$

$$g_{lm} = \begin{Vmatrix} 1 & 0 & 0 \\ 0 & r^2 & 0 \\ 0 & 0 & 1 \end{Vmatrix}. \qquad (7.5)$$

Therefore

$$G_{2,21} = -G_{1,22} = \frac{1}{r} \qquad (7.6)$$

with all other components vanishing, and

$$dx_r = dr, \quad dx_\theta = r\, d\theta, \quad dx_z = dz, \qquad (7.7)$$

$$2\omega_{r\theta} = \frac{\partial u_r}{\partial \theta} - \frac{\partial u_\theta}{\partial r}; \quad 2\omega_{\theta z} = \frac{\partial u_\theta}{\partial z} - \frac{\partial u_z}{\partial \theta}; \quad 2\omega_{zr} = \frac{\partial u_z}{\partial r} - \frac{\partial u_r}{\partial z}, \qquad (7.8)$$

$$\left. \begin{array}{c} d_{rr} = \dfrac{\partial u_r}{\partial r}; \quad d_{\theta\theta} = \dfrac{1}{r}\dfrac{\partial u_\theta}{\partial \theta} + \dfrac{u_r}{r}; \quad d_{zz} = \dfrac{\partial u_z}{\partial z}; \\[4pt] 2d_{r\theta} = \dfrac{1}{r}\dfrac{\partial u_r}{\partial \theta} + \dfrac{\partial u_\theta}{\partial r} - \dfrac{u_\theta}{r}; \quad 2d_{\theta z} = \dfrac{\partial u_\theta}{\partial z} + \dfrac{1}{r}\dfrac{\partial u_z}{\partial \theta}; \\[4pt] 2d_{zr} = \dfrac{\partial u_r}{\partial z} + \dfrac{\partial u_z}{\partial r}. \end{array} \right\} \qquad (7.9)$$

The dynamical equations (4.4) become

$$\left. \begin{array}{c} \dfrac{\partial s_{rr}}{\partial r} + \dfrac{1}{r}\dfrac{\partial s_{r\theta}}{\partial \theta} + \dfrac{\partial s_{rz}}{\partial z} + \dfrac{(s_{rr} - s_{\theta\theta})}{r} + \varrho(B_r - a_r) = 0, \\[4pt] \dfrac{\partial s_{\theta r}}{\partial r} + \dfrac{1}{r}\dfrac{\partial s_{\theta\theta}}{\partial \theta} + \dfrac{\partial s_{\theta z}}{\partial z} + \dfrac{2 s_{r\theta}}{r} + \varrho(B_\theta - a_\theta) = 0, \\[4pt] \dfrac{\partial s_{zr}}{\partial r} + \dfrac{1}{r}\dfrac{\partial s_{z\theta}}{\partial \theta} + \dfrac{\partial s_{zz}}{\partial z} + \dfrac{s_{rz}}{r} + \varrho(B_z - a_z) = 0, \end{array} \right\} \qquad (7.10)$$

$$\left. \begin{array}{c} a_r = \ddot{r} - r\dot{\theta}^2, \\ a_\theta = r\ddot{\theta} + 2\dot{r}\dot{\theta}, \\ a_z = \ddot{z}. \end{array} \right\} \qquad (7.11)$$

β) *Tensor-invariants.* The *proper* (or *principal*) *values* of the tensors will be denoted by $t(i), t(j), t(k)$, where i, j, k are mutually perpendicular principal directions. The principal invariants of the symmetrical tensor are

$$\left. \begin{array}{l} I_t = t_{xx} + t_{yy} + t_{zz} = t(i) + t(j) + t(k), \\ II_t = t_{xx} t_{yy} + t_{yy} t_{zz} + t_{zz} t_{xx} - t_{xy}^2 - t_{yz}^2 - t_{zx}^2 \\ \quad = t(i)\, t(j) + t(j)\, t(k) + t(k)\, t(i), \\ III_t = t_{xx} t_{yy} t_{zz} + 2 t_{xy} t_{yz} t_{zx} - t_{xx} t_{yz}^2 - t_{yy} t_{zx}^2 - t_{zz} t_{xy}^2 = t(i)\, t(j)\, t(k). \end{array} \right\} \qquad (7.12)$$

When the tensor is resolved into an isotropic component and a deviator in accordance with

$$t_{lm} = \frac{t_{\alpha\alpha}}{3}\delta_{lm} + t_{lm(0)}, \qquad (7.13)$$

the first invariant of the deviator vanishes or

$$I_{t(0)} = t_{(0)}(i) + t_{(0)}(j) + t_{(0)}(k) = 0, \qquad (7.14)$$

while the second is

$$\begin{aligned}II_{t(0)} &= -\tfrac{1}{2}[t_{xx(0)}^2 + t_{yy(0)}^2 + t_{zz(0)}^2 + 2t_{xy(0)}^2 + 2t_{yz(0)}^2 + 2t_{zx(0)}^2] \\ &= -\tfrac{1}{2}[t_{(0)}^2(i) + t_{(0)}^2(j) + t_{(0)}^2(k)].\end{aligned} \qquad (7.15)$$

When the stress deviator is two-dimensional, or $t_{(0)}(k) = 0$, we have from (7.14)

$$t_{(0)}(j) = -t_{(0)}(k) \qquad (7.16)$$

and from (7.15)

$$II_{(0)} = -t_{(0)}^2(i), \qquad (7.17)$$

while from the third Eq. (7.12)

$$III_{t(0)} \equiv 0. \qquad (7.18)$$

γ) *Isotropic tensor functions.* As said in Sect. 6, the fulcrum of theoretical rheology is the rheological equation (6.1) through which the stress tensor s is defined as a function of the tensor of deformation D. Both are symmetric tensors of rank two. As said in Sect. 1, we assume in the present article that the materials to which the different rheological equations apply are isotropic or quasi-isotropic. The relation between both tensors will therefore also be isotropic. This implies certain conditions which are common to all isotropic tensor functions.

Considering the simplest case in which s_{lm} is an explicit function of D_{no}, writing y_{lm} for s_{lm} and x_{no} for D_{no}, the relation

$$y_{lm} = F(x_{no}), \qquad (7.19)$$

can be expressed as a sum of symmetric tensors of second rank

$$y_{lm} = \mathfrak{F}_0 \delta_{lm} + \mathfrak{F}_1 x_{lm} + \mathfrak{F}_2 x_{l\alpha} x_{\alpha m} + \mathfrak{F}_3 x_{l\alpha} x_{\alpha\beta} x_{\beta m} + \cdots \qquad (7.20)$$

which we may write as

$$y = \sum_{n=0}^{n=\infty} \mathfrak{F}_n x^n. \qquad (7.21)$$

Because of the isotropy of the material, the $\mathfrak{F}$ must be scalars. They are in general functions of the principal invariants I_x, II_x, III_x of x, or in a power expansion

$$\mathfrak{F}_n = \sum_{p,q,r} k_{p,q,r}^{(n)} I^p II^q III^r. \qquad (7.22)$$

Eq. (7.21) can be further developed by means of the Cayley-Hamilton equation, which can be written as

$$x^n = I_x x^{n-1} - II_x x^{n-2} + III_x x^{n-3} \qquad (7.23)$$

or

$$x^3 = I_x x^2 - II_x x^1 + III_x x^0. \qquad (7.24)$$

Therefore

$$x^n = A_n x^2 + B_n x^1 + C_n x^0, \qquad (7.25)$$

where the A_n, B_n, C_n, are also power-series in I_x, II_x, III_x. This makes (7.21)

$$y = \mathfrak{F}_0 x^0 + \mathfrak{F}_1 x^1 + \mathfrak{F}_2 x^2 + \sum_{n=3}^{\infty} \mathfrak{F}_n (A_n x^2 + B_n x^1 + C_n x^0) \qquad (7.26)$$

or

$$y = \left(\mathfrak{F}_0 + \sum_{n=3}^{\infty} \mathfrak{F}_n C_n\right) x^0 + \left(\mathfrak{F}_1 + \sum_{n=3}^{\infty} \mathfrak{F}_n B_n\right) x^1 + \left(\mathfrak{F}_2 + \sum_{n=3}^{\infty} \mathfrak{F}_n A_n\right) x^2, \qquad (7.27)$$

for which we can write with Reiner (1945, 1948)

$$y_{lm} = F_0 \delta_{lm} + F_1 x_{lm} + F_2 x_{l\alpha} x_{\alpha m}, \tag{7.28}$$

where the scalars F_0, F_1, F_2 are in general power series in the principal invariants of x_{lm}. The Eq. (7.28) is valid in general, i.e. is valid also outside the fields of elasticity and hydrodynamics for which it was originally derived; and it is thus shown that the most general isotropic relation of symmetric tensors of second rank is of second order in tensors and not higher. However, the F are power series of the components of x_{lm} to any degree through the invariants, and the relation (7.28) can be of *any degree* in tensor-components.

The problem then arises how to express in a general manner the F through the A_n, B_n, C_n, and th $\mathfrak{F}$, in terms of the invariants I, II, III. As shown by Hanin and Reiner[1] the expressions are as follows:

$$A_n = \sum_{\substack{a+2b+3c \\ =n-2}} (-1)^b \frac{(a+b+c)(a+b+c-1)!}{a!\,b!\,c!} I^a II^b III^c, \tag{7.29}$$

$$B_n = \sum_{\substack{a+2b+3c \\ =n-1}} (-1)^b \frac{(b+c)(a+b+c-1)!}{a!\,b!\,c!} I^a II^b III^c, \tag{7.30}$$

$$C_n = \sum_{\substack{a+2b+3c \\ =n}} (-1)^b \frac{c(a+b+c-1)!}{a!\,b!\,c!} I^a II^b III^c. \tag{7.31}$$

The next step is the determination of the constants k of (7.22). This can only be done by means of suitable experiments in which related *components* of y and x are measured, and the component y is expressed as a function of the component of x. The simplest experiment will be one-dimensional in x. Let $x(i), x(j), x(k)$ be the proper values of x with reference to the principal axes i, j, k. Then we can put $x(j) = x(k) = 0$ and

$$I = x(i), \quad II = III = 0. \tag{7.32}$$

In this case from (7.29) to (7.31) with $b=c=0$

$$A_n = x(i)^{n-2}; \quad B_n = C_n = 0, \tag{7.33}$$

while

$$\mathfrak{F}_n = \sum_p k^{(n)}_{p,0,0} x(i)^p. \tag{7.34}$$

This would serve for the determination of the constants $k^{(n)}_{p,0,0}$. Similarly one can use another experiment in which

$$x(j) = -x(i), \quad x(k) = 0 \tag{7.35}$$

and

$$I = 0, \quad II = -x(i)^2, \quad III = 0 \tag{7.36}$$

when $a=c=0$ and

$$A_n = (-1)^{n-2} x(i)^{n-2}, \quad B_n = (-1)^{n-1} x(i)^{n-1}, \quad C_n = 0 \tag{7.37}$$

for the determination of $k^{(n)}_{0,q,0}$ in

$$\mathfrak{F}_n = \sum_q k^{(n)}_{0,q,0} (-x(i))^{2q}. \tag{7.38}$$

---

[1] M. Hanin and M. Reiner: Z. angew. Math. Phys. **7**, 377–393 (1956).

The most important case arises when the $\mathfrak{F}_n$ are constants, say K_n, and therefore

$$F_0 = K_0 + \sum_{n=3}^{\infty} K_n C_n,$$
$$F_1 = K_1 + \sum_{n=3}^{\infty} K_n B_n, \qquad (7.39)$$
$$F_2 = K_2 + \sum_{n=3}^{\infty} K_n A_n.$$

Then in the first experiment

$$y(i) = \sum_{n=0}^{\infty} K_n x(i)^n, \quad y(j) = y(k) = K_0, \quad (7.40)$$

while in the second experiment

$$\left.\begin{array}{l} y(i) = K_0 + K_1 x(i) + K_2 x(i)^2, \\ y(j) = K_0 - K_1 x(i) + K_2 x(i)^2, \\ y(k) = K_0. \end{array}\right\} \quad (7.41)$$

We shall have occasion to make use of these expansions in Sect. 29.

δ) *Mohr mapping.* A symmetrical physical tensor can be mapped in a plane Cartesian system by means of the "Mohr-circles" as follows:

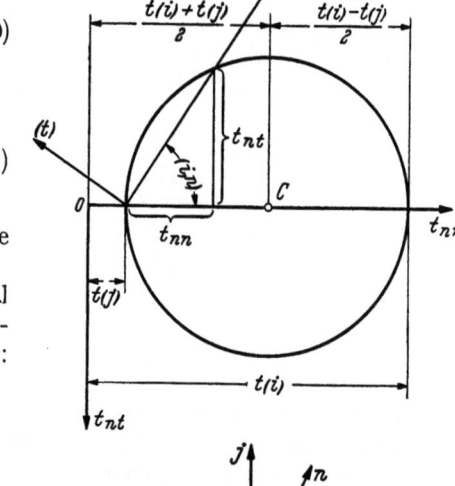

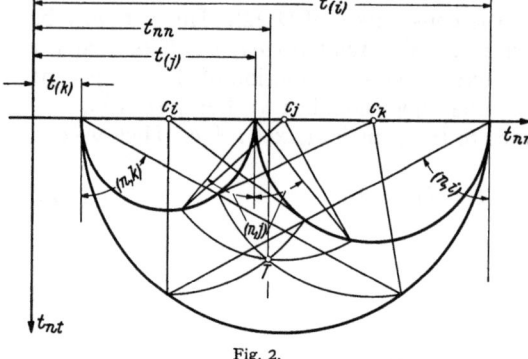

Fig. 2.

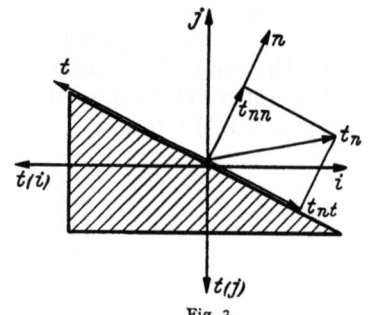

Fig. 3.

Fig. 2. Mohr-mapping. — Tensor plane. $t(i) > t(j) > t(k)$ principal values of tensor t_{lm}, c_i, c_j, c_k centres of Mohr-circles.

Fig. 3. Mohr-circle for plane tensor. $t(i) > t(j)$ principal values; $\vec{t}_n$ vector related to surface element with orientation $\vec{n}$; t_{nn} normal component of t_n, t_{nt} tangential component of t_n; (n) parallel to $\vec{n}$ through point $(t(j), 0)$. If $\vec{n}$ in direction x, $\vec{t}$ is in direction y and $t_{nt} = t_{xy}$.

Let $\vec{t}_n$ denote the vector, connected with the normal $\vec{n}$, which is one of the components of the tensor[1] t_{lm}. Select two other mutually perpendicular directions t ("tangential") and c ("cross") such that t_{nc} vanishes. We then have

$$\left.\begin{array}{l} t_{nn} = t_{(\iota)} [n \cdot \iota]^2, \\ t_{nn}^2 + t_{nt}^2 = \{t(\iota) [n \cdot \iota]\}^2, \end{array}\right\} \quad (7.42)$$

where ι (Greek letter iota) stands for the directions i, j, k respectively, and is to be treated as a summation index. The tensor t_{lm} can accordingly be mapped in a plane rectangular Cartesian sytem, the *tensor-plane*, as shown in Fig. 2,

---
[1] In the same manner as that in which the traction $\vec{s}_n$ is a component of the stress-tensor s_{lm} (compare Sect. 4).

with t_{nn} as the abscissa, and t_{nt} the ordinate of a point P defined by the angles (n, i), (n, j) and (n, k). The curved triangle between three semi-circles corresponding to $[n \cdot i] = 0$, $[n \cdot j] = 0$, $[n \cdot k] = 0$, represents the octant between axes i, j, k. In the two-dimensional case when e.g. $[n \cdot k] = 0$, conditions are much simplified, and are as shown in Fig. 3. For an isotropic tensor all three Mohr-circles are reduced to one single point on the t_{nn}-axis. Shifting the origin on the abscissa axis therefore means the superposition of an isotropic component. Fig. 4 shows a simple construction[1] for determining an origin O so that the tensor has the property of a deviator

$$t_{(0)}(i) + t_{(0)}(j) + t_{(0)}(k) = 0. \quad (7.43)$$

ε) *Convected axes.* It is sometimes convenient to use a system of reference which is formed by material points of the body, and takes part in its deformation. In this case the coordinates of a material point do not change in value, or $x_l = {}_l x$, and the deformation is defined by the metric tensor g_{lm} as a function of time. This method was introduced by HENCKY[2] and used by OLDROYD[3] and GREEN and ZERNA[4]. It is called the method of convected axes. In the present article no use is made of this method.

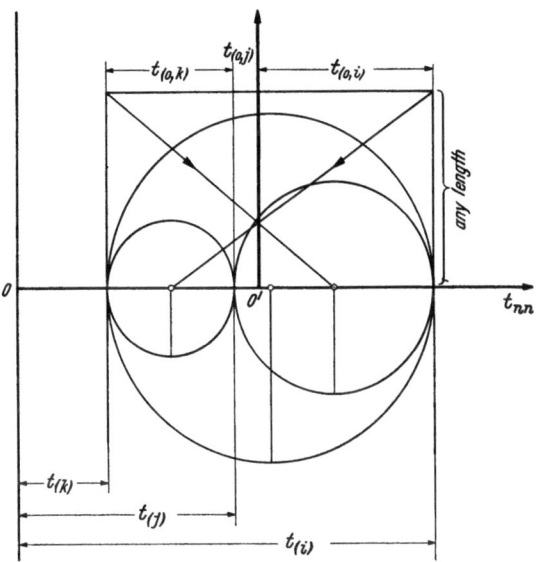

Fig. 4. Construction of deviator-origin. O origin of Mohr-reference system for tensor with principal values $t(i) > t(j) > t(k)$. O' origin of reference system for which Eq. (7.44) is valid.

8. Rheological properties.

Rheological properties are of two kinds.

α) *Essential rheological properties* are those the investigation of which has reached the stage when they can be exactly defined as parameters in a postulated rheological equation. Essential properties are either (a) *fundamental* or (b) *complex*. Fundamental properties are (I) elasticity, (II) viscosity and (III) plasticity (internal solid friction). Complex properties result from combinations of the fundamental properties, such as represented by complex models, and are theoretically unlimited, but in practice limited in number. Special designations have been coined for some of them, namely delayed elasticity or firmoviscosity, elasticoviscosity, dynamical plasticity, anelasticity and viscoelasticity. Certain other properties for which designations exist, such as "fore-effect", "after-effect", "retardation", "relaxation" etc. are not independent properties but are *derived* properties which can be expressed in terms of fundamental properties. When one comes to higher degrees of complexity, such language fails, and we must use the terminology described in Sect. 6, speaking for instance, of the Burgers-body etc. It may be advisable to drop all special designation for the essential rheological properties, and use only the terminology of ideal bodies. There would then simply be different elasticities, namely those of the Hooke-, the Kelvin- and the Maxwell-bodies etc. Much confusion has, for

---

[1] Given by NADAI (1931), p. 76.
[2] H. HENCKY: Z. angew. Math. Mech. **5**, 144 (1925).
[3] J. G. OLDROYD: Proc. Roy. Soc. Lond., Ser. A **200**, 523 (1950).
[4] A. E. GREEN and W. ZERNA: Theoretical Elasticity. Oxford 1954.

instance, been caused by the use of the term "visco-elasticity"[1], to give only one example.

To the three fundamental properties mentioned above, (IV) strength must be added as a fourth.

β) **Technological properties** are those for which a method of measurement has been devised, but of which the theoretical investigation has not yet reached a stage where the property is shown either to be fundamental or to be expressible in terms of known fundamental properties. The result of the measurement is an "index" which may be of relative significance or may serve only as an identification number. Examples of such properties are "penetration", "ductility", "tack", "thixotropy". The investigation of other technological properties has not even reached the stage where they can be measured in any apparatus. They are mentioned briefly below in Sect. 46. No attempt can be made at such properties as "seepage", „covering power of a pigment" and similar.

II. The classical bodies.

9. The volumetric equation. To a first approximation the *volumetric rheological equation* is for *all materials*

$$-p = \varkappa \varepsilon_v \qquad (9.1)$$

where p is the "hydrostatic" or "thermodynamic static" pressure and $\varkappa$ is the *bulk modulus*. This has the dimension of stress and is measured in bars (1 bar = 1 dyne cm^{-2}). Some values of $\varkappa$ are listed in Table 1[2].

Table 1. *Bulk moduli* $\varkappa$.

Material	ether	alcohol	water	tuff	clay	concerte	glass	mercury	iron	steel	incompressible
$\varkappa$ in 10^{11} bars	0.08	0.1	0.2	0.5	1	2	4	5.4	15	18	∞

In classical mechanics of continua p is identified with p_m. The second Eq. (5.7) now becomes

$$\dot{w}_v = \varkappa \varepsilon_v \dot{\varepsilon}_v \qquad (9.2)$$

and by integration

$$w_v = \tfrac{1}{2} \varkappa \varepsilon_v^2 = \tfrac{1}{2} p^2/\varkappa, \qquad (9.3)$$

and w_v is thus entirely conserved, vanishing in a closed cycle. Such specific work is named *strain-work*, and in accordance with what has been said in Sect. 7 above, to a first approximation all volumetric stress-work is strain-work. For the Euclid solid and for "incompressible" materials in general ε_v vanishes and $\varkappa = \infty$. In this case p can be determined from the dynamical boundary-conditions.

10. The Hooke-solid. The rheological equation[3] of the Hooke-solid is

$$s_{lm(0)} = 2\mu \varepsilon_{lm(0)}, \qquad (10.1)$$

where μ is the *shear modulus* or *modulus of rigidity*. Some values of μ are listed in Table 2.

---

[1] Compare JEFFREYS (1952), p. 10, who complains of those physiologists and engineers who confuse the terms.

[2] The purpose of the figures contained in Tables 1 to 3 is only to indicate orders of magnitude.

[3] By this we mean here and later the "distortional" equation.

Eq. (10.1) is an expression for HOOKE's law of proportionality of stress and strain. Considering (4.5), (3.13), (4.6), the total stress is related to the total strain by the linear equation

$$s_{lm} = \lambda \varepsilon_{\alpha\alpha} \delta_{lm} + 2\mu \varepsilon_{lm}, \tag{10.2}$$

where

$$\lambda = \varkappa - \tfrac{2}{3}\mu \tag{10.3}$$

is *Lamé's constant*.

Inversion of (10.2) gives the linear equation

$$\varepsilon_{lm} = -\frac{\lambda}{2\mu(3\lambda+2\mu)} s_{\alpha\alpha} \delta_{lm} + s_{lm}/2\mu. \tag{10.4}$$

The moduli $\varkappa$ and μ have physical significance, which LAMÉ's constant does not possess[1].

Table 2. *Shear moduli μ.*

Material	silk	lead	tin	glass	aluminium quartz silver	zinc copper	wrought iron platinum	nickel steel
μ in 10^{11} bars	0.1	0.2	1.5	2	3	4	7	8

In simple pull when s_{lm} is reduced to, say, $\sigma_{(i)}$ with $\sigma_{(j)} = \sigma_{(k)} = 0$, and therefore $s_{\alpha\alpha} = \sigma_{(i)}$, we have from (10.4)

$$\left.\begin{array}{l} \varepsilon_{(i)} = \dfrac{\lambda+\mu}{\mu(3\lambda+2\mu)} \sigma_{(i)} = \dfrac{\sigma_{(i)}}{E}, \\[6pt] \varepsilon_{(j)} = \varepsilon_{(k)} = -\dfrac{\lambda}{2\mu(3\lambda+2\mu)} \sigma_{(i)} = -\dfrac{\lambda}{2(\lambda+\mu)} \varepsilon_{(i)} = \nu \varepsilon_i \end{array}\right\} \tag{10.5}$$

where

$$E = \frac{\mu(3\lambda+2\mu)}{\lambda+\mu} = \frac{9\varkappa\mu}{3\varkappa+\mu} \tag{10.6}$$

is *Young's modulus*, and

$$\nu = \frac{\lambda}{2(\lambda+\mu)} = \frac{3\varkappa - 2\mu}{6\varkappa + 2\mu} \tag{10.7}$$

is *Poisson's ratio*.

For an incompressible material for which $\varkappa = \infty$, $E^* = 3\mu$, $\nu^* = \tfrac{1}{2}$. If $\mu = \infty$, the Hooke-body degenerates to the Euclid-body. While the moduli $\varkappa$, μ, E, λ each have the dimension force/area, ν is a dimensionless number. Handbook Tables often show poor agreement with the relations expressed by Eqs. (10.6) and (10.7), because the moduli were measured on different samples etc. For cork and sponge rubber POISSON's ratio approaches zero.

By introducing s_{lm} from (10.2) into the dynamical equations (4.4), we get the *momentum equations*

$$\mu u_{l,\alpha\alpha} + (\lambda+\mu) u_{\alpha,\alpha l} + \varrho(B_l - \ddot{x}_l) = 0. \tag{10.8}$$

By introducing $s_{lm(0)}$ from (10.1) into the third Eq. (5.7), we find

$$\dot{w}_{(0)} = 2\mu \varepsilon_{\alpha\beta(0)} \dot{\varepsilon}_{\alpha\beta(0)}, \tag{10.9}$$

and by integration

$$w_{(0)} = \mu \varepsilon_{\alpha\beta(0)} \varepsilon_{\alpha\beta(0)} = s_{\alpha\beta(0)} s_{\alpha\beta(0)}/4\mu. \tag{10.10}$$

---

[1] J. MEIXNER: Z. Physik **130**, 30–43 (1954).

This is therefore entirely conserved strain work, vanishing in a closed cycle when both ε and s return to zero. The rheological equation of the Hooke-body can accordingly be derived by postulating the existence of a *strain-energy* function φ, while ψ vanishes identically.

Due to non-vanishing rotations, the *relative displacements* which a Hooke-body may suffer will in general be finite, and when the body possesses an "open structure" may be quite large. Typical of such large relative displacements in infinitesimal strain is the deformation of a close-coiled helical spring (e.g. as used in a spring balance). If we do not know what the internal structure of a material is, these relative displacements may be mistaken for finite strains. In the first attempts to explain the large extensibility of rubber, it was attributed to a hidden internal open structure. Later it was found that this explanation was not correct (compare Sect. 27), but there may be other materials where it does apply, as for instance blown bitumen. The extension $\varepsilon(i)$ of (10.5) will then be large with small E, and macroscopically this is a case of finite strain. However, microscopically speaking, the physical strain of the material, i.e., the strain which causes stress, is actually infinitesimal, and for this reason HOOKE'S law in its simple form is applicable. From this point of view, the strain with which we deal in such cases is "pseudofinite".—When the helical spring is so much extended that the coils open up, the linear Hooke-relation ceases to apply. This is also so in other open structures.

11. The Newtonian liquid. Its rheological equation is

$$s_{lm(0)} = 2\eta f_{lm}, \tag{11.1}[1]$$

where η is the *coefficient of (shear) viscosity*. It has the dimension of a stress times time, and is measured in poises (1 poise = 1 dyne cm^{-2} sec). Some values of η for 25° C are listed in Table 3. Sometimes it is more convenient to use the *coefficient of fluidity* $\varphi = 1/\eta$.

Table 3. *Coefficients of viscosity η at 25° C.*

Material	ideal liquid	toluene	water	alcohol	mercury	milk	sugar solution 60% conc.	olive oil	castor oil
η in centipoises	0	0.6	1	1.2	1.6	2	57	100	1000

Eq. (11.1) is an expression of NEWTON'S law that "the resistance ($s_{lm(0)}$) which arises from the lack of slipperiness of the parts of the liquid ... is proportional to (the gradient of) the velocity with which the parts ... are separated (f_{lm})". The analogy between Eqs. (11.1) and (10.1) should be noted. This permits the translation of a problem of elasticity into an analogous problem of slow viscous flow (secondary creep), substituting $f_{lm} = \dot{\varepsilon}_{lm}$ for ε_{lm} and η for μ.

The rheological equation for the total stress is

$$s_{lm} = \varkappa \varepsilon_v \delta_{lm} + 2\eta f_{lm}. \tag{11.2}$$

For an incompressible liquid ε_v vanishes, but $\varkappa = \infty$, and the first term on the right of Eq. (11.2) is therefore rheologically indeterminate. One writes in this case, but also in general

$$s_{lm} = -p\delta_{lm} + 2\eta \dot{d}_{lm(0)}, \tag{11.3}$$

and finds p from the dynamical boundary conditions of the problem. For very viscous materials, such as bitumen, the viscosity can be determined by extending

---
[1] Taking account of (3.15).

a prismatic rod under a constant traction, say $\sigma(i)$. Accordingly a *coefficient of viscous traction* λ_T has been defined by TROUTON[1] from

$$\sigma(i) = \lambda_T f(i), \qquad (11.4)$$

which is the viscous analogy to YOUNG's modulus E. Because $f_{\alpha\alpha}$ is assumed to vanish [compare Eq. (3.15)]

$$\lambda_T^* = 3\eta. \qquad (11.5)$$

Introducing (11.2) into (4.4), considering at the same time (3.10), results in the *Navier-Stokes equations*

$$p_{,l} = \eta v_{l,\alpha\alpha} + \eta/3 \cdot \dot{\varepsilon}_{v,l} + \varrho(B_l - \ddot{x}_l) \qquad (11.6)$$

which form the basis of the classical hydrodynamics of viscous fluids identifying $\dot{d}_v$ with $\dot{\varepsilon}_v$. By introducing $s_{lm(0)}$ from (11.1) into the third Eq. (5.7), we find, considering (3.10)

$$\dot{w} = 2\eta f_{\alpha\beta} f_{\alpha\beta}. \qquad (11.7)$$

While the volumetric stress-work of a Newtonian-liquid is conserved in accordance with Eq. (9.3), its distortional stress power is therefore continually and completely dissipated. The rheological equation of the Newtonian liquid can accordingly be derived by postulating the existence of a *dissipation function* ψ, while Φ vanishes identically[2].

12. The St. Venant- and the Prandtl-body. For the St. Venant body it is postulated that it is rigid up to a certain stress $\vartheta_{lm(0)}$, the *yield stress*, and that the stress than remains constant, or

$$s_{lm(0)} = \vartheta_{lm(0)}, \qquad (12.1)$$

while the material flows plastically. In this flow it is assumed that the principal axes of the rate of deformation coincide with the principal axes of stress. In analogy with (11.1), LÉVY[3] accordingly postulated

$$s_{lm(0)} = \vartheta_{lm(0)} = 2\overline{\dot{D}_{lm(0)}}/\lambda_M, \qquad (12.2)[3]$$

where λ_M is not a material parameter but a factor of proportionality[4] which will vary from place to place, and in the course of the deformation, in order to keep $s_{lm(0)}$ constant.

If we assume that the body follows HOOKE's law as long as the stress is below the yield stress, and that the total deformation is the sum of the elastic strain and the plastic deformation, there results the *Prandtl-body* in which

$$2\overline{\dot{D}_{lm(0)}} = \frac{\overline{\dot{s}_{lm(0)}}}{\mu} + \lambda_M s_{lm(0)}, \qquad (12.3)$$

which is reduced to (12.2) for $\mu = \infty$. The strain at yielding or *yield strain* which in LÉVY's equation vanishes, is for the Prandtl body given by

$$2\Theta_{lm(0)} = \vartheta_{lm(0)}/\mu. \qquad (12.4)$$

Yield stress and yield strain together define the yield point[5].

---

[1] F. TROUTON: Proc. Phys. Soc. Lond. **19**, 47 (1905). — Proc. Roy. Soc. Lond., Ser. A **77**, 326 (1906).
[2] As said before, this refers only to the deviatoric part of the work.
[3] M. LÉVY: C. R. Acad. Sci. Paris **70**, 1323—1325 (1870).
[4] We use the subscript M in λ_M to indicate "MISES"; compare v. MISES, l. c.
[5] The rheological equation (12.3) forms the basis of the developments in R. H. HILL, Plasticity, Oxford 1950. HILL writes $d\lambda$ for λ_M, using the increment of deformation instead of the rate of deformation. As increment of deformation he introduces $d(d_{l,m}) = \frac{1}{2}[(du)_{l,m} + (du)_{m,l}]$, compare Eq. (3.8).

The strain work expended by the internal forces up to the yield point is conserved as strain energy, and can be regained on de-loading. The stress-work in excess of this energy is dissipated by solid *friction*[1]. In some materials (notably mild steel) there exists an analogy to the two kinds of solid friction, namely static and kinetic friction, manifesting itself in an "upper" and "lower" yield point.

The classical theory of plasticity is applied to metals in two very different fields: in structures and machine parts where plastic deformations must be small, and in technological forming processes such as rolling, drawing and extruding where they are very large. In the first case plastic zones exist which are restrained by the adjacent elastic material; this is the range of *contained plastic deformation*. When the load intensity is further increased *unrestricted plastic flow* sets in. In the first range the Prandtl theory, in the second range the St. Venant-Lévy theory is appropriate[2].

B. Macrorheology.

I. First-order phenomena.

13. The rheological tree. Essential rheological properties can be systematically expressed in a hierarchy of ideal bodies. In this, the three bodies which are treated in the classical mechanics of continua, namely the Hooke solid (denoted by the symbol H), the St. Venant plastic (symbol $St\ V$) and the Newtonian liquid (symbol N) can be regarded as the "primitive" bodies from which the "complex" bodies can be derived by suitable combinations.

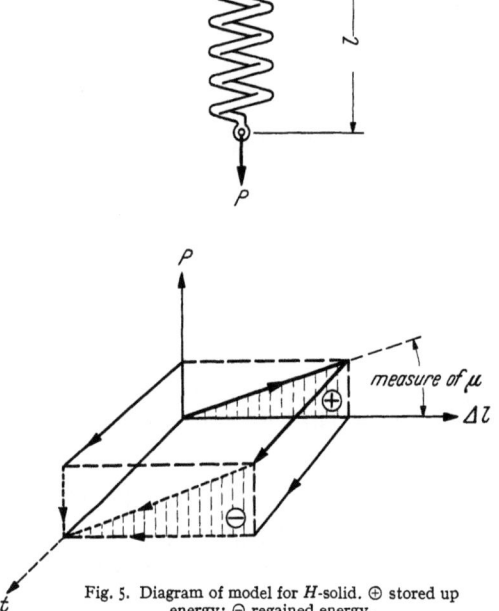

Fig. 5. Diagram of model for H-solid. ⊕ stored up energy; ⊖ regained energy.

As has been said in Sect. 1 above, in the task of establishing the rheological equations of the complex bodies from experimental observations, it is of great heuristic help to build, if only in imagination, a mechanical model which will behave qualitatively in a manner similar to that of the actual material, to some degree of approximation, and to describe that behaviour in terms of forces and elongations. If these terms are translated into stresses and deformations the result easily leads to a rheological equation, as will presently be shown.

The *primitive* bodies, typifying the three fundamental properties of elasticity, plasticity and viscosity, can be conveniently represented by the following mechanical models:

(i) a helical steel spring as shown in Fig. 5, forms a suitable model for the H-body. As mentioned in Sect. 9, its elongation Δl, even if finite, will not divert from the Hookean behaviour

$$\Delta l = \frac{P}{(E)}, \tag{13.1}$$

where (E) is a "gross" elastic modulus.

---

[1] Viscosity is sometimes spoken of (wrongly) as "internal friction".

[2] W. Prager and P. G. Hodge: Perfectly Elastic Solids. New York and London 1951. The Prandtl-theory is also known as Prandtl-Reuss theory.

Sect. 13. The rheological tree 453

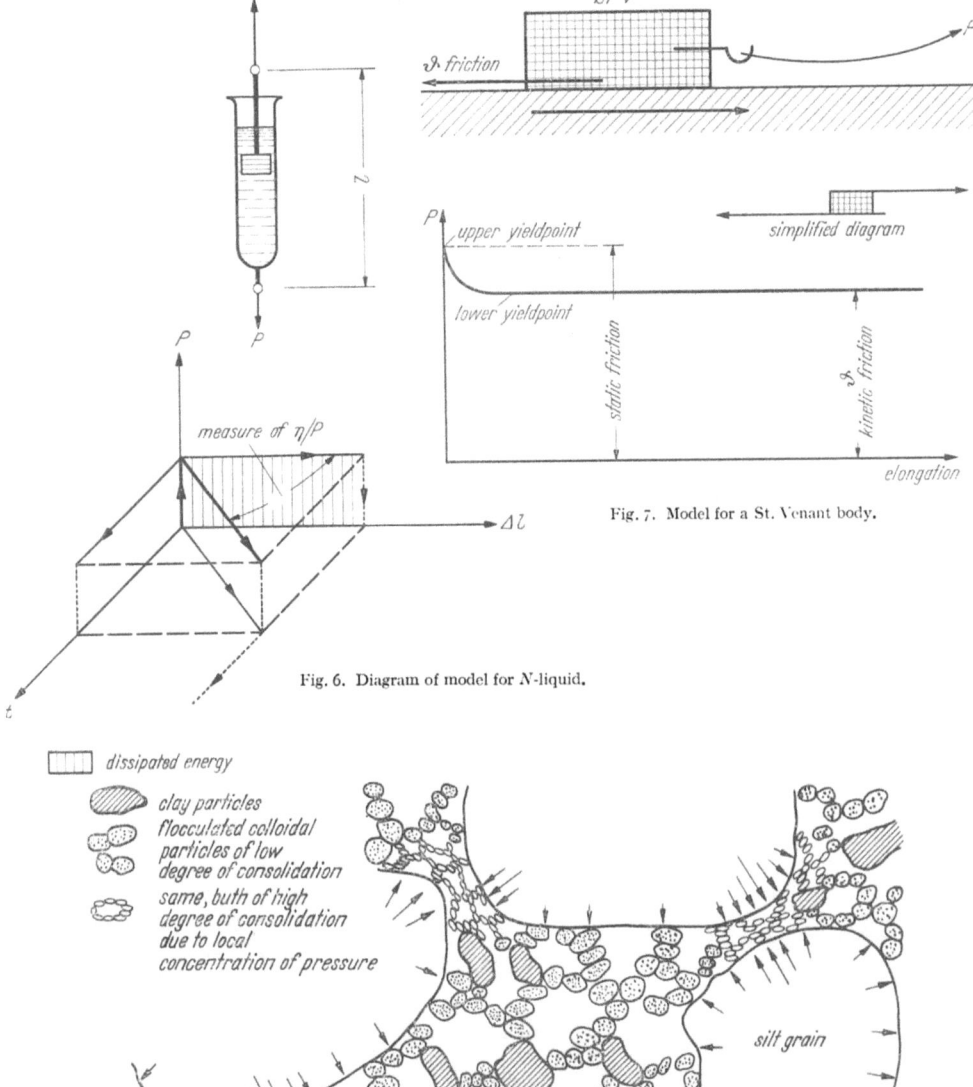

Fig. 6. Diagram of model for N-liquid.

Fig. 7. Model for a St. Venant body.

Fig. 8. Structure of loam.

Handbuch der Physik, Bd. VI. 29a

(ii) a dashpot, e.g. in the form of a test tube filled with a viscous oil in which a loosely fitted "stopper" moves as a piston, can serve to represent the N-body,

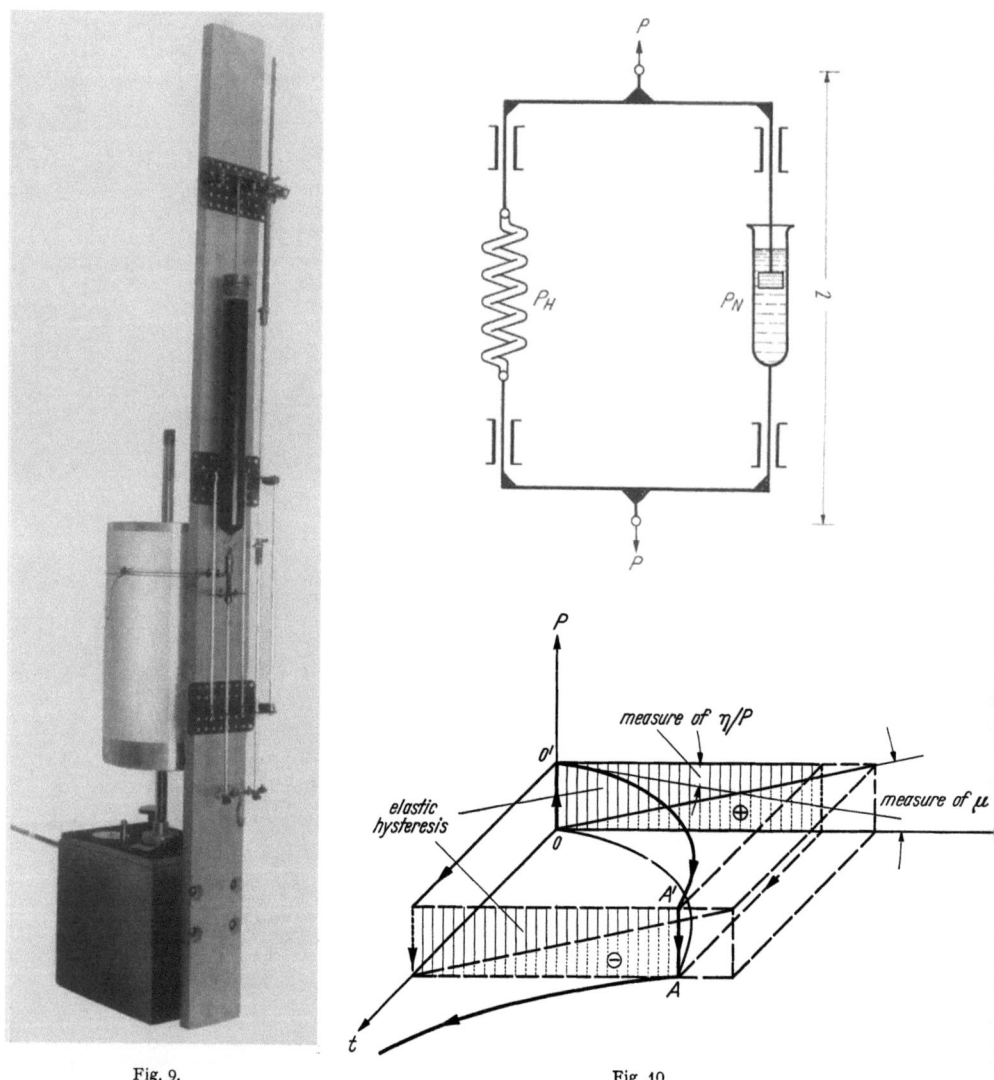

Fig. 9. Fig. 10.

Fig. 9. Mechanical model for K-body. A close-coiled steel spring (on the right) and a testtube filled with a very viscous oil (on the left) acting as a dashpot are coupled in parallel. The deformation versus time curve is automatically registered.

Fig. 10. Diagram of model for K-body $K = H|N$. On the Δl-versus-t plane, curves for the elastic fore- and after-effect are shown, in accordance with the loading diagram shown on the P-versus-t plane.

in accordance with
$$\overline{\dot{\Delta l}} = \frac{P}{(\eta)}. \tag{13.2}$$
It will behave qualitatively as shown in Fig. 6.

(iii) a weight resting on a table top with solid friction between them, as shown in Fig. 7, forms a suitable model for St V-behaviour with
$$P = (\vartheta). \tag{13.3}$$

These three elements can be coupled either in parallel (|) or in series (—). When coupled in parallel, the loads taken by each one of the elements are additive, while the rates of elongation of both are the same. When coupled in series, the rates of elongation are additive, while each takes the same total load. The models work with elongations under pulls, but they can serve to represent not only linear dilatation, but also shear (or distortion in general) and cubical dilatation.

As an example, consider the problem of postulating the rheological equation for the description of the rheological behaviour of a material such as shown in Fig. 8. This pictures a piece of loam magnified about 10^4 times. Silt particles are held together by chains of colloidal clay, with the interstices filled with water; the whole forming a gel. This is the kind of material for which KELVIN's description as follows, is appropriate[1]. "Consider a perfectly elastic vesicular solid, whether like a sponge with communications between the vesicles, or with each vesicle separately enclosed in elastic solid: imagine its pores and interstices filled up with viscous liquid, such as oil. Static experiments on such a solid will show perfect elasticity of bulk and shape; kinetic experiments will show losses of energy such as are really shown by vibrators of ... elastic homogeneous solids ... According to STOKES' law of viscosity of fluids[2], our supposed vesicular vibrator should follow the law of subsidence of a simple vibrator experiencing a resistance simply proportional to the velocity of its motion".

A mechanical model which will reproduce the rheological behaviour as described by KELVIN is shown in Fig. 9. Diagrammatically it can be represented as in Fig. 10. By combining (13.1) and (13.2),

$$P_K = P_H + P_N = (E)\, \Delta l + (\eta)\, \overline{\dot{\Delta l}}. \tag{13.4}$$

In a further simplification the diagram of Fig. 10 can be written in "shorthand" as

$$K = H|N. \tag{13.5}$$

This is called a *rheological formula*.

From Eq. (13.4) we are led to combining Eqs. (10.1) and (11.1) into

$$s_{lm(0)} = 2\mu\, \varepsilon_{lm(0)} + 2\eta_s\, \dot{\varepsilon}_{lm(0)}, \tag{13.6}$$

where we have used the subscript "s" in η_s to indicate "solid" viscosity, for reasons explained in Sect. 15.

On this example we shall now build up a "tree" of rheological bodies. For short we write s, where $s_{lm(0)}$ is meant, and ε and f for $\varepsilon_{lm(0)}$ and $f_{lm(0)}$.

Consulting Table 4, we see how a system of rheological bodies can be formed in accordance with the number of fundamental structural elements making up the body. Coupling in series is indicated by one branch-line, coupling in parallel by two parallel branch-lines. Some possible combinations have been postulated either on theoretical grounds or from empirical considerations. They are, besides the K-body mentioned before:

$M = N - H$ named after MAXWELL (1868).

$B = H - (St\,V|N)$ after BINGHAM[3]

$J = N|M$ after JEFFREYS (1929). This is, as a mechanical

---
[1] Lord KELVIN (1875).
[2] This refers to Eq. (11.6).
[3] E. C. BINGHAM and H. GREEN (1919).

model, equivalent to $L = K - N$ as postulated by Lethersich[1].

$Schw = H - (StV|M)$ after Schwedoff[2].

$PTh = H|M$ after Poynting and Thomson (1902).

$Bu = M - K$ after Burgers[3]. This is, as a mechanical model, equivalent with $TR = N - PTh$ as postulated by Trouton and Rankine[4].

$Sch\,ScB = Schw - K$ after Schofield and Scott Blair[5].

Table 4. *Tree of rheological bodies.*

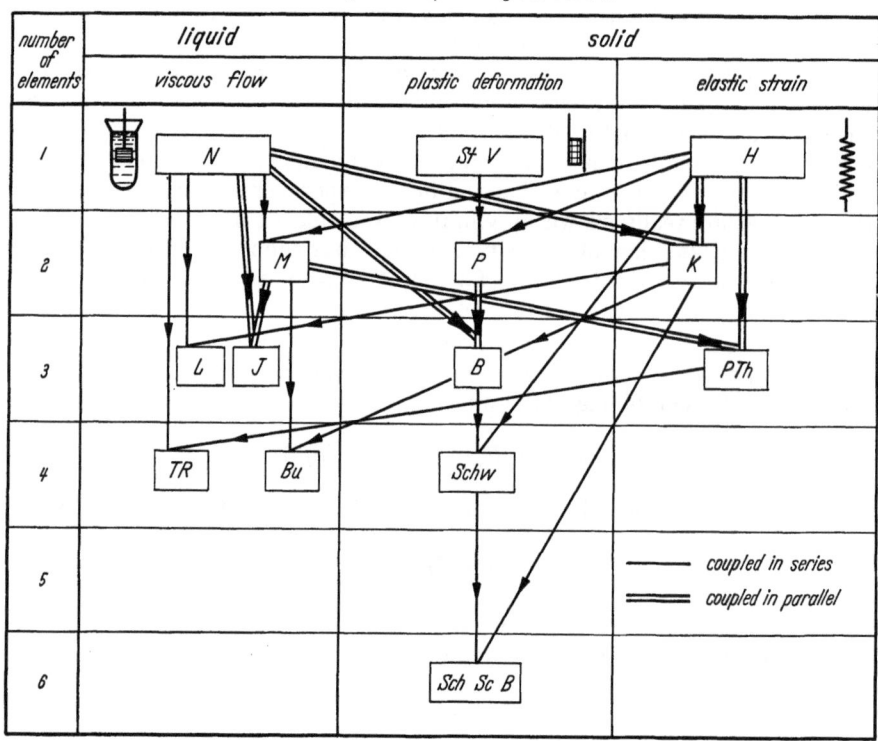

Other combinations are possible and may arise for the description of the rheological behaviour of materials not investigated so far. However, it may be considered as an axiom of rheology that every *real* material possesses *all rheologica-properties*, albeit in varying degrees.

As has also been said in Sect. 1 above, the materials with which rheology has to deal are mostly dispersed systems composed of solid and liquid phases. In such solid-liquid systems the H-element may stand for the solid, the N-element for the liquid phase. For instance a material in transition from sol to gel, where gel-"sponges", themselves imbued with liquid, are suspended in a liquid, might be represented by the structural formula $(H|N) - N$. When

---

[1] W. Lethersich: J. Soc. Chem. Ind. **61**, 101 (1942).
[2] T. Schwedoff: J. de Phys. (2) **9**, 34 (1890). — Rapp. pres. congr. internat. phys. **1**, 478 (1900).
[3] J. M. Burgers: In First Report 1935.
[4] F. Trouton and Rankine: Phil. Mag. (6) **8**, 555 (1904).
[5] R. K. Schofield and G. W. Scott Blair: Proc. Roy. Soc. Lond., Ser. A **138**, 707 (1932); **139**, 557 (1933); **141**, 72 (1933); **160**, 87 (1937).

explaining solid viscosity, KELVIN (as described before) considered a jelly, but VOIGT[1] derived the concept of solid viscosity for a homogeneous one-phase material. While the designation Voigt body is sometimes used for a material represented by the model of Fig. 10, this is a case where the symbolic character of the models becomes prominent. They should not be taken literally[2]. The mechanism assumed by MAXWELL when deriving the concept of relaxation which lies at the basis of the M-body, likewise refers to a homogeneous one-phase system. — The models will also fail when the elastic or viscous responses are not linear, but even in such cases they retain their heuristic value.

14. Plastico-dynamics or the Bingham body. Rheology started when BINGHAM[3] investigated concentrated clay-suspensions, and BINGHAM and GREEN (1919) investigated oilpaints. As an introduction to the method employed in the following pages, it will be useful to describe the second investigation in some detail showing why the departure from classical hydrodynamics became necessary, with which rheology emerged as a new branch of physics.

An oilpaint consists mainly of an oil as "vehicle", in which "pigments" are suspended. The oils used in the preparation of paints are simple viscous liquids. It was therefore natural to assume that oil paint is a liquid, the same as oil, which is its main component, and that one could determine its "viscosity" as for a Newtonian liquid. When brushing out a paint on a surface, it is requisite that this should be possible without great effort, and that the paint should be able to flow enough to cause the brush marks to disappear. Paint technologists tried to relate "brushing" and "leveling" properties with the viscosity of paint. The greater the fluidity, the more easily the paint could be brushed out and the better would the brush marks disappear, and one could therefore think that the rheological properties of paint would be sufficiently described by the value of η.

Paint, however, is required to have a *third* rheological property. When applied to a vertical surface, it must not run. This requirement seemed to be in conflict with the two just mentioned. It looked as if paint should have a low viscosity in order to satisfy the first two conditions, and a high viscosity in order to satisfy the third one. Then, when the "viscosity" of paint was determined at different rates of flow, it turned out not to be a constant. Ultimately BINGHAM and GREEN found the solution as indicated in the title of their paper: "Paint, a plastic material and not a viscous liquid". They arrived at the concept of what we may call in our more developed language the property of plastico-dynamics typified by the *Bingham body*.

This ideal material is a plastic solid which resists plastic flow not only with plastic friction (ϑ), as do the St. Venant or the Prandtl body, but also with a viscosity which may be named *plastic viscosity* (η_{pl}). Its rheological formula can be expressed as

$$B = H - (N \mid StV) \tag{14.1}$$

shown diagrammatically in Fig. 11. By combining Eqs. (13.2) and (13.3), we find for its mechanism, adding forces,

$$P_B = P_N + P_{StV} = \eta_{pl} \overline{\dot{\Delta l}} + \vartheta. \tag{14.2}$$

---

[1] VOIGT: Abh. Ges. Wiss. Göttingen **36** (1890). — Ann. d. Phys. **47**, 671 (1892), but the priority of the concept belongs to KELVIN.
[2] Compare discussion ROSCOE-REINER: Brit. J. Appl. Phys. **1**, 332 (1950).
[3] E. C. BINGHAM: U.S. Bur. Stand. Bull. **13**, 309 (1916).

Translating forces and elongations into stresses and deformations, we are led to the rheological equation[1]

$$s = 2\mu\varepsilon \quad \text{for} \quad |s| < |\vartheta|, \\ s - \vartheta = 2\eta_{pl}\dot{d} \quad \text{for} \quad |s| \gtreqless |\vartheta|. \quad (14.3)$$

The yield point is given by the transition conditions

$$\dot{d} = 0, \quad \varepsilon = 0, \quad s = \vartheta = 2\mu\Theta, \quad (14.3,1)$$

where Θ is the *yield strain* and ϑ the *yield stress*.

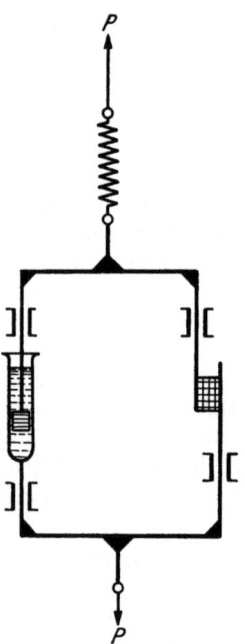

Fig. 11. Diagram of model for B-body, $B = H - (N|StV)$.

The two parameters of a Bingham material can best be determined in one of the two "plastometers" which operate on the model of the two viscosity prototypes, namely the *pressure tube* and the *co-axial rotating cylinder* apparatus.

α) *The pressure tube.* In this instrument, the material is forced by means of the pressure gradient $\Delta p/l$ to flow through a cylindrical tube of radius R and length l. Using cylindrical co-ordinates we assume in a semi-inverse method that $v_z = f(r)$, while v_r and v_ϑ vanish. We then find from (7.9), replacing u by v, and d by $\dot{d}$, that all components of the flow tensor vanish, except

$$2\dot{d}_{zr} = dv/dr = f'(r). \quad (14.4)$$

This makes (14.3)

$$\tau_{zr} = \vartheta_t + \eta_{pl} f'(r), \quad (14.5)$$

where ϑ_t is the tangential component of the yield stress. For the complete stress we must add the isotropic term $-p\delta_{lm}$. In the absence of mass-forces ($\varrho = 0$), the first two stress-equations (7.10) yield

$$\frac{\partial p}{\partial r} = \frac{\partial p}{\partial \Theta} = 0, \quad (14.6)$$

while the third gives

$$\eta_{pl}[rf''(r) + f'(r)] + \vartheta_t = r\,\partial p/\partial z = r\Delta p/l. \quad (14.7)$$

Integrating twice with respect to r, and introducing again v for $f(r)$, gives

$$\eta_{pl}v + r\vartheta_t = \frac{r^2}{4}\frac{\Delta p}{l} + C_1 \log r + C_2. \quad (14.8)$$

As v must be finite for $r = 0$ and whatever ϑ, C_1 must vanish. For the determination of C_2, we assume the kinematical boundary condition that v vanishes for $r = R$, i.e. that the liquid adheres to the wall. We thus find

$$\eta_{pl}v = \frac{(R^2 - r^2)\Delta p}{4l} - \vartheta(R - r). \quad (14.9)$$

In accordance with the second Eq. (14.3), this relation is valid in the region of flow only, i.e. where the shearing stress

$$\tau = r\Delta p/2l \quad (14.10)$$

is equal to or exceeds the yield stress ϑ. For $\tau = \vartheta$ we have

$$r = r_0 = \frac{2l\vartheta}{\Delta p}. \quad (14.11)$$

---

[1] J. G. OLDROYD: Proc. Cambridge Phil. Soc. **43**, 100 (1947).

For all $r < r_0$, we have $\tau < \vartheta$, and in this region therefore the material does not flow but moves as a whole, i.e. as a solid plug (compare Fig. 12). Stresses in the solid plug are purely elastic. From Eq. (14.9) the quantity Q of material passing through the tube in time t is

$$\frac{Q}{t} = \frac{\pi R^4 \Delta p}{8\eta_{pl} l}\left[1 - \frac{4}{3}\frac{2l\vartheta}{R\Delta p} + \frac{1}{3}\left(\frac{2l\vartheta}{R\Delta p}\right)^4\right]. \tag{14.12}$$

This is the *Buckingham-Reiner equation*. For $\vartheta = 0$, it yields the *Hagen-Poiseuille equation* of Newtonian viscosity

$$\frac{Q}{t} = \frac{\pi R^4 \Delta p}{8\eta l}, \tag{14.12,1}$$

while (14.9) is reduced to the parabolic distribution

$v = (R^2 - r^2)\Delta p/4l\eta$ (14.9,1)

with vanishing plug.

Eq. (14.12) was found to be applicable by SCOTT BLAIR and CROWTHER[1] to clay and soil pastes, and by WOLAROWITCH, KULAKOFF and ROMANSKY[2] to soil suspensions.

β) *The co-axial rotating cylinder apparatus.* In this instrument the material is forced into circular flow in the annulus between two co-axial cylinders of radii R_i and R_e, the inner cylinder being kept at rest through the torque M and the outer cylinder rotating with the angular velocity Ω. In this case we assume $v_r = v_z = 0$, while

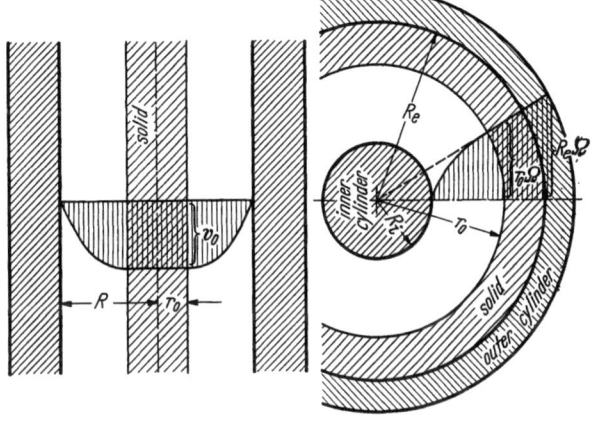

Fig. 12. Plastic flow in the tube (left) and rotating-cylinder (right) instrument. R radius of tube. R_i radius of internal. R_e of external cylinder.

$$v_\Theta = r\dot\theta = rf(r). \tag{14.13}$$

Then from (7.9)

$$2 d_{r\Theta} = r\, d\dot\theta/dr = r f'(r), \tag{14.14}$$

and from (14.3)

$$\tau_{r\Theta} = \vartheta_t + \eta_{pl} r f'(r). \tag{14.15}$$

From (7.10), neglecting mass-forces, p is constant and equal to the air pressure, while from the second equation

$$\frac{d\tau}{dr} + 2\frac{\tau}{r} = 0. \tag{14.16}$$

Integration yields

$$\tau = \frac{C_1}{r^2}. \tag{14.17}$$

Introducing τ from (14.5) this gives

$$\frac{C_1}{r^3} = \frac{\vartheta}{r} t + \eta_{pl} f'(r), \tag{14.18}$$

and through integration

$$\eta_{pl}\dot\theta = \frac{C_1}{2r^2} - \vartheta \log r + C_2. \tag{14.19}$$

---

[1] G. W. SCOTT BLAIR and E. M. CROWTHER: J. Phys. Chem. **33**, 321 (1921).
[2] M. P. WOLAROWITCH, N. N. KULAKOFF and A. N. ROMANSKY: Kolloid-Z. **71**, 267 (1935).

Again assuming that there is no slippage at $r = R_i$

$$C_2 = \frac{C_1}{2R_i^2} + \vartheta \log R_i, \tag{14.20}$$

and therefore

$$\eta_{pl}\dot{\theta} = \frac{C_1}{2}\left(\frac{1}{R_i^2} - \frac{1}{r^2}\right) - \vartheta \log \frac{r}{R_i}. \tag{14.21}$$

The integration constant C_1 can be determined from the dynamical boundary condition that the torque of the surface traction acting upon one of the two cylinders is given as M_z, and is thus a constant for every r. Accordingly

$$M_z = \tau \cdot 2\pi r h \cdot r \tag{14.22}$$

where h is the wetted length of the internal cylinder. Therefore from (14.17)

$$C_1 = \frac{M}{2\pi h}, \tag{14.23}$$

which makes (14.21)

$$\Omega = \frac{1}{\eta_{pl}}\left[\frac{M}{4\pi h}\left(\frac{1}{R_i^2} - \frac{1}{R_e^2}\right) - \vartheta \log \frac{R_e}{R_i}\right]. \tag{14.24}$$

This is the *Reiner and Riwlin-equation*. For $\vartheta = 0$ it yields the *Margules equation* of Newtonian viscosity

$$\dot{\Theta} = \frac{M}{4\pi h \eta}\left(\frac{1}{R_i^2} - \frac{1}{R_e^2}\right). \tag{14.24,1}$$

We see from (14.22) that the shearing traction τ is largest where r is sallestm, i.e. on the surface of the internal cylinder when $r = R_i$. Therefore when $M_z < 2\vartheta \pi R_i^2 = M_0$, the material between both cylinders will be strained, but there will be no flow. When M_z exceeds the value M_0, the material will start to flow. However, as long as the stress near the inner surface of the external cylinder is smaller than ϑ, there will be a shell of solid material adjacent to the external cylinder which will rotate together with the external cylinder as a whole (compare right side of Fig. 12). The stress at the wall of the external cylinder is $= M_z/2\pi h R_e^2$. When M_z increases beyond $M_1 = 2\pi h R_e^2 \vartheta$, all material between both cylinders flows. This constitutes an essential difference between the plastic flow in the rotating cylinder apparatus and the plastic flow in the tube.

Eq. (14.24) was found applicable to clay-suspensions by WOLAROWITCH and TOLSTOI (1935)[1] and to flour-dough by WOLAROWITCH and SAMARINA (1935)[2]. GREEN (1949) has investigated different materials in a high speed rotational plastometer[3] of wide range and found complete confirmation of the equation.

It is convenient to introduce in the case of the tube instrument as variables in Eq. (14.12) the quantities

$$\left.\begin{array}{l} V = \dfrac{4Q/t}{R^3 \pi}, \\[4pt] P = \dfrac{R\Delta p}{2l} \end{array}\right\} \tag{14.25}$$

---

[1] M. P. WOLAROWITCH and D. M. TOLSTOI: Kolloid-Z. **70**, 165 (1935).
[2] M. P. WOLAROWITCH and K. I. SAMARINA: Kolloid-Z. **70**, 280 (1935).
[3] This is the instrument used in thixotropy investigations, compare Sect. 46.

and in the case of the co-axial rotating cylinder instrument in Eq. (14.24)

$$V = 2\frac{\Omega}{1-a},$$
$$P = \frac{M}{2R_i^2 \pi h},$$
$$a = \left(\frac{R_i}{R_e}\right)^2$$
(14.26)

Eqs. (14.12) and (14.24) then become

$$V = \frac{P}{\eta_{pl}}\left[1 - \frac{3}{4}\frac{\vartheta}{P} + \frac{1}{3}\left(\frac{\vartheta}{P}\right)^4\right], \quad (14.27)$$

$$V = 2\frac{P}{\eta_{pl}}\left(1 + \frac{\vartheta}{P}\frac{\log a}{2(1-a)}\right), \quad (14.28)$$

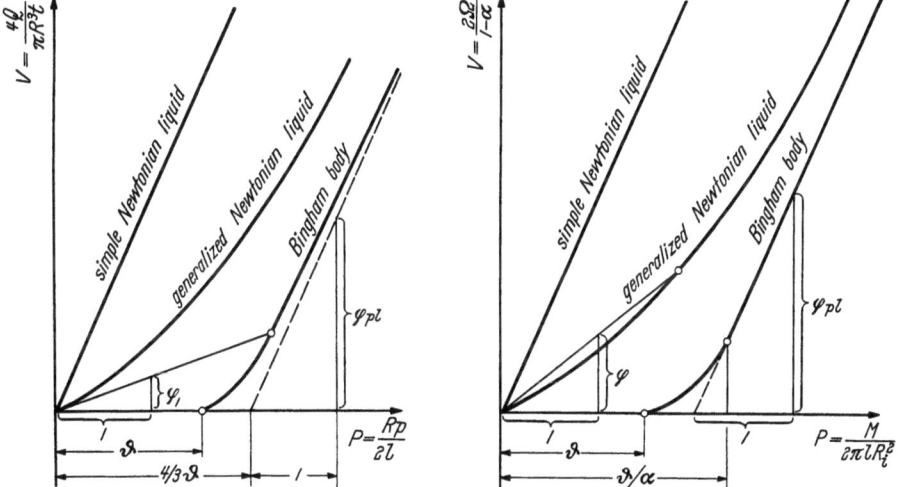

Fig. 13. Consistency curves of three types of materials, in the tube (left) and rotating-cylinder instrument (right). Apparent fluidity φ, varies between 0 and φ_{pl}.

and the V versus P curves are independent of the dimensions of the apparatus and therefore represent a *property* of the material. This property is called *consistency*, defined by the American Society of Testing Materials as "that property of a material by which it resists permanent change of shape ... defined by the complete flow-force relation". The V and P are accordingly called *consistency variables* and the V versus P-curves (14.27) and (14.28) are *consistency curves* (compare Fig. 13). As can be seen, P is the shearing stress τ in the material near the wall of the instrument which is at rest, while V is the mean velocity gradient.

It should be noted that in the derivation of Eqs. (14.12) and (14.24), and therefore also of Eqs. (14.27) and (14.28), it is assumed that there is no "slippage" or that the material adheres to the wall of the tube or cylinder. When this is not the case, "wall-effects" become manifest. If there are such effects, the V versus P curves will not be independent of the dimensions of the apparatus. For such cases "slippage terms" have been introduced with $v = v_R$ when $r = R$, instead of equal to zero. Scott Blair and collaborators[1] have examined regions of very small shearing velocities. For instance, the consistency curve of clay

---

[1] G. W. Scott Blair (1938), pp. 32—41.

and soil pastes was found to divide itself into four regions. In region I there is no flow, region II is a straight line, region III is markedly curved and region IV becomes asymtotically straight. Region II is due to slippage. It was Scott Blair's contribution to find region I, i.e. the fact that a definite stress is necessary to *start* slippage.

Another set of phenomena which causes deviations from the R^4 relation of the Poiseuille equation was described by Schofield and Scott Blair[1] and named σ-effect. When a positive σ-effect is present, the material will flow relatively faster in a narrow than in a wider tube. Negative σ-effects have also been described. The σ-effect is sometimes of a much higher order of magnitude than could be accounted for by flow in region II mentioned before.

If an apparent coefficient of viscosity η_1 is defined in accordance with (11.1) from $\eta_1 = s/2\dot{d}$, we find

$$\eta_1 = \eta_{pl} + \frac{\vartheta}{2\dot{d}}. \tag{14.29}$$

Table 5. *Yield stresses ϑ and plastic viscosities η_{pl} (orders of magnitude).*

Material	ϑ in bars	η_{pl} in poises	Material	ϑ in bars	η_{pl} in poises
oilpaint	10^2	10^0	sand-bitumen mixture	10^6	10^{11}
dough	10^3	10^3	copper aluminum	10^9	
cement paste	10^4	10^2	iron steel	10^{10}	
clay	10^5				

This is the viscosity which paint technologists determined, as described above. It is infinite when $s = \vartheta$, and $\dot{d}$ therefore vanishes, and it approaches asymptotically η_{pl} for $\dot{d} \to \infty$ (compare Fig. 13).

Oldroyd[2] has pointed out that dynamical similarity in geometrically similar systems is governed in the case of the Bingham body by two dimensionless numbers, namely in addition to the *Reynolds number*

$$R = V \varrho d / \eta_1 \tag{14.30}$$

by a number which may be named the Oldroyd number

$$O = \vartheta d / \eta_1 V. \tag{14.31}$$

where d is a typical distance and V is a typical velocity. If R is small so that inertia terms can be neglected, and O is large, i.e. in sufficiently slow plastic flow, a *plastic boundary layer* will be formed, at the edge of which a *yield surface* occurs, similar to the cylindrical surfaces in the two cases treated above in detail. The yield surface separates the layer in which plastic flow occurs from the rest of the material which is elastically strained, and may move, but does not flow. Introducing a stream-function, Oldroyd has solved the momentum equations for some cases of plane plastic flow. In further papers[3] he treated axial rectilinear flow under zero pressure gradient between two eccentric circular cylinders and two confocal elliptic cylinders which are in relative motion.

---

[1] Scott Blair l. c.
[2] J. G. Oldroyd: Proc. Cambridge Phil. Soc. **43**, 383 (1947).
[3] J. G. Oldroyd: Proc. Cambridge Phil. Soc. **43**, 396, 521 (1947).

With $\vartheta \to 0$, the Bingham body degenerates to the Newtonian liquid. With $\eta_{pl} \to 0$ it degenerates to the Prandtl body. In this manner every solution for the Bingham body can be specialized for these two classical bodies.

Some values for ϑ and η_{pl} are listed in Table 5 in their order of magnitudes.

15. Delayed elasticity or the Kelvin-solid. The property of delayed elasticity is typified by the *Kelvin body* with the rheological formula

$$K = H|N \qquad (15.1)$$

from which, by the method indicated in the preceding section, we arrive at the rheological equation

$$s = 2\mu\varepsilon + 2\eta_s\dot{\varepsilon}. \qquad (15.2)$$

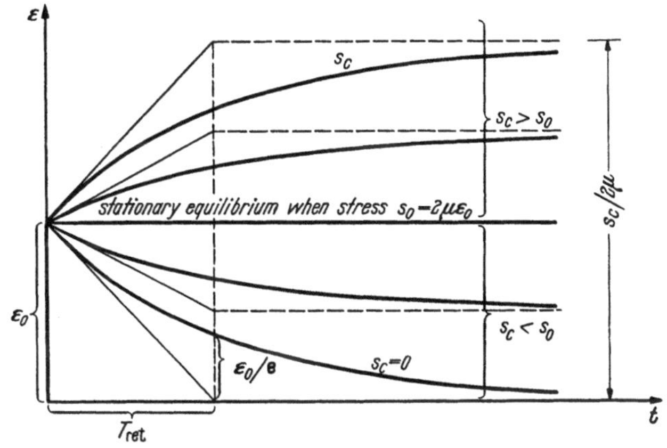

Fig. 14. Strain-time curves of a Kelvin-body for constant stresses s_c. The testpiece is first ($t=0$) instantaneously strained to the amount ε_0 und then subjected to a constant stress s_c. There exists for every initial strain ε_0 a certain stress s_0 such that stationary equilibrium is maintained.

This is a linear differential equation in ε, the integral of which is

$$\varepsilon = e^{-\frac{\mu}{\eta_s}t}\left\{\varepsilon_0 + \frac{1}{2\eta_s}\int s e^{\frac{\mu}{\eta_s}t}dt\right\}, \qquad (15.3)$$

where ε_0 is the initial strain. Let the stress s be const $= s_c$. Eq. (15.3) then gives

$$\varepsilon = s_c/2\mu + (\varepsilon_0 - s_c/2\mu)\exp(-t/T_{\text{ret}}) \qquad (15.4)$$

where we have introduced

$$T_{\text{ret}} = \eta/\mu. \qquad (15.5)$$

Eq. (15.4) gives a series of strain-time curves as shown in Fig. 14.

If $s_c = s_0 = 2\mu\varepsilon_0$, there will be stationary equilibrium as with a Hooke body. The Kelvin-body is therefore a solid, and its viscosity is "solid" viscosity, indicated by the index s in η_s. If $s_c < s_0$ the strain gradually decreases; if $s_c > s_0$ it gradually increases, albeit at a decreasing rate. This, when T_{ret} is very large, may simulate slow flow. It is therefore sometimes spoken of as (primary) *creep*. Finally (for $t = \infty$) the strain $s_c/2\mu$ is reached. The strain therefore does not appear instantaneously but is delayed in an elastic *fore-effect* with T_{ret} as *time of retardation*. If the stress is removed ($s_c = 0$), the shear vanishes in an elastic *after-effect* (or *creep-recovery*) theoretically at $t=\infty$, but, if T_{ret} is not too large, practically at finite time. — Both elastic fore- and after-effect constitute *delayed* or *retarded elasticity*. Fig. 15 shows the strain-time curves when the body is loaded and when the load has been removed. Both curves are taken from Fig. 14, and can also be seen on the $\Delta l - t$ plane of Fig. 10. As pointed out by Jeffreys[1], the Kelvin body shows a resistance *additional* to the elastic one. He therefore proposed for the property represented by it the name *firmo-viscosity*.

From the quotation from Kelvin in Sect. 12, it can be seen that Kelvin postulated this body in order to explain the damping of elastic oscillations.

It is clear that the same concepts as developed here for stress- and strain-deviator can also be used for the isotropic components of both tensors. We accordingly arrive at a *volume viscosity* of the Kelvin or "solid" type. Le ζ denote volume viscosity and ζ_s *solid volume viscosity*, then the damping of voluminal oscillations will be due, inter alia, to ζ_s, and Eq. (9.1) must be replaced by

$$-p_m = \varkappa \varepsilon_v + \zeta_s \dot{\varepsilon}_v \tag{15.6}$$

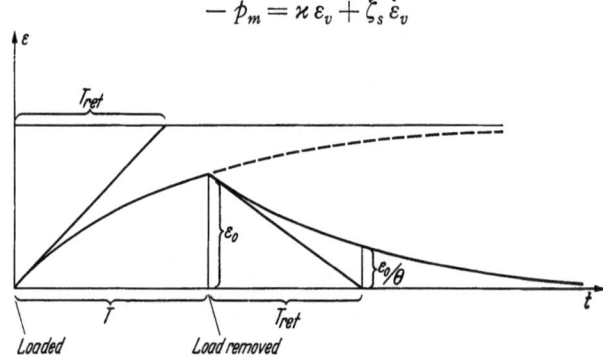

Fig. 15. Elastic fore- and after-effect. It takes infinite time to recover a strain ε_0 attained in the finite time-interval T.

where p_m, the mean pressure is defined by Eq. (4.6), and is now different from p. Combining Eqs. (15.2)[2] and (15.6), we get, considering (3.13) and (10.3),

$$s_{lm} = (\lambda \varepsilon_v + \lambda_v \dot{\varepsilon}_v) \delta_{lm} + 2\mu \varepsilon_{lm} + 2\eta_s \dot{\varepsilon}_{lm} \tag{15.7}$$

where

$$\lambda_v = \zeta_s - \frac{2\eta_s}{3}. \tag{15.8}$$

In classical hydrodynamics the rheological equation of the viscous liquid is written in the form

$$s_{lm} = (-p + \lambda_v \dot{\varepsilon}_v) \delta_{lm} + 2\eta f_{lm}. \tag{15.9}$$

Then f_{lm} is resolved on the model of (3.11) into

$$f_{lm} = \tfrac{1}{3} f_{\alpha\alpha} \delta_{lm} + f_{lm(0)}, \tag{15.10}$$

and $f_{\alpha\alpha}$ is identified with $\dot{\varepsilon}_v$, thus Eq. (15.9) yielding

$$s_{lm} = -p \delta_{lm} + \tfrac{1}{3}(3\lambda_v + 2\eta) \dot{\varepsilon}_v \delta_{lm} + 2\eta f_{lm(0)}. \tag{15.11}$$

It is then usually assumed that

$$3\lambda_v + 2\eta = 0 \tag{15.12}$$

which reduces Eq. (15.11) to the classical (11.3). This is called the *Stokes relation* about which more is said in Sect. 24.

---
[1] l. c. p. 10.
[2] Note that s stands there for $s_{lm(0)}$ and ε for $\varepsilon_{lm(0)}$.

16. Elastico-viscosity, or the Maxwell body. The property of elastico-viscosity is typified by the *Maxwell body*

$$M = H - N \tag{16.1}$$

shown as model in Fig. 16. Its rheological equation results from the Eqs. (9.1) of the Hooke solid and (10.1) of the Newtonian liquid respectively, by adding rates of deformation and is accordingly

$$\dot{d} = \dot{s}/2\mu_l + s/2\eta, \tag{16.2}$$

where we have written $\dot{d}$ for both $\dot{\varepsilon}$ in the first and f in the second case, as it comprises both a rate of strain and flow.

This is a linear differential equation in s. Its integral is

$$s = e^{-\frac{\mu_l}{\eta}t}\left\{s_0 + 2u_l\int \dot{d}\,e^{\frac{\mu_l}{\eta}t}\,dt\right\}, \tag{16.3}$$

where s_0 is the initial stress. Let the rate of deformation $\dot{d}$ be const $= \dot{d}_c$, Eq. (16.3) then gives

$$s = 2\dot{d}_c\eta + (s_0 - 2\dot{d}_c\eta)\exp(-t/T_{\rm rel}) \tag{16.4}$$

where we have introduced

$$T_{\rm rel} = \eta/\mu_l. \tag{16.5}$$

Eq. (16.4) represents a series of stress-time curves as shown in Fig. 17. If $\dot{d}_c = \dot{d}_0 = s_0/2\eta$, there will be steady flow with the internal stress in equilibrium with the load, as with a Newtonian liquid. The Maxwell-body, even if elastic, is therefore a liquid which is indicated by the index l in μ_l. When it is deformed by some stress s_0, and the deformation is kept constant ($\dot{d}_0 = 0$), the stress diminishes in time in accordance with the relaxation curve

$$s = s_0 \exp(-t/T_{\rm rel}) \tag{16.6}$$

with $T_{\rm rel}$ as *time of relaxation*.

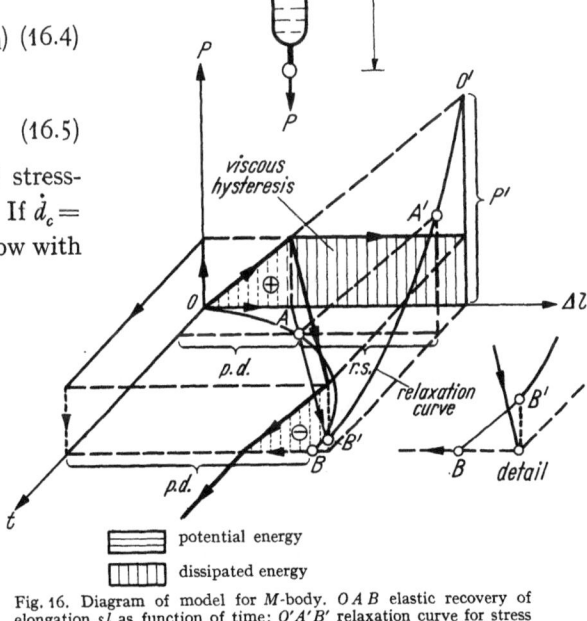

Fig. 16. Diagram of model for M-body. OAB elastic recovery of elongation sl as function of time; $O'A'B'$ relaxation curve for stress P'; p.d. permanent deformation; r.s. recovered strain. ⊕ stored; ⊖ regained.

If $\dot{d}_c > s_0/2\eta$ the stress increases, if $\dot{d}_c < s_0/2\eta$ it decreases, in both cases until a stress appropriate to $\dot{d}_c$, i.e. $=2\eta\dot{d}_c$, is reached, when it remains constant. If on the contrary, the stress increases at a constant rate, i.e. $\dot{s} = \dot{s}_c$, and therefore $s = \dot{s}_c t$, the deformation increases as

$$2d(t) = \frac{\dot{s}_c}{\mu}\left(t + \frac{t^2}{2T_{\rm rel}}\right) = \frac{s}{\mu}\left(1 + \frac{t}{2T_{\rm rel}}\right). \tag{16.7}[1]$$

---

[1] The index l in μ_l is omitted for simplicity.

The Maxwell body behaves in accordance with the magnitude of $T_{\rm rel}$, which constitutes its intrinsic time-scale. Writing (16.2) in the form

$$2d = \left(s + \frac{1}{T_{\rm rel}} \int s\, dt\right)\!/\mu,$$
$$s = 2\eta \dot{d} - \dot{s}\, T_{\rm rel},$$
(16.8)

we see that if the duration of the experiment is very short ($t \ll T_{\rm rel}$), $d = s/2\mu$, and the material impresses as an elastic Hooke solid; while if the duration of the experiment is very long ($T_{\rm rel} \ll t$), $s = 2\eta \dot{d}$, and the material impresses as a viscous, Newtonian liquid. As an example we may name concrete. To loadings lasting hours or days it responds as an elastic solid, and this makes it a suitable material for the structural engineer, but if it is observed for years, flow is discernible[1].

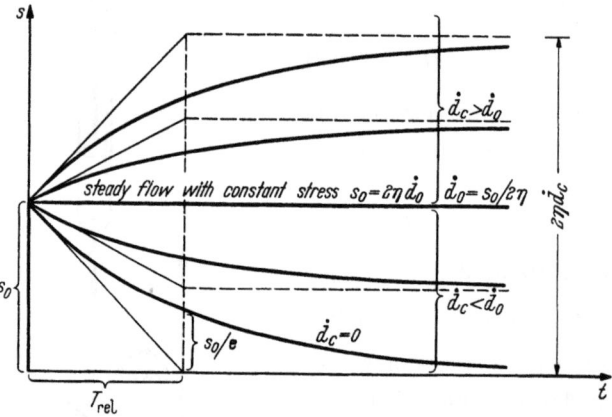

Fig. 17. Stress-time curves of a Maxwell-body for constant rates of deformation $\dot{d}_c$. The testpiece is first ($t = 0$) instantaneously stressed to the amount σ_0 and then deformed at the constant rate $\dot{d}_c$. There exists for every initial stress s_0 a certain rate of deformation $\dot{d}_0$ such that the stress is not changed.

As seen from $2\dot{d}_0 = s_0/\mu\, T_{\rm rel}$, one speaks of (secondary) *creep*, but in contradistinction to the primary creep, this creep is not recovered; on removal of the load, the deformation remains.

JEFFREYS has named the property represented by Eq. (16.2) *elastico-viscosity*. The model and its rheological equation have been widely used for the representation of the rheological behaviour of such dissimilar substances as a 1.5% starch solution[2] and concrete[3], as mentioned before, but it must be pointed out that Maxwell himself derived the equation in order to explain the viscosity of a gas from the relaxation of an elastic response. While the orthodox theory of viscosity follows STOKES' approach, as embodied in Eq. (11.1), Maxwell considered the deformation of an elastic "body" of liquid in equilibrium. The deformation causes a deviation of the pressure from the isotropic state of stress. This deviation tends to disappear, but actually disappears only when the state of equilibrium is completely re-established. The transition from the original to the final state of equilibrium requires a certain period of time which differs for various kinds of fluids: the relaxation time. He calculated from considerations of the kinetic theory of gases that for air $T_{\rm rel} = 1.961 \times 10^{-10}$ sec. MAXWELL's idea of viscous

---

[1] For data see Sect. 20.
[2] W. R. HESS: Kolloid-Z. **27**, 154—163 (1920).
[3] A. M. FREUDENTHAL: in REINER (Editor) 1954.

17. Elastic sols, or the Lethersich body.

For the description of the rheological behaviour of bitumen sols Lethersich postulated a material typified by the rheological formula

$$L = N - K. \tag{17.1}$$

Its rheological equation will be found by adding rates of deformations due to the N and K element respectively. We have from (11.1), introducing d for f,

$$\dot{d}_N = \frac{s}{2\eta_N}. \tag{17.2}$$

On the other hand from (15.3) introducing now d for ε

$$\dot{d}_K = \frac{s}{2\eta_K} - \frac{\mu}{\eta_K} e^{-\frac{\mu}{\eta_K}t} \left(d_{0K} + \frac{1}{2\eta_K} \int s\, e^{\frac{\mu}{\eta_K}t}\, dt \right), \tag{17.3}$$

where d_{0K} is the strain in the Kelvin element at the time $t=0$. Therefore

$$\dot{d}_L = \dot{d}_N + \dot{d}_K = \frac{s}{2} \frac{\eta_N + \eta_K}{\eta_N \eta_K} - \frac{\mu}{\eta_K} e^{-\frac{\mu}{\eta_K}t} \left(d_{0K} + \frac{1}{2\eta_K} \int s\, e^{\frac{\mu}{\eta_K}t}\, dt \right), \tag{17.4}$$

where we distinguish between the viscosities of the two elements by the subscripts N and K respectively. If the stress is const $= s_c$, we can solve the integral in (17.4), and find

$$\dot{d} = \frac{s_c}{2\eta_N} + \frac{s_c - 2\mu d_{0K}}{2\eta_K} e^{-\frac{\mu}{\eta_K}t}, \tag{17.5}$$

and the rate of deformation thus decreases from $\dot{d}_0 = \frac{s_c}{2}\left(\frac{1}{\eta_N} + \frac{1}{\eta_K}\right) - \frac{\mu}{\eta_K} d_{0K}$ at $t=0$ to $\dot{d}_\infty = \frac{s_c}{2\eta_N}$ at $t=\infty$. At the same time the deformation is, from (17.5) by integration

$$d = d_0 + \frac{s_c}{2\eta_N} t + \left(1 - e^{-\frac{\mu}{\eta_K}t}\right)\left(\frac{s_c}{2\mu} - d_{0K}\right) \tag{17.6}$$

and increases from d_0 to $d_\infty = \infty$.

If we unload the specimen ($s_c = 0$), when the deformation has reached a certain magnitude d_0, there will be a recovery in accordance with

$$d = d_0 - d_{0K}\left(1 - e^{-\frac{\mu}{\eta_K}\cdot t}\right) \tag{17.7}$$

from d_0 to $d_0 - d_{0K}$. The viscous part of the extension, namely

$$d_0 - d_{0K} = \frac{s_c}{\eta_N} T \tag{17.8}$$

where T is the period of loading, will not be recovered, however long we may wait.

Eq. (17.4) can be transformed by differentiation and elimination of the second term on the right (in brackets), resulting in

$$s + \dot{s}\,\frac{\eta_N + \eta_K}{\mu} = 2\eta_N \left(\dot{d} + \frac{\eta_K}{\mu}\ddot{d}\right). \tag{17.9}$$

This equation can be further transformed as follows: Let there be an initial deformation d_0, and the body be released of stresses at the time $t=0$, so that

$s = \dot{s} = 0$. Eq. (17.9) is then reduced to a differential equation in d, which on integration yields

$$d - d_\infty = (d_0 - d_\infty)\, e^{-\frac{\mu}{\eta_K} t}. \qquad (17.10)$$

The $(d_0 - d_\infty)$ part of the deformation is accordingly recovered with a retardation time

$$T_{\text{ret}} = \eta_K/\mu. \qquad (17.11)$$

On the other hand, let the deformation be kept constant by adjusting the stress, so that $\dot{d} = \ddot{d} = 0$. Then Eq. (17.9) yields

$$s = s_0\, e^{-\frac{\mu}{\eta_N + \eta_K} t}, \qquad (17.12)$$

and the stress relaxes with a relaxation time

$$T_{\text{rel}} = \frac{\eta_N + \eta_K}{\mu}. \qquad (17.13)$$

We can therefore re-write Eq. (17.9) in the more general form

$$s + \dot{s}\, T_{\text{rel}} = 2\eta_L(\dot{d} + \ddot{d}\, T_{\text{ret}}). \qquad (17.14)$$

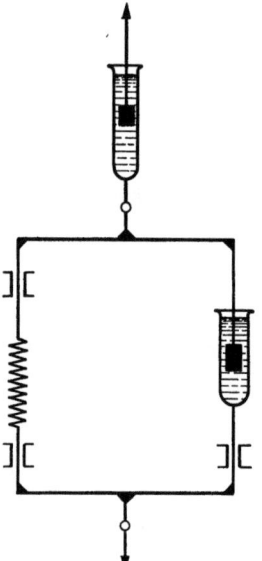

Fig. 18. Diagram of model for L-body.

The Lethersich body is suitable for the description of the rheological behaviour of elastic sols in general. However, in this case there is no question of a relaxation of elastic stresses or a gradual dissipation of elastic potential energy, as suggested by the appearance of a relaxation time in accordance with Eq. (17.13). A rough representation of the structure of an elastic sol will show minute springs or other elastic particles suspended in the solvent, e.g. water, which itself is not elastic. When the solvent is subjected to continuous deformation or flow in which the suspended particles participate, forces are exerted by the solvent upon the "springs" which are extended or compressed, and which they can resist elastically. The elasticity of the dispersed system accordingly resides *in the solute*; when the flow stops, extended springs contract etc., and the internal stresses of the system disappear. Because of the viscous resistance of the solvent this takes time, and thus simulates a relaxation time. In such a sol-system the stress is transmitted by the liquid phase to the solid phase, the external surface forces acting directly upon the liquid phase. This mechanism is reflected by the rheological formula (17.1), but better visualized in the diagram Fig. 18.

18. Relaxing gels, or the Jeffreys body. For the description of the rheological behaviour of the earth's crust, JEFFREYS postulated a material typified by the rheological formula

$$J = N|M. \qquad (18.1)$$

This formula represents essentially a gel. In a gel, which is a dispersion of liquid in solid, the external surface forces act directly upon the solid phase, and the stress is transmitted by the solid to the liquid phase. In such systems flow or creep is due to the relaxation of the elastic forces set up by the straining of the solid phase. At the same time the elastic response of the solid phase is retarded by the liquid phase in the manner described by KELVIN (compare Sect. 13).

Sect. 18. Relaxing gels, or the Jeffreys body.

The rheological equation corresponding to (18.1) is

$$s = 2\eta_N \dot{d} + e^{-\frac{\mu}{\eta_M}t}\left(s_{0M} + 2\mu \int \dot{d}\, e^{\frac{\mu}{\eta_M}t}\, dt\right). \tag{18.2}$$

If we keep the stress a const $=s_c$, differentiation of (18.2), elimination of the integral and integration yields

$$2d = 2d_0 + \frac{s_c}{\eta_M + \eta_N}t + \frac{\eta_M \eta_N}{\mu(\eta_M + \eta_N)}\left[2\dot{d}_0 - \frac{s_c}{\eta_M + \eta_N}\right] \times \\ \times \left[1 - \exp\left(-\mu\frac{\eta_M + \eta_N}{\eta_M \eta_N}t\right)\right]. \tag{18.3}$$

But from (18.2)

$$\dot{d}_0 = (s_c - s_{0M})/2\eta_N \tag{18.4}$$

which makes (18.3)

$$2d = 2d_0 + \frac{s_c}{\eta_M + \eta_N}t + \frac{\eta_M}{\mu(\eta_M + \eta_M)^2}[s_c \eta_M - s_{0M}(\eta_M + \eta_N)] \times \\ \times \left[1 - \exp\left(-\mu\frac{\eta_M + \eta_N}{\eta_M \eta_N}t\right)\right]. \tag{18.5}$$

If we load an unstrained specimen ($d_0 = 0$), in which internal stresses have entirely relaxed ($s_{0M} = 0$), the deformation increases at a rate which changes in accordance with

$$2\dot{d} = \frac{s_c}{\eta_M + \eta_N} + s_c \frac{\eta_M}{\eta_N(\eta_M + \eta_N)} \exp\left(-\mu\frac{\eta_M + \eta_N}{\eta_M \eta_N}\right)t \tag{18.6}$$

from

$$\dot{d}_0 = s_c/2\eta_N \tag{18.7}$$

to

$$\dot{d}_\infty = \frac{s_c}{2(\eta_M + \eta_N)}. \tag{18.8}$$

If we unload the specimen when the deformation has reached a certain magnitude d_0, there will be a recovery in accordance with

$$2d = 2d_0 + \frac{s_{0M}\eta_N}{\mu(\eta_M + \eta_N)}\left[1 - \exp\left(-\mu\frac{\eta_M + \eta_N}{\eta_M \eta_N}t\right)\right] \tag{18.9}$$

from d_0 to $d_0 - s_c \frac{\eta_M}{2\mu(\eta_M + \eta_N)}$, where the internal stress in the M-element, namely s_{0M}, is the stress s_M attained at the end of the period of loading T. This can be calculated from

$$s_{0M} = s_c - 2\eta_N \dot{d} = s_c \frac{\eta_M}{\eta_M + \eta_N}\left[1 - \exp\left(-2\frac{\eta_M + \eta_N}{\eta_M \eta_N}T\right)\right]. \tag{18.10}$$

Therefore

$$d_\infty = \frac{s_c}{2(\eta_M + \eta_N)}t. \tag{18.11}$$

The viscous part of the extension is accordingly not recovered, however long we may wait.

Eq. (18.2) can be simplified by differentiation in respect of time, and eliminating the expression within brackets. Introducing at the same time

$$T_{\text{rel}} = \eta_M/\mu, \tag{18.12}$$

we find

$$s + \dot{s}\, T_{\text{rel}} = 2(\eta_N + \eta_M)\left(\dot{d} + \frac{\eta_N}{\eta_M + \eta_N}T_{\text{rel}}\,\ddot{d}\right). \tag{18.13}$$

Imposing a deformation d_c, and keeping it constant ($\dot{d} = \ddot{d} = 0$), we find that the stress relaxes in accordance with T_{rel}. On the other hand, if we release the deformed body from stresses so that $s = \dot{s} = 0$, we find

$$d - d_\infty = (d_0 - d_\infty) \exp\left(-\frac{\eta_N + \eta_M}{\eta_N} \frac{t}{T_{\text{rel}}}\right), \quad (18.14)$$

and we see that the elastic response is delayed in accordance with

$$T_{\text{ret}} = \frac{\eta_N}{\eta_N + \eta_M} T_{\text{rel}}. \quad (18.15)$$

Introducing the term T_{ret} into Eq. (18.13), we have

$$s + \dot{s} T_{\text{rel}} = 2\eta_J (\dot{d} + T_{\text{ret}} \ddot{d}), \quad (18.16)$$

where

$$\eta_J = \eta_N + \eta_M. \quad (18.17)$$

We thus again arrive at a rheological equation of the type Eq. (17.14) with however, a different structural interpretation of the parameters. The diagram for the model of the J-body is shown in Fig. 19. Its mechanical action is not different from the one of Fig. 18. Accordingly, when the flow curves corresponding to the L- and the J-bodies are plotted, as in Fig. 20, both graphs fall on the same curve and are thus undistinguishable. This is a case of the shortcoming of the mechanical models mentioned at the end of Sect. 13.

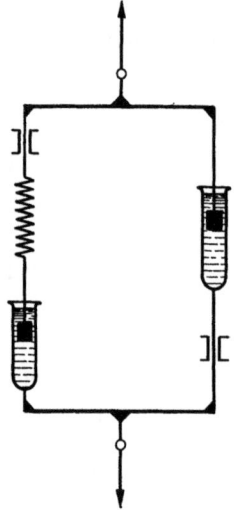

Fig. 19. Diagram of model for J-body.

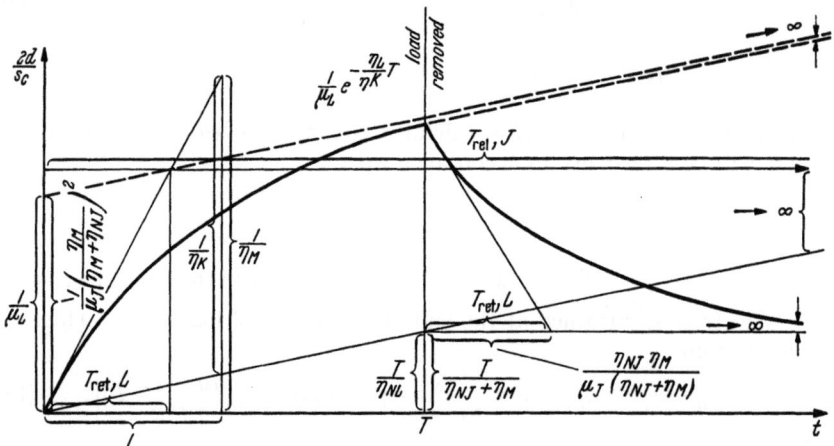

Fig. 20. Deformation per unit stress versus time for constant stress s_c. It is assumed that there is no initial deformation at $t = 0$. The L-body is characterised by $\mu_L, \eta_{NL}, \eta_K, T_{\text{ret}, L} = \eta_K/\mu_L$. The J-body is characterised by μ_J, η_{NJ}, η_M, $T_{\text{rel}, J} = \eta_M/\mu_J$.

The question then arises whether we can by means of rheological observations gain any insight into the structure of such a material, in order to determine in a specific case whether the material is of the sol type of the L-body, or the gel-type of the J-body. Two ways are open. There is firstly the order of magnitude of the relaxation time, which for the J-body will be much higher than for the L-body. A striking example of a material which exists in both states is bitumen. The order of magnitude for a bitumen-sol, or a Lethersich-body, is about 10^3 sec or less; for a bitumen-gel or a Jeffreys-body it is 10^5 sec or more, while in a region of transition both exist side by side. When JEFFREYS (1929) postulated Eq. (18.16) in the form

$$2\mu (d + T_{\text{ret}} \dot{d}) = s + \frac{1}{T_{\text{rel}}} \int s\, dt \qquad (18.18)$$

for the description of the rheological behaviour of the earth's rocky shell, he was considering times of relaxation of the order of 10^8 sec or more. Going back to Eq. (18.12), and assuming μ of the order of 10^{12} bars, he calculated a viscosity of $\eta_M = 5 \times 10^{20}$ poises. For the time of retardation he found $T_{\text{ret}} = 4 \times 10^{-3}$ sec. The second way is provided by the examination of the material in simple shear when the principal axes rotate. In this case a sol in which, as has been said before, the stress is transmitted from the liquid phase to the solid phase, behaves very differently from a gel, in which, on the contrary, the stress is transmitted from the solid to the liquid phase. This could be represented in a model diagram as shown in Fig. 21. As mechanisms both models will, of course, behave in the same manner, and the diagrams have only symbolical meaning.

19. Anelasticity, or the Poynting-Thomson body. ZENER[1] has coined the term *anelasticity* to denote the property typified by the rheological formula

$$PTh = H\,|\,M \qquad (19.1)$$

but, as was already emphasized by KELVIN, the elasticity of such a system is "perfect" in the sense that all deformation is ultimately recovered, and is therefore "strain". The mechanical model represented by (19.1), was first proposed by POYNTING and THOMSON[2] in order to explain the behaviour of glass-fibres[3].

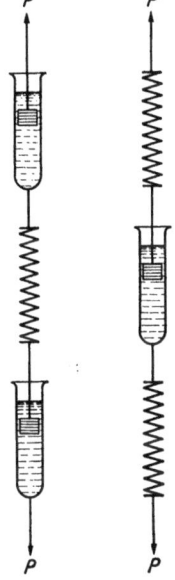

Fig. 21. Model diagrams for Maxwell-sol and Maxwell-gel.

The system represented by the model will have an instantaneous displacement when a force is suddenly applied. The magnitude of this instantaneous displacement is determined solely by the two spring constants. As the force is maintained, the force across the dashpot is gradually relaxed by movement of the piston, resulting in a gradual increase in the overall deformation. Conversely, when the force is suddenly removed, the springs will suddenly release some of their stored energy, resulting in a partical instantaneous recovery. The complete release of all energy in the springs must await the gradual relaxation of force across the dashpot. Such a system therefore manifests the features of both instantaneous and delayed elasticity.

In order to find the rheological equation of the ideal material corresponding to the formula (19.1), we must add the stresses taken by the H and M element respectively. These are given by Eqs. (9.1) and (16.3), and we thus obtain

$$s = 2\mu_H \varepsilon + e^{-\frac{\mu_M}{\eta} t} \left\{ s_0 + 2\mu_M \int \dot{\varepsilon}\, e^{\frac{\mu_M}{\eta} t}\, dt \right\}, \qquad (19.2)$$

where we distinguish between the elasticities of the two elements by the subscripts H and M respectively. If the rate of strain is const $= \dot{\varepsilon}_c$, we can solve the

---

[1] C. ZENER: Elasticity and Anelasticity of Metals. Chicago 1948.

[2] (1902). They were the first to introduce a mechanical model for the representation of rheological behaviour.

[3] They mentioned that the effects exist also in metals and are absent or nearly absent in quartz fibres.

integral in (19.2) and find

$$s = 2\mu \dot{\varepsilon}_c t + 2\eta \dot{\varepsilon}_c + (s_0 - 2\eta \dot{\varepsilon}_c)\, e^{-\frac{\mu_M}{\eta}t}. \qquad (19.3)$$

If the body is strained to some value ε_0 which is kept constant so that $\dot{\varepsilon}_c = \dot{\varepsilon}_0 = 0$, we find that the stress relaxes in accordance with

$$s = s_0\, e^{-t/T_{\text{rel}}} \qquad (19.4)$$

where the relaxation time is

$$T_{\text{rel}} = \frac{\eta}{\mu_M}. \qquad (19.5)$$

Eq. (19.2) can be transformed by differentiation and elimination of the second term on the right resulting in

$$s + \dot{s}\, \frac{\eta}{\mu_M} = 2\mu_H \varepsilon + 2\frac{\mu_H + \mu_M}{\mu_M}\eta\dot{\varepsilon}. \qquad (19.6)$$

Now let the body be deformed to some strain ε_0, and let it be released at the time $t = 0$ from all external force or stress, so that $s = \dot{s} \equiv 0$. We then find from (19.6)

$$\varepsilon = \varepsilon_0 \exp\left(-\frac{\mu_H \mu_M}{\eta(\mu_H + \mu_M)}t\right), \qquad (19.7)$$

and the recovery is delayed with a time of retardation

$$T_{\text{ret}} = \frac{(\mu_H + \mu_M)\eta}{\mu_H \mu_M} \qquad (19.8)$$

but is *complete* at $t = \infty$.

We can therefore give Eq. (19.6) the form

$$s + \dot{s} T_{\text{rel}} = 2\mu_H (\varepsilon + \dot{\varepsilon} T_{\text{ret}}). \qquad (19.9)$$

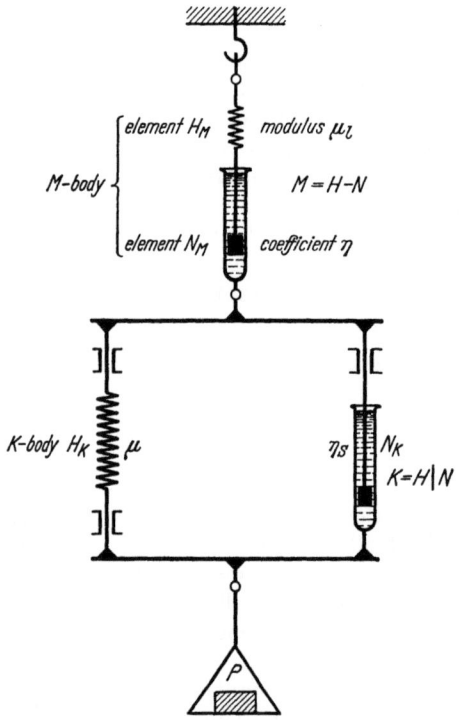

Fig. 22. Model for a Burgers-body; H elastic Hooke-element, N visccus Newtonian element, M visco-elastic Maxwell-body, K firmo-viscous Kelvin-body.

20. Visco-elasticity, or the Burgers body. The rheological formula of the Burgers body[1] is

$$Bu = M\text{-}K, \qquad (20.1)$$

shown as model in Fig. 12. The property which it represents has been named *visco-elasticity*. It exhibits both primary and secondary creep. Its rheological equation is formed by adding rates of deformation due to the M and K bodies respectively. We have from (16.2)

$$\dot{d}_M = \frac{\dot{s}}{2\mu_M} + \frac{s}{2\eta_M}. \qquad (20.2)$$

On the other hand from Eq. (15.3), writing d for ε, we arrive at expression (17.3). Therefore

$$\dot{d}_{Bu} = \dot{d}_M + \dot{d}_K = \frac{s}{2}\frac{\eta_M + \eta_K}{\eta_M \eta_K} + \frac{\dot{s}}{2\mu_M} - \frac{\mu_K}{\eta_K} e^{-\frac{\mu_K}{\eta_K}t}\left(d_{0K} + \frac{1}{2\eta_K}\int_0^t s\, e^{\frac{\mu_K}{\eta_K}t}\, dt\right). \qquad (20.3)$$

If the stress is const $= s_c$, we find the same rate of deformation as for the Lethersich body, Eq. (17.5).

---

[1] First report (1935), pp. 5—72.

Eq. (20.3) can be transformed by differentiation and elimination of the third term on the right (in brackets), resulting in

$$\dot{d} + \frac{\eta_K}{\mu_K}\ddot{d} = \frac{s}{2\eta_M} + \frac{\dot{s}}{2}\frac{\eta_M\mu_K + \eta_M\mu_M + \eta_K\mu_M}{\eta_M\mu_K\mu_M} + \frac{\ddot{s}}{2}\frac{\eta_K}{\mu_M\mu_K}. \qquad (20.4)$$

If $\mu_M \to \infty$, Eq. (20.2) is reduced to that of the Newtonian liquid, and Eq. (20.4) is reduced to Eq. (17.9). The Lethersich body can therefore be regarded as a degeneracy of the Burgers body.

On loading, the Burgers body shows instantaneous and delayed elasticity and viscous flow (i.e. primary and secondary creep); on un-loading it shows instantaneous and delayed recovery with an unrecovered rest from the viscous flow. The Burgers body has both a relaxation time and a retardation time.

In order to find an expression for the relaxation time, let there be a deformation which is kept constant by adjusting the stress so that $\dot{d} = \ddot{d} = 0$. Eq. (20.4) then yields

$$s + \dot{s}\frac{\eta_M\mu_K + \eta_M\mu_M + \eta_K\mu_M}{\mu_K\mu_M} + \ddot{s}\frac{\eta_M\eta_K}{\mu_M\mu_K} = 0 \qquad (20.5)$$

from which

$$s = C_1 e^{-t/T_1} + C_2 e^{-t/T_2}, \qquad (20.6)$$

where T_1 and T_2 are two relaxation times given by

$$\frac{1}{T_{1,2}} = \frac{\eta_M\mu_M + \eta_K\mu_M + \eta_M\mu_K \pm \sqrt{\Delta}}{2\eta_K\eta_M} \qquad (20.7)$$

with

$$\Delta = (\eta_M\mu_M + \eta_K\mu_M + \eta_M\mu_K)^2 - 4\eta_M\eta_K\mu_M\mu_K. \qquad (20.8)$$

On the other hand, let there be an initial deformation d_0, and the body be released of stresses at the time $t=0$, so that $s = \dot{s} = 0$. Eq. (20.4) now yields

$$\dot{d} + \frac{\eta_K}{\mu_K}\ddot{d} = 0 \qquad (20.9)$$

from which

$$d + \frac{\eta_K}{\mu_K}\dot{d} = C. \qquad (20.10)$$

But introducing $s = \dot{s} = 0$ into Eq. (20.3), we find

$$\dot{d} = -\frac{\mu_K}{\eta_K} e^{-\frac{\mu_K}{\eta_K}t} d_{0K} \qquad (20.11)$$

from which

$$d = \left(e^{-\frac{\mu_K}{\eta_K}t} - 1\right) d_{0K} + d_0. \qquad (20.12)$$

The constant C of Eq. (20.10) is therefore

$$C = d_0 - d_{0K}. \qquad (20.13)$$

The deformation d_0 at $t=0$ is accordingly reduced to $d_0 - d_{0K}$ at $t = \infty$ with a retardation time

$$T_{\text{ret}} = \frac{\eta_K}{\mu_K}. \qquad (20.14)$$

The Burgers body is suitable for the description of the rheological behaviour of cement stone. As an example we may note that the rheological coefficients of this material are of the following orders of magnitude: μ_M, modulus of instantaneous elasticity, 10^{11} bar; μ_K, modulus of delayed elasticity, 2×10^{11} bar; η_M, secondary creep viscosity, 6×10^{17} poises; η_K, primary creep viscosity, 3×10^{12} poises. Therefore $T_{\text{ret}}=$ about 2 hrs, $T_{\text{rel}}=$ about 30 days.

When TROUTON and RANKINE examined the rheological behaviour of a lead wire beyond the elastic limit, they represented it by a model with the rheological formula

$$TR = PTh - N = (H|M) - N. \tag{20.15}$$

As a mechanism this is equivalent to

$$TR = (H|M) - N = (H|N) - (H-N) = K - M = Bu, \tag{20.16}$$

but, as in the case of the L- and J-bodies, the internal structure of lead is very different from that of a cement stone, and while, again, their "consistency curves" will have the same shape, the interpretation of their parameters will be quite different.

They found that the model correctly represents the rheological behaviour of lead in a qualitative manner, but did not find quantitative agreement with an exponential law of relaxation. This is typical also in other cases and is due to the presence of a "spectrum" of bodies (compare Sect. 23γ).

21. Plastic gels, or the Schwedoff body. SCHWEDOFF examined the rheological behaviour of a 0.5% gelatine solution 24 hours old in a co-axial cylinder apparatus. The internal cylinder is suspended upon a wire. Let this wire be given a twist through the angle Ω. If the material between both cylinders is an elastic solid, the internal cylinder will be rotated through another angle, say θ. Making use of the elasticity-viscosity analogy, the angle θ can be calculated from the first Eq. (14.24) with the result that

$$\theta = M_z/4\mu\pi h \cdot (1/R_i^2 - 1/R_e^2) \tag{21.1}$$

where M_z is the torque exerted upon the wire.

On the other hand, if the material between the cylinder is a Newtonian liquid, the inner cylinder will follow the wire immediately with decreasing velocity until it has moved through the angle Ω, i.e. until no torsion is left in the wire. Thirdly, if the material is a Maxwell liquid, the cylinder will not follow immediately, but will in the first instance rotate through the angle $\theta (<\Omega)$, the elastic resistance of liquid balancing the torque M_z of the wire, which is

$$M_z = (\Omega - \theta)\frac{D}{l} \tag{21.2}$$

where D is the torsional resistance of the wire and l its length.

Introducing $\Omega - \theta = \varphi$, Eq. (21.1) becomes

$$\theta = \frac{D}{l}\varphi/4\mu_l\pi h \cdot (1/R_i^2 - 1/R_e^2). \tag{21.3}$$

From Eq. (21.3) the rigidity of the elastic liquid can be determined, and is found to be

$$\mu_l = \frac{D}{l}\varphi(1/R_i^2 - 1/R_e^2)/4\pi h\theta. \tag{21.4}$$

The internal cylinder does not remain at the deflection θ but relaxation sets in and the cylinder gradually follows the wire. The relaxation of the stresses required to maintain the elastic strain θ can be studied by reducing the torsion of the wire (Ω) from time to time, so as to maintain the cylinder at the deflection θ and by plotting the angles which measure the torques $\varphi_1, \varphi_2, \varphi_3 \ldots$ against the times $t_1, t_2, t_3 \ldots$. If the material is a Maxwell liquid, the φ-t curve is of the exponential type and the stress vanishes for $t = \infty$. SCHWEDOFF, however, found that in his gelatine solution the relaxation was not of this type, but that the material apparently maintained permanently a small residual deformation permanently; i.e. if Ω was maintained at a constant value, the inner cylinder did not follow until $\theta = \Omega$ or $\varphi = 0$, but stopped at a finite value of φ. Instead of Eq. (16.6) he found the equation

$$s = \vartheta + (s_0 - \vartheta)\,e^{-t/T_{\mathrm{rel}}} \tag{21.5}$$

to be applicable, where ϑ is the residual stress which does not relax after practically infinite time. Similar observations were made by HATSCHEK and JANE[1] on a number of other solutions. This would mean that the materials in question were in fact not liquids, as was thought from their appearance, but solids having a yield value. They can thus be considered as plastic gels.

To represent such materials, we postulate the Schwedoff body, built up of three elements in accordance with the structural formula

$$Schw = (M \mid StV) - H. \tag{21.6}$$

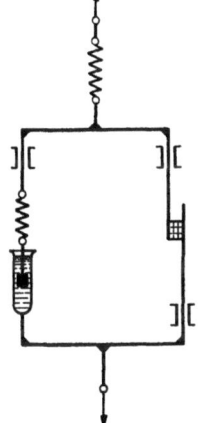

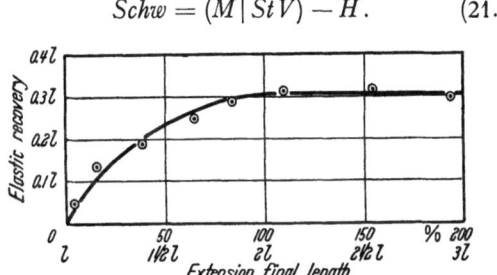

Fig. 23. Diagram of model for *Schw*-body. $Schw = H - (M \mid StV)$.

Fig. 24. Elastic recovery of dough as function of extension.

The model is shown in Fig. 23. The rheological equation is

$$2\dot{d} = (s - \vartheta)/\eta_{pl} + \dot{s}/\mu. \tag{21.7}$$

Recently Soviet scientists have claimed for SCHWEDOFF priority over BINGHAM[2].

22. Flour-dough as an example for high complexity. As an example of the method by which one arrives at the rheological equation of a material through a mechanical model which can reproduce qualitatively its rheological behaviour, we may treat flour-dough which appears to require the most complex combination of elements proposed so far.

The rheological behaviour of dough was extensively investigated by SCHOFIELD and SCOTT BLAIR[3]. These workers first stretched out long cylindrical pieces of dough to different extents for one minute, at the end of which time the cylinders were cut loose. It was found that part of the extension was recovered, but part was permanent (see Fig. 24). This showed that flour-dough (*FD*), is not a perfect Hooke body although it has a Hooke component

---

[1] E. HATSCHEK and R. S. JANE: Kolloid-Z. **40**, 53—58 (1926).
[2] N. V. TABIJN: Kolloidnij Ž. **11**, 438—441 (1949).
[3] See footnote 5, p. 456.

and because of the permanent extension of the cylinders we may write as a first approximation, $FD = H - X$, where X represents another element or elements to be inferred from further experiments.

In the next series of experiments it was found that the elastic recovery decreases with time under stress, in accordance with an exponential law (see Fig. 25). This, however, is what happens in the relaxation of a Maxwell body, and as a second approximation we tentatively write: $FD = H\text{-}N = M$. The relaxation of stresses in accordance with this model was confirmed by further experiments in which the decay of internal stresses was followed in pieces of dough which had been stretched and held stretched (see Fig. 26).

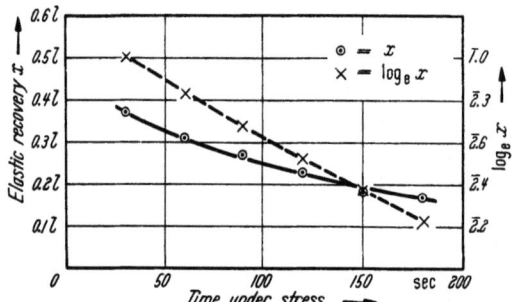

Fig. 25. Elastic recovery of dough as function of time under stress.

SCHOFIELD and SCOTT BLAIR'S second paper describes observations in which the extension of cylinders of dough hung vertically and allowed to elongate under the action of gravity for different periods of time, is related to the stress. The stress at any point depends on the weight of the sample below that point. It was found that while the rate of extension generally decreased with decreasing stress, there was a *finite* stress at which the rate of extension vanishes, in fact a yield-point. This shows that a St. Venant element must also be added; giving as a third approximation $FD = M \mid StV$. It was noticed, however, that a "considerable time often elapses between the release of stress and the cessation of contraction". This indicated an elastic after-effect to the investigation of which the third paper is devoted. For an elastic after-effect a K body must be added. As the structural formula FD contains a StV element the question arises as to which end of the StV element the K body is to be connected. Experiments (see Fig. 27) showed that the elastic after-effect makes its appearance with strains *below the yield point*. This means that the K body must be connected at the spring end. It could be formed there by parallel coupling of the spring with an N element. The same figure, however, illustrates that besides the delayed elastic recovery there is also an instantaneous recovery i.e. the spring of the StV element is not impaired in its working, and the K body is therefore connected to it in series. We accordingly obtain, as a fourth approximation the formula

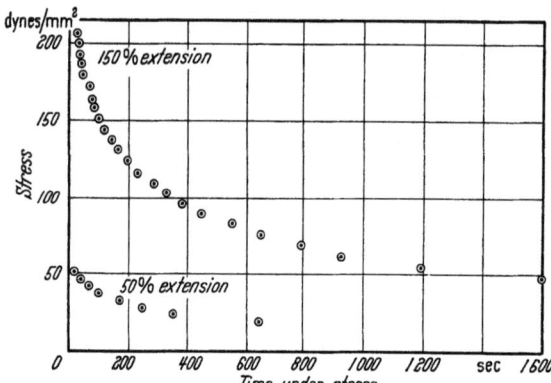

Fig. 26. Relaxation of stress in dough.

$$Sch\, ScB = (M \mid StV) - K = Schw \mid N. \qquad (22.1)$$

Fig. 28 confirms the formula for deformations above the yield-point. The formula embodies all observations made by SCHOFIELD and SCOTT BLAIR.

Sect. 22. Flour-dough as an example for high complexity.

With the help of the rheological formula (21.1) we arrive at the rheological equation of the SCHOFIELD-SCOTT BLAIR body by writing in Eq. (20.3) $s-\vartheta$ for s and introducing η_{pl} for η_M

$$2\dot{d} = (s-\vartheta)\frac{\eta_{pl}+\eta_K}{\eta_{pl}\eta_K} + \frac{\dot{s}}{\mu_M} - \frac{\mu_K}{\eta_K} e^{-\frac{\mu_K}{\eta_K}t}\left[2d_{0K} + \frac{1}{\eta_K}\int_0^t (s-\vartheta) e^{\frac{\mu_K}{\eta_K}t} dt\right]. \quad (22.2)$$

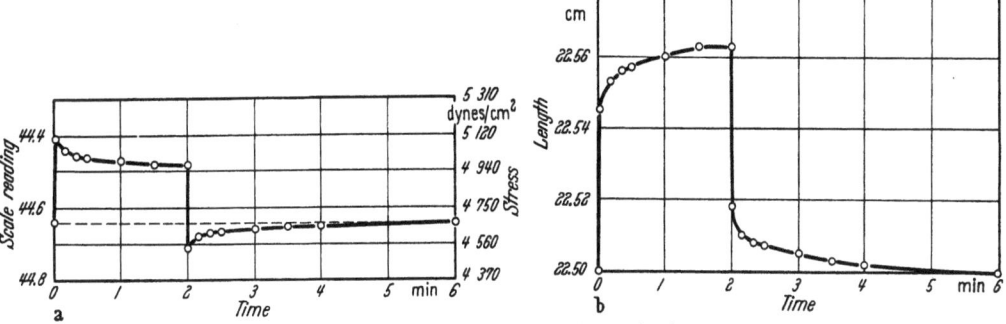

Fig. 27. Delayed elasticity effects in dough.

Eq. (22.2) can again be simplified by differentiation and elimination of the expression within [] brackets resulting in

$$2\dot{d} + 2\ddot{d}\frac{\eta_K}{\mu_K} = \frac{s-\vartheta}{\eta_{pl}} + \dot{s}\frac{\eta_{pl}\mu_K + \eta_{pl}\mu_M + \eta_K\mu_M}{\eta_{pl}\mu_K\mu_M} + \ddot{s}\frac{\eta_K}{\mu_M\mu_K}. \quad (22.3)$$

Eq. (22.1) can also be written as

$$\left.\begin{array}{l}FD = Schw|N \\ = (H-B)|N,\end{array}\right\} \quad (22.4)$$

and flour dough accordingly contains a Bingham element. In certain experiments, other than those performed by SCHOFIELD and SCOTT BLAIR, this Bingham element may be more conspicuous than the H and K elements connected with it. For instance VOLARO-

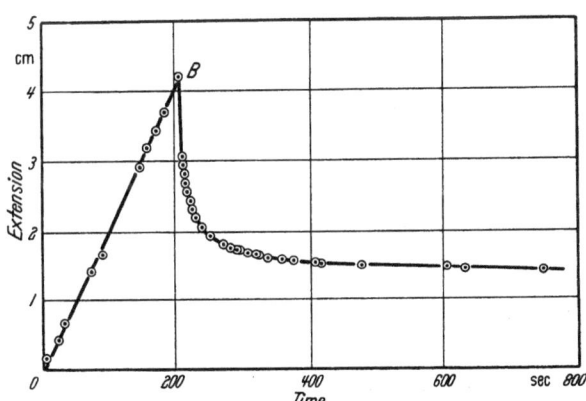

Fig. 28. Deformations above the yield point in dough.

VITCH and SAMARINA examined flour-dough in a co-axial rotating cylinder plastometer where it was subjected to steady flow, and found that Eq. (14.24) was applicable. In their instrument the elastic response of the material could not be determined.

One should not conclude that flour-dough is rheologically the most complicated material ever investigated. This impression results only from the fact that flour-dough was investigated in so many different aspects which at the same time are easily observed. As stated in Sect. 13, one is safe in saying as an axiom of rheology that every material possesses, to some extent, all rheological properties, although in varying degrees. The concepts of Hookean elasticity, Newtonian viscosity etc. are abstractions, useful when *one* distinct aspect of the material only is under consideration. In the *analytical method* used here, the ideal bodies H, N, M, K, etc., are postulated, and the real material is represented as being made up of some combinations of these elements.

It will be noted that it was possible to denote the K, M, L, J, PTh, B and *Schw* bodies by means of terms which imply certain attributes such as "delayed elasticity", "elastico viscosity" etc. This terminology fails us entirely when we reach the complexity of the *Sch ScB* body, and we are thrown back upon the nomenclature introduced in Sect. 6. Even when a "connotative" term is available, it will be preferable to speak of the M, the Bu and the K bodies than to try to distinguish between elasticoviscosity, viscoelasticity and firmo-viscosity.

23. More complex bodies. α) *A general linear body.* HOHENEMSER and PRAGER[1] have postulated a general linear body with the rheological equation

$$a_0 + a_1 s + a_2 \dot{s} + a_3 d + a_4 \dot{d} = 0 \tag{23.1}$$

which we can re-write as

$$(s - \vartheta) + T_{\text{rel}} \dot{s} = 2\mu d + 2\eta \dot{d} = 2\mu (d + T_{\text{ret}}) \dot{d}). \tag{23.2}[2]$$

Through making some coefficients vanish, we arrive at the H, N, StV, M, K, B bodies as special cases.

IRMAI[3] introduces a time-factor T as numerical parameter

$$T = T_{\text{rel}}/T_{\text{ret}} \tag{23.3}$$

and (disregarding ϑ) classifies the linear bodies as follows:

(i) the *homothermal* bodies with $T = 1$, which show no dissipation of energy in a cycle, thus resembling the Hooke-body,

(ii) the *exothermal* bodies with $T < 1$, which are strained less than the homothermal bodies, with part of the energy disspated. These include the N, K and M bodies.

In addition he allows for

(iii) *endous*[4] bodies with $T > 1$, which are strained more than the homothermal bodies with *gain* of energy. In order to comply with the second law of thermodynamics, he assumes that this gain of energy is compensated by changes in the internal structure of the body and suggests that complex poly-electrolytes may behave endothermally. This would permit a rheological equation

$$\dot{s} = 2d/\eta \tag{23.4}$$

when $T_{\text{ret}} \to 0$ and $T_{\text{rel}} \to \infty$.

β) *Superposition of individual bodies.* More complex bodies result from combinations of the individual bodies which constitute the "family" listed in Table 4. Even in the simplest cases there will be combinations of the bodies of the same species. It is obvious that coupling H bodies among themselves either in series or in parallel does not change their rheological character. The same applies to the N- and StV-bodies. The position changes when we come to the second "generation" of M-, P- and K-bodies. These bodies behave differently when coupled in series or in parallel, as will presently be shown:

---

[1] K. HOHENEMSER and W. PRAGER: J. Rheology **3**, 16—22 (1932).
[2] One should note that the equation is not general enough to include the *Sch ScB*-body.
[3] S. IRMAI: Quart. J. Mech. Appl. Math. **7**, 399—409 (1954).
[4] This term has been coined in analogy to "endothermal reactions" to indicate a kind of negative viscosity, named *endosity*.

Examining the M-complex
$$M = M_1 - M_2, \tag{23.5}$$

we have to add rates of deformation. But from (16.2)

$$\dot{d} = \dot{s}\left(\frac{1}{2\mu_1} + \frac{1}{2\mu_2}\right) + s\left(\frac{1}{2\eta_1} + \frac{1}{2\eta_2}\right) = \frac{\dot{s}}{2\mu} + \frac{\dot{s}}{2\eta}, \tag{23.6}$$

where

$$\frac{1}{\mu} = \frac{1}{\mu_1} + \frac{1}{\mu_2}; \quad \frac{1}{\eta} = \frac{1}{\eta_1} + \frac{1}{\eta_2}. \tag{23.7}$$

Therefore several M-bodies coupled in series are equivalent to one M-complex with coefficients of elasticity[1] and coefficients of fluidity adding up.

It is very different with the case

$$M = M_1 | M_2. \tag{23.8}$$

In accordance with the model Eq. (23.8) the initial deformation of the two members is the same d_0 imposed upon the body at zero time. From this

$$s_{01} = 2\mu_1 d_0, \quad s_{02} = 2\mu_2 d_0. \tag{23.9}$$

Adding stresses we find from (16.4) and (23.9)

$$s = 2\dot{d}_c(\eta_1 + \eta_2) + 2e^{-\frac{\mu_1}{\eta_1}t}(\mu_1 d_0 - \eta_1 \dot{d}_c) + 2e^{-\frac{\mu_2}{\eta_2}t}(\mu_2 d_0 - \eta_2 \dot{d}_c). \tag{23.10}$$

Now, we can select a rate of deformation

$$\dot{d}_c = \frac{\mu}{\eta} d_0 \tag{23.11}$$

such that either the second or the third term on the right of (23.10) vanishes, but both these terms will vanish only if both constituent M-complexes have the *same relaxation time* $T_{\rm rel} = \eta_1/\mu_1 = \eta_2/\mu_2$. Generally, therefore, we cannot deform the body at a constant rate such that the stress remains constant, as we can with an individual M-body. Conversely under a constant stress the material will not flow at a constant rate of deformation. In other words, there will be no state of simple viscous flow as is possible with a Maxwell liquid. We therefore find that several M-bodies connected in series behave as one single M-body, while when connected in parallel they do not follow the e-law of relaxation, but a more complicated one, and behave in fact as a M-body with a non-constant time of relaxation[2]. Similarly several K-bodes coupled in parallel behave as one single K-body, but when connected in series they follow not the e-law of retardation but a more complicated one, and behave as one K-body with a non-constant time of retardation. We treat variability of the rheological parameters in Sect. 26. Here we only mention that earlier attempts to describe the rheological behaviour of high polymeric and other substances by assuming their rheological parameters to be variable, have largely given place to theories based upon a *principle of superposition* formulated originally as a heuristic surmise, and now recognized as the mathematical formulation of the basic assumption of the theory, which is the hypothesis of the linearity of viscoelastic effects[3].

---

[1] The *coefficient of elasticity* is the reciprocal of the modulus of elasticity.
[2] This, possibly, provides the explanation to TROUTON and RANKINE's findings, compare end of Sect. 20, but compare also KUBÁT, Kolloid-Z. **134**, 205 (1953).
[3] GROSS (1953).

γ) *Spectra of individual bodies.* In these theories of viscoelasticity, effects are classified in two groups: group I, referring to creep effects made under given stress, and group II, referring to relaxation effects made under given deformation. Then two types of relationship appear in the theory, one involving only functions belonging to the same group of phenomena, and the other involving functions belonging to different groups. Interpretation proceeds in terms of a *continuous set* of exponential functions with time-constants distributed continuously over a finite or infinite interval. Thus two distribution functions are introduced, namely, the *spectrum of retardation times* belonging to group I, and the *spectrum of relaxation times* belonging to group II. The form of the distribution function is a characteristic of the rheological properties of the material. A general relationship between the two distributions was established independently by B. GROSS[1] and W. KUHN[2].

These theories, which are important in the description and the understanding of the rheological behaviour of high-polymers, are treated in detail in Vol. XIII of this Encyclopaedia[3].

δ) *The integrative method.* When the number of elements making up the model increases as much as in the case of the *Sch ScB* body, and the material even then does not follow quantitatively the rheological equation postulated to represent the ideal body, so that a whole spectrum, i.e. an infinite number of individual bodies must be assumed, then SCOTT BLAIR (1949) has advocated a different approach, which he calls the *integrative method*[4]. To describe a behaviour intermediate between the Hookean solid $(s = 2\mu\varepsilon)$ in accordance with Eq. (10.1), the Newtonian liquid $(s = 2\eta f)$ in accordance with Eq. (11.1) and the Pascalian liquid, he defines a quantity Ψ by

$$\Psi = s^\beta d^{-1} t^k \qquad (23.12)$$

in which Ψ can stand for a coefficient of viscosity when $\beta = 1$ and $k = 1$, and for a modulus of elasticity when $\beta = 1$ and $k = 0$, but is also suitable for describing more complex behaviour even when β and k have these special values. Equation (23.12) was proposed by NUTTING and independently by SCOTT BLAIR, and is known as the *Nutting-Scott Blair equation*. As it stands, its application is limited to the case of either constant stress or constant deformation. But while it contains the time explicitly, by differentiating with respect to time, t can be eliminated between the original and the diffentiated equation. The result for constant Ψ is

$$\dot{d}/d - \beta \dot{s}/s = k(s^\beta/d\Psi)^{1/k}. \qquad (23.13)$$

This equation can be used when both stress and deformation vary.

The Nutting-Scott Blair equation does not claim to express a physical law, and the properties which it represents are "quasi-properties". The parameters β and k are sometimes indicative of certain changes in the rheological properties which might occur in a material when subjected to certain processes. In this function they have proved invaluable in many branches of technology. Taking logarithms of Eq. (23.12), we find what has been called the *logarithmic homologue* of the curve representing it. This will be a straight line. It is an empirical fact found in many laboratories all over the world that observational results when

---

[1] B. GROSS: Phys. Rev. **71**, 144 (1947).
[2] W. KUHN, O. KUNZLE and H. PREISSMANN: Helv. chim. Acta **30**, 307, 464 (1947).
[3] See also Sect. 20 of the article by FREUDENTHAL and GEIRINGER in this volume and chapters by STAVERMAN and SCHWARZL in Stuart (Editor), 1956.
[4] In contradistinction to the analytical method, compare Sect. 22.

plotted in log: log scale give straight lines over certain limited ranges. The Nutting-Scott Blair equation is an analytical expression of this empirical fact which manifests itself in the graphical representation. For instance when at a certain laboratory the flow curves of different filler bitumen mixtures were plotted in log: log scale, a series of straight lines were found, which up to a certain filler content has a slope of 1 in 1, and which above that content increased in slope. This indicated that there is a breakdown in the filler-binder system at a certain definite filler content, revealing itself in the slopes of the log: log lines or, correspondingly, in the power of a Nutting-Scott Blair equation representing them, and one can therefore use the magnitude of that power as an "indicator" for the structureal change.

It would be incorrect to regard the Nutting-Scott Blair equation as a "rheological equation". This has been wrongly claimed for certain "power laws" about which more is said in Sect. 26. However, an extension of the integrative method has been developed by Scott Blair using fractional differentiation by which Eq. (23.12) is expanded so as to give it a theoretical basis.

24. Volume-viscosity. As said in Sect. 3, every deformation can be resolved into a change of volume or *volumetric dilatation* and a change of shape or *distortion*. With regard to the volumetric part, classical mechanics of continua, as said in Sect. 7, postulates that all materials, solids and liquids alike, behave in the same highly degenerate manner, namely as perfectly elastic solids with the (elastic) bluk modulus $\varkappa$ as the only rheological coefficient.

Rheology has on the whole taken a similar view, concentrating upon the investigation of distortion and identifying flow with a distortion proceeding in time. This view is too narrow. More detailed observations have shown that while the different rheological properties are more spectacular in distortion, they are all also present in volumetric dilatation. The properties in question are characterized by four coefficients or parameters, which may or may not be constants. These are (i) the *elastic modulus*, (ii) the *retardation time*, (iii) the *relaxation time* and (iv) the *plastic strength* (ϑ). The two "proper" times of a material are connected with two coefficients of viscosity, namely the retardation time with the *solid viscosity* and the relaxation time with *the liquid viscosity*. All these coefficients exist in pairs, one class refers to distortion; the other to volumetric deformation. With the first class we have dealt in the preceding Sects. 13 to 23; the second class is the subject of the present Section. With regard to the elastic modulus, this fact of existing in independent pairs is well known, the modulus of distortion being the modulus of rigidity (μ), the other the bulk modulus ($\varkappa$)[1]. The plastic strength in shear is operative in the St. Venant and Bingham bodies, but in general there will also be a volumetric plastic strength which we may denote by ϑ_v.

That the volumetric straining of a real material must be accompanied by a viscous resistance is obvious from the Second Law of Thermodynamics. Let a sphere of some material be put into free volumetric vibrations by the sudden release of an isotropic pressure. These are connected with ordered radial movements of all particles in unison; a highly improbable state. The ordered movements will be disturbed by internal mechanisms, and ultimately converted into random movements, a process which phenomenologically is a *viscous damping*. We thus arrive at the notion of *solid volume viscosity* ζ_s.

---

[1] Attempts by Cauchy and Poisson in the early stages of the development of the classical theory of elasticity to derive a relation between both and to reduce them to one, failed.

Considering that every liquid has volume elasticity, this must be connected with solid volume viscosity. On this view the rheological equation of the Newtonian liquid should be derived from a combination of Eqs. (11.1) and (15.6). We then find, instead of (11.2), the following relation:

$$s_{lm} = (\varkappa \varepsilon_v + \zeta_s \dot{\varepsilon}_v) \delta_{lm} + 2\eta f_{lm}. \tag{24.1}$$

If no distinction is made between strain and flow, expressing both in terms of deformation through d_{lm} in accordance with Eqs. (3.8) and (3.10), and thus replacing (22.1) by

$$s_{lm} = (\varkappa d_{\alpha\alpha} + \zeta_s \overline{\dot{d}_{\alpha\alpha}}) \delta_{lm} + 2\eta \overline{\dot{d}_{lm}}, \tag{24.2}$$

one gets

$$s_{\alpha\alpha} = -3p_m = 3\varkappa d_{\alpha\alpha} + (3\zeta_s + 2\eta) \overline{\dot{d}_{\alpha\alpha}}. \tag{24.3}$$

Then if the mean pressure p_m is identified with p, the static pressure $3\zeta_s + 2\eta = 0$. This is named the "Stokes relation", but is evidently due to a misunderstanding because it presumes that $f_{\alpha\alpha}$ does not vanish. But non-vanishing of $f_{\alpha\alpha} = f_v$ implies the existence of volume-*flow*, i.e. a continuous cubical dilatation (positive or negative) progressing in time under the action of a constant isotropic stress. Such volume-flow was not considered by STOKES. Actually, therefore, the classical concept of the Newtonian liquid implies $\zeta_s = 0$. This is the point of view accepted in most textbooks of hydrodynamics. It has, however, become evident, mostly through observations on the damping of ultrasonic waves, that ζ_s must be quite large and KARIM and ROSENHEAD[1] have reviewed the reasons for the existence of a finite "second" coefficient of viscosity in liquids and gases. This second coefficient of viscosity is not connected with flow, but with elastic strain, and accepting its non-vanishing existence simply means that the model of the Kelvin body can represent the volumetric behaviour of a Newtonian liquid. The developments of Sect. 15 are therefore applicable, writing $\varkappa$ for μ and ζ_s for η. More especially, there exists a (volumetric) time of retardation

$$T_{\text{ret}} = \zeta_s/\varkappa. \tag{24.4}$$

The elastic response from the action of an isotropic pressure is accordingly delayed, and free volumetric (symmetrical) oscillations are damped in the manner described there.

However, REINER[2] has pointed out that there will in general be a second kind of second coefficient of viscosity connected with volume *flow*. It may be named the coefficient of *liquid volume-viscosity*, denoted by ζ_l and defined in analogy to (9.1) by

$$f_v = f_{\alpha\alpha} = -p_m/\zeta_l. \tag{24.5}$$

Volume-flow (f_v) is the continuous cubical dilatation resulting in a change of density of the material progressing in time under constant isotropic stress. Positive volume flow was found by LEE, REINER and RIGDEN[3] in asphalt and observed quantitatively by REINER, RIGDEN and THROWER[4]; negative volume-flow by GLANVILLE and THOMAS[5] in concrete (compare also REINER[6]). BOSWORTH[7] has

---

[1] S. M. KARIM and L. ROSENHEAD: Rev. Mod. Phys. **24**, 108 (1952).
[2] M. REINER: Appl. Sci. Res. A **1**, 475 (1949).
[3] A. R. LEE, M. REINER and P. RIGDEN: Nature **158**, 706 (1946).
[4] M. REINER, P. RIGDEN and E. N. THROWER: J. Soc. Chem. Ind. **69**, 257 (1950).
[5] W. H. GLANVILLE and F. C. THOMAS: Building Research Technical Papers, No. 21, 1939.
[6] M. REINER: Appl. Sci. Res. A **1**, 475 (1949).
[7] R. C. L. BOSWORTH: Austral. J. Res. A **2**, 394—404 (1949).

observed volume-flow in solidified carbon-dioxide under isotropic pressure, in tension and in torsion. It should be kept in mind that, rheologically, these materials must be considered as liquids (even if possessing elasticity of shape), the order of magnitude of their coefficient of shear-viscosity being respectively- 10^{12}, 10^{17} and 10^{10} poises. Conservation of mass requires, of course, that positive volume-flow is connected with an increase, negative volume-flow with a decrease of voids in the material. However, in accordance with EYRING[1] "a liquid is a binary mixture of molecules and holes", and the materials cited are special only because of the size of their holes which makes volume-flow so pronounced. Gases in explosions may also show a volume-flow effect. When the isotropic load on the body is kept constant, both positive and negative volume-flow must at some time come to an end, the first through rupture or an ultimate rarefication of the material, the second through its ultimate compaction—this requires that ζ_l is not a constant, but does not invalidate their character as flow and the process may take geological times.

For reasons similar to those leading to Eq. (10.6), there will be a relation between λ_T, TROUTON's coefficient of viscous traction, ζ_l and η, namely

$$\lambda_T = 9\zeta_l \eta / (3\zeta_l + \eta). \tag{24.6}$$

We must therefore distinguish between two kinds of volume-viscosity, one (ζ_s) connected with volume-strain, the other (ζ_l) with volume-flow. They have sometimes been confused[2].

When a material is elastically "incompressible", $\varkappa = \infty$, which makes $e_v = 0$. Analogously, for volume-flow to be absent in every case, ζ_l must be equal to ∞, from which $f_v = 0$. In this case from (24.6)

$$\left. \begin{aligned} \lambda_T^* &= 3\eta, \\ \nu_l^* &= \tfrac{1}{2}. \end{aligned} \right\} \tag{24.6'}$$

For solid viscosity to be absent in every case, ζ_s must be equal to 0. However, for volume-flow and solid viscosity to be absent, these are not necessary conditions. For instance, in laminar deformation or flow the volume of the material does not change, whatever the magnitude of either $\varkappa$ or ζ_l. Therefore, as STOKES put it, "in most cases in which it would be interesting to apply the theory of friction of fluids the density of the fluid is either constant or may without sensible error be regarded as constant, or else changes slowly with the time. In the first two cases the results would be the same and in the third nearly the same whether ζ_s were equal to zero or not. Consequently, ... in such cases ... the experiments must not be regarded as confirming ... the theory which relates to supposing ζ_s to be equal to zero" (or ζ_l equal to ∞ as we may add).

Considering that in negative volume-flow the "matter" of the material (EYRING's molecules) flows into the holes, it should be possible to express ζ_l in terms of η, resulting in a uni-constant theory of liquid viscosity, similar to that attempted by CAUCHY and POISSON in the case of elasticity[3].

When there is both volume-strain and volume-flow, the total rate of cubical dilatation in a loaded body is the sum of two, one recoverable and the other irrecoverable. In the unloaded body there may be a recovering volume-strain,

---

[1] J. EYRING: Chem. Phys. **4**, 283 (1936).
[2] Also by REINER in earlier demonstrations in Amer. J. Math. **68**, 672 (1946) and Quart. Appl. Math. **8**, 341 (1951).
[3] REINER: Appl. Sci. Res. A **1**, 475 (1949).

but there will be no volume-flow. We have

$$-\dot{e}_v = \frac{p_m}{\zeta_s} - \frac{\varkappa}{\zeta_s} e^{-\varkappa t/\zeta_s} \left(e_{v,0} + \int \frac{p_m}{\zeta_s} e^{\varkappa t/\zeta_s} dt \right). \tag{24.7}$$

Now let $\dot{d}_{l_m}$ be the tensor of the instantaneous rate of deformation (whether receoverable or not), then

$$-\dot{d}_v = -(\dot{e}_v + \dot{f}_v) = p_m \frac{\zeta_s + \zeta_l}{\zeta_s \zeta_l} - \frac{\varkappa}{\zeta_s} e^{-\varkappa t/\zeta_s} \left(e_{v,0} + \int \frac{p_m}{\zeta_s} e^{\varkappa t/\zeta_s} dt \right). \tag{24.8}$$

Differentiating with respect to time, and eliminating the second expression within brackets on the right side of (24.8) yields

$$p_m + \dot{p}_m \frac{\zeta_s + \zeta_l}{\varkappa} = -\zeta_l \dot{d}_v + \frac{\zeta_s}{\varkappa} \ddot{d}_v, \tag{24.9}$$

which is the volumetric rheological equation of the general viscous liquid.

In order to understand its behaviour, let $\varkappa$, ζ_s and ζ_l be constants so that explicit integration can be carried out. We now perform a number of experiments in imagination:

(i) Let the body be compressed to such extent that its (negative) cubical dilatation is $-d_{v,0}$, and then be released at the time $t=0$, so that $p_m = \dot{p}_m = 0$. Eq. (24.9) is then reduced to a linear differential equation in d, which twice integrated yields

$$d_v - d_{v,\infty} = (d_{v,0} - d_{v,\infty}) e^{-\varkappa t/\zeta_s}. \tag{24.10}$$

The part $d_{v,0} - d_{v,\infty}$ of the dilatation is therefore asymptotically ($t = \infty$) recoverable and is the elastic strain. The recovery is delayed, the retardation time being the same as in the absence of volume-flow. There is a permanent dilatation $d_{v,\infty}$, which is not recovered.

(ii) Let the dilatation $-d_{v,0}$ be kept constant by gradually reducing p_m as required. Now $\dot{d}_v = \ddot{d}_v = 0$, and (24.9) is a linear differential equation in p_m, which on integration yields

$$p_m = p_{m,0} e^{-\frac{\varkappa t}{\zeta_s + \zeta_l}}. \tag{24.11}$$

The stress $p_{m,0}$ applied to produce $d_{v,0}$, therefore relaxes with a relaxation-time

$$T_{\text{rel}} = \frac{\zeta_s + \zeta_l}{\varkappa}. \tag{24.12}$$

Another form for (24.9) is accordingly

$$p_m + \dot{p}_m T_{\text{rel}} = -(\dot{d}_v + \ddot{d}_v T_{\text{ret}}). \tag{24.13}$$

In the Stokesian liquid $\zeta_s = 0$ and $\zeta_l = \infty$ and then $T_{\text{ret}} = 0$ and $T_{\text{rel}} = \infty$.

It appears that the BURGERS body of Sect. 20 offers a suitable representation of the general volumetric behaviour of the viscous liquid[1].

25. The shortcomings of the classical approach. The classical approach on which the preceding review is based, with the rheological equations summarized in Table 6, can only account for first order phenomens. It should not be rashly assumed that first order effects are always more conspicious than those of the second order. When in an expression such as $y = ax + bx^2 = ax(1 + \frac{a}{b}x)$, $\frac{a}{b}x \ll 1$, y can be approximated by ax, but when $\frac{a}{b}x \approx 1$, a change-over of the

---

[1] Compare also J. G. OLDROYD: Proc. Roy. Soc. Lond., Ser. A **226**, 57—58 (1954).

order of magnitude takes place and second order phenomena come to the fore, with $y \sim bx^2$, when $\frac{a}{b}x \gg 1$. Nor should it be forgotten that this "number one" which forms the criterion must be dimensionless. Finally, one cannot neglect bx^2, however small x may be, if the first order term is absent or $a = 0$[1]. On this account the classical approach has four shortcomings, all resulting from the wish to linearize the relevant equations. Two of these shortcomings refer to kinematics, the other two to dynamics.

Table 6. *System of rheological equations.*

number of elements	liquid	solid	
	viscous flow or creep	plastic flow	elastic strain
1	N $2f = s/\eta$	St V $s - \vartheta = 0$ $2\dot{d} = \lambda_M s$	H $s = 2\mu\varepsilon$
2	$M = N - H$ $2\dot{\varepsilon} = s/\eta + \dot{s}/\mu_1$	$P = St V - H$ $s - \vartheta = 0$ $2\dot{d} = \dot{s}/\mu + \lambda_M s$	$K = H \mid N$ $s = 2\mu\varepsilon - 2\eta_s\dot{\varepsilon}$
3	$L = N - K$ $J = N \mid M$ $s + \dot{s}T_{rel} = 2\eta(\dot{d} + \ddot{d}T_{rel})$ $2\dot{d} = s \frac{\eta_N + \eta_K}{\eta_N \eta_K} - \frac{\mu_K}{\eta_K} e^{\frac{\mu_K}{\eta_K}t}(2d_{0K} + \frac{1}{\eta_K}\int s e^{\frac{\mu_K}{\eta_K}t}dt)$	$B = (N \mid StV) - H$ $2\dot{d} = (s - \vartheta)\eta_{pl}$	$PTh = H \mid M$ $s + \dot{s}T_{rel} = 2\mu(\varepsilon + \dot{\varepsilon}T_{rel})$ $s/2 = \mu_H \varepsilon + e^{\frac{\mu_K}{\eta}t}(\varepsilon_0 + \mu_M \int \dot{\varepsilon} e^{\frac{\mu_K}{\eta}t}dt)$
4	$Bu = M - K$ $TR = N - PTh$ $2\dot{d} = s \frac{\eta_M + \eta_K}{\eta_M \eta_K} - \frac{\dot{s}}{\mu_M} \frac{\mu_K}{\eta_K} e^{\frac{\mu_K}{\eta_K}t}(2d_{0K} + \frac{1}{\eta_K}\int s e^{\frac{\mu_K}{\eta_K}t}dt)$	$Schw = (M \mid StV) - H$ $2\dot{d} = (s - \vartheta)/\eta_{pl} + \dot{s}/\mu$	$K \mid M$
6		$SchwScB = (M \mid StV) - K$ $2\dot{d} = (s - \vartheta)\frac{\eta_{pl} + \eta_K}{\eta_{pl}\eta_K} + \frac{\dot{s}}{\mu_M} - \frac{\mu_K}{\eta_K} e^{-\frac{\mu_K}{\eta_K}t}[2d_{0K} - \frac{1}{\eta_K}\int(s-\vartheta)e^{\frac{\mu_K}{\eta_K}t}dt]$	— coupled in series | coupled in parallel

The kinematical shortcomings result from the linearized Eqs. (3.8) and (3.10) for the measures of relative deformation, strain and flow. Dealing first with deformation and strain, Eq. (3.8) is connected with the assumption that the displacement gradient γ_{lm} is infinitesimal. One might think that the classical theory is consistent and complete within this self-imposed limitation. However, it is clear that when the displacement gradient is finite (which we indicate by writing for it Γ_{lm}), we cannot define the relative deformation analogously to (3.8) by $D_{lm} = \frac{1}{2}(\Gamma_{lm} + \Gamma_{ml})$ because such expression for D_{lm} will imply, in general, a rotation. As will be seen, any correct expression for D_{lm} involves higher powers of Γ_{lm}, or products $\Gamma_{l\alpha}\Gamma_{\alpha m}$. When reducing the finite D_{lm} to the infinitesimal d_{lm} by falling back from the finite Γ_{lm} to the infinitesimal γ_{lm}, it is customary to discard their products as "infinitesimal of higher order". However, as pointed out before, this procedure is admissible only if the first order terms are present. When these are absent, the products $\gamma_{l\alpha}\gamma_{\alpha m}$ cannot be neglected. As is shown in Sect. 29, this is the case when there is rotation of the principal axes, as for instance in simple shear. In this manner second order phenomena appear even in infinitesimal elasticity, and the classical theory is limited to "pure" (i.e. irrotational) strains.

---

[1] The mathematically inclined reader will excuse such trivialities; they have often been overlooked in rheological literature.

The second kinematical shortcoming refers to Eq. (3.10), and is of similar kind. In accordance with (3.10) $f_{lm} = \dot{d}_{lm}$, and such second order effects as appear when one reduces D_{lm} to d_{lm} will therefore also be present in f_{lm}. But when it is desired to express *flow* as the *rate of relative deformation* we can only consider *finite* deformations. To consider infinitesimal deformations of a liquid means restricting oneself to infinitesimal oscillations, excluding real flow. When there is flow, the deformation is finite and while (3.10) is correct within this meaning, it is not correct in general. It is important to have this point quite clear.

Therefore consider dilatational flow. Consider a cylindrical tube of increasing length l, surrounding a straight streamline of a viscous liquid. Let the origin of the coordinate system permanently coincide with one end of the tube. Let the other end move with the velocity v_l, and let the flow be homogeneous so that $v_x = \frac{v_l}{l} x$, where $v_l = \frac{dl}{dt}$. In accordance with the arguments of classical hydrodynamics the longitudinal rate of deformation or "flow" which causes and determines the viscous resistance is $f_l = \frac{v_l}{l}$. Introducing the above expression for v_l, we find $f_l = \left(\frac{dl}{dt}\right)\big/l = \left(\frac{dl}{l}\right)\big/dt = \frac{d}{dt} \log(l/l_0)$, where l_0 is the "original" length of the tube at some arbitrary time $t=0$. If the flow is not homogeneous, we must consider an *element* of length, say dx, the two ends of which move with the velocities v_x and $v_{x+dx} = v_x + \frac{dv_x}{dx} dx$ respectively, so that the relative velocity (our previous v_l) is $\frac{dv_x}{dx} dx$ and $f_{xx} = \frac{dv_x}{dx} = v_{x,x}$. We then find as before $f_{xx} = d\left[\log \frac{\delta x_1}{\delta x_0}\right]\big/dt$, where x_0 and x_1 are the values of x for $t=0$ and $t=t$ respectively. But $\delta x_1 = \delta x_0 + u_x(x_0 + \delta x_0, dt) - u_x(x_0, dt)$, where $u_x(x_0, t)$ is the relative displacement as a function of the initial coordinates and time. Neglecting terms of second and higher order ($u_{x,x}$ cannot vanish identically) against the first order term, there results $\delta x_1 = \delta x_0 + u_{x,x}(x_0, 0) \delta x_0$, and therefore $f_{xx} = \frac{d}{dt}[\log(1 + u_{x,x})] = \frac{d}{dt}\log(1 + \Gamma_{xx})$. Only when the displacement gradient is infinitesimal, can we, by developing

$$d_{xx} = \log(1 + \gamma_{xx}) = \gamma_{xx} - \gamma_{xx}^2/2 + \cdots \tag{25.1}$$

use the Cauchy deformation, Eq. (3.8). When the displacement gradient is finite, as it will be in every case of real flow, whether viscous or plastic, we must measure the deformation in the logarithmic measure introduced systematically by HENCKY[1] in which

$$\overset{H}{D} = \log(l/l_0) \tag{25.2}$$

replaces

$$\overset{C}{D} = l/l_0 - 1. \tag{25.3}$$

We find then

$$f_l = \frac{d}{dt}\left(\overset{H}{D}_l\right), \tag{25.4}$$

and the designation "rate of deformation" for f_{lm} is justified. The adoption of the definition (25.4) has certain consequences which are treated in Sects. 34 and 45.

The other two shortcomings have their source in the linearization of the rheological Eq. (6.1) in the manner of Eq. (22.1). In this equation the coefficients

---

[1] More about this measure is said in Sect. 29 below.

are assumed to be constants. Stress deformation or stress rate-of-deformation graphs are then straight lines. When this does not conform with experimental results, e.g. in the case of high polymers, two procedures are open. In the first, the coefficient involved is assumed not to be a constant. This has been named *physical non-linearity*. In the second procedure a different measure of deformation is assumed, namely, one which, while non-linear in the displacement gradients, permits the coefficient to be a constant. This has been named *geometrical non-linearity*.

The second shortcoming in the dynamical stress-deformation relation results from the fact that in Eq. (22.1), which is typical for all rheological equations treated in Part B I, tensors appear in their first power t_{lm} (which stands for d_{lm} etc.). This may be named tensorial linearity or quasi-linearity. It has, however, been shown in Sect. 7 that any isotropic relation between tensors of second rank involves the appearance of second powers $t_{l\alpha} t_{\alpha m}$ of the tensor t_{lm}. When these are introduced in the rheological equation the result is *tensorial non-linearity*.

The repair of the shortcomings of the classical approach effected by physical, geometrical and tensorial non-linearity make second and higher phenomena manifest. These are treated in the following Part B II.

II. Higher order phenomena.

26. Physical non-linearity in elasticity.
α) *Physical non-linearity in general*. In the preceding sections, the rheological coefficients—moduli of elasticity, coefficients of viscosity, times of relaxation etc.—have been assumed to be "constants". While their numerical value will, of course, in each case depend upon physical parameters which are outside the field of rheology, such as temperature, electrical charge, etc., they were assumed to be independent of the rheological variables themselves. This view is too-narrow. First of all, it is known that the rheological coefficients appearing in the distortional equations (μ and η etc.) depend upon the isotropic pressure (or tension) under which the material stands, or, what is the same, its density or its cubical dilatation. The cubical dilatation is an invariant of the deformation. If the deformation is infinitesimal, it is as mentioned in Sect. 3, its *first* invariant. In isotropic and quasi-isotropic materials the rheological coefficients appearing as parameters in the rheological equations, are *scalars*. Therefore, if they are variable, they can be functions of stress or deformation only through their three invariants, which are themselves scalars. Expressions for the principal invariants are given in (7.12) and (7.15). When s, d, f or D is written for t, they become expressions for the invariants of the stress, deformation- and flow-tensor respectively. One can say that generally every rheological coefficient will be a function of the first invariant of either the stress or strain tensor through the change of density of the material, and of the second and third of their deviators. Taking account of the first dependence from tables or empirical equations relating the coefficient to either isotropic (hydrostatic) pressure or density, we are left with the problem of its dependence upon the other two.

In a purely formal mathematical approach we can use the invariants of the deviators of either stress or deformation as arguments in the functional dependence of the coefficient. In a physical approach, which will facilitate a better understanding of the problem, we may use what TRUESDELL (1952) has named the joint invariant of both

$$J(s, \dot{d}) = s_{\alpha\beta} \dot{d}_{\alpha\beta}; \quad J(s, f) = s_{\alpha\beta} f_{\alpha\beta} \tag{26.1}$$

which is the stress-power, and therefore has immediate physical significance. For the physical interpretation it may be assumed that a change of any of the rheological coefficients in the course of the deformation or through deformation must be due to a change of microstructure of the material. Such change of structure, dislodging the system from a state of equilibrium or of minimum internal energy, will require work: since that work has to be performed in a certain time against the time of relaxation, the governing quantity is power. Assuming that all volumetric stress work is conserved as elastic potential energy, it will be the distortional stress power which causes a change of structure, and thereby of the rheological coefficients. We may denote the latter in a general manner by M. In its variations we shall naturally distinguish between two values: one when the material is in the "natural" state, unstrained in the case of elastic solid, at rest in a viscous liquid. This we may denote by M_0. The other will be its value when changed to the maximum extent by the expenditure of extreme stress power: let this be M_∞. Finally M will be changed from M_0 to M_∞ less readily or more readily in accordance with the magnitude of a stability coefficient χ.

The structural state at a certain stage characterized by a definite value of $\dot{w}_{(0)}$ is determined by $M_\infty - M$ and χ can therefore be defined by

$$\chi = \frac{M_\infty - M}{dM/d\dot{w}_0}. \tag{26.2}$$

The coefficient of stability χ is then that power input through which $M_\infty - M_0$ is reduced to its e-th part. Here again the presence of a spectrum of structures will cause the experimental curve to deviate from the exponential curve of (26.2), but the definition of χ given before may be retained.

When the parameters are thus made variable, the rheological equations, while not losing their tensorial linearity, turn out to be non-linear in the components of the tensor. This kind of non-linearity has been named physical non-linearity[1] and the rheological equation is then *quasi-linear*. The stress-strain or flow-curve will cease to be a straight line: the ideal body represented by the rheological equation is thereby generalized, and we may speak of a generalized or non-linear HOOKE solid, generalized or non-linear Newtonian liquid, etc. For the non-linear Newtonian liquid the term *non-Newtonian liquid* is in use, and for its variable viscosity Wo. OSTWALD has introduced the term *structural viscosity*.

It must, however, be realized that the physical non-linearity may in some cases not be of intrinsic physical significance, notwithstanding its name. It is always presupposed that in the stress-strain or stress-flow relation the kinematic quantities are defined as by Eqs. (3.8) and (3.10). There are cases in which this procedure, upon which the considerations of the preceding part B I are based, is open to objections. The first has been indicated at the end of Sect. 10. To illustrate it, we may go back to an experiment performed by BERNOULLI[2] who, without apparently knowing of HOOKE's work, attempted to establish the law of elasticity. While HOOKE had experimented with metal wire strings, BERNOULLI took a gut string and found a parabolic stress-strain curve. Here the non-linearity is due to the body under examination possessing an open structure, while composed of a material which itself may be "linear".

In a second case, the non-linearity of the stress-strain relation vanishes when the expression for the strain is itself made non-linear (geometric non-linearity); the rheological coefficient (modulus etc.) may thus change to a constant. Finally

---
[1] W. OLSZAK: Bull. Acad. Polon. Sci., Cl. IV **2**, 107–111 (1954).
[2] D. BERNOULLI: Mém Acad. Sci. **1705**, 176–185.

the non-linearity of a stress-strain (etc.) relation may point to the presence of a spectrum of individual "bodies" making up the material in the manner described in Sect. 23, with a spectrum of coefficients, each one a constant by itself.

β) *Variability of the bulk modulus.* The bulk modulus occupies a special position. While the shear modulus μ, if defined as by Eq. (10.1), could *in principle* be a constant, the bulk modulus as defined by Eq. (9.1) cannot be so without coming into conflict with a basic physical law. Writing (9.1) in the form

$$V = V_0 + \Delta(V_0) = V_0\left(1 + \frac{\Delta(V_0)}{V_0}\right) = V_0(1 + \varepsilon_v) = V_0\left(1 - \frac{p}{\varkappa}\right), \qquad (26.3)$$

the volume of a body would be compressed to nothing[1] by a finite hydrostatic pressure $p = \varkappa$. Therefore $\varkappa$ must increase with increasing p. To avoid this result, a correction can be made to Eq. (9.1) without abandoning it in principle. We must take into account that, strictly speaking, the stress is not a tensor, but a *tensor-density*. In general, there is no necessity in rheology to consider stress under this aspect because in rheological phenomena the density of the material does not change to any appreciable extent, and can be considered as a parameter[2]. We therefore only indicate HENCKY's[3] procedure by which he arrives at constant rheological coefficients in elastic compressibility. Substituting first

$$dp = -\varkappa \frac{dV}{V} \qquad (26.4)$$

and defining

$$e_v = \log(V/V_0) \qquad (26.5)$$

he finds

$$-p = \varkappa e_v \qquad (26.6)$$

with

$$V = V_0 e^{-p/\varkappa}. \qquad (26.7)$$

This avoids a vanishing volume at finite pressure. Then, taking account of the character of tensor-density of p, he introduces the *reduced stress*

$$p_{\text{red}} = \frac{\varrho_0}{\varrho} p = \frac{V}{V_0} p \qquad (26.8)$$

and thus replaces (26.6) by

$$-p = \varkappa e_v \exp(-e_v). \qquad (26.9)$$

Finally he introduces a *limiting relative volume* ψ, which cannot be reached at finite pressure. He thus arrives at a formula containing the two coefficients $\varkappa$ and ψ. Using BRIDGMAN's[4] observations, he claimed that his final formula was applicable for high stresses, while near atmospheric pressure Eq. (26.6) may be applied. For ordinary rheological phenomena, this provides a sufficiently close approximation.

---

[1] This was used by BERNOULLI, l.c., in an argument against the assumption of a constant YOUNG's modulus E. He went even farther. Having argued that a prism cannot be compressed by 100 percent of its length because then it would be compressed to nothing, he argued that, as an extension is only a negative compression, it follows that it also cannot be extended by 100 percent. From this we learn not only that in 1705 rubberbands were unknown, but also that physicists had exaggerated ideas about the power of mathematical reasoning.

[2] It may not be out of place to quote the definition of the term as "a controllable or variable *constant* in a function which may be varied in order to distinguish the various specific cases".

[3] H. HENCKY: J. Rheology 2, 169—176 (1931).

[4] P. W. BRIDGMAN: Handbuch der Experimentalphysik, Vol. VIII/2, p. 247—395.

These considerations show how by the introduction of the geometrical non-linearity, Eq. (26.3), the physical non-linearity (variability of $\varkappa$) is dismissed.

γ) *Variability of Young's modulus.* The extensive literature on physical non-linearity in elasticity is mainly concerned with the variability of YOUNG's modulus E, defined as the ratio between a normal traction $\sigma = P/A_0$, and a Cauchy extension $e_l = \Delta l/l_0$ of a cylindrical testpiece of original length l_0 and cross-sectional area A_0 in the tensile test, namely

$$E = \sigma/e_l. \tag{26.10}$$

When it is found that the "stress-strain" diagram is not a straight line, this is interpreted as due to a variability of E.

From the second expression in Eq. (10.6) it is clear that the variability of $\varkappa$ must bring with it a variability of E, even if μ is a constant. We may, however, disregard this side of the problem; while in the course of the tensile test the isotropic component of the stress increases together with the pull P, this increase is not sufficient to make the variability of $\varkappa$ felt as a variability of E. A variability of E must therefore be interpreted as being mainly due to a variability of μ. That this constant is not examined directly, is due to the fact that the tensile test is connected with a homogeneous deformation, while the deformation connected with the torsion of a cylindrical rod, the typical test for the determination of μ, is heterogeneous.

It must be said at once that in spite of the great amount of experimental work with which it has been attempted to find a "law" governing the variability of E, this is of no theoretical significance, and has at best resulted in empirical formulae useful within a certain limited range, but more often entirely unsound. The historically interested reader will find references in v. KÁRMÁN (1913) and TIMOSHENKO[1]. We can quote here as typical the BACH-SCHUELE "power-law"

$$e = \alpha \sigma^m, \tag{26.11}$$

which was widely used in Strength of Materials a generation ago, and can be occasionally met even today in engineering literature. Comparison with (26.8) gives

$$E = \sigma^{1-m}/\alpha. \tag{26.12}$$

In extensive experimental work BACH attempted to determine values for m and α for different materials with $m < 1$ for weld-iron, cast-iron, copper, granite, cement and concrete and $m > 1$ for leather, going to such absurd precisions as $m = 1.13713$ for one concrete and $m = 1.20677$ for another one without apparently noticing that on the basis of his equations both kinds of concrete could not be compared at all, E being of different dimension in both cases. The objections to power-laws have been discussed in detail by REINER[2]. The "dimensional" objection just mentioned can be overcome by writing

$$\sigma = a\, e^n \tag{26.13}$$

or

$$E = a\, e^{n-1} \tag{26.14}$$

because e is a dimensionless number. There remains the objection that for very small strains, E would either vanish or become infinitely great in accordance

---

[1] S. P. TIMOSHENKO: History of Strength of Materials, New York: McGraw Hill Publ. Co. 1953.

[2] M. REINER: Naturwiss. **21**, 294—299 (1933).

with whether $n>1$ or $n<1$. This can easily be amended in the first case by adding on the right of Eq. (26.14) a constant term, and in the second case by adding a constant term to the reciprocal. While we cannot thus arrive at a physical "law", we possibly obtain a useful interpolation formula. It must, however, be said, that in all probability the E thus arrived at does not represent an elastic modulus, but a quantity S, which has been introduced by van der Poel[1] as *stiffness modulus*, the reason being that e thus observed does not represent a strain, but a deformation d composed of recoverable ind irrecoverable parts. It appears that most investigators advocating the power-formula plotted the stress-deformation curve from conditions when the specimen was *loaded*, while, as was emphasized in Sect. 1 above, the strain can only be determined from conditions on unloading.

Two properties will give a variable S, even should E be a constant. They are (i) the permanent set, and/or (ii) viscoelasticity.

(i) The permanent set can be considered as a plastic deformation without noticeable yield point. To explain this property, we must distinguish between macro- and micro-stresses. The macro-stress is the gross stress as usually calculated in accordance with the theory of elasticity or, in a simplified manner, in "strength of materials". However, every material body will have notches on its surface and voids or cracks in its interior. These cause local stress *concentrations*, and these micro-stresses will much exceed the gross stress which is recorded in the stress-strain diagram. The stress concentration causes local ruptures in brittle materials such as concrete, stone or timber, and plastic deformation in plastic materials such as metals and moist earth, with asphalt in an intermediate position. Therefore, there will be an irrecoverable macro-deformation, which is the permanent *set*, even at very low and practically infinitesimal macro-stress, with a corresponding absence of yield point. This set is most pronounced on *first loading*. It is usually smaller at each further loading *which does not exceed the first*, and the elasticity of the material is accordingly improved through a series of loadings and un-loadings with consequent appearance of a yield point. The phenomenon is prominent in wet clay, where another micromechanism is operative[2], and concrete, but is noticeable also in metals, especially in cast iron.

The usual definition of the plastic state, connecting it with the existence of a yield value, is therefore not applicable to the "virgin" state of such a material; but if the set has once taken place, it creates a yield value up to the last stressing.

It is clear that the stress-deformation diagram of a material which exhibits a noticeable permanent set will be of parabolic shape on loading. The unloading diagram, on the contrary, is a straight line cutting the deformation axis at a distance from the origin. This distance is a measure of the sum total of the permanent set. The rest which has been recovered is the strain, and the constant E must be calculated from the slope of the straight unloading diagram.

(ii) When the material is pronouncedly viscoelastic, as is the case, for instance, with cement- or concrete-stone, the usual testing procedure of gradually increasing loads in the loading process, and likewise gradually decreasing them in the unloading process, taking readings at the same time, results in a hysteresis loop, both branches being bent upwards. Bingham and Reiner[3] tested a cement-mortar beam in this manner, and also by loading the beam with the full weight, so that the loading did not take time, and recorded the related deflection (as a measure of deformation). This was repeated with increasing loads until a maxi-

---

[1] C. van der Poel: Proc. 2nd Internat. Congr. Rheology 1953.
[2] Compare § 1.4 of Chapter by H. H. Macey in Reiner (Editor), 1954.
[3] E. C. Bingham and M. Reiner: Physics **4**, 88 (1933).

mum was reached. Fig. 29 shows the deflection-load (or deformation-stress) diagram. In the second kind of experiment, the curve was near enough a straight line, its slope being the same as the slope of the tangent at the origin of the curve obtained in the first procedure. The conclusion can be drawn that the instantaneous strain of cement-morter-stone (the one represented by the H_M-spring in the BURGERS model, compare Sect. 20) follows HOOKE's law with constant E, the curvature in the first procedure being due to the delayed elasticity. Contrary to commonly held opinions, this seems also to apply to a coarse-aggregate concrete. For instance BLAKEY and BERESFORD[1] state that "there is little evidence of any marked deviation from a linear tensile stress-strain relationship for concrete below the cracking stress".

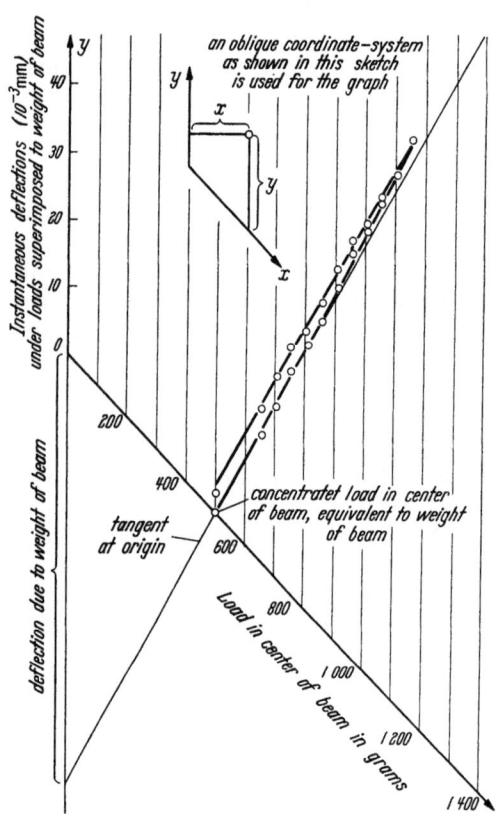

Fig. 29. Stress-strain diagram for 1:3 cement mortar far beam.

27. Rubberlike elasticity. We have said in Sect. 1 that in this chapter rheology is treated from the phenomenological point of view, excluding non-mechanical phenomena such as heat, and thus considering temperature as a parameter. We have therefore not distinguished so far between the two essentially different kinds of elasticity, namely that due to the potential energy of the internal structure, and that due to the kinetic energy of the ultimate particles. For the first kind the elasticity of metals is typical, and for the second the elasticity of gases. The first is the one treated in classical theory of elasticity[2], the second is the subject of the kinetic (or dynamical) theory of gases[3]. It is surprising how long it took before the essential dissimilarity of the second kind from the first was generally recognized. The second edition of the Encyclopaedia Britannica (1779) contains in the article "elasticity" the statement that "The cause of elasticity of the atmosphere hath been commonly ascribed to a repulsion between its particles", and goes on to describe the repulsive force as similar to the one in a compressed spring. In a similar manner a third kind of elasticity, namely rubber like elasticity has only recently come to be distinguished from potential elasticity. For a long time it was thought to be of the same kind, with the ultimate particles likewise pictured as little springs. The subject is treated in this Encyclopaedia in detail by STAVERMAN in

---

[1] F. A. BLAKEY and F. D. BERESFORD: Report C 2, 2—1. Commonwealth Scient. Ind. Res. Org. Melbourne, 1953.

[2] See I. N. SNEDDON and D. S. BERRY, this volume.

[3] Treated in this Encyclopaedia in Vol. XII.

Vol. XIII. For the sake of selfcontainedness of the present chapter, we must, however, briefly mention this kind of elasticity from the phenomenological point of view, even if it is finite, possessing an elastic extensibility of up to 500 to 1000% with a YOUNG modulus of not more than of the order at 10^7 bar.

From the phenomenological point of view the distinction between potential, kinetic and rubberlike elasticity does not arise. True, materials possessing elasticity of the first kind suffer only infinitesimal strains when they are in a compact state, and then follow HOOKE's law. But as mentioned in Sect. 26, and before in Sect. 10, when the same material has an open structure, it can be subjected to large gross strains which, while actually pseudo-finite need not be distinguished from the phenomenological point of view, from the finite strains. This may point to a weakness of the phenomenological point of view, but one arrives in every branch of physics, when dealing with certain problems at regions where its concepts become ill-defined and must be abandoned; this does not, however, invalidate their usefulness in their own field.

Actually the difference between potential and rubberlike elasticity was only discovered about 25 years ago when their entirely different dependence upon the temperature parameter was discovered, similarly to the difference between the temperature dependence of the potential and the kinetic kind. Roughly speaking, in potential elasticity the material becomes "softer" with rise of temperature, while in both kinetic and rubberlike elasticity it becomes "stiffer". In this, rubber-like elasticity is therefore similar to the volume-elasticity of a gas.

There are many materials which either exist in nature or can be prepared artificially, that consist of long-chain "macro"-molecules or are high-polymers. The flexible molecules are either linear or branched, and in anyone material not all of the same length; there is a statistical distribution of chain lengths.

The molecules may also be *cross-linked*, and then form *net*-works, as vulcanized rubbers. The chains have irregularly kinked configurations which are in random Brownian motion, and are thus statistically coiled. The materials are usually amorphous, becoming crystalline at lower temperatures and when highly deformed. The usual theory of rubberlike elasticity refers to the amorphous state where general statements can be made without reference to the chemical nature of the material.

To derive the elasticity law of high-polymers two methods have been used. In the statistical approach of the kinetic theory one calculates the probability that the distance between the ends of a chain at any instant has a certain length. This leads to certain stress-strain relations. The newest approach to the problem of rubberlike elasticity, however, follows the methods of rheology. Unlike the statistical approach, no assumption is made about the molecular structure. Instead a reasonable rheological equation is postulated from which the elastic behaviour is predicted under given conditions. This approach is due to MOONEY[1]. Pure homogeneous deformation with the principal extension ratios $\lambda(\iota)$ is considered. These define a tensor, the third invariant of which

$$\mathrm{III} = \lambda(i) \cdot \lambda(j) \cdot \lambda(k) \qquad (27.1)$$

defines the relative strained volume. It is assumed that the material is incompressible or

$$\mathrm{III} = 1. \qquad (27.2)$$

---

[1] M. MOONEY: J. Appl. Phys. **11**, 582 (1940).

In the kinetic energy theory it is shown that the free stored energy Φ (compare Sect. 5) can be expressed as

$$\Phi = \frac{\mu}{2}[\lambda^2(i) + \lambda^2(j) + \lambda^2(k) - 3] \tag{27.3}$$

with μ proportional to the absolute temperature T. When the stored-energy function Φ is known, the stress-strain relation for any type of strain can be determined.

In the second approach one starts directly from a rheological stress-strain equation. Thus MOONEY postulates for the uni-axial case

$$\sigma = 2\left(C_1 + \frac{C_2}{\lambda}\right)\left(\lambda - \frac{1}{\lambda^2}\right). \tag{27.4}$$

MOONEY's equation takes account of certain deviations from the statistical theory. From (27.4) it follows that

$$\Phi = C_1[\lambda^2(i) + \lambda^2(j) + \lambda^2(k) - 3] + C_2\left[\frac{1}{\lambda^2(i)} + \frac{1}{\lambda^2(j)} + \frac{1}{\lambda^2(k)} - 3\right]. \tag{27.5}$$

It is seen that (27.3) is a special case of MOONEY's more general theory, when C_2 vanishes. RIVLIN[1] has maintained that considerations of symmetry require that λ should appear as λ^2.

From experiments it was found that for a certain swollen rubber the order of magnitude of MOONEY's constants is $C_1 \approx 10^6$ bar, $C_2 \approx 10^5$ bar. For a dry rubber no constant C_1/C_2 ratio could be found.

28. Structural viscosity. The discovery in the United States of the fact that there are true liquids of which the viscosity is not constant, but decreases with increasing shear, *after* the discovery by BINGHAM of the very soft plastic solids which show variability of an "apparent viscosity", led to naming the former "pseudo-plastics". However, BINGHAM himself found that the consistency curve of a gelatine-in-water sol or of a nitro-cellulose solution does not follow his equation and, as he remarked, "the difference seems to be not one of degree, but to constitute a different class entirely". In Europe the order of discovery was reverse. Following GRAHAM, the founder of colloid chemistry who in the middle of the last century spoke of the viscometer as a "colloidoscope", viscosity measurements—generally by means of the Ostwald viscometer which does not permit of the controlled variation of pressure—became one of the most widely used methods of investigating colloidal solutions. However, much of the material thus accumulated was difficult to interpret, possibly due to the "anomalous" behaviour in colloidal solutions, i.e. the variability of the coefficient of viscosity. The variability was observed at the beginning of this century. Later Wo. OSTWALD and pupils[2] carried out a great number of investigations by means of a capillary instrument of special construction. He attributed the anomalous behaviour to the presence of a "structure" in the liquid and its change through flow, and coined the term "Strukturviskosität" which was translated into English as "structural viscosity". This term is now in general use. For the liquids showing this kind of variable viscosity the designation "non-Newtonian liquids" was introduced by REINER[3], and is also in general use, even if open to the objection that it might be thought to include Maxwell-liquids, which is

---

[1] RIVLIN: Proc. Ind. Rubber Techn. Conf. London, 1948.
[2] Wo. OSTWALD: Kolloid.-Z. **36**, 99 (1925). — Wo. OSTWALD and R. AVERBACH: Kolloid-Z. **38**, 261—280 (1926).
[3] M. REINER: J. Rheology **1**, 5 (1929).

not the intention. While therefore such a term as "generalized" or "non-linear" Newtonian liquid might be more appropriate, we shall speak of non-Newtonian liquids possessing structural viscosity.

The consideration of the different structural mechanisms which might cause variability of the coefficient of viscosity belongs to micro-rheology and is dealt with there (Part C). From the phenomenological point of view of macro-rheology it was first pointed out by OSTWALD that a consistency curve will generally have the following features:

The curve starts at the origin (this makes the material a liquid) and the V-ordinates increase regularly with the P-abcissae i.e. there exists no minimum V. At the origin, the curve has a definite slope. For large V/P values, the curve approaches a straight line radiating from the origin and possessing a steeper slope than the slope of the curve at the origin. In consequence thereof, the curve first turns its convex side towards the P-axis and after passing through an inflection point, reverses its curvature (see Fig. 30). The first discoverers of nonlinear consistency curves observed the unshaded parts of the figure only, and were led by this to translate the curves into "power-laws",

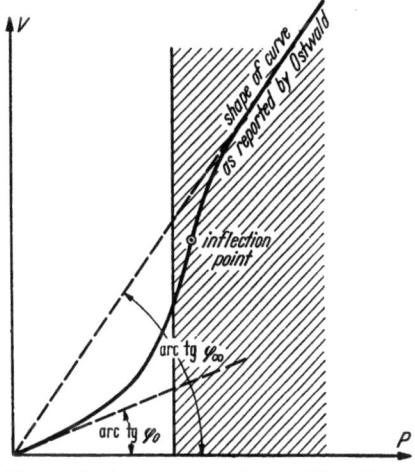

Fig. 30. Consistency curve of an on-Newtonian liquid. φ_0 zero fluidity = fluidity of the liquid at rest. φ_∞ maximum fluidity which the liquid can attain. The unshaded part shows the shape of the curve as reported by many investigators.

which were mentioned in Sect. 26. Wo. OSTWALD was the first to observe and describe the shaded part, and a curve of such form has therefore been called an *Ostwald-curve*. Apart from observations which show that the curve approaches a definite viscosity η_∞, we can say, *a priori*, that while the characteristic property of a non-Newtonian liquid is the decrease of its viscosity with increasing shear, the viscosity naturally cannot be reduced to zero; in fact the liquid being a dispersed system, its viscosity cannot be less than the viscosity of the dispersion medium.

The problem of structural viscosity can be formulated as follows: We must replace the linear Eq. (11.1) by a more general one $s=F(f)$, and the problem is to find the form of the function F.

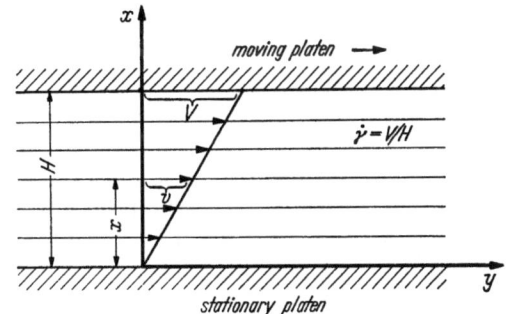

Fig. 31. Plane laminar flow.

For greater simplicity we consider plane laminar flow as shown in Fig. 31 where the stress component is τ and write $\dot{\gamma}$ for the velocity gradient. We then have

$$\dot{\gamma} = \varphi \tau, \qquad (28.1)$$

where φ is the coefficient of fluidity, itself a function of the stress τ, so that our problem is further reduced to finding $\varphi(\tau)$.

If we do not know the form of $\varphi(\tau)$, it seems natural to develop it into a power series,

$$\varphi = a_0 + a_1 \tau + a_2 \tau^2 + \cdots = \sum_{0=n}^{n} a_n \tau^n. \qquad (28.2)$$

From this equation we find for $\tau=0$, $\varphi=a_0$. We call it the *fluidity at rest* and write for it φ_0, and so

$$\varphi = \varphi_0 + \sum_1^n a_n \tau^n. \qquad (28.3)$$

This expression can be specialized in the following way from elementary considerations: If we determine the viscosity by means of a rotation instrument as mentioned in Sect. 14, it is obvious that the viscosity does not depend on whether the outer cylinder is rotated clockwise or anti-clockwise. The sign of τ, however, is changed if the rotation is reversed. If this reversal is to be of no consequence, the power of τ in (28.3) must be an *even* power or

$$\varphi = \varphi_0 + \sum_{n=1}^n \alpha_{2n} \tau^{2n} \qquad (28.4)$$

and therefore

$$\dot{\gamma} = \varphi_0 \tau + \sum_{n=1}^n \alpha_{2n} \tau^{2n+1}. \qquad (28.5)$$

This equation can easily be integrated for the two standard types of viscometers mentioned in Sect. 14, as follows:

| Capillary tube | Co-axial rotating cylinders. |

We introduce the appropriate expressions for the velocity gradient $\dot{\gamma}$ and the shearing traction τ.

[Compare (14.4) and (14.10)] | [Compare (14.14) and (14.22)]

$$\dot{\gamma} = dv/dr, \quad \tau = r\Delta p/2l \quad \Big| \quad \dot{\gamma} = r\frac{d\theta}{dr}, \quad \tau = M/2\pi r^2 h \qquad (28.6)$$

into the rheological equation (27.5) and get (assuming that φ_0 vanishes)

$$\frac{dv}{dr} = \sum_{n=0} \alpha_{2n}(r\Delta p/2l)^{2n+1}, \quad \Big| \quad r\,d\dot{\theta}/dr = \sum_{n=0} \alpha_{2n}(M/2\pi r^2 h)^{2n+2}. \qquad (28.7)$$

Integration with respect to r gives

$$v = \frac{2l}{\Delta p}\sum_0 \frac{\alpha_{2n}}{2n+2}\left(\frac{r\Delta p}{2l}\right)^{2n+2} + C. \quad \Big| \quad \dot{\theta} = -\sum_{n=0}\frac{\alpha_{2n}}{4n+2}\left(\frac{M}{2\pi h r^2}\right)^{2n+1} + C. \qquad (28.8)$$

The integration constant is determined from the boundary condition that the velocity of the liquid in the layer adjacent to a solid wall is identical with the velocity of the wall. We then find

$$-v = \frac{2l}{\Delta p}\sum_{n=0}\frac{\alpha_{2n}}{2n+2}\times \quad \Big| \quad \theta = \sum_{n=0}\frac{\alpha_{2n}}{4n+2}\times$$

$$\times\left[\left(\frac{R\Delta p}{2l}\right)^{2n+2} - \left(\frac{r\Delta p}{2l}\right)^{2n+2}\right]. \quad \Big| \quad \times\left[\left(\frac{M}{2\pi h R_i^2}\right)^{2n+1} - \left(\frac{M}{2\pi h r^2}\right)^{2n+1}\right]. \qquad (28.9)$$

The quantity of flow in unit time is accordingly | The angular velocity of the external cylinder is accordingly

$$\frac{Q}{t} = \int_{r=0}^{r=R} v \cdot 2\pi r \cdot dr \quad \Big| \quad \dot{\Omega} = \sum_{n=0}\frac{\alpha_{2n}}{4n+2}\left(\frac{M}{2\pi h}\right)^{2n+1}\times$$

$$= \sum_{n=0}\frac{\alpha_{2n}}{2n+4}R^3\pi\left(\frac{R\Delta p}{2l}\right)^{2n+1}. \quad \Big| \quad \times\left(\frac{1}{R_i^{4n+2}} - \frac{1}{R_e^{4n+2}}\right). \qquad (28.10)$$

For $n=0$, this equation is reduced with $1/\alpha_0 = \eta$, to

Poiseuille's law, first Eq. (14.12) | Margules' law, first Eq. (14.24).

Introducing into Eqs. (28.10) the consistency variables Eqs. (14.25) and (14.26) we find expressions which are independent of the dimensions of geometrically similar apparatus as follows

$$V = \sum_{n=0} \frac{4\alpha_{2n}}{2n+4} P^{2n+1}. \qquad \left. \begin{array}{l} V = \dfrac{2}{1-a} \sum_{n=0} \dfrac{\alpha_{2n}}{4n+2} \times \\ \times (1-a^{2n+1})\, P^{2n+1}. \end{array} \right\} \qquad (28.11)$$

These power-series equations obviously cover only the unshaded part of Fig. 30. No equations have so far been definitely established which would cover the whole range. Reiner (1949, b), pp. 113—119 has applied Eq. (26.2), which for the present case assumes the form

$$\chi = \frac{\varphi_\infty - \varphi}{d\varphi/d(\tau^2)}. \qquad (28.12)$$

By integration, when χ is constant,

$$\dot{\gamma} = [\varphi_\infty - (\varphi_\infty - \varphi_0)\exp(-\tau^2/\chi)]\,\tau. \qquad (28.13)$$

Postulating Eq. (28.13) as the rheological equation of the non-Newtonian liquid, he has applied it do observations by Reiner and Schoenfeld-Reiner[1] on rubber-toluene solutions in tube and rotating cylinder instruments of various dimensions. While the equation was shown to be adequate in a qualitative manner, the dependence was not in accordance with a simple exponential function. The reason must be sought in the existence of a spectrum of dispersed rubber particles.

The observations by Reiner and Schoenfeld-Reiner revealed a generalization of the Ostwald curve for very small shearing stresses, as shown in Fig. 30. The curve starts at the origin as a straight line; after this it follows a curved part, first convex and then concave towards the P-axis; the end is again formed by a straight line which can be extrapolated towards the origin. The curve therefore has four characteristic points, viz. (i) the origin, (ii) the point where the first straight part ends (a), (iii) the inflexion point (b), (iv) the point where it again becomes straight (c). There are two regions in which the liquid is *simple* Newtonian, viz. (I) from the rest to point a, (II) from c to turbulence. The points a and c are the lower and upper limit of structural viscosity. The points a, b, c define the following coefficients of viscosity; the viscosity at rest η_0, the minimal viscosity η_∞ and the apparent viscosity at the inflexion point η_b'. The first two, applying as they do to simple Newtonian states, are *real* viscosities characteristic of the material. Between the points 0 and a on one hand and c to turbulence on the other, there is *the same viscosity* in every lamina of the flowing liquid, and one can therefore speak of the viscosity of the liquid. In contradistinction, between a and c the viscosity varies from lamina to lamina; one cannot speak of the viscosity of the liquid and the calculated viscosities are therefore *apparent* viscosities.

These observations are typical for non-Newtonian liquids. It is convenient to plot the *logarithmic homologues* of the consistency curves. The logarithmic homologue of a straight line is a straight line sloped at 45°, therefore the log V versus log P curve starts and finishes as such straight lines. The whole subject

---
[1] M. Reiner and R. Schoenfeld-Reiner: Kolloid-Z. **65**, 44 (1933).

was extensively treated by PHILIPPOFF (1942), to whose book the reader is referred. Fig. 32 shows the logarithmic homologues of consistency curves of a 0.25% nitrocellulose solution after PHILIPPOFF.

The equations derived in this section for a non-linear Newtonian liquid can easily be generalized for a non-linear Bingham body by introducing $\tau - \vartheta$ for τ.

29. Geometrical non-linearity. α) *Measures of finite deformation.* We have mentioned in Sect. 26 that behind a non-linear stress-strain relation there may lie not a variability of the rheological coefficient—in this case the elastic modulus—but a different definition of the strain. The definition of the straintensor given in Sect. 3 is linear in displacement gradient components. This can only be so when these are infinitesimal, and the strain ε then also is infinitesimal. We have also said in Sect. 6 that we deal in the present chapter with infinitesimal strain only, but for what follows in subsection β we must briefly mention the subject of measures of finite deformation.

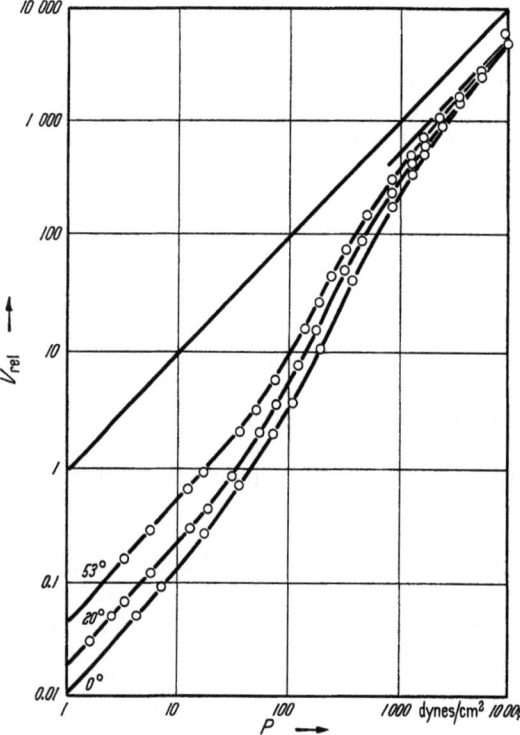

Fig. 32. Consistency curves of 0.25% nitrocellulose solutions (after PHILIPPOFF). The logarithmic homologues have been plotted.

When the strain is finite, and when we replace in (10.2) ε by e, writing

$$s_{lm} = \lambda_L e_{\alpha\alpha} \delta_{lm} + 2\mu e_{lm} \quad (29.1)$$

the equation looks no different, and is still linear in tensors; but as will presently be shown, this is misleading. Actually second-order effects in the strain components and therefore in the stresses appear at once.

The magnitude of these effects depends upon the measure of strain which is used for e_{lm}.

There is no unique measure of strain. A prismatic element of length l_0 oriented in the direction of a principal axis will generally be deformed into a prismatic element of length

$$l = l_0 + \Delta l. \quad (29.2)$$

If the strain is infinitesimal, we have in accordance with the Cauchy measure, Eq. (3.8),

$$\varepsilon = \frac{\Delta l}{l_0} = \frac{l - l_0}{l_0} = \frac{l}{l_0} - 1 = \lambda - 1 \quad (29.2')$$

where λ, the *principal extension ratio*, is

$$\lambda = \frac{l}{l_0}. \quad (29.3)$$

We can generalize (29.2) for finite strain, writing

$$e^C = \lambda - 1. \quad (29.4)$$

But any function of λ can be used as a measure of strain (or, in general, of deformation) provided (a) it vanishes for $\lambda=1$, when $l=l_0$, and there is no change of length, (b) it is reduced to Eq. (29.2), when $\lambda-1$ is infinitesimal; and (c) it is a dimensionless figure[1]. This is so in the case of the following measures of deformation or strain which have been proposed, among others, by SWAINGER[2], GREEN, ALMANSI and HENCKY[3]. They are in this order

$$e^S = 1 - \frac{1}{\lambda} = (\lambda - 1) - (\lambda - 1)^2 + \cdots, \tag{29.5}$$

$$e^G = \tfrac{1}{2}(\lambda^2 - 1) = (\lambda - 1) + \tfrac{1}{2}(\lambda - 1)^2, \tag{29.6}$$

$$e^A = \frac{1}{2}\left(1 - \frac{1}{\lambda^2}\right) = (\lambda - 1) - \frac{3}{2}(\lambda - 1)^2 + \cdots, \tag{29.7}$$

$$e^H = \log\frac{l}{l_0} = \log[1 + (\lambda - 1)] = (\lambda - 1) - \frac{(\lambda - 1)^2}{2} + \cdots. \tag{29.8}$$

SWAINGER proposed Eq. (29.5) in connection with his investigations on large deformation of metals, following (but independently of) KOERBER, to describe plastic deformation and not recoverable strain, but such distinction is immaterial for the purpose of the following considerations. In order to appreciate the differences between these measures, let the length of a rod be doubled; then the different measures will give

$$e^C = 100\%, \quad e^G = 150\%, \quad e^H = 66\%, \quad e^A = 37.5\%, \quad e^S = 50\%,$$

while when its length is halved, they give

$$e^C = -50\%, \quad e^G = -37.5\%, \quad e^H = -66\%, \quad e^A = -150\%, \quad e^S = -100\%.$$

The tensor expressions for two measures have been derived by MURNAGHAN (1937). They are the Green measure

$$2_{lm}\overset{G}{e} = g_{\alpha\beta}(_l x^\alpha)(_m x^\beta) - {}_{lm}g \tag{29.9a}$$

and the Almansi measure

$$2\overset{A}{e}_{lm} = g_{lm} - {}_{\alpha\beta}g(^\alpha x,_l)(^\beta x,_m), \tag{29.10a}$$

or introducing the displacement gradient defined in Sect. 25,

$$2_{lm}\overset{G}{e} = {}_{lm}\Gamma + {}_{ml}\Gamma + (_{\alpha l}\Gamma)(_{\alpha m}\Gamma) \tag{29.9b}$$

$$2\overset{A}{e}_{lm} = \Gamma_{lm} + \Gamma_{ml} - (\Gamma_{\alpha l})(\Gamma_{\alpha m}). \tag{29.10b}$$

The expression for the volume-strain in the Almansi measure is

$$\overset{A}{e}_v = 1 - (1 - 2I + 4II - 8III)^{\frac{1}{2}}. \tag{29.10c}$$

These two measures are suitable for describing elastic strains. They are not suitable, as explained in Sect. 25, for plastic deformations and deformations due to flow. For these we must use the Hencky measure.

---

[1] This is required in order that the strain should not depend upon the dimensions of the body. For example, a steel rod of 1 cm² cross-sectional area if pulled by means of a force of 1000 kg is elongated approximately by $\frac{1}{2}$ mm, if its length is 1 m; by 1 mm, if its length is 2 m, etc. Physical considerations, however, require that the physical quantity characterizing the process, i.e., the strain, should be independent of the accidental length, and therefore should be the same in both cases.

[2] K. H. SWAINGER: Phil. Mag. [7] **36**, 443 (1945).

[3] For references cf. TRUESDELL (1952). Eq. (29.6) is the measure given in LOVE (l. c) from theoretical considerations as "the" measure of finite strain without mentioning its special character. It was used by KIRCHHOFF in connection with his investigations on very thin rods.

A tensorial expression for the Hencky measure can be derived by the method of Sect. 7, as follows:

What has been described there as the "first experiment" is given in Eqs. (29.7) and (29.8), from which we have

$$2\overset{H}{e}(i) = -\log\left(1 - 2\overset{A}{e}(i)\right) = \sum_{n=1}^{n=\infty} \frac{1}{n}\left(2\overset{A}{e}(i)\right)^n. \tag{29.11}$$

We now introduce $2\overset{H}{e}(i)$ for $y(i)$ and $2\overset{A}{e}(i)$ for $x(i)$, while $x(j) = x(k) = 0$. Comparison with Eq. (7.36) now shows that K_0 vanishes in every case. From $n=1$ upwards the expressions for K_n are

$$K_n = \frac{1}{n}. \tag{29.12}$$

It then follows from (7.35) that

$$\left.\begin{aligned}
F_0 &= \sum_{n=3}^{\infty} \frac{1}{n} \sum_{\substack{a+2b+3c \\ =n}} (-1)^b \frac{c(a+b+c-1)!}{a!\,b!\,c!} I_x^a II_x^b III_x^c \\
&= \sum_{\substack{a,b,c=0 \\ (a+2b+3c \geq 3)}}^{\infty} \frac{c(a+b+c-1)!}{(a+2b+3c)\,a!\,b!\,c!} I_x^a (-II_x)^b III_x^c \\
&= \sum_{a,b,c=0}^{\infty} \frac{c(a+b+c-1)!}{(a+2b+3c)\,a!\,b!\,c!} I_x^a (-II_x)^b III_x^c,
\end{aligned}\right\} \tag{29.13}$$

$$\left.\begin{aligned}
F_1 &= 1 + \sum_{n=3}^{\infty} \frac{1}{n} \sum_{\substack{a+2b+3c \\ =n-1}} (-1)^b \frac{(b+c)(a+b+c-1)!}{a!\,b!\,c!} I_x^a II_x^b III_x^c \\
&= 1 + \sum_{\substack{a,b,c=0 \\ (a+2b+3c \geq 2)}}^{\infty} \frac{(b+c)(a+b+c-1)!}{(a+2b+3c+1)\,a!\,b!\,c!} I_x^a (-II_x)^b III_x^c \\
&= 1 + \sum_{a,b,c=0}^{\infty} \frac{(b+c)(a+b+c-1)!}{(a+2b+3c+1)\,a!\,b!\,c!} I_x^a (-II_x)^b III_x^c,
\end{aligned}\right\} \tag{29.14}$$

$$\left.\begin{aligned}
F_2 &= \frac{1}{2} + \sum_{n=3}^{\infty} \frac{1}{n} \sum_{\substack{a+2b+3c \\ =n-2}} (-1)^b \frac{(a+b+c)(a+b+c-1)!}{a!\,b!\,c!} I_x^a II_x^b III_x^c \\
&= \frac{1}{2} + \sum_{\substack{a,b,c=0 \\ (a+2b+3c \geq 1)}}^{\infty} \frac{(a+b+c)(a+b+c-1)!}{(a+2b+3c+2)\,a!\,b!\,c!} I_x^a (-II_x)^b III_x^c \\
&= \frac{1}{2} + \sum_{a,b,c=0}^{\infty} \frac{(a+b+c)(a+b+c-1)!}{(a+2b+3c+2)\,a!\,b!\,c!} I_x^a (-II_x)^b III_x^c.
\end{aligned}\right\} \tag{29.15}$$

After some calculations we find

$$F_0(I_x, II_x, III_x) = III_x \int_0^1 \frac{\sigma^2\, d\sigma}{1 - I_x\sigma + II_x\sigma^2 - III_x\sigma^3}, \tag{29.16}$$

$$F_1(I_x, II_x, III_x) = \int_0^1 \frac{(1 - I_x\sigma)\, d\sigma}{1 - I_x\sigma + II_x\sigma^2 - III_x\sigma^3}, \tag{29.17}$$

$$F_2(I_x, II_x, III_x) = \int_0^1 \frac{\sigma\, d\sigma}{1 - I_x\sigma + II_x\sigma^2 - III_x\sigma^3}, \tag{29.18}$$

Sect. 29. Geometrical non-linearity.

so that finally

$$2\overset{H}{e^l_m} = \left[III\int_0^1 \frac{\sigma^2\,d\sigma}{1-I\sigma+II\sigma^2-III\sigma^3}\right]\delta^l_m + \left[\int_0^1 \frac{(1-I\sigma)\,d\sigma}{1-I\sigma+II\sigma^2-III\sigma^3}\right]2\overset{A_l}{e^l_m} + \left.\left[\int_0^1 \frac{\sigma\,d\sigma}{1-I\sigma+II\sigma^2-III\sigma^3}\right]4\overset{A_l}{e^{A_l}_\alpha}\overset{A}{e^\alpha_m}\right\} \quad (29.19)$$

As an example we treat homogeneous simple shear in Cartesian coordinates in accordance with

$$x_0 = x, \quad y_0 = y - Sz, \quad z_0 = z. \qquad (29.20)$$

We have

$$\Gamma_{lm} = \begin{Vmatrix} 0 & 0 & 0 \\ 0 & 0 & S \\ 0 & 0 & 0 \end{Vmatrix} = {}_{lm}\Gamma \qquad (29.21)$$

and therefore

$$2\overset{A}{e}_{lm} = S\begin{Vmatrix} 0 & 0 & 0 \\ 0 & 0 & 1 \\ 0 & 1 & -S \end{Vmatrix}, \quad 2\overset{G}{{}_{lm}e} = S\begin{Vmatrix} 0 & 0 & 0 \\ 0 & 0 & 1 \\ 0 & 1 & S \end{Vmatrix} \qquad (29.22\text{a, b})^1$$

with the invariants [compare Eqs. (7.12)]

$$\begin{aligned} I_A &= -S^2/2; & II_A &= -S^2/4; & III_A &= 0 \\ I_G &= S^2/2; & II_G &= -S^2/4; & III_G &= 0. \end{aligned} \right\} \qquad (29.23)$$

Furthermore

$$4\overset{A}{e}_{lm}\overset{A}{e}_{\alpha m} = S^2\begin{Vmatrix} 0 & 0 & 0 \\ 0 & 1 & -S \\ 0 & -S & 1+S^2 \end{Vmatrix} \qquad (29.24)$$

so that

$$2\overset{H}{e}_{lm} = S\begin{Vmatrix} 0 & 0 & 0 \\ 0 & SF_2 & F_1 - S^2F_2 \\ 0 & F_1 - S^2F_2 & S(F_2 - F_1) + S^3F_2 \end{Vmatrix}. \qquad (29.25)$$

Now

$$F_0 = 0, \qquad (29.26)$$

$$F_1 = \int_0^1 \frac{(1+S^2\sigma)\,d\sigma}{1+S^2\sigma - S^2\sigma^2} = \frac{S^2+2}{S^2C}\log\left|\frac{C+1}{C-1}\right|, \qquad (29.27)$$

$$F_2 = \int_0^1 \frac{\sigma\,d\sigma}{1+S^2\sigma - S^2\sigma^2} = \frac{1}{S^2C}\log\left|\frac{C+1}{C-1}\right|, \qquad (29.28)$$

where

$$C = \sqrt{1+4/S^2}. \qquad (29.29)$$

Therefore finally,

$$\overset{H}{e}_{lm} = \frac{1}{CS}\log\left|\frac{C+1}{C-1}\right|\begin{Vmatrix} 0 & 0 & 0 \\ 0 & S/2 & 1 \\ 0 & 1 & -S/2 \end{Vmatrix} \qquad (29.30)$$

with

$$I_H = 0; \quad II_H = -\frac{1}{4}\left(\log\frac{C+1}{C-1}\right); \quad III_H = 0. \qquad (29.31)$$

---

[1] In accordance with Eq. (3.8) we would have $d_{yz} = d_{zy} = S/2$ in both cases with all other components vanishing.

β) *Cross-elasticity in infinitesimal strain.* Having thus found expressions for finite strain, we calculate the corresponding stresses from Eq. (29.1). This becomes non-linear in the displacement-gradient component $\Gamma_{yz} = S$, while not loosing its *tensorial* linearity. The equation has for such cases been termed "quasi-linear" (compare TRUESDELL l. c.).

The stress must be referred to the final state, which is the state of equilibrium. The Almansi and Hencky measures refer to this state, but when the Green measure is used, which refers to the initial state, we must carry out a rotation which brings it into the final state.

The rotation matrix is

$$R_{lm} = \begin{Vmatrix} 1 & 0 & 0 \\ 0 & \cos\alpha & \sin\alpha \\ 0 & -\sin\alpha & \cos\alpha \end{Vmatrix}, \tag{29.32}$$

where [compare LOVE (1927), Art. 3]

$$\tan\alpha = S/2. \tag{29.33}$$

This makes

$$2\overset{G}{e}_{lm} = S \begin{Vmatrix} 0 & 0 & 0 \\ 0 & S & 1 \\ 0 & 1 & 0 \end{Vmatrix} \tag{29.34}$$

We do not follow up these developments further for finite strain. As said in Sect. 6, we deal in this chapter with infinitesimal strain only[1]. We therefore reduce the expressions for $\overset{G}{e}_{lm}, \overset{A}{e}_{lm}, \overset{H}{e}_{lm}$ to $\overset{G}{\varepsilon}_{lm}, \overset{A}{\varepsilon}_{lm}, \overset{H}{\varepsilon}_{lm}$ by neglecting S or S^2 against 1 or a greater number, but not neglecting S^2 where it is the first term in a series. The expressions for $\overset{G}{\varepsilon}_{lm}$ and $\overset{A}{\varepsilon}_{lm}$ are not changed, but

$$2\overset{H}{\varepsilon}_{lm} = S \begin{Vmatrix} 0 & 0 & 0 \\ 0 & S/2 & 1 \\ 0 & 1 & -S/2 \end{Vmatrix}. \tag{29.35}$$

We see that when the gradient is infinitesimal, the first order terms are the same in every measure, while the second order terms vary considerably from measure to measure.

Introducing the expressions for the infinitesimal strain from Eqs. (29.22a), (29.34) and (29.35) into (29.1), we find

$$\overset{G}{S}_{lm} = \frac{\lambda s^2}{2}\delta_{lm} + \mu S \begin{Vmatrix} 0 & 0 & 0 \\ 0 & S & 1 \\ 0 & 1 & 0 \end{Vmatrix}, \tag{29.36}$$

$$\overset{A}{S}_{lm} = -\frac{\lambda S^2}{2}\delta_{lm} + \mu S \begin{Vmatrix} 0 & 0 & 0 \\ 0 & 0 & 1 \\ 0 & 1 & -S \end{Vmatrix}, \tag{29.37}$$

$$\overset{H}{S}_{lm} = \mu S \begin{Vmatrix} 0 & 0 & 0 \\ 0 & S/2 & 1 \\ 0 & 1 & -S/2 \end{Vmatrix}. \tag{29.38}$$

Contrary to the linear theory of classical elasticity, simple shear therefore cannot be supported by shearing tractions only. In addition to these, it is necessary

---

[1] Finite strain is treated in this Encyclopaedia by W. NOLL in Vol. X.

to apply an isotropic tension in the first and an isotropic pressure in the second case. Furthermore tractions are required in the direction of the displacement and in the direction normal to it. These form a system of *cross-stresses*. These cross-stresses have the following components

	G	A	H
in the direction of the displacement	+	0	+
normal to the direction of the displacement	0	−	−,

where + indicates tension and − compression. If these second order stresses and tractions are absent, a simple shearing stress

$$s_{lm} = \tau \begin{Vmatrix} 0 & 0 & 0 \\ 0 & 0 & 1 \\ 0 & 1 & 0 \end{Vmatrix} \tag{29.39}$$

will produce, in addition to the shearing strain

$$2\varepsilon_{lm} = \frac{\tau}{\mu} \begin{Vmatrix} 0 & 0 & 0 \\ 0 & 0 & 1 \\ 0 & 1 & 0 \end{Vmatrix} \tag{29.40}$$

the following strains:

	G	A	H
volumetric	shrinkage	expansion	nil
in the direction of the displacement	contraction	nil	contraction
normal to the direction of the displacement	nil	expansion	expansion

The volumetric change due to simple shear has been named *dilatancy*. "Dilatancy" was first observed in sand by REYNOLDS[1] who coined the term to distinguish it from "dilatation". There will be elastic, plastic and viscous dilatancy. Plastic dilatancy is known in soils; positive in sandy soils and negative in clayey soils. It was observed in asphalt by LEE and MARWICK[2] and by WOOD[3] in single metal crystals. We are concerned here with *elastic* dilatancy, a phenomenon predicted by Lord KELVIN (1890) with the words "It is possible that a shearing stress may produce in a truly isotropic solid condensation or dilatation in proportion to the square of its value". A modulus of elastic dilatancy has been defined by REINER (1948) by

$$\delta = \left(\varkappa e_v - \frac{s_{\alpha\alpha}}{3}\right) \Big/ 4 II_{(0)e}. \tag{29.40'}[4]$$

Where II_e denotes the second invariant of the strain tensor, while the subscript (0) indicates the deviator.

It should be noted that in the terms which define these phenomena, the first power of S is absent, and the second powers therefore *cannot be neglected* even

---

[1] A. REYNOLDS: Phil. Mag. (2) **20**, 469 (1885).
[2] A. R. LEE and A. H. D. MARKWICK: J. Soc. Chem. Ind. **56**, 1461 (1937).
[3] W. A. WOOD: Proc. Roy. Soc. Lond., Ser. A **172**, 231 (1939).
[4] This δ should not be confounded with KRONECKER's delta.

when S is infinitesimal. There is, of course, nothing to prevent an investigator from *disregarding* them—provided he knows of their existence.

γ) *Expressions in cylindrical coordinates.* Homogeneous simple shear cannot easily be realized experimentally. If it is desired to test the theory expounded here by means of experiments, the laminar shear of a right circular cylinder either simple or combined with pure extension, is more suitable. Laminar shear of a cylinder is shown in Fig. 33. Simple shear as shown in Fig. 31 is a laminar distortion in which parallel plane slices slide over each other as in a pack of cards, each card as if it were rigid. (The cards should, of course be imagined as infinitely thin.) When the laminae are hollow cylinders, they can either slide

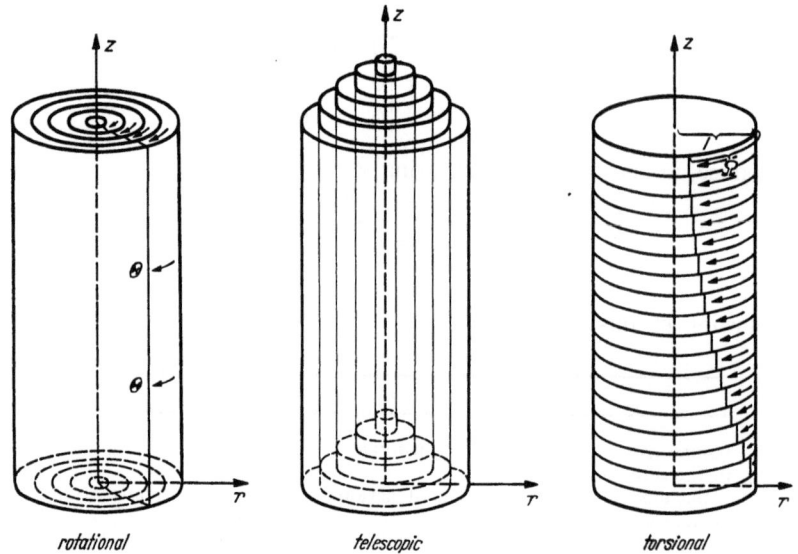

Fig. 33. Laminar distortion of cylinders.

in the direction of their common axis which is a "telescopic" distortion; or they can rotate around the axis, which gives a "rotational" distortion. For the case of visco-plastic flow, these two have been treated in Sect. 4. In the third case, the laminae are plane slices and the distortion is "torsional". In pure extension of a cylinder its axis is elongated (or shortened) and its radius contracts or expands.

To treat these cases, we need expressions in the cylindrical coordinates of Eqs. (7.4). The initial coordinates of a particle are

$$_l x = r_0, \theta, x_0 \tag{29.41}$$

and the coordinate differences

$$\Delta r = r - r_0; \quad \Delta \theta = \theta - \theta_0; \quad \Delta z = z - z_0. \tag{29.42}$$

In terms of coordinate differences, Eq. (29.10) becomes

$$2 \overset{A}{e}_{lm} = g_{lm} - {}_l{}_m g + {}_{\alpha m} g [(\Delta x)^\alpha_{,l}] + {}_{l\beta} g [(\Delta x)^\beta_{,m}] - {}_{\alpha\beta} g [(\Delta x)^\alpha_{,l}] [(\Delta x)^\beta_{,m}]. \tag{29.43}$$

Introducing

$$_{lm}g = \begin{Vmatrix} 1 & 0 & 0 \\ 0 & r_0^2 & 0 \\ 0 & 0 & 1 \end{Vmatrix}, \quad g_{lm} = \begin{Vmatrix} 1 & 0 & 0 \\ 0 & r^2 & 0 \\ 0 & 0 & 1 \end{Vmatrix} \tag{29.44}$$

Geometrical non-linearity.

we then find from (29.43)[1]

$$\begin{aligned}
2e_{rr} &= 2\frac{\partial \Delta r}{\partial r} - \left[\left(\frac{\partial \Delta r}{\partial r}\right)^2 + r^2\left(\frac{\partial \Delta \theta}{\partial r}\right)^2 + \left(\frac{\partial \Delta z}{\partial r}\right)^2\right] + 2r\Delta r\left(\frac{\partial \Delta \theta}{\partial r}\right)^2 - \\
&\quad - (\Delta r)^2\left(\frac{\partial \Delta \theta}{\partial r}\right)^2, \\
2e_{\theta\theta} &= 2\frac{\Delta r}{r} + 2\frac{\partial \Delta \theta}{\partial \theta} - \left(\frac{\Delta r}{r}\right)^2 - 4\frac{\Delta r}{r}\frac{\partial \Delta \theta}{\partial \theta} - \\
&\quad - \left[\frac{1}{r^2}\left(\frac{\partial \Delta r}{\partial \theta}\right)^2 + \left(\frac{\partial \Delta \theta}{\partial \theta}\right)^2 + \frac{1}{r^2}\left(\frac{\partial \Delta z}{\partial \theta}\right)^2\right] + \\
&\quad + 2\left(\frac{\Delta r}{r}\right)^2\frac{\partial \Delta \theta}{\partial \theta} - 2\frac{\Delta r}{r}\left(\frac{\partial \Delta \theta}{\partial \theta}\right)^2 + \left(\frac{\Delta r}{r}\right)^2\left(\frac{\partial \Delta \theta}{\partial \theta}\right)^2, \\
2e_{zz} &= 2\frac{\partial \Delta z}{\partial z} - \left[\left(\frac{\partial \Delta r}{\partial z}\right)^2 + r^2\left(\frac{\partial \Delta \theta}{\partial z}\right)^2 + \left(\frac{\partial \Delta z}{\partial z}\right)^2\right] + 2r\Delta r\left(\frac{\partial \Delta \theta}{\partial z}\right)^2 - \\
&\quad - (\Delta r)^2\left(\frac{\partial \Delta \theta}{\partial z}\right)^2, \\
2e_{\theta z} &= \frac{1}{r}\frac{\partial \Delta z}{\partial \theta} + r\frac{\partial \Delta \theta}{\partial z} - 2\Delta r\frac{\partial \Delta \theta}{\partial z} - \\
&\quad - \left[\frac{1}{r}\frac{\partial \Delta r}{\partial \theta}\frac{\partial \Delta r}{\partial z} + r\frac{\partial \Delta \theta}{\partial \theta}\frac{\partial \Delta \theta}{\partial z} + \frac{1}{r}\frac{\partial \Delta z}{\partial \theta}\frac{\partial \Delta z}{\partial z}\right] + \\
&\quad + \frac{(\Delta r)^2}{r}\frac{\partial \Delta \theta}{\partial z} + 2\Delta r\frac{\partial \Delta \theta}{\partial \theta}\frac{\partial \Delta \theta}{\partial z} - \frac{(\Delta r)^2}{r}\frac{\partial \Delta \theta}{\partial \theta}\frac{\partial \Delta \theta}{\partial z}, \\
2e_{zr} &= \frac{\partial \Delta r}{\partial z} + \frac{\partial \Delta z}{\partial r} - \left[\frac{\partial \Delta r}{\partial z}\frac{\partial \Delta r}{\partial r} + r^2\frac{\partial \Delta \theta}{\partial z}\frac{\partial \Delta \theta}{\partial r} + \frac{\partial \Delta z}{\partial z}\frac{\partial \Delta z}{\partial r}\right] + \\
&\quad + 2r\Delta r\frac{\partial \Delta \theta}{\partial z}\frac{\partial \Delta \theta}{\partial r} - (\Delta r)^2\frac{\partial \Delta \theta}{\partial z}\frac{\partial \Delta \theta}{\partial r}, \\
2e_{r\theta} &= r\frac{\partial \Delta \theta}{\partial r} + \frac{1}{r}\frac{\partial \Delta r}{\partial \theta} - 2\Delta r\frac{\partial \Delta \theta}{\partial r} - \frac{1}{r} \times \\
&\quad \times \left[\frac{\partial \Delta r}{\partial r}\frac{\partial \Delta r}{\partial \theta} + r\frac{\partial \Delta \theta}{\partial r}\frac{\partial \Delta \theta}{\partial \theta} + \frac{1}{r}\frac{\partial \Delta z}{\partial r}\frac{\partial \Delta z}{\partial \theta}\right] + \\
&\quad + \frac{(\Delta r)^2}{r}\frac{\partial \Delta \theta}{\partial r} + 2\Delta r\frac{\partial \Delta \theta}{\partial r}\frac{\partial \Delta \theta}{\partial \theta} - \frac{(\Delta r)^2}{r}\frac{\partial \Delta \theta}{\partial r}\frac{\partial \Delta \theta}{\partial \theta}.
\end{aligned} \qquad (29.45)$$

When neglecting all second order terms, these equations are reduced to the expressions of infinitesimal theory, Eqs. (7.9).

To write down expressions in the Green measure, one alters those of Eqs. (29.45) by replacing r, θ, z by r_0, θ_0, z_0, and changing all negative signs into positive ones.

In the cases mentioned before there will be axial symmetry, i.e. all derivates in respect of θ will vanish. This reduces some of the expressions (29.45) as follows:

$$\begin{aligned}
2e_{\theta\theta} &= 2\frac{\Delta r}{r} - \left(\frac{\Delta r}{r}\right)^2, \\
2e_{\theta z} &= r\frac{\partial \Delta \theta}{\partial z} - 2\Delta r\frac{\partial \Delta \theta}{\partial z} + \frac{(\Delta r)^2}{r}\frac{\partial \Delta \theta}{\partial z}, \\
2e_{r\theta} &= r\frac{\partial \Delta \theta}{\partial r} - 2\Delta r\frac{\partial \Delta \theta}{\partial r} + \frac{(\Delta r)^2}{r}\frac{\partial \Delta \theta}{\partial r}.
\end{aligned} \qquad (29.46)$$

---

[1] We are using in the following the Almansi measure, but drop the index A over e for simpler writing. — Intermediate step between (29.43) and (29.45) is

$$e_{(lm)} = (g^{ll}g^{mm})^{\frac{1}{2}} e_{lm}.$$

The expressions are still more reduced if we consider laminar displacement. In this case $\Delta r = 0$, and we have

$$\begin{aligned}
2e_{rr} &= -r^2\left(\frac{\partial \Delta \theta}{\partial r}\right)^2 - \left(\frac{\partial \Delta z}{\partial r}\right)^2, \\
2e_{\theta\theta} &= 0, \\
2e_{zz} &= 2\frac{\partial \Delta z}{\partial z} - r^2\left(\frac{\partial \Delta \theta}{\partial z}\right)^2 - \left(\frac{\partial \Delta z}{\partial z}\right)^2, \\
2e_{\theta z} &= r\frac{\partial \Delta \theta}{\partial z}, \\
2e_{zr} &= \frac{\partial \Delta z}{\partial r} - r^2\frac{\partial \Delta \theta}{\partial z}\frac{\partial \Delta \theta}{\partial r} - \frac{\partial \Delta z}{\partial z}\frac{\partial \Delta z}{\partial r}, \\
2e_{r\theta} &= r\frac{\partial \Delta \theta}{\partial r}.
\end{aligned} \quad (29.47)$$

Finally we specialize these equations for three cases in which Δz vanishes. We have

(i) *Simple torsion*

$$\begin{aligned}
2e_{rr} &= -r^2\left(\frac{\partial \Delta \theta}{\partial r}\right)^2; & 2e_{zz} &= -r^2\left(\frac{\partial \Delta \theta}{\partial z}\right)^2, \\
2e_{\theta z} &= r\frac{\partial \Delta \theta}{\partial z}; & 2e_{zr} &= -r^2\frac{\partial \Delta \theta}{\partial z}\frac{\partial \Delta \theta}{\partial r}, \\
2e_{r\theta} &= r\frac{\partial \Delta \theta}{\partial r}; & e_{\theta\theta} &= 0.
\end{aligned} \quad (29.48)$$

Let the cylinder be of length l and let Ω/l be the twist per unit length, then

$$\Delta \theta = \frac{\Omega}{l} z. \quad (29.49)$$

Going over to infinitesimal strains we now have

$$2\overset{A}{\varepsilon}_{lm} = \frac{r\Omega}{l}\begin{Vmatrix} 0 & 0 & 0 \\ 0 & 0 & 1 \\ 0 & 1 & -\frac{r\Omega}{l} \end{Vmatrix}. \quad (29.50)$$

Comparing Eqs. (29.50) with (29.22), we see that the former can be derived from the latter simply by replacing S by $\frac{r\Omega}{l}$. We can therefore write from (29.34) and (29.35)

$$2\overset{G}{\varepsilon}_{lm} = \frac{r\Omega}{l}\begin{Vmatrix} 0 & 0 & 0 \\ 0 & \frac{r\Omega}{l} & 1 \\ 0 & 1 & 0 \end{Vmatrix}, \quad (29.51)$$

$$2\overset{H}{\varepsilon}_{lm} = \frac{r\Omega}{l}\begin{Vmatrix} 0 & 0 & 0 \\ 0 & \frac{r\Omega}{2l} & 1 \\ 0 & 1 & -\frac{r\Omega}{2l} \end{Vmatrix}. \quad (29.52)$$

(ii) *Torsion combined with pure extension.* Going back to Eqs. (29.45), we introduce
$$\Delta r = -\varepsilon_t r; \quad \Delta\theta = \Omega z/l; \quad \Delta z = \varepsilon_l z \tag{29.53}[1]$$
where Ω is the twist of the cylinder, l its length, Δl the elongation and
$$\varepsilon_l = \frac{\Delta l}{l}, \quad \varepsilon_t = \frac{\Delta R}{R} \tag{29.54}[1]$$
with R the radius of the cylinder and ΔR its lateral contraction.

The strain components are accordingly
$$\left.\begin{array}{l} 2e_{rr} = 2e_{\theta\theta} = -\varepsilon_t(2+\varepsilon_t) \\ 2e_{zz} = \varepsilon_l(2-\varepsilon_l) - (r\Omega/l)^2(1+\varepsilon_t)^2 \\ 2e_{\theta z} = \dfrac{r\Omega}{l}(1+\varepsilon_t)^2 \\ 2e_{zr} = 2e_{r\theta} = 0. \end{array}\right\} \tag{29.55}$$

For infinitesimal strains we can neglect ε_l and ε_t against unity and thus find
$$\left.\begin{array}{l} \varepsilon_{rr} = \varepsilon_{\theta\theta} = -\varepsilon_t \\ \varepsilon_{zz} = \varepsilon_l - \dfrac{1}{2}\left(\dfrac{r\Omega}{l}\right)^2; \quad \varepsilon_{\theta z} = \dfrac{r\Omega}{2l} \\ \varepsilon_{zr} = \varepsilon_{r\theta} = 0. \end{array}\right\} \tag{29.56}$$

We see that in the case of infinitesimal strain the strains can be superposed by addition.

(iii) *Rotation of laminae.* For the case of rotation we specialize Eqs.(29.48) by letting all derivatives with respect to z vanish. We find
$$\left.\begin{array}{l} 2e_{rr} = -r^2\left(\dfrac{\partial\Delta\theta}{\partial r}\right)^2, \quad 2e_{r\theta} = r\dfrac{\partial\Delta\theta}{\partial r} \\ e_{zz} = e_{\theta\theta} = e_{\theta z} = e_{zr} = 0. \end{array}\right\} \tag{29.57}$$

30. Tensorial non-linearity. α) *General considerations.* The results found in Sect. 29, namely that simple shear can only be maintained by cross-stresses, which are *different for different measures of strain*, is rather disturbing. Note that λ and μ, or K and μ, whatever their magnitude and dependence upon invariants, are necessarily positive. Therefore the situation cannot be met by physical non-linearity, i.e. making λ and μ depend upon e_{lm} through its invariants. Accordingly the measure of strain chosen would prejudice the experimental results. We are therefore forced to conclude that the classical equation is not general enough, even for infinitesimal strain. It has been generalized by REINER (1948) to
$$s_{lm} = F_0 \delta_{lm} + 2\mu e_{lm} + 4\mu_c e_{l\alpha} e_{\alpha m} \tag{30.1}$$
when the introduction of μ_c, an independent *modulus of cross-elasticity*, solves this difficulty. The equation can be derived by applying the consideration of Sect. 7γ, as follows: We postulate a general stress-strain relation
$$s_{lm} = F(e_{n0}), \tag{30.2}$$
and applying Eq. (7.24) find (30.1). Eq. (29.1) is thus a degenerative form of (30.1) when μ_c vanishes.

It should be emphasized that the considerations of Sect. 7 show that by virtue of the Cayley-Hamilton equation, Eq. (30.1) cannot be further generalized *by the addition of other tensor-terms*. There is only *one* possible way of obtaining

---
[1] The index "*l*" in ε_l means "longitudinal", the index "*t*" in ε_t, "transversal".

results more general than those issuing from this equation when μ and μ_c are assumed to be rheological constants; it is by means of physical non-linearity, when there is practically no limit to the number of terms in terms of deformation-*components*. For the same reason, however, the physical significance of such method is slight.

β) *Non-linear elasticity.* We now give examples of non-linear infinitesimal elasticity when the non-linearity is *tensorial* only, and the moduli of elasticity are therefore constants. REINER[1] has used observations by POYNTING (1909, 1912), which will be reported upon in Sect. 31, to calculate the moduli for certain metals, assuming *finite strain*. The results showed that μ_c, the modulus of cross-elasticity, is in these cases of the same order of magnitude as the shear modulus μ. In the following we assume this to be generally so. The examples are the three cases mentioned in Sect. 29. We use the Almansi measure, and derive first the complete cross-terms and stress-components. Going over to infinitesimal strains, we neglect terms constituting infinitesimal numbers, against the number 1 and numbers >1.

(i) *Torsion combined with pure extension.* We have from (29.55)

$$\left. \begin{aligned} & 4\,(e_{r\alpha})\,(e_{\alpha r}) = \varepsilon_t^2 (2+\varepsilon_t)^2\,, \\ & 4\,(e_{\theta\alpha})\,(e_{\alpha\theta}) = \varepsilon_t^2 (2+\varepsilon_t)^2 + \left(\frac{r\Omega}{l}\right)^2 (1+\varepsilon_t)^4\,, \\ & 4\,(e_{z\alpha})\,(e_{\alpha z}) = \left(\frac{r\Omega}{l}\right)^2 (1+\varepsilon_t)^4 + \left[\varepsilon_l(2-\varepsilon_l) - \left(\frac{r\Omega}{l}\right)^2 (1+\varepsilon_t)^2\right]^2\,, \\ & 4\,(e_{\theta\alpha})\,(e_{\alpha z}) = -\left[\varepsilon_t(2+\varepsilon_t) - \varepsilon_l(2-\varepsilon_l) + \left(\frac{r\Omega}{l}\right)^2 (1+\varepsilon_t)^2\right] \frac{r\Omega}{l} (1+\varepsilon_t)^2\,, \\ & 4\,(e_{r\alpha})\,(e_{\alpha\theta}) = 4\,(e_{z\alpha})\,(e_{\alpha r}) = 0\,. \end{aligned} \right\} \quad (30.3)$$

This makes Eq. (30.1)

$$\left. \begin{aligned} s_{rr} &= F_0 + \varepsilon_t(2+\varepsilon_t)\left[-\mu + \mu_c\,\varepsilon_t(2+\varepsilon_t)\right], \\ s_{\theta\theta} &= F_0 + \varepsilon_t(2+\varepsilon_t)\left[-\mu + \mu_c\,\varepsilon_t(2+\varepsilon_t)\right] + \mu_c\left(\frac{r\Omega}{l}\right)^2 (1+\varepsilon)^4, \\ s_{zz} &= F_0 + \mu\left[\varepsilon_l(2-\varepsilon_l) - \left(\frac{r\Omega}{l}\right)^2 (1+\varepsilon_t)^2\right] + \\ &\quad + \mu_c\left\{\left(\frac{r\Omega}{l}\right)^2 (1+\varepsilon_t)^4 + \left[\varepsilon_l(2-\varepsilon_l) - \left(\frac{r\Omega}{l}\right)^2 (1+\varepsilon_t)^2\right]^2\right\}, \\ s_{r\theta} &= s_{zr} = 0, \\ s_{\theta z} &= \mu \frac{r\Omega}{l}(1+\varepsilon_t)^2 - \mu_c\left[\varepsilon_t(2+\varepsilon_t) - \varepsilon_l(2-\varepsilon_l) + \left(\frac{r\Omega}{l}\right)^2 (1+\varepsilon_t)^2\right] \times \\ &\quad \times \frac{r\Omega}{l}(1+\varepsilon_t)\,. \end{aligned} \right\} \quad (30.4)$$

In the case of infinitesimal strain we note the ε_l and ε_t are of the same order as $\left(\frac{r\Omega}{l}\right)^2$, and find the stress-components

$$\left. \begin{aligned} s_{rr} &= F_0 - 2\mu\,\varepsilon_t\,, \\ s_{\theta\theta} &= F_0 - 2\mu\,\varepsilon_t + 4\mu_c\left(\frac{r\Omega}{2l}\right)^2, \\ s_{zz} &= F_0 + 2\mu\left[\varepsilon_l - 2\left(\frac{r\Omega}{2l}\right)^2\right] + 4\mu_c\left(\frac{r\Omega}{2l}\right)^2, \\ s_{r\theta} &= s_{zr} = 0\,, \qquad s_{\theta z} = \mu\frac{r\Omega}{l}\,. \end{aligned} \right\} \quad (30.5)$$

---

[1] M. REINER: Appl. Sci. Res. A **5**, 281–295 (1955).

(ii) Simple torsion. The cross-terms for simple torsion result from (30.3) by making ε_l vanish, and similarly the stress-components from (30.5).

$$\left. \begin{aligned} s_{rr} &= F_0, \\ s_{\theta\theta} &= F_0 + 4\mu_c \left(\frac{r\Omega}{2l}\right)^2, \\ s_{zz} &= F_0 - \mu \left(\frac{r\Omega}{l}\right)^2 + 4\mu_c \left(\frac{r\Omega}{2l}\right)^2, \\ s_{r\theta} &= s_{zr} = 0, \\ s_{\theta z} &= \mu \frac{r\Omega}{l}. \end{aligned} \right\} \quad (30.6)$$

(iii) Rotation of laminae. The complete cross-terms are

$$\left. \begin{aligned} 4(e_{r\alpha})(e_{\alpha r}) &= r^2 \left(\frac{\partial \Delta\theta}{\partial r}\right)^2 \left[1 + r^2 \left(\frac{\partial \Delta\theta}{\partial r}\right)^2\right], \\ 4(e_{\theta\alpha})(e_{\alpha\theta}) &= r^2 \left(\frac{\partial \Delta\theta}{\partial r}\right)^2, \\ (e_{z\alpha})(e_{\alpha z}) &= 0; \quad 4(e_{r\alpha})(e_{\alpha\theta}) = -r^3 \left(\frac{\partial \Delta\theta}{\partial r}\right)^3, \\ (e_{\theta\alpha})(e_{\alpha z}) &= (e_{z\alpha})(e_{\alpha r}) = 0. \end{aligned} \right\} \quad (30.7)$$

Going over to infinitesimal strains, the stresses are

$$\left. \begin{aligned} s_{rr} &= F_0 - r^2 \left(\frac{d\Delta\theta}{dr}\right)^2 (\mu - \mu_c), \\ s_{\theta\theta} &= F_0 + \mu_c r^2 \left(\frac{d\Delta\theta}{dr}\right)^2, \\ s_{zz} &= F_0, \\ s_{r\theta} &= \mu_r \frac{d\Delta\theta}{dr}, \\ s_{\theta z} &= s_{zr} = 0. \end{aligned} \right\} \quad (30.8)$$

31. The Poynting effect. The considerations of the preceding Section find their main application in an analysis of what has been named the Poynting effect.

The standard test for the determination of the magnitude of the shear-modulus μ is the torsion of a long, comparatively thin right circular cylinder. Despite the innumerable tests that have been and are being carried out in the testing laboratories all over the world on the torsion of metal wires, it has never occurred to any investigator, with one exception to be presently mentioned, to measure changes in length l and radius R of the wires[1]. This is only partly due to the experimental difficulties; it is mainly due to the fact that the classical theory of elasticity of the Hooke body predicts that there are no such changes when the strain is infinitesimal.

The exception refers to POYNTING (1909, 1912). Nearly half a century ago he observed that cylinders of steel and copper in the form of long wires, when acted upon by forces equivalent to a torsional torque, are not only twisted as predicted by the classical theory of elasticity, but also elastically elongated and

---

[1] Very recently such measurements have been made.

increased in volume. These observations were never repeated and fell into oblivion; we shall therefore treat them in some detail.

The experiments were made on several wires hung vertically from a fixed support. They were first lightly loaded until kinks were taken out. It was ascertained that this stretching has no influence upon the results. The lower end of the wire was twisted within the elastic limit. The temperature was kept fairly constant, and the changes were nearly isothermal. Four steel piano wires, two copper wires and one brass wire were tested. The main results of the experiment were, in POYNTING's words: "There was always a lengthening on twisting of the same order whether the twist was clockwise or counterclockwise. The lengthening was nearly proportional to the square of the twist put on. The

Table 7. POYNTING's *observations. Lenght of wires* $= 160.5\ cm$

1	material	steel		hard copper
2	$2R$ in cm	0.0986	0.1210	0.1219
3	$\bar{\varepsilon}_l = \Delta l/l$ for one turn	1.71×10^{-6}	2.90×10^{-6}	4.25×10^{-6}
4	$\nu_c \bar{\varepsilon} = \Delta R/R$ for one turn	0.319×10^{-6}	0.524×10^{-6}	1.75×10^{-6}
5	ε_v for one turn	1.07×10^{-6}	1.85×10^{-6}	0.75×10^{-6}
6	E in dyne cm^{-2}	2.12×10^{12}	2.12×10^{12}	1.31×10^{12}
7	μ in dyne cm^{-2}	0.835×10^{12}	0.825×10^{12}	0.493×10^{12}
8	ν	0.270	0.287	0.331
9	$\nu_c = \bar{\varepsilon}_t/\bar{\varepsilon}_l$	0.187	0.181	0.41
10	maximum number of turns of wire	4	3	1
11	$\varkappa$ in dyne cm^{-2}	1.53×10^{12}	1.65×10^{12}	1.29×10^{12}
12	$2\pi R/l$	0.00193	0.00237	0.00239
13	$(2\pi R/l)^2$	3.7×10^{-6}	5.6×10^{-6}	5.7×10^{-6}
14	$c = \mu_c/\mu$	-2.40	-2.90	-6.40
15	μ_c in dyne cm^{-2}	-2.000×10^{12}	-2.393×10^{12}	-3.155×10^{12}
16	δ in dyne cm^{-2}	-0.885×10^{12}	-1.108×10^{12}	-0.346×10^{12}

lengthening for a given twist was proportional to the square of the radius. The change in diameter of a wire when twisted is a contraction, also proportional to the square of the angle of twist." POYNTING made sure by a subsidiary experiment that the lengthening on twisting is not due to a change in YOUNG's modulus. The Table 7 summarizes the results of his observations on two steel wires and one copper wire, recorded in the second paper, when he also measured the lateral contraction which he had not done in the first series.

As can be seen from the value for $2\pi R/l$, which is negligible against unity, the strain was infinitesimal. The values for the extension $\bar{\varepsilon}_l$ and lateral contraction $\nu_c \bar{\varepsilon}_l$ for one turn of the wire can also be neglected against unity and against $R\Omega/l$, but are comparable with and cannot be neglected against $(R\Omega/l)^2$. Accordingly, in contradiction to classical theory second order effects were noticeable. POYNTING's theoretical attempts in dealing with his observations must be considered as inadequate. His experimental technique was supreme, but for the theoretical analysis he lacked the appropriate mathematical instrument, namely the tensor calculus. This is the case $\beta(i)$ of Sect. 30. Applying the stress-equations (7.10) in the absence of mass-forces, we have

$$\left. \begin{array}{l} \dfrac{dF_0}{dr} = \mu_c r \left(\dfrac{\Omega}{l}\right)^2 \\ \dfrac{\partial F_0}{\partial \theta} = \dfrac{\partial F_0}{\partial z} = 0 \end{array} \right\} \quad (31.1)$$

from which
$$F_0 = \frac{\mu_c}{2}\left(\frac{r\Omega}{l}\right)^2 + C \tag{31.2}$$
and
$$S_{rr} = 2\mu\,\varepsilon_t + \frac{\mu_c}{2}\left(\frac{r\Omega}{l}\right)^2 + C. \tag{31.3}$$

Assuming that the sides of the cylinder ($r = R$) are stressfree, we find
$$C = 2\mu\,\varepsilon_t - \frac{\mu_c}{2}\left(\frac{R\Omega}{l}\right)^2. \tag{31.4}$$

Therefore
$$F_0 = 2\mu\,\varepsilon_t - \frac{\mu_c}{2}\left(\frac{\Omega}{l}\right)^2 (R^2 - r^2). \tag{31.5}$$

This makes
$$S_{zz} = \mu\left[2\varepsilon_t + 2\varepsilon_l - \left(\frac{r\Omega}{l}\right)^2\right] + \mu_c\left[\frac{1}{2}\cdot\left(\frac{\Omega}{l}\right)^2\cdot(3r^2 - R^2)\right]. \tag{31.6}$$

From Eq. (31.6) we find
$$P_z = 2\pi\int_0^R S_{zz}\,r\,dr = R^2\pi\left\{\mu\left[2\varepsilon_t + 2\varepsilon_l - \frac{1}{2}\cdot\left(\frac{R\Omega}{l}\right)^2\right] + \mu_c\left(\frac{R\Omega}{2l}\right)^2\right\}. \tag{31.7}$$

However, for simple torsion $P_z = 0$. Therefore in this case
$$\mu_c = 2\mu\left[1 - \frac{\varepsilon_l + \varepsilon_t}{\left(\frac{R\Omega}{2l}\right)^2}\right]. \tag{31.8}$$

Taking into account the values for $\bar{\varepsilon}$ add ν_c from Table 7, this gives for the ration $c = \mu_c/\mu$ the values entered in line 14 of the table, which also shows in line 15 the values for μ_c.

Considering the change in volume, we find
$$e_v = \varepsilon_l(1 - 2\nu_c) \tag{31.9}$$

as in the infinitesimal theory. As ε_l is proportional to the square of the twist, so is e_v. However, POISSON's ratio ν as calculated from E and μ in accordance with the wellknown formula [compare (10.6) and (10.7)]
$$\mu = \frac{E}{2(1+\nu)} \tag{31.10}$$

entered in line 8 of the table, is much larger than ν_c, which points to the fact that the observed cubical dilatation is greater than the cubical dilatation ε_v caused by the mean tension s_m, the isotropic component of s_{lm}, namely
$$s_m = \tfrac{1}{3} s_{\alpha\alpha} \tag{31.11}$$

in accordance with the formula
$$\varepsilon_v = s_m/\varkappa. \tag{31.12}$$

The difference
$$e_d = e_v - \varepsilon_v \tag{31.13}$$

is due to elastic dilatancy. Dilatancy has been mentioned above in Sect. 29. The modulus of elastic dilatancy defined in Eq. (29.40') was calculated by REINER[1]. Its values are entered in line 16 of the table. They turn out to be negative.

---
[1] M. REINER: Appl. Sci. Res. A **5**, 281—295 (1955).

A negative modulus of dilatancy means that an isotropic pressure is required, in addition to shearing stresses, in order to produce simple shear, or that simple shear stresses will be accompanied by an *increase* of volume.

Poynting's observations thus show that in the case of two metals the complete elasticity law may be written in the form (30.1), with non-vanishing modulus of cross-elasticity μ_c and constant moduli μ and μ_c. This law does not prejudice experimental results. A torsional torque will cause in a right-circular cylinder either (a) simple torsion with no elongation, if $\mu_c = 2\mu$, (b) an elongation of the cylinder, if $\mu_c < 2\mu$, or (c) a shortening of the cylinder if $\mu_c > 2\mu$. In the case of the two metals examined by Poynting, an elongation was observed, and no other experimental observations are known. However, Swift[1] when subjecting metal cylinders to large plastic torsions observed that while most metals showed a lengthening, lead which does not work-harden showed a shortening of the cylinder. Insofar as the plastic deformation of a metal can be considered as a "freezing-in" of successive elastic strains, Swift's experiments suggest that similar conditions may prevail in the elastic range.

The theory similarly predicts in general positive or negative dilatancy. In the case of the two metals investigated by Poynting, a positive dilatancy was observed; shearing stresses causing an increase in volume. It might be expected that in lead the volume would decrease.

The theory has been applied by using a definite measure of strain, namely the one due to Almansi. The application of any other measure will only affect the numerical values of μ_c without changing its character as a constant.

While the theory expanded in Sect. 30 is proved by Poynting's observations as correct thus far in the case of two metals only, it may be expected that it is generally valid for the complete description of the elasticity of polycrystalline metals.

32. Second order effects in viscous flow. Considering that the flow-tensor f_{lm} is unequivocally defined by Eq. (3.10) in strict analogy to d_{lm} of Eq. (3.8), second order effects in viscous flow can only arise when Eq. (15.11), which we may write in the form,

$$s_m = F_0 \delta_{lm} + 2\eta f_{lm} \tag{32.1}$$

is enlarged to

$$s_m = F_0 \delta_{lm} + 2\eta f_{lm} + 4\eta_c f_{l\alpha} f_{\alpha m} \tag{32.2}$$

following Reiner (1945) in analogy to the generalization which yields Eq. (30.1).

Truesdell (1952) has named a fluid with (32.2) as the rheological equation, Reiner-Rivlin fluid. Rivlin[2] and Braun and Reiner[3] have derived formulae describing the behaviour of such a fluid in various apparatus. The subject is treated in more detail elsewhere in this Encyclopaedia. For our purpose we mention briefly two cases.

α) *Simple laminar shearing flow.*

$$v_x = G y, \quad v_y = v_z = 0 \tag{32.3}$$

yields

$$s_{lm} = F_0 \delta_{lm} + \eta G \begin{Vmatrix} 0 & 1 & 0 \\ 1 & 0 & 0 \\ 0 & 0 & 0 \end{Vmatrix} + \eta_c G^2 \begin{Vmatrix} 1 & 0 & 0 \\ 0 & 1 & 0 \\ 0 & 0 & 0 \end{Vmatrix}. \tag{32.4}$$

---

[1] H. W. Swift: Engineering 1947.
[2] R. S. Rivlin: Proc. Roy. Soc. Lond., Ser. A **193**, 260—281 (1948).
[3] I. Braun and M. Reiner: Quart. J. Mech. Appl. Math. **5**, 42—53 (1952).

Sect. 32. Second order effects in viscous flow.

If the liquid is exposed to air pressure Π in the z-direction

$$s_{zz} = F_0 = -\Pi. \tag{32.5}$$

This determines the unknown function F_0 as a constant.

The cross-stresses are dependent upon the coefficient of cross-viscosity η_c, and vanish with the latter. It is important to note that in contradistinction of all cases of cross-elasticity treated in Sects. 29 and 30, both cross-stresses (in the x- and the y-direction) have the *same sign* depending upon the sign of η_c: they are both either tensions or compressions.

β) *Torsional flow.* Let the liquid be contained between a plate rotating around a vertical axis with angular velocity Ω and another parallel one at a distance H which is stationary. We assume

$$v_\theta = r\omega, \tag{32.6}$$

where ω is a function of z only, while v_r and v_z vanish. Then

$$f_{lm} = A \begin{Vmatrix} 0 & 0 & 0 \\ 0 & 0 & 1 \\ 0 & 1 & 0 \end{Vmatrix}, \tag{32.7}[1]$$

where

$$A = \frac{r}{2}\frac{d\omega}{dz}. \tag{32.8}$$

From (32.7) there follows

$$f_{l\alpha}f_{\alpha m} = A^2 \begin{Vmatrix} 0 & 0 & 0 \\ 0 & 1 & 0 \\ 0 & 0 & 1 \end{Vmatrix}. \tag{32.9}$$

We now have

$$s_{lm} = \begin{Vmatrix} F_0 & 0 & 0 \\ 0 & F_0 + A^2 4\eta_c & 2\eta A \\ 0 & 2\eta A & F_0 + A^2 4\eta_c \end{Vmatrix}. \tag{32.10}$$

From the second stress-equation, neglecting inertia-terms, we find

$$\omega = \Omega\left(1 - \frac{z}{H}\right). \tag{32.11}$$

The other two equations yield

$$s_{rr} = F_0 = \frac{r^2\Omega^2}{2H^2}\eta_c + K. \tag{32.12}$$

At the edge of the plates where $r = R$, the liquid has a free surface, more or less vertical, where s_{rr} can be put as equal to $-\Pi$. This makes

$$K = -\frac{R^2\Omega^2}{2H^2}\eta_c - \Pi \tag{32.13}$$

and

$$\Pi + F_0 = -\frac{\Omega^2}{2H^2}(R^2 - r^2)\eta_c \tag{32.14}$$

therefore

$$\Pi + s_{zz} = F_0 + 4A^2\eta_c = -\frac{\Omega^2}{2H^2}(R^2 - 3r^2)\eta_c. \tag{32.15}$$

---

[1] Here and in the following, subscripts have been used in physical components for simpler writing, as mentioned in Sect. 7α.

Handbuch der Physik, Bd. VI.

This shows that s_{zz} is distributed parabolically over the width of the plate with pressure at the centre, changing to tension at $r = R/\sqrt{3}$.

33. The Weissenberg effect. At the 1946 meeting of the British Rheologists' Club, K. WEISSENBERG first demonstrated very strange phenomena in the hydrodynamic behaviour of certain very viscous liquids. These were subsequently described by WEISSENBERG in a short account[1]. When such a liquid is sheared between a rotating vessel and some co-axial inner member which is held against

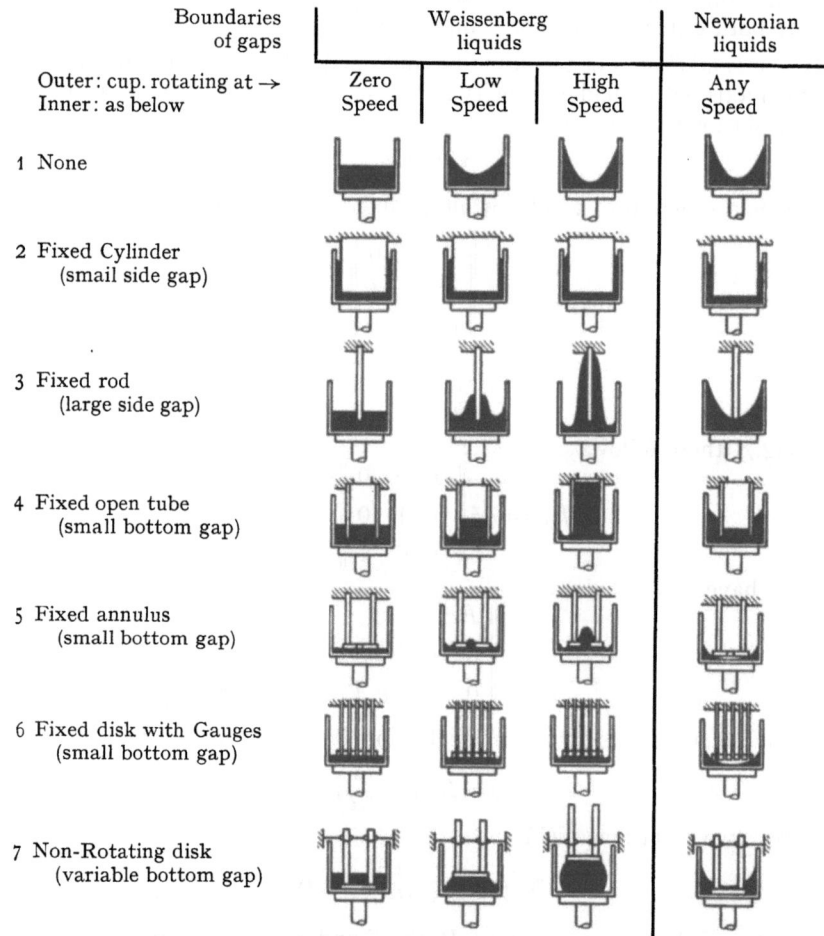

Fig. 34. Weissenberg-effects.

rotation, and is either fixed in position or is free to move up and down the axis, the liquid is drawn *inward* against the action of the centrifugal forces, and *upward* against the forces of gravity, the whole arrangement forming a sort of "centripetal pump" (compare Fig. 34).

These phenomena can be comprised under the designation of the "Weissenberg effect". In the simplest form the Weissenberg effect is exhibited when a rod is rotated in a "Weissenberg liquid"; the latter then "climbs" up the rod. Fig. 35 shows this for a denaturized sweetened condensed milk[2]. A similar effect

---

[1] K. WEISSENBERG: Nature, Lond. **159**, 310 (1947).
[2] M. REINER, G. W. SCOTT BLAIR and H. B. HAWLEY: J. Soc. Chem. Ind. **68**, 327 (1949).

has been observed in blown bitumen[1]. All liquids for which the Weissenberg effect has been demonstrated so far, as for instance rubber-, cellulose-, and starch-solutions or, in general, concentrated colloidal and high polymer solutions with

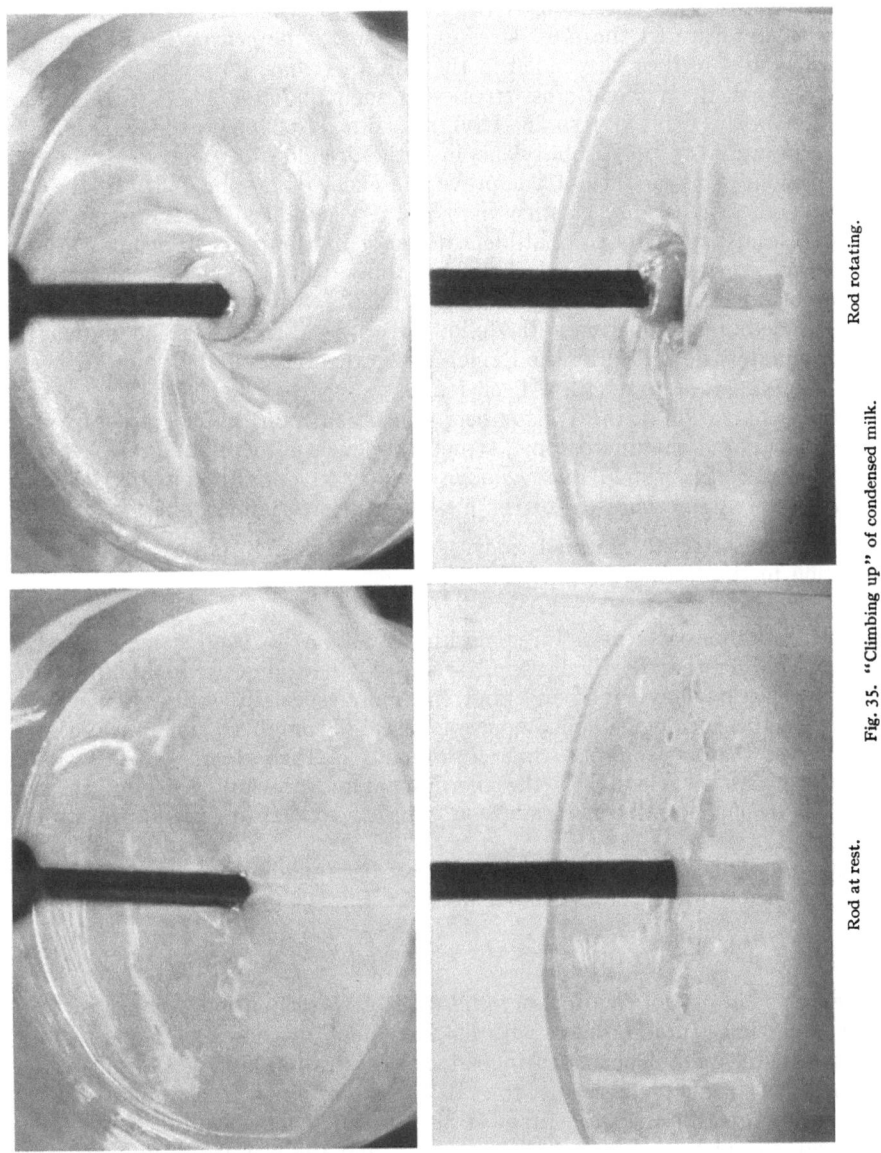

Fig. 35. "Climbing up" of condensed milk.

a more or less complex structure, exhibit also elasticity of *shape* under the action of transient forces or are elastic sols of the L-type. It stands to reason that when elastic elements are suspended in a sol, which as a result of velocity gradients existing in the liquid are subjected to elastic strains, cross-stresses should arise from these strains. Elastic cross-stresses will cause a centripetal pump action, and this may be the explanation for the mechanism of the Weissenberg

[1] K. Lax-Weiner and R. Schoenfeld-Reiner: Bull. Res. Council. Israel **2**, 66 (1952).

effect. This explanation has been advanced by Weissenberg[1] himself, by Burgers[2] and by Mooney[3]. Appropriate theories have been developed by Oldroyd[4] and DeWitt[5]. The same would apply if the material was an elastic gel on the model of a Maxwell body. On the other hand the effect might be due to cross-viscosity as described in the preceding section. Rivlin[6] who adopted this idea, interpreted the cross-stresses as due to the *orientation* of long chain molecules dissolved or suspended in the liquid. It has, however, been shown in Sect. 32, that the viscous cross-stresses in such liquid *must be distributed in a definite manner*. Greensmith and Rivlin[7] claimed to have proved experimentally that a solution of poly-isobutylene in ortho-di-chloro-benzene does show this distribution. This would not yet prove the exclusive applicability of the cross-viscosity theory because elastico-viscosity, if generalized by the inclusion of a cross-elasticity term, is compatible with *any* distribution. It would, however, show that cross-viscosity *might* be a real thing. But Roberts[8] claims to have repeated Greensmith's and Rivlin's experiments with negative results. Mooney has summed up the position in the following words: "There may be liquids which are adequately described by the Reiner-Rivlin theories; but I do not believe any such liquids have so far (1954) been found."

The explanation of the Weissenberg effects, both on the macroscopic, phenomenological, and the microscopic, structural, level, is therefore *an open problem*. It may well be that *different mechanisms* are operative in different cases, similarly to the different mechanisms of the elasticity of steel and rubber.

34. The perfected Maxwell body. When Maxwell (1868) postulated his equation in the form

$$\dot{s} = E\dot{e} - s/T_{\text{rel}}, \tag{34.1}$$

where e is an elastic strain "of some kind", and E "a coefficient ($=$ modulus) of elasticity for that particular kind of strain", he did so with the intention to describe elastic behaviour of any kind, but more especially with a view of applying it to the rheological behaviour of gases. From his "dynamical theory" he reasoned that air would be what we now call "elastico-viscous" with a modulus of shear-elasticity μ equal to the thermodynamic pressure p, i.e. of the order of 10^6 dyne cm^{-2}, and consequently a time of relaxation T_{rel} of the order of 10^{-10} sec.

It does not seem that his theory has ever been made use of in aerodynamics. The reason for this becomes clear if we write Eq. (34.1) in the form

$$E e = s + \frac{1}{T_{\text{rel}}} \int s\, dt. \tag{34.2}$$

An experimental confirmation or otherwise of this equation seemed to be out of the question. In accordance with (34.2) the deformation e is composed of two parts, one recoverable constituting an elastic strain, the other resulting from viscous flow (compare also Fig. 16). If T_{rel} is very small, the second part is incomparably greater than the first. If at any time t, however small, we remove the load, the viscous deformation stays put, and the elastic strain is recovered.

---

[1] K. Weissenberg: Proc. 1st Int. Rheological Congr. Amsterdam, 1948.
[2] J. M. Burgers: Proc. Acad. Sci. Amsterd. **51**, 787 (1948).
[3] M. Mooney: J. Coll. Sci. **6**, 96 (1951).
[4] J. G. Oldroyd: Proc. Roy. Soc. Lond., Ser. A **202**, 345 (1950).
[5] T. W. DeWitt: J. Appl. Phys. **20**, 889 (1955).
[6] R. S. Rivlin: Trans. Faraday Soc. **45**, 739 (1949).
[7] H. W. Greensmith and R. S. Rivlin: Phil. Trans. Roy. Soc. Lond., Ser. A **245**, 899, 399 (1953).
[8] J. E. Roberts: Nature, Lond. **179**, 487 (1957).

However, the latter will be so small in comparison with the former that it cannot be observed, coming within experimental error. This view has been expressed by JEFFREYS (1952) for the case of water with the following words: "If water has a rigidity of order 10^{10}, T will be of order 10^{-12} sec. Thus in any experiment that tests the behaviour over intervals large compared with 10^{-12} sec, the flow will be much greater than the elastic deformation, and *the insertion of an elasticity term in the stress-strain relation will affect nothing observable*". It has recently been shown experimentally by REINER[1] and by POPPER and REINER[2], that this view is not correct, the first by describing an observed centripetal pump effect in air, the other two by building a centripetal airpump. The main features of these two experimental set-ups will be described in the following Sect. 35; the present section is devoted to the theoretical analysis which shows how a Maxwell body must necessarily exhibit a Weissenberg effect.

We have already mentioned in the preceding section that an elasticoviscous material would show a Weissenberg effect. This is not evident from the rheological equation of the Maxwell body as formulated in (34.1).

In the reasoning which lead MAXWELL to this equation there are three imperfections. Firstly, no distinction is made between the total strain, its isotropic component (i.e. the cubical dilatation) and its deviator (the distortion or shear). However, MAXWELL certainly did not want to say that the isotropic stress relaxes, but had in mind the deviator of stress and strain. This imperfection was amended by REIGER[3], and has been taken care of in Eq. (16.2) which is written in terms of deviators. Secondly, MAXWELL did not say how e should be measured. He identified $\dot{e}$ with f. This, however, presupposes, as was shown in Sect. 25, that e must be expressed in the Hencky measure. He might have reasoned that, as the elastic strain in the case considered by him is certainly infinitesimal, it would not matter how the strain is defined—all possible measures being reduced to the infinitesimal Cauchy measure, Eq. (3.8), the rate of which is, in accordance with Eq. (3.10), identical with that measure of flow on which STOKES based his theory of viscous liquids, to which MAXWELL referred. This, however, as shown in Sect. 29 above, is true only if we disregard cross-stresses. This brings us to the third imperfection in MAXWELL's theory. Not possessing tensor-concepts, he naturally considered only the one-dimensional case and JEFFREY's reasoning is on these lines. If we rewrite Eq. (34.2) in the form

$$\overset{H}{\varepsilon}_{lm(0)} = \left[s_{lm(0)} + \frac{1}{T_{\text{rel}}} \int s_{lm(0)} \, dt \right] \Big/ 2\mu \qquad (34.3)$$

it is clear from considerations of Sect. 30 that in certain modes of flow the second term within [ ] brackets may vanish, leaving the first term only, which now is not negligible. Let there be e.g. laminar shear and let the flow take place in the x-direction. There is then no flow in the y-direction, but there is an elastic cross-stress in that direction, and this may be observable in a suitable experimental arrangement as will be shown below.

Such situation was foreseen by TRUESDELL (1952). As suggested by him, a non-dimensional number, which we may name the Truesdell-number, can be defined by

$$T = \eta \dot{\gamma}/p, \qquad (34.4)$$

where η is the viscosity of the liquid, $\dot{\gamma}$ the velocity gradient and p the air pressure. TRUESDELL reasons that if $T \gg 1$ a "Weissenberg effect" might be present.

---

[1] M. REINER: Proc. Roy. Soc. Lond., Ser. A **240**, 173 (1957).
[2] B. POPPER and M. REINER: Brit. J. Appl. Phys. **7**, 452 (1956), **8**, 493 (1957).
[3] R. REIGER: Ber. dtsch. Phys. Res. **1919**, 421—434.

For water and normal atmospheric pressure he calculates that $\dot{\gamma}$ must be about 7×10^5 sec^{-1} in order that $T \sim 1$. We may add that for air this would require that $\dot{\gamma}$ be 2×10^7 sec^{-1}. TRUESDELL adds "By ascent into the atmosphere (i.e. reducing p), however, T can be made arbitrarily large, and thus ... is particularly appropriate to problems of high altitude aerodynamics". The experimental difficulty for taking this course is obvious. With regard to the second course which is open, namely making $\dot{\gamma}$ very large, we must have a high velocity gradient. The latter increases as the relative velocity (V) is increased, and/or as the distance between the sheared surfaces (d) is decreased. In a centripetal pump arrangement the streamlines are circles, which gives rise to centri*fugal* forces. In order to produce centri*petal* forces, the former must be overcome. One therefore cannot increase V very much, and must turn to decreasing d. One must expect that this decrease will have to be of an extreme degree, with an accuracy of the distance between the shearing surfaces hard to obtain. However, when occupied with an investigation in a rather elementary rheological field, J. F. T. BLOTT discovered by chance an arrangement where d automatically decreases to such an extreme degree as to make cross-stresses in air observable.

35. A centripetal pump effect in air. When examining the behaviour of grease in ball and roller bearings, one finds that airfilms carried by the rotating surfaces prevent contact between the grease and the moving surface in certain parts of the bearing. It could be expected that the vibration of the bearing would cause sufficient slumping of the grease to bring it against the moving surface, and that once contact had been established the grease in the covers would be churned continuously. This however, does not occur, and it therefore appears that some "agency" actively prevents contact between the grease and the moving surfaces. In order to investigate the effect more closely, a brass disc 6 cm in diameter and 0.4 cm thick was put into rotation at speeds varying between 3000 and 8000 R.P.M., and brought slowly into contact edgewise with a thick layer of grease. While a trace of grease was picked up on the edge of the disc, further penetration causes grease to be carried forward, leaving a "slot" within which the disc rotates without making contact. When the same experiment was tried with a very viscous cylinder oil, a depression was made in the oil, and no wetting of the disc occurred. However, with thinner oils or water the disc was wetted immediately. Similar experiments were performed with a hollow cylinder. The cylinder was mounted on a powerful motor, but when the cylinder was in contact with the oil which wetted it, the cylinder could not be turned at more than a very slow speed. However, when the cylinder was *first* given rotation at about 5000 R.P.M., it would penetrate the surface of the oil without wetting. If the speed was then reduced to somewhat about 2500 R.P.M., contact occurred suddenly, stopping the rotation.

The experiment was then varied by closing the cylinder at the top, and fitting the oil bath with a tube which entered at the bottom, and passed up the centre to a height above the maximum level of the surface of the oil inside the cylinder. This tube was connected outside the oil bath to a water manometer so that pressure inside the cylinder could be measured. It was found possible to immerse the cylinder as before without wetting. Observing the air pressure inside the cylinder, it was found that it was negative at speeds of about 6000 R.P.M., but surprisingly became positive *on reducing the speed* to about 4000 R.P.M. On further reduction of the speed, wetting of the cylinder by the oil occurred as before.

Here then was the desired experimental arrangement for investigating a centripetal-pump effect in air. As can be observed in an experiment as described, the oil due to its high viscosity, moves so slowly that its velocity is negligible in comparison with the velocity of the cylinder-wall. There is, therefore, a high relative velocity V from cylinder to oil. The thickness d of the air gap is certainly very small, and the velocity gradient therefore great, and possibly so great as to imply a high Truesdell-number even at normal air pressure of 1 atm. One does not know, but need not care, what the distance d is, because it is maintained automatically at the right value. There must be a cross-pressure which prevents the contact between the oil and the moving surface, balancing the hydrostatic pressure of oil. This is the "agency" mentioned above. When the distance d is smaller than the right distance, the velocity gradient increases, and therefore also the cross-pressure. The latter then pushes the liquid away, decreasing the velocity-gradient and the cross-pressure until it balances the hydrostatic pressure. One can understand why wetting occurs with an oil of low viscosity; such oil is easily brought into rotational movement, and the velocity difference V reduced too much. Wetting also occurs when the rotational velocity of the cylinder is reduced, for the same reason.

The air gap between the metal cylinder and the co-axial cylinder formed by the bitumen is thus maintained by one cross-stress which in the case of the Hencky measure is a pressure. The other cross-stress acting in the direction of flow, which is a tension, produces the strangulation pressure which causes the pumping.

On the basis of this experience, POPPER and REINER designed and built an all-metal centripetal airpump. This consists in the main of two circular metal plates of about 5 cm diameter, one stationary, the other rotating opposite it with an adjustable narrow gap between them. When the gap is reduced to about 0.02 mm and the rotor has a velocity of about 10000 R.P.M. the air is drawn in a centripetal direction into the gap. They have also built a centripetal vacuum pump[1].

III. Strength.

We have said in Sect. 6 that the deformations which any body may suffer, are limited by the strength of the material of which that body is composed. Strength must therefore be included in rheology.

36. The classical criteria. Qualitatively speaking, strength is the resistance of a material against either *plastic deformation or fracture*. In order to find a quantitative definition, we reason that to deform a body plastically or to break it, the external forces must perform work. We thus arrive at a theory first advanced by BELTRAMI in accordance with which a material fails when the stress-work w_s exceeds a certain limit. This theory was easily discredited when applied to the case of isotropic compression. BELTRAMI had overlooked the radical difference in the behaviour of all materials in cubical dilatation (positive or negative) and distortion. Allowing for this, HENCKY in his theory of plastic strength assumed that there is no volume-plasticity. This is normally taken for granted, but as can be seen from Lord KELVIN's example quoted in the footnote of Sect. 1, is valid only as a first approximation. Disregarding such cases, HENCKY[2] postulated that a material is made to yield plastically when the distortional stress-work per unit volume $w_{(0)}$ reaches or exceeds a certain limit. HUBER[3] applied a similar consideration to fracture. He argued that the volume strength against isotropic pressure is theoretically unlimited, even if in practice a material may fail through

---

[1] Proc. Symposium on boundary layer research. 1958.
[2] H. HENCKY: Proc. 1st Internat. Congr. Mech. Delft, 1924 and Z. angew. Math. Mech. **4**, 323—334 (1924).
[3] M. T. HUBER: Czasapismo Techniczne, Lemberg **22**, 81 (1904).

local crushing around pores or holes. In cubical dilatation under isotropic tension failure will be due to brittle fracture when the true cohesion of the material is exceeded. Turning to distortion, he postulated (prior to HENCKY) that a material will break when the distortional stress-work per unit volume $w_{(0)}$ reaches or exceeds a certain limit. Both theories can thus be regarded as modifications of BELTRAMI's theory, which preceded them. However, both overlooked a further point. To make a material flow plastically or to break it, work must be *expended*. If part of the stress-work is dissipated in the course of deformation and converted into heat, this part is not available to produce either fracture or plastic flow. Going back to Eq. (5.6) we can accordingly say with REINER and WEISSENBERG[1] that in both, HUBER's and HENCKY's theory, the specific work in question must be *strain-work*. In other words *strength is the limit to which a material can be strained*. In accordance with the resolution of the specific work Eq. (5.5), there will be two such limits, one volumetric, the other distortional. The maximum strain-work is called *resilience* (E) and we can thus formulate

$$\Phi = w_e \leq E, \tag{36.1}$$

where we have to distinguish E_v and $E_{(0)}$ and, in addition, plastic E_{pl} and fracture or breaking resilience E_b, the symbol E standing for elastic energy. With E_{pl} we get HENCKY's *flow-condition*, with E_b HUBER's *fracture-condition*.

The distinction between stress- and strain-work does not arise in the case of the three classical bodies particularized in Part A II (pp. 448—450). Both the Hooke and the St. Venant-Prandtl solid are assumed to be perfectly elastic up to failure, and the Newtonian liquid is not supposed to have any strength in distortion. Accordingly the Huber-Hencky criteria suffice in these cases.

On the basis of rheological equation (9.1) we find now from (9.3) as the volumetric condition of fracture, common to all three classical types,

$$\overline{w}_v = \frac{\varkappa}{2} \overline{\varepsilon}_v^2 = \frac{1}{2\varkappa} \vartheta_v^2 = E_v, \tag{36.2}$$

where $\overline{w}_v$ is the maximum cubical strain-work, and $\overline{\varepsilon}_v$ is the maximum cubical dilatation attainable at the maximum isotropic tensile traction ϑ_v. From (36.2)

$$\vartheta_v = +\sqrt{2\varkappa E_v}; \quad \overline{\varepsilon}_v = +\sqrt{2E_v/\varkappa}. \tag{36.3}$$

We use in Eq. (36.3) the positive sign to take account of the fact that in most materials $-\vartheta_v$ will be immeasurably great.

Then, considering the Hooke-body, we have from (5.7) and (10.10)

$$\overline{w}_0 = \vartheta_{\alpha\beta(0)} \vartheta_{\alpha\beta(0)} / 4\mu = E_{(0)}. \tag{36.4}$$

Writing explicitly, we find in Cartesian coordinates

$$\vartheta_{\alpha\beta(0)} \vartheta_{\alpha\beta(0)} = \vartheta_{xx(0)}^2 + \vartheta_{yy(0)}^2 + \vartheta_{zz(0)}^2 + 2\vartheta_{yz(0)}^2 + 2\vartheta_{zx(0)}^2 + \vartheta_{xy(0)}^2. \tag{36.5}$$

Comparison with (7.15) shows that Eq. (36.4) can also be written in the form

$$-\Pi_{s(0)} = 2\mu E_{(0)}. \tag{36.6}$$

This, with E_{pl} for E, is called *Mises' flow condition*. MISES[2] derived his equation (before HENCKY) not from energy-considerations. He reasoned that strength must be measured by a scalar quantity, a scalar expression of stress, and therefore

---
[1] M. REINER and K. WEISSENBERG: Rheolog. Leaflet **1939**, No. 10, 12.
[2] R. v. MISES: Göttinger Nachr. math.-phys. Kl. **1913**, 528.

a function of the stress invariants. Considering the deviator, its first invariant vanishes, and distortional strength is accordingly determined in the first instance by the second invariant. One often speaks of the Mises-Hencky flow condition, but it should be noted that these two conditions are identical only in the case of the Hooke body. In this case the strength conditions can also be expressed in terms of ultimate strain instead of ultimate stress, there being a one-to-one relation between them. The resilience is the joint invariant of both ultimate stress and ultimate strain. This joint invariant forms the criterion in the Hencky theory, while the Mises theory is based on the second invariant of either stress or strain. When the material is not represented by a Hooke body, these three invariants give different results.

37. Rheological criteria. When we transcend the range of the Hooke body and consider the strength of a material which shows viscous deformation or flow, we must distinguish between stress-work and strain-work.

α) While the stress work of the Hooke solid is entirely conserved, the cubical stress-work only of the Newtonian liquid is conserved, the distortional stress power being continually and completely dissipated.

A Newtonian liquid accordingly has isotropic strength, practically infinite in compression and considerable in tension, but no shear strength. The isotropic tension strength appears only if the liquid is truly homogeneous i.e., in the absence of air bubbles. The phenomenon of cavitation in which small parts of a solid (e.g., cast iron) are broken off, when in contact with streaming water is partly due to the strength of water exceeding that of the solid[1].

β) For the Kelvin solid we find from Eq. (15.2)

$$\dot{w}_{(0)} = s\dot{\varepsilon} = 2\mu\varepsilon\dot{\varepsilon} + 2\eta_s(\dot{\varepsilon})^2. \tag{37.1}$$

The condition for failure through either plastic flow or rupture in distortion is therefore

$$\int (\dot{w}_{(0)} - 2\eta_s \dot{\varepsilon}^2)\, dt = 2\mu \int_0^{\varepsilon_f} \varepsilon\dot{\varepsilon}\, dt = \mu \varepsilon_f^2 = E_{(0)}, \tag{37.2}$$

where ε_f is the strain at which failure occurs.

From Eq. (37.2)

$$\varepsilon_f = \sqrt{\frac{E_0}{\mu}} \tag{37.3}$$

so that from Eq. (15.2)

$$s_{sf} = 2\sqrt{E_0 \mu} + 2\eta_s \dot{\varepsilon} = s_{sf} + 2\eta_s \dot{\varepsilon}, \tag{37.4}$$

where s_{sf} is the ultimate stress in static failure.

A Kelvin solid therefore fails in distortion when the strain reaches a definite limit (ε_f) in accordance with (37.3), while the stress at which the material fails (s_f) increases with the rate of strain. This, for instance, is the case with mild steel; the strength of steel is thus connected with its "anelasticity". There is competition between voluminal and distortional resilience. In the "cup and cone" fracture of a mild steel rod, both are present, the former being exceeded first at the axis of the rod where the isotropic component of stress is greatest.

It should be noted that simple pull causing longitudinal extension has an isotropic component which may cause fracture by separation when volume-strength is exceeded.

---

[1] M. REINER: Engineering **1943**, 454.

γ) For the Maxwell-body we find from Eq. (16.2)

$$\dot{w}_{(0)} = s\dot{d} = \frac{s\dot{s}}{2\mu} + \frac{s^2}{2\eta}. \tag{37.5}$$

The failure condition therefore is

$$\int \left(\dot{w}_{(0)} - \frac{s^2}{2\eta}\right) dt = \frac{1}{2\mu} \cdot \int_0^{s_f} s\dot{s}\, dt = \frac{s_f^2}{4\mu} = E_{(0)}, \tag{37.6}$$

where s_f is the stress at which failure occurs. From Eq. (37.6)

$$s_f = 2\sqrt{E_0 \mu}, \tag{37.7}$$

so that from Eq. (16.2)

$$\dot{d}_{sf} = \frac{\dot{s}}{2\mu} + \sqrt{\frac{E_0 \mu}{\eta^2}} = \frac{\dot{s}}{2\mu} + \varepsilon_{sf}, \tag{37.8}$$

where $\dot{d}_{sf}$ is the rate of deformation at which the material fails under a constant static load.

A Maxwell body therefore fails when the stress reaches a definite limit (s_f) in accordance with (37.7), while the rate of deformation at which the material fails ($\dot{d}_f$) increases with the rate of stress. This has been confirmed for certain "plastics" (compare REINER and FREUDENTHAL[1]).

C. Microrheology[2].

38. General considerations. Most materials with which rheology has to deal are *dispersed systems* consisting of two or more rheologically different materials or *phases*, of which one (considering that every material has voids or pores) will be air. Even a polycrystalline metal possesses an amorphous thin layer between the single crystals which may be considered as a separate phase. In macrorheology such materials are treated as if they were homogeneous. The aim of microrheology is to determine the relation between the rheological behaviour of the composite system and that of its constituent parts. When this aim is attained, it should be possible to determine the structure of a composite material from rheological measurements. This procedure has been named "rheological analysis".

One of the phases will be *continuous*. This is also called the *medium*. The dispersed phase can sometimes be considered as consisting of "particles". The phases can be solid or liquid or gaseous. The dispersion can be any combination of such phases. A solid in liquid is called a *sol*, a liquid in solid is called a *gel*. There often is a continuous transition from sols to gels with increasing concentration of the solid phase. When the solid phase forms one continuous structure throughout the body, American terminology speaks of a *jelly*.

This terminology originated from colloidchemistry where it is, however, much more involved. Rheology uses a simplified version.

As an example we may name bitumen which we have already mentioned in Sects. 17 and 18. PFEIFFER and SAAL[3] have shown that from the rheological point of view two kinds of bitumen can be distinguished. Bitumen consists of a dispersion of micelles in heavy viscous oils. The micelles are high-molecular-weight hydrocarbons, which adsorb oil on the surface and also absorb it interiorly. When the micelles are fully peptized and able to move through the bitumen as

---

[1] M. REINER and A. FREUDENTHAL: Proc. 5th Internat. Congr. Appl. Mechanics, 1938.
[2] In the preparation of Sects. 38 to 42 I have had the help of Dr. Z. HASHIN.
[3] T. PH. PFEIFFER and R. N. SAAL: J. Phys. Chem. **44**, 139 (1940).

freely as the viscosity of the oily phase permits, the system is a sol, showing viscous flow with intramicellar elasticity (compare Fig. 36). If, however, the micelles attract each other, a gel-structure is formed, with irregular open packing of the micelles, the spaces being filled by the viscous liquid (compare Fig. 37). The intermicellar forces will be subject to relaxation, and the structure thus forms a Maxwell body. Owing to the presence of the viscous oil between the micelles, however, the gel-structure cannot work "instantaneously" at acoustic speed but will be delayed by the Kelvin-type of solid viscosity. In this case the different phases cannot be distinguished by the naked eye. As extreme cases of what in rheology may be considered as a dispersed system we may mention loam as pictured in Fig. 8, and still more so concrete. In concrete a solid skeleton is formed by the "coarse aggregate", the interstices being filled with "mortar".

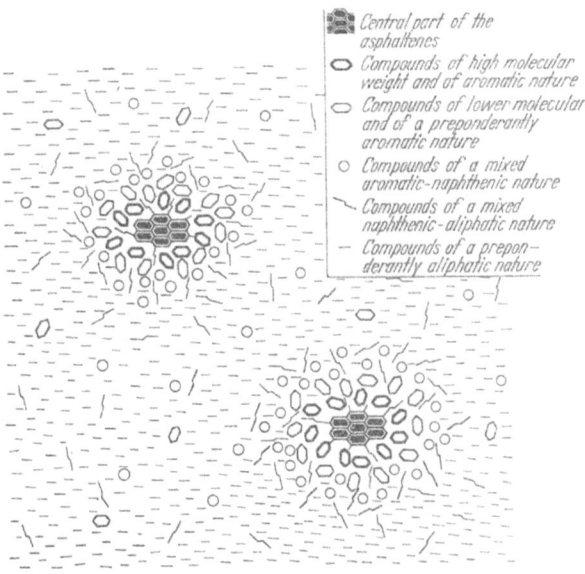

Fig. 36. Bitumen-sol.

The mortar itself is a dispersed system of sand in cement, and the latter can be considered as a very viscous liquid in relation to the concrete. The concrete as a whole can be accordingly considered as a gel. In a very good concrete the mortar will completely fill the interstices of the coarse aggregate skeleton which, in turn will form one solid structure throughout the body of the concrete. In this case, rheologically speaking, the concrete can be treated as a "jelly".

Besides solid-liquid dispersions, we have to distinguish solid-solid and liquid-liquid dispersions. The latter are named *emulsions*.

The schematic pictures of dispersed systems as shown in Figs. 8, 36 and 37 makes it obvious that the mathematical treatment of the basic problem of microrheology must meet great difficulties.

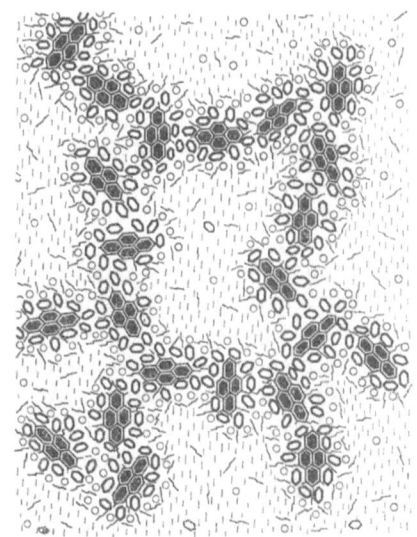

Fig. 37. Bitumen-gel.

In order for mathematical analysis to be at all possible the following conditions must be realized:
(a) the rheological equations of the constituent materials involved are known;
(b) the dispersed phase has mathematically defined boundaries;

(c) conditions on interfaces are known, and may be expressed mathematically;
(d) functions for stress-work and/or stress-work per unit time of the constituent materials exist and are known.

As a rheological body the dispersed system may be assumed to be quasi-homogeneous, i.e. statistically homogeneous. This presupposes that even at the lowest concentration the number of particles is great, and their size and distribution are reasonably constant throughout the medium.

When the particles are elongated or similarly non-symmetrical the dispersed system will nevertheless be quasi-isotropic provided the axes of the particles are statistically oriented in all directions of space. This state may be assumed to exist in a liquid at rest or a plastic material in the virgin state. Such material, however, may become anisotropic through flow.

The volume concentration of particles, which is a parameter of the material is defined by

$$C_v = \frac{\text{volume of particles in a region of the dispersed system}}{\text{volume of region}}. \tag{38.1}$$

Let a rheological constant of the medium be denoted by M_m, that of the particle by M_p, while M^* denotes the corresponding rheological constant of the dispersed system. Then the ratio

$$\frac{M^*}{M_m} = M_{\text{rel}} \tag{38.2}$$

is defined as the relative rheological constant.

The problem is to calculate M_{rel} from the given data of M_m, M_p, C_v and the geometry of structure.

39. Dilute sols. The pioneering work in this field is due to EINSTEIN as part of his doctor's thesis[1], when he calculated the coefficient of shear viscosity of a very dilute solution from the equations of hydrodynamics.

EINSTEIN's method can be applied for the determination of the material constants of any dispersion of dilute concentration, provided a certain boundary value problem can be solved. We shall presently describe the general method.

Rheological constants are determined from experiments in which specimens are submitted to certain simple states and/or applications of stress and strain. This may be called a "primitive state". (For example—simple tension for determining YOUNG's modulus of elasticity, E, simple bending for measuring TROUTON's coefficient of viscosity, λ_T, simple torsion for measuring either the shear modulus of elasticity, μ, or shear viscosity, η, etc.)

Let a homogeneous isotropic material be submitted to such a primitive state of stress producing a uniform state of principal deformations $d_i^{(0)}$, and/or principal rates of deformation $\dot{d}_i^{(0)}$. Let this material constitute the medium for a dispersion of dilute volume concentration formed by the introduction of particles of another material. The dispersed system is submitted to the same primitive state of stress which produces on this occasion deformations d_i^* and/or rates of deformation $\dot{d}_i^*$.

By dilute dispersion or low volume concentration we mean the following:

(a) It will in general be possible to find between any two particles a region, however small, for which it may be assumed, with sufficient accuracy, that the primitive state $d_i^{(0)}$ and/or $\dot{d}_i^{(0)}$ is not perturbed by the particles (Fig. 38).

---

[1] A. EINSTEIN (1906), (1911).

(b) At any point in the dispersed system the effect of all particles may be determined by summation of all their individual effects at that point, hence for the determination of the influence of one specific particle the presence of all others is ignored.

(c) The direction of the principal axes is the same for the $d^*_{(i)}$; $\dot{d}^*_{(i)}$ as for the $d^{(0)}_{(i)}$; $\dot{d}^{(0)}_{(i)}$ respectively.

(d) Squares and higher powers of volume concentration c_v may be neglected. It has thus been assumed that interaction between particles may be neglected.

In the attempt to determine the generally complex state of stress, deformation and/or rate of deformation around a particle the following boundary value problem arises: to solve the differential equations for medium and particle in such a way that at their common boundary i.e. the surface of the particle, certain conditions be satisfied (such as continuity of stress vector, vanishing of displacement and velocities etc.), and that at an infinite distance from the particle the uniform state $d^{(0)}_{(i)}$; $\dot{d}^{(0)}_{(i)}$ prevails.

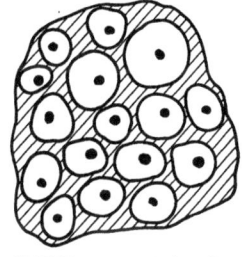

Fig. 38. Dilute dispersion.

In order to have a more definite picture in mind we shall choose the case when medium and particles are solids showing no time effects.

In imagination we cut off from the dispersed system an element containing one particle. Let the surface of the cut lie wholly in an unperturbed region (Fig. 39), and let it be geometrically similar to the particle boundary. Such an element may be called the "composite element". Having solved the boundary value problem, the strain energy $W(r_s, r_p, M, d_i^{(0)})$ of the composite element can be calculated, where r_s denotes parameters of the composite element surface, r_p parameters of the particle surface, and M-material constants of medium or particle.

When the particle is absent, and its place is taken by the medium, we can easily calculate the strain energy, assuming the uniform state. This strain energy may be denoted by $W_0(r_s, M, d_i^{(0)})$.

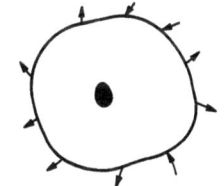

Fig. 39. Element containing one particle.

The change in strain energy caused by the presence of the particle is therefore

$$\Delta W = W - W_0. \tag{39.1}$$

We now cut off from the dispersed system a region large in comparison with the composite element. When the particle is absent the strain energy $W_0(R, M, d_i^{(0)})$, stored in it can be calculated from the uniform state. Here the R are parameters of the surface of the "region".

Considering the same region in the dispersed system, the strain energy stored in it is

$$W = W_0 + \sum_1^n \Delta W, \tag{39.2}$$

where n is the number of particles in the region.

The same region is now considered as consisting of a different homogeneous material with material constants M^*, with $W^*(R, M^*, d_i^*)$ the strain energy. This must be equal to W, and we get

$$W^*(R, M^*, d_i^*) = W_0(R, M, d_i^{(0)}) + \sum_1^n \Delta W(r_s, r_p, R, M, d_i^{(0)}). \tag{39.3}$$

There remains to express the d_i^* in terms of known quantities. To this end we use the solution of the boundary value problem. We choose an arbitrary point in space within the dispersed system, and determine the perturbation caused by any particle upon the uniform state as a function of its distance from this point. If this be denoted by $u'_{i(k)}$ for the k-th particle, then we have

$$u_i^* = u_i^{(0)} + \sum_1^k u'_{i(k)}, \qquad (39.4)$$

where $u_i^{(0)}$ is the displacement in the absence of the particle, and u_i^* the displacement in its presence.

The d_i^* can be calculated by differentiation, and in general

$$d_i^* = d_i^* (d_i^{(0)}, c_v, r_p). \qquad (39.5)$$

Introducing this expression into Eq. (39.3) the only remaining unknown is M^*.

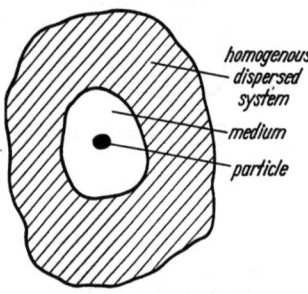

Fig. 40. Composite element.

When dealing with liquids the method is exactly the same. It is sufficient to replace displacements by velocities, strains by rates of deformation, energy by power. For materials showing time effects and for dispersed systems consisting of solids and liquids we may equate energies and powers separately.

Another method which is not less general has been developed by FROEHLICH and SACK[1]. In this method displacements or velocities are equated instead of energies.

Here the composite element is imagined to be surrounded by a thick shell of the dispersed system (Fig. 40), which possesses the unknown material constants which it is desired to determine. A uniform state of stress is applied to the boundary. It is then necessary to solve the differential equations of the materials fulfilling conditions at the common boundaries. The displacements at the last boundary are next calculated. It is then assumed that the whole region consists of the dispersed system with unknown constants, and the displacements at the boundary are again calculated. Equating the displacements, one obtains an expression for the material constants of the dispersion.

40. Dispersions in liquids. EINSTEIN obtained his solution under the following assumptions:

(a) The solvent, i.e. the medium, is a viscous liquid of known coefficient of shear viscosity η. The motion of this liquid is so slow that the linearized Navier-Stokes equations describe conditions with sufficient accuracy.

(b) The molecules of the solute, i.e. the particles, are regarded as small rigid spheres.

(c) There is no slippage at interfaces.

The boundary value problem to be solved is the perturbation caused by a rigid sphere in a field of flow having the velocities

$$u_i^{(0)} = \dot{d}_i^{(0)} x_i.$$

This is solved by generalizing STOKES' method for the flow of a viscous liquid of constant velocity around a rigid sphere. The result is

$$\eta_{\text{rel}} = \frac{\eta^*}{\eta_m} = 1 + 2.5\, c_v. \qquad (40.1)$$

---

[1] H. FROEHLICH and R. SACK: Proc. Roy. Soc. Lond., Ser. A **185**, 415—430 (1946).

EINSTEIN'S work was first generalized for different shapes of particles.

The problem of a dispersion of rigid ellipsoids and rigid sticks has been treated under different assumptions regarding the orientation of axes by JEFFERY, EISENSCHITZ and BURGERS[1]. BURGERS has developed a special method for calculating viscosities of dispersions of rigid particles by using OSEEN's method[2] for viscous flow instead of the linearized Navier-Stokes equations. Accounts of the work of JEFFERY, EISENSCHITZ and BURGERS, as well as the question of the influence of Brownian movement may be found in HERMANS (1953)[3].

With a view of determining interaction effects, SIMHA[4] has calculated the viscosity of a dispersion of doublets of rigid spheres (dumbells), using SMOLUCHOWSKI's theory[5] for viscous flow around two spherical obstacles.

Other work is concerned with particles consisting of various materials, their form remaining spherical.

Treating the case when the medium is a viscous liquid, GUTH and MARK[6] have calculated the viscosity of a dispersion of bubbles and found, neglecting surface tension,

$$\eta_{rel} = 1 - c_v. \tag{40.2}$$

The case of an emulsion of spherical droplets of one viscous liquid in another has been treated by TAYLOR[7] taking into account surface tension. He found

$$\eta_{rel} = 1 + 2.5 \frac{2\eta_m + 5\eta_p}{5\eta_m + 5\eta_p} \cdot c_v, \tag{40.3}$$

where η_m is the coefficient of shear viscosity of the liquid medium and η_p of the droplets.

Further generalizations of this case, such as taking into account slippage between the droplets and the medium fluid when their mutual coefficient of friction is known, have been carried out by OLDROYD[8] using the Froehlich and Sack method.

The case of a dispersion of elastic spheres in a viscous liquid was treated by FROEHLICH and SACK by their method described above. The medium which is a Newtonian liquid obeys the relation

$$\tau = \eta_m \dot{\gamma}, \tag{40.4}$$

where τ is a pure shearing stress and γ the change in a right angle due to it.

The dispersed system will be viscoelastic, and Eq. (40.4) is replaced by

$$\left(1 + T_1 \frac{d}{dt}\right) \tau = \eta^* \left(1 + T_2 \frac{d}{dt}\right) \gamma, \tag{40.5}$$

where the time of relaxation T_1 and the time of retardation T_2 are given by

$$T_1 = \frac{3\eta_m}{2\mu_p}\left(1 + \frac{5}{3} c_v\right), \tag{40.6}$$

$$T_2 = \frac{3\eta_m}{2\mu_p}\left(1 - \frac{5}{2} c_v\right), \tag{40.7}$$

where μ is the shear modulus.

---

[1] G. B. JEFFERY: Proc. Roy. Soc. Lond., Ser. A **102**. 161–180 (1923). — R. EISENSCHITZ: Z. phys. Chem. A **163**, 133–141 (1933). — J. M. BURGERS: Second Report on viscosity and plasticity, Amsterdam 1938.
[2] C. W. OSEEN: Hydrodynamik. Leipzig 1927.
[3] HERMANS: Flow properties of dispersed systems. Amsterdam 1953.
[4] R. SIMHA: J. Res. Nat. Bur. Stand. **42**, 409–418 (1949).
[5] M. SMOLUCHOWSKI: Bull. Acad. Sci. Cracovie A **1**, 28 (1911).
[6] E. GUTH and H. MARK: Ergebn. exakt. Naturw. **12**, 115–162 (1933).
[7] G. I. TAYLOR: Proc. Roy. Soc. Lond., Ser. A **138**, 41–48 (1932).
[8] J. G. OLDROYD: Proc. Roy. Soc. Lond. Ser. A **218**, 122 (1953).

41. Dispersions in solids. The case of a solid medium was treated by MACKENZIE[1]. He determined the bulk modulus $\varkappa$ and the shear modulus μ of a dispersion of spherical holes in an elastic medium. The method used was that of FROEHLICH and SACK. As it was necessary to determine two different independent constants, two uniform states, one of hydrostatic pressure and the other of homogeneous shear, were considered. The more difficult boundary value problem is that for shear, which was solved in terms of spherical harmonics [compare LOVE (1927), p. 250].

MACKENZIE's results are re-written here to conform with equations for relative viscosity. Neglecting also some higher order terms, we have

$$\varkappa_{\text{rel}} = 1 - \frac{3}{2} \frac{1 - \nu_m}{1 - 2\nu_m} c_v, \tag{41.1}$$

$$\mu_{\text{rel}} = 1 - 15 \frac{1 - \nu_m}{7 - 5\nu_m} c_v, \tag{41.2}$$

where ν_m is POISSON's ratio for the medium.

In the same paper the case of pressure on the surfaces of the holes has also been investigated.

A second case of dispersion of rigid spherical particles in an elastic medium, has been treated by HASHIN[2] using EINSTEIN's method. The problem of the perturbation caused by a rigid sphere in an infinite elastic medium in a state of uniform strain was solved by an explicit method due to Lord KELVIN[3]. By this method the equations of elasticity can be solved within or outside a spherical boundary under condition that the displacements attain arbitrary values on the boundary. An account of this method is given in LOVE (1927), Chap. XI.

The solution expressed in terms of spherical harmonics of negative integral degree is applied to the case of hydrostatic and pure shearing strain. It is found that

$$\varkappa_{\text{rel}} = 1 + 3 \frac{1 - \nu_m}{1 + \nu_m} c_v, \tag{41.3}$$

$$\mu_{\text{rel}} = 1 + \frac{15}{2} \frac{1 - \nu_m}{4 - 5\nu_m} c_v, \tag{41.4}$$

when E is YOUNG's modulus.

It should be noted that in both MACKENZIE's and HASHIN's papers it has not been assumed that the particles are of equal size.

A mathematical check on formulas (41.2) and (41.4) is provided by the analogy between the slow motion of a viscous liquid and the strain of an incompressible medium. In this analogy, velocity is analogous to elastic displacement, rate of deformation to strain, and coefficient of shear viscosity to modulus of shear. Thus introducing $\nu_m = \frac{1}{2}$ into (41.2) we obtain $\mu_{\text{rel}} = 1 - c_v$, which is analogous to the Guth and Mark result (40.4). Introducing $\nu_m = \frac{1}{2}$ into (41.4), we find $\mu_{\text{rel}} = 1 + 2.5 c_v$, which is analogous to EINSTEIN's result (40.1).

Investigation of the case of a dispersion of spherical particles of one elastic material in another one has also been undertaken. HASHIN found

$$\varkappa_{\text{rel}} = 1 - 3(1 - \nu_m) \frac{1 - \varkappa_p/\varkappa_m}{2(1 - 2\nu_m) + (1 + \nu_m)\varkappa_p/\varkappa_m} c_v, \tag{41.5}$$

$$\mu_{\text{rel}} = 1 - 15(1 - \nu_m) \frac{1 - \mu_p/\mu_m}{7 - 5\nu_m + 2(4 - 5\nu_m)\mu_p/\mu_m} c_v \tag{41.6}$$

---

[1] J. K. MACKENZIE: Proc. Phys. Soc. Lond. B **63**, 361 B, 2—11 (1950).
[2] Z. HASHIN: Bull. Res. Council. Israel **5** c, 46 (1955).
[3] THOMSON and TAIT: Natural Philosophy Pt. II. Cambridge 1879—1883.

where the index p refers to the particles. In the case of holes $\varkappa_p = \mu_p = 0$ and (41.1) and (41.2) are obtained. In the case of rigid spheres $\varkappa_p \to \infty$, $\mu_p \to \infty$ and (41.3) and (41.4) result.

The case of spherical inclusions, filled with viscous fluid, in an elastic solid has been treated by OLDROYD[1].

This is the inverse of the case of FROEHLICH and SACK, and has been treated by their method. Instead of the relation

$$\tau = \mu_m \gamma, \qquad (41.7)$$

where τ is a pure shearing stress, the relation

$$\left(1 + T_1 \frac{d}{dt}\right) \tau = \mu^* \left(1 + T_2 \frac{d}{dt}\right) \qquad (41.8)$$

will hold for the dispersed system.

For an incompressible medium it is found that

$$\mu^* = \mu_m \left(1 - \frac{5}{3} c_v\right). \qquad (41.9)$$

Explicit formulae for $\varkappa^*$ and μ^* are given in terms of the rheological constants of the elastic medium and the viscous fluid in the inclusions.

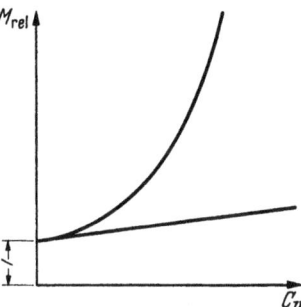

Fig. 41. Dependence of relative constant upon concentration.

42. Dispersed systems of high volume concentration. α) *General considerations.* It can not be expected that linear formulae in c_v for the relative material constants such as given in Sect. 41, will hold for high volume concentrations. Experiments show that EINSTEIN's formula is not valid when the volume concentration exceeds 2 to 3%.

From the practical point of view the case of large volume concentration is of major importance. Thus the volume concentration of sand particles in mortar reaches 60%, that of rubber fillers 50%.

When dealing with large volume concentrations one can no longer neglect interaction effects, nor are effects at any point additive. Moreover in theories of dilute concentration volume integrations are carried out in which it is tacitly assumed that the volume of the particle is negligible within the region of integration.

We should in principle expect $M_{\rm rel}$ to be represented as a function of c_v by a curve (Fig. 41), such that its tangent at the origin represents conditions for dilute concentrations. $M_{\rm rel}(c_v)$ will be represented by a curve convex towards the abscissa when the medium is "strengthened" by the particles (such as is done by rigid spheres), and by a concave curve if the medium is "weakened" (such as is done by holes).

We may assume that it is possible to represent $M_{\rm rel}$ as a power series in c_v such as

$$M_{\rm rel} = 1 + \alpha_1 c_v + \alpha_2 c_v^2 + \alpha_3 c_v^3 + \cdots, \qquad (42.1)$$

where we have taken care that for $c_v = 0$, $M_{\rm rel} = 1$, and by choosing α_1 as the coefficient of c_v in the equations of Sect. 41, that for small c_v Eq. (42.1) degenerates into the results for dilute concentration.

Following KRAEMER, we write

$$\alpha_1 = M_i \qquad (42.2)$$

naming it the *intrinsic material constant*. From Fig. 46 or differentiation of (42.1) it follows that

$$M_i = \left.\frac{dM_{\rm rel}}{dc_v}\right|_{c_v=0}. \qquad (42.3)$$

---

[1] J. G. OLDROYD: In GRAMMEL (Editor) l.c.

β) *Viscosity of concentrated dispersions.* In the field of high concentrations the only work hitherto carried out consists in attempts to find the relative viscosity of a dispersion of rigid spheres in a viscous liquid, thus generalizing EINSTEIN'S formula.

Attempts have been made to determine interaction effects between particles when these approach each other. Calculations become very complicated, and results are not consistent.

It is universally agreed that $\alpha_1 = \eta_i = 2.5$. However, calculations of α_2 resulted in, among others, $\alpha_2 = 14.1$ [1] and $\alpha_2 = 7.349$ [2].

The question arises whether it is at all possible to obtain by an interaction method, starting from $c_v = 0$, a formula which would be valid for any c_v.

The lower limit of such a formula is well defined by (42.3), and the results of Sect. 41, but it is questionable what the upper limit might be. Clearly, the concentrations cannot increase after the spheres touch each other in close packing. At such a stage no flow will be possible, and the viscosity would attain infinite value. But there are different ways of packing spheres. Assuming equal size for a cubical packing which produces the least density. $c_v = \pi/6 = 0.52$, while for hexagonal packing, which is the densest, $c_v = 0.74$. It is difficult to imagine that by starting from $c_v = 0$ with increasing c_v, a formula can be derived which would predict the correct packing and yield η_{rel} for its appropriate c_v.

Assuming even that this can be done, another serious difficulty consists in the assumption of the spherical shape of the particles. This offers no serious drawback in theories of dilute concentration for there the immediate vicinity of the particles does not matter, energies being calculated from conditions at the surfaces of the composite elements, and it may well be assumed that conditions at some distance from the particle are not seriously changed when substituting for an approximately spherical particle a perfectly spherical one.

In contradistinction, when spheres are close, any deviation from perfect spherical shape will be of great influence.

It appears that a more rational approach would consist of demanding *a priori* that η_{rel} should satisfy the condition that $\frac{d\eta_{rel}}{dc_v} = 2.5$ for $c_v \to 0$ while approaching infinity for some c_v associated with a certain kind of packing. A very elegant approach in this direction has been made by MOONEY[3], who determines the functional form of η_{rel} by the simple consideration that the order of adding volume fractions of particles to the liquid cannot influence the viscosity attained for a certain c_v. This leads to the formula

$$\eta_{rel} = \exp\left(\frac{2.5 c_v}{1 - k c_v}\right), \tag{42.4}$$

where it is assumed that interaction and "crowding" effects of particles may all be expressed by one constant k, which is called the *self crowding factor*.

It may be seen that the first previously mentioned condition is satisfied. In order to satisfy the second one it is required that the denominator $1 - k c_v$ should vanish between the limits $c_v = \pi/6$ and $c_v = 0.74$, corresponding to least and densest packings. It is thus found that $1.35 < k < 1.91$.

Fitting an appropriate value for k, good agreement with experiments is obtained. The theory is also extended to particles of different sizes.

---

[1] E. GUTH and R. SIMHA: Kolloid-Z. **74**, 266 (1936).
[2] V. VAND: J. Phys. Colloid Chem. **52**, 277—299 (1948).
[3] M. MOONEY: J. Coll. Sci. **6**, 162—170 (1951).

Another approach for high concentrations has been developed by SIMHA[1]. It is assumed that at high concentrations the only interactions of importance are those of a particle with its nearest neighbours. As the dispersed system is considered to be isotropic, a hypothetical spherical surface may be imagined to exist around any particle such that the interactions of the particle at the centre with others lying on the surface vanish. Accordingly, a hydrodynamic boundary value problem is formulated and solved. The result is

$$\eta_{\text{rel}} = 1 + 2.5\,\lambda\,c_v, \tag{42.5}$$

where λ is a function of a/b only, with a the radius of the particle and b the radius of the surface.

The factor λ tends to 1 for small c_v, and to infinity for a closed packing of spheres. It may be regarded as a semi-empirical function[2].

Although we here are concerned only with the theoretical determination of the material constants of a dispersed system, some interesting experimental results for mortar due to ARNSTEIN and REINER[3] may be mentioned. Investigating the creep of a mortar proceeding with a viscosity of the order of 10^{17} poises, these authors found that EINSTEIN's equation (42.1) remains valid for volume concentrations of sand up to 60%.

More particularly they found experimentally that

$$\eta_{\text{rel}} = \frac{\eta_{\text{mortar}}}{\eta_{\text{cement}}} = \frac{1+WC}{1+\overline{WC}}(1+2.5\,c_v), \tag{42.6}$$

where WC is the volume water-cement ratio of neat cement and $\overline{WC}$ is the volume water-cement ratio of the cement base used in the mortar.

This result may be due to the high coefficient of viscosity of the cement base so that a steady state of flow (creep) of the mortar is not reached even in an experiment taking some years. The experiment which took four years, must therefore be regarded as remaining within a transient state.

43. Rheological analysis of structural viscosity. When approaching the problem of the generalised Newtonian liquid of physical non-linearity for which η or φ varies (see Sect. 26), we are at the outset confronted with the fact that these liquids are as far as we know, all dispersed systems, and that η decreases (or increases) with increased rate of shear (or increased shearing stress). The suspended particle *interferes* with the mobility of part of the dispersion medium. If it can be shown that this interference decreases with increased rate of shear, this would explain non-linear behaviour. The problem is therefore to show why and in what manner an increased rate of shear should decrease the interference.

We can distinguish two sorts of interference, which we may name (1) disturbance and (2) immobilization[4].

α) *Disturbance* is the mechanism which has been assumed in the preceding sections. It can lead in four different ways to non-linear behaviour, as follows:

α1) *Competition between orientation and Brownian movement.* As shown in Sect. 3, simple shear, and therefore flow in general, is accompanied by rotation. If the suspended particles are not spheres but elongated, say, ellipsoidal, the magnitude of the interference depends on the orientation of the particle in the streaming liquid, i.e. on the angle φ between the main axis of the ellipsoid and the direction of flow. At rest all values of this angle are equally probable. This

---
[1] R. SIMHA: J. Appl. Phys. **23**, 1020–1024 (1952).
[2] See also Chapter by FRISCH and SIMHA in EIRICH (Editor).
[3] A. ARNSTEIN and M. REINER: Civ. Engng., Lond. **40**, 198–202 (1945).
[4] Compare Table 8.

determines a certain viscosity η_0. Through flow this angle is constantly changed by the rotation of the particle[1]. If the temperature is low, and there is no Brownian movement, the particles may ultimately become completely oriented or directed due to the flow of liquid. This determines another value η_∞. In a random orientation the particle will interfere with the laminar flow of a greater volume of liquid, while, when oriented in the direction of flow, the volume will be less. Therefore $\eta_0 > \eta_\infty$.

The orientation is counteracted by the rotational diffusion of the Brownian movement. Where the Brownian movement is negligible, η_∞ is established as soon as a steady state of flow is reached, and the system is therefore *simple linear* of that viscosity η_∞. Where the influence of the Brownian movement prevails over the orientation forces, η_0 is maintained, and in this case also the flow is simple linear. Generally there will be at every temperature and every rate of shear a dynamical equilibrium between the forces of diffusion and orientation, with corresponding η, where $\eta_0 > \eta > \eta_\infty$.

In this case, *the relative viscosity η_{rel}[2], must depend upon the temperature*. An increase of temperature not only changes the viscosity of the medium but also brings with it increasing Brownian movement, and accordingly shifts η towards η_0. Therefore, in this case η_{rel} increases with increasing temperature.

α2) Elastic deformation. The interference with the deformation of the liquid produces by reaction stresses in the particle. Through these stresses the particle is deformed, acting as a "spring". The deformation may be considerable. The tension and extension of the particle is at its highest when its axis is inclined by an angle of 45° towards the direction of flow. However, the particle is not only deformed but also rotates. Through rotation, the spring is unloaded and recovers. This must result in an oscillation of the particle, which oscillation is transmitted to the liquid, and the amount of dissipated energy is thereby increased. If such a mechanism exists, the result must therefore be an *increase* of the relative viscosity with increased shear.

α3) Straightening of curled threads. If the particles are long threads without rigidity the result is entirely different. The tension produces a straightening of the thread. This decreases the interference of the particle with the flow of the liquid, and therefore the viscosity also decreases. Rotation of the particle tends to restore coiling and curling. Such mechanism may be independent of temperature.

α4) Shortening. If the tension is very great, or the breaking strength of the particle very small, the particle may break as soon as its axis coincides with the direction of maximum tension ($\varphi = 45°$). The particle may also be broken through bending. Shortening of particles decreases the ratio between the larger and the smaller axis of the particle, makes it more sphere-like and therefore decreases η. Whenever the fragments touch each other *in a suitable* position, the particle may be reformed. The probability for this is, however, small. In this case the consistency curve produced in decreasing shear will be different from the curve produced in increasing shear, i.e. the consistency curve will show a hysteresis loop.

β) Immobilization can cause non-linear behaviour in two ways:

β1) Adsorption. The adsorption layer may be sheared off through the flow of the liquid. Such a mechanism was suggested by HATSCHEK[3] and mathematically

---

[1] If the particle is a sphere, all axes of the particle are geometrically equivalent, and the rotation is of no influence. There is, therefore, no reason why a suspension of rigid spheres in a simple Newtonian liquid, as treated by EINSTEIN (compare Sect. 40) should not behave in the simple Newtonian manner, as regards disturbance.

[2] Defined by Eq. (40.1).

[3] E. HATSCHEK and R. JANE: Kolloid-Z. **40**, 53 (1926).

formulated by REINER and RIWLIN[1]. In this case there must be an influence of change of temperature. Adsorption is an exothermal process and should decrease with increasing temperature.

$\beta 2$) *Steric immobilization* has been discussed at great length by KRAEMER and WILLIAMSON[2]. They imagined the particle to be an "ultramicroscopic body which is generally permeated in gel-like fashion with the liquid medium. The union of the structural elements forming the particle may involve any or every type of bond which can operate between atoms or molecules". The particle may be either (i) a giant molecule immobilizing solvent between the branches and in its ring, (ii) a micelle (crystalite), (iii) a group of micelles with a regular arrangement, (iv) an aggregate or unordered assemblage of smaller particles.

Non-linear behaviour can be brought about in two ways: (a) through deformation; (b) through rupture.

Regarding (a), it should be noted that, in general, the particle can be much deformed. It will therefore partially participate in the deformation of the liquid. By deformation of the particle, solvent is squeezed out of it as from a sponge and part of the liquid is therefore mobilized.

Regarding (b), the deformation is accompanied by stresses. The particle can resist these stresses to a certain limit only. If the limit is exceeded, the particle breaks, and this also releases part of the immobilized liquid. If the fragments of the particle come to touch each other in the course of flow, the particles may be re-formed. "For a particular rate of shear, a dynamic equilibrium would be established between the rate of disaggregation and the rate of re-aggregation." However, the probability of reaggregation is generally not the same as that of disaggregation. Both can be equal in case (iv) only, i.e. in the case of unordered aggregates. Only in this case can every meeting of the fragments in any position result in a re-aggregation. In the case of molecules, micelles and ordered aggregates, a binding of fragments can take place only if the fragments are in a certain definite position in respect to each other. Not every position is therefore equally favourable. If consistency measurements are made both ways, i.e. the rate of shear is increased from zero to a maximum and then again decreased to zero, there will result not one single curve, but two such curves forming a hysteresis loop.

With regard to an influence of change of temperature, it may be assumed that the binding forces in unordered aggregates are so small that the aggregate will partially be destroyed by an increase of temperature.

γ) *Structural criteria.* These considerations lead to the conclusion that there may, and probably do, exist different types of variable viscosity. Table 6 gives a summary of the different mechanisms and their properties. There are four criteria which enable us to decide whether a given generalized Newtonian liquid belongs to one type or another. They are (I) the influence of a variation of the parameter of temperature on η_{rel}; (II) the reversibility or irreversibility of a variation of rate of shear, or existence or absence of a hysteresis loop; (III) the existence of a lower limit of non-linear behaviour; (IV) the relation between η_0 and η_∞.

Regarding (I), it is obvious that η_{rel} cannot show a dependence on temperature as long as the phenomenon of non-linearity is entirely geometrico-mechanical in nature. In this case the variation of the viscosity of the solution with temperature is solely due to the variation of the viscosity of the solvent. "Shortening" (1, iv of the table) represents such a mechanism, because the breaking of the

---
[1] M. REINER and R. RIWLIN: Kolloid-Z. **43**, 72 (1927); **44**, 9 (1928).
[2] E. O. KRAEMER and R. V. WILLIAMSON: J. Rheology **1**, 76 (1929).

particles changes their geometrical shape only. In case 1, iii the influence of temperature will probably also be absent. On the other hand, one can say that there must be a dependence on temperature in the cases 1, i and 1, iv. Nothing can be said *a priori* in this respect, of the other mechanisms.

With regard to (II), we have already mentioned that an instantaneous change of viscosity with changing rate of shear, i.e. without any lag, is possible only if all positions of the particles are equally favourable for a re-aggregation. This is not the case if the particles are molecules or micelles.

Regarding (III) it should be noted that whether the non-linear viscosity is due to a partial destruction of the structure (2B, ii, iii, iv) or a shortening or reduction of the primary particle (1, iv; 2A) there must be a lower limit of non-linearity. The forces which form the primary particles or the secondary structures can be overcome only when the stresses produced by the flow reach and exceed a certain value. Up to that value the behaviour is normal, i.e. simple linear. Such a limit cannot exist in the cases (1, i, ii, iii) where the smallest stress must be of influence.

Regarding (IV), if the structure of the dispersed phase is the same, *whatever the concentration*, η_0/η_∞ should be independent of c_v. For instance, if the viscosity anomaly was due to mechanism 1, i—competition between orientation and Brownian movement—the effect upon the value of η_0/η_∞ should be the same whether there was one particle only present in the dispersion or very many. On the other hand, in the case of 2B, iv—unordered aggregates—secondary structures will possibly appear at higher concentrations only, while at low concentrations the dispersed phase will consist of stable primary articles. In this case η_0/η_∞ would increase with c_v.

All mechanisms require that when η_∞ is reached, the behaviour is simply linear. This value of η_∞ may be reached either at a finite rate of shear or be asymptotically approached at an infinite rate of shear.

The temperature dependence of the "structural mechanism" becomes conspicious from *relative consistency curves* at different temperatures, when the

Table 8. *Types of structural viscosity.*

Nature of interference with free flow of solvent	(I) Temperature dependence of η_{rel}	(II) Reversibility of consistency curve	(III) Existence of lower limit		(IV) Dependence of η_0/η_∞ upon c_v
1. Disturbance					
(i) Competition between orientation and Brownian movement	increases	yes	no		no
(ii) Elastic deformation . . .	increases	yes	no		no
(iii) Straightening of curled threads	nil	yes	no		no
(iv) Shortening	nil	no	yes		no
2. Immobilization					
(A) Adsorption	decreases	yes	yes		no
(B) Steric immobilization . . .			(a) deformation	(b) rupture	
(i) Macromolecule	?	no	no	yes	no
(ii) Micelle.	?	no	no	yes	no
(iii) Ordered group	?	no	no	yes	no
(iv) Unordered aggregate . .	decreases	yes	no	yes	yes

relative consistency variable
$$V_{\text{rel}} = V \cdot \eta_{\text{solv}}, \tag{43.1}$$

where η_{solv} is the viscosity of the medium (solvent), is plotted against P. When there is no influence of temperature, they all coincide in a single curve.

D. Rheometry.

44. The method. Rheological properties are determined quantitatively in *rheometers* of which the various *viscometers* are the prototype for the more liquid kind and the *tensile test* of metals for the more solid kind of material. The property is determined either with its "absolute" value or relatively to the magnitude of the same property in a standard material. The instruments are of three types: In type I the material under test is subjected to pure homogeneous deformation; in type II it is subjected to laminar semi-homogeneous shear; in type III it is deformed in streamlined flow of a more complicated kind.

Examples for type I are the tensile test of mild steel, the compression test of concrete and Trouton's viscous traction[1]. "End-effects" must be taken into account. Second order effects are hard or impossible to detect by this method.

Examples for type II are the capillary tube and rotating co-axial cylinder viscometers. Because of non-homogeneity of deformation, these require either *integration* of a postulated rheological equation or *differentiation* of empirical results[2]. Sometimes the laminar shear may cause the simulation of a spurious effect. For instance observations in a rotating cylinder instrument may simulate non-Newtonian viscosity due to the rotation of the principal axes of deformation. However, for the same reason, this method is especially suited for the detection of second order effects.

An example for type III is the free fall of a small heavy sphere governed by Stokes' law.

In instruments of types I and II the method of testing may be either "static" under conditions of equilibrium, or "dynamic" with either loads or deformations alternating in a given manner. When the measured quantities are plotted in graphs which represent their mutual dependence, the result is a *technical test curve*. This does not give direct information about the rheological properties. To obtain this, a *rheological test curve* must be derived in which "consistency variables" are plotted. By comparing the general form of such a curve with curves obtained from models in imagined experiments, the appropriate rheological equation for the material can be found.

The application of this method may be illustrated by the example of concrete. When a concrete or reinforced slab is loaded, a maximum deflection δ is produced[3]. A certain part of the deflection occurs instantaneously (δ_i). When the load is removed immediately after loading[4] part of the deflection ($\delta_{e,i}$) is *recovered* instantaneously; this is due to the stresses induced by the instantaneous strain. A rest (δ_s) is not recovered, so that

$$\delta_i = \delta_{e,i} + \delta_s. \tag{44.1}$$

On the other hand, if we leave the slab under load for some time T, δ increases from δ_i to δ_T, and if we now remove the load we shall find, besides an instantaneous recovery of the same magnitude as before, that after a day or two still more

---

[1] F. Trouton: Proc. Roy. Soc. Lond., Ser. A **77**, 326 (1906).

[2] M. D. Hersey: J. Rheology **3**, 196 (1932). — M. Reiner and G. W. Scott Blair: J. Appl. Phys. **21**, 1195 (1950); **22**, 236 (1951).

[3] A. Arnan, M. Reiner and M. Teinowitz: Loading Test Reinf. Concr. Struct. Res. Counc. Israel, Jerusalem, 1950.

[4] Some time must elapse for reading observations, unloading, etc.

is recovered and that this process goes on for days and weeks up to a maximum of additional recovery δ_d. This is an indication of *delayed elasticity*. At the same time, the non-recovered permanent deformation has also increased by, say, δ_v. If we double the time of loading we shall find that δ_v has increased to roughly $2\delta_v$ and δ_d to $n\delta_d$, where $1 < n < 2$. If we leave the load on for "infinite" time, δ_d will not increase beyond a certain maximum $\delta_{e,d}$ which, therefore, is called the *delayed elastic strain*. As δ_v increases both with time and with the load roughly in direct proportion, as in liquids, this is a case of *viscous deformation*[1].

We can group the deformation just described in three different manners:

Firstly, we can distinguish the recoverable deflections $\delta_{e,i}$ and $\delta_{e,d}$ from the non-recoverable deflection δ_s and δ_v. The first two indicate *elastic strains*, the other two *permanent deformations*. Secondly, we can distinguish the instantaneous deflections $\delta_{e,i}$ and δ_s from the time-dependent deflections $\delta_{e,d}$ and δ_v. As the latter usually proceed with low speed, they indicate *creep*. They can be separated on unloading only, and are then distinguished as "primary" or recoverable creep which is a strain, and "secondary" or irrecoverable creep which is a deformation resulting from slow viscous flow.

As said before, δ_s does not appear in a second loading.

Disregarding δ_s, we connect in a third manner of grouping $\delta_{e,i}$ with δ_v, leaving $\delta_{e,d}$ separate.

It can then be seen that when $\delta_{e,i}$ and δ_v are grouped together, the rheological behaviour can be represented by an M-body. On the other hand the $\delta_{e,d}$ can be represented by a K-body. When all deflections are considered together, the M and K bodies must be coupled in series resulting in a Bu-body. We thus arrive at the result that the Bu-body is suitable for the representation of the rheological behaviour of concrete[2], but that in addition, a permanent set δ_s must be taken into account.

45. Simple pull and push. Tension and compression tests are the most common tests because they are the simplest tests to carry out, even though the material tested is in practice rarely subjected to simple tension or compression.

The tension test is most commonly used with both ferrous and non-ferrous metals and plastics. The materials most often tested under compression are those having a low tensile strength compared to the compressive strength, and thus are the materials that are usually employed to resist compressive loadings. Tension or compression tests are not only made to determine the properties of the material but are also used for tests of the manufactured product. For example performance tests of wires, rods, tubing, fabrics and fibres are made in tension, while tests for tile, masonry blocks, brick, cast iron and concrete are made in compression.

When the rheological properties of a material are determined by means of the extension of a prismatic specimen, its length l_0 is gradually increased by Δl, and this causes its cross-sectional area A_0 gradually to decrease to A. Therefore, if the load P is kept constant, the tensile traction σ

$$\sigma = \frac{P}{A} \tag{45.1}$$

---

[1] When first discovered in concrete the phenomenon was called "plastic" deformation, but this rheological behaviour described here on the example of concrete is typical for all materials which are *not plastic*. Concrete, while it can be moulded and is therefore "a plastic", is not a plastic substance. When an attempt is made to force upon it a high rate of deformation it breaks in a brittle manner. Mild steel which is a plastic substance can be deformed by impact, but concret cannot.

[2] M. REINER: Chap. I in REINER (Editor) 1954.

gradually increases. The traction is very often referred to the "original" cross-sectional area A_0. For instance in the standard-test of mild steel in the so called "stress-strain diagram", the "nominal stress" is the load divided by the original area

$$\sigma_0 = \frac{P}{A_0}. \qquad (45.2)$$

Keeping the load P constant, and thereby varying the traction during the experiment would seem to complicate the theoretical evaluation. Different devices for keeping the traction constant by gradually reducing the load have therefore been developed by ANDRADE[1], ANDRADE and CHALMERS[2] and SCOTT BLAIR and collaborators[3]. However, keeping the load constant has its advantages. Not only is the apparatus considerably simplified, but at the same time there is the advantage that one *single experiment* fulfills the task of a series of experiments with varying σ; and diversion from simple Hookian or simple Newtonian behaviour can be detected in *one* experiment.

For these reasons LETHERSICH[4] and REINER[5] have used constant *load* extensometers in their rheological investigations of bitumen. When recording his experiments, LETHERSICH plotted the percentage longitudinal deformation

$$D_l^C = \frac{\Delta l}{l_0} \qquad (45.3)$$

against time, the slope of the curve being a measure of the rate of extension $\dot D$. Because of the gradually increasing traction, $\dot D$ also gradually increases, and even when the material is a simple Newtonian liquid with constant viscosity η, the D_l^C versus t curve is not a straight line, but convex towards the t axis. The theoretical evaluation of such a curve is complicated. It was therefore fortunate that LETHERSICH[6] later discovered that when the longitudinal deformation is plotted in a measure proposed by SWAINGER[7], namely related to the extended length

$$l = l_0 + \Delta l \qquad (45.4)$$

and measured by

$$D_l^S = \frac{\Delta l}{l} = \frac{l - l_0}{l} = 1 - \frac{l_0}{l} \qquad (45.5)$$

the curve is a straight line for a simple Newtonian liquid. This relation can be derived from the law of extensional flow of a simple Newtonian liquid as follows:

During the extension of the prismatic specimen one end is fixed and the other travels with the velocity

$$v_l = \frac{dl}{dt} = \dot l. \qquad (45.6)$$

The extensional flow is

$$f_l = \frac{v_l}{l} = \frac{\dot l}{l} = \frac{d}{dt}(\log l). \qquad (45.7)[8]$$

For a Newtonian liquid this is related to the tensile traction σ by

$$\sigma = \lambda f_l, \qquad (45.8)$$

---

[1] E. N. DA C. ANDRADE: Proc. Roy. Soc. Lond., Ser. A **84**, 1 (1911).
[2] E. N. DA C. ANDRADE and B. CHALMERS: Proc. Roy. Soc. Lond., Ser. A **138**, 348 (1932).
[3] G. W. SCOTT BLAIR: Survey of General and Applied Rheology, London 1949, p. 159.
[4] W. LETHERSICH: Brit. Electr. Applied Ind. Res. Assoc. Report A 1—83, 1941.
[5] M. REINER: Research on Rheological Properties of Bitumen. Pal. Board Scient. Ind. Res. 1948.
[6] W. LETHERSICH: J. Sci. Instrum. **21**, 27, 111 (1944).
[7] Compare Eq. (29.5).
[8] Compare Eq. (25.5).

where λ is Trouton's coefficient of viscous traction. Introducing (45.7) into (45.8) gives

$$\sigma = \lambda \frac{\dot{l}}{l}. \tag{45.9}$$

As said before, if the load P is constant, the traction σ varies during the experiment with varying length l. Considering that the magnitude of σ depends upon the cross-sectional area, and the latter is connected with l through the volume V of the specimen, the variation of σ with l can be calculated as follows: We may assume that the volume of the deformed material is constant[1] or

$$V = A\,l = A_0\,l_0. \tag{45.10}$$

This gives

$$\sigma = P/A = P\,l/A_0\,l_0 = \sigma_0\,l/l_0. \tag{45.11}$$

Eq. (45.9) now yields

$$\sigma_0 = \lambda \frac{l_0}{l^2} \dot{l}. \tag{45.12}$$

But from (45.5)

$$\frac{l_0}{l^2} \dot{l} = D_i^S, \tag{45.13}$$

and therefore by integration

$$D_i^S/\sigma_0 = t/\lambda. \tag{45.14}$$

If the material is a *simple* Newtonian liquid, i.e. λ is constant and therefore independent of σ_0, Eq. (45.14) represents a straight line passing through the origin with $1/\lambda$ as slope. This confirms Lethersich's method.

Lethersich applied the same method upon the large extension of a *simple Hookian* elastic solid, i.e. one in which the extensional strain e_i^C is defined in the usual (Cauchy) way, namely by

$$e_i^C = \frac{\Delta l}{l_0} = \frac{l}{l_0} - 1. \tag{45.15}$$

This is analogous to (45.3) but now Δl is the *recoverable* elongation. Let E be Young's modulus, then in analogy with (45.8)

$$\sigma = E\,e_i^C. \tag{45.16}$$

Now, from (45.11) and (45.15)

$$\sigma_0 = E\,D_i^S, \tag{45.17}$$

where D_i^S is defined by (45.5). If $\sigma_0 = E$, (45.17) yields $D_i^S = 1$ and $l = \infty$. Equilibrium therefore is possible only if the initial traction σ_0 does not exceed the Young-modulus E. Introducing

$$\sigma_0 = a\,E \tag{45.18}$$

the factor a cannot be larger than unity.

Eq. (45.17) can also be written in the form

$$D_i^S/\sigma_0 = 1/E. \tag{45.19}$$

Therefore, from (45.14) and (45.19), if we plot a series of constant load experiments D_i^S/σ_0, i.e. the Swainger extension per unit "nominal" traction, σ_0, against time, t, and all observations are represented by *one single curve*, this proves (i) in the case of a viscous liquid that it is simple Newtonian, (ii) in the case of an elastic solid that it is simple Hookian.

It should be kept in mind that D_i^S has in our case a formal meaning only resulting from (45.11), and is of no physical significance. Lethersich regarded the method as the application of a "correction".

---

[1] Lethersich has taken into account that there may be a change of volume.

For the Kelvin body, we start from Eq. (15.2), which we write as

$$\sigma = E(e_i^C + T_{\text{ret}} \dot{f}_i), \qquad (45.20)$$

where

$$T_{\text{ret}} = \lambda_j / E \qquad (45.21)$$

is the time of retardation.

We now take account of (45.11), (45.7) and (45.18) and find from (45.20)

$$dt = T \frac{l_0}{al - e l_0} \cdot \frac{dl}{l}, \qquad (45.22)$$

where we have dropped indices for simpler writing.

If the elasticity is "simple" Hookian, Eq. (45.22) becomes

$$dt = T \, de/[e^2(a-1) + e(2a-1) + a]. \qquad (45.23)$$

Integration yields

$$e/a = (e^{t/T} - 1)/[a + (1-a) e^{t/T}]. \qquad (45.24)$$

Introducing e from (45.15), Eq. (45.24) becomes

$$\left(1 - \frac{l_0}{l}\right)\sigma_0 = (1 - e^{t/T})/E. \qquad (45.25)$$

We may write for $1 - l_0/l$ in accordance with (45.5) the symbol D_i^S. It should, however, be kept in mind that this is not the strain of the material, the latter being measured by e^C.

We find

$$D_i^S/\sigma_0 = (1 - e^{t/T})/E. \qquad (45.26)$$

Therefore *if we plot observations in the Swainger measure per unit nominal traction* (σ_0) *against time*, all observations lie on *one single exponential curve* provided the elasticity of the K-body is simple Hookian. This was confirmed by BRAUN, SCHOENFELD-REINER and TRAUM[1] for an "oxidized bitumen, 20 to 30 pen"[2], which showed elasticity to an extraordinary degree. Its rheological coefficients were found to be: YOUNG's modulus of elastic phase $E = 1.331 \times 10^6$ dyne cm^{-2}, time of retardation $T = 372$ sec, TROUTON's coefficient of viscous traction of liquid phase $\lambda = 4.96 \times 10^8$ poises.

46. Technological properties. As said in Sect. 8, we can only briefly mention technological properties for some of which instrumental methods of measurements have been devised, while the investigation of others has not reached such stage. In this latter case the craftsman assesses the property by feel and sight. This is where psycho-physiological considerations enter rheology, as mentioned in Sect. 1. With the one exception of thixotropy, we need not give here more than a glossary, so that the physicist who comes across the term in the literature will not get lost. He will find more information in SCOTT BLAIR (1949) which contains extensive references.

Dealing with *thixotropy*, this term was introduced by FREUNDLICH and coworkers[3] to denote an isothermal reversible gel-sol transformation brought about by mechanical "working". It has later been more widely defined as "that property of a body by virtue of which the ratio of shear stress to rate of deformation is temporarily reduced by previous deformation". The time factor is essential, distinguishing thixotropy from structural viscosity. GREEN (1949) has devised an instrument by which thixotropy is measured and expressed in a quantitative manner.

---

[1] I. BRAUN, R. SCHOENFELD-REINER and E. TRAUM: Bull. Res. Council Israel **2**, 89—98 (1952).
[2] "pen" is short for "penetration", explained in the following section.
[3] H. FREUNDLICH: Thixotropy. Paris: Hermann & Co. 1935.

L'HERMITE[1] has coined the term anti-thixotropy to denote an *increase* in the above mentioned ratio. Thixotropy and anti-thixotropy should be distinguished from structural viscosity. Both structural viscosity and thixotropy denote a reversible fall in viscosity induced by a rise in shearing stress, with recovery of the viscosity in a state of rest. However, while in the case of structural viscosity the recovery takes place instantaneously, without any time-lag, in the case of thixotropy the recovery requires some time which often is considerable. Anti-thixotropy is distinguished from thixotropy by the fact that the rise in shearing stress causes a rise in viscosity. From structural viscosity it is distinguished in that, as in the case of thixotropy, the recovery is not instantaneous. While many cases of thixotropy are known, anti-thixotropy seems to be a comparatively rare phenomenon.

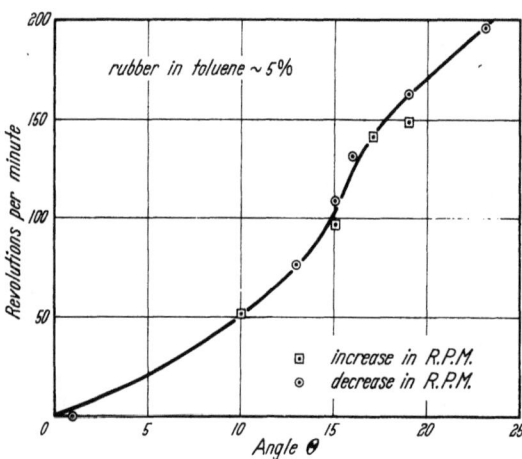

Fig. 42. Ostwald flow-curve of a rubber-toluene solution.

A striking example of anti-thixotropy was discovered by KATCHALSKY[2], and collaborators a 5% solution of polymetacrylic acid in water showing a 350 fold increase in viscosity after subjection for one minute to a rate of shear of 10 sec^{-1}.

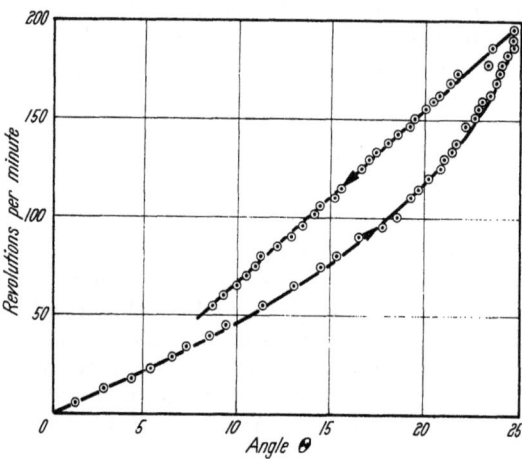

Fig. 43. Thixotropie flow curve of a soft bitumen.

The difference between the three types of variable viscosity can be shown by means of curves of the following substances[3], namely a 1.5% natural rubber solution in toluene which shows structural viscosity (Fig. 42), a very soft bitumen exhibiting thixotropic breakdown (Fig. 43), and a 5% P.I.B. solution in tetralene (Fig. 44). In the figures the rotational velocity in a coaxial cylinder instrument is plotted against the torque of the wire on which the internal cylinder is suspended (measured in degrees of the torsional angle θ) and which is a measure of the shearing stress.

It can be seen that the anti-thixotropy loop is not apparent at the second test while it shows again, although a little narrower, after a 24 hour rest. It appears that the effect of shearing vanishes when the solution is left undisturbed.

So much for thixotropy. Going over to a glossary as mentioned before, here are explanations for some of the more often used terms.

---

[1] L'HERMITE: Ann. Institut Technique du Bâtiment et des Travaux Publics, No. 92, 1949.

[2] J. ELIASSAF, A. KATCHALSKY and A. SILBERBERG: Nature **176**, 1119 (1955).

[3] J. CRANE and D. SCHIFFER: Int. Symp. Macromol. Chem. Rehovoth. 1956.

Roughness and *smoothness* are tactual perceptions experienced in movement over a surface. In the absence of movement the experience is labelled *eveness* and *uneveness*. In the rubber-industry *rugosity* has been introduced as a measurable property for roughness. However, the same term is also used to describe the reciprocal of the sphericity of sand, i.e. the "ruggedness" of its particles. In the latter case alternatively a *coefficient of angularity* is used.

When a portion of a certain clay is held in the fingers of both hands, and is alternatively pulled and compressed, there is a range of easy movement culminating in a sudden resistance. This property has been called *back-lash*.

When certain materials, notable sulphur melts and gum arabic, are made to flow in a capillary, and the pressure is sharply released, the material apparently flows back along the capillary tube. This has been named *flow-elasticity*. There is probably a close connection between flow-elasticity and the capacity of some materials to work-harden into long elastic strands or fibres, which has been named in German *Spinnbarkeit*, a term for which in English the word *leptogenesis* has been proposed. Researches in Spinnbarkeit led to the invention of nylon.

Micromeritics is concerned with the rheology of fine particles.

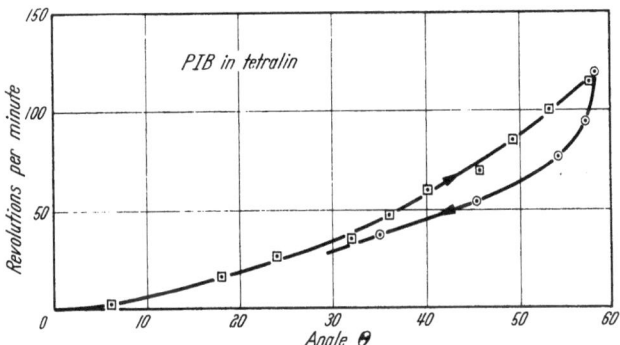

Fig. 44. Autothixotropic flow-curve of a poly-iso-butylene solution.

Redress is a recovery of deformation which takes place under the influence of external stimuli; thus "thermo-redress" is used when heat is the activation.

Rheodestruction is an irreversible mechanical breaking up of gels in flow processes.

Rheopexy is the phenomenon by which the setting of a thixotropic gel is accelerated by gentle streaming.

Stickiness is defined for soils as the force which is required to pull away a plate which has been pressed against the soil. This "property" is not much different from what is called *adhesivity* or *tack*. The latter is important in the printing industry and has been defined as "the surface property of a plastic solid by virtue of which two pieces will stick together when brought in contact and will thereafter resist separation". Tackiness has been dealt with by GREEN[1]. There exists a simple machine called the tackmeter for measuring it in printing inks.

Workhardening[2] is mostly used to describe the rise of the yield stress due to excessive plastic deformation, but also includes a rise of viscosity. Some authors have mentioned *strain-softening*.

Structural turbulence appears in liquids, possessing structural viscosity, well below the critical velocity of REYNOLDS' turbulence. REINER[3] has argued that it might be due to a breaking up of the internal structure to which its variable viscosity is due, when the shearing stress exceeds a certain limit.

Penetration and *ductility* are two terms used in bitumen technology. Penetration is determined from the depth to which a weighted needle sinks into a

---
[1] H. GREEN: Industr. Engng. Chem. **13**, 632—639 (1941).
[2] Also named strain-hardening.
[3] M. REINER: Kolloid-Z. **39**, 314 (1926).

dish of bitumen.—Ductility is measured by an "extension" type of test. A briquette made of bitumen is extended at a specified speed until the thread connecting the two ends breaks. The length at which the thread breaks is designed as its ductility.

As an *envoi* to this chapter we may mention that while *solid* and *liquid* are important technological distinctions, they have no place in rheology. While the term "rheology" brings to the mind HERACLITOS' παντα ῥει we may with as much justification say that "everything is solid". The criterion for the distinction between solid and liquid is not qualitative but quantitative: it has been stated in Sect. 16. In the present chapter we have treated concrete with a time of relaxation of the order of 10^6 sec as a fluid, and air with a time of relaxation of the order of 10^{-10} sec as a solid. We can even say that when concrete behaves as a solid and air as a fluid, they are of no interest to the rheologist.

E. Addenda.

To Sects 1 and 3. In soil-mechanics, the soil is usually considered as a continuous medium with the displacements continuous functions of the coordinates, and phenomenological macrorheology as treated here is applied. Such treatment is inadequate for the solution of certain problems encountered in dynamic geology and mining practice, when the earth crust has the character of a macrorubble composed of a large number of elements. LITWINISZYN[1] has applied the equations of stochastic processes to such problems.

To Sect 11. The experimental basis of the Newtonian liquid is formed nearly exclusively by POISEUILLE's experiments leading to an equation of the form (14.12, 1), derived before theoretically by HAGEN from (11.1). POISEUILLE was interested in the flow of blood through the arteries, but found the equation when experimenting with water. Blood is a suspension showing "structural viscosity". In addition, JEFFORD and KNISELY[2] have establishes that "segments of arteries are actually portions of truncated cones rather than cylinders". The flow of blood through the blood vessels has not yet found adequate treatment.

To Sect. 13. SCOTT BLAIR (1949, p. 187) uses as a model for the St. Venant body a piston moving in a tube filled with a clay-paste which has a yield value.

Another model has been proposed by PAEZ[3] for concrete. In order to represent a pronounced curvature in the stress-straindiagram, and also a residual deformation when the load has been removed, with a limiting value of stress equal to the maximum failure load, he proposes a large number of pistons of the Scott-Blair type connected by springs. He then substitutes for the pistons springs which transmit their load to the enclosing cylinder wall by friction. By connecting the cylinder with a reservoir of viscous liquid so that liquid is drawn in or expelled when the piston moves up or down, he is able to represent deformations due to shrinkage and swelling, and also those due to absorption or loss of water during curing etc.

OLSZAK and LITWINISZYN[4] have replaced the tube by a vessel of varying width in order to represent non-linear flow.

To Sect. 14. A biography of SCHWEDOFF in which priority for Eq. (14.24) is claimed for his pupil WEINBERG is contained in M. P. WOLAZOWITCH and S. M. LEVI: Kolloidnij Ž. **18**, 129—134 (1956).

---

[1] J. LITWINISZYN: Nadbitka z Arch. Mech. Stos. **8**, 393—411 (1956) (in English).
[2] J. V. JEFFORD and M. H. KNISELY: Angiology **7**, 105—136 (1956).
[3] E. TORROJA and A. PAEZ: Set and reinforced concrete. Chap. VIII in REINER 1954.
[4] W. OLSZAK and J. LITWINISZYN: Nadbitka z Arch. Mech. Stos. **5**, 557—583 (1953).

The rotating co-axial cylinder instrument is named after Couette. It is described by Hatschek[1], who designed an improved type.

Using an instrument of this kind Papadakis[2] has found that concentrated cement-suspensions can be considered as Bingham materials.

Applying the Mises flow condition (36.6) Paslay and Slibar[3] write the second of Eqs. (14.3) in the form $s\left(1 - \dfrac{\vartheta_t}{\sqrt{-II_{s(0)}}}\right) = 2\eta_{pl}\dot{d}$, and solve the case of a Bingham-material in a plane thrust bearing.

To Sect. 16. Broer[4] has solved the equations for two cases of periodic laminar flow in a cylindrical tube and, approximately, for flow in a film of variable thickness.

To Sect. 18. The difference in behaviour of an elastic sol, such as the L-body, and a relaxing gel, such as the J-body, in simple shear, has been discussed by Reiner[5]. In a sol, the principal axis i of the flow tensor is inclined at 45° towards the direction of flow, and so is the stress, being produced by the flow, and co-axial with it. In a gel, the material is sheared to a certain extent, with the principal axes of strain at an angle *smaller* than 45°. Only when such elastic strain is reached, does flow arise through relaxation. The stress is in this case co-axial with the strain and not with the flow tensor. This simulates a "decrease" of viscosity, even when the viscosity of the liquid phase is constant.

To Sect. 25. The assertion that only the Hencky-measure is suitable for the description of a finite deformation due to viscous or plastic flow is important enough to warrant support by an additional argument. Consider the plastic elongation of a bar of "original" length l_0 *in stages*. If we elongate the bar first by a small increment Δl, the extension will be $\Delta l/l_0$. Now elongate the extended bar of length $l_0 + \Delta l$ by another increment Δl. The question then arises how to express its extension. In accordance with the Cauchy measure, if the second elongation follows the first immediately, so that both can be considered as forming one single operation upon the original length l_0, the second extension will again be $\Delta l/l_0$ and the total $2\Delta l/l_0$.

But if the second extension was carried out after some time as an entirely new operation, the original length l_0 not being known, the second extension would be $\Delta l/(l_0 + \Delta l)$ and the total $\Delta l[1/l_0 + 1/(l_0 + \Delta l)]$, which is different from $2\Delta l/l_0$.

Now let us consider the progressive elongation in the light of the Hencky measure. We have $l_1 = l_0 + \Delta l$, $l_2 = l_1 + \Delta l = l_0 + 2\Delta l$ and the final extension is

$$e^H = \ln(l_2/l_0). \tag{1}$$

The expression on the right side can, however, be developed as follows:

$$\ln(l_2/l_0) = \ln\left(\frac{l_2}{l_1} \cdot \frac{l_1}{l_0}\right) = \ln(l_2/l_1) + \ln(l_1/l_0). \tag{2}$$

Therefore, if a bar of length l_0 is extended by Δl in a first operation and by another Δl in a second operation, the result is the same as if it were extended in one single operation by $2\Delta l$, provided extensions are defined as by Hencky, but no so, if defined as by Cauchy. Mathematically, this is expressed by the statement that Hencky-extensions form a group, while Cauchy extensions do not form a group.

---

[1] E. Hatschek: The viscosity of liquids. London 1928.
[2] M. Papadakis: Rec. des Matériaux de construction. Mai 1955.
[3] P. R. Paslay and A. Slibar: Öst. Ing. Arch. **10**, 328 (1956).
[4] L. J. F. Broer: Appl. Sci. Res. A **6**, 226—236 (1956).
[5] M. Reiner: Bull. Res. Counc. Israel **1**, 5—25 (1951).

To Sect. 26. For non-linear behaviour of Newtonian liquids the term *pseudo-plastics* is also in use. It originated at a time when, after BINGHAM's success in explaining "viscosity anomalies" by plasticity, it was thought that this was the correct explanation in every case when a supposed liquid showed variability of an "apparent" coefficient of viscosity. KRAKAUER[1] has described a method which by application of Eq. (14.24) determines whether flow curves obtained in a co-axial rotating cylinder instrument indicate structural viscosity of a non-linear Newtonian liquid or a Bingham type of flow.

To Sect. 28. DE WAELE and, independently, OSTWALD proposed a "power-law" $\dot{\gamma} = k\tau^n$ to describe structural viscosity. This must be regarded rather as an interpolation formula of sometimes great practical usefulness, but SCHEELE and TIMM[2] have claimed that the number n is of physical significance.

SWEENY and GECKLER[3] have examined concentrated suspensions of glass spheres in an equal density medium with the roational viscometer, and found confirmation of an equation tentatively proposed by REINER, namely

$$\varphi = \frac{\varphi_0 + \varphi_\infty \tau^2/c^2}{1 + \tau^2/c^2}$$ rather than Eq. (28.13).

SCHULTZ-GRUNOW[4] has proposed to replace in non-Newtonian liquids Eq. (28.1) $\dot{\gamma} = f(\tau)$ by $\frac{\dot{\gamma}}{C} = f\left(\frac{\tau}{A}\right)$, where A and C are constants of the dimensions of τ and $\dot{\gamma}$. He found for certain liquids agreement with an equation $\frac{\dot{\gamma}}{C} = \sinh \frac{\tau}{A}$ first proposed by PRANDTL[5].

PAWLOWSKI[6] has based structural viscosity on thermodynamical considerations. In another paper[7] he has discussed the meaning of "viscosity" in non-Newtonian liquids, and maintains that the form proposed by SCHULTZ-GRUNOW is not suitable as a general rheological equation.

To Sect. 29. The developments of this section, leading to a general expression for HENCKY's measure of deformation—and by the same method also suitable for CAUCHY's, SWAINGER's and others—are rather involved. If expressions for laminar shear only, in Cartesian or cylindrical coordinates, are desired, two simpler methods are available, one indicated by LOVE, Art. 37 and 49, the other based on MOHR's circle explained in Sect. 7, δ. An exposition of them follows here:

(i) Consider the simple shear of Sect. 3 given by

$$x_1 = x_0 + S y, \quad y_1 = y_0, \quad z_1 = z_0. \tag{1}$$

LOVE (l.c. Art. 37) calculates from purely geometrical considerations

$$\lambda_{(i)} = (1 - \sin)/\cos, \quad \lambda_{(j)} = (1 + \sin)/\cos, \quad \lambda_{(k)} = 1, \tag{2}$$

and proves that the directions of the principal axes of strain are the bisectors of the angle $\pi/2 + \alpha$ with the x-axis, and the angle through which the principal axes are turned is the angle α. The stress caused by strain will have the principal components from (29.1)

$$s(i) = \lambda I_e + 2\mu e(i). \tag{3}$$

---

[1] V. O. KRAKAUER: J. Appl. Phys. **21**, 850—852 (1950).
[2] W. SCHEELE and T. TIMM: Kolloid-Z. **120**, 103—119 (1951); **121**, 140—143 (1951).
[3] K. H. SWEENY and R. D. GECKLER: J. Appl. Phys. **25**, 1135—1145 (1954).
[4] F. SCHULTZ-GRUNOW: Kolloid-Z. **138**, 167 (1954).
[5] L. PRANDTL and FR. VANDEY: Z. angew. Math. Mech. **30**, 169 (1950).
[6] J. PAWLOWSKI: Kolloid-Z. **131**, 11 (1953); **138**, 6 (1954).
[7] J. PAWLOWSKI: Kolloid-Z. **143**, 92—97 (1955).

The components of stress with respect to the system x, y, z will be from Love's equations, Art. 49:

$$\begin{aligned}
s_{xx} &= \tfrac{1}{2} \cdot [s(i) + s(j)] - \tfrac{1}{2} \cdot [s(i) - s(j)] \sin \alpha \\
s_{yy} &= \tfrac{1}{2} \cdot [s(i) + s(j)] + \tfrac{1}{2} \cdot [s(i) - s(j)] \sin \alpha \\
s_{zz} &= s(k) \\
s_{xy} &= \tfrac{1}{2} \cdot [s(i) - s(j)] \cos \alpha, \quad s_{yz} = s_{zx} = 0.
\end{aligned} \quad (4)$$

Introducing the expressions for the principal stresses from (3) into (4) gives

$$\begin{aligned}
s_{xx} &= \lambda I_e + \mu \{e(i) + e(j) - [e(i) - e(j)] \sin \alpha\} \\
s_{yy} &= \lambda I_e + \mu \{e(i) + e(j) + [e(i) - e(j)] \sin \alpha\} \\
s_{xy} &= -\mu [e(i) - e(j)] \cos \alpha, \quad s_{yz} = s_{zx} = 0.
\end{aligned} \quad (5)$$

Now from (1), $e(k) = 0$, and therefore $s_{zz} = \lambda_L I_e$. We then introduce the five different measures of (29.4) to (29.8), and find the stress components as entered in the following table[1] [expressing the goniometrical functions in terms of S of Eq. (1)].

Comparison of Eq. (29.22) and (29.30) shows agreement. In addition the table also shows the cross-effects for the Cauchy and Swainger measures.

Table 8. *Stresses in finite simple shear.*

stress	$e^C = \lambda - 1$	$e^G = \dfrac{\lambda^2 - 1}{2}$	$e^H = \ln \lambda$	$e^A = \dfrac{1 - 1/\lambda^2}{2}$	$e^S = 1 - 1/\lambda$
$(s_{xx} - s_{zz})/2\mu$	$\dfrac{2 + S^2}{CS} - 1$	$\dfrac{S^2}{2}$	$\dfrac{1}{2C} \ln \dfrac{C+1}{C-1}$	0	$1 - \dfrac{2}{CS}$
$(s_{yy} - s_{zz})/2\mu$	$\dfrac{2}{CS} - 1$	0	$-\dfrac{1}{2C} \ln \dfrac{C+1}{C-1}$	$-\dfrac{S^2}{2}$	$1 - \dfrac{2 + S^2}{CS}$
$s_{xy}/2\mu$	$\dfrac{1}{C}$	$\dfrac{S}{2}$	$\dfrac{1}{CS} \ln \dfrac{C+1}{C-1}$	$\dfrac{S}{2}$	$\dfrac{1}{C}$
$(s_m - s_{zz})/2\mu$	$\dfrac{CS - 2}{3}$	$\dfrac{S^2}{6}$	0	$-\dfrac{S^2}{6}$	$\dfrac{2 - CS}{3}$

$C = \sqrt{1 + 4/S^2}$

(ii) Let a sheet consisting of an elastic material, say, of rubber, which can suffer large deformations be fixed between two parallel runners as shown in Fig. 45. Let a circle of unit radius bet drawn on the sheet. Let one runner be fixed, while the other is displaced parallel to the first at constant distance H. In this case of simple homogeneous shear the circle is deformed into an ellipse. Two stages of such deformation are shown. The construction of the ellipse is a purely geometrical problem which has nothing to do with rheology. It is indicated in Stage 2. Two radii of the circle normal to each other in the directions $\bar{i}$ and $\bar{j}$ are extended and concentrated, respectively into the half-axes l_i and l_j of the ellipse in the directions i and j, which are also normal to each other. These are the principal directions of the deformation. Note that the axes $\bar{i}$ and $\bar{j}$ are rotated by the angle $\alpha = \arctan(\gamma/2)$ into the positions i and j. This is the known rotation connected with simple shear. The Mohr-circles for Stages 2 and 3 are also shown. Note that the abscissa axis of the Mohr circle, indicated by (i) must be

---

[1] M. REINER: Proc. Intern. Congr. Theoret. Appl. Mech., Istanbul 1954.

parallel to the direction i. In order to construct the circle we must know the principal extensions D_i and D_j. Assuming the Hencky measure we have

$$D_i^H = \ln l_i. \tag{6}^1$$

This can be determined by measuring l_i from the figure. As the area of the circle is not changed

$$D_a^H = D_i^H + D_j^H = 0 = D_{xx}^H + D_{yy}^H = I_D^H \tag{7}$$

from which

$$D_j^H = -D_i^H. \tag{8}$$

From the Mohr circle we determine D_{xy}^H and $D_{xx}^H = -D_{yy}^H$. These are entered in the figure as graphs of D versus γ. These graphs represent the formulae in the centre column of the table.

To Sect. 32. PADDEN and DEWITT[2] have made measurements on concentrated polyisobutylene solutions, and have interpreted them on the basis of a rheological equation due to DEWITT which includes the vorticity ω_{lm} of fluid motion.

GIESEKUS[3] has generalized Eq. (32.2) by considering in addition to the symmetrical flow tensor f_{lm}, the anti-symmetrical tensor ω_{lm} related to a coordinate system connected with the principal axes of f_{lm}. In other words, ω_{lm} defines the rotation of the principal axes of f_{lm} in a convective system (compare Sect. 7ε above). The result is a rheological equation as follows (in simplified writing):

$$s = \eta_0 \delta + 2\eta_1 f + 4\eta_c f^2 + 2\zeta_1(\omega f - f\omega) + 4\zeta_2(\omega f^2 - f^2\omega) +$$
$$+ 8\zeta_3 f(\omega f - f\omega) + 4\xi_0 \omega^2 + 4\xi_1(\omega^2 f + f\omega^2) + 8\xi_2(\omega^2 f^2 + f^2\omega^2) +$$
$$+ 8\psi_1 \omega(\omega f - f\omega)\omega + 16\psi_2 \omega(\omega f^2 - f^2\omega)\omega + 32\psi_3 f\omega(\omega f - f\omega)\omega f.$$

The twelve viscosity-parameters η_0 to ψ_3 will in general be functions of the six invariants of the tensors f and ω. The symmetrical tensors s and f are now not co-axial.

As can be seen from an inspection of Fig. 45, material particles which at a certain stage lie on a principal axis, and are therefore not sheared at this moment, will leave the axis at the next moment. This will influence the rheological behaviour of an elastico-viscous liquid, and/or one forming a dispersed system, but probably not of a homogeneous viscous liquid devoid of elasticity. The foregoing equation is equivalent with one derived by OLDROYD[4].

BRAUN[5] has derived the momentum-equations for the liquid defined by the rheological equation (32.2), which replace the Navier-Stokes equations (11.6).

To Sect. 33. SCHULTZ-GRUNOW[6] maintains that the opening in the stator of WEISSENBERG'S (Fig. 35) and similar instruments, leading to manometers, affects the flow and thus the pressure reading. He has shown this optically in a liquid streaming between co-axial rotating cylinders by making in the stator an indent of a width of about half the width of the gap between the cylinders. The pressure measurements would therefore not be significant.

To Sect. 36. NADAI[7] has interpreted $\sqrt{-II_{s(0)}}$ as an octahedral shearing stress and the Mises-criterion can accordingly be expressed in terms of a limiting octahedral stress.

To Sect. 37. TORDELLA[8] suggests that the strength criterion Eq. (37.7) applies in the case of the fracture of amorphous polymers when extruded through capillaries.

---

[1] But we can similarly use any other measure.
[2] F. J. PADDEN and T. W. DEWITT: J. Appl. Phys. **25**, 1086—1091 (1954).
[3] H. GIESEKUS: Kolloid-Z. **147**, 29—45 (1956).
[4] J. G. OLDROYD: l. c. p. 447, note 3.
[5] I. BRAUN: Rend. Circ. Mat. Palermo, (2) **2**, 1—8 (1954).
[6] F. SCHULTZ-GRUNOW: Z. VDI **9**, 409—416 (1955).
[7] A. NADAI: J. Appl. Phys. **8**, 205 (1937).
[8] J. P. TORDELLA: J. Appl. Phys. **27**, 454—458 (1956).

Addenda. 547

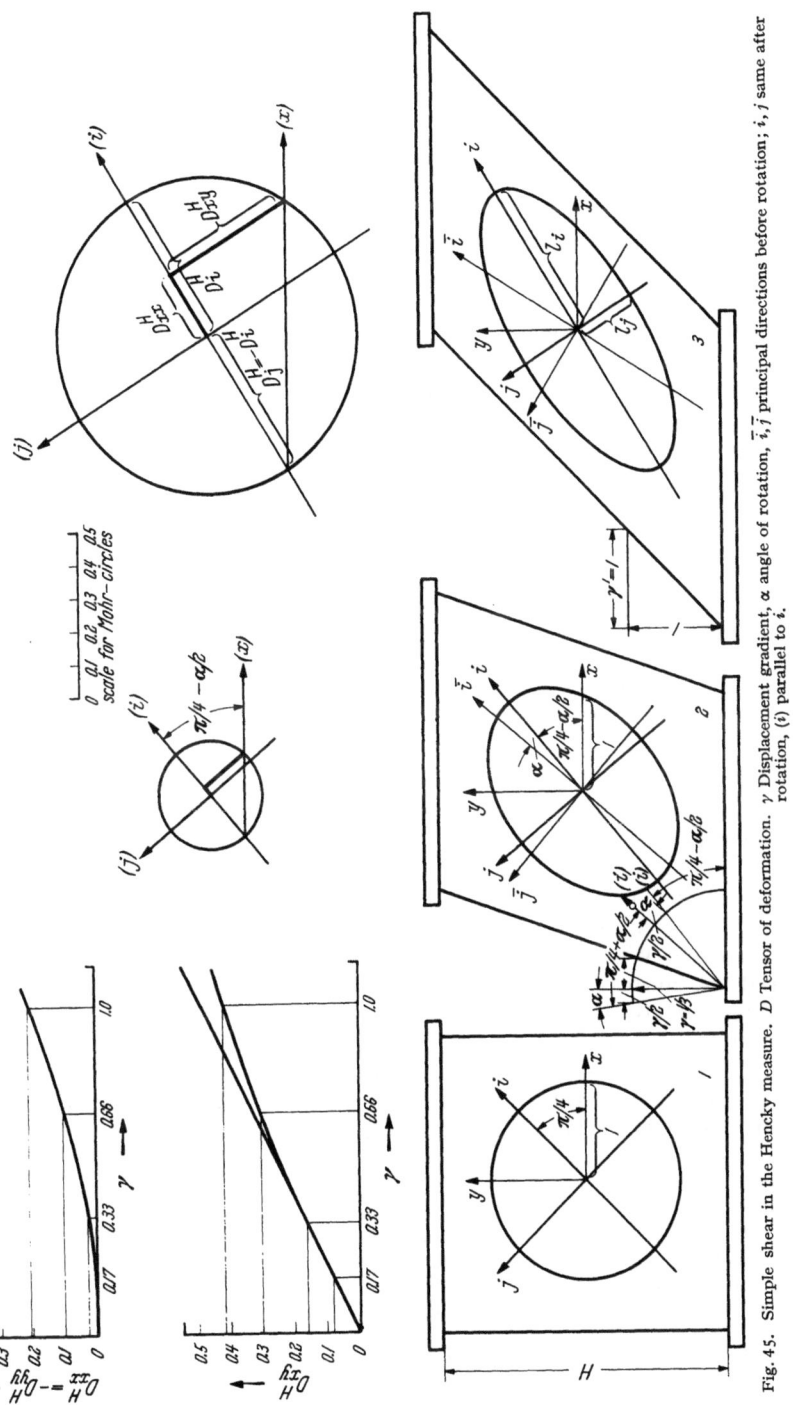

Fig. 45. Simple shear in the Hencky measure. D Tensor of deformation. γ Displacement gradient, α angle of rotation, $\bar{i}, \bar{j}$ principal directions before rotation; $\bar{\bar{i}}, \bar{\bar{j}}$ same after rotation, (i) parallel to $\bar{i}$.

To Sect. 37 β. BEN-ARIE[1] has determined the tensile strength of Napalm gels by extruding the gel vertically through a nozzle, and weighing the gel-column falling by its own weight. Assuming that the weight of the column

---
[1] M. BEN-ARIE: J. Polymer Sci. **17**, 179—190 (1955).

divided by the area of the nozzle gives the value of the tensile strength, he found that the latter depends linearly upon the velocity of extension.

To Sects. 39 and 40. LIFSON[1] has examined theoretically the flow of a solution through ultrafine capillaries, and found an increase of apparent viscosity as a result of the inhomogeneity of the flow. This increase is dependent on the ratio between the dimensions of the macromolecules and the capillary.

To Sect. 40. GIESEKUS[2] has taken into account Brownian movement. Starting from a theory similar to the one of RISEMAN and KIRKWOOD[3], he considers a suspension of elastic dumbbells. He shows that all twelve parameters defined in the same paper (compare addendum to Sect. 32) will in general be present, depending upon whether the dumbbells are rigid, hard-elastic or soft-elastic. A suspension of rigid dumbbells represents an elastic liquid when the elasticity due to the entropy decreases with the orientation of the particles.

To Sect. 43. RIGDEN[4] has examined the rheological behaviour of non-aqueous suspensions through rheological analysis using consistency curves. MYERS, MILLER and ZETTLEMOYER[5] applied the method to the study of dispersions of calcium carbonate and clays, and established the existence of a critical concentration for the occurence of structural viscosity.

To Sect. 44. NEVILLE[6] has reviewed the different theories on the creep in concrete, some of which express a different view.

Viscometers and plastometers are comprehensively treated by MESKAT[7].

A Couette-instrument with electrostatic restoring torque for viscosity measurements in the centipoise range is described by EISENBERG and FREI[8].

WELTMANN and KUHNS[9] have described a Couette-instrument which automatically records flow curves for all kinds of non-Newtonian materials.

WELTMANN[10] has calculated dimensionless parameters from consistency diagrams of non-Newtonian flowing materials (both plastics and liquids), and used them for the design of friction diagrams to determine the flow characteristics of such materials in piping systems.

To Sect. 46. THORNTON[11] has measured the time-dependence of the viscosity of the thixotropic materials in a co-axial rotating cylinder viscometer.

GREEN and WELTMANN[12] published extensive investigations on thixotropic materials by means of GREEN's apparatus. DAHLGREN[13] has treated their observations analytically.

BESTUL and BRYANT[14] have recently examined polyisobutulenes and related substances with results "consistent with REINER's theoretical analysis of the early turbulence mechanism of OSTWALD or with a variation of it involving the energetics of molecular disentanglement".

---

[1] S. LIFSON: J. Polymer Sci. **20**, 1 (1956).
[2] H. GIESEKUS: l. c. p. 546, note 3.
[3] J. RISEMAN and J. G. KIRKWOOD in EIRICH (Editor), Vol. I.
[4] P. J. RIGDEN: Road Research Technical Paper No. 28, London 1954.
[5] R. R. MYERS, J. C. MILLER and A. C. ZETTLEMOYER: J. Appl. Phys. **27**, 468—471 (1956).
[6] A. M. NEVILLE: J. Amer. Concr. Inst. **27**, 47—60 (1955).
[7] W. MESKAT: Messen und Regeln in der chemischen Technik, pp. 698—781. Berlin: Springer 1957.
[8] H. EISENBERG and E. H. FREI: Bull. Res. Counc. Israel **3** (1954). — J. Polymer Sci. **14**, 417—426 (1954).
[9] R. N. WELTMANN and P. W. KUHNS: Nat. Adv. Com. Aeronautics, Technical Note 3510, Washington 1955.
[10] R. N. WELTMANN: Nat. Adv. Com. Aeronautics, Technical Note 3397, Washington 1955.
[11] S. THORNTON: Proc. Phys. Soc. Lond. B **46**, 115—119 (1953).
[12] H. GREEN and R. N. WELTMANN: Industr. Engng. Chem., Anal. Ed. **15**, 201—206 (1943); **18**, 167—172 (1946).
[13] S. E. DAHLGREN: Trans. Chalmers Univ. **1955**, No. 159.
[14] A. B. BESTUL and C. B. BRYANT: J. Polymer Sci. **19**, 255 (1956).

Symbols.

A	Area.	u, u_l	Displacement.
a	Acceleration.	v	Index indicating "volumetric".
B, B_l	Body force.	v, v_l	Velocity.
C	Constant.	w	Specific work.
D, D_{lm}	Deformation.	$\dot{w}$	Specific power
d	Distance, also differential.	$\dot{w}_s$	Specific stress power
d_{lm}	Infinitesimal Cauchy deformation tensor.	${}_lx$	Initial coordinate.
		x_l	Final coordinate.
d_v	Cubical dilatation.	y	Coordinate.
$\vec{ds}$	Line element vector.	z	Coordinate.
E	Young's modulus.	α	Dummy index.
e, e_{lm}	Strain.	Γ_{lm}	Finite displacement gradient
$\dot{e}_k$	Kinetic energy per unit volume.	γ, γ_{lm}	Infinitesimal displacement gradient.
f, f_{lm}	Flow.	$\dot{\gamma}$	Velocity gradient.
f_v	Volume flow.	Δ	Difference, increment.
g	Acceleration of gravity.	δ_{lm}	Kronecker's delta.
g_{lm}	Metric tensor.	ε_{lm}	Infinitesimal strain tensor.
i, j, k	Principal directions.	ε_v	Volumetric strain.
l	Tensor index.	ζ	Volume viscosity.
M, M_z	Torque.	Θ	Angular coordinate.
m	Tensor index.	ϑ	Yield stress.
n	Indicating "normal".	η	Shear viscosity.
$\vec{n}$	Normal direction vector.	$\varkappa$	Bulk modulus.
(0)	Subscript to indicate deviator.	λ	Lamé's constant.
O	Oldroyd number.	λ_T	Trouton's coefficient.
P, P_l	Load.	μ	Shear modulus.
p	Pressure.	ν	Poisson's ratio.
p_m	Mean pressure.	ϱ	Density.
R	Radius.	Σ	The sum of.
r	Radius, also coordinate.	$\sigma_n = s_{nn}$	normal component of traction.
$\vec{s}$	Direction vector.	$\tau_n = s_{nt}$	Tangential component of traction.
s_{lm}	Stress tensor.	φ	Intrinsic free energy density.
$\vec{s}_n$	Traction.	ψ	Bound energy density.
T	Truesdell number.	ω_{lm}	Infinitesimal rotation tensor.
T_{rel}	Relaxation time.	$\cdot$	Material time derivative.
T_{ret}	Retardation time.	I	First invariant.
t	Time, also index indicating "tangential".	II	Second invariant.
		III	Third invariant.
$\vec{t}$	Tangential direction vector.	$\vert$	Coupled in parallel.
t_{lm}	Tensor.	$-$	Coupled in series.

Bibliography.

Books.

The classical books for Hookian elasticity and Newtonian hydrodynamics are still A. E. H. Love, Elasticity, Cambridge 1906—1927, and H. Lamb, Hydrodynamics, Cambridge 1879—1932.

Hill, R.: Plasticity, Oxford 1950, has attempted to achieve the same for "classical" plasticity, without, however, covering the field to such extent. In addition the following may therefore be consulted:

Nadai, A.: Plasticity. New York and London 1931.

Prager, W., and P. G. Hodge: Perfectly Elastic Solids. New York and London 1951.

An exhaustive survey of new developments in elasticity and hydrodynamics has been given by C. Truesdell, J. Rat. Mech. a. Analysis **1**, 125—300 (1952); **2**, 593—616 (1953).

Phenomenological rheology is treated in two textbooks by M. Reiner: "Twelve Lectures on Theoretical Rheology", North Holland Publ. Co., Amsterdam, 1949, and "Deformation and Flow", H. K. Lewis, London, 1949. Older literature was reviewed by v. Kármán in Art. 31 of vol. IV. 4 of Encyclopaedie der Math. Wiss. in 1913, under the title ,,Physikalische Grundlagen der Festigkeitslehre", before the term "rheology" was known.

Other basic books in the field of rheology are in alphabetical order.

Alfrey, T. jr.: Mechanical Behaviour of High Polymers, Interscience, New York 1948.

Committee for the study of viscosity. First report on viscosity and plasticity. Amsterdam 1935.

Ditto: Second report. Amsterdam 1938.

EIRICH, F. R. (Editor): Rheology, Theory and Applications. New York 1956. This work in three volumes is planned to give in 45 chapters by different authorities an exhaustive survey.

FREUDENTHAL, A. M.: The Inelastic Behavior of Engineering Materials and Structures. New York 1950.

GREEN, H.: Industrial Rheology and Rheological Structures. New York and London 1949. Mainly for thixotropy.

GROSS, B.: Theories of Viscosity. Paris 1953.

HOUWINK, R.: Elasticity, Plasticity and the Structure of Matter. Cambridge 1937. — (Elastizität, Plastizität und Struktur der Materie, Dresden und Leipzig 1938.) Dealing with glass, resisns, asphalt, rubber, gut-percha, balata, cellulose, starch, proteins, dough, paints, lacquers, clay and sulphur.

JEFFREYS, H.: The Earth. Cambridge 1929 and 1952. — For the rheologist the earlier edition is more interesting. JEFFREYS coined the term "elasticoviscosity" and "firmoviscosity".

LEADERMAN, H.: Physics of High Polymers. Utrecht 1951.

NADAI, A.: Theory of Flow and Fracture of Solids. New York: McGraw-Hill Book Co. 1950.

PHILIPPOFF, W.: Viscosität der Kolloide. Dresden u. Leipzig 1942.

SCOTT-BLAIR, G. W.: A survey of general and applied rheology. London 1949. — Mainly for the integrative approach and psychophysical considerations in rheology.

STUART, H. A. (Editor): Die Physik der Hochpolymeren. Berlin: Springer 1956. Vol. IV with Chapters by A. J. STAVERMAN, F. SCHWARZL, L. R. G. TRELOAR and J. D. FERRY.

TRELOAR, L. R. G.: The Physics of Rubber Elasticity. Oxford 1949.

When planning rheological experiments, it is advisable to consult:

BURGERS, J. M., and G. W. SCOTT BLAIR: Report on the principles of rheological nomenclature. Amsterdam 1949.

A series of monographs "on the rheological behaviour of natural and synthetic products" has been edited by J. M. BURGERS, J. J. HERMANS and G. W. SCOTT BLAIR, and published by the North Holland Publishing Co., Amsterdam. They comprise:

FREY-WYSSLING, A. (Editor): Deformation and Flow in Biological Systems, (1952), dealing with protoplasm, muscle, plant cell walls, latex, blood, lymph, cerebrospinal fluids, intra-ocular fluids, intra-ocular fluid and secretions.

HERMANS, J. J. (Editor): Flow Properties of Dispersed Systems, 1953, dealing with suspensions, emulsions, gels, dilute solutions of impenetrable rigid particles, dilute solutions of flexible chain molecules, liquid sprays, atomization of liquids, foams, smoke, powders.

MEREDITH, R. (Editor): Mechanical Properties of Wood and Paper. 1953.

REINER, M. (Editor): Building Materials, their Elasticity and Inelasticity, 1954, dealing with metals, wood, plaster and mortar, fresh concrete, set and reinforced concrete, road asphalt, soils, earth walling, clay products and minor materials of building construction.

Papers of major historical importance.

Lord KELVIN (Sir W. THOMSON): "Elasticity", Encyclopedia Britannica. 9th ed., 1875; also Papers, 3. London 1890. This is an article which can still today be read with great profit.

EINSTEIN, A.: Ann. Physik **19**, 289 (1906); **34**, 591 (1911). — The first microrheological analysis.

MAXWELL, J. C.: Phil. Mag. (4) **35**, 129, 185 (1868). — Introduced the relaxation of stress.

BINGHAM, E. C., and H. GREEN: Proc. Amer. Assoc. Testing Materials, II **19**, 640 (1919). — This paper started rheology.

POYNTING, J. H., and J. J. THOMSON: Properties of Matter. London 1902. These authors first used the picture of a mechanical model.

POYNTING, J. H.: Proc. Roy. Soc. Lond., Ser. A **82**, 546 (1909); **86**, 534 (1912). — The first experimental observation of second order effects in elasticity.

MURNAGHAN, F. D.: Amer. J. Math. **59**, 235 (1937). — The first to use systematically tensor notation in elasticity.

REINER, M.: Amer. J. Math. **67**, 350 (1945); **70**, 433 (1948). — Established the isotropic relation between two tensors of second rank and the resulting second order effects in viscous and elastic media.

Fracture.

By

GEORGE R. IRWIN.

With 10 Figures.

1. Introduction. In 1776 COULOMB [1] expressed the view that fracture of a solid would occur if the maximum shear strain at some point surpassed a critical value characterizing the mechanical strength of the material. Although the suggestion is of no more than historic interest it may be noted as the oldest of a number of empirical "critical stress" or "critical strain" relations [2].

In 1920 A. A. GRIFFITH [3] proposed an explanation of fracture strength of glass based on the idea that glass contains crack-like flaws. GRIFFITH reasoned that the largest crack-like flaw would become self-propagating when the rate of release of strain energy became greater than the rate of increase of surface energy of the extending crack. Subsequently, during the interval to 1938, a statistical concept of mechanical strength was introduced. A widely known exposition of this viewpoint, sometimes referred to as a "worst flaw" theory is due to WEIBULL [4]. WEIBULL showed that fracture strength might be expected to depend upon the extensiveness of the region of greatest stress as well as upon the stress magnitude.

It will be seen that the strain energy release rate idea basic to GRIFFITH'S proposal, and the statistical viewpoint inherent in the "worst-flaw" idea are essential for development of a basis for understanding fracturing and fracture strength of real materials. The additional concepts which are needed are best shown through illustrative examples and discussion.

I. Tensile strength of liquids.

2. Growth of a cavity to unstable size. A cavity in a large volume of liquid subjected to a hydrostatic tension p will expand or collapse depending upon whether p is greater or less than the value

$$p = \frac{2\gamma}{r} - p_v$$

where γ is the surface tension, r is the radius of a spherical cavity, and p_v is the vapor pressure. A rough estimate of the limiting tensile strength of a pure liquid may be obtained by neglecting p_v and by making the size of the cavity correspond to the specific volume per single molecule of liquid. Thus for carbon tetrachloride (CCl_4) at room temperature

$$p = \frac{2 \times 27 \text{ ergs/cm}^2}{7.3 \times 10^{-8} \text{ cm}} = 740 \times 10^6 \frac{\text{ergs}}{\text{cm}^3}.$$

Experimental values of tensile strength for CCl_4 of nearly half this magnitude have been measured by BRIGGS[1]. In contrast to this near agreement, similar

---

[1] L. J. BRIGGS: J. Chem. Phys. **19**, 970 (1951).

estimation procedures applied to pure solids lead to calculated strengths which are two or three orders of magnitude greater than the strengths found by actual measurement. Evidently the tensile strength of pure liquids is more easily related to fundamental theoretical considerations than is the case for solids. For this reason the tensile strength of pure liquids will be considered in sufficient detail to bring out features which pertain to fracturing generally.

The procedure used in this discussion extends somewhat an analysis of this problem by FISHER[1].

Assume at first that the liquid has no impurities, that it adheres perfectly to the walls of its container, and that the negative pressure is uniform throughout. One may imagine the liquid is in a rigid cylindrical container fitted with a piston. Suppose, then, that the effect of piston motion in exerting an initial tension or negative pressure p_0 has been to increase the liquid volume from V to $(V+w)$. If the modulus of compression is k_0

$$p_0 = \frac{k_0}{V} w. \tag{2.1}$$

We suppose next that the tension is relaxed by growth of one spherical cavity whose volume v is smaller than w. Then the pressure is

$$p = p_0(1 - v/w) \tag{2.2}$$

and the strain energy in the liquid is

$$U = \tfrac{1}{2} p(w - v). \tag{2.3}$$

In addition to affecting U the growth of the cavity increases the energy in the liquid by the surface energy term γA, where A is the surface area. As the cavity grows in volume with the walls and piston position fixed

$$\frac{d}{dv}(U + \gamma A) = -p + \frac{2\gamma}{r}. \tag{2.4}$$

The loss in strain energy is just offset by the increase in surface energy when

$$p = \frac{2\gamma}{r}. \tag{2.5}$$

When the test volume is large compared to the value

$$\omega = \frac{16\pi}{3} \left(\frac{8\gamma}{3p_0}\right)^3 \frac{k_0}{p_0} \tag{2.6}$$

Eq. (2.5) is satisfied at two values of v, an unstable equilibrium cavity size v_1 much smaller than w and a stable cavity size v_2 nearly as large as w. At the condition such that the cavity size is v_1, the loss in strain energy due to relaxation of pressure is $p_0 v_1 (1 - \tfrac{1}{2} u)$ where $u = v_1/w$. The gain in surface energy is $\tfrac{2}{3} p_0 v_1 (1 - u)$. Thus there is a net energy increase of

$$\Delta H_1 = \tfrac{1}{3} p_0 v_1 (1 - 2u). \tag{2.7}$$

When the hydrostatic tension is large, say 3×10^8 ergs/cm³, the magnitude of expression (2.6) is about 10^{-17} cm³. In a volume of liquid equal to or less than ω, unstable cavity growth would be impossible without the aid of strain energy from neighboring volumes. In a volume of liquid of, say, 10ω sudden growth of a cavity of size v_1 would relax the pressure by only 1% even without

---

[1] J. FISHER: J. Appl. Phys. **19**, 1062 (1948).

motion of particles at the outer boundaries of that volume. Thus we may consider the whole test volume of the liquid to consist of a large number of cavity growth sites, each surrounded by a liquid volume of ample size, say, 10ω, so that the development of an unstable cavity might, conceptually, occur within any growth site independent of neighboring elements. At each cavity growth site thermal energy fluctuations are, from time to time, momentarily converted by various degrees into surface energy through cavity dilations. Since a considerable general movement of molecules must occur during the opening of a cavity to critical size, such an event is unlikely to occur in a time shorter than

$$\tau = \frac{(10\omega)^{\frac{1}{3}}}{c} \tag{2.8}$$

where c is the velocity of sound in the liquid. Thus we are concerned with thermal energy fluctuations integrated over time periods roughly equal to τ and over volumes roughly equal to 10ω. It will be assumed that when a thermal energy fluctuation meeting these conditions occurs then the probability of cavity growth to unstable size can without appreciable error be taken to be unity. It may be seen later that taking this probability to be 10^{-3} would increase the computed strength by only 4%.

3. The probability of unstable cavity growth. The local thermal energy fluctuations constitute a statistical population. It is a property of this population that the a priori probability of any given magnitude of integrated energy fluctuation for the time period, τ, and the volume, 10ω, is not dependent upon the sizes of τ and of 10ω. For the temperature range of interest the a priori probability of occurrence of ΔH_1 is given by the expression, $\exp(-\Delta H_1/kT)$.

If the liquid test volume contains N molecules, there are, at most, N growth sites for cavities. From one viewpoint these must be counted as operating independently even though the regions, 10ω, surrounding adjacent sites would overlap. In any case it would appear safe to assume a number of independent cavity growth sites, N_1, such that

$$N > N_1 > \frac{V}{10\omega}. \tag{3.1}$$

Similarly the restriction of one's consideration to t/τ time elements, the testing time being t, would overlook the fact that a larger number of positions in time might act as the mid-points of the intervals τ. The frequency of fluctuations of magnitude kT or more must be about kT/h for consistency with HEISENBERG'S uncertainty principle. We will therefore assume a there are $\nu\,dt$ significant time sites in the interval dt where

$$\frac{kT}{h} > \nu > \frac{1}{\tau}. \tag{3.2}$$

Thus the probability for occurrence of cavity growth to unstable size in the interval of testing time dt becomes

$$f(t)\,dt = N_1 \nu\,dt \exp\left(-\frac{\Delta H_1}{kT}\right) \tag{3.3}$$

where ΔH_1 with the aid of (2.5) and (2.7) is given by

$$\Delta H_1 = \frac{16\pi}{3}\frac{\gamma^3}{p_0^2}\frac{1}{(1-u)}. \tag{3.4}$$

The tension p_0 is regarded as a function of time t obtainable from knowledge of the test procedure under consideration. In steps leading to expression (3.4) advantage was taken of the fact that u is very small to make a substitution of $(1-u)(1-2u)$ for $(1-u)^3$.

Define, next, a quantity F such that

$$F(t) = \int_0^t f(t)\,dt. \tag{3.5}$$

Then by the usual methods of extreme value statistics, the expected average time to fracture is given by

$$\bar{t} = \int_0^\infty t f(t)\,dt \exp[-F(t)]. \tag{3.6}$$

If (dp_0/dt) is greater than zero one may replace $f(t)\,dt$ by $g(p_0)\,dp_0$ so that with

$$g(p_0)\,dp_0 = N_1 v \frac{dt}{dp_0} dp_0 \exp\left[-\frac{\Delta H_1}{kT}\right] \tag{3.7}$$

and

$$G(p_0) = \int_0^{p_0} g(p_0)\,dp_0 \tag{3.8}$$

one obtains the average tension for fracture of the liquid $\bar{p}_0$ by the equation

$$(\bar{p}_0) = \int_0^\infty p_0 g(p_0)\,dp_0 \exp[-G(p_0)]. \tag{3.9}$$

The coefficient of p_0 in the integrand of (3.9) differs appreciably from zero over so small a range that the average value of p_0 will not differ significantly from the value at which $g(p_0)\exp(-G)$ attains maximum size. Thus the expected average strength, $(\bar{p}_0)$, is closely approximated by the value of p_0 which satisfies the relation

$$\frac{dg}{dp_0} - g^2 = 0. \tag{3.10}$$

Neglecting variation of ΔH_1 with p_0 through the negligible volume ratio u,

$$\frac{dg}{dp_0} = g\left(\frac{dN_1}{N_1 dp_0} + \frac{dv}{v\,dp_0} + \frac{2\Delta H_1}{p_0 kT}\right). \tag{3.11}$$

Thus

$$g(p_0) = \frac{dN_1}{N_1 dp_0} + \frac{dv}{v\,dp_0} + \frac{2\Delta H_1}{p_0 kT}. \tag{3.12}$$

For purposes of specific calculation it will be assumed that one may take

$$t = \beta p_0. \tag{3.13}$$

Where β is constant, as the relation of p_0 to time in experiments such as those of BRIGGS. The magnitude of the right side of (3.12) has so little influence upon the calculation that one may drop the terms involving N_1 and v. In fact it may be shown that

$$0 < \frac{dN_1/N_1}{dp_0/p_0} + \frac{dv/v}{dp_0/p_0} < \frac{1}{20}\frac{\Delta H_1}{kT}. \tag{3.14}$$

Using (3.7), (3.13) and (3.12) one obtains

$$\log(N_1 v\,t) - \frac{\Delta H_1}{kT} = \log\left(2\frac{\Delta H_1}{kT}\right) \tag{3.15}$$

where t is the duration of the test. Taking t as one second, N_1 as the number of molecules in one cubic centimeter, and ν as (kT/h), the strength should be such that ΔH_1 is about 80 times kT or about 2 electron volts.

4. Pure liquid tensile strength measurements. In the experiments which have given the highest values of liquid fracture strength a strong capillary glass tube filled with a pure liquid is rotated about a transverse axis[1]. The speed of rotation is rapidly increased until the liquid column breaks. Calculation of the negative pressure at the axis of spin is based upon knowledge of the rotational speed at the instant the column separates. The assumption made above that p_0 increases linearly with time may be applied to this experiment without significant error. However, since the tension is not uniform but decreases as a quadratic function of distance from the axis of spin some consideration should be given to estimating the equivalence of the liquid volume used in this experiment with a liquid volume which would have the same strength in a uniform tension experiment. This estimate can be made with sufficient precision by a further application of extreme value statistics. The details need not be given here. If the length of the liquid column between points of zero tension lies between 5 and 10 cm and the diameter of the liquid column is 0.6 to 0.8 mm, the experiment may be taken as equivalent to applying uniform tension to a volume of 2.5×10^{-3} cm^3. Taking the time of rise of the tension as 5 sec as a further adaptation to the BRIGGS type of experiment, calculations were made of theoretical strengths for a number of pure liquids. The extremes allowed by the inequalities (3.1) and (3.2) may be represented by

$$\log\left(\frac{kT}{h}\frac{75 \times 10^{20}}{V_m} \text{cm}^3 \text{sec}\right) = \frac{\Delta H_1}{kT} + \log\left(\frac{2\Delta H_1}{kT}\right) \tag{4.1}$$

and

$$\log\left(\frac{c}{(10\omega)^{\frac{4}{3}}} \times 6.25 \text{ cm}^3 \text{sec}\right) = \frac{\Delta H_1}{kT} + \log\frac{2\Delta H}{kT} \tag{4.2}$$

where V_m is the molar volume, c is the velocity of sound, and 10ω is given by

$$10\omega = 10\left(\frac{8}{3}\right)^3 \frac{k_0}{p_0^2} \Delta H_1. \tag{4.3}$$

Values of p_0 were obtained from equation (3.4) after ΔH_1 had been obtained from (4.1) and (4.2). The results are shown in Table 1 along with the highest strength values observed by BRIGGS for these same liquids. Any non-wetting areas of container wall, non-wetting impurities, or gas bubbles of about 10^{-5} cm size would cause the observed strengths to be considerably less than those given in the table. Experimental uncertainties of this kind are discussed by DONOGHUE, VOLLRATH, and GERGOUY[2]. The high strength of 150 atmospheres for pure benzene reported by BRIGGS was attained by DONAGHUE, VOLLRATH, and GERGOUY in only two out of a large number of trials. It is apparent that the multiplicity of factors which normally limit the tensile strength of a liquid to 1% or less of its theoretical value, calculated as above, are difficult to control experimentally Thus the theoretical strength calculations for pure liquids are of doubtful practical utility. Nevertheless, the degree of completeness permitted in the theoretical considerations make the pure liquid tensile strength analysis of importance.

In the foregoing analysis it was observed that removal of flaws from the liquid reduced the fracture starting situation to one of cavity growth governed

---

[1] L. J. BRIGGS: J. Chem. Phys. **19**, 970 (1951).
[2] J. J. DONAGHUE, R. E. VOLLRATH and E. GERGOUY: J. Chem. Phys. **19** (1951).

Table 1. *Theoretical and measured tensile strengths of liquids*[1].

Liquid	Molecular weight	Temp. range °C	Surface tension ergs/cm²	Tensile Strength, p_0 (bars)		
				from (4.2)	from (4.1)	measured
H_2O	18	$+5$ to $+20$	74	1528	1640	275[2]
Acetic acid CH_3COOH	60	$+17$ to $+30$	27.6	352	385	288[3]
Aniline $C_6H_5NH_2$ . .	93	-2 to $+5$	43	685	744	300[3]
Benzene C_6H_6	78	$+15$ to $+20$	28.9	377	412	150[3]
Carbon Tet. CCl_4 . .	154	-15 to $+20$	26.8	336	368	275[3]
Chloroform $CHCl_3$ . .	119	-15 to $+10$	27.2	344	377	315[3]

by thermal fluctuations. The cavities thus generated from time to time constituted the strength controlling flaw population. Elimination of flaws of the impurity and non-wetting kinds from the theoretical analysis did not eliminate influence of test volume and testing time. Although changes by a factor of 10^7 in volume or in testing time would have been necessary in order to change the calculated strength by 10% nevertheless no estimate of strength could have been made without inserting specific values for the test volume and the testing time.

The tendency for conversion from potential or strain energy to thermal energy was restricted by an activation energy requirement ΔH_1 which was proportional to the inverse square of the applied tension. The size of ΔH_1 and its relation to the tension was obtained from the conditions describing the point of onset of unstable cavity growth. Relation (2.5) which expressed these conditions is a simple example of a strain energy release rate consideration which is of basic importance in every kind of macroscopic fracturing.

II. Stress and force relations in fracture.

5. Stresses near a flat internal free surface and the crack-extension-force. When a fracture develops in a solid from minute origins, these origins are embedded in a larger field of inelastic or plastic strains. Knowledge of the complex patterns of distortion which characterize such early stages of fracturing is for the most part qualitative. As a definite crack develops and extends it is a characteristic of the fracturing process that the growth of crack length tends to overtake expansion of the field of plastic strains. Eventually, with continued crack extension, plastic strains are limited to a distance from the crack which is small relative to crack size and may be neglected in the elastic strain field analysis. Obviously no single descriptive model of fracturing could be applied to the entire range of the phenomenon. The condition of fracturing which closely resembles a thin internal cut is conceptually possible in any solid material[4]; its simplicities assist theoretical analysis and it is basic to development of a fracture strength concept of practical usefulness. The discussion of fracture in solids will therefore begin with extension of a well developed crack and with a "macro" rather than "micro" point of view.

A. A. GRIFFITH [3] presented a theory of fracture strength of a solid in terms of its surface energy. For this theory one assumes cracks exist or form in the solid as it is subjected to tensile forces. The theory then states that whenever elevation of the tensile forces produces a strain-energy release rate with crack extension which is larger than the rate of energy gain by formation of new free

---
[1] Table prepared by H. L. SMITH.
[2] L. J. BRIGGS: J. Appl. Phys. **21**, 721 (1950).
[3] L. J. BRIGGS: J. Chem. Phys. **19**, 970 (1951).
[4] This statement is discussed in Sect. 15.

surface area, rapid crack extension occurs and the solid breaks. Presumably the balance of energy rates for fracture strength determination would refer to the pre-existent crack with largest dimensions normal to the applied tensile load. It must also be assumed that the pattern of bonding forces characteristic of the solid internally are not easily re-established when an internal free surface closes so that any crack extensions are irreversible.

It was suggested by IRWIN [6] that the Griffith theory could be made generally applicable by substitution of energy spent in localized plastic strain for surface tension as a measure of resistance to crack extension. A similar proposal restricted to brittle fracture was advanced by OROWAN [7]. Direct applicability of the modified Griffith theory of fracture strength to the various large scale fracture propagation problems was shown by KIES and IRWIN[1]. Recognition of the role of GRIFFITH'S strain energy release rate as the force conjugate to time rate of crack extension and as a measure of elevation of stresses near the end of a crack[2] contributed further to the body of theory now available for analysis of the fracturing process.

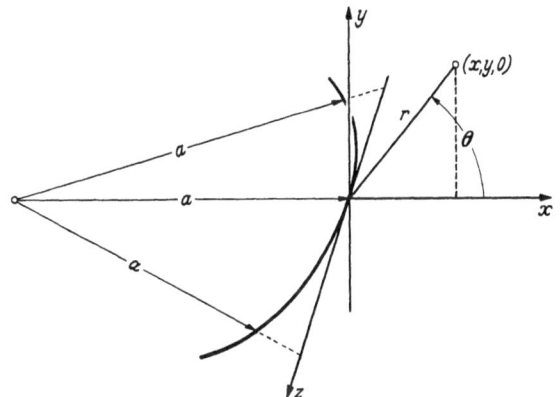

Fig. 1. Rectangular coordinates, x, y, z and polar coordinates r, ϑ, at the edge of a penny-shaped interior crack.

A paper by SNEDDON[3] gave a linear elastic theory analysis of the stresses and strains near a "penny-shaped" crack in an infinite solid[4]. The analysis assumed uniform tension at large distances from the crack. Superposition of uniform extensional stresses parallel to the plane of the crack would not alter the character of the results. SNEDDON used his analysis to obtain a value for the strain energy release rate appropriate to a Griffith theory [3] for predicting an unstable crack size. At one point in his paper SNEDDON gave approximate equations for the stress components in a region very close to the outer edge of the circular disc-shaped crack.

Consider a rectangular coordinate system with its y axis parallel to the axis of symmetry of SNEDDON's penny-shaped crack and with the x and z axes in the plane containing the crack. A uniform tension, σ, assumed to be applied at infinity causes a symmetrical pattern of displacement discontinuities which constitute the crack opening.

Assume next that the origin of coordinates is located at any point on the outer circular edge of the crack with the z axis tangent to the circle (see Fig. 1).

For points in the x, y plane and at distances from the origin which are small compared to the radius of the crack, a, the appropriate stress relations are those approached in the limit as r approaches zero and may be obtained from SNEDDON's paper. These are most simply written in terms of the coordinates r, ϑ shown

---

[1] G. R. IRWIN and J. A. KIES: Welding J. Res. Suppl. **31**, 95 (1952).
[2] G. R. IRWIN: IXth Int. Congr. of Appl. Mech., paper No. 101—II, Brussels, 1956.
[3] I. N. SNEDDON: Proc. Phys. Soc. Lond. **187**, 229 (1946).
[4] See also R. A. SACK: Proc. Phys. Soc. **58**, 729 (1946).

in Fig. 1 and may be expressed as follows

$$\sigma_x = K \frac{\cos \vartheta/2}{\sqrt{2r}} \left\{ 1 - \sin \frac{\vartheta}{2} \sin \frac{3\vartheta}{2} \right\}, \tag{5.1}$$

$$\sigma_y = K \frac{\cos \vartheta/2}{\sqrt{2r}} \left\{ 1 + \sin \frac{\vartheta}{2} \sin \frac{3\vartheta}{2} \right\}, \tag{5.2}$$

$$\sigma_z = \nu (\sigma_x + \sigma_y) = 2\nu K \frac{\cos \vartheta/2}{\sqrt{2r}}, \tag{5.3}$$

$$\tau_{xy} = K \frac{\cos \vartheta/2}{\sqrt{2r}} \sin \frac{\vartheta}{2} \cos \frac{3\vartheta}{2}, \tag{5.4}$$

$$\tau_{yz} = \tau_{xz} = 0 \tag{5.5}$$

where ν is Poisson's ratio.

Comparison with Sneddon's paper shows that the factor K in these equations is $\frac{2}{\pi}(\sigma^2 a)^{\frac{1}{2}}$.

To the same approximation as Eqs. (5.1) through (5.5) the y-direction displacements at the crack surface for small values of r/a are given by

$$v = \frac{(1-\nu)}{\mu} K \sqrt{2r} \tag{5.6}$$

where μ is the modulus of rigidity and r extends in the negative x direction.

Consider the edge region of a crack in plane strain corresponding to Eqs. (5.1) through (5.5) and assume the crack edge to be straight rather than a section of a circle. It is mathematically possible to close a short segment of length α to the left of the origin by superposition of tensile forces per unit area on the crack surfaces given by

$$S_y(p) = p K \frac{1}{\sqrt{2(\alpha - r)}}. \tag{5.7}$$

The factor, p, may be regarded as a proportional loading parameter. When p has been increased from 0 to 1 the crack will be closed along the segment α.

During the closure operation the y-direction displacements at the crack surfaces within the closing segment are given by

$$v(p) = (1-p) \frac{(1-\nu) K}{\mu} \sqrt{2r}. \tag{5.8}$$

Since the displacements, Eq. (5.8), are a linear function of the applied forces, Eq. (5.7), it is clear that the work to close up a length α at the edge of a crack is given by

$$\int_0^\alpha v(0) S_y(1) dr = \frac{(1-\nu)}{\mu} K^2 \int_0^\alpha \left(\frac{r}{\alpha - r} \right)^{\frac{1}{2}} dr = \frac{(1-\nu)\pi}{2\mu} K^2. \tag{5.9}$$

The calculation is on the basis of unit thickness of material in the z-direction. If the system is isolated from receiving energy, say by a fixed-grip situation, Eq. (5.9) represents the amount of energy which disappears from the strain energy field when the crack extends a small distance α. It is, therefore, equal to $\alpha \mathscr{G}$ where $\mathscr{G}$ is the strain energy release rate applicable to a Griffith theory analysis.

Sect. 5. Stresses near a flat internal free surface and the crack-extension-force.

Eq. (5.9) was presented as pertaining to a crack whose edge was straight and coincided with the z axis. However, since α can be taken very small, since the calculation is per unit of length in the z direction, and since the stress equations might apply to any point on the circular boundary of the penny-shaped crack, one may insert for K SNEDDON's value

$$K = \frac{2}{\pi}(\sigma^2 a)^{\frac{1}{2}}. \tag{5.10}$$

One then obtains

$$\mathscr{G} = \frac{(1-\nu)\pi}{2\mu} K^2 = \frac{2(1-\nu)\sigma^2 a}{\pi\mu} \tag{5.11}$$

which verifies the value for $\mathscr{G}$ obtained by SNEDDON for this specific problem.

In agreement with precedent in other discussions of dislocation mechanics, $\mathscr{G}$ is the force per unit of edge length tending to cause extension of the crack.

One might question the applicability of the above discussion to a real crack, since linear elasticity relations were assumed valid to the edge of the crack, and, for cracks in real materials, substantial plastic distortion is found near the crack surfaces even when the fracture appears to be quite brittle. In this connection it is clear from physical considerations and may also be shown mathematically that the calculation of Eq. (5.9) is the equivalent of an overall integral of the strain energy loss rate due to crack extension. If Eqs. (5.1) through (5.5) are used to form an expression for strain energy density, ψ, the contribution to $\mathscr{G}$ from a region of radius r_1 around the edge of the crack is

$$\int_0^{r_1}\int_{-\pi}^{\pi} \frac{\partial \psi}{\partial a} r\, dr\, d\vartheta = \left(\frac{5-8\nu}{8-8\nu}\right) r_1 \frac{\partial \mathscr{G}}{\partial a}. \tag{5.12}$$

As a first approximation it may be assumed that the preceding integral represents the order of magnitude of error due to neglect of plastic strains when r_1 includes the major portion of the plastic strain zone.

For a central crack of length $2a$ in a large plate, or for the interior penny-shaped crack subjected to uniform tensile stress at a large distance from the crack, the value of $\mathscr{G}$ is proportional to a, so that

$$r_1 \frac{\partial \mathscr{G}}{\partial a} = \left(\frac{r_1}{a}\right) \mathscr{G}. \tag{5.13}$$

For cracks similarly located but opening under the action of wedge forces applied at the crack centers, $\mathscr{G}$ is inversely proportional to a, so that

$$r_1 \frac{\partial \mathscr{G}}{\partial a} = -\left(\frac{r_1}{a}\right) \mathscr{G}. \tag{5.14}$$

When $\mathscr{G}$ is passing through a maximum or a minimum value as a function of crack extension the integral equivalent to Eq. (5.12) is zero. Under the assumed conditions of this analysis the plastic strains are confined to distances from the crack surfaces small in comparison to the crack size. Thus stress relaxation by plastic flow near the crack surfaces may be estimated as having an influence upon the calculation of $\mathscr{G}$ which varies with $\partial \mathscr{G}/\partial a$ but is relatively small.

For a crack traversing a plate under conditions such that r_1 can be considered to be a few plate thickness in magnitude without significant loss of accuracy, a generalized plane stress analysis can be used. The relations for stresses near the crack tip are as in Eqs. (5.1), (5.2), and (5.4). However, because of the increase in the displacement magnitudes when σ_z is put equal to zero, the value of $\mathscr{G}$ for a crack in plane stress is $\frac{1}{1-\nu^2}$ times the value of $\mathscr{G}$ for this crack subjected to similar σ_y values in a situation of plane strain. Thus for plane stress

$$K = \left[\frac{E\mathscr{G}}{\pi}\right]^{\frac{1}{2}} \qquad (5.15)$$

where E is YOUNG's modulus. With this value of K, and remembering that a uniform stress can be added to the value of σ_x as well as σ_z, Eqs. (5.1), (5.2), and (5.4) provide the stress relations for the crack tip region of a separational crack dislocation in plane stress.

With simplifying approximations similar to those in the preceding discussion it is possible to write general expressions for elastic stresses and displacements near the edge of a crack. One assumes the internal cut, representing the crack, has an edge boundary which along any short segment can be replaced by a straight line, and that the locus of the crack near such a segment of its edge can be represented approximately as a plane. The portion of crack edge unter consideration may then be represented as on Fig. 1. The system of elastic stresses and displacements near the crack edge separates naturally into the sum of three systems. These refer to the three sets of displacement discontinuities which relax the three stresses $\sigma_y, \tau_{xy}, \tau_{yz}$, to constant or zero values at the locus of the crack.

The stress system related to σ_y is a special case of the set of problems solved by WESTERGAARD [5] starting from the assumed Airy stress function

$$F = \operatorname{Re} \bar{\bar{Z}} + y \operatorname{Im} \bar{Z}. \qquad (5.16)$$

In WESTERGAARD's notation $\bar{Z}, Z$, and Z' are successive derivatives of a function, $\bar{\bar{Z}}(\zeta)$ where ζ is $(x+iy)$ or, when polar coordinates are more convenient, ζ is $re^{i\vartheta}$.

For the stresses, one has

$$\sigma_x = \frac{\partial^2 F}{\partial y^2} = \operatorname{Re} Z - y \operatorname{Im} Z', \qquad (5.17)$$

$$\sigma_y = \frac{\partial^2 F}{\partial x^2} = \operatorname{Re} Z + y \operatorname{Im} Z', \qquad (5.18)$$

$$\sigma_z = \nu(\sigma_x + \sigma_y) = 2\nu \operatorname{Re} Z, \qquad (5.19)$$

$$\tau_{xy} = -\frac{\partial^2 F}{\partial x \partial y} = -y \operatorname{Re} Z', \qquad (5.20)$$

$$\tau_{yz} = \tau_{xz} = 0. \qquad (5.21)$$

By taking

$$\bar{Z} = K_1 \sqrt{2\zeta} \qquad (5.22)$$

and $re^{i\vartheta}$ for ζ, one obtains Eqs. (5.1) through (5.4). For the displacements, u, v, w in the x, y, z directions one has

$$u = \frac{K_1 \sqrt{2r}}{2\mu} \cos\frac{\vartheta}{2} \left\{1 - 2\nu + \sin^2\frac{\vartheta}{2}\right\}, \qquad (5.23)$$

$$v = \frac{K_1 \sqrt{2r}}{2\mu} \sin\frac{\vartheta}{2} \left\{2(1-\nu) - \cos^2\frac{\vartheta}{2}\right\}, \qquad (5.24)$$

$$w = 0.$$

As in Eq. (5.11)

$$K_1^2 = \frac{2\mu \mathscr{G}_1}{(1-\nu)} \qquad (5.25)$$

where $\mathscr{G}_1$ is the crack-extension-force for the opening mode of crack extension.

Sect. 5. Stresses near a flat internal free surface and the crack-extension-force.

A similar procedure to that given above solves the problem for the stress system related to τ_{xy}. For this one assumes

$$F = -y \operatorname{Re} \bar{Z}. \tag{5.26}$$

By taking

$$\bar{Z} = K_2 \sqrt{2\zeta} \tag{5.27}$$

one obtains

$$\sigma_x = -\frac{K_2}{\sqrt{2r}} \sin\frac{\vartheta}{2}\left\{2 + \cos\frac{\vartheta}{2}\cos\frac{3\vartheta}{2}\right\}, \tag{5.28}$$

$$\sigma_y = \frac{K_2}{\sqrt{2r}} \sin\frac{\vartheta}{2}\cos\frac{\vartheta}{2}\cos\frac{3\vartheta}{2}, \tag{5.29}$$

$$\sigma_z = -2\nu \frac{K_2}{\sqrt{2r}} \sin\frac{\vartheta}{2}, \tag{5.30}$$

$$\tau_{xy} = \frac{K_2}{\sqrt{2r}} \cos\frac{\vartheta}{2}\left\{1 - \sin\frac{\vartheta}{2}\sin\frac{3\vartheta}{2}\right\}, \tag{5.31}$$

$$\tau_{yz} = \tau_{xz} = 0. \tag{5.32}$$

For the displacements

$$u = \frac{K_2\sqrt{2r}}{2\mu} \sin\frac{\vartheta}{2}\left\{2(1-\nu) + \cos^2\frac{\vartheta}{2}\right\}, \tag{5.33}$$

$$v = -\frac{K_2\sqrt{2r}}{2\mu} \cos\frac{\vartheta}{2}\left\{(1-2\nu) - \sin^2\frac{\vartheta}{2}\right\}, \tag{5.34}$$

$$w = 0.$$

By a calculation similar to that of Eq. (5.9)

$$K_2^2 = \frac{2\mu \mathscr{G}_2}{\pi(1-\nu)} \tag{5.35}$$

where $\mathscr{G}_2$ is the crack-extension-force for the first sliding or shear mode of crack extension.

For the stresses related to τ_{yz} one may assume

$$u = v = 0, \tag{5.36}$$

$$\mu w = K_3 \operatorname{Im}\left(\sqrt{2\zeta}\right) = K_3\sqrt{2r}\sin\frac{\vartheta}{2}. \tag{5.37}$$

Thus

$$\sigma_x = \sigma_y = \sigma_z = \tau_{xy} = 0, \tag{5.38}$$

$$\tau_{xz} = -\frac{K_3}{\sqrt{2r}}\sin\frac{\vartheta}{2}, \tag{5.39}$$

$$\tau_{yz} = \frac{K_3}{\sqrt{2r}}\cos\frac{\vartheta}{2}. \tag{5.40}$$

By procedures similar to those used above

$$K_3^2 = \frac{2\mu \mathscr{G}_3}{\pi} \tag{5.41}$$

where $\mathscr{G}_3$ is the crack-extension-force for the second sliding or shear mode of crack extension.

Any constant terms representing superimposed uniform stress fields can be added to the stress components of the above stress systems. Otherwise linear combinations of the above three stress groups represent the stress fields possible at the edge of an internal cut representing a crack.

The three quantities $\mathscr{G}_1$, $\mathscr{G}_2$, and $\mathscr{G}_3$ are rates of transfer with crack extension of energy from the surrounding elastic strain field into other forms. In the case of metals the released strain energy is converted primarily to heat through local plastic deformation. In the case of all solids so far investigated the most rapid crack extension rates are much less than the velocity of sound. Thus the speed of the process is primarily limited by the reaction rates characteristic of local inelastic deformations rather than by inertia considerations.

For fracturing of a solid whose strength properties from point to point are homogeneous, the average time rate of crack extension is an increasing function of the crack-extension-force. The detailed relationship of crack extension velocity to crack-extension-forces is a complex rate theory problem. Knowledge of this relationship is currently a matter of experimental observation as is true also of the analogous problem for plastic yielding.

The sliding processes associated with $\mathscr{G}_2$ and $\mathscr{G}_3$ if not accompanied by the first mode to provide actual separation of the crack surfaces, are not crack-like deformations in the usual interpretation. Solids can be severed, for example, in compression tests of ceramics, by a localized sliding action. However, because the conditions of restricted plastic deformation are poorly met in shear fracturing of metals and because the attention of investigations has been primarily on fractures of the opening mode type, discussion of the sliding modes of fracture will be quite limited. The symbol $\mathscr{G}$ with no subscript will refer to the crack-extension-force for the opening mode.

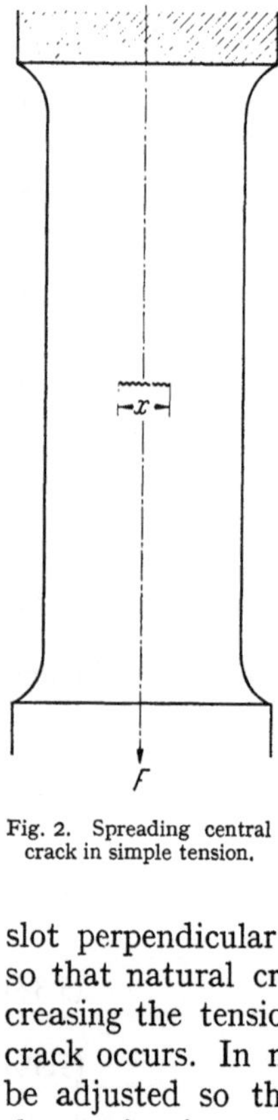

6. Calculation and measurement of the crack-extension-force. Consider next the energy changes which occur when segments of a solid material are separated by fracture. Assume the solid material under consideration is a plate in simple tension with a centrally located crack as in Fig. 2. Such a situation can be developed experimentally in various ways, for example, one may saw or cut a narrow slot perpendicular to the direction of the tension and apply wedging actions so that natural cracks are produced outward at either end of the slot. By increasing the tension, a load F can be found such that lengthening of the central crack occurs. In materials of moderate ductility the experimental situation can be adjusted so that the average time rate of crack extension increases from slow to fast in a manner which, at least from trend analysis viewpoint, can be considered as continuous. Consider an increment δx of crack extension during a period in which the process is slow enough so that kinetic energy may be omitted from the energy considerations. During this increment we assume the force F extended the plate by amount δl. A certain part l_e of the extension of the plate is recoverable by unloading. The balance which is not recoverable by unloading will be referred to as l_p. Then the total extension from the beginning of the experiment is

Fig. 2. Spreading central crack in simple tension.

and
$$\left.\begin{array}{c} l = l_e + l_p \\ \delta l = \delta l_e + \delta l_p. \end{array}\right\} \tag{6.1}$$

For the purpose of this elementary illustration it may be assumed
$$F = M\, l_e \tag{6.2}$$
where the spring constant, M, is a decreasing function of the length of the crack. Corresponding to this linear relation the strain energy U in the plate is
$$U = \tfrac{1}{2} F\, l_e \tag{6.3}$$
and
$$\delta U = F\, \delta l_e - \tfrac{1}{2} F^2 \delta\!\left(\tfrac{1}{M}\right). \tag{6.4}$$

It is useful to define a quantity δW by the relation,
$$F\, \delta l = \delta U + \delta W. \tag{6.5}$$
Using (6.1) and (6.4) one observes that
$$\delta W = F\, \delta l_p + \tfrac{1}{2} F^2 \delta\!\left(\tfrac{1}{M}\right). \tag{6.6}$$

Since the fracturing process with its accompanying plastic deformations is unaffected if, at any instant, the grips shown in Fig. 2 are considered to be fixed, one may write
$$\frac{dW}{dx}\,\delta x = -\,\delta U_f = F\, \delta l_p + \tfrac{1}{2} F^2 \frac{d}{dx}\!\left(\tfrac{1}{M}\right)\delta x \tag{6.7}$$
where $(-\delta U_f)$ is the loss of strain energy for fixed grip conditions. Assuming a situation in which crack extension is accomplished by negligible plastic extensions δl_p, it is clear that the coefficient of δx in the last term of Eq. (6.7) is the crack-extension-force. Thus, on a unit thickness basis
$$\mathscr{G} = \tfrac{1}{2} F^2 \frac{d}{dx}\!\left(\tfrac{1}{M}\right). \tag{6.8}$$

Consider a tensile test such as that of Fig. 2 but with no crack present. For this situation the last term of Eq. (6.7) is zero and the force acting to produce plastic extension, δl_p, is the longitudinal force, F. On a unit area basis one would say the force driving the deformation process is the longitudinal stress. A special relationship between time rate of plastic extension and longitudinal stress is usually found. For most materials the time rate is in the creep range and considered negligible for structural purposes so long as the stress is below the yield strength of the material. When, as the stress is increased, an abrupt increase occurs in the time rate of deformation, the material is said to have a sharp or well defined yield strength.

It is useful to define the fracture strength of a solid material in a way analogous to that of the yield strength. Since the change from slow to fast fracturing of a developing crack is usually abrupt, fracture strength may be described as the critical value, $\mathscr{G}_c$, of the crack-extension-force, $\mathscr{G}$, necessary for onset of rapid crack extension.

Eq. (6.8) suggests several practical procedures for measurement of $\mathscr{G}$. The elastic deflection of specimens containing cuts, representing cracks, of various length may be measured. From these measurements M as a function of crack length must be found experimentally with sufficient precision to permit computation of the slope of the curve representing $(1/M)$ as a function of crack length. With this information at hand one needs only to observe the applied force and the crack length for onset of rapid fracturing. Eq. (6.8) then permits calculation of $\mathscr{G}_c$. The method has been applied with reasonable success to tests of bars of steel and aluminum alloys broken in bending. However, a high degree of precision is required in the spring constant measurements for good accuracy.

Historically the second experimental method for measurement of the crack-extension-force was that used by Wells[1] for cracks extending rapidly in steel plates. The method assumes the crack is in rapid motion and that the conversion from strain energy to heat is confined to a region close to the path of the crack. By means of thermocouples pressed into small holes near the path of the crack Wells was able to measure the rise and fall of local temperature after rapid fracturing had occured. Assuming the path of the crack to be the source of the observed heat wave, Wells was able to calculate the energy converted to heat per unit of fracture area. In contrast with the method based upon Eq. (6.8) which is applicable only to stationary cracks, the thermocouple method is applicable only to rapidly moving cracks. Satisfactory agreement of results from the thermocouple method with those expected from theoretical considerations was found.

For both of the experimental procedures discussed above a knowledge of the stress distribution is unnecessary. However, because use of the thermocouple method is limited to special materials and because of inconveniences inherent in both procedures, an approximate theoretical stress distribution is normally employed. An example of such a procedure using a photoelasticity technique is given in Sect. 10.

The relation of the stresses near a flat crack to the applied loads may be determined theoretically in several ways. Either semi-inverse procedures such as those of Neuber[2] and Westergaard [5] or integral equation procedures such as those of Sneddon[3] and Muskhelishvili[4] may be used. The Westergaard procedure was described in Sect. 5 of this article. The integral equation methods are potentially more powerful particularly when the desired answers require use of computing machines but will not be discussed in this article. Neuber's methods of exact stress analysis have been widely used as a means of determining the largest stress at the root of a notch. For purposes of elastic theory analysis, a flat crack can be considered to be a notch, external or internal, having zero flank angle and an edge of nearly zero radius of curvature. Inspection of Neuber's equations for the maximum stress at the notch σ_m shows that the product $\sigma_m \sqrt{q}$, where q is the radius of curvature of the notch, approaches a non-zero finite limit as q approaches zero. This product is related to the stress intensity factor K of this article by the equation

$$K = \lim \left(\tfrac{1}{2} \sigma_m \sqrt{q}\right) \quad \text{as} \quad q \to 0. \tag{6.9}$$

In his consideration of the influence of plastic yielding at a notch Neuber suggested the effective stress concentration might be related to the average stress from the root of the notch across a "plastic particle" distance, ε. From Eq. (5.1) with ϑ equal to zero

$$\sigma_y = \frac{K}{\sqrt{2r}}.$$

The average stress from $r=0$ to $r=\varepsilon$ is given by

$$\bar{\sigma} = \frac{1}{\varepsilon} \int_0^\varepsilon \sigma_y \, dr = K \sqrt{\frac{2}{\varepsilon}}. \tag{6.10}$$

---

[1] A. A. Wells; Welding Res. **7**, No. 2, 34-r (1953).

[2] H. Neuber: Kerbspannungslehre: Grundlagen für genaue Spannungsrechnung. Berlin: Springer 1937.

[3] I. N. Sneddon: Proc. Phys. Soc. Lond. **187**, 229 (1946).

[4] N. I. Muskhelishvili: Some Basic Problems of the Mathematical Theory of Elasticity. Groningen, Holland: P. Noordhoff 1953.

Fracture experiments do not determine ε and $\bar{\sigma}$ separately. However, by arbitrarily choosing $\bar{\sigma}$ to equal, say, the ultimate tensile strength found by standard tensile bars, a characteristic ε for onset of rapid crack extension can be determined. This procedure has been employed for prediction of applied loads which will cause crack propagation in steel and aluminum alloys[1]. The equivalence of a characteristic value of the product $\bar{\sigma}\sqrt{\varepsilon}$ to a characteristic value of K is obvious from Eq. (6.10).

Knowledge of the stresses near a flat crack in tension for a variety of situations is needed not only to assist fracture strength measurements but also to permit estimates of the crack-extension-force generally in the different laboratory or practical situations in which knowledge of this force tendency assists understanding of observed events. For two-dimensional stress fields associated with a straight crack the semi-inverse procedure suggested by WESTERGAARD has provided solutions for a number of problems. Three of these which have applications of general value are stated next in terms of the Westergaard stress function, Z.

Series of co-linear two-dimensional cracks. Consider the following stress function

$$Z = \sigma \prod_{i=1}^{N} \left\{ 1 - \frac{a_i^2}{(\zeta - b_i)^2} \right\}^{-\frac{1}{2}} \tag{6.11}$$

where b_i is an increasing series of real numbers, the values of a_i are positive, ζ is $(x+iy)$, and

$$b_i + a_i + a_{i+1} < b_{i+1}.$$

At infinity both σ_y and σ_x approach σ. Along the x axis there are N regions such that

$$|x - b_i| < a_i$$

where free surface boundary conditions are met. At the right end of the j-th crack, the stress intensity factor is

$$K_j = \sigma \sqrt{a_j} \prod_{i \neq j}^{N} \left\{ 1 - \frac{a_i^2}{(a_j + b_j - b_i)^2} \right\}^{-\frac{1}{2}}. \tag{6.12}$$

Periodically repeated two-dimensional crack. The limit of the stress function of Eq. (6.11) as N approaches infinity for equal length and equally spaced cracks is a stress function suggested by WESTERGAARD [5],

$$Z = \sigma \left\{ 1 - \left(\frac{\sin(\pi a/2A)}{\sin(\pi \zeta/2A)} \right)^2 \right\}^{-\frac{1}{2}}. \tag{6.13}$$

Internal free surfaces representing cracks occur along the x axis whenever

$$\left(\sin \frac{\pi x}{2A} \right)^2 < \left(\sin \frac{\pi a}{2A} \right)^2.$$

The length of each crack is $2a$ and the period of repetition is $2A$. The stress intensity, K, near a crack end is

$$K = \sigma \left(\frac{2A}{\pi} \tan \frac{\pi a}{2A} \right)^{\frac{1}{2}}. \tag{6.14}$$

Eq. (6.14) is a useful approximation to the stress intensity for plates of width $2A$, containing either a central crack of length $2a$, or two colinear cracks of length a, extending inward from the side boundaries.

---
[1] P. KUHN: Stockholm Colloquium on Fatigue, Int. Union of Theor. and Appl. Mech., pp. 131—140. Berlin: Springer 1956.

Localized pressure in a two-dimensional straight crack. Consider the stress function

$$Z(\zeta) = \frac{P\sqrt{a^2 - b^2}}{\pi \zeta (\zeta - b)\sqrt{1 - (a/\zeta)^2}}. \tag{6.15}$$

The stress field corresponding to this function represents a crack of length $2a$, in which splitting or wedge forces of strength $\pm P$ per unit of length in the z direction act at $x = b$. The stress intensity factor for stresses surrounding $x = a$ is

$$K = \frac{P}{\pi\sqrt{a}}\sqrt{\frac{a+b}{a-b}}. \tag{6.16}$$

Since the Z and K of component stress fields are additive, values of Z and K appropriate to any prescribed pressure relation,

$$P = p(b)\, db$$

can be obtained from appropriate integrals of Eqs. (6.15) and (6.16).

The values of $\mathscr{G}_c$ given in Table 2 were for the most part obtained using a plate in tension with a central cut or slot as a starting crack. A stress distribution corresponding to the stress function of Eq. (6.13) was assumed. The corresponding $\mathscr{G}$ value is

$$\mathscr{G} = \frac{\pi K^2}{E} = \frac{2A\sigma^2}{E}\tan\frac{\pi a}{2A}. \tag{6.17}$$

Table 2. *Typical values of the crack-extension-force $\mathscr{G}_c$ necessary for onset of unstable fast fracturing*[1].

Material	Plate or sheet thickness cm	Method	Temperature deg. Cent.	$\mathscr{G}_c$ 10^5 ergs/cm²
Steel, ship plate cleavage	1.9	a	−20	175
Steel, ship plate ductile	1.9	d	+25	3500
Aluminum alloy 24 ST 4	2.5	b	+25	525
Aluminum alloy 24 ST 6	0.1	a	+25	1050 [2]
Aluminum alloy 75 ST 6	0.1	a	+25	525 [2]
Aluminum alloy 78 ST 6	0.1	a	+25	260 [2]
Polymethylmethacrylate, as cast plates	0.3 to 1.2	a	+25	8.7
Polyesters, plates	0.3 to 1.2	a	+25	2.0
Polymethyl-alpha chloracrylate plates	0.3 to 0.6	a	+25	4.9
Vulcanized natural rubber	0.09	c	+25	120
Glass, lantern slide covers, moist	0.05	a	+25	0.07
Glass, lantern slide, in 2% RH air	0.05	a	+25	0.14
Cellulose acetate	0.06	a	+25	25
Cellulose acetate	0.005	a	+25	50

a) Central crack type tension specimens in which the unstable crack length was about one quarter of the plate width.

b) Slow bend tests of notched bars.

c) Tear tests by Rivlin: J. Polymer Sci. **10**, 291—318 (1953). Various compositions give values of $\mathscr{G}_c$ from 10^6 to 2×10^8 ergs/cm².

d) Deduced from side-notched tear test fracture work rate results with the assumption that unstable fast ductile fracture would occur without significant decrease of work rate beyond the slow fracture value listed.

[1] Table prepared by J. A. Kies and H. L. Smith.

[2] These values are increased by a factor of 2 when restraints are used to prevent buckling across the span of the initial crack.

III. Forming and spreading of cracks.

7. Illustrative models of fracture development. The processes by means of which small openings form, grow, and join to produce a visible crack are primarily understood only in qualitative terms. ZENER [6] and HOLLOMON [6] discussed several mechanisms for micro-crack formation in terms of the arrangement and movement of crystalline dislocations. They pointed out that the intensified stress field near the leading edge of a slip band contains regions of large tensile stress. Opportunities for openings to develop in this zone of increased tensions may be expected when continued slip is delayed, for example by arrest of a slip band at a grain boundary or inclusion.

If a uniform shear stress equal to the relaxed resolved shear stress at the slip plane is added, the equations for stresses pertaining to sliding modes of separation as discussed in Sect. 5 are applicable. From Eqs. (5.28) through (5.31) one finds the greatest tension is nearly four times the maximum shear stress at a ϑ value of $-\pi/2$. ZENER noted that large tensile stresses with a high ratio of tension to shear might also develop at grain corners due to relaxation of stress in the grain boundary. His discussion emphasized the close association of micro-crack development with plastic strain.

A similar viewpoint was developed in somewhat greater detail by PETCH[1] and by STROH[2]. Evidence was presented by PETCH[3] that fracture stress of polycrystalline pure iron has a linear relation to the inverse-square-root of grain size over a wide temperature range. The similarly of this result to the relation between yield stress and grain size suggested that both crack formation and slip propagation were related to attainment of certain critical stress intensities near the boundary of a blocked array of dislocations. The appearance of experimental evidence for a relation which can be justified in terms of a theoretical model in this instance, as in others to which reference will be made, provides a helpful view of one aspect.

The growth of a crack to macroscopic size results from a series of events ranging to submicroscopic scale. Proper conditions for formation of holes or cracks at smallest scale must exist as prerequisites for coalescence of these in groups to form fracture elements of next larger size. However, the events at large as well as small scale must be considered. A description of this whole complex series of processes which is both concise and general in application is not at hand. It is believed the broad features of fracture development can be seen in terms of the selected illustrative models which will presently be discussed.

Very small fracture origins in metals and plastics can often be identified on a fracture surface. Microscopic studies of these invariably show an appearance which suggests a rounded cavity rather than a thin crack. Although openings of more crack-like nature such as segments of cleavage may be present both in earlier and later stages of fracture development there are rather general grounds for considering that fracture development also includes openings which more nearly resemble holes than cracks.

Consider a small pennyshaped crack of radius r embedded in a polycrystalline rod. If the rod is subjected to a tensile stress of value S at its ends and if the plane of the crack is normal to the direction of the tensile stress, the enlargement of the volume of the crack by elastic strains is very nearly.

$$V_1 = \frac{4\pi r^3 S}{3E}. \qquad (7.1)$$

---

[1] N. J. PETCH: J. Iron Steel Inst. **173**, 25 (1953).
[2] A. N. STROH: Proc. Roy. Soc. Lond., Ser A **223**, 404 (1954).
[3] N. J. PETCH: Phil. Mag., VIII. Ser. **1**, 186 (1956).

A factor of approximately 1.15 is neglected in this equation. If the area of the crack is extended by adding a row of vacancies all around the outer edge, the volume of vacancies added is

$$V_2 = 2\pi r \lambda^2 \qquad (7.2)$$

where λ is the average atomic separation distance.

The new crack volume enlargement due to the tension S now exceeds V_1 by the amount

$$\lambda \frac{\partial V_1}{\partial r} = 4\pi r^2 \frac{S}{E} \lambda. \qquad (7.3)$$

The ratio of this volume increment to V_2 is $2rS/\lambda E$. If one takes E/S to be about 10^3 as would be true for strong metals, the above ratio approaches unity when r is about 10^{-5} cm. The volume enlargement of a crack-like opening during a tensile strength test is, therefore, comparable to the volume of the added vacancies and somewhat insensitive as to their influence upon cavity shape when either the opening or the stress is extremely small.

Examination of polycrystalline metals usually discloses a wide variety of flaws and defects whose linear dimensions are very much larger than 10^{-5} cm. Normally, cracks started by separations of these larger magnitudes would be expected to extend by plastic deformation processes localized in edge regions of largest stress. The corresponding crack volume enlargement then results primarily from contractions in the elastic strain field and the shape of the opening becomes more crack-like. As the crack size increases, flaw containing regions approached by the leading edge are subjected to an environment of increasing tensile stress. The extension of the crack to encompass regions of this nature when they separate near a leading edge of the crack becomes the process contributing most to extension velocity as the dominant crack becomes large in comparison with the individual flaws. Two easily reproduced model type experiments are of value in bringing out these and other features characteristic of growth of cracks.

One model of this nature may be constructed by forming a raft of small equal sized bubbles[1]. Experimentally one finds that the opposite sides of the bubble raft will cling tenaciously to glass rods and that attempts to fracture the bubble raft can be performed by separating the glass rods. When the raft is essentially a single crystal and contains none or very few dislocations, all attempts to produce a fracture result in separation by local width reduction with no obvious crack. This action resembles the pinching or sliding apart action of a coarse crystalline cadmium rod pulled slowly in tension. However, if a polycrystalline raft of bubbles is formed, such a raft can be made to fracture by development and extension of interior openings. These interior openings always occur at the crystalline boundaries. This behavior appears to be due to the mobility of the bubble raft dislocations which prevents transcrystalline separation, and to the tendencies for dislocation movements to center in the neighborhood of crystalline boundaries, thus facilitating enlargement of grain boundary openings by addition of vacancies. It has been generally accepted that bubble raft models reproduce metal behavior at elevated rather than at low temperatures. The tendency of fracture to follow grain boundaries at elevated temperatures is illustrated by the polycrystalline bubble raft in an interesting fashion.

Consider next another experimental model of a separation process. This model consists of a 0.002" thick zinc foil pulled in simple tension by a dead

---

[1] W. L. BRAGG: Proc. Roy. Soc. Lond., Ser. A **190**, 474 (1947).

weight[1]. If a central crack is created in such a foil by a razor blade cut and the subsequent extension of this crack observed, several significant characteristics of fracturing may be noted. A continuing steady general creep deformation of the sheet results in a continuing movement of the weight. Extensive local plastic

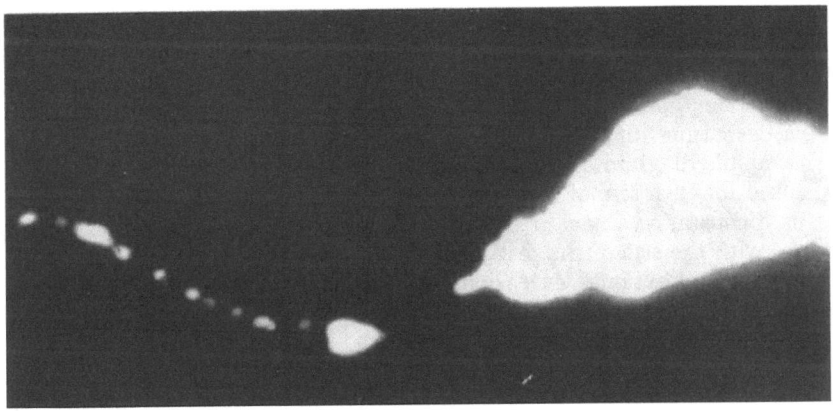

Fig. 3. Enlarged view of the end of a crack extending slowly in zinc foil. Fracture origins beyond end of crack are about to join together forming a quick extension of the main crack.

deformations near the ends of the crack result in thickness reductions which are unequal in magnitude in small neighboring regions. In part because of local thinning of the sheet and in part because of unequal strengths of different grains, holes appear near the ends of the extending crack. In the zinc foil used by McLean and George for these experiments, the grain size was approximately equal to the original sheet thickness. The number of such holes was counted and might be said to be either one to three per original grain diameter, or one to three per original sheet thickness. A preference for hole formation at grain corners would account for the observed spacing of "pin hole" fracture origins. Fig. 3 shows the appearance of a group of holes just prior to a joining up process which will result in a relatively fast advance of the crack.

When results of this experiment are studied by methods of trend analysis, the general crack extension behavior observed may be represented as shown

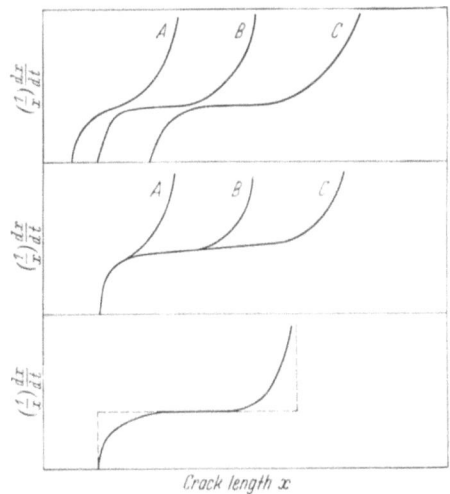

Fig. 4. Crack growth rate curves as a function of crack length for specimen sizes A, B, and C proportional respectively to $\frac{1}{2}$, 1, and 2.

on Fig. 4. In this schematic figure, crack length is indicated by x and time by t. In the upper diagram, the fractional time rate of extension is shown as a function of x for a size effect experiment. The arrangement of this experiment is such that the crack extension in a medium size sheet is compared to that in a half size and a double size sheet. In each case, the starting crack length and

---

[1] E. A. McLean, and T. W. George: Phys. Rev. **94**, 761 (1954) (abstract).

the load are proportional to the width of the test sheet. The velocity dx/dt changes from a slow "creeping" extension rate to a very fast rate during the course of the experiment, typically several hours. Only the approach to unstable fast fracturing is shown on the diagram. Midway between the starting crack length and the crack length for fast fracturing the fractional time rate of crack extension is about the same for each test size. The crack length for onset of unstable fast fracturing is not proportional to the width of test sheet. If this fast fracturing crack length is thought of as representing the end of the experiment, the ratio of the final to the initial crack length is, in this set of size effect experiment, a decreasing function of the size of the test sheet.

The middle diagram of Fig. 4 shows the result when the same starting crack size is used for each size of test. Again, the load is adjusted so that the nominal stress is a constant value. In this case the ratio of final to initial crack length increases with the size of the test sheet.

If the slow fractional extension rate period is represented by a constant average value as shown by the dashed lines on the lower diagram of Fig. 4, then the total time to fracture may be represented as follows:

$$\frac{1}{x}\frac{dx}{dt} = \frac{1}{t_0}, \tag{7.4}$$

$$t_f = t_0 \log \frac{x_2}{x_1}. \tag{7.5}$$

In these equations, t_0 is a constant, t_f is the total time to fracture, x_2 is the final crack length, and x_1 is the initial crack length.

One observes that, for the size effect experiment in which the starting crack was proportional to the test sheet size, t_f decreases with increasing test sheet size. This is a normal size effect in terms of time to fracture failure. However, the second set of tests discussed above in which the starting crack was the same for each sheet size possesses the opposite tendency in terms of time to failure as affected by sheet size. When these same tests are performed with no initial starting crack, the time to final fracture decreases with increasing sheet size.

8. Growth of damage and onset of rapid separation. For some purposes a fatigue fracture experiment may also be regarded as an illustrative model of a fracture developed during a single load application. When the nominal stress is held well below yield stress levels as in a fatigue test inelastic strains are detectable during the initial cycles only by internal friction measurements. However, some regions are weaker than others, some shifting of internal load distribution occurs, and configurations for load relaxation by localized plastic slip eventually form. It then becomes possible to develop cracks which grow to sufficient size for self-fracturing without the development of measurable general yielding. Because of the special conditions of long times, strain reversals, and local temperature fluctuation the balance of factors contributing to crack formation during fatigue differ from the case of fracturing under single load application, as discussed by Freudenthal[1]. However, the suppression of general yielding assists observation of cracks during early stages of their growth and studies of development and growth of cracks during fatigue are of definite general interest.

Careful microscopic observations have been made by Thompson[2] and his associates of the appearance of minute fatigue cracks in polycrystalline pure

---

[1] A. M. Freudenthal: "Fatigue", this volume.
[2] N. Thompson, N. Wadsworth and N. Louat: Phil. Mag. 1, 113—126 (1956).

copper. The first detectable cracks appeared to form directly from slip bands and their lengthening was accompanied by additional local deformation. As was shown in reference [11] the macroscopic extension process of large cracks consists in the formation, spreading, and joining to the main crack of new crack-like elements. Microscopic studies of fatigue fracture origins reveal an equally complex pattern of events. These are not discussed in further detail since the crack forming actions are somewhat specialized to surface cracks and to fatigue. However, one may note that new fracture origins, whether located in the elevated stress zone of an advancing large crack or developed within a zone of localized deformation during fatgiue, are consistently associated with substantial amounts of plastic strain.

Observable stages of crack lengthening are practically confined to the final half of the fatigue life. At the end of the fatigue life rapid fracturing occurs during one load cycle. $\mathscr{G}_c$ values have been calculated by KIES and IRWIN[1] from estimates of the crack size and load at instability using illustrations accompanying papers on fatigue of steel and aluminium alloys. The values obtained in this way compare will with $\mathscr{G}_c$ values measured on similar material not subjected to fatigue damage.

Large amounts of experimental information exist relating the creep rate of metal rods in tension to the tensile stress. Such studies are simplified by the fact that the force motivating creep, the longitudinal stress, can be held to a constant value as the deformation proceeds. It would be desirable at this point to discuss similar information relating time rate of crack extension to the crack-extension-force. However, the force definition appropriate when the crack length is large compared to the accompanying zone of plastic strains may not be appropriate for a small crack embedded in a region in which there is general plastic yielding. Furthermore, continual readjustment of the loads to hold the crack-extension-force constant while conceptually possible is experimentally difficult and little data based upon such tests is available[2]. Nondestructive measurement of the spreading of cracks within thick solids is also experimentally difficult and again the experimental information at hand is meager.

The first of the above difficulties may be resolved somewhat arbitrarily by computing the crack-extension-force as if no plastic strains were present throughout the entire range of crack lengths. This procedure has the virtue of uniformity and one is less at fault in ignoring influence of plastic strains than in ignoring influence of crack opening size. The second and third difficulties restrict this discussion to what can be inferred from illustrative models as discussed in Sect. 7 coupled with overall consistency considerations.

Such observations as have been reported indicate that the zinc foil results can be considered a good model of slow extension of cracks in glassy and polycrystalline materials. These results can be explained if one assumes that the development of fracture origins in each flawed region of a test specimen follows the same trend as do the fractional time rates of extension shown in Fig. 4. Thus, if a number of microscopic separations are growing in a competitive way with fractional or specific time rates of extension which tend to increase, then in terms of averages, the ratio of the length of the largest to that of the smallest cracks will also increase.

Consider two similarly stressed regions of the same material whose sizes are in the ratio, say, of one to two. Assume that each contains a normal distribution

---
[1] Unpublished notes.
[2] For an isolated exception see J. J. BENBOW and F. C. ROESLER: Proc. Phys. Soc. B **70**, 201 (1957).

of flaws. The larger region may be thought of as slightly weaker than the smaller one because of its greater chance of containing a more serious flaw. With growth of separations under the influence of tensile stress and time, the strength disadvantage of the larger region relative to the smaller region may be expected to increase steadily.

Amplification of an initially small "flaw probability" type of size effect by growth of fracture origins is a prominent characteristic of fatigue tests. A similarly large size effect is the normal expectancy of creep-rupture tests. The end point of slow fracture extension in fatigue, creep-rupture, and tensile tests is the onset of a fast fracture using locally released strain energy for propagation.

Judging by the curves of Fig. 4 the time rate of crack extension may be represented as a linear increasing function of $\mathscr{G}$ during stages of crack growth well removed from onset of fast fracturing. However, toward the last of the growth period the time rate of crack extension increases quite rapidly with increase of crack-extension-force. The general features of the relationship between time rate and force for crack extension are thus similar to the relationship which has been found to apply for creep rates as a function of stress.

In the case of strong metals, glass, and hard plastics, the onset of final self-fracturing is often observed to be abrupt and sudden. The reason for this is two-fold. In the first place, the resistance of the test specimen to fracture extension decreases in regions of largest stress or largest flaws due to growth of fracture origins. In the second place, the average fracture extension rate prior to fast fracture is composed of discontinuous segments of relatively fast fracture extension. The crack extension process waits for an accumulation of damage in material near the end of the crack sufficient to lower the resistance to crack extension within range of the existing crack-extension-force. Then the crack extends quickly through this damaged region. Finally, one such quick extension provides a sufficient increase of force so that the separation process no longer has to wait for time and stress to produce additional fracture strength damage. In extreme cases, the first visible crack extension process starts the final self-fracturing separation.

Characteristically, high temperature grain boundary fractures under creep-rupture conditions occur with little or no local reduction of area of the test specimen at the location of the fracture. The separation is of sudden brittle kind starting from one or more surface cracks of relatively small depth. Clearly, a general loss of strength is of at least comparable importance to lengthening of the starting crack in causing onset of rapid fracture in such instances.

In many situations of practical importance, there are residual stresses within fabricated metallic alloys, which are not removed by removal of the external loads. In such instances, crack growth, assisted by hydrogen embrittlement, corrosion, thermal fatigue, or by other causes, continues in unloaded structural members and may produce a substantial loss of fracture strength over a long period of time.

9. Conditions for fracture in terms of stress. The law of constant resolved shear stress represents rather well the influence of orientation upon yielding of ductile single crystals. NABARRO[1] has pointed out that this result corresponds to a characteristic critical force per unit dislocation line length necessary for expansion of dislocation loops in the acting slip planes. A similar inference would pertain to the passing of obstacles on the slip planes by crowded arrays or "pile-ups" of dislocations. Grain boundaries, interstitial atoms, and undissolved

---

[1] F. R. N. NABARRO: Phil. Mag., Ser. VII **12**, 213 (1951).

constituents add to the complexity of analysis of yielding of real materials. Whereas the force on a segment of a single dislocation loop depends only upon the resolved shear stress and not upon the loop size, the force on a dislocation "pile-up" is dependent upon spacing and extent of these arrays. Of the factors influencing force on a dislocation array only the average stress on the yielding region is directly observable. Thus, although the idea of relating forces on dislocations to their time rates of movement is of fundamental importance in dislocation mechanics, results of yield strength measurements are always given in terms of average stress.

Somewhat the reverse of this occurs in the application of a dislocation-type force concept to crack extension. As the crack becomes larger, restriction of

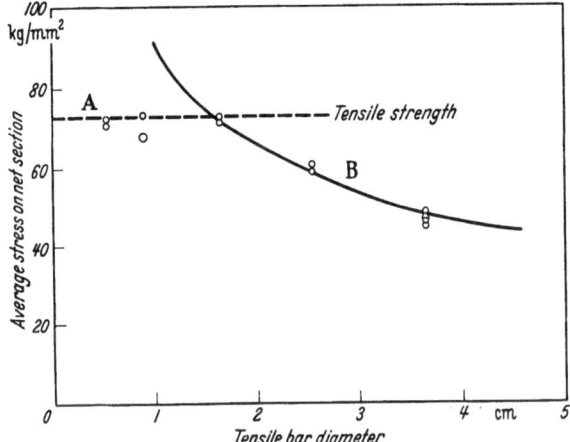

Fig. 5. Influence of size upon fracture stress for notched round bars of 7075-T6 Aluminum alloy. Notch reduced section area by one-half and had root radius of 0.25 mm. Curve B corresponds to a $\mathscr{G}_c$ value of 177×10^3 ergs/cm². Data from WEISS, reported by LUBAHN.

plastic strains on a relative scale to the zone of overstress near the crack edges leads finally to a clear physical situation well adapted for application of a crack-extension-force concept. Furthermore, fracture test data expressed in terms of the characteristic value of crack-extension-force for onset of rapid crack extension are directly useful in engineering strength calculations.

On the other hand, applications of the force idea to extension of very small cracks, while of assistance to qualitative understanding of their development and growth, are of little practical value. When a solid is tested to rupture under conditions such that the unstable crack size is not observed or under conditions such that a force calculation based upon applied loads and crack configuration is not feasible, then the conditions for fracturing can only be stated in terms of the average stresses and strains descriptive of the region within which the fracture instability developed.

Consider for example the test results discussed by LUBAHN[1] shown on Fig. 5. A somewhat brittle aluminum alloy was subjected to tests at various sizes in the form of round bars sharply side notched to a depth which reduced the net section-area by one half. The results shown are typical of those from a wide variety of fracture tests in which a notch or saw cut is used to assist the starting

---

[1] J. D. LUBAHN: Proc. of 1955 Sagamore Conference on Strength Limitations of Metals, Syracuse University, p. 159, 1956.

of a crack. When the test piece is sufficiently small, the zone of plastic deformation extends across the net section. The average stress on this section at the time of fracture then cannot exceed the yield stress by more than the moderate amounts due to work hardening and the influence of biaxiality of stresses upon the maximum shear stress. Thus line A on Fig. 5 is only about ten percent above the yield stress and coincides, in this case, with the ultimate tensile strength as measured on unnotched bars of the same material.

When the test piece size is large enough so that the plastic strains accompanying onset of rapid fracture are confined to the zone near the root of the sharp notch, the net section stress is less and reduces toward zero as the test piece size is increased. For points on curve B (Fig. 5) the product of the square root of the specimen size times the average net section stress has a constant value. Curve B corresponds to development and propagation of a crack from the vicinity of the notch root when the stress environment of that region reaches a critical magnitude and the result can be stated in terms of a characteristic crack extension force.

When a specimen with no notch is broken, as in the case of the most commonly used tests, the situation more or less resembles that of Fig. 5, curve A, depending upon the size and ductility of the specimen. Although the specimen has no intentional notch, as the load is increased, a starting crack of some type eventually must form. If at onset of fast extension of the starting crack the entire net section is yielding, then the average tensile stress for such a fracture, like that for the smallest test specimens of Fig. 5, can be estimated from the known resistance to plastic deformation of the material.

When the material is quite brittle, plastic strains accompanying development of the starting crack may be restricted to the close neighborhood of largest flaws to such an extent that the entire strain prior to fracture appears to be elastic and recoverable by unloading. All degrees of this condition occur in various crystalline and amorphous solids. Cleavage surfaces apparently devoid of plastic strains can be produced in some materials; in mica primarily because of the weakness of bonding across the cleavage plane, and in diamond primarily because of the absence of dislocation mobility. On the other hand, evidence of non-recoverable strain is found near fracture origins in glass and in all metals in which brittleness is brought about by lowering of temperature.

When the object of measurement is the determination of stress conditions for development of a small starting crack, the pertinent stresses are those averaged across the entire region within which crack development occurs. The results of such tests are often stated in terms of the average tensile stress normal to the plane of the section eventually severed by the fracture. This is usually the direction of greatest tension. Due to developed or inherent directional weakness the fracture may not have this orientation. For example, single crystal fractures under brittle conditions occur preferentially upon cleavage planes, usually the set of planes of widest spacing. In such cases the average tension resolved normal to the cleavage plane provides a fracture stress component which is essentially constant for fracture of the crystal in various orientations.

Cleavage of single crystals was discussed by BARRETT[1] who illustrated constancy of the critical normal tensile stress with results on bismuth. Results with zinc crystals in the temperature range $-253°$ C to $-80°$ C and with bismuth in the range $-80°$ C to $20°$ C substantiate the relative insensitivity of the critical normal tensile stress to temperature changes. According to BARRETT cleavage

---

[1] C. S. BARRETT: Structure of Metals, p. 317—320. New York: McGraw-Hill 1943.

in most crystals is facilitated by cooling to liquid air temperature and by driving a sharp blade into the crystal along the cleavage plane.

The constancy of critical normal tensile stress corresponds, in terms of crack-extension-force, to a characteristic value for $\mathscr{G}_1$ as defined in Sect. 5. The influences of temperature, strain rate, and biaxiality are common to many solids and are discussed in Sect. 15.

Fracture stress conditions for aluminum and magnesium alloys and steel were discussed by DORN [6] who concluded the results corresponded more closely to a critical maximum shear stress law than to invariance of the greatest tensile stress. A less ductile material, cast iron follows a critical maximum shear stress law only when fracture occurs by the sliding modes. Similar conclusions were reached by PARKER[1].

The general behavior pattern of materials relative to conditions for fracture in terms of average stresses is quite clear. For test specimens in which development of the starting crack is accompanied by general plastic yielding one would expect to predict the average stresses from known conditions for plastic yielding. In extremes of brittleness such that a condition of general yielding is not attained prior to propagation of the starting crack the tensile stress normal to the plane of separation dominates. When fracture occurs in compression tests of materials such as cast iron and ceramics, a shear fracture normally develops from sliding on a plane of greatest shear stress and results of tests are predictable in terms of a critical maximum shear stress.

For a wide variety of materials including single crystals experiments have established that both the yield stress and the fracture stress are dependent upon the specimen size. The size dependency is more noticeable for fracture stress than for yield stress. Reasons for this will be discussed more fully in Sects. 12 and 13.

When the object of measurement is the determination of stress conditions for extension of a well developed and somewhat brittle crack, the stress conditions in the region of the advancing edge of the crack may be given in terms of crack-extension-force as was brought out in Sect. 5. The description of fracture strength in terms of a $\mathscr{G}_c$ value while attractive because of its close association with the crack extension process is nevertheless a considerable oversimplification. Essentially it is a description of the average stress conditions for growth and joining of new fracture elements modified in a way appropriate for the region enclosing the advancing edge of a crack. Representation of fracture strength in terms of average stress is convenient and appropriate when no well formed crack can be observed prior to rupture. Primarily one needs to bear in mind (1) that fracture stress is not easily distinguished from plastic flow strength except for materials so brittle that yielding is not observed, (2) that fracture stress is substantially reduced by increase of specimen size, and (3) that both fracture stress and characteristic crack-extension-force are approximate descriptive concepts dependent in a complex way upon time rates of response to forces at a much smaller scale.

IV. Stress field, velocity, and division of a running crack.

10. Stresses near a running crack. A relatively complete photoelastic investigation of stresses near a rapidly moving crack was made by WELLS and POST [14]. The transparent material used was Columbia resin (CR-39) in the form of flat plates 3.2 mm thick and 12.7 cm wide. The length of plate in the direction

---

[1] E. R. PARKER: Brittle Behavior of Engineering Materials, p. 66. New York: John Wiley & Sons 1957.

of applied tension was 38 cm. The load, applied manually in a small fraction of a second, caused extension to occur from a pre-cracked notch at one of the plate side boundaries. Four spark light sources fired in sequence furnished four views of the stress field as shown in Fig. 6. The isochromatic loops close to the end of

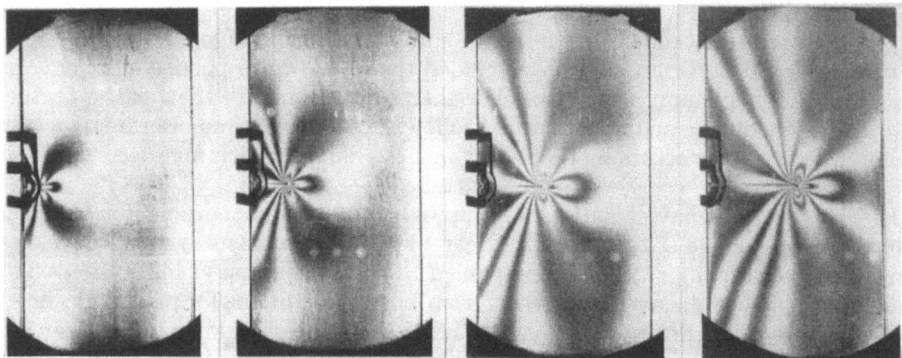

Fig. 6. Stress field near end of running crack from WELLS and POST [14]. Flash photographs show isochromatic fringes for four positions of the same crack traversing plate of Columbia resin.

the crack agree in appearance with calculated theoretical isochromatics based upon Eqs. (5.1), (5.2), (5 4), and an added uniform stress, $-\sigma_{0x}$, parallel to the crack. Close to the crack tip each isochromatic loop leans toward the direction

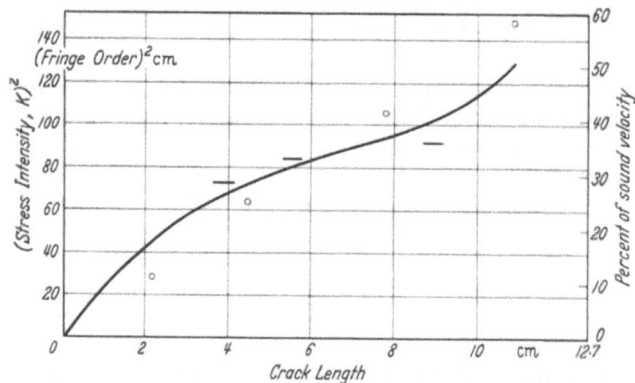

Fig. 7. Circles show average values of K^2 from isochromatics using photographs by WELLS and POST [14]. The curve is from Eqs. (6.14) and (10.4) with L proportional to time. Horizontal bars show average velocity measurements by WELLS and POST relative to a velocity of sound of 1.53 km/sec. One fringe order is 0.0092 kg/mm².

of crack motion by an amount which increases with σ_{0x} and has a size which increases with K. By measuring the angle, ϑ_m, from the x-axis (the extended path of the crack) to the position of greatest separation of the fringe from the crack tip and by measuring this greatest separation distance, r_m, it was possible, to obtain values of the two stress field parameters, K and σ_{0x}, for each of the WELLS-POST photographs[1]. The equations employed were

$$K = (2\tau_m)\sqrt{2r_m}\,f(\vartheta_m) \tag{10.1}$$

---

[1] See discussion by G. R. IRWIN published with reference [14].

and
$$\sigma_{0x} = (2\tau_m) \sec \frac{3\vartheta_m}{2} \left\{1 + \left(\frac{3}{2}\tan\vartheta_m\right)^2\right\}^{-\frac{1}{2}} \tag{10.2}$$
where
$$f(\vartheta_m) = \operatorname{cosec}\vartheta_m \left(1 + \frac{2}{3}\tan\frac{3\vartheta_m}{2}\cot\vartheta_m\right)\left\{1 + \left(\frac{2}{3}\cot\vartheta_m\right)^2\right\}^{-\frac{1}{2}}. \tag{10.3}$$

The results for K^2 are shown on Fig. 7. Each isochromatic loop corresponds to a constant maximum shearing stress, τ_m, in the plane of the plate. In the calculations the value of τ_m was taken equal to the fringe order and was not converted to stress units. The average values of K^2 obtained were nearly in simple proportionality to the crack length. The lateral compression factor, σ_{0x}, tended to be least in the central portion of the plate and was about half of the initial longitudinal stress, σ_0.

To obtain a theoretical calculation for purposes of comparison, use was made of a modified static stress analysis. Eq. (6.14) was employed with the value of A set equal to the plate width and σ adjusted for dynamic unloading due to rapid motion of the crack. To make the dynamic unloading adjustment, a study was made of the influence of the crack length upon the spring constant of a long plate. The result was
$$\frac{\sigma_0}{\sigma} = 1 + \left(\frac{8A}{\pi L}\right) \log\left(\sec\frac{\pi a}{2A}\right) \tag{10.4}$$
where L is the plate length and σ_0 is the initial tensile stress applied to the plate.

For the curve shown on Fig. 7 it was assumed the applicable value of L increased in proportion to time from the start of rapid crack extension and had the value $1.5 A$ when the end of the crack reached the midpoint of the plate width. The values of crack-extension-force predicted in this way are within the scatter range of the observations. An influence of bending in the plane of the plate not considered in the calculations would tend to increase the observed values for the longest crack lengths.

It is evident that, since the crack velocity is substantially below the velocity of elastic disturbances, the stresses in the end region of the running crack must resemble the stress field characteristic of the end region of a stationary crack. This was verified by WELLS and POST [14] by appropriate comparisons with the isochromatic fringes near the ends of stationary cracks. Additional support for this conclusion is provided by the mathematical studies of YOFFE[1].

Table 3. *Relative values of largest tensile stress and largest shear stress near the edge of an opening mode crack for $\sigma_{0x}=0$ and $\nu=0.3$.*

ϑ	$\sigma_1\left(\frac{\sqrt{2r}}{K}\right)$	$\tau_m\left(\frac{\sqrt{2r}}{K}\right)$
0°	1.00	0.200
30°	1.18	0.306
60°	1.30	0.433
90°	1.21	0.500
120°	0.93	0.433

11. Texture, velocity, and division of fractures. Examination of the stress equations for the opening mode crack dislocation, Eqs. (5.1) to (5.4), shows that the locus of greatest tensile stress does not lie in the x, z plane which contains the crack. Table 3 gives the relative values of the largest tensile stress, σ_1, and of the largest shear stress, τ_m, for various angles, ϑ. For simplicity it is assumed the components of added constant stress are zero. The ϑ value of largest tensile stress at a fixed small distance, r, from the edge of the crack then becomes $\pm\pi/3$.

Previous discussion has brought out that plastic strains assist the development of the smallest elements of separation. Thus, in the elevated stress zone

---
[1] E. A. YOFFE: Phil. Mag. **42**, 739 (1951).

surrounding the crack edge, flaws develop into components of fracture most rapidly at positions near a ϑ value of $\pm \pi/3$ where substantial shearing stress as well as maximum tensile stress exists. As new fracture elements form away from the idealized locus plane of the main crack and attain sufficient size, sliding modes of separation develop joining them with the main crack and with each other thus advancing the composite edge of the main crack. The magnitude of the roughness of the fracture surface depends in part upon ductility factors which handicap new openings in attaining a crack-like shape and in part upon the numerosity of serious flaws which determine the largest r values at which the stress elevation is effective. Due to the nature of the opening mode stress field, fractures can avoid roughness only in exceptional situations, for example the cleavage of brittle single crystals.

At the position r, $\pi/3$ the direction of σ_1 is normal to the plane of the main crack. However, the character of new openings forming in this region depends upon ductility, directional weakness, and the nature of the local flaws as well as upon the stress field. In low carbon steel at room temperature a flat fracture may be primarily composed of small elements of sliding mode separation as discussed by Parker[1]. These show a preference for planes of greatest shear stress although the locus of the main crack is a plane normal to the greatest tension of the general stress field. In the same material at low temperature, opening mode separations by cleavage dominate. The fracture surface then contains many small elements of separation in planes nearly parallel to the plane of the main crack.

In materials for which resistance to plastic flow increases with time rate of strain the velocity of crack extension has a marked influence upon the texture of the fracture surface near onset of rapid crack extension. As crack velocity increases, the development of outlying flaws is at first suppressed and the fracture surface becomes more smooth. However as will be clear from later discussion, the velocity increase lags behind increase of crack-extension-force. Thus the effectiveness of the stress field in development of flaws away from the crack plane soon dominates and increased roughening occurs.

The stress field situation for the sliding modes of fracture is not analogous to that for opening mode. For such fractures the role of new elements ahead of the main crack in the extension process has not been observed. Judging by position and direction of the largest shear stress, formation of new elements would be expected to occur most rapidly in the plane of expected extension of the crack.

As an aid to understanding the fracture strength of a material careful study of the fracture surface is always rewarding. Guiding principles to assist such studies are given in references [10] and [11].

Under the influence of a steadily increasing crack-extension-force, the velocity of a running crack approaches a limiting magnitude. Smith and Kies[2] investigated the effect of the tensile stress at crack initiation upon the velocity of cracks traversing plates of cellulose acetate. It was found the velocity increased with the initial tension until the crack began to divide. Trials at successively greater magnitudes of initial tension produced, first, the occurrence of division at shorter crack lengths, then sub-division of the branches, and finally a condition which one might term shattering.

---

[1] E. R. Parker: Brittle Behavior of Structural Materials, p. 67—70. New York: John Wiley & Sons 1957.

[2] H. L. Smith, J. A. Kies and G. R. Irwin: Phys. Rev. **83**, 872 (1951) (abstract).

At tensile loads sufficient to cause crack division the velocity was estimated in terms of movement of the ends of the most advanced cracks. This velocity appeared to be independent of the initial tensile stress. At tensile loads just less than that necessary for crack division the velocity across more than half of the plate width was equal within experimental error to the limiting velocity found after development of shattering.

It will be assumed negligible error is incurred in referring to the limiting crack velocity in a material either as the velocity of the locus of most advanced cracking or as the velocity of a single crack when the crack-extension-force is near the magnitude required for crack division. Table 4 gives a summary of limiting velocities on this basis.

Table 4. *Limiting velocities of running cracks*[1].

Material	Values of $0.5\,c_2$ km/sec	Experimental velocities km/sec
Glass (soda-lime-silica)	1.57	1.54
Glass (silica)	1.85	2.19
Cellulose acetate	0.34	0.30
Steel (0.18 C-annealed)	1.58	1.50
Columbia resin (CR-39)	0.47	0.55

$$\mu = \varrho\, c_2^2 = \frac{E}{2(1+\nu)}; \quad \mu = \text{shear modulus}; \quad \varrho = \text{density}.$$

Note: Glass data — H. RAWSON, Soc. of Glass Tech. **36**, 173 (1952). Cellulose acetate data — H. L. SMITH and W. J. FERGUSON, Naval Research Laboratory Progress Report, April 1950. Steel data — OSRD Report No. 6452, Jan. 1946. Columbia resin data — WELLS and POST [14].

Returning to the WELLS-POST experiment, on Fig. 7 the horizontal bars indicate average observed values of crack velocity as a function of crack length. One may note that as the crack extends the increase of velocity becomes much less than the increase with crack length of the measured crack-extension-force. A similar remark applies to comparison of measured velocities with the theoretical estimates of crack-extension-force although the margin of difference in this case is not so great.

The fact that a limiting velocity is at least nearly reached prior to branching of a crack suggests a calculation of limiting velocity may be allowable within the general frame work of comments on this topic by MOTT[2]. Exploratory calculations on this basis were made[3] which suggested the limiting velocity should be about half the velocity of elastic shear waves, c_2. Both the theoretical estimates and limiting velocity measurements are segments of incomplete work. The measurements at hand correspond closely to the value of one-half c_2 which suggests an inertial type velocity limitation whereas the appearance features of the experiments suggest the velocity is limited by onset of crack division.

Along the borders of long single cracks resulting from fast fractures of polymethylmetacrylate plates SMITH and KIES observed numerous small pairs of cracks suggestive of unsuccessful efforts to divide the main crack. There was an increased roughening of the texture of the fracture surface with distance from the region of onset of fast fracturing. This behavior, which is typical of most solids,

---

[1] Table prepared by H. L. SMITH.
[2] N. F. MOTT: Engineering **165**, 16 (1948).
[3] Correspondence between A. A. WELLS and G. R. IRWIN.

is apparently due to the expansion of the field of elevated stresses and, correspondingly, the increase of the crack-extension-force.

It was noted, however, that a doubling of the force, while sufficient to produce occasional small pairs of cracks, did not cause crack division. For plates of polymethylmetacrylate, cellulose acetate, and Columbia resin approximately a ten fold increase of crack-extension-force above the characteristic value, $\mathscr{G}_c$, was necessary to cause crack division.

V. Effects of size upon fracturing.

12. Analysis of strength by extreme value statistics. It was shown in Division I that, even for a pure liquid, an influence of size upon tensile strength is predicted. In glasses and metals, processes such as slip, twinning, cleavage, and growth of holes or micro-cracks which, in the aggregate, compose plastic flow and fracture depend in various ways upon pre-existing flaws or strength inequalities. The existence of strength inequalities implies existence of size effects. Thus, it would be self contradictory to assume that plastic deformation and separation processes occur and at the same time that no dimensional effects exist. Influences of size are fundamental to the nature of the deformation and separation processes. It is only the magnitude and character of the size effects which are in question. Larger size effects for fracture than for plastic flow are anticipated because fracturing tends to accentuate pre-existent strength inequalities to a greater degree than plastic flow. In fact, in a fracture test, the largest flaw developed by accentuation of strength inequalities is the crack, extension of which severs the test piece.

If one considers classes of fracture in which the volume element whose failure controls the measured strength is small compared to the total volume tested, statistical methods sometimes called "extreme value statistics" are useful. Fractures generally described as brittle and fractures of slender filaments and wires are of this class.

The fact that fracture origins grow at various rates under the influence of applied load and temperature still permits application of a statistical theory of fracture strength based on a flaw probability function. One merely assumes that the flaw probability function refers, not to the flaws in the material prior to the test, but to those which exist just prior to onset of unstable fast fracturing. The significant flaws in each small volume element are a function of the load-time history of that region. To a fair approximation, one may think of the effects of the load-time history as merely altering the values of the parameters in the flaw probability distribution function. These parameters, then, may be different for the same material under different conditions of testing.

For the statistical analysis it will be sufficient to use concepts developed by WEIBULL [4] with some modifications to show the influence of flaw growth. Assume, at first, the test volume is in uniform uniaxial tension. The influence of each weakening defect is assumed independent of other defects and such that it would by itself limit the supportable load on the test bar to a tensile stress S. Let P_S be the probability that a fixed small volume ω will contain an "S-flaw" in the range S to $(S+dS)$. The probability of no S-flaw of this kind in a large volume V is $(1-P_S)^{V/\omega}$. Assume the range of S values sub-divided and numbered so that

$$\delta S_K = S_{K+1} - S_K > 0. \tag{12.1}$$

The probability for failure in the range, δS_K, for the whole test volume is the probability that V contains no flaw as serious as S_K diminished by the probability that V contains no flaw as serious as S_{K+1}. This difference is

$$(1-G_{K-1})^{V/\omega}-(1-G_K)^{V/\omega} \tag{12.2}$$

where

$$G_K=\sum_{i=0}^{K} P_{S_i}. \tag{12.3}$$

For test specimen sizes of practical interest it is reasonable to limit attention to a group of most serious flaws which are a small fraction of the total number. This means that, for S values of primary interest, $\frac{V}{\omega} P_S$ is small compared with unity. The difference expression, (12.2), may then be replaced by

$$\frac{V}{\omega} P_{S_K} (1-G_{K-1})^{\left(\frac{V}{\omega}-1\right)}. \tag{12.4}$$

For mathematical simplicity it will also be assumed G_K can be approximated as a continuous function of S, namely $G(S)$, and that

$$\frac{V}{\omega} P_{S_K} (1-G_{K-1})^{\left(\frac{V}{\omega}-1\right)} \approx -\frac{\partial}{\partial G}\left[\exp\left(-\frac{V}{\omega}G\right)\right] dG \tag{12.5}$$

where the right side is evaluated at S_K.

It is clear that $G(S)$ must be a positive increasing function of S which is zero when S is zero. Consequently, S, considered as a function of G, is a monotonically increasing function of that argument. From these comments, one can see that the assumptions made above, general as they are, have certain necessary implications regarding the effect of the size of the volume subjected to fracture testing. For example, the expected average failure stress S_A from a large number of trials is

$$S_A=\int_{G=0}^{\infty} S\, e^{-\frac{V}{\omega}G}\, \frac{V}{\omega}\, dG \tag{12.6}$$

and

$$\frac{\partial S_A}{\partial V}=\frac{1}{V}\int_{0}^{\infty} S\left(1-\frac{V}{\omega}G\right) e^{-\frac{V}{\omega}G}\, \frac{V}{\omega}\, dG. \tag{12.7}$$

When plotted as a function of G, the expression

$$\left(1-\frac{V}{\omega}G\right) e^{-\frac{V}{\omega}G} \tag{12.8}$$

has the same area under the positive portion, G less than ω/V, as is enclosed by the negative portion, G greater than ω/V. Since S always increases with G, the result of the integration by which $\partial S_A/\partial V$ is calculated must be different from zero and negative under all permissible choices of V, ω, and $G(S)$.

For illustration, consider the choice of $G(S)$ made by WEIBULL [4].

$$\left.\begin{array}{ll} G=0 & \text{for } S<\sigma_0, \\ G=\left(\dfrac{S-\sigma_0}{\sigma}\right)^n & \text{for } S>\sigma_0. \end{array}\right\} \tag{12.9}$$

This corresponds to having σ_0 as a lower threshold of strength and to a flaw probability P_S of

$$P_S=\left(\frac{S-\sigma_0}{\sigma}\right)^{n-1}\frac{dS}{\sigma}. \tag{12.10}$$

With n greater than unity, this implies an ever greater probability as the supposed flaw becomes less serious in its weakening effect, an assumption which is plausible in its gross aspects. A frequency plot of test results falling within small, equal intervals of S is expected to resemble the function

$$f(S) = n\left(\frac{S - \sigma_0}{\sigma}\right)^{n-1} e^{-\frac{V}{\omega}\left(\frac{S-\sigma_0}{\sigma}\right)^n}. \tag{12.11}$$

If n is large, this function has negligible magnitude, except near the value of S for which it is maximum.

For the assumed case of uniform tension, we obtain for the average strength,

$$S_A = \sigma_0 + \sigma\left(\frac{\omega}{V}\right)^{1/n} \Gamma\left(1 + \frac{1}{n}\right) \tag{12.12}$$

and for the mean square relative deviation (relative variance)

$$\eta^2 = \left(\frac{S - S_A}{S_A}\right)^2_{\text{average}} = \left(1 - \frac{\sigma_0}{S_A}\right)^2 \frac{1.5}{n^2}. \tag{12.13}$$

If σ_0/S_A is negligible, the value of η is about $1.2/n$ and can be computed readily for comparison with observed scatter of experimental results when only n is known.

To attribute to a flaw population the lowering of fracture stress from theoretical estimates by a factor of 100 to 1000 has been criticized by some as inferring less uniformity in the results of fracture stress measurements than is actually found. However, measurements of the effect of size upon fracture stress provide a direct indication of the magnitude of expected scatter due to the flaws. Relation (12.13) gives the predicted theoretical scatter when P_S is represented by relation (12.10). The scatter actually found is invariably greater rather than less than the magnitude predicted by Eq. (12.13).

This point was discussed from a different aspect by FISHER and HOLLOMON[1]. Assuming a distribution of micro-cracks which is equivalent to a more realistic P_S than that of Eq. (12.10), FISHER and HOLLOMON concluded that the degree of uniformity of fracture stress experimentally found corresponded in theoretical terms to having a realistically large number of flaws in the test specimen sizes under consideration. The increase of η with decrease in number of flaws predicted by their analysis may help to explain the large variation of results which has been found in measurements of the strength of single crystal filaments or "whiskers" of metal[2].

In addition it was shown that this model gave essentially correct predictions of the influences upon fracture stress of flaw orientation during deformation of metals.

To assume σ_0/S_A negligible permits considerable simplification of mathematical analysis. It may be noted that one cost of this simplification is the fact that the corresponding value of η is insensitive to size of test volume whereas a review of strength measurement data shows a definite trend toward decrease of η with increase in size of the specimen tested as predicted by the Fisher-Hollomon statistical model and by Eq. (12.13).

13. Influences of growth of crack-life flaws. When a statistical analysis of fracture strength is employed with a view toward obtaining general trends of fundamental significance, it is helpful to keep in mind certain limitations of this

---
[1] J. C. FISHER and V. H. HOLLOMON: Metals Techn. **14**, No. 5 (1947).
[2] S. S. BRENNER: J. Appl. Phys. **27**, 1484 (1956).

approach. For example, to obtain determinations of a parameter such as σ_0 requires determination of S_A for scaled sets of strength measurements extending over an extremely large size range. Although in principle it is possible to do this without alteration of the material flaws, in practice it is difficult to avoid introduction of new classes of more serious flaws or increased seriousness of existing flaws in the largest sizes tested. Intuitively, one would expect some lower threshold to the weakening influence of the various possible material defects. However, if one studies graphs of log S_A as a function of log V from past experimental work, no convincing evidence for a σ_0 term of significant size is found. The previous comment may assist in reconciling this fact with one's intuitive idea that a limit of weakening influence nevertheless exists. In each set of experimental results on a particular material, a limited range and variety of largest flaws dominate the results. For consistency with general trends, it is necessary to choose for the flaw probability function a form which permits small flaws to be more numerous than large ones to a degree controllable by adjustment of the parameters of the function. The Weibull flaw probability function satisfies this requirement and permits convenient simplifications in the consideration of flaw growth and of non-uniform stress fields. For reasons apparent from the above comment, it is believed mathematical convenience also justifies dropping the σ_0 term of the Weibull formula in the analyses which follow.

Consider next the influence of flaw growth upon the parameter n. For illustration, it will be assumed that the controlling S-flaw of each volume element ω can be thought of as an embedded penny-shaped crack of radius r and that the growth rate of r under tensile stress S follows the trends shown in Fig. 4. The resistance to fracture extension at onset of unstable fast fracturing is assumed to be a constant. Since dG/dA is proportional to the product $S^2 r$, one has

$$S^2 r = S_0^2 r_0 \tag{13.1}$$

where S_0 and r_0 are any selected reference pair of values of stress and critical crack size. The S value of a volume element containing a crack of radius r then becomes

$$S = S_0 \left(\frac{r_0}{r}\right)^{\frac{1}{2}}. \tag{13.2}$$

Thus

$$\frac{dS}{S\,dt} = -\frac{1}{2}\left(\frac{dr}{r\,dt}\right). \tag{13.3}$$

The number of volume elements ω in V containing S-flaws in the range S to $(S+dS)$ is, with σ_0 omitted from equation (12.10)

$$N_s = \frac{V}{\omega}\left(\frac{S}{\sigma}\right)^{n-1}\frac{dS}{\sigma}. \tag{13.4}$$

This relation is represented by curve A of Fig. 8. If the flaw growth effect corresponds to a constant fractional change in diameter per unit time of each penny-shaped crack, equation (13.3) shows that each point of curve A may be thought of as moving to the left with constant velocity, so that, after some period of time under tensile loading, the frequency graph of S-flaws would be represented by a parallel curve such as curve C. However, this situation is most unlikely. Since our interest is primarily in representing the behavior of the largest flaws, it is expected that the fractional time rate of extension will be an increasing function of their length or a decreasing function of their S-value.

Thus the influence of flaw growth is expected to be represented by a curve of smaller slope, such as curve B of Fig. 8. There is no reason to draw curve B other than as a straight line since no more is attempted than an approximate representation of the largest flaws as allowed by a Weibull type probability function. A decrease of the parameter n with time under load is, therefore, to be expected. Results of size effect tests with varied loading time have consistently shown this trend. From the above considerations, one anticipates that damage to the strength properties of a material will cause an increase in the scatter of test data, a lowering of the average measured strength, and an increased sensitivity of the strength to size. These trends are in general agreement with engineering experience.

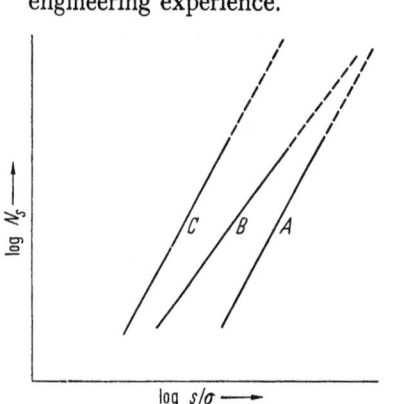
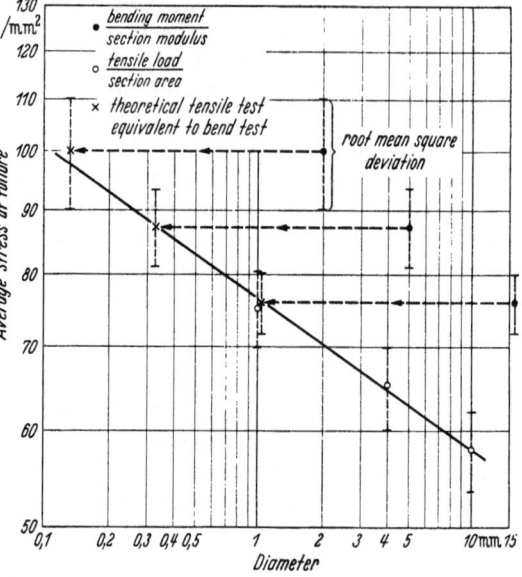

Fig. 8. Schematic graphs of the relation between $\log N_s$ and $\log(S/\sigma)$ from Eq. (13.4) showing influence of two hypothetical trends of flaw growth.

Fig. 9. Tensile and bend strength results on a low ductility mild steel at liquid air temperature. Replotted from data of DAVIDENKOV, SHEVANDIN, and WITTMAN.

14. Measurements of effects of size upon fracturing. When the statistical approach is applied to a non-uniformly stressed solid, for example, a bar subjected to bending moments, some modifications of the above analysis are required. It is, at the outset, necessary to consider whether the influence of stress gradient over regions comparable to the instability size of a crack may be neglected. With this condition satisfied, one may make a computation of the expected average value, S'_A, for a set of trials in which the strength value of each trial is given as the computed largest tensile stress, S_m, in the test piece corresponding to the loads just before onset of sudden fracture. The result is

$$S'_A = \left(\frac{1}{R}\right)^{1/n} S_A \tag{14.1}$$

where S_A is as given by Eq. (12.12) and

$$R = \frac{1}{V} \iiint \left(\frac{S(x, y, z)}{S_m}\right)^n dx\, dy\, dz. \tag{14.2}$$

Here the integral is taken to extend throughout that portion of the test piece subjected to tension and $S(x, y, z)$ is the maximum tensile stress at the point (x, y, z).

In the experiment to be discussed next, test conditions were selected for which very little flaw growth during time under load would be expected. The experiment provides the best known comparison of results of statistical fracture strength analysis for bend tests with those for tensile tests. Fig. 9 shows results replotted from DAVIDENKOV, SHEVANDIN, and WITTMAN[1]. It was the purpose of this work to investigate fracture size effect of a commercial metal under conditions of extreme brittleness. Thus the steel chosen had a high phosphorous content and the tests were done at liquid air temperature, the area reduction at this temperature being insignificant. The room temperature ultimate tensile strength was 62 kg/mm². Small specimens were machined from the ends of larger specimens after they had been broken. The tests were made in tension in one set of tests and in bending in a second set of tests. For each specimen size, sixteen to thirty results were obtained.

The results show areas of agreement with the statistical analyses discussed above. When the logarithm of the failure stress is plotted against the logarithm of specimen diameter as in Fig. 9, the slopes of the curves through the data are about the same for the bend tests as for the tension tests and suggest an average value of 25 for the constant n. Estimates of η were made from the published data and were found to vary in the range of 1.1 to 1.9 times the theoretical value. Each group of bend tests provides an average failure stress which is theoretically equal to the failure stress for a group of tensile bars which are of smaller size by an amount which can be calculated by computing R for the bend tests from Eq. (14.2). The close agreement with the tensile test results of the shifted bend test points as shown in Fig. 9 is somewhat surprising in view of uncertainties of calculation and measurement. The computation of R for the bend tests assumed the stress to be proportional to distance from a neutral plane through the axis of the cylindrical specimen and to fall off linearly with distance along the rod from the central point of load application. Due to the large value of n, the results are determined primarily by stresses in a small region at the position of greatest tension. Since the stress drop with distance along the bar from the midpoint is undoubtedly not linear in the region of the greatest stress, the computation of R using that stress gradient may be questioned. Possibly changes in the magnitude of the greatest tensile stress from that predicted by the simple beam formula compensate in some way so that a correct computation of R would not produce greatly different results.

Notched bar size effect studies in tension using a good quality structural low carbon steel were made by BROWN, LUBAHN, and EBERT[2]. The test conditions permitted considerable amounts of plastic deformation at the root of the notch. The bars broke suddenly, essentially at the maximum recorded load. The trend with size of the measured nominal stresses corresponded to $n = 50$. Accurately scaled bend tests of notched bars using tough heat-treated Ni—Cr alloy steels were reported by SHEARIN, RUARK, and TRIMBLE [6]. The values of n, again based upon maximum load, showed little consistency from one plate to another but were all in the range 60 to 100.

From experiments by ANDEREGG[3] on small diameter glass rods in tension, a value of n of about 8 may be deduced as characterizing the effect of changes in specimen length.

McVICKER and IRWIN[4] studied the effect of specimen length upon strength of glass fibers on a basis similar to that of ANDEREGG. A comparison was made of

---

[1] N. DAVIDENKOV, E. SHEVANDIN and F. WITTMAN: Trans. Amer. Soc. Mech. Engrs. **69**, A 63 (1947).
[2] W. F. BROWN, J. P. LUBAHN and L. J. EBERT: Welding J. Res. **26**, S. 554 (1947).
[3] F. O. ANDEREGG: Ind. and Eng. Chem. **31**, 290 (1939).
[4] NRL Laboratory Report in preparation.

fibers damaged by handling with others carefully protected from surface abrasion. The decrease of average strength for the damaged fibers was much greater for long than for short specimens suggesting the damage did not greatly alter the number of relatively small flaws. The corresponding n values were 3 for the damaged fibers and 12 for the undamaged fibers. Despite use of careful testing methods the measured η values exceeded the value $1.2/n$ by somewhat more than a factor of three. Use of a flaw distribution function more like that of the Fisher-Hollomon model discussed in Sect. 12 would have improved agreement of theoretical and measured η values.

In the report of their experiments by SHEARIN, RUARK, and TRIMBLE [6], attention was primarily upon total work done by the loading forces in causing fracture rather than on the value of the maximum load. Fig. 10 is typical of tests of this nature on steels which are tough in their resistance to crack initiation. Unstable fast fractures occurred in some, though not in all, of the largest size bars. The load-strain curve for the largest size bar of Fig. 10 was selected to show an example of a test which ended with a sudden fracture. Three prominent size effects were noted: as the size of the test piece was increased, the total work per unit volume to complete fracture became less, the initial crack at the root of the notch tended to become well developed at a smaller total strain, and the chance of sudden fracturing was increased. With respect to the first two effects, one may observe that the notch indentation is, in each specimen, the largest flaw and that for equal nominal stress the force tendency acting to extend the notch is proportional to the specimen size. For a more detailed view, one may observe that, during early stages of yielding, flaws in similar positions relative to the notch would have the same stress environment in the different test sizes. Individually, similar flaws in similar positions but in different size bars would develop into extending cracks at the same rate. However, the region subjected to large stresses increases with the specimen size. Thus, because of flaw probability considerations, a notch develops into a natural crack the more readily the larger the bar size. Even the largest bar size of Fig. 10 is too small for a pronounced influence of $\mathscr{G}_c$ upon the nominal or average stress. Thus for these tests the nominal stress at which the crack develops is controlled by plastic yielding and is reduced in the larger bars only by the relatively small amounts typical of size effects associated with plastic deformation.

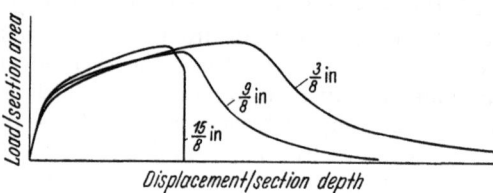

Fig. 10. Typical load-deformation curves for three sizes of notched bar bend tests of a steel. From SHEARIN, RUARK, and TRIMBLE.

15. Factors which control degree of brittleness. The zone of plastic strains in the region near the advancing edge of a running crack is influenced by the fact that propagation velocities for plastic strains are normally much less than the propagation velocities of elastic strains. For example, it was pointed out by VON KARMAN[1] and by TAYLOR[2] that the velocity, $V(\varepsilon)$, of a plastic strain, ε, under conditions of rapidly applied uniaxial tension or compression might be estimated from

$$V(\varepsilon) = \sqrt{S'/\varrho}$$

---

[1] T. v. KARMAN and P. DUVEZ: J. Appl. Phys. **10**, 987 (1950).

[2] G. I. TAYLOR: J. Inst. Civ. Engrs. **26**, 486 (1956).

where S' is the slope, $dS/d\varepsilon$, of a curve relating the force per unit original area, S, sometimes called "engineering stress", to the average sectional strain, ε; and where ϱ is the density of the material.

At the strain for which S' is zero, the plastic reduction of area of a tensile bar becomes locally concentrated because strengthening due to work hardening becomes at this point less than weakening due to reduction of section area. In consistency with this interpretation $V(\varepsilon)$ is zero when S' is zero.

In the tearing of a ductile metal foil the separations which compose the advance of the crack form in a zone of large plastic strains. A considerable fraction of the total deformation work is expended in this region which does not expand in proportion to crack length because the characteristic propagation speed for large plastic strains is less than even a slow time rate of motion of the crack. Thus the zinc foil fracture development model discussed in Sect. 7 corresponds roughly to the conditions of the modified Griffith theory even though lengthening of the zinc sheets by creep was observed along with the crack extension.

Consider next a cleavage crack rapidly traversing a large plate of structural grade low carbon steel, several centimeters thick. Crack velocities ranging from 0.9 to 1.5 km/sec have been observed. For thickness reduction by plastic strain to occur in advance of such cracks the slope, S', would need to be in the range of 3 to 9% or more of Young's modulus. One would estimate from inspection of static stress-strain curves that plastic strains greater than 2% would not occur, a result which agrees with what has been commonly observed. However, a dynamic rather than a static stress-strain relation should be used for this estimate. Reference [15] gives a brief summary of dynamic measurements for two steels.

In the first suggestions of a relation between plastic instability and brittleness[1] local adiabatic temperature rise was of equal importance to rate of strain hardening. This consideration as well as influence of strain rate upon plastic flow stress should be taken into account. A simple association of strain hardening with brittleness under various conditions of crack extension cannot therefore be expected. However, the general features of all these considerations are at least qualitatively known.

When a crack is subject only to two-dimensional considerations as for the severing of a thin ductile foil, the fact that plane strain conditions are necessary at the advancing edge is not relevant. However, in the case of cracks traversing thick sections the regions of the crack well removed from free surfaces must conform to conditions of plane strain. Calculations using Eqs. (5.1) through (5.4) show that, for fixed r and K, the ratio of greatest shearing stress for plane stress to greatest shearing stress for plane strain has values of approximately 1, 1.4, and 3 at the three ϑ values of $\pi/2$, $\pi/3$, and zero. Thus the zone of plastic strains near the advancing edge of a crack is substantially altered and restricted when the situation changes from one of plane stress to one of plane strain. The magnitude of this influence upon values of $\mathscr{G}_c$ can be estimated by comparing the plane strain $\mathscr{G}_c$ value shown on Fig. 5, 1.8×10^7 erg/cm, with the measurement for the same aluminum alloy under conditions of plane stress shown in Table 2, 5.2×10^7 erg/cm. The yield and fracture strengths of high strength aluminum alloys are only slightly influenced by temperature and strain rate for temperatures within 100° C of room temperature.

Observations on several steels indicate a condition which is predominately one of plane strain will prevail when

$$3E\mathscr{G}_c \leq \sigma_Y^2 B \qquad (15.1)$$

---

[1] See articles by ZENER and by READ in Ref. [6].

where σ_Y is yield stress, B is the span of the crack between free surfaces, and E is Young's modulus. When this condition is met, a predominately cleavage crack can be made to progress slowly through a section of low carbon steel. Otherwise rapid motion of the crack is necessary and plane stress conditions accompanied by large plastic strains occur if the motion drops below a critical speed.

From the preceding discussion it is clear that tendencies toward embrittlement from lowering of temperature and from increase of strain rate should be abrupt in those situations where plastic flow stress changes abruptly with temperature and strain rate. In addition a pronounced decrease in the plastic strains near a crack results from the addition of tensile stress parallel to the leading edge.

These factors are of a kind which might be influenced beneficially by planned variations in composition and microstructure of the materials. They are superimposed upon the cracks, porosity, and unwanted constituents which normally occur due to manufacturing conditions in most commercial materials.

Other causes for embrittlement exist frequently in forms which are also influenced by temperature, strain rate, and state of stress. For example it has been known for many years that crack extension is assisted in the case of brass by slight amounts of mercury and in the case of glass by water[1]. Kies[2] has measured values of $\mathscr{G}_c$ for commercial lantern slide glass as influenced by water vapor. He found little strengthening of the glass with reduction of humidity except at a relative humidity of 2% or less. The results, which are included in Table 2 (end of Sect. 6), include a substantial segment of slow crack extension as a portion of the crack length used in the $\mathscr{G}_c$ computation. The observations by Kies are difficult to explain unless one assumes that $\mathscr{G}_c$, regarded as a function of time rate of crack extension, can have a maximum in the region of relatively low velocity. A relationship of this kind has in fact been found by Benbow and Roesler[3] for polymethylmethacrylate. Assuming such a maximum exists, it then appears the influence of moisture in the case of glass can be described as essentially a lowering of the $\mathscr{G}_c$ barrier to rapid crack extension. The strong affinity of water molecules for a clean surface of glass is well known. Presumably the increase of crack extension velocity at a given stress and crack length may be ascribed to a lowering of the surface tension of glass. This interpretation would be consistent with the fact that under rapid load application the strength of glass is not improved by removal of moisture.

An ambiguity of interpretation occurs in discussions of the embrittling influence of hydrogen in steel and other metals. It is known that molecular hydrogen films form on clean metal surfaces. Petch and Stables[4] suggested that the embrittling influence of hydrogen should be interpreted as lowering the effective surface tension of the metal. However, the older internal pressure theory appears to have at least equal justification. A large number of papers on hydrogen embrittlement have appeared[5]. As Baldwin has suggested the major features can be simply understood in terms of internal pressure even though an additional surface tension type of influence is simultaneously present.

It is well known that very large pressures develop from diffusion of atomic hydrogen through steel and the entrapment of molecular hydrogen in internal cavities. Eq. (5.11) gives the crack-extension force for a penny-shaped internal

---

[1] See, for example, T. C. Baker and F. W. Preston: J. Appl. Phys. **17**, 179 (1946).
[2] Unpublished laboratory notes.
[3] J. J. Benbow and F. C. Roesler: Proc. Phys. Soc. B **70**, 201 (1957).
[4] N. J. Petch and P. Stables: Nature, Lond. **169**, 842 (1952).
[5] e.g. see W. M. Baldwin and J. T. Brown: J. of Metals, Trans. Sect. **6**, 298 (1954).

crack subjected to remotely applied tension σ. For internal pressure p, the expression for $\mathscr{G}$ has the same form with σ replaced by p. Remembering that when stress systems are combined, the K values are additive, the crack-extension force for combined tension and internal pressure becomes

$$\mathscr{G} = \frac{2(1-\nu)}{\pi\mu}(\sigma + p)^2 a. \tag{15.2}$$

Thus if p is an appreciable fraction of σ, say one-half, the presence of the internal pressure may increase the value of $\mathscr{G}$ by more than a factor of two.

Whether $\mathscr{G}$ increases or decreases when the crack extends quickly depends upon the ratio of p to σ and upon the initial non-elastic volume of the crack. A self-arresting behavior with crack extension is necessary in order to explain appearance features of the fracture and the influence of time upon stress rupture life. One finds that minimum internal pressures, p, ranging from $\sigma/4$ (for negligible non-recoverable crack volume) to $5\sigma/4$ (for relatively large non-recoverable crack volume) are required for a self-arresting behavior. These values p are theoretically possible at room temperature and at hydrogen contents commonly observed under conditions known to produce embrittlement. For rapidly applied loads no influence of hydrogen is expected because there is no time for restoration of pressure during the large cavity expansion which occurs as the starting crack develops to critical size. Influence of hydrogen pressure upon the deepening of a crack open to a low pressure region is not excluded because the new fracture elements, whose growth and joining compose the crack extension process, are internal. The uncertain effect of regions of large plastic strain upon collection of hydrogen in a cavity as well as the surface energy effect proposed by Petch and Stables provide ample substance for additional study.

16. Closing comments. The plan of this article has been to present the principal concepts needed for explanation of experimental observations. The experiments discussed were limited to those necessary for purposes of illustration. Among significant experiments which were not discussed are those using essentially single crystal whiskers[1]. The condition of this area of work currently lies intermediate between that of tensile strength of pure liquids and fracture strength of polycrystalline and otherwise flawed solids.

With regard to solids of this latter kind published experimental observations of fracturing and fracture strength are numerous, as might be expected for a topic of considerable practical interest. Among such papers, those which describe the processes associated with fracture and those which report measurements of fracture size effects are of most value. Noteworthy investigations of these features have been made by Bridgman [8], Tipper [9], de Leiris [10], Kies, Sullivan and Irwin [11], Stanton and Batson [12], and Docherty [13]. The fracture mechanics concepts now at hand were stimulated by the experimental fact that fracture size effects of large magnitude were found in spite of painstaking elimination of superficial causes for differences in results, such as specimen shape, surface finish, selection of material, and the testing equipment.

Bibliography references.

[1] Timoshenko, S. P.: History of the Strength of Materials, p. 51. New York: McGraw-Hill 1953.
[2] Nadai, A.: Theory of Flow and Fracture of Solids, pp. 207–228. New York: McGraw-Hill 1950.
[3] Griffith, A. A.: The Phenomenon of Rupture and Flow in Solids. Phil. Trans. Roy. Soc. Lond., Ser. A **221**, 163–198 (1920).

---
[1] e.g. see S. S. Brenner: J. Appl. Phys. **27**, 1484 (1956).

[4] WEIBULL, W.: A Statistical Theory of the Strength of Metals. Proc. Roy. Swed. Inst. Eng. Res. **1939**, No. 151.
[5] WESTERGAARD, H. M.: Bearing Pressure and Cracks. J. Appl. Mech. **6** (2) (1939).
[6] Fracturing of Metals (A.S.M.S., Oct. 1947), A.S.M., Cleveland, 1948. Contains, among other papers: C. ZENER: The Micro-Mechanism of Fracture; J. E. DORN: The Effect of Stress State on the Fracture Strength of Metals; G. R. IRWIN: Fracture Dynamics; P. E. SHEARIN, A. E. RUARK, R. M. TRIMBLE: Size Effects in Steels and other Metals from Slow Bend Tests; T. A. READ: Plastic Flow and Rupture of Steel at High Hardness Levels; J. H. HOLLOMON: Fracture and the Structure of Metals.
[7] Fatigue and Fracture of Metals (M.I.T. Symposium, June 1950), p. 139. New York: Wiley 1950. Contains, among other papers: E. OROWAN: Fundamentals of Brittle Behavior in Metals; W. WEIBULL: The Statistical Aspects of Fatigue Failures and its Consequences; P. L. TEED: The Influence of Metallographic Structure on Fatigue.
[8] BRIDGMAN, P. W.: Studies in Large Plastic Flow and Fracture. New York: McGraw-Hill 1952. Chapter 12 on volume changes in simple compression is of special interest in reference to the nature of fracture origins.
[9] TIPPER, C. F.: Dimensions in Testing (Conference on Brittle Fracture in Steel). J. West Scotland Iron a. Steel Inst. **60** (1953).
[10] LEIRIS, H. DE: L'Analyse Morphologique des Cassures. Métaux, Corrosion, Industries, No. 316, Dec. 1951.
[11] KIES, J. A., A. M. SULLIVAN and G. R. IRWIN: Interpretation of Fracture Markings. J. Appl. Phys. **21**, 716 (1950).
[12] STANTON, T. E., and R. G. C. BATSON: On the Characteristics of Notched-Bar Impact Tests. Proc. Inst. Civ. Eng. **211**, 67 (1920/21). The published discussions of this paper are also of interest.
[13] DOCHERTY, J. G.: Slow Bending Tests on Large Notched Bars, Engineering, Vol. 139, p. 211. 1935.
[14] WELLS, A. A., and D. POST: The Dynamic Stress Distribution Surrounding a Running Crack. Proc. Soc. for Exper. Stress Analysis **15**, No. 2 (1958).
[15] KRAFFT, J. M., A. M. SULLIVAN and G. R. IRWIN: J. Appl. Phys. **28**, 379 (1957).

Fatigue.

By

ALFRED M. FREUDENTHAL.

With 9 Figures.

I. The fatigue phenomenon.

1. Features of fatigue fracture. Fracture produced by a monotonically increasing force represents the terminal point of a process of inelastic deformation at which the rate of such deformation is no longer sufficient to prevent the rapid spreading and coalescence of existing submicroscopic cracks. Under a repeatedly applied cyclic force, however, fracture is finally produced by a force-amplitude which is far below the force associated with fracture under a single load application; this amplitude decreases with increasing number of repetitions. If the number is relatively large, fracture occurs well within the range of stresses and deformations which, at least phenomenologically, can be considered elastic. Such fractures are designated as *fatigue fractures*. They occur without any noticeable overall permanent deformation, and the fracture surfaces in metals have a characteristic appearance which indicates the progressive character of the fatigue damage (Fig. 1): A smooth zone of brittle separation, showing "clam shell" markings which represent the outline of the fatigue crack at various periods in its growth, starts from one or several nuclei (stress concentrations) indicating the immediate "cause" or causes of the fatigue fracture, and spreads gradually; when the remaining cross-section can no longer withstand even a single further application of the cyclic force, fracture by momentary overload suddenly starts, producing a surface of coarse crystalline appearance. It is only at this stage that the crack-propagation mechanism associated with fracture under a single rapid load application becomes relevant with respect to fatigue fracture.

The fatigue crack progresses quickly under a high intensity of the cyclic force, but may take years to develop under a low force-amplitude; in the latter case the crack surface is frequently discolored as a result of surface friction and oxidation. An inverse relation exists, therefore, between the force- or deformation-amplitude or, rather, their specific intensities expressed in terms of nominal stress S or strain ε, and the *fatigue life* N, usually defined as the number of repetitions of the stress- or strain-cycle producing fracture at the applied amplitude. (When progressive damage proceeds under conditions of resonant vibration the fatigue life is, however, usually defined in terms of "fatigue failure" rather than fatigue fracture, by the number of cycles that produce a fatigue crack of a depth sufficient to eliminate the resonant vibration by its damping effect and consequent change of natural frequency.) Within the range of essentially elastic strains the trend of this relation is represented in the form of $S-N$ diagrams. Beyond the elastic range the lack of a simple relation between stress and strain prevents the establishment of an $S-N$ relation from the $\varepsilon-N$ relation observed under conditions of controlled deformation-amplitude; the phenomenon can therefore only be described directly in the form of an $\varepsilon-N$ diagram.

The repeated stress-amplitude $\pm S$ producing fatigue fracture or failure within a certain range of fatigue lives N depends on the applied mean stress S_m roughly in a way indicated in Fig. 2. The maximum reversed stress-amplitude $\pm S$ which may be superimposed on the steady (mean) stress S_m and repeated on the average N times before causing fracture decreases with increasing level of S_m.

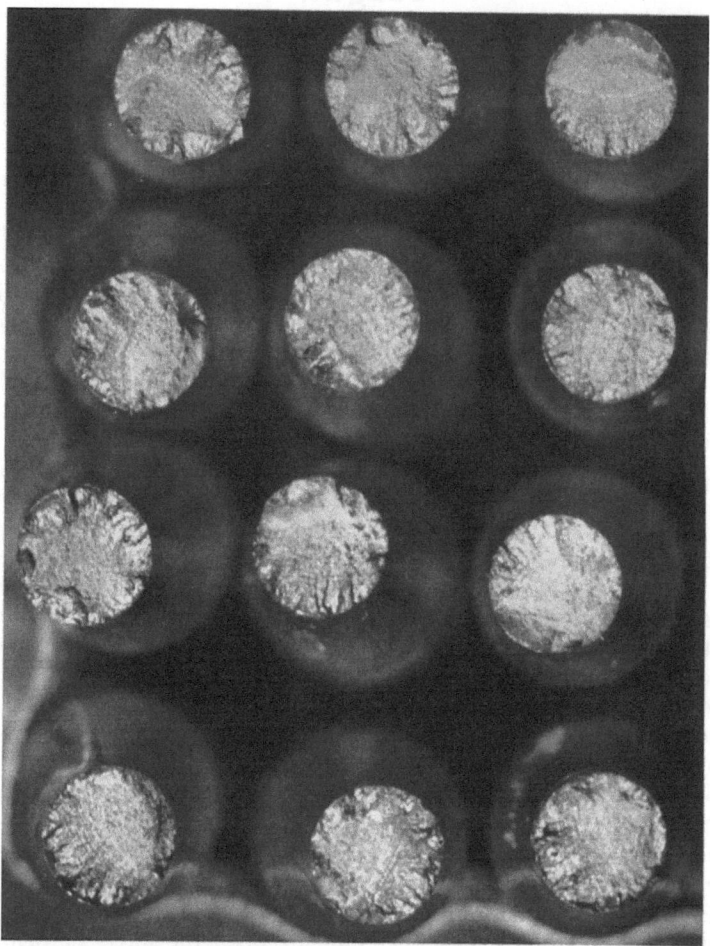

Fig. 1. Characteristic fatigue fracture surfaces in bending of structural aluminum alloy under random loading.

The actual shape of the function $S = f(S_m)$ depends essentially on the ductility of the material; while for ductile metals a relation of the form $S = S_r [1 - (S_m/S_u)^2]$ represents the test results fairly well, the behavior of brittle metals is better fitted by the linear relation $S = S_r \cdot [1 - (S_m/S_u)]$ or by relations falling below this line [1]. The maximum stress-amplitude at zero mean stress, representing conditions of complete stress-reversal, is usually designated as the "*fatigue stress*" or "*fatigue limit*" $S = S_r$, while S_u is an experimental constant of the order of magnitude of, but not equal to, the fracture strength under a single load application.

With increasing life the fatigue limit decreases asymptotically towards a stress-amplitude S_0, the "*endurance limit*", which can be repeated indefinitely without ever producing fatigue failure or fracture. For certain materials, such as aluminum and copper, with recovery temperatures relatively close to room temperature, as well as under conditions of pronounced time-sensitivity of mechanical behavior produced by elevated temperatures and by corrosive effects of the environment, the endurance limit may be so low as to be practically indistinguishable from zero; for other materials, particularly ferrous metals under conditions of moderate or low temperatures and non-corrosive environment, the endurance limit may be high enough to eliminate fatigue as a phenomenon except at the high stress-amplitudes associated with considerable local plastic deformation.

It should be noted, however, that the *endurance limit*, where it exists, *is a characteristic feature of the conventional fatigue test*, in which only constant stress-amplitudes are applied to failure. Under variable stress-amplitudes a definite endurance limit can hardly exist unless all stress-amplitudes are below the constant stress-amplitude limit. Otherwise the fatigue damage produced by stress-cycles above this limit will tend to reduce the limit in relation to the number and intensity of such stress-cycles applied.

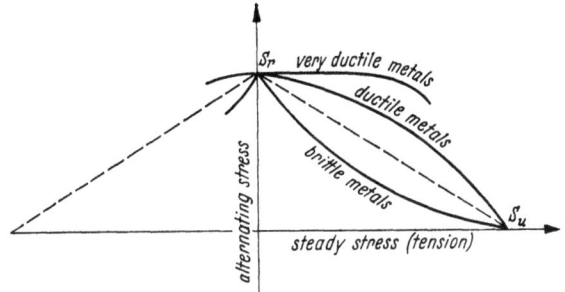

Fig. 2. Schematic diagram of effect of mean stress on strength under stress reversal.

2. Cycle- and time-effects. Most metals occupy intermediate positions, with finite fatigue life ranges extending towards $N = 10^7$ or 10^8 cycles; beyond this limit for all practical purposes fatigue failures no longer occur. In certain very brittle materials, such as glasses and ceramics, as well as in most materials at very high temperatures, fatigue damage defined as the progressively destructive effect of repeated load-cycles does not exist, or is of such minor significance that it can not be observed independently; progressive damage under such conditions is essentially due to time-effects, such as creep, corrosion, oxidation or chemical reaction. However, in partly crystalline high polymers, such as nylon, definite cycle-dependent fatigue effects exist jointly with creep-fracture effects. It appears, therefore, that fatigue is a phenomenon associated only with materials and conditions for which individual stress-cycles produce localized minute, but distinct permanent changes in the sub-microscopic or microscopic structure of the material. Where such changes are absent or indistinguishable because of the quasi-isotropy of the structure, as in highly brittle or in glassy substances, or are overshadowed by recovery and recrystallization processes, as at high temperatures, fatigue effects are insignificant or do not exist. The dependence of the fracture stress on the number N of stress-cycles is therefore replaced by its dependence on the time t under stress. In the intermediate range in which both cycle-dependent and time-dependent progressive damage exist and interact, the fracture stress S is a function of N and t. An $S-N-t$ surface (Fig. 3) indicates the trend of the phenomenon over the whole range; its intersection with the plane $t = 0$ represents the fatigue-relation $S-N$, its intersection with the plane $N = 0$ the creep-fracture-relation $S-t$. The joint effects on the fracture strength of N and t will appear as an apparent effect of the frequency of application of the cyclic stress if only the $S-N$ relation is considered; in this case it might also appear as an effect of increasing temperature reducing the fatigue strength, while the reduction may, in fact, be due only to the temperature-controlled intensification of the creep-damage. Under conditions of vanishing time-effects, represented by absence of measurable creep during the total period of the repeated stressing, the fatigue strength is largely independent of the frequency of the cyclic stress. A relatively slight increase of the fatigue strength with frequency, however, is observed in the range of very high frequencies (>500 cps.), probably as the result of blocking, at high strain rates, of certain inelastic deformation mechanisms [2]. Similar blocking effects at low temperatures are probably responsible for the observed moderate increase of fatigue life with decreasing temperature. On the other hand, it appears that even in the absence of significant creep-damage the fatigue strength tends to decrease at very low frequencies of the applied stress cycle ($\ll 1$ cps). This may be due to the fact that the characteristic deformation within the crystal structure associated with fatigue will occur only above a certain order of magnitude of the rate of straining, while below this range the type of deformation produced is practically identical with that occuring under slow unidirectional straining [7c], [8a], leading therefore to damage by "*alternating plastic deformation*" rather than to actual fatigue damage. Test results in this range are, however, scarce and not quite conclusive.

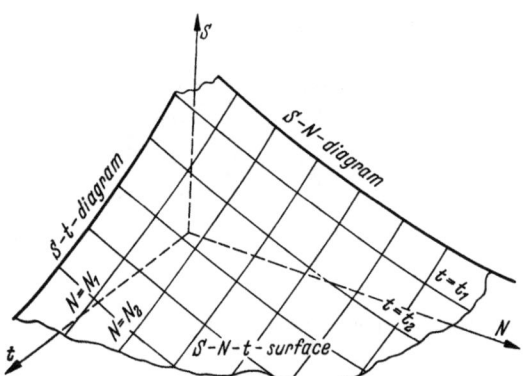

Fig. 3. Schematic S–N–t surface.

3. Fatigue relation. The simplest form in which the inverse trend of the relation between S and N can be expressed is

$$(N - N_{0S}) = \left(\frac{k}{S - S_0}\right)^\varrho \tag{3.1}$$

where $\varrho > 1$ and k is a scale factor of the dimension of stress. The introduction of the constant N_{0S}, respressenting the "*minimum life*" at a given stress-level S reflects the fact that neither physically nor by definition can the fatigue relation be extrapolated to $N = 1$. The phenomenon of fracture under the single application of a force ($N = \frac{1}{4}$) can physically not be construed as a limiting case of fatigue because of the basic dissimilarity of the mechanisms; when damage is "progressive" a certain number of load applications is required by definition to produce final separation. The value of N_{0S} may be so small in relation to the range of N as to be practically negligible; but, theoretically, it will always be an integer $N_{0S} \gg 1$.

Since any monotonically decreasing function can, over a certain limited range, be fairly well approximated by a single-term power-law of the type of Eq. (3.1), results of fatigue tests can usually be represented in double-logarithmic scale by the equivalent relation

$$\log(N - N_{0S}) = \varrho [\log k - \log(S - S_0)] \tag{3.2}$$

the expediency of which is in its linearity, permitting easy plotting and extrapolation. The difference between the double-logarithmic form of Eq. (3.2) and the occasionally used semi-logarithmic form $(S - S_0) = C_1 - C_2 \log(N - N_{0S})$

(usually with $S_0 = 0$ and $N_{0S} = 0$) is frequently not significant enough within the range of the observations for a distinction to be made between the two relations.

The relation between S and N must not be considered as a single-valued function, but as an expression of the trend of a statistical relationship. This reflects the fact that fatigue lives under nominally identical conditions of service or of testing show a characteristic wide scatter, frequently exceeding one order of magnitude even in carefully controlled tests. A complete relation between S and N should therefore specify the probability of failure or of survival for any combination (S, N); thus a family of relations between S and N associated with

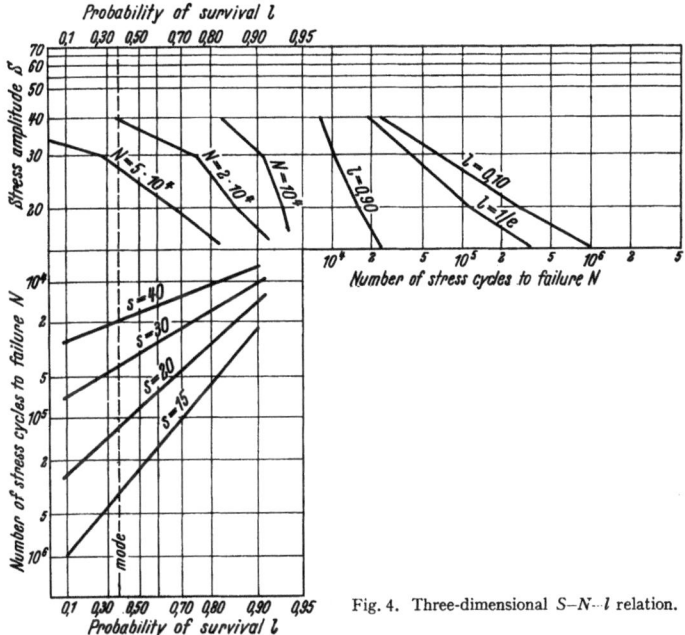

Fig. 4. Three-dimensional S–N–l relation.

different values of the probability of failure $P(N, S)$ or of survival $l(N, S)$ as parameters can be established, and the trend presented for a specific statistical measure of central tendency, such as the mean, the mode or the median, according to Eq. (3.1). The complete relation is 3-dimensional; it can be represented 2-dimensionally, as shown in Fig. 4, in its three different aspects: $S(N)_l$-relations for various probabilities of survival, $l(N)_S$-distributions, indicating probability of survival at N cycles for various applied stress-amplitudes, and $l(S)_N$-distributions, indicating probabilities of survival at stress-amplitudes S for various numbers of cycles N [3]. When both cycle- and time-dependent progressive damage is present and its trend represented by a S–N–t surface, the consideration of the scatter requires the introduction of the probability of survival l at a certain combination (S, N, t) as a parameter defining a family of $(S$–N–$t)_l$-surfaces. No test results exist of a scope sufficient for the establishment even of the trend of such a 4-dimensional relation.

In any specified situation of application of alternating loads, the magnitude of the stress may be represented by a convenient *nominal* stress parameter S, and certain analyses, statistical or physical, of strength measurements may be considered independently of what may be the mechanism of local damage leading to failure. However, for comparison of results, particularly between situations of differing degrees of biaxial or triaxial constraint, consideration should be given to the establishment of an effective stress S_e, reflecting in some rational way the mechanism tending to produce local fatigue damage. As will be shown, the damage mechanisms are most closely related to local plastic deformation and heating. Thus the maximum shear stress or a stress based upon a characteristic distortional strain energy might be selected; S_e may be expressed conveniently in an invariant form [1]. However, the use of this form, in this application, must not be considered to imply that size effects of large magnitude do not exist in fatigue. A representation of S_e which is somewhat more general than the distortional strain energy yield criterion is

$$S_e = \bar{\sigma} + f(p) \tag{3.3}$$

where $p = \tfrac{1}{3}\sigma_{ii}$ is the hydrostatic stress (tension positive) and $\bar{\sigma} = \sqrt{3 J_2'}$; J_2' denotes the second invariant of the stress deviation (see p. 240). The function $f(p)$ can be expanded into a polynomial; in first approximation therefore $f(p) = a p$. For conditions of plane stress or of plane strain of a nearly incompressible material ($\nu \sim \tfrac{1}{2}$) Eq. (3.3) may be written in the form

$$S_e = \frac{\sqrt{3}}{2}(\sigma_1 - \sigma_2) + f\left(\frac{\sigma_1 + \sigma_2}{2}\right). \tag{3.4}$$

II. Micromechanism of progressive damage.

4. General considerations. Two stages can be distinguished in the progress of fatigue: (1) the stage of initiation of submicroscopic fatigue damage, and (2) the stage of microscopic and macroscopic propagation of fatigue cracks. Such distinction does not imply a sharp discontinuity in the damage progress, but only attempts to delimit the stage of propagation of a visible crack from the stage preceding its appearance and recognition. This delimitation is necessarily somewhat arbitrary, since it depends on the size of the smallest crack that can be optically resolved, and thus on the momentary state of micrographic technique.

The essential features of the micromechanism of the *propagation* of fatigue cracks have been established in the late twenties and thirties by H. F. MOORE [4], F. OSHIBA [5], F. KOERBER [6], and, particularly, by H. J. GOUGH and co-workers [7], whose classical investigations of the mechanisms of slip and propagation of the fatigue crack represent the major fatigue research contribution of that period. It is only rather recently, however, that as a result of improved micrographic and X-ray techniques and the development of the electron microscope a more detailed concept of the micromechanism of *initiation* of fatigue damage is gradually emerging from the investigations of W. A. WOOD [8], P. H. LAURENT [9], M. HEMPEL [10], G. C. SMITH [11], N. J. WADSWORTH and N. THOMPSON [12] and, particularly, P. J. E. FORSYTH [13], as well as from recent advances in the understanding of the mechanism of slip and heterogeneous glide summarized by A. F. BROWN [14]. The following discussion is based primarily upon the concepts established in references [8], [13] and [14].

In a discussion of the mechanism of fatigue crack initiation distinctions should be made between the annealed and the workhardened state of a metal, as well as between pure metals or stable alloys and unstable strain-aging alloys. The micromechanism of fatigue is closely related to that of slip and heterogeneous glide, particularly during the very early stages; it now appears that the changes in crystal structure produced during these early stages decisively affect the subsequent progress of fatigue. The basic mechanism of slip is obviously the same for unidirectional and for cyclic stressing; however, the distribution of glide-bands under reversed stressing differs significantly from that produced by unidirectional stressing and depends, moreover, on grain-size and the metallurgical structure of the metal.

5. Significant features of slip.

The slip bands that can be observed on the polished surface of metal crystals after unidirectional plastic straining are the traces, on this surface, of packets of finely spaced slip lamellae. These dark bands are really steps of a height of several tenth of a micron, made up of terraces of smaller steps, formed by avalanche-like slip over a distance of about 2000 Å on several parallel slip planes, spaced some hundred Å apart (Fig. 5). The surface between slip-bands appears slightly crumpled, probably due to micro-slip that can not be clearly resolved even by electron microscope; at the ends of the slip band the steps cease rather abruptly.

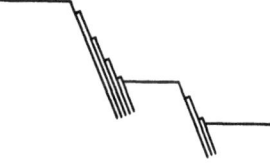

Fig. 5. Glide-step formation.

The spacing of slip bands decreases with increasing stress or strain from a mean of several tens of microns for very small strains, to a few microns for large strains, the actual values differing for different metals. The statistical distribution of the spacing about the mean at a constant stress or constant strain is rather wide and significantly skew towards the small spacings: no spacings are found below a minimum which depends on strain, but can not be smaller than the critical block-size of the substructure. Cumulative distribution functions of slip band spacings in aluminum single crystals as determined by Brown [14] for 0.7 and 2.5% strain have been derived and plotted on extreme-value probability paper in Fig. 6.

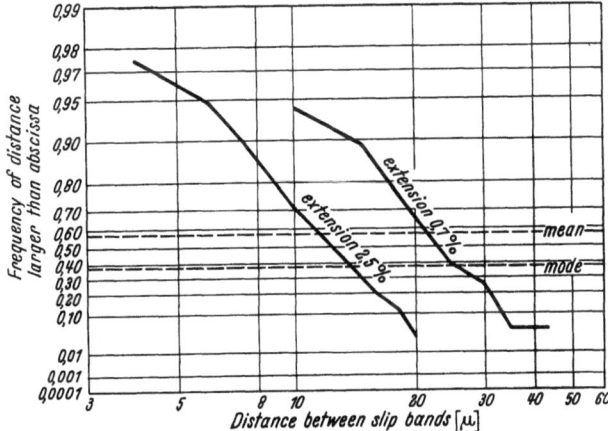

Fig. 6. Frequency function of slip-band spacing in pure aluminum.

The relatively good fit of the observations by the distribution function of extreme values [34] might be related to the fact that the observed slip-line spacings represent the largest distance over which deformation is either elastic or by unresolvable micro-slip, without avalanche-slip. Extrapolation of the observations indicates that for 0.7% strain with a mean spacing of some 22 microns the 0.01 and 0.99 probability limits are associated with spacings varying over one order of magnitude, or roughly between 50 and 5 microns.

It appears that the number n of slip bands per unit length is not linearly proportional to the shear strain γ as would be the case if each new band contributed a fixed amount of slip to the strain, but rather of the form (for small strains) $n = \text{const } \sqrt{\gamma}$, which fits the observations fairly well [15]. While new bands are formed in positions approximately mid-way between existing bands, fresh slip avalanches occur at the same time within the existing bands, forming new lamellae and broadening the bands. The possibility of slip within existing glide zones, presumably hardened by previous slip, is attributed to a process of self-annealing, initiated as soon as the local stresses around the active slip plane attain a critical magnitude, which depends on temperatures. Since self-annealing is sharply reduced at very low temperatures while it is intensified by elevated temperatures, the tendency is to form new slip bands with increasing strain at low temperatures, and new avalanches of slip on existing bands at elevated temperatures. The effect of low strain rates is similar to that of elevated temperatures, that of high strain rates to that of low temperatures, although a change of the strain rate over several orders of magnitude is required to produce a significant effect.

For large strains the deformation process is no longer limited to simple glide, but is accompanied by kinking and wide-spread fragmentation and disorientation of crystal fragments. This, in turn, reduces the thermal stability of the deformed structure and intensifies the recovery rates, leading to grain-boundary migration and polygonization.

A linear relation has been found to exist [16] between the density of slip bands and the level of workhardening of an annealed metal, particularly within the range of small and moderate strains that is of significance in fatigue. For n slip bands per unit length the excess of the workhardening shear-stress τ over the elastic limit shear-stress τ_0 at which the first slip band occurs is

$$\tau - \tau_0 = k \cdot n \tag{5.1}$$

where k is a constant of dimension of stress. This relation would indicate that workhardening is related to the density of slip bands only, not to the amount of slip per band; it leads to a plausible stress-strain relation of the form $(\tau - \tau_0) = $ const $\sqrt{\gamma}$. The essential contribution to hardness is thus made by slip processes by which new bands are formed; subsequent slip *within* the bands does not contribute to hardness, but only increases the strain without increasing the stress.

The spacing of slip bands at the surface is considerably affected by the character of this surface [17]. The slip process may have a different appearance on surfaces prepared in different ways. At small strains the usual wide-spaced slip bands do not appear unless the surface is rubbed with abrasive before electropolishing; on an electro-polished non-abraded surface diffuse markings are usually observed, which suggest micro-slip that can not be resolved. Only at relatively high strains is the dissimilarity between the two types of surface eliminated by the break-through of slip bands even in the unabraded surface. It appears therefore that an undistorted electropolished surface acts as a barrier against slip reaching the surface, at least as long as the strains are small; this barrier can be destroyed either by mechanical work or by chemical action (corrosion) [18].

The spacing of visible slip bands is also affected by surface scratches, however light; even scratches 10 micron deep have been shown to produce a slight increase of the density of slip bands in their vicinity [19]. Macroscopic notches and other stress-concentractions must therefore be expected to cause an increase of slip-band density.

6. Fatigue striations. The effect of reversed cyclic stressing on the annealed crystal structure differs from that of unidirectional stress by the sharp localization of the glide processes in widely spaced slip-bands, usually referred to as "*striations*" (Fig. 7) which tend to grow in width and "intensity" with continuous cyclic stressing, as well as by the apparent suppression of disorientation between the crystal fragments. The striations have usually grooved contours rather than the step-form characteristic of unidirectional stressing (Fig. 8a); less frequently they form stiles or ridges rising above the surface, indicating that

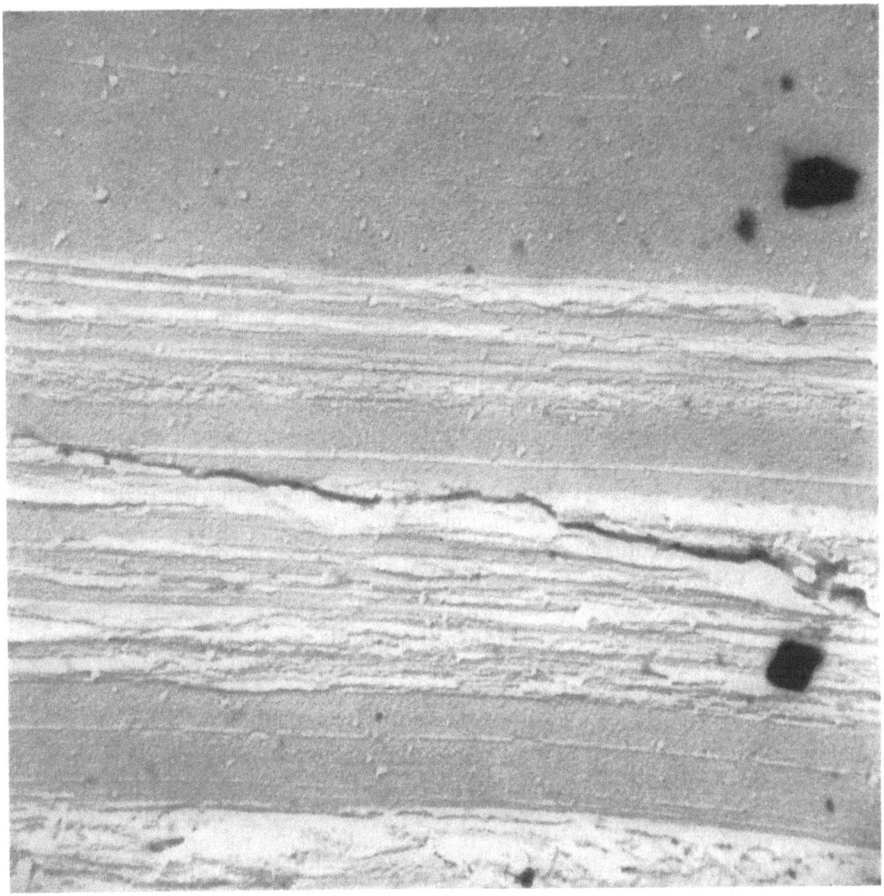

Fig. 7. Striations and fatigue crack in aluminium. (Courtesy Dept. Theor. and Appl. Mech., University of Illinois, and W. J. CRAIG.) (Magnification original photo 11000×.)

slip clusters of "rising" and "falling" lamellea have been combined (Fig. 8b). Wide ridges are pushed up slightly above the surrounding surface; under certain conditions narrow ridges, several micron high, appear as "extrusions" [13].

Fig. 8a and b. Contours of striations. (a) Grooves, (b) ridges and stiles.

The spacing of the striations seems to be determined by the slip band distribution produced by the first few stress cycles, and by the frequency range of the cyclic stress. According to Eq. (5.1) the number of striations within a certain region would be roughly proportional to the local value of $(\tau - \tau_0)$. The sharp localization of slip into striations appears distinctly only at frequencies above a critical frequency, which is probably different for different materials and different stress-levels; at lower frequencies the slip markings produced by cyclic stressing do not differ significantly from those under unidirectional stress. At the frequencies of at least a few hundred cycles per minute usually applied in fatigue testing the formation of striations is, in general, well developed; it becomes especially pronounced at frequencies in the range of 10000 to 30000 cycles per minute. At these frequencies slight stains and oxide surface films of granular structure appear in the striations, suggesting that in these regions a considerable temperature increase has occured in the course of reversed slip, accompanied by extensive thermal softening. This heating could be due either to the energy dissipation in the course of the multitude of slip processes preceding crack-formation, or to friction-heating within cracks of submicroscopic size, or to a combination of causes. The relatively high frequency with which incomplete slip is observed under conditions of reversed rapid stressing suggests that the energy of the slip motion is dissipated within a distance shorter than the grain-size; this would only be possible if the speed of the dislocations were reduced by thermally induced drag-forces to a fraction of the speed of sound. Such a phenomenon is necessarily associated with sharply localized rapid temperature increases [20]. The resulting steep temperature gradients and localized temperature stresses, combined with a reduction of cohesive strength due to the temperature increase might conceivably be responsible for the formation of submicroscopic cracks within the striations (see Fig. 7). Since melting temperatures are attained in relatively slow friction processes under low stresses between clean metal surfaces it is likely that friction-heating within the initial cracks will further raise the temperature within the striations, thus extending the regions of thermal softening and further localizing the slip under subsequent stress cycles, with resulting increase of the heat input into these regions (see Sect. 10).

Under reversed stressing the heat production is intensified by the large number of stress-cycles, while under unidirectional slip heat-production on a slip plane is limited both by hardening in the plane and by transfer of slip to distant planes and the consequent rapid loss of heat to surrounding cold regions. Reversed slip within the striations is thus associated with large numbers of minute "heat flashes" of very short duration and of statistically distributed intensity; their number per cycle must be expected to increase with increasing strain-amplitude, their number per unit time, which determines the heat imput, with increasing frequency of cyclic stressing. The accumulation, particularly at high frequencies, of large numbers of heat flashes within the striations will gradually increase the temperature over regions of macroscopic size and produce phenomena like heat-staining, granular surface films, as well as accelerated local precipitation followed by extrusion of depleted material [13], [21], indicating that the development of localized high temperatures and steep thermal stress gradients in the course of sharply localized slip may represent the basic phenomenon of fatigue initiation. The difference between the rates of heat imput and heat loss by conduction and radiation necessarily increases with the frequency of the cyclic stressing and, at the same stress amplitude, with ratio of volume to surface, or of cross-section to circumference, that is with specimen size.

Cracks in the striations, being so closely related to slip, would necessarily form rather early in the fatigue life. Recent observations [12] support this assumption, showing that significant fatigue damage in the form of persistent striations, the traces of which can be removed neither by electro-polishing (which removes all surface glide-steps that have not been concentrated in striations) nor by annealing and repolishing, and which could already contain cracks of unresolvable magnitude, may occur within the first tenth of the expected fatigue life, although definite cracks can not be clearly identified before one-half of the fatigue life has been expended. Thus the complete course of fatigue damage may be largely predetermined by the first few hundred stress-cycles, by which time the geometric pattern of the distribution of striations has been well established, although no actual damage may have occured.

The most severe cracks are likely to form at points where striations meet grain boundaries which block their path. These are the locations of the most severe disturbances in the crystalline structure; they are associated with severe fragmentation of crystals, and severe thermal gradients and textural stresses, creating particularly favorable conditions for crack initiation, unless grain-boundary migration and polygonization by the thermal agitation relaxes the microstresses within the disturbed region before a micro-crack has had time to form.

7. Effects of workhardening. While workhardening of annealed metal is associated with decreasing spacing and crossing patterns of slip bands and thus with the formation of new bands, fatigue appears to be associated with intensification of slip bands into striations, by localization of new glide processes in the regions of the initially formed bands; the fatigue process apparently sets in when workhardening has been largely completed. The extent of workhardening may thus predetermine the subsequent process of formation and propagation of fatigue cracks; the two processes are, however, neither parallel nor even simultaneous. Only to the extent to which workhardening affects the spacing of the subsequent striations is there an interrelation between the two processes. The resulting conclusion that workhardening as such does not appear to be a cause of fatigue cracking is supported on the one hand by the observation [7c], [8d] that similar hardening may result either from fragmentation associated with grain-distortion produced by unidirectional stressing, or from sharply localized slip without significant lattice distortion during the initial stage of reversed stressing; only the latter process is associated with fatigue. On the other hand, workhardening during the initial stage of cyclic stressing is a characteristic of metals in the annealed condition only; numerous observations indicate [8e], [22] that initially cold-worked metals are softened rather than hardened by cyclic stressing, thus suggesting that fatigue in initially workhardened materials is accompanied by recovery or even recrystallization within the fragmented crystal structure. The softening is hardly noticeable as long as the reversed stress is considerably below the static yield stress of the metal; it increases significantly when the stress-amplitude approaches this level.

While these observations indicate that the workhardening and the fatigue mechanisms can not be causally related, the indirect effect of prior workhardening on subsequent fatigue performance is well illustrated by the observed difference in the change of crystal structure produced by cyclic stressing of a metal in annealed and in work-hardened condition [9], [13]. This difference is due to the fact that slip in a workhardened metal is modified by the presence of crystallite sub-boundaries produced by prior workhardening. While slip under conditions of uniaxial stressing is not particular sensitive to the presence of such sub-boundaries, new slip planes being formed when the local stress is high enough to produce slip in the fragmented structure, slip under reversed stressing is localized within the sub-boundaries in a similar way and for the same reason that it is localized within the striations in the annealed metal. As a result of the cyclic stressing the boundaries of the work-hardened sub-structure, particularly those lying in the direction of the operating slip planes following the maximum resolved shear stress, are heavily delineated by slip bands, which appear to spread into striations from these boundaries, frequently producing on the surface a regular network of crossing, heavily ridged contours. It is reasonable to assume that this localization of slip in the sub-boundaries under reversed stressing is accompanied by a local temperature increase similar to that presumed to occur within the fatigue striations in an annealed metal. In the workhardened metal structure, however, the temperature increase appears to intensify the natural tendency of grain-boundary migration and polygonization to such an extent that significant growth of sub-grains may occur within the regions of heaviest localized slip. Obviously, these regions are considerably softened, since not only is the overall hardness of a work-hardened metal noticeably reduced by cyclic stressing, but it has been observed that by cyclic stressing of cold rolled aluminum alloy [13] at room temperature the softened regions are forced up above the general surface-level of the surrounding unaffected matrix. It is only natural that the first cracks are found in these regions. That localized temperature increase is a dominant factor in this process may be concluded from the fact that repeated stressing of the same cold-rolled aluminum at very low temperature produced no evidence of grain-growth, softening and rising striations, though the localization of slip around grain-boundaries was well developed.

Since the stress required to initiate slip in a single crystal is known to increase with decreasing crystal size [23], the total number of slip processes that can be produced in a polycrystalline aggregate by an applied stress necessarily decreases with the mean grain-size. Since the fatigue life can be assumed to increase with decreasing number of fatigue striations in which cracks could be initiated, and since the number of striations formed in the course of cyclic stressing appears to be roughly proportional to the number of initially operative slip-planes, the fatigue life of a workhardened metal can be expected to increase with decreasing mean grain size and therefore with increasing extent of hardening. This tendency has been confirmed by observations on brass, cold drawn to various extents between 20 and 60% as well as annealed to various grain sizes between 0.012 and 0.13 mm. [24].

8. Extrusion cracks. Crack propagation. The effect of alloying elements on the fatigue mechanism depends on the stability of the alloy formed. Stable solid solution alloys under cyclic stressing have been found to deform in striations in essentially the same manner as the pure metal. Under conditions under which the alloy is metastable, and age-hardening or strain-ageing occurs within the time scale of the cyclic stressing, the fatigue mechanism appears to be affected by strain-accelerated precipitation along striations during the early stages of their formation [21]. It is under those conditions that thin, ribbon-shaped extrusions have been observed to form suddenly along operative slip planes in aluminium alloys, leaving deep crevasses in the crystal surface which develop into fatigue cracks [13]; the observed extrusion bands are several μ high, of a thickness of the order of 0.1 μ. Since the appearance of such extrusions can be suppressed at sufficiently low temperatures, at which the fatigue mechanism appears to revert to normal, it is reasonable to conclude that their formation is due to the development, under cyclic stressing, of very narrow soft regions at elevated temperature, as a result of the especially pronounced concentration of the deformation in these regions, possibly due to the blocking of large numbers of potential slip planes by segregation of precipitates. At low stress amplitudes the same strain-accelerated blocking process of potential slip planes may raise the over-all resistance to slip at higher stress amplitudes, thus increasing the fatigue life at those amplitudes, a phenomenon known as "training" or "coaxing".

The characteristic type of *propagation* of visible fatigue cracks and of the development of fracture surfaces is necessarily determined by the process of deformation associated with crack initiation. In the case of fracture under unidirectional stressing, following extensive crystal-fragmentation and disorientation, an incipient crack has to propagate through a matrix of disoriented crystallites; the rate of energy-dissipation in such a process, in the course of which the crack has to twist and turn, is necessarily high. The suppression of the disorientation under conditions of sufficiently rapid reversed stressing, however, permits the fatigue crack to propagate along roughly parallel striations or sub-boundaries in the over-all direction of maximum resolved shear stress [7]; less energy is required for this process which, therefore, is initiated at a lower stress level than unidirectional fracture.

Thus, it appears that the sharp reduction of the fracture strength in fatigue is essentially the consequence of the concentration of slip into striations produced by reversed stressing, as a result of which both the initiation and the propagation of a crack proceed at stresses far below the static yield stress.

III. Fatigue theories.

9. Phenomenological and structural theories. Various attempts have been made to correlate the significant variables by a theoretical concept on the basis of phenomenological, micro-structural or statistical considerations. The inconclusiveness of the proposed quantitative fatigue theories is largely due to the fact that almost any not too implausible concept can form the basis of a theory reproducing fairly well the principal empirical fatigue relation Eq. (3.1). Since the fitting of this relation can therefore not be considered as a criterion for acceptance or rejection of a proposed theory, the reality of the assumed underlying fatigue mechanism itself must be considered as the principal criterion. On this basis and in the light of recent experimental studies of the initiation of fatigue damage summarized in Part II, none of the quantitative theories so far proposed appear completely acceptable.

The phenomenological theories based upon the assumption of a causal relation between mechanical damping, as an expression of "internal friction" and progressive fatigue damage [25], implying or directly postulating the existence of a critical amount of dissipated energy associated with final fatigue failure, appear to be as far from representing all significant aspects of fatigue as are the structural theories based upon the assumption that workhardening can be considered a measure of accumulation of fatigue damage [26]. The damping theories have the merit that the assumed interrelation in fact exists, although it is certainly not single-valued; the idea of a fixed critical limit of dissipated energy is not only a priori untenable, but has been experimentally disproved [27]. The workhardening theories owe their existence to the ad hoc identification of the rate of fatigue damage under reversed stress with the rate of workhardening under unidirectional stress, and of the static workhardening limit with the fracture-stress in fatigue, thus disregarding the basic difference in the changes of crystal structure under the two conditions (see Sect. 7). Fatigue theories based on the concept of dislocations moving foreward and backward under reversed stressing and producing microcracks by the concentration of vacancies resulting from such motion have been proposed [12], [13], [28]; at present, however, none of these theories has been formulated in sufficient detail to account for the differences in the deformation pattern (distribution of slip) observed under reversed and uni-directional stressing.

10. Relaxation and temperature effects. An acceptable fatigue theory must account for this difference and for the fact that localized softening phenomena are developed in the course of the reversed slip processes. Since the development of temperatures sufficiently high to cause softening or localized thermal cracking or both requires necessarily more than one stress cycle, it is reasonable to assume that localization of slip in reversed stressing is the cause, not the effect of the temperature rise. Such localization must therefore be a characteristic feature of the stress-reversal.

This is, in fact, the case; because of the viscous nature of a newly formed slip band [29], the shear stresses in and around this band are relieved almost immediately after its formation. If the stress, after having produced the slip band, is sustained or further increased, the stress-relaxation along the first slip band is accompanied by rapid transfer of stress to adjacent potential slip planes, followed by slip on those planes, subsequent stress-relaxation and further spreading of slip to new planes. However, if the stress, after having produced slip along the first band, is rapidly decreased, shear stresses of opposite sign are built up on the nearly stress-free slip band, while the surrounding elastic matrix is being unloaded; hence, when the applied stress reverses its sign, the region of the first slip band will yield again under a stress-amplitude reduced by the residual stress built-up in the unloading part of the previous half-cycle. This yielding is followed by relaxation of the reversed stress and the build-up of new residual stresses in the unloading part of the same cycle, and so on. Slip under stress reversal is thus sharply concentrated within regions of initial slip by the same process of "viscous" stress-relaxation along newly formed slip bands that causes spreading of slip under uni-directional stressing.

The concentration of slip into striations is accompanied and intensified by local temperature flashes associated with the individual slip processes. That the local temperature on a slip plane may be quite high can be shown by a rough estimate of the temperature level that can be attained on the slip planes of a crystal cube in the course of the dissipation into heat of the stored elastic energy and the subsequent slip-motion.

Assuming that slip over a distance δ is the result of the motion at velocity V of a trail of piled-up dislocations across the elastically strained crystal, considered as a uniformly moving "heat source" of length λ, the range of momentary temperatures in °C that may be produced by such dissipation can be roughly estimated from the equations of quasi-stationary temperature distribution around a moving heat source either in a infinite medium (single slip plane) or between insulating planes spaced at a distance equal to the thickness $2l$ of the glide lamellae (multiple slip on parallel planes) [30]

$$\frac{\tau}{mk}\left(\frac{\varkappa \lambda V}{\pi}\right)^{\frac{1}{2}} < T < \frac{\tau \delta}{2l\varrho c} \tag{10.1}$$

where k denotes the thermal conductivity, ϱ the density, c the specific heat, $\varkappa = k/\varrho c$ the diffusivity, m the distance between consecutive dislocations in interatomic distances ($5 < m < 10$) and τ the shear stress on the slip plane.

Assuming $0.1\ V_0 < V < 0.9\ V_0$, where V_0 denotes the velocity of sound, for a standard structural aluminium alloy

$$15 \text{ to } 50°\ C < T < 500°\ C.$$

The expected temperature flash range for steel is much higher because of the lower conductivity and the higher yield stress.

The estimated levels of temperature support the assumption that high temperatures and steep temperature- and stress-gradients might develop in the course of the repetition of stress cycles. This fact can easily account for the initiation of fatigue cracks even at low stress levels, provided at least some localized slip occurs at this level. The width of the statistical distribution of crystal-sizes in a polycrystalline metal may therefore have a significant influence on its fatigue life, particularly at low stress-levels. A characteristic frequency function of grain-sizes in a high-purity annealed aluminum is shown in Fig. 9.

The effect of a relatively small number of large grains in which slip is produced under a low over-all stress [23] in reducing the fatigue life or the endurance limit

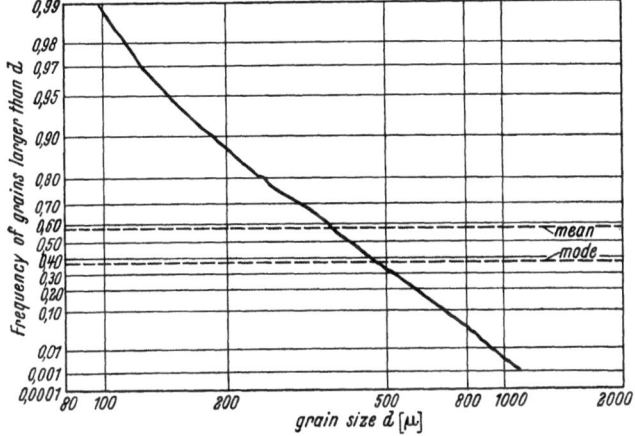

Fig. 9. Typical frequency function of grain-size in high purity annealed aluminum.

is probably out of proportion to their actual volume in the polycrystalline aggregate. The probability of encountering one or a number of grains exceeding the critical size for a certain stress-amplitude will necessarily determine the probable fatigue life at that amplitude. Since, obviously, this probability decreases with decreasing volume for a specimen under uniform stress or, at the same peak stress, with decreasing surface and increasing stress-gradient for a specimen under non-uniform stress distribution where peak stress and stress gradient are at the surface, a distinct size effect with respect to fatigue life at constant stress-amplitude must be expected: the smaller the volume or the surface of the specimen and the steeper the stress gradient at constant peak stress, the longer its expected fatigue life at the given stress-amplitude. Assuming that the probability of encountering a certain size of crystal grain in a given volume can be expressed in the same manner as the probability of encountering a certain size of flaw, the statistical approach developed in the preceding article of G. R. IRWIN, Part V, and its results can be directly applied to the problem of fatigue.

However, the statistical size effect on probable fatigue life at constant stress amplitude, and thus on probable fatigue strength at specified life, is not the only size effect. An additional effect necessarily arises from the interrelation with the local temperature rise within the specimen due to the heat-flashes produced by stress-reversal. Since the local temperature- and associated thermal stress-gradients depend on the heat-transfer characteristics of the specimen, the ratio of surface to volume, which determines the heat-loss by surface-radiation, is likely to affect the temperature level, particularly near the surface. The increase of this ratio with decreasing specimen size may therefore be expected to reduce the rate of progressive damage with decreasing specimen size, and thus to lengthen the fatigue life at a given stress amplitude in a manner similar to that of an externally applied cooling process.

Finally, at low stresses, a third type of size effect producing a parallel trend, also depending on the ratio of surface to volume, is likely to arise as a result of the sensitivity of surface-slip initiation with respect to the character of the surface. Probably this effect also accounts for the particularly high surface-sensitivity of the endurance limit and of the fatigue life at low stresses, as well as for the disproportionately severe effects of even very shallow surface scratches, producing a slight local increase in the density of slip bands (see Sect. 5).

The trend of the S–N relation can be obtained from the simple physical assumption that the expected number of cycles producing fatigue fracture $(N - N_{0s})$ is an inverse function of the density of areas of striations per unit volume, which is $(1/s)$, where s denotes the mean spacing of striations:

$$(N - N_{0s}) = \text{const } (1/s)^{-\varrho}. \tag{10.2}$$

Since the number of striations per unit length $n \approx 1/s$ and since, roughly, $n \approx (S - S_0)/k$ according to Eq. (5.1), Eq. (10.2) is transformed into Eq. (3.1).

11. Statistical theories. The wide scatter of fatigue lives at constant stress-amplitude appears to be the consequence of the above fatigue mechanism: The fatigue life of a polycrystal depends on the statistical distribution of both the size of crystal-grains in the aggregate and of the initial spacing of slip bands under the applied stress. Both are subject to a statistical variation over at least one order of magnitude (see Fig. 6 and 9). The scatter associated with relations (10.2) or (3.1) at constant stress-amplitude might be obtained by computing the scatter of the area-density of striations per unit volume from a convolution of the frequency function of grain-size with the frequency function of slip band spacings.

Considering the pronounced statistical character of the process of initiation and of propagation of fatigue damage within the inhomogeneous polycrystalline structure, it is understandable that a purely statistical theory based on a rather vague mechanism of progressive destruction of submicroscopic cohesive bonds [31] reproduces the significant aspects of fatigue fairly well; the very vagueness of the underlying mechanism makes such a theory easily adaptable to the changing physical concepts of fatigue, without loss of generality.

If the existence is assumed of a distribution function $\varphi(r)$ of the strength r of cohesive bonds in the aggregate, and of a distribution function $\psi(s)$ of bond forces s induced by an applied stress amplitude S, the probability function $P(S)$ of the strength of the aggregate in terms of the applied stress might be obtained from the convolution

$$\pi(S) = \int_0^\infty \psi(s) \left[\int_0^s \varphi(r) \, dr \right] ds,$$

evaluated with $\psi(s)$ taken at different levels S. At any stress-amplitude S the probability of surviving N stress-cycles under the assumption of independent probabilities is therefore

$$l(S)_N = [1 - \pi(S)]^{mN} \doteq [1 - P(S)]^N \tag{11.1}$$

where m denotes the number of bonds the destruction of which would actually produce fracture. The relation $[1 - \pi(S)]^m = [1 - P(S)]$ defines the effect of specimen size (number of bonds) on $P(S)$; since the substitution $m = V/\omega$ transforms it into the basic equation underlying the analysis of size effects in Part V of Dr. Irwin's preceding article the resulting size effect is necessarily of similar type and therefore also identical with the first type of size effect discussed in Sect. 10. The distribution function associated with $l(S)_N$

$$p(S)_N = -dl(S)_N/dS = N[1 - P(S)]^{N-1} dP(S) \qquad (11.2)$$

has the typical form of the distribution of extreme (smallest) values S for N stress-cycles, every one of which is considered as an independent "experiment".

The relation between the mode $\tilde{S}$ of $p(S)_N$ and N determines the trend of a theoretical S–N-diagram, derived independently of any specific fatigue mechanism. Its form is necessarily determined by the assumption made concerning the shape of $P(S)$; there hardly exists any basis for assumptions concerning the form of $\varphi(r)$ und $\psi(s)$. It has been shown [32] that if $P(S)$ is assumed as the integral of a Laplace (exponential) distribution, the mode decreases as a multiple of $(\log N)$, if it is assumed as a cumulative Gaussian distribution, the mode decreases as a multiple of $(\sqrt{\log N})$, leading to the alternative relations $S = C_1 - C_2 (\log N)$ and $S = C_3 - C_4 \sqrt{\log N}$, respectively. These relations provide theoretical support for the empirical assumption that $\log N$ rather than N is a measure of fatigue life.

The arbitrariness of the assumption of $P(S)$ can be avoided by considering that for the large values of N, characteristic of the fatigue problem, the type of the probability (survivorship)-function Eq. (11.1) is determined independently of the specific form of $P(S)$: it has been shown [33] that as N tends towards infinity only three limiting probability functions of extreme values exist, of which the third alone is relevant to the fatigue problem [34]. It permits the consideration of an *endurance limit* and can be written in the form

$$l(S)_N = \exp\left[-\left(\frac{S - S_{0N}}{S_V - S_{0N}}\right)^{\beta_N}\right] \qquad (11.3)$$

where S_{0N} denotes the endurance limit for any particular (large) number N of stress-cycles, and the value of β_N varies with N. The "characteristic" stress-amplitude $S = S_V$ represents a measure of central tendency such that $l(S_V) = 1/e$, while $1/\beta_N$ is a dimensionless scale parameter which determines the width and skewness of the distribution. The limiting function for $N = \infty$, for which $S_{0N} = S_0$, represents the probability of permanent survival or the probability function of the *true endurance limit* S_0 [35].

Because of the fact that fatigue tests are most expediently performed by subjecting specimens of characteristic shapes to repeated stress cycles of constant amplitude, recording the number of cycles sustained to fracture or otherwise defined failure, the survivorship function $l(S)_N$ must be derived from the actually observed survivorship functions $l(N)_S$ at constant stress-amplitudes; it could be directly observed only by a very wasteful and cumbersome experimental procedure.

Theoretical reasoning and experimental evidence support the conclusion that the function $l(N)_S$ has also the character of an extremal distribution of the form

$$l(N)_S = \exp\left[-\left(\frac{N - N_{0S}}{V_S - N_{0S}}\right)^{\alpha_S}\right] \qquad (11.4)$$

where N_{0S} represents the minimum life at any particular stress-amplitude [36]. The "characteristic life" $N = V_S$ is the measure of central tendency for which $l(V_S) = 1/e$, while $1/\alpha_S$ is a dimensionless scale parameter varying with S.

Eqs. (11.3) and (11.4) are compatible only if at every point along the family of curves $l(N, S) = $ const the relation holds

$$l(N)_S = l(S)_N = \text{const}.$$

Hence

$$\left(\frac{N - N_{0S}}{V_S - N_{0S}}\right)^{\alpha_S} = \left(\frac{S - S_{0N}}{S_V - S_{0N}}\right)^{\beta_N} = \text{const}. \tag{11.5}$$

Assuming, for instance, for $l = 1/e$ a reference level (S', V'_S) or (S'_V, N') the relations are obtained

$$(V_S - N_{0S})^{\alpha_S} = \text{const} \, (S - S_{0N})^{-\beta_N} \tag{11.6}$$

or

$$(N - N_{0S})^{\alpha_S} = \text{const} \, (S_V - S_{0N})^{-\beta_N} \tag{11.7}$$

where the "constants" depend on $\alpha_{S'}, \beta_{N'}$, and the reference values selected.

Eqs. (11.6) and (11.7) are made equivalent by the assumption that $\alpha_S = \alpha$ and $\beta_N = \beta$ are constants, independent of S and N respectively. In this case the equations are also equivalent with Eq. (3.1), provided that the variation with N of S_{0N} can, in first approximation, be neglected by introducing $S_{0N} = S_0$, and that $\varrho = \beta/\alpha$. The conventional straight-line relation $\log V_S = \text{const} - \varrho (\log S)$ is obtained by the further approximation of setting both $N_{0S} = 0$ and $S_0 = 0$. It is therefore hardly surprising that the trend of fatigue test results is not too well represented by this doubly approximate relation; the relation $\log V_S = \text{const} - \varrho \log (S - S_0)$ is usually preferable [3a].

Since the distribution of N at constant S is usually much wider than that of S at constant N, so that $\alpha < \beta$, the value of $\varrho > 1$. Values of α_S determined from fatigue tests at constant stress vary roughly between $2 < \alpha_S < 6$, depending on material and stress-level; values of ϱ obtained from test-evaluation vary roughly between $1 < \varrho < 3$ for soft copper and $6 < \varrho < 12$ for aluminum and structural steel. The resulting large values of $\beta = \alpha \varrho$ suggest rather narrow ranges of variation of S at constant N.

The established interrelation between the two-dimensional statistical distributions $l(N)_S$ and $l(S)_N$ and the fatigue equation $(S-N)_l$ permits the derivation of any one of the relations if the two others have been specified. While the reproduction of the *trend* of the $S-N$-relation does not in itself provide a sufficient criterion for discrimination between possible fatigue theories, the complete 3-dimensional $S-N-l$ relation might represent a better criterion.

IV. Cumulative damage.

The problem of the prediction of fatigue life under repeated stress cycles of varying amplitude, generally referred to as "cumulative damage", can be approached from three different points of view resulting, respectively, in a physical, a probabilistic, and a statistical formulation [37].

12. The physical approach. The physical formulation is based on the concept of progressive damage accumulating in every stress cycle. Under a constant stress-amplitude S the increment dD of damage per cycle is necessarily a function of the number N of prior load applications. Hence,

$$\frac{dD}{dN} = f(N)_S \tag{12.1}$$

where the parameters of $f(N)_S$ are functions of S. After N stress cycles the total damage

$$D = \int_{N_{0S}}^{N} \frac{dD}{dN} dN = \int_{N_{0S}}^{N} f(N)_S \, dN. \tag{12.2}$$

The damage increases from $D=0$ during the "incubation period" of fatigue when $N < N_{0S}$ to $D=1.0$ at fracture, when the expected value of $N \approx V_S$.

The form of $f(N)_S$ should represent the fact that during the period of crack-initiation, which may generally exceed half the expected fatigue life, the damage rate is small but increases rapidly as N approaches the expected life V_S. Defining the "*cycle-ratio*" x by

$$x = \frac{N - N_{0S}}{V_S - N_{0S}} \tag{12.3}$$

the simplest function fulfilling this condition is

$$\frac{dD}{dN} = \frac{\alpha}{V_S - N_{0S}} x^{\alpha-1} \quad \text{and} \quad D = x^\alpha \tag{12.4}$$

with $\alpha = \alpha(S) > 1$. For $\alpha = 1$ Eq. (12.4) degenerates into the straight-line relation $D = x$, with the constant damage rate $\frac{1}{V_S - N_{0S}}$.

If it is assumed that under stress cycles of varying amplitudes the damage rate at each stress-amplitude is independent of that at any other stress-amplitude, the total fatigue damage from N_i cycles at i stress-amplitudes S_i would accumulate according to the relation

$$D = \sum_i x_i (N_i, S_i)^{\alpha_i(S_i)}, \tag{12.5}$$

fracture being associated with $D=1.0$. When the stress-amplitudes vary in a distinct sequence, the sum in Eq. (12.5) necessarily reflects this sequence because of the non-linearity of the damage rate, and can therefore not give a stable value of $D=1.0$. For random loading, however, which is defined by a histogram of relative frequencies of the stress-amplitudes $p_i(S_i)$ or by a continuous distribution function $p(S) = \frac{d}{dS} l(S)$, where $l(S)$ denotes the frequency of a stress level $> S$, and for which, therefore, no sequence-effect exists, this sum tends towards a stable value [38]. Hence an equivalent or "reduced" constant stress-amplitude S_R which would produce the same total damage as the random stress spectrum after a total number of stress cycles $V_{SR} = \sum N_i$ can be computed by trial and error from the equation

$$D = \sum_i x_i (p_i V_{SR}, S_i)^{\alpha_i(S_i)} = 1 \tag{12.6}$$

or the integral

$$D = \int_S x(p V_{SR}, S)^{\alpha(S)} dS = 1 \tag{12.7}$$

and from the general $S - V_S$-relation Eq. (3.1), introducing $V_S = V_{SR}$. For $\alpha = 1$ the procedure is considerably simplified, leading to the conventional linear damage law $\sum_i x_i = 1$. The expected fatigue life for random loading with $N_{0S} = 0$ is therefore

$$V_{SR} = \frac{1}{\sum_i (p_i/V_{S_i})}. \tag{12.8}$$

It is well known that Eq. (12.8) is quite unreliable for the prediction of the expected fatigue life under random loading [39]; this is frequently assumed to be due to its linearity. Attempts have been made to use Eq. (12.6) or equivalent equations based on alternative non-linear damage laws, retaining the assumption of the independence of the damage rates under different stress-amplitudes. It appears, however, that the error introduced by the linearization of the damage functions may be less significant than the assumption of the independence of damage rates.

Assuming that the number of fatigue cracks initiated is roughly proportional to the density of striations and inversely proportional to the number $(N - N_{0S})$ of stress-cycles leading to fracture [see Eq. (10.2)], it is evident that the density of slip bands at stress-amplitude S_i will be increased by intermittent stress-amplitudes $S_{i+k} > S_i$, but will be much less affected by amplitudes $S_{i-k} < S_i$. Hence the expected life V_{S_i} will be reduced by all stress-amplitudes $S > S_i$, the more so the larger the excess of S over S_i, while it is largely unaffected by amplitude $S < S_i$. The effect of such stress-interaction may be introduced in the form of a simple factor ω_i indicating the reduction of V_{S_i} by the stress amplitudes $S > S_i$; this factor is a function of the intervals $(S - S_i)$, of the form of the frequency function of stress-amplitudes and, particularly, of its ordinate $p(S_{\max})$. The larger the interval the more pronounced the reduction of V_{S_i} and the larger therefore the contribution to ω_i.

Assuming that one of the effects of intermittent high stress-amplitudes is to eliminate the "incubation period" N_{0S_i} at stress-amplitude S_i, the linear damage law with expected lives adjusted for stress-interaction can be written in the form:

$$V_{SR} = \frac{1}{\sum_i (\omega_i p_i / V_{S_i})}. \qquad (12.9)$$

Test results indicate that the expected fatigue lives at low stress-amplitudes may easily be reduced by relatively infrequent high stress amplitudes by more than two orders of magnitude, leading to large *"interaction factors"* ω_i for stress-amplitudes producing fatigue lives exceeding 10^5 cycles, while $\omega_i \approx 1$ at stress-amplitudes S_i close to the static yield-stress [39]. It is obvious that stress-interaction effects of such magnitude reduce the influence of the non-linearity of the damage-rate to practical insignificance.

13. The probability approach. A purely formal approach to the problem, unrelated to any physical mechanism, can be based on the assumption of a constant mean probability of failure p in every cycle. The actual probability of failure in each cycle need not be constant, but may vary considerably between cycles, provided p is related to the probabilities of failure p_m in the m-th out of an expected total of $N = V_S$ stress cycles of amplitude S by the relation $\sum_m p_m = p V_S$. Because of the relatively large number of stress-cycles sustained in fatigue, p_m and p are very small. This creates the conditions for the application of POISSON's law governing the probability of "rare events"

$$p(n) = \frac{1}{n!} \frac{N}{V_S} \exp\left(-\frac{N}{V_S}\right). \qquad (13.1)$$

Eq. (13.1) with $n = 1$ indicates the probability of occurence of failure after N stress-cycles; with $n = 0$ it defines the probability of surviving N cycles. Hence

$$p(1) = \frac{N}{V_S} \exp\left(-\frac{N}{V_S}\right) = x e^{-x} \qquad (13.2)$$

may be considered to represent the distribution of lives of fatigue specimens at a constant stress-amplitude S as a function of the "cycle ratio" x with $N_{0S} = 0$, while

$$p(0) = \exp\left(-\frac{N}{V_S}\right) = e^{-x} = l(S) \qquad (13.3)$$

is the survivorship function at this amplitude.

Obviously, the probability of survival under i stress amplitudes S_i, each applied N_i times is

$$l(\textstyle\sum N_i) = \Pi_i \left[\exp\left(-\frac{N}{V_S}\right)\right] = \exp\left(-\sum_i \frac{N_i}{V_{S_i}}\right). \tag{13.4}$$

The equivalent expected constant stress-amplitude S_R that can be repeated $N = \sum N_i$ times is obtained by computing V_{SR} from Eq. (3.1) and the equality

$$l(V)_{S_R} = e^{-1} = \exp\left(-\sum_i \frac{N_i}{V_{S_i}}\right) \tag{13.5}$$

which, with $N_i = p_i V_{SR}$ leads to Eq. (12.8). This identity illustrates the fact that the assumption of a constant damage rate and of a constant mean probability of failure in a stress cycle are equivalent. If it is assumed that the mean probability of failure in a cycle of constant stress amplitudes S_i is increased by intermittent cycles of amplitudes $S > S_i$ by a factor ω_i from $1/V_{S_i}$ to (ω_i/V_{S_i}) Eq. (12.9) is obtained.

14. The statistical aspect. The statistical approach is based on observations of the distribution of fatigue lives N under constant stress amplitudes. Since the observed survivorship functions at constant stress-amplitudes S_i are fairly well represented by Eq. (11.4) the probability of surviving a stress spectrum of i stress-amplitudes S_i each applied N_i times can be expressed by:

$$l(\textstyle\sum N_i) = \Pi_i l(N_i)_{S_i} = \Pi_i \exp(-x_i^{\alpha s_i}) = \exp\left[-\sum x_i^{\alpha s_i}\right]. \tag{14.1}$$

The equivalent constant stress-amplitudes S_R with $N_i = p_i V_{SR}$, is determined by the condition $l(N)_{S_R} = l(\sum N_i)$ or,

$$\sum_i x_i^{\alpha s_i} = \sum_i \left(\frac{p_i V_{SR} - N_{0 S_i}}{V_{S_i} - N_{0 S_i}}\right)^{\alpha s_i} = 1. \tag{14.2}$$

With $N_{0 S_i} = 0$ this relation is transformed into Eq. (12.6).

This transformation suggests the identification of the parameters of dispersion α_{S_i} with the exponents α_i defining the non-linearity of the physical damage rate at constant stress amplitudes S_i. Since analysis of fatigue test results indicates that higher stress levels are usually associated with larger values of α_S, the non-linearity of the specific damage rate would appear to increase with increasing stress-amplitudes.

In order to consider stress-interaction effects the probability of survival according to Eq. (11.4) could again be adjusted by a factor ω_i indicating the reduction of constant-stress-amplitude-lives resulting from such interaction. Eq. (14.2) would then be replaced by

$$\sum_i (\omega_i x_i)^{\alpha s_i} = 1. \tag{14.3}$$

With $\alpha_{S_i} = 1$, Eq. (11.4) degenerates into the exponential distribution associated with the linear damage law. It is interesting to note that values of $\alpha_S \approx 1$ are very infrequently obtained in tests; they are characteristic of fatigue tests performed under conditions of inadequate experimental control. It appears that only under such conditions would fatigue failure become a probability phenomenon that can be dealt with by Poisson's law as well as by considering a physical mechanism.

References.

[1] a) SMITH, J. O.: Univ. Illinois Engng. Exp. Stat. Bull. No. 334 (1942).
 b) GOUGH, H. J., H. V. POLLARD and W. J. CLENSHAW: Aeron. Res. Council Rep. and Mem. No. 2522 (1951).
 c) FINDLEY, W. N.: Nat. Adv. Comm. Aeron. Tech. Note No. 2924 (1952).
 d) MARIN, J.: Int. Conference on Fatigue of Metals, London Session 2, Paper 12 (1956).
[2] CLARK, D. S., and D. S. WOOD: Proc. Amer. Soc. Test. Mater. **49**, 717 (1949); **53**, 755 (1953). — Trans. Amer. Soc. Metals **43**, 571 (1951); **45**, 620 (1953).
[3] a) WEIBULL, W.: Trans. Roy. Inst. Tech., Stockholm **10**, 29 (1949).
 b) FREUDENTHAL, A. M.: Proc. Amer. Soc. Test. Mater. **51**, 583 (1951).
[4] MOORE, H. F., and T. VER: Univ. Illinois Engng. Exp. Stat. Bull. No. 208 (1930).
[5] OSHIBA, F.: Sci. Rep. Tôhoku Univ. **23**, 589 (1934).
[6] KOERBER, F.: Proc. Fifth Internat. Congress Appl. Mech., Cambridge, Mass., 20, 1938.
[7] a) GOUGH, H. J.: Proc. Amer. Soc. Test. Mater. **33**, 3 (1933).
 b) GOUGH, H. J., and W. A. WOOD: Proc. Roy. Soc. Lond., Ser. A **154**, 510 (1936).
 c) GOUGH, H. J., and W. A. WOOD: Proc. Inst. Mech. Engrs., Lond. **141**, 175 (1939).
[8] a) WOOD, W. A., and P. L. THORPE: Proc. Roy. Soc. Lond., Ser. A **174**, 310 (1940).
 b) WOOD, W. A., and A. K. HEAD: J. Inst. Met. **79**, 89 (1951).
 c) WOOD, W. A., A. K. HEAD and F. P. BULLEN: Proc. Roy. Soc. Lond., Ser. A **216**, 332 (1953).
 d) WOOD, W. A., and R. B. DAVIS: Proc. Roy. Soc. Lond., Ser. A **220**, 255 (1954).
 e) WOOD, W. A.: Fatigue in Aircraft Structures, 1. New York: Academic Press 1956.
 f) WOOD, W. A., and R. L. SEGALL: Proc. Roy. Soc. Lond., Ser. A **242**, 180 (1957).
[9] LAURENT, P. H.: Publ. sci. et techn. Min. de l'Air No. 256 (1952).
[10] HEMPEL, M.: Fatigue in Aircraft Structures, p. 83. New York: Academic Press 1956. — Int. Conf. on Fatigue of Metals London, Session 6, Paper 7 (1956).
[11] SMITH, G. C.: Proc. Roy. Soc. Lond., Ser. A **242**, 189 (1957).
[12] a) THOMPSON, N., and N. J. WADSWORTH: Phil. Mag. (7) **45**, 223 (1954); (8) **1**, 113 (1956). — Phil. Mag. Suppl. **7**, 72 (1958).
 b) THOMPSON, N.: Fatigue in Aircraft Structures, p. 43. New York: Academic Press 1956.
[13] a) FORSYTH, P. J. E.: J. Inst. Met. **80**, 181 (1951/52); **82**, 449 (1953/54); **83**, 395 (1955). — Phil. Mag. **2**, 437 (1957). — Proc. Roy. Soc. Lond., Ser. A **242**, 198 (1957).
 b) FORSYTH, P. J. E.: Fatigue in Aircraft Structures, p. 20. New York: Academic Press 1956.
 c) FORSYTH, P. J. E.: Int. Conference on Fatigue in Metals, London, Session 6, Paper 5 (1956).
 d) FORSYTH, P. J. E., and C. A. STUBBINGTON: J. Inst. Met. **83**, 173 (1954); **85**, 339 (1957).
[14] BROWN, A. F.: Adv. Physics **1**, 427 (1952).
[15] BROWN, A. F.: J. Inst. Metals **80**, 115 (1952).
[16] a) YAMAGUCHI, K.: Sci. Pap. Inst. Phys. Chem. Res., Tokyo **8**, 289 (1928); **11**, 223 (1929).
 b) COX, H. L., and W. J. CLENSHAW: Proc. Roy. Soc. Lond., Ser. A **149**, 312 (1935).
[17] BROWN, A. F., and R. W. K. HONEYCOMBE: Phil. Mag. **42**, 1146 (1951).
[18] GOULD, A. J.: Int. Conference on Fatigue of Metals, London, Session 4, Paper 2 (1956).
[19] NISHIMURA, H., and J. TAKAMURA: Mem. Kyoto Univ. **23**, 1 (1951).
[20] LEIBFRIED, G.: Z. Physik **127**, 344 (1950).
[21] WEVER, F.: Proc. IUTAM Colloquium on Fatigue, Stockholm 1955, p. 299. Berlin: Springer 1956.
[22] a) DEHLINGER, U.: Naturwiss. **17**, 545 (1929). — Metallwirtsch. **10**, 26 (1931).
 b) PFARR, B.: Z. techn. Phys. **14**, 220 (1933).
 c) KOERBER, F., and M. HEMPEL: Mitt. K.-Wilh.-Inst. Eisenforsch. **17**, 255 (1935).
 d) DAVIES, R. B., J. Y. MANN and D. S. KENSLEY: Int. Conference on Fatigue of Metals, London, Session 6, Paper 9 (1956).
[23] a) BRAGG, W. L.: Nature, Lond. **149**, 511 (1942). — Trans. N-E. Coast Instn. Engrs. Shipb. **62**, 25 (1945).
 b) FREUDENTHAL, A. M.: J. Franklin Inst. **248**, 523 (1949).
[24] SINCLAIR, G. M., and W. J. CRAIG: Trans. Amer. Soc. Metals **44**, 929 (1952).
[25] a) HAIGH, B. P.: Trans. Faraday Soc. (II) **24**, 125 (1928).
 b) LEHR, E.: Z. Metallkde. **20**, 78 (1928).
 c) HANSTOCK, R. F.: Proc. Phys. Soc. Lond. **59**, 275 (1947).
 d) GEROLD, E., and A. KARIUS: Arch. Eisenhüttenw. **21**, 191 (1950).
[26] a) GOUGH, H. J., and D. HANSON: Proc. Roy. Soc. Lond., Ser. A **104**, 539 (1923).
 b) OROWAN, E.: Proc. Roy. Soc. Lond., Ser. A **171**, 79 (1939).
 c) AFANASEV, N. N.: J. techn. Physics USSR. **19**, 1553 (1940).

[27] FOEPPL, O.: J. Iron Steel Inst. (II) **134**, 393 (1936).
[28] ODING, I. A.: Proc. IUTAM Colloquium on Fatigue, Stockholm 1955, p. 178. Berlin: Springer 1956. — MOTT, N. F.: Proc. IUTAM Colloqiuum on Deformation and Flow of Solids, Madrid 1955, p. 57. Berlin: Springer 1956. — MASON, W. P.: J. Acoust. Soc. Amer. **28**, 1207 (1956). — COTTRELL, A. H., and D. HULL: Proc. Roy. Soc. Lond., Ser A **242**, 211 (1957).
[29] a) ROSENHAIN, W.: J. Iron Steel Inst. (II) **70**, 189 (1906).
 b) SNOEK, J. L.: Physica, Haag **8**, 711 (1941).
 c) ZENER, C.: Trans. Amer. Inst. Min. Metallurg. Engrs. **167**, 155 (1946).
[30] FREUDENTHAL, A. M., and J. H. WEINER: J. Appl. Phys. **27**, 44 (1956).
[31] FREUDENTHAL, A. M.: Proc. Roy. Soc. Lond., Ser. A **187**, 416 (1946).
[32] EPSTEIN, B.: J. Appl. Phys. **19**, 140 (1948).
[33] MISES, R. v.: Rév. Math. Univ. Interbalkanique **1**, 1 (1936).
[34] FREUDENTHAL, A. M., and E. J. GUMBEL: Adv. Appl. Mech. **4**, 116 (1955).
[35] FREUDENTHAL, A. M., and E. J. GUMBEL: J. Appl. Phys. **25**, 1435 (1954).
[36] FREUDENTHAL, A. M., and E. J. GUMBEL: Trans. Roy. Soc. Lond., Ser. A **216**, 309 (1953). — J Amer. Statist. Assoc. **49**, 575 (1954).
[37] FREUDENTHAL, A. M.: Proc. IUTAM Colloquium on Fatigue, Stockholm 1955, Berlin: Springer 1956.
[38] FREUDENTHAL, A. M.: Proc. Amer. Soc. Test. Mater. **53**, 896 (1953).
[39] FREUDENTHAL, A. M., and R. A. HELLER: Fatigue in Aircraft Structures, p. 146. New York: Academic Press 1956.

Sachverzeichnis.

(Deutsch-Englisch.)

Bei gleicher Schreibweise in beiden Sprachen sind die Stichwörter nur einmal aufgeführt.

Abbildung von Gebieten auf eine Halbebene, *mapping of regions onto a half-plane* 66—71.
— — auf einen Kreis, *onto a circle* 62—64.
Abmessungen, Einfluß auf die Zerreißfestigkeit, *size-effects, fracture* 580—586, 589.
absolute Unstetigkeit, *absolute discontinuity* 342.
— — der Spannung, *of stress* 343, 366.
— — der Tangentialgeschwindigkeit, *of tangential velocity* 342, 366.
adiabatische Bedingungen im unelastischen Kontinuum, *adiabatic conditions of inelastic continuum* 231, 233, 236, 321.
adiabatische Deformation, *adiabatic deformation* 14.
adiabatisches elastisches Potential, *adiabatic elastic potential* 247, 250.
Aelotropie s. Anisotropie, *aelotropy see anisotropy*.
Äquipotentiallinien, *contour or stress lines* 400f.
äußere Ableitungen, *exterior derivatives* 341.
Airysche Spannungsfunktion, *Airy stress function* 43—47, 50, 353, 424.
— — in Polarkoordinaten, *in polar coordinates* 45—46.
Aktivierungsenergie, Zerreißen reiner Flüssigkeit, *activation energy, pure liquid fracture* 552, 555, 556.
Almansisches Deformationsmaß, *Almansi measure of deformation* 499.
α-Kurve s. Charakteristik, *α-curve see characteristic*.
Alterung, *age-hardening* 602.
anelastische Medien, Ansprechen auf periodische Kräfte, *anelastic media, response to periodic forces* 268.
anelastisches Medium, generalisiertes, *generalized anelastic medium* 292.
Anelastizität (Poynting-Thomsonscher Körper), *anelasticity (Poynting-Thomson body)* 263—269, 471—472.
Anfangskurve, *initial curve* 333.
Anfangswertproblem, charakteristisches, *initial-value problem, characteristic* 361, 396, 397.
— des Cauchyschen Typs, *of Cauchy type* 333, 336f., 340, 360f., 365, 396.
—, gemischtes, *mixed* 362, 397.
anisotrope Substanz, *non-isotropic material* 370.
anisotroper Körper, elastische Konstanten, *anisotropic body, elastic constants* 21.

Anlaufvorgänge bei plastischem Fließen, *incipient plastic flow* 427.
Antithixotropie, *anti-thixotropy* 540—541.
Aufreißen von Flüssigkeiten, *flaws in liquids* 555—556, 580—584.
Ausbreitung von Ermüdungsrissen, *propagation of fatigue cracks* 596, 601, 603, 606.
— des Lichtes in einem Kristall, *of light in a crystal* 132—135, 151—153.
Ausnahmerichtung, *exceptional direction* 385, 389—390.

Babinetscher Kompensator, *Babinet compensator* 136.
Bach-Schuelesches Gesetz, *Bach-Schuele law* 490.
Balken, s. auch Stab.
Balken, elastisch-plastisches Verhalten, *beam, elastic-plastic behaviour* 314.
— von L-förmigem Querschnitt, *L-shaped beam* 400, 406.
— aus viscoelastischem Material, *of viscoelastic material* 311, 312.
—, vorgespannter, *prestressed beam* 223.
Balkenbelastung nach der Airyschen Methode behandelt, *beam loading treated by Airy's method* 44—45.
Balkenbiegung, elastische, *elastic beam bending* 38—39.
Bauschinger-Effekt, *Bauschinger effect* 283.
Bearbeiten spannungsoptischer Materialien, *machining of photoelastic materials* 164.
Belastung, zufällige, bei Ermüdungserscheinungen, *random loading in fatigue* 609.
Belastungsfunktion, *loading function* 280, 283.
Berührungsproblem, viscoelastische Lösung, *contact problem, viscoelastic solution* 296.
β-Kurve s. Charakteristik, *β-curve see characteristic*.
Beton, rheologisches Verhalten, *concrete, rheological behaviour* 435, 466, 536.
Bettische Integrationsmethode, *Betti's integration method* 95.
Bewegungsgleichungen, *equations of motion* 10—12, 139, 235, 323.
Beziehungen zwischen den elastischen Konstanten, *relations between elastic constants* 26.
Biegungswellen in Zylindern, *flexural waves in cylinders* 115—116.
biharmonische Gleichung, *biharmonic equation* 43, 50.

Binghamscher Körper, *Bingham body* 455, 457—463.
Biotsche Entropiegleichung, *Biot's entropy equation* 123.
Bitumen (Sol oder Gel), *bitumen (sol or gel)* 522—523.
B-Körper s. Binghamscher Körper, *B-body see Bingham body*.
Bolzen, spannungsoptische Streifenbilder, *bolts, photoelastic fringe patterns* 211—215.
Böschungsfläche, *surface of constant slope* 401, 402.
Boussinesqsche Lösung zweiter Art, *Boussinesq's second type solution* 91.
Boussinesq-Papkowitsch-Lösung, *Boussinesq-Papkovich solution* 89—91, 267.
—, einfache Beispiele, *simple examples* 91.
—, Vollständigkeit, *completeness* 94.
Brewster (Einheit), *Brewster (unit)* 149.
Brewsterscher Doppelbrechungseffekt, *Brewster effect of double refraction* 128.
Brewsterscher Winkel, *polarising angle* 135.
Brotteig, rheologisches Verhalten, *flour dough, rheological behaviour* 475—478.
Bruch, Rauhigkeit und Faserstruktur, *fracture, roughness and texture* 577, 578.
Bruchbedingung von HUBER, *fracture condition of Huber* 520.
Bruchentstehung, *fracture origins* 567, 568, 569.
Bruchfestigkeit, Definition, *fracture strength, definition* 563.
— s. auch Festigkeit, *see also strength*.
—, Frequenzabhängigkeit, *frequency dependence of strength* 593f., 600.
—, Messungen (Tabelle), *fracture strength measurements (table)* 566.
Bruchspannung, *fracture stress* 567, 573, 574, 575.
— bei Ermüdungsversuchen, *in fatigue experiments* 594.
Buckingham-Reinersche Gleichung, *Buckingham-Reiner equation* 459.
Burgersscher Körper (= viscoelastische Substanz), *Burgers body (= visco-elastic material)* 456, 472—474.

Cauchysche Anfangsdaten (Anfangsstreifen), *Cauchy initial data* 333, 336f., 340, 360f., 365, 396.
Cauchysches Deformationsmaß, *Cauchy measure of deformation* 438, 486, 498, 544.
Cauchysches Integral, *Cauchy integral* 58, 59 bis 62.
Cayley-Hamiltonscher Satz, *Cayley-Hamilton theorem* 258, 261, 444.
Cerrutische Lösung für den Halbraum, *Cerruti's solution for the half-space* 97—98.
Charakteristik s. auch Gleitlinie, *characteristic see also slip line* 355, 357f., 399.
Charakteristiken bei ebener Torsion, *characteristics in plane torsion* 402.
— für ein plastisches Rohr, *for the plastic tube* 414.
—, reelle, *real* 339.

charakteristische Bedingung, *characteristic condition* 341.
charakteristische Determinante, *characteristic determinant* 372.
charakteristische Ebene, *characteristic plane* 337.
charakteristische Fläche, *characteristic surface* 333, 340—341.
charakteristische Koordinaten, *characteristic coordinates* 377.
charakteristische Linie (Charakteristik), *characteristic line* 334.
Christoffelsches Symbol erster Art, *Christoffel symbol of the first kind* 7, 85.
— —, physikalisches, *physical* 442.
— — zweiter Art, *of the second kind* 4, 11, 85.
Coulombsche Bedingung, *Coulomb's condition* 372, 380.

Dämpfungskonstante viscoelastischer Wellen, *attenuation coefficient of viscoelastic waves* 300.
Dämpfungstheorien der Ermüdungserscheinungen, *damping theories of fatigue* 603.
Darcysches Gesetz, *Darcy's law* 265.
Deformation, elastische, in der Rheologie, *strain in rheology* 434.
Deformationsarbeit, reversible, *strain-work* 440, 448, 520.
Deformationsgeschwindigkeitstensor, *rate-of-deformation tensor* 239.
Deformationstensor, *strain tensor* 3—5.
—, Invarianten, *strain invariants* 6.
Deformations-Vorgeschichte, *strain history* 280, 287.
Deformationszustand, *strain* 140—141.
—, ebener, *plane strain* 40—41.
Dehnung, *normal strain* 140.
—, logarithmische, *logarithmic strain* 237.
Deviator 438.
Dickenmesser von COKER, *lateral extensometer of Coker* 173.
Diffusion 229, 231, 233.
Diffusionsgleichung für Druckrelaxation, *diffusion equation for pressure relaxation* 266.
— für die Entropie, *for the entropy* 264.
Diffusions-Relaxation, *diffusion-relaxation* 263.
Dilatanz (= Volumänderung durch Scherung), *dilatancy (= volume change by shear)* 261, 503.
Dilatation, 24, 143, 438, 481.
—, Bestimmung nach der Bettischen Methode, *determination by Betti's method* 95 bis 97.
disperses System, *dispersed system* 435.
Dispersion hoher Konzentration, *dispersion of high concentration* 529.
Dispersionen in festen Körpern, *dispersions in solids* 528—529.
— in Flüssigkeiten, *in liquids* 526—527.
Dissipation der inneren Energie, *dissipation on internal energy* 244.

Dissipationsfunktion, *dissipation function* 451.
— einer zähen Flüssigkeit, *of a viscous fluid* 250, 251, 252.
Dissipationsmechanismus, *dissipation mechanism* 230, 231.
Dissipationspotential, *dissipation potential* 250, 252, 256, 259, 264, 285, 291, 296, 305 bis 306.
Dissipationspotentiale, quasilineare Beziehungen, *dissipation potentials, quasilinear relations* 260.
Doppelbrechung unter Druck, *double refraction under pressure* 128.
doppelter Grenzpunkt, *double limit point* 387.
Drehgeschwindigkeitstensor, *vorticity tensor* 239.
Drehschwingungsrahmen, *tilting stage* 187f.
Drehung, infinitesimale, *infinitesimal rotation* 438.
Drehungstensor, *rotation tensor* 238.
dreidimensionaler Spannungszustand, spannungsoptische Untersuchung, *three-dimensional stress, photoelastic investigation* 130, 177—202.
Dreieckslösung für elastisch-plastische Torsion, *triangular elastic-plastic torsion solution* 407.
Drillung s. Torsion.
Druck, *pressure* 323, 439, 448.
—, innerer, der eine Kugelschale dehnt, *internal, expanding a spherical shell* 422.
—, —, der ein Rohr dehnt, *expanding a tube* 365, 408—418.
— als Ursache freier plastischer Strömung, *pressure originating free plastic flow* 425.
Druckkraft, *pressure force* 323.
Druckrelaxation, *pressure-relaxation* 265, 266.
Duhamel-Neumannsches Gesetz der Thermoelastizität, *Duhamel-Neumann law of thermoelasticity* 123.
Durchbiegung eines Balkens, *bending of a beam* 38—39, 311—314.
Durchlässigkeitskoeffizient in einem porösen Medium, *permeability coefficient in a porous medium* 266.
dynamische Grundgleichungen der Rheologie, *dynamical equations of rheology* 439.
dynamische Probleme der Elastizitätstheorie, *dynamical problems of elasticity* 107—122.
dynamische Spannungen, spannungsoptische Untersuchung, *dynamic stress, photoelastic investigation* 227.
dynamische Viscosität, *dynamic viscosity* 277.
dynamischer Modul, *dynamic modulus* 268 bis 269, 277—278.

ebener Deformationszustand, *plane strain* 40 bis 41, 339, 347, 349—350, 353—367, 409—410.
— — bei allgemeiner Fließbedingung, *under general yield condition* 350f.
— — in plastischem Material, vollständiger Gleichungssatz, *in plasticity, complete set of equations* 349.
— — eines Rohres, *of a tube* 409, 410, 418.

ebener Deformations- und Spannungszustand für Viscoelastizität, *plane strain and stress in visco-elasticity* 272.
ebener Spannungszustand, *plane stress* 40—42.
— —, generalisierter, *generalised* 129, 143 bis 145.
— — bei quadratischer Fließbedingung, *with quadratic yield condition* 351.
ebenes Problem der Plastizitätstheorie, vollständiges, *plane problem of plasticity, complete* 352.
ebenes viscoplastisches Fließen, *plane viscoplastic flow* 302.
Effekte zweiter Ordnung bei zähem Fließen, *second order effects in viscous flow* 512 bis 514.
Eichung spannungsoptischer Materialien, *calibration of photoelectric materials* 164—166 180.
einachsiger Spannungszustand, *uniaxial stress state* 272.
Eindeutigkeit, *uniqueness* 253, 254, 255, 284, 316.
einfache Scherung, *simple shear* 437.
einfache Welle, *simple wave* 359, 390—396.
Einfrieren des Spannungszustandes, *stress freezing* 130, 177—180.
— —, geeignete Substanzen, *materials* 178, 180.
Einfrierverfahren, *freezing procedure* 179.
Einleitung von Ermüdungsschäden, *initiation of fatigue damage* 596, 606.
Einschnitt, *indentation* 101, 432.
Einsteinsche Theorie der verdünnten Sole (Dispersionen in Flüssigkeiten), *Einstein's theory of dilute sols (dispersions in liquids)* 524, 526.
Einzelkraft, *point-force* 76, 88—89.
elastische Konstanten der klassischen Theorie, *elastic constants of the classical theory* 20 bis 26.
elastische Substanz, *Hooke solid (= elastic material)* 441, 448—450.
elastische Volumänderung, *volumetric strain* 438.
elastische Wellen, *elastic waves* 107—118.
— —, Erzeugung durch Volumkräfte, *generation by body forces* 116—118.
elastischer Körper, mathematische Definition, *elastic body, mathematical definition* 14.
elastisches Potential, *elastic potential* 13—14, 244, 256.
— —, adiabatisches, *adiabatic* 247, 250.
— —, isothermes, *isothermal* 247, 250.
— —, kombiniert aus fester und flüssiger Phase, *combined of solid and liquid phases* 265.
— —, Variation, *variation* 252.
elastisches Sol (Lethersichscher Körper) *elastic sol (Lethersich body)* 467—468.
elastisches Verzerrungspotential, *elastic distortional potential* 291.
elastisch-plastische Grenzzone, *elastic-plastic boundary* 400, 403.

elastisch-plastische Randwertprobleme, *elastic-plastic boundary-value problems* 399 bis 425.
elastisch-plastisches Verhalten, *elastic-plastic behaviour* 313—322.
Elastizität zweiter Ordnung, *cross-elasticity* 502f., 507.
—, *second order elasticity* 502f., 507—509.
Elastizitätskoeffizienten, generalisierte, *generalized compliances* 248, 258.
Elastizitätsmodul, *Young's modulus* 24—25, 142, 329, 449, 451.
—, Veränderlichkeit, *variability* 490.
Elastoviscosität (Maxwellscher Körper), *elastico-viscosity (Maxwell body)* 465 bis 467.
elliptischer Zylinder, Torsion, *elliptical cylinder, torsion* 30, 34.
elliptisches Differentialgleichungsproblem, *elliptic problem of differential equations* 339, 369, 371, 384.
elliptisches Gebiet, *elliptic region* 384.
elliptisches Loch in einer unendlich ausgedehnten Platte, *elliptic hole in an infinite plate* 57—58, 64.
Emulsion 435, 523.
Enden des Rohres offen, *open-end conditions of a tube* 409, 410, 418.
— — verschlossen, *closed-end conditions of a tube* 409, 410, 414, 418.
endliche Deformation, *finite deformation or strain* 414, 489—501.
endliche Dehnung, *finite extension* 237.
endliche Dicke einer Platte s. generalisierter ebener Spannungszustand, *finite thickness of a plate, see general plane stress.*
Endosität (= negative Viscosität), *endosity (= negative viscosity)* 478.
Energiedissipation, *energy dissipation* 249, 260.
entartete Lösung, *degenerate solution* 359, 391.
Entlastung, *unloading* 280, 285.
Entropiebilanz, *entropy balance* 245, 246, 294.
Entropieerzeugung, *entropy production* 245, 246, 248, 251, 264.
—, minimale Zunahme, *minimum rate* 253.
Entropiefluß, *entropy flux* 245, 264.
Enveloppe Mohrscher Kreise, *envelope of Mohr circles* 352, 376.
Epizykloiden als Gleitlinien, *epicycloids as slip lines* 363.
Erhaltung der Energie, *conservation of energy* 243, 293.
— des Impulses, *of momentum* 235.
— der Masse, *of mass* 234.
Erholung als Begleiterscheinung der Ermüdung, *recovery accompanying fatigue* 601.
— vom Kriechen, *from creep* 464.
Erinnerungsvermögen von Spannung und Verformung, *memory functions of stress and strain* 273.
Ermüdungserscheinungen in Hochpolymeren, *fatigue effects in high polymers* 593.
— in Keramik und Gläsern, *in ceramics and glasses* 593.

Ermüdungsgrenze, *fatigue limit* 592.
—, untere, *endurance limit* 592—593, 606.
Ermüdungsriß, *fatigue crack* 591.
Ermüdungsrisse, Ausbreitung, *fatigue cracks, propagation* 596, 601, 603, 606.
Ermüdungsschäden, *fatigue damage* 571, 572.
—, Einleitung, *initiation* 596, 606.
Eshelbysche Lösung, *Eshelby's solution* 121.
Extremwert-Statistik der Festigkeit, *extreme-value-statistics of strength* 580, 581, 582.

Farbe als Binghamscher Körper, *paint as a Bingham body* 457, 462.
Festigkeit, *strength* 448, 519—522.
—, Einfluß der mittleren Spannung darauf, *effect of mean stress upon it* 593.
Filonsche Theorie des generalisierten ebenen Spannungszustandes, *Filon theory of generalised plane stress* 129, 143—145.
flacher Ring in elastisch-plastischem ebenen Spannungszustand, *flat ring in elastic-plastic plane stress* 418—421.
Fließen, Definition, *flow, definition* 434.
— mit Volumänderung, *volume flow* 482.
Fließbedingung, *yield condition* 278, 281 bis 283, 370—372.
— von HENCKY, *flow condition of Hencky* 520.
— für einen isotropen, inkompressiblen, vollkommen plastischen Körper, *yield condition for an isotropic, incompressible, perfectly plastic body* 327.
— von MISES, *flow condition of Mises* 521.
—, parabolische Form, *yield condition, parabolic form* 282.
—, Parameterdarstellung, *yield condition, parametric representation* 368.
— von SOKOLOVSKY, *yield condition of Sokolovsky* 381.
— von TRESCA, *yield condition of Tresca* 328, 339, 350, 371, 411f., 417.
Fließfunktion, *yield function* 327, 350, 352.
Fließgeschwindigkeit, *strain rate* 328, 426.
—, gemittelt über die Plattenstärke, *averaged through plate thickness* 353.
—, Variationsbreite, *variation* 233.
Fließgrenze, *yield point* 434.
Fließspannung, *yield stress* 230, 451, 458, 462.
— bei einfacher Dehnung, *in simple tension* 324.
— bei einfacher Scherung, *in simple shear* 324.
Fließspannungsfläche, *yield surface* 278, 283.
Fließspannungsort, *yield locus* 279, 283.
Flüssigkeit, Newtonsche, *Newtonian liquid* 441, 450—452.
—, nicht Newtonsche, *non-Newtonian liquid* 488, 497, 544.
Fluidität, *fluidity coefficient* 450.
— in Ruhe, *at rest* 496.
Flußtensor, *flow-tensor* 438.
Formänderung s. Deformation.
Fortpflanzungsgeschwindigkeit, *propagation velocity* 343.
Fouriertransformationen in der Elastizitätstheorie, *Fourier transforms in elasticity* 72—78.

freie Energie, *free energy* 16—17, 247.
— — ohne kinetischen Anteil, *intrinsic free energy* 440.
Frequenz der periodisch veränderlichen Spannung, Einfluß auf Bruchfestigkeit, *frequency of cyclic stress, effect on strength* 593f., 600.
Fresnelsches Ellipsoid, *Fresnel's ellipsoid* 133, 146, 186.
Frochtsche Methode der Spannungsbestimmung, *Frocht's method of stress determination* 184—186.

Galinsche Lösung für die bewegte Stanzlinie, *Galin's solution for the moving punch* 121.
Gangunterschied, *relative retardation of optical paths* 149, 153—155, 171.
gebietsweise holomorphe Funktion, *sectionally holomorphic function* 60.
Gel 435, 522.
gemischtes Problem, *mixed problem* 362, 381, 397, 430.
generalisierte Fließgeschwindigkeiten, *generalized strain-rates* 255.
generalisierte Kraft, *generalized force* 248, 249, 267.
generalisierte Spannung, *generalized stress* 255, 327.
generalisierter ebener Spannungszustand, *generalized plane stress* 129, 143—145, 352—353.
generalisierter Fluß, *generalized flux* 248, 249.
geometrische Nichtlinearität, *geometrical nonlinearity* 487, 488, 498—507.
geradlinige Charakteristiken, *rectilinear characteristics* 377, 392.
Geschwindigkeit der Verrückung einer Wellenfläche, *velocity of displacement of a wave surface* 343.
Geschwindigkeitscharakteristiken, *velocity characteristics* 354.
Geschwindigkeitsebene, *velocity plane* 357, 375.
Geschwindigkeitsfeld bei ebener plastischer Strömung, *velocity field in plane plastic flow* 364.
Geschwindigkeits-Kompatibilitätsrelationen, *velocity compatibility relations* 356—357.
Gesetz vom Deformationstyp, *deformation-type law* 332.
— vom Strömungstyp, *flow-type law* 332.
Gibbs-Helmholtzsche Gleichung, *Gibbs-Helmholtz equation* 440.
Gläser, Ermüdungserscheinungen, *glasses, fatigue effects* 593.
Gleichgewichtsbedingungen, *equilibrium equations* 235, 293.
—, Kelvinsche Lösung, *Kelvin's solution* 87 bis 88.
—, kovariante Schreibweise, *covariant form* 11, 15, 28.
—, vektorielle Schreibweise, *vector form* 87.
Gleichgewichtsviscosität, *equilibrium viscosity* 293.
Gleitbänder, *glide bands* 232, 597.

Gleitbandabstände, *slip-band spacing* 597, 606.
Gleitbandverteilung, *slip-band distribution* 597.
Gleitebenenblockierung, *slip-plane blocking* 602.
Gleitlawine, *slip avalanche* 597—598.
Gleitlinie, *slip line* 355, 357f., 399.
Gleitlinienfeld, *slip line field* 357—360, 361, 366, 431.
— der Telegraphengleichung, *of the telegraph equation* 363.
Gleitmodul, *rigidity modulus or shear modulus* 25—26, 142, 448.
Gleitung, *slip* 230, 321.
— bei Ermüdungserscheinungen, *in fatigue* 597.
Goodiersches thermoelastisches Potential, *Goodier's thermoelastic potential* 125.
Greenscher Satz, *Green's theorem* 95.
Greensches Deformationsmaß, *Green measure of deformation* 499.
Grenzlinie (Enveloppe von Charakteristiken), *limit line (envelope of characteristics)* 384 bis 385, 408.
— zwischen plastischer und starrer Substanz, *between plastic and rigid material* 383.
Grenzlinien-Singularitäten, *limit line singularities* 384—390.
Grenzpunkt, *limit point* 385.
Grenzschicht bei viscoplastischer Strömung, *boundary layer in visco-plastic flow* 305.
Griffithsche Theorie, *Griffith theory* 557.
Gummi s. Kautschuk.

Haar-Kármánsche Hypothese, *Haar-Kármán hypothesis* 427.
Hagen-Poiseuillesches Gesetz, *Hagen-Poiseuille equation* 459, 497.
Halbkreiseinschnitt, *semicircular groove* 423.
Hankelsche Integraltransformation, *Hankel transform* 100, 101.
Hauptspannungen, *components of stress, principal* 13, 137.
Hauptspannungs-Differenz, *principal stress-difference* 149, 166.
Hauptspannungslinie, *stress trajectory* 139, 168—170.
Hauptspannungsrichtung, *principal stress axis* 13, 137, 327, 354.
—, sekundäre, *secondary* 138.
Helmholtzsche freie Energie s. freie Energie, *Helmholtz' free energy see free energy*.
Henckysche Fließbedingung, *Hencky's flow condition* 520.
Henckysches Deformationsmaß, *Hencky measure of deformation* 486, 499—500, 544.
hexagonale Symmetrie, elastische Konstanten, *hexagonal symmetry, elastic constants* 22—23.
H-Körper s. elastische Substanz, *H-body see Hooke solid*.
Hohlraum-Wachstum in Flüssigkeiten, *cavity growth in liquids* 551, 555.
holomorph, *holomorphic* 60.

homogener isotroper Körper, *homogeneous isotropic body* 17—19, 21, 23—26.
Hookesches Gesetz, *Hooke's law* 15, 17, 24, 142, 329, 400, 401, 409.
Hubersche Bruchbedingung, *Huber's fracture condition* 520.
hydrostatischer Druck, *hydrostatic tension* 143.
hyperbolischer Punkt, *hyperbolic point* 369, 372, 377.
hyperbolisches Differentialgleichungsproblem, *hyperbolic problem of differential equations* 339, 341, 352, 360, 362, 369, 371, 384.
hyperbolisches Gebiet, *hyperbolic region* 384 bis 385.
hypoelastisches Verhalten, *hypo-elasticity* 262.
Hypozykloiden als Gleitlinien, *hypocycloids as slip lines* 363.

ideale Flüssigkeit, *ideal liquid* 441.
ideale Substanz, *ideal material* 440.
inelastisches Verhalten, physikalische Grundlagen, *inelasticity, physical basis* 229—231.
infinitesimale Deformation, *infinitesimal strain or deformation*, 6—7, 12, 438.
inkompressibles, ebenes viscoplastisches Fließen, *incompressible plane visco-plastic flow* 302.
inkompressibles Medium, Deformationsinvarianten, *incompressible matter, strain invariants* 6.
— —, Spannungs-Dehnungs-Relationen, *stress-strain relations* 18.
innere Ableitungen, *interior derivatives* 341, 345.
innere Energie, *internal energy* 243, 244.
innere Reibung, *internal friction* 372.
Integraltransformationen in der dreidimensionalen Elastizitätstheorie, *integral transforms in three-dimensional elasticity* 98 bis 102.
— in der Elastodynamik, *in dynamic elasticity* 121—122.
— in der zweidimensionalen Elastizitätstheorie, *in two-dimensional elasticity* 72 bis 78.
Integration der zweidimensionalen Spannungsgleichungen, numerische Methoden, *integration of two-dimensional stress equations, numerical methods* 173—176.
Integrativmethode von SCOTT BLAIR, *integrative method of Scott Blair* 480.
integrierender Faktor, *integrating factor* 382.
Intensitäten von Dehnung und Spannung, *intensities of strain and stress* 240, 241, 284, 292.
Instabilität, zeitabhängige, *time dependent instability* 310, 311.
invariante Form der Theorie, *invariant form of theory* 335, 337, 338.
Invarianten des Deformationstensors, *invariants of strain tensor* 6.
— eines Tensors, *of a tensor* 239—243, 258, 443f.

Impedanz, komplexe mechanische, *complex mechanical impedance* 276f.
Isobaren, *isobars* 384, 387, 390.
isochromatische Streifen, *isochromatic fringes* 149.
isokline Streifen, *isoclinic fringes* 148, 166.
Isoklinen, *isoclines* 384, 387, 390.
isotherme Bedingungen im unelastischen Kontinuum, *isothermal conditions of inelastic continuum* 233, 236.
isotherme Deformation, *isothermal deformation* 14, 16—17.
isothermes elastisches Potential, *isothermal elastic potential* 247, 250.
isotrope Tensorfunktion, *isotropic tensor function* 444.
isotroper homogener Körper, *isotropic homogeneous body* 17—19, 21, 23—26.
isotroper Körper, elastische Konstanten, *isotropic body, elastic constants* 23—26.
isotroper Punkt, *isotropic point* 166, 168.

Jeffreysscher Körper (s. auch Lethersichscher Körper), *Jeffreys body (= elastic sol) (see also Lethersich body)* 455, 468—471.

Kanteneffekte bei spannungsoptischen Messungen, *edge effects on photoelastic measurements* 162, 164.
Kármánsche Beziehungen der Spannungskomponenten, *Kármán's notation of stress components* 26.
Kautschukelastizität, *rubber elasticity* 16, 18, 492—494.
Kautschuk-Struktur, *rubber structure* 18, 450, 493.
Keil unter einseitiger Belastung, Anlaufvorgang des plastischen Fließens, *wedge unilaterally loaded, incipient plastic flow* 427—429.
Kelvin-Effekt (Dilatanz), *Kelvin effect (dilatancy)* 261.
Kelvinsche elastische Kugel, *Kelvin's elastic sphere* 102—104.
Kelvinsche Lösung der Gleichgewichtsbedingungen, *Kelvin's solution of equilibrium equations* 87—88.
Kelvinscher Körper (= verzögerte Elastizität), *Kelvin body (= delayed elasticity)* 312—313, 454f., 463—464.
Keramik, Ermüdungserscheinungen, *ceramics, fatigue effects* 593.
Kerbprobleme, *indentation problems* 101, 432.
Kettenmoleküle, *chain molecules* 18.
kinematische Unstetigkeitsbedingungen, *kinematical discontinuity conditions* 344.
K-Körper, s. Kelvinscher Körper, *K-body see Kelvin body*.
Kohäsion, *cohesion* 372.
Kollimation des Lichtes, *collimation of light* 156.
Kompatibilitätsrelationen, *compatibility equations* 7—8, 238, 239, 341, 345f., 356f., 366, 373f., 398.
— für ein viscoelastisches Medium, *for the visco-elastic medium* 298.

Komplementärenergie, *complementary energy* 253.
komplexe Potentiale für den ebenen Spannungszustand, *complex potentials for plane stress* 48—58.
komplexe Variable in der Elastodynamik, *complex variables in dynamic elasticity* 118—121.
kompressible Strömung, *compressible flow* 369.
Kompressionsmodul, *bulk modulus* 25, 143, 266.
—, Veränderlichkeit, *variability of bulk modulus* 489.
konkave Ecken, lokal plastisches Verhalten, *concave corners, local plastic behavior* 400.
Konsistenzparameter, *consistency variables* 461.
Konsistenzkurve, *consistency curve* 461, 494, 497.
Konsistenz einer Substanz, *consistency of a material* 461.
Konstanten der klassischen Elastizitätstheorie, *constants of classical elasticity* 20—26.
Kontaktunstetigkeit, s. substantielle Unstetigkeit, *contact discontinuity see material discontinuity*.
kontinuierlicher Übergang vom elastischen zum plastischen Zustande, *continuous transition from elastic to plastic state* 331 f.
Kontinuitätsgleichung, *continuity equation* 234, 323.
Koordinaten des abgeplatteten Rotationsellipsoids in der Elastizitätstheorie, *spheroidal coordinates, oblate, in elasticity* 106 bis 107.
—, mitbewegte, *convected coordinates* 447.
—, orthogonale krummlinige, *orthogonal curvilinear coordinates* 84—86.
Korngrenzen-Diffusion oder -Wanderung, *grain-boundary diffusion or migration* 267, 598.
Korngröße, Einfluß auf Gleitprozesse, *grain size effect on slip processes* 602, 606.
Korrosionseinflüsse auf die Ermüdungsgrenze, *corrosive effects upon endurance limit* 593.
kovariante Ableitung, *covariant derivative* 4.
kreisförmiges Loch in elastisch-plastischem ebenen Spannungszustand, *circular hole in elastic-plastic plane stress* 422.
— — in plastischer Substanz, *in plastic material* 365.
— —, spannungsoptische Streifenbilder, *photoelastic fringe patterns* 205—206.
Kreisquerschnitt eines Stabes, *circular cross section of a bar* 405.
Kreiszylinder, Torsion, *circular cylinder, torsion* 27—28, 36—38.
Kriechen, *creep* 271, 292, 310, 463, 466, 472.
— bei spannungsoptischen Messungen, *effects on photoelastic measurements* 162, 177.

Kriechfunktion, *creep function* 273.
Kriechschäden, *creep-damage* 594.
Kristallgitter, *crystal lattice* 18.
Kristalloptik, *crystal optics* 132—136.
Krümmung der Gleitlinien, *curvature of slip lines* 358, 364, 378.
kugelförmiger Hohlraum, visco-elastische Lösung, *spherical cavity, viscoelastic solution* 295.
Kugelkoordinaten in der Elastizitätstheorie, *spherical polar coordinates in elasticity* 86, 102—104.
Kugelschale in elastisch-plastischer Dehnung unter Innendruck, *spherical shell in elastic-plastic expansion by internal pressure* 422.
Kuhn-Treloarsche Theorie kautschukähnlicher Substanzen, *Kuhn-Treloar theory of rubberlike substances* 18.

Längswellen in Zylindern, *longitudinal waves in cylinders* 113—114.
Lagrange-Funktion elastischer Spannungen, *Lagrangian of elastic stress* 15.
$\lambda/4$-Plättchen, *quarter-wave plate* 135, 151, 153, 157—158.
Lamésche elastische Konstanten, *Lamé's elastic constants* 24, 449.
Lamé-Maxwellsche Gleichungen, *Lamé-Maxwell equations* 140, 174, 182.
Latenzzeit bei Ermüdungserscheinungen, *incubation period of fatigue* 609, 610.
Lebensdauer, minimale, bei Ermüdungsversuchen, *minimum life in fatigue experiments* 593, 607.
Leerstellenkonzentration, *vacancy concentration* 603.
Legendresche Transformation, *Legendre transformation* 370.
Lehm als Kelvinscher Körper, *loam as a Kelvin body* 453—455.
Leistung der äußeren Kräfte, *power input, specific* 439.
Lethersichscher Körper (=elastisches Sol), *Lethersich body (elastic sol)* 456, 467 bis 468.
Linearisierung der ebenen Plastizitätstheorie, *linearization of plane plasticity* 367—370.
linear polarisiertes Licht, Herstellungsmethoden, *plane polarised light, methods of production* 135, 158.
Linsensystem für spannungsoptische Ausrüstung, *lens system in photoelastic equipment* 155—157.
Loch in einer unendlich ausgedehnten Platte, *hole in an infinite plate* 58.
Löcher verschiedener Form, spannungsoptische Streifenbilder, *holes of different shapes, photoelastic fringe patterns* 205 bis 211.
logarithmische Dehnung, *logarithmic strain* 237.
logarithmische Homologen der Konsistenzkurven, *logarithmic homologues of consistency curves* 497.

logarithmische Spiralen als Gleitlinien, *logarithmic spirals as slip lines* 363—365, 380.
— — als Stromlinien, *as streamlines* 364 bis 365.
Lovesche Wellen in geschichteten Medien, *Love waves in stratified media* 111—112.
Lüderssche Bänder (Gleitbänder), *Lueders bands* 232.

Makrorheologie, *macro-rheology* 435.
Margulessches Gesetz, *Margules equation* 460, 497.
Maße für endliche Deformationen, *measures of finite deformation* 498—501.
Maxwellscher Körper (= elastisch-viscose Substanz, *Maxwell body* (= *elastico-viscous material)* 290, 301, 312—313, 455, 465—467, 516—518.
— —, Ausbreitung einer Scherungswelle, *propagation of a shear wave* 301.
mechanische Modelle für Substanzen, *mechanical models of materials* 452f.
mehrdeutige Verrückung, *multi-valued displacement* 53.
Membrangleichnis von PRANDTL, *membrane model of Prandtl* 401, 424.
Metarheologie, *metarheology* 436.
Mikromechanismus der Ermüdungserscheinungen, *micro-mechanism of fatigue* 596 bis 597.
Mikrorheologie, *micro-rheology* 436, 522—535.
Mikrorisse, *microcracks* 567.
Minuswelle, *minus wave* 391f.
Misessche Fließbedingung, *Mises yield condition* 282, 289, 320, 322, 324, 328, 370, 521.
Misessche Plastizitätsgleichungen, *Mises plasticity equations* 323, 404.
— —, Charakteristiken, *characteristics* 337 bis 339.
Misessche Strömungsgleichungen, *Mises flow equations* 339.
Mises-Lévysche Gleichungen, *Mises-Lévy equations* 284, 286, 323.
Mittellinie (neutrale Faser) eines Balkens, *central line af a beam* 38.
M-Körper s. Maxwellscher Körper, *M-body see Maxwell body*.
Modelle für Substanzen, *models of materials* 452f.
Modul der Elastizität zweiter Ordnung, *modulus of cross-elasticity* 507.
Moduln, generalisierte, *generalized moduli* 248, 258.
Mohrsche Kreise für die Spannungen auf beiden Seiten einer Unstetigkeitslinie, *Mohr circles of stresses on both sides of a discontinuity* 367.
Mohrscher Kreis, *Mohr circle* 241, 292, 325f., 352, 375—376, 446.
Monge-Ampèresche Gleichung für die Airysche Spannungsfunktion, *Monge-Ampère equation for Airy function* 353.
Mooneysche Gleichung für das elastische Kautschukpotential, *Mooney's equation for elastic rubber potential* 19.

Muschchelischwilische Potentiale, *Mushkhelishvili potentials* 48—58.
Muschelbruch bei Ermüdung, *clam shell markings of fatigue crack* 591.

Nachwirkung, elastische, *elastic after-effect* 464.
Näherungslösungen der ebenen Plastizitätstheorie, *approximate solutions in plane plasticity* 362, 396—398.
Navier-Stokessche Gleichungen, *Navier-Stokes equations* 451.
neutrale Belastung, *neutral loading* 290.
neutrale Faser, *central line* 38.
neutrale Spannungsänderung, *neutral stress change* 280, 288.
Newtonsche Flüssigkeit (= ideale zähe Flüssigkeit), *Newtonian liquid* (= *ideal viscous liquid)* 441, 450—451, 452.
nichtlineare Elastizität, *non-linear elasticity* 508—509.
nichtlineare dissipative Substanzen, *non-linear dissipative media* 255.
nichtlineare Scherung, *non-linear shear* 19.
nichtlineare zähe Strömung, *non-linear viscous flow* 289.
Nichtlinearität der Ermüdungsverluste, *non-linearity of damage rate in fatigue* 609 bis 611.
— in der Rheologie, *in rheology* 487.
nicht-Newtonsche Flüssigkeit, *non-Newtonian liquid* 488, 497, 544.
nichtstationäres Fließen bei ebener plastischer Verformung, *non-steady flow in plane plastic strain* 427, 432.
Nicolsches Prisma, *Nicol's prism* 135.
Niete s. Bolzen, *rivets see bolts*.
N-Körper s. Newtonsche Flüssigkeit, *N-body see Newtonian liquid*.
Normalsystem von Differentialgleichungen, *normal system of differential equations* 366f., 340.
Nutting-Scott-Blairsche Gleichung, *Nutting-Scott-Blair equation* 480.

Oberflächenkratzer, *surface scratches* 598.
Oberflächenwellen Rayleighscher Art, *surface waves of the Rayleigh type* 109—111.
Ölfarbe s. Farbe, *oil paint see paint*.
Oldroydsche Zahl, *Oldroyd number* 462.
Onsagersche Symmetriebeziehungen, *Onsager symmetry relations* 249.
Ordnung der Unstetigkeit, *order of discontinuity* 342.
orthogonale krummlinige Koordinaten, *orthogonal curvilinear coordinates* 84—86.
orthogonales Gleitliniennetz, *orthogonal net of slip lines* 355, 357 bis 360, 397.
Ort für inkompressibles Fließen, *locus of incompressible yield* 279.
Orthotropie, elastische Konstanten, *orthotropy, elastic constants* 21—22.
Oszillatormodell der Relaxation, *oscillator model of relaxation* 275f.

ovaler Querschnitt, Stab, *oval cross section bar* 406.
Oxydschichten in Ermüdungsstriemen, *oxide films in fatigue striations* 600.

Parabelgrenze, *parabola limit* 371, 379—380, 394—395.
parabolischer Punkt, *parabolic point* 388.
parabolisches Differentialgleichungsproblem, *parabolic problem of differential equations* 369, 384, 422.
Papkowitsch-Lösung s. Boussinesq-Papkowitsch-Lösung, *Papkovich solution see Boussinesq-Papkovich solution*.
Paynesche Potentialfunktionen, *Payne's potential functions* 78—80.
Phasengeschwindigkeit viscoelastischer Wellen, *phase velocity of viscoelastic waves* 300, 301.
Photozelle bei spannungsoptischen Beobachtungen, *photoelectric cell in photoelastic observations* 161.
physikalische Komponenten des Spannungstensors, *physical components of stress tensor* 10, 12.
— — des Verschiebungsvektors, *of displacement vector* 5—6.
physikalische Nichtlinearität, *physical nonlinearity* 487—488.
plastische Arbeit, *plastic work* 325.
plastische Festigkeit, *plastic strength* 481.
plastisches Kriechen, *creep-buckling* 311.
plastische Masse zwischen zwei gekrümmten Platten, *plastic mass between two curved plates* 383.
— — zwischen zwei starren, rauhen Platten, *between two rigid rough plates* 366, 429—432.
plastische Torsion, *plastic torsion* 339, 399—408.
plastische Viscosität, *plastic viscosity* 457, 462.
plastische Volumänderung, *plastic volumetric flow* 438.
plastische Volumdilatation, *plastic cubic dilatation* 286.
plastischer Fließgeschwindigkeitstensor, *plastic flow velocity tensor* 325, 326.
plastisches Fließen, *plastic flow* 279, 425, 434.
— — in spannungsoptischem Material, *yield in photoelastic material* 176, 225—226.
plastisches Gel (Schwedoffscher Körper), *plastic gel (Schwedoff body)* 474—475.
plastisches Gleichgewicht, *plastic equilibrium* 418f., 425.
plastisches Potential, *plastic potential* 254, 284, 285, 326, 350, 373.
— —, stückweise stetig, *piece-wise continuous* 254.
plastisch-elastische Grenzzone, *plastic-elastic boundary* 400, 403.
plastisch-starre Grenze, *plastic-rigid boundary* 366, 383, 399, 422.
Plastizitätskoeffizient, *plastic compliance* 290.

Platte zwischen zwei Stempeln gedrückt (plastischer Spannungszustand), *plate pressed betwen two pistons (plastic stress)* 362.
Platten, elastisch-plastisches Verhalten, *plates, elastic-plastic behaviour* 320.
— aus viscoelastischem Material, *of viscoelastic material* 310, 311.
Plattengleichnis, *plate analogy* 424.
Plattengleichung, viscoelastische, *plate equation, visco-elastic* 310.
Poissonsche Formel der Wahrscheinlichkeitsrechnung in Anwendung auf Ermüdungsbrüche, *Poisson's probability law applied on fatigue fractures* 610—611.
Poiseuillesches Gesetz s. Hagen-Poiseuillesches Gesetz, *Poiseuille equation see Hagen-Poiseuille equation*.
Polarisationsebene des Lichtes, *polarisation plane of light* 131.
polarisiertes Licht, Durchgang durch einen Kristall, *polarised light, passage through a crystal* 133—135.
— —, Durchgang durch eine Platte unter Spannung, *passage through a stressed plate* 147—150.
Polariskop für zirkulares Licht, *polariscope, circular* 153.
polaroides Material, *polaroid material* 135, 158.
Polygonisierung, *polygonization* 598, 602.
Polymere, Ermüdungserscheinungen, *polymers, fatigue effects* 593.
Porenflüssigkeit, *pore-fluid* 231, 265.
Porenflüssigkeitsanelastizität, *pore-fluid anelasticity* 263, 265—267.
Porenvolumen, *pore volume* 230.
Potentialfunktion von AIRY, *potential function of Airy* 43—47, 50.
— von BOUSSINESQ und PAPKOWITSCH, *of Boussinesq-Papkovich* 91—93.
— von KELVIN, *of Kelvin* 87—88.
—, komplexe, für den ebenen Spannungszustand, *complex, for plane stress* 48—58.
— von MUSCHCHELISCHWILI, *of Mushkhelishvili* 48—50.
— von PAYNE, *of Payne* 78—80.
—, reelle, für den ebenen Spannungszustand, *real, for plane stress* 78—84.
— der Torsion, *of torsion* 29—33.
Potentialgleichung für ebene Spannungen, *Laplace equation for plane stress* 144, 173.
—, zweidimensionale, in der Theorie der Torsion, *two-dimensional, in the theory of torsion* 29.
potentielle Energie elastischer Spannungen, *potential energy of elastic stress* 13—14.
Poynting-Effekt, *Poynting effect* 261, 509 bis 512.
Poynting-Thomsonscher Körper (= anelastische Substanz), *Poynting-Thomson body (= anelastic material)* 456, 471—472.
Prandtlscher Körper (= elastisch-plastische Substanz), *Prandtl body (= elastic-plastic body)* 285, 290, 451.
Prandtl-Reußsche Gleichungen, *Prandtl-Reuss equations* 330, 331, 340, 404, 412, 416.

Prandtl-Reußscher Körper, *Prandtl-Reuss body* 285, 290, 451.
Prandtl-Reußsche Theorie, *Prandtl-Reuss theory* 329f.
Primärwelle (P-Welle), *primary wave (P-wave)* 107—109.
Problem mit freier Grenzlinie, *free-boundary problem* 404.
proportionale Belastung, *proportional loading* 294
P-Wellen (Primärwellen), *P-waves (primary waves)* 107—109.

quadratische Fließbedingung s. Misessche Fließbedingung, *quadratic yield condition see Mises yield condition.*
quadratische Form des Deformationstensors, *strain quadric* 141.
— — des Spannungstensors, *stress quadric* 137, 146.
quasi-homogenes Material, *quasi-homogeneous material* 435.
quasi-isotropes Material, *quasi-isotropic material* 436.
Querkontraktionszahl, *Poisson ratio* 25, 142, 177, 212, 329, 449.
—, modifizierte, *modified* 50.

Radialbelastung, *radial loading* 281, 287.
Radoksche Lösung, *Radok's solution* 120.
Randbedingungen, dynamische, *boundary conditions, dynamical* 439, 448.
Randspannung, *boundary stress* 170, 182.
Randveränderung während spannungsoptischer Messungen, *time-edge effect in photoelastic measurements* 162, 164.
Randwertaufgabe, erste, der Potentialtheorie, *Dirichlet's boundary problem of potential theory* 32.
—, zweite, der Potentialtheorie, *Neumann's boundary problem of potential theory* 28.
Randwertproblem, erstes, für ebenen Spannungszustand, *boundary value problem, first, of plane stress* 62—64.
—, gemischtes, *mixed* 58, 70, 81—84, 101 bis 102.
—, homogenes, *homogeneous* 80.
—, zweites, für ebenen Deformationszustand, *second, of plane strain* 64 .
Randwertprobleme in elastisch-plastischer Substanz, *boundary-value problems in elastic-plastic material* 399—425.
— der Elastodynamik, *of dynamic elasticity* 118—122.
— der Viscoelastizität, *in visco-elasticity* 294f.
— der Viscoplastizität, *in visco-plasticity* 305—308.
Rayleighsche Wellen, *Rayleigh waves* 109 bis 111.
rechteckiges Loch, spannungsoptisches Streifenbild, *rectangular hole, photoelastic fringe pattern* 209.
reduzierte Spannung nach HENCKY, *reduced stress of Hencky* 489.

reelle Potentiale für den ebenen Spannungszustand, *real potentials for plane stress* 78—84.
Reibungserwärmung, *friction heating* 600.
Reiner-Rivlinsche Flüssigkeit, *Reiner-Rivlin fluid* 512.
Reiner-Rivlinsche Gleichung, *Reiner-Rivlin equation* 460.
Rekristallisation, *recrystallization* 601.
Relaxation im Standard-Festkörper, *relaxation in the standard solid* 268.
Relaxationsfrequenz, *relaxation frequency* 275.
Relaxationsfunktion, *relaxation function* 273, 274f.
Relaxationsgleichung, *relaxation equation* 291, 292
Relaxationslösungen, *relaxation modes* 266.
Relaxationsspektren, *relaxation spectra* 273 bis 278, 312.
Relaxationszeit, *relaxation time* 465, 479, 481.
—, variable, *variable* 291.
relaxierende Körper, *relaxing media* 269.
relaxierendes Gel (Jeffreysscher Körper), *relaxing gel (Jeffreys body)* 468—471.
remanentes Streifensystem, *residual stress pattern* 130.
Resonanzerschütterung in Ermüdungsversuchen, *resonant vibration in fatigue experiments* 591.
retardierte elastische Körper, *retarded elastic media* 269f.
Retardierung der Verformung s. auch Kriechen, *retardation of strain see also creep* 463.
Retardierungszeit, *retardation time* 269, 463, 481.
Reußsche Gleichungen, Theorie usw. s. Prandtl-Reußsche Gleichungen, Theorie usw., *Reuss equations, theory etc. see Prandtl-Reuss equations, theory etc.*
reversible elastische Erscheinungen, *reversible elastic phenomena* 14, 15.
Reynoldssche Zahl, *Reynolds number* 462.
rheologische Formel einer Substanz, *rheological formula of a material* 455.
rheologische Gleichung, quasilineare, *rheological equation, quasi-linear* 488, 502.
— — für eine Substanz, *of a material* 440.
rheologische Grundeigenschaften, *rheological fundamental properties* 447.
rheologische Körper, Definitionen, *rheological bodies, definitions* 454—456.
— —, Superposition, *superposition* 478f.
— —, zusammengesetzte, *complex* 478f.
rheologischer Stammbaum, *rheological tree* 456.
Rheologie, phänomenologische, *phenomenological rheology* 435.
Rheometrie, *rheometry* 535—542.
Riemannsche Funktion, *Riemann function* 360, 361.
Riemannsche Integrationsmethode, *Riemann integration method* 360f.
Riemann-Christoffelscher Tensor, *Riemann-Christoffel tensor* 7.
Ring, ebener, *plane annulus* 56—57.

Riß, spannungsoptische Untersuchung, *crack, photoelastic investigation* 207—208.
Risse, Grenzgeschwindigkeit, *cracks, limiting velocity* 579.
—, kolineare, zweidimensionale, *co-linear two-dimensional* 565, 566.
—, laufende, *running* 479, 575, 576.
—, scheibenförmige, *disc-shaped* 557, 558.
Riß-Ausweitungskraft, *crack-extension-force* 559—561, 573, 574, 575
Rißproblem in der dreidimensionalen Elastizitätstheorie, *crack problem in three-dimensional elasticity* 102, 106.
— in der ebenen Elastizitätstheorie, *in plane elasticity* 76—78.
Rißteilung, *crack division* 578, 579, 580.
Rohr s. zylindrisches Rohr, *tube see cylindrical tube*.
Rotationskörper, symmetrischer Deformationszustand, *revolution solid, symmetrical strain* 105.
rotierende Scheibe, *rotating disk* 423.
rotierender Körper, spannungsoptisches Verhalten, *rotating body, photoelastic behaviour* 227.
rotierender Zylinder, *rotating cylinder* 423.

SAINT VENANT s. ST. VENANT.
Sandhaufen-Gleichnis von NADAI, *sand hill analog of Nadai* 402, 424.
Schäden, Bruchfestigkeit, *damage, fracture strength* 570, 571, 572.
Scheibe, rotierende, *rotating disk* 423.
Scherung, *shear strain* 141 437.
— in Oktaederfläche, *octahedral shear strain* 240.
—, Volumänderung (= Dilatanz), *volume change by shear (= dilatancy)* 261, 503.
Scherungsviscosität, *shear viscosity* 450.
Schofield-Scott Blairscher Körper, *Schofield-Scott Blair body* 475—478.
Schublinien, *shearing stress lines* 32.
Schubmodul s. Gleitmodul.
Schubspannung, *shear stress* 137.
—, kritische maximale, *critical maximum shear stress* 575.
— in Oktaederfläche, *octahedral shear stress* 240, 241, 326.
—, Richtung maximaler, *maximum direction* 354, 373.
schwache Unstetigkeit, *weak discontinuity* 342, 347.
Schwedoffscher Körper (= plastisches Gel), *Schwedoff body (= plastic gel)* 456, 474 bis 475.
Schwingungsebene, *plane of vibrations* 131.
Sechseck von TRESCA, *hexagon of Tresca* 371, 379—380.
sechsseitiges Prisma der Trescaschen Fließbedingung, *hexagonal prism of Tresca yield condition* 328, 371.
Seifenhaut-Gleichnis von PRANDTL, *soap film analog of Prandtl* 401, 424.
seismische Wellen (Hinweis), *seismic waves (reference)* 107.

sekundäre Hauptspannungsrichtung, *secondary principal stress axis* 138.
Sekundärwelle (S-Welle), *secondary wave (S-wave)* 107—109.
Selbst-Temperung, *self-annealing* 598.
Sénarmontsche Methode zur Messung von Gangunterschieden, *Sénarmont method of measuring retardations* 154—155, 160 bis 161, 172.
Sneddonsche Lösung, *Sneddon's solution* 120.
Sokolovskysche Fließbedingung, *Sokolovsky yield condition* 381.
Sol 435, 522.
—, verdünntes, *dilute* 524.
Soleil-Babinetscher Kompensator, *Soleil-Babinet compensator* 136.
Spaltfläche, *cleavage surface* 574.
Spannungen in einer Symmetrieebene, *stresses in a plane of symmetry* 180.
Spannungsänderung, neutrale, *neutral stress-change* 280.
Spannungsbestimmung in dreidimensionalen Modellen, *stress determination in three-dimensional models* 180—186.
Spannungscharakteristiken, *stress characteristics* 354, 373.
Spannungs-Dehnungs-Relationen im isotropen, elastischen Medium, *stress-strain relations in an isotropic elastic material* 15—26, 142—143.
Spannungsdeviator, *stress deviation tensor* 323.
Spannungsdiagramm (s. auch Spannungsebene), *stress graph (see also stress plane)* 357, 368—369, 388—389.
Spannungsdifferenz, *stress-difference* 149, 166.
—, Bestimmung aus Streifenbild, *determination from fringe pattern* 170—171.
Spannungsebene (s. auch Spannungsdiagramm), *stress plane (see also stress graph)* 357, 368—369, 388—389.
Spannungs-Formänderungs-Beziehungen inelastischer Substanzen, *constitutive equations of the inelastic continuum* 236, 294.
Spannungsfunktion, *stress function* 31—33, 400.
— für ein viscoelastisches Medium, *for the viscoelastic medium* 298.
— für ein viscoplastisches Medium, *for the viscoplastic medium* 303, 304, 306.
Spannungshauptachse, *stress axis, principal* 13, 137, 327.
spannungslose Deformation, *stressless deformation* 434.
spannungslose Volumänderung, *stressless volume change* 434.
Spannungskegelschnitt, *stress conic* 138, 166.
Spannungskonzentration, *stress concentration* 202.
Spannungsoptik, Entdeckung und Geschichte, *photoelasticity, detection and history* 128—130.
— in einem rotierenden Körper, *in a rotating body* 227.
— auf Stoßwellen angewandt, *applied to shock waves* 227.
—, Theorie, *theory* 145—155.

spannungsoptische Bank, *photoelastic bench* 159—161.
spannungsoptische Versuchsmaterialien, Eigenschaften, *photoelastic materials, properties* 163.
— — zum Einfrieren des Spannungszustandes, *for stress freezing* 178, 180.
spannungsoptischer Koeffizient, *stress-optical coefficient* 149.
Spannungsrelaxation, *stress-relaxation* 270, 271.
— längs der Gleitbänder, *along slip bands* 604.
Spannungsstreifenbilder, dreidimensionale Beispiele, *stress fringe patterns, three-dimensional examples* 215—224.
—, zweidimensionale Beispiele, *two-dimensional examples* 203—215.
Spannungstensor, *stress tensor* 8—10, 234, 323.
—, physikalische Komponenten, *physical components* 10, 12.
Spannungsunstetigkeit, *stress discontinuity* 366, 399.
Spannungszustand, ebener, *plane stress* 40 bis 42.
Spitzpunkt einer Linie, *cusp* 384, 385, 387, 389, 393.
Sprödigkeit, *brittleness* 574, 586, 589.
Stab von kreisförmigem Querschnitt, *bar of circular cross section* 405.
— von ovalem Querschnitt, *of oval cross section* 406.
Stabilität viscoplastischer Strömung, *stability of visco-plastic flow* 307.
Stäbe aus viscoelastischem Material, *rods of visco-elastic material* 313.
Standard-Festkörper, *standard solid* 268, 271.
Standard-Substanz, *standard material* 440.
Stanzloch, kreisförmiges, in plastischer Substanz, *circular punch in plastic material* 365.
Stanzproblem in der ebenen Elastizitätstheorie, *punch problem in plane elasticity* 70, 71—72, 76—78, 81.
starke Unstetigkeit, *strong discontinuity* 342, 347.
starrer Kern einer Scheibe zwischen rauhen Platten, *rigid kernel of slab between rough plates* 431—432.
starrer Körper, *rigid body* 441.
Starrkörper-Verschiebung, infinitesimale, *infinitesimal rigid-body displacement* 7.
starr-plastische Näherung, *rigid-plastic approximation* 281, 316, 319f.
stationäre Unstetigkeit, *stationary discontinuity* 343.
stationäres Fließen bei ebener plastischer Verformung, *steady flow in plane plastic strain* 427.
Steifigkeitsmodul, *stiffness modulus* 491.
stereographische Projektion der Spannungshauptachsen, *stereographic projection of stress axes* 189.
Störung viscoplastischer Strömung, *perturbation of visco-plastic flow* 306—308.

Stokessche Beziehung, *Stokes relation* 300, 464, 482.
Stoßausbreitung, *shock propagation* 333.
Stoßbedingungen, *shock conditions* 346.
Stoßfortpflanzung in einem viscoelastischen Medium, *pulse propagation in the visco-elastic medium* 300, 313.
Stoßwellen, spannungsoptische Untersuchung, *shock waves, photoelastic investigation* 227.
— in der Strömung kompressibler Flüssigkeiten, *shock of compressible fluid flow* 342, 346.
Streifen, isochromatische, *isochromatic fringes* 149.
—, isokline, *isoclinic* 148, 166.
—, unendlich langer, elastische Lösung, *infinite strip, elastic solution* 84.
—, — —, viscoplastische Lösung, *visco-plastic solution* 303, 306—307.
Streifenbild, spannungsoptisches, *stress fringe pattern* 149, 166f.
Streulichtmethode zur Spannungsbestimmung, *scattered light method of stress determination* 193—202.
Streuung bei Ermüdungsbruch, *scatter in fatigue fracture* 595.
Striemenbildung bei Ermüdungserscheinungen, *striations in fatigue* 598—601, 606, 610.
Strömungsfunktion für ebene viscoplastische Strömung, *stream function for plane viscoplastic flow* 303, 306.
Strömungsgeschwindigkeitstensor, plastischer, *plastic flow velocity tensor* 325, 326.
Stromlinie, *stream line* 386, 400, 426.
Strukturviscosität, *structural viscosity* 488, 494—498, 531—535.
St. Venantscher Körper (= starrplastische Substanz), *St. Venant body* (= *rigid-plastic body*) 441, 451—452.
St. Venantsches Prinzip, *St. Venant's principle* 203.
St. Venant-Lévy-Misessche Beziehungen, *St. Venant-Lévy-Mises relations* 323.
St. Venant-Trescasche Fließbedingung, *St. Venant-Tresca yield condition* 283, 286f., 320, 328.
StV-Körper s. St. Venantscher Körper, *StV-body see St. Venant body*.
substantielle Unstetigkeit, *material discontinuity* 343.
Superposition rheologischer Körper, *superposition of rheological bodies* 478f.
Suspension 435.
Swaingersches Deformationsmaß, *Swainger measure of deformation* 499.
S-Wellen (Sekundärwellen), *S-waves (secondary waves)* 107—109.

Tardysche Methode zur Messung von Gangunterschieden, *Tardy method of measuring retardations* 153—154, 160—161, 172.
technologische Materialeigenschaften, *technological properties of materials* 540—542.
Teig s. Brotteig, *dough see flour dough*.

teilweise plastisches Rohr, *partly plastic tube* 411.
Telegraphengleichung, *telegraph equation* 363, 369f.
Telegraphengleichungen viscoelastischer Wellen, *telegraph equations of viscoelastic waves* 301, 313.
Temperaturanstieg, lokaler, *local temperature increase* 602, 604.
Temperaturänderung während der Belastung, *temperature change under load* 130, 177–180.
Temperatureffekte in unelastischen Körpern, *temperature effects in inelastic media* 292, 293.
Temperatureinfluß auf die (untere) Ermüdungsgrenze, *temperature effect upon endurance limit* 593.
Temperaturempfindlichkeit der Viscosität, *temperature sensitivity of viscosity* 296.
Temperaturfeld, nicht homogenes, *temperature field, non-homogeneous* 293.
Tempern spannungsoptischer Materialien, *annealing of photoelastic materials* 163–164.
tensorielle Nichtlinearität, *tensorial non-linearity* 487, 507–509.
Tensorinvarianten, *tensor invariants* 6, 239 bis 243, 258, 443f.
Tensorkomponenten, physikalische, *physical tensor components* 441f.
Terezawasche Lösung für den Halbraum, *Terezawa's solution for the half-space* 99 bis 101.
thermische Erweichung, *thermal softening* 600.
thermische Schwankungen, *thermal fluctuations* 553.
thermische Spannungen, spannungsoptische Untersuchung, *thermal stresses, photoelastic investigation* 228.
thermisches Dissipationspotential, *thermal dissipation potential* 264.
Thermoanelastizität, *thermo-anelasticity* 263–264.
thermoelastisches Potential von GOODIER, *thermoelastic potential of Goodier* 125.
Thermoelastizität, *thermoelasticity* 123–126
Thixotropie, *thixotropy* 540–541.
Thomsonscher Körper s. Poynting-Thomsonscher Körper, *Thomson body see Poynting-Thomson body*.
Ton, rheologisches Verhalten, *clay, rheological behaviour* 461–462.
Torsion mit Dehnung, *torsion with extension* 507, 508.
—, einfache, *simple* 506, 509.
—, elastisch-plastische, *elastic-plastic* 403 bis 408.
— eines elliptischen Zylinders, *of an elliptic cylinder* 30, 34.
—, endliche, *finite* 36–38.
— eines Hohlzylinders, *of a hollow cylinder* 33.
— eines kreisrunden Stabes mit kreiszylindrischer Vertiefung, *of a circular shaft with circular groove* 35–36.
— eines Kreiszylinders, *of a circular cylinder* 27–28, 36–38.

Torsion, nicht ebene, *warping in torsion* 401, 405–408.
—, völlig elastische, *fully elastic* 399.
—, völlig plastische, *fully plastic* 399.
— eines Zylinders allgemeinen Querschnitts, *of a non-circular cylinder (general)* 28–31.
— — von Dreiecksquerschnitt, *of triangular cross section* 34.
— — von rechteckigem Querschnitt, *of rectangular cross section* 35.
Torsionsfließen, *torsional flow* 513.
Torsionsfunktion, *torsion function* 28.
— für einen elliptischen Zylinder, *for an elliptic cylinder* 30.
—, komplexe, *complex* 31.
Torsionsmodul, *torsional rigidity* 28.
Torsionswellen in Zylindern, *torsional waves in cylinders* 112–113.
totales Differential, *total differential* 382.
Traglast, *collapse load* 315–318.
Traglastverfahren, *limit design* 315.
Trennungsfläche, *separation surface* 341.
Trescasche Fließbedingung, *Tresca's yield condition* 328, 339, 350, 371, 411f., 417.
Truesdellsche Zahl, *Truesdell number* 517.

unendlich langer Streifen, *infinite strip* 84, 303, 306–307.
unstetige Lösung, *discontinuous solution* 341 bis 343.
Unstetigkeit, die sich bewegt, *moving discontinuity* 342, 346.
—, substantielle, *material discontinuity* 343.
Unstetigkeitsbedingungen, kinematische, *kinematical discontinuity conditions* 344.
Unstetigkeitsfläche, *discontinuity surface* 333, 343.
Unstetigkeitslinie (im ebenen Deformationszustand), *discontinuity line (in plane strain)* 348, 428.
— als Übergangszone, *as a transition zone* 348.
Unstetigkeitsordnung, *discontinuity order* 342.
Unstetigkeitswechselwirkung, *discontinuity interaction* 367.

Variationsprinzipien in der Elastizitätstheorie, *variational principles in elasticity* 15.
— des unelastischen Kontinuums, *of the inelastic continuum* 251–256.
verdünntes Sol, *dilute sol* 524.
Verfestigung, *consolidation* 263.
—, *workhardening* 280, 285, 598, 603.
Verformung s. Deformation.
Verformung, generalisierte, *generalized strain* 327.
Verlustanteil bei Ermüdungserscheinungen, *damage rate of fatigue* 609, 611.
Verlustwinkel, *loss angle* 268, 277–278.
Verrückung s. Verschiebung.
Verschiebungstensor, *displacement gradienttensor* 437.
Verschiebungsvektor, *displacement vector* 2, 437.
—, physikalische Komponenten, *physical components* 5.

Sachverzeichnis.

Versetzung, *dislocation* 53, 230, 263, 603, 604.
Verteilung von Relaxationszeiten, *distribution of relaxation times* 274f.
Verwindung s. Torsion, *twisting see torsion*.
Verzerrung, *distortion* 438, 481.
— s. Deformation.
Verzweigungslinie, *branch line* 389.
Viertelwellenlängenplättchen, *quarter-wave plate* 135, 151, 153, 157—158.
virtuelle Arbeit, *virtual work* 15.
viscoelastische Modulu, *visco-elastic moduli* 309—310.
viscoelastische Spannungs-Dehnungs-Relationen in einer oder zwei Dimensionen, *visco-elastic stress-strain relations in one or two dimensions* 272.
viscoelastische Wellengleichungen, *visco-elastic wave equations* 299—301.
viscoelastisches Randwertproblem, *visco-elastic boundary value problem* 294f.
Viscoelastizität, *visco-elasticity* 269—278, 293 bis 301, 308—313, 448.
— (Burgersscher Körper), *(Burgers body)* 472—474, 491.
viscoplastische Grenzschicht, *visco-plastic boundary layer* 305.
viscoplastische Randwertprobleme, *visco-plastic boundary value problems* 305—308.
Viscoplastizität, *visco-plasticity* 289, 301—308.
viscoser Dehnungskoeffizient, *viscous traction coefficient* 451.
viscoses Fließen, *viscous flow* 434.
Viscosität, *viscosity coefficient* 450, 459.
—, Temperaturempfindlichkeit, *temperature sensitivity* 296.
vollkommene Plastizität, *perfect plasticity* 278, 323, 327, 425.
vollständige Welle, *complete wave* 393.
Volumänderung durch Scherung (= Dilatanz), *volume change by shear (dilatancy)* 261, 503.
Volumdilatation, *volume dilatation* 24, 143, 438, 481.
Volumkraft, *body force* 10, 15, 72, 98.
Volumkräfte, die elastische Wellen erzeugen, *body forces generating elastic waves* 116 bis 118.
Volumviscosität, *volume-viscosity* 231, 249, 267, 269.
Volumen-Anelastizität, *volume-anelasticity* 267.
vorgespannter Balken, *pre-stressed beam* 223.
Vorzeichenfestsetzung für die Krümmung *sign conventions of curvature* 358, 378.
Vorzeichenkonventionen der Spannungsoptik, *sign conventions in photoelasticity* 169—170.

Wärmeblitze, die Striemenbildung verursachen, *heat-flashes causing striation* 600, 604f.
Wahrscheinlichkeitsgesichtspunkte beim Ermüdungsbruch, *probability viewpoints in fatigue fracture* 595, 607f.
Wandeffekte in der Rheologie, *wall effects in rheology* 461.

Handbuch der Physik, Bd. VI.

Wandgleitung, *slippage along the wall* 461.
Wasser als rheologischer Gegenstand, *water as a rheological object* 435.
Wechselverfestigung, *shake down* 317—318.
Wechselwirkung von Unstetigkeiten, *interaction of discontinuities* 367.
Weissenberg-Effekt, *Weissenberg effect* 261, 514—517.
Wellen, elastische, *elastic waves* 107—118.
Wellenfläche, *wave surface* 343.
Wellenfront, *wave front* 132.
Wellengleichungen, viscoelastische, *wave equations, visco-elastic* 299, 301.
Werfung (nicht-ebene Torsion), *warping (in torsion)* 401, 405—408.

Yoffesche Lösung für den bewegten Griffithschen Riß, *Yoffe's solution for the moving Griffith crack* 121.

Zähigkeit s. Viscosität.
Zementstein, rheologisches Verhalten, *cement stone, rheological behaviour* 473.
zentrierte Welle, *centred wave* 391.
Zerreißfestigkeit von Flüssigkeiten, *tensile strength, liquids* 551, 555.
—, Messungen an reinen Flüssigkeiten, *measurements, pure liquids* 555, 556.
Zerreißfestigkeitsmessungen, Einfluß der Abmessungen, *fracture size-effect, measurements* 584, 585, 586.
Zerreißspannung, kritische normale, *critical normal tensile stress* 575.
zirkulare Polarisation, *circular polarisation* 134, 151—153.
Zirkularpolariskop, *circular polariscope* 153.
Zugspannung, *normal stress* 137.
Zugversuch, *straining frame experiment* 158—159.
Zusatzviscosität im Kelvinschen Körper, *firmo-viscosity of the Kelvin body* 464.
Zustandsfunktionen des unelastischen Kontinuums, *state functions of the inelastic continuum* 244, 245, 248—251.
Zustandsgleichung, *equation of state* 236.
Zykloiden als Gleitlinien, *cycloids as slip lines* 363, 366, 381, 384, 429.
Zylinder unter axial wirkendem Moment, *cylinder under axial moment* 399.
— der Misesschen Fließbedingung, *of Mises yield condition* 328.
—, rotierender, *rotating* 423.
— verschiedenen Querschnitts, Torsion, *of different cross sections, torsion* 27—38.
Zylinderkoordinaten in der Elastizitätstheorie, *cylindrical polar coordinates in elasticity* 85—86, 105.
Zylinderwellen, *cylinder waves* 112—116.
zylindrische Deformation, *cylindrical distortion* 504—507.
zylindrisches Rohr, elastisch-plastische Dehnung, *cylindrical tube, elastic-plastic expansion* 408—418.
— —, völlig plastische Dehnung, *fully plastic expansion* 365.

Subject Index.

(English-German.)

Where English and German spelling of a word is identical the German version is omitted.

Absolute discontinuity, *absolute Unstetigkeit* 342.
— — of stress, *der Spannung* 343, 366.
— — of tangential velocity, *der Tangentialgeschwindigkeit* 342, 366.
Activation energy, pure liquid fracture, *Aktivierungsenergie, Zerreißen reiner Flüssigkeit* 552, 555, 556.
Adiabatic conditions of inelastic continuum, *adiabatische Bedingungen im unelastischen Kontinuum* 231, 233, 236, 321.
Adiabatic deformation, *adiabatische Deformation* 14.
Adiabatic elastic potential, *adiabatisches elastisches Potential* 247, 250.
Aelotropy see anisotropy, *Aelotropie s. Anisotropie*.
After-effect, elastic, *elastische Nachwirkung* 464.
Age-hardening, *Alterung* 602.
Airy stress function, *Airysche Spannungsfunktion* 43—47. 50, 353, 424.
— — — in polar coordinates, *in Polarkoordinaten* 45—46.
Almansi measure of deformation, *Almansisches Deformationsmaß* 499.
α-curve see characteristic, *α-Kurve s. Charakteristik*.
Anelasticity (Poynting-Thomson body), *Anelastizität (Poynting-Thomsonscher Körper)* 263—269, 471—472.
Anelastic media, response to periodic forces, *anelastische Medien, Ansprechen auf periodische Kräfte* 268.
Anelastic medium, generalized, *generalisiertes anelastisches Medium* 292.
Anisotropic body, elastic constants, *anisotroper Körper, elastische Konstanten* 21.
Annealing of photoelastic materials, *Tempern spannungsoptischer Materialien* 163—164.
Annulus, plane, *ebener Ring* 56—57.
Anti-thixotropy, *Antithixotropie* 540—541.
Approximate solution in plane plasticity, *Näherungslösungen der ebenen Plastizitätstheorie* 362, 396—398.
Attenuation coefficient of viscoelastic waves, *Dämpfungskonstante viskoelastischer Wellen* 300.
Axis of principal stress, *Hauptspannungsrichtung* 13, 137.
— — —, secondary, *sekundäre* 138.

Babinet compensator, *Babinetscher Kompensator* 136.
Bach-Schuele law, *Bach-Schuelesches Gesetz* 490.
Bar see also beam and rod.
Bar of circular cross section, *Stab von kreisförmigem Querschnitt* 405.
— of oval cross section, *von ovalem Querschnitt* 406.
Bauschinger effect, *Bauschinger-Effekt* 283.
B-body see Bingham body, *B-Körper s. Binghamscher Körper*.
Beam see also bar and rod.
Beam bending, elastic, *elastische Balkenbiegung* 38—39.
Beam, elastic-plastic behaviour, *Balken, elastisch-plastisches Verhalten* 314.
Beam loading treated by AIRY's method, *Balkenbelastung nach der Airyschen Methode behandelt* 44—45.
Beam, pre-stressed, *vorgespannter Balken* 223.
Beam of visco-elastic material, *Balken aus viscoelastischem Material* 311, 312.
Bending of a beam, *Durchbiegung eines Balkens* 38—39, 311—314.
β-curve see characteristic, *β-Kurve s. Charakteristik*.
BETTI's integration method, *Bettische Integrationsmethode* 95.
Biharmonic equation, *biharmonische Gleichung* 43, 50.
Bingham body, *Binghamscher Körper* 455, 457—463.
BIOT's entropy equation, *Biotsche Entropiegleichung* 123.
Bitumen (sol or gel), *Bitumen (Sol oder Gel)* 522—523.
Body force, *Volumkraft* 10, 15, 72, 98.
Body-force equations see equations of motion.
Body forces generating elastic waves, *Volumkräfte, die elastische Wellen erzeugen* 116 to 118.
Bolts, photoelastic fringe patterns, *Bolzen, spannungsoptische Streifenbilder* 211 bis 215.
Boundary conditions, dynamical, *dynamische Randbedingungen* 439, 448.
Boundary layer in visco-plastic flow, *Grenzschicht bei viscoplastischer Strömung* 305.
Boundary stress, *Randspannung* 170, 182.

Boundary value problem, first, of plane stress, *Randwertproblem, erstes, für ebenen Spannungszustand* 62—64.
— — —, homogeneous, *homogenes* 80.
— — —, second, of plane strain, *zweites, für ebenen Deformationszustand* 64.
Boundary value problems of dynamic elasticity, *Randwertprobleme der Elastodynamik* 188—122.
— — — in elastic-plastic material, *in elastisch-plastischer Substanz* 399—425.
— — — in visco-elasticity, *der Viscoelastizität* 294 seq.
— — — in visco-plasticity, *der Viscoplastizität* 305—308.
Boussinesq-Papkovich solution, *Boussinesq-Papkowitsch-Lösung* 89—91, 267.
— —, completeness, *Vollständigkeit* 94.
— —, simple examples, *einfache Beispiele* 91.
BOUSSINESQ's second type solution, *Boussinesqsche Lösung zweiter Art* 91.
Branch line, *Verzweigungslinie* 389.
Brewster (unit), *Brewster (Einheit)* 149.
Brewster effect of double refraction, *Brewsterscher Doppelbrechungseffekt* 128.
Brittleness, *Sprödigkeit* 574, 586, 589.
Buckingham-Reiner equation, *Buckingham-Reinersche Gleichung* 459.
Bulk modulus, *Kompressionsmodul* 25, 143, 266.
Bulk modulus, variability, *Veränderlichkeit des Kompressionsmoduls* 489.
Burgers body (= visco-elastic material), see also visco-elasticity, *Burgersscher Körper (= visco-elastische Substanz) s. auch Viscoelastizität* 456, 472—474.

Calibration of photoelectric materials, *Eichung spannungsoptischer Materialien* 164—166, 180.
Cauchy initial data, *Cauchysche Anfangsdaten (Anfangsstreifen)* 333, 336 seq., 340, 360 seq., 365, 396.
Cauchy integral, *Cauchysches Integral* 58, 59 to 62.
Cauchy measure of deformation, *Cauchysches Deformationsmaß* 438, 486, 498, 544.
Cavity growth in liquids, *Hohlraum-Wachstum in Flüssigkeiten* 551, 555.
Cayley-Hamilton theorem, *Cayley-Hamiltonscher Satz* 258, 261, 444.
Cement stone, rheological behaviour, *Zementstein, rheologisches Verhalten* 473.
Central line of a beam, *Mittellinie (neutrale Faser) eines Balkens* 38.
Centred wave, *zentrierte Welle* 391.
Ceramics, fatigue effects, *Keramik, Ermüdungserscheinungen* 593.
CERRUTI's solution for the half-space, *Cerrutische Lösung für den Halbraum* 97—98.
Chain molecules, *Kettenmoleküle* 18.
Characteristic see also slip line, *Charakteristik s. auch Gleitlinie* 355, 357 seq., 399.
Characteristic condition, *charakteristische Bedingung* 341.

Characteristic coordinates, *charakteristische Koordinaten* 377.
Characteristic determinant, *charakteristische Determinante* 372.
Characteristic line, *charakteristische Linie (Charakteristik)* 334.
Characteristic plane, *charakteristische Ebene* 337.
Characteristic surface, *charakteristische Fläche* 333, 340—341.
Characteristics in plane torsion, *Charakteristiken bei ebener Torsion* 402.
— for the plastic tube, *für ein plastisches Rohr* 414.
—, real, *reelle* 339.
Christoffel symbol of the first kind, *Christoffelsches Symbol erster Art* 7, 85.
— — —, physical, *physikalisches* 442.
— — — of the second kind, *zweiter Art* 4, 11, 85.
Circular cross section of a bar, *Kreisquerschnitt eines Stabes* 405.
Circular cylinder, torsion, *Kreiszylinder, Torsion* 27—28, 36—38.
Circular hole in elastic-plastic plane stress, *kreisförmiges Loch in elastisch-plastischem ebenen Spannungszustand* 422.
— —, photoelastic fringe patterns, *spannungsoptische Streifenbilder* 205—206.
— — in plastic material, *in plastischer Substanz* 365.
Circular polarisation, *zirkulare Polarisation* 134, 151—153.
Circular polariscope, *Zirkularpolariskop* 153.
Clam shell markings of fatigue crack, *muschelförmige Kennzeichen des Ermüdungsrisses* 591.
Clay, rheological behaviour, *Ton, rheologisches Verhalten* 461—462.
Cleavage surface, *Spaltfläche* 574.
Closed-end conditions of a tube, *Enden des Rohres verschlossen* 409, 410, 414, 418.
Cohesion, *Kohäsion* 372.
Collapse load, *Traglast* 315—318.
Collimation of light, *Kollimation des Lichtes* 156.
Compatibility equations, *Kompatibilitätsrelationen* 7—8, 238, 239, 298, 341, 345 seq., 356 seq., 366, 373 seq., 398.
Complementary energy, *Komplementärenergie* 253.
Complete wave, *vollständige Welle* 393.
Complex potentials for plane stress, *komplexe Potentiale für den ebenen Spannungszustand* 48—58.
Complex variables in dynamic elasticity, *komplexe Variable in der Elastodynamik* 118—121.
Compliances, generalized, *generalisierte Elastizitätskoeffizienten* 248, 258.
Components of stress, principal, *Hauptspannungen* 13, 137.
Compressible flow, *kompressible Strömung* 369.
Concave corners, local plastic behaviour, *konkave Ecken, lokal plastisches Verhalten* 400.

Concrete, rheological behaviour, *Beton, rheologisches Verhalten* 435, 466, 536.
Conservation of energy, *Erhaltung der Energie* 243, 293.
— of mass, *der Masse* 234.
— of momentum, *des Impulses* 235.
Consistency curve, *Konsistenzkurve* 461, 494, 497.
Consistency of a material, *Konsistenz einer Substanz* 461.
Consistency variables, *Konsistenzparameter* 461.
Consolidation, *Verfestigung* 263.
Constants of classical elasticity, *Konstanten der klassischen Elastizitätstheorie* 20—26.
Constitutive equations of inelastic continuum, *Spannungs-Formänderungsbeziehungen inelastischer Substanzen* 236, 294.
Coordinates, convected, *Koordinaten, mitbewegte* 447.
—, orthogonal curvilinear, *orthogonale krummlinige* 84—86.
Contact discontinuity see material discontinuity, *Kontaktunstetigkeit, s. substantielle Unstetigkeit.*
Contact problem, viscoelastic solution, *Berührungsproblem, viscoelastische Lösung* 296.
Continuity equation, *Kontinuitätsgleichung* 234, 323.
Continuous transition from elastic to plastic state, *kontinuierlicher Übergang vom elastischen zum plastischen Zustande* 331 seq.
Contour lines, *Äquipotentiallinien* 400 seq.
Corrosive effects upon endurance limit, *Korrosionseinflüsse auf die (untere) Ermüdungsgrenze* 593.
COULOMB'S condition, *Coulombsche Bedingung* 372, 380.
Covariant derivative, *kovariante Ableitung* 4.
Crack division, *Rißteilung* 578, 579, 580.
Crack, photoelastic investigation, *Riß, spannungsoptische Untersuchung* 207—208.
Crack problem in plane elasticity, *Rißproblem in der ebenen Elastizitätstheorie* 76 to 78.
— — in three-dimensional elasticity, *in der dreidimensionalen Elastizitätstheorie* 102, 106.
Cracks, co-linear two-dimensional, *Risse, kolineare, zweidimensionale* 565, 566.
—, disc-shaped, *scheibenförmige* 557, 558.
—, limiting velocity, *Grenzgeschwindigkeit* 579.
—, running, *laufende* 479, 575, 576.
Crack-extension-force, *Riß-Ausweitungskraft* 559—561, 573, 574, 575.
Creep, *Kriechen* 271, 292, 310, 463, 466, 472.
Creep-buckling, *plastisches Knicken* 311.
Creep-damage, *Kriechschäden* 594.
Creep effects on photoelastic measurements, *Kriechen bei spannungsoptischen Messungen* 162, 177.
Creep function, *Kriechfunktion* 273.

Critical crack-extension-force, *kritische Riß-Ausweitungskraft* 559—561, 573, 574, 575.
Critical maximum shear stress, *kritische maximale Schubspannung* 575.
Critical normal tensile stress, *kritische normale Zerreißspannung* 575.
Cross-elasticity, *Elastizität zweiter Ordnung* 502 seq., 507.
Crystal lattice, *Kristallgitter* 18.
Crystal optics, *Kristalloptik* 132—136.
Curvature of slip lines, *Krümmung der Gleitlinien* 358, 364, 378.
Cusp, *Spitzpunkt einer Linie* 384, 385, 387, 389, 393.
Cycloids as slip lines, *Zykloiden als Gleitlinien* 363, 366, 381, 384, 429.
Cylinder under axial moment, *Zylinder unter axial wirkendem Moment* 399.
— of Mises yield condition, *der Misesschen Fließbedingung* 328.
—, rotating, *rotierender* 423.
Cylinder waves, *Zylinderwellen* 112—116.
Cylinders of different cross sections, torsion, *Zylinder verschiedenen Querschnitts, Torsion* 27—38.
Cylindrical distortion, *zylindrische Deformation* 504—507.
Cylindrical polar coordinates in elasticity, *Zylinderkoordinaten in der Elastizitätstheorie* 85—86, 105.
Cylindrical tube, elastic-plastic expansion, *zylindrisches Rohr, elastisch-plastische Dehnung* 408—418.
— —, fully plastic expansion, *völlig plastische Dehnung* 365.

Damage, fracture strength, *Schäden, Bruchfestigkeit* 570, 571, 572.
Damage rate of fatigue, *Verlustanteil bei Ermüdungserscheinungen* 609, 611.
Damping theories of fatigue, *Dämpfungstheorien der Ermüdungserscheinungen* 603.
DARCY's law, *Darcysches Gesetz* 265.
Deformation-type law, *Gesetz vom Deformationstyp* 332.
Degenerate solution, *entartete Lösung* 359, 391.
Deviator 438.
Diffusion 229, 231, 233.
Diffusion equation for the entropy, *Diffusionsgleichung für die Entropie* 264.
— — for pressure relaxation, *für Druckrelaxation* 266.
Diffusion-relaxation, *Diffusions-Relaxation* 263.
Dilatancy (= volume change by shear), *Dilatanz (= Volumänderung durch Scherung)* 261, 503.
Dilatation 24, 143, 438, 481.
—, determination by BETTI's method, *Bestimmung nach der Bettischen Methode* 95 to 97.
Dilute sol, *verdünntes Sol* 524.
Direction condition see characteristic condition.

DIRICHLET's boundary problem of potential theory, *erste Randwertaufgabe der Potentialtheorie* 32.
Discontinuity conditions, kinematical, *kinematische Unstetigkeitsbedingungen* 344.
Discontinuity interaction, *Unstetigkeitswechselwirkung* 367.
Discontinuity line (in plane strain), *Unstetigkeitslinie (im ebenen Deformationszustand)* 348, 428.
— — as a transition zone, *als Übergangszone* 348.
Discontinuity, material or stationary, *substantielle oder stationäre Unstetigkeit* 343.
Discontinuity order, *Unstetigkeitsordnung* 342.
Discontinuity surface, *Unstetigkeitsfläche* 333, 343.
Discontinuous solution, *unstetige Lösung* 341 to 343.
Disk, rotating, *rotierende Scheibe* 423.
Dislocation, *Versetzung* 53, 230, 263, 603, 604.
Dispersed system, *disperses System* 435.
Dispersion of high concentration, *Dispersion hoher Konzentration* 529.
Dispersions in liquids, *Dispersionen in Flüssigkeiten* 526—527.
— in solids, *in festen Körpern* 528—529.
Displacement gradient-tensor, *Verschiebungstensor* 437.
Displacement vector, *Verschiebungsvektor* 2, 437.
— —, physical components, *physikalische Komponenten* 5.
Dissipation function, *Dissipationsfunktion* 451.
— — of a viscous fluid, *einer zähen Flüssigkeit* 250, 251, 252.
Dissipation of internal energy, *Dissipation der inneren Energie* 244.
Dissipation mechanism, *Dissipationsmechanismus* 230, 231.
Dissipation potential, *Dissipationspotential* 250, 252, 256, 259, 264, 285, 291, 296, 305—306.
Dissipation potentials, quasi-linear relations, *Dissipationspotentiale, quasilineare Beziehungen* 260.
Distortion, *Verzerrung* 438, 481.
Distribution of relaxation times, *Verteilung von Relaxationszeiten* 274 seq.
Double limit point, *doppelter Grenzpunkt* 387.
Double refraction under pressure, *Doppelbrechung unter Druck* 128.
Dough see flour dough, *Teig s. Brotteig*.
Drilled crack see crack.
Duhamel-Neumann law of thermo-elasticity, *Duhamel-Neumannsches Gesetz der Thermoelastizität* 123.
Dynamical equations of rheology, *dynamische Grundgleichungen der Rheologie* 439.
Dynamical problems of elasticity, *dynamische Probleme der Elastizitätstheorie* 107—122.
Dynamic modulus, *dynamischer Modul* 268 to 269, 277—278.

Dynamic stress, photoelastic investigation, *dynamische Spannungen, spannungsoptische Untersuchung* 227.
Dynamic viscosity, *dynamische Viscosität* 277.

Edge effects on photoelastic measurements, *Kanteneffekte bei spannungsoptischen Messungen* 162, 164.
EINSTEIN's theory of dilute sols (dispersions in liquids), *Einsteinsche Theorie der verdünnten Sole (Dispersionen in Flüssigkeiten)* 524, 526.
Elastic body, mathematical definition, *elastischer Körper, mathematische Definition* 14.
Elastic constants of the classical theory, *elastische Konstanten der klassischen Theorie* 20—26.
Elastic distortional potential, *elastisches Verzerrungspotential* 291.
Elastic-plastic behaviour, *elastisch-plastisches Verhalten* 313—322.
Elastic-plastic boundary, *elastisch-plastische Grenzzone* 400, 403.
Elastic-plastic boundary-value problems, *elastisch-plastische Randwertprobleme* 399 to 425.
Elastic potential, *elastisches Potential* 13—14, 244, 256.
— —, adiabatic, *adiabatisches* 247, 250.
— —, combined of solid and liquid phases, *kombiniert aus fester und flüssiger Phase* 265.
— —, isothermal, *isothermes* 247, 250.
— —, variation, *Variation* 252.
Elastic sol (Lethersich body), *elastisches Sol (Lethersichscher Körper)* 467—468.
Elastic waves, *elastische Wellen* 107—118.
— —, generation by body forces, *Erzeugung durch Volumkräfte* 116—118.
Elastico-viscosity (Maxwell body), *Elasticoviscosität (Maxwellscher Körper)* 465 to 467.
Elliptical cylinder, torsion, *elliptischer Zylinder, Torsion* 30, 34.
Elliptic hole in an infinite plate, *elliptisches Loch in einer unendlich ausgedehnten Platte* 57—58, 64.
Elliptic problem of differential equations, *elliptisches Differentialgleichungsproblem* 339, 369, 371, 384.
Elliptic region, *elliptisches Gebiet* 384.
Emulsion 435, 523.
Endosity (= negative viscosity), *Endosität (= negative Viscosität)* 478.
Endurance limit, *untere Ermüdungsgrenze* 592 to 593, 606.
Energy dissipation, *Energiedissipation* 249, 260.
Entropy balance, *Entropiebilanz* 245, 246, 294.
Entropy flux, *Entropiefluß* 245, 264.
Entropy production, *Entropieerzeugung* 245, 246, 248, 251, 264.
— —, minimum rate, *minimale Zunahme* 253.

Envelope of Mohr circles, *Enveloppe Mohrscher Kreise* 352, 376.
Epicycloids as slip lines, *Epizykloiden als Gleitlinien* 363.
Equation of state, *Zustandsgleichung* 236.
Equations of motion, *Bewegungsgleichungen* 10—12, 139, 235, 323.
Equilibrium equations, *Gleichgewichtsbedingungen* 235, 293.
— —, covariant form, *kovariante Schreibweise* 11, 15, 28.
— —, KELVIN's solution, *Kelvinsche Lösung* 87—88.
— —, vector form, *vektorielle Schreibweise* 87.
Equilibrium viscosity, *Gleichgewichtsviscosität* 293.
ESHELBY's solution, *Eshelbysche Lösung* 121.
Exceptional direction, *Ausnahmerichtung* 385, 389—390.
Exceptional line, plane etc. see characteristic.
Exterior derivatives, *äußere Ableitungen* 341.
Extreme-value statistics of strength, *Extremwert-Statistik der Festigkeit* 580, 581, 582.

Fan see simple wave.
Fatigue crack, *Ermüdungsriß* 591.
Fatigue cracks, propagation, *Ermüdungsrisse, Ausbreitung* 596, 601, 603, 606.
Fatigue damage, *Ermüdungsschäden* 571, 752.
— —, initiation, *Einleitung* 596, 606.
Fatigue effects in ceramics and glasses, *Ermüdungserscheinungen in Keramik und Gläsern* 593.
— — in high polymers, *in Hochpolymeren* 593.
Fatigue limit, *Ermüdungsgrenze* 592.
Filon theory of generalised plane stress, *Filonsche Theorie des generalisierten ebenen Spannungszustandes* 129, 143—145.
Finite deformation or strain, *endliche Deformation* 414, 489—501.
Finite extension, *endliche Dehnung* 237.
Finite thickness of a plate, see general plane stress, *endliche Dicke einer Platte s. generalisierter ebener Spannungszustand*.
Firmo-viscosity of the KELVIN body, *Zusatzviscosität im Kelvinschen Körper* 464.
Flat ring in elastic-plastic plane stress, *flacher Ring in elastisch-plastischem ebenen Spannungszustand* 418—421.
Flaws in liquids, *Aufreißen von Flüssigkeiten* 555—556, 580—584.
Flexural waves in cylinders, *Biegungswellen in Zylindern* 115—116.
Flour dough, rheological behaviour, *Brotteig, rheologisches Verhalten* 475—478.
Flow condition see also yield condition.
Flow condition of HENCKY, *Fließbedingung von Hencky* 520.
— —, of MISES, *von Mises* 521.
Flow, definition, *Fließen, Definition* 434.
Flow-tensor, *Flußtensor* 438.
Flow-type law, *Gesetz vom Strömungstyp* 332.

Flow velocity tensor, plastic, *plastischer Strömungsgeschwindigkeitstensor*, 325, 326.
Fluidity coefficient, *Fluidität* 450.
Fluidity at rest, *Fluidität in Ruhe* 496.
Fore-effect, elastic, see creep.
Fourier transforms in elasticity, *Fouriertransformationen in der Elastizitätstheorie* 72—78.
Fracture condition of HUBER, *Bruchbedingung von Huber* 520.
Fracture origins, *Bruchentstehung* 567, 685, 569.
Fracture, roughness and texture, *Bruch, Rauhigkeit und Faserstruktur* 577, 578.
Fracture size-effect, measurements, *Zerreißfestigkeitsmessungen, Einfluß der Abmessungen* 584, 585, 586.
Fracture strength see also strength, *Bruchfestigkeit s. auch Festigkeit*.
— —, definition, *Definition* 563.
— —, frequency dependence, *Frequenzabhängigkeit* 593 seq., 600.
— —, measurements (table), *Messungen (Tabelle)* 566.
Fracture stress, *Bruchspannung* 567, 573, 574, 575.
— —, in fatigue experiments, *bei Ermüdungsversuchen* 594.
Free-boundary problem, *Problem mit freier Grenzlinie* 404.
Free energy, *freie Energie* 16—17, 247.
— —, intrinsic, *freie Energie ohne kinetischen Anteil* 440.
Freezing procedure, *Einfrierverfahren* 179.
Frequency of cyclic stress, effect on strength, *Frequenz der periodisch veränderlichen Spannung, Einfluß auf Bruchfestigkeit* 593 seq., 600.
FRESNEL's ellipsoid, *Fresnelsches Ellipsoid* 133, 146, 186.
Friction heating, *Reibungserwärmung* 600.
Fringes, isochromatic, *Streifen, isochromatische* 149.
—, isoclinic, *isokline* 148, 166.
Fringe value (of a material) 149, 164.
FROCHT's method of stress determination, *Frochtsche Methode der Spannungsbestimmung* 184—186.
Frozen stress (see also stress freezing) *eingefrorener Spannungszustand* 130, 177 to 180.

GALIN's solution for the moving punch, *Galinsche Lösung für die bewegte Stanzlinie* 121.
Gel 435, 522.
Generalized flux, *generalisierter Fluß* 248, 249.
Generalized force, *generalisierte Kraft* 248, 249, 267.
Generalized plane stress, *generalisierter ebener Spannungszustand* 129, 143—145, 352 to 353.
Generalized strain-rates, *generalisierte Fließgeschwindigkeiten* 255.

Generalized stress, *generalisierte Spannung* 255, 327.
Geometrical non-linearity, *geometrische Nichtlinearität* 487, 488, 498—507.
Gibbs-Helmholtz equation, *Gibbs-Helmholtzsche Gleichung* 440.
Glasses, fatigue effects, *Gläser, Ermüdungserscheinungen* 593.
Glide bands, *Gleitbänder* 232, 597.
GOODIER's thermoelastic potential, *Goodiersches thermoelastisches Potential* 125.
Grain-boundary diffusion or migration, *Korngrenzen-Diffusion oder -Wanderung* 267, 598.
Grain size effect on slip processes, *Korngröße, Einfluß auf Gleitprozesse* 602, 606.
Green measure of deformation, *Greensches Deformationsmaß* 499.
GREEN's theorem, *Greenscher Satz* 95.
Griffith theory, *Griffithsche Theorie* 557.

Haar-Kármán hypothesis, *Haar-Kármánsche Hypothese* 427.
Hagen-Poiseuille equation, *Hagen-Poiseuillesches Gesetz* 459, 497.
Hankel transform, *Hankelsche Integraltransformation* 100, 101.
H-body see Hooke solid, *H-Körper s. elastische Substanz.*
Heat-flashes causing striation, *Wärmeblitze, die Striemenbildung verursachen* 600, 604 seq.
Helmholtz' free energy see free energy, *Helmholtzsche freie Energie s. freie Energie.*
Hencky measure of deformation, *Henckysches Deformationsmaß* 486, 499—500, 544.
HENCKY's flow condition, *Henckysche Fließbedingung* 520.
Hexagonal prism of TRESCA yield condition, *sechsseitiges Prisma der Trescaschen Fließbedingung* 328, 371.
Hexagonal symmetry, elastic constants, *hexagonale Symmetrie, elastische Konstanten* 22—23.
Hexagon of TRESCA, *Sechseck von Tresca* 371, 379—380.
Hole in an infinite plate, *Loch in einer unendlich ausgedehnten Platte* 58.
Holes of different shapes, photoelastic fringe patterns, *Löcher verschiedener Form, spannungsoptische Streifenbilder* 205—211.
Holomorphic, *holomorph* 60.
Homogeneous isotropic body, *homogener isotroper Körper* 17—19, 21, 23—26.
Hooke solid (= elastic material) *elastische Substanz* 441, 448—450.
HOOKE's law, *Hookesches Gesetz* 15, 17, 24, 142, 329, 400, 401, 409.
HUBER's fracture condition, *Hubersche Bruchbedingung* 520.
Hydrostatic tension, *hydrostatischer Druck* 143.
Hyperbolic point, *hyperbolischer Punkt* 369, 372, 377.
Hyperbolic problem of differential equations, *hyperbolisches Differentialgleichungsproblem* 339, 341, 352, 360, 362, 369, 371, 384.

Hyperbolic region, *hyperbolisches Gebiet* 384 to 385.
Hypocycloids as slip lines, *Hypozykloiden als Gleitlinien* 363.
Hypo-elasticity, *hypoelastisches Verhalten* 262.

Ideal liquid, *ideale Flüssigkeit* 441.
Ideal material, *ideale Substanz* 440.
Impedance, complex mechanical, *komplexe mechanische Impedanz* 276 seq.
Incipient plastic flow, *Anlaufvorgänge bei plastischem Fließen* 427.
Incompressible matter, strain invariants, *inkompressibles Medium, Deformationsinvarianten* 6.
— —, stress-strain relations, *Spannungs-Dehnungs-Relationen* 18.
Incompressible plane visco-plastic flow, *inkompressibles, ebenes viscoplastisches Fließen* 302.
Incubation period of fatigue, *Latenzzeit bei Ermüdungserscheinungen* 609, 610.
Indentation, *Einschnitt* 101, 432.
Inelasticity, physical basis, *inelastisches Verhalten, physikalische Grundlagen* 229 to 231.
Infinite strip, *unendlich langer Streifen* 84, 303, 306—307.
Infinitesimal deformation (or strain) *infinitesimale Deformation* 6—7, 12, 438.
Initial curve, *Anfangskurve* 333.
Initial-value problem of Cauchy type, *Anfangswertproblem des Cauchyschen Typs* 333, 336 seq., 340, 360 seq., 365, 396.
— —, characteristic, *charakteristisches* 361, 396, 397.
— —, mixed, *gemischtes* 362, 397.
Initiation of fatigue damage, *Einleitung von Ermüdungsschäden* 596, 606.
Instability, time dependent, *zeitabhängige Instabilität* 310, 311.
Integral transforms in dynamic elasticity, *Integraltransformationen in der Elastodynamik* 121—122.
— —, in three-dimensional elasticity, *in der dreidimensionalen Elastizitätstheorie* 98 to 102.
— —, in two-dimensional elasticity, *in der zweidimensionalen Elastizitätstheorie* 72 to 78.
Integrating factor, *integrierender Faktor* 382.
Integration of two-dimensional stress equations numerical methods, *Integration der zweidimensionalen Spannungsgleichungen, numerische Methoden* 173—176.
Integrative method of SCOTT BLAIR, *Integrativmethode von Scott Blair* 480.
Intensities of strain and stress, *Intensitäten von Dehnung und Spannung* 240, 241, 284, 292.
Interaction of discontinuities, *Wechselwirkung von Unstetigkeiten* 367.
Interior derivatives, *innere Ableitungen* 341, 345.
Internal energy, *innere Energie* 243, 244.

Internal friction, *innere Reibung* 372.
Invariant form of theory, *invariante Form der Theorie* 335, 337, 338.
Invariants of strain tensor, *Invarianten des Deformationstensors* 6.
— of a tensor, *eines Tensors* 239—243, 258, 443 seq.
Isobars, *Isobaren* 384, 387, 390.
Isochromatic fringes, *isochromatische Streifen* 149.
Isoclines, *Isoklinen* 384, 387, 390.
Isoclinic fringes, *isokline Streifen* 148, 166.
Isothermal conditions of inelastic continuum, *isotherme Bedingungen im unelastischen Kontinuum* 233, 236.
Isothermal deformation, *isotherme Deformation* 14, 16—17.
Isothermal elastic potential, *isothermes elastisches Potential* 247, 250.
Isotropic body, elastic constants, *isotroper Körper, elastische Konstanten* 23—26.
Isotropic homogeneous body, *isotroper homogener Körper* 17—19, 21, 23—26.
Isotropic point, *isotroper Punkt* 166, 168.
Isotropic tensor function, *isotrope Tensorfunktion* 444.

Jeffreys body (= elastic sol) (see also Lethersich body), *Jeffreysscher Körper (s. auch Lethersichscher Körper)* 455, 468—471.
Jelly, see also gel 435, 522.

KÁRMÁN's notation of stress components, *Kármánsche Beziehungen der Spannungskomponenten* 26.
K-body see Kelvin body, *K-Körper, s. Kelvinscher Körper*.
Kelvin body (= delayed elasticity), *Kelvinscher Körper (= verzögerte Elastizität)* 312 to 313, 454 seq., 463—464.
Kelvin effect (dilatancy), *Kelvin-Effekt (Dilatanz)* 261.
KELVIN's elastic sphere, *Kelvinsche elastische Kugel* 102—104.
KELVIN's solution of equilibrium equations, *Kelvinsche Lösung der Gleichgewichtsbedingungen* 87—88.
Kinematical discontinuity conditions, *kinematische Unstetigkeitsbedingungen* 344.
Kuhn-Treloar theory of rubber-like substances, *Kuhn-Treloarsche Theorie kautschukähnlicher Substanzen* 18.

Lagrangian of elastic stress, *Lagrange-Funktion elastischer Spannungen* 15.
LAMÉ's elastic constants, *Lamésche elastische Konstanten* 24, 449.
Lamé-Maxwell equations, *Lamé-Maxwellsche Gleichungen* 140, 174, 182.
Laplace equation for plane stress, *Potentialgleichung für ebene Spannungen* 144, 173.
— —, two-dimensional, in the theory of torsion, *zweidimensionale, in der Theorie der Torsion* 29.

Lateral extensometer of COKER, *Dickenmesser von Coker* 173.
Legendre transformation, *Legendresche Transformation* 370.
Lens system in photoelastic equipment, *Linsensystem für spannungsoptische Ausrüstung* 155—157.
Lethersich body (= elastic sol), *Lethersichscher Körper (= elastisches Sol)* 456, 467 to 468.
Limit design, *Traglastverfahren* 315.
Limit line (envelope of characteristics), *Grenzlinie (Enveloppe von Charakteristiken)* 384—385, 408.
— — between plastic and rigid material, *zwischen plastischer und starrer Substanz* 383.
Limit line singularities, *Grenzlinien-Singularitäten* 384—390.
Limit point, *Grenzpunkt* 385.
Linearization of plane plasticity, *Linearisierung der ebenen Plastizitätstheorie* 367 to 370.
Line of rupture see stress discontinuity line.
Liquid, Newtonian, *Flüssigkeit, Newtonsche* 441, 450—452.
—, non-Newtonian, *nicht Newtonsche* 488, 497, 544.
Loading function, *Belastungsfunktion* 280, 283.
Loam as a Kelvin body, *Lehm als Kelvinscher Körper* 453—455.
Locus of incompressible yield, *Ort für inkompressibles Fließen* 279.
Logarithmic homologues of consistency curves, *logarithmische Homologen der Konsistenzkurven* 497.
Logarithmic spirals as slip lines, *logarithmische Spiralen als Gleitlinien* 363—365, 380.
— — as streamlines, *als Stromlinien* 364 to 365.
Logarithmic strain, *logarithmische Dehnung* 237.
Longitudinal waves in cylinders, *Längswellen in Zylindern* 113—114.
Loss angle, *Verlustwinkel* 268, 277—278.
Love waves in stratified media, *Lovesche Wellen in geschichteten Medien* 111—112.
L-shaped beam, *Balken von L-förmigem Querschnitt* 400, 406.
Lueders bands, *Lüdersche Bänder (Gleitbänder)* 232.

Macro-rheology, *Makrorheologie* 435.
Machining of photoelastic materials, *Bearbeiten spannungsoptischer Materialien* 164.
Mapping of regions onto a circle, *Abbildung von Gebieten auf einen Kreis* 62—64.
— — onto a half-plane, *auf eine Halbebene* 66—71.
Margules equation, *Margulessches Gesetz* 460, 497.
Material discontinuity, *substantielle Unstetigkeit* 343.

Maxwell body (= elastico-viscous material), *Maxwellscher Körper (= elastisch-viscose Substanz)* 290, 301, 312—313, 455, 465 to 467, 516—518.
— —, propagation of a shear wave, *Ausbreitung einer Scherungswelle* 301.
M-body see Maxwell body, *M-Körper s. Maxwellscher Körper*.
Measures of finite deformation, *Maße für endliche Deformationen* 498—501.
Mechanical models of materials, *mechanische Modelle für Substanzen* 452seq.
Membrane model of PRANDTL, *Membrangleichnis von Prandtl* 401, 424.
Memory functions of stress and strain, *Erinnerungsvermögen von Spannung und Verformung* 273.
Metarheology, *Metarheologie* 436.
Microcracks, *Mikrorisse* 567.
Micro-mechanism of fatigue, *Mikromechanismus der Ermüdungserscheinungen* 596 to 597.
Micro-rheology, *Mikrorheologie* 436, 522 to 535.
Minimum life in fatigue experiments, *minimale Lebensdauer bei Ermüdungsversuchen* 593, 607.
Minus wave, *Minuswelle* 391 seq.
Mises flow equations, *Misessche Strömungsgleichungen* 339.
Mises plasticity equations, *Misessche Plastizitätsgleichungen* 323, 404.
— —, characteristics, *Charakteristiken* 337—339.
Mises yield condition, *Misessche Fließbedingung* 282, 289, 320, 322, 324, 328, 370, 521.
Mises-Lévy equations, *Mises-Lévysche Gleichungen* 284, 286, 323.
Mixed boundary value problem, *gemischtes Randwertproblem* 58, 70, 81—84, 101 to 102.
Mixed problem, *gemischtes Problem* 362, 381, 397, 430.
Models of materials, *Modelle für Substanzen* 452 seq.
Moduli, generalized, *generalisierte Moduln* 248, 258.
Modulus of cross-elasticity, *Modul der Elastizität zweiter Ordnung* 507.
— of rigidity see shear modulus.
Mohr circle, *Mohrscher Kreis* 241, 292, 325 seq., 352, 375—376, 446.
Mohr circles of stresses on both sides of a discontinuity line, *Mohrsche Kreise für die Spannungen auf beiden Seiten einer Unstetigkeitslinie* 367.
Monge-Ampère equation for Airy function, *Monge-Ampèresche Gleichung für die Airysche Spannungsfunktion* 353.
MOONEY'S equation for elastic rubber potential, *Mooneysche Gleichung für das elastische Kautschukpotential* 19.
Moving discontinuity, *Unstetigkeit, die sich bewegt* 342, 346.

Multi-valued displacement, *mehrdeutige Verrückung* 53.
Mushkhelishvili potentials, *Muschchelischwilische Potentiale* 48—58.

Navier-Stokes equations, *Navier-Stokessche Gleichungen* 451.
N-body see Newtonian liquid, *N-Körper s. Newtonsche Flüssigkeit*.
NEUMANN'S boundary problem of potential theory, *zweite Randwertaufgabe der Potentialthoerie* 28.
Neutral loading, *neutrale Belastung* 290.
Neutral stress change, *neutrale Spannungsänderung* 280, 288.
Newtonian liquid (= ideal viscous liquid), *Newtonsche Flüssigkeit (= ideale zähe Flüssigkeit)* 441, 450—451, 452.
NICOL'S prism, *Nicolsches Prisma* 135.
Non-isotropic material, *anisotrope Substanz* 370.
Non-linear dissipative media, *nichtlineare dissipative Substanzen* 255.
Non-linear elasticity, *nichtlineare Elastizität* 508—509.
Non-linear viscous flow, *nichtlineare zähe Strömung* 289.
Non-linearity of damage rate in fatigue, *Nichtlinearität der Ermüdungsverluste* 609—611.
— in rheology, *in der Rheologie* 487.
Non-linear shear, *nicht-lineare Scherung* 19.
Non-Newtonian liquid, *nicht-Newtonsche Flüssigkeit* 488, 497, 544.
Non-steady flow in plane plastic strain, *nichtstationäres Fließen bei ebener plastischer Verformung* 427, 432.
Normal strain, *Dehnung* 140.
Normal stress, *Zugspannung* 137.
Normal stress, maximum direction, *Hauptspannungsrichtung* 354.
Normal system of differential equations, *Normalsystem von Differentialgleichungen* 336seq., 340.
Nutting-Scott Blair equation, *Nutting-Scott-Blairsche Gleichung* 480.

Octahedral shear strain, *Scherung in Oktaederfläche* 240.
Octahedral shear stress, *Schubspannung in Oktaederfläche* 240, 241, 326.
Oil paint see paint, *Ölfarbe s. Farbe*.
Oldroyd number, *Oldroydsche Zahl* 462.
Onsager symmetry relations, *Onsagersche Symmetriebeziehungen* 249.
Open-end conditions of a tube, *Enden des Rohres offen* 409, 410, 418.
Order of discontinuity, *Ordnung der Unstetigkeit* 342.
Orthogonal curvilinear coordinates, *orthogonale krummlinige Koordinaten* 84—86.
Orthogonal net of slip lines, *orthogonales Gleitliniennetz* 355, 357—360, 397.
Orthotropy, elastic constants, *Orthotropie, elastische Konstanten* 21—22.

Oscillator model of relaxation, *Oszillatormodell der Relaxation* 275 seq.
Oval cross section bar, *ovaler Querschnitt, Stab* 406.
Oxide films in fatigue striations, *Oxydschichten in Ermüdungsstriemen* 600.

Paint as a Bingham body, *Farbe als Binghamscher Körper* 457, 462.
Papkovich solution see Boussinesq-Papkovich solution, *Papkowitsch-Lösung s. Boussinesq-Papkowitsch-Lösung*.
Parabola limit, *Parabelgrenze* 371, 379—380, 394—395.
Parabolic point, *parabolischer Punkt* 388.
Parabolic problem of differential equations, *parabolisches Differentialgleichungsproblem* 369, 384, 422.
Partly plastic tube, *teilweise plastisches Rohr* 411.
Pascalian liquid see ideal liquid.
PAYNE's potential functions, *Paynesche Potentialfunktionen* 78—80.
Perfect plasticity, *vollkommene Plastizität* 278, 323, 327, 425.
Permeability coefficient in a porous medium, *Durchlässigkeitskoeffizient in einem porösen Medium* 266.
Perturbation of visco-plastic flow, *Störung viscoplastischer Strömung* 306—308.
Phase velocity of viscoelastic waves, *Phasengeschwindigkeit viscoelastischer Wellen* 300, 301.
Photoelastic bench, *spannungsoptische Bank* 159—161.
Photoelastic materials, properties, *spannungsoptische Versuchsmaterialien, Eigenschaften* 163.
— — for stress freezing, *zum Einfrieren des Spannungszustandes* 178, 180.
Photoelasticity applied to shock waves, *Spannungsoptik auf Stoßwellen angewandt* 227.
—, detection and history, *Entdeckung und Geschichte* 128—130.
— in a rotating body, *in einem rotierenden Körper* 227.
—, theory, *Theorie* 145—155.
Photoelectric cell in photoelastic observations, *Photozelle bei spannungsoptischen Beobachtungen* 161.
Physical components of displacement vector, *physikalische Komponenten des Verschiebungsvektors* 5—6.
— — of stress tensor, *des Spannungstensors* 10, 12.
Physical non-linearity, *physikalische Nichtlinearität* 487—488.
Plane of polarisation, *Polarisationsebene* 131.
Plane polarised light, methods of production, *linear polarisiertes Licht, Herstellungsmethoden* 135, 158.
Plane problem of plasticity, complete, *ebenes Problem der Plastizitätstheorie, vollständiges* 352.

Plane strain, *ebener Deformationszustand* 40 to 41, 339, 347, 349—350, 353—367, 409 to 410.
— — under general yield condition, *bei allgemeiner Fließbedingung* 350 seq.
— — in plasticity, complete set of equations, *in plastischem Material, vollständiger Gleichungssatz* 349.
— — of a tube, *eines Rohres* 409, 410, 418.
Plane strain and stress in visco-elasticity, *ebener Deformations- und Spannungszustand für Viscoelastizität* 272.
Plane stress, *ebener Spannungszustand* 40 to 42.
— —, generalised, *generalisierter* 129, 143 to 145.
— — with quadratic yield condition, *bei quadratischer Fließbedingung* 351.
Plane of vibrations, *Schwingungsebene* 131.
Plane visco-plastic flow, *ebenes viscoplastisches Fließen* 302.
Plastic compliance, *Plastizitätskoeffizient* 290.
Plastic cubic dilatation, *plastische Volumdilatation* 286.
Plastic-elastic boundary, *plastisch-elastische Grenzzone* 400, 403.
Plastic equilibrium, *plastisches Gleichgewicht* 418 seq., 425.
Plastic flow, *plastisches Fließen* 279, 425, 434.
Plastic flow velocity tensor, *plastischer Fließgeschwindigkeitstensor* 325, 326.
Plastic gel (Schwedoff body), *plastisches Gel (Schwedoffscher Körper)* 474—475.
Plastic mass between two curved plates, *plastische Masse zwischen zwei gekrümmten Platten* 383.
— — between two rigid rough plates, *zwischen zwei starren, rauhen Platten* 366, 429—432.
Plastic potential, *plastisches Potential* 254, 284, 285, 326, 350, 373.
— —, piece-wise continuous, *stückweise stetig* 254.
Plastic-rigid boundary, *plastisch-starre Grenze* 366, 383, 399, 422.
Plastic strength, *plastische Festigkeit* 481.
Plastic torsion, *plastische Torsion* 339, 399 to 408.
Plastic viscosity, *plastische Viscosität* 457, 462.
Plastic work, *plastische Arbeit* 325.
Plastic yield in photoelastic materials, *plastisches Fließen in spannungsoptischem Material* 176, 225—226.
Plate analogy, *Plattengleichnis* 424.
Plate equation, visco-elastic, *viscoelastische Plattengleichung* 310.
Plate pressed between two pistons (plastic stress), *Platte zwischen zwei Stempeln gedrückt (plastischer Spannungszustand)* 362.
Plates, elastic-plastic behaviour, *Platten, elastisch-plastisches Verhalten* 320.
— of visco-elastic material, *aus viscoelastischem Material* 310, 311.

Point-force, *Einzelkraft* 76, 88—89.
Poiseuille equation see Hagen-Poiseuille equation, *Poiseuillesches Gesetz s. Hagen-Poiseuillesches Gesetz*.
Poisson ratio, *Querkontraktionszahl* 25, 142, 177, 212, 329, 449.
— —, modified, *modifizierte* 50.
POISSON's probability law applied on fatigue fractures, *Poissonsche Formel der Wahrscheinlichkeitsrechnung in Anwendung auf Ermüdungsbrüche* 610—611.
Polarisation plane of light, *Polarisationsebene des Lichtes* 131.
Polariscope, circular, *Polariskop für zirkulares Licht* 153.
Polarised light, passage through a crystal, *polarisiertes Licht, Durchgang durch einen Kristall* 133—135.
— —, passage through a stressed plate, *Durchgang durch eine Platte unter Spannung* 147—150.
Polarising angle, *Brewsterscher Winkel* 135.
Polaroid material, *polaroides Material* 135, 158.
Polygonization, *Polygonisierung* 598, 602.
Polymers, fatigue effects, *Polymere, Ermüdungserscheinungen* 593.
Pore-fluid, *Porenflüssigkeit* 231, 265.
Pore-fluid anelasticity, *Porenflüssigkeitsanelastizität* 263, 265—267.
Pore volume, *Porenvolumen* 230.
Potential energy of elastic stress, *potentielle Energie elastischer Spannungen* 13—14.
Potential function of AIRY, *Potentialfunktion von Airy* 43—47, 50.
— — of BOUSSINESQ-PAPKOVICH, *von Boussinesq und Papkowitsch* 91—93.
— —, complex, for plane stress, *komplexe, für den ebenen Spannungszustand* 48—58.
— — of KELVIN, *von Kelvin* 87—88.
— — of MUSHKHELISHVILI, *von Muschchelischwili* 48—58.
— — of PAYNE, *von Payne* 78—80.
— —, real, for plane stress, *reelle, für den ebenen Spannungszustand* 78—84.
— — of torsion, *der Torsion* 29—33.
Power input, specific, *Leistung der äußeren Kräfte* 439.
Poynting effect, *Poynting-Effekt* 261, 509 to 512.
Poynting-Thomson body (= anelastic material), *Poynting-Thomsonscher Körper (= anelastische Substanz)* 456, 471—472.
Prandtl body (= elastic-plastic body), *Prandtlscher Körper (= elastisch-plastische Substanz)* 285, 290, 451.
Prandtl-Reuss body, *Prandtl-Reußscher Körper* 285, 290, 451.
Prandtl-Reuss equations, *Prandtl-Reußsche Gleichungen* 330, 331, 340, 404, 412, 416.
Prandtl-Reuss theory, *Prandtl-Reußsche Theorie* 329 seq.
Pressure, *Druck* 323, 439, 448.
Pressure force, *Druckkraft* 323.

Pressure, internal, expanding a spherical shell, *innerer Druck der eine Kugelschale dehnt* 422.
—, —, expanding a tube, *der ein Rohr dehnt* 365, 408—418.
— originating free plastic flow, *als Ursache freier plastischer Strömung* 425.
Pressure-relaxation, *Druckrelaxation* 265, 266.
Pre-stressed beam, *vorgespannter Balken* 223.
Primary wave (P-wave), *Primärwelle (P-Welle)* 107—109.
Principal stress axis, *Hauptspannungsrichtung* 13, 137, 327, 354.
— — —, secondary, *sekundäre* 138.
Principal stress-difference, *Hauptspannungs-Differenz* 149, 166.
Probability viewpoints in fatigue fracture, *Wahrscheinlichkeitsgesichtspunkte beim Ermüdungsbruch* 595, 607 seq.
Propagation of fatigue cracks, *Ausbreitung von Ermüdungsrissen* 596, 601, 603, 606.
— of light in a crystal, *des Lichtes in einem Kristall* 132—135, 151—153.
Propagation velocity, *Fortpflanzungsgeschwindigkeit* 343.
Proportional loading, *proportionale Belastung* 294.
Pulse propagation in the visco-elastic medium, *Stoßfortpflanzung in einem viscoelastischen Medium* 300, 313.
Punch, circular, in plastic material, *kreisförmiges Stanzloch in plastischer Substanz* 365.
Punch problem in plane elasticity, *Stanzproblem in der ebenen Elastizitätstheorie* 70, 71—72, 76—78, 81.
P-waves (primary waves), *P-Wellen (Primärwellen)* 107—109.

Quadratic yield condition see Mises yield condition, *quadratische Fließbedingung s. Misessche Fließbedingung*.
Quarter-wave plate, *Viertelwellenlängenplättchen* 135, 151, 153, 157—158.
Quasi-homogeneous material, *quasi-homogenes Material* 435.
Quasi-isotropic material, *quasi-isotropes Material* 436.

Radial loading, *Radialbelastung* 281, 287.
RADOK's solution, *Radoksche Lösung* 120.
Random loading in fatigue, *zufällige Belastung bei Ermüdungserscheinungen* 609.
Rate-of-deformation tensor, *Deformationsgeschwindigkeitstensor* 239.
Rayleigh waves, *Rayleighsche Wellen* 109 to 111.
Real potentials for plane stress, *reelle Potentiale für den ebenen Spannungszustand* 78 to 84.
Recovery accompanying fatigue, *Erholung als Begleiterscheinung der Ermüdung* 601.
— from creep, *vom Kriechen* 464.

Recrystallization, *Rekristallisation* 601.
Rectangular hole, photoelastic fringe pattern, *rechteckiges Loch, spannungsoptisches Streifenbild* 209.
Rectilinear characteristics, *geradlinige Charakteristiken* 377, 392.
Reduced stress of HENCKY, *reduzierte Spannung nach Hencky* 489.
Reiner-Rivlin equation, *Reiner-Rivlinsche Gleichung* 460.
Reiner-Rivlin fluid, *Reiner-Rivlinsche Flüssigkeit* 512.
Relations between elastic constants, *Beziehungen zwischen den elastischen Konstanten* 26.
Relaxation equation, *Relaxationsgleichung* 291, 292.
Relaxation frequency, *Relaxationsfrequenz* 275.
Relaxation function, *Relaxationsfunktion* 273, 274 seq.
Relaxation modes, *Relaxationslösung* 266.
Relaxation spectra, *Relaxationsspektren* 273 to 278, 312.
Relaxation in the standard solid, *Relaxation im Standard-Festkörper* 268.
Relaxation time, *Relaxationszeit* 465, 479, 481.
— —, variable, *variable* 291.
Relaxing gel (Jeffreys body), *relaxierendes Gel (Jeffreysscher Körper)* 468—471.
Relaxing media, *relaxierende Körper* 269.
Residual stress pattern, *remanentes Streifensystem* 130.
Resonant vibration in fatigue experiments, *Resonanzerschütterung in Ermüdungsversuchen* 591.
Retardation, relative, of optical paths, *Gangunterschied* 149, 153—155, 171.
Retardation of strain, see also creep, *Retardierung der Verformung, s. auch Kriechen* 463.
Retardation time, *Retardierungszeit* 269, 463, 481.
Retarded elastic media, *retardierte elastische Körper* 269 seq.
Reuss equations, theory etc. see Prandtl-Reuss equations, *Reußsche Gleichungen, Theorie usw. s. Prandtl-Reußsche Gleichungen, Theorie, usw.*
Reversible elastic phenomena, *reversible elastische Erscheinungen* 14, 15.
Revolution solid, symmetrical strain, *Rotationskörper, symmetrischer Deformationszustand* 105.
Reynolds number, *Reynoldssche Zahl* 462.
Rheological bodies, complex, *zusammengesetzte rheologische Körper* 478 seq.
— —, definitions, *Definitionen* 454—456.
— —, superposition, *Superposition* 478 seq.
Rheological equation of a material, *rheologische Gleichung für eine Substanz* 440.
— —, quasi-linear, *quasilineare* 488, 502.
Rheological formula of a material, *rheologische Formel einer Substanz* 455.

Rheological fundamental properties, *rheologische Grundeigenschaften* 447.
Rheological tree, *rheologischer Stammbaum* 456.
Rheology, phenomenological, *phänomenologische Rheologie* 435.
Rheometry, *Rheometrie* 535—542.
Riemann-Christoffel tensor, *Riemann-Christoffelscher Tensor* 7.
Riemann function, *Riemannsche Funktion* 360, 361.
Riemann integration method, *Riemannsche Integrationsmethode* 360 seq.
Rigid body, *starrer Körper* 441.
Rigid-body displacement, infinitesimal, *infinitesimale Starrkörper-Verschiebung* 7.
Rigidity modulus, *Gleitmodul* 25—26, 142, 448.
Rigid kernel of slab between rough plates, *starrer Kern einer Scheibe zwischen rauhen Platten* 431—432.
Rigid-plastic approximation, *starr-plastische Näherung* 281, 316, 319 seq.
Rivets see bolts, *Niete s. Bolzen.*
Rod see also bar and beam.
Rods of visco-elastic material, *Stäbe aus viscoelastischem Material* 313.
Rotating body, photoelastic behaviour, *rotierender Körper, spannungsoptisches Verhalten* 227.
Rotating cylinder, *rotierender Zylinder* 423.
Rotating disk, *rotierende Scheibe* 423.
Rotation, infinitesimal, *infinitesimale Drehung* 438.
Rotation tensor, *Drehungstensor* 238.
Rubber elasticity, *Kautschukelastizität* 16, 18, 492—494.
Rubber structure, *Kautschuk-Struktur* 18, 450, 493.

SAINT VENANT see ST. VENANT.
Sand hill analog of NADAI, *Sandhaufen-Gleichnis von Nadai* 402, 424.
Scatter in fatigue fracture, *Streuung bei Ermüdungsbruch* 595.
Scattered light method of stress determination, *Streulichtmethode zur Spannungsbestimmung* 193—202.
Schofield-Scott Blair body, *Schofield-Scott Blairscher Körper* 475—478.
Schwedoff body (= plastic gel), *Schwedoffscher Körper (= plastisches Gel)* 456, 474 to 475.
Secondary principal stress axis, *sekundäre Hauptspannungsrichtung* 138.
Secondary wave (S-wave), *Sekundärwelle (S-Welle)* 107—109.
Second order effects in viscous flow, *Effekte zweiter Ordnung bei zähem Fließen* 512 to 514.
Second order elasticity, *Elastizität zweiter Ordnung* 502 seq., 507—509.
Sectionally holomorphic function, *gebietsweise holomorphe Funktion* 60.

Seismic waves (reference), *seismische Wellen (Hinweis)* 107.
Self-annealing, *Selbst-Temperung* 598.
Semicircular groove, *Halbkreiseinschnitt* 423.
Sénarmont method of measuring retardations, *Sénarmontsche Methode zur Messung von Gangunterschieden* 154—155, 160—161, 172.
Separation surface, *Trennungsfläche* 341.
Shake down, *Wechselverfestigung* 317—318.
Shattering 578, 579.
Shear, *Scherung* 437.
Shear modulus, *Gleitmodul* 25—26, 142, 448.
Shear strain, *Scherung* 141.
— —, octahedral, *in Oktaederfläche* 240.
Shear stress, *Schubspannung* 137.
— —, maximum direction, *Richtung maximaler Schubspannung* 354, 373.
— —, octahedral, *in Oktaederfläche* 240, 241, 326.
Shear viscosity, *Scherungsviscosität* 450.
Shearing stress lines, *Schublinien* 32.
Shock conditions, *Stoßbedingungen* 346.
Shock of compressible fluid flow, *Stoßwellen in der Strömung kompressibler Flüssigkeiten* 342, 346.
Shock propagation, *Stoßausbreitung* 333.
Shock waves, photoelastic investigation, *Stoßwellen, spannungsoptische Untersuchung* 227.
Sign conventions of curvature, *Vorzeichenfestsetzung für die Krümmung* 358, 378.
Sign conventions in photoelasticity, *Vorzeichenkonventionen der Spannungsoptik* 169—170.
Simple shear, *einfache Scherung* 437.
Simple wave, *einfache Welle* 359, 390 bis 396.
Size-effects, fracture, *Abmessungen, Einfluß auf die Zerreißfestigkeit* 580—586, 589.
Slip, *Gleitung* 230, 321.
Slip avalanche, *Gleitlawine* 597—598.
Slip-band distribution, *Gleitbandverteilung* 597.
Slip-band spacing, *Gleitbandabstände* 597, 606.
Slip in fatigue, *Gleitung bei Ermüdungserscheinungen* 597.
Slip line, *Gleitlinie* 355, 357 seq., 399.
Slip line field, *Gleitlinienfeld* 357—360, 361, 366, 431.
— — — of the telegraph equation, *der Telegraphengleichung* 363.
Slippage along the wall, *Wandgleitung* 461.
Slip-plane blocking, *Gleitebenenblockierung* 602.
SNEDDON's solution, *Sneddonsche Lösung* 120.
Soap film analog of PRANDTL, *Seifenhaut-Gleichnis von Prandtl* 401, 424.
Sokolovsky yield condition, *Sokolovskysche Fließbedingung* 381.
Sol 435, 522.
—, dilute, *verdünntes* 524.
Soleil-Babinet compensator, *Soleil-Babinetscher Kompensator* 136.
Sonic line 387—388.

Spherical cavity, viscoelastic solution, *kugelförmiger Hohlraum, viscoelastische Lösung* 295.
Spherical polar coordinates, *Kugelkoordinaten in der Elastizitätstheroie* 86, 102—104.
Spherical shell in elastic-plastic expansion by internal pressure, *Kugelschale in elastisch-plastischer Dehnung unter Innendruck* 422.
Spheroidal coordinates, oblate, in elasticity, *Koordinaten des abgeplatteten Rotationsellipsoids in der Elastizitätstheorie* 106 to 107.
Stability of visco-plastic flow, *Stabilität viscoplastischer Strömung* 307.
Standard material, *Standard-Substanz* 440.
Standard solid, *Standard-Festkörper* 268, 271.
State functions of the inelastic continuum, *Zustandsfunktionen des unelastischen Kontinuums* 244, 245, 248—251.
Stationary discontinuity, *stationäre Unstetigkeit* 343.
Steady flow in plane plastic strain, *stationäres Fließen bei ebener plastischer Verformung* 427.
Stereographic projection of stress axes, *stereographische Projektion der Spannungshauptachsen* 189.
Stiffness modulus, *Steifigkeitsmodul* 491.
Stokes relation, *Stokessche Beziehung* 300, 464, 482.
Strain, generalized, *generalisierte Verformung* 327.
Strain history, *Deformations-Vorgeschichte* 280, 287.
Strain invariants, *Deformationstensor, Invarianten* 6.
Strain, logarithmic, *logarithmische Dehnung* 237.
Strain quadric, *quadratische Form des Deformationstensors* 141.
Strain rate, *Fließgeschwindigkeit* 328, 426.
— —, averaged through plate thickness, *gemittelt über die Plattenstärke* 353.
— — variation, *Variationsbreite* 233.
Strain in rheology, *elastische Deformation in der Rheologie* 434.
Strain tensor, *Deformationstensor* 3—5.
Strain-work, reversible *Deformationsarbeit* 440, 448, 520.
Straining frame experiment, *Zugversuch* 158 to 159.
Strains, *Deformationszustand* 140—141.
Stream function for plane viscoplastic flow, *Strömungsfunktion für ebene viscoplastische Strömung* 303, 306.
Stream line, *Stromlinie* 386, 400, 426.
Strength, *Festigkeit* 448, 519—522.
—, effect of mean stress upon it, *Einfluß der mittleren Spannung darauf* 593.
Stress axis, principal, *Spannungshauptachse* 13, 137, 327.
Stress-change, neutral, *neutrale Spannungsänderung* 280.
Stress characteristics, *Spannungscharakteristiken* 354, 373.

Stress concentration, *Spannungskonzentration* 202.
Stress conic, *Spannungskegelschnitt* 138, 166.
Stress determination in three-dimensional models, *Spannungsbestimmung in dreidimensionalen Modellen* 180—186.
Stress deviation tensor, *Spannungsdeviator* 323.
Stress-difference, *Spannungsdifferenz* 149, 166.
—, determination from fringe pattern, *Bestimmung aus Streifenbild* 170—171.
Stress discontinuity, *Spannungsunstetigkeit* 366, 399.
Stress on a free boundary see boundary stress.
Stress freezing, *Einfrieren des Spannungszustandes* 130, 177—180.
— — materials, *geeignete Substanzen zum Einfrieren des Spannungszustandes* 178, 180.
Stress fringe pattern, *spannungsoptisches Streifenbild* 149, 166 seq.
Stress fringe patterns, three-dimensional examples, *Spannungsstreifenbilder, dreidimensionale Beispiele* 215—224.
— — —, two-dimensional examples, *zweidimensionale Beispiele* 203—215.
Stress function, *Spannungsfunktion* 31—33, 400.
— — for the viscoelastic medium, *für ein viscoelastisches Medium* 298.
— — for the viscoplastic medium, *für ein viscoplastisches Medium* 303, 304, 306.
Stress graph (see also stress plane), *Spannungsdiagramm (s. auch Spannungsebene)* 357, 368—369, 388—389.
Stress lines, *Äquipotentiallinien* 400.
Stress-optical coefficient, *spannungsoptischer Koeffizient* 149.
Stress plane (see also stress graph), *Spannungsebene (s. auch Spannungsdiagramm)* 357, 368—369, 388—389.
Stress quadric, *quadratische Form des Spannungstensors* 137, 146.
Stress-relaxation, *Spannungsrelaxation* 270, 271.
— along slip bands, *längs der Gleitbänder* 604.
Stress state, uniaxial, *einachsiger Spannungszustand* 272.
Stress-strain relations in an isotropic elastic material, *Spannungs-Dehnungs-Relationen im isotropen, elastischen Medium* 15—26, 142—143.
Stress tensor, *Spannungstensor* 8—10, 234, 323.
— —, physical components, *physikalische Komponenten* 10, 12.
Stress trajectory, *Hauptspannungslinie* 139, 168—170.
Stresses in a plane of symmetry, *Spannungen in einer Symmetrieebene* 180.
Stressless deformation, *spannungslose Deformation* 434.
Stressless volume change, *spannungslose Volumänderung* 434.

Striations in fatigue, *Striemenbildung bei Ermüdungserscheinungen* 598—601, 606, 610.
Strip, infinite, elastic solution, *unendlich langer Streifen, elastische Lösung* 84.
—, —, visco-plastic solution, *viscoplastische Lösung* 303, 306—307.
Strong discontinuity, *starke Unstetigkeit* 342, 347.
Structural breakdown, see yielding 230, 231.
Structural viscosity, *Strukturviscosität* 488, 494—498, 531—535.
StV-body see St. Venant body, *StV-Körper s. St. Venantscher Körper.*
St. Venant body (= rigid-plastic body), *St. Venantscher Körper (= starr-plastische Substanz)* 441, 451—452.
St. VENANT's principle, *St. Venantsches Prinzip* 203.
St. Venant-Lévy-Mises relations, *St. Venant-Lévy-Misessche Beziehungen* 323.
St. Venant-Tresca yield condition, *St. Venant Trescasche Fließbedingung* 283, 286 seq., 320, 328.
Superposition of rheological bodies, *Superposition rheologischer Körper* 478 seq.
Surface of constant slope, *Böschungsfläche* 401, 402.
Surface scratches, *Oberflächenkratzer* 598.
Surface waves of the Rayleigh type, *Oberflächenwellen Rayleighscher Art* 109—111.
Suspension 435.
Swainger measure of deformation, *Swaingersches Deformationsmaß* 499.
S-waves (secondary waves), *S-Wellen (Sekundärwellen)* 107—109.

Tardy method of measuring retardations, *Tardysche Methode zur Messung von Gangunterschieden* 153—154, 160—161, 172.
Technological properties of materials, *technologische Materialeigenschaften* 540—542.
Telegraph equation, *Telegraphengleichung* 363, 369 seq.
Telegraph equations of viscoelastic waves, *Telegraphengleichungen viscoelastischer Wellen* 301, 313.
Temperature change under load, *Temperaturänderung während der Belastung* 130, 177 to 180.
Temperature effect upon endurance limit, *Temperatureinfluß auf die (untere) Ermüdungsgrenze* 593.
Temperature effects in inelastic media, *Temperatureffekte in unelastischen Körpern* 292, 293.
Temperature field, non-homogeneous, *nicht homogenes Temperaturfeld* 293.
Temperature increase, local, *lokaler Temperaturanstieg* 602, 604.
Temperature sensitivity of viscosity, *Temperaturempfindlichkeit der Viscosität* 296.
Tensile strength, liquids, *Zerreißfestigkeit von Flüssigkeiten* 551, 555.
— — measurements, pure liquids, *Messungen an reinen Flüssigkeiten* 555, 556.

Tensor components, physical, *physikalische Tensorkomponenten* 441 seq.
Tensor invariants, *Tensorinvarianten* 6, 239 to 243, 258, 443 seq.
Tensorial non-linearity, *tensorielle Nichtlinearität* 487, 507—509.
TEREZAWA's solution for the half-space, *Terezawasche Lösung für den Halbraum* 99—101.
Thermal dissipation potential, *thermisches Dissipationspotential* 264.
Thermal fluctuations, *thermische Schwankungen* 553.
Thermal softening, *thermische Erweichung* 600.
Thermal stresses, photoelastic investigation, *thermische Spannungen, spannungsoptische Untersuchung* 228.
Thermo-anelasticity, *Thermoanelastizität* 263 to 264.
Thermoelasticity, *Thermoelastizität* 123—126.
Thermoelastic potential of GOODIER, *thermoelastisches Potential von Goodier* 125.
Thixotropy, *Thixotropie* 540—541.
Thomson body see Poynting-Thomson body, *Thomsonscher Körper s. Poynting-Thomsonscher Körper.*
Three-dimensional stress, photoelastic investigation, *dreidimensionaler Spannungszustand, spannungsoptische Untersuchung* 130, 177—202.
Tilting stage, *Drehschwingrahmen* 187 seq.
Time-creep see creep.
Time-edge effect in photoelastic measurements, *Randveränderung während spannungsoptischer Messungen* 162, 164.
Torsion of a circular cylinder, *Torsion eines Kreiszylinders* 27—28, 36—38.
— of a circular shaft with circular groove, *eines kreisrunden Stabes mit kreiszylindrischer Vertiefung* 35—36.
— of a cylinder of rectangular cross section, *eines Zylinders von rechteckigem Querschnitt* 35.
— of a cylinder of triangular cross section, *eines Zylinders von Dreiecksquerschnitt* 34.
—, elastic-plastic, *elastisch-plastische* 403 to 408.
— of an elliptic cylinder, *eines elliptischen Zylinders* 30, 34.
— with extension, *mit Dehnung* 507, 508.
—, finite, *endliche* 36—38.
—, fully elastic, *völlig elastische* 399.
—, fully plastic, *völlig plastische* 399.
Torsion function, *Torsionsfunktion* 28.
— —, complex, *komplexe* 31.
— — for an elliptic cylinder, *für einen elliptischen Zylinder* 30.
Torsion of a hollow cylinder, *Torsion eines Hohlzylinders* 33.
— of a non-circular cylinder (general), *eines Zylinders allgemeinen Querschnitts* 28—31.
—, simple, *einfache* 506, 509.
Torsional flow, *Torsionsfließen* 513.
Torsional rigidity, *Torsionsmodul* 28.

Torsional waves in cylinders, *Torsionswellen in Zylindern* 112—113.
Total differential, *totales Differential* 382.
TRESCA's yield condition, *Trescasche Fließbedingung* 328, 339, 350, 371, 411 seq., 417.
Triangular elastic-plastic torsion solution, *Dreieckslösung für elastisch-plastische Torsion* 407.
Truesdell number, *Truesdellsche Zahl* 517.
Tube see cylindrical tube, *Rohr s. zylindrisches Rohr.*
Twisting see torsion, *Verwindung s. Torsion.*

Uniaxial stress state, *einachsiger Spannungszustand* 272.
Uniqueness, *Eindeutigkeit* 253, 254, 255, 284, 316.
Unloading, *Entlastung* 280, 285.

Vacancy concentration, *Leerstellenkonzentration* 603.
Variational principles in elasticity, *Variationsprinzipien in der Elastizitätstheorie* 15.
— — of the inelastic continuum, *des unelastischen Kontinuums* 251—256.
Velocity characteristics, *Geschwindigkeitscharakteristiken* 354.
Velocity compatibility relations, *Geschwindigkeits-Kompatibilitätsrelationen* 356—357.
Velocity of displacement of a wave surface, *Geschwindigkeit der Verrückung einer Wellenfläche* 343.
Velocity field in plane plastic flow, *Geschwindigkeitsfeld bei ebener plastischer Strömung* 364.
Velocity plane, *Geschwindigkeitsebene* 357, 375.
Virtual work, *virtuelle Arbeit* 15.
Visco-elastic boundary value problem, *viscoelastisches Randwertproblem* 294 seq.
Visco-elastic moduli, *viscoelastische Moduln* 309—310.
Visco-elastic stress-strain relations in one or two dimensions, *viscoelastische Spannungs-Dehnungs-Relationen in einer oder zwei Dimensionen* 272.
Visco-elastic wave equations, *viscoelastische Wellengleichungen* 299—301.
Visco-elasticity, *Viscoelastizität* 269—278, 293—301, 308—313, 448.
— (Burgers body), *(Burgersscher Körper)* 472—474, 491.
Visco-plastic boundary layer, *viscoplastische Grenzschicht* 305.
Visco-plastic boundary value problems, *viscoplastische Randwertprobleme* 305 to 308.
Visco-plasticity, *Viscoplastizität* 289, 301 to 308.
Viscosity coefficient, *Viscosität* 450, 459.
—, temperature sensitivity, *Temperaturempfindlichkeit* 296.
Viscous flow, *viscoses Fließen* 434.
Viscous traction coefficient, *viscoser Dehnungskoeffizient* 451.

Volume-anelasticity, *Volumen-Anelastizität* 267.
Volume change by shear (= dilatancy), *Volumänderung durch Scherung (= Dilatanz)* 261, 503.
Volume dilatation, *Volumdilatation* 24, 143, 438, 481.
Volume flow, *Fließen mit Volumänderung* 482.
Volume-viscosity, *Volumviscosität* 231, 249, 267, 269.
Volumetric flow, *plastische Volumänderung* 438.
Volumetric strain, *elastische Volumänderung* 438.
Vorticity tensor, *Drehgeschwindigkeitstensor* 239.

Wall effects in rheology, *Wandeffekte in der Rheologie* 461.
Warping (in torsion), *Werfung (nicht-ebene Torsion)* 401, 405—408.
Water as a rheological object, *Wasser als rheologischer Gegenstand* 435.
Wave equations, visco-elastic, *viscoelastische Wellengleichungen* 299—301.
Wave front, *Wellenfront* 132.
Wave surface, *Wellenfläche* 343.
Waves, elastic, *elastische Wellen* 107—118.
Weak discontinuity, *schwache Unstetigkeit* 342, 347.
Wedge unilaterally loaded, incipient plastic flow, *Keil unter einseitiger Belastung, Anlaufvorgang des plastischen Fließens* 427 to 429.

Weissenberg effect, *Weissenberg-Effekt* 261, 514—517.
Workhardening, *Verfestigung* 280, 285, 598, 603.

Yield condition, *Fließbedingung* 278, 281 to 283, 370—372.
— — for an isotropic, incompressible, perfectly plastic body, *für einen isotropen, inkompressiblen, vollkommen plastischen Körper* 327.
— —, parametric representation, *Parameterdarstellung* 368.
— —, parabolic form, *parabolische Form* 282.
— — of Sokolovsky, *von Sokolovsky* 381.
— — of Tresca, *von Tresca* 328, 339, 350, 371, 411 seq., 417.
Yield function, *Fließfunktion* 327, 350, 352.
Yield locus, *Fließspannungsort* 279, 283.
Yield point, *Fließgrenze* 434.
Yield stress, *Fließspannung* 230, 451, 458, 462.
— — in simple shear, *bei einfacher Scherung* 324.
— — in simple tension, *bei einfacher Dehnung* 324.
Yield surface, *Fließspannungsfläche* 278, 283.
Yoffe's solution for the moving Griffith crack, *Yoffesche Lösung für den bewegten Griffithschen Riß* 121.
Young's modulus, *Elastizitätsmodul* 24—25, 142, 329, 449, 451.
— —, variability, *Veränderlichkeit* 490.

MIX
Papier aus verantwortungsvollen Quellen
Paper from responsible sources
FSC® C105338

If you have any concerns about our products,
you can contact us on
ProductSafety@springernature.com

In case Publisher is established outside the EU,
the EU authorized representative is:
**Springer Nature Customer Service Center GmbH
Europaplatz 3, 69115 Heidelberg, Germany**

Printed by Libri Plureos GmbH
in Hamburg, Germany